Centennial Special Volume 2

Geomorphic Systems
of
North America

Edited by

William L. Graf
Department of Geography
Arizona State University
Tempe, Arizona 85287

1987

Acknowledgment

Publication of this volume, one of the Special Volumes of *The Decade of North American Geology Project* series, has been made possible by members and friends of the Geological Society of America, corporations, and government agencies through contributions to the Decade of North American Geology fund of the Geological Society of America Foundation.

Following is a list of individuals, corporations, and government agencies giving and/or pledging more than $50,000 in support of the DNAG Project:

ARCO Exploration Company
Chevron Corporation
Cities Service Company
Conoco, Inc.
Diamond Shamrock Exploration
    Corporation
Exxon Production Research Company
Getty Oil Company
Gulf Oil Exploration and Production
    Company
Paul V. Hoovler
Kennecott Minerals Company
Kerr McGee Corporation
Marathon Oil Company
McMoRan Oil and Gas Company
Mobil Oil Corporation
Pennzoil Exploration and Production
    Company

Phillips Petroleum Company
Shell Oil Company
Caswell Silver
Sohio Petroleum Corporation
Standard Oil Company of Indiana
Sun Exploration and Production Company
Superior Oil Company
Tenneco Oil Company
Texaco, Inc.
Union Oil Company of California
Union Pacific Corporation and
    its operating companies:
        Champlin Petroleum Company
        Missouri Pacific Railroad Companies
        Rocky Mountain Energy Company
        Union Pacific Railroad Companies
        Upland Industries Corporation
U.S. Department of Energy

Published by the Geological Society of America, Inc.
3300 Penrose Place, P.O. Box 9140, Boulder, Colorado 80301

Printed in U.S.A.

Cover Photo: The Goose Neck on the Colorado River, from Dead Horse Point, Utah.

Library of Congress Cataloging-in-Publication Data

Geomorphic systems of North America.

(Centennial special volume ; 2)
Includes bibliographies and index.
1. Geomorphology—North America. 2. Decade of North
American Geology Project. I. Graf, William L.,
1947–   . II. Geological Society of America.
III. Series.
GB427.G46   1987   551.4'097   87-8411
ISBN 0-8137-5302-3

Contents

Contents

Preface

This volume is a contribution from the Quaternary Geology and Geomorphology Division of the Geological Society of America to the Decade of North American Geology (DNAG) Project as a part of the celebration of the Centennial of the Society. The able editorship of William L. Graf of the Department of Geography of Arizona State University at Tempe, Arizona, and the support of the officers and membership of the Division are greatly appreciated.

In addition to four Centennial Special Volumes such as this, the DNAG Project includes a 29 volume set of syntheses that constitute *The Geology of North America,* six Centennial Field Guides that highlight 100 of the best geologic sites in the area of each of the six regional Sections of the Society, 23 Continent/Ocean Transects, and eight wall maps at a scale of 1:5,000,000 that summarize the geology, tectonics, magnetic and gravity anomaly patterns, regional stress fields, thermal aspects, seismicity and neotectonics of North America and its surroundings. Together, the books and maps of the DNAG Project are the first coordinated effort to integrate all available knowledge about the geology and geophysics of a crustal plate on a regional scale.

The products of the DNAG Project present the state of knowledge of the geology and geophysics of North America in the 1980s, and they point the way toward work to be done in the decades ahead.

Allison R. Palmer
Centennial Science Program Coordinator

Foreword

Because I am writing this foreword in the field, I am reminded that the field experience has always been at the heart of geomorphology. Laboratory and computational experiments make valuable and lasting contributions, but it all starts in places like this. Behind me rise mountains with basalt caps and ringed by debris flows. I sit on a pediment that sweeps down to an alluvial valley floor dotted with saguaro cactus and carpeted with creosote bush. In the distance the crags of the faulted and eroded Eagle Tail Mountains shimmer in the Arizona heat. How old is this landscape? How did it come to be? What processes are even now shaping its forms? Here there are secrets and puzzles enough for a lifetime.

This is a marvelous time to be a gemorphologist. Formally the science is not yet a century old, but it is in one of its most exciting periods. New ideas, new techniques, new data, even new extraterrestrial worlds are in greater abundance than ever before. The emergence of geomorphology to take its rightful place among the global earth sciences was evident a year ago at the First International Conference on Geomorphology in Manchester, England. This book, also a first, celebrates the centennial anniversary of one of the world's foremost earth-science organizations, the Geological Society of America, by collecting together recent geomorphological research from North America.

A small group of geomorphologists in the Quaternary Geology and Geomorphology Division of The Geological Society of America proposed in 1983 that a book be assembled to demonstrate the range and types of research developments in the science. The division appointed Rich Madole, Marie Morisawa, Nat Rutter (chair), and me as a committee to explore alternative structures for the volume. After numerous consultations among ourselves and with our colleagues, we elected to use a regional framework to insure compatibility with other volumes being planned and published by the Society that would emphasize regional geology.

I invited chapter leaders to oversee the production of individual chapters, either by writing entire chapters themselves or by inviting other contributing authors. The committee, Madole, Morisawa, and Rutter provided expertise and sage advice. As editor I established a general policy and performed the role of a gate tender in an irrigation project by channeling the flow of information and ideas. If this volume furnishes the crops of the science with too much of the wrong kinds of intellectual water or not enough of the right kinds, I am solely to blame.

The real credit for this effort belongs to the authors who were the source watersheds of ideas that appear here. They were invited to chronicle the development of geomorphologic theory that was growing from field research in the various geomorphic provinces. In some cases, such development was clearly taking place, while in other areas the science was still collecting the basic pre-theory data. I suspect that most users of this work will read individual chapters rather than read the volume cover-to-cover, so I have not attempted to impose uniformity on the various chapters, authors, and regions. The result is a collection of individual elements, each with the capability to stand alone. The whole is greater than the sum of its parts, however, because of the continental-scale picture of the science that it provides.

The work presented here has been rigorously reviewed and revised, taking into account the input of referees. Two external referees and at least one member of the editorial board reviewed each chapter. The manuscripts benefited from the input of several anonymous reviewers as well as the following.

Charles S. Alexander
Department of Geography
University of Illinois
Urbana, Illinois 61801

William B. Bull
Department of Geosciences
University of Arizona
Tucson, Arizona 85721

Robert W. Blair, Jr.
Department of Geology
Fort Lewis College
Durango, Colorado 81301

Arthur L. Bloom
Department of Geological Sciences
Cornell University
Ithaca, New York 14853

John J. Clague
Geological Survey of Canada
Vancouver, British Columbia V6B 1R8

Richard G. Craig
Department of Geology
Kent State University
Kent, Ohio 44242

William E. Dietrich
Department of Geology and Geophysics
University of California
Berkeley, California 94720

Ronald I. Dorn
Department of Geography
Texas Tech University
Lubbock, Texas 79409

Robert J. Fulton
Geological Survey of Canada
601 Booth Street
Ottawa, Ontario K1A OE8

James S. Gardner
Department of Geography
University of Waterloo
Waterloo, Ontario N2L 3G1

Richard F. Hadley
Department of Geography and Geology
University of Denver
Denver, Colorado 80208

Bernhard Hallet
Quaternary Research Center
University of Washington—AK60
Seattle, Washington 98195

R. Craig Kochel
Department of Geology
Southern Illinois University
Carbondale, Illinois 62901

Harold E. Malde
U.S. Geological Survey
Box 25046, Mail Stop 913
Denver Federal Center
Denver, Colorado 80225

Helaine Walsh Markewich
U.S. Geological Survey
6481 Peachtree Industrial Blvd.
Doraville, Georgia 30360

Marie Morisawa
Department of Geological Sciences
and Environmental Studies
State University of New York
Binghamton, New York 13901

Troy L. Péwé
Department of Geology
Arizona State University
Tempe, Arizona 85287

Norbert P. Psuty
Center for Coastal and Environmental Studies
Rutgers University
New Brunswick, New Jersey 08903

John B. Ritter
Department of Geosciences
Pennsylvania State University
University Park, Pennsylvania 16802

Olav Slaymaker
Department of Geography
University of British Columbia
Vancouver, British Columbia
V6T 1W5

Tom Spencer
School of Geography
University of Manchester
Manchester MI3 9PL
United Kingdom

David L. Weide
Department of Geoscience
University of Nevada, Las Vegas
Las Vegas, Nevada 89154

Garnett P. Williams
U.S. Geological Survey
Box 25046, Mail Stop 413
Denver Federal Center
Denver, Colorado 80225

All of these professionals contributed generously of their time and talent. A. R. (Pete) Palmer and his staff at the headquarters of the Geological Society of America literally made the volume possible with their expert management and editing. To all of these contributors I extend my personal sincere thanks.

In viewing the stark basin and range landscape around me and considering this volume, I have two sets of thoughts. The first is an immense sense of relief and satisfaction that the project is finished. The second is a feeling of excitment and anticipation that the development of geomorphology has just begun.

William L. Graf

In the Gila Bend Mountains
October 1986

Geological Society of America
Centennial Special Volume 2
1987

Chapter 1

Regional geomorphology of North America

William L. Graf
Department of Geography, Arizona State University, Tempe, Arizona 85287

INTRODUCTION

Geomorphology is the study of earth-surface processes and forms. As a component of the fields of geology and geography, geomorphology is particularly concerned with change: progressive changes through time or changing spatial distributions at the earth's surface. Geomorphology as a science views the objects of its study as collections of elements in complex systems that are groupings of forms and related processes. Because these groupings have regional manifestations strongly controlled by geologic materials and structures, the geomorphic systems of North America have a definable regional expression that serves as a method of organizing knowledge. The Geological Society of America's publication series celebrating the Decade of North American Geology and the Centennial of the founding of the organization is generally organized according to geographic regions, so it is appropriate that this companion volume to the series also adopts a regional framework (Figs. 1 and 2).

The geomorphologic divisions of the continent depicted in Figure 2 are broad generalizations. Discussions in subsequent chapters that more specifically define the extent of each division use boundaries suggested by workers several decades ago. In the United States, for example, the modern geomorphologic divisions derive from the efforts of Neville M. Fenneman, who in 1913 proposed to the Association of American Geographers that geomorphic divisions would be useful in research (Fenneman, 1914). Later, a committee chaired by Fenneman and consisting of M. R. Campbell, F. E. Matthes, Douglas Johnson, and Eliot Blackwelder produced a map with three orders of geomorphic divisions for the United States, which was adopted by the U.S. Geological Survey. Its use has continued with only slight modifications. Fenneman (1928) provided an accompanying paper, which remains the definitive work on regional classification of geomorphic systems in the United States.

The emphasis in the following chapters is not on definition and description of regions, but rather on analyses of modern processes and forms. It is not possible, however, to completely separate geomorphology from Quaternary geology because some research questions addressing processes do so on time scales that extend into the middle and late Tertiary. While this volume recognizes the symbiotic relationship of "process geomorphology" and Quaternary geology, the emphasis in the subsequent pages is generally on questions of modern processes. The value of Quaternary studies in this framework is that they inform on the operations of present-day systems.

PURPOSE

The most important purpose of this volume is to bring to the earth science community a review of selected research problems from each of the geomorphic provinces of North America, and to outline the progress of recent efforts at their solutions. A simple accounting of the subjects of research efforts in the 1980s provides a cross section of the field of North American geomorphology, while more detailed discussion of methods and results provides a review of the state of development of the science.

Another major purpose of the volume is to assess the development of geomorphologic theory in the various provinces of the continent. The object of any science is the creation of generalizations, but in geomorphology, theory development is not the same everywhere. In some provinces, the construction of theory using empirical evidence for hypothesis testing is highly advanced, while in other areas, the collection of additional basic observations is required before further progress toward theory is possible.

A reading of *Geomorphic Systems of North America* provides a general review of the strengths and weaknesses in present geomorphologic knowledge about the continent. Users of the volume will find that what we know appears to be a tiny fraction of what we are on the verge of learning, and that many gaps in our knowledge remain to be addressed through research. In other cases, we, as a discipline, seem to have relentlessly pursued some questions well beyond the point of diminishing return for our investment of expertise. It is hoped that the volume can provide some guidance in the selection of future research topics.

Finally, the volume may serve as a reference work by indicating published sources of information about the recent major questions in each province. By limiting discussion to a few significant themes, it is possible to provide a full exploration of the

Graf, W. L., 1987, Regional geomorphology of North America, *in* Graf, W. L., ed., Geomorphic systems of North America: Boulder, Colorado, Geological Society of America, Centennial Special Volume 2.

Figure 1. Landform map-diagram of North America by Erwin Raisz. Published with the permission of Kate Raisz.

published work addressing a few subjects and to relate some material that may have appeared in sources difficult to obtain or with limited distribution. The volume is not a region-by-region literature review because such a task now seems beyond the scope of a single volume. The shear enormity of the general and specific literature precludes a complete review.

The volume is also not a detailed regional description of geomorphic systems of the continent. The only descriptions provided are those necessary to appreciate the scientific questions addressed in each province. References to descriptive literature provide the reader with keys to other sources that have done for restricted regions what was impossible for this volume to do for the entire continent.

PREVIOUS REGIONAL WORKS

Previously published volumes provide a historical perspective on the development of regional geomorphology and descriptive material that is lacking in the present volume. One of the earliest efforts at a regional approach was *The Physiography of the United States* by John Wesley Powell (1896). Representing an attempt to capitalize on the rapid, though embryonic, developments in the field later known as geomorphology, the book includes introductory essays on processes, forms, and geomorphic regions by Powell. His map of physiographic regions is generally similar to most subsequent renditions. The book also contains essays on specific research problems in particular areas. I. C. Russell provided a chapter on present and extinct lakes in Nevada, N. S. Shaler wrote about beaches and marshes on the Atlantic coast, Baily Willis and C. Willard Hayes expounded on the Appalachian region, and W. M. Davis interpreted southern New England. G. K. Gilbert deciphered the dynamics and history of Niagara Falls.

The first comprehensive regional physiography of the United States that included all the major regions in a descriptive and explanatory effort was *Forest Physiography* by Isaiah Bowman (1911). Bowman, director of the American Geographical Society, sought to provide students of forestry with a brief review of basic principles followed by a detailed region-by-region description of the geologic, geomorphic, pedologic, and hydrologic systems that controlled forest development. He relied heavily on the publications from previous surveys and was able to condense information from hundreds of sources into a single lengthy volume that presented a clear statement of the character of the geomorphic provinces. Bowman did not emphasize interpretations or explanations, apparently because he saw his work as a source of information for forestry management decisions. As a result, the book is still useful 75 years after its initial publication as an introduction to the various provinces and subdivisions.

In the *Physiography of the Western United States*, Fenneman (1931) offered a regional description of the geomorphology of the United States, but he also included interpretations and explanations. As a geologist, Fenneman viewed landforms as an end product of geologic evolution. Although some of his interpre-

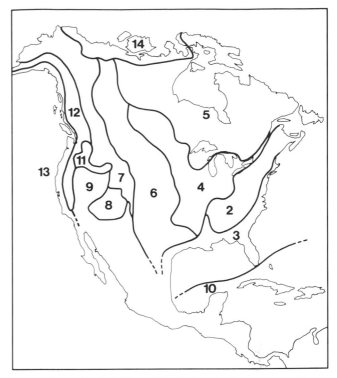

Figure 2. Generalized division of North America into geomorphic provinces. Approximate boundaries outline the geographic scope of each of the subsequent chapters in this book. Numbers correspond to chapter numbers: 2, Appalachian Mountains and Plateaus; 3, Atlantic Coastal Plain; 4, Central Lowland; 5, Canadian Shield; 6, Great Plains; 7, Rocky Mountains; 8, Colorado Plateau; 9, Basin and Range; 10, Central America; 11, Columbia Plateau; 12, Interior Mountains and Plateaus; 13, Pacific Rim; 14, Arctic Lowlands.

tations based on a strict application of W. M. Davis' principles of erosion cycles now appear outdated, many of his reviews of the connections between geologic structures and landforms provide succinct summaries of large bodies of literature yet to be contradicted. The majority of the volume is descriptive, which remains useful. A companion volume, *The Physiography of the Eastern United States,* appeared in 1938, and as a matched set the books became a fixture in the libraries of geologists and geographers for several decades. A shorter rendition of the subject by Loomis (1937) emphasized discussions of origin of landforms, but was less widely read than Fenneman's work.

In 1940, Wallace W. Atwood, a geographer at Clark University, produced *The Physiographic Provinces of North America,* also a shorter version of Fenneman's two-volume effort. Atwood's explanations reflected the continued dominance of the Davisian view of landscape evolution. Atwood's volume is notable for at least two reasons: first, he demonstrated an appreciation of the human consequences of landforms and landform processes, and second, the volume was accompanied by a remarkable map. Tucked in the back cover of the book was a comprehensive map by Erwin Raisz, a talented young cartographer. Using three-dimensional sketch techniques largely of his own design, Raisz produced a map-diagram depicting the surface of the United States and southern Canada that provided a view of landscape impossible to obtain from the ground or even an airplane (for example, see Fig. 1). No more effective visual tool for the continent-wide perspective was available until the advent of photography from orbital spacecraft. Atwood's book went through twelve printings and 24 years of publication, while Raisz's map is still in production in 1986.

The general regional works, from Bowman (1911) through Atwood (1940), developed from physiography, a combined study of geomorphology, soils, vegetation, and climate. William D. Thornbury's (1965) *Regional Geomorphology of the United States* departed from the tradition by limiting discussion to the geomorphic systems. As a geologist, Thornbury was most concerned with the influence of geologic structures and materials in landscape development, but like the authors before him he relied heavily on Davisian concepts for interpretation of the surface forms of various regions. His work includes many post-Davisian concepts, however, and represents the beginning of a transition to a more process-oriented perspective.

Appearing at about the same time as Thornbury's work, Charles B. Hunt's (1967) *Physiography of the United States* continued the tradition of the multi-faceted physiographic approach, but his work represented a much stronger orientation to process and showed less reliance on evolutionary explanations for landscapes. Hunt's work also included a resource perspective and his text contained less description than previous volumes. He later expanded the work into a more comprehensive volume, *Natural Regions of the United States and Canada* (Hunt, 1974), which remains the most comprehensive volume with general coverage.

Descriptive and interpretive volumes provide more detail for some limited areas. Published volumes on physiography and geomorphology appeared between 1896 and 1925 for Missouri, New Jersey, Maryland, Texas, New York, Wisconsin, Oklahoma, Indiana, and Georgia. Additionally, subsequent publications for various geomorphic regions have addressed the educated lay as well as the professional audience.

CONCLUSION

The present volume is not a direct descendant of these earlier works because it seeks to focus on a limited number of research themes within each geomorphic region. The object in the present volume is not extensive coverage, but rather to identify the emerging theoretical developments in each region. As such, it represents a statement about the intellectual development of the science seen against the backdrop of research in specific regions.

REFERENCES CITED

Atwood, W. W., 1940, The physiographic provinces of North America: New York, Blaisdell, 536 p.

Bowman, I., 1911, Forest physiography; Physiography of the United States and principles of soils in relation to forestry: New York, Wiley and Sons, 759 p.

Fenneman, N. M., 1914, Physiographic boundaries within the United States: Annals of the Association of American Geographers, v. 4, p. 84–134.

——— , 1928, Physiographic divisions of the United States: Annals of the Association of American Geographers, v. 18, p. 261–353.

——— , 1931, Physiography of the western United States: New York, McGraw-Hill, 534 p.

——— , 1938, Physiography of the eastern United States: New York, McGraw-Hill, 714 p.

Hunt, C. B., 1967, Physiography of the United States: San Francisco, W. H. Freeman, 480 p.

——— , 1974, Natural regions of the United States and Canada: San Francisco, W. H. Freeman, 725 p.

Loomis, F. K., 1937, Physiography of the United States: Doubleday, Doran and Company, 350 p.

Powell, J. W., 1896, The physiography of the United States: New York, American Book Company, 345 p.

Thornbury, W. D., 1965, Regional geomorphology of the United States: New York, Wiley and Sons, 609 p.

MANUSCRIPT ACCEPTED BY THE SOCIETY NOVEMBER 20, 1986

Geological Society of America
Centennial Special Volume 2
1987

Chapter 2

Appalachian mountains and plateaus

Hugh H. Mills
Department of Earth Sciences, Tennessee Technological University, Cookeville, Tennessee 38505
G. Robert Brakenridge
Department of Geological Sciences, Wright State University, Dayton, Ohio 45435
Robert B. Jacobson*
Department of Geography and Environmental Engineering, The Johns Hopkins University, Baltimore, Maryland 21218
Wayne L. Newell
U.S. Geological Survey, 925 National Center, Reston, Virginia 22902
Milan J. Pavich and John S. Pomeroy
U.S. Geological Survey, 928 National Center, Reston, Virginia 22092

INTRODUCTION: AN OUTLINE OF THE HISTORY OF APPALACHIAN GEOMORPHOLOGY

H. H. Mills

William Morris Davis (1888) considered the Appalachians to have had an important role in the development of geomorphology. He saw the systematic study of topography as largely of American origin, and suggested that there were two steps in its development. The first began about 1840 with the founding of the Eastern state surveys, during which geologists such as Lesley (1856) established the intimate relationship between topography and structure (in the broad sense) in the Appalachians. During this first step, according to Davis, topographic form was regarded as a

completed product of extinct processes. Topography revealed structure, but it did not then reveal the long history that the structure has passed through. . . . The systematic relation of forms to structure, base level, and time; the change of drainage areas by contest of headwaters at divides; the revival of exhausted rivers by massive elevations of their drainage areas; all these consequences of slow adjustments were then unperceived (Davis, 1888, p. 14).

The awareness of such principles, the second step in the advancement of the study of topography, came about largely as a result of the geological exploration of the western U.S., exemplified by the work of Powell (1875, 1876) and Gilbert (1877). Once this awareness came about, however, these principles were applied to the Appalachians; Davis (1888, p. 15) stressed the importance of the Appalachians in this second phase as well as the first:

If it be true that the greater part of this second advance is American like the first, it must be ascribed to the natural opportunities allowed us. The topographers of the Appalachians had a field in which one great lesson was repeated over and over again and forced on their attention. The patchwork structures of Europe gave no such wide opportunities.

McGee (1888a, 1888b) was the first to apply the idea that sequences of erosional and depositional phases of landform development could be applied to the Appalachians, as well as the first to apply the concept of base level. However, Davis was soon to become the dominant influence on Appalachian geomorphology. His first article on the Appalachians, "The Rivers and Valleys of Pennsylvania" (1889), was also the first to treat in detail his concepts of the geographical cycle and peneplains. Despite the brilliance of the article, this early association of the Appalachians with the cycle probably augured poorly for the future of Appalachian geomorphology. As Beckinsale and Chorley (1968) have noted, under the influence of Davis, the admirable aim of elucidating the latest parts of geological history became fused with the cyclical approach.

The stage was soon set for decades to come. The Davisian Cycle proved immensely popular with most geologists, who began to apply it with a vengeance. Only nine years after Davis's seminal paper, Tarr (1898, p. 41–42) complained that peneplains ". . . are being found nearly everywhere. Indeed, they are announced upon the most meager evidence, and oftentimes with no statement of evidence whatever . . . The literature of geology is becoming overburdened with peneplains . . ." This "dark age" of geomorphology lasted in the Appalachians until the 1940s. As

*Present address: U.S. Geological Survey, 926 National Center, Reston, Virginia 22092.

Mills, H. H., Brakenridge, G. R., Jacobson, R. B., Newell, W. L., Pavich, M. J., and Pomeroy, J. S., 1987, Appalachian mountains and plateaus, *in* Graf, W. L., ed., Geomorphic systems of North America: Boulder, Colorado Geological Society of America, Centennial Special Volume 2.

Flemal (1971, p. 4) has remarked, this period was "strikingly sterile," as few new concepts entered geomorphology and landscape processes were almost ignored. Nowhere was the influence of the peneplain concept more profound than in the Appalachians.

The effect of this development on geomorphic research in the Appalachians can be demonstrated by comparing the number of papers on Appalachian peneplains with the number on any phase of Appalachian geomorphology, by decade (Fig. 1). Papers on peneplains peaked in the 1930s, declining rapidly thereafter (Sevon and others, 1983). Corresponding to this decline, the total number of geomorphic papers dropped in the 1940s and 1950s relative to the 1930s. Apparently, geomorphology in this region had become so closely identified with the Davisian Cycle that when the latter fell from favor, so did the Appalachians. Only in the 1960s did the number of papers exceed that of the 1930s, and, in fact, relative to the total growth in the field of geomorphology, the level of Appalachian research of the 1930s has only recently been matched.

To burden Davis with sole responsibility for this long unproductive period is unjust, however. First, Davis's papers on the Appalachians were confined to the period 1889 to 1903, several decades before the peak in Appalachian peneplain studies. Hence, Davis's disciples, rather than he himself, were to blame. Second, although many of his followers were obsessed with peneplains, Davis's papers were brimming with many ideas other than the geographical cycle. Hack (1976), for example, has noted that many of Davis's concepts, concerning such subjects as the formation of wind and water gaps, headward migration of divides, and stream piracies, fit in perfectly well with an equilibrium model of landscape development. It was Davis's followers who chose to give so much emphasis to the cycle while paying scant attention to Davis's other ideas.

Although peneplains and the cycle were pervasive in Appalachian geology during this period, it should be noted that other, more productive, work was also done. Large-scale drainage evolution was a favorite subject; although it is true that such studies often made reference to erosion cycles. Perhaps more importantly, advances in the understanding of the adjustment of landforms to structure continued to be made, to some extent laying the foundation for Hack's later work. Understanding of the nonglacial surficial mantle remained rudimentary, however, and work on geomorphic processes was almost nonexistent.

The decline in Appalachian peneplain studies that began in the 1940s paralleled that elsewhere, of course, and basically was due to a changing outlook in geomorphology. First, even by the 1930s, the cyclical concept was beginning to fall into disrepute, as discussed by Beckinsale and Chorley (1968). Second, geomorphologists began to acquire new interests. Attention turned from the effect of time on landscape to that of process and structure. As Mark (1980) pointed out, the interest in process was frequently accompanied by work on small-scale landforms, as opposed to landscape-scale landforms that occupied the attention of earlier workers. An interest in quantification developed, inspired in part

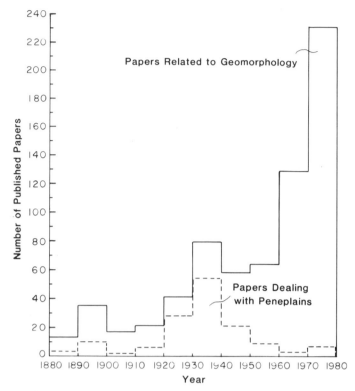

Figure 1. Histograms comparing number of papers dealing with peneplains in the Appalachians with number dealing with any aspect of Appalachian geomorphology, by decade, from 1890 to 1980. Data on peneplain papers is from Sevon and others (1983). Data on total geomorphology papers was compiled by the use of the Bibliography of North American Geology and GEOREF, plus the use of bibliographies of papers found in this manner. These numbers are to some extent arbitrary, because many papers are only partly concerned with geomorphology, and many are on the periphery of geomorphology. The admittedly subjective criterion for inclusion was that at least part of the report must deal "substantially" with geomorphic questions. Dissertations, theses, and abstracts are not included. Routine compilations, such as landslide maps, were not included unless accompanied by interpretative reports.

by Horton (1945). Finally, the concept of equilibrium, as exemplified by Strahler (1950), became influential.

Post-1950 research in the Appalachians at first incorporated only some of these new directions. Although quantification increased, with a few exceptions (e.g., Wolman, 1955), little work on processes was done, the emphasis remaining on landscape-scale forms. Basically, research from the 1950s to the early 1970s was dominated by reinterpretations of landscapes previously explained by the cyclic approach in terms of "dynamic equilibrium" (Hack, 1960). That is, efforts were made to show that landforms are adjusted to the bedrock underlying them and the processes acting on them. Although the "processes" in these studies were mainly of an inferred nature, much new interest developed in the relationship between structure and landforms during this period.

Although he attributed the concept of "dynamic equilibrium" to Gilbert (1877), Hack (1960) was one of the chief advocates of this concept during this period, and did much to popularize it (e.g., Hack, 1960, 1965, 1966; Hack and Goodlett, 1960). It is probably no coincidence that the strongest proponent of the equilibrium concept in landform analysis during this period did much of his work in the central Appalachians. Landscapes affected by glaciation, highly active tectonics, or Quaternary sea-level changes, for example, commonly undergo dramatic modifications that obscure any tendency to equilibrium. In the old landscapes of the Appalachians, however, topography commonly has had time to make a close approximation to dynamic equilibrium.

Geomorphic research in the Appalachians during the last decade or so has been dominated by practical applications. Geohazards, especially earthquakes and landslides, have been of particular interest. Study of surficial sediments and landforms in order to detect Cenozoic faulting has been undertaken on an intensive level by the U.S. Geological Survey (e.g., Mixon and Newell, 1977; Reinhardt and others, 1984), and the Survey's ground-failure hazards reduction program (U.S. Geological Survey, 1982) has resulted in extensive mapping and investigation of slope failures, particularly in the Appalachian Plateau (e.g., Pomeroy, 1980, 1982a). Similar investigations have been pursued by state surveys.

Floods, of course, have been a perennial concern to mankind, and research on this topic continues (e.g., Costa, 1974; Renwick, 1977; Moss and Kochel, 1978). Of more recent interest is the effect of land use on erosion and sedimentation (e.g., Trimble, 1974, 1977; Costa, 1975; Arnold and others, 1982).

Caves and other karst landforms have long been of interest to geomorphologists as well as speleologists, and this area of research now has considerable practical application. Karst subsidence has become a particular problem in areas where groundwater tables are being lowered (e.g., Newton, 1976). Karst terrains are also particularly prone to local flooding (e.g., Crawford, 1981; Kemmerly, 1981). In addition, the unpredictable drainage pathways in karst areas also make such terrains of environmental concern (e.g., Quinlan and Rowe, 1977, 1978).

More academic topics have not been neglected, however. Archaeological investigations have provided the main opportunity for terrace-chronology studies in the region (e.g., Delcourt, 1980; Brakenridge, 1984). Other areas that have received attention include relict periglacial deposits (e.g., Sevon and others, 1975; Walters, 1978), debris slides and resulting deposits (e.g., Williams and Guy, 1973; Kochel and Johnson, 1984), geochemical weathering studies (e.g., Cleaves and others, 1970; Cleaves and others, 1974; Velbel, 1985; Pavich, 1986), karst studies (e.g., Crawford, 1984; Kemmerly, 1982, 1986; White and White, 1979, 1983), structural geomorphology (e.g., Hack, 1980, 1982), and long-term evolution of landforms (e.g., Costa and Cleaves, 1984).

This chapter presents selected research that is thought to be of importance for Appalachian geomorphology, either because of historical significance or current interest. Much omitted research is equally important, and the chapter leader apologizes to readers whose fields of interest have been slighted. In particular, the decision was made to emphasize the unglaciated part of the Appalachians. Although glacial geomorphology represents a major field of research in the Appalachian region as defined here, it has been treated in a number of symposia volumes, unlike nonglacial topics, which have received little such coverage. The present chapter partially corrects this imbalance.

THE ENDOGENIC IMPRINT ON LANDSCAPE IN THE VALLEY AND RIDGE PROVINCE

H. H. Mills

INTRODUCTION

Endogenic influences include structure, lithology, and past and present tectonics. These effects may be subsumed under the term "structural geomorphology", where structure is used in the broadest sense. Forced for decades to take second place to the cycle, this field now lingers in the shadow of process geomorphology. Nevertheless, structural geomorphology has played such an important role in the development of Appalachian geomorphology, and is so well illustrated in this province, that no review of geomorphology in this region could justifiably omit it.

Davis (1888), as mentioned above, thought the Appalachians important in the development of geomorphology because of the insights they allowed into the close relationship between structure and topography. It is ironic that having so recognized the importance of structural geomorphology, Davis (1889) almost immediately launched a view of landscape development that would greatly subordinate the role of structure to stage. And it is doubly ironic that the adjustment of landforms to structure would eventually be used by Hack and others as a cudgel to attack the cycle. As might be expected, we find research on structural geomorphology in the Appalachians concentrated in the pre- and post-cycle periods, although many insightful studies were made during the barren cycle years, as will be mentioned below. Structural control of landforms is most obvious in the Valley and Ridge, and has also received the most attention here. The discussion of structural geomorphology will therefore concentrate on this province.

RELATION OF RELIEF TO LITHOLOGY AND STRUCTURE

Geologists recognized very early the effect of resistance and thickness of a bedrock unit on relief, and the effect of dip was also soon realized. Lesley (1892, p. 696, but formulated decades earlier) referred to the latter effect as a "law of topography," that ". . . the flatter the rocks the higher the mountain; the steeper the dip the lower the mountain." Using such rules, Lesley (1892, p. 685), for example, was able to explain the following in the Valley and Ridge of central Pennsylvania:

first, why these mountains sometimes run in straight lines parallel to each other for many miles; secondly, why these parallel lines sometimes come

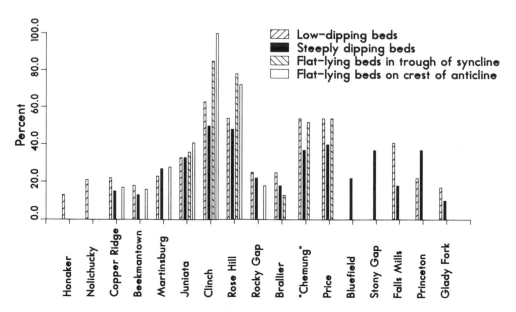

Figure 2. Cooper's "topography potency" of the various ridge makers in the Burkes Garden Quadrangle, Virginia. Both lithology and structure are incorporated in this scheme. Stratigraphic names do not necessarily correspond to present usage (after Cooper, 1944, Fig. 6).

together at both ends, as do the gunwales of a boat at the prow and stern; thirdly, why in other districts they unite in a series of zigzags; fourthly, why the opposite points of such a series of zigzags have two totally different characteristics, one long and sloping gradually into the plain, the other high, sharp, and abrupt, projecting like a knob into the air. . . .

As detailed knowledge of elevations became readily available, additional relationships were realized. Ashley (1935, p. 1422) formulated seven rules for predicting ridge-crest elevations: (1) a low-dipping monocline is higher than a steeply dipping monocline; (2) an anticline is higher than a monocline; (3) a broad anticline is higher than a narrow anticline; (4) a syncline is higher than an anticline; (5) two monoclinal ridges close together are higher than the same ridges well separated; (6) the point of juncture of two monoclinal ridges of a syncline or of a breached anticline is higher than either ridge elsewhere; and (7) any change in the structure of a ridge is registered in either higher or lower elevations.

A number of attempts to express relationships between relief and bedrock characteristics in a semiquantitative manner have been made, based for the most part on the relative elevations of landforms underlain by the various rock types and structures. Hayes (1899, Fig. 1) for example, presented a curve illustrating the relation of relief to lithologic composition in the Chattanooga district. Cooper (1944, Fig. 6) rated the "topographic potency" of the various ridge makers in the Burkes Garden Quadrangle, Virginia, in a manner that reflects structure as well as lithology (Fig. 2). Thompson (1939, 1941) used area-altitude distributions as an index of the resistance of rock formations. Such efforts

presaged later, more quantitative efforts to relate topography to bedrock (e.g., Chorley, 1969), very few of which, unfortunately, have been done in the Appalachians. Dinga (197) quantified the relationship between dip angles of homoclinal ridges and their height above the local base level. For seven areas in the Valley and Ridge, he found correlation coefficients ranging from 0.68 to 0.97.

As discussed below, water and wind gaps in the ridges were of great interest in arguments concerning the evolution of drainage in the Appalachians. One important controversy was over how consistently the gaps were located at points of weakness along the ridges. Earlier investigators such as Rogers (1858) attributed the gaps to transverse faults. Strahler (1945) insisted that this was rarely the case. More recent studies show that while the cause might not be as simple as a transverse fault, gaps commonly are localized by particular structural or lithological features. Epstein (1966), from work in the Stroudsburg area of eastern Pennsylvania, for example, found that each of the six major gaps in this area possessed some combination of the following features: (1) dying out of folds over a short distance, and associated abrupt changes in strike of bedding; (2) steep dip and narrow width of outcrop of resistant strata; and (3) folding that was more intense than in the surrounding rocks. None of these features was present along the ridges where there were no gaps. Epstein and Epstein (1969) noted additional correlations between structure and gaps along Blue Mountain, and Clark (1967) found that at the Wills Mountain anticlinorium, West Virginia, four out of five water gaps in which the Tuscarora Formation is exposed show faulting or tight monoclinal flexuring. Based on these studies, it seems probable that many gaps are localized by structure.

RELATION OF STREAM CHARACTERISTICS TO LITHOLOGY AND STRUCTURE

Although earlier investigators were well aware that stream characteristics were related to bedrock characteristics, Hack (1957) was the first to study this relationship in a systematic fashion.

Longitudinal gradients of streams flowing on alluvium reflect, in large part, the size of the stream and the size of bedload particles. In a study of rivers in several Appalachian provinces, Hack (1957) found that stream slope can be predicted fairly well by the equation

$$S = 18 \, (M/A)^{0.6}, \qquad (1)$$

where S is channel slope in feet/mile, M is median particle size of the bed material in millimeters, and A is drainage basin in square miles. Hack made a detailed effort to analyze the factors that control the changes in the size of bed material along the stream, using detailed size-distribution analyses in which stream-bed samples were separated into lithologic components. He found that coarse material enters the stream wherever the valley walls are steep and composed of bedrock. The size of bed material at any point along a stream is determined by the distance from such a source, the initial size of the material, and the relative resistance of the material to abrasion and breakage. Because streams flowing on a given rock type are likely to flow through similar topography and to have similar bed material, they are likely to have similar profiles where stream size is held constant. In the Shenandoah Valley, for example, for a given stream length, streams on sandstones have slopes about seven times as steep as those of streams on Martinsburg Shale (Fig. 3). Streams in carbonate areas show intermediate slope values, with slopes on cherty dolomites generally being somewhat steeper than those on limestones.

Hack (1957, 1980) has pointed out that as longitudinal profiles of streams are indicative of the relief of an area and are closely related to its topography, a geomorphological analysis of a region based upon a comparison of stream profiles is of value. Consider, for example, the variation of regional relief. Although, as described above, local relief can be explained by local variations in lithology and structure, regional relief is heavily dependent upon the profiles of master streams that traverse the area. Surveying the Appalachian Highlands as a whole, Hack (1980) noted that generally the higher areas are on the most resistant rocks, but that anomalies do exist. One of the largest anomalies occurs in the Great Valley of southwest Virginia, where the crest of the Appalachians crosses the nonresistant Cambrian and Ordovician sequence at an altitude of more than 730 m. At the Hudson River, for comparison, the Cambrian-Ordovician belt is almost at sea level.

At least part of this anomaly can be explained by adjustments of the major streams crossing the Cambrian-Ordovician belt to bedrock downstream from the belt. Figure 4 shows a

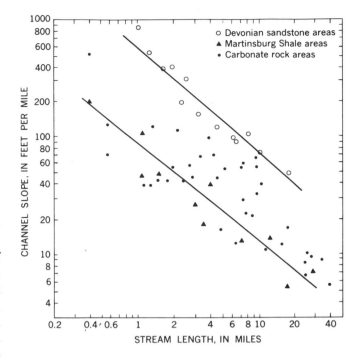

Figure 3. Relation of channel slope to stream length (measured from the source) on different kinds of bedrock at various localities for streams in the Shenandoah Valley (Hack, 1965, Fig. 5).

profile along a series of stream reaches that flow along strike in the belt. Note that the profile is divided into concave-upward segments, each of which is part of the drainage basin of a large stream that exits in the belt. The high point is near Rural Retreat between headwaters of the Tennessee and New Rivers. This striking divide can be explained by differences in the geology of the Tennessee River drainage basin and the New River basin (Hack, 1973). The Tennessee system flows from the divide all the way to Chattanooga through a broad valley in shale and carbonate rocks. Although it crosses resistant sandstone beds in a short reach below Chattanooga, almost its entire coarse is in nonresistant rocks. The New River, in contrast, heads in the Blue Ridge and crosses the Cambrian and Ordovician belt at a narrow point. It then flows northward toward the Ohio River, crossing thick sequences of Pennsylvanian sandstones. This barrier accounts for its high elevation near Rural Retreat. Northeast of the New River, the regional drainage is southeast to the Atlantic. Although these streams must cross hard-rock barriers to reach the Piedmont lowland, the hard-rock outcrop widths are narrow compared to those crossed by the New River. The barriers between the Cambrian-Ordovician belt and the Piedmont generally decrease to the northeast, and the stream profiles become correspondingly lower. Thus, as Hack (1980) noted, the seemingly anomalous topography of this belt can be explained without recourse to differential uplift.

Channel pattern can also be affected by bedrock. Several authors have described the spectacular elongate incised meanders

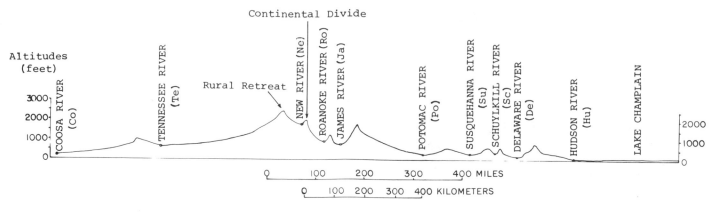

Figure 4. Profile of the Cambrian and Ordovician belt from Coosa River, Alabama, to Lake Champlain, New York (from Hack, 1980, Fig. 4).

that occur along some rivers in the Valley and Ridge, especially those flowing on shale (e.g., Strahler, 1946; Fisher, 1955; Hack and Young, 1959). The most comprehensive study, however, is that by Braun (1983). Braun investigated 78 reaches in the Valley and Ridge Province from Pennsylvania to Tennessee and found that meander lengths increase with bedrock erodibility, where stream size is held constant. Specifically, he found that meanders cut in shaly lithologies (ML = 105 $Q_f^{0.50}$, where ML is meander length and Q_f is most probable annual flow) are about twice the length of meanders cut in thick-bedded to massive lithologies, typically carbonates (ML = 39.3 $Q_f^{0.56}$; Fig. 5). In addition, the valley-floor width of the meanders cut in shaly bedrock (VF = 28 $Q_f^{0.43}$) is two to three times wider than the valley-floor width of the meanders cut in nonshaly bedrock (VF = 8 $Q_f^{0.50}$). No other factors besides the lithologies appeared to help explain these differences.

Drainage-basin geometry and stream-network topology have been investigated in this province. Whereas the flat-bedded Appalachian Plateau has been used as an area likely to minimize the influence of structure on such studies, the Valley and Ridge Province has been investigated with the intent of demonstrating the effect of structure. Miller (1953) made one of the first basin-geometry studies in this province on Clinch Mountain, Virginia and Tennessee. The mountain is a long homoclinal ridge, capped by the Clinch Sandstone, which overlies less-resistant sandstones, shales, and limestones. The dip angle varies along strike from less than 20° to greater than 60°. Hence, Clinch Mountain provides a setting in which, as lithology remains constant, the effect of dip angle alone can be ascertained. Miller found that, on the southeast (dip) flank, stream lengths and basin area show good inverse correlation with dip (Fig. 6, and that hypsometric curves yield significantly lower mean integrals where dips are higher. Drainage density and basin circularity, however, show little relation to dip. Second-order basins were more sensitive to dip than first-order basins. Flank-slope profiles also varied as a function of dip, being concave upward or straight where dip was steep, close to straight where medium, slightly convex where low to medium,

and very convex upward where low. On the northwest (outcrop) slope, however, dip shows little control of basin morphology.

Several studies have investigated network topology of trellis drainage in the province. Surprisingly, earlier studies found that the topology of these striking patterns is not significantly different from those predicted by the random topology model (Smart, 1969; Mock, 1971) or were only subtly different (Mock, 1976). Abrahams and Flint (1983), however, have demonstrated that the topological properties of trellis networks in the Valley and Ridge of central Pennsylvania are indeed strongly influenced by the underlying geology. Perhaps most significantly for structural geomorphology, the authors were able to show preferential development of tributaries flowing downplunge or downdip.

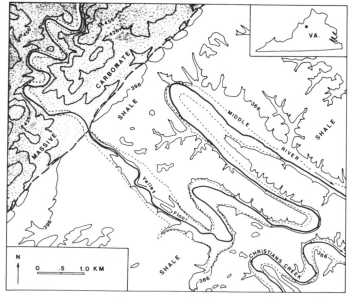

Figure 5. Map of Middle River near Staunton, Virginia, showing the effect of bedrock on meander and flood-plain size. The meanders cut into shaly units are about twice the size of the meanders cut into thick-bedded to massive carbonates. The flood plain (outlined by fine dashed line) cut in the shale is typically about two or three times as wide as that cut in the massive carbonates (Braun, 1983, Fig. 1).

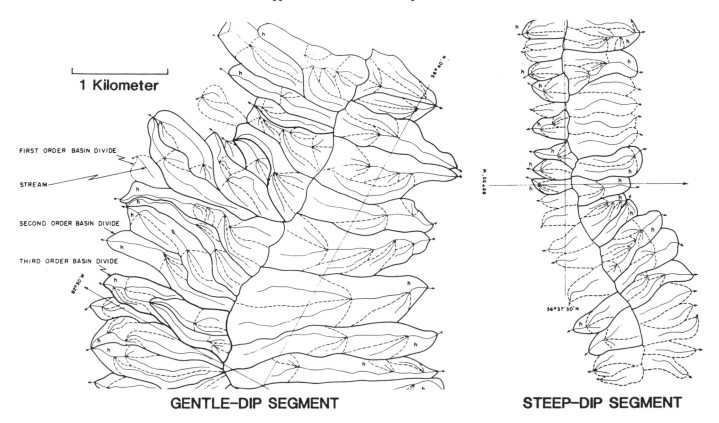

1 Kilometer

FIRST ORDER BASIN DIVIDE

STREAM

SECOND ORDER BASIN DIVIDE

THIRD ORDER BASIN DIVIDE

GENTLE–DIP SEGMENT   STEEP–DIP SEGMENT

Figure 6. Drainage map of part of Clinch Mountain, Virginia and Tennessee, demonstrating the effect of dip on drainage-basin geometry (from Miller, 1953, Plate 2). Dip direction is to the right.

INDIRECT EFFECTS OF LITHOLOGY AND STRUCTURE: THE INFLUENCE OF TRANSPORTED REGOLITH ON TOPOGRAPHY

Where surficial deposits are substantially more resistant to erosion than is the bedrock they overlie, they can greatly affect the course of erosion in the vicinity. If such deposits result from continental glaciation or from an invasion of the sea, their effect on erosion, of course, has nothing to do with structural geomorphology. Where the deposits are locally derived, however, their effect on topographic evolution represents an extension of the effect of resistant rocks to neighboring areas. Although not previously included as a part of structural geomorphology, such effects logically seem to belong in this field. Examples of these effects are abundant in the Valley and Ridge, owing to the dramatic variability in bedrock erodibility in this province.

"Piedmont alluvial aprons" (Hack, 1965, p. 53) are excellent examples. These pedimentlike features occur along the margins of lowland areas (underlain commonly by shale and limestone) where they abut mountains of sandstone or quartzite (Fig. 7). According to Hack (1965, p. 53–54), a piedmont alluvial apron

consists of a broad plain cut on bedrock covered with a thin sheet of gravel arranged with other sheets of various ages in a fanlike complex of

dissected and undissected plains at the mountain front . . . [the deposit] tends to grow laterally because the mountain cobbles are more resistant to corrasion and corrosion in the stream than are the softer rocks of the valley. Thus the stream has a tendency to widen its valley. . . . The gradient of the alluvial surface is adjusted to the discharge and load of the mountain stream. Because of the size of the load, this gradient is generally steeper than the gradient of a smaller adjacent lowland stream, which erodes a narrower valley at the faster rate. Eventually a piracy occurs, and the lowland stream receives the load of coarse debris formerly deposited on the adjacent apron. The lowland valley is now alluviated but of course is never filled to the same level as the abandoned alluvial apron. As the process of degradation by the valley streams, piracy, and alluviation continues, the piedmont apron is gradually spread and becomes a complex of gravelly flood plains, terraces, and dissected terraces.

An analogous phenomenon on a smaller scale is seen on the flanks of many of the major ridges in this province. For the most part, these ridges are capped with resistant sandstones or quartzites, underlain by relatively nonresistant shale and limestone. Where the ridges are sufficiently high, their flanks are corrugated by small streamless valleys. Hack and Goodlett (1960) called these valleys *hollows* and the divides between them, *noses*.

Observation suggests that the form of these hollows is greatly affected by the lithology and structure of the caprock. All caprocks are of course resistant to some degree, but they can be classified into two types, according to whether or not they break

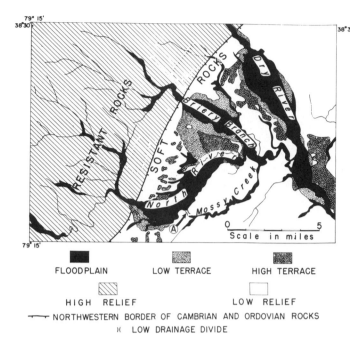

FLOODPLAIN   LOW TERRACE   HIGH TERRACE

HIGH RELIEF   LOW RELIEF

NORTHWESTERN BORDER OF CAMBRIAN AND ORDOVIAN ROCKS
)(LOW DRAINAGE DIVIDE

Figure 7. Simplified map showing areas of resistant and nonresistant (soft) rocks and their relation to alluvial terraces on the northwest side of the Shenandoah Valley, Virginia, to illustrate the formation of piedmont alluvial aprons (Hack, 1960, Fig. 4).

down into large boulders that remain intact as they move downslope. Thin-bedded sandstones, for example, do not produce boulders. Thick-bedded, poorly cemented sandstones produce large boulders, but these boulders rarely move far from the outcrop without disintegrating. Structure is also a factor, of course, for boulders generally are much more plentiful on outcrop than on dip slopes.

The northwest slopes of Brush and Sinking Creek mountains, Virginia, provide a good example of this effect (Fig. 8). Brush Mountain is capped with the McCrady-Price Formation, which at this location does not produce resistant boulders. The hollows on the northwest flank of this mountain are deep, V-shaped, and symmetrical. Sinking Creek Mountain, in contrast, is capped with the Tuscarora Sandstone, which produces large, resistant boulders. These boulders move down the ridge flanks and into the hollows, whose floors they commonly cover. Note (Fig. 8) that these hollows, for the most part in the shale of the Martinsburg Formation, are shallow and commonly asymmetric. They also appear to be somewhat less regularly spaced. The cross-sectional form of these hollows apparently arises from the armoring effect of the boulders, which prevents vertical incision during the rare flows of water in the hollows. Instead, erosion from these flows is directed at the nonresistant shale sides of the hollows, which results in the migration of the hollow to one or both sides, rather than in its deepening. This process has been discussed in more detail elsewhere (Mills, 1981a).

Miller (1953) has suggested that large boulders may also affect the gradients of hollows and thereby the hypsometric integral of drainage basins on the mountain flank. From his observations on Clinch Mountain, he inferred that the protective role of the boulders may decrease sapping of the weaker rocks from beneath the cap rock, and thereby decrease the concavity of slope that would otherwise result.

ORIGIN OF TRANSVERSE DRAINAGE IN THE APPALACHIANS: SOME OLD AND NEW PERSPECTIVES

A classic geomorphic problem in the Appalachians is the origin of the major streams that drain across Appalachian structures to the Atlantic. There are two basic problems. First, the sedimentary sequences laid down during the Appalachian orogenies testify to a westerly or northwesterly drainage, originating in the Piedmont or farther east. The question arises, then, of just how the drainage reversed to its present southeasterly direction. Second, many master streams flow in deep gorges through ridges of resistant rock, with the Valley and Ridge of Pennsylvania having the most dramatic examples. The problem of how the streams were able to cut through such obstacles has fascinated many geomorphologists. The purpose here is to review briefly the hypotheses that have been proposed to explain these phenomena, and to reevaluate them in light of recent advances made in the understanding of the earth's crust.

Hypotheses of transverse drainage development have been classified in different ways (Strahler, 1945; Thornbury, 1965; Oberlander, 1965), but for discussion purposes can be subsumed under the categories consequent, subsequent, and superposed. No purely consequent hypothesis has been proposed, but those of Davis (1889) and Meyerhoff and Olmsted (1936) are "modified consequent" hypotheses (Strahler, 1945). Davis's theory, although one of the oldest, is also the most intricate, and cannot be characterized in a few words. Its main aspects, however, include partial drainage reversal by warping and faulting during Triassic time, additional captures of headwaters portions of the Anthracite River (the hypothetical original master stream that flowed to the northwest) by steeper-graded streams flowing into the downfaulted Triassic lowland, and establishment of the lower portion of reversed streams on a Cretaceous coastal-plain cover hypothesized to extend inland farther than today. Problems of this theory include the absence of any vestige of the Anthracite River and the attribution of drainage reversal to Triassic faulting. Davis recognized the latter as a problem because the northwestward dip of the Newark-age rocks seemed to imply an uptilting from the southeast, which would have opposed the development of southeastern drainage. Oberlander (1965) noted that the theory also fails to account for transverse cuts through the noses of synclinal mountains.

Meyerhoff and Olmsted (1936) modified Davis's theory by postulating that the drainage divide in Triassic time lay northwest of the Great Valley, owing to thrust sheets and overturned folds

Figure 8. Topographic map demonstrating the effect of caprock type on shape of ridge-flank hollows. McCrady-Price Formation capping Brush Mountain supplies few large boulders and northwest flank of ridge A shows deep, V-shaped, symmetrical hollows. Tuscarora Sandstone capping Sinking Creek Mountain supplies abundant large boulders and northwest flank of ridge B displays shallow, commonly asymmetrical hollows. From U.S. Geological Survey Newport, Virginia, 7½-minute Quadrangle.

that have since been eroded away, and that consequent drainage developed on the surfaces of these features. As Mackin (1938) has pointed out, however, reconstruction of the proposed topography from the present structures is purely speculative, and field evidence does not support certain implications of the theory.

Subsequent-drainage hypotheses involve exploitation of weak rocks and/or westward shifting of drainage divides. The earliest of these is that of Rogers (1858), who assumed that southeast-flowing streams are subsequent along transverse fault lines. This hypothesis fails to account for the reversal of drainage, however, as well as for water gaps that are not along fault lines. The major subsequent hypothesis is the "progressive piracy" one of Thompson (1939). He assumed that the original divide was along the Blue Ridge and Reading Prong, and has been migrating westward as a result of the steeper gradients of streams draining to the Atlantic, aided by repeated eastward tilting of the Atlantic slope. He noted that the amount of migration increases to the north, and attributed this to greater tilting and narrower belts of crystalline rocks to the north. Thompson stressed the exploitation of favorable locations in this migration, noting that water gaps commonly occur where the ridge-forming strata are thin or weak. The major problem is how streams were able to effect capture through ridges of resistant rock, a problem Strahler (1945) thought insuperable, for it would require a great elevation difference on the two sides of the ridge. Thompson's (1939) example of stream capture along the high escarpment of the southern Blue Ridge is convincing, but it is questionable whether such asymmetry existed elsewhere along the migrating divide. Obviously the previous discussion indicating that gaps are commonly controlled by structural factors is favorable to Thompson's hypothesis, but nevertheless probably does not outweigh Strahler's objections.

Superposed drainage in the Appalachians has been proposed by several authors. Thompson (1939) proposed local superposition of some streams by downcutting from a weaker formation, and Meyerhoff and Olmsted's (1936) theory involved superposition from now-removed folds and thrust sheets. Johnson (1931) proposed regional superposition from an unconformable cover mass. According to this concept, an invasion of the Cretaceous sea extended westward over an old-age surface and deposited coastal plain sediments over what is now the Valley and Ridge, obliterating all former drainage lines. After withdrawal of the sea, a system of southeast-flowing consequent streams was initiated on the Cretaceous cover. These streams subsequently were superposed on the folds of this province. The basic problem with this theory is that there is no evidence that such an extensive invasion of the Cretaceous sea took place. Groot (1955), in fact, noted that the petrology of Cretaceous sediments in Delaware indicates a Piedmont source, suggesting subaerial exposure of this province during the Cretaceous. Thompson (1939) has raised other objections based on a contrast between actual topography and that which might be expected from the theory.

The advent of plate-tectonics research has greatly increased our understanding of orogens, providing new criteria for evalua-

14　　　　　　　　　　　　　　　*H. H. Mills and Others*

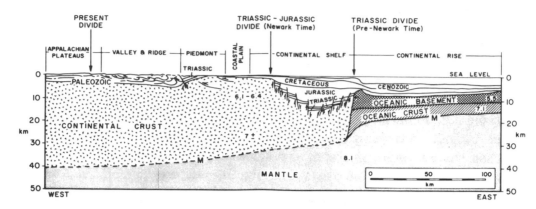

Figure 9. Geologic section of the eastern margin of the North American continent and adjacent oceanic crust in the New Jersey–Pennsylvania area. The presumed westward migration of the main east-west divide is shown (Judson, 1975, Fig. 2).

ting the above-mentioned hypotheses. Judson (1975), for example, points out that Davis's (1889) difficulty with the northwestward dip of the Newark rocks seems to be resolved by our present knowledge of divergent plate boundaries. By analogy with the East Africa and Red Sea rifts, there was *uplift* during the Triassic rather than subsidence, as Davis thought. Downward motion did occur along rift valleys, but the heights flanking these valleys serve as divides. This thermally induced arching was probably superimposed on whatever Paleozoic elevations may have been present. The reversal of stream flow to a southeasterly direction probably began in the Jurassic as the Atlantic began to open, and resulted from subsidence of the continental margin. Before the subsidence, the east-west divide lay somewhere along the crest of the thermal ridge. With initial rifting the divide moved toward the western edge of the rift valley. With subsidence it moved farther and farther to the west (Fig. 9). Hence, there is an endogenous mechanism for westward shifting of the drainage divide, although, of course, the divide may have been pushed still farther west by erosion.

A common approach to geomorphic problems involving vast time intervals is that of substituting space for time. For the problem of Appalachian drainage this approach would consist of studying drainage development in an orogen similar to the Appalachian orogen, but in an earlier stage of development. Oberlander (1965, 1985) has done just this for a very recent orogen that is similar in many ways to the Appalachians, the Zagros Mountains of Iran. Because erosion on fold surfaces here ranges from almost nil to advanced, Oberlander was able to ascertain the manner in which drainage evolves on a fold belt.

Oberlander found that none of the classical drainage hypotheses of transverse drainage could account for the anomalous relationships between drainage and geology in the Zagros, but that every anomaly in the region could be explained by some variant of one or another of the theories. Regarding the theory of modified consequent drainage, he reported that shifts of consequent drainage predicted by Davis (1889) to accompany relief

inversion did not occur in the manner suggested. As for superposition from an unconformable cover mass, there is no evidence of such a cover, and unlike in the ancient Appalachians, it is much more difficult to claim that the cover mass has been removed by erosion.

Oberlander's most exciting insight involves superposition from conformable covers of nonresistant clastics. Although such superposition has been appealed to before (Meyerhoff and Olmsted, 1936), Oberlander demonstrated that it may involve wide areas and, in addition, may be a general phenomenon in orogens. His reasoning is as follows. Thick masses of erodible flyschlike deposits typify all continental-margin orogens. This flysch is likely to overlie, and to be overlain by, resistant units. As folding and uplift proceeds, the anticlines will be the first to have their cap rocks breached, owing to their greater height and fracturing of their crests. Unroofing leads to relief inversion with growth of subsequent valleys in the flysch. Whether such valleys are confined to individual anticlines, or can link together adjacent anticlines, largely depends on the thickness of the flysch relative to the amplitudes of the folds. If the latter is greater, then the subsequent lowlands of individual anticlines will rarely merge (Fig. 10, top), whereas if flysch thickness exceeds the fold amplitude, such merging will be extensive (Fig. 10, bottom). Merging is greatest along transverse highs, and it is here that subsequent streams on the flysch have the best chance of establishing themselves across the fold belt. Eventually these streams are let down onto the resistant cores of anticlines, producing transverse gorges. Oberlander reported that half the transverse gorges in the central Zagros originated in this manner.

Although sequences of erodible clastics are not as thick in the Appalachians as in the Zagros, an analogous interpretation can be made for parts of the Appalachians (Fig. 11). Oberlander (1985) pointed out that despite transecting the anticlinal axes north of Harrisburg, flowing against synclinal plunges, and oscillating in and out of single anticlinal structures, the Susquehanna River is in fact closely attuned to geologic structures, if past outcrop patterns are considered. He noted that where the West Branch and the trunk stream are transverse to the structural grain, they lie between—and conspicuously parallel to—the transverse axes of the major anticlinorium and synclinorium in the fold belt. This course suggests that the river may have evolved in a conformably folded erodible Paleozoic cover mass. Unlike the Zagros, this unit was not thick enough to produce an uninterrupted strip across the nearby anticlinorium, but produced a broad lowland between resistant outcrops on the east-plunging flanks of the anticlinorium. Oberlander suggested that the most appropriate unit for transverse stream development would have been the relatively nonresistant Mauch Chunk Formation (Upper Mississippian), between the resistant Pottsville (Lower Pennsylvanian) and Pocono (Lower Mississippian) sandstones and conglomerates. The Mauch Chunk in central Pennsylvania has a thickness exceeding local fold amplitudes in many areas, although not in the anticlinorium west of the Susquehanna River.

Oberlander's study also sheds light on Thompson's (1939) concept of divide migration by headward-eroding streams. Oberlander noted that such migration indeed takes place, but not in the way Thompson envisaged it. Whereas Thompson thought streams extended headward on landscapes with fully integrated drainage systems, in the Zagros, migration consists of the headward extension of consequent streams on the initial deformational surface. Such extension is not along preexisting erodible outcrops, but is simply the expansion of consequent gullies into axial basins, where growth is controlled by scarp retreat following breaching of the anticline.

NEOTECTONICS AND APPALACHIAN GEOMORPHOLOGY

The effect of neotectonics on geomorphology of the Appalachians is problematic. Neotectonics has received a great deal of attention in the eastern United States, particularly the study of Cenozoic faults by the U.S. Geological Survey (e.g., Mixon and Newell, 1977; Prowell and O'Connor, 1978; Reinhardt and others, 1984). For the most part, such faults have been located near the Piedmont–Coastal Plain boundary, and no Holocene or even well-dated Quaternary faulting has been demonstrated. No such faults have been found in the Valley and Ridge.

Apparent vertical crustal movements have been inferred from releveling data (Brown and Oliver, 1976; Brown, 1978; Brown and others, 1980). These data suggest that the Appalachian Highlands are rising relative to the Atlantic Coast at rates up to 6 mm/yr. The rate maxima correlate strongly with topographic highs, suggesting that, if the uplift is indeed real, that at least some of the Appalachian topography owes its existence to modern tectonics rather than to structure and lithology. However, the rates of uplift are much too high to be sustained for long, making their interpretation difficult. Brown and Oliver (1976) suggested that because the rates are so much higher than averages over geological time, the movements are either episodic or oscillating about a long-term trend.

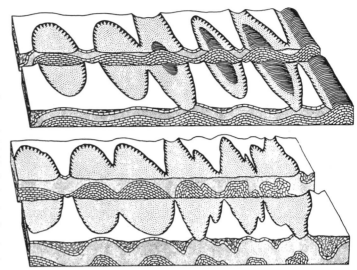

Figure 10. Basin coalescence related to amplitude of folding and depth of nonresistant unit. Top: Gentle folding with a thin layer of erodible materials interposed between resistant formations produces exposure of fold cores prior to basin coalescence. The development of capacious transverse lowlands (and thus transverse drainage) is precluded. Bottom: Although fold amplitudes are greater, the presence of a compensatory depth of erodible materials allows the development of extensive transverse subsequent lowlands prior to exposure of the second resistant stratum, thereby permitting development of transverse drainage (Oberlander, 1965, Fig. 64).

Considerations by Hack (1980, 1982), however, suggest that if modern tectonics does indeed shape the Appalachians, the effect is much subordinate to that of rock control, and indeed, in most places is difficult to detect. Most landforms owe their existence to erosion of rocks of differential resistance rather than to tectonic processes. This would be particularly true of the Valley and Ridge Province, where rocks differ so greatly in resistance. Some exceptions do occur, however. The southern Blue Ridge escarpment cannot be completely explained by differential erosion alone, and thus appears to reflect differential uplift. The Piedmont northeast of the Potomac River appears to have been recently uplifted relative to the rest of the Piedmont, based on higher relief and streams with unusually narrow valleys and convex longitudinal profiles (Hack, 1982).

No similar anomalous areas exist in the Valley and Ridge Province, despite the presence of a seismic zone inferred to have been the locus of the second largest earthquake in the southeastern United States (MMI VIII; Bollinger and Wheeler, 1982, 1983). This northeast-striking zone in Giles County, Virginia, appears to be related to features below the Appalachian overthrust belt, and most likely resulted from compressional reactivation of an Iapetan normal fault. This location within the basement rocks may explain why no surface faults related to the zone have been detected. The only possible evidence of differential uplift

Figure 11. Geologic setting of the Susquehanna River in the Appalachian fold belt. Anticlinorium exposing early Silurian and older rocks at left (crosshatched); synclinorium of Pennsylvanian-Permian rocks at right (dotted). Solid black indicates ridge-forming sandstones: St, Tuscarora (Silurian); Do, Oriskany (Devonian); Dm, marine beds of Susquehanna Group (Devonian) Mp, Pocono (Mississippian); Pp, Pottsville (Pennsylvanian). Other formations: Su, Upper Silurian; Dho, Hamilton Group (Devonian); Dck, Catskill (Devonian); Mmc, Mauch Chunk (Mississippian. From Oberlander, 1985, Fig. 9.)

associated with the zone is the finding that broad, high New River terraces in the vicinity of the zone are roughly 30 m higher than comparable terraces southeast of the zone, a displacement whose sense corresponds to the motion inferred from seismic data (Mills, 1986). However, correlation of these undated terraces is uncertain, owing to their discontinuous and poorly preserved character (Mills and Wagner, 1985).

Further geomorphic research related to the Giles County seismic zone involves a search for evidence of seismic ground shaking. No soft-sediment deformation features likely to be associated with shaking have been found (Mills, 1985). Seismic-induced slope failure, however, appears to be a fertile field for future investigation. The presence of slide-rock dammed Mountain Lake near the zone is suggestive. Recently, Schultz (1986) has identified what appear to be enormous relict rockslides off Sinking Creek anticline southwest of the zone. If these or other anomalously large failures can be shown to be geographically restricted to areas near the Giles County seismic zone, the importance of this zone for earthquakes would be established.

SUMMARY

The effect of structure and lithology on landforms is superbly illustrated in the Valley and Ridge Province. Many fundamental relationships between bedrock character and relief were first discovered here. More-recent research has demonstrated other relationships. For example, the longitudinal profiles of streams are strongly affected by the lithologies over which the streams flow, especially by the caliber of the bed material supplied to them by the bedrock. Incised valley meanders have been shown to be longer and wider in shaly than nonshaly meanders. Drainage-basin geometry and stream-network topology are affected by dip direction and angle. Many water and wind gaps appear to be localized by particular structural and lithologic features.

Lithology can affect topography indirectly via transported regolith: clasts weathered from resistant bedrock may be moved downstream or downslope onto adjacent areas underlain by nonresistant bedrock, where they affect topographic evolution. Plate tectonics and the study of young orogens elsewhere have shed new light on the classic problem of transverse drainage in the Appalachians. Modern tectonic movement may account for some geomorphic features in the Appalachians, but is subordinate to rock control in shaping the Appalachian landscape.

THE SURFICIAL MANTLE AND LANDSCAPE EVOLUTION IN THE PIEDMONT AND INNER COASTAL PLAIN

M. J. Pavich and W. L. Newell

INTRODUCTION

The upland surface of the Piedmont has long interested geomorphologists because of its low relief over areas encompassing many different types of crystalline rocks. As Thornbury (1965, p. 94) has noted, it "probably comes as near as any area in the eastern United States to looking like a former peneplain surface." Costa and Cleaves (1984) interpreted the upland surface of Piedmont Maryland to be a relict feature dating to the Late Miocene. Implicit in many arguments for the antiquity of the Piedmont landscape is the assumption of great age for the thick saprolite that mantles this surface. However, recent studies discussed below indicate that most saprolite is Quaternary in age, and thus do not support the concept of a relict erosion surface.

Many recent geomorphic studies of the Piedmont landscape have been carried out on the outer Piedmont near its junction with the Atlantic Coastal Plain at the Fall Line. The dramatically different geological substrates juxtaposed at the Fall Line allow the effect of substrate on geomorphology to be determined while holding climatic and tectonic history essentially constant. Contrasting modes of weathering, erosion, and evolution of Piedmont and inner–Coastal Plain landscapes provide insight concerning rates of Piedmont surficial processes. Therefore, although the subject of this chapter is the Appalachians, a discussion of the inner–Coastal Plain landscape is included.

REGOLITH PRODUCTION

Regolith, including residual or transported surficial materials, occurs above unweathered bedrock. Distributions and characteristics of regolith types on the Piedmont and Coastal Plain are indicative of chemical and physical processes that form them. Estimation of long-term rates of landscape reduction can be based on relations of regolith characteristics to age-dated parent materials and to relative topographic positions.

The untransported regolith of the Piedmont is predominantly saprolite, the isovolumetric residuum of chemical weathering. The altered mineralogy, petrography, and structural fabric of saprolite reflects a variety of original crystalline rock types. Hydrochemical reactions in groundwater remove mass and alter the mineralogy of igneous and metamorphic rocks that contain minerals unstable in dilute, low-temperature solutions at the earth's surface. Although the solubility and instability of metamorphic minerals varies greatly, the movement of water through the rock structure is a primary factor in promoting chemical reactions and determining rates of mass removal in solution. Comparing regional Piedmont drainage divides (i.e., on maps with scale of 1:500,000), areas of low altitude are generally underlain by finer-grained metasediments with steeply dipping (e.g., >60°) foliation (Hack, 1982). These rocks contain numerous open structural planes (secondary porosity) that facilitate the rapid flushing of groundwater. By contrast, relatively high areas are generally underlain by monomineralic or structurally massive rocks without well developed, steeply dipping foliation, and with widely spaced joints. Even in those massive rocks that contain labile minerals (e.g., massive granitic to mafic rocks containing iron silicates and plagioclase feldspar), depth of weathering is limited by the lack of structurally controlled permeability.

The typical upland mantle of Piedmont regolith (Fig. 12) on a steeply dipping, well-foliated rock includes 1 m of soil, 0.5 to 1 m of massive subsoil, and 10 to 20 m of saprolite over fresh rock (Pavich and others, 1987). Beneath relatively broad, subhorizontal (e.g., <5° slope) drainage divides, numerous observations in quarries indicate that the regolith thickness varies with the spacing of open, vertical fractures in the bedrock, but averages 12 to 15 m (Pavich and others, 1987). Variation with latitude cannot be systematically documented (J. T. Hack, oral communication, 1985).

As regolith is formed, progressive mass loss occurs in the sequence: rock → saprolite + dissolved solids → soil + dissolved solids. The dissolved solids are removed by groundwater flow through the regolith and discharge to surface streams. Residence time of solutions is not sufficient to cause cementation even though the solutions are generally supersaturated with silica relative to quartz and amorphous-silica solubilities (Stumm and Morgan, 1970). Soil formed on saprolite undergoes volume reduction with respect to saprolite due to compaction and mechanical reorganization of parent fabric. Within a few meters or less of the surface, the relict saprolite structure is destroyed in the soil zone by freeze/thaw, wetting/drying, burrowing animals, and other biological activity.

By contrast with foliated, structurally permeable rocks, massive rocks such as quartzites and granitic plutons are typically covered with thin, spotty regolith. On these rocks regolith is produced more slowly and the rate of removal exceeds or equals the rate of production. It is probable, though unproven, that processes of landscape lowering are, or have formerly been, slower than on well-fractured foliated rocks.

Regionally, the inner Coastal Plain exhibits a much more variable regolith than the Piedmont (Newell and others, 1980). Along the inner Coastal Plain margin, north to south, marine and nonmarine deposits of different Cenozoic ages are juxtaposed against the older Piedmont rocks. Where the innermost Coastal Plain deposits are old (early to middle Tertiary) and permeable, as in parts of Georgia and South Carolina, the high Coastal Plain interfluves are underlain by thick zones of residuum that com-

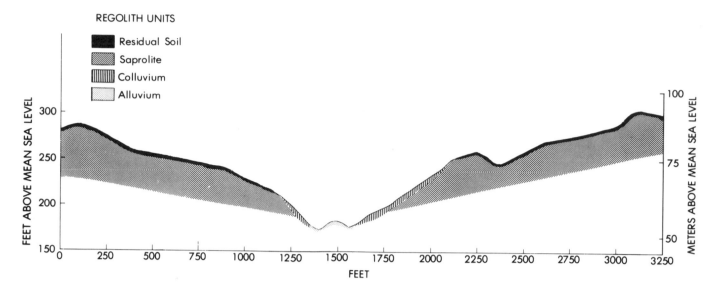

Figure 12. Schematic cross section of Piedmont valley showing typical broad sloping interfluves and steep valley side walls. Regolith is thickest beneath the divides and thins beneath the valley sides.

monly include remnants of surficial deposits that developed on now-eroded ancient surfaces (Fig. 13). These deposits provide evidence of multiple episodes of colluviation and local alluviation. Colluvium and alluvial fans are extensive and complex, and soils are extremely thick and mineralogically mature. Elsewhere, as in parts of North Carolina and Virginia, where younger (Neogene) Coastal Plain deposits abut older Coastal Plain and Piedmont rocks, the erosional history, accumulation of residuum, and soil development are much less advanced.

REGOLITH MOVEMENT

Slope Processes

The dominant processes of sediment transport on slopes in the Piedmont and inner Coastal Plain areas are sheet wash, creep, slumping, and debris flows (Newell and others, 1980). Little evidence for widespread aeolian transport exists, although locally, thin veneers of loess and scattered dune fields do occur.

The dominant physical process on relatively flat Piedmont interfluves underlain by saprolite appears to be washing of fines from the surface, leaving a typical pavement of coarse, angular to subrounded vein-quartz pebbles and cobbles on the Piedmont upland. This process appears to be less effective on Coastal Plain surfaces, perhaps because those soils that contain gravel are underlain by more-permeable deposits and infiltration capacities are much higher.

On Piedmont valley-side slopes (generally >12°) the dominant process is creep. Plastic deformation due to creep extending below 1 m on steep hillslopes is indicated by downslope reorientation of steep foliation dips in saprolite. Six-meter-thick accumulations of colluvium have been reported from the western

Piedmont of South Carolina (Eargle, 1940, 1977; Cain, 1944). In contrast to other parts of the Appalachians (see following section), mass wastage in the form of landslides is rarely reported in the Piedmont (Costa, 1974; Moss and Kochel, 1978), except where hillslopes have been modified by construction.

On the Coastal Plain, earth flows, landslides, and slumps are locally effective means of scarp retreat in headward-eroding drainages (Newell and others, 1980). Failures are usually concentrated along permeable/impermeable contacts where groundwater intersects slopes, and shear strength is minimal during wet periods.

Fluvial Processes

On the Piedmont, materials delivered to stream valleys are alternately stored for short periods (10^1 to 10^3 years) and removed by fluvial transport. Although eroded soils have produced Holocene fills in Piedmont streams (Trimble, 1974), there is no widespread pre-Holocene terrace stratigraphy indicating long-term sediment storage on the Piedmont. Stream gradients and capacities are adjusted to transport even the coarsest material delivered to the stream. In areas of fine- to medium-grained metamorphic rocks, coarse debris other than vein quartz is rare. In contrast, Coastal Plain stream valleys commonly contain numerous levels of Pleistocene, and even Pliocene, terraces (Newell, 1985; Soller, 1984). The terraces may include suites of fluvial to estuarine deposits, and sediment storage on the terraces can exceed 10^6 years (Newell, 1985). Alluvium is a complex assemblage of fresh, labile constituents from the Piedmont and resistant constituents from the Coastal Plain.

It is unlikely that periglacial processes had a substantial influence on erosional or depositional systems of the outer Pied-

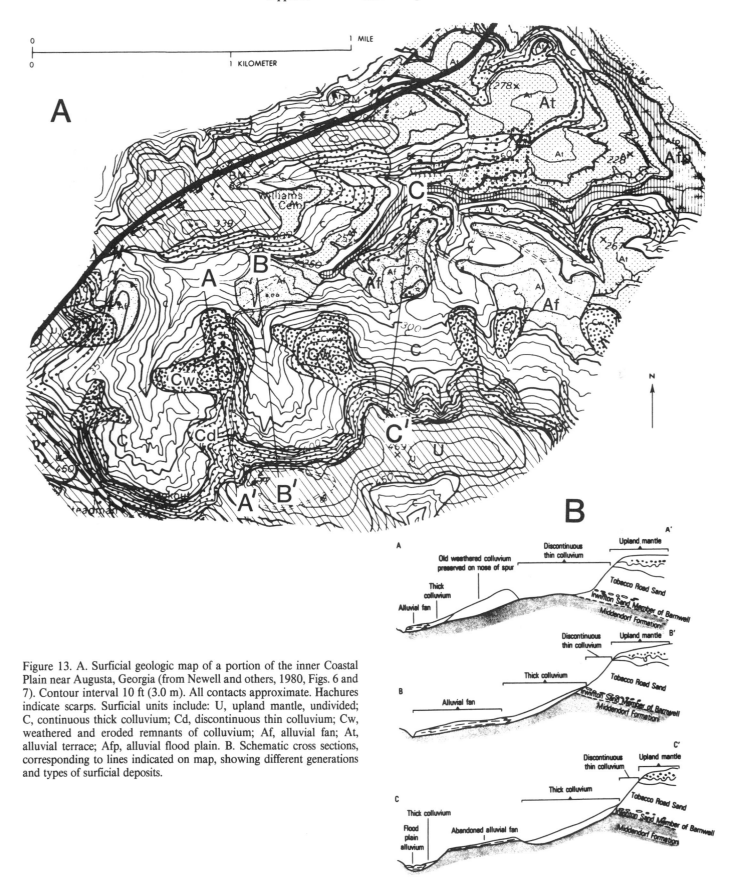

Figure 13. A. Surficial geologic map of a portion of the inner Coastal Plain near Augusta, Georgia (from Newell and others, 1980, Figs. 6 and 7). Contour interval 10 ft (3.0 m). All contacts approximate. Hachures indicate scarps. Surficial units include: U, upland mantle, undivided; C, continuous thick colluvium; Cd, discontinuous thin colluvium; Cw, weathered and eroded remnants of colluvium; Af, alluvial fan; At, alluvial terrace; Afp, alluvial flood plain. B. Schematic cross sections, corresponding to lines indicated on map, showing different generations and types of surficial deposits.

mont and Coastal Plain at distances greater than 160 km from the late Wisconsin border. Although block fields have been documented by Clark (1968), Godfrey (1975), and Hupp (1983) in higher elevations of the Blue Ridge and Valley and Ridge in Maryland and Virginia, and Whitehead and Barghoorn (1962) have presented palynologic evidence for pre–late Pleistocene colluviation of the western Piedmont of South Carolina, no evidence of widespread periglacial slope movement has been documented for the outer Piedmont beyond this limit. Congeliturbationlike structures (pots) are present in silty deposits of old gravels near Washington, D.C. (Conant and others, 1976). However, extensive silty soils on terraces flanking the Potomac and Rappahannock rivers show no congeliturbation structures or fossil frost wedges, suggesting that periglacial processes were absent at low altitudes.

LANDSCAPE/REGOLITH RELATIONS

The differences in surficial stratigraphy between the Piedmont and Coastal Plain apparently reflect differences in rates of mass loss. In places, the long-term result of faster rates of denudational mass loss from the Piedmont has been to lower the Piedmont relative to the inner edge of the Coastal Plain (Newell and others, 1980). Topographic inversion, with the formerly higher piedmont surface now standing lower than the adjacent inner–Coastal Plain surface, occurs in northern Virginia where Coastal Plain sediments are Miocene in age (Pavich and Obermeier, 1985). Inversion is even more pronounced where the Piedmont abuts Eocene-age Coastal Plain sediments along the South Carolina–Georgia border (Newell and others, 1980).

The evidence for continued denudation of the Piedmont through the late Cenozoic fits Hack's (1960) concept of dynamic equilibrium in the Appalachians. Geomorphically, topographic expression of granite balds, massive greenstones, and quartzites can be explained in terms of their resistance to mass loss (both chemical and mechanical) relative to the nonresistant, well-foliated metasedimentary rocks that are more common on the Piedmont. Over short time scales (e.g., 10^6 yr), all parts of the landscape are not necessarily lowered at the same rate. Denudation rates, for example, may not be constant under varying climatic conditions, and the relative effectiveness of physical versus chemical denudation has probably changed from time to time as a function of independent variables such as climatic change, particularly in higher-relief areas (Godfrey, 1975). Over longer time scales (e.g., $\geq 10^8$ yr), however, relief processes and rates appear to be balanced, with all parts of the landscape being lowered at approximately the same rate.

Costa and Cleaves (1984) have discussed evidence for cyclic erosion in the Maryland Piedmont. In our opinion, however, until new stratigraphic evidence indicates otherwise, the Piedmont landscape is most reasonably understood as the result of continuous mineral alteration at depth in the regolith, continual erosion or stripping events, and limited short-term sediment storage in a continuously downwasting region, which maintains constant re-

lief, slope angles, and regolith configuration for uniform rock types.

Inner–Coastal Plain topography, by contrast, is generated from initially broad, low-gradient depositional surfaces at or below the lower elevations of the adjacent Piedmont. As major rivers traversed the Coastal Plain and cut through the nearly flat-lying strata, a tributary drainage net developed at the expense of the flat upland surface. As the drainage net develops and valleys intersect permeable and impermeable strata, the hydrologic parameters change; weathered residuum on upland areas is eroded and incorporated in surficial deposits accumulating in and moving through tributary and major river valleys. Because the materials are generally more permeable and gradients are lower, colluvium and alluvium on these landscapes are weathered, eroded, and redeposited many times before ultimately being transported out of the region (Newell and others, 1980). Reworking of deposits and long residence times on the landscape typify the denudation of the Coastal Plain.

RESIDENCE TIMES OF REGOLITH AND RATES OF REMOVAL

Geomorphic evidence suggests that denudation rates are faster on the Piedmont than on the inner Coastal Plain, and that residence times of regolith units are significantly longer on the latter (Newell and others, 1980). Obtaining absolute ages for regolith is more difficult. Terraces and associated soils can be used to construct a chronostratigraphic framework for the Coastal Plain. Study of soils and weathering zones along the major Coastal Plain rivers (Newell, 1985; McCartan and others, 1984; Markewich, 1985; Soller, 1984) shows that fluvial terraces range in age from late Pleistocene to Pliocene. Older deposits exist but are extensively dissected, and original surfaces are probably not preserved. Markewich and others (1987) have demonstrated that soil profiles (especially argillic-B horizons) continue to change through this time period on Rappahannock River terraces studied by Newell (1985).

Surficial deposits on the Piedmont cannot be morpho-, bio-, or chronostratigraphically correlated to the Coastal Plain deposits. However, at isolated, well-protected locations, soil development relative to Coastal Plain terrace soils of approximately known age suggests a maximum age of about 1 Ma for the Piedmont upland soils. This is also a maximum age, therefore, for upland saprolite. Minimum rates of saprolite development based on stream-baseflow chemistry indicate that a time interval consistent with 1 m.y. would be required to develop the commonly observed thickness (15 m) of saprolite (Pavich, 1986). Further support for this age estimate is furnished by recent ^{10}Be "age" measurements (Pavich and others, 1985), which indicate a minimum age of 0.8 Ma for upland soil and saprolite.

These age estimates (Pavich and Obermeier, 1985) and the observed topographic inversion indicate that saprolite production and removal are continuous processes that lower the Piedmont upland surface at a rate of 10 to 20 m/m.y. over broad areas near the Fall Line.

SUMMARY

The production of Piedmont saprolite is strongly dependent upon rock structure. Foliated, steeply dipping metasediments weather much more rapidly than massive igneous rocks. Saprolite thickness on the former averages about 15 m near drainage divides. Thickness shows no apparent variation with latitude.

The juxtaposition of Piedmont and Coastal Plain geological units along the Fall Line allows the effect of geological substrate on surficial processes and regolith to be ascertained. The inner Coastal Plain is underlain primarily by fluvial and marine sediments, whereas the Piedmont is underlain by metamorphic rocks. Differences are most dramatic where the Coastal Plain units are early-to-middle Tertiary in age and have been subject to subaerial erosion since that time. In such settings, the extensive and complex surficial deposits of the Coastal Plain show evidence of multiple episodes of colluviation and local alluviation. Soils are extremely thick and mineralogically mature. Regolith on the Coastal Plain appears to be weathered, eroded, and redeposited many times before ultimately being transported out of the region.

In contrast, on the adjacent Piedmont, the regolith consists chiefly of saprolite, with only thin, discontinuous patches of colluvium present. Soils are much thinner and more poorly developed than those on Coastal Plain regolith. Long-term storage of eroded sediments appears to be absent. These observations, plus the fact that upland-surface elevations commonly are lower on the Piedmont than on the adjacent Coastal Plain, suggest that eroded sediments on the Piedmont have residence times an order of magnitude lower than those on the Piedmont. This difference may well stem from the more-permeable materials and the lower initial gradients on the Coastal Plain.

Several dating methods indicate that Piedmont saprolite has an age on the order of 1 Ma. A Quaternary age means that the existence of thick saprolite can no longer be used as evidence for great age of the Piedmont surface. Rather than a cyclic interpretation of the Piedmont landscape, more appropriate is a dynamic-equilibrium model, in which continuous weathering, continual erosion, and limited sediment storage result in a landscape that maintains constant relief, slope angles, and regolith configuration as long as bedrock remains uniform.

SLOPE FAILURES IN THE APPALACHIAN PLATEAU

R. B. Jacobson and J. S. Pomeroy

INTRODUCTION

The ground-failure hazards reduction program of the U.S. Geological Survey has provided a strong impetus for research on landslides in the Appalachians. The bulk of this research has been carried out in the Appalachian Plateau, for compared to other Appalachian physiographic provinces, the Appalachian Plateau has the greatest abundance of conspicuous slope failures (Fig. 14;

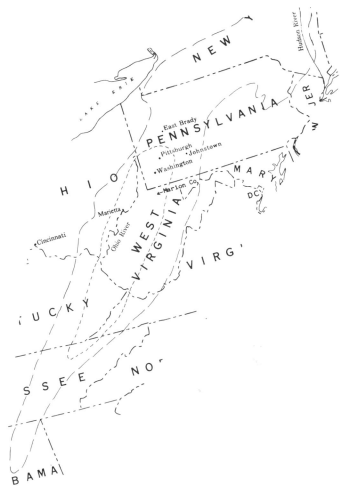

Figure 14. Map showing Appalachian Plateau and locations referred to in text. Long dashes show approximate boundary of plateau and short dashes enclose that portion of plateau with an especially high incidence of slope failures (after Radbruch-Hall and others, 1982).

Radbruch-Hall and others, 1982, p. 13). This abundance may be more apparent than real, because our knowledge of the distribution of slope failures in time and space is biased by (1) a short sampling interval, (2) a sampling scale that limits studied features and processes to those recognizable on air photos with scales of approximately 1:24,000, and (3) an emphasis on areas of human disturbance, particularly large industrial and population centers. It appears, however, that the greater prominence of slope failures in the plateau may arise in part from the high-frequency, low-magnitude nature of failures compared to those of (say) the Blue Ridge and Valley and Ridge provinces.

Most slope failures in the Appalachian Plateau are slump-earth flows (Fig. 15a) and earth flows (Figs. 15b and 16), but earth slumps, debris slides, debris flows, topples and rockfalls are also common. Bedrock slumps, although much less frequent than other types, have been documented along major river valleys from Pennsylvania to northern Alabama (Pomeroy, 1982b, 1984a, 1984b; Pomeroy and Thomas, 1985).

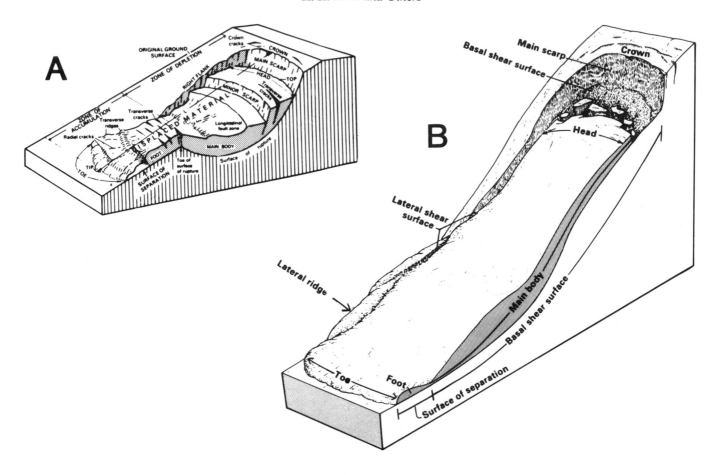

Figure 15. A. Block diagram of slump-earth flow (from Varnes, 1978). B. Idealized earth flow (modified from Keefer and Johnson, 1983).

REGIONAL DISTRIBUTION OF FAILURES

The greatest abundance of failures has been mapped in the region underlain by subhorizontal cyclothemic sedimentary rocks (shale, mudstone, claystone, siltstone, sandstone, and coal) of Pennsylvanian to Permian age. Fewer slides have been mapped on slopes underlain by shale, siltstone, and sandstone of Devonian and Mississippian age. Presumably, lower percentages of weak rock in these units result in less frequent failures, or alternatively, erosional processes may be dominated by high-frequency, low-magnitude creep.

Different parts of the plateau are characterized by distinct kinds of slope failures. The northeastern edge of the plateau along the Hudson River in New York displays slumps and earth flows in Pleistocene marine clays (Dunn and Banino, 1977). North-central Pennsylvania shows slumps and slump-earth flows in glaciolacustrine sediments, till, and colluvium as well as debris slides along steep-walled valleys underlain by shales and sandstones of Devonian to Pennsylvanian age (Delano and Wilshusen, 1985). Southwestern Pennsylvania (Pomeroy, 1982a), southeastern Ohio (Fisher and others, 1968; Pomeroy, 1984a), central and

western West Virginia (Lessing and others, 1976), and northeastern Kentucky (Outerbridge, 1986) commonly have slow-moving earth flows in addition to slump-earth flows and slumps. The susceptible surficial mantle in this area is largely derived from fine-grained clastic rocks of the Conemaugh, Monongahela, and Dunkard groups (Pennsylvanian and Permian). Large parts of eastern Kentucky and southwestern West Virginia are characterized by debris slides and flows because of the steep slopes and large amount of coarse clastic rock material in this area (Outerbridge, 1986). Commonly, sandstone and shale of the Breathitt and equivalent formations of Pennsylvanian age underlie these slopes.

LITHOLOGIC CONTROLS ON SLOPE FAILURE

Most recognizable slope failures in the Appalachian Plateau occur in colluvium and weathered fine-grained rocks. Failures on sandstones are less conspicuous, presumably because they are less frequent, but also because hillslopes underlain by sandstone are more likely to be forested than those underlain by fine-grained rocks, thus making failures more difficult to map.

Figure 16. Photograph of earth-flow terrain, Washington County, southwestern Pennsylvania.

Rock type and structure are important determinants of failure mechanisms, sizes, and rates. Sandstone commonly weathers to a residuum with low cohesion and a high angle of internal friction that is most likely to fail as planar slabs along joint planes. If triggering conditions result in flow, this regolith is likely to produce cohesionless debris slides. Mudstone, in contrast, may fail as either planar or rotational slides, either of which may progress to a muddy debris flow or earth flow under favorable conditions. On hillslopes where weak mudstones and claystones are interbedded with sandstones, the stronger sandstones limit total hillslope erosion over long time scales. Over shorter time scales, many failures occur in the weaker rocks, but their size and shape generally is limited by adjacent sandstone layers. For example, in an area of concentrated slope failures 10 km southeast of Marietta, Ohio, heads of landslides occurring in a red mudstone generally lie below a sandstone ledge. The ledge not only limits headward progression of the failures, but, as numerous springs at the base of the ledge indicate, the sandstone/mudstone contact is a zone of concentrated pore-water pressure where failures may be initiated.

Several argillaceous rocks of the Appalachian Plateau have reputations as particularly weak, failure-prone lithologies. Among these, the Pennsylvanian and Permian rocks of the Dunkard Basin have received the most attention. The red mudstones of the Conemaugh and Monongahela formations and the Dunkard Group account for nearly 95 percent of slope failures mapped in eastern Ohio (Fisher and others, 1968, p. 71), and for the majority of failures in West Virginia (Lessing and Erwin,

1977, p. 250). Instability of redbeds stems from their fine particle size, mineralogy, and diagenetic history. Red mudstones slake readily in water, falling apart into constituent grains in minutes to hours (Stollar, 1976, p. 37). Slaking has been attributed to unstable, swelling illite mineral phases in which ferric oxyhydroxide species are substituted for interlayer potassium (Fisher and others, 1968). High-angle slickensided surfaces are common in the red mudstones and have been attributed to compaction, stress release, and Paleozoic pedogenesis. Whatever their origin, these discontinuities significantly reduce rock strength and enhance permeability and weathering.

Although generally more resistant to weathering than red mudstones, gray and green shales and underclays in the Pennsylvanian-Permian rocks are also relatively weak lithologies. Some underclays will slake almost as rapidly as red mudstones. Gray and green lithologies may also have abundant, easily weathered sulfide minerals. Oxidation and dissolution of pyrite and marcasite produce acidic solutions that can intensely weather silicate minerals. Precipitation of sulfate minerals from these solutions can contribute to physical breakdown by expansion during crystallization (Dougherty and Barsotti, 1972).

Figure 17 shows a trench wall through a typical natural slope failure in the Washington Formation of the Dunkard Group in Marion County, West Virginia. This section illustrates several important features common to natural slope failures in this area. The upper 1 to 2 m is a yellowish brown to reddish brown diamicton composed of sandstone and siltstone clasts in a clayey silt matrix. In the center part of the failure the diamicton

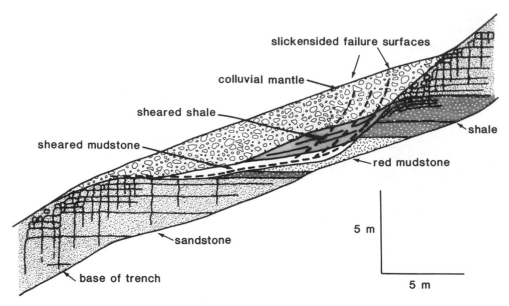

Figure 17. Cross section of a typical rotational-planar slope failure, upper Washington Formation, Dunkard Group, Marion County, West Virginia.

layer is transitional to a sheared red mudstone unit 5 to 75 cm thick. Discontinuous slickensided shear surfaces occur both in the sheared unit and in the overlying diamicton. The most recently active, main shear surface occurs at the base of the sheared zone on top of relatively unweathered red mudstone and shale. In many trenched slope failures, the slickensided shear surfaces stand out as bleached, light gray to white, 1- to 10-mm-thick planes. Typically, the head scarp occurs at a sandstone layer, the main body of the failure is in the mudstone layer, and the toe is deflected out by the next-lower sandstone layer.

Almost all mudstone slope segments have prominent slickensided shear surfaces subparallel to the slope at depths of 1 to 4 m, similar to those shown in Figure 17, even where no failure scarp is apparent at the surface. Residual strengths in Appalachian Plateau mudstones are believed to be mobilized on slickensided surfaces (Richardson, 1979; Hamel and Adams, 1981; Elnagger and Flint, 1976). Although natural slope stability may in some cases be controlled by residual strength (Skempton, 1964), other factors commonly may be more important. For example, the lowest shear surface on these slopes commonly extends into unweathered, unsheared rock. This indicates that near-peak strengths are mobilized along some parts of these failure surfaces. Also, additional strength may be added by roots (Riestenberg and Sovonick-Dunford, 1983), "healing" of slickensided shear surfaces (Elnagger and Flint, 1976, p. 36), or shear-surface roughness (Patton, 1969). Some combination of these factors allows many slopes underlain by mudstone to stand at greater inclinations than residual strengths alone should allow under common soil-moisture conditions. Typical peak and residual Mohr-Coulomb shear-strength parameters for Pennsylvanian-Permian mudstones are listed in Table 1.

The size and shape of this type of natural slope failure is largely controlled by the thickness of mudstone layers. Thicker mudstones tend to fail as rotational slides, whereas thinner mudstone units tend to show shallow, planar slides (Jacobson, 1985, p. 312). Movement on the failure surface creates diamicton colluvium by shearing mudstone and mixing in sandstone and siltstone clasts at the toe and head of the slide. Shear strength of colluvium and mudstone are both mobilized; however, slope stability analyses of this type of failure indicate that 70 to 90 percent of the mobilized shear strength is contributed by the argillaceous rock (Jacobson, 1985). Natural slope failures in the Appalachian Plateau thus can be controlled by complex relations of bedrock and colluvial lithologies.

Natural slopes in the mudstone-sandstone cyclothems are benched; sandstones support slopes of 25° to greater than 50°, whereas slopes on mudstones range from 10° to 30°. Slope angles on sandstones are higher because these lithologies possess higher shear strengths and are less susceptible to weathering. Declivity of sandstone slopes is highly variable, however, because of variation in joint density and in thickness of the surficial cover. Most landslide research has focused on the argillaceous rocks, and, in the short term, variations in the strength of these rocks probably is an important control on where and when failure occurs. On longer time scales, however, the rate of sandstone weathering limits failure initiation because sandstone layers ultimately control failure susceptibility.

Pleistocene glacial till, outwash, and lacustrine deposits mantle bedrock in parts of New York, Pennsylvania, Ohio, West Virginia, and Kentucky. Of these deposits, much attention has been given to very fine, poorly consolidated lacustrine silts and clays left as remnants from repeated dammings of major river

TABLE 1. EFFECTIVE MOHR-COULOMB PARAMETER TEST VALUES, PEAK AND RESIDUAL, FOR UPPER OHIO VALLEY PENNSYLVANIAN-PERMIAN RED MUDSTONES

$\bar{\phi}$ peak (degrees)	$\bar{c}$ peak (kN/m^2)	$\bar{\phi}$ residual (degrees)	$\bar{c}$ residual (kN/m^2)	Source
13–24	7–14	8–18	0	Hamel and Adams, 1981
12–13	0	6–12	6–19	Richardson, 1979
18	0	15	0	Elnagger and Flint, 1976
27.5	12.7	9.0	0.98	Hooper, 1969

systems during Pleistocene ice advances. Although these deposits cover relatively little area, they have attracted much slope-stability research because of their instability. For example, in Vinton County, Ohio, equilibrium natural slopes on the Minford Silt are 6.0°; liquid limits range from 50 to 60 percent and plasticity indices range from 22 to 39 percent, indicating that the material can have very low strength (Webb and Collins, 1967, p. 72). Fleming and others (1981) showed that Illinoian-age proglacial lake clays in the Cincinnati area are high in clay (50 to 70 percent) and have high liquid limits (39 to 77 percent) and plasticity indices (14 to 42 percent). Natural slopes on these lacustrine materials are 7° to 8°, and back-calculated residual angles of internal friction range from 6.3° to 15°.

COLLUVIUM ON HILLSLOPES—MODERN AND ANCIENT

Most Appalachian Plateau hillslopes have a mantle of poorly sorted debris that has moved downslope by a variety of mechanical processes including seasonal creep, tree-throw, faunal turbation, sliding, and flowing. This mantle is referred to as colluvium regardless of exact origin. In the field, most workers define colluvium as material that overlies residual or unweathered bedrock and has clasts of upslope bedrock mixed in with it. Colluvium is important to slope failure processes because it commonly is unstable and because it influences subsurface water flow and bedrock weathering. Figure 18 shows examples of colluvial slope movements that occur on benched hillslopes along the Allegheny River in western Pennsylvania (Pomeroy, 1984b). Along these steep, relatively undissected slopes, colluvium tends to collect on benches and in shallow downslope-trending swales. Similar slopes occur along other major river valleys in the Appalachian Plateau.

Thin colluvium similar to that shown in Figure 17 mantles most West Virginia Dunkard Group hillslopes to thicknesses as great as 2 m. Contemporary failure rates indicate that this colluvium is being produced by present-day weathering and slope failure (Jacobson, 1985). Other colluvial deposits, however, appear to be relict. Figure 19 shows thick remnant diamicton deposits that formerly filled first- and zero-order drainages on a

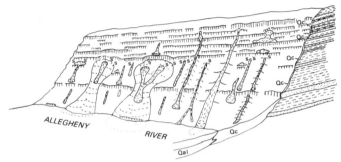

Figure 18. Schematic diagram showing colluvial slope movements along swales and above benches. Relative distribution of lithologies is typical of those along slope opposite East Brady, Pennsylvania (Pomeroy, 1984b, Fig. 12). S, horizon of abundant seeps; Qc, Quaternary colluvial deposits; Qal, Quaternary alluvial deposits. Diagram suggested by Gray and Gardner, 1977, Fig. 2.

Dunkard Group landscape about 125 km south of the Wisconsin glacial border. (Zero-order basins are hillslope depressions with concave-outward contours that are tributary to first-order basins, but lack stream channels themselves.) These deposits are up to 15 m thick. Most of the old diamicton deposits have been correlated with Wisconsin glaciation, and presumably are the product of paleoclimatic influence (Jacobson, 1985). Similar thick colluvial deposits in the upper Ohio River valley have been radiocarbon dated as early Holocene (8940 ± 350 and 9750 ± 200 yr B.P.; Gray and others, 1979, p. 466), greater than 20,400 yr B.P. (Gray and Gardner, 1977, p. 31), and greater than 40,000 yr B.P. (D'Appolonia and others, 1967, p. 452).

Ancient debris flows also have been identified in northwestern and north-central Pennsylvania (Pomeroy, 1983). The flows, commonly 600 to 900 m long and 200 to 500 m wide, occur on forested concave-upward slopes ranging from 11° to 27°. The deposits are bouldery with hummocky, lobate feet; the lower one-third to one-half of each flow frequently shows a convex-transverse profile. Large ancient debris flows have also been found in northeastern Alabama (Pomeroy and Thomas, 1985). No radiocarbon dates have been reported for these flows, however.

Thick accumulations of old colluvium indicate that rate of

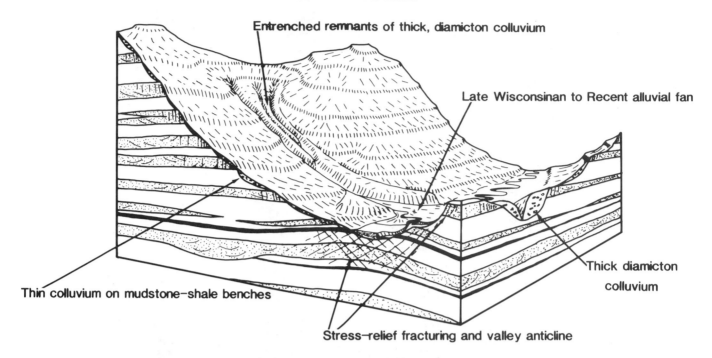

Entrenched remnants of thick, diamicton colluvium

Late Wisconsinan to Recent alluvial fan

Thick diamicton colluvium

Thin colluvium on mudstone-shale benches

Stress-relief fracturing and valley anticline

Figure 19. Diagrammatic sketch showing geomorphic and surficial stratigraphic features of a typical landscape underlain by Dunkard Group rocks in Marion County, West Virginia. Hillslopes are strongly benched with steep slopes on sandstones and gentle slopes on mudstones and shales. Entrenched remnants of thick, diamicton colluvium occur in zero-order drainage basins. Thin colluvium occurs on shale and mudstone benches. Late Wisconsin and younger alluvial fans stratigraphically crosscut thick diamicton colluvium and are graded to valley-bottom alluvial deposits. Valley anticlines and near-surface fracture zones in sandstones have resulted from stress relief produced by valley erosion. Sandstones are stippled, coals are black, and shales and mudstones are unpatterned.

production of colluvium has exceeded the rate of removal by fluvial processes during intervals in the late Pleistocene and early Holocene. Increased colluvial transport during seasonal thawing under periglacial climatic conditions has been inferred to explain old, thick colluvium in northern Pennsylvania (Denny, 1956; Pomeroy, 1983; Delano and Wilshusen, 1985). Farther south of the glacial border and at lower elevations, paleoclimatic data are lacking and the former existence of periglacial processes is unclear. In these areas, discrete intervals of accumulation of colluvium may have been caused by climatically controlled periods of accelerated physical and/or chemical weathering, climatically induced vegetation changes, increased rates of seasonal creep, or increased frequency of severe, triggering storms. Because evidence for these processes is limited and their interactions probably are complex, it is difficult to generalize about the exact process of accumulation that may have operated in a given area.

Colluvium produced by older mass movements commonly is weaker than residual soils and bedrock, and therefore is more subject to failure, especially when it is steepened beyond its original depositional slope. However, colluvium can also be quite stable depending upon its physical properties and position on the landscape. Colluvium deposited on low-angle alluvial fans by debris flows, for example, is likely to be very stable relative to bedrock and may persist on the landscape for long periods of

time. Jacobson (1985) attributes persistence of early Wisconsin (55,000 to 70,000+ yr B.P.) debris-flow colluvium in zero-order drainage basins in West Virginia (Fig. 19) to entrenchment following deposition. This sequence of events left colluvium as long, well-drained fingers that are at least as stable as surrounding bedrock slopes.

Although these examples of old colluvium illustrate that past processes are at least locally important in determining present-day slope stability, the relative importance of past and present processes in the formation of colluvial deposits remains a topic for debate. One position is that most colluvial deposits can be explained by rare meteorological events, and that paleoclimatic influences are not required (e.g., Hack and Goodlett, 1960, p. 63). Whatever its origin, thick colluvium is common on much of the Appalachian Plateau and may mantle a dissected bedrock topography giving highly variable, unpredictable thicknesses (Royster, 1973, p. 255).

Mechanical properties of colluvium vary mainly with the nature of the source rock. For example, data from southeastern Ohio (Hooper, 1969, p. 64-68) show that index properties of colluvium derived from red mudstone are very similar to those of red-mudstone residuum, with the exception of a slight increase in plasticity index. Typical Mohr-Coulomb shear-strength-parameter values for claystone colluvium are shown in Table 2. Collu-

TABLE 2. EFFECTIVE MOHR-COULOMB PARAMETER TEST VALUES, PEAK AND RESIDUAL,
FOR UPPER OHIO VALLEY COLLUVIAL DEPOSITS

$\bar\phi$ peak (degrees)	$\bar c$ peak (kN/m^2)	$\bar\phi$ residual (degrees)	$\bar c$ residual (kN/m^2)	Source
31–32	0	---	---	unpublished ASCS data
20–25	35	11–16	0	Gray and others, 1979
27	0	18	0	Gray and Gardner, 1977
20	7.7	16	0	D'Appolonia and others, 1967
13–24	7–14	8–18	0	Hamel and Adams, 1981
20–25	7–21	8–18	0–10	Hamel, 1980

vium derived from sandier and interbedded lithologies tends to have greater effective angles of internal friction and zero effective cohesion. For example, diamicton colluvium formed from Dunkard Group cyclothemic rocks has typical peak Mohr-Coulomb values of $\bar\phi$ = 31° and $\bar c$ = 0 kN/m^2 (unpublished data, Agricultural Stabilization and Conservation Service).

Hamel (1980, p. 15) stated that engineering behavior of colluvium is largely controlled by preexisting failure surfaces and shear zones. Residual strength attained by large-strain laboratory tests are much less than peak strengths, as shown by comparisons in Table 2. In addition, D'Appolonia and others (1967, p. 454) showed that water contents and plasticity indices of colluvial material above and below shear surfaces are lower than those of the shear-surface material.

Residual-strength values probably are reliable conservative values for design purposes and for remobilized failures. For many natural colluvial slopes, however, slope stability may not be controlled by residual strength, for two reasons. First, many thick deposits of colluvium have been emplaced by debris flow rather than by sliding along a single surface (Jacobson, 1983; Pomeroy, 1983). In these cases, preexisting shear surfaces do not necessarily exist and peak strengths may be mobilized. Second, as mentioned previously, where preexisting surfaces do exist, shear strength may be supplemented by root strength and surface roughness.

HILLSLOPE HYDROLOGY AND MORPHOLOGY

Pore-water pressures decrease effective normal stresses on potential failure planes and therefore reduce shear strength. In the Appalachian Plateau, spatial distributions of soil moisture have been attributed to the effects of colluvial mantles, microclimatic influences (especially slope aspect), slope morphology, and down-dip flow of groundwater.

Colluvium and bedrock commonly differ in permeability. In landscapes underlain by the Dunkard Group, the colluvial mantle is more permeable than nonfractured mudstone and less permeable than fractured argillaceous rocks and sandstone. At a Marion County, West Virginia, Agricultural Stabilization and Conservation Service dam site, for example, colluvial material has per-

meabilities ranging from 4×10^{-7} to 5×10^{-8} cm/sec (as measured in the laboratory by falling head permeability apparatus). Packer pump tests in boreholes show that total bedrock permeability is highly variable, ranging from impervious in unfractured shale and mudstone to nearly 4×10^{-2} cm/sec in fractured sandstones and coals. Where colluvium drapes over the more permeable lithologies it can act as a confining layer so that excess pore-water pressures can develop. During spring rains it is not uncommon for crayfish burrows (3 to 4 cm in diameter) in Dunkard Group colluvium to have spouts of water up to 10 cm high issuing from them. Similarly, at a monitored landslide site in the Cincinnati area, Fleming and others (1981, p. 551, 556) recorded groundwater uplift pressures of 1.5 m at the contact between colluvium and a fractured limestone unit.

In contrast, colluvium with permeability greater than underlying bedrock and residuum appears to increase slope-failure susceptibility in Tennessee (Royster, 1973). Here, colluvium on slopes underlain by the Pennington Formation is highly permeable because of additions of sandy materials from sandstone and conglomerate upslope. Royster (1973, p. 261) described common situations in which groundwater flow is concentrated in the colluvial mantle overlying less-permeable shales. Concentration of flow at the contact accelerates weathering and produces a wet, relatively unstable zone.

Many workers have tested hypotheses about the influences of hillslope-scale microclimate (topoclimate), groundwater flow down geologic structure, and topographic concentration of moisture on the spatial distribution of slope failures. The combined effect of these three interdependent factors on moisture distribution is complex, and the importance of any one of them is difficult to isolate.

Lessing and others (1983) suggested that in areas studied in West Virginia, failures show little relation to slope aspect. Detailed mapping in the Marietta, Ohio, area (Pomeroy, 1984a) and in an area south of Washington, Pennsylvania (Pomeroy, 1982a), however, has revealed that the majority of recently active earth flows are on northwest- to east-facing slopes (Fig. 20). In contrast, most south-facing slopes in the Marietta area have well-developed drainage networks and show relatively few failures.

The greater wetness and steepness of north-facing slopes appear to provide the optimum conditions for creep and slope failure.

Hall (1974) and Briggs and others (1975) suggested that dip slopes may be more susceptible to failure because of preferential groundwater flow down geologic structure, but Pomeroy (1982b), Lessing and others (1983), and Jacobson (1985) found no significant relationship between failure-slope aspect and bedrock-dip direction. Such findings may seem surprising since any dip on contacts between rocks of differing permeabilities should influence groundwater flow direction. However, Wyrick and Borchers (1981) have shown that in Wyoming County, West Virginia, a layer of fractured rock, produced by stress relief, parallels the topography. Hydrologic studies in the area have shown that the shell of fractured rock is more transmissive than other surficial or bedrock hydrologic units. Similar stress-relief fracturing has been noted elsewhere in the Appalachian Plateau (Ferguson, 1967; Hamel and Adams, 1981). Because such fracturing is produced by unloading due to valley erosion, it probably occurs throughout the Appalachian Plateau. To the extent that groundwater is concentrated in the stress-relief zone where secondary permeability is high, bedrock-dip influence on groundwater flow direction will be minimized.

Slope morphology influences pore-water pressure distribution by guiding surface and shallow subsurface flow. In homogeneous, isotropic regolith, concave slope forms should be moister than convex forms. Lessing and others (1976) and Pomeroy (1982a) have shown that about 65 percent of recently active landslides in West Virginia, Pennsylvania, and Ohio have taken place on concave-upward slope profiles. In addition, studies by Lessing and others (1983, p. 98), Pomeroy (1982a, p. 13; 1982b, p. 36), and Jacobson (1985) show that more mappable failures occur on laterally-concave slopes (i.e., slopes with concave-outward contours) than on laterally-convex slopes.

Failures might also be expected to be concentrated at points where drainage area above the failure is at a maximum for a given slope. However, in Marion County, West Virginia, Jacobson (1985) found only a weak increase in slope-failure volume with increase in drainage area above the failure.

METEOROLOGICAL TRIGGERS AND FAILURE FREQUENCY

The distribution of natural slope failures over time is highly dependent on the frequency of meteorological events that can trigger failures by increasing pore-water pressures to failure conditions. Triggering meteorological events usually are a combination of precipitation, evapotranspiration, and runoff that allows soil moisture to accumulate. Failure thresholds at a site, however, also change with time, usually decreasing as weathering decreases the shear strength of the material. As shear strength is reduced, the frequency of the meteorological events sufficient for failure is increased. High-frequency, low-magnitude processes, such as seasonal creep, faunal turbation, and tree throw, may also reduce the

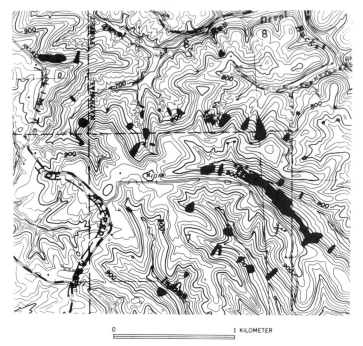

Figure 20. Preferred slope failures on northwest- to east-facing colluvium-covered, forested hillslopes near Marietta, Ohio. Bedrock is shale, mudstone, and sandstone, with regional dip to south of less than 1°. Dark areas show active or recently active slope failures, mostly earth flows.

failure threshold of a slope over time by building up colluvial thicknesses and steepening slopes.

Most of the work on failure frequency in the Appalachian Plateau has concentrated on the effects of rare, severe storms. For example, two storms in August and September, 1969, one of which was associated with Hurricane Camille, accounted for 1,534 debris flows and avalanches in a 94-km² drainage basin in the eastern part of the Appalachian Plateau (Schneider, 1973). Several hundred storm-induced landslides took place during a 9-hour period as a result of up to 30 cm of rain that fell in the Johnstown, Pennsylvania, area in 1977 (Pomeroy, 1980). The recurrence interval of this event is estimated to be 5,000 to 10,000 yr. Less but probably more-intense rain triggered failures on about 2.5 percent of the slope surface in two areas adjacent to East Brady, Pennsylvania, in 1980 (Pomeroy, 1984a). The number of failures ranged from 56 to 85 per km² and totalled at least 50,000 tons of regolith. A July 1976 rainstorm that produced 6.6 cm in 2 hr near Ithaca, New York, resulted in 500 m³ of landslide erosion in a 1.0-km² drainage basin (Renwick, 1977). This storm had an estimated recurrence interval of 70 yr; the flood produced from the runoff, however, had an estimated recurrence interval of only 35 yr, indicating that much of the precipitation was extracted to vegetation or groundwater storage.

A 33-yr record of tree-ring-dated slope failures in Marion County, West Virginia, shows that failures occur in almost any given year, and that total yearly volume is best predicted by 30-day-duration runoff events (Jacobson, 1985). High-magnitude 30-day runoff events invariably occur in the late winter and spring when evapotranspiration is low and soil moisture is high. Frequency analysis of the annual series of maximum 30-day runoff events and a regression model that predicts aggregate yearly failure volume as a function of 30-day runoff indicate that yearly expected failure volume is 220.2 $m^3/km^2/yr$. The product of the magnitude and frequency models shows the trade-off between frequent, small events and rare, large events; the peak of the product curve occurs at the events that do the most cumulative geomorphic work (Wolman and Miller, 1960). The peak of the magnitude-frequency product curve for the Marion County data occurs at 30-day runoff events that recur, on average, every 4 years.

The Marion County data support three ideas that contrast with previous studies. First, the timing of slope failures is best explained by runoff data rather than precipitation data, a result that is consistent with hydrologic models that attribute runoff to the combined effects of precipitation intensity, infiltration rates, evapotranspiration, and filling of available soil-moisture reservoirs. Second, the Marion County data show that long wet periods rather than intense storms are responsible for triggering the most failure volume. This result probably is primarily due to the influence of fine-grained argillaceous rocks in this area, which require long, slow infiltration to build up soil moisture at failure depths, whereas more intense rainfall tends to run off before it can infiltrate deeply enough. Third, these data show that in terms of sediment flux and total geomorphic work, low-magnitude, relatively frequent slope-failure events can be more important over long time periods than large, rare events. While this finding may hold for sediment flux and geomorphic work, it may not hold for slope-failure hazards because costs of failures do not necessarily increase linearly with failure volume.

SUMMARY AND CONCLUSIONS

Natural slope failures in the Appalachian Plateau are distributed in time and space by complex interactions of geology and climate. Weak mudstones and shales have attracted the most engineering and design research but are not necessarily the critical factor over longer time scales in natural settings where competent sandstones are present. Pleistocene sediments and colluvium mantle parts of the Appalachian Plateau; these unconsolidated surficial deposits are relatively weak materials and often influence subsurface water flow, pore-water pressures, and location of weathering.

It is generally thought that soil-moisture distribution on slopes is controlled by a complex combination of climatic input, down-dip groundwater flow, and topographic concentration. In the Appalachian Plateau, some studies show slope-failure aspect distributions that conform with expected topoclimatic influences; in other studies, lack of aspect correlations may be caused by bedrock influences on slope morphologies and materials or other factors, such as land use, that vary among failures. Down-dip

groundwater flow does not appear to have a general, important influence on the spatial distribution of natural slope failures in the Plateau because hillslope hydrology can be dominated by zones of secondary permeability created by stress-relief fracturing. More failures are found on laterally and longitudinally concave slope segments, indicating that topographically influenced flow concentration is an important factor in determining relative slope stability.

There are too few data on slope failure frequencies to generalize for the Appalachian Plateau. Events such as the 1977 Johnstown debris slides indicate that a great deal of geomorphic work can be accomplished during rare events. However, the contribution of more moderate events has not been assessed at that site. In contrast, the Marion County, West Virginia, data indicate that events with more moderate recurrence intervals (in this case, 4 yr) are responsible for the greatest cumulative geomorphic work on hillslopes in that area.

DEBRIS SLIDES AND FOOT-SLOPE DEPOSITS IN THE BLUE RIDGE PROVINCE

H. H. Mills

INTRODUCTION

Mass wasting in the Blue Ridge has received somewhat less attention than in the Appalachian Plateau. An exception is the spectacular debris slides that appear to be by far the most common form of rapid mass wasting in this province. This dominance derives in part from the relative lack of other forms of rapid mass wasting that are common in the plateau, probably due to the absence of weak shales. In addition, however, the absolute number of debris slides appears to be greater in the Blue Ridge (Fig. 21). This abundance may stem in part from the greater permeability of regolith derived from crystalline rocks relative to that of regolith derived from the shales of the plateau. In contrast to the fine-grained colluvium of the plateau, such permeable colluvium may be very insensitive to prolonged, slow rainfall, but sensitive to rainfall that is so intense as to infiltrate faster than it drains. The large number of debris slides in the Blue Ridge may also reflect the intense rainfalls that characterize the province, particularly the southern part. Hack and Goodlett (1960) have pointed out that, according to Yarnell's (1935, Fig. 59) data, the 100-yr 24-hr rainfall increases from 6 in (152 mm) in northern Virginia to 13 in (330 mm) in the small area near the border of Georgia, North Carolina, and South Carolina. The susceptibility of crystalline rocks to saprolitization, which provides a ready source of material for slides, may be another contributing factor.

Vast areas covered by bouldery, generally unsorted and unstratified debris lie along mountain foot slopes in many parts of the Blue Ridge. The origin of these "foot-slope" deposits has been considered problematic. Study of modern debris slides in this region, however, makes these deposits somewhat more understandable.

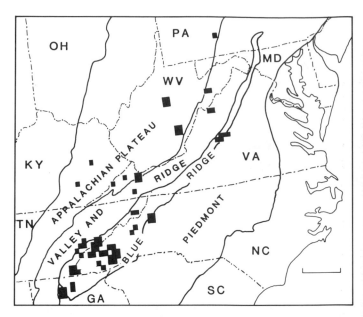

Figure 21. Location of Appalachian debris-slide areas south of the glacial border (after Schneider, 1973, Plate 1).

Although placed by Fenneman (1938) in the Piedmont Province, the Nelson County area of Virginia—site of hundreds of debris slides induced by Hurricane Camille in 1969—possesses topography much more characteristic of the Blue Ridge than of the Piedmont Province, and therefore is herein considered as part of the Blue Ridge Province.

CHARACTERISTICS OF BLUE RIDGE DEBRIS SLIDES

Clark's (1973, 1984) terminology for debris-slide features is used in a modified form herein. "Debris slides" are rotational or translational movements of coherent masses of debris. Typically, these slides rapidly turn into "debris flows," with flowage by internal deformation. For brevity, "debris slides" is used below to refer to both these phenomena, except where specified. "Chute" is a general term for hillslope areas stripped of vegetation by slides/flows (Fig. 22). Clark subdivided chutes into the upper "slide scar" and the lower "flow track," the latter being that portion of the chute modified by debris-flow activity and accompanied by fluvial erosion, transportation, and deposition. Because this distinction requires field study of individual chutes, however, often it is not made.

Chutes generally are much longer than they are wide, with lengths ranging from 6 m (Williams and Guy, 1973) to greater than 700 m (Neary and Swift, 1984), although typical values are 50 to 300 m. Widths range from 3 m (Koch, 1974) to 60 m (Williams and Guy, 1973), with 8 to 24 m being typical. Thickness of debris removed commonly is less than 1 m, but may reach as much as 6 m (Williams and Guy, 1973). Commonly slides

Figure 22. Photograph of typical debris-slide chute in Blue Ridge of Virginia, produced by Camille storm, August 19 to 20, 1969. Located about 7.2 km south of Fairfield, Rockbridge County. (Photo supplied by Virginia Division of Mineral Resources.)

occur in troughlike depressions down the hillslope, and typically the failure surface is the contact between colluvium and bedrock.

Commonly only a small part of the material in a debris slide is deposited at the base of the chute, the greater part being carried away as a debris flow, or washed away by water flows subsequent to the slide. Deposits from a slide depend upon the topographic setting of the hillslope on which the slide occurs, as well as the amount of runoff affecting the deposits. Where a stream is undercutting the base of the slope, virtually no deposits occur. Where the stream is distant from the base, debris fans typically form (Fig. 23). Where a narrow mountain channel emerges into a relatively broad valley, alluvial fans may result. The deposits include enormous boulders and log jams (Fig. 23). Kochel and Johnson (1984) found inverse grading in the 1969 Nelson County deposits, but otherwise, sediments show little sorting or stratification and only a very weak orientation. They also found

Figure 23. Photograph of debris fan (produced by Camille storm) containing boulders and tree trunks at base of east slope of Woods Mountain; Muddy Creek visible at toe of fan. Located about 9.3 km north of Lovingston, Virginia. (Photo supplied by Virginia Division of Mineral Resources.)

that particle-size distributions are coarser than those reported for alluvial fans in the southwestern U.S. (Bull, 1972). On CM diagrams, these distributions plot within the mudflow envelope defined by Bull (1972). They observed no downstream changes in sedimentary characteristics, although lateral variations are common owing to the lobate nature of debris deposition on the fans.

LOCALIZING FACTORS

Many attempts have been made to explain why debris slides occur in particular locations at particular times. Rainfall is certainly the most important control. Scott (1972) found that twentieth-century debris flows in the Blue Ridge were associated with three types of storms. The most destructive and widespread slides, such as those of Nelson County, Virginia, in 1969 (Williams and Guy, 1973), have been caused by tropical hurricanes (see also Kochel, 1986). Extra-tropical low-pressure systems have also produced slides, as have localized severe thunderstorms.

Scott also discovered that all known debris slides in the Blue Ridge occurred in the months of May through October, with the majority taking place during July and August. This finding probably reflects the fact that only warm air can hold sufficient water vapor to produce the intense rainfalls required for debris slides.

Precipitation amounts associated with slides range from 710 mm in 8 hr for Hurricane Camille in Nelson County, Virginia (Williams and Guy, 1973) to as little as 102 mm in 1 hr at Mount LeConte, Tennessee, in 1951 (Bogucki, 1970, 1976). Another example of slides induced by relatively low amounts of precipitation is the slides in Pisgah National Forest, North Carolina, in 1977 (Neary and Swift, 1984). These slides also have the best-controlled precipitation estimates. In the general area, peak intensities at 15 rain gages ranged from 21 to 102 mm/hr, with nearly half exceeding 76 mm/hr. One gage near a slide showed 69 mm in 1 hr, 137 in 3 hr, and 159 mm in 12 hr.

Studies such as the above fail to explain why one slope fails and the adjacent one does not. Possibly there are major differ-

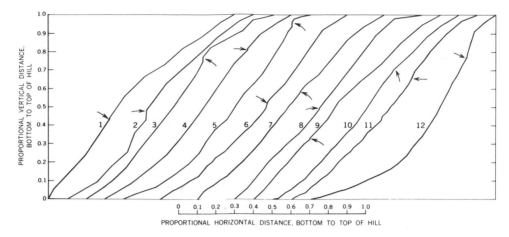

Figure 24. Longitudinal profiles of hillslopes, including avalanche scars, in Nelson County, Virginia. Vertical and horizontal distances have been standardized to extend from 0 to 1.0. Arrow indicates position of head of debris-slide scar (Williams and Guy, 1973, Fig. 16).

ences in the precipitation amounts on adjacent slopes, in which case regional precipitation studies are of limited value. Much effort has been expended in attempting to find other factors that may explain local differences in slope behavior.

Hillslope characteristics are obvious candidates for causal factors, with slope angle being the most investigated. Bogucki (1970, 1976) reported slope angles of 35° to 44°, with an average of 40°, for slides on Mount LeConte, Tennessee. Koch (1974) reported slopes of 32° to 43° (average of 37°) for Webb Mountain, Tennessee, and Neary and Swift (1984) reported that the largest slides in Pisgah National Forest occurred on slopes steeper than 35°. Williams and Guy (1973), in contrast, reported a range of 16° to 39°, with 26° to 31° being most common. These relatively low angles perhaps reflect the nearly unprecedented intensity of the Nelson County storm. Williams and Guy also noted, however, that average slope gradients appear to mean little, for most chutes head at the steepest part of the hillslope profile (Fig. 24).

Length and form of hillslope profiles have also been investigated, but predictive ability of these variables seems to be poor. Slide occurrence shows some association with slope aspect, commonly because dip slope is more favorable to slides (Bogucki, 1976; Koch, 1974), although Williams and Guy (1973) suggested that more rain may have been driven onto slopes in the northeast quadrant. By far the most important facet of hillslope morphology involves the cross-slope profile. Williams and Guy (1973) reported that at least 85 percent of debris avalanches occurred along previously existing depressions in hillslopes, and Bogucki (1976) and Moneymaker (1939) made similar observations (Fig. 25). Dietrich and Dorn (1984) have pointed out that subsurface flow convergence in such hollows makes these sites the most probable locations on a hillslope for sliding.

Aside from the fact that slides are more likely on dip slopes, other structural and lithological factors seem to be of little conse-

quence, except that Bogucki (1976) found slides to be more common on phyllite than on metasandstone. Depth of colluvium has been found to have little predictive ability, as has clay mineralogy of the colluvium. Because most steep slopes (where most slides occur) in the Blue Ridge are forested, ascertaining the effect of vegetation and land use is difficult.

RECURRENCE INTERVALS AND IMPORTANCE FOR DENUDATION

Recurrence intervals of debris slides are of great interest, both because of the importance for geological hazards assessment, and for evaluating the importance of this phenomenon for landscape denudation. One approach is to consider the recurrence intervals of rainfalls that set off debris slides. Neary and Swift (1984) estimated a return interval of 50 to 200+ yr for the intensities of 76 to 102 mm/hr associated with the Pisgah slides of 1977. As these rainfalls are among the lightest associated with Blue Ridge slides, rainfalls responsible for most other slides in this province must have still longer recurrence intervals. Thompson (1969), for example, estimated the return interval for the Nelson County rainfall at greater than 1,000 yr.

Estimating return times of rare precipitation events is subject to large error, however, and other methods may yield more precise estimates of debris-flow recurrence intervals. Kochel and others (1982) and Kochel and Johnson (1984) have attempted to determine recurrence intervals in Nelson County, Virginia, by correlating and dating superposed debris flows and flood deposits in alluvial fans and associated flood plains. Individual units were recognized on the basis of buried soil horizons, changes in matrix mineralogy, bedding character, and clast weathering. Six radiocarbon dates on organics in soils indicate that at least three events of a magnitude similar to the 1969 Camille storm have occurred during the past 10,700 yr.

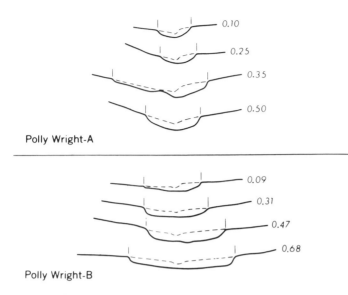

Polly Wright-A

Polly Wright-B

Figure 25. Typical cross profiles of two debris-slide scars (A and B), drawn looking downslope. Number at edge of each profile indicates proportional distance of profile section down hillslope from scar head. Vertical tick marks show edge of scar. Dashed line shows prestorm ground surface. Located in Polly Wright Cove, Nelson County, Virginia (from Williams and Guy, 1973, Fig. 18).

A third approach, not yet utilized in the Blue Ridge, involves colluvial fills in the hillslope hollows. Scott (1972) suggested that most hillslope erosion in the Blue Ridge is accomplished by debris slides. This is true because the line of slope on most hillslopes leads not to the base of the hillslope, but into one of the numerous hollows that corrugate the mountain slopes. Debris thus moves by creep and other slow mass-wasting processes into hollows, where it accumulates until at rare intervals it is flushed from the hollows by debris slides. Dietrich and Dunne (1978) and Dietrich and Dorn (1984) have further developed this model for mountains of the U.S. Pacific coast, and propose that the hollow deposits can be used to estimate debris-slide recurrence intervals. The rationale is that if debris-slide chutes are scoured to bedrock by debris slides, colluvium in hollows must postdate the latest debris slide in that hollow. Hence, by radiocarbon dating basal sediments in the hollow, one can establish a minimum age for the slide. It is difficult to use this method in the Appalachians because hollow fills almost never contain organics. If an alternative method for dating hollow sediments could be found, however, this approach would be extremely useful. The recent application of thermoluminescence dating to colluvium in the Blue Ridge (Shafer, 1984) offers promise.

A factor that might affect recurrence intervals is the length of time required to refill a hollow after a debris slide. However, in the Blue Ridge, the "healing" of slide chutes seems to proceed so rapidly relative to the long recurrence intervals of the slides that this factor probably is not very important. Although erosion may continue on the bare chute for years subsequent to a slide (Scott,

1972), healing is taking place even during this interval (Bogucki, 1976). After 36 yr, Koch (1974) found the slide scars on Webb Mountain to have an average soil thickness of 18 cm, versus 28.5 cm on the sides of the scars, and an average humus-layer thickness of 5.6 cm versus 7.8 cm. Refilling appears to take place by slumping from the chute sides and by erosion in the scar head. The narrowness of the chutes greatly aids the recovery. Revegetation proceeds from the sides, and from the lower part of the chutes upward.

Debris slides obviously are very important agents of denudation in the Blue Ridge. Williams and Guy (1973), for example, calculated the mean denudation in small Nelson County drainage basins resulting from the Camille slides. The means ranged from 35.6 to 50.8 mm. As the long-term denudation rate in the Appalachians is about 40 mm/1000 yr (Hack, 1980), this one storm alone produced 1,000 yr of erosion. Estimates of the long-term importance of debris slides, of course, must await more precise estimates of recurrence intervals. If debris slides can be assumed to be the chief means of catastrophic erosion, however, a gross estimate can be made by comparing "normal" erosion rates to long-term rates. Velbel (1985), for example, noted that in small basins of the Coweeta Lab in North Carolina, sediment-export rates accounted for only about 5 to 45 percent of the long-term rate of 40 mm/1000 yr. Thus, 50 to 95 percent of physical erosion must occur as high-magnitude, low-frequency events. Most such events probably derive a considerable portion of their sediment from debris slides. Williams and Guy (1973), for example, estimated that in the 1969 Nelson County flood, almost half the sediment contributed to the stream system came from this source.

It thus appears that rare events of large magnitude may be somewhat more important for long-term denudation in the Blue Ridge than in the Appalachian Plateau. In contrast to the Plateau, however, events of moderate frequency and size have received almost no attention in the Blue Ridge, so that the assumption that such events are not very important in this province cannot yet be demonstrated.

MORPHOMETRY AND SEDIMENTOLOGY OF FOOT-SLOPE DEPOSITS

Blue Ridge foot-slope deposits may be divided on the basis of morphology into continuous and discrete types. The former occur as aprons that may extend for many kilometers along mountain piedmonts. Large examples occur in Virginia, where resistant rocks of the Blue Ridge abut the carbonates of the Great Valley (e.g., Hack, 1965, plate 2). A smaller example occurs along the northwest flank of Grandfather Mountain, North Carolina (Mills, 1981b). Discrete types commonly have been referred to as "fans." Although many of these features are not at all fan shaped (Fig. 26), this term will be used for brevity.

Unlike the classical alluvial fans that occur below relatively straight fault scarps in the Basin and Range Province of the southwestern United States, the size and shape of the Blue Ridge

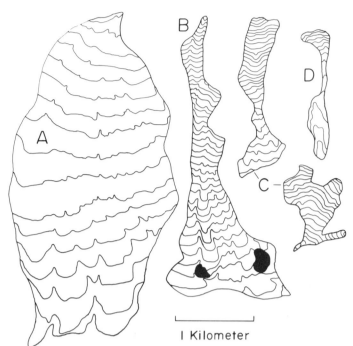

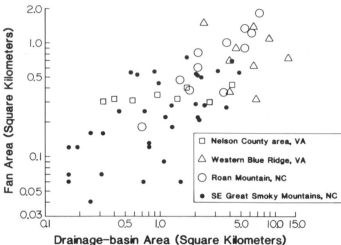

Figure 27. Log-log plot of fan area versus drainage-basin area for fans in Blue Ridge Province. Sources of data: Nelson County area and western Blue Ridge, Virginia—Kochel and Johnson, 1984; Roan Mountain—Mills, 1983; southeast Great Smoky Mountains—Mills, 1982.

Figure 26. Examples of discrete foot-slope deposits, or "fans," in Blue Ridge Province. Downslope direction is to bottom of page. Contour interval is 40 ft (12.2 m) for A to C and 20 ft (6.1 m) for D. Fans are from the following quadrangles: A, Big Levels and Stuarts Draft quadrangles, western Blue Ridge, Virginia; B, Bakersville Quadrangle, Roan Mountain, North Carolina; C, Dellwood Quadrangle, southeast Great Smoky Mountains, North Carolina; D, Horseshoe Mountain Quadrangle, Nelson County, Virginia.

fans commonly are somewhat more constrained by the surrounding topography. In arid regions, log-log correlations between alluvial-fan and drainage-basin areas generally are high, and regression equations of the form $A_f = c\,A_d^n$ (where A_f is fan area and A_d is drainage-basin area) yield values of the exponent n that are generally between 0.8 and 1.0 (Bull, 1962, 1964; Denny, 1965; Hooke, 1968). Similar plots for Blue Ridge fans show somewhat different results (Fig. 27). The results can be related to the degree of topographic constraint. The most unconstrained fans on which measurements have been made are those produced by streams debouching into the Great Valley from the Virginia Blue Ridge (A in Fig. 26). These fans are the most similar to those of the Southwest. They are the largest in the Blue Ridge, and are composed of clast-supported gravels that appear to be braided-stream deposits. Thickness ranges from 30 to 110 m (Kochel and Johnson, 1984). Kochel and Johnson attributed the poor correlation between fan and basin areas to problems in determining the areal extent of the coalescing fans. However, it also appears that a relatively small size range (Fig. 27) may have made discerning a relationship difficult.

The next less constrained fans are those at Roan Mountain, North Carolina, which occur in large embayments in the moun-

tain front known in this region as "coves." (B in Fig. 26). These are smaller than the above fans, and although shapes are grossly fan shaped, they are somewhat less regular than classical fans. Although fluvial facies occasionally are seen in the distal parts of the fans, the fans are composed primarily of bouldery diamictons. Average thickness is 18.5 m. A fairly good relationship between fan and basin areas exists (r = 0.83). In the equation $A_f = c\,A_d^n$, n is 0.76, only slightly less than the 0.8 to 1.0 range typical of Southwest fans.

Still more constrained are the fans of Dellwood Quadrangle in the southeastern Great Smoky Mountains (C of Fig. 26), which occur in coves that generally are smaller and narrower than those of Roan Mountain. The fans generally are smaller than the Roan Mountain fans, and many are not fan shaped. Most exposures reveal bouldery diamictons. Thicknesses are poorly known, but appear to be less than those at Roan Mountain. The relationship between fan and basin areas (Fig. 27) shows an r value of 0.63 and an n value of only 0.53, much less than that shown by Southwest fans. The irregular shapes and low n values probably reflect the limiting effect of topography. No matter how much basin area is increased, there simply is not room for fan size to increase a corresponding amount.

The most topographically constrained fans are those of Nelson County, Virginia, most of which are constricted between narrow basin interfluves on steep hillslopes (D in Fig. 26). Fan shapes typically are elongate and irregular. Fan material consists chiefly of diamictons, and thickness ranges from 5 to 20 m. Although the log-log correlation between fan and basin areas is 0.60, the n value is only 0.09, not significantly different from zero. In other words, topographic restraint is so great that fan area is relatively constant regardless of basin area.

Gradients of Southwest alluvial fans decrease as an inverse log-log function of drainage-basin area (Bull, 1962), and Dellwood fans show a similar relationship. However, the regression line for the latter plots far above that for the former, indicating somewhat steeper gradients for a comparable-sized basin (Mills, 1982, Fig. 3e). Gradients for Roan Mountain fans, though not as steep, also plot well above those of Bull (Mills, 1983). These steeper gradients probably reflect the paucity of water-laid sediment on many of the Blue Ridge fans, as well as the coarse, bouldery texture of the fan debris. Kochel and Johnson (1984), on the other hand, reported gradients for Nelson County fans that are more comparable to those reported by Bull. The reason for this discrepancy is not clear.

Although clast-supported gravels of the western Blue Ridge fans in Virginia are obviously water-laid deposits, the origin of the diamictons on other fans has been viewed as problematic. Michalek (1968), for example, studying Blue Ridge fans south of Virginia, thought much of this material is relict from glacial ages, having been emplaced by solifluction. Detailed work on modern debris slides in the Blue Ridge, however, makes it appear very likely that foot-slope deposits were emplaced by debris flows. Sediments laid down by modern debris flows appear very similar to those in older foot-slope deposits, and there seems little reason to invoke solifluction. To be sure, distinguishing Pleistocene debris-flow deposits from solifluction deposits of a similar age may be difficult. Characteristics of solifluction tend to be either not very distinctive or else are unlikely to survive very long after solifluction ceases (Benedict, 1976). This problem is compounded by the difficulty of dating the deposits. One apparent sedimentary difference between debris-flow and solifluction deposits, however, is consistency of clast long-axis orientation. Although fabric patterns may be similar, debris flows generally have weak fabrics (Mills, 1984), whereas solifluction fabrics are quite strong (Nelson, 1985). The weak fabric observed in most of the fan deposits thus supports a debris-flow origin.

ON THE ORIGIN OF FOOT-SLOPE DEPOSITS

From their investigation of fans in the Blue Ridge of Virginia, Kochel and Johnson (1984) proposed a geomorphological model for alluvial fans in their study area. Essentially, the basis for this model is the extrapolation into the past of processes and features associated with the 1969 Hurricane Camille storm, using fan stratigraphy and radiocarbon dating to establish recurrence intervals. The main features of the model are that rainfalls with recurrence intervals of 3,000 to 6,000 yr initiate debris slides, and the resulting debris flows cover between 20 and 70 percent of the downslope fan surface. Commonly these rare storms erode deep fan-head trenches and deposit debris on the lower part of the fan. During the long intervals between events, soils form and depressions may be eroded into the fans. Successive events show a shifting of the locus of deposition across the fan, induced by deposition of sediments into fan depressions. Such shifting results in the smoothing of fan topography.

Although this model is satisfactory for the small, Holocene fans studied by Kochel and Johnson (1984), it has two limitations that prevent its applicability to the Blue Ridge Province as a whole. First, it ignores the effect of Quaternary climatic change. Because deposits during a given storm are relatively thin, with a recurrence interval of 3,000 to 6,000 yr, even a very small fan takes many tens of thousands of years to form. Consequently, the building of most fans probably spans at least several Pleistocene stadials and interstadials, if not glaciations and interglaciations. During glacial stadia, large areas of the Appalachian Highlands were above tree line (Delcourt and Delcourt, 1981). Because of the lack of trees and the accelerated frost action, the volume of material available for debris slides probably was greatly increased in many basins.

Hence, although Michalek's (1968) attribution of fan sediments to solifluction is probably erroneous, his suggestion that fans were related to former periglacial climates may be at least partly correct. As evidence for a periglacial nature, Michalek pointed out that (for fans south of Virginia) the upper ends of many fans are covered with block fields, that no fans are found below about 700 m, and that fans diminish in both number and development southward, becoming uncommon south of Asheville, North Carolina. Such observations require confirmation, but they are suggestive.

That glacial climates provided optimum conditions for fan growth cannot be assumed with certainty, however. Although the available debris may have been greater, it is probable that the drier air associated with Arctic airmasses (Delcourt and Delcourt, 1984) made intense rainstorms somewhat less common than in the present climatic regime, although to be sure the absence of forest cover probably decreased the rainfall intensity necessary to initiate debris slides. Based on his studies of terraces along the Little Tennessee River, Delcourt (1980) suggested that episodes of massive alluviation (and by implication, colluviation) in the southern Appalachians are confined primarily to lateglacial/interglacial (and stadial/interstadial) transitions. Knowledge of exactly which Quaternary climates were optimum for debris slides must await extensive dating of fan deposits.

The effect of climate, of course, probably varied with fan and basin elevations. The Nelson County fans, for example, occur at elevations of only 300 m or so, and the highest peaks in their basins seldom exceed 700 m. These fan systems may well have remained largely forested during glacial climates, making Kochel and Johnson's model more applicable to them than to fans at much higher elevations farther south in the Blue Ridge Province. Fans in the southeastern Great Smokies, for example, are at elevations of 900 to 1100 m, and peaks in the basins rise above 1600 m.

The second limitation of the model stems from the small size and constricting topography of the Nelson County fans. Larger, less-constrained fans commonly show large areas of permanently abandoned depositional surfaces. These persist as old, high-level remnants of fan surfaces (Fig. 28). Soil development and weathering of sediments on these surfaces increase in proportion to their

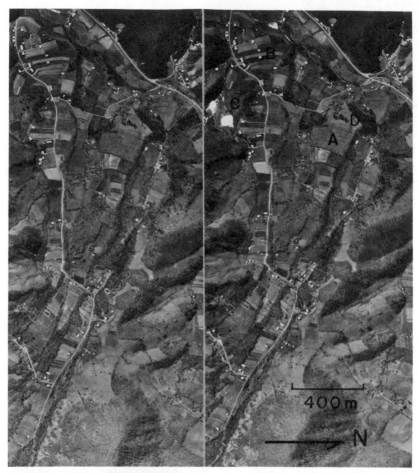

Figure 28. Stereo aerial photograph of high-level remnants of fan surfaces at Roan Mountain, North Carolina (Mills, 1983). Highest remnant (A) is surrounded by young surfaces. A lower, but still relict, surface is at B. C and D indicate bedrock knobs.

elevation above the modern depositional surface. The process by which these surfaces is abandoned has been described by Hack (1960, 1965) and Mills (1983), and is similar to the process described for pediments by Rich (1935). Basically, the abandonment simply reflects the fact that although accumulation on a fan surface may continue for a long interval of time, the Blue Ridge is, after all, an area characterized by long-term erosion, not deposition. Eventually, because the surrounding landscape is gradually being lowered by erosion, streams on the depositional surface are captured by streams flowing at lower elevations. Remnant surfaces are rare on small, constricted fans such as those in Nelson County and in Dellwood Quadrangle, probably because they are quickly eroded away.

SUMMARY

Whereas mass wasting in the Appalachian Plateau seems to be dominated by low-magnitude, high-frequency events, the reverse appears true for the Blue Ridge, where debris slides seem to dominate. The difference may result in part from the more

permeable colluvium in the Blue Ridge, as well as from the more intense rainfall that occurs in this province, particularly the southern part.

Geologic and topographic factors are at best only fair predictors of debris-slide locations, except that most slides occur where hillslopes are laterally concave. Recurrence intervals of slides probably are measured in thousands of years and are best estimated by correlating and dating superposed debris flows and flood deposits in fans and associated flood plains. Although they occur only at long intervals, debris slides may account for more than half of landscape denudation in the Blue Ridge.

Fans that occur on mountain foot slopes in the Blue Ridge are much less regular than alluvial fans of the southwestern U.S., owing to greater constraint by surrounding topography. Generally, areas with smaller fans show greater control by the adjacent topography than areas with larger fans, and correlations between fan and basin areas tend to be lower. Gradients of Blue Ridge fans tend to be somewhat steeper than those of fans in the Southwest, perhaps because of a greater prevalence of debris-flow deposits on the former. Blue Ridge fans probably are formed by repeated episodes of debris slides set off by rare storms. However, particularly for larger fans at higher elevations, paleoclimate may have had a substantial effect on long-term rates of accumulation. Old fan surfaces are much more likely to be preserved on large fans than on small ones.

FLUVIAL SYSTEMS IN THE APPALACHIANS

G. R. Brakenridge

INTRODUCTION

Previous sections of this chapter discuss examples of the major impact of streams on landform development and of the strong influence of nonhomogeneous bedrock and structure on the localization and morphology of streams. The focus in this section is on fluvial sedimentary and erosional processes in their own right, how they operate and how they respond to external and internal changes, viewed at a variety of time scales.

Specific goals of Appalachian work on fluvial systems have included determining the effects of human impacts on modern-day fluvial hydrology and sedimentation, understanding the process implications of documented fluvial sedimentary histories ("alluvial chronologies"), and quantifying local fluvial erosion rates and the processes responsible for bedrock-valley evolution. These topics will be considered below.

HUMAN IMPACTS

European geomorphologists often visit North America to enjoy their first observations of unchannelized and unregulated rivers. By their standards, many Appalachian rivers are wild and represent a chance to study fluvial systems that are not as affected by human impacts as are rivers in Europe. However, a number of studies have emphasized the strong impact of recent environmental change, including human land use, on many watersheds in this region (Arnold and others, 1982; Costa, 1975; Faye and others, 1980; Trimble, 1975; Wolman, 1967). Several workers have also demonstrated that complex processes and significant lag times occur during fluvial responses to environmental changes (Schumm and Parker, 1973; Knox, 1975). Thus, many apparently "natural" undisturbed streams and rivers in the Appalachians may actually be in various stages of responding to earlier, and profound, changes in the environment.

Working in the Piedmont, Trimble (1975, 1977) described a post-Euroamerican-settlement watershed history characterized by high rates of soil erosion and fluvial sedimentation due to forest clearing, burning, grazing, row-crop agriculture, and other activities. This major change, which may have affected many other fluvial systems in the Appalachian region, began diachronously in the late 1700s to early 1800s in different areas. In the 1900s, this change was followed in many areas by a reversed trend: widespread regrowth of Appalachian forests, probably coupled with similarly widespread reductions in hillslope sedi-

Figure 29. Photograph of modern flood plain (TO) (right foreground), and immediately prehistoric flood plain (T1) (left background), along the Duck River, Tennessee.

ment yield. However, at least some streams and rivers in the region are still in the process of reworking the early inputs of large sediment volumes, as most of the sediment is still in storage within the system (Trimble, 1975, 1977).

Even in the mountainous upland areas, extensive "natural" areas do not exist. Patches of old-growth forests remain only in a few relatively inaccessible areas. Extensive logging in the 1800s bared most of even the highest Appalachian peaks, and the present, heavily forested appearance of much of the Appalachians is thus a very recent phenomenon.

Recent alterations of fluvial geomorphology in this humid and now forested region are not as obvious, or as well studied, as the changes experienced in the semi-arid western U.S. However, the famous "arroyo-cutting" episodes at the beginning of the twentieth century in the West (Cooke and Reeves, 1976) may actually be no more profound than the changes that occurred along Appalachian streams somewhat earlier. For example, along the Duck River in the Interior Low Plateau of Tennessee, a postsettlement alluvial fill is adjacent to the immediately prehistoric floodplain, which is now a terrace (Fig. 29). The river bank of immediately presettlement age is marked by a dipping paleosol (Fig. 30); since this time, the bank has moved about 60 m to the east. At this location, neither surface is forested, but at others in the same valley, approximately 100-yr-old forest stands are present on the T0 surface and obscure the dramatic, and very recent, change in flood-plain morphology.

Elsewhere along this river, tributary gullies are graded to the modern channel and expose prehistoric flood-plain sediments in their walls. The stage was set for gullying when the T1 surface was abandoned by the river as an active locus of sediment accumulation. Thus, a change initiated, as best can be determined, early in the nineteenth century (Brakenridge, 1984), is still strongly affecting this fluvial system's operation.

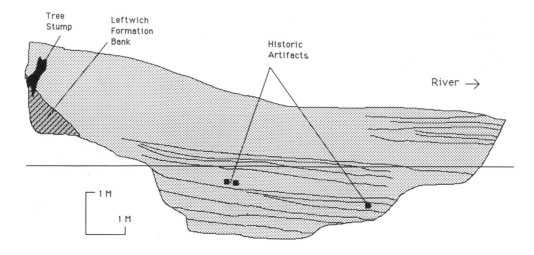

Figure 30. Cross section of trench exposure of T1/T0 contact shown in Figure 29.

Despite the modern trend toward reforestation of Appalachian hillslopes, more recent, and more localized, human impacts continue to occur. In the 1950s through the present, a number of Appalachian watersheds witnessed the rapid growth of urban, suburban, and highway development, resulting in a second widespread episode of extrinsic fluvial perturbations. In studies assessing such changes, increases in the frequencies and magnitudes of small- to intermediate-sized flood events were found (Wolman, 1967; Arnold and others, 1982). In most of these cases, the rapid readjustment of the streams involved is manifested by increased channel dimensions, rapid sedimentation and erosion, and by increased flooding-related engineering problems. These changes may, in some cases, be superimposed on systems already perturbed by the previous, larger-scale land-use changes noted above. Thus, fluvial geomorphologists working in this region must understand the recent landscape and river history in their study areas before interpreting modern-day process measurements.

PROCESS IMPLICATIONS OF ALLUVIAL CHRONOLOGIES

Fortunately, the operation of fluvial systems through time in much of this region has been recorded by geomorphic landforms and alluvial terrace sequences. Recent years have seen the development of the first detailed studies of the chronology of river terraces in and near the Appalachian region (Haynes, 1985; Scully and Arnold, 1981; Brakenridge, 1981, 1983, 1984). Three examples of studied alluvial stratigraphies are shown in Figure 31. Such studies are of interest here for their implications concerning (1) the episodic operation of fluvial sedimentary systems at relatively long time scales, and (2) the processes responsible for the morphology of some classes of Appalachian bedrock valleys.

Holocene Episodes of Fluvial Activity

Stratigraphic-geochronological investigations along rivers in the Appalachian region have documented episodes of fluvial-system changes during prehistoric time that are similar in nature to those that occurred in early historic time. Features common to the Holocene deposits of these rivers (Fig. 31) are (1) buried soils that mark old flood-plain surfaces, (2) the presence, side by side, of flood-plain deposits of quite different ages ("row terraces" of Schirmer, 1983), and (3) the existence of both conformable contacts (marked by dipping paleosols that mark former riverbank positions), and unconformable contacts (marked by truncation of stratification) at terrace boundaries. According to this evidence and radiocarbon dates, the Holocene fluvial history along all three streams is characterized by periods of relative morphologic stability, during which soil profiles develop, alternating with intervals of fluvial activity, during which rapid fluvial sedimentation, erosion, and lateral channel migration occur (Fig. 32). A pressing question is: what caused the prehistoric episodes?

Currently, the alternatives suggested are (1) the inherent episodicity of fluvial processes (Schumm and Parker, 1973), (2) the effects of Holocene climatic changes (Knox, 1975; Brakenridge, 1980), or (3) the combined operation of (1) and (2). A variety of ultimate causes could result in the destabilization of channels with respect to their channel-forming dominant discharge, including changes in the magnitude of that discharge itself. Any such changes, whether of local or regional extent, could result in a period of readjustment such as is modeled in Figure 32, and documented for Holocene time along actual rivers as shown in Figure 31. One cannot interpret causation, therefore, from even the most accurate determination of the river's sedimentary response.

A useful approach to this problem is to compare independent records of environmental change with the alluvial record.

This is unfortunately still difficult to do for the Appalachians because of a lack of sufficient proxy paleoclimatic data of the appropriate time resolution. However, there seems to be plausible evidence for a causal relationship between Holocene climatic change and episodes of fluvial activity on the basis of synchroneity of timing. Recent work along rivers in Missouri, Tennessee, and Vermont suggests a widespread episode of fluvial activity at about 8000 to 7000 ^{14}C yr B.P., which interrupted (e.g., Brakenridge, 1984) an earlier period of stability and flood-plain soil development. This was followed by closely spaced periods of activity and erosion in late Holocene time (Fig. 31). It is, at the least, clear that Holocene time witnessed important climatic changes in many regions, including the northern and southern Appalachians (Davis, 1983; Delcourt and Delcourt, 1984). The impact such climatic changes should have had on hydrologic systems (e.g., Schumm, 1965, 1968) argues that some of the documented prehistoric episodes of fluvial activity may have had a climatic cause.

It should be emphasized that many fluvial sedimentary systems, including those in the Appalachians, are strongly "buffered"; the dominant processes and self-dampening feedback adjustments in such systems are resilient and persist even in the face of strong outside perturbations. Environmental changes are more likely to change the system kinetics (e.g., *rates* of vertical or lateral accretion or erosion) than the actual nature of the processes themselves (a fact recognized by Knox, 1975). Appalachian valleys contain several specific examples of this effect. Consider, for example, the phase of prehistoric flood-plain deposition, following an interval of soil development immediately after 6,885 yr B.P. for the Duck River (Fig. 31). Renewed deposition was associated with lateral channel migration as well, and was very similar to the historic phase of activity (compare the relationship between depositional units II and III and units V and VI in Figure 31C. In both cases, process rates were affected, and the construction of new flood-plain accumulations occurred simply by the acceleration of normal processes. The similarities between prehistoric and historic fluvial events in the Appalachian region underscores the observation that, although fluvial systems are affected by external changes, they impose the strong imprint of their own dominant processes on the resulting sedimentary record.

Bedrock-valley Morphologies

Many rivers in the Appalachian region can be classified as one of three types: (1) freely meandering rivers that swing back and forth across the valley floor, (2) ingrown meandering rivers that are relatively confined within asymmetrical bedrock walls, and (3) straight rivers that flow within symmetrical bedrock walls, with little or no net valley-floor alluviation. Figure 33 illustrates these three river and valley morphologies as a continuum developed along a hypothetical drainage of a mature orogenic belt such as the Appalachians. Symmetrically incised meandering valleys are a special class and are not considered here.

The downstream cross section (A in Fig. 33) illustrates a

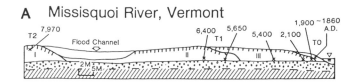

A   Missisquoi River, Vermont

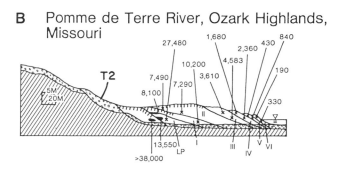

B   Pomme de Terre River, Ozark Highlands, Missouri

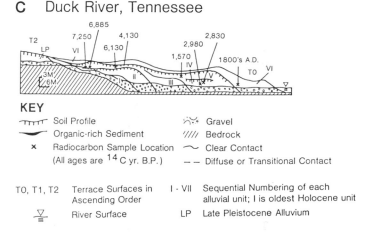

C   Duck River, Tennessee

KEY

⊤⊤⊤⊤ Soil Profile	⋏⋏ Gravel
⌣ Organic-rich Sediment	//// Bedrock
× Radiocarbon Sample Location	⌒ Clear Contact
(All ages are ^{14}C yr. B.P.)	— — Diffuse or Transitional Contact

T0, T1, T2   Terrace Surfaces in Ascending Order   I - VII   Sequential Numbering of each alluvial unit; I is oldest Holocene unit

▽ / ═   River Surface   LP   Late Pleistocene Alluvium

Figure 31. Examples of cross sections of alluvial-terrace sequences. A. Missisquoi River, Vermont, which drains the west slopes of the Green Mountains (see also Brakenridge and others, 1986). B. Pomme de Terre River, Missouri, in the southern Ozark Plateau (modified from Brakenridge, 1983); T2 surface indicated. C. Duck River, Tennessee, in the Interior Low Plateau (modified from Brakenridge, 1984).

typical morphology for freely meandering rivers. Such rivers may be approximately in equilibrium with respect to sediment if channel beds are at or near bedrock (as shown). If, however, channel beds are underlain by Holocene alluvium, they may be undergoing net aggradation. An example from the Appalachians is the upper reaches of the Susquehanna River in northern Pennsylvania and southern New York.

The cross section somewhat farther upstream (B in Fig. 33) illustrates a typical morphology for ingrown meandering rivers dissecting regions such as the Appalachian and Ozark plateaus. Actual examples are shown in Figures 34 and 35. Such rivers

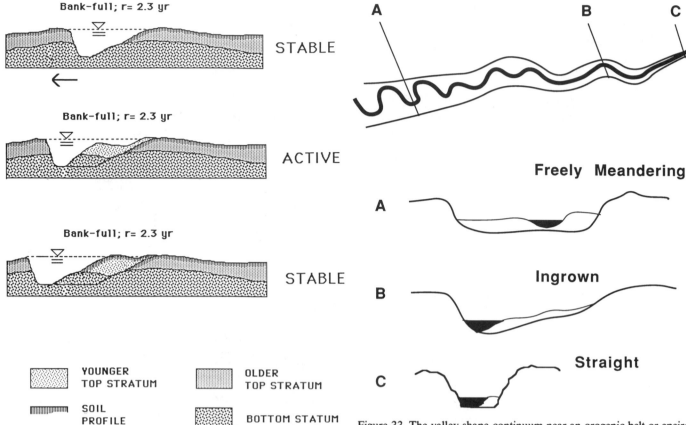

Figure 32. Accumulation of a new flood-plain deposit consisting of fine-grained bank and overbank (top stratum) facies overlying coarse-grained channel-bed (bottom stratum) facies. The channel-forming dominant discharge is indicated as the flow with a 2.3-yr recurrence interval. During an active phase of channel motion, new deposits are rapidly accreted on convex banks until approximate net sedimentary equilibrium is attained. During the stable phase, slow overbank-deposition rates prevail and soil profiles develop. Note that two row terraces, separated by a conformable contact, have been created while coeval cut-bank erosion has occurred on the opposite bank (from Brakenridge, 1986).

Figure 33. The valley-shape continuum near an orogenic belt or epeirogenic uplift. Cross section A illustrates a typical morphology for freely meandering rivers, B illustrates a typical morphology for ingrown meandering rivers, and C illustrates typical low-order, actively incising straight rivers.

exhibit steep cliffs on the outsides of valley bends and more gentle slip-off slopes on the insides of bends.

The most-upstream reach illustrated (C in Fig. 33) exhibits the morphology of straight rivers with little or no flood-plain development. In the Appalachians, these typically are vertically incising tributary streams of low order (see examples in Fig. 35), but also include trunk rivers flowing through canyons and gorges (e.g., middle portions of the Susquehanna River in eastern Pennsylvania).

The ingrown-valley morphology is of special interest because its origin has been controversial and recently has received study. The sinuous form of many ingrown bedrock valleys suggested to earlier investigators an "underfit stream" hypothesis for their origin (Dury, 1977). This hypothesis invoked paleohydrological changes (which, as noted previously, undoubtedly did

occur) to explain the meandering valley forms, the asymmetrical cross-valley profiles, and other characteristics. In brief, discharges many times larger than present discharges would have filled the valley floors and allowed valley erosion to occur in a manner analogous to channel erosion. The large-scale slip-off slopes and steep cliffs opposing them were thus inferred by Dury to be analogous to point-bar and cut-bank topography developed along present-day meandering rivers.

This hypothesis viewed many modern rivers in the Appalachians as occupying relict valleys carved during late Pleistocene time. The rivers of today were thought capable of only extremely slow valley modification because of their much-reduced competence and size. However, Tinkler (1971) and Brakenridge (1984, 1985) later concluded that several supposedly underfit rivers, without any changes in their overall discharge regimes, could have eroded the valleys they now occupy. In particular, alluvial stratigraphic studies along valleys of the Pomme de Terre River in the Ozark Plateau of Missouri and the Duck River in Tennessee clearly demonstrate that the late-Quaternary rivers occupying these valleys differed little in size from the modern rivers (Haynes,

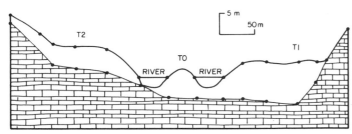

Figure 34. Cross section of an ingrown meandering valley (as determined from drill-core data) of the Duck River, Tennessee. Along this relatively straight reach, a reversal of channel migration has occurred in Holocene (T1) time; lateral valley widening is underway.

1985; Brakenridge, 1981, 1985). The same studies also indicate that significant bedrock erosion rates have prevailed through Holocene time, so that the stream channels are actively modifying their valleys today (see following section). Thus, instead of viewing such meandering valleys as relict features, it now appears more reasonable to consider them as normal transitional forms between valleys enlarging by vertical incision and those enlarging by lateral erosion.

FLUVIAL EROSION RATES AND THE EVOLUTION OF BEDROCK VALLEYS

The above discussion suggests that, other factors being equal, bedrock-valley morphology in upland regions is controlled by the relative rates of lateral and vertical erosion. Alluvial chronological studies in the Appalachians have allowed estimates of such erosion rates averaged over intervals of hundreds or thousands of years. Such long-term averages are appropriate figures to use in modeling landform evolution, and constitute "baseline" information useful for comparisons with modern process observations. The following description of valley-evolution processes and rates attempts to seek a middle path between Dury's (1977) interpretation of many Appalachian valleys as relict features and a too-simplistic reliance on modern process measurements as reliable keys to fluvial erosion over long time spans.

Working in the Ozark Plateau, Haynes (1985) and Brakenridge (1981, 1983) described in detail a local sequence of fluvial erosion for the time span accessible to radiocarbon dating. The summary diagram Figure 31B illustrates (in much-abbreviated form) the chronology; the ingrown valley is shown in map view in Figure 35. Note the two abandoned valley meanders shown in this figure. The available radiocarbon and subsurface information indicates that the more-northern meander was abandoned before

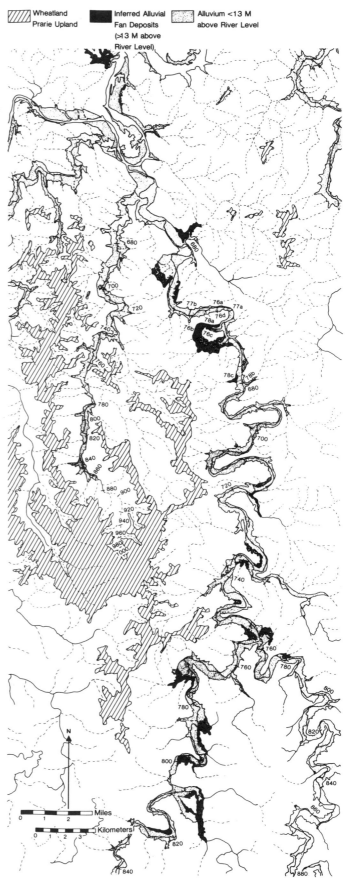

Figure 35. Geomorphic map of the Pomme de Terre drainage in the Ozark Plateau Province, showing heavily dissected topography near the ingrown meandering valley of the river, remnants of the Wheatland Prairie Upland at greater distances from the river, and two abandoned meanders that are presently filling with alluvial-fan sediments.

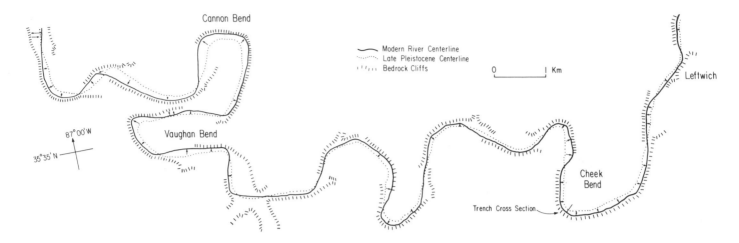

Figure 36. Locations of modern Duck River channel and bedrock cliffs, and inferred position of the late Pleistocene (ca. 14,000 yr B.P.) river. Flow is from east to west (from Brakenridge, 1985).

about 49,000 ^{14}C yr B.P. The bedrock valley floor then incised about 5 m sometime between about 49,000 and 45,000 ^{14}C yr B.P. (Brakenridge, 1981). Subsequently, the abandoned valley meander has been filling slowly with colluvial and alluvial fan sediments, whereas lateral erosion has been widening the modern valley.

Abandoned valleys such as shown in Figure 35 have not yet been well studied, even though they are common and their relative ages can be estimated by quantitative geomorphological methods (e.g., see Leopold and Bull, 1979). They are also significant modern sedimentary storage sites, capable of trapping portions of a watershed's sediment yield for long periods of time; important paleontological records are often preserved in such locations.

The higher bedrock-valley floor beneath the dated meander, the coeval bedrock step (below the T2 surface, Fig. 31B) along the modern valley, and the still higher strath terraces, confirm a history of episodic incision along the Pomme de Terre River. However, Figure 31B and the detailed work of Haynes (1985, see especially Fig. 3 therein) both show that the studied river reach has not undergone significant vertical erosion since T2 time. The channel floor today is no lower than it was at various times during the past 45,000 yr (see also Brakenridge, 1981, Fig. 5). Such observations raise important, and still unresolved, questions regarding valley-erosion processes. Is vertical bedrock erosion inherently episodic along this river (the natural imprint of the fluvial system, *sensu* Shepherd and Schumm, 1974)? Or do the bedrock-valley steps actually reflect pulses of isostatic or tectonic uplift? Because bedrock-terrace levels may be our best information on Quaternary tectonism (Mike, 1975), and because they are found along other valleys in the region (see below), this question is an especially important one for future workers to address.

The valley of the Duck River, Tennessee, exhibits a similar ingrown morphology and alluvial stratigraphy (Figs. 31C and 34). Figure 36 shows a map view of this valley and the modern channel, as well as the inferred position of the late Pleistocene (ca. 14,000 ^{14}C yr B.P.) channel. The relict channel position was reconstructed on the basis of river-facing scarps of the T2 terrace mapped along this river (Brakenridge, 1985). Comparison of the two river positions suggests similarities between the style of long-term fluvial bedrock erosion and that of freely meandering rivers, whose much more rapid changes have been measured: valley meander bends enlarge outward and bend apices move downstream through time. A stepped bedrock topography was also mapped along this river (Brakenridge, 1984), again indicating the long-term episodic nature of vertical incision.

Lateral erosion, also, was demonstrably not a continuous process along the river. Instead, such erosion was the result of episodic lateral channel migration, as indicated by the emplacement of an alluvial sequence burying relatively stable, convex-bank positions, separated by more rapidly deposited lateral-accretion sediments. Along one river bend, the preserved and dated fossil soils mark former bank and flood-plain positions (Fig. 37) and allow estimation of Holocene rates of lateral bank accretion. The estimates accord with independent lateral bedrock-erosion rates obtained by comparing the positions of bedrock cliffs and terrace scarp positions (Fig. 36), indicating that bank accretion and flood-plain extension, at the observed rates, may occur coevally with lateral bedrock erosion. Average rates of lateral bedrock erosion obtained in this manner for the Duck River, underlain by Paleozoic limestone, vary from 0.6 to 1.9 m/100 yr.

Such high rates suggest that, given a lack of further vertical incision, the residence time of a particular ingrown valley bend along this river reach will be limited. Within, at most, tens of thousands of years, the bend will either be cut off, or the focus of

Figure 37. Photograph of a buried early Holocene flood-plain surface along the Duck River. Surface corresponds to unit II in Figure 31C and is marked by a paleosol and concentrations of cultural artifacts, charcoal, and pebbles (flagging). Middle Holocene flood-plain alluvium (unit III in Fig. 31C) has buried this surface, which still exhibits typical levee (background) and swale (foreground) topography. A period of fluvial activity, here dated at ca. 7200 to 6400 ^{14}C yr B.P., rearranged this river's morphology in a manner similar to that which occurred at about 1820 A.D.

active erosion will shift downstream. Especially through the latter process, valley floors are gradually broadened, and, barring renewed vertical incision, the rates of lateral erosion greatly decline, because channel contact with valley margins becomes less frequent as the valley widens (compare profiles A and B in Fig. 33; also see Fig. 34).

EROSION RATES IN THE NORTHERN APPALACHIANS

Thus far, the fluvial sequences examined for their long-term morphogenetic-process implications have been in nonglaciated portions of the Appalachians. In glaciated portions, similar processes occur, but the substrate on which they work includes various types of Pleistocene glacial deposits as well as consolidated bedrock. Also, glacio-eustatic tectonism has played an important role in river-valley evolution.

Figure 38 shows a meander bend along the middle Missis-

quoi River, which drains the western flanks of the Green Mountains in northern Vermont. To the left of the channel are the flood plain and low alluvial terraces left by the Holocene river; their subsurface stratigraphy and chronology is given in Figure 31A. To the right of the channel are forested, late-Pleistocene marine clay deposits dating from the immediately postglacial incursion of the ocean into this region. The surface of these marine deposits stands 37 m above the modern channel; the river incised the deposits in an asymmetric manner (Fig. 39) as the landscape slowly rose above sea level by isostatic crustal rebound.

The radiocarbon dates on logs contained in this river's Holocene deposits has allowed a partial description of the history of this river bend (Brakenridge and others, 1986). These data, when combined with information regarding geomorphic surfaces and valley topography, illustrate the roles played by both lateral and vertical processes in this youthful valley's genesis. Figure 40 shows summary plots of the Missisquoi River channel position during Holocene time. In response to glacio-eustatic uplift, the

Figure 38. Oblique aerial photograph of a meander bend of the Missisquoi River in northern Vermont. Note trench location (cross section shown in Fig. 31A).

river initially incised late Pleistocene marine deposits very rapidly, at rates averaging at least 1 m/100 yr. Then, following about 7,000 to 8,000 ^{14}C yr B.P., vertical incision slowed to nearly zero and episodic lateral channel migration (and concave-bank erosion of the Pleistocene substrate) began to occur, at rates varying from 0 to 4 m/100 yr. As discussed above for other rivers, local fluvial processes have also been affected recently by human impacts. Figure 40 shows that the lateral channel-migration rate increased from near zero to 1.5 m/100 yr about A.D. 1860 (see also Fig. 31A and Brakenridge and others, 1986).

SUMMARY

This discussion of Appalachian fluvial systems has examined the influence of human and climatic-change effects, the process implications of recent alluvial chronological studies, and the relative roles of lateral and vertical erosion in carving valleys. Ingrown-valley morphologies were stressed, partly because of the controversy over their origin and partly because they constitute an important transitional link between the freely meandering and the incised-valley types of rivers.

A major unanswered question concerns the relative influence of system-external and system-internal effects in causing episodes of fluvial activity, especially those related to prehistoric river terracing in the Appalachians. Fluvial geomorphologists need more detailed chronologies of Holocene environmental changes in order to better interpret fluvial histories and understand their process implications. On the positive side, much new knowledge recently has been accumulated along several Appalachian valleys concerning long-term sedimentary and erosional process dynamics. This knowledge is useful in comparing styles of fluvial responses. For example, episodes of enhanced fluvial activity are also documented for the ephemeral stream valleys of the southwestern U.S. In these streams, periods of prehistoric activity were characterized by headcut migration upstream, resulting "arroyo cutting," and, later, backfilling of the enlarged channels (e.g., Patton and Schumm, 1975). In contrast, no such alluvial cut-and-fill cycles must have been associated with active phases along the Appalachian streams described. Instead, episodically enhanced rates of lateral channel migration, cut-bank erosion, and convex-bend sedimentation resulted in the formation of row terraces. Therefore, to infer cut-and-fill cycles along Appalachian rivers solely on the basis of the presence of terraces is no longer valid. The new work suggests that it is time for the cut-and-fill model, developed along western streams and rivers, to assume its proper role as only one of several alternative hypotheses for terrace formation in the Appalachians.

As described, streams flowing in unglaciated Appalachian upland regions periodically undergo intervals of enhanced bedrock-valley erosion. The alluvial deposits left on the older, higher bedrock levels form step terraces. Again, the external or internal causation for episodes of bedrock incision is not certain, and this problem constitutes another unsolved question in Appalachian fluvial geomorphology. In contrast, along a river in the glaciated northern Appalachians, incision into late Pleistocene marine deposits apparently occurred in direct response to postglacial crustal rebound. The average erosion rate was estimated at 1 m/100 yr based on radiocarbon dates, but the actual short-term rate may have been much higher.

In closing, another area of research on fluvial systems in this region is currently gaining momentum and merits attention due to its possible importance for fluvial geomorphology. The effect of infrequent, large-magnitude floods on fluvial systems is now being intensely restudied. Both hydrogeomorphic methods of flood-hazard evaluation (Baker, 1976) and "paleoflood hydrology" (Kochel and Baker, 1982) represent opportunities for future research that will engage both theoretically minded geomorphologists (who may, following Wolman and Miller, 1960, reassess the importance of rare events in landform evolution) and applied workers (who are already using paleoflood techniques to assist hydrologists and engineers). Although much of the recent work is being conducted in arid and semiarid regions of the U.S., geomorphologists in the Appalachian region have for some time also been active in such studies (e.g., Mansfield, 1938; Jahns, 1947; Wolman and Eiler, 1958; Hack and Goodlett, 1960; Williams and Guy, 1973; Renwick, 1977; Costa, 1974; and Moss and Kochel, 1978). A large number of Appalachian valleys are heavily populated and flood prone; hurricane incursions bring infrequent torrential rainfalls, and most discharge records are less than 50 yr in duration. As data-acquisition methods become more sophisticated, and reproducibility and efficiency of geomorphic flood-hazard techniques improve, many more such studies in this region are likely, suggesting that Appalachian fluvial geomorphology may soon enter a golden age of theoretical and applied advances based on understanding the long-term hydrologic history of watersheds and the impact of flood events.

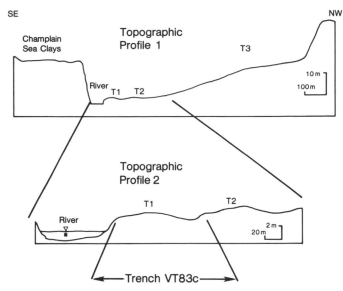

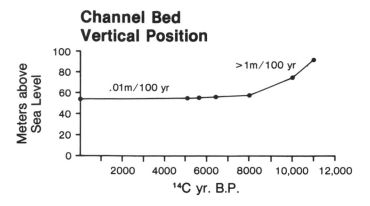

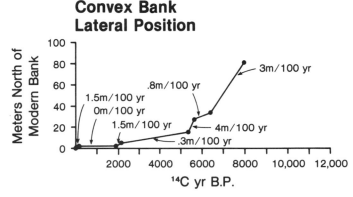

Figure 39. Topographic profiles across the Missisquoi River valley near the trench site. Note asymmetrical cross-valley profile of Holocene age. Location of cross section presented in Figure 31A is shown in profile 2.

Figure 40. Inferred position of the Missisquoi River channel through late Quaternary time. Following vertical incision in latest Pleistocene and earliest Holocene times, the river has engaged mainly in widening of its valley.

CONCLUSIONS

The rapid increase during the 1970s and 1980s in the amount of research related to Appalachian geomorphology is encouraging. Nevertheless, there are problems that make further progress more difficult than in many other regions. In the first place, there are inherent natural problems. For example, surficial deposits, required for dating geomorphic surfaces, are thin, discontinuous, poorly exposed, and very difficult to date in the Appalachians. Radiocarbon-datable organics are rarely found except where they have been preserved below the water table, and currently there is no reliable method of dating older deposits.

Problems created by people are perhaps more serious. Despite the increase in geomorphic work, there is too little basic research, because of the scarcity of institutions supporting such work. In the past several decades, geomorphic research in the Appalachians has been carried out largely by the U.S. Geological Survey. As illustrated herein, this work has greatly increased understanding of structural geomorphology, geological controls on slope failures, and the relation between landform evolution and Cenozoic tectonics. However, one institution must necessarily concentrate on a relatively small number of topics, and as a result many areas of geomorphology have received little or no attention in the Appalachians. In other regions, academic institutions or state surveys commonly supply the diversity that helps to fill in the gaps left by the focused projects of the U.S. Geological Survey. However, there are presently very few geomorphologists at research-oriented universities who are engaged in active research in the Appalachians. Likewise, with one or two excep-

tions, state surveys have contributed little in the past decade or so to Appalachian geomorphology, aside from the mapping of Pleistocene glacial deposits in northern states. Until this situation improves, it will be difficult for geomorphological knowledge of the Appalachians to advance on a broad front.

A related problem is that many parts of the Appalachian region are deficient in public perception of environmental problems that have stimulated geomorphic research elsewhere, such as water quality, adverse effects of logging, destruction of commercial fish habitat, and seismic hazards. In addition, although floods and slope failures certainly constitute costly hazards in this region, the potential of geomorphic research for mitigating these hazards has not yet been fully applied.

Despite the above-mentioned handicaps, the expansion in Appalachian research seems likely to continue. Although the best topics for future research undoubtedly are those not yet conceived, certain directions seem promising. One is the relationship between structure and landforms. It would be of interest to extend Hack's work on this subject in a more quantitative fashion, perhaps attempting to predict landforms from independent measures of rock resistance, and integrating such work with our increasing knowledge of Appalachian neotectonics. A related topic is the evolution of very old landforms and landscapes. Are there really remnants of erosion surfaces? Answers to such questions will require the development and refinement of new dating techniques, as well as advances in understanding denudational processes on land surfaces of low relief.

Another direction is determining the recurrence intervals of erosional events that control the form of hillslope and fluvial landforms. As suggested previously, the type of dominant events and their recurrence intervals may well differ among provinces. Geomorphic response to climatic change is also of great interest. Over the past decade, understanding of Appalachian paleoclimate has increased greatly, but knowledge of the effect of climatic change on geomorphology is still primitive. Dating of hillslope colluvium, foot-slope deposits, and alluvial terraces will help resolve this question.

Other promising directions for research on hillslopes include the development of models that predict the location of slope failures in space and time as a function of geologic, topographic,

and meteorological factors. An understanding of the variation in regolith thickness and other properties as a function of geology, topography, and climate is also very desirable. In addition, the dearth of Appalachian studies on hillslope processes not involving slope failure certainly needs to be corrected.

Research on streams should continue to address the question of the effect of rare, high-magnitude floods on stream landforms. Another question of interest is that of which model best explains the formation of stream terraces in the Appalachians. The answers to both these questions may differ for different provinces. A third desirable line of study concerns high terraces. In the unglaciated Appalachians, most terrace studies have been concerned with terraces that are Holocene to late Wisconsin in age. Commonly these terraces are only a few meters above modern river level, yet many terraces in the Appalachians are hundreds of meters above modern river level. Clearly, many such terraces must be Tertiary in age, and may contain information regarding tectonic history and long-term climatic change.

REFERENCES CITED

Abrahams, A. D., and Flint, J., 1983, Geological controls on the topological properties of some trellis channel networks: Geological Society of America Bulletin, v. 94, p. 80–91.

Arnold, C. L., Boison, P. J., and Patton, P. C., 1982, Sawmill Brook; An example of rapid geomorphic change relating to urbanization: Journal of Geology, v. 90, p. 155–166.

Ashley, G. H., 1935, Studies in Appalachian Mountain sculpture: Geological Society of America Bulletin, v. 46, p. 1395–1436.

Baker, V. R., 1976, Hydrogeomorphic methods for the regional evaluation of flood hazards: Environmental Geology, v. 1, p. 261–281.

Beckinsale, R. P., and Chorley, R. J., 1968, History of geomorphology, in Fairbridge, R. W., ed., The encyclopedia of geomorphology: Stroudsburg, Pennsylvania, Dowden, Hutchinson, and Ross, p. 410–416.

Benedict, J. B., 1976, Frost creep and gelifluction features; A review: Quaternary Research, v. 6, p. 55–76.

Bogucki, D. J., 1970, Debris slides and related flood damage associated with the September 1, 1951, cloudburst in the Mt. LeConte–Sugarland Mountain area, Great Smoky Mountains National Park [Ph.D. thesis]: Knoxville, University of Tennessee, 164 p.

—— , 1976, Debris slides in the Mt. LeConte area, Great Smoky Mountains National Park, U.S.A.: Geografiska Annaler, v. 58A, p. 179–191.

Bollinger, G. A., and Wheeler, R. L., 1982, The Giles County, Virginia, seismogenic zone; Seismological results and geological interpretations: U.S. Geological Survey Open-File report 82-585, 136 p.

—— , 1983, The Giles County, Virginia, seismic zone: Science, v. 219, p. 1063–1065.

Brakenridge, G. R., 1980, Widespread episodes of stream erosion during the Holocene and their climatic cause: Nature, v. 283, p. 655–656.

—— , 1981, Late Quaternary floodplain sedimentation along the Pomme de Terre River, southern Missouri: Quaternary Research, v. 15, p. 62–76.

—— , 1983, Late Quaternary floodplain sedimentation along the Pomme de Terre River, southern Missouri; Part II, notes on sedimentology and pedogenesis: Geologisches Jahrbuch, Series A, v. 71, p. 265–283.

—— , 1984, Alluvial stratigraphy and radiocarbon dating along the Duck River, Tennessee; Implications regarding flood-plain origin: Geological Society of America Bulletin, v. 95, p. 9–25.

—— , 1985, Rate estimates for lateral bedrock erosion based on radiocarbon ages, Duck River, Tennessee: Geology, v. 13, p. 111–114.

—— , 1987, Floodplain stratigraphy, in Baker, V. R., and others, eds., Flood geomorphology, New York, John Wiley (in press).

Brakenridge, G. R., Thomas, P. A., Schiferle, J. S., and Conkey, L. E., 1986, Floodplain sedimentation, postglacial uplift, and environmental change, Missisquoi River, Vermont [abs.]: American Quaternary Association Program and Abstracts, p. 119.

Braun, D. D., 1983, Lithologic control of bedrock meander dimensions in the Appalachian Valley and Ridge Province: Earth Surfaces Processes and Landforms, v. 8, p. 223–237.

Briggs, R. P., Pomeroy, J. S., and Davies, W. E., 1975, Landsliding in Allegheny County, Pennsylvania: U.S. Geological Survey Circular 728, 18 p.

Brown, L. D., 1978, Recent vertical crustal movement along the East Coast of the United States: Tectonophysics, v. 44, p. 205–231.

Brown, L. D., and Oliver, J. E., 1976, Vertical crustal movements from leveling data and their relation to geologic structure in the eastern United States: Reviews of Geophysics and Space Physics, v. 14, p. 13–35.

Brown, L. D., Reilinger, R. E., and Citron, G. P., 1980, Recent vertical crustal movements in the U.S.; Evidence from precise leveling, in Morner, N., ed., Proceedings of a symposium, Earth rheology, isostasy, and eustasy, Stockholm, 1977: New York, John Wiley, p. 389–405.

Bull, W. B., 1962, Relations of alluvial-fan size and slope to drainage-basin size and lithology in western Fresno County, California: U.S. Geological Survey Professional Paper 450-B, p. 51–53.

—— , 1964, Geomorphology of segmented alluvial fans in western Fresno County, California: U.S. Geological Survey Professional Paper 352-E, p. 89–129.

—— , 1972, Recognition of alluvial-fan deposits in the stratigraphic record, in Rigby, J. K., ed., Recognition of ancient sedimentary environments: Society of Economic Palaeontologists and Mineralogists Special Publication 16, p. 63–83.

Cain, S. A., 1944, Pollen analysis of some buried soils, Spartanburg County, South Carolina: Bulletin of the Torrey Botanical Club, v. 71, p. 11–22.

Chorley, R. J., 1969, The elevation of the Lower Greensand Ridge, southeast England: Geological Magazine, v. 106, p. 231–248.

Clark, G. M., 1967, Structural geomorphology of a portion of the Wills Mountain anticlinorium, Mineral and Grant counties, West Virginia [Ph.D. thesis]: University Park, Pennsylvania State University, 201 p.

—— , 1968, Sorted patterned ground; New Appalachian localities south of the glacial border: Science, v. 161, no. 3839, p. 355–356.

——, 1973, Remote sensor utilization for environmental systems studies: final report: Bedford, Massachusetts, Air Force Cambridge Research Laboratories, v. 3, Contract no. F19628-69-C-0016, Proj. no. 7259, p. 303–375.

——, 1984, Debris slide/debris flow events in the Appalachians south of the glacial border; Processes, site factors, and hillslope development in selected historical events: Geological Society of America Abstracts with Programs, v. 16, p. 472.

Cleaves, E. T., Fisher, D. W., and Bricker, O. P., 1974, Chemical weathering of serpentine in the eastern Piedmont of Maryland: Geological Society of America Bulletin, v. 85, p. 437–444.

Cleaves, E. T., Godfrey, A. E., and Bricker, O. P., 1970, Geochemical balance of a small watershed and its geomorphic implications: Geological Society of America Bulletin, v. 81, p. 3015–3032.

Conant, L. C., Black, R. F., and Hosterman, 1976, Sediment-filled pots in upland gravels of Maryland and Virginia: U.S. Geological Survey Journal of Research, v. 4, no. 3, p. 353–358.

Cooke, R. U., and Reeves, R. W., 1976, Arroyos and environmental change in the American Southwest: Oxford, Clarendon Press, 213 p.

Cooper, B. N., 1944, Geology and mineral resources of the Burkes Garden Quadrangle, Virginia: Virginia Geological Survey Bulletin, v. 60, 299 p.

Costa, J. E., 1974, Response and recovery of a Piedmont watershed from tropical storm Agnes, June, 1972: Water Resources Research, v. 10, no. 1, p. 106–112.

——, 1975, Effects of agriculture on erosion and sedimentation in the Piedmont Province, Maryland: Geological Society of America Bulletin, v. 86, p. 1281–1286.

Costa, J. E., and Cleaves, E. T., 1984, The Piedmont landscape of Maryland; A new look at an old problem: Earth Surface Processes and Landforms, v. 9, p. 59–74.

Crawford, N. C., 1981, Karst flooding in urban areas; Bowling Green, Kentucky, *in* Beck, B. F., ed., 8th International Congress of Speleology: Western Kentucky University Press, p. 763–765.

——, 1984, Karst landform development along the Cumberland Plateau escarpment of Tennessee, *in* LaFleur, R. G., ed., Groundwater as a geomorphic agent: Boston, Allen and Unwin, p. 294–339.

D'Appolonia, E., Alperstein, R., and D'Appolonia, D. J., 1967, Behavior of a colluvial slope: Journal of Soil Mechanics, Foundation Engineering Division, American Society of Civil Engineers, v. 93 (sm 4), p. 447–473.

Davis, M. B., 1983, Holocene vegetational history of the eastern United States, *in* Wright, H. E., Jr., ed., The Holocene, Volume 2, Late Quaternary environments of the United States: Minneapolis, University of Minnesota Press, p. 166–181.

Davis, W. M., 1888, Geographic methods in geologic investigations: National Geographic Magazine, v. 1, p. 11–26.

——, 1889, The rivers and valleys of Pennsylvania: National Geographic Magazine, v. 1, p. 183–253.

Delano, H. L., and Wilshusen, J. P., 1985, Landslides in the Williamsport 1° × 2° quadrangle, north-central Pennsylvania: Geological Society of America Abstracts with Programs, v. 17, p. 15.

Delcourt, P. A., 1980, Quaternary alluvial terraces of the Little Tennessee River valley, east Tennessee, *in* Chapman, J., ed., The 1979 archaeological and geological investigations in the Tellico Reservoir: University of Tennessee Department of Anthropology, Report of Investigations 29, p. 110–121, 175–212.

Delcourt, P. A., and Delcourt, H. R., 1981, Vegetation maps for eastern North America; 40,000 yr B.P. to the present, *in* Romans, R. C., ed., Geobotany II: Plenum, p. 123–165.

——, 1984, Late Quaternary paleoclimates and biotic responses in eastern North America and the western North Atlantic Ocean: Palaeogeography, Palaeoclimatology, Palaeoecology, v. 48, p. 263–284.

Denny, C. S., 1956, Surficial geology and geomorphology of Potter County, Pennsylvania: U.S. Geological Survey Professional Paper 288, 72 p.

——, 1965, Alluvial fans in the Death Valley region, California and Nevada:

U.S. Geological Survey Professional Paper 466, 62 p.

Dietrich, W. E., and Dorn, R., 1984, Significance of thick deposits of colluvium on hillslopes; A case study involving the use of pollen analysis in the coastal mountains of northern California: Journal of Geology, v. 92, p. 147–158.

Dietrich, W. E., and Dunne, T., 1978, Sediment budget for a small catchment in mountainous terrain: Zeitschrift für Geomorphologie, Supplementband 29, p. 191–206.

Dinga, C. F., 1971, An analysis of the relationship between geologic structure and the geometric surface form of homoclinal ridges [Ph.D. thesis]: Terre Haute, Indiana State University, 219 p.

Dougherty, M. T., and Barsotti, N. J., 1972, Structural damage and potentially expansive sulfide minerals: Bulletin of the Association of Engineering Geologists, v. 9, p. 105–125.

Dunn, J. R., and Banino, G. M., 1977, Problems with Lake Albany "clavs," *in* Coates, D. R., ed., Landslides: Geological Society of America Reviews in Engineering Geology, v. 3, p. 133–136.

Dury, G. H., 1977, Underfit streams; Retrospect, perspect, and prospect, *in* Gregory, K. J., ed., River channel changes: New York, Wiley, p. 181–196.

Eargle, D. H., 1940, The relations of soils and surface in the South Carolina Piedmont: Science, v. 91, p. 337–338.

——, 1977, Piedmont Pleistocene soils of the Spartanburg area, South Carolina: South Carolina Division of Geology, State Development Board, Geologic Notes, v. 21, p. 57–74.

Elnagger, H. A., and Flint, N. K., 1976, Analysis and design of highway cuts in rock: Pennsylvania Department of Transportation Research Report EHWA-PA-71-688, 49 p.

Epstein, J. B., 1966, Structural control of wind gaps and water gaps and of stream capture in the Stroudsburg area, Pennsylvania and New Jersey: U.S. Geological Survey Professional Paper 550-B, p. B80–B86.

Epstein, J. B., and Epstein, A. G., 1969, Geology of the Valley and Ridge Province between Delaware water gap and Lehigh gap, Pennsylvania, *in* Subitsky, S., ed., Geology of selected areas in New Jersey and eastern Pennsylvania: New Brunswick, Rutgers University Press, p. 132–205.

Faye, R. E., Carey, W. P., Stammer, J. K., and Kleckner, R. L., 1980, Erosion, sediment discharge, and channel morphology in the upper Chattahoochee River basin, Georgia: U.S. Geological Survey Professional Paper 1107, 85 p.

Fenneman, N. M., 1938, Physiography of eastern United States: New York, McGraw-Hill, 691 p.

Ferguson, H. F., 1967, Valley stress relief in the Allegheny Plateau: Bulletin of the Association of Engineering Geologists, v. 4, p. 63–68.

Fisher, C. C., 1955, Elongate meanders of the North Fork of the Shenandoah River [abs.]: Geological Society of America Bulletin, v. 66, p. 1687.

Fisher, S. P., Fanoff, A. S., and Picking, L. W., 1968, Landslides of southeastern Ohio: Ohio Journal of Science, v. 68, p. 67–80.

Flemal, R. C., 1971, The attack on the Davisian system of geomorphology; A synopsis: Journal of Geological Education, v. 19, p. 3–13.

Fleming, R. W., Johnson, A. M., Hough, J. E., Gokce, A. O., and Lion, T., 1981, Engineering geology of the Cincinnati area, *in* Roberts, T. G., ed., Cincinnati 1981 field trip guidebook, Volume 3: Geological Society of America, p. 543–570.

Gilbert, G. K., 1877, Geology of the Henry Mountains (Utah); U.S. Geographical and Geological Survey of the Rocky Mountain Region: Washington, D.C., U.S. Government Printing Office, 160 p.

Godfrey, A. E., 1975, Chemical and physical erosion in the South Mountain anticlinorium, Maryland: Maryland Geological Survey Information Circular 19, 35 p.

Gray, R. E., and Gardner, G. D., 1977, Processes of colluvial slope development at McMechen, West Virginia: Bulletin of the International Association of Engineering Geologists, no. 16, p. 29–32.

Gray, R. E., Ferguson, H. E., and Hamel, J. V., 1979, Slope stability in the Appalachian Plateaus of Pennsylvania and West Virginia; Rock slides and avalanches, Part 2, Engineering sites, *in* Voight, B., ed., Developments in geotechnical engineering, v. 14B: New York, Elsevier, p. 447–471.

Groot, J. J., 1955, Sedimentary petrology of the Cretaceous sediments of northern

New Jersey in relation to paleogeographic problems: Delaware Geological Survey Bulletin 5, 157 p.

Hack, J. T., 1957, Studies of longitudinal stream profiles in Virginia and Maryland: U.S. Geological Survey Professional Paper 294-B, p.45–97.

——, 1960, Interpretation of erosional topography in humid temperate regions: American Journal of Science, v. 258A, p. 80–97.

——, 1965, Geomorphology of the Shenandoah Valley, Virginia and West Virginia, and origin of the residual ore deposits: U.S. Geological Survey Professional Paper 484, 83 p.

——, 1966, Interpretations of the Cumberland Escarpment and Highland Rim, south-central Tennessee and northeast Alabama: U.S. Geological Survey Professional Paper 524-C, 16 p.

——, 1973, Drainage adjustment in the Appalachians, *in* Morisawa, M., ed., Fluvial geomorphology: Boston, Allen and Unwin, p. 51–69.

——, 1976, Dynamic equilibrium and landscape evolution, *in* Melhorn, W. N., and Flemal, R. C., eds., Theories of landform development: Boston, Allen and Unwin, p. 87–102.

——, 1980, Rock control and tectonism; Their importance in shaping the Appalachian Highlands: U.S. Geological Survey Professional Paper 1126-B, 17 p.

——, 1982, Physiographic divisions and differential uplift in the Piedmont and Blue Ridge: U.S. Geological Survey Professional Paper 1265, 49 p.

Hack, J. T., and Goodlett, J. C., 1960, Geomorphology and forest ecology of a mountain region in the central Appalachians: U.S. Geological Survey Professional Paper 347, 66 p.

Hack, J. T., and Young, R. S., 1959, Intrenched meanders of the North Fork of the Shenandoah River, Virginia: U.S. Geological Survey Professional Paper 354-A, p. 1–10.

Hall, G. A., 1974, The development of design criteria for soil slopes on West Virginia highways [Ph.D. thesis]: Morgantown, University of West Virginia, 271 p.

Hamel, J. V., 1980, Geology and slope stability in western Pennsylvania: Association of Engineering Geologists Bulletin, v. 17, p. 1–26.

Hamel, J. V., and Adams, W. R., 1981, Claystone slides, Interstate Route 79, Pittsburgh, Pennsylvania, U.S.A.: Tokyo, Proceedings, International Symposium on Weak Rocks, p. 549–553.

Hayes, C. W., 1899, Physiography of the Chattanooga district: U.S. Geological Survey 19th Annual Report, pt. 2, p. 1–58.

Haynes, C. V., 1985, Mastodon-bearing springs and late Quaternary geochronology of the lower Pomme de Terre valley, Missouri: Geological Society of America Special Paper 204, 35 p.

Hooke, R. LeB., 1968, Steady-state relationships on arid-region alluvial fans in closed basins: American Journal of Science, v. 266, p. 609–629.

Hooper, J. R., 1969, Slope stability investigation of residual clay soils of southeastern Ohio: Ohio State University, Transportation Research Center Report EES 312, 91 p.

Horton, R. E., 1945, Erosional development of streams and their drainage basins: Geological Society of America Bulletin, v. 56, p. 275–370.

Hupp, C. R., 1983, Geobotanical evidence of late Quaternary mass wasting in block field areas of Virginia: Earth Surface Processes and Landforms, v. 8, p. 439–450.

Jacobson, R. B., 1983, Sedimentation in a small arm of proglacial Lake Monongahela, Buffalo Creek, West Virginia: Geological Society of America Abstracts with Programs, v. 15, p. 125.

——, 1985, Spatial and temporal variations in slope activity, Upper Buffalo Creek watershed, Marion County, West Virginia [Ph.D. thesis]: Baltimore, The Johns Hopkins University, 485 p.

Jahns, R. H., 1947, Geologic features of the Connecticut Valley, Massachusetts, as related to recent floods: U.S. Geological Survey Water-Supply Paper 996, 158 p.

Johnson, D. W., 1931, Stream sculpture on the Atlantic slope, a study in the evolution of Appalachian rivers: New York, Columbia University Press, 142 p.

Judson, S., 1975, Evolution of Appalachian topography, *in* Melhorn, W. N., and

Flemal, R. C., eds., Theories of landform development: Boston, Allen and Unwin, p. 29–44.

Keefer, D. K., and Johnson, A. M., 1983, Earth flows; Morphology, mobilization, and movement: U.S. Geological Survey Professional Paper 1264, 56 p.

Kemmerly, P., 1981, The need for recognition and implementation of a sinkhole-floodplain hazard designation in urban karst terrains: Environmental Geology, v. 3, p. 281–292.

——, 1982, Spatial analysis of a karst depression population; Clues to genesis: Geological Society of America Bulletin, v. 93, p. 1078–1086.

——, 1986, Exploring a model for karst-terrane evolution: Geological Society of America Bulletin, v. 97, p. 619–625.

Knox, J. C., 1975, Concept of the graded stream, *in* Melhorn, W. N., and Flemal, R. C., eds., Theories of landform development: Boston, Allen and Unwin, p. 169–198.

Koch, C. A., 1974, Debris slides and related flood effects in the 4–5 August, 1938, Webb Mountain cloudburst; Some past and present environmental geomorphic implications [M.Sc. thesis]: Knoxville, University of Tennessee, 112 p.

Kochel, R. C., 1986, Holocene debris flows in central Virginia, *in* Costa, J. E., and Wieczorek, G. F., eds., Debris flows/avalanches; Process, recognition, and mitigation: Geological Society of America, Reviews in Engineering Geology, v. VII (in press).

Kochel, R. C., and Baker, V. R., 1982, Paleoflood hydrology: Science, v. 215, p. 353–361.

Kochel, R. C., and Johnson, R. A., 1984, Geomorphology and sedimentology of humid temperate alluvial fans in central Virginia, *in* Koster, E., and Steel, R., eds., Memoir on gravels and conglomerates: Canadian Society of Petroleum Geologists, p. 109–122.

Kochel, R. C., Johnson, R. A., and Valastro, S., Jr., 1982, Repeated episodes of Holocene debris avalanching in central Virginia: Geological Society of America Abstracts with Programs, v. 14, p. 31.

Leopold, L. B., and Bull, W. B., 1979, Base level, aggradation, and grade: Proceedings of the American Philosophical Society, v. 123, p. 168–202.

Lesley, J. P., 1856, Manual of coal and its topography: Philadelphia, J. B. Lippincott and Company, 216 p.

——, 1892, Summary description of the geology of Pennsylvania: Harrisburg, Pennsylvania Geological Survey, 3 volumes, 2,588 p.

Lessing, P., and Erwin, R. B., 1977, Landslides in West Virginia, *in* Coates, D. R., ed., Landslides: Geological Society of America Reviews in Engineering, v. 3, p. 245–254.

Lessing, P., Kulander, B. R., Wilson, B. D., Dean, S. L., and Woodring, S. M., 1976, West Virginia landslides and slide-prone areas: West Virginia Geological and Economic Survey, Environmental Geology Bulletin 15, 64 p.

Lessing, P., Messina, C. P., and Fonner, R. F., 1983, Landslide risk assessment: Environmental Geology, v. 5, p. 93–99.

Mackin, J. H., 1938, The origin of Appalachian drainage; A reply: American Journal of Science, v. 236, p. 27–53.

Mansfield, G. R., 1938, Flood deposits of the Ohio River, January–February, 1937; A study of sedimentation: U.S. Geological Survey Water-Supply Paper 838, p. 693–733.

Mark, D. M., 1980, On scales of investigation in geomorphology: Canadian Geographer, v. 24, p. 81–82.

Markewich, H. W., 1985, Geomorphic evidence for Pliocene-Pleistocene uplift in the area of the Cape Fear arch, North Carolina, *in* Morisawa, M., and Hack, J. T., eds., Tectonic geomorphology: Boston, Allen and Unwin, p. 279–297.

Markewich, H. W., Pavich, M. J., Mausbach, M. J., Hall, R. L., Johnson, R. G., and Hearn, P. P., 1987, Age relations between soils and geology in the Coastal Plain of Maryland and Virginia: U.S. Geological Survey Bulletin 1589-A (in press).

McCartan, L., Lemon, E. M., and Weems, R. E., 1984, Geological map of the area between Charleston and Orangeburg, South Carolina: U.S. Geological Survey Miscellaneous Geologic Investigations Map I-1472, scale 1:250,000.

McGee, W. J., 1888a, The geology of the head of Chesapeake Bay: U.S. Geological Survey, 7th Annual Report, p. 537–646.

——, 1888b, Three formations of the middle Atlantic slope: American Journal of Science, v. 35 (3rd ser.), p. 120–143, 328–330, 367–388, 448–466.

Meyerhoff, H. A., and Olmsted, E. W., 1936, The origins of Appalachian drainage: American Journal of Science, v. 232, p. 21–42.

Michalek, D. D., 1968, Fanlike features and related periglacial phenomena of the southern Blue Ridge [Ph.D. thesis]: Chapel Hill, University of North Carolina, 198 p.

Mike, K., 1975, Utilization of ancient river beds for the detection of Holocene crustal movements: Tectonophysics, v. 29, p. 359–368.

Miller, V. C., 1953, A quantitative geomorphic study of drainage basin characteristics in the Clinch Mountain area, Virginia and Tennessee [Ph.D. thesis]: New York, Columbia University, 135 p.

Mills, H. H., 1981a, Boulder deposits and the retreat of mountain slopes, or, "gully gravure" revisited: Journal of Geology, v. 89, p. 649–660.

—— , 1981b, Some observations on slope deposits in the vicinity of Grandfather Mountain, North Carolina: Southeastern Geology, v. 22, p. 209–222.

—— , 1982, Piedmont-cove deposits of the Dellwood Quadrangle, Great Smoky Mountains, U.S.A.: Zeitschrift für Geomorphologie, v. 26, p. 163–178.

—— , 1983, Pediment evolution at Roan Mountain, North Carolina, U.S.A.: Geografiska Annaler, v. 65A, p. 111–126.

—— , 1984, Clast orientation in Mount St. Helens debris-flow deposits, North Fork Toutle River, Washington: Journal of Sedimentary Petrology, v. 54, p. 626–634.

—— , 1985, Descriptions of backhoe trenches dug on New River terraces between Radford and Pearisburg, Virginia, June, 1981: U.S. Geological Survey Open-File Report 85-474, 67 p.

—— , 1986, Possible differential uplift of New River terraces in southwest Virginia: Neotectonics, v. 1, p. 75–86.

Mills, H. H., and Wagner, J. R., 1985, Long-term change in regime of the New River indicated by vertical variation in extent and weathering intensity of alluvium: Journal of Geology, v. 93, p. 131–142.

Mixon, R. B., and Newell, W. L., 1977, Stafford fault system; Structures documenting Cretaceous and Tertiary deformation along the Fall Line in northeastern Virginia: Geology, v. 5, p. 437–440.

Mock, S. J., 1971, A classification of channel links in stream networks: Water Resources Research, v. 7, p. 1558–1566.

—— , 1976, Topological properties of some trellis pattern channel networks: Cold Regions Research and Engineering Laboratory Report 76-46, 54 p.

Moneymaker, B. C., 1939, Erosional effects of the Webb Mountain (Tennessee) cloudburst of August 5, 1938: Tennessee Academy of Science Journal, v. 14, p. 190–196.

Moss, J. H., and Kochel, R. C., 1978, Unexpected geomorphic effects of the Hurricane Agnes storm and flood, Conestoga drainage basin, southeastern Pennsylvania: Journal of Geology, v. 86, no. 1, p. 1–11.

Neary, D. G., and Swift, L. W., 1984, Rainfall thresholds for triggering a debris avalanching event in the southern Appalachians: Geological Society of America Abstracts with Programs, v. 16, p. 609.

Nelson, F. E., 1985, A preliminary investigation of solifluction macrofabrics: Catena, v. 12, p. 23–33.

Newell, W. L., 1985, Architecture of the Rappahannock estuary; Neotectonics in Virginia, *in* Morisawa, M., and Hack, J. T., eds., Tectonic geomorphology: Boston, Allen and Unwin, p. 321–342.

Newell, W. L., Pavich, M. J., Prowell, D. C., and Markewich, H. W., 1980, Surficial deposits, weathering processes, and evolution of an inner Coastal Plain landscape, Augusta, Georgia, *in* Frey, R. W., ed., Excursions in southeastern geology, Volume 2: American Geological Institute, p. 527–544.

Newton, J. G., 1976, Early detection and correction of sinkhole problems in Alabama, with a preliminary evaluation of remote sensing applications: Alabama Highway Department, Bureau of Research and Development, Research Report HPR-76.

Oberlander, T. M., 1965, The Zagros streams; A new interpretation of transverse drainage in an orogenic zone: Syracuse University Geographical Series, no. 1, 168 p.

—— , 1985, Origin of drainage transverse to structure in orogens, *in* Morisawa, M., and Hack, J. T., eds., Tectonic geomorphology: Boston, Allen and Unwin, p. 155–182.

Outerbridge, W. F., 1986, The Logan Plateau, a young physiographic region in West Virginia, Kentucky, Virginia, and Tennessee: U.S. Geological Survey Bulletin 1620 (in press).

Patton, F. D., 1969, The determination of shear strength of rock masses, *in* Proceedings of conference on engineering geology in Appalachian shales: West Virginia University, 19 p.

Patton, P. C., and Schumm, S. A., 1975, Gully erosion, northern Colorado; A threshold phenomenon: Geology, v. 3, p. 88–90.

Pavich, M. J., 1986, Processes and rates of saprolite production and erosion on a foliated granitic rock of the Virginia Piedmont, *in* Colman, S. M., and Dethier, D. P., eds., Rates of chemical weathering of rocks and minerals: New York, Academic Press, p. 551–590.

Pavich, M. J., and Obermeier, S. F., 1985, Saprolite formation beneath Coastal Plain sediments near Washington, D.C.: Geological Society of America Bulletin, v. 96, p. 886–900.

Pavich, M. J., Brown, L., Valette-Silver, J. N., Klein, J., and Middleton, R., 1985, ^{10}Be analysis of a Quaternary weathering profile in the Virginia Piedmont: Geology, v. 13, no. 1, p. 39–41.

Pavich, M. J., Leo, G. W., Obermeier, S. F., and Estabrook, J. R., 1987, Investigations of the characteristics, origin, and residence time of the upland residual mantle of the Piedmont of Fairfax County, Virginia: U.S. Geological Survey Professional Paper (in press).

Pomeroy, J. S., 1980, Storm-induced debris avalanching and related phenomena in the Johnstown area, Pennsylvania, with references to other studies in the Appalachians: U.S. Geological Survey Professional Paper 1191, 23 p.

—— , 1982a, Mass movement in two selected areas of western Washington County, Pennsylvania: U.S. Geological Survey Professional Paper 1170-B, 17 p.

—— , 1982b, Landsliding in the greater Pittsburgh region, Pennsylvania: U.S. Geological Survey Professional Paper 1229, 48 p.

—— , 1983, Relict debris flows in northwestern Pennsylvania: Northeastern Geology, v. 5, p. 1–7.

—— , 1984a, Preliminary map showing recently active landslides in the Marietta area, Washington County, southeastern Ohio: U.S. Geological Survey Open-File Report 85-4, scale 1:24,000, with text.

—— , 1984b, Storm-induced slope movements at East Brady, northwestern Pennsylvania: U.S. Geological Survey Bulletin 1618, 16 p.

Pomeroy, J. S., and Thomas, R. E., 1985, Geologic relationships of slope movement in northern Alabama: U.S. Geological Survey Bulletin 1649, 13 p.

Powell, J. W., 1875, Exploration of the Colorado River of the West (1869–1872): Washington, D.C., U.S. Government Printing Office, 291 p.

—— , 1876, Report on the geology of the eastern portion of the Uinta Mountains: Washington, D.C., U.S. Government Printing Office, 218 p.

Prowell, D. C., and O'Connor, B. J., 1978, Belair fault zone; Evidence of Tertiary fault displacement in eastern Georgia: Geology, v. 6, p. 681–684.

Quinlan, J. F., and Rowe, D. R., 1977, Hydrology and water quality in the central Kentucky karst, Phase I: University of Kentucky, Water Resources Research Institute, Research Paper no. 101, 89 p.

—— , 1978, Hydrology and water quality in the central Kentucky karst, Phase II, Part A; Preliminary summary of the hydrogeology of the Mill Hole sub-basin of the Turnhole Spring groundwater basin: University of Kentucky, Water Resources Research Institute, Research Report no. 109, 42 p.

Radbruch-Hall, D. H., Colton, R. B., Davies, W. E., Skipp, B. A., Luschitta, I., and Varnes, D. J., 1982, Landslide overview map of the conterminous United States: U.S. Geological Survey Professional Paper 1183, 25 p.

Reinhardt, J., Prowell, D. C., and Christopher, R. A., 1984, Evidence for Cenozoic tectonism in the southwest Georgia Piedmont: Geological Society of America Bulletin, v. 95, p. 1176–1187.

Renwick, W. H., 1977, Erosion caused by intense rainfall in a small catchment in New York State: Geology, v. 5, p. 361–364.

Rich, J. L., 1935, Origin and evolution of rock fans and pediments: Geological Society of America Bulletin, v. 46, p. 999–1024.

Richardson, A. M., 1979, Landslide in claystone derived soil: Journal of the

Geotechnical Engineering Division, Proceedings American Society of Civil Engineering, v. 105, no. GT7, p. 857–869.

Riestenburg, M. M., and Sovonick-Dunford, S., 1983, The role of woody vegetation in stabilizing slopes in the Cincinnati area, Ohio: Geological Society of America Bulletin, v. 94, p. 506–518.

Rogers, H. D., 1858, The geology of Pennsylvania: First Pennsylvanian Geological Survey, v. 2, 1015 p.

Royster, D. L., 1973, Highway landslide problems along the Cumberland Plateau in Tennessee: Association of Engineering Geologists Bulletin, v. 10, p. 255–287.

Schirmer, W. 1983, Criteria for the differentiation of late Quaternary river terraces: Warsaw, Polish Academy of Sciences, Quaternary Studies in Poland, v. 4, p. 199–204.

Schneider, R. H., 1973, Debris slides and related flood damage resulting from Hurricane Camille, 19–20 August, and subsequent storm, 5–6 September, 1969, in the Spring Creek drainage basin, Greenbrier County, West Virginia [Ph.D. thesis]: Knoxville, University of Tennessee, 131 p.

Schultz, A. P., 1986, Ancient, giant rockslides, Sinking Creek Mountain, southern Appalachians, Virginia: Geology, v. 14, p. 11–14.

Schumm, S. A., 1965, Quaternary paleohydrology, *in* Wright, W. E., Jr., and Frey, D. G., eds., The Quaternary of the United States: Princeton, New Jersey, Princeton University Press, p. 783–794.

——, 1968, River adjustment to altered hydrologic regime; Murrumbidgee River and paleochannels, Australia: U.S. Geological Survey Professional Paper 598, 65 p.

Schumm, S. A., and Parker, R. S., 1973, Implications of complex response of drainage systems for Quaternary alluvial stratigraphy: Nature, v. 243, p. 99–100.

Scott, R. C., Jr., 1972, The geomorphic significance of debris avalanching in the Appalachian Blue Ridge Mountains [Ph.D. thesis]: Athens, University of Georgia, 185 p.

Scully, R. W., and Arnold, R. W., 1981, Holocene alluvial stratigraphy in the upper Susquehanna River basin, New York: Quaternary Research, v. 15, p. 327–344.

Sevon, W. D., Crowl, G. H., and Berg, T. M., 1975, The late Wisconsinan drift border in northeastern Pennsylvania; Guidebook for the 40th annual field conference of Pennsylvania Geologists: Pennsylvania Topographic and Geologic Survey, 108 p.

Sevon, W. D., Potter, N., Jr., and Crowl, G. H., 1983, Appalachian peneplains; An historical review: Earth Sciences History, v. 2, p. 156–164.

Shafer, D. S., 1984, Late Quaternary paleoecologic, geomorphic, and paleoclimatic history of Flat Laurel Gap, Blue Ridge Mountains, North Carolina [M.Sc. thesis]: Knoxville, University of Tennessee, 148 p.

Shepherd, R. G., and Schumm, S. A., 1974, Experimental study of river incision: Geological Society of America Bulletin, v. 85, p. 257–268.

Skempton, A. W., 1964, Long-term stability of clay slopes: Geotechnique, v. 14, p. 77–102.

Smart, J. S., 1969, Topological properties of channel networks: Geological Society of America Bulletin, v. 80, p. 1757–1773.

Soller, D. R., 1984, The Quaternary history and stratigraphy of the Cape Fear Valley, North Carolina [Ph.D. thesis]: Washington, D.C., George Washington University, 198 p.

Stollar, R. L., 1976, Geology and some engineering properties of near-surface Pennsylvanian shales in northeastern Ohio [M.Sc. thesis]: Kent, Ohio, Kent State University, 52 p.

Strahler, A. N., 1945, Hypotheses of stream development in the folded Appalachians of Pennsylvania: Geological Society of America Bulletin, v. 56, p. 45–88.

——, 1946, Elongate intrenched meanders of Conodoguinet Creek, Pennsylvania, American Journal of Science, v. 244, p. 31–40.

——, 1950, Equilibrium theory of slopes approached by frequency distribution analysis: American Journal of Science, v. 248, p. 800–814.

Stumm, W., and Morgan, J. J., 1970, Aquatic chemistry: New York, Wiley-Interscience, 583 p.

Tarr, R. S., 1898, The peneplain: American Geologist, v. 21, p. 351–370.

Thompson, H. D., 1939, Drainage evolution in the southern Appalachians: Geological Society of America Bulletin, v. 50, p. 1323–1356.

——, 1941, Topographic analysis of the Monterey, Staunton, and Harrisonburg quadrangles: Journal of Geology, v. 49, p. 521–549.

Thompson, H. J., 1969, The James River flood of August 1969 in Virginia: Weatherwise, v. 22, October, p. 180–183.

Thornbury, W. D., 1965, Regional geomorphology of the United States: New York, John Wiley and Sons, 609 p.

Tinkler, K. J., 1971, Active valley meanders in south-central Texas and their wider implications: Geological Society of America Bulletin, v. 82, p. 1783–1800.

Trimble, S. W., 1974, Man-induced soil erosion on the southern Piedmont, 1700–1970: Soil Conservation Society of America, 180 p.

——, 1975, Denudation studies; Can we assume stream steady state?: Science, v. 188, p. 1207–1208.

——, 1977, The fallacy of stream equilibrium in contemporary denudation studies: American Journal of Science, v. 277, p. 876–887.

U.S. Geological Survey, 1982, Goals and tasks of the landslide part of a ground-failure hazards reduction program: U.S. Geological Survey Circular 880, 49 p.

Varnes, D. J., 1978, Slope movement types and processes, *in* Schuster, R. L., and Krizek, R. J., eds., Landslides; Analysis and control: National Research Council, Highway Research Board Special Report 176, p. 11–33.

Velbel, M. A., 1985, Geochemical mass balances and weathering rates in forested watersheds of the southern Blue Ridge: American Journal of Science, v. 285, p. 904–930.

Walters, J. C., 1978, Polygonal patterned ground in central New Jersey: Quaternary Research, v. 10, p. 42–54.

Webb, D. K., Jr., and Collins, H. R., 1967, Geologic aspects of a recent landslide in Vinton County, Ohio: Ohio Journal of Science, v. 67, p. 65–74.

White, E. L., and White, W. B., 1979, Quantitative morphology of landforms in carbonate rock basins in the Appalachian Highlands: Geological Society of America Bulletin, v. 90, p. 385–396.

——, 1983, Karst landforms and drainage basin evolution in the Obey River basin, north-central Tennessee, U.S.A.: Journal of Hydrology, v. 61, p. 69–82.

Whitehead, D. R., and Barghoorn, E. S., 1962, Pollen analytical investigation of Pleistocene deposits from western North Carolina and South Carolina: Ecological Monographs, v. 32, no. 4, p. 347–369.

Williams, G. P., and Guy, H. P., 1973, Erosional and depositional aspects of Hurricane Camille in Virginia, 1969: U.S. Geological Survey Professional Paper 804, 80 p.

Wolman, M. G., 1955, The natural channel of Brandywine Creek, Pennsylvania: U.S. Geological Survey Professional Paper 271, 56 p.

——, 1967, A cycle of sedimentation and erosion in urban river channels: Geografiska Annaler, v. 49A, p. 385–395.

Wolman, M. G., and Eiler, J. P., 1958, Reconnaissance study of erosion and deposition produced by the flood of August, 1955, in Connecticut: Transactions of the American Geophysical Union, v. 39, p. 1–14.

Wolman, M. G., and Miller, J. P., 1960, Magnitude and frequency of forces in geomorphic processes: Journal of Geology, v. 68, p. 54–74.

Wyrick, G. G., and Borchers, J. W., 1981, Hydrologic effects of stress-relief fracturing in an Appalachian valley: U.S. Geological Survey Water Supply Paper 2177, 51 p.

Yarnell, D. L., 1935, Rainfall intensity-frequency data: U.S. Department of Agriculture Miscellaneous Publication 204, 68 p.

MANUSCRIPT ACCEPTED BY THE SOCIETY NOVEMBER 4, 1986

Geological Society of America
Centennial Special Volume 2
1987

Chapter 3

Atlantic and Gulf Coastal Province

H. Jesse Walker
Department of Geography, Louisiana State University, Baton Rouge, Louisiana 70803
James M. Coleman
Coastal Studies Institute, School of Geoscience, Louisiana State University, Baton Rouge, Louisiana 70803

INTRODUCTION

The Atlantic and Gulf Coastal Province differs from most other geomorphic provinces in North America in that it has a large suboceanic area (685×10^3 km^2) in addition to its subaerial segment ($1,166 \times 10^3$ km^2; Table 1). The subaerial portion, usually referred to as the Coastal Plain*, extends from Long Island in New York (with outliers in Martha's Vineyard and Cape Cod) to the Mexico border in Texas and includes all or part of 19 states. The suboceanic portion, which composes part of North America's Continental Shelf, extends from the Canadian border south and west to the border with Mexico.

The unifying character of this province stems mainly from a geologic history that has recorded alternating periods of submergence and emergence. Thus, all parts of the province have experienced coastal processes at one time or another, and most areas several times. However, over many parts of the province the surface expression of these processes has been destroyed or highly modified by erosion or burial. Most of the suboceanic portion of the province was subaerial as recently as 18,000 years ago, whereas parts of the subaerial portion of the Coastal Plain have not been submerged since the Cretaceous. The interface between the two divisions has been near its present elevational position for only some 5,000 to 6,000 years.

During most of the province's history, a number of geomorphic processes have been operative upon a relatively gently sloping surface consisting of rocks that generally increase in age upslope. In addition to slope and rock type, the effectiveness of processes within the province has varied mainly with tectonic activity, climate, time, and, more recently, human activity.

REGIONAL SETTING

Boundary Conditions

The three main boundaries that characterize the province

differ markedly from each other. The outer and inner boundaries, in their smoothed form, are about the same length (Table 1). The outer boundary, here taken as the Continental Slope, is relatively steep and descends over 2,000 m from the Continental Shelf to the Continental Rise. In places, especially when indented by submarine canyons, it is very rugged. The inner boundary has elevations that vary irregularly along its entire length. The elevation of the boundary ranges from sea level at the province's contact with New England to a maximum of more than 300 m at its southern limit in Texas. In between, representative elevations include 130 m in Maryland, 250 m in Georgia, 110 m near Little Rock, Arkansas, and 210 m at Austin, Texas. The province shares its inner boundary with nine other provinces (Fig. 1) ranging in common lengths from about 100 km with the Valley and Ridge Province to over 1,700 km with the Piedmont Province (Table 2). In contrast to the Continental Slope, the geologic types found at the boundary between the Coastal Plain and adjacent provinces differ markedly (Fig. 1).

The Coastal Plain side of the boundary is represented by Cretaceous sedimentary rocks along most of its length, although in some locations Tertiary and even Quaternary sedimentary rocks occur at the juncture. Facing them are rocks that range from Precambrian to Cretaceous in age, including sedimentary, igneous, and metamorphic types (Table 2).

The third boundary, the one that separates the province into its two most distinctive segments, is the present-day interface between the Coastal Plain and the Continental Shelf. Even in its smoothed form, it is about half again as long as the other two boundaries and is the one presently undergoing the most rapid change. As such, it is appropriate to consider the coast a transitional zone with varying widths and varying proportions of oceanic and terrestrial characteristics. Quaternary (especially Holocene) sediment dominates the interface. However, in a few locations, especially along the shorelines of Chesapeake Bay and western Florida, upper Tertiary and even lower Tertiary sedimentary rocks are present (Fig. 1).

---

*In some classifications the term Coastal Plain also includes the subaqueous Continental Shelf. However, in this paper it is used to designate only the subaerial portion of the province.

Walker, H. J., and Coleman, J. M., 1987, Atlantic and Gulf Coastal Province, *in* Graf, W. L., ed., Geomorphic systems of North America: Boulder, Colorado, Geological Society of America, Centennial Special Volume 2.

52   *H. J. Walker and J. M. Coleman*

TABLE 1. THE ATLANTIC AND GULF COASTAL PROVINCE*

	Length (km)	Area (km²)	Area (km²)	(%)
Boundaries				
1. Continental Shelf/Slope	5,150			
2. Continental Shelf/Coastal Plain	7,365			
3. Coastal Plain/Adjacent Provinces	4,665			
Area				
1. Continental Shelf			685,000	100
a. Atlantic portion		354,000		52
b. Gulf portion		331,000		48
2. Coastal Plain			1,166,000	100
a. Cretaceous		196,000		17
b. Lower Tertiary		315,000		27
c. Upper Tertiary		302,000		26
d. Quaternary		353,000		30
Total--Atlantic and Gulf Coast Province			1,851,000	

*Lengths and areas were calculated from the Geology Map (scale 1:7,500,000) in the U.S. Geological Survey (1970) National Atlas. For purposes of comparison, the General Coastline length (Table 7) determined by the U.S. Department of Commerce, NOAA (1975) is 6.8% longer than in this table.

Figure 1. Coastal Plain geology. From several sources including the U.S. Geological Survey, National Atlas (1970, p. 74–75).

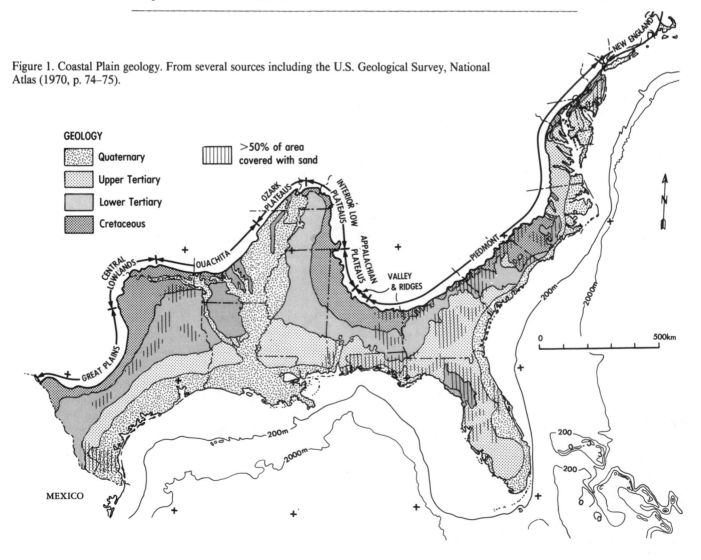

TABLE 2. THE COASTAL PLAIN AND ITS NEIGHBORING PROVINCES

Province Name	Common Boundary (km)	(%)	Geologic Age and Predominant Rock Types
1. New England	350	7.5	Precambrian and Paleozoic sedimentary and igneous--gneisses, granites
2. Piedmont Plateaus	1,710	37.0	Precambrian and lower Paleozoic metamorphic and sedimentary; Triassic sediments to the north--schists, gneisses, slates
3. Valley and Ridge	100	2.2	Upper and lower Paleozoic sedimentary
4. Appalachian Plateaus	225	4.8	Upper Paleozoic sedimentary--sandstones, conglomerates, shales
5. Interior Low Plateaus	420	8.8	Upper Paleozoic sedimentary--both marine and continental sandstones, shales, limestones
6. Ozark Plateaus	300	6.5	Lower Paleozoic sedimentary--limestones, dolomites
7. Ouachita	570	12.2	Lower to upper Paleozoic sedimentary--sandstones, shales, some limestones and conglomerates
8. Central Lowlands	325	6.7	Cretaceous sedimentary--sandstones, shales, limestones, conglomerates
9. Great Plains	665	14.3	Lower Cretaceous sedimentary--limestone
TOTAL	4,665	100	

Geology: Lithology and Structure

The Atlantic and Gulf Coastal Province is one of the least complicated provinces in the United States insofar as its geology is concerned. Essentially it is a series of seaward-dipping layers of marine sediments that accumulated during periods of submergence, although vast areas are covered with fluvial and some with eolian deposits. The distributions of the four main formations in the Coastal Plain are Cretaceous (17%), lower Tertiary (27%), upper Tertiary (26%), and Quaternary (30%; Fig. 1 and Table 1). Nearly all are sedimentary in origin and represent various combinations of onshore, nearshore, and offshore deposits.

Although there is some Lower Cretaceous clay, sand, and gravel near the heads of the Potomac, Chesapeake, and Delaware estuaries and some sand, clay, shale, limestone, chalk, and marl adjacent to the Ouachita and Central Lowlands boundaries, most Cretaceous deposits are younger in age. The Cretaceous is especially broad in Texas, Alabama, and along the South Carolina and North Carolina boundary, where it approaches within a few kilometers of the shoreline. Upper Cretaceous deposits are varied and include gravel, sand, clay, limestone, chalk, marl, and even volcanic ash.

Lower Tertiary sediments are present in a somewhat wider band (over 250 km wide in east Texas) than the Cretaceous. This band is nearly continuous except across the Mississippi River flood plain and north of Virginia. The sediment is mostly marine limestone and sand although some nonmarine clay and sand are also present.

The upper Tertiary, although essentially parallel to the lower Tertiary, does not reach northward up the Mississippi Valley, but

does extend nearly the full length of Florida. From Virginia north, where the lower Tertiary is only represented by a narrow band bordering the Fall Zone, the upper Tertiary is quite extensive. The upper Tertiary includes marine sand, clay, and marl and much of the limestone of Florida.

The Quaternary deposits of the Coastal Plain, which are extensive in the Mississippi Valley and along nearly all of the coast, consist mainly of riverine gravel, sand, and clay, marine limestone, and eolian loess. They also cover nearly all of the Continental Shelf and Slope. Sand and sand with gravel predominate in the Atlantic and eastern Gulf of Mexico sectors, whereas off the Louisiana and Texas coasts, mud, sometimes mixed with sand, is the major type of surface deposit (Fig. 2).

Structurally, the Atlantic and Gulf Coastal Province is dominated by seaward-dipping beds. During and subsequent to deposition, these beds were affected by warping (folding), faulting, and diapirism. The resulting structures divide the Coastal Plain into subprovinces with unique depositional and erosional histories.

Although the general dip is eastward along the Atlantic and southward along the Gulf of Mexico, warping has locally modified directions in both segments (Fig. 3). Upwarping created five major arches and uplifts, and downwarping resulted in seven major basins and embayments (Fig. 3; Murray, 1961, p. 79).

Surface Configuration: Relief and Drainage Patterns

The Atlantic and Gulf Coastal Province is noted for its generally low local relief. However, when the relief of the province as a whole is considered, it is at least 2,500 m, with elevations ranging from over 300 m above sea level on the inner edge of the

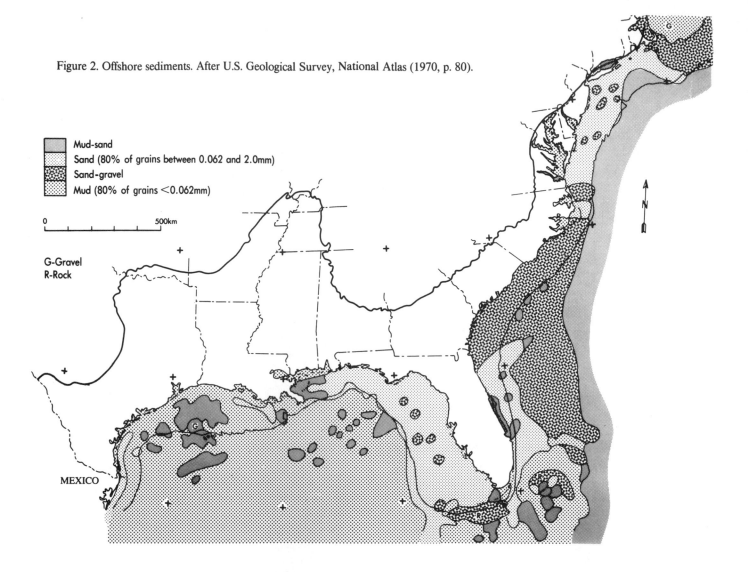

Figure 2. Offshore sediments. After U.S. Geological Survey, National Atlas (1970, p. 80).

Coastal Plain in south Texas to more than 2,500 m below sea level at the bottom of some of the submarine canyons at the outer edge of the Continental Shelf.

The Continental Shelf, except in the vicinity of submarine canyons, has local relief that seldom averages more than a few meters, although in places, such as on the subaqueous portion of the Mississippi Delta and formerly glaciated subaqueous area off New England, surfaces may be quite irregular. The Coastal Plain, however, has a fairly large range of local relief, one that tends to increase away from the present-day shoreline and the Mississippi River alluvial plain, that is, in a direction consistent with increasing age of the rocks. Areas with local relief of over 90 m are generally confined to the Cretaceous in Texas and Tennessee-Alabama-Georgia and to small areas in the lower Tertiary rocks in Texas and Alabama (Fig. 4). In these same areas the province's most impressive hills are common.

Hills with lesser relief (50 to 75 m) are found in a number of other widely distributed locations within the province, including the morainal (Pleistocene) areas of Long Island and northern New Jersey, a crescentic area of the lower Tertiary in Mississippi, the upper Tertiary of central and northwest Florida, and especially those locations where rivers cut through cuestas in Alabama, Mississippi, and Texas.

Although there is a degree of correlation between age and relief, a major exception occurs in the South Carolina–North Carolina area where the Cretaceous nearly reaches the shoreline (Fig. 1) and the bulk of it has local relief of less than 30 m.

Even though local relief within the province is considered low, the position of the relatively smooth surfaces has a regional pattern. Areas where local relief is more than 30 m and where the gently-sloping portions occur in the upland are especially concentrated on the inner part of the Atlantic Coastal Plain, to the east of the Mississippi River bluff, and in the upper Tertiary of Texas and Alabama (Fig. 4). Most of the rest of the Coastal Plain with local relief over 30 m has the bulk of its gently-sloping portions in the lowland.

Figure 3. Structural features of the Atlantic and Gulf Coastal Province. After Murray (1961, p. 80).

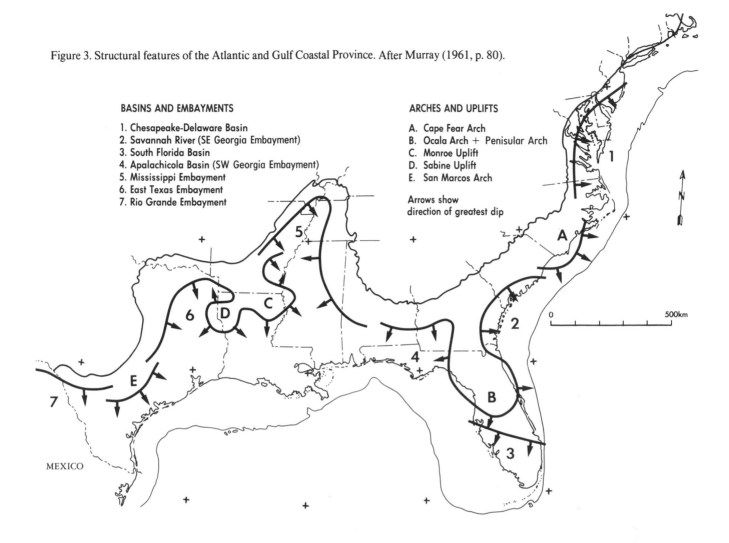

BASINS AND EMBAYMENTS

1. Chesapeake-Delaware Basin
2. Savannah River (SE Georgia Embayment)
3. South Florida Basin
4. Apalachicola Basin (SW Georgia Embayment)
5. Mississippi Embayment
6. East Texas Embayment
7. Rio Grande Embayment

ARCHES AND UPLIFTS

A. Cape Fear Arch
B. Ocala Arch + Penisular Arch
C. Monroe Uplift
D. Sabine Uplift
E. San Marcos Arch

Arrows show
direction of greatest dip

500km

MEXICO

River courses in the Coastal Plain reflect the downward slope, toward the sea, of the province's formations and are generally perpendicular to the shoreline. Thus, the trend of stream courses along the Atlantic seaboard is from northwest to southeast, whereas those rivers flowing into the Gulf of Mexico vary in direction from north to south between Georgia and Texas and northwest to southwest in Texas (Fig. 5). Although some rivers originate in the Coastal Plain, most major rivers flow into and across it from adjacent provinces (Fig. 6 and Table 3).

The total drainage area of all rivers affecting the Coastal Plain is 5.5×10^6 km^2 of which less than 1.2×10^6 km^2 (about 21%) is within the Coastal Plain itself. The Mississippi River drainage basin, with a total area of 3.3×10^6 km^2, occupies over 60% of the total drainage area of all rivers affecting the Coastal Plain (Table 3). Within the Coastal Plain, the Mississippi River and its tributaries occupy 26% of the total area. These two rivers, including their tributaries, cross numerous provinces and, along with some of the rivers of the northeast, originate in formerly glaciated sections of the continent (Fig. 6). The area of the present-day Mississippi River drainage basin that drains glaciated areas is 1.2×10^6 km^2, of which about 53,000 km^2 is in the Rocky Mountains.

Most of the rivers and streams that flow across and that originate within the Coastal Plain are perennial. The major exception occurs in south Texas where precipitation amounts are relatively low (less than 50 cm/year) and intermittent flow is common (Fig. 5). Surface flow is also reduced or eliminated in parts of Florida because of the dominance of subsurface drainage. The discharge and sediment load of the rivers of the province vary greatly (Table 4).

The total discharge into the ocean is about 35,000 m^3/sec, of which just over half (52%) is from the Mississippi River. The rivers flowing into the Atlantic Ocean contribute 29%, whereas those flowing into the Gulf of Mexico, excluding that of the Mississippi River, have a discharge proportion of 19% (Table 4). Over 83% of the total sediment delivered to the Continental Shelf is carried by the Mississippi River; the Rio Grande contributes another 10%. The dominance of these two rivers is also reflected in the relative concentration of suspended sediment (Table 4).

Figure 4. Relief and land-surface form of the Coastal Plain. After U.S. Geological Survey, National Atlas (1970, p. 62–63).

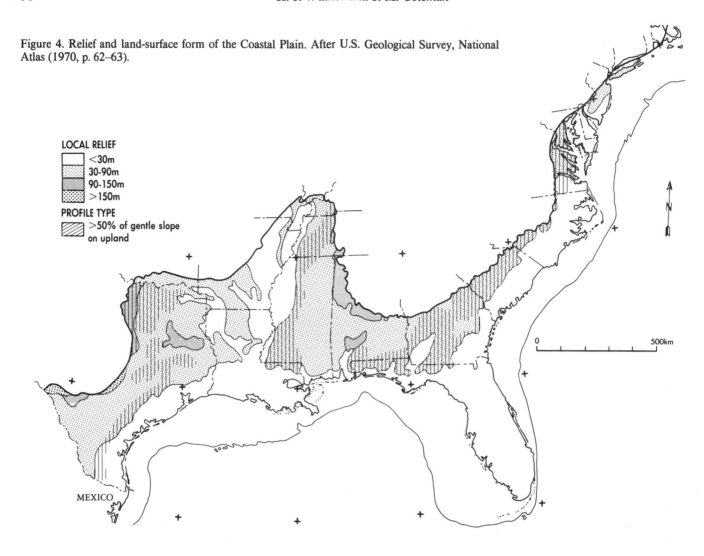

With only four exceptions (drainage areas 5, 15, 16, and 19), the total contributed quantity of sediment to the ocean per year is less than 0.5% of that of the Mississippi River. This stems from the fact that in all but those four cases, low discharge (ranging from 0.38% for area 11 to 6% for area 2) occurs in combination with low concentrations (ranging from 3% to 17%). The most extreme case is the area that includes nearly all of the Texas rivers. The sediment concentration of these Texas rivers is 820 mg/ℓ, which is some 60% higher than the 510 mg/ℓ of the Mississippi River. The total sediment contribution to coastal lagoons and to the Gulf of Mexico by Texas rivers is high.

Landforms and Processes

Within the Atlantic and Gulf Coastal Province, it is convenient to consider all processes and forms under three basic divisions: those predominantly subaerial, those predominantly suboceanic, and those that are transitional. Only a few processes and forms are common to all three zones; most of those that are subaerial are not present (or, if present, are highly modified) in the suboceanic zone. As mentioned in the introduction, virtually all of the province has experienced all three situations—that is, subaerial, suboceanic, and transitional. The present distribution of these three zones is epochal. At the present time, the subaerial and suboceanic portions of the province are nearly equal in size.

The subaerial surface tends to be dominated by erosional processes and forms, the suboceanic surface by depositional processes, and the transitional surface by combinations of the two. The subaerial zone supports depositional forms (e.g., point bars and eolian dunes) and the suboceanic has erosional forms (e.g., mudflow gullies and submarine canyons), but such forms are generally superimposed on erosional surfaces in the case of the subaerial zone and "carved" into depositional surface in the case of suboceanic environments.

Because all of the province was suboceanic in the past, it can be theorized that those portions that are now subaerial had, at one time, processes and forms similar to those of the present Continental Shelf. Likewise, the present Continental Shelf was character-

TABLE 3. MAJOR RIVERS AND THE COASTAL PLAIN*

| River | | Drainage Basin Area | | |
Name	Designation (see Fig. 5)	Total (km²)	Coastal Plain (km²)	(%)
Connecticut[†]	a	28,000	0	0
Hudson	b	35,400	0	0
Delaware	c	29,700	5,600	19
Susquehanna	d	73,000	0	0
Potomac	e	37,800	5,600	15
James	f	26,500	5,200	20
Roanoke	g	25,600	3,500	14
Cape Fear	h	23,100	11,800	51
Pee Dee-Yadkin	i	40,700	17,100	42
Santee	j	42,900	9,200	21
Savannah	k	27,100	7,900	29
Altamaha-Ocmulgee	l	36,300	20,700	57
Sawannee	m	25,100	25,100	100
Apalachicola	n	46,700	29,700	64
Tombigbee-Alabama	o	112,900	62,500	55
Pearl	p	21,700	21,700	100
Mississippi[§]	q	3,360,000	302,000	9
Sabine-Neches	r	54,400	54,400	100
Trinity	s	44,000	40,400	92
Brazos	t	120,600	33,500	28
Colorado	u	103,000	11,700	11
Nueces	v	40,700	36,900	84
Rio Grande[†]	w	590,000	21,000 (U.S. only)	6 (U.S. only)

*All areas calculated from the 1:3,168,000 equal area Base Map of the United States, 1965 edition, U.S. Geological Survey, U.S. Department of the Interior, Washington, D.C., except those of the Mississippi and Rio Grande, which were calculated from Curtis and others (1973).

[†]Includes: Rio Grande--265,000 km² in Mexico and 29,000 km² of closed basins; Mississippi--59,000 km² in Canada; Connecticut--4,000 km² in Canada.

[§]Includes the Atchafalaya.

Figure 5. Water bodies of the Coastal Plain. See Table 3 for river names. From several sources including the U.S. Geological Survey 1:3,168,000 Base Map (1965). Compare with Figure 6.

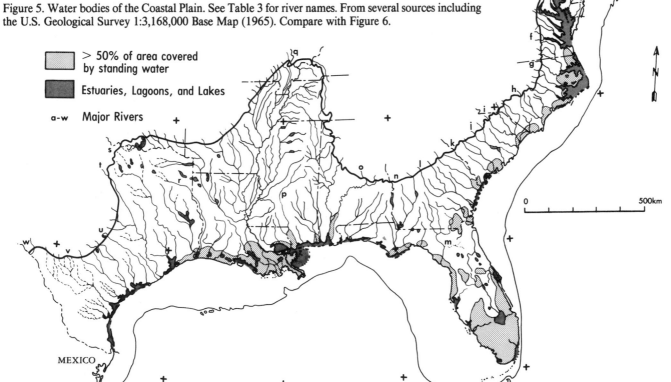

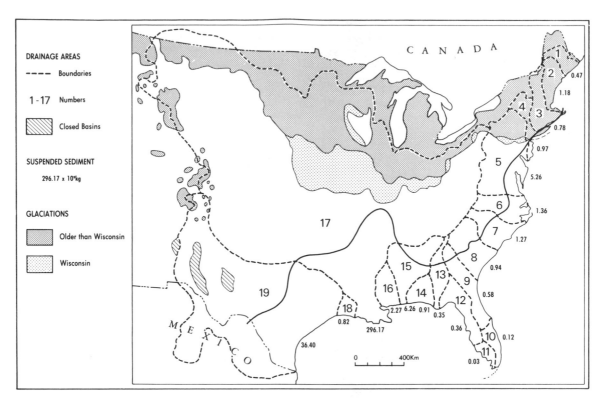

Figure 6. Drainage areas, annual suspended sediment load, and glacial limits. See Table 4. Mainly after Curtis and others (1973, p. 2–3). Note: Many of the drainage areas include a number of river basins; compare with Figure 5 and Table 3.

ized by erosional forms, many of which have been buried by sediment if not previously obliterated by waves and currents during sea-level fluctuation.

After discussions of the history of the development of geomorphological research on and the geological history of the province, selected landforms and processes are examined under the suboceanic, transitional, and subaerial divisions. Because of the relatively distinct character of each of these divisions, the approach to their presentation varies. The forms and processes within the suboceanic portion are treated within the three regional subheadings of the Atlantic Margin, the Florida Platform, and the Gulf of Mexico Basin. The transitional division is examined systematically and is treated in order of the shoreface, barrier islands, deltas, lagoons, marshes, and dunes and their related forms. The subaerial division is discussed from the standpoint of present-day processes, including those of tectonic, gravitational, solutional, and fluvial origin.

THE DEVELOPMENT OF GEOMORPHIC KNOWLEDGE

Among the first portions of the New World to be explored after Columbus rediscovered the Americas were the Atlantic and Gulf coasts of North America. From the beginning of these contacts, geomorphic information began to accumulate and even

geological processes were enunciated. Explorers seeking a route through the American barrier along the North Atlantic Coast were stopped by the rapids and falls present at the heads of the bays and estuaries they entered. Later, after some successes and failures near the coast, English settlers pushed inland up rivers such as the James, where they encountered "that cataract or fall of water, which the Indians call Paquachowng" (cited by Bakeless, 1961, p. 186).

The recognition that such rapids and falls occur along a line that separates a wide flat coastal margin (the Coastal Plain) from a hard-rock hilly region (the Piedmont) was documented as early as 1588 in a book by Thomas Hariot, a naturalist, who was with Sir Walter Raleigh. Less than a quarter-century later (in 1612), Captain John Smith published a map depicting the same information (see White, 1953, p. 135).

The significance of the line of falls, to become known as the Fall Line, was impressed on all who ventured up the rivers, at least from New York to central Alabama. William Bartram exemplified this fact well when, in 1775, he wrote that from Augusta, where the Savannah River

. . . falls over the cataracts, . . . downwards to the ocean, . . . [it] uninterruptedly flows with a gentle meandering course, and is navigable for vessels of twenty or thirty tons burthern to Savannah, where ships of three hundred tons lie in a capacious and secure harbor (cited by Cruickshank, 1961, p. 112).

TABLE 4. DISCHARGE AND SUSPENDED LOAD*

Drainage Area (No.)	Discharge		Suspended Sediment			
	m³/sec (x10²)	% of Miss. R.†	mg/l (Avg)	% of Miss. R.	Metric tons/yr (x10⁶)	% of Miss. R.†
A. Mississippi River						
17	184.00	100	510	100	296.17	100
1	6.65	4	20	40	0.47	0.2
2	11.62	6	32	6	1.18	0.4
3	9.86	5	25	5	0.78	0.3
4	9.28	5	33	6	0.97	0.3
5	28.66	16	58	11	5.26	1.8
6	7.79	4	54	11	1.36	0.5
7	8.05	4	50	10	1.27	0.5
8	9.99	5	30	6	0.94	0.3
9	7.32	4	25	5	0.58	0.2
10	2.55	1	15	3	0.12	---
TOTAL	101.77	54			12.94	
B. Atlantic Ocean						
11	0.71	---	15	3	0.03	---
12	7.70	4	15	3	0.36	0.1
13	7.56	4	15	3	0.35	0.1
14	7.11	4	40	8	0.91	0.3
15	18.18	10	109	21	6.26	2.1
16	8.83	5	81	16	2.27	0.8
--	(see Mississippi River, above)					
18	3.06	2	85	17	0.82	0.3
19	14.07	8	820	161	36.40	12.3
TOTAL	67.22	37			47.40	
GRAND TOTAL	352.99				356.51	

*Data from Curtis and others (1973); all percentages have been rounded.

†To convert discharge to % of total across the Coastal Plain, multiply by 0.52. To convert the annual suspended load to a % of the total, multiply by 0.83.

Many of the early descriptions of this new land were contained in colonization or promotional tracts intended to attract settlers. Although basically propaganda, they nonetheless provided some geomorphic information. Much of this information dealt with bays, lagoons, and sounds and their function as harbors. Because of the smallness of the vessels of the time, excellent anchorage was available all along the coast. Whereas early descriptions and maps depict the nature of barrier islands, tidal inlets, and lagoons along this coast, they also frequently imply the dangers associated with navigation in their vicinity (Fig. 7), and, by extension, something of the shifting of inlets and severity of storms.

Thoughts of sea-level change must have existed early. For example, in 1606, G. Percy in his *Discourse of Virginia* wrote that

... within the shoares of our rivers, whole bancks of oysters and scallopps, which lye unopened and thick together, as if there had bene their naturall bedd before the sea left them (cited by Bakeless, 1961, p. 189).

One of the most perceptive of the early writers was Lewis Evans. In 1755 he wrote that

... a Rief or Vein of Rocks ... some two or three, or Half a Dozen Miles broad ... extending from New York City South Westerly ... was the antient maritime Boundary of America ...

The Land between this Rief and the Sea, ... Westward as far as this map extends, and probably to the Extremity of Georgia ... consists of Soil washed from above, and Sand accumulated from the Ocean (cited by Mather and Mason, 1970, p. 55).

The notion that the "Region of Sea Sand" (the Coastal Plain) was originally sea bottom and that the Fall Line had been a shoreline was accepted by other naturalists of the time despite its conflict with Biblical teachings (Brown, 1948, p. 98).

The quote above is from Evan's analysis that accompanied a map upon which he divided the eastern United States into physiographic and geologic provinces. Both his map and analysis served as a basis for much of the descriptive work that followed.

Observations of the Gulf Coast from Florida " ... with vast lakes and marshes of stagnated fresh water ..." (Bakeless, 1961, p. 34) to Texas came early (Fig. 8). The Mississippi River Delta appears on maps as early as 1513 and qualitative descriptions of

H. J. Walker and J. M. Coleman

Figure 7. Sixteenth century map of the coast of Virginia, showing the island of Roanoke. From Bakeless (1961, p. 191).

Figure 8. Florida swamp in 1872. From Bakeless (1961, p. 30).

the river's width, volume, and sediment load appeared in accounts of De Soto's 1539–42 expedition.

Without much doubt the Mississippi River attracted most of the early attention of the naturalists who studied the Gulf Coast Region. Darby, in 1816, writing about Louisiana, explained the fact that the Mississippi River below the Ohio flows along the eastern bluffs because " . . . all the large tributary rivers entering from the west have filled that side of the great valley with their deltas . . ." (See Lyell, 1856, p. 265). Thirty years after Darby, Lyell made his study of the Mississippi River, which many considered a prime exhibit for the fluvialists molding geological theory at the time. Lyell was especially intrigued with the Mississippi River Delta and calculated that it was 67,000 years old and had an area of 22,000 km^2 (1856, p. 273). Upon learning that its growth rate was decreasing, he theorized that as the delta's "tongues" extended into the gulf, stronger currents carried sediment away. Lyell discussed many of the forms (and processes) associated with the river, including crevasses, meanders, cut-offs, semicircular lakes (oxbows), and log rafts. Impressed with the river's organic-matter load, he wrote:

The prodigious quantity of wood annually drifted down by the Mississippi and its tributaries, is a subject of geological interest, not merely as illustrating the manner in which abundance of vegetable matter becomes . . . imbedded in submarine and estuary deposits, but as attesting the constant destruction of soil and transportation of matter to lower levels by the tendency of rivers to shift their courses (1856, p. 268).

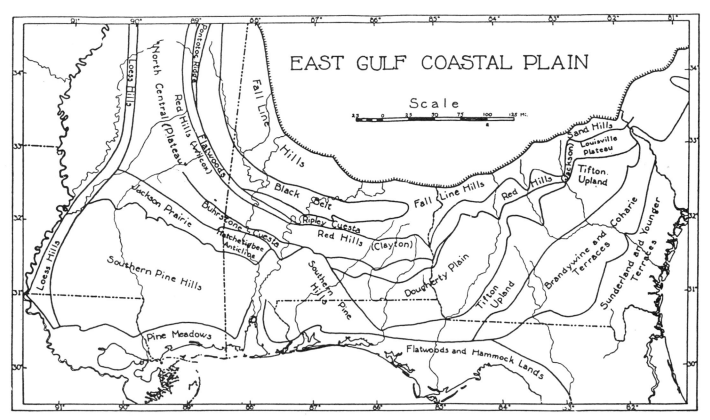

Figure 9. Map of physiographic regions in the eastern Gulf Coastal Plain. From Fenneman (1938, p. 68).

Lyell's work on the Mississippi River was soon followed by a study of the river for the purpose of assisting flood control. This study, authorized by the U.S. Congress in 1850, led to what Chorley and others (1964, p. 436) labeled "phenomenal progress" in hydrological studies. A. A. Humphries and H. J. Abbot, in their definitive study on hydrology published in 1861, demonstrated the amount of work performed by the Mississippi River.

Although a number of maps (including that of Evans) were produced before the beginning of the nineteenth century, King and Beikman considered that " . . . the first geologic map of the United States is that published by William Maclure in 1809 . . ." (1974, p. 1). Maclure traveled much of the Coastal Plain as far west as the Mississippi and included Cape Cod and Long Island in his "alluvial rock" region, which is representative of the present Coastal Plain at least to central Alabama. Lyell, not surprisingly, included much more detail on his 1845 map than had Maclure in 1809. Using conventional systems and series, Lyell's map showed relatively accurately the distribution of the Cretaceous, Eocene, Miocene, and post-Pliocene and alluvium deposits. From the Mississippi River east, distributions differ little from those used today, and by 1855 (Marcou's map), even the Coastal Plain in Texas was relatively well delineated (King and Beikman, 1974, p. 5).

Thus, for over 125 years the outline of the province has been

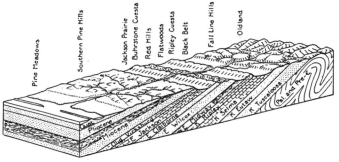

Figure 10. Three-dimensional section. From Fenneman (1938, p. 69).

well established, even though the maps prior to that time were not necessarily intended to be topographic or geomorphic. However, during the mid-nineteenth century such maps began to appear and led eventually to that produced by the Fenneman Committee in 1916 (Fenneman, 1916, p. 21). In 1938, Fenneman published *Physiography of eastern United States* in which Chapter 1 (p. 1–120) is entitled "Coastal Plain Province." In it, Fenneman included numerous regional maps and cross sections (Figs. 9 and 10), most, if not all, of which have been used in standard works such as Thornbury (1965) and specific studies such as Murray (1961).

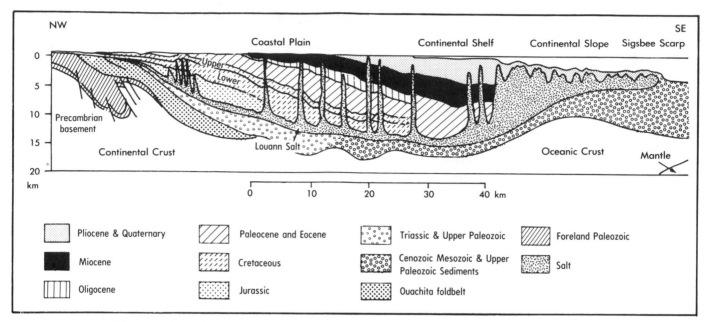

Figure 11. Generalized section of the Gulf Coast Geosyncline from central Texas to the Sigsbee Scarp. After King (1977, p. 83).

THE GEOMORPHIC HISTORY OF THE ATLANTIC AND GULF COASTAL PROVINCE

Pre-Quaternary Geology

The Atlantic and Gulf Coastal Province formed along the trailing edge of the North American Plate during Mesozoic and Cenozoic times. Formation occurred in an actively subsiding geosyncline that contains numerous basins, all of which have had somewhat different geologic histories. Although the province has a surface with relatively low relief, it is, nonetheless, a three-dimensional structural-stratigraphic entity (Murray, 1961, p. 493) that rests upon a basement formed mostly in pre-Mesozoic times.

The basement beneath the area that now constitutes the Atlantic and Gulf Coastal Province consists of a variety of rock types including metamorphic and igneous crystalline rocks as well as Paleozoic and lower Mesozoic (Triassic) sedimentary rocks (Fig. 11). Many of these rocks have their counterparts in the highlands that border the province to the west and north. Like the present Coastal Plain, the basement surface slopes seaward, although, as pointed out by Murray (1961, p. 76), variations in its slope occur for several reasons, including orogenic modification, isostatic adjustment, igneous intrusion, and preexisting surface irregularity. This basement slopes variously beneath the Coastal Plain reaching depths of 6 to 8 km and 11 to 13 km beneath the North Atlantic and Gulf of Mexico shorelines respectively.

The major variations in the basement surface consist of a series of alternating positive and negative features, the arches and embayments shown on Figure 3. These features have been a major factor in determining not only the present position of the province

as a whole but also of much of the nature of present-day surface topography. For the most part, they have continued to be active to the present, that is, through Mesozoic and Cenozoic times.

Folding, faulting, and intruding igneous and salt masses have interrupted the overall homoclinal dip of these Mesozoic-Cenozoic deposits and have also influenced: (1) the width of the plain, (2) the rates of erosion and deposition, (3) the development of drainage patterns, and (4) the nature and character of the shoreline.

The nine bordering provinces, as noted in the introduction, are very diverse and include highlands such as the Appalachian structural belt, the Ozark Uplift, the Ouachita structural belt and the Wichita and Arbuckle mountains, and the Llano and Marathon uplifts. These highlands have had considerable influence on the Atlantic and Gulf Coastal Province, especially through regulation of the supply of sediment during the Cenozoic.

The Atlantic and Gulf Coastal Province, when considered in terms of the structural elements that have influenced its development, is divisible into: (1) the Atlantic Coastal Margin, (2) the central Georgia-Florida Platform, and (3) the Gulf of Mexico Basin. The Atlantic Coastal Margin can be characterized as a region with relatively broad, slowly subsiding basins or negative areas that are separated by broad arches or positive areas. The central Georgia-Florida Platform has been controlled primarily by the Ocala Arch (Fig. 3), which was the site of carbonate sediment accumulation during the Cenozoic. The Gulf of Mexico Basin is complex. During the Tertiary, extremely large amounts of clastic sediments were deposited, large salt masses were mobilized, and highly active penecontemporaneous (growth) faults were formed (Fig. 12).

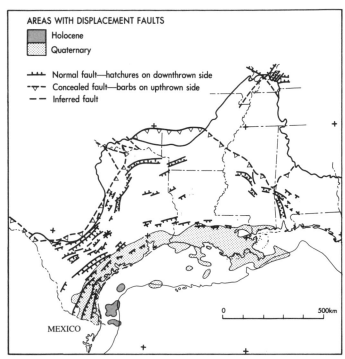

Figure 12. Tectonic features of the Gulf Coastal Plain and Continental Shelf. From several sources including the U.S. Geological Survey, National Atlas (1970, p. 70–71) and Murray (1961, p. 168).

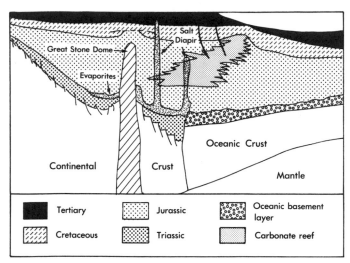

Figure 13. Interpretive crustal section across Baltimore Canyon Trough. Section from NW (left) to SE. Modified from Grow (1981, p. III-20).

The Atlantic Coastal Margin. The Atlantic Coastal Margin has a Tertiary layer that, although generally thin across the subaerial plain, thickens rapidly in a seaward direction and especially across the Continental Shelf. This layer, as shown in an interpretive crustal section across the Baltimore Canyon Trough (Fig. 13), overlies Cretaceous and Jurassic sequences, which are interrupted by carbonate reefs and banks. During Late Jurassic and Early Cretaceous times, carbonate reefs and banks dominated the sedimentary sequence and controlled the seaward margin of the Coastal Plain sediments.

Along this coastal margin are the Chesapeake-Delaware and Savannah (southeast Georgia) basins and the Cape Fear and Ocala arches (Fig. 3). Both basins or embayments are large, with the Chesapeake-Delaware Basin extending from North Carolina past Cape Cod. Its continuation as a negative area to the present is reflected in the presence of extensive Pleistocene deposits, riverine and coastal terraces, and numerous bays and estuaries that line the present shoreline (Murray, 1961, p. 94). The Savannah Basin, situated along the South Carolina–Georgia Coast, is smaller and shallower than its counterpart to the north. Its offshore profile is unique in the province in that it includes a break (the Blake Plateau) in the Continental Slope between the depths of about 600 and 1,200 m.

Separating the Chesapeake-Delaware and Savannah basins is Cape Fear Arch (Fig. 14), the crest of which nearly coincides with the North Carolina–South Carolina boundary. Although, at

the shoreline, the basement is within about 600 m of the surface, at its crest (near Wilmington, North Carolina) it is only about half that depth. Uplift during the Cenozoic resulted in the outcrops of Cretaceous deposits that characterize this portion of the Atlantic Margin (Fig. 1). The shoreline, with its prominent capes and with the shoals that extend seaward from them separated by smooth cuspate bays, is undoubtedly related to the uplift of Cape Fear Arch (Maher and Applin, 1971, p. 9). In addition, drainage patterns, deeply incised river courses, and nonparallelism of river terraces reflect a long-term regional deformation associated with the arch (Markewich, 1984, p. 289).

The Georgia-Florida Platform. The Georgia-Florida Platform, which has been rather stable, is dominated by the Ocala Arch (also known as the Peninsular Arch), which extends from south Georgia south to central Florida (Fig. 3). The arch is a complex structure with early Paleozoic strata at its core. It is thought that it stood sufficiently high during Early and early Late Cretaceous time so that deposition occurred around but not on top of it (Maher and Applin, 1971, p. 24). Subsequently, that is, during the Upper Cretaceous and the Cenozoic, deposition produced only a thin sedimentary sequence. The thick Jurassic and Cretaceous sediments extend eastward across the Blake Plateau to the Continental Slope.

The South Florida Basin, which dominates the area south and southwest of the Ocala Arch, was the depository for vast amounts of sediment during the Cretaceous, and since the Tertiary has alternated between being a submerged shallow shelf and a low-lying emerged coastal plain.

The Gulf of Mexico Basin. The Gulf of Mexico Basin is very large and extremely complex. It has been dominated, from the beginning of the Tertiary to the present, by sediments from the Mississippi River (Martin and Bouma, 1978; Bouma and others, 1978). Mesozoic and Cenozoic deposits, which are much thicker

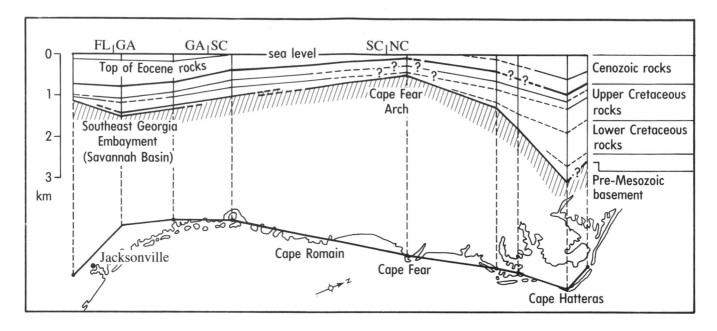

Figure 14. Geologic section, Florida to Virginia. After Maher and Applin (1971, Plate 9).

in the Gulf of Mexico Basin than along the Atlantic, are estimated to have attained a total thickness in excess of 15 km (Fig. 11). The zone of maximum thickness trends east-west near the present-day Coastal Plain of Louisiana and west Texas. Rapid subsidence associated with sediment loading has been responsible for such thick localized sedimentary accumulations.

In the Coastal Plain, especially west of Alabama, deep-seated fault structures are numerous (Fig. 12). Some extend over great distances, whereas others, such as those associated with salt domes, are highly localized. The extensive fault systems, as is true of most structures in the province, tend to be parallel to the coastline and also are generally parallel to the strike of the strata (Fig. 12). Murray (1961, p. 167) noted that normal faults predominate, that most are steepest near the surface, and that those near the coast tend to be steeper (up to 70°) than those inland (averaging 45° to 50°).

The modern depositional pattern in the Gulf Basin, where terrigenous deposits dominate on the northern and western shelf areas and carbonates occur on the broad platforms of the eastern gulf, has persisted since the Upper Cretaceous. At that time, the terrigenous clastic influx from tectonically elevated northern and western continental interiors began to overwhelm the mainly carbonate environments that encircled the gulf during earlier Cretaceous times (Garrison and Martin, 1973).

The bulk of the sediment was delivered to the northern margin of the gulf during the Cenozoic and prograded the shelf as much as 300 km from the margin of the Cretaceous platform deposits to the present shelf edge. This sedimentation rate provides a magnitude of shelf-edge progradation of 5 to 6 km per million years, which contrasts sharply with that of the eastern margin of

the United States, where the average shelf-edge progradation is measured in fractions of a kilometer per million years.

Along the northern half of the gulf, the province is outlined by a semicircular arc of alternating major embayments or basins and arches or uplifts (Fig. 3). In addition, there are many other tectonic depressions or uplifts of lesser importance. The four major arches and five major basins that comprise this arc, which are similar in many ways to their counterparts along the Atlantic, have all been associated with the major geosynclinal development of the Gulf of Mexico Basin as a whole.

The central portion of the northern part of the Gulf Basin is dominated by the Mississippi Embayment, which extends over 800 km north from south Louisiana to south Illinois (Fig. 3). This exceptionally large downwarp, which influences the Coastal Plain from Alabama to east Texas, consists of Cretaceous and Tertiary rocks. The downwarp is asymmetric at various locations and has an axis that varies in alignment. It has been proposed that variations in alignment correlate with basement fractures, which also control the form and shape of the Mississippi Valley and River (Murray, 1961, p. 108 after Fisk, 1947). However, Saucier (1974, p. 4) suggested that the configuration of the lower Mississippi valley is more related to preglacial drainage than to fault control. The Mississippi Embayment, like most of the rest of the western part of the Gulf Basin, has been subject to vertical movement through the formation of salt domes and development of near-surface faults. The major fault systems are deep-seated and originated in Mesozoic and even earlier times; faulting is a process that continues to the present and influences the development of specific surface forms.

Usually, changes in sea level are considered the hallmark of

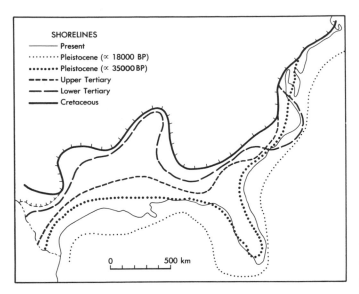

Figure 15. Shoreline positions and Coastal Plain extent at various times. Modified from Hunt (1967, p. 163).

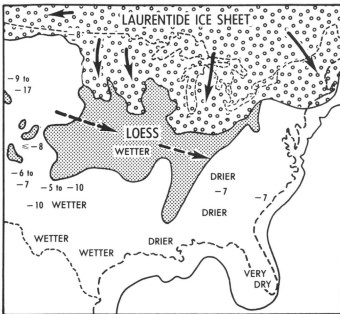

Figure 16. Environmental conditions during Laurentide ice sheet maximum. After Barry (1983, p. 401).

the Quaternary. However, the Atlantic and Gulf Coastal Province was subjected to sea-level changes often during pre-Quaternary times (Fig. 15) and much of its morphology today reflects those changes. One of its distinctive characteristics is that, at one time or another during the past 100×10^6 years, most, if not all, of it experienced both submergence and emergence, some of it several times. It has been affected to a much greater extent than any other province by changing sea levels. On a temporal basis, sea-level fluctuations were not so rapid or cyclic as in the Quaternary when glacial controls were of critical importance.

Between the Cretaceous and the end of the Tertiary, there was a cumulative (albeit at variable rates) lowering of relative level to near the position of the present shoreline (Fig. 15). The entire area of the present province (and even beyond in the case of west Texas) was suboceanic during the Cretaceous. During early Tertiary times virtually all of the same area was covered. The major exceptions were two crescentic belts, one southeast of the Central Lowlands and one southwest of the Appalachian Plateau, and a large region centered over what is now the middle Atlantic shoreline.

Sea level during the Tertiary appears to have been controlled mainly by the rate of crustal spreading in the Atlantic. During most of the Miocene, crustal spreading was rapid and sea level was generally high. Toward its end (approx. 5.5×10^6 years B.P., crustal spreading slowed and sea level dropped slowly (Chorley and others, 1984, p. 398). About the same time, the Antarctic ice sheet probably began to form and, since then, has helped maintain a sea level lower than it otherwise would have been.

The Quaternary

The Quaternary, characterized mainly by cyclic glaciation, witnessed major impact on all parts of the province, even though

the ice sheets directly affected only a small area in the northeast. The province's morphology was influenced by drastic and relatively frequently changes in climate (temperature, precipitation, wind), vegetation, drainage, and sea level throughout the epoch.

Although there is still debate as to the precise history of the Quaternary, the general nature of the effects of advancing and retreating ice on climate and vegetation are becoming clearer through research in palynology and other sciences. Although the results are far from precise, there is a fair correlation between advancing ice and decreasing temperatures and retreating ice and increasing temperatures within the Atlantic and Gulf Coastal Province. These variations are also reflected, although again imprecisely, in vegetation variability. Whatever correlation exists between ice advance or retreat and precipitation is somewhat less clear. For example, at the time of maximum ice advance, precipitation appears to have been lower in the southeast and higher in the southwest and midwest than at present (Fig. 16). Wind direction, persistence, and velocity were apparently affected greatly (even if, to a large extent, indirectly) by ice-sheet position. The effects of variations in precipitation on runoff, of temperature on weathering and vegetation, and of wind on deflation have all been important in landform development within the province.

The lowering of sea level with waxing glaciers exposed vast areas of former Continental Shelf and lowered the base level of streams draining into the ocean; this, in turn, led to entrenchment. With the rise in sea level that occurred as glaciers waned, vast areas of former Coastal Plain were submerged, the base level of streams was raised, and river valleys aggraded.

There is much debate as to just how many times and at what

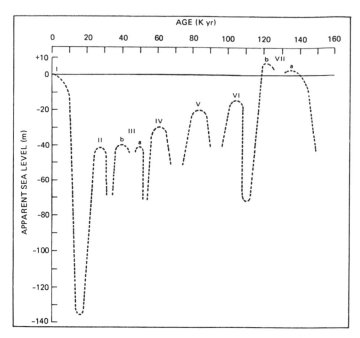

Figure 17. High stands of sea level during the past 150,000 years. From Revelle (1983, p. 434).

rates ice advanced and retreated during the Quaternary. Equally as elusive is the determination of the quantities of ice, and thus seawater, involved in these transfers between ocean and land and therefore in frequencies, rates, and positional (both vertical and horizontal) limits of sea-level change. Early events, some now believed to be Pliocene, were of sufficient duration and height relative to present-day sea level to have created the marine and associated riverine terraces that are so numerous in the province.

Cronin and others, (1981), using multiple paleoenvironmental criteria, documented at least five intervals during the past 500,000 years that were sufficiently warm to raise sea level above that of the present. The last four of these intervals provided sea levels of 7 ±5 m at 188,000 years ago, 7 ± 1.5 m at 120,000 years ago, 6.5 ± 3.5 m at 94,000 years ago, and 7 ± 3 m at 72,000 years ago (Cronin and others, 1981, p. 233). They assumed that during other middle and late Pleistocene warm intervals, sea level remained below that of the present. Such extreme oscillations in sea level are now recognized as the rule rather than the exception (Fig. 17).

Cronin (1981, p. 830) concluded that the coastal plain record contains a tectonic uplift component of 1 to 3 cm/1,000 yr as a result of lithospheric flexure, sediment loading, and possibly hydro-isostatic response during deglaciation.

Most interest in the Quaternary centers around the time since the beginning of the most recent deglaciation, about the last 18,000 years. Just where sea level was at that time is also subject to much debate. Despite the fact that the history of the late Wisconsin is much more clear than that of any other glaciation, Bloom suggested that:

the current range of estimates of 120 ± 60 m for the Late Wisconsin sea-level change is an accurate expression of the extreme complexity of a variable that each scientist previously and naively expected to be unique and easily measured (1983b, p. 219).

It has become clear that to the relatively straightforward changes in level that result from changes in the temperature and salinity of the oceans, the quantity of ice on the continents, and even continental drift and mid-ocean ridge growth must be added the continuously and highly variable changing nature of the geoid (Mörner, 1983, p. 87) as well as all of the changes due to regional and local subsidence, compaction, tectonic movement, and isostatic adjustment due to variations in load by ice, water, and sediment. Thus, it is not surprising that it is now considered that sea-level change along every coast has probably been unique (Bloom, 1983b, p. 219).

The East Coast of the United States would seem to illustrate just such a premise. Clark (1981, p. 438), who along with many others considered the average glacio-eustatic rise in sea level to be only about 80 m, developed models to illustrate relative sea-level changes along the mid-Atlantic coast (Fig. 18A) and the deformation of its coast by glacial isostasy (Fig. 18B). His models suggest that 16,000 years ago the coastline in New Jersey near the ice margin was depressed by as much as 40 m in comparison with the coastline of Florida, and further, that the New Jersey and Virginia coasts actually experienced emergence during the first 6,000 years of sea-level rise. Figure 18C is a model (utilizing the interpolated data from Clark's two models) to graphically show the predicted shoreline depth below present sea level through time along the Atlantic Coast.

Although a eustatic lowering of sea level by only 80 to 90 m is probably too conservative, Clark's models nonetheless suggest how variable the actual areal exposure might have been, whatever the true value.

Based upon the nature of the modifications that occurred during the Quaternary, the province is divisible into: (1) the small region in the northeast modified directly by glacial ice; (2) that portion, mostly the present-day shelf, that alternated between suboceanic and subaerial conditions and therefore was frequently subjected to coastal processes; and (3) that portion of the present Coastal Plain that was neither ice- nor ocean-covered during the Quaternary.

The Glaciated Northeast. Continental glaciers extended seaward past the present shoreline nearly to the edge of the Continental Shelf (Fig. 19). In the process, they eroded and stripped off previously deposited sediment, eroded channels into the shelf, and deposited coarse detritus as moraines and outwash. Sandy, braided outwash plains were probably good sources of sediment for dune formation (Bloom, 1983b, p. 220). Meltwater rivers cut valleys, built deltas, interrupted the longshore drift of sediments, and trapped detritus in estuaries on what is now the Continental Shelf. The Hudson Canyon, one of the ancestral river channels, has 3 to 7 m of fill, which is characteristic of modern shelf sediments that cap fluvial-deltaic deposits (Knebel, 1981).

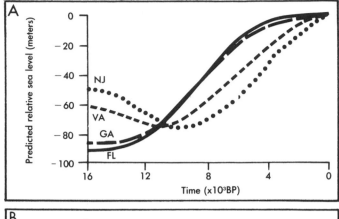

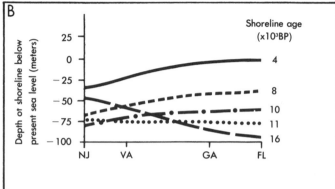

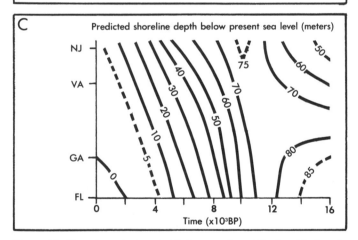

Figure 18. Shorelines during the Holocene. A and B after Clark as in Bloom (1983b, p. 219 and 220).

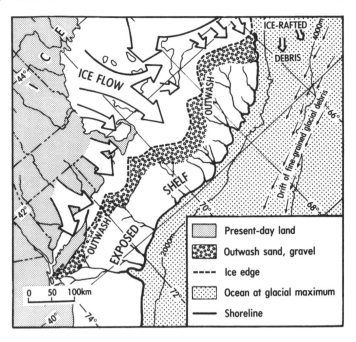

Figure 19. Schematic map of the northeastern U.S. Continental Shelf during Pleistocene time. After Knebel (1981, p. 352).

The only subaerial portions of the formerly glaciated area that are part of the present Coastal Plain are Cape Cod and Long Island and the islands between, such as Martha's Vineyard and Nantucket. They are part of the large northeast-trending belt that includes Georges Bank and extends northeast into the Gulf of Maine (Denny, 1982, p. 6). These subaerial portions are capped by materials deposited by glacial ice and glacially-fed rivers and in glacial lakes. Most types of glacial landforms (e.g., hummocky topography, moraines, kames, outwash plains, and proglacial val-

leys) are present. Much of the present size, shape, and local relief of Cape Cod and especially Long Island is determined by the positioning and size of moraines and glacial outwash. Cape Cod, which is tied to the mainland by glacial outwash deposits, also has outwash deposits along the open Atlantic shoreline that have been eroded into cliffs 60 m high (Fisher, 1985, p. 223). In the case of Long Island, which is 190 km long, glacial materials have resulted in an increase of its maximum elevation by some 60%, to 120 m, and of its size by four times (Thornbury, 1965, p. 37).

The Continental Shelf and Slope. Although a part of the present Continental Shelf was directly modified by glacial ice (Fig. 19), most of it was affected only indirectly and for highly variable lengths of time. Estimates differ on how much of the present shelf (and slope) was actually subaerial during various times in the Quaternary, even during late Wisconsin time, depending on the extent of sea-level lowering accepted. Another major factor affecting shelf forms was the rate of sea-level change. A vertical rise on the order of 5 m per century may well have occurred during the early stages of the last Wisconsin deglaciation (Revelle, 1983, p. 433). Such rapid vertical change could well have translated into a distance normal to the shoreline of as much as 5 km per century had it been over the gently sloping shelf. However, the period of most rapid sea-level rise (i.e., immediately after deglaciation began) was probably when sea level was below the shelf break.

During its subaerial stage, the Continental Shelf was modified in much the same way as the present Coastal Plain. As sea level was lowering, formerly established rivers extended their courses seaward and new drainage patterns were established on

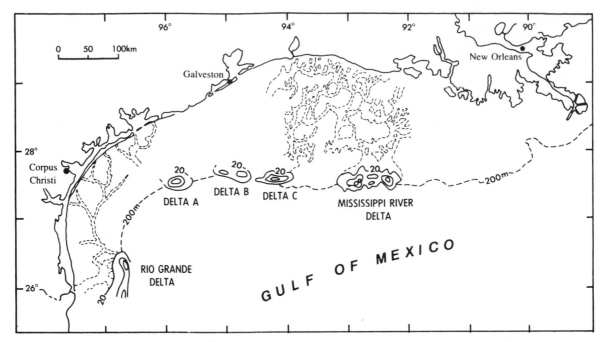

Figure 20. Map of northwest Gulf of Mexico showing locations of shelf-margin deltas and courses of ancient streams across the Continental Shelf during last lowstand of sea level. Thickness of deltaic sediments in intervals of 40 m beginning at 20 m. After Suter and Berryhill (1985, p. 78).

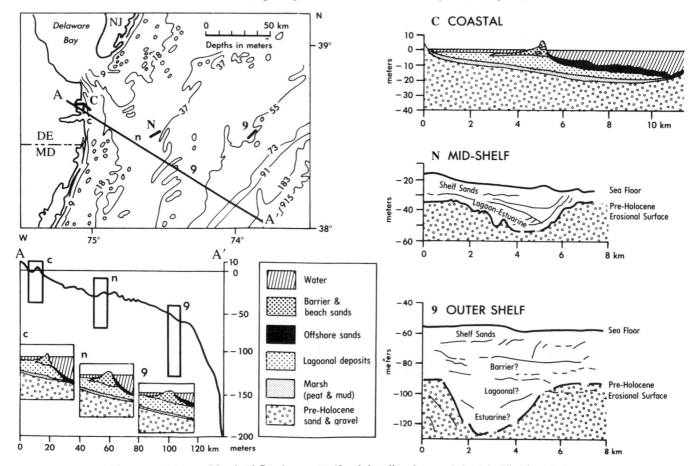

Figure 21. Delaware-Maryland Continental Shelf and shoreline characteristics. Modified from Belknap and Kraft (1981, p. 432, 433, and 439).

newly exposed surfaces. They carried their sediment seaward, much of which was deposited as deltas (Fig. 20). Many of these newly-created deltas in the Gulf of Mexico formed at the edge of the shelf (Suter and Berryhill, 1985, p. 78). If the changes of sea level presented by Revelle (1983, p. 434) are valid, the most seaward interface forms were created in a relatively short period of time (Fig. 17). The suboceanic deltas shown on Figure 20 with sediment thicknesses of generally less than 100 m would seem to substantiate this conclusion.

Considering the period of time since sea level was last above that of the present (approximately 120×10^3 years) and using the graph of Figure 17, it appears that the sea frequently traversed the zone between depths of about –20 m and –80 m (relative to the present), many times creating coastal features at many locations. During such fluctuations of sea level, including especially the most recent rise (between 18,000 and 6,000 B.P.), erosion and redeposition altered to varying degrees the preexisting topography. Belknap and Kraft noted that:

> preservation of fractions of the sediment column in a transgressive sequence depends on the depth of erosion. This, in turn, depends on impinging wave energy, sediment supply, resistance to erosion, preexisting topography, tidal ranges, and rate of relative sea-level change. It is hypothesized that the preservation of coastal lithosomes will vary from a high proportion of preservation on the outer shelf to a low proportion of preservation on the inner shelf . . . (1981, p. 430).

Although little is actually known about former shoreline forms on the Continental Shelf, some submerged shorelines have been preserved sufficiently for recognition. These, such as the Nichols, Franklin, and Fortune shorelines, occur in the deeper waters of the shelf where transgression was rapid. In shallower waters, on the other hand, where sea level rose relatively slowly, destruction was much more complete. However, as shown by Belknap and Kraft, cores from these shallower waters allow the identification of former lagoon, estuary, and inlet environments (Fig. 21). Sand dunes and barrier islands, subjected directly to wave action or a rising sea, are more likely destroyed (Belknap and Kraft, 1981, p. 430).

On the shelf from Cape Hatteras, North Carolina, south to at least the latitude of Miami Beach, Florida, there is a series of cemented ridges that apparently represent late-Pleistocene nearshore environments. Along this stretch of the shelf the biotic component of the sediments increases in a southward direction. South of Miami Beach, the shelf, which is composed of carbonates, has karst features that formed during lower sea level (Bloom, 1983b, p. 222).

The Coastal Plain. Possibly the most conspicuous Pleistocene forms present on the Coastal Plain today are those features associated with former shorelines, including terraces, seaward facing scarps, beach ridges, ridge-and-swale lineations, and aligned karst features (Winker and Howard, 1977, p. 123). Hoyt and Hails (1974, p. 205) noted that barrier-island and lagoonal-marsh facies are present along all Pleistocene shorelines, although in places

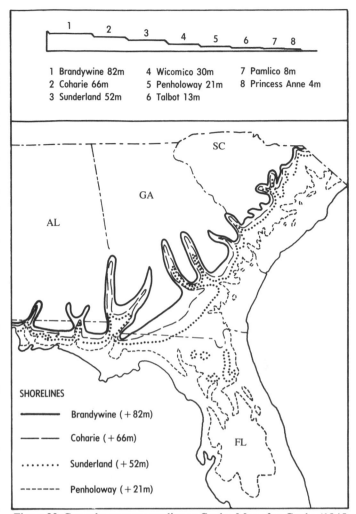

1 Brandywine 82m	4 Wicomico 30m	7 Pamlico 8m
2 Coharie 66m	5 Penholoway 21m	8 Princess Anne 4m
3 Sunderland 52m	6 Talbot 13m	

SHORELINES

———— Brandywine (+82m)

– – – – Coharie (+66m)

.... Sunderland (+52m)

- - - - - Penholoway (+21m)

Figure 22. Coastal terraces according to Cooke. Map after Cooke (1945, p. 275) who believed that individual terraces could be correlated on the basis of elevation alone (Cronin, 1981, p. 813). Cross section from Fenneman (1938, p. 24).

extensive erosion during the late stage of a stillstand removed segments of the barriers.

These features are so conspicuous that they easily attracted the attention of geologists, including Lyell when he made his way through Georgia in 1844. Attempts at correlating observations between different locations and with time also came early. These attempts, as exemplified in 1945 by Cooke (Fig. 22), were based on elevation. It was assumed that warping had not occurred, and therefore individual sea stands formed terraces that today would be at the same elevation wherever preserved along the Atlantic and Gulf coasts (Fig. 23). Research since the early 1970s has shown that such uniformity is not the case. Indeed, the extension of the 30 m terrace in Georgia (designated Wicomico by Cooke, Fig. 22) is 50 m high in Florida and 65 m high in central South Carolina. It has been labeled the Trail Ridge–Orangeburg Sequence (Fig. 24) by Winker and Howard (1977, p. 123).

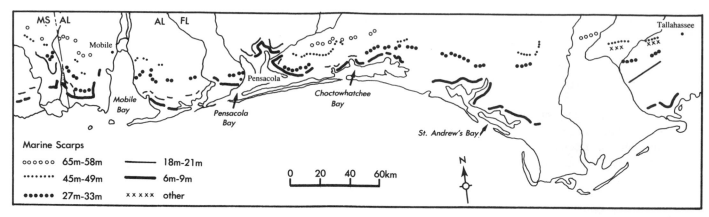

Figure 23. Distribution of marine scarps on the east Gulf Coast. After Carlston (1950, p. 1122).

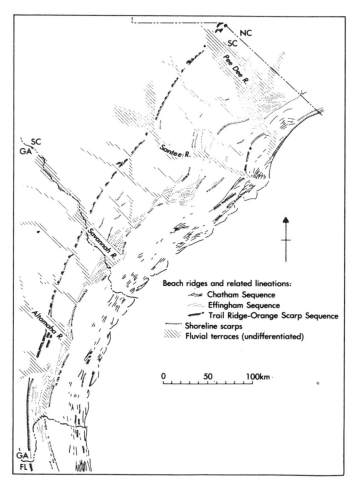

Figure 24. Geomorphic features of the Atlantic Coastal Plain. After Winker and Howard (1977, p. 125).

Likewise, Cooke's Penholoway terrace, when extended, varies from 23 to 33 m (as the Effingham Sequence) and his Pamlico varies from 7 to 15 m (as the Chatham Sequence; Fig. 24; Winker and Howard, 1977, p. 126). Such research indicates that the Cape Fear and Ocala uplifts continued upward movement during the Pliocene and Quaternary.

Thom (1967), during his examination of a section of the South Carolina Coast, found evidence of five phases of coastal progradation ranging in elevations from 33 m to present sea level. During each phase, a barrier bar and backbarrier flat formed. Superimposed on these five depositional surfaces are numerous other forms including entrenched streams, fluvial terraces, sinkholes, Carolina bays, and sand dunes that developed as specific geomorphic processes with sea-level fluctuation (Thom, 1967, p. 50).

One of the enigmas of Atlantic Coastal Plain morphology is the Carolina Bay. Douglas Johnson, (1942, p. 341), for example, considered that it is not only "one of the most remarkable geomorphic features on the face of the earth" but also one of the "most difficult of earth forms to explain." Numbering more than a half-million, the Carolina bays are found from Maryland to northern Florida (Fig. 25) and, although not limited to Quaternary terraces, they are especially abundant on them. They are most numerous and best developed in the region between central North Carolina and northern Georgia.

Today, few of the Carolina Bays contain water. Those that have not been converted into agricultural fields (Fig. 25) usually are densely vegetated (Kaczorowski, 1976, p. 16). Carolina Bays are usually elliptical in form, with the long axis oriented in a general northwest-southeast direction, and usually are rimmed with grey or white sand.

The origin of these bays has been debated for over 100 years. Tuomey, apparently the first to describe these features, explained them as being initiated by what was known locally as "boiling springs," and suggested that later their surface form was modified by pond wave action. He noted that "they have quite an artificial appearance" and "are not deep and conical like 'lime-sinks'"

Figure 25. Carolina bays. Photo courtesy T. Kana. Note conversion of bays for agriculture.

Figure 26. Loess deposit near Vicksburg, Mississippi. Photo by H. J. Walker.

(1848, p. 143–4). Although dozens of theories (including meteor showers, lagoon segmentation, and solution weathering) have been offered, the theory advocating the modification of basin form by the waves and currents set up by a unidirectional wind system appears to be in most favor today (see Kaczorowski, 1976; Thom, 1970).

The rivers of the province were also affected greatly in various ways during the late Pleistocene. Those fed with glacial meltwater included the Susquehanna, Delaware, and other rivers of the northeast, the Mississippi River and most of its major tributaries, and the Rio Grande (Figs. 5 and 6). These rivers and all the others of the province were also affected by climatic changes. At the time when the last major deglaciation began (approximately 18,000 B.P.), the climate of the eastern and southern Coastal Plain was cooler and drier than at present, whereas, in those drainage areas to the west, wetter and cooler conditions prevailed (Fig. 16). Such conditions led to an increase in the discharge of the Mississippi River; a discharge that was already high because of the vast volume of meltwater that it carried.

Most, if not all, of the rivers entrenched themselves as sea level lowered and as their courses extended across the newly exposed continental shelf. Many of them, such as the Susquehanna and Delaware, developed glaciofluvial terraces. In addition, new and geologically ephemeral streams developed on the newly exposed surfaces.

Although all rivers were affected, the system that was affected most was that of the Mississippi. It drained a melting ice front that extended over 3,000 km from New York to Montana (Fig. 6). During ice-sheet retreat, the amount of water it carried was exceptionally large. Just when this amount reached its maximum is unknown, although it may well have been when the transgression phase was at a maximum, around 11,500 B.P. (Emiliani cited in Baker, 1983, p. 117).

Continental glaciers reached to within about 200 km of the lower Mississippi River valley at the maximum of the Flandrian (approximately 18,000 B.P.). However, during earlier glaciations it had reached to within about 50 km (Fig. 6). Although ice never directly affected this part of the Coastal Plain, its meltwater was responsible for extensive modification of the preglacial drainage system and for the transport of vast quantities of sediment to and through the system (Saucier, 1974, p. 2).

During lowering sea levels prior to this date, entrenchment of the lower Mississippi River probably extended upstream only to middle Louisiana, although during earlier glacials it may have reached as far upriver as Vicksburg, Mississippi (Saucier, 1974, p. 2).

Upstream from the latitude of Alexandria, Louisiana, braided streams widened and aggraded the valley with glacial debris. Such braided streams were abundant sources of meltwater silts, which were subsequently transported eastward by wind for deposition as loess (Figs. 16 and 26).

The great variations in meltwater volume during periods of ice-sheet advance and retreat combined with accompanying changes in climate and sea level to produce extreme variations in discharge, sediment load, and morphologic modification. Saucier wrote that:

. . . during each glacial cycle, a stage was reached when stream discharge and sediment load declined to a point where the ancestral Mississippi River changed from a braided to a meandering regimen. This apparently first occurred nearest the Gulf Coast and proceeded upvalley. In the case of the last cycle, the Mississippi River is known to have changed from braided to meandering quite abruptly about 12,000 years ago south of the latitude of Baton Rouge, Louisiana (1974, p. 4).

Such change results in the burial of the coarse-grained sediments deposited during braided drainage by fine sediments deposited during meandering (Baker, 1983, p. 118).

Although most deposits of early glaciations were either de-

TABLE 5. LATE PLEISTOCENE FLUVIAL EVENTS IN THE
MISSISSIPPI RIVER DRAINAGE SYSTEM*

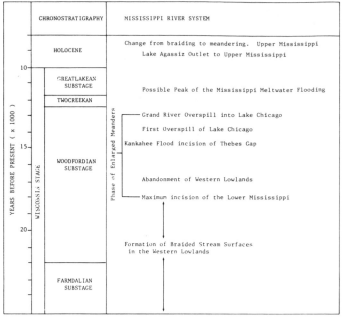

*From Baker (1983).

stroyed by erosion or buried, there are exceptions. For example, Macon Ridge, Arkansas, a prominent feature about 150 km long, is believed to have formed some 35,000 to 40,000 years ago as a result of deposition in the Mississippi River valley by the ancestral Arkansas River (Saucier, 1974, p. 18).

During the waning stages of the Pleistocene, there were many changes in the course of the Mississippi River along virtually all of its route. The northern quarter of its alluvial valley, which is about 100 km wide, is nearly bisected by Crowley's Ridge. This ridge, which has an Eocene and Cretaceous core covered by Pleistocene gravels and Pleistocene loess, is about 300 km long and rises in places some 60 m above the alluvial plain. About 18,000 years ago, the Mississippi River, which had been flowing down the west side of the alluvial plain west of Crowley's Ridge, established a new course through the ridge and thus began a move toward the eastern side of the valley. Around 9,000 years ago the last major diversion in this part of the upper valley occurred when the Mississippi River cut through Thebes Gap. At this time the juncture with the Ohio River near Cairo, Illinois, was established 650 km north of where it had been 10,000 years earlier.

Although the Mississippi River is by far the most spectacular of the rivers of the Coastal Plain from the standpoint of Pleistocene influence (Table 5), other rivers, including those not subjected to meltwater flow, were also affected mainly by climatic and base-level changes. The Colorado River of Texas, studied by Baker and others, is an example of late-Pleistocene river metamorphosis (Baker, 1983, p. 119). He wrote that " . . . the late Quaternary was characterized by adjustment from low sinuosity, sand- and silt-transporting streams . . . The low-sinuosity streams

probably drained the basin during more arid phases and the high-sinuosity streams characterized humid phases, . . ." Such adjustments have caused major changes in fluvial channel patterns and the development of terraces (Fig. 27).

Although many of the river terraces in the Coastal Plain are quite small, those of the Mississippi River are large. Those along the lower Mississippi River are upvalley equivalents of coastwise terraces. These river terraces combine with coastal terraces to provide the main evidence of what happened in Louisiana during the Pleistocene. During lowering sea level, the river cut into the former flood plain. With sea-level rise, deposition occurs. At any one location the vertical sequence grades upward from coarser (often gravels) to finer (silts and clays) sediments. Four major terraces—the Prairie, Montgomery, Bentley, and Williana—have been named (Fig. 28). Because of isostatic adjustments, these terraces are tilted upward in an upriver direction; the slope steepens with increasing age.

From Louisiana north, parallel to the Mississippi River, is a band of Wisconsin loess (see Fig. 29). The loess deposits have basal ages that range from 29×10^3 to 17×10^3 years B.P., with a termination date of about 12×10^3 years B.P. (Ruhe, 1983, p. 136). The deposits, which are about 20 m thick just east of the Mississippi River, thin rapidly with distance from the river (Fig. 29). Within the loess deposits are incipient soil layers indicating that loess deposition must have been intermittent during its 17×10^3 years of formation.

THE SUBOCEANIC ENVIRONMENT

The suboceanic portion of the province (i.e., its present Continental Shelf and Continental Slope) is separated here into the three relatively distinct segments that comprise the suboceanic portions of the Atlantic Margin, the Florida Platform, and the Gulf of Mexico Basin.

Atlantic Margin Continental Shelf and Slope

Continental Shelf. The Continental Shelf varies considerably in width, gradient, and morphologic complexity over the more than 3,000 km it extends along the east coast of the United States. Almost all of it is covered by a surficial sand sheet, often with some gravel (Hollister, 1973; Fig. 2). In the area of Georges Bank in the north, coarse gravel left by glaciers is common, and surface forms include remnant river channels cut by glacial meltwater and those left from glacial deposition, which form an irregular surface.

South of the former glaciated area, the shelf is characterized by fields of relatively well-oriented, linear, northeast-trending shoals (Duane and others, 1972). These shoals form a small angle with the coast (usually less than 35°), display complex bathymetric topography, have up to 10 m of local relief, and have side slopes of a few degrees (Fig. 30). As well-recognized morphologic features, they extend from water depths of only a few meters (see section on the shoreface) out to depths of approximately 60 m.

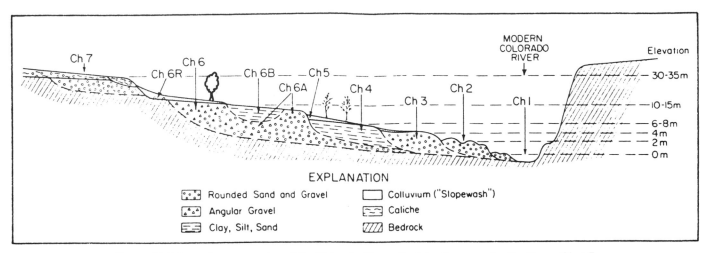

Figure 27. Schematic cross section of the Colorado River alluvial valley near Austin, Texas. Ch 1-7 represent assemblages of channels of various ages: 1 is modern; 2-4 are Holocene; 6, 6A, 6B are late Pleistocene. From Baker (1983, p. 120).

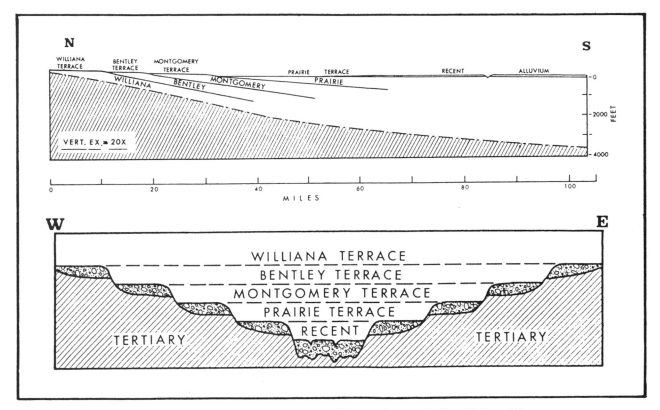

Figure 28. Louisiana terraces based on the Fisk model. From Kniffen (1968, p. 40).

Although they show high variation in size, complexity, and distribution along the eastern seaboard, they nonetheless can be grouped into arcuate (inlet and cape-associated) and linear types (Duane and others, 1972). The shoals, composed of Holocene sands, rarely attain thicknesses greater than 10 m and generally rest upon horizontal strata of marsh, lagoon, and estuarine deposits. Radiocarbon dating of this underlying material indicates that the shoals postdate the last transgression. Therefore, most of them are less than 11,000 years old. Shoal sands are generally well-sorted, medium-grained sands that are similar in lithology to present shoreline beaches; all bear evidence of recent modification by current and wave activity. Although there are numerous

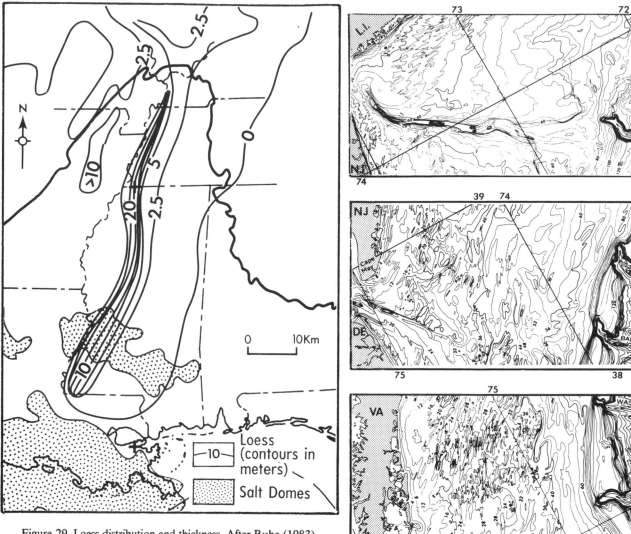

Figure 29. Loess distribution and thickness. After Ruhe (1983).

Figure 30. Bathymetric fabric of Continental Shelf off New York, New Jersey, Delaware, and Virginia. Contour interval is 4 m. After Duane and others (1972, p. 450, 451, and 452).

theories of their origin, it is generally accepted that they were formed by nearshore processes. Shoals that are presently isolated on the shelf are judged to have been formerly shoreface-connected and subsequently detached during the coastal retreat that accompanied the last rise in sea level.

Continental Slope. The continental slope off the east coast of the United States ranges in depth from 40 to 2,500 m and extends southward from Georges Bank to south Florida. Over the past few years, process-oriented and geophysical studies have added considerably to our knowledge of this submarine margin (Knebel, 1984; McGregor and others, 1984; Folger and others, 1979; Robb and others, 1981).

Between Georges Bank and Cape Lookout, North Carolina, the Continental Slope is very rugged, dissected by numerous submarine canyons and gullies. The shelf break varies in depth from 120 to 140 m, and slopes vary from 1° to 2° at upper levels and from 3° to 8° at lower levels. South of Cape Lookout, there are three distinct physiographic units: (1) the Florida-Hatteras

Slope, (2) the Blake Plateau, and (3) the Blake Escarpment (Knebel, 1984). The shelf-edge break at the outer edge of the Florida-Hatteras Slope occurs at depths of 40 to 80 m, is generally featureless, and has a gradient that is usually less than 3°. The major break in slope occurs at depths of 800 to 1,400 m and is associated with the Blake Escarpment; it also is rather featureless and has a gradient of 2° to 6°. Southward, the average gradient increases abruptly to an overall slope greater than 20°.

The slope is dissected by large canyons, which in turn are cut by smaller canyons and gullies (Fig. 31). The floors of the canyons average about 500 m in width and are generally flat. In contrast, canyon sidewalls are complex. They are marked by outcrop-controlled benches. Canyon morphology, which varies

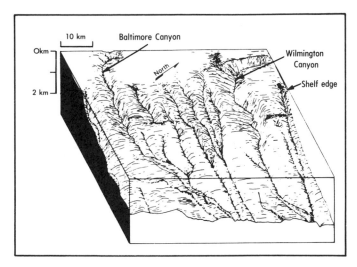

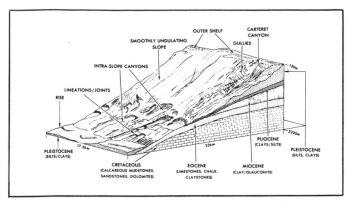

Figure 32. Schematic of Continental Slope morphology near Cateret Canyon. After Prior and Doyle (1986).

Figure 31. Schematic of submarine canyons. After McGregor (1984, p. 276).

considerably from one canyon to another, is controlled mainly by the underlying bedrock. The morphology of Carteret Canyon (Prior and Doyle, 1986) is illustrative of the most significant aspects of those canyons that indent the Continental Slope from Georges Bank to Cape Lookout (Fig. 32).

Pleistocene sediments, which overlie Tertiary sediments, are thickest and most continuous on the outer shelf and upper slope. Unconformities attest that there have been major erosional episodes during the Tertiary and Quaternary. In the upper and middle slopes, the canyons are being cut primarily into Pleistocene and late Tertiary sediments. Wide, smooth-topped, convex-crested ridges separate the relatively narrow canyons. On the lower slope, where Eocene rocks outcrop, slope morphology is much more complex.

Intraslope canyons (Prior and Doyle, 1986), lineations, joint patterns, subparallel stripes (Robb and others, 1981), and localized angular steep scarps are present (Fig. 32). The intraslope canyons (labeled valleys by Robb and others, 1981) are generally common and begin at some depth below the shelf break. At the base of the slope, the Continental Rise is rather flat or gently undulating with extensive smooth areas of Pleistocene deposits. Locally, there are outcrop benches and blocky or hummocky areas, resulting in highly irregular microrelief.

The processes responsible for the formation of the canyons have long been debated; the two most commonly cited are subaerial cutting by rivers during low sea levels and submarine cutting by density currents. Much of the new data tend to indicate that river scouring during low sea level was only a minor process and only locally important. Density currents, which obviously play a role in the evolution of these canyons, is not the only process responsible for their formation. Mass-wasting processes—including slumping, rockfall, biological erosion, undersea discharge of groundwater, and sediment resuspension—are also involved in shaping them. It is highly probable that the canyons are the cumulative result of long periods of gradual slope degradation, rather than the products of short-term catastrophic processes (Prior and Doyle, 1986). Most of the processes responsible for the long-term evolution and shaping of the Continental Slope are probably more active during periods of lowered sea level. Radiocarbon dating of sediments on the slope and in the canyons tends to indicate that accumulation rates were relatively slow during the Holocene. Some evidence, however, suggests that the modern slope is not totally static (Knebel, 1984). Sharp, angular boundaries on erosional furrows, rockfall debris on canyon wall ledges, finely-textured erosional gullies, observed biologic erosion, and localized areas of abnormally high accumulation rates all attest to some activity during the Holocene. However, it is thought that these presently active processes are of relatively small magnitude in comparison with those that were active during periods of lowered sea level.

Around the head of Hudson Canyon in depths of 120 to 500 m is an area of 800 km[2] that has a rough surface. A relief of 1 to 10 m has been formed in semilithified, silty clay since the beginning of the Holocene. Twichell and others (1985, p. 712) attribute the erosion that is forming this rough surface to tilefish (*Lopholatilus chamaeleonticeps*) and associated species of crustaceans. They calculate that an erosion rate of 13 cm/1,000 years has been necessary to provide the relief measured.

In the region of Cape Lookout to south Florida, the Continental Slope is much less complex. Evidence of abundant mass-wasting processes and significant canyon incisement is not present. Other than for a few slope-instability features (Popenoe and others, 1981; Cashman and others, 1981), the slope is relatively smooth and erosional in nature, having only a thin covering of Pleistocene and Holocene sediments.

Florida Platform Continental Shelf and Slope

The Florida Platform consists primarily of carbonate sediments, there is virtually no major source of terrigenous clastics.

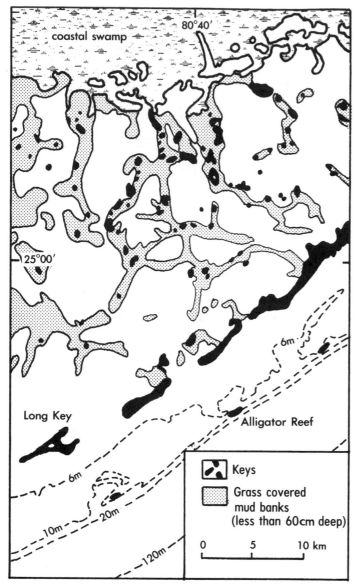

Figure 33. Florida Bay and reef. Modified from Ginsburg (1956).

production is again active, resulting in a wide range and distribution of environments and morphologic expressions. The Florida Platform consists of the narrow eastern Florida Shelf, the Bahamas Bank, the southern Florida reef tract, and the broad western shelf.

The eastern Florida Shelf is extremely narrow and merges southward into the Florida reef tract. It is flanked on its seaward side by a rather steep, abrupt slope that forms the edge of the Straits of Florida. It, and the shallow lagoon of Florida Bay, has a complicated morphology (Fig. 33). Polychaete worm reefs, grass-covered mud banks, and algal flats form the major morphologic elements of the back-reef environment. The active reef track, which faces the Florida Straits, is composed of a series of complex reefal facies. These reefs are actively growing today in response to the Holocene sea-level rise. During the last low sea level, many of the reefal facies were exposed, and dissolution under subaerial conditions resulted in formation of an irregular topography upon which modern reefal facies became established.

The Bahamas Bank, separated from the Florida Peninsula by the narrow Florida Straits, consists of two major regions, the Great Bahama Bank and Little Bahama Bank. Although the Continental Shelf is extremely broad and low-sloping, its edge is abrupt and quite steep. The bank is composed mostly of shallow Continental Shelf and lagoonal sediments. Active reefs are present along its windward side, but account for only a small percentage of the total area of the region. The greatest relief elements are found on the outer margin of the shelf where reefs are actively growing. In water depths of less than 20 m, the greatest irregularity occurs in the areas of reef pinnacles, which can display local relief of up to 10 m (Fig. 34).

Across the vast majority of the bank, the sediment surface is covered with grass flats of carbonate sands. The relict sand bodies cover a much larger area than that covered by presently active oolite shoals, suggesting that sand transport on the bank was once much more active than it is today (Hine, 1983). Apparently, as water depth increased over the bank with rising sea level, the frequency of grain movement due to waves, storms, and tides decreased until the benthic marine flora became established and completely or partially stabilized the substrate (Hine, 1983). Today, active sand movement is occurring only in those areas that have maintained vertical aggradation with rising sea level. Tides and storm surges are the primary forces that maintain sediment movement.

The west Florida Shelf is extremely broad, often exhibiting widths of nearly 200 km. The surface is relatively smooth and displays little morphologic variability. Most of the sediment zones are oriented parallel to the shelf contours and thus parallel the present shelf edge (Fig. 35). The most pronounced relief occurs in the coralline algae ridges (Fig. 35).

Continental Slope. The southeastern, southern, and southwestern Continental Slope off the Florida Platform is extremely steep and faces the Florida Straits, where the floor is at a general depth of 1,500 to 2,000 m. The slope is marked by a shelf-edge reef at a depth of approximately 230 to 250 m. This

The platform consists of approximately 10 km of Jurassic to Holocene carbonates and evaporites resting on a patchwork of Precambrian to Jurassic igneous, metamorphic, and sedimentary rocks (Doyle and Sparks, 1980).

Continental Shelf. The emergent portions of the Florida Keys and Bahamas Bank are shallow platforms capped with indurated Pleistocene carbonate bedrock. Holocene sediments are variable in thickness and in morphologic expression. Fluctuations in sea level alternately exposed suboceanic environments, resulting in the dissolution of carbonates and submerged subaerial landscapes, which allowed active carbonate production. These processes caused the major morphologic expressions present today. With a rising sea level during the Holocene, carbonate reef is built upon a thick (70 m) Pleistocene progradational wedge of sediments that extends downslope to water depths of 400 to 500 m. One of the more prominent morphologic elements is a Miocene outcrop that forms a terrace at a water depth of 400 m. The surface of the outcrop has only a thin cover of younger sediments (Holmes, 1985). Several small submarine channels or canyons cut the Continental Slope and extend from the shelf-edge reef to the base of the slope, at which point small carbonate fans are present on the floor of the Florida Straits. Small compressional folds (interpreted as sediment creep) and displaced slabs of sediment (indicating sediment instability on the steep slope) have been reported (Holmes, 1985).

The western Continental Slope margin is defined by a double reef trend; the shallower reef lies between 130 and 150 m, whereas the deeper reef trend ranges between 210 and 300 m in depth (Doyle and Holmes, 1985). Holocene and late Pleistocene sediments composed of a foraminifera-coccolith ooze accumulated at a very rapid rate, averaging 30 cm/1,000 yrs (Doyle and Holmes, 1985). At a depth between 1,000 and 2,000 m, the slope increases significantly, in places exceeding 20°, and forms the West Florida Escarpment. The escarpment is erosional in nature and is composed of sediments deposited in a shallow-water, back-reef and lagoonal facies (Freeman-Lynde, 1983). A series of gullies and small canyons create the middle and upper slope. Some of the most striking morphology of the west Florida upper Continental Slope are the irregular features associated with mass-movement processes. The mass-wasting features range in size from creep forms to massive slides and gravity-induced folds tens of kilometers long (Doyle and Holmes, 1985). The northern portion of the Continental Slope is relatively unbroken all the way to the northern limit of the carbonate margin at about the De Soto Canyon.

Gulf of Mexico Basin Continental Shelf and Slope

The emerged and submerged platforms of the Gulf of Mexico form the bulk of the Coastal Plain and Continental Shelf of the southern margin of the United States (Fig. 36). Known as the Gulf Basin, it has been dominated by the sediment yield of the Mississippi River (Martin and Bouma, 1978; Bouma and others, 1978).

Continental Shelf. The depositional environments on the Continental Shelf are primarily terrigenous in the north, with carbonates on the broad platforms to the east.

The shelf bordering the west coast of Florida is part of an extensive system of carbonate banks that dominates the southeastern Gulf of Mexico. The surface is low in relief and consists primarily of fine-grained carbonate sediments and worm-reef accumulations. The only prominent bathymetric expression is the Florida Middle Ground (Fig. 36), a living reef with an area of more than 750 km² that has a relief of 18 m. Poorly-developed karst topography dominates the innermost shelf between the Florida Keys and Tampa. This karst topography is the result of subaerial exposure and resulting dissolution of carbonate sediments

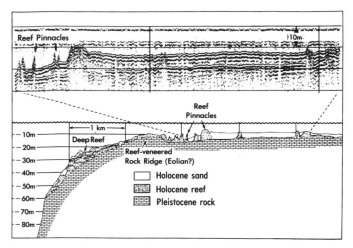

Figure 34. Profile across northern margin of Little Bahama Bank. After Hine (1983, p. 53).

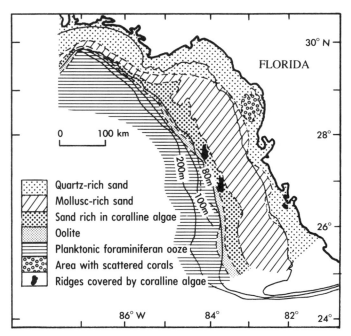

Figure 35. Bathymetry and facies off west Florida. After Hine (1983, p. 89).

during the last lowering of sea level. A 100 km zone of sand shoals, typical of those off the eastern Atlantic Shelf, is present off the cuspate foreland of Cape San Blas at the northern end of the province.

The shelf break off westernmost Florida, Alabama, and Mississippi ranges in depth from 60 to 100 m and is characterized by a relict topography covered by a thin sand and mud sheet that was deposited during the last sea-level rise. The shelf is narrow because of a large reentrant known as De Soto Canyon, which is

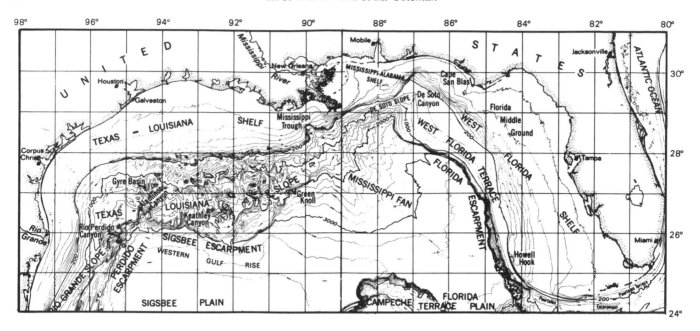

Figure 36. Bathymetric map of the northern Gulf of Mexico. Contour intervals: 0 to 200 m in 20 m isobaths; below 200 m in 200 m isobaths. After Martin and Bouma (1978, p. 6 and 7).

not a true submarine canyon. The canyon occurs at the point where Tertiary clastic sediments lap up against the carbonate facies of the northeastern Gulf. The sediments underlying the thin Holocene sediment cover consist of fluvial sands and gravels deposited during the last low sea-level stand. Most of the inner shelf consists of a clean quartz-rich multicycle sand (Fig. 37).

Linear shoals, probably representing relict nearshore topography, are present within this sand sheet. Small bed forms are presently actively migrating on the shelf, indicating modern-day reworking of these relict topographic features. The outer rim of the shelf consists of a lime-mud facies; a mixture of calcium carbonate, quartz, and clay. Offshore of Mississippi and Alabama, where the same lime-mud facies exists, there are also several zones of reefs and interreef facies (Fig. 37). The reefs occur in two depth zones, one at 65 to 80 m, the other at 97 to 110 m. The reef zone is characterized by algal limestone pinnacles, some attaining relief in excess of 15 m. In the inner-reef areas, mollusc shell and reef debris embedded in a muddy matrix is the most common lithology. Separating the sand from the lime-mud facies is a transition zone consisting of calcareous muddy sands and silts.

The shelf off Louisiana is highly variable in width, ranging from generally less than 20 km off the active mouths of the Mississippi River delta to more than 180 km off the western Louisiana coast. The narrow shelf in front of the Mississippi River is probably the most dynamic area in the submerged portion of the Atlantic and Gulf margins. Sedimentation rates are extremely rapid, generally in excess of 1 m per year immediately off the mouths of the Mississippi River. The high rate of sediment accumulation results in underconsolidated sediments and inherent

sediment instability even though the shelf gradient is generally less than 0.5°. The most common type of instability feature is the mudflow gully. Submarine failures result in a radiating pattern of channels that cut the narrow shelf all the way to the shelf edge, providing the sea floor with a relief in excess of 5 m. It is estimated that as much as 50% of the sediment annually delivered to the mouth of the Mississippi River is displaced seaward to the shelf edge by these instability processes.

West of the active Mississippi River Delta, off central Louisiana, the shelf is to a large extent mud covered (Fig. 2), extremely broad, and displays little topographic relief. The Holocene muds vary in thickness from a few meters to 10 m and were derived from the Mississippi River by the slow westward drift that characterizes this portion of the northern Gulf. Toward the inner edge of the shelf, the surface sediment is reworked sand (Mazzullo, 1986, p. 639).

The shelf offshore of Texas is relatively broad and displays little relief. Most of it has only a thin cover of Holocene sediments and, in many locations, late Pleistocene sediments crop out on the sea floor. The fine quartzose sand that composes much of the late Quaternary sediment in the East Texas–Louisiana Shelf is derived from two distinct environmental situations. Based on a grain-shape analysis, Mazzullo (1986, p. 641) was able to divide the shelf into five sedimentary petrologic provinces, three of which (Guadalupe, Trinity, and Mississippi) consist of sands from the Coastal Plain and two of which (Brazos-Colorado and Red River) have sands from the mid-continent.

The carbonate banks off Texas are much like those off Louisiana except that they are not covered to the same extent by

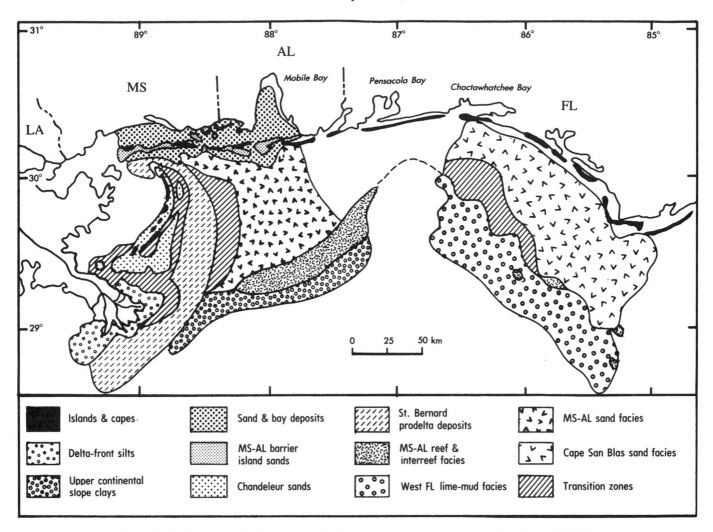

Figure 37. Sediment distribution map compiled from sediment-type occurrences. After Ludwick (1964, p. 218).

Holocene sediments. The most prominent of these carbonate banks are the West and East Flower Garden Banks off Galveston, Texas. Active faunal and floral growth on them is capping diapiric structures near the shelf edge. Although many of the carbonate banks that cap diapiric highs were actively growing during the time of lowered late Pleistocene sea level, today only a few are active because of the hinderance to growth by smothering sediments and a thick water layer that limits sunlight penetration.

Continental Slope. The Continental Slope of the Gulf Basin gently descends from the shelf edge, at about the 200 m isobath, to the upper limit of the Continental Rise, at a depth of 2,800 m. The slope (Fig. 36) occupies more than 500,000 km^2 (including the portion off Mexico) of prominent escarpments, smooth and gently sloping surfaces, knolls, intraslope basins, ridge and valley topography, and submarine channels (Martin and Bouma, 1978; Bouma and others, 1978). Martin and Bouma (1978) described nine distinctive subprovinces and many indi-

vidual features. The factors that have controlled the present-day morphology of the northern portions of the slope include reef building on the Florida carbonate platform, erosion, nondeposition and faulting in the Straits of Florida, diapirism and differential sedimentation on the slopes off Texas and Louisiana, and rapid accumulation of terrigenous sediment during the Pliocene and Pleistocene in offshore Louisiana.

The Rio Grande Slope is a region of linear and irregular hills underlain by large massifs interspersed with deep basins and troughs that contain thick sections of sedimentary material. The massifs and smaller domes and anticlines usually have cores of salt (Fig. 38). With the exception of the domes and isolated basins, the topography is relatively smooth in comparison with the rest of the Gulf Basin.

The most complex unit is the Texas-Louisiana Slope, comprising 120,000 km^2 of hill and basin sea floor (Fig. 36). The average gradient of the slope is slightly less than 1°, but gradients

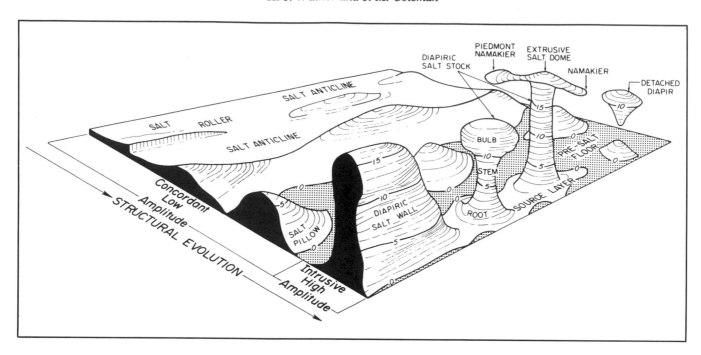

Figure 38. The main types of large salt structures. After Jackson and Talbot (1986, p. 305).

greater than 20° are not uncommon around the many knolls and basins. Steep-sided knolls, enclosed intraslope basins, and canyonlike topography characterizes the eastern two-thirds of the slope, whereas occasional knolls and low-relief noses mark the otherwise featureless slope of the western sector (Martin and Bouma, 1978; Bouma and others, 1978). The extreme topographic relief of the Texas-Louisiana Slope is associated with salt diapirism and salt withdrawal beneath the basins. Intraslope basins, such as the Gyre Basin (Fig. 36), are flanked and commonly surrounded by salt domes and contain exceptionally thick sections of Tertiary sediments (Bouma and others, 1975). The basins are directly related to the growth of the adjacent salt spines, which blocked active submarine channels or coalesced to create sea floor expression in noncanyon areas. The Sigsbee Escarpment is the most prominent feature at the base of the slope (Fig. 39). This escarpment is nearly continuous along the entire base of the slope from the western Gulf to De Soto Canyon. The scarp is the expression of the lobate frontal edge of the northern Gulf Diapiric Province and is underlain throughout its length by a complex system of salt ridges, overthrust tongues, and steep-sided massifs (Humphries, 1978; Martin and Bouma, 1978; Martin, 1984). The continuity of the escarpment is broken locally by diapiric outliers and large pronounced reentrants of several large interlobal canyons such as the Alaminos and Keathley canyons (Fig. 36).

The Mississippi Canyon and Fan (described in the section on Quaternary History) mark the eastern boundary of the Diapiric Province and form one of the more prominent physiographic features in the northern Gulf of Mexico. The De Soto Slope lies between the eastern limit of the upper Mississippi Fan and the West Florida Terrace (Fig. 36). This section of the slope is rela-

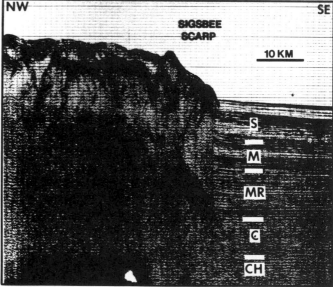

Figure 39. The Sigsbee scarp. Seismic-stratigraphic units are: S, Sigsbee; M, Cinco de Mayo; MR, Mexican Ridges; C, Campeche; CH, Challenger. From Worzel and Burke (1978, p. 2299).

tively smooth and is underlain by a thick sequence of conformably bedded sediment that is deformed by minor monoclinal folds and isolated small salt domes. The most conspicuous physiographic element in this area is the broad valley formed by the depositional convergence of the terrigenous slope with the northernmost exposure of the Florida Escarpment.

TABLE 6. SHORELINE CHARACTERISTICS

	Tidal Shoreline Length* (km)	Total Length (km)	National Shoreline[†] Sheltered Shoreline (%)	Beach Length (km)	(%)
MA	2,445	1,931		1,513	78
NY	2,977	1,027		533	52
NJ	2,884	755		346	46
DE	413	364	45	122	34
MD	5,134	3,120		74	2
VA	5,335	1,598		473	30
NC	5,431	5,892	91	2,042	35
SC	4,628	4,929	94	315	6
GA	3,772	328	55	164	50
FL	13,560	10,082	80	2,185	22
AL	977	566	87	365	64
MI	577	397	87	156	39
LA	12,425	3,127	58	1,343	43
TX	5,406	4,020	85	606	15
Total	65,964	38,136	70[§]	10,237	27

*After U.S. Department of Commerce, NOAA (1975), The Coastline of the United States. In Ringold and Clark (1980).

[†]U.S. Army Corps of Engineers (1973).

[§]Includes states of ME, NH, RI, and CT.

THE PRIMARY INTERFACE: THE COASTAL ZONE

Throughout the history of the province there has been a highly mobile transition zone between ocean and land. Sea-level variation accompanied by erosion and deposition and often by local and regional tectonic activity has insured continuous morphological alteration. The forms present along this zone, and the processes operating on it today, although generally similar to those of the past, differ from them in some critical respects and likely will differ from those that occur in the future as well. One of the major differences between the present shoreline and those occurring during much of the Quaternary results from the relatively long period of present-day stillstand that has allowed processes to operate within relatively low vertical and horizontal limits along most of the province's coastline.

Many calculations of the length of the shoreline have been made. The results are highly variable and depend to a large extent on the objectives of the determination. The values produced by U.S. Department of Commerce, NOAA (1975) give a tidal shoreline length of 66×10^3 km (Table 6), which is nearly 12 times the length it uses for the general shoreline (Table 7).

The interface between the Coastal Plain and the Continental Shelf is dominated by nonresistant sediments that have several origins. The bulk of these sediments originate inland and, in some cases, are transported several thousand kilometers before reaching the coast. All of the provinces within the United States east of the Continental Divide serve as source areas (Fig. 6). Most of the sediments are clastics, although organics are common along some shorelines, such as those in southern Florida, and in sections with protected swamps and marshes.

Despite the general similarity of the geology of the province's interface, major differences in the types of coast exist (Fig. 40). Although most of the coast is dominated by marine processes, some sections, such as the Mississippi River Delta and the province's large estuaries, are influenced most by fluvial processes, and others, such as the south Florida coast, are affected by biotic and chemical processes.

The relative importance of many of these processes depends, to a large extent, on spatial and temporal variations in climate and weather. Among the most important variables are wind speed and direction primarily because of control over the nature of waves, which are the most important source of energy in coastal modification. Surges that result from hurricanes or other large storms are extreme events that have a major impact along nearly all parts of the province's coastline. Precipitation and temperature are also important, especially because of the effect they have on streamflow and vegetation and therefore sediment load and deflation protection.

The coastal zone as discussed here is the zone extending from the inner edge of the Continental Shelf (where it joins the shoreface), past the shoreline, inland to a distance that includes all estuaries and bordering wetlands. It is a zone composed of varying proportions of water (salt, brackish, and fresh), sediment (sand, silt, clay, and organic matter), and biota (flora and fauna). These properties are modified continuously by natural processes and increasingly by human processes as well.

Various combinations of material, process, and form provide the basis for dividing the coastal zone into a number of geomorphic systems. Each system has its own suite of landforms that are affected in specific ways by a variety of processes operat-

TABLE 7. COASTAL CHARACTERISTICS

State		General Coastline Length* (km)	Major Barrier Islands				Estuaries Area* (km2)
			No.*	Length* (km)	(%)	Area[†] (km2)	
MA		309	2	29	9	152	838
NY		204	4	150	73	123	1,524
NJ		209	10	161	77	194	3,150
DE		45	1	10	21	41	1,601
MD		50	2	50	98	58	569
VA		180	9	108	60	279	6,758
NC		484	20	459	95	592	8,930
SC		301	18	154	51	583	1,732
GA		161	12	143	89	670	691
FL	(Atlantic)	933	17	455	49		
FL	(Gulf)	1,239	32	397	32	1,893	4,254
AL		85	3	31	36	114	2,145
MI		71	4	48	70	38	1,017
LA		639	11	95	15	166	14,347
TX		591	6	349	59	1,552	5,439
Total	(Atlantic)	2,876	95	1,719	60		
Total	(Gulf)	2,625	56	920	35		
Total		5,501	151[§]	2,639	48	6,455	52,995

*After U.S. Department of commerce, NOAA (1975), in Ringold and Clark (1980).

[†]Pilkey and Evans (1981).

[§]The number of major islands given by the U.S. Department of the Interior (1979, p.7) is 268.

ing individually and jointly. The systems discussed below are: (1) shoreface, (2) delta, (3) barrier-beach-strandplain-chenier, (4) bay-sound-lagoon-estuary, (5) marsh-swamp, and (6) coastal dune.

Shoreface

The zone extending seaward from low tide level, known as the shoreface, has a character that varies with sediment (type, texture, mobility, and input), wave climate, and biological activity and often reflects the nature of the shore itself. Along most of the coast of the Atlantic and Gulf Coastal Plain, the shoreface is relatively uniform because it is a continuation seaward of the extensive retrograding barrier islands that line the coast. The shoreface extends out to depths of as much as 15 m, that is, to depths beyond the zone where wave energy has lost its importance in shaping the sea floor under normal wave climates. Even at these depths, however, the sea floor is affected during hurricanes and other strong storms.

The barrier island coasts of Virginia and North Carolina tend to have bands of two distinct grain sizes (Swift, 1976, p. 267). The upper shoreface consists mainly of sands that become finer in a seaward direction until the lower shoreface is reached where the sediments are more variable and coarser grains become important. Along the Texas coast the texture varies from sand in the upper shoreface to muds in the lower shoreface. In both areas the modern sediments form a thin veneer overlying

relict (especially Holocene and Pleistocene) deposits. The thickness of these veneers varies with season and is greatest in winter. Thickness also varies with sedimentation rates, which are low along much of the coastline and in some locations so low that the Holocene and Pleistocene deposits are exposed on the sea floor. The bottom materials of the lower shoreface and inner Continental Shelf, although little disturbed by physical processes, are constantly mixed by burrowing organisms along the Texas Coast.

Although the barrier island type of shoreface is the most common in the province, some locations have a shoreface dominated by muds. The most extensive area with such textures is that to the west of the Atchafalaya Bay in Louisiana, the area seaward of the Chenier Plain. Fine sediments are contributed by the Mississippi and Atchafalaya rivers and are transported westward at least as far as the Sabine River (Gosselink and others, 1979). In addition, some sections of the shoreline are eroding back over marshland (Fig. 41), supplying additional fines to the system.

The shoreface of much of the Florida coast is unique in that biological products such as coral reefs in the keys, sepellariid worm reefs along the east coast, and vermetid gastropod reefs in the southwest all contribute sediments to the subtidal zones. However, biotic reefs are not limited to Florida. Oyster reefs are common along the shores of Louisiana and Texas. Equally as unique are the shoreface sediments of the Long Island–Cape Cod segment. Submerged glacial deposits offshore and eroding sea cliffs of Pleistocene deposits, both of which have a wide range in texture, serve as major sources of sediment.

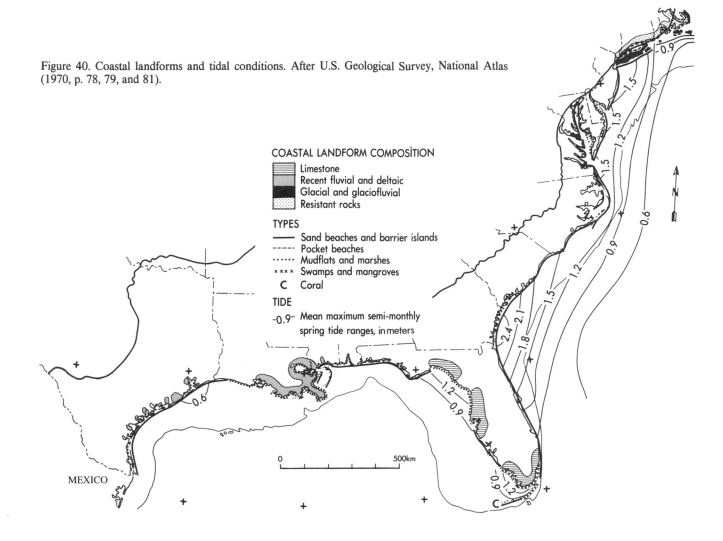

Figure 40. Coastal landforms and tidal conditions. After U.S. Geological Survey, National Atlas (1970, p. 78, 79, and 81).

The gradient of the shoreface in the province varies from as low as 1 minute in the case of the "Big Bend" area of northwest Florida (Tanner, 1985, p. 163) to over 1° along some of the barrier coasts. Normal widths range between 1 and 6 km. In Texas, there is a general decrease in width of the shoreface from north to south. The average is less than 2 km, with the area of the entire shoreface being about 1,000 km². Along the Texas coast, low tidal ranges and relatively uniform wind-driven waves (about 0.7 to 1.2 m high) help account for uniformity in the width of the shoreface. In contrast, along the Georgia coast, with its higher tidal range, the transition zone between the upper shoreface with its fine sand and the lower shoreface with its coarser sediments is some 14 km from shore.

The surface configuration of the shoreface is also variable. Although quite flat and uniform in some locations, it is very uneven in others. Brown (1976, p. 10) constructed 70 profiles off the South Carolina coast to record the changes that occur to depths of 15 m (Fig. 42). The profiles seaward of arcuate strands are concave, steep near shore, and relatively featureless beyond

Figure 41. Chenier plain sediment being eroded at Holly Beach, Louisiana. Photo by H. J. Walker.

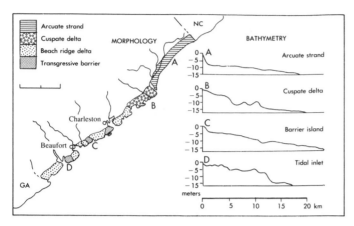

Figure 42. South Carolina coastal morphology and bathymetric trends. After Brown (1976, p. II-3 and II-11).

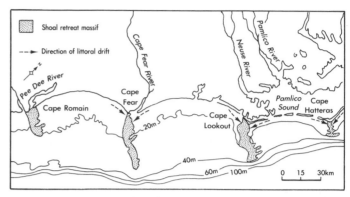

Figure 43. Arcuate shoreline of the middle Atlantic Coast. After Swift (1976, p. 286).

an 8 m depth; those off the cuspate delta are highly irregular; those off the central part of barrier islands are fairly uniform in slope and nearly featureless; whereas those off tidal inlets are irregular and flat to a great distance from shore before dropoff occurs.

The shoreface may support a variety of other forms. Offshore bars including breaker-point bars that are parallel or subparallel to the shoreline are the most common. Their development depends to a large extent on the supply of sand. Offshore bars are generally missing along the Chenier Plain of Louisiana but do occur toward the west along Bolivar Peninsula where sand becomes more plentiful.

Some of the most distinctive shoreface forms are the shore-retreat massifs (Fig. 43) that extend seaward from the four major capes (Hatteras, Lookout, Fear, and Romain) of the Carolinas. These forms, which extend out across the shoreface into deeper waters, were thought by Swift (1976, p. 284) to have originated as shelf-edge deltas during low sea levels. Today, littoral drifts from both directions feed them. It is likely that the Apalachicola cuspate foreland in northwest Florida originated in the same way.

In addition to these large, normal-trending forms are the many shoreface-connected ridges or shoals that obliquely extend into deeper water. Some are large and long (over 10 km) and may even have a sinuous pattern, as occurs off the Delmarva shoreline. Although best developed along the middle Atlantic Coast, shoreface-connected ridges are also present in Florida, Alabama, and Texas.

Delta

Deltas are possibly the paramount example of forms that are transitional between the Coastal Plain and the Continental Shelf, depending on sediment delivered by a river to a shoreline in greater quantities than can be removed by marine processes. The specific form that develops depends on the nature of the interactions among the physical, chemical, and biological processes op-

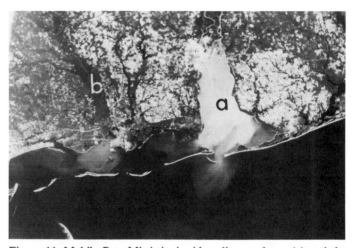

Figure 44. Mobile Bay, Mississippi with sediment plume (a) and the Pascagoula River, a former estuary that is now almost completely filled (b). Photo courtesy NASA.

erating within the riverine and marine portions of the deltaic environment. These processes lead to the formation of a deltaic plain that includes subaerial and subaqueous segments.

Within the province, deltas occur in many different settings and vary greatly in size. Many are very small, especially those forming on the landward sides of lagoons. Some lagoonal deltas, for example, that of the Colorado River in Texas, have grown to such an extent that they have bridged their lagoon and continued to grow into the open ocean. A very common type of delta in the province is the one that forms at the head of an estuary. The creation of estuaries along most of the coastline of the province (with most of the Florida coast excepted) as a result of rising sea level left an ideal situation for estuarine delta formation. In estuaries, partly because of reduced marine activity, delta formation proceeded relatively rapidly. Some, such as the Mississippi River estuary, completely filled and were eliminated as surface forms. Others, for example, Mobile Bay, have only partially filled since relative stillstand was reached (Fig. 44).

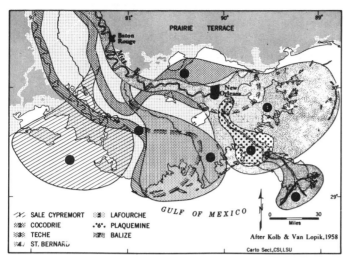

Figure 45. Deltaic lobes of the Mississippi River. From Coleman (1976, p. 26).

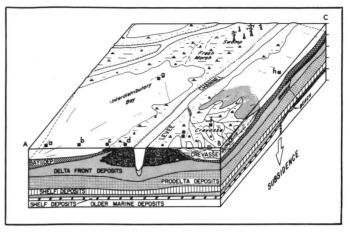

Figure 46. Block diagram of deltaic facies. Letters a-h are borehole locations. After Coleman (1976, p. 45).

The Mississippi River Delta is not only one of the most studied but also one of the most often used examples when deltaic processes and forms are described. It is one of the world's largest deltas, with an area of 28,600 km², of which 4,700 km² (16%) is subaqueous. The bulk of the subaerial portion is composed of marshes and open bays. Because shoreline wave power is extremely low and river discharge is high, riverine processes dominate. Although fine-grained sediments are spread far into the Gulf of Mexico, the suspended load is sufficiently high that deposition occurs. It is accompanied by methane gas production and is also responsible for excess pore water pressures, all of which lead to the creation of various deformational and slump forms. Subaqueous deltaic growth is largely through lengthening of the delta's distributaries in fingerlike extensions over subaqueous deposits.

The sediments that have been continuously delivered to the Gulf of Mexico since at least Cretaceous times by the Mississippi River and its ancestral system compose the bulk of the deposits in the Gulf Coast geosyncline (Fig. 11).

However, the present deltaic plain is mainly the result of development occurring since relative stillstand was reached. Modern Mississippi River Delta deposits overlie the thin sandy shell horizon that developed as sea level was rising from its 70 to 80 m position below present sea level about 15,000 years ago.

Approximately 7,000 years ago a series of progradational lobes was begun (Kolb and Van Lopik, 1966; Morgan, 1970). It is generally accepted that there are 7 major deltaic lobes, and that they normally develop over a period of 1,000 to 1,500 years (Fig. 45). Such a reconstruction masks many finer developments. Frazier (1967), for example, identified 17 individual delta lobes within those determined by Kolb and Van Lopik (Morgan, 1977, p. 36).

The most recent of these lobes, known as the Balize Delta, is almost a textbook example of the birdfoot type delta. It has an area of about 600 km², which is less than one-fourth the average of 2,700 km² for the older lobes. In contrast, it is five to six times as thick (100 to 120 m versus 20 m). Progradation rates vary among specific distributaries from less than 50 m/year to over 100 m/year. Slopes are also variable; near the shelf break they average 1.7 to 2.2°, whereas in the interdistributary bays they are seldom more than 0.2°.

Subsidence in the Mississippi River Delta generally ranges between 30 and 100 cm/100 years, although in the vicinity of distributary mouths, where sands are being deposited on weak clays, local subsidence may reach 200 cm/year (Coleman and Prior, 1980, p. 37). Also of significance is the consolidation that results from dewatering and degassing the subsurface sediment flowage because of sediment loading.

Although deltaic plains are transitional between true subaerial and suboceanic environments, their subenvironments fall conveniently into subaerial and subaqueous units. The subaerial unit includes distributary channels, levees, interdistributary bays, crevasses, and marshes (Fig. 46); the subaqueous unit includes levees, distributary-mouth bars, distal bars, prodelta depositional forms, and offshore slumps.

Subaerial Environment. The distributary channels in the Mississippi River Delta range in width from a few meters to 1 km and in depth from 1 or 2 m to over 30 m. Because distributary channels often scour through their mouth-bar deposits into the clays beneath, they become stable, which reduces meandering (Russell, 1967, p. 77).

Distributary abandonment may be caused by a number of processes, including storm surges, channel alterations upstream, loss of gradient advantages, and log jams. Infilling, following abandonment, is from both upstream and downstream. The sediment is from local sources and is usually clay. Thus the infilled portion contrasts in texture with the coarser-grained sediments of the natural levees that bordered the former distributary.

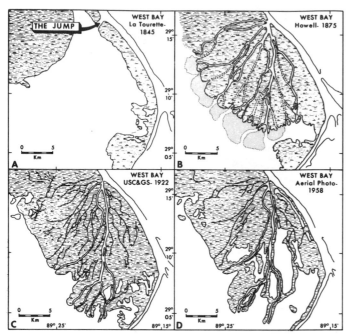

Figure 47. Historic development of West Bay crevasse. After Coleman (1976, p. 41).

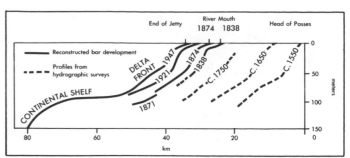

Figure 48. Advance of the Mississippi River Delta during the past four centuries. After Fisk and others (1954).

The interdistributary bays of the Mississippi River Delta are partially open to the sea (the usual case), connected to the sea by small tidal channels, or surrounded by distributary levees or marsh. These bays range in area to over 200 km^2 and have irregular shapes. Deposition, which usually occurs only at times of high river flooding and high tides, consists mainly of fine-grained sediments, which are usually bioturbatious.

Crevassing was of frequent occurrence along the distributaries of the developing delta (Fig. 47), especially prior to the construction of artificial levees. The formation of crevasses leads to the infilling of interdistributary bays and the creation of some of the major land areas of the lower delta. Such crevassing and bay filling tend to be cyclic as explained by Coleman and Prior.

Each bay fill forms initially as a break in the major distributary channel during flood stage, gradually increases in flow through successive floods, reaches a peak of deposition, wanes, and becomes inactive. As a result of subsidence, the crevasse system is inundated by marine waters, reverting to a bay environment, and thus completing a sedimentary cycle (1980, p. 53).

Such a sequence has been cartographically documented in the case of the West Bay Complex (Fig. 47). The 1958 map shows that much of the marsh present in 1922 had been destroyed. This destruction, which was the result of subsidence coupled with wave action, continued to such an extent that by 1980 nearly all of the former bay (as of 1838) was once again marine.

Subaqueous Environment. The subaqueous portion of the Mississippi River deltaic plain consists of the area of the Gulf of Mexico below low-tide level that is actively receiving riverborne sediments. The major subenvironments radiate generally out from the distributary mouth and its subaqueous natural levees, past the distributary-mouth bar and distal bar, to the prodelta. Generally, the sediment fines in the same direction, ranging from sands in the natural levees and distributary mouth bar to clays on the prodelta. Once river water debouches through the distributary mouth, it spreads out on top of the denser sea water, decelerates, and rapidly drops the coarser clastics it is carrying. The deposition of these coarser sediments produces a distributary-mouth bar which, being in relatively shallow water, is subject to reworking by marine processes. Because of rapid deposition, pore pressures are often high. These deposits also contain large quantities of organic matter including water-saturated logs, as noted previously in the quote from Lyell. These distributary-mouth bars, although varying in width from one distributary to another, average between 3 and 6 km.

Deposition of silts and some of the clays produces a distal bar, whereas the zone where fine silts and the bulk of the clay are deposited is the prodelta. Because of the great temporal variation in river discharge, the distances at which specific textural groups are deposited also vary. With delta-front progradation (Fig. 48) there is a concomitant seaward advance of subaqueous forms (Kenyon and Turcotte, 1985, p. 1462).

Subaqueous slumping and downslope mass movement of sediments, although only recently recognized as normal processes, are very important to subaqueous morphology off the Mississippi River Delta. Recent research has shown that low-angle slopes at the delta front are unstable and that large amounts of sediment are transported from shallow to deep water in a variety of ways (Fig. 49). Factors influencing the stability of bottom sediments include rapid deposition and sedimentary loading, deposition of coarse-grained sediments on top of fine-grained sediments, rapid biochemical degradation and gas production, underconsolidation of fine-grained deposits, and winter storm and hurricane wave induced loading on the bottom (Coleman and Prior, 1980, p. 89).

The instability that occurs in the distributary-mouth bar area because of the deposition of dense sand on weak clays often leads to diapiric intrusions of clay into and through the sands. Such

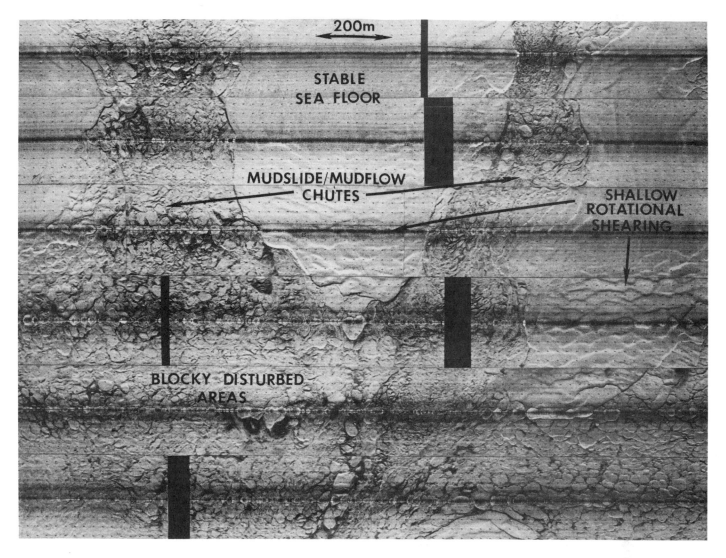

Figure 49. Subaqueous forms on the Mississippi River Delta, shown on a side-scan sonar mosaic (courtesy D. Prior).

forms are usually called mud lumps or mud diapirs. Other instability forms include collapse depressions, peripheral rotational slides, mudflow gullies, depositional lobes, and shelf-edge slumps (Fig. 50).

Barrier, Beach, Strandplain, Chenier

Barrier islands are accumulations of coarse-grained (especially sand) sediments that stand above the level of high tide. Strandplains are accumulations of sediment left behind as the shoreline shifts seaward. Cheniers are ridges that develop in environments that have high quantities of mud and undergo alternating periods of progradation and erosion.

Barrier Island. The coastal conditions that favor barrier island formation (a low gradient plain and shelf, adequate sediment supply, and low tidal range) are present along most of the province's shoreline. Barrier islands and associated lagoons flank nearly half of the coast (Table 7). Most of these barriers, some of which are among the world's longest (Shepard, 1982, p. 593), have sand beaches along their ocean front. They are frequently joined texturally with barrierless shorelines, most of which also support sand beaches. Except for such sand beaches and open-ended estuaries, nearly all of the Atlantic seaboard belongs in the barrier island category. In contrast, the Gulf Coast is much more diverse. It has not only lengthy barrier islands but also sections with mudflats, open sand beaches, marshes, mangroves, and deltas.

Barrier islands, barrier spits, and bay barriers (Leatherman, 1979) form the three major types of barriers common along the province's coastline. The latter two types are anchored to the mainland or an island at one or both ends, whereas the barrier island is not attached. Its island nature comes from tidal inlets that

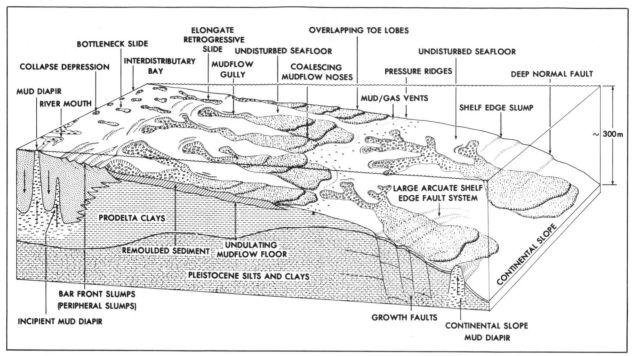

Figure 50. Illustration of major types of submarine landslides, diapirs, and contemporary faults in the Mississippi River Delta. After Coleman and Prior (1982, p. 168).

isolate it from adjacent islands or spits. Barriers are composed of sand and other sediments that are easily transported by longshore and tidal currents, waves, and wind.

The number, length, width, area, stratigraphy, and associated forms of barriers vary greatly along the coast. Whereas only about 15% of the coast of Louisiana has barrier islands, some 95% of North Carolina's coast is lined with them (Table 7). The major barrier islands have an area of about 6,500 km² or somewhat more than that of Delaware.

Fisher (1982, p. 131) separated the barrier island coastlines of the Coastal Plain into four major groups: (A) the Middle Atlantic states, (B) South Carolina and Georgia, (C) the Florida east coast, and (D) Texas. The barrier islands of the Middle Atlantic states, which extend from Long Island to North Carolina, are long and relatively straight and backed by major estuaries such as Pamlico Sound. The South Carolina and Georgia islands, usually referred to as the Sea Islands, are short and separated by numerous tidal inlets. Along eastern Florida the barriers are again long but, unlike most of those of the Middle Atlantic states, they are backed by long, narrow lagoons. The Texas coast barriers are long with few tidal inlets and are backed by lagoons of variable widths. Smaller, more or less isolated, barrier islands are scattered along other parts of the Gulf Coast from western Florida to Louisiana.

Such great variability in the character of the province's barriers suggests that several modes of formation and modification must be at work, modes that are dependent on such factors as

local topography, sediment type and supply, onshore, offshore, and longshore currents, wave energy, sea-level change, tidal range, and stage of development among others. Thus, it is not surprising that numerous theories have been proposed to account for their origin (for literature reviews see Schwartz, 1973; Swift, 1975; Leatherman, 1979; and Fisher, 1982). The three theories that have gained most attention are: (1) spit elongation and segmentation, (2) vertical growth of offshore bars, and (3) drowning of coastal ridges. Because many barriers are probably the result of a combination of processes, Schwartz (1971) proposed that a multiple causality hypothesis be adopted.

Fisher (1982, p. 130) suggested that spit elongation may be the most important factor in the formation of the barriers in the Middle Atlantic states. This section of coast divides into five subsections each with sequentially repetitive segments (Fig. 51). The northern end of each subsection is a spit or cape, which is followed in order southward by a baymouth barrier, a long spit and long barrier islands with few inlets, and finally by short barrier islands with numerous inlets. The spits and capes (such as Sandy Hook, Cape Henlopen, and Cape Henry) were formed by northerly littoral sand transport from eroding Quaternary coastal highlands. The barrier islands in the southern part of each subsection are formed by southerly littoral transport (Kraft, 1985, p. 215).

The islands along the coast of South Carolina and Georgia consist of at least three different types: (1) erosional remnants of older islands (Sea Islands), (2) sandy islands with dune ridges

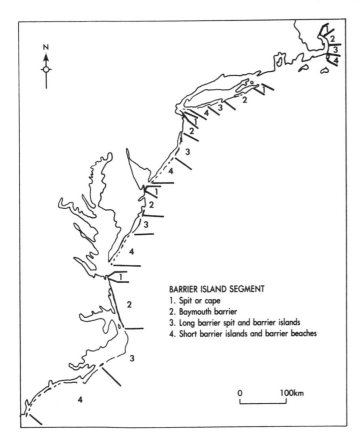

Figure 51. Middle Atlantic barrier island coastlines. After Fisher (1982, p. 131).

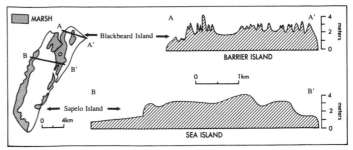

Figure 52. Topographic profiles across Blackbeard Island (a barrier island) and Sapelo Island (a sea island). After Mathews and others (1980, p. 65).

(barrier islands), and (3) marsh islands (Mathews and others, 1980). The older islands are Pleistocene in age, fringed by marsh and dune ridges and generally do not front directly on the open ocean; the other two types are Holocene in age and normally face the sea. The Sea Islands proper are usually rectangular in form and have a subdued relief as illustrated by Sapelo Island (Fig. 52), whereas the sandy barrier islands generally support a ridge and swale topography with steep slopes as in the case of Blackbeard Island (Fig. 52). Maximum elevations of both types are usually less than 10 m, although some of the dune crests on the barriers may extend over 15 m in height. Marsh islands are generally flat and floodable by tidal waters. Scattered through these marsh islands are small sand ridges.

Mathews and others (1980, p. 72) noted that the islands have parallel ridges that are often organized into distinct sets and argued that, because seaward progradation and landward erosion have sufficiently alternated through time, there has not been a general migration landward. Overwash processes, such as those along the barriers of the Middle Atlantic states, are not important in the sea island section, although spit elongation does occur.

Whatever the origin of the barriers, they may be reshaped by waves and currents and may migrate landward or accrete seaward. Hayes (Hayes and Kana, 1976, p. I-96), emphasizing the importance of tidal range (but also implying wave energy), divided barrier morphology into microtidal (<2 m), mesotidal (2 to 4 m), and macrotidal (>4 m) types. Nearly all of this province's coastline has mean maximum spring tidal ranges of less than 2 m (Fig. 40); therefore, it is a coastline along which microtidal barriers dominate. Mesotidal barriers are limited to the South Carolina–Georgia and Cape Cod shorelines.

Microtidal barriers are long, with what Hayes and Kana (1976, p. I-96) called a "hot dog" shape, and tend to have numerous washover features, few inlets, large flood-tidal deltas, and small ebb-tidal deltas. Mesotidal barriers, on the other hand, are short, often with a "drumstick" shape, and usually have few washover features, numerous tidal inlets, moderate-sized flood-tidal deltas, and large ebb-tidal deltas. Combining the rate of sand supply with Hayes' energy groupings provides the basis for four barrier types: (1) microtidal transgressive, (2) microtidal regressive, (3) mesotidal transgressive, and (4) mesotidal regressive (Leatherman, 1979, p. 11).

Transgressive barriers—those where erosion dominates—are generally unstable and are more numerous than regressive barriers partly because of a rising sea level. Those barriers occurring in the lee of prominent capes (such as Cape Lookout, North Carolina) may be regressive (Hayes, 1985). The long barriers of North Carolina and southern Texas, as well as some of the shorter barriers of the Sea Islands and northern Gulf of Mexico groups, are examples of the transgressive type (Fig. 53). Regressive barriers, although less common than the transgressive type, are well represented by Kiawah Island, South Carolina, and Galveston Island, Texas (Fig. 54). The rate of growth seaward of Galveston Island was determined by Bernard and LeBlanc (1965) to average about 1.3 m/year during the last 3,500 years. However, Kraft (1978, p. 374) suggested that total growth actually took 5,300 years, thus reducing the average growth rate to 0.85 m/year.

Although some barrier islands consist mainly of a single ridge of sand, others are complex and in addition to the beach that fronts most of them, consist of various combinations of ridges, dunes, washover fans (Fig. 55), marshes and tidal flats—exemplified by Kiawah Island, South Carolina (Figs. 56 and 57).

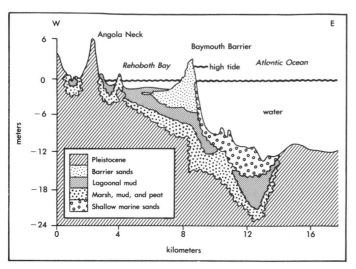

Figure 53. A geological cross section of transgressive environments in the vicinity of Rehoboth Bay, Delaware. Modified from Kraft (1985, p. 217).

Beach. A beach, as defined by the U.S. Army Coastal Engineering Research Center (1977, v. 3, p. A-3), is "the zone of unconsolidated material that extends landward from the low water line to the place where there is marked change in material or physiographic form, or to the line of permanent vegetation (usually the effective limit of storm waves)."

The length of individual beaches varies greatly from very small highly arcuate pocket beaches to long, straight, or only slightly arced beaches that may extend for several hundred kilometers, interrupted only by an occasional tidal inlet or river mouth (Fig. 57). Pocket beaches are rare in the province; their only major occurrence is on the sound side of Long Island and in some estuaries. Cross-sectional profiles throughout the province are generally similar, although width and slope vary with such

factors as texture, composition, wave action, and whether the shoreline is in an erosional, progradational, or stable stage.

The beaches on Cape Cod, Martha's Vineyard Island, Nantucket Island, and Long Island contain sediments that are derived from eroding morainal material and outwash plain sands from offshore. South of the formerly glaciated area, beaches are composed of sediments brought to the coast by the longer rivers that drain the Piedmont and Blue Ridge Mountains, by the shorter rivers of the Coastal Plain that transport sediments of already reworked materials, by current and waves from offshore, and of a biotic origin. Quartz sands predominate.

The beaches of east Florida are also mainly quartz sands transported southward from the rivers of Georgia and South Carolina, although in numerous localities coquina fragments are abundant. Quartz sands dominate along the western coast of Florida, except in the Keys and the 150-km section north to Sanibel Island, where beach sands are mainly carbonates. In the Florida Panhandle, clays are important in the marsh areas and in estuarine situations. The beaches that are found along the Alabama and Mississippi shorelines are nearly pure quartz sand that had its origin in the Appalachians. The predominant longshore currents are from the east in this section of the Gulf of Mexico. Therefore, the Mississippi River has little influence on its depositional characteristics.

In contrast, most Louisiana beaches are dominated by sediments from the Mississippi River. The coarser fractions are either carried seaward or transported west along the shore, usually for relatively short distances. The finer fractions, however, are carried farther west, with many of them becoming important components of the Chenier Plain. Sand beaches are present along nearly all of the Texas coastline and are generally composed of mostly fine quartz sands. However, locally (as at Big Shell Beach and Little Shell Beach) the shell content may be more than 50% of the total by volume. In other locations along the Texas coast, calcareous fragments are important; in contrast to the other beaches of the province, those of Central Padre Island contain some basal-

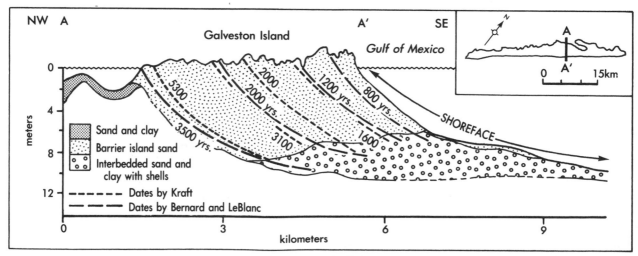

Figure 54. Regressive barrier of Galveston Island. Modified from Bernard and LeBlanc (1965, p. 158), Bloom (1983a, p. 47), and Kraft and Chrzastowski (1985, p. 646).

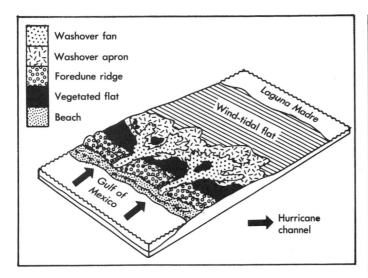

Figure 55. Model for washover channels on barrier islands. After Hayes and Kana (1976, p. I-98).

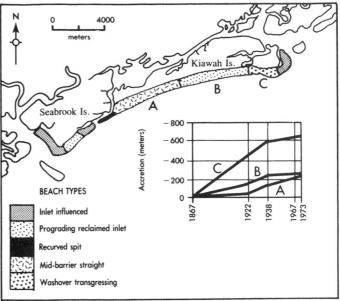

Figure 56. Morphology and accretion for central Kiawah Island, South Carolina. After Hayes and Kana (1976, p. II-83).

tic hornblende and pyroxene sands brought to the shore by the Rio Grande.

Beach widths in the province range from less than 10 m to over 100 m, with the narrowest generally occurring along eroding shorelines. The average slope of beaches varies between 2° and 4°, although many examples outside these values occur. Those beaches that have shell contents of more than 50%, as in Florida and Texas, are as steep as 6°. All beaches within the province are subject to hurricane or other large storm-generated waves. At such times entire beaches may be temporarily lost and the others will be reduced in width and steepened.

Longitudinally, beaches also vary in character, often within short distances. For example, the 25-km-long prograding beach of Kiawah Island, South Carolina, has been divided into five distinct morphological units (Fig. 56). In contrast, beaches along some barrier islands—such as those of the Outer Banks, North Carolina, and Padre Island, Texas—stretch for great distances with little variation.

Strandplain and Chenier. Beach ridges are deposits that generally parallel the shoreline and are formed primarily by wave action, especially the action of storm waves. They often occur in groups with a number of alternating ridges and swales. While normally linear, those that form on a recurving spit or near a migrating inlet (the east end of Kiawah Island, Fig. 57) usually curve and often bifurcate.

In Florida they are best developed at Cape Canaveral on the east coast and at Marco, Sanibel, Dog, and St. Vincent islands on the Gulf coast (Tanner, 1985, p. 164). St. Vincent Island has an especially well developed but complicated beach-ridge field with over 150 ridges that Tanner (1985, p. 166) has grouped into 12 separate sets (Figs. 58 and 59).

A special category of beach ridge is that found in the Chenier Plain of Louisiana west of Vermillion Bay. This plain is a 25

Figure 57. Kiawah Island with tidal inlet. Seabrook Island in lower left. Compare with Figure 56. Photo courtesy T. Kana.

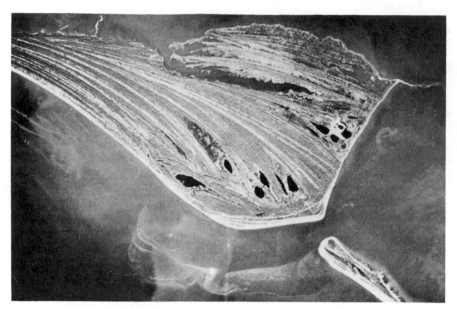

Figure 58. St. Vincent Island beach ridges. Photo courtesy W. Tanner.

to 30 km wide band, about 200 km long, that contains basically only two types of landscape. Areally dominant is a low (nearly all of it less than 1 m above sea level) coastal marsh with its associated streams and lakes. Within the marsh are numerous long, narrow sand and shell ridges that generally parallel the coast. The ridges, which rise above the marsh surface by only 2 to 4 m, are normally covered with oak (chêne in French) trees. The generally accepted explanation for the formation of cheniers is reflective of the dominance of the Mississippi River along this part of the Gulf of Mexico shoreline (Russell and Howe, 1935). When the river discharges on the western side of its plain, fine sediments accumulate, coastal progradation occurs, and marshes develop. When, however, the river discharges at an easterly location, sediment supply is decreased and shoreline erosion occurs. In the process, coarse sediments (sand and shell hash) form beaches that are pushed landward and eventually become ridges on top of the marsh. Presently much of the shoreline along the Chenier Plain is eroding and is exposing former marsh surface (Fig. 41). However, in the eastern section of the plain, progradation is occurring, probably in response to the mud being delivered into the system by the Atchafalaya River.

Although cheniers, which are erosional features, are the ridges that dominate in the Chenier Plain, two other ridge types are also present (Gosselink and others, 1979): the barrier islands (spits) that develop seaward of coastal embayments, and accretion ridges that develop at river mouths (Fig. 60). Both are created by progradation and differ mainly in that the barrier islands are relatively long and straight, whereas the river mouth accretion forms are multiple ridges that have a fan shape, with the handle forming a single ridge away from the river mouth. Some of the river mouths have shifted westward with time, stranding ridges formed earlier.

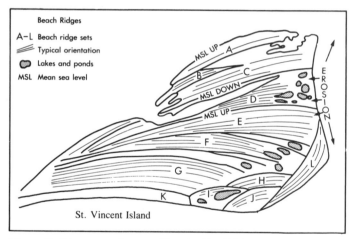

Figure 59. Diagrammatic map of beach-ridge field on St. Vincent Island, Florida. See Figure 58. After Tanner (1985, p. 166).

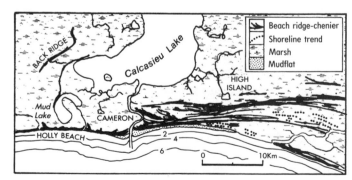

Figure 60. Landforms of the western Chenier Plain, Louisiana. After Gosselink and others (1979, p. 12).

Bay, Sound, Lagoon, Estuary

The waterbodies occurring in the transitional zone between the Coastal Plain and the Continental Shelf are extremely varied in size, shape, depth, chemical and biological composition, freshwater input, degree of protection from oceanic forces, developmental history, and designation. Within the province, such terms as bay, sound, and lagoon form part of the proper name of most of the transitional zone waterbodies. Although there is much overlap in most characteristics among them, there are some general distinctions that are commonly accepted. Bays usually have relatively large mouths opening either directly to the sea or to another bay or lagoon. Sounds are fairly broad bodies of water generally open at both ends, such as Pamlico Sound, North Carolina, and Breton Sound, Louisiana. Lagoons, although often containing bays, are normally linear bodies of water parallel to the coast, having limited contact with the ocean. Because all of these transitional waterbodies are semi-enclosed with a free connection to the sea and contain a mixture of fresh water and seawater (even if highly seasonal in relative proportion), they all have estuarine characteristics (Walker and Prior, 1986). Although the term "estuary" has historically been considered to be the lower tidal reaches of rivers (Pritchard, 1967, p. 3), contemporary usage often broadens it to also include bays, sounds, inlets, and lagoons (Geophysics of Estuaries Panel, 1977, p. 1).

Estuaries and lagoons dominate the transitional zone of the province (Fig. 5); they are present along over 75% of the coast (Emery, 1967, p. 9). This dominance reflects the physical processes of: (1) the drowning of river valleys with sea-level rise, and (2) the formation of barriers along the shore as discussed above. The estuaries and lagoons of today are but the most recent expression of a series of systems that formed as sea level began to rise and progressed upslope across the Continental Shelf.

With rapid sea-level rise, the rate of estuarine development also proceeded rapidly; river valley drowning dominated over sediment infilling. From the standpoint of the Atlantic and Gulf Coastal Plain, estuaries were probably at their maximum development about the time (or just prior to the time) stillstand was reached, about 3,000 to 4,000 years B.P. By that time, not only were the main river valleys drowned, but the lower courses of many of their tributaries (such as the Potomac which flows into the Chesapeake) were drowned in turn.

As the rate of sea-level rise began to decrease (especially after about 6,000 B.P.), those estuaries that have large sediment inputs began to fill. The classic example of the rate at which such filling can occur is the Mississippi River estuary, which was soon "overfilled" once stillstand was achieved. Mobile Bay, Mississippi, is the remnant of an estuary that was over twice as large as it is at present; the Pascagoula River estuary (Fig. 44) has been almost completely filled and converted to marsh.

The extent of estuarine fill affects drastically the nature and length of the shoreline. The deep indentation that the Pascagola estuary possessed until recently has been eliminated, shortening the shoreline in the process. Coasts with estuarine shorelines tend

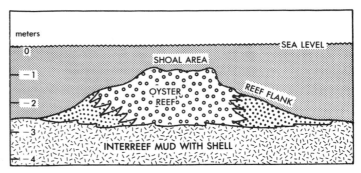

Figure 61. Schematic cross section of relict oyster reef in a Texas estuary. After Brown and others (1976, p. 72).

to be long. The length of the inner shoreline of the Texas and North Carolina coasts, with their unfilled estuaries, contrasts greatly with the outer shoreline of Padre Island and the Outer Banks.

Nonetheless, estuaries are ephemeral features, mainly because of the rapidity with which they fill. This fill is usually heterogeneous. Basal sediments (at depths between 40 and 50 m below present sea level in the case of both the Potomac estuary, Maryland and Baffin Bay estuary, Texas) are coarse-grained and of fluvial origin. As a subaerial river is converted into an estuary, riverine sediments become finer and sediments of marine and organic origin are added. The shallow portions of filling estuaries usually support fresh-brackish-saline water marsh and often oyster reefs as well (Fig. 61). In the deeper parts (i.e., the unfilled or slowly filling channels) of these estuaries, mud often dominates. In the lower Potomac more than 14 m of mud has accumulated (Knebel and others, 1981, p. 588).

The depositional history of the province's estuaries depends on many factors in addition to sea-level change. Once sea level reached its present position along the south Texas coast, the tidal inlet or pass that connected Baffin Bay estuary to the ocean was cut off by the formation of Padre Island. In turn, marine sediment no longer reached the estuary, a circumstance that, when combined with the fact that the streams draining into it carry little sediment, has produced "a sediment-starved bay-estuary system" (Brown and others, 1977, p. 22).

Just as the vertical sediment column in an estuary may possess great textural variation, so may the present-day bottom sediments. Mobile Bay, for example, has bottom sediments that range from relatively pure sand to relatively pure clay, with zones showing various combinations of sand, silt, and clay (Fig. 62). Areal distribution of sediment, to a large extent, reflects depth and circulation patterns. Sand is most abundant in the shallow areas near the head of the bay and around its edges where currents are strongest. Clay, on the other hand, is concentrated not only in the deeper parts of the bay but also where energy is low and eddy currents predominate.

Because sea level in estuaries and lagoons has been virtually the same for over 3,000 years, the original valleys and inner

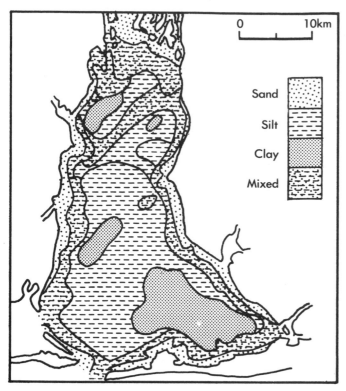

Figure 62. Sediment distribution in Mobile Bay. After Loyacano and Smith (1979, p. 10).

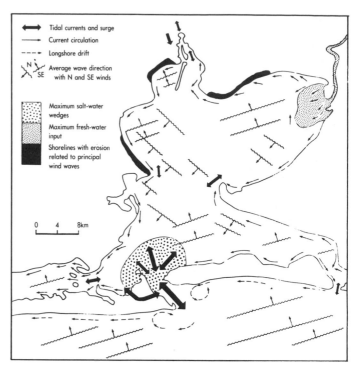

Figure 63. Galveston Bay and physical processes emphasizing wind direction and shoreline erosion. After Fisher and others (1972, p. 19).

shorelines have been subject to a relatively long period of erosion. Thus, much of the sediment being incorporated in estuaries, and especially those protected estuaries of Texas, is from local shoreline erosion. The extent of such erosion depends in large measure on wind direction and strength and fetch. In the case of Galveston Bay, northwest and southeast winds are responsible for much of the shoreline erosion (Fig. 63).

One of the major agents modifying estuaries and lagoons (including the barriers to which lagoons owe their existence) is the hurricane. It is probable that all of the province's lagoons have been affected by hurricanes at some time or other (Fig. 64). Along some sections, especially the Gulf Coast and southern Atlantic Coast, they are of major importance. A number of factors influence the effect of hurricanes including tidal conditions, the nature of the shoreface, the type of barrier (microtidal barriers are more affected than mesotidal barriers), vegetation, and the amount of rainfall, among others. Storm surges, after producing washover channels through barriers, carry sediment from the shoreface and beach into the lagoon (Fig. 55). On some of the barriers of Texas, washover channels have been eroded to depths of 5 m below sea level and have remained open for months (Brown and others, 1976, p. 68). They seldom become permanent because longshore drift plugs them. However, small ponds may remain leeward of these plugs for years.

The heavy rainfall that is associated with hurricanes is im-

portant in that it (at least in central and south Texas) is usually responsible for the transport of the majority of the fluvial sediment that reaches the estuaries and lagoons.

The offshore is a major source of sediment for many estuarine environments. Although the quantity of sediment available varies greatly along the foreshore, in many locations amounts are finite. Along the New Jersey coast, the contribution of sediment by rivers has been minimal. Yet infilling has occurred; the source was offshore. This offshore source has supplied the bulk of the sediment that has infilled the state's estuaries. However, this infilling has required so much sediment that the offshore source is now depleted and it is proposed that the accretionary trend of the past will now be reversed (Psuty, 1986).

Marsh and Swamp

A marsh ecosystem in the coastal estuarine region includes both the emergent, grassy and woody land zones and the tidal creeks that exchange salt water. Often organisms form a controlling interface and their activities are an integral component that controls the geomorphic landscape. In the Atlantic Coastal Plain, salt marshes and tidal creeks dominate; on the Florida Peninsula, tidal mangrove swamps form the most important coastal vegetation; and in the Gulf Coastal Plain, narrow salt marshes and broad brackish marshes, separated by broad estuarine sounds and lagoons, form the dominant landscape morphology.

Salt marshes are beds of intertidal rooted vegetation that are

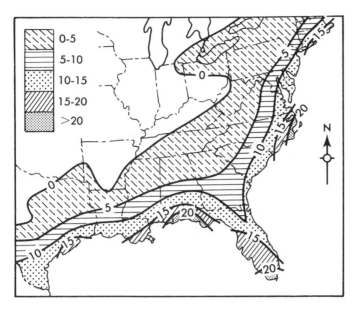

Figure 64. Number of times destruction was caused by tropical storms for the period 1901 to 1955. From Mathews and others (1980, p. 59).

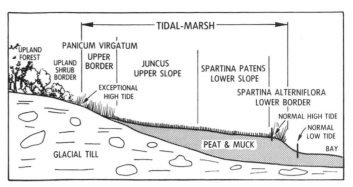

Figure 65. Cross-sectional diagram of New England–type salt marsh. From Miller and Egler (1950).

alternately inundated and drained by the rise and fall of the tide (Cooper, 1974). Salt marshes become established in areas where salt tolerant plant species have invaded shallow, relatively protected tidal flats and on the landward side of regressive barrier island systems. Most of the present salt marshes of the Atlantic Coastal Plain have developed during the past 5,000 to 6,000 years, in response to the outbuilding of regressive barrier island plains. Salt marshes, which can be exposed to salinities that vary up to 40% during flood tides, may also be inundated by fresh water during heavy rainfall and river flooding. Only a relatively small number of plants and animals can tolerate such extremes. Grasses of the genus *Spartina* and species of *Juncus* and *Salicornia* are almost universal in their occurrence, as are certain kinds of animals, such as fiddler crabs and mussels.

Along the New England coast, the salt marshes display a rather clear-cut zonation pattern (Fig. 65). *Spartina alterniflora* occurs in pure stands within the intertidal region and along creek banks. *Spartina patens* and *Distichlis spicata* stands are generally found landward of the intertidal zone; the vegetation is often rooted in firm clays or peaty zones. *Juncus* species form relatively pure stands in the zones above normal high-tide levels. Around the fringes of the salt marsh is a narrow belt of *Panicum virgatum* and *Spartina pectinate,* as well as a number of other species. The marsh surface is usually traversed by numerous tidal creeks, which often form a complex dendritic drainage pattern. The smaller tidal creeks are essentially drained on each ebbing tide and refilled on the flooding tide. The larger tidal channels are rarely completely drained. Flood tides act as a barrier in the lower reaches of the streams, holding fresh water in the upper reaches of the channel system. This damming effect often results

in overbank inundation of relatively fresh water into an otherwise saline environment.

Some of the most extensive salt marshes in the United States occur between Chesapeake Bay and Jacksonville, Florida. They are formed primarily in estuaries where major rivers, draining large areas of uplands, deposit large quantities of highly nutrient-bearing inorganic sediment. There are usually two major plant communities in these marshes, *Juncus roemerianus* and *Spartina patens.*

The marsh characteristics of the South Atlantic disappear between Jacksonville and St. Augustine, Florida, where mangroves first appear (Egler, 1952). Mangrove swamps cover about 1,750 km² of peninsular Florida (Craighead, 1964). These swamps are best developed in the Ten Thousand Island region and around Cape Sable. Three species of mangrove—red (*Rhizophora mangle*), black (*Avicennia germinans*), and white (*Laguncularia racemosa*)—dominate these swamps. In some instances, these mangroves often have a tree canopy that is in excess of 30 m tall. Often salt marshes form large flat expanses within the mangrove swamps. These salt-tolerant species are relatively intolerant of major variations in temperature; westward, along the Gulf of Mexico shoreline, the mangrove swamps diminish in size in response to winter freezes. Those in Louisiana consist primarily of dwarfed black mangrove.

Marsh types along the Gulf coast are composed essentially of the same plant species that occur along the South Atlantic coast. Often only a small thin zone of *Spartina alterniflora* occurs along the back sides of the barrier islands and along small tidal creeks. Landward there are large expanses of *Juncus roemerianus.* The largest expanses of marsh in the province are associated with the broad plains of the Mississippi River Delta and its adjacent downdrift (westward) region of western Louisiana and eastern Texas. *Spartina alterniflora* forms a relatively thin seaward zone that is regularly flooded by low tides. The brackish marsh is the most extensive of marsh types and contains dense stands of *Spartina patens, Distichlis spicata,* and *Juncus roemerianus.* Tidal channel patterns are not well developed and generally consist of a high density of relatively small channels that do not tend to show the pronounced dendritic pattern so characteristic of the tidal creeks along the Atlantic Coastal Plain. Tidal ranges are low, rarely exceeding 0.5 m. Within the Mississippi River Delta region, the marshes are subjected to high rates of subsidence and compaction, which results in a constantly changing marsh landscape. In many instances, biomass production (marsh accretion) is barely able to maintain pace with rising sea level and subsidence. In the most landward part of the delta plain, fresh water marshes are found. In many instances, these fresh to slightly brackish marshes consist of a relatively thin floating mat of vegetation overlying a highly turbid water and organic ooze. These marshes are referred to locally as flotant (Russell, 1942). Flotant results when the rooted living marsh mat separates from the substrate (often former delta deposits) because of rapid subsidence. These marshes are prone to rapid deterioration because of the extreme water levels that occur during the passage of hurricanes and tropical storms and the variety of destructional processes caused by organisms such as nutria, muskrat, and waterfowl.

In general, the marshes of the Atlantic Coastal Plain are relatively stable in comparison to the constantly changing river regimes found in the central and western Gulf of Mexico Coastal Plain. Flooding, changes in delta distributaries, rapid compaction, and subsidence along the Gulf Coast cause the marshes to be subjected to a wide range of stresses over a relatively short period of time. In the Florida Peninsula, minor climatic changes place significant stresses on mangrove swamps. In most marshes, modifications to drainage patterns, damming of stream channels, diking and land reclamation projects have had adverse effects on the marshes in both coastal plains. Only in the past decade have scientists realized the significant impacts to which these marshes are being subjected.

Coastal Dune

Much of the shoreline of the province is bordered by sand dunes, partly because they form an integral part of nearly all barriers. Those portions of barriers that have grown to elevations higher than storm surges can reach owe their existence to wind. However, dune morphology is influenced by other factors such as topographic setting, shoreline trend, sand availability, and vegetation, which is probably the most important (Hayes and Kana, 1976, p. I-51).

Thom (1967, p. 17) has noted that the foredunes along parts of the South Carolina coast are generally low (1 to 3 m high), but where they have become extensive they are typically 3 to 7 m high. These more extensive dunes are covered by dense scrub brush and are stabilized today. However, "within living man's memory, they have migrated a few hundred feet across the gently sloping Myrtle barrier before being stabilized."

In Florida, coastal dunes occur where the coastline has been stable for some time and longshore drift of sand occurs, or where present-day beach ridges are undergoing erosion (Tanner, 1985, p. 167). Dune fields are rare, with most of them concentrated in the Florida Panhandle.

Onshore winds favor the construction of large dunes such as those on Padre Island, Texas. Such winds also favor dune migration and the formation of blowouts. Much of Padre Island has two dune systems, a fore-island dune ridge and a back-island dune field. The fore-island dune ridge, which rises to as much as 12 m high, is composed of steeply dipping, well-sorted sand that is derived from the backbeach area. Although it protects the back of the barrier island from hurricane surges, its face is often eroded back. Normally, it takes years after such erosion for this dune ridge to grow gulfward to its former position (Brown and others, 1977, p. 57). Blowouts frequently occur along the dune ridge. Within the blowouts, various types of dunes (barchan, parabolic, and elongate dunes) may develop. The life of blowouts is variable—alternating between periods of activity and stability.

Back-island dune fields are large systems that develop on the lagoon side of the barrier. In the Kingsville, Texas area, bare, shifting dunes, extend over an area of more than 75 km^2. The sand for these fields is supplied through hurricane washover channels and fore-island blowouts. Barchan, longitudinal, and transverse dunes are all represented. Because the migration direction of these dunes is westerly, most of their sand eventually ends up in Laguna Madre (Brown and others, 1977, p. 58).

THE SUBAERIAL ENVIRONMENT

The subaerial portion of the province is nearly synonymous with what is generally considered to be the Coastal Plain; excluded are those areas discussed above as part of the province's transitional belt. It comprises a variety of landforms, despite the fact that the total relief of the more than 1.1×10^6 km^2 area is less than 300 m. Although the gross character of most Coastal Plain forms (especially such as cuestas, plains, hills, and terraces—Figs. 9, 10, and 24), stems from pre-Holocene times, present-day processes are responsible not only for alterations in the morphology of these "stable" forms but also for the creation of additional features. Neotectonic, gravitational, solutional, fluvial, eolian, and biologic processes, which often work in concert, are all important.

Neotectonic, Gravitational, and Solutional Processes

Although the province has not experienced the extreme tectonic activity that characterizes other provinces, much of its morphology is the result of the tilting that accompanied the development of the province's geosynclines, arches, and basins (Fig. 3). Such tilting accounts for the variation in the slope of the Mississippi River terraces and the variation in elevation along the marine terraces of the Coastal Plain facing the Atlantic Ocean.

Movement, most of which occurred in pre-Holocene times, continues at present. For example, although the downwarping and lateral movement at the inner edge of the Coastal Plain that caused the southwestward deflections of the Potomac, Susquehanna, and Delaware rivers mainly occurred in the Cretaceous and Tertiary, lateral movement continued in the Quaternary (Mixon and Newell, 1977, p. 437; Mixon, 1985, p. 65).

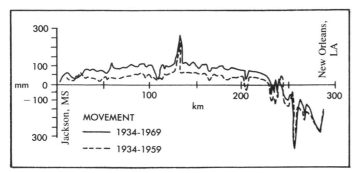

Figure 66. Vertical bench mark movement between Jackson, Mississippi, and New Orleans, Louisiana. After Burnett and Schumm (1983, p. 49).

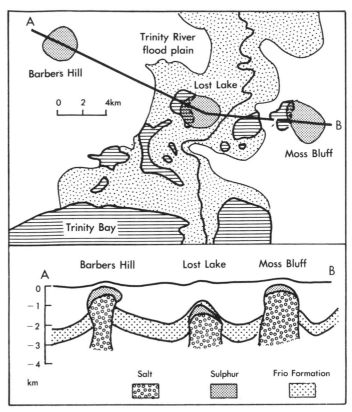

Figure 67. Three salt domes and their topographic expression in southeast Texas. After Fisher and others (1972, p. 74).

In the Rio Grande Valley, repeated geodetic surveys have shown that uplift is on the order of 10 mm/year and that of the Monroe Uplift (Fig. 3) and other parts of Louisiana and Mississippi is some 5 mm/year (Burnett and Schumm, 1983, p. 49; Fig. 66).

Although the long-term effect of uplift on drainage patterns, as illustrated in the case of Cape Fear Arch (Fig. 14), is often obvious, the recognition of present-day responses to continued uplift has been slow in coming, especially in such presumed relatively stable areas as most of the Coastal Plain. It has been demonstrated by Burnett and Schumm (1983, p. 50) that some of the rivers in Louisiana and Mississippi are presently adjusting to neotectonics. In the case of one of the smaller creeks (namely, the Big Colewa), they write that:

> above the axis of the Monroe uplift, the reduction of valley slope and channel gradient has caused a reduction of sinuosity and an anastomosing pattern has developed; gradient and channel depth is less than that below the axis, and overbank flooding is frequent. Downstream of the axis, there is bank erosion and the channel is meandering. In a downstream direction, as the axis of the uplift is crossed, channel and valley slope, channel depth, and sinuosity increase (1983, p. 49).

Even large rivers, such as the Pearl (Table 3), are affected. Although the Pearl River is not anastomosing above the uplift axis that occurs in the smaller rivers, it nonetheless has "developed a new floodplain below the axis of uplift and the former floodplain is now a low terrace" (Burnett and Schumm, 1983, p. 50).

Neotectonics is apparently also responsible for the formation of asymmetrical cross-section profiles and for lateral stream migration in southeast Alabama and southwest Georgia. A renewal of subsidence in the Apalachicola Basin (Fig. 3) is causing a southward migration of those streams that have a general east-west trend. Relatively broad terraces are forming on the gentle northern slopes, whereas little or no terracing is occurring on the steeper southern slopes (Price and Whetstone, 1976, p. 136).

Surface uplift is also caused by the growth of salt domes, which are numerous across the Gulf Coastal Plain (Fig. 67). Although most salt domes are very old, apparently some are still

growing (Murray, 1961, p. 274). As they grow upward and approach the surface, they dome up the overlying sediments. The dip of these beds is away from the dome, which is usually partially or even completely surrounded by a structural depression. These depressions are commonly filled with either water (Fig. 67) or marsh vegetation.

Although most of the neotectonic deformation occurring in the province is aseismic, earthquakes are not uncommon. The recent seismic record of the Coastal Plain is dominated by the earthquakes that occurred in the New Madrid Seismic Zone (NMSZ) in 1811–12 and at Charleston, South Carolina, in 1886. Nonetheless, every state within the Coastal Plain has suffered damage (mostly minor) from earthquakes in historic times (Hays, 1981, p. B4).

The NMSZ is the most seismically active zone east of the Rocky Mountains; the New Madrid earthquakes affected more area (over 620×10^3 km^2) than any other series of quakes recorded in the United States (Crone and others, 1985, p. 547). The NMSZ, which is over 300 km long and some 70 km wide, nearly straddles the Mississippi River at the upper end of the Mississippi Embayment. The three major quakes that made up the 1911–12 series were responsible for major topographic and hydrologic changes in the NMSZ. Reelfoot Lake, Tennessee, which is more than 30 km long and over 6 m deep, is the most

Figure 68. Landslide incidence and susceptibility. After Radburch-Hall and others (1982).

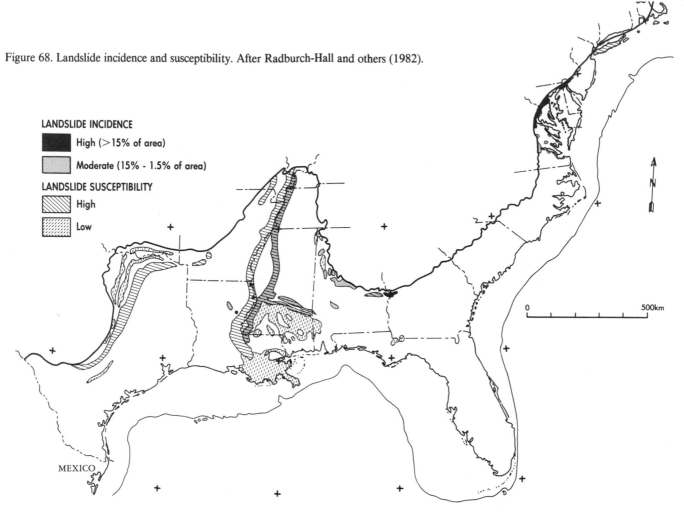

LANDSLIDE INCIDENCE
■ High (>15% of area)
▨ Moderate (15% - 1.5% of area)
LANDSLIDE SUSCEPTIBILITY
▨ High
▨ Low

MEXICO

conspicuous feature caused by the earthquake. However, at the same time numerous other lakes were also formed.

Although not so intense, the Charleston, South Carolina, earthquake of 1886, was, nonetheless, felt in Virginia and Florida. It apparently caused no major topographic modification. However, it was responsible for the formation of liquefaction-flowage features called "sand flows" and for the craters they formed (Obermeier and others, 1985, p. 408). Work by Talwani and Cox (1985, p. 379) showed that at least two other earthquakes prior to 1886 but later than 3,740 B.P. caused similar forms.

Seismic activity of low magnitude is common, especially in the Gulf Coastal Plain (Fig. 12). The localized nature of some of these seismic-generated surface forms was described by Russell who, in 1943, witnessed one in formation west of New Orleans, Louisiana. He wrote that

the throw at the surface increased gradually during a 10-day period to a maximum of 35 centimeters . . . Borings by the U.S. Army Corps of Engineers demonstrated that at various places along the 2-mile surface trace of the fault there were increasingly greater offsets in strata of greater depth (Sternberg and Russell, 1952, p. 381).

Gravity, even in the generally low-sloping terrain that char-

acterizes most of the province, is the force that is ultimately responsible for most of the mass movement that occurs. However, it usually works in concert with, or is triggered by, other processes such as flooding, wave action, groundwater sapping, and earthquakes. It operates in the suboceanic (Fig. 32) and transitional (Figs. 49 and 50) portions of the province as well as in the subaerial portion. As in the case of surface faulting, the incidence of and susceptibility to landslides is much less than in most mountainous areas (Hays, 1981, p. B62).

The areas of the Coastal Plain where landslides have occurred (incidence) and where the strength of earth materials is such that landslides are likely to occur (susceptible) are shown in Figure 68. Among the most susceptible area are the clay-rich Cretaceous deposits in Texas, especially along the Balcones Escarpment, and in Alabama (Radburch-Hall and others, 1982, p. 13). Texas and Alabama, along with North Carolina, Virginia, and Maryland, are the only coastal plain states ranked by the Committee on Ground Failure Hazards (1985, p. 8) as having moderately severe landsliding problems. However, most of these problems are in the non-coastal portions of each state.

In the northeast, along the western shore of Chesapeake Bay and on Long Island, Nantucket Island, and Martha's Vineyard, slumping, aggravated by groundwater sapping and wave erosion, is common. Within the Mississippi River alluvial plain, erosion initiates frequent riverbank failures (Brunsden and Kesel, 1973). From Baton Rouge, Louisiana, and to the north, such failure is more common than in the delta area to the south

because fine-grained deposits in the upper valley are underlain by coarse, easily eroded sand at depths to which the river can scour; . . . in the lower delta area, the fine-grained deposits are thicker, the river runs wholly within them, and bank failures are much less frequent (Radburch-Hall and others, 1982, p. 14).

Subsidence, "the lowering or collapse of the land surface either locally or over broad regional areas" (Hays, 1981, p. B73), is a common phenomenon in the Coastal Plain. Natural subsidence, which is emphasized here, is caused by earthquakes, the accumulation of a load (be it ice, water, or sediment), compaction, and the dissolution of limestone or other soluble material.

The formation of Reelfoot Lake, Tennessee, as discussed above, is an excellent example of earthquake-induced subsidence. However, the ground shaking caused by the New Madrid quakes was so intense that much subsurface sediment and water were ejected to the surface, leaving subsurface voids which prompted extensive subsidence.

Because of the great load of sediment (nearly five times as much as all of the other rivers of the province combined) transported to the Gulf by the Mississippi River (Table 4), the central Gulf Coast is the most actively subsiding large area in the United States. This subsidence is critical in the formation of the Gulf Coast geosyncline. Landward of this subsiding zone is a compensatory zone of uplift as evidenced by the tilted terraces. The hinge line between the zone of subsidence and zone of uplift shifts seaward with continued deposition. Although by far the greatest amount of natural subsidence is occurring along the Louisiana coast, nearly all other Gulf and Atlantic Coastal Plain coasts are also subsiding. Estimates of this coastal subsidence vary from a low of about 10 cm in eastern Florida to over 90 cm for Louisiana during the next century (Titus, 1985).

The accumulation of sediment also causes compaction, which is added to the downwarping that occurs. Compactional subsidence tends to be more localized than geosynclinal subsidence (Morgan, 1977, p. 39). The rate of compaction varies with the composition (mineral or/and organic) and texture of the sediment as well as with the amount of water in them.

Much more conspicuous than either the geosynclinal or compactional subsidence is that associated with dissolution. The Coastal Plain has a number of areas—especially in Florida, Alabama, and Georgia—where solution weathering of limestone has created a karst terrane (Fig. 69) with thousands of caves and sinkholes (Fig. 70).

Solution weathering in Florida is responsible for the loss of 1.2×10^6 m³/yr of limestone through its systems of springs. This is the equivalent of 1 m loss from the surface of Florida every 38,000 years (Opdyke and others, 1984, p. 227). Such a loss has been considered of sufficient importance to be used as an expla-

nation for the 36 to 41 m of height differential between the terraces in south Georgia and north-central Florida (Opdyke and others, 1984). The isostatic uplift generated by solution weathering, begun during or before the early Pleistocene, is continuing.

Within karst areas, such as in northern Florida, lakes are numerous. Generally they tend to be circular (Fig. 70) and frequently occur in clusters rather than in lineal chains. Thus, the elongate lakes that occur along the St. Johns River are believed to be relicts of a former estuary rather than limestone depressions (White, 1970, p. 103).

Not all areas underlain by carbonate strata, including the Florida Panhandle, the Apalachicola Embayment, and the low-lying Atlantic coastal zone of Florida, are subject to sinkhole development because they may be overlain by a thick layer of unconsolidated sand and clay, have a poorly developed cavernous system, or have a potentiometric surface near the ground surface (Windham and Campbell, 1981, p. 20).

Although most of the sinkholes tend to be located on plains or in valley floors, some also develop on ridges (Fig. 69). Most of the ridge sinkholes are located in the Cretaceous rocks of south Texas, southwest Tennessee, and Alabama.

Fluvial Processes

The Atlantic and Gulf coastal plains owe their origins to the sediment introduced via alluvial channels throughout the Cretaceous, Tertiary, and Quaternary. This addition of sediment to the shoreline and regional uplift have resulted in a net outbuilding of the coast. In the Atlantic Coastal Plain, there was a relatively constant transport of sediment by the large number of rivers that drain the Appalachian Mountains. In the Gulf of Mexico, however, the major supply of sediment was through the drainage basin of the ancestral Mississippi River system. In both cases, broad coastal plains developed, but the volume of sediment is vastly different, being much higher in the Gulf Coastal Plain.

During the Quaternary, changing sea levels strongly influenced the morphology of the coastal plain alluvial systems. In times of low sea level, coastal rivers, both large and small, incised and carved out erosional valleys, which, with rising sea level, were infilled and formed broad alluvial terraces.

The stream channels of the Coastal Plain basically have a meandering pattern today. However, during periods of low and rising sea level, channels with increased stream gradients and higher sediment loads had braided channel patterns. These patterns have largely been obscured by subsequent sedimentation. Most of the coastal fluvial channels have displayed a strong tendency toward meandering since the beginning of the Holocene. Channel morphology is controlled by a large number of variables, including discharge (volume and variability), sediment yield (volume and size), channel width, depth, velocity, bed roughness, and stream gradient. Braiding channels are favored by high sediment yield (especially large volumes of coarse-grained sediment), high variability in discharge, and relatively high channel gradients. These conditions are generally present during periods of

Figure 69. Karst terrane in the Coastal Plain. After Hays (1981, p. B76).

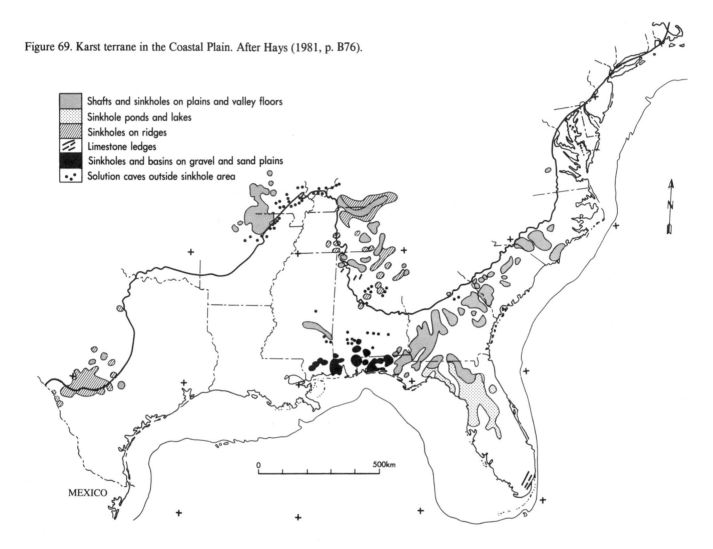

	Shafts and sinkholes on plains and valley floors
	Sinkhole ponds and lakes
	Sinkholes on ridges
	Limestone ledges
	Sinkholes and basins on gravel and sand plains
	Solution caves outside sinkhole area

MEXICO

0   500km

Figure 70. Sinkhole in central Florida. Photo by H. J. Walker.

Figure 71. Oxbow lake and ridge and swale topography. Photo courtesy F. Beatty.

falling and low sea level. This is the time that the coastal streams eroded their alluvial valleys in the upper parts of the river system. As sea level rose, channel gradients decreased, discharge became less erratic, and the volume of fine-grained sediment in the lower parts of fluvial systems increased in quantity. Channel infilling, which was intensive, favored the development of meandering rivers.

Each channel in the Coastal Plain displays various degrees of the meander-belt development including such forms as oxbow lakes, ridge and swale topography, and backswamps (Fig. 71). The size of meander belts, as well as of individual point bars, is controlled by a variety of factors, including fluid and sediment discharge, variability of flood levels, stream gradients, and bank erodibility. Some of the best-developed meander-belt sequences, and by far the largest in the Coastal Plain, are found in the valley of the Mississippi River. These meander belt landforms and deposits were described by Fisk (1947) in his classic paper on the alluvial deposits in the lower Mississippi Valley. Fisk referred to the lower initial fill of the valley system as the substratum. This fill is extremely coarse and was deposited during low sea level as well as during the early stages of sea level rise. It is highly probable that many streams, attempting to respond to changing gradients and excessive sediment discharge from the melting continental glaciers, displayed braided patterns. As stream gradients became established, channels gradually assumed meandering patterns. Although most of the alluvial channels in the Altantic and Gulf Coastal Plain have not had as intensive study as the Mississippi, it is highly probable that most of them underwent similar changes.

Flooding in the province's channels is frequent and the production of natural levees common. The sediment deposited is usually clay and silt, although some fine sand may be included. The height of the natural levee, which is controlled by the maximum height of flood level, tends to decrease in height downstream, as does the width. Breaches through natural levees may develop into crevasses and lead to the development of overbank splays, to the spread of sediment into interlevee basins, and to the infilling of fresh-water lakes. Crevasses frequently maintain their discharge for several years, enabling the deposition of considerable amounts of sediment within the flood-plain basin. Eventually, deposits will seal the channel breach, and vegetation will quickly occupy the slightly higher topographic elevations created by the crevasse splays.

Large areas of most flood plains are occupied by fresh-water lakes or dense stands of water-tolerant woody vegetation. Fisk (1952) referred to these areas as the backswamp and to the resulting deposits as backswamp deposits. They represent principally the long-continued accumulation of deposits in the low-lying areas that flank the meander belts, and consist primarily of highly organic clayey materials. Often sluggish drainage channels will be found in this area, carrying primarily rain water and flood overbank discharge down the valley. In many of the flood basins of the Coastal Plain, these backswamp areas form the bulk of the surface expression of the alluvial valley and are characterized by

dense stands of cypress, willow, gum, and cottonwood trees with a thick understory of small shrubs.

As population increased within the drainage basins and on the flood plains of the river systems, man-made modifications changed the natural conditions of most of the alluvial streams. Dams in the upper reaches, leveeing for flood protection, and clearing of flood plains for development have all taken their toll on this natural environment and most of the streams are nearing total control by man.

Although differential erosion has been responsible for the relatively high relief present in the numerous cuestas that nearly parallel the Gulf Coast and in the hilly sections that occur elsewhere in the province, it appears that in some locations surface reduction by running water has been minimal. This generalization is especially applicable to the Atlantic Coastal Plain where "stratigraphic, geomorphic, paleobotanical, and climatic evidence indicate that the drainage divides in most of the middle part of the upper Coastal Plain have not been eroded since the surficial sediments were deposited" (Daniels and others, 1971, p. 61).

THE HUMAN ELEMENT

In the preceeding sections, emphasis was placed on natural processes, materials, and forms. However, the province today is very different from what it would have been had not man, an ever increasingly important geomorphic agent, entered on the scene.

Prehistoric Impact

Just when humans first began to modify the province is unknown. Although estimates have Indians in the area more than 15,000 years ago (Haag, 1978, p. 2), it was not until prehistoric agriculture began to spread across the Coastal Plain around A.D. 1,000 that significant morphological modifications began to occur. With agriculture, even on a limited scale and with a hill- or mound-type planting, some deforestation, alteration in runoff, and soil erosion occurred. Neitzel (1981, p. 57) demonstrated that Indian occupation in the loess hills near Natchez, Mississippi, was responsible for extensive colluviation. Functional as well as abandoned clearings (which became known as "Indian old field") were numerous along the Atlantic seaboard at the time the first Europeans arrived. Intentional burning, which was used to help make these clearings, was more extensively used to control the undergrowth within forests (Brown, 1948, p. 13). This practice was probably more significant ecologically than geomorphologically.

An intentional alteration of the landscape by Indians is evidenced by the many temple, burial, and living mounds that dot the Coastal Plain. Mounds were constructed on cheniers and even in the marsh (Fig. 72); many of them have been destroyed by coastal erosion or have subsided beneath the marsh surface. Numerous larger mounds were constructed on higher ground, for example, those at Poverty Point, Louisiana (Fig. 73). Construction at the Poverty Point site, which covers about 160 ha, began

Figure 72. Indian mounds in coastal marsh of south Louisiana. Courtesy R. Neuman.

about 1800 B.C. and apparently reached its maximum development about 1000 B.C. (Haag, 1978, p. 4). In addition to a number of mounds, the Indians constructed other types of earthworks such as the six concentric elevated ridges illustrated in Figure 73. The diameter of the outer ridge is 3,745 ft (1,208 m; Webb, 1982, p. 9). This site, which contains the largest early earthen constructions in the Coastal Plain, illustrates that the Indian, even with a limited technology, was an effective direct geomorphic agent.

Historic Impact

In the early days of European settlement, modifications of the landscape proceeded along somewhat similar lines as those of the Indian. However, when cattle, which numbered about half that of the human population in 1650, and field crops became important, the geomorphic impact of man was increased. Per Kalm, a Swedish traveller, noted in 1749 that cattle would wander through the woods "where they are half-starved, having long ago extirpated the annual grasses by cropping them too closely in the Spring" (as quoted in Brown, 1948, p. 30). It is unlikely that the forest floor suffered much increase in erosion because of this practice. However, the opposite conclusion holds for agriculture. The increased size of clearings and the introduction of the plow led to sizable increases in the rate of soil erosion.

Although agricultural practices may be considered the first major human factor affecting the morphology of the province, they were soon joined by others. The growth of population, improvement in transportation, development of industry, and expansion of technology all expanded the geomorphic options of man.

Today these options are being exercised in all three of the divisions of the province, although they were introduced only recently to its suboceanic portion.

The Subaerial Division. Virtually all of the subaerial portion of the province has been affected to some degree by man. In addition to the widespread impact of agriculture, there has been

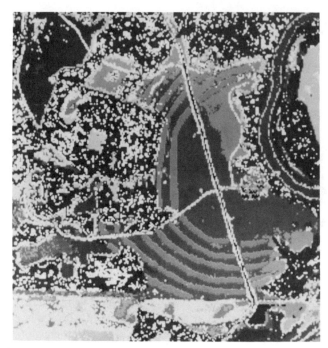

Figure 73. Classified image using TMS Bands 2, 3, 4, 5, and 7 of Poverty Point, Louisiana; NASA/NSTL/ERL, 10-83-011. Courtesy T. Sever.

an almost equally important alteration of the province's rivers. Many river channels have been straightened, shortened, steepened, leveed, and dammed. All of these practices affect the discharge of both water and sediment. Damming, for example, results in sediment trapping and downstream sediment starvation (Fig. 74). Artificial levees (Fig. 75), which now line many of the rivers of the Coastal Plain, prevent floodwater from flowing across floodplains.

Within the Coastal Plain, natural lakes tend to be few in number except for the Florida sinkholes and Carolina bays. Two major modifications have occurred; some lakes have been eliminated, and many lakes and ponds have been created (Fig. 76). Among the Coastal Plain wetlands that have been reclaimed are many of the floodplain swamps. North Carolina's freshwater wetlands have an area of 18,000 km², of which about half are evergreen forested and scrub-shrub wetlands, known locally as "pocosins" (Tiner, 1984, p. 49). The conversion of pocosins to agricultural land has been rapid since the 1950s (Fig. 77).

The Transitional Division. Some of the most significant modifications by man in the province are to be found in the environs of the shoreline. Marshes, estuaries, lagoons, barrier islands, and beaches have all been altered. Canal construction has been extensive, ranging from narrow trapper trails through the marsh to broad waterways such as the Intracoastal Canal. The spoil produced by dredging is frequently piled along the dredged channel, although it may be transported to other locations, such as marsh surfaces, lagoon bottoms, and offshore (Fig. 78). The area covered by spoil in the coastal zone is large; in Texas alone it

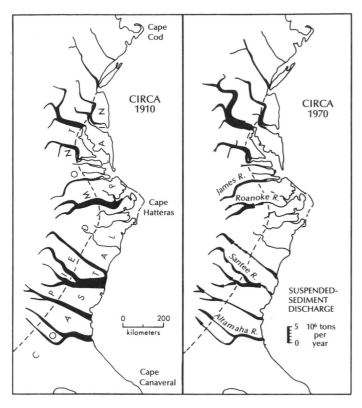

Figure 74. Average annual load of suspended material of some Atlantic Coast rivers before and after the construction of dams. After Meade and Trimble (1974, p. 100).

Figure 75. Artificial levee south of Baton Rouge, Louisiana. Photo by H. J. Walker.

Figure 76. A portion of the shoreline of Toledo Bend Reservoir, Texas and Louisiana. Photo courtesy F. Beatty.

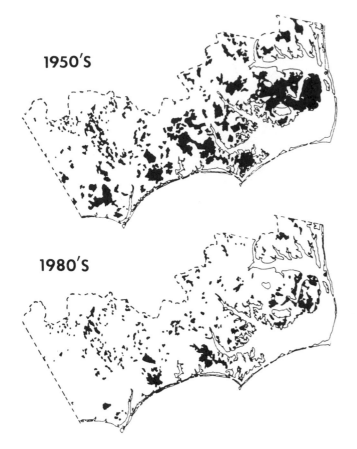

Figure 77. Comparison of the extent of natural or only slightly modified pocosins in North Carolina. From Tiner (1984, p. 50).

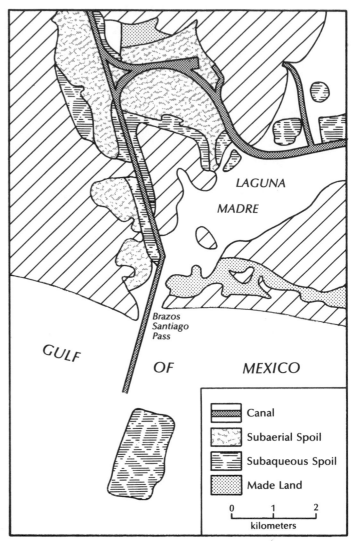

Figure 78. Canals and spoil deposition locations. Modified from Brown and others (1980).

amounts to over 500 km². In some areas, such as at Brownsville, Texas, there is more sediment supplied by dredging than by natural erosion (Brown and others, 1980, p. 24).

Pressures from many directions (economic, industrial, residential) on the coastal zone have led to extensive reclamation. Most reclamation involves the conversion of wetland into dry land by drainage and filling. However, recently the development of marinas, with their combination of water and dry land, has brought a new form to the shoreline. Recreation has been a major factor in the modification of many coastal areas. Estero Island, Florida, is an example of the nearly complete conversion from the natural to the cultural state (Fig. 79). Recreation is also largely responsible for the conversion of many coastal dunes into dwelling platforms (Fig. 80) and for upsetting a delicate balance that characterizes many dunes.

The susceptibility of much of the coast to erosion has led to the building of protective structures along the shoreline. Seawalls (Fig. 81) now border much of the shoreline, especially in populated locations such as Atlantic City, New Jersey. In addition to seawalls, groins and jetties are numerous. These structures alter coastal circulation patterns and affect sediment transport and deposition; many aggravate the problems they were designed to solve.

Of major concern in many coastal locations is subsidence. Although subsidence is a natural process, it can be aggravated by man. Human-induced subsidence, mainly through the withdrawal of groundwater from subsurface aquifers, has been responsible for the submergence of low-lying areas such as Bay Town, Texas.

The Suboceanic Division. The Continental Shelf and Continental Slope are the parts of the province least affected by man. Direct modification is only of recent origin. It is true that sedimentation rates have changed because of the modification of the rivers that discharge into the sea. For example, the levees and jetties of the Mississippi River help direct the sediment across the shelf and onto the slope.

Direct modification of the Continental Shelf began with offshore drilling in connection with the oil and gas industry. Such activity has, at least to date, been mainly concentrated off the

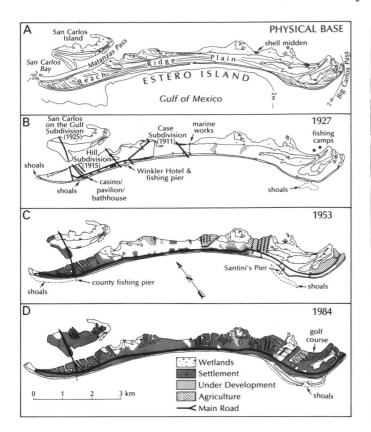

Figure 79. Estero Island, Florida, illustrating the development of a coastal recreational area. Maps supplied by K. Meyer-Arendt.

Figure 80. Sand dune along the coast of New Jersey with sand fence and planted grass. Photo by H. J. Walker.

Figure 81. Sea wall at Galveston, Texas. Photo by H. J. Walker.

Louisiana and Texas coasts and has been responsible for the construction of numerous platforms and the laying of the oil and gas pipelines that connect the wells to shore (Fig. 82).

FUTURE NEEDS

The suboceanic, transitional, and subaerial provinces of the Atlantic and Gulf Coastal Plain have been the site of numerous investigations. Much of the early morphologic work in the United States was conducted along the Atlantic Coast, followed by studies in the Gulf Coastal Plain, especially the alluvial valley system of the Mississippi River. Shoreline environments, including estuaries, have also received intensive study, particularly along the Atlantic Coast. Coastal terraces, for example, have attracted the attention of numerous workers throughout the twentieth century. Early investigators tended to concentrate on the processes responsible for terrace formation and their response to changing Quaternary sea levels. The suboceanic province was one of the least-studied areas; not until the petroleum industry began drilling offshore on the Continental Shelf in the Gulf of Mexico did any significant information concerning processes, subaqueous landforms, and time relationships become available. Off the east coast

such information has only been acquired within the past few years.

Although a considerable amount of literature has been published on the Coastal Plain, there remains a considerable amount of future work to be completed. Some of the more important areas of investigation include the following topics.

Subaerial Province

1. Regional analysis of landform distribution and its response to Quaternary events such as sea-level fluctuation and neo-tectonics. Although certain areas of the subaerial Coastal Plain have been intensively studied, there is a need to have a regional analysis completed based on similar types of landform analysis. With the advent of satellite photography, this type of study could be completed with less field checking than was

needed in the past. Such a compilation could have a significant impact on the future utilization of this important province.

2. As the drainage basins of the numerous rivers that feed onto the Coastal Plain became more populated and modified by man, significant changes occurred in the natural hydraulics of these rivers. Channel migration, frequency of flooding, sediment load, water quality, and navigability have been modified from their natural state. Specific studies, many of them local in nature, have been completed. There remains, however, a need for a systematic study of man's effect on these critical arteries of the Coastal Plain.

3. Alluvial channels and their morphologic responses have been studied in some detail, but details are still lacking in correlating channel response to specific processes. For example, bedload volumes and characteristics are poorly known for many of the coastal streams. Channel response and timing to changing base levels associated with variations in sea level, both in the past and in the future, are poorly known. With the exception of the early Corps of Engineers studies on the Mississippi Alluvial Valley, there has been no systematic study of the other major fluvial systems in the Coastal Plain Province.

4. Individual landforms within the subaerial province have received specific investigations, but several types of landforms have not had major investigations oriented toward their origins. Pimple mounds, generally present on the terraced surfaces of the coastal lowlands, have been the subject of numerous papers, but little intensive detailed study of this feature has been completed. Carolina bays, also the subject of numerous papers, have lacked a rigorous systematic study. Drainage valley incisement processes have also escaped detailed study, except in rather local and isolated areas.

Transitional Province

5. The shoreface regions along the Gulf and Atlantic margins have been studied in considerable detail within the past few years, but the response of these coastal features to long-term regional processes such as variations in sea level, isostatic uplift, and sediment loading has not been fully investigated. With the advent of modern computational methods such as three-dimensional modeling, significant studies are capable of being conducted if the proper scientific personnel are assembled. Quantitative geomorphic methods, common in some areas of study, have not been applied in any great detail to the coastal zone. In addition, the differences in response to short-term catastrophic stresses versus long-term steady state stresses have not been systematically evaluated.

6. With the realization that the shoreface and its adjacent coastal marshes and swamps are valuable resources, studies concerning variations in rates of landform change within this region could be significant for the future of many of the coastal states along the Atlantic and Gulf margins. Individual regions have been the sites of specific studies, but quantification of the effects of man as compared to changes resulting from natural processes

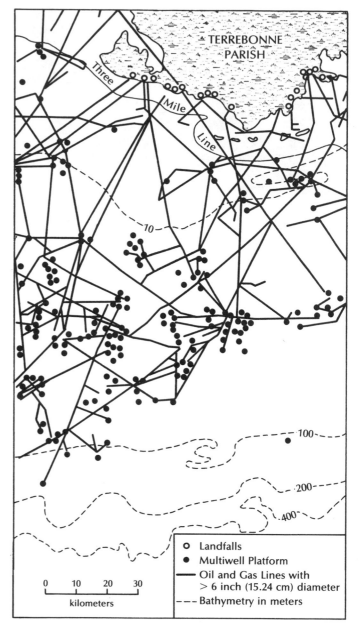

Figure 82. Platforms, and oil and gas pipelines south of Terrebonne Parish, Louisiana, in 1983. Oil and gas pipelines with a diameter of less than 15.24 cm (6 in) are not shown. From U.S. Department of the Interior (1983, visual 7).

have received only minimal attention. In many cases, this lack of quantification stems from our inability to establish an adequate natural base in the first place. For example, coastal erosion rates vary with many factors such as the rate of present-day sea-level rise, the rate of subsidence, and the emplacement of artificial structures. With the exception of a few localized areas, quantitative subsidence rates are poorly known. Separating relative sea-level rise from eustatic sea-level rise is a major problem that will require intensive study in the near future.

7. Sediment transport volumes in the nearshore are virtually unknown except in a few intensively studied areas along the extremely long shoreline of the Coastal Plain. Sediment sources, especially from adjacent shelf and complex estuaries, are poorly known.

Suboceanic Province

8. The evolution of the continental margin throughout the Quaternary has been studied in a few localities, especially along some portions of the Gulf of Mexico related to petroleum exploration and along some parts of the Atlantic margin where drilling by the U.S. Geological Survey and the Deep Sea Drilling Project have produced data. Other areas of the continental margin remain virtually unknown. Although the continental margin has had a history of continual basinward progradation since the Tertiary, details concerning the extremely variable rates, both in time and space, are still to be determined.

9. Subaqueous morphology of the continental margin is also poorly known. The presence of submarine canyons that crease this margin have been known since the early 1900s, but it has only been in the past few years that the detailed configuration of some of these complex canyons has been recorded. With the advent of high resolution geophysical and side-scan sonar techniques, the methodology exists to study these deep submarine landforms, but only small localized regions have been studied in any great detail to date. Sediment distribution along the outer Continental Shelf and upper Continental Slope is virtually unknown in a regional sense. Intensive sediment sampling is concentrated in only a few regions.

10. The Continental Shelf has been the site of numerous recent investigations, primarily associated with exploration for hydrocarbons. Although there have been several detailed studies, these are often confined to and associated with offshore leasing practices. These broad continental margins were the sites of subaerial coastal plains during the last low sea level, yet we know very little of the morphology that existed during these periods of exposure. The relationship between the relict topography that has escaped the erosive nature of the last sea-level rise and the landforms that were created during the rise in sea level are virtually unknown in a regional sense. During periods of low sea level, many coastal streams eroded channels across the modern shelf and large deltas were built out near the shelf margin. Even in an area such as the Gulf of Mexico, where petroleum exploration has been active since the 1950s, it has only been in the past few years that we have learned of the complexity of these drainage patterns that crossed the shelf during subaerial exposure. During periods of low sea levels and with a constantly changing climate, carbonate biomasses were more extensive than at present. The distribution, morphology, and biologic nature of these low sea-level reefs are poorly known.

11. In areas of rapid sediment loading, such as along the present continental margin of the northern Gulf Coastal Plain, rapid subsidence and salt- and shale-related diapirs severely deform the continental shelf sediments. Even with dense grids of drilling activity and seismic data acquisition, details concerning present-day rates of salt dome growth and variable subsidence rates are poorly known.

The major topics of concern outlined above are obviously not inclusive, but hopefully suffice to indicate that significant problems still exist, despite the fact that the Atlantic and Gulf Coastal Plain has been the site of a long period of geomorphic investigation. In a few decades, it is certain that another series of volumes will be published outlining the state of knowledge of this part of North America; we hope that some of the questions posed above will have been answered; certainly new problems will have been identified and the study of one of the most dynamic and interesting geomorphic areas will continue.

REFERENCES

Bakeless, J., 1961, The eyes of discovery: New York, Dover Publications, 439 p.

Baker, V. R., 1983, Late Pleistocene fluvial systems, *in* Wright, H. E., Jr., ed., Late-Quaternary environments of the United States: Minneapolis, University of Minnesota, v. 1, p. 115–129.

Barry, R. G., 1983, Late-Pleistocene climatology, *in* Wright, H. E., Jr., ed., Late-Quaternary environments of the United States: Minneapolis, University of Minnesota, v. 1, p. 390–407.

Belknap, D. F., and Kraft, J. C., 1981, Preservation potential of transgressive coastal lithosomes on the U.S. Atlantic Shelf: Marine Geology, v. 42, p. 429–442.

Bernard, H. A., and LeBlanc, R. J., 1965, Resume of the Quaternary geology of the northwestern Gulf of Mexico province, *in* Wright, H. E., Jr., and Frey, D. G., eds., The Quaternary of the United States: Princeton, Princeton University Press, p. 137–185.

Bloom, A. L., 1983a, Sea level and coastal changes, *in* Wright, H. E., Jr., ed., Late-Quaternary environments of the United States: Minneapolis, University of Minnesota, v. 2, p. 42–51.

—— , 1983b, Sea level and coastal morphology of the United States through the late Wisconsin glacial maximum, *in* Wright, H. E., Jr., ed., Late-Quaternary environments of the United States: Minneapolis, University of Minnesota,

v. 1, p. 215–229.

Bouma, A. H., Moore, G. T., and Coleman, J. M., eds., 1978, Framework, facies, and oil-trapping characteristics of the upper continental margin: American Association Petroleum Geologists, Studies in Geology no. 7, 326 p.

Bouma, A. H., Smith, L. B., Sidner, B. R., and McKee, T. R., 1975, Submarine geomorphology and sedimentation patterns of the Gyre intraslope basin, northwest Gulf of Mexico: Texas A&M University, Dept. of Oceanography Technical Report 75-9-T, 163 p.

Brown, L. F., Jr., Brewton, J. L., McGowen, J. H., Evans, T. J., Fisher, W. L., and Groat, C. G., 1976, Environmental geologic atlas of the Texas coastal zone; Corpus Christi area: Austin, The University of Texas, Bureau of Economic Geology, 123 p.

Brown, L. F., Jr., McGowen, J. H., Evans, T. J., Groat, C. G., and Fisher, W. L., 1977, Environmental geologic atlas of the Texas coastal zone; Kingsville area: Austin, The University of Texas, Bureau of Economic Geology, 131 p.

Brown, L. F., Jr., Brewton, J. L., Evans, T. J., McGowen, J. H., White, W. A., Groat, C. G., and Fisher, W. L., 1980, Environmental geologic atlas of the Texas coastal zone; Brownsville-Harlingen area: Austin, The University of Texas, Bureau of Economic Geology, 140 p.

Brown, P. J., 1976, Variations in South Carolina coastal geomorphology, *in*

Hayes, M. O., and Kana, T. W., eds., Terrigenous clastic depositional environments: Columbia, University of South Carolina, Part II, p. 2–15.

Brown, R. H., 1948, Historical geography of the United States: New York, Harcourt, Brace and Company, 596 p.

Brunsden, D., and Kesel, R., 1973, Slope development on a Mississippi River bluff in historic time: Journal of Geology, v. 81, p. 576–597.

Burnett, A. W., and Schumm, S. A., 1983, Alluvial-river response to neotectonic deformation in Louisiana and Mississippi: Science, v. 222, p. 49–50.

Carlston, C. W., 1950, Pleistocene history of coastal Alabama: Geological Society of America Bulletin, v. 61, p. 1119–1130.

Cashman, K. V., Popenoe, P., and Chayes, D., 1981, Sidescan sonar depiction of slump features associated with diapirism on the Continental Slope off southeastern United States: American Association of Petroleum Geologists Bulletin, v. 65, p. 909.

Chorley, R. J., Dunn, A. J., and Beckinsale, R. P., 1964, The history of the study of landforms: London, Methuen and Company, v. I, 678 p.

Chorley, R. J., Schumm, S. A., and Sugden, D. E., 1984, Geomorphology: New York, Methuen, 605 p.

Clark, J. A., 1981, Comment on late Wisconsin and Holocene tectonic stability of the United States mid-Atlantic coastal region: Geology, v. 9, p. 438.

Coleman, J. M., 1976, Deltas; Processes of deposition and models for explanation: Champaign, Continuing Education Publishing Company, Incorporated, 102 p.

Coleman, J. M., and Prior, D. B., 1980, Deltaic sand bodies, Continuing Education Course Note Series: American Association of Petroleum Geologists, no. 15, 171 p.

—— , 1982, Deltaic environments of deposition: American Association of Petroleum Geologists Memoirs 31, p. 139–178.

Committee of Ground Failure Hazards, 1985, Reducing losses from landsliding in the United States: Washington, D.C., National Academy Press, 41 p.

Cooke, C. W., 1945, Geology of Florida: Tallahassee, Florida Geological Survey Bulletin no. 29, 339 p.

Cooper, A. W., 1974, Salt marshes, *in* Odum, H. T., Cooper, B. J., and McMahan, E. A., eds., Coastal Ecological Systems of the United States: Washington, D.C., The Conservation Foundation, p. 55–98.

Craighead, F. C., 1964, Land, mangroves, and hurricanes: Fairchild Tropical Gardens Bulletin, v. 19, no. 4, p. 5–32.

Crone, A. J., McKeown, F. A., Harding, S. T., Hamilton, R. M., Russ, D. P., and Zoback, M. D., 1985, Structure of the New Madrid seismic source zone in southwestern Missouri and northeastern Arkansas: Geology, v. 13, p. 547–550.

Cronin, T. M., 1981, Rates and possible causes of neotectonic vertical crustal movements of the submerged southeastern United States Atlantic Coastal Plain: Geological Society of America Bulletin, v. 92, p. 812–833.

Cronin, T. M., Szabo, B. J., Ager, T. A., Hazel, J. E., and Owens, J. P., 1981, Quaternary climates and sea levels of the U.S. Atlantic Coastal Plain: Science, v. 211, p. 233–240.

Cruickshank, H. G., 1961, John and William Bartram's America: New York, The American Museum of Natural History, 378 p.

Curtis, W. F., Culbertson, J. K., and Chase, E. B., 1973, Fluvial-sediment discharge to the oceans from the conterminous United States: U.S. Geological Survey Circular 670, 17 p.

Daniels, R. B., Gamble, E. E., and Wheeler, W. H., 1971, Stability of Coastal Plain surfaces: Southeastern Geology, v. 13, p. 61–75.

Denny, C. S., 1982, Geomorphology of New England: U.S. Geological Survey Professional Paper 1208, 18 p.

Doyle, L. J., and Holmes, C. W., 1985, Shallow structure, stratigraphy, and carbonate sedimentary processes of west Florida upper Continental Slope: American Association Petroleum Geologists Bulletin, v. 69, p. 1133–1144.

Doyle, L. J., and Sparks, T. N., 1980, Sediments of Mississippi, Alabama, and Florida (MAFLA) Continental Shelf: Journal of Sedimentary Petrology, v. 50, p. 905–916.

Duane, D. B., Field, M. E., Meisburger, E. P., Swift, D.J.P., and Williams, S. J., 1972, Linear shoals on the Atlantic inner Continental Shelf, Florida to Long

Island, *in* Swift, D.J.P., Duane, D. B., and Pilkey, O. H., eds., Shelf sediment transport: Stroudsburg, Dowden, Hutchinson and Ross, Incorporated, p. 447–498.

Egler, F. E., 1952, Southeast saline Everglades vegetation, Florida, and its management: Vegetation, v. 3, p. 213–265.

Emery, K. O., 1967, Estuaries and lagoons in relation to continental shelves, *in* Lauff, G. H., ed., Estuaries: Washington, D.C., American Association for the Advancement of Science Publication 83, p. 9–11.

Fenneman, N. M., 1916, Physiographic divisions of the United States: Annals of the Association of American Geographers, v. 6, p. 3–18.

Fenneman, N. M., 1938, Physiography of eastern United States: New York, McGraw-Hill Book Company, Incorporated, 691 p.

Fisher, J. J., 1982, Barrier islands, *in* Schwartz, M. L., ed., The encyclopedia of beaches and coastal environments: Stroudsburg, Hutchinson Ross Publishing Company, p. 124–133.

—— , 1985, Atlantic USA; North, *in* Bird, E.C.F., and Schwartz, M. L., eds., The world's coastline: New York, Van Nostrand Reinhold Company, p. 223–234.

Fisher, W. L., McGowen, J. H., Brown, L. F., Jr., and Groat, C. G., 1972, Environmental geologic atlas of the Texas coastal zone; Galveston-Houston area: Austin, The University of Texas, Bureau of Economic Geology, 91 p.

Fisk, H. N., 1947, Fine-grained alluvial deposits and their effects on Mississippi River activity: Vicksburg, U.S. Waterways Experiment Station, v. 1 and 2.

—— , 1952, Geological investigation of the Atchafalaya Basin and the problem of Mississippi River diversion: Waterways Experiment Station, U.S. Army Corps of Engineers, v. 1, 145 p.

Fisk, H. N., McFarlen, E., Jr., Kolb, C. R., and Wilbert, L. J., Jr., 1954, Sedimentary framework of the modern Mississippi Delta: Journal of Sedimentary Petrology, v. 24, p. 76–99.

Folger, D. W., Dillon, W. P., Grow, J. W., Klitgord, K. D., and Schlee, J. S., 1979, Evolution of the Atlantic continental margin of the United States, *in* Talwani, M., Hay, W., and Ryan, W.B.F., eds., Continental margins and paleoenvironments, Volume 3: American Geophysical Union, p. 87–108.

Frazier, D. E., 1967, Recent deltaic deposits of the Mississippi River, their development and chronology: Gulf Coast Association Geological Society Transactions, v. 17, p. 287–315.

Freeman-Lynde, R. P., 1983, Cretaceous and Tertiary samples dredged from the Florida escarpment, eastern Gulf of Mexico: Gulf Coast Association of Geological Societies Transactions, v. 33, p. 91–99.

Garrison, L. E., and Martin, R. G., 1973, Geologic structures in the Gulf of Mexico basin: U.S. Geological Survey Professional Paper 773, 85 p.

Geophysics of Estuaries Panel, 1977, Estuaries, geophysics, and the environment: Washington, D.C., National Academy of Sciences, 127 p.

Ginsburg, R. N., 1956, Environmental relationships of grain size and constituent particles in south Florida carbonate sediments: American Association of Petroleum Geologists Bulletin, v. 40, p. 2384–2427.

Gosselink, J. G., Cordes, C. L., and Parsons, J. W., 1979, An ecological characterization study of the Chenier Plain coastal ecosystem of Louisiana and Texas: Slidell, U.S. Fish and Wildlife Service, 305 p.

Grow, J. A., 1981, The Atlantic margin of the United States, *in* Bally, A. W., Watts, A. B., Grow, J. A., Manspeizer, W., Bermoulli, D., Schreiber, C., and Hunt, J. M., eds., Geology of passive continental margins: Tulsa, Association of American Petroleum Geologists, Part 3, p. 1–41.

Haag, W. G., 1978, A prehistory of the lower Mississippi River valley: Geoscience and Man, v. 19, p. 1–8.

Hayes, M. O., 1985, Atlantic USA; South, *in* Bird, E.C.F., and Schwartz, M. L., eds., The world's coastline: New York, Van Nostrand Reinhold Company, p. 207–211.

Hayes, M. O., and Kana, T. W., eds., 1976, Terrigenous clastic depositional environments; Some modern examples: Columbia, University of South Carolina, Technical Report No. 11-CRD, Part I, 131 p. and Part II, 171 p.

Hays, W. W., 1981, Facing geologic and hydrologic hazards; Earth-science considerations: U.S. Geological Survey Professional Paper 1240-B, 108 p.

Hine, A. C., 1983, Modern shallow water carbonate platform margins, *in* Cook,

H. E., Hine, A. C., and Mullins, H. T., eds., Platform margin and deep water carbonate: Tulsa, Society of Economic Paleontologists and Mineralogists Short Course No. 12, Part 3, p. 1–100.

Hollister, C. D., 1973, Atlantic Continental Shelf and Slope of the United States; Texture of surface sediments from New Jersey to southern Florida: U.S. Geological Survey Professional Paper 529-m, 23 p.

Holmes, C. W., 1985, Accretion of the south Florida platform, Late Quaternary development: American Association of Petroleum Geologists Bulletin, v. 69, p. 149–160.

Hoyt, J. H., and Hails, J. R., 1974, Pleistocene stratigraphy of southeastern Georgia, in Oaks, R. Q., and DuBar, J. R., eds., Post Miocene stratigraphy central and southern Atlantic Coastal Plain: Utah State University Press, p. 232–245.

Humphreys, A. A., and Abbot, H. L., 1861, Report on the physics and hydraulics of the Mississippi River Corps of Topographical Engineers: Washington, D.C., U.S. Government Printing Office, Professional Paper no. 4, 691 p.

Humphries, C. C., Jr., 1978, Salt movement on Continental Slope, northern Gulf of Mexico, in Bouma, A. H., Moore, G. T., and Coleman, J. M., eds., Framework, facies, and oil-trapping characteristics of the upper continental margin: American Association Petroleum Geologists, Studies in Geology no. 7, p. 69–85.

Hunt, C. B., 1967, Physiography of the United States: San Francisco, W. H. Freeman and Company, 480 p.

Jackson, M.P.A., and Talbot, C. J., 1986, External shapes, strain rates, and dynamics of salt structures: Geological Society of America Bulletin, v. 97, p. 305–323.

Johnson, D. W., 1942, The origin of the Carolina bays: New York, Columbia University Press, 341 p.

Kaczorowski, R. T., 1976, Origin of Carolina bays, in Hayes, M. O., and Kana, T. W., eds., Terrigenous clastic depositional environments: Columbia, University of South Carolina, Coastal Research Division, Technical Report no. II-CRD, p. II-16–II-36.

Kenyon, P. M., and Turcotte, D. L., 1985, Morphology of a delta prograding by bulk sediment transport: Geological Society of America Bulletin, v. 96, p. 1457–1465.

King, P. B., 1977, The evolution of North America: Princeton, Princeton University Press, 197 p.

King, P. B., and Beikman, H. M., 1974, Explanatory text to accompany the geologic map of the United States: U.S. Geological Survey Professional Paper 901, 40 p.

Knebel, H. J., 1981, Processes controlling the characteristics of the surficial sand sheet, U.S. Atlantic outer Continental Shelf: Marine Geology, v. 1–4, p. 349–368.

Knebel, H. J., 1984, Sedimentary processes on the Atlantic Continental Slope of the United States: Marine Geology, v. 61, p. 453–74.

Knebel, H. J., Martin, E. A., Glenn, J. L., and Weedelle, S. W., 1981, Sedimentary framework of the Potomac River estuary, Maryland: Geological Society of America Bulletin, v. 92, p. 578–589.

Kniffen, F. B., 1968, Louisiana, its land and people: Baton Rouge, Louisiana State University Press, 196 p.

Kolb, C. R., and Van Lopik, J. R., 1966, Depositional environments of the Mississippi River deltaic plain, southeastern Louisiana, in Shirley, M. L., and Ragsdale, J. A., eds., Deltas in their geologic framework: Houston, Houston Geological Society, p. 17–61.

Kraft, J. C., 1978, Coastal stratigraphic sequences, in Davis, R. A., ed., Coastal sedimentary environments: New York, Springer-Verlag, p. 361–683.

—— , 1985, Atlantic USA; Central, in Bird, E.C.F., and Schwartz, M. L., eds., The world's coastline: New York, Van Nostrand Reinhold Company, p. 213–219.

Kraft, J. C., and Chrzastowski, M. J., 1985, Coastal stratigraphic sequences, in Davis, R. A., ed., Coastal sedimentary environments (second revised, expanded edition): New York, Springer-Verlag, p. 625–663.

Leatherman, S. P., 1979, Barrier island handbook: College Park, University of Maryland, 109 p.

Loyacano, H. A., Jr., and Smith, J. P., 1979, Symposium on the natural resources of the Mobile Estuary: Alabama, U.S. Army Corps of Engineers, 290 p.

Ludwick, J. C., 1964, Sediments in northeastern Gulf of Mexico, in Miller, R. L., ed., Papers in marine geology, Shepard commemorative volume: New York, The Macmillan Company, p. 204–238.

Lyell, Sir C., 1856, Principles of geology (sixth edition): New York, D. Appleton and Company, 816 p.

Maher, J. C., and Applin, E. P., 1971, Geologic framework and petroleum potential of the Atlantic Coastal Plain and Continental Shelf: U.S. Geological Survey Professional Paper 659, 98 p.

Markewich, H. W., 1984, Geomorphic evidence for Pliocene-Pleistocene uplift in the area of the Cape Fear Arch, North Carolina, in Morisawa, M., and Hack, J. T., eds., Tectonic geomorphology: Boston, Allen and Unwin, p. 279–297.

Martin, R. G., 1984, Diapiric trends in the deepwater Gulf Basin: Austin, Gulf Coast Section of Society of Economic Paleontologists and Mineralogists Foundation Research Conference, p. 60–62.

Martin, R. G., and Bouma, A. H., 1978, Physiography of Gulf of Mexico, in Bouma, A. H., Moore, G. T., and Coleman, J. M., eds., Framework, facies, and oil-trapping characteristics of the upper continental margin: American Association of Petroleum Geologists, Studies in Geology no. 7, p. 3–19.

Mather, K. F., and Mason, J. L., 1970, A source book in geology: Cambridge, Harvard University Press, 702 p.

Mathews, T. D., Stapor, F. W., Jr., Richter, C. R., Miglarese, J. V., McKenzie, M. D., and Barclay, L. A., 1980, Ecological characterization of the sea island coastal region of south Carolina and Georgia: U.S. Fish and Wildlife Service, Department of the Interior, v. I, 321 p.

Mazzullo, J., 1986, Sources and provinces of late Quaternary sand on the east Texas–Louisiana Continental Shelf: Geological Society of America Bulletin, v. 97, p. 638–647.

McGregor, B. A., 1984, The submerged continental margin: American Scientist, v. 72, p. 275–281.

McGregor, B. A., Nelson, T. A., Stubblefield, W. L., and Merrill, G. F., 1984, The role of canyons in late Quaternary deposition of the United States Atlantic Continental Rise, in Stow, D.A.V., and Piper, D. J., eds., Fine-grained sediments; Deep-water processes and environments: Geological Society of London, Special Publication Number 15, p. 319–330.

Meade, R. H., and Trimble, S. W., 1974, Changes in sediment loads in rivers of the Atlantic drainage of the United States since 1900: International Association of Hydrological Sciences Publication no. 113, p. 99–104.

Miller, W. R., and Egler, F. E., 1950, Vegetation of the Wequetequock-Pawcatuck tidal marshes, Connecticut: Ecological Monographs, v. 20, p. 143–192.

Mixon, R. B., 1985, Stratigraphic and geomorphic framework of uppermost Cenozoic deposits in the southern Delmarva Peninsula, Virginia and Maryland: U.S. Geological Survey Professional Paper 1067-G, 53 p.

Mixon, R. B., and Newell, W. L., 1977, Stafford fault system; Structures documenting Cretaceous and Tertiary deformation along the Fall Line in northeastern Virginia: Geology, v. 5, p. 437–440.

Morgan, D. J., 1977, The Mississippi River Delta: Geoscience and Man, v. 16, 196 p.

Morgan, J. P., 1970, Deltas; A résumé: Journal of Geological Education, v. 18, p. 107–117.

Mörner, N-A., 1983, Sea levels, in Gardner, R., and Scoging, H., eds., Mega-Geomorphology: Oxford, Clarendon Press, p. 73–91.

Murray, G. E., 1961, Geology of the Atlantic and Gulf Coastal Province of North America: New York, Harper and Brothers, 692 p.

Neitzel, R. S., 1981, A suggested technique for measuring landscape changes through archaeology: Geoscience and Man, v. 22, p. 57–70.

Obermeier, S. F., Gohn, G. S. Weems, R. E., Gelinas, R. L., and Rubin, M., 1985, Geologic evidence for recurrent moderate to large earthquakes near Charleston, South Carolina: Science, v. 227, p. 408–410.

Opdyke, N. D., Spangler, D. P., Smith, D. L., Jones, D. S., and Lindquist, R. C., 1984, Origin of the epeirogenic uplift of Pliocene–Pleistocene beach ridges in Florida and development of the Florida karst: Geology, v. 12, p. 226–228.

Pilkey, O. H., and Evans, M., 1981, Rising sea, shifting shores, *in* Jackson, T. C., and Reische, D., eds., Coast alert; Scientists speak out: San Francisco, Friends of the Earth, p. 13–47.

Pepenoe, P., Cashman, K. V., and Coward, E. L., 1981, Environmental considerations for OCS development, lease sale 78 call area: U.S. Geological Survey Open-File Report 81-749, p. 59–108.

Price, R. C., and Whetstone, K. N., 1976, Lateral stream migration as evidence for regional geologic structures in the eastern Gulf Coastal Plain: Southeastern Geology, v. 18, p. 129–147.

Prior, D. B., and Doyle, E. H., 1985, Intra-slope canyon morphology and its modification by rock-fall processes: U.S. Atlantic continental margin: Marine Geology, v. 67, p. 177–196.

Pritchard, D. W., 1967, What is an estuary; Physical viewpoint, *in* Lauff, G. H., ed., Estuaries: American Association for the Advancement of Science Publication, v. 83, p. 3–5.

Psuty, N. P., 1986, Holocene sea level in New Jersey: Physical Geography, v. 7, p. 154–165.

Radburch-Hall, D. H., Colton, R. B., Davies, W. E., Lucchitta, I., Skipp, B. A., and Varnes, D. J., 1982, Landslide overview map of the conterminous United States; U.S. Geological Survey Professional Paper 1183, 25 p.

Revelle, R. R., 1983, Probable future changes in sea level resulting from increased atmospheric carbon dioxide, *in* National Academy of Science, Changing climate: Washington, D.C., National Academy Press, p. 433–448.

Ringold, P. L., and Clark, J., 1980, The coastal almanac for 1980; The year of the coast: San Francisco, The Conservation Foundation, W. H. Freeman and Company, 172 p.

Robb, J. M., Hampson, J. C., Jr., Kirby, J. R., and Twichell, D. C., 1981, Geology and potential hazards of the Continental Slope between Linkenkohl and South Toms Canyon, offshore mid-Atlantic United States: U.S. Geological Survey Open-File Report 81-600, 33 p.

Robb, J. M., Hampson, J. C., Jr., Kirby, J. R., Gibson, P. C., and Hecker, B., 1983, Furrowed outcrops of Eocene chalk on the lower Continental Slope offshore New Jersey: Geology, v. 11, p. 182–186.

Ruhe, R. V., 1983, Depositional environment of late Wisconsin loess in the midcontinental United States, *in* Wright, H. E., Jr., ed., Late-Quaternary environments of the United States: Minneapolis, University of Minnesota, v. 1, p. 130–137.

Russell, R. J., 1942, Flotant: The Geographical Review, v. 32, p. 74–98.

——, 1967, Origins of estuaries, *in* Lauff, G. H., ed., Estuaries: American Association for the Advancement of Science Publication no. 83, p. 93–99.

Russell, R. J., and Howe, H. V., 1935, Cheniers of southwestern Louisiana: Geographical Review, v. 25, p. 449–461.

Saucier, R. T., 1974, Quaternary geology of the lower Mississippi valley: Arkansas Archaeological Survey Research Series no. 6, 25 p.

Schwartz, M. L., 1971, The multiple causality of barrier islands: Journal of Geology, v. 79, p. 91–94.

——, ed., 1973, Barrier islands: Stroudsburg, Dowden, Hutchinson and Ross, 451 p.

Sever, T., and Wiseman, J., 1985, Remote sensing and archaeology; Potential for the future: National Aeronautics and Space Administration, 76 p.

Shepard, F. P., 1982, North America, coastal morphology, *in* Schwartz, M. L., ed., The encyclopedia of beaches and coastal environments: Stroudsburg, Hutchinson Ross Publishing Company, p. 593–602.

Sternberg, H. O., and Russell, R. J., 1952, Fracture patterns in the Amazon and Mississippi Valleys, *in* Proceedings, XVIIth Congress: Washington, D.C., International Geographical Union, p. 380–385.

Suter, J. R., and Berryhill, H. L., Jr., 1985, Late Quaternary shelf-margin deltas, northwest Gulf of Mexico: American Association Petroleum Geologists Bulletin, v. 69, p. 77–91.

Swift, D.J.P., 1975, Barrier island genesis; Evidence from the central Atlantic Shelf, eastern U.S.A.: Sedimentary Geology, v. 14, p. 1–43.

——, 1976, Coastal sedimentation, *in* Stanley, D. J., and Swift, D.J.P., eds., Marine sediment transport and environmental management: New York, John Wiley and Sons, p. 255–310.

Talwani, P., and Cox, J., 1985, Paleoseismic evidence for recurrence of earthquakes near Charleston, South Carolina: Science, v. 229, p. 379–381.

Tanner, W. F., 1985, Florida, *in* Bird, E.C.F., and Schwartz, M. L., eds., The world's coastline: New York, Van Nostrand Reinhold Company, 163–167.

Thom, B. G., 1967, Coastal and fluvial landforms: Horry and Marion counties, South Carolina: Baton Rouge, Coastal Studies Institute, 75 p.

——, 1970, Carolina bays in Horry and Marion counties, South Carolina: Geological Society of America Bulletin, v. 81, p. 783–814.

Thornbury, W. D., 1965, Regional geomorphology of the United States: New York, John Wiley and Sons, 609 p.

Tiner, R. W., Jr., 1984, Wetlands of the United States; Current status and recent trends: Washington, D.C., U.S. Government Printing Office, 59 p.

Titus, J. G., 1985, How to estimate future sea level rise in particular communities: Environmental Protection Agency, Mimeograph Table, 1 p.

Tuomey, M., 1848, Report on the geology of South Carolina: Columbia, South Carolina, A. S. Johnston, 293 p.

Twichell, D. C., Grimes, C. B., Jones, R. S., and Able, K. W., 1985, The role of erosion by fish in shaping topography around Hudson submarine canyon: Journal of Sedimentary Petrology, v. 55, p. 712–719.

U.S. Army Corps of Engineers, 1973, National shoreline study, (5 volumes): Washington, D.C., U.S. Government Printing Office.

U.S. Army Coastal Engineering Research Center, 1977, Shore protection manual (3 volumes): Fort Belvair, Virginia.

U.S. Department of Commerce, National Oceanic and Atmospheric Administration, 1975, The coastline of the United States map: Washington, D.C., scale 1:3,168,000.

U.S. Department of the Interior, 1978, Report of the Barrier Island Work Group: Heritage Conservation and Recreation Service, Washington, D.C., 80 p.

U.S. Department of the Interior, 1983, Environmental impact statement, proposed oil and gas lease sales 104 and 105: Mineral Management Service, 636 p.

U.S. Geological Survey, 1965, Base map of the United States: Washington, D.C., scale 1:3,168,000.

U.S. Geological Survey, 1970, The national atlas: Washington, D.C., 417 p.

Walker, H. J., and Prior, D. B., 1986, Estuarine environments, *in* Fookes, P. G., and Vaughan, P. R., eds., A handbook of engineering geomorphology: New York, Chapman and Hall, p. 180–192.

Webb, C. H., 1982, The Poverty Point culture: Geoscience and Man, School of Geoscience, Louisiana State University, v. 17, 85 p.

White, G. W., 1953, Early American geology: The Scientific Monthly, v. 76, p. 134–141.

White, W. A., 1970, The geomorphology of the Florida Peninsula: Florida Geological Survey Bulletin 51, 164 p.

Windham, S. R., and Campbell, K. M., 1981, Sinkholes follow the pattern: Geotimes, v. 26, p. 20–21.

Winker, C. D., and Howard, J. D., 1977, Correlation of tectonically deformed shorelines on the southern Atlantic Coastal Plain: Geology, v. 5, p. 123–127.

Worzel, J. L., and Burke, C. A., 1978, Margins of Gulf of Mexico: American Association of Petroleum Geologists Bulletin, v. 62, p. 2290–2307.

MANUSCRIPT ACCEPTED BY THE SOCIETY SEPTEMBER 19, 1986

Geological Society of America
Centennial Special Volume 2
1987

Chapter 4

*Central Lowlands**

David M. Mickelson
Department of Geology and Geophysics, University of Wisconsin at Madison, Madison, Wisconsin 53706

STUDIES BEFORE 1960

"The craton was the focal point for the development of glacial geology in North America." So begins a summary of nineteenth century contributions to the study of glacial geology on the North American craton by Totten and White (1985). In that paper they trace the development of early ideas about the origin of till from its association with drift ice forward in time to our present concept of till being deposited by an ice sheet that moved into the Central Lowland from the north, producing many of the landforms we see today. Familiar names such as Hitchcock, Whittlesey, Newberry, Chamberlin, Gilbert, and many others attest to the importance of studies of glacial deposits in the Central Lowland in the development of early ideas in glacial geomorphology.

In this chapter I discuss only the development of glacial geomorphological concepts in the glaciated region. Because of space considerations, the very important developments in stratigraphic interpretation, dating, and Quaternary history of the Central Lowland are not covered. Because Totten and White (1985) give a detailed discussion of the development of ideas concerning the glacial geology of the North American craton, I will only briefly summarize contributions of the nineteenth century before discussing the more recent development of ideas in more detail. Much of the following discussion is summarized from Totten and White (1985).

The effectiveness of glacial erosion was widely debated in the latter half of the nineteenth century. Newberry (1862, 1870) and Whittlesey (1866) were instrumental in the development of the concept that glaciers were responsible for the enlargement of river valleys to form the Great Lakes basins and the Finger Lakes of New York. In addition, it was recognized that numerous smaller lakes exist in depressions carved by the debris-laden ice. T. C. Chamberlin, a major contributor to concepts of glacial geology in the midcontinent region, also believed that flowing ice had created the Finger Lakes, the Great Lakes, and numerous

grooves and striations found across the Central Lowland (Chamberlin, 1883, 1888). Others, including Spencer (1890), Davis (1882), and Wright (1891, 1892), argued in favor of a combination of tectonic movement and fluvial valley development as the cause of the Great Lakes basins. By the turn of the century the concept that ice sheets substantially erode their beds was accepted (Leverett, 1899, 1901) and became common knowledge among geologists working in the Central Lowland (Totten and White, 1985).

T. C. Chamberlin and Frank Leverett, Chamberlin's successor in the U.S. Geological Survey, made major contributions to our knowledge of the Quaternary history of the central lowland as well as the development of landforms. Besides developing the concept of multiple glaciation in the midcontinent region, naming most of the glacial and interglacial stages (Totten and White, 1985), Chamberlin and Leverett made major contributions to concepts of landform development. The extensive program announced by Chamberlin in the early 1800s to map "the terminal moraine" from the Atlantic Ocean westward to the Rockies (Totten and White, 1985) was instrumental in allowing Chamberlin, Leverett, and numerous other workers to make detailed maps and observations of glacial landforms and glacial materials. Observations on the pattern of moraines and the lobate nature of the southern part of the ice sheet, the origin of end moraines, interlobate moraines, kettle holes, kames, eskers, and other landforms all developed at this time. Although ideas on the origin of many of these features came from Europe or other parts of the United States, certainly the mapping program begun by Chamberlin in the 1880s had a major impact on the development of geomorphic ideas in the Central Lowland.

By the turn of the century, topographic maps were becoming available for many parts of the Central Lowland (White, 1973). Classic end moraines of the Midwest were mapped by Leverett and his associates at the U.S. Geological Survey, and Leverett's monographs on the Illinois Glacial Lobe (1899), the glacial and drainage features of the Erie and Ohio basins (1901), the moraines and shorelines of the Lake Superior region (1929), and one with F. B. Taylor on the Pleistocene of Indiana and

---

*Additional material planned for this chapter, covering landscape evolution and stream processes in the upper Mississippi River basin, was not available at publication time.

Mickelson, D. M., 1987, Central Lowlands, *in* Graf, W. L., ed., Geomorphic systems of North America: Boulder, Colorado, Geological Society of America, Centennial Special Volume 2.

Michigan and the history of the Great Lakes (1915) are an amazing contribution to our knowledge of the areal glacial geology in the Central Lowland.

From that time on, most of the glacial studies provided elaboration of concepts developed earlier and refinement of mapping as more and more topographic maps became available (White, 1973). Interpretation of landforms took another major leap forward as air photos began to be used just prior to 1940. With the photos, much more detailed mapping of soils and small landforms could be done, and there was further refinement of the relative age and significance of moraines. Still, the concept of each moraine marking a readvance and being associated with a single till unit dominated thinking in the Midwest.

MORE RECENT CONCEPTS

Although some studies relating subsurface geology to groundwater supply were done soon after World War II, especially in the dry western areas, White (1973) placed the real beginnings of stratigraphic and subsurface studies at about 1960. At this time, large excavations for buildings and major highways were being made, better drill rigs became available, and geophysics provided much more information on the internal composition of various landforms. In particular, the significance of end moraines, each of which had been considered by Leverett and others to represent at least a minor glacial advance of the last glaciation, was questioned. Totten (1969) suggested that some moraines were overridden features of earlier advances. The concept of palimpsest landscapes became more widely recognized. Studies by White (1974) and his students in Ohio, and studies in Illinois at about the same time, demonstrate that a number of the moraines mapped by Leverett, refined by later workers, and assumed to be ice-front positions of a retreating ice sheet, can be demonstrated to be palimpsest.

In the 1960s, and in particular in the 1970s, glacial geologists in the Central Lowland began to think more critically about the genesis of glacial landforms, and they began to study modern glaciers to an extent not known since Chamberlin. Along the west edge of the Central Lowland and in the Great Plains to the west, deposits of hummocky till and flowed sediment are components of a landscape that is very different than the classic glacial landscape of end moraines further east. Gravenor and Kupsch (1959), Clayton and Freers (1967), Stalker (1960), and Parizek (1969) are among the important contributions to the interpretation of this landscape. Based on detailed descriptions of landforms from air photos, combined with information on stratigraphy and the age of logs buried beneath the surface, it became apparent that extensive areas of hummocky topography resulted from very thick accumulations of supraglacial sediment. Along the edges of the Missouri Coteau, intense compressive flow accelerated the movement of debris toward the ice surface and the accumulation of a thick sediment cover. The hummocky topography formed slowly as buried ice took perhaps thousands of years to melt out.

In addition to the observations of Clayton (1966) on the development of hummocky topography on stagnant modern glaciers, Boulton (1970a, 1970b, 1971, and later papers), and somewhat later work by Lawson (1979a, 1979b, 1981), and others working on modern glaciers had a major influence on the direction of research in the Central Lowland. It was clear by then that without detailed study of the sedimentology of deposits, an understanding of the genesis of landforms or till and till-like sediment is impossible. Detailed studies of single till units, particularly in a single moraine or in a small area (e.g., Kemmis, 1981; Wickham and Johnson, 1981; Johnson and others, 1982), were attempts to adapt the methodology and criteria for recognition of different sediment types developed around modern glaciers. Since the early 1970s, the INQUA Commission on the Genesis and Lithology of Quaternary Deposits exerted an influence on the terminology used to describe glacial deposits and the interpretation of glacial deposits that were studied in the Central Lowland.

Another important development during the 1970s was the attempt to reconstruct subglacial and periglacial conditions during Pleistocene time in the Central Lowland. Using observations by glaciologists on modern ice sheets, glacial geologists working in the Central Lowland attempted to deduce the character of the ice sheet from geomorphic features. Wright (1973) combined a knowledge of periglacial conditions based on the pollen record and stratigraphy, with observations of large valleys in the area covered by the Superior Lobe. These tunnel valleys, he concluded, formed during catastrophic flows of water from beneath the Superior Lobe. He hypothesized a zone around the margin of the ice where the glacier was frozen to its bed and, under thicker ice, a zone where the bed was at the pressure melting point and sliding. The concept of these zones beneath the ice sheet developed in several parts of the Central Lowland at about the same time, and has been included in many more recent papers to explain the distribution of glacial landforms.

A glacial process-form model for the western part of the Central Lowland was developed by Clayton and Moran (1974). In this model they identify four elements: preadvance, subglacial, supraglacial, and postglacial. These four elements combine with different relative importance in several zones behind the former ice margin. Much of their attempt to reconstruct glacier characteristics was concentrated on the temperature regime at the bed. They proposed in their model that there were three zones, a proglacial permafrost zone, a frozen-bed zone, and, under thicker ice, a thawed-bed zone. They also pointed out the effect of excess pore pressure on the movement of sediment beneath the ice, especially in the development of transverse compressional features on the bed. Because the model is based on observations in the western part of the Central Lowland, it concentrates on the importance of supraglacial deposits. The model is clearly applicable, however, to large parts of the Central Lowland. A similar model with a frozen bed zone was used by Whittecar and Mickelson (1977, 1979) and Stanford and Mickelson (1985) to help explain features of till deposition and erosion and deformation of sand and gravel in drumlins formed behind the margin.

In the late 1970s and early 1980s, actual ice-sheet models

were developed. Most of these attempt the reconstruction of the whole Laurentide ice sheet, and thus are of a different scale than studies of landform distribution within the Central Lowland. They have, however, had a great impact on the thinking of glacial geologists in the Central Lowland because they appear to place limits on ice thickness, bed conditions, and flow directions that would otherwise be unavailable.

Andrews (1982), in a review paper on the reconstruction of Pleistocene ice sheets, pointed out that there have been three separate approaches to ice-sheet reconstruction. The first of these is based on evidence in the geologic record such as crossing striations, till fabric, and the overall reconstruction of glacial history. These studies have had a considerable impact on our knowledge of the overall flow pattern of the ice sheet, and, in fact, because of models of this type it appears that much of the ice moving into the Central Lowland was derived not from an ice center over Hudson Bay as was thought by many workers since Tyrrell (1898), but that a major component of ice was derived from Labrador. Only in its northwestern parts was the Central Lowland covered by ice coming from a flow center in Keewatin, west of Hudson Bay (Shilts, 1980, 1982). Because of the lobate nature of the ice mass in its marginal areas, most flow directions within the Central Lowland were fairly well established before the recent provenance studies of Shilts and others.

An early attempt to use geological data and glaciological constraints to model ice-sheet morphology and flow in the central lowland was made by Hook and others (1976). They used flow reconstructions based on the distribution of shale particles in the till of the Des Moines lobe determined by Matsch (1972) to develop a model that would explain these observations. The model produced very low ice-surface profiles, nearly plug flow in the central part of the lobe, and areas of ice accumulation on the Des Moines lobe itself, with flow in all directions from an ice dome. The low ice-surface gradients are similar to the low ice-surface gradients estimated by Mathews (1974) along much of the west edge of the Laurentide ice sheet in the Central Lowland and the adjacent Great Plains.

One early model that was developed to compare basal thermal regimes with the large-scale distribution of landforms was that by Sugden (1977, 1978). Because of the scale and simplicity of the model, fine distinctions cannot be made within the Central Lowland; in general, however, it predicts a zone of basal melting out near the edge of the ice sheet, and a zone with a frozen bed 100 to 200 km back. Because the model was developed for one point in time, the time of the ice sheet maximum, it is difficult to relate results of the model with detailed distribution of landforms (Mickelson and others, 1983).

Glaciological models and reconstructions based on rebound or sea-level rise during postglacial time are based on a number of assumptions exclusive of the landform record of the ice sheet. Although primarily designed to further our understanding of the dynamics and chronology of the disappearance of the last ice sheet, a model by Denton and Hughes (1981) provides the most detail available on predicted conditions at the bed of the ice in the Central Lowland. The Denton and Hughes model (Fig. 1) predicts a wet bed in the Des Moines lobe of Iowa and across the southern Great Lakes area in Illinois, Indiana, and Ohio. To the north, a zone of freezing is predicted to exist across the Great Lakes and westward into Manitoba.

The development of ice-sheet models should probably be an iterative process. Thus, glacial geologists armed with some constraints on ice sheet conditions produced by glaciological models should now be attempting to interpret landforms and glacial materials in terms of subglacial conditions and ice dynamics.

Since Wright's 1973 paper suggesting surging of the Superior lobe, several papers have been published suggesting the possibility that the southern margin of the Laurentide ice sheet was dominated by surges. The Denton and Hughes (1981) glaciological model suggests surges or ice streams in the valleys of the Great Lakes, supporting field observations by glacial geologists. Evidence of surging in the western part of the Central Lowland is documented by Clayton and others (1985). Mickelson and others (1981) suggested that during the retreat of ice, rapid advances and retreats of ice in the Lake Michigan basin may have been caused by the presence of a floating ice shelf that was unstable because of the shape of the basin and because of large fluctuations in water depth.

This mechanism seems likely because of the rapid rates of advance and retreat necessary to fit the stratigraphic story presented by Acomb and others (1982). The possibility of ice shelves in the Great Lakes basins was also considered likely by Mayewski and others (1981). They suggested that the lowlands of the Great Lakes basins were occupied by ice streams during the latter part of late Wisconsin time. In particular they suggest that Port Huron and Two Rivers advances may have been surges triggered by minor climatic events or possibly changes in water level. A final advance into the northern edge of the central lowland at 9,900 B.P. (Hughes, 1978), out of the Lake Superior basin may also have been a surge.

Boulton and others (1985) used a variety of transverse and longitudinal features to document the pattern of deglaciation in the Scandinavian and Laurentide ice sheets. Again, the scale of the model is such that detailed reconstruction of glaciological conditions of the Central Lowland cannot be made. An important contribution of this paper, however, is the suggestion that parts of the ice sheet underlain by a moderately thick sediment layer, as opposed to bare bedrock, had a deforming bed (Boulton and Jones, 1979; Boulton and others, 1985). There is geophysical evidence that a deforming sediment bed is now present beneath ice streams leading to the Ross ice shelf (Alley and others, 1986), perhaps providing a modern analog for basal conditions in parts of the Laurentide ice sheet.

Observations of sediments along the edges of the Lake Michigan basin do not suggest that there was a deforming bed during much of late glacial time. Although debris-rich ice was certainly deforming, lake sediment present between thin till units is for the most part undeformed and is relatively uniform in thickness and distribution. However, in the southern parts of the

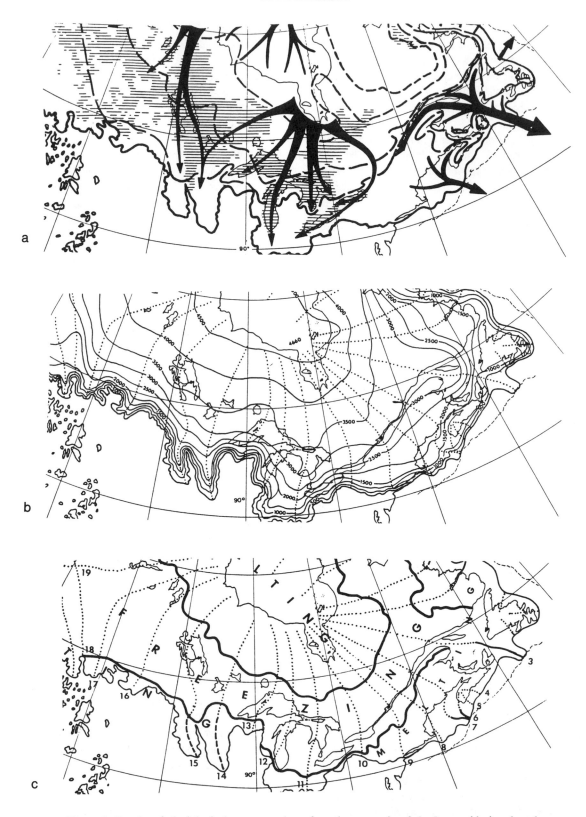

Figure 1. Results of glaciological reconstruction of southern margin of the Laurentide ice sheet by Denton and Hughes (1981). a, Location of ice streams (Denton and Hughes, Fig. 8-1); b, Ice thickness (m) for maximum late Wisconsin (Denton and Hughes, Fig. 6-15); c, Zones of melting and freezing (Denton and Hughes, 6-6). Reprinted by permission of the authors and John Wiley & Sons, Inc., from Denton, G., and Hughes, T., The Last Great Ice Age, copyright 1981, G. Denton.

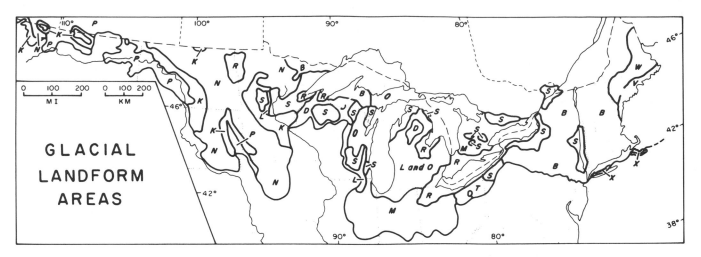

Figure 2. Glacial landform areas in the United States and part of southern Canada. From Mickelson and others, 1983. B, generally thin but locally thick till over bedrock topography and older drift; D, extensive pitted outwash separated by low till uplands; K, thick, clayey supraglacial till forming high-relief hummocky topography; L, thick, sandy, and loamy supraglacial till forming high-relief hummocky topography; M, active-ice moraine topography with moraines of low local relief and nearly flat till plain between composed mostly of subglacial till with thin superglacial till; N, thin supraglacial clayey or loamy till forming very low-relief hummocky topography, washboard, or minor moraines and drumlins; O, low-relief subglacial till with some drumlins and extensive pitted outwash; P, drumlinized thin till draped over older topography washboard moraines but with few end moraines; R, plains formed by glaciolacustrine sediment or with thin till on surface; S, drumlins and streamlined forms of subglacial till with only thin superglacial deposits; T, thin till over thick pre–late Wisconsin drift; V, glaciomarine sediment in lowlands surrounding till and bedrock hills and washboard moraines; W, low-relief till-and-outwash surface with very long esker systems; X, thin till over glaciotectonically thrust and stacked stratified sediments.

ice sheet, there is evidence of a deforming bed, probably indicating very wet conditions. Bleuer and others (1983) exhibit excellent examples of this at the Adams Mill locality in central Indiana. Bleuer has suggested that wet-bed, surging conditions occurred in Indiana during much of late Wisconsin time (Bleuer, 1983).

Mickelson and others (1983), in addition to outlining the chronology of events across the southern part of the Laurentide ice sheet, attempted to use the distribution of landforms as an indicator of glacier-bed conditions and of the dynamics of the last ice sheet (Fig. 2). The record left by the retreating ice in the southern and eastern part of the central lowland is one of flat or gently undulating till plain with classic end moraines (Frye and Willman, 1973). Although all of these moraines may not represent still stands of the retreating late Wisconsin ice margin, it is clear that many of them do. Unlike moraines in the northern part of the Central Lowland, the moraines in Illinois, Indiana, and Ohio appear to be composed mostly of basal till, and they have a very low-relief hummocky surface. Much of the sediment in these moraines may have been transported from the Great Lakes as a debris-rich basal layer without being moved up into the ice. It seems likely that the only englacial or supraglacial load in the ice is what appears now as a scatter of Precambrian boulders on

some of the moraine surfaces (Goldthwait, 1952; Goldthwait and Forsyth, 1955; Forsyth and Goldthwait, 1962). Similar landforms are present in the central part of the Des Moines lobe in Iowa. Mickelson and Clayton (1981) suggested that the ice sheet in this southern area had a wet bed and was sliding. A map indicating the likely zones of frozen and unfrozen bed based on glacial features is shown in Figure 3.

It appears that there was a zone of frozen bed near the ice margin in areas to the north (Fig. 3) before 13,000 B.P. Drumlins, tunnel channels, ice-thrust features, and high-relief hummocky moraine all indicate a different temperature regime than to the south. It seems likely that equilibrium conditions at the bed of the ice were never reached, and that ice advanced onto a permafrost surface. Permafrost then slowly decayed beneath the ice sheet, and large amounts of erosion and streamlining took place in a zone of drumlins behind the margin, where there was a patchy distribution of frozen and unfrozen bed.

Tunnel channels are very common along the outermost margin of the ice from Chicago westward through Minnesota, but they are generally absent to the south in Illinois, Indiana, and Ohio. As suggested by Wright (1973), these may represent catastrophic flows of water that were dammed behind a frozen-bed zone at the margin. In areas to the south, where the bed was

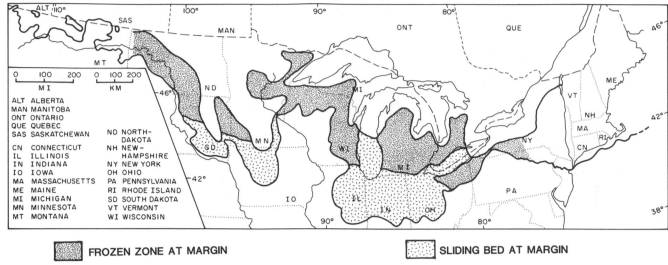

Figure 3. Areas of frozen and unfrozen bed based on landform distribution. Modified from Mickelson and Clayton, 1981.

thawed, it was possible for water to move in small tunnels or sheets along the bed instead of being concentrated into large catastrophic drainages.

The distribution of large thrust masses is probably also an indicator of bed conditions (Clayton and Moran, 1974; Moran and others, 1980; Bluemle and Clayton, 1984). It seems unlikely that these features can form without a frozen bed where the strength of the ice-sediment contact is stronger than contacts in the rock or sediment beneath and where pore builds up behind a permafrost zone near the margin.

FUTURE RESEARCH

It appears that we are on the threshold of another major surge of interest in glacial landform genesis in the Central Lowland and adjacent physiographic regions. New topographic maps at a scale of 1:24,000 enable much better interpretation of landforms than was possible earlier. Just as the availability of the first topographic maps around the turn of the century created a surge in the mapping and understanding of glacial landforms in the Central Lowland, the availability of these detailed maps should also allow much better resolution of landform distributions. Satel-

lite imagery may also aid in the understanding of landform distribution on a regional basis. Mathematical models are becoming increasingly sophisticated, and more complex ice-sheet models will demand more and more detail on the limits put on glaciological models by the landforms themselves.

Thus, it seems likely there will be an emphasis in the next ten years on the detailed interpretation and mapping of landforms and, therefore, the refinement of our understanding of the genesis of landforms and the behavior of the southern sector of the Laurentide ice sheet. Glacial geologists can provide boundary conditions for future glaciological models, and glaciologists can improve our understanding of landform genesis by putting limits on what physical conditions are possible.

Another area of future research that is critical to the understanding of glacial landform genesis in the Central Lowland is the understanding of the genesis of glacial sediments and landforms in the modern glacier environment. Only by careful observation of processes in the modern glacier environment can we make sound interpretations on the genesis of features in the Central Lowland. Likewise, although modern glaciers are not always good analogs, they should provide a testing ground for models of sediment deposition or landform genesis that are deduced from observations of Pleistocene deposits in the Central Lowland.

REFERENCES

Acomb, L. J., Mickelson, D. M., and Evenson, E. B., 1982, Till stratigraphy and late glacial events in the Lake Michigan lobe of eastern Wisconsin: Geological Society of America Bulletin, v. 93, p. 289–296.

Alley, R. B., Blankenship, D. D., Bentley, C. R., and Rooney, S. T., 1986, Deformation of till beneath ice stream B, West Antarctica: Nature, v. 322, no. 6074, p. 57–59.

Andrews, J. T., 1982, On the reconstruction of Pleistocene ice sheets: Quaternary

Science Reviews, v. 1, p. 1–30.

Bleuer, N. K., 1983, Load and slide deformation of Wisconsinan substrates in the central Indiana till plain: Geological Society of America Abstracts with Programs, v. 15, no. 4, p. 250.

Bleuer, N. K., Melhorn, W. N., and Pavey, R. R., eds., 1983, Interlobate stratigraphy of the Wabash Valley, Indiana: Midwest Friends of the Pleistocene, 30th Field Conference, 136 p.

Bluemle, J. P., and Clayton, L., 1984, Large scale glacial thrusting and related processes in North Dakota: Boreas, v. 13, p. 279–299.

Boulton, G. S., 1970a, On the origin and transport of englacial debris in Svalbard glaciers: Journal of Glaciology, v. 9, no. 56, p. 213–229.

—— , 1970b, The deposition of subglacial and meltout tills at the margins of certain Svalbard glaciers: Journal of Glaciology, v. 9, no. 56, p. 231–245.

—— , 1971, Till genesis and fabric in Svalbard, Spitsbergen, *in* Goldthwait, R. P., ed., Till: A symposium: Ohio State University Press, p. 41–72.

Boulton, G. S., and Jones, A. S., 1979, Stability of temperate ice caps and ice sheets resting on beds of deformable sediment: Journal of Glaciology, v. 24, p. 29–43.

Boulton, G. S., Smith, G. D., Jones, A. S., and Newsome, J., 1985, Glacial geology and glaciology of the last mid-latitude ice sheets: Journal of the Geological Society of London, v. 142, p. 447–474.

Chamberlin, T. C., 1883, Preliminary paper on the terminal moraine of the second glacial epoch: U.S. Geological Survey, Annual Report, v. 3, p. 17–21.

—— , 1888, The rock-scorings of the great ice invasions: U.S. Geological Survey, Annual Report, v. 7, p. 147–248.

Clayton, L., 1966, Karst topography on stagnant glaciers: Journal of Glaciology, v. 5, p. 107–112.

Clayton, L. and Freers, T. F., eds., 1967, Glacial geology of the Missouri Coteau and adjacent areas: North Dakota Geological Survey, Miscellaneous Series 30, 170 p.

Clayton, L. and Moran, S. R., 1974, A glacial process-form model, *in* Coates, D. R., ed., Glacial geomorphology: State University of New York at Binghamton, p. 89–119.

Clayton, L., Teller, J. T., and Attig, J. W., 1985, Surging of the southwestern part of the Laurentide ice sheet: Boreas, v. 15, p. 235–241.

Davis, W. M., 1982, On the classification of lake basins: Proceedings of the Boston Society of Natural History, v. 21, p. 315–381.

Denton, G. H., and Hughes, T. J., eds., 1981, The last great ice sheet: New York, John Wiley and Sons, 484 p.

Forsyth, J. L., and Goldthwait, R. P., 1962, Midwest Friends of the Pleistocene Field Guide: Friends of the Pleistocene, Midwest Section, Guidebook, 26 p.

Frye, J. C., and Willman, H. B., 1973, Wisconsinan climatic history interpreted from the Lake Michigan Lobe deposits and soils, *in* Black, R. F., Goldthwait, R. P., and Willman, H. B., eds., The Wisconsinan stage: Geological Society of America Memoir 136, p. 135–152.

Goldthwait, R. P., 1952, The 1952 field conference of the Friends of the Pleistocene, Midwest Section, Guidebook, 14 p.

Goldthwait, R. P., and Forsyth, J. L., 1955, Pleistocene chronology of southwestern Ohio: Indiana and Ohio Geological Surveys Fifth Biennial Field Conference Guidebook, p. 35–72.

Gravenor, C. P., and Kupsch, W. O., 1959, Ice disintegration features in western Canada: Journal of Geology, v. 67, p. 48–64.

Hook, R. L., Matsch, C. L., and Gasser, M. M., 1976, Implications of computer reconstructions of the Pleistocene Des Moines Lobe, Minnesota and Iowa: Geological Society of America Abstracts with Programs, v. 8, p. 924.

Hughes, J. D., 1978, A post Two Creeks buried forest in Michigan's northern peninsula: Proceedings, Institute on Lake Superior Geology, 5th, p. 16.

Johnson, M. D., Mickelson, D. M., Attig, J. W., Stanford, S. D., Boley-May, S., and Socha, B., 1982, Lodgement till, meltout till, and supraglacial debris; Can they be distinguished in sequences of Pleistocene age?: Geological Society of America Abstracts with Programs, v. 14, no. 5, p. 263.

Kemmis, T. J., 1981, Importance of the regulation process to certain properties of basal tills deposited by the Laurentide ice sheet in Iowa and Illinois, U.S.A.: Annals of Glaciology, v. 2, p. 147–152.

Lawson, D. E., 1979a, A comparison of the pebble orientations in ice and deposits of The Matanuska Glacier, Alaska: Journal of Geology, v. 87, p. 629–645.

—— , 1979b, Sedimentological analysis of the western terminus region of the Matanuska Glacier, Alaska: Cold Regions Research Lab Report 79-9, 122 p.

—— , 1981, Distinguishing characteristics of diamictons at the margin of the Matanuska Glacier, Alaska: Annals of Glaciology, v. 2, p. 78–84.

Leverett, F., 1899, The Illinois glacial lobe: U.S. Geological Survey Monograph 38, 818 p.

—— , 1901, Glacial formations and drainage features of the Erie and Ohio basins: U.S. Geological Survey Monograph 41, 802 p.

—— , 1929, Moraines and shorelines of the Lake Superior region: U.S. Geological Survey Professional Paper, 154, 72 p.

Leverett, F., and Taylor, F. B., 1915, The Pleistocene of Indiana and Michigan and the history of the Great Lakes: U.S. Geological Survey Monograph 53, 529 p.

Mathews, W. H., 1974, Surface profile of the Laurentide ice sheet in its marginal areas: Journal of Glaciology, v. 13, p. 37–43.

Matsch, C. L., 1972, Quaternary geology of southwestern Minnesota, *in* Sims, P. K., and Morey, G. B., eds., Geology of Minnesota; A centennial volume: Minnesota Geological Survey, p. 548–560.

Mayewski, P. A., Denton, G. H., and Hughes, T. J., 1981, Late Wisconsin ice sheets of North America, *in* Denton, G. H., and Hughes, T. J., eds., The last great ice sheets: New York, John Wiley and Sons, p. 67–179.

Mickelson, D. M., and Clayton, L., 1981, Subglacial conditions and processes during middle Woodfordian time in Wisconsin: Geological Society of America Abstracts with Programs, v. 13, p. 310.

Mickelson, D. M., Acomb, L. A., and Bentley, C., 1981, Possible mechanisms for the rapid advance and retreat of the Lake Michigan Lobe between 13,000 and 12,000 years B.P.: Annals of Glaciology, v. 2, p. 185–186.

Mickelson, D. M., and Clayton, L., 1981, Subglacial conditions and processes during middle Woodfordian time in Wisconsin: Geological Society of America Abstracts with Programs, v. 13, p. 310.

Moran, S. R., Clayton, L., Hooke, R. LeB., Fenton, M. M., and Andriashek, L. D., 1980, Glacier-bed landforms of the prairie region of North America: Journal of Glaciology, v. 25, no. 93, p. 457–476.

Newberry, J. S., 1862, Notes on the surface geology of the basins of the Great Lakes: Boston Society of Natural History Proceedings, v. 9, p. 42–46.

—— , 1870, On the surface geology of the basin of the Great Lakes and the valley of the Mississippi: Annals Lyceam Natural History New York, v. 9, p. 213–234.

Parizek, R. R., 1969, Glacial ice-contact rings and ridges: Geological Society of America Special Paper 123, p. 49–102.

Shilts, W. W., 1980, Flow patterns in the central North American ice sheet: Nature, v. 286, p. 213–218.

—— , 1982, Quaternary evolution of the Hudson/James Bay region: Le Naturaliste Canadien, v. 109, p. 309–332.

Spencer, J. W., 1890, Origin of the basins of the Great Lakes of America: Quarterly Journal of the Geological Society of London, v. 46, p. 523–531.

Stalker, A. M., 1960, Ice-pressed drift forms and associated deposits in Alberta: Canada Geological Survey Bulletin 57, 38 p.

Stanford, S. D., and Mickelson, D. M., 1985, Till fabric and deformational structures in drumlins near Waukesha, Wisconsin, U.S.A.: Journal of Glaciology, v. 31, no. 109, p. 220–228.

Sugden, D. E., 1977, Reconstruction of the morphology, dynamics, and thermal characteristics of the Laurentide ice sheet at its maximum: Arctic and Alpine Research, v. 9, p. 21–47.

—— , 1978, Glacial erosion by the Laurentide ice sheet: Journal of Glaciology, v. 20, p. 367–392.

Totten, S. M., 1969, Overridden recessional moraines of north-central Ohio: Geological Society of America Bulletin, v. 80, p. 1931–1946.

Totten, S. M., and White, G. W., 1985, Glacial geology and the North American craton; Significant concepts and contributions of the nineteenth century: Geological Society of America, Centennial Special Volume 1, p. 125–141.

Tyrrell, J. B., 1898, The glaciation of north-central Canada: Journal of Geology, v. 6, p. 147–160.

Whittlesey, C., 1866, On the fresh-water glacial drift to the northwestern states: Smithsonian Contributions to Knowledge 197, 32 p.

White, G. W., 1973, History of investigation and classification of Wisconsinan drift in north-central United States, *in* Black, R. F., Goldthwait, R. P., and

Willman, H. B., eds., The Wisconsinan stage: Geological Society of America Memoir 136, p. 3–64.

—— , 1974, Buried glacial geomorphology, *in* Coates, D. R., ed., Glacial geomorphology: State University of New York at Binghamton, p. 331–350.

Whittecar, G. R., and Mickelson, D. M., 1977, Sequence of till deposition and erosion in drumlins: Boreas, v. 6, p. 213–217.

—— , 1979, Composition, internal structures, and an hypothesis of formation for drumlins, Waukesha County, Wisconsin, U.S.A.: Journal of Glaciology, v. 22, no. 87, p. 357–371.

Wickham, S. S., and Johnson, W. H., 1981, The Tiskilwa Till, a regional view of its origin and depositional process: Annals of Glaciology, v. 2, p. 176–182.

Wright, G. F., 1891, The ice age in North America and its bearing on the antiquity of man (third edition): New York, D. Appleton and Company, 648 p.

—— , 1892, Man and the glacial period: New York, D. Appleton and Company, 385 p.

Wright, H. E., Jr., 1973, Tunnel valleys, glacial surges, and subglacial hydrology of the Superior Lobe, Minnesota, *in* Black, R. F., Goldthwait, R. P., and Willman, H. B., eds., The Wisconsinan stage: Geological Society of America Memoir 136, p. 251–276.

MANUSCRIPT ACCEPTED BY THE SOCIETY NOVEMBER 20, 1986

Geological Society of America
Centennial Special Volume 2
1987

Chapter 5

Canadian Shield

William W. Shilts, Janice M. Aylsworth, Christine A. Kaszycki, and Rodney A. Klassen
Geological Survey of Canada, 601 Booth Street, Ottawa, Ontario K1A 0E8, Canada

INTRODUCTION

W. W. Shilts and J. M. Aylsworth

GENERAL COMMENTS

The Canadian Shield* (Fig. 1) is a geologically complex terrain that makes up about one-third of the North American landmass. Its rocks were deposited, formed, and deformed over a time span encompassing about three-quarters of the earth's known geologic history. Consequently, it includes lithologies, structural elements, and topography as varied as those found on the North American continent as a whole. Because of the complexity of geomorphic features that reflect this geologic diversity, we have decided to focus this chapter on the most obvious geomorphologic elements of its modern landscape, the various landforms produced by the latest geologic events to affect the area, the Pleistocene glaciation. Virtually all of the distinctive landscape elements commonly associated with the Canadian Shield were formed or noticeably modified as a result of the passage of glaciers that coalesced to form the last great continental ice mass, the Laurentide Ice Sheet (Prest, 1970).

Before discussing the glacial geomorphology of the Shield, it is appropriate to make some general observations of the geologic factors that have influenced the nature and distribution of glacial landforms and sediments. Because of the antiquity of the rocks that comprise this "stable" core of the continent, original lithologies have been modified by numerous orogenic and metamorphic events as the earth's crust evolved and as crustal elements shifted through the agency of plate tectonics. The composition of the oldest rocks, in fact, reflects both sedimentation in a primitive, pre-oxygen atmosphere and igneous events that took place as the crust was still differentiating, giving rise to lithological assemblages that are in some cases markedly different from those associated with later geologic processes.

---

*In this chapter the Canadian Shield, or, simply, Shield, is defined as the mainland area of outcrop of Precambrian rocks and excludes Precambrian terranes of Baffin Island, Adirondack Mountains, and the high Arctic Islands. For convenience it also includes the Paleozoic and Mesozoic terranes of the Hudson Bay basin.

Superimposed on this 3 to 4 billion years of sedimentation and igneous activity is almost universal metamorphism of the Precambrian rocks, giving rise to one of the most important characteristics of the Shield: recrystallization related to metamorphism has produced rocks that are physically hard and often coarsely crystalline—many originally lithologically diverse units being converted through repeated and/or intense metamorphism to vast tracts of granitoid gneiss. Thus, the glaciers that passed over most of the Canadian Shield traversed a hard substrate that yielded relatively little debris through glacial erosion.

Notwithstanding the above generalization, some areas of the Canadian Shield, as defined in this chapter, yielded abundant glacial debris and probably played major roles in influencing the dynamic behavior of the continental ice sheets. A central, long-lived depression in the Shield, now largely occupied by Hudson Bay, is filled by up to 2,000 m of unmetamorphosed, flat-lying Paleozoic and Mesozoic sediments. These rocks were easily eroded by Laurentide ice, and products of this erosion were dispersed widely over crystalline terrane south and west of Hudson Bay (Shilts, 1980, 1982). Within the Precambrian terrane itself, the Athabaska and Thelon sedimentary basins, filled with Late Proterozoic, unmetamorphosed, flat-lying sedimentary and volcanogenic rocks, likewise produced large volumes of clastic sediment through glacial erosion of their generally poorly consolidated outcrops. These sediments were also dispersed widely over adjacent crystalline outcrops. The significance of these major sources of sediment, as well as that of older metamorphosed volcanic and sedimentary terranes that served as preferential sources of glacial sediment, will become apparent later in this discussion.

The Canadian Shield is typically a region of low, rolling hills, rounded by glacial erosion of the hard crystalline bedrock of which it is typically composed. In areas of unmetamorphosed sedimentary bedrock, such as the Thelon and Athabaska sedimentary basins and the Hudson Bay basin and adjacent Paleozoic-Mesozoic lowlands, local relief is negligible. The plains in these areas are broken only by ridges formed by inliers of crystalline rock and by post-glacial trenches cut by rivers. In other regions, particularly in northern Quebec and Ungava, relief may exceed 1,000 m. In the rugged uplands of Ungava, cirques are developed.

Shilts, W. W., Aylsworth, J. M., Kaszycki, C. A., and Klassen, R. A., 1987, Canadian Shield, *in* Graf, W. L., ed., Geomorphic systems of North America: Boulder, Colorado, Geological Society of America, Centennial Special Volume 2.

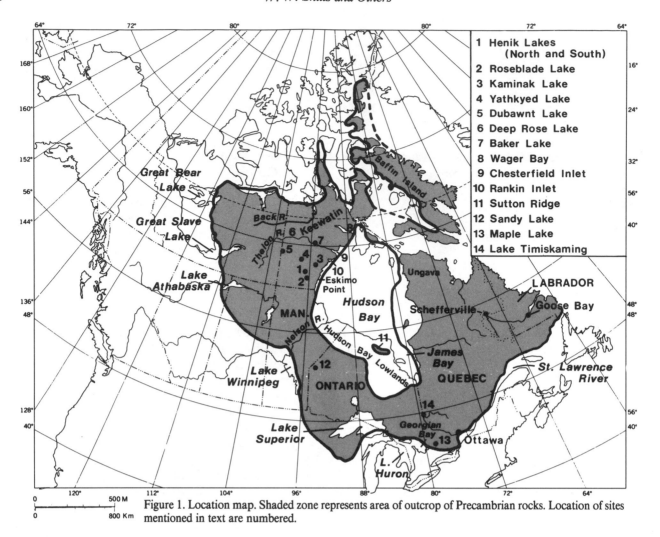

Figure 1. Location map. Shaded zone represents area of outcrop of Precambrian rocks. Location of sites mentioned in text are numbered.

Drainage of the Shield has been extensively disrupted by glacial processes, giving rise to its most characteristic landscape features, the myriads of lakes that dot its surface. Although some of these lake basins may predate glaciation, particularly those which occupy ancient astroblemes, most occupy depressions scoured by glacial erosion or formed by melting of ice blocks in areas of thicker drift. The ice-scoured depressions are often structurally controlled. Many smaller lakes occur in areas formerly submerged beneath postglacial seas, or as proglacial lakes behind dams consisting of ridges of nearshore sediment deposited during offlap of these water bodies, or in areas of permafrost where they occupy thermokarst depressions.

Both patterns of glacier dynamics and postglacial geomorphological processes are related to climatic zones over the Canadian Shield. Although there are many local subenvironments, the climatic zones of the Shield can be broken into two major divisions in which climatic effects are expressed by vegetation: tundra and forest. In the more continental, central part of the Shield and in its northern regions, tundra vegetation prevails and permafrost is continuous. The tundra biomass decreases markedly north of 66° latitude and at higher altitudes east of Hudson Bay in northern Quebec and Labrador. Except on its northern edge, the outer part of the Shield is covered, largely by boreal forest. The northern tree line is near the southern limit of continuous permafrost (Fig. 2); southward from this limit permafrost becomes discontinuous, then sporadic, and finally disappears completely. The southern fringe of the Shield in southern Ontario, Minnesota, Wisconsin, and Michigan experiences a much more temperate climate as expressed by its deciduous forests.

MAJOR PHYSIOGRAPHIC REGIONS OF THE SHIELD

The physiography of the Canadian Shield is the product of a combination of Precambrian peneplanation and Pleistocene glaciation. The overall surface of the Shield is an exhumed pre-Paleozoic erosion surface (Ambrose, 1964), which slopes gently from mean elevations of 400 to 600 m near its outer margins toward a central depression occupied by Hudson Bay. The major elements of the landscape are parts of the ancient erosion surface and as such are controlled by geological structure and bedrock lithology. The Shield is made up of broad sloping or undulating uplands and plateaus, rolling or flat lowlands, subdued flat-topped or rounded hills, and a few deeply incised highland regions. Much of the surface is characterized by a subdued, nearly level terrane, dotted with innumerable lakes, ponds, and swamps and interrupted by flat-topped or rounded monadnocks and ranges of hills (Bostock, 1970). The summits of some of the ranges are not significantly higher than their neighboring uplands, and the uniformity of summit levels within these ranges demonstrates that these are also remnants of the old erosion surface. Relief rarely exceeds 95 m except in the highlands of eastern Canada where mean elevations exceed 600 m and the rugged surface is deeply incised by river valleys. Maximum relief is attained along the northeastern coast where summits rise 1,500 m above sea level. Superimposed on the major features of the Precambrian erosion surface is the minor relief produced by glacial erosion and deposition during the Pleistocene.

Bostock (1970) divided the Canadian Shield into five major physiographic regions: Kazan, James, Hudson, Davis, and Laurentian; he further subdivided these regions into seven categories: lowlands, plains, uplands, plateaus, hills, highlands, and mountains (Fig. 3). Four of the major physiographic regions coincide with one or more of the geological structural provinces of the Precambrian Shield of Canada depicted in Figure 4; Hudson Region coincides with the Hudson Platform, which, although largely underlain by unmetamorphosed Paleozoic sediments, also includes Belcher Fold Belt of Churchill Province (Stockwell and others, 1970).

Characteristics of the physiographic subregions generally coincide with distinctive geological characteristics. (1) "Plains" are formed on areas of relatively flat-lying unmetamorphosed rock. The largest plains, Athabaska Plain and Thelon Plain (Fig. 3), are flat to gently rolling surfaces underlain predominantly by flat-lying clastic sediments, as is Cobalt Plain. Nipigon Plain is underlain by flat-lying Proterozoic gabbro sills. (2) "Lowlands," (3) "Uplands," and (4) "Plateaus," distinguished in roughly ascending mean elevation, are underlain by massive, mainly crystalline rock. These three categories cover much of the Canadian Shield and, for the most part, they are characterized by gently sloping or rolling terrain or undulating terrain in the form of low broad hills. Relief is generally subdued. In contrast to the above generality, in eastern Canada Kaniapiskau Plateau, Mecatina Plateau, Hamilton Upland, and Hamilton Plateau (Fig. 3) have been incised deeply by rivers, producing a more rugged appearance. (5) "Hills" are composed of low-grade metamorphic rocks. Ranges of hills are the result of differential erosion of faulted or gently folded strata. (6) "Highlands" are broad, uplifted areas of deeply incised, highly metamorphosed, massive rocks. The deep dissection produces a mountainous appearance, although the level nature of the summits reflects the undulating surface of ancient peneplains. The Labrador Highlands include the highest ranges in eastern Canada (1,500 m elevation) and are rugged and deeply incised with U-shaped valleys and fiords along the coast. Cirques occur on the summits. (7) "Mountains" are composed of particularly resistant massive rock. Bostock (1970) restricts the category on the Canadian Shield to Mealy Mountains

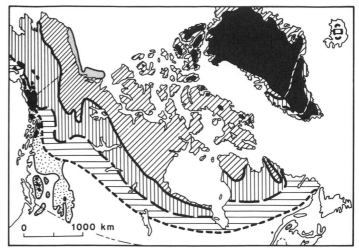

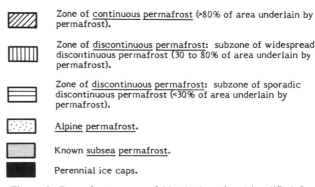

	Zone of <u>continuous</u> permafrost (>80% of area underlain by permafrost).
	Zone of <u>discontinuous</u> permafrost: subzone of widespread discontinuous permafrost (30 to 80% of area underlain by permafrost).
	Zone of <u>discontinuous</u> permafrost: subzone of sporadic discontinuous permafrost (<30% of area underlain by permafrost).
	<u>Alpine</u> permafrost.
	Known <u>subsea</u> permafrost.
	Perennial ice caps.

Figure 2. Permafrost zones of North America (simplified from data provided by J. A. Heginbottom, personal communication, 1986).

in eastern Canada (Fig. 3). Mealy Mountains are formed of resistant anorthosites that rise abruptly above the surrounding terrane and reach altitudes over 1,100 m.

For a complete description of the physiographic regions of the Canadian Shield in Canada the reader is referred to Bostock (1970). Part of James Region extends southward into the Lake Superior district of the United States. This region has been named Superior Uplands by Fenneman (1938).

PERMAFROST AND RELATED FEATURES

The principal geomorphic features related to the climate of the Shield are forms of patterned ground, most of which are clearly related to permafrost (Fig. 2) or to the active layer developed in perennially frozen terrain. In the shallow lakes and bogs of the boreal forest, patterns comprising straight to sinuous, subparallel stripes of vegetation begin to become common north of approximately 50° north latitude. These patterned fens, or bogs, are particularly well developed in the flat, muskeg-covered terrane of the Hudson Bay Lowlands and in southern Labrador–Nouveau Quebec. North of about 60° these patterns disappear. In the zone of discontinuous permafrost, especially in the Hudson

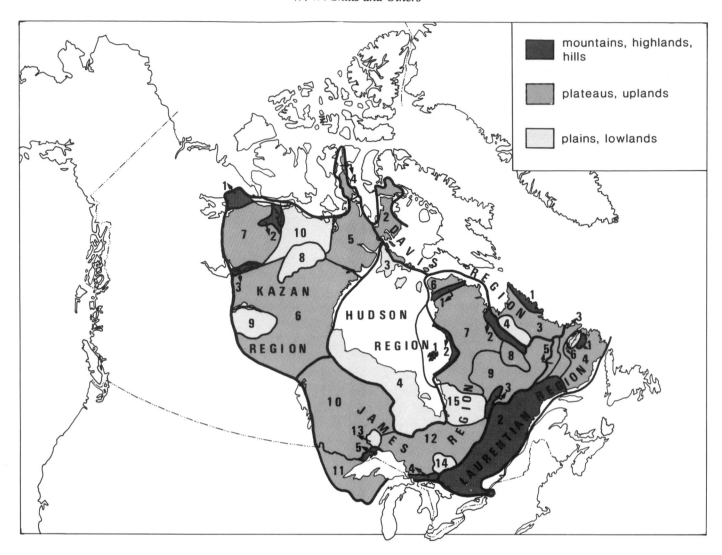

Figure 3. Major physiographic subdivisions of Canadian Shield (after Bostock, 1970).

KAZAN REGION
1. Coronation Hills
2. Bathurst Hills
3. East Arm Hills
4. Boothia Plateau
5. Wager Plateau
6. Kazan Upland
7. Bear-Slave Upland
8. Thelon Plain
9. Athabaska Plain
10. Back Lowland

HUDSON REGION
1. Belcher Islands
2. Richmond Hills
3. Southampton Plain
4. Hudson Bay Lowlands

JAMES REGION
1. Povungnituk Hills
2. Labrador Hills
3. Mistassini Hills
4. Penokean Hills
5. Port Arthur Hills
6. Sugluk Plateau
7. Larch Plateau
8. Kaniapiskau Plateau
9. Lake Plateau
10. Severn Uplands
11. Superior Uplands
12. Abitibi Uplands
13. Nipigon Plain
14. Cobalt Plain
15. Eastman Lowland

LAURENTIAN REGION
1. Mealy Mountains
2. Laurentian Highlands
3. Hamilton Plateau
4. Mecatina Plateau
5. Hamilton Upland
6. Melville Plain

DAVIS REGION
1. Labrador Highlands
2. Melville Plateau
3. George Plateau
4. Whale Lowland

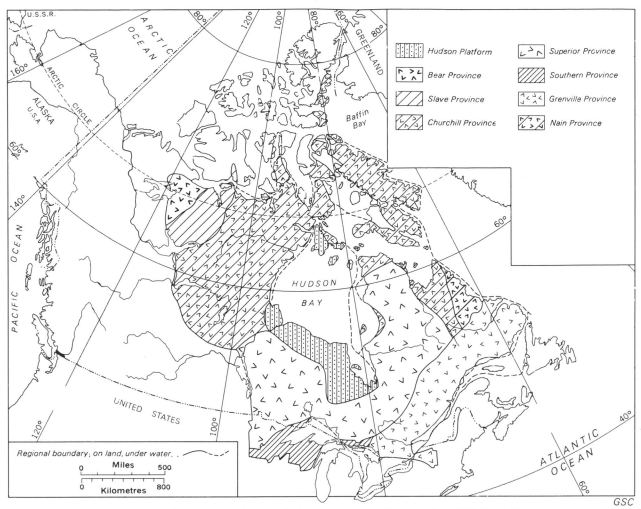

Figure 4. Structural provinces of the Canadian Shield (after Douglas, 1970, Fig. 1-1).

Bay Lowlands west of Hudson Bay, permafrost features such as palsen are associated with patterned fens and peatlands. These low hummocks are formed over accreting ice lenses that form in water-saturated organic deposits.

North of the treeline, which extends south of 60° latitude on the west side of Hudson Bay, reflecting the extreme continentality of the central part of the Shield, a large variety of patterned ground features is evident. Most of these features may be classified into three main categories including: features resulting from (1) soft sediment deformation, (2) cracking of frozen soil, and (3) frost heaving of bedrock.

The first category encompasses circular to linear features 1–5 m in diameter that form largely as a result of soft sediment deformation processes taking place in the active layer. These features occur preferentially in glacial or post-glacial sediments that have a high content of fine particles, particularly silt and finer sizes, and become relatively less common northward as the annual depth of the active layer decreases. They are largely products of deformation that takes place under conditions of saturation or near saturation during the thaw season. Among the forms in this category are various types of sorted and unsorted circles (Washburn, 1956) or mudboils (Fig. 5; Shilts, 1978); subparallel straight to sinuous rib and trough structures (Fig. 6; Shilts and Dean, 1975) that occur at water depths of less than 3 m in nearly all lakes that occupy depressions in suitably fine-grained substrate; and large and small-scale solifluction stripes (Washburn, 1956) of uncertain genesis. The latter features are particularly well-developed on finer grained tills and muds.

Polygonal or orthogonal patterns are formed by repetitive frost cracking caused by shrinkage of frozen soil subjected to extremely cold temperatures (Fig. 5). These features, which dominate all surficial sediment types where the active layer is shallow, especially in the northernmost parts of the Shield, are formed in more southerly areas on nonplastic glacial and postglacial sediments, such as gravels and sands, as well as on organic terrains where vegetative insulation causes the depth of the active layer to be negligible. Where the active layer is thicker than about 1 m, these features form and persist only in materials that are not de-

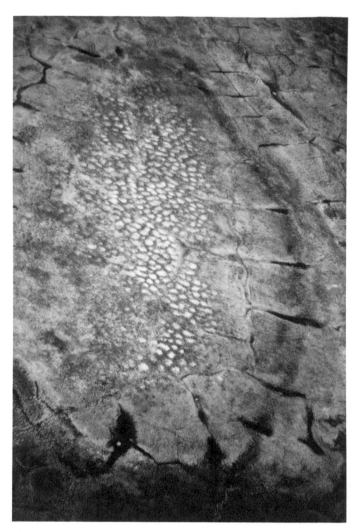

Figure 5. Mudboils formed on 1 to 1.5 m deep active layer developed in marine silty clay draping a drumlin, Kaminak Lake area, District of Keewatin. Frost polygons are developed in adjacent, poorly drained organic terrain where active layer depth is less than 50 cm. Photo by W. W. Shilts (GSC-201696W).

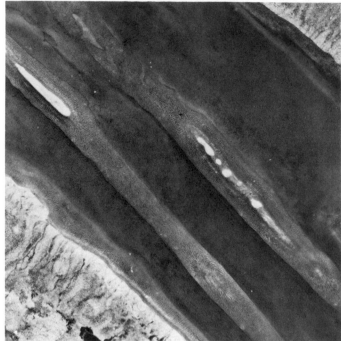

Figure 6. Rib and trough structures developed on till-forming drumlins submerged in a lake just southeast of Kaminak Lake, District of Keewatin. (Portion of NAPL photo A 19396-226.)

GLACIAL GEOMORPHOLOGY

In the following discussion the authors have concentrated on areas or processes with which they and their colleagues are most familiar, with the objective of providing information that may be extrapolated to parts of the Shield for which data are more sparse. Because of its size and the logistical difficulties of working in the climatically hostile, virtually uninhabited regions that comprise most of the Shield, there are still vast areas for which there is no detailed geomorphological information. For this reason, the discussion of some areas will be necessarily brief or vague, while others will be discussed in some detail.

The Laurentide Ice Sheet was responsible for virtually all of the glacial erosion and deposition that affected the Shield during the last glaciation. Although similar ice sheets must have grown and traversed the Shield several times during the Pleistocene, as evidenced by the multiple glacial units exposed in the Hudson Bay Lowlands (McDonald, 1969; Skinner, 1973; Andrews and others, 1983) and encountered here and there in boreholes or excavations elsewhere on the Shield, their effects have been largely erased or masked by erosional and depositional processes that took place during the last glaciation. As this chapter is entering its final revision, P. Wyatt and H. Thorleifson (personal communication, 1986) have reported to the authors overwhelming evidence of northwestward ice flow in the form of striations, rôches moutonnées, and whalebacks, preserved on the leeside or down-ice from outcrops and islands shaped by southwestward flowing ice near Big Trout Lake, Ontario (latitude 53°45′N;

formable enough to "heal" the cracks caused by seasonal extremes in temperature.

The third category comprises bedrock frost-heave features including: fields of felsenmeer composed of individual frost heaved blocks bounded by primary outcrop fractures such as joint planes, and rock "bursts," which, although relatively rare, are often spectacular, having the aspect of a small crater (DiLabio, 1978). These latter features form in much the same way as palsen, with ice lensing causing heave along subhorizontal fractures within the outcrop.

Pingos, common in the shallow lakes and fine-grained drift northwest of the Shield, are rare on the Shield; one well-known exception is the Thelon River Pingo described by Craig (1964).

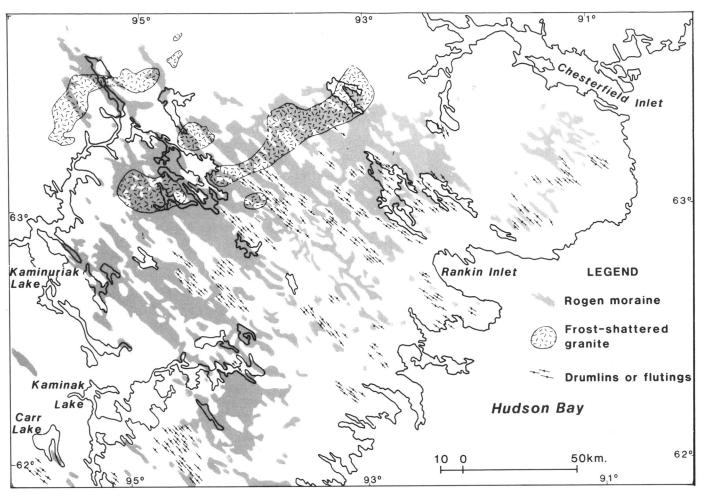

Figure 7. Trains of rogen moraine and drumlins, northwest side of Hudson Bay. Note close association
of rogen pattern with frost-susceptible granite and drumlin train southeast of Kaminak Lake.

longitude 90°W). These features, predating the last major glacial phase recorded in the Hudson Bay Lowlands, suggest that, at least once, a major ice flow centre existed south of Hudson Bay. Tyrrell first suggested such a Patrician Ice Centre in 1913.

We have chosen as a unifying theme for our discussion the regular arrangement of glacial landforms* about the Laurentide Ice Sheet's major ice dispersal centers in Keewatin and in Quebec/ Labrador. Spatially related groups of landforms also exist on the southern Shield but they have less-clear temporal relationships and seem to be related to major late glacial lobations and surges from ice masses in or near southern Hudson Bay.

In following sections, we shall refer to these regular sediment/landform assemblages as *zones,* which in the case of the

Keewatin Ice Divide, wrap around the south half of the divide like a series of layers. Less is known of their overall extent around the Quebec/Labrador Divide, and higher topographic relief to some extent distorts the zone, but they are present nevertheless. The zone concept is also used for the southern Shield, but the landform assemblages are not comparable in area or origin to those around the major ice divides.

Although most modern literature emphasizes the role of ice *dynamics* in promoting distinctive glacial landform/sediment associations (i.e., Hughes and others, 1985), the *geology* of the glacier bed (lithology, topography, deformability) is increasingly being recognized as having played a significant, if not dominant, role in localizing sediment/landform patterns (Boulton and others, 1985; Shilts, 1985; Fisher and others, 1985).

The zones that we recognize around the ice divides are probably related both to geologic and dynamic factors. Because dynamic conditions at the base of glaciers vary as a function of ice-flow velocity, degree of ice stagnation, ice thickness (thicker ice retarding loss of geothermal heat), local nature of basal flow

---

*In this chapter glacial landforms are defined as both *erosional,* caused by glacial sculpting of bedrock or previously deposited drift or other unconsolidated deposits, and *constructional,* built by deposition of glacial, glaciofluvial, glaciolacustrine, or glaciomarine sediments.

(compressive or extensive), and other factors—all parameters that tend to vary radially away from a center of outflow—it is likely that some of the zones are at least partially related to ice dynamics. However, some of the zones are related to amount of sediment load in the glacier and, within the zones themselves, landform types may be related to the physical nature of the basal load, which is in turn a function of the geology of the glacier's bed.

For instance, the generally close lateral association of drumlins and rogen moraine near the ice divides can be assumed to be related to similar dynamic conditions responsible for their formation. Which one of these two features developed or, indeed, if any recognizable constructional landform developed at all, is apparently a function of the amount and physical properties of debris entrained in or lying below the sole of the glacier. The nature and quantity of debris is in turn closely related to the "geology" of the source and dispersal areas, including the structure and degree of frost shattering of the source outcrops, their lithology, and the topography of source outcrops and dispersal area. Because debris was entrained and dispersed in the form of linear glacial trains, the landforms themselves tend to occur in trains, extending from specific outcrops or source areas.

Examples of trains of glacial landforms extending from distinct source areas can be found in the low-relief terrain between the central part of the Keewatin Ice Divide and Hudson Bay (Fig. 7). In this region, trains of rogen moraine, composed of stony till, typical of the red, clay-rich till of the region, are overlain by a dense cover of coarse boulders of granite, which can be traced to outcrops of a distinctive late Archean granite. These outcrops are presently characterized by advanced development of felsenmeer that has formed subsequent to the last glaciation and postglacial marine inundation, over a time span of less than 5,000 years. Therefore, it is reasonable to assume that these same outcrops were covered with an easily entrained layer of felsenmeer boulders prior to the last glacial event, resulting in a dense load of boulders in basal ice. This bouldery load in turn may have created the basal conditions necessary to form rogen moraine instead of the drumlins that commonly occur in intervening boulder-free areas. Elsewhere in this region, rogen moraine is characteristically composed of boulder-covered till (matrix of the till is indistinguishable from till that forms intervening drumlins), or of reworked sand and gravel derived from proglacial outwash, or from friable Proterozoic sandstones and conglomerates of the Thelon sedimentary basin.

In the same region, drumlins can also often be seen to occur in trains trailing away from specific source areas. In some cases the source areas are underlain by easily eroded rocks that do not produce abundant boulders, as in the case of the spectacular drumlin fields extending down-ice from the friable sandstones of the Thelon basin, or from lake basins excavated in various rock types that for one reason or another produced abundant debris. Drumlin trains extending eastward or southeastward from Kaminak Lake (Fig. 7) and Henik Lakes basins are good examples of this latter type.

Interspersed with drumlin and rogen moraine trains are areas where no organized glacial constructional landforms occur. It is the authors' opinion that when these areas occur within a region where rogen moraine and drumlins are common, they simply reflect a sediment-poor part of the glacier bed; the lack of sediment reflecting either the natural resistance of underlying lithologies to erosion, or, less likely, some dynamic condition at the base of the glacier which precluded significant glacial erosion or deposition of already entrained debris. Whatever the cause, a paucity of glacial sediment would have resulted inthere not being enough sediment to be shaped or formed into recognizable glacial landforms.

The above examples, drawn from areas in which the authors have carried out extensive, detailed ground and airphoto studies, demonstrate the importance of geological factors, at least at a local scale, in influencing the patterns of occurrence of glacial bedforms on the Canadian Shield. In the following section it will be seen that other, larger scale patterns of these and associated glaciofluvial landforms are distributed in discrete zones that surround the ice divides. These concentric zones, which are similar to those that other authors* have alluded to on a continental scale, are probably a result of a combination of (1) changing ice sheet dynamics along radii extending from the ice divides, (2) the effects of local topography, and (3) the geometry of the various major structural and lithological terranes of the Shield.

GLACIAL GEOMORPHOLOGY OF NORTHWESTERN CANADIAN SHIELD

W. W. Shilts and J. M. Aylsworth

INTRODUCTION

Detailed surficial geological mapping carried out near Hudson Bay by the authors and colleagues has been combined with reconnaissance airphoto interpretation of selected glacial features in a large area of the northwestern part of the Canadian Shield (Fig. 8). The resulting map, based largely on morphology of the terrain and inferred sediment associations, reveals distinctive patterns of glacial deposition and erosion that may have considerable significance in interpretation of the history and dynamics of the western part of the Laurentide Ice Sheet. Although the genesis of the landforms described in this chapter will not be discussed to any extent, the regional and local patterns of their occurrence place many constraints on theories of their origin published elsewhere.

Most of the area discussed in this section was mapped at a small-scale, general-reconnaissance level by Geological Survey of

---

*Other authors have, however, generally attributed these zones strictly to variations in ice-sheet dynamics—specifically to the presence and amount of liquid water at the glacier's base (Sugden, 1977, 1978). The concentricity of the zones is generally described about Hudson Bay, rather than about the ice divides.

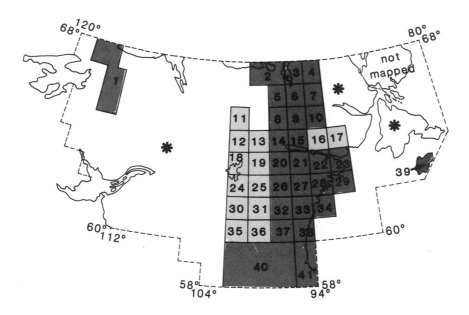

Figure 8. Sources of glacial geomorphological information for northwestern Canadian Shield.

SOURCES

  Surficial geology maps, 1:125 000 and 1:250 000 scale, based on detailed airphoto interpretation and extensive ground observation.

  Surficial geology maps, 1:125 000 and 1:250 000 scale, based on detailed airphoto interpretation and minimal or no ground observation.

[✳] Maps of major glacial features, 1:250 000 scale, unpublished, based on airphoto interpretation.

1. St-Onge, in press.
2. Helie, 1984.
3, 4, 6, 7, 9, 10. Thomas and Dyke, 1981a, b, c, d, e, f.
5, 8. Thomas, 1981a, b.
11. Clarke, 1983.
15, 26, 38. Aylsworth, Cunningham, and Shilts, 1986, 1981a, b.
21. Aylsworth, Boydell, Cunningham, and Shilts, 1981.
22, 23, 28, 29. Aylsworth, Boydell, and Shilts, 1985, 1986, 1981a, b.
24, 35, 36, 31. Aylsworth, 1986a, b, c, in press.
27. Arsenault, Aylsworth, Kettles, and Shilts, 1981.
32. Aylsworth, Cunningham, Kettles, and Shilts, 1986.
33. Arsenault, Aylsworth, Cunningham, Kettles and Shilts, 1982.
34. Aylsworth, Kettles, and Shilts, 1981.
39. Aylsworth and Shilts, 1986.
12-14, 16-20, 25, 30, 37. Aylsworth and others, unpublished maps.
40. Dredge, Nixon, and Richardson, 1985.
41. Dredge and Nixon, 1986.

Canada personnel in the late 1950s and early 1960s (Craig, 1964, 1965; Fyles, 1955; Lee, 1959). These early maps were restricted to depicting or symbolizing the most prominent geomorphic features, such as major esker ridges and large drumlin fields, and only recorded rogen (ribbed) moraine in a fraction of the area actually covered by it. No attempt was made to differentiate drift-covered from drift-free areas. These works, supplemented by further air photo reconnaissance mapping, were compiled onto the Glacial Map of Canada (Prest and others, 1968, 1:5,000,000 scale).

Since the Glacial Map of Canada was published, more detailed surficial geology maps (1:125,000 or 1:250,000) have been produced for much of the eastern part of the region. These maps, in published or unpublished form, cover approximately 40% of our map of the northwestern Shield. For the remaining 60%, major glacial features were compiled from interpretation of 1:60,000 scale air photographs plotted on 1:250,000 scale maps.

KEEWATIN ICE DIVIDE

The Keewatin Ice Divide (Lee and others, 1957; Fig. 9) is defined in a broad elongate area stretching from near the Manitoba-Keewatin border to the vicinity of Wager Bay—a distance of about 700 km. It is an area generally devoid of distinctive glacial landforms and represents the region toward which the last Laurentide Ice Sheet shrank west of Hudson Bay. Ice still occupied this area as recently as approximately 6,000 to 6,500 years ago, based on ^{14}C dates on shells marking the onset of marine incursion about 100 km east of the divide, in the Kaminak Lake area (GSC-1434; 6,600 ± 230 yr B.P.). From disper-

W. W. Shilts and Others

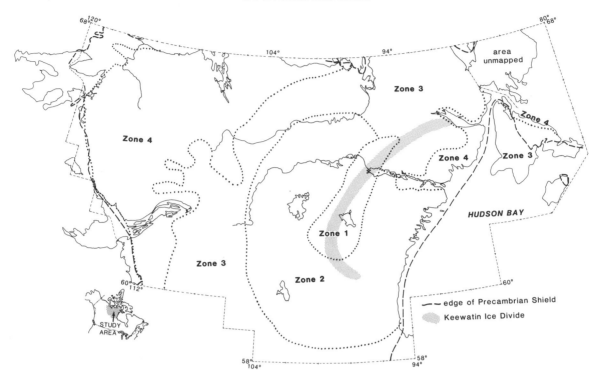

Figure 9. Glacial geomorphological zones around the Keewatin Ice Divide, northwestern Canadian Shield.

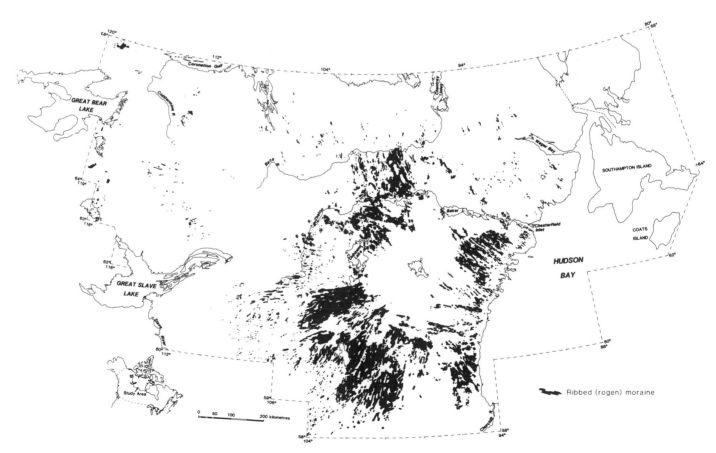

Figure 10. Distribution of rogen moraine, northwestern Canadian Shield.

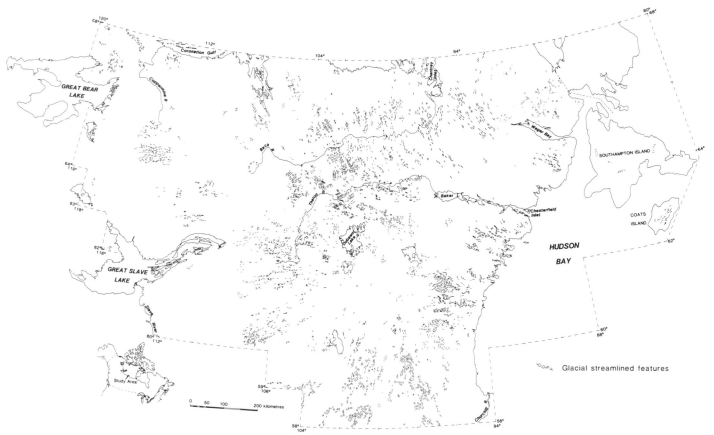

Figure 11. Distribution and orientation of drumlins and other streamlined constructional features, northwestern Canadian Shield.

sal patterns of distinctive rock types, ice-inscribed forms, such as striations, and the general pattern of landforms around it, the Ice Divide can be seen to have served also as the dispersal center or highest part of the western half of the Laurentide Ice Sheet. Although the elongate dispersal center migrated eastward and southward during its existence, and undoubtedly had many more offshoots and irregularities than are apparent in its final mapped form, it migrated no more than 100 km, and its final location is a reasonable approximation of its general position throughout the last glaciation.

The Keewatin Ice Divide, like the Labrador–Nouveau Quebec Ice Divide east of Hudson Bay, serves as the focus for nearly all of the glacial landforms created in the sector of the Laurentide Ice Sheet that it influenced. These forms can be seen to be arranged in an organized way in a crudely radial pattern around this feature, not, as implied in earlier literature, around and radially away from the Hudson Bay basin (Flint, 1943; Sugden, 1978). In this chapter, we will endeavor to show that the predictable and regular zonation of glacial landforms described in the introduction can be mapped clearly around the major ice divide located west of Hudson Bay.

Four glacial geomorphological zones (Fig. 9) that are roughly concentric about the Keewatin Ice Divide are derived from the distribution patterns of glacial landforms (Fig. 10, 11, 12, 13).

Zone 1

The innermost zone, characterized by the absence of eskers and rogen moraine, extends as a broad ellipse through southern Keewatin and extends about 50 km on either side of the axis of the Keewatin Ice Divide (Lee and others, 1957). The characteristic landscape of Zone 1 comprises till plains with areas of low till hummocks (Fig. 14) and virtually no oriented depositional features. It is almost completely devoid of any glaciofluvial deposits, except, perhaps, for minor outwash in some valleys. The location of the Ice Divide itself (Fig. 9), which follows the axis of the region, is clearly defined by striation orientations.

Zone 2

A second horseshoe-shaped zone, 200 to 250 km wide, bounds all but the northern part of the central ellipse and is characterized by the presence of well-developed rogen moraine and esker-outwash systems.

Rogen (ribbed) moraine typically comprises areas of short,

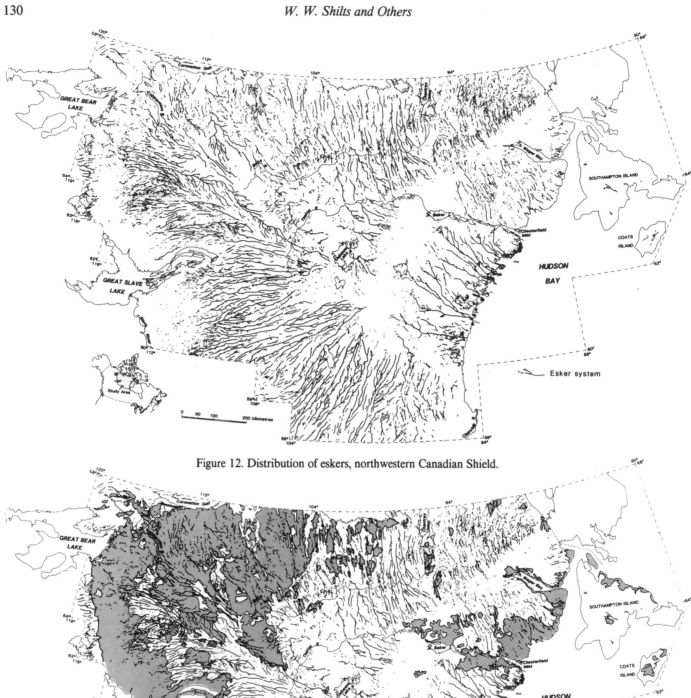

Figure 12. Distribution of eskers, northwestern Canadian Shield.

Figure 13. Distribution of drift-free areas and eskers, northwestern Canadian Shield.

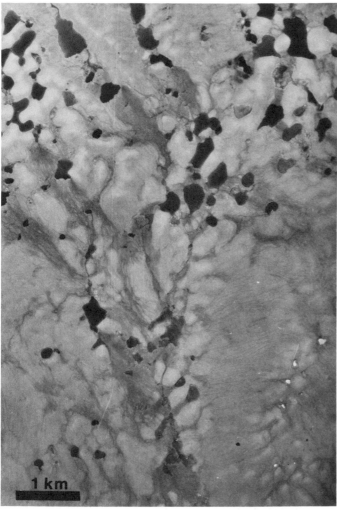

Figure 14. Disorganized, hummocky till plain west of Yathkyed Lake, in vicinity of Keewatin Ice Divide. (Portion of NAPL photo A14378-132.)

Figure 15. Sharp contact between drumlinized till plain (to north) and large train of classical rogen moraine southwest of Dubawnt Lake. (Portion of NAPL airphoto mosaic no. RE 11662-6, east ½.)

sinuous ridges, trending roughly perpendicular to ice flow (Cowan, 1968; Lundqvist, 1969; Shilts, 1977; Bouchard, 1980). Fields of rogen moraine form linear belts or trains parallel to major directions of ice movement. These trains radiate from the Keewatin Ice Divide like the spokes of a wheel (Fig. 10). Featureless or drumlinized (Fig. 11) till plains with minor bedrock outcrop occur between the rogen trains with almost no transition from one landscape to the other. Individual drumlins may occur within the trains and individual rogen ridges may be fluted. Where best developed, rogen moraine distribution is little influenced by topography, but where the belts become more linear and separate down-ice from the Ice Divide, rogen moraine generally occurs preferentially in depressions. Near the Ice Divide, clusters of rogen ridges have been observed on the tops of hills.

The radial distribution of rogen moraine about the Keewatin

Ice Divide is not continuous. Rogen moraine is rare west of Dubawnt Lake and east of Baker Lake, where large areas are largely devoid of glacial deposits or are covered by a featureless till veneer. Northwest of Dubawnt Lake an extensive area is covered exclusively by drumlins (Fig. 11). The northern boundary of the southwestern-most field of rogen moraine is remarkably distinct against the generally featureless till plain west of Dubawnt Lake (Fig. 15). In general, the down-ice margin of the zone of rogen moraine is abrupt; although isolated, small, linear trains occur in the inner half of Zone 3.

Eskers also begin near the inner edge of Zone 2 and, like the rogen moraine trains, radiate from the Keewatin Ice Divide (Fig. 12). A typical esker system begins as a series of hummocks or short, flat-topped segments (Fig. 16) that pass downstream into continuous large eskers joined by areas of outwash or meltwater

Figure 16. Flat-topped esker deposited in open channel in thin ice at upstream termination of tributary of Maguse River esker system, near Keewatin Ice Divide. Note till stripped by meltwater from bedrock beside esker ridge. (Photo by W. W. Shilts, GSC-203539-Y.)

channels (Fig. 17). Along a section measured across the southwestern part of Zone 2, eskers are spaced approximately 13 km apart, with spacing varying from 2 to 27 km. Throughout most of the area eskers are sharp ridged and up to 40 m high, with occasional conical knobs projecting well above the average elevation of the esker crest (Fig. 18). Along their length they may be interrupted periodically by bulges where the single ridge splits into multiple ridges that coalesce downstream (Fig. 19). These bulges may be the result of greatly increased sedimentation in the esker tunnel, perhaps due to slowing retreat of the ice front due to climatic deterioration. When the tunnel was blocked by sediment, a bypass tunnel formed; this process may have occurred repeatedly until the ice front retreated upstream (Shilts, 1984a). Above marine limit and above the shorelines of proglacial lakes, eskers are commonly flanked by outwash in terraces disrupted by kettle lakes. North of the Thelon River, eskers are associated with prominent outwash terraces (Fig. 19), and the tops of the eskers are in some places flat, planed off by subaerial meltwater flowing on a stagnant ice floor into which the esker ridge was temporarily frozen after the ice front had retreated (Shilts, 1984a).

Below marine limit, eskers commonly were reworked by wave action during the offlap phase of the Tyrrell Sea. This has had the effect of subduing their relief to the extent that they rarely project more than 10 m above the adjacent terrain. Sonar profiles show them to retain their relief and sharp crest where they are submerged in deep modern lakes.

Zone 3

Zone 3 is characterized by an intricate dendritic pattern of eskers, fairly continuous drift cover, drumlin fields, and widely scattered fields of rogen moraine. It forms a 200- to 300-km-wide ring of unconsolidated sediment, commonly till, draped over the rolling bedrock surface. Drift thins to a discontinuous veneer in the outer portion of the ring. This zone, if developed east of the Keewatin Ice Divide, lies beneath Hudson Bay. Large esker systems are known to extend into Hudson Bay, and wave-reworked eskers bearing erratics common to the Keewatin mainland occur on Coats Island in northern Hudson Bay (Shilts, 1982), suggesting that an area similar to Zone 3 may lie east of Zone 2, submerged in Hudson Bay. In the northwesternmost part of the Shield, an area similar to Zone 3, with eskers and a well-developed mantle of till, is developed in a region of relatively flat-lying, unmetamorphosed Proterozoic sedimentary rocks. The presence of Zone 3 features here and on the Paleozoic terrane that bounds the Shield to the west suggests that this zone is related to relatively high production of glacial debris from noncrystalline outcrops. The concentricity of this zone about the ice divide may reflect as much the presence of easily eroded outcrops between it and the divide as it reflects regular change of glaciodynamic conditions away from the divide. The intricate esker patterns and lack of well-developed rogen moraine do suggest, however, that ice dynamics in this zone differ from adjacent zones.

Within Zone 3, rogen moraine occurs only as isolated, short, narrow, linear trains, lying in depressions and closely associated, in space, if not in time or genesis, with eskers. Within the rogen areas, individual ridges are much smaller and depart considerably in shape from those in Zone 2 and from "classical" rogen moraines described by Lundqvist (1969; S. Paradis, personal communication, 1986).

Compared to the 13 km spacing of eskers in Zone 2, spacing between eskers decreases to approximately 8 km (3 to 15 km range) along a section 200 km downstream from the section that was measured in Zone 2. Although numbers of eskers and their tributaries increase in Zone 3, the average size (height and width) of individual eskers decreases in this zone.

In the outer part of Zone 3, esker systems become more discontinuous and disorganized, and esker distributaries are common. Short esker segments are commonly oriented in various directions and areas of crevasse fillings are associated with the disruption of the esker system.

Within the northwestern part of this zone, prominent ice recessional positions are marked by aprons formed by coalescing ice-contact deltas or fans, outwash trains, and closely spaced short

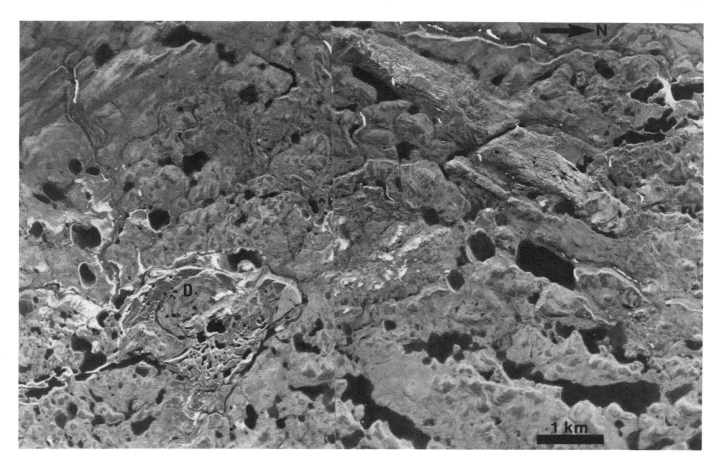

Figure 17. Esker, esker delta (D), and subglacial meltwater channel; northward flowing tributary of Deep Rose Lake esker system, District of Keewatin. Channel is cut through orthoquartzite ridge, which also dammed lake into which delta was built. Overflow from lake cut channel through till plain at left (west) side of figure. Note hummocky and asymmetric moraine in northeast quadrant of photo. Collapse features and abandoned distributary channels are well preserved on delta. (Portion of NAPL airphotos A15390-28, 29.)

a
b

Figure 18. Sharp-crested eskers built above marine limit, District of Keewatin. Note in 18b protruberance flattened by later meltwater erosion of streams flowing on stagnant ice into which esker was frozen. Lateral terraces are still younger outwash deposits (arrows). Esker ridges are 30 to 40 m high. (18a: arrow points to figure for scale, photo by J. M. Aylsworth, 1980, GSC 203540P; 18b, photo by B. C. McDonald, 1968, GSC 10-27-68.)

134   W. W. Shilts and Others

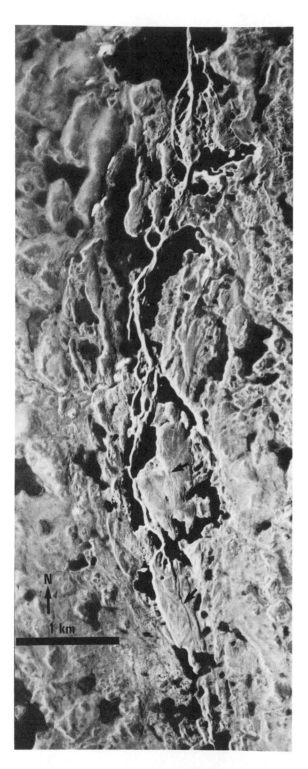

Figure 19. Bifurcating esker ridges forming nodes on Deep Rose Lake esker tributary, about 20 km north (downstream) from esker delta complex of Figure 17. Note *outwash terrace* remnants superimposed on esker and valley sides; those in the south part of photo show scars of braided meltwater stream channels (arrows). The esker splits into *multiple ridges* in several places, particularly at the north end. Most lakes were filled with ice blocks when the outwash terraces were deposited. (Photo enlarged from a portion of NAPL A15390-33.)

esker segments with distributaries, all of which form a line parallel to and north of the Back River (Fig. 20). In the vicinity of this feature, short esker segments abound between the major systems, and major outwash trains parallel the inferred ice front.

Zone 4

The outermost zone of the northwestern Canadian Shield is characterized by extensive bedrock outcrops that are nearly bare of drift (Fig. 13). Although the transition from Zone 3 to 4 is abrupt in the southern part of the northwestern Shield, long trains of drift project northwestward into major lowlands in the northern part. The abrupt transition in the south corresponds roughly to the eastern boundary of the Fort Smith Belt, a northward-trending zone of gneisses and intrusive rocks (Charbonneau, 1980; Fig. 21). The transition zone north of Great Slave Lake corresponds with the eastern (up-glacier) edge of the Slave structural province (Fig. 4), a region of massive crystalline rock. The projections of relatively continuous drift cover that extend northwestward along lowlands through the crystalline bedrock of the Slave Province probably represent trains of debris eroded from the poorly consolidated sedimentary rocks of the Dubawnt Group of the Thelon Basin (Fig. 21) and transported preferentially along the lower elements of the landscape. On a smaller scale, Paleozoic carbonate debris has been eroded from the Ottawa Valley in southeastern Ontario and transported preferentially up valleys in the Precambrian terrane to the southwest (Shilts, 1985). This depositional model is repeated in many glacial terrains in North America where heavy loads of easily eroded debris in the basal part of a glacier are blocked by highlands but transported through lowlands.

ESKERS

Eskers radiating outward from the Keewatin Ice Divide have an intricate framework of tributaries that form a dendritic pattern. The largest of the systems can be traced, with gaps, as far as 600 km and unbroken lengths of individual eskers can be traced for up to 75 km. In many places, discontinuous segments are linked by meltwater channels (Fig. 17), usually scoured through drift to bedrock or filled by outwash.

Although the pattern is by no means universal, there is a tendency for tribuary eskers to join main eskers preferentially from the left; that is, from the north for those eskers deposited by eastward-flowing meltwater streams east of the Keewatin Ice Divide and from the south for those eskers deposited by westward-flowing meltwater west of the Divide. This observation is best illustrated by the esker system trending northwestward from Dubawnt Lake (Fig. 12) where, in a distance of 350 km, 12 tributaries join from the south and one from the north. Although there are segments in which tributaries join from the north, throughout the length of this system, the majority of esker tributaries approach from the south; these tributaries are generally the largest and best developed, often possessing tributaries them-

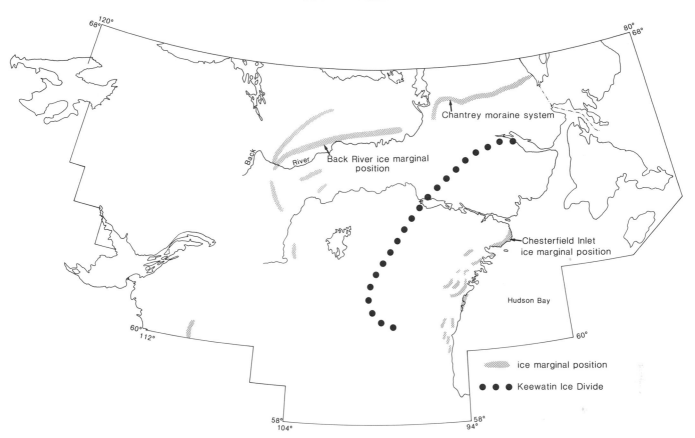

Figure 20. Locations of deposits suggesting halts of ice margin during retreat toward Keewatin Ice Divide.

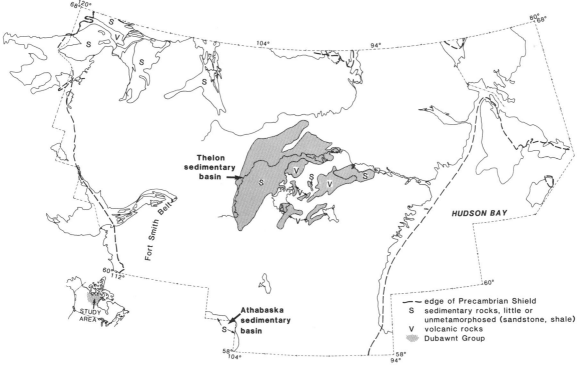

Figure 21. Major Precambrian sedimentary basins of northwestern Canadian Shield. Outcrops in these basins were particularly susceptible to glacial erosion because of their generally poorly consolidated nature or lack of metamorphism and deformation.

Figure 22. Esker superimposed on drumlin field at angles approaching 90° suggesting sluggish or no ice flow at time of deposition, west of Rankin Inlet, District of Keewatin. (Portion of airphoto NAPL A-14302-17.)

selves. Similarly, between its origin near Yathkyed and Henik lakes and Hudson Bay, most tributaries of the Maguse River esker system make a sharp turn to the south a few kilometres before joining the trunk stream on its north side. In most cases the orientation of the tributary esker changes abruptly as it approaches the main esker, and the tributary joins the main esker at right angles. Downstream from where the eskers turn abruptly to the south, they often cut local drumlin fields at sharp angles (Fig. 22).

In some locations, clusters of short, subparallel esker ridges occur between trunk eskers in 1 to 10 km wide zones that trend at right angles to general esker direction. In these zones, trunk eskers may be interrupted by major subaqueous fans or deltas, and distributary ridges fan out from them. Several of these clusters and sedimentary bulges occur along the Hudson Bay coast from

Chesterfield Inlet southward to Eskimo Point. Somewhat similar features, although less regular and associated with crevasse fillings, occur near the western edge of Zone 3. The eastern clusters may mark the ice front during a halt or slowdown in retreat and thus may correlate with the clusters of esker segments and outwash features that mark retreat positions northwest of the Keewatin Ice Divide (see above, Zone 3 and Fig. 20).

Esker systems appear to have been fully integrated Horton systems (Horton, 1945; Scheidegger, 1969) with tributaries, including meltwater channels, and trunk streams regularly bifurcating upstream into lower order tributaries (Figs. 23a and b), until their deposits disappear near the Keewatin Ice Divide (Shilts, 1984a). This pattern may be interpreted in at least three ways. (1) The whole system may have functioned in sub-ice tunnels extending from the centre of a thin, relatively stagnant glacier to

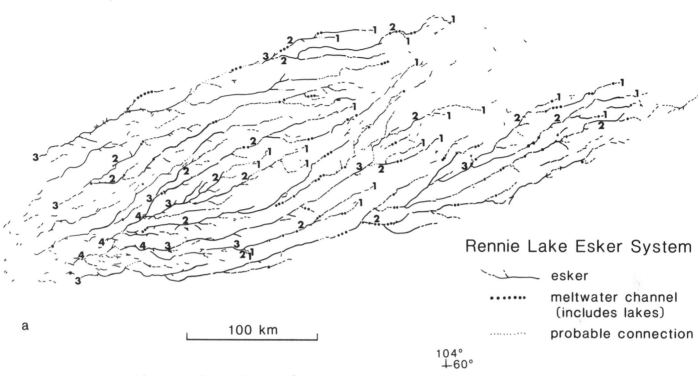

+63°
104°

Rennie Lake Esker System

⌒⌐ esker

•••••• meltwater channel
(includes lakes)

·········· probable connection

104°
+60°

a

|—— 100 km ——|

Figure 23. Examples of esker systems as fully integrated fourth-order Horton drainage systems. Numbers 1-4 refer to first- to fourth-order streams. Abrupt disappearance of Rennie Lake system (23a) downstream probably due to lack of basal sediment in ice overriding hard, crystalline terrain.

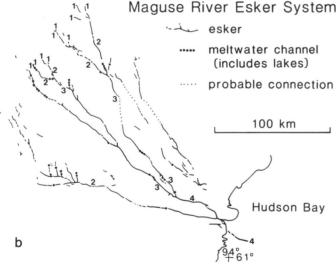

Maguse River Esker System

⌒⌐ esker

••••• meltwater channel
(includes lakes)

····· probable connection

|—— 100 km ——|

Hudson Bay

b

94°
+61°

its retreating margins, which lay at one time some 300 to 500 km away. Although this model has some merit, the very size of the ice sheet at the inception of esker deposition would seem to argue against it; the thicker ice near the divide is too plastic at the base to maintain open tunnels, even with thermal erosion of their walls. In addition, it is probable that the Horton system could not have developed so fully within a solid mass of ice; topographic irregularities at the base of the ice would have exercised greater influence on esker trends than is evident from the Horton pattern.

(2) Eskers may have been deposited by streams flowing in short tunnels near the margin of the glacier, continuity being maintained by up-ice migration of the heads of the tunnels by melting (St-Onge, 1984a, p. 274). Although this model is more compatible with observed sedimentation features and probable dynamic conditions in the retreating ice, it does not explain well the Horton pattern of tributaries, particularly the regular bifurcation of eskers upstream. It is hard to imagine how a subglacial tunnel would bifurcate regularly as it melted up-ice without some external control. (3) A more satisfactory model for esker formation and development of internal meltwater drainage is an integrated system of drainage channels developed on the surface of the glacier, with meltwater plunging to the base of the glacier to flow in a subglacial tunnel the last few kilometres of its course before issuing from the retreating glacial front. Such a system would have developed quite late in the glacial cycle, when most of the glacier was below the equilibrium line. The tunnels near the ice edge would have extended themselves headward by melting, as in

the preceding model, but their headward migration would have followed roughly the traces of the surface drainage, thus accounting for regular bifurcation upstream and the common lack of control by subglacial topography. This hybrid model best explains both the Horton drainage pattern and the manifest evidence of subglacial deposition of most esker sediments.

As meltwater streams descended to the base of the ice near the outer margin of an ice sheet, sediment for esker deposition would have been largely limited to debris in the basal ice. Therefore, the size of the esker can be used as a rough indication of the amount of basal debris. Within the outer three zones of the western Canadian Shield, the size of esker ridges commonly diminishes with distance from the ice centre. The small size of eskers and the paucity of drift cover within Zone 4 indicate that the ice in that area may have been relatively clean. This implies that dispersal trains of sediment from the east had been depleted and little erosion occurred within this zone.

Although eskers are a geomorphic feature that is typical of the Canadian Shield, they are by no means confined to the Shield. Their scarcity in Zone 4 is probably due to a lack of debris in the ice rather than to some difference in ice dynamics in this zone. This is evident from the fact that they appear again on the Paleozoic terrain that bounds Zone 4 to the west and in the extension of Zone 3 to the extreme northwest, where the glacial debris load must have been high because of the easily eroded substrate. In the northeast, on the northwest side of Hudson Bay, north of the U-shaped Zone 2, eskers do exist, but are very small and discontinuous compared to eskers in the more southerly parts of Zones 2 and 3. Thus, within this portion of the Shield, the presence or absence of eskers is principally a function of the amount of debris in the ice.

If this is the case for eskers on and adjacent to the Canadian Shield, why are eskers relatively rare throughout most of the rest of the area covered by the Laurentide Ice Sheet? Thick till cover on topographically similar Paleozoic and younger terranes attests to the fact that the ice was carrying abundant basal debris that could have been reworked to form eskers. The answer must lie in a contrast in ice-sheet dynamics between ice retreating from the outer portions of the Laurentide Ice Sheet and those parts that covered the Shield. Our assessment of the meaning of this contrast is that, for most of the terrain surrounding the Shield, ice was flowing so actively during retreat that the integrity of basal meltwater tunnels required for esker deposition could not be maintained long enough to allow esker sediments to accumulate and be preserved. By the time the ice sheet had shrunk approximately to the edge of the Shield, the velocity of flow had diminished to the point where tunnels could be maintained. Thus, the confinement of esker systems roughly to terrane underlain by the Canadian Shield is thought by the authors to be a geographical coincidence rather than a result of some geological or physiographic condition peculiar to the Shield compared to younger terranes. The edge of the Shield, therefore, may mark roughly the edge of the Laurentide Ice Sheet when it ceased to receive enough nourishment to flow vigorously and began to dissipate largely by downwasting. Alternatively, the deformable nature of the bedrock in Mesozoic and Paleozoic beds of the Laurentide Ice Sheet may have promoted rapid flow compared to the rigid Shield terrane, causing tunnel integrity to be difficult to maintain (suggestion by R. J. Fulton; see also Fisher and others, 1985; Clayton and others, 1985).

The role of thermal erosion of the sides of subglacial meltwater channels has not been studied well on the Canadian Shield. Major eskers and interlobate glaciofluvial sediments in Zone 3 in Manitoba are sites of convergent ice flow, which may well reflect drawdown of ice toward sites of thermal erosion. The very bulk of the esker ridges, commonly comprising over 30 m of coarse glacial sediment in areas where the till itself rarely exceeds 5 m thickness, suggests that basal glacial sediment is being fed into esker tunnels from areas adjacent to the esker. This would be accomplished by ice flow toward the tunnel, a process documented in Finland (Repo, 1954) and suspected from the Manitoba data. This local activity within the glacier would be caused by special conditions related to maintaining an optimum tunnel size in the face of thermal erosion, and cannot be considered to reflect activity in the ice sheet as a whole. The trends of ice flow convergent on esker or interlobate complexes and the trends of esker tributaries are commonly at sharp angles to regional trends of drumlins and other bedforms created when the ice sheet was active, further suggesting that the ice sheet which retreated from large parts of the northwestern Shield was sluggishly flowing or stagnant. The dendritic pattern of the eskers provides additional evidence of the relatively stagnant nature of the ice sheet. It is difficult to conceive of a widespread, integrated dendritic meltwater system developing and maintaining itself on the surface of an active ice sheet; it is similarly difficult to conceive of tributary eskers joining the main esker at right angles within vigorously moving ice. The reason for the preferred direction of approach of tributaries (from the left) is unknown. Perhaps it reflects the influence of coriolis force on the movement of water on a large featureless plain—the surface of the ice sheet (a hypothesis that requires most of the ice cover to be below the equilibrium line).

ROGEN MORAINE

Rogen moraine (Lundqvist, 1969) describes fields of short, sinuous ridges oriented more or less at right angles to the local direction of glacial flow. Although introduced formally into the literature as late as 1969 by Jan Lundqvist, the term has been widely used in Europe since the 1930s to describe features similar to those that occur around Lake Rogen in Sweden. In Canada the features are referred to as ribbed moraine by some authors (Elson, 1969; Cowan, 1968; Hughes, 1964) and features described in this section can equally well be described by that name.

On the northwestern Shield, rogen moraine occurs largely in a "U"-shaped belt, Zone 2, around the south end of the Keewatin Ice Divide, but smaller individual belts of rogen moraine are found in several places in Zone 3 (Figs. 10, 11).

Classic rogen moraine comprises nested series of short, sinuous ridges, generally less than 2 km long and less than 10 m high

Figure 24. Typical rogen (ribbed) moraine ridges of Lake Rogen, Sweden type, southern District of Keewatin. (Photo by C. C. Cunningham, 1977, GSC 203315-A.)

(Fig. 24), which are in many places asymmetric in cross section, having a gentle stoss side and steep lee side (Fig. 25). The asymmetry is not everywhere obvious, but where it is developed, the ridges have the aspect of inclined plates of sediment thrust one on top of another (Fig. 25). These forms are associated in space with less well-defined ridges or hummocks of similar composition and relief (Fig. 17). Individual ridges may be fluted (Fig. 26). Larger lakes in areas of rogen moraine have characteristic dentate shorelines where the ridges project into the water, and are typically full of islands where ridges rise above the lake's surface (Fig. 15). Dense concentrations of small linear lakes occupy the depressions between rogen ridges in other areas. In some areas, the classical rogen pattern has smaller rogen ridges superimposed on it, the smaller ridges being less than 100 m long and under 5 m high. In the area east of Roseblade Lake (Fig. 1), trains of asymmetric rogen ridges are fairly widely spaced and assume a crescentic or barchan-like form, concave down-ice.

Fields of rogen moraine and individual ridges commonly

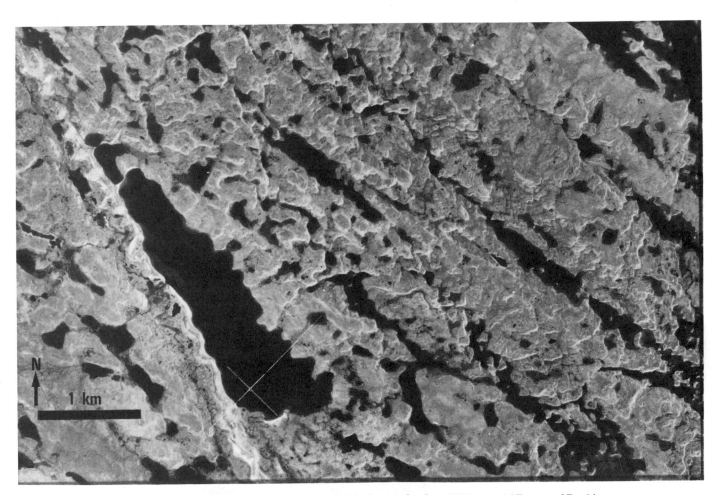

Figure 25. Asymmetric rogen moraine deposited by ice moving from NW toward SE, west of Rankin Inlet, Hudson Bay. Rogen ridges have aspect of plates of boulder-covered till thrust one on the back of another, plate edges facing down-ice. Note beaded, meandering esker superimposed at an angle of about 20° on the rogen train. (Portion of airphoto NAPL A14887-2.)

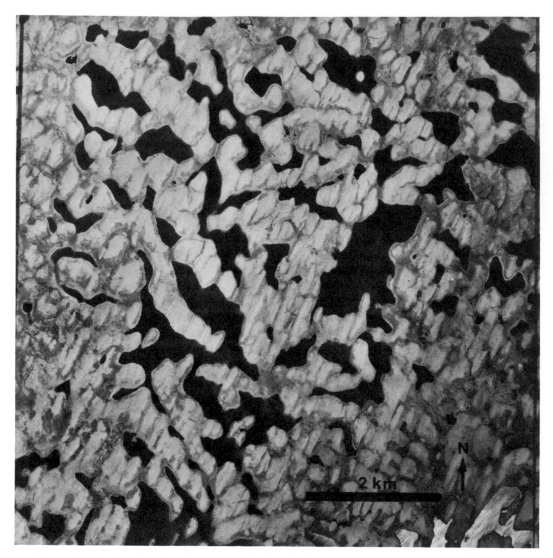

Figure 26. Fluted or drumlinized rogen moraine deposited by ice flowing southwestward, southwestern District of Keewatin. (Portion of airphoto NAPL-A17181-76.)

pass laterally into drumlins, but equally commonly the contact between the two types of terrain may be abrupt with no sign of a transitional zone. Individual drumlin ridges are observed to be broken up into rogen-like ridges where a transition zone does exist (Fig. 27).

The up-ice terminations of fields of rogen moraine are very sharp and linear against the featureless terrain of Zone 1, the Keewatin Ice Divide. There is virtually no transition from featureless till plain or bedrock outcrop to rogen moraine. Unlike the pattern of distribution of rogen moraine in Sweden (Lundqvist, 1969) where rogen moraine is developed only above the postglacial limit of marine inundation, rogen moraine is equally well-developed above and below marine limit around the Keewatin Ice Divide.

DRUMLINS

Fields of drumlins or fluted till occur throughout the area around the Keewatin Ice Divide (Fig. 22). Their form is typically much more elongate than those commonly described in text books, and they rarely have the "upside-down–spoon" form. Individual drumlin or flute ridges can attain lengths of over a kilometre, but are rarely more than 100 m wide. Although individual drumlins are usually straight, some are curved where ice flow directions changed along their length. A field of curved drumlins occurs near the south end of the Keewatin Ice Divide.

Drumlins usually occur in fields that are noticeably elongated in the direction of ice flow (Fig. 11), similar in shape and dimensions to fields of rogen moraine. In most instances these

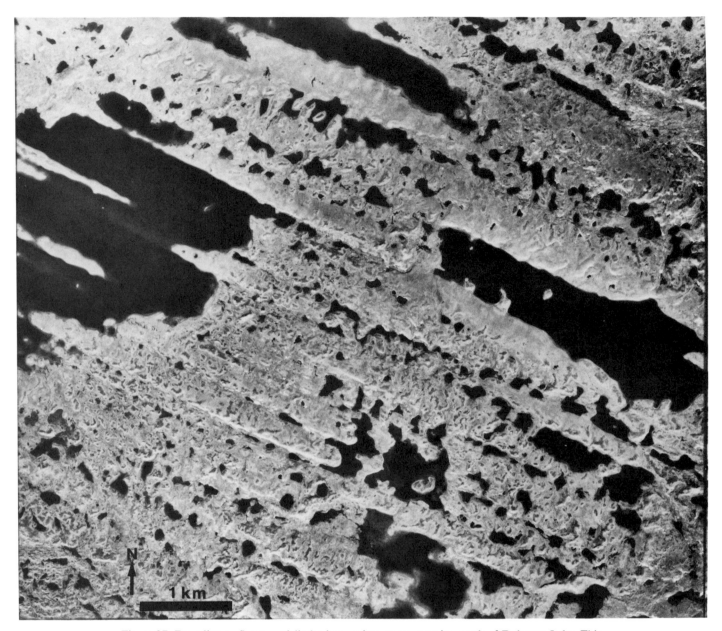

Figure 27. Drumlins or flutes partially broken up into rogen moraine north of Dubawnt Lake. This terrain passes down-ice (northwest) into long flutes or drumlins. (Portion of airphoto NAPL A15066-71.)

fields trend away from specific geological units within the Shield terrane. The most spectacular drumlin fields on the Shield are developed within and down-ice from the Athabaska and Thelon sedimentary basins where poorly consolidated, unmetamorphosed clastic sedimentary rocks have been subjected to intense glacial erosion, producing large quantities of glacial sediment (Fig. 21).

COMPOSITION AND ORIGIN OF DRUMLINS AND ROGEN MORAINE

Because of their close and mutually exclusive spatial rela-

tionship in Zone 2, drumlins and rogen moraine are considered to be formed under similar glaciodynamic conditions. Because both groups of landforms contain elements that reflect regional flow patterns of active ice, they are considered to have been formed at the base of the ice. The composition of the drift that has been shaped into one or the other of these features reflects the source rocks from which the landform trains extend. Where observed in the field, rogen moraine seems to be developed in gravelly sediments or to be formed of till associated with dense concentrations of boulders, which may mantle the till or be intermixed with it. Drumlins, on the other hand, are rarely formed in drift with high boulder concentrations and are underlain by sandy or gravelly sediments only where previously deposited glaciofluvial sediments have been fluted by glacial erosion, or where the source rocks are poorly consolidated sandstones, as in and adjacent to the Thelon and Athabaska sedimentary basins.

Given the general characteristics of the composition and distribution of the two groups of landforms, some explanation for their origin must be sought that explains not only their morphology, but also their close spatial relationship in Zone 2 and the general lack of rogen features in Zone 3 and elsewhere beneath the Laurentide Ice Sheet away from ice divides. Although basal stress conditions have been cited as responsible for formation of one or the other type of feature in Scandinavia (Lundqvist, 1969; Sugden and John, 1976)—rogen moraine reflecting compressive ice flow and drumlins reflecting extensive ice flow—these arguments obviously cannot account for the paucity of rogen features outside of Zone 2. It is the authors' opinion that, although basal stress conditions are important, whether one landform or the other was developed was not so much a function of compression or noncompression as it was a function of the physical character of the debris the ice was carrying. More dilatant material—in the absence of heavy boulder concentrations—formed drumlins or flutes; less dilatant—gravel, crushed bedrock, bouldery debris from outcrops covered by felsenmeer—tended to form rogen and related features. Perhaps drumlinized or fluted terraine was preserved no matter what the conditions of deglaciation, whereas rogen features were destroyed under all but those conditions associated with melting of a largely stagnant ice sheet.

OTHER GLACIAL CONSTRUCTIONAL LANDFORMS

Hummocky Moraine

Small, low hummocks of till, less than 5 m high and 100 m across cover much of the plain in the vicinity of the Keewatin Ice Divide (Fig. 14). Similar clusters of hummocks also have been noted immediately adjacent to and within areas of rogen moraine, but they are rare elsewhere outside of Zone 1. They generally appear to be underlain by till similar in composition to material that is formed into drumlins in nearby areas and rarely have the dense surface cover of boulders typical of rogen ridges formed of till.

We interpret these features to be deposits of the basal portions of the glacier, released from ice that was almost totally stagnant. In the vicinity of the final position of the Ice Divide, flow velocities would have been low even when the ice sheet was active. Before the divide migrated southeastward to this region, the region would have been an area of higher ice flow velocities and therefore an area where sediment was effectively eroded and entrained. The low flow velocities that existed just before Keewatin ice finally melted may have prevented these sediments from being formed into organized landforms, such as the drumlins or flutes typical of similar deposits in zones down-ice from Zone 1. The hummocky surface is probably a reflection of release of debris from the lower parts of an essentially stagnant ice mass.

De Geer Moraine

Short, straight, low ridges of classical De Geer moraine (De Geer, 1940; Elson, 1969) are developed along the coast of Hudson Bay in the vicinity of Rankin Inlet. They are spaced from 150 m to 250 m apart and are thought to represent annual ice front positions of an actively flowing glacier retreating in the sea (Lee, 1959). They are only developed in a zone about 10 km inland from the present coast of Hudson Bay and are superimposed across drumlins.

Very large areas of De Geer moraine also occur southeast of Hudson Bay in Quebec (Fig. 28). Vincent (1977) has noted that these ridges, which are similar in size and spacing to those near Rankin Inlet, are formed only east of the Sakami Moraine, which represents the grounding line of the Labrador ice sheet at the time when the postglacial Tyrrell Sea replaced the high level proglacial lakes (Barlow-Ojibway) dammed to the south (Hillaire-Marcel and others, 1981).

Based on average spacing between individual ridges of the well-developed system of De Geer moraines east of Richmond Gulf, Vincent (1977) estimated an ice retreat rate of about 217 m per year. De Geer moraines in the vicinity of Rankin Inlet and spacing of supposedly annual "beads" in beaded eskers in the same region yield retreat rates of 250 to 290 m/year.

GLACIAL GEOMORPHOLOGY OF LABRADOR AND EASTERN QUEBEC

R. A. Klassen

INTRODUCTION

The area of eastern Canada covered in this section is bounded by longitudes 56° and 68°W, and by latitudes 52° and 57°N; it represents most of central and southern Labrador, and some parts of adjacent Quebec (Fig. 29). Principal glacial and glaciofluvial landforms have been mapped in the area largely by interpretation of aerial photographs at 1:60,000 scale, compiled initially at 1:250,000 scale and finally at 1:1,000,000 scale. Reference to drift composition, and to evidence of ice flow recorded by striae are based, for the most part, on recent field studies in the area and to a lesser extent on published work.

The area, which straddles the southeastern part of the Labrador Ice Divide near Schefferville and extends over 500 km to the east and southeast, occupies only a part of the area associated with ice originating from the divide. Its boundaries are based on lines of latitude and longitude and do not encompass the full extent of the Labradorean Sector of the Laurentide Ice Sheet. Reference has been made to the Glacial Map of Canada to extend interpretations to other areas of Labrador and Quebec included within the Labradorean Sector.

Studies of surficial geology in Labrador and eastern Quebec have been done by the Geological Survey of Canada and by the McGill Sub-Arctic research group since the early 1950s. The work has varied in style of mapping and scale of investigation, and large parts of the area remain to be studied. The most de-

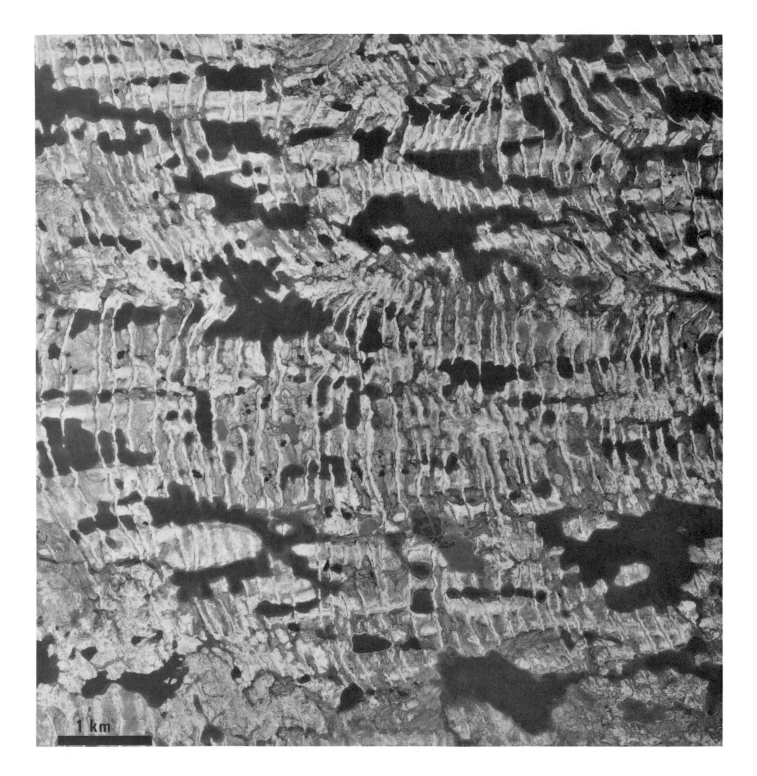

Figure 28. Well-developed field of De Geer moraine formed above marine limit east of south part of Hudson Bay. Ridges are 10 to 15 m high. (From Prest, 1983; portion of airphoto NAPL A14882-91.)

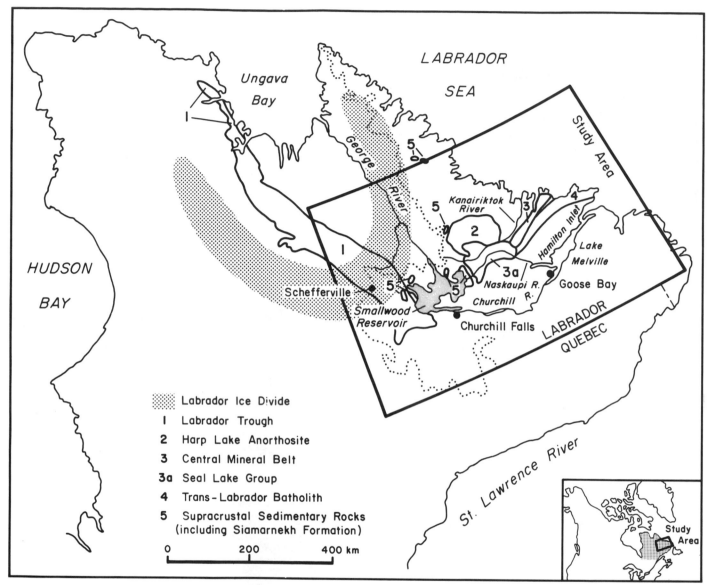

Figure 29. Location map of Quebec and Labrador part of Canadian Shield.

tailed work has been done in the area of Schefferville, due to the logistical support base available there (Ives, 1956, 1960; Henderson, 1959; Kirby, 1961; Derbyshire, 1959). The Glacial Map of Canada at 1:5,000,000 scale (Prest and others, 1968) represents a summary compilation of much of the early work. Surficial geological maps of most of southern Labrador at 1:250,000 scale have been done by Fulton and Hodgson (1979, 1980) and Fulton and others (1975, 1979, 1980a to f, 1981a to d), based largely on interpretation of aerial photographs. The physiography of the region was described first by Douglas and Drummond (1955) and then by Bostock (1970); both physiography and geology were covered by Greene (1974).

Labrador occupies the eastern sector of the Canadian Shield and includes parts of the Superior, Churchill, Grenville, and Nain

geological provinces (Fig. 4). The area is underlain, for the most part, by a basement complex of crystalline intrusive and metamorphic rocks of Archean-Aphebian ages and, to a lesser degree, by less-deformed supracrustal rocks of Aphebian age that are sedimentary and volcanic origin. The supracrustal assemblages outcrop mainly in the Labrador Trough in the west and in the Central Mineral Belt, which extends eastward across central Labrador (Fig. 29).

LABRADOR ICE DIVIDE

Summaries of the glacial history of the Labrador sector of the Laurentide Ice Sheet are given by Prest (1970, 1984), Ives (1978), and by Mayewski and others (1981). Regional patterns

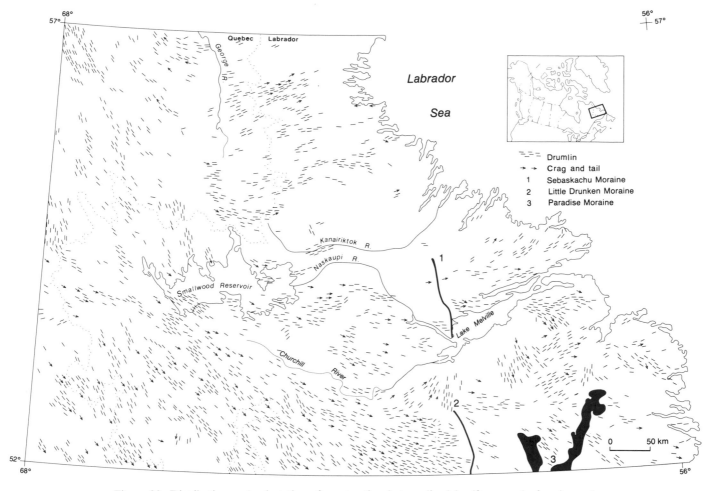

Figure 30. Distribution and orientation of constructional streamlined landforms and of major end moraines, Labrador and eastern Quebec.

of ice flow, recorded on the Glacial Map of Canada, indicate northward flow from central Labrador-Quebec convergent into Ungava Bay, and divergent flow outward toward the south, east, and northeast across Labrador. The boundary between the convergent northward and the divergent southward flow trends describes a large, northward-opening U-shaped area around Ungava Bay; this area defines the Labrador Ice Divide (Fig. 29). The divide extends from northwestern Quebec, southward through Schefferville, and northward along the axis of the Labrador Peninsula.

The glacial history of the Ice Divide is not well-known, and it has not been demonstrated whether the patterns of ice flow associated with the divide occurred throughout late Wisconsinan time or are the result of a late glacial phase of flow associated with ablation of the Laurentide Ice Sheet (Ives and others, 1975). The extent of ice during late Wisconsinan also remains problematic; 'maximum' and 'minimum' portrayals that show ice sheet margins beyond the modern Labrador Coast and well inland of the coast, respectively, have been summarized by Prest (1984). If Prest's (1984) "minimum" late Wisconsin configuration of the Labrador ice sheet is accepted, it is possible that some landforms in eastern Labrador that appear to be related to the divide actually may have remained unglaciated during the late Wisconsinan. Glacial landforms may be time transgressive; thus, the regional patterns evident on Figure 30 may not necessarily reflect the overall pattern of ice flow during any single time or phase of ice flow.

The northeastern margin of Labrador, in the Labrador Highlands, has been proposed as an area where growth of the Laurentide Ice Sheet could have been initiated (Flint, 1943; Ives, 1978). As part of the first geological survey of Labrador, Low (1896) recorded evidence of glacial dispersal centers in central Quebec and of change in their location during the course of glaciation. More recent work has shown an 'early' northeast-trending pattern of ice flow across central Labrador that is associated with glacial dispersal from a center south and west of Labrador prior to flow associated with the U-shaped ice divide (Klassen and Bolduc, 1984). The evidence suggests a more dynamic history to the Laurentide Ice Sheet than is represented by the U-shaped Labrador Ice Divide alone.

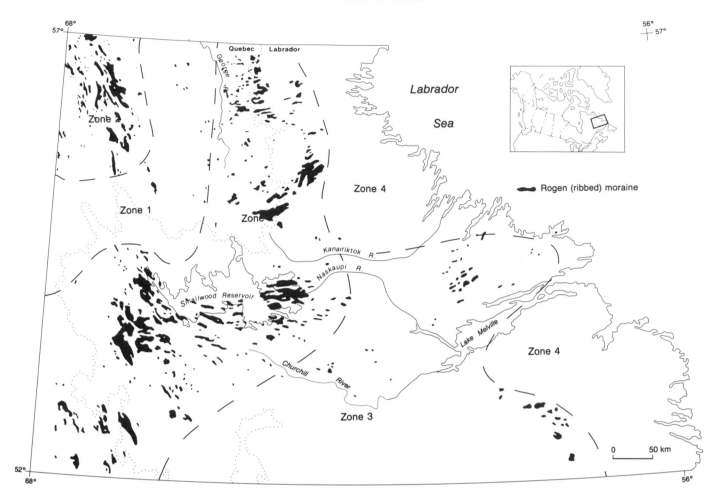

Figure 31. Glacial landform zones and distribution of rogen moraine, Labrador and eastern Quebec.

The significance of ice flow outward from the Schefferville area, which is associated with the Labrador Ice Divide, can be shown by study of the glacial transport patterns of distinctive erratics, particularly iron formation derived from the Labrador Trough. Iron formation erratics have been found at least 400 km from their source outcrops. They are common in the upper Churchill River valley and east of the Smallwood Reservoir, and they are present, but rare, near the western end of Lake Melville. To the north of Schefferville, they have been identified on the southwestern coast of Ungava Bay. The eastward and southeastward dispersal of most of the known erratics of iron formation accords well with regional orientations of glacier bedforms presently associated with the Labrador Ice Divide. The long distance of transport (>400 km) they have undergone indicates they were transported by ice from the Schefferville region during a significant period of time. The dispersal evidence suggests that the southeastern part of the Labrador Ice Divide near Schefferville served as an important ice-dispersal center, as well as being the region toward which the last ice sheet shrank. On this basis, the spatial arrangement of most of the glacial landforms around the southern part of the divide, between Schefferville and Lake Melville, can be discussed with regard to changing dynamics of the ice sheet away from a long-lived center of ice dispersal. Until further work is done, the maximum extent of ice eastward from this center must remain problematic (e.g., Prest, 1984), although it appears to have occupied Zone 3.

Erratics of iron formation have also been found in the area between Smallwood Reservoir and the head of George River, although they are extremely rare. The occurrence of Labrador Trough erratics in that area cannot be readily explained by glacial transport along flow trends associated with the Labrador Ice Divide, although the net distance of transport (>300 km) indicates that they are the result of prolonged ice flow. Their distribution is associated with an ice sheet configuration distinct from the Labrador Ice Divide and possibly with the regional pattern of 'old' northeastern flow identified by Klassen and Bolduc (1984).

The distribution of glacial landforms (Figs. 30, 31, 32) relative to the Labrador Ice Divide is broadly similar to that described for the District of Keewatin, and sedimentation patterns are described according to the system of zones used for that area.

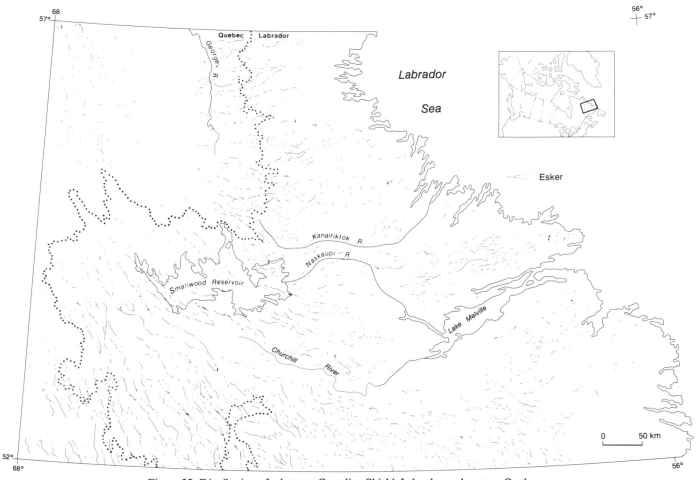

Figure 32. Distribution of eskers on Canadian Shield, Labrador and eastern Quebec.

As used here, the zones describe an assemblage and spatial organization of glacial landforms that are considered to be the result of large-scale ice sheet dynamics. They appear to form discrete zones about the Labrador Ice Divide. Although Sugden and John (1976) relate such patterns only to glaciological conditions prevailing at a glacial maximum, the role of bedrock geology and physiography also appears to be regionally important in formation of the patterns. It should be noted, however, that although the overall landform patterns are similar, there is at present no geological basis for direct comparison of either the dynamic or temporal histories of the ice divides in the District of Keewatin and in Labrador.

Zone 1

The innermost zone (Zone 1, Fig. 31) occupies the area of the Labrador Ice Divide near Schefferville and forms a belt 50 to 100 km in width on either side of the axis of the divide. Its eastern margin is bounded by the George River basin, and describes a large arc curving generally southwestward through the Schefferville area. Zone 1 is characterized by an absence of rogen moraine

(Fig. 31) and eskers (Fig. 32), and such glacially streamlined landforms (Fig. 30) that are present display marked local change in flow direction, much of which is at variance with overall regional flow patterns. According to Henderson (1959), and Kirby (1961), drift cover near Schefferville is variable and generally less than 2 m thick; bedrock controls surface morphology throughout most of the area.

Zone 2

The second zone, Zone 2, forms a belt 100–300 km in width to the east and to the south of Zone 1. It is characterized by well-developed streamlined landforms, rogen moraine and esker systems (Fig. 30, 31, 32). With some exceptions, noted below, the orientation and expression of glacial landforms all reflect the last-recorded trends of regional ice flow outwards from the Labrador Ice Divide, as indicated by detailed examination of striae. Drift cover is continuous to discontinuous throughout the zone, and is thickest in areas lying south and east of the Labrador Trough and east of George River. In the area of Churchill Falls, drift is characterized by large (1 to 2 m) boulders that form an extensive, and in some areas continuous, surface mantle. Glacial landforms appear to be most abundant and best-developed down-ice of the Labrador Trough where they record a divergent pattern of southward-to-eastward ice flow.

Streamlined landforms, including drumlins and crag-and-tail features (Fig. 30), are not uniformly distributed, but preferentially occur in elongate belts that are oriented along regional ice flow trends and that commonly occur between areas of rogen moraine.

Rogen moraine was recognized by early workers to characterize large areas of central Labrador, and has been variously referred to as 'ribble till' (Douglas and Drummond, 1955), 'rippled till' (Ives, 1956), 'cyclical moraines' (Henderson, 1959), and 'ribbed moraine' (Hughes, 1964). The term ribbed moraine was used by Prest (1970). Detailed studies of ribbed (rogen) moraine in the area of Schefferville have been made by Cowan (1968). In Zone 2, rogen moraine is concentrated in patches of irregular outline that are generally elongate in the direction of ice flow. Cowan (1968) noted that in the area south of Schefferville it occurs preferentially in topographical lows. Commonly, major eskers cross through the centre of areas of rogen moraine and occupy erosional channels that cut through moraine ridges.

As in the District of Keewatin, the moraine ridges are most commonly sinuous, less commonly straight, and they are oriented transverse to regional ice flow trends. The ridges may be asymmetric, having gentle up-ice slopes and steeper down-ice slopes, although the moraine can vary in shape and become 'mound-like' and irregular in outline. Areas of rogen moraine pass laterally into areas of streamlined landforms, and Cowan (1968) has noted that the ridges themselves may have a glacially fluted aspect suggesting that they formed within an active ice sheet. The overall distribution of rogen moraine is not uniform and the landform appears to be preferentially developed within the paths of northward and of southward-to-eastward flow out of Labrador Trough, and in the area north of Smallwood Reservoir, east of George River (Fig. 31).

Within Zone 2, eskers define major, glaciofluvial drainage systems with 'trunk' eskers that extend nearly unbroken over distances of 100 to 300 km and numerous, smaller tributary eskers (Fig. 32). The overall orientation of the esker ridges reflects patterns of regional ice flow; an exception is noted northwest of Smallwood Reservoir where eskers are oblique to the trend of glacially streamlined landforms. Tributary eskers commonly connect to the larger trunk eskers with a pronounced left (clockwise) curve, although there are numerous exceptions.

In the north, eskers originate on the western side of George River basin and can be traced eastward through several hundred metres of change in elevation as they extend downslope across the George River and up over the modern watershed (marked by the Quebec-Labrador political boundary). They terminate at the eastern margin of Zone 2 where they reach the approximate inland extent of major valleys transecting Labrador Highlands and southeastern George Plateau (Fig. 3) and connecting directly with Labrador Sea. South of George River, the integrated character of esker systems changes to a segmented one east of Small-

wood Reservoir, at the approximate head of large east-flowing rivers (Kanairiktok, Naskaupi, and Churchill rivers). In valleys heading at the points of esker termination, there are extensive, thick deposits of outwash and proglacial deposits. Although on a regional scale eskers appear continuous, where they cross the modern watershed in the southwest they commonly broaden to form large, flat-topped kames. Formation of the kames may have occurred subsequent to esker deposition as the ice sheet margin retreated north of the watershed and ponded glacial lakes against the drainage divide.

Zone 3

Zone 3 is characterized by continuous to discontinuous drift cover, widely scattered fields of rogen moraine, and a complex pattern of esker systems. It comprises much of southeastern Labrador, extending to the Labrador coast northeast of Lake Melville. There appears to be a direct transition from Zone 2 to Zone 4 in the north. Zone 3 in Labrador is characterized by esker systems that are better developed, with numerous tributary branches in the west, but become segmented and discontinuous, following the general trend of valleys, to the east and southeast. South of Churchill River and in the area west of Lake Melville basin, eskers are notably absent. Ice flow and esker trends are generally parallel within Zone 3. The northern boundary with Zone 4 to the north is marked by an abrupt transition from eastward to northeastward ice flow, by decrease in the number of eskers, and streamlined landforms, and by a dramatic increase in large (>2 m diameter) boulders within drift. The boundary is approximately coincident with the granitic trans-Labrador batholith and the Harp Lake anorthosite.

Zone 4

Zone 4 is characterized in Labrador by a general absence of rogen moraine and either an absence of eskers or discontinuous esker segments that do not define an integrated or continuous pattern of drainage. Surficial cover is largely felsenmeer and locally-derived bedrock rubble, and glacial sediments are thin to absent. The transition from Zone 2 to Zone 4 is coincident with geological and physiographical changes marked to the north by George Plateau, and to the south by Mealy Mountains (Fig. 3). Both physiographic areas are high, and regional ice flow patterns include an element of divergent flow around them.

RELATIONSHIP OF GLACIAL LANDFORMS TO BEDROCK GEOLOGY AND TOPOGRAPHY

Throughout central Labrador, drift cover is generally thin (2–3 m), thickening locally in areas of drumlins and rogen moraine. Drift cover is discontinuous to absent in the east, particularly in the central Labrador uplands and in topographically high areas underlain by intrusive rock (Kaniapiskau Plateau and Mealy Mountains, Fig. 3, and Harp Lake Anorthosite, Fig. 29).

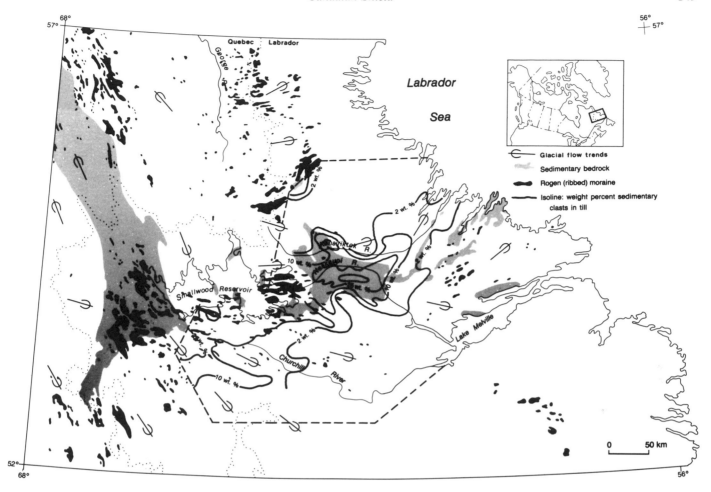

Figure 33. Relationship between rogen moraine trains and major Precambrian metasedimentary basins, Labrador and eastern Quebec. Contours indicate dispersal patterns of metasedimentary clasts based on weight percent of clasts in 4 to 6 mm fraction of till.

In low-lying areas down-ice from the Labrador Trough (Whale Lowland, central Lake Plateau, Fig. 3), drift is relatively thick and continuous. Although glacial deposition appears to have been most significant in the central parts of the ice sheet (Zones 2 and 3), drift thicknesses can be related in part to high sediment production in particularly erodable bedrock terranes. Large moraines in southeastern Labrador, mapped by Fulton and Hodgson (1979), include recessional moraines (Little Drunken Moraine) formed about 9,500 B.P., and a moraine interpreted as a possible limit of late Wisconsinan ice (Paradise Moraine; Fig. 30). Paradise Moraine has been more recently interpreted as a recessional moraine (King, 1985). Little Drunken Moraine marks an apparent change from relatively thick, continuous drift characterized by numerous streamlined landforms aligned with regional flow, to thinner drift and fewer streamlined landforms having more varied orientations.

Regional variations in susceptibility of various bedrock terranes to glacial erosion can be related in part to patterns of landform development. For instance, large intrusive bodies, such as Harp Lake Anorthosite, are relatively resistant, producing little glacial sediment and, consequently, few constructional landforms. Supracrustal sedimentary and volcanic sequences, conversely, produced much more debris, which is expressed by continuous drift cover and by well-developed constructional landforms on and down-ice from their outcrops. These relationships have been supported by lithological analyses of the drift that comprises constructional landforms. There is, for instance, a positive correlation between high concentrations of sedimentary rock fragments and presence of thick drift, fields of rogen moraine, and development of eskers. In particular, the distribution of drift with more than 2% sedimentary clasts in the 4–6 mm fraction is nearly coincident with the distribution of rogen moraine and major esker systems, at least in central Labrador: compositional analysis of till to the north and to the south of the areas where contours are shown has not been done (Fig. 33).

Although sample data are not presently available from Whale Lowland, north of the Schefferville part of the Ice Divide, the northward-trending belt of rogen moraine lies along the principal path of ice flow northward from the Labrador Trough. East of George River, however, belts of rogen moraine are not so easily related to specific bedrock terranes. Their distribution appears to be associated with an arcuate belt of migmatitic and agmatitic bedrock, which elsewhere is relatively resistant to glacial erosion. Alternatively, they could be formed on debris trains extending down-ice from the sedimentary facies of the Siamarnekh Formation (Fig. 29) in the south and from adamellite outcrops in the north. Substantial amounts of glacial debris have been dispersed northward from the adamellite according to Batterson and others (1985).

The association between the occurrence of rogen moraine and erodable bedrock in Zone 2 would suggest that in Labrador the availability of debris to the ice is an important factor in its formation, in addition to large-scale glacial dynamics. The close association between rogen moraine and streamlined landforms, the preferred orientation of moraine ridges transverse to flow trends, and their partially fluted aspect (Henderson, 1959; Cowan, 1968; Hughes, 1964) indicate that rogen moraine is not a product of stagnant ice flow conditions. Shaw (1979) has suggested that rogen moraine reflects high concentrations of basal debris in ice and local zones of compressional flow producing folding. Compressional flow associated with ice flowing eastward, up the regional slope of George Plateau may in part be the cause of the problematic belt of rogen moraine east of George River. Although drift over large areas of sedimentary bedrock east of Smallwood Reservoir contains significant amounts (>10 wt.%) of sedimentary rock debris, rogen moraine is notably absent there, as it is elsewhere in Zone 3 (Fig. 31). Its absence from this area of Zone 3, despite apparently high debris loads, could indicate extensional flow conditions, in comparison with Zone 2.

In addition to supplying debris, bedrock may also exert an influence, which is presently unknown, on the thermal regime at the base of the ice sheet and on discharge of basal meltwater. This might be caused by variations in geothermal flux as well as by variations in permeability of the glacial bed.

Orientations of glacial landforms diverge east and south from the Schefferville area, indicating that the ice sheet thinned in these directions during flow associated with the Labrador Ice Divide. Because overall orientations of eskers are subparallel to these trends, the ice sheet during deglaciation appears to have had a similar configuration. Distinctive physiographic features also appear to have influenced both ice-flow patterns and subglacial discharge of meltwater and associated esker formation. Although striae commonly show ice directly crossing ridges and hills without apparent deflection, there are some areas where ice appears to have been diverted around topographic obstacles and toward basins. Evidence of topographically influenced flow can be seen by the orientation of streamlined landforms around the Mealy Mountains and in the arc around the northern margin of Hamilton Upland, east of Smallwood Reservoir. Northeast of the reservoir, flow trends also indicate flow around the massif underlain by Harp Lake Anorthosite. Topographic influence is limited within the regional context and would appear to be related to late

glacial flow of a thin ice sheet. Patterns of convergent flow into the western end of Lake Melville may be the result of late-glacial drawdown into a low-lying area, possibly by a glacier calving mechanism.

Meltwater discharge at the base of the ice sheet has been influenced by eastward-sloping bedrock valleys that are incised up to several hundred metres into the westward-sloping George Plateau and by large valleys that extend well into the Lake Plateau. Esker systems terminate near the heads of the major valleys, marking the Zone 2–4 boundary north of Smallwood Reservoir, and appear less well developed adjacent to Churchill River and the head of Lake Melville. East of the reservoir, eskers follow the floors of large valleys, curving around the northern end of Hamilton Upland.

Although compositional analysis of the esker systems has not been carried out, their preferential association with rogen moraine in Zone 2 suggests that the supply of abundant glacial debris could be an important factor in their development. An example of this may be seen where poorly developed eskers that drain the Kaniapiskau Plateau, which is underlain by acidic intrusive rocks and pyroxene granulite, become much better developed upon crossing onto sedimentary bedrock terrane of the Labrador Trough. Conversely, the relative rarity of eskers in the northern Zone 4 may result from the limited supply of debris, which is suggested by the general lack of drift cover characteristic of the region.

GLACIAL GEOMORPHOLOGY OF THE SOUTHERN CANADIAN SHIELD

C. A. Kaszycki

INTRODUCTION

The southern and western parts of the Canadian Shield (collectively termed the southern Shield, Fig. 34) are characterized, for the most part, by low to moderate relief (<100 m), gently to moderately rolling topography, and a thick cover of boreal forest. In the region north of the Hudson Bay drainage divide, retreating ice was fronted by immense proglacial lakes, into which large quantities of fine-grained sediment were deposited. These laminated silty clays form a blanket over much of the inundated area. This, coupled with extensive vegetation, conspire to make production of small-scale air photo compilations difficult, without extensive ground checking. The only glacial landforms that can be identified consistently on a regional scale in the southern Shield are glaciofluvial systems and major end moraines. Even these features are sometimes completely buried in the Ontario "clay belt," being frequently encountered in bore holes.

The interior portion of the southern Shield (Zone A, Fig. 34) is characterized by extensive esker systems. In the west and south, eskers tend to terminate at major end moraines such as the Cree Lake (Schreiner, 1984), Lac Seul (Prest, 1963), and

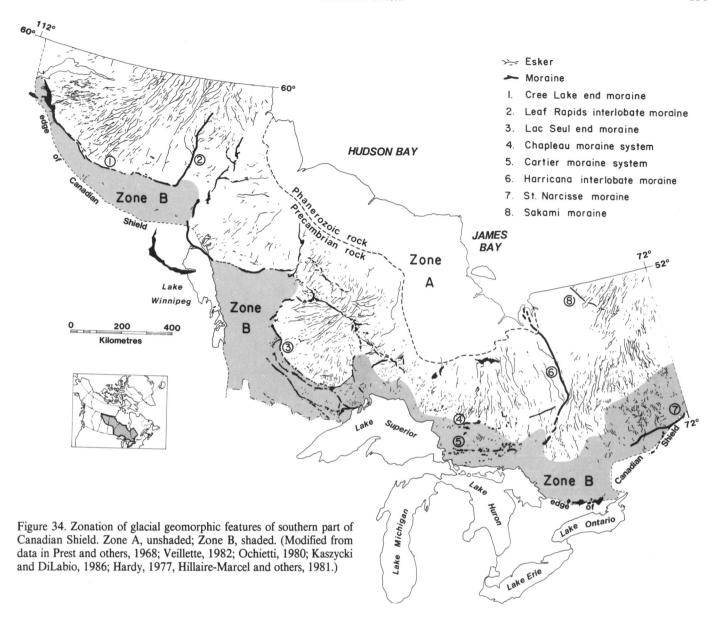

Figure 34. Zonation of glacial geomorphic features of southern part of Canadian Shield. Zone A, unshaded; Zone B, shaded. (Modified from data in Prest and others, 1968; Veillette, 1982; Ochietti, 1980; Kaszycki and DiLabio, 1986; Hardy, 1977, Hillaire-Marcel and others, 1981.)

Legend:
- Esker
- Moraine
1. Cree Lake end moraine
2. Leaf Rapids interlobate moraine
3. Lac Seul end moraine
4. Chapleau moraine system
5. Cartier moraine system
6. Harricana interlobate moraine
7. St. Narcisse moraine
8. Sakami moraine

Chapleau morainic systems (Saarnisto, 1974), producing an exterior esker-free zone (Zone B, Fig. 34) along the outer margin of the southern Shield. In the southern and eastern parts of this region (Zone B), in southeastern Ontario and southwestern Quebec, esker systems become discontinuous and less abundant to absent near the periphery of the southern Shield. This zonation of major landforms may reflect a change in the style of deglaciation from zonal stagnation and active retreat in Zone B (exterior; Kaszycki, 1985), to regional stagnation and downwasting of large ice masses in Zone A (interior). In the east, this zonation may also be a partial function of enhanced relief and topographic control on deglaciation in the Laurentian Highlands (Ochietti, 1980).

ZONE A

Eskers and Moraines

Zone A constitutes by far the major portion of the southern Shield and is similar in topography to the northwestern, and parts of the northeastern Shield. Distinct differences exist, however, between the type and distribution of glacial landforms in the northern Shield (Figs. 10, 11, 12) and those of the southern Shield (Fig. 34). The most striking morphologic aspect of the southern Shield, not observed in Shield terranes to the north, is the pronounced overlapping lobate pattern produced by the con-

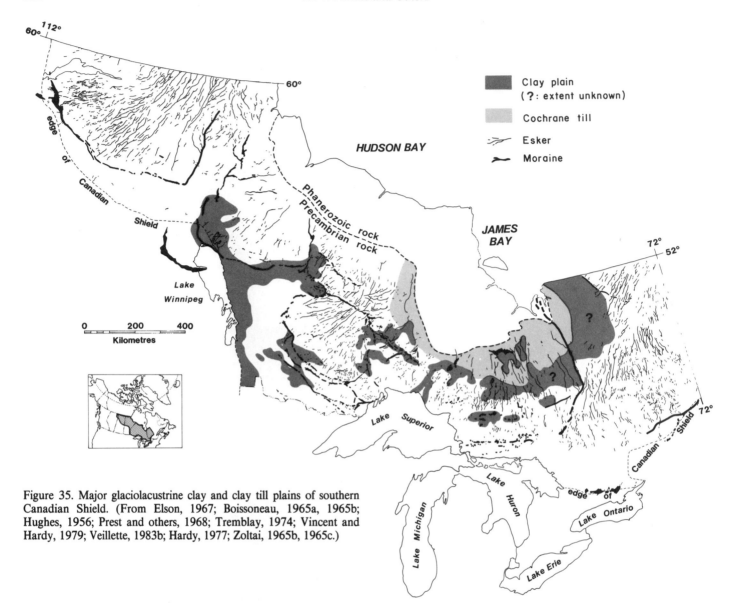

Figure 35. Major glaciolacustrine clay and clay till plains of southern Canadian Shield. (From Elson, 1967; Boissoneau, 1965a, 1965b; Hughes, 1956; Prest and others, 1968; Tremblay, 1974; Vincent and Hardy, 1979; Veillette, 1983b; Hardy, 1977; Zoltai, 1965b, 1965c.)

figuration of end and interlobate moraines (Fig. 34), and the distribution of esker systems related to these features. As a general rule, this pattern is most pronounced where the ice sheet was flowing out of the Paleozoic limestone terrane flooring Hudson Bay, James Bay, and adjacent lowlands to the south. The approximate western limit of carbonate dispersal onto the Shield is defined by the Leaf Rapids interlobate moraine (Kaszycki and DiLabio, 1986). To the south and east this limit is marked by the Harricana interlobate moraine (Wilson, 1938; Veillette, 1982, 1983a, 1983b; Hardy, 1977). These two moraines also mark the western and eastern extent of pronounced multiple lobations of the ice margin. These relationships suggest that where flat-lying, relatively impermeable fine-grained Paleozoic rocks and their derived tills formed the glacier bed, water trapped at the glacier sole caused a decrease in yield strength of the basal ice, resulting in localized readvances or surges of the ice margin (Clayton and others, 1985). In the area north of the Hudson Bay

drainage divide, large proglacial lakes fronted the retreating ice (Fig. 35). Lobation of the ice margin in this region also may have been generated by active calving into proglacial lakes, which would have locally increased ice surface gradient, producing instabilities at the ice margin that resulted in readvance or surging (Prest, 1970; Dyke and others, 1982).

Integrated esker systems in Zone A commonly are well developed within subregions defined by enclosing end and interlobate moraines (Fig. 34). In some instances these systems terminate at end moraines that represent the maximum extent of readvance of an individual ice lobe (i.e., the Lac Seul moraine; Prest, 1963). In the vicinity of large interlobate moraines, eskers commonly have been deflected toward the moraine, producing strongly convergent patterns along the lateral margins of confluent ice lobes (Kaszycki and DiLabio, 1986; Veillette, 1982, 1983a, 1983b). Well-integrated esker systems reflecting the configuration of individual ice lobes suggest that glacial readvance or surging was followed by regional downwasting and stagnation. Periodic resurgence of the ice margin possibly was related to localized glaciological conditions such as an increase in ice surface gradient, or development of an extensive water layer beneath the ice sheet, either near the margin, or up-ice where the glacier bed rested on impermeable fine-grained substrate. These events, therefore, do not necessarily reflect renewed or increased activity of the Laurentide Ice Sheet in this region, but rather represent inherent instability of the ice sheet during deglaciation (see Dyke and others, 1982).

Proglacial Lakes

As previously stated, north of the Hudson Bay–St. Lawrence River drainage divide, and north of the Hudson Bay–Mississippi River divide, ice retreated down gradient, damming northward drainage and producing large proglacial lakes. Fine-grained laminated sediments were deposited within these basins and formed a blanket over much of the area inundated, ranging in thickness from a thin drape (1-2 m) to several tens of meters. The thicker sediments formed featureless plains beneath which glacial and bedrock topography was completely obscured (Fig. 35).

Two major freshwater bodies covered the southern Shield: Lake Agassiz in the western part of the region was impounded between retreating ice and the Mississippi River drainage divide (Teller and Clayton, 1983); Lake Barlow-Ojibway, in the central and eastern part of the region was dammed by retreating ice north of the St. Lawrence River drainage divide (Zoltai, 1965a, 1967; Boisonneau, 1966, 1968; Hughes, 1965; Prest, 1963; Vincent and Hardy, 1979). As Hudson Bay became ice free approximately 7,800 yr B.P. (Skinner, 1973; Vincent and Hardy, 1979), these lakes drained, and marine water inundated isostatically depressed central portions of the former ice sheet. Marine inundation in this region did not extend over vast expanses of the southern Shield, but was confined primarily to the Hudson Bay–James Bay lowlands (Craig, 1968; Hardy, 1982; Dredge and Cowan, 1987).

The Ontario clay belt is probably the most important of the clay plains in the Canadian Shield because of its economic significance. These fine-grained sediments form an agricultural oasis in the normally sterile Shield, but they also present a serious impediment to mineral exploration because they overlie some of the richest mineralized Precambrian terrane in Canada. Thick sequences of laminated silt and clay were deposited in Lake Barlow-Ojibway both before and after the late glacial Cochrane readvances, which reworked part of these sediments into the clay tills covering the terrain north of the clay plain proper (Hughes, 1965; Fig. 35). Large quantities of silt and clay derived from glacial erosion of fine-grained Paleozoic lithologies flooring the Hudson Bay–James Bay lowlands were deposited by both density under flow mechanisms near the ice margin, and by rain-out from suspension in deep water distal environments and near the ice margin during periods when sediment yield was low (i.e., winter months). Where sediment yield was large and density underflow mechanisms dominated sedimentation, thick se-

quences of laminated sediment filled depressions on the till-bedrock surface, producing extensive clay plains. For this reason the distribution of clay plain tends to mimic the stable positions of former ice margins as indicated by moraines and the distribution of the Cochrane clay till (Fig. 35). In areas between these ice retreat positions where deposition from suspension predominated, laminated sediment draped the topography, and the underlying morphology was preserved beneath a thin, 1–2 m, veneer of lacustrine clay (not represented in Fig. 35).

In the Shield area inundated by Lake Agassiz, extensive clay cover occurs within the Red River and Lake Winnipeg basins, and extends northward along the Nelson River system, and eastward into northern Ontario in the Sandy Lake basin (Elson, 1967; Fig. 35). Thick lacustrine sequences (7–20 m) have also been observed within the Wabigoon–Lac Seul basins (Zoltai, 1965c; Elson, 1967; Fig. 35). Sediment deposited within the ice-proximal environment commonly contains a larger proportion of debris carried in the ice and hence is enriched in carbonate and non-expandable clay minerals derived from the Hudson Bay basin and adjacent terranes to the east. Except for the basins described above, most of the Shield terrane forming the Lake Agassiz basin, is draped by a thin veneer of laminated clay, 1 to 2 m thick. Clay mineral studies on lacustrine sediment throughout the Agassiz basin indicate that a large proportion of clay deposited in the distal environment is composed of smectite. This suggests that much of the Lake Agassiz clay was derived from inflow channels draining Cretaceous terranes to the west (Teller, 1976; Kaszycki and DiLabio, 1986).

Zone B

The exterior part of the southern Shield is marked by an absence of well-integrated esker systems. In the southeastern part of the area, relatively small, discontinuous eskers are abundant, but decrease in frequency toward a southern limit marked, in part, by the St. Narcisse moraine (Fig. 34) in southern Quebec, and the Cartier-Chapleau morainic systems in central Ontario. The difference in disposition of eskers and major moraines between zones A and B suggests that the style of deglaciation and ice sheet disintegration in each of these areas was dramatically different. Well-integrated esker systems reflect the configuration of independent ice lobes in Zone A (Fig. 34), as well as the configuration of disintegrating ice masses in the northwestern and northeastern parts of the Shield (Figs. 12, 20, 31, 32), and suggest that final deglaciation was characterized by large-scale regional downwasting and stagnation of the interior portions of the ice sheet. The absence of major esker systems in the exterior part of the southern Shield, may indicate that deglaciation was characterized by active retreat, such that large continuous surface and subglacial drainage systems could not remain open for prolonged periods of time. Consequently, eskers that do exist tend to be short and discontinuous. This pattern is particularly pronounced in the southeastern part of Zone B, where enhanced topographic relief in the Laurentian Uplands also may have played a signifi-

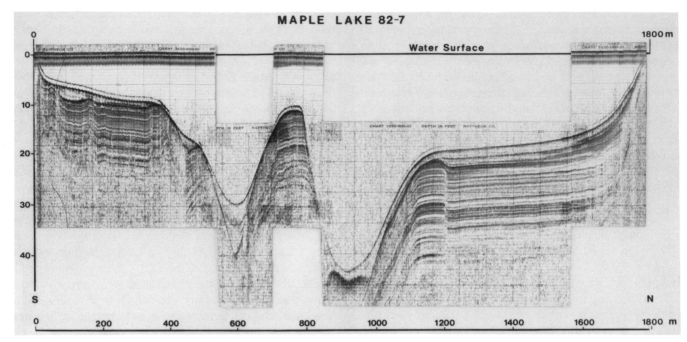

Figure 36. Subbottom acoustic profile across Maple Lake, southern Ontario. Deformation and truncation of the structure of the thick proglacial clay underlying skin of modern organic sediment indicates deposition around stranded blocks of glacier ice. "Pillar" of laminated sediments was deposited between two ice blocks (VE ~15×; from Shilts, 1984b; Kaszycki, 1985.)

cant role in localizing subglacial drainage, and eskers, where present, tend to follow the trend of structural valleys.

Most detailed work relating to the glacial geomorphology of the southern Shield has focused on the description and distribution of surficial sediments, and deglacial history of specific areas. Until recently, little emphasis has been placed on interpretation of glacial processes as related to the distribution, morphology, and sedimentology of glacial deposits (Bouchard, 1980; Bouchard and others, 1984).

In southern Ontario, sediment/landform assemblages have been described that suggest the final phase of deglaciation was characterized by zonal stagnation near the periphery of the ice sheet (Kaszycki, 1985). Terraced glaciofluvial sediment is observed perched along valley walls, marking the location of former high-level ice-marginal drainage routes. Glaciofluvial sediment also forms valley train deposits which both bypass and emerge from lakes that should have acted as sediment traps. These features require the presence of ice blocks in depressions to provide a temporary floor for the through flow of water and sediment. In addition, many lakes within the southern Ontario region are rimmed by glaciolacustrine sediment, indicating that ice blocks were present within modern lake basins at least during the initial phases of proglacial sedimentation (Kaszycki, 1985; Shilts, 1983). Subbottom seismic profiling of lakes within the Shield terrane of southern Ontario and Quebec (Kaszycki, 1985; Shilts, 1983, 1984b) provides additional evidence for local stagnation. At

Maple Lake, in southern Ontario, the stagnation process is documented by the spectacular morphology of lacustrine sediment forming the lake floor (Fig. 36). Prominent ridges of laminated sand, silt, and clay were produced by deposition on and between remnant ice blocks. In places, lamination is truncated abruptly by ice-block depressions. Elsewhere, faulting and slumping are pronounced, as evidenced by the rotational slump block in Figure 36.

In general, the dominent depositional landforms in southern Ontario are ice-contact terraces that flank valley walls wherever sediment accumulation was localized (Kaszycki, 1985). Sediment comprising terraces is variable depending upon sediment supply and the depositional processes operative in the ice-marginal environment. In upland areas sediment cover is thin and discontinuous. Resistant bedrock lithologies minimized glacial erosion and debris production, and the landscape is sediment poor. The most common glacigenic sediment sequences in these areas are composed of complex sediment flow assemblages, often located on the lee side of bedrock knobs, and also forming ice-contact terraces flanking rock knobs and valley sides. Individual sediment flows or 'flow tills' may be complexly interbedded with glaciofluvial and/or glaciolacustrine sediment, depending upon ice-marginal and subglacial drainage conditions.

Basal till, although relatively rare, occurs as two distinct lithofacies. Where present as a thin (<1 m) discontinuous mantle, till is commonly loose to compact with few to abundant stringers of sand, and clasts of contorted fine sand and silt. There is little to no evidence of shear deformation, although clast fabric reflects ice transport orientation. This till facies is thought to represent deposition by basal meltout beneath wholly stagnant ice.

Thick till sequences (>5 m) may occur as rare isolated deposits filling or flanking narrow troughs in the bedrock. This till facies is extremely compact, may exhibit pronounced partings or fissility, and rarely contains inclusions of sorted material. Elongated boulders are commonly striated parallel to the long axis, and these, as well as clast fabric, reflect ice transport orientation. Intratill fluvial deposits have planar upper contacts indicating deposition within subglacial meltwater channels, flowing beneath an active shear plane (Eyles and others, 1982; Shaw, 1985). Likewise, the development of fissility, which may be pronounced in this till facies, has been attributed to deposition beneath active ice. Under conditions of compressive ice flow, the basal shear zone shifts upward, successively stacking thin slabs of debris-rich ice and producing a substrate composed of alternating zones of debris-rich and debris-poor ice. Subsequent melting of interstital ice then produces a fissility inherited from structure at the base of the glacier (Virkalla, 1952). Prolonged compressive flow within and around bedrock obstructions results in locally thick accumulations of till. Streamlined features, primarily flutes, although rare, are located invariably in flat upland areas, and commonly down-ice from major structural valleys. The proximity of flutes to structural valleys suggests that less resistant lithologies forming these valleys, and/or pre-glacial valley-fill sediment may have provided an abundant supply of sediment to be reworked by overriding ice.

The overall pattern and morphology of sedimentation suggests that most deposition occurred during deglaciation, within the proglacial and ice-marginal environment. The prevalence of ice-contact terraces as the dominant depositional landform indicates that localized stagnation occurred near the ice margin, characterized by entrapment of ice blocks within enclosed bedrock basins. Although these observations are limited in areal extent, it is the authors' opinion that similar processes were operative throughout large portions of the exterior southern Shield. It appears that zonal stagnation during active retreat is an important element in the regional style of deglaciation, particularly in areas of moderate to high local bedrock relief (>50–100 m). Stagnation is thought to have occurred, in part, subglacially as the ice sheet thinned, and entrapped basal ice became increasingly isolated from its source. In areas of pronounced topographic relief, however, it is possible that large independent ice lobes may have remained active for a short time after detachment from the Laurentide Ice Sheet (Ochietti, 1980).

This discussion has focused primarily on aspects of glacial geomorphology as observed in southern Ontario. Due to the fact that few previous studies in the exterior portion of the southern Shield have attempted to interpret glacial processes and dynamics, the foregoing discussion must serve as a preliminary interpretation of the style of deglaciation over the entire region. This discussion also has not focused on that part of the 'Canadian' Shield that extends into the north-central United States. A com-

prehensive review of glacial stratigraphy and correlation problems in this area is given by Mayewski and others (1981) to which the reader is referred for additional information.

GLACIAL EROSION OF THE CANADIAN SHIELD

C. A. Kaszycki and W. W. Shilts

INTRODUCTION

Studies of the magnitude of Pleistocene glacial erosion of the Canadian Shield have yielded disparate results, with estimates ranging from a minimum of less than 10 m (e.g., Flint, 1971), to a maximum of 100's to 1,000 m (White, 1972). Various techniques have been used to estimate the thickness of rock removed by glacial erosion. Most of these methods are based on indirect evidence of glacial erosion, such as the configuration of the Precambrian paleoplain (Ambrose, 1964; Flint, 1971), the volume of Pleistocene sediment deposited in the ocean basins (Mathews, 1975; Laine, 1980; Bell and Laine, 1985), and regional distribution of morphologic features in the area covered by the Laurentide Ice Sheet (White, 1972; Sugden, 1978). A more direct method for estimating the average thickness of rock eroded during a single glacial event involves calculating the volume of lithologically distinctive debris dispersed from a known source area and dividing the volume by the area of outcrop (Kaszycki and Shilts, 1979, 1980).

Two basic schools of thought have developed concerning the effects of glacial erosion on the Shield. In the first, proponents of deep erosion by continental ice sheets adhere to the concept that maximum erosional intensity occurred beneath the central portion of the ice sheet, over a region now encompassed by the Canadian Shield and Hudson Bay (White, 1972; Laine, 1980; Bell and Laine, 1985). These researchers advocate glacial exhumation of the Precambrian paleoplain from beneath a substantial cover of Paleozoic strata. White (1972) initially suggested that thick sequences of Precambrian rock have also been stripped from the interior portion of the Shield both adjacent to and underlying Hudson Bay. Obvious geological constraints, such as the presence of thick sequences (>2,000 m) of Paleozoic rock flooring Hudson Bay (Stockwell and others, 1970) and structural control of even small-scale bedrock relief near Hudson Bay, place severe limitations on White's model for deep erosion of Precambrian crystalline rocks in this region.

The second school of thought, to which the authors adhere, advocates pre-glacial exhumation of the Precambrian paleoplain with little subsequent modification of the Precambrian surface by Pleistocene glaciers (Ambrose, 1964; Flint, 1971; Gravenor, 1975; Sugden, 1976). Ambrose (1964) summarized the following early field observations on the nature of the Precambrian surface that support this conclusion: (1) continuation of typical Precambrian topography beneath Paleozoic strata; (2) preservation of Paleozoic outliers within small, deep Precambrian basins; (3) inliers of Precambrian hills piercing Paleozoic strata; and

(4) drainage systems adjusted to structure abruptly truncated along contacts with younger rocks. Cretaceous and Tertiary outliers that rest unconformably on Precambrian rocks have been documented at several localities in the Shield (Higgs, 1978; Andrews and others, 1972; Stockwell and others, 1970; Poole and others, 1970). These observations suggest that most of the Shield was exhumed during the late Paleozoic/early Mesozoic, with sporadic, areally restricted episodes of sedimentation occurring during the late Mesozoic/early Cenozoic.

Gravenor (1975) documented the presence of fresh igneous and metamorphic materials in the earliest known glacial deposits south of the maximum extent of the Illinoian ice margin in the north-central United States. The fact that tills of this age can be differentiated on the basis of Precambrian indicator lithologies outcropping near the present Precambrian-Paleozoic border also suggests that the Shield was exhumed near its present position before the earliest glacial events in the terrestrial record. Preglacial exhumation of the Precambrian Shield and post-glacial preservation of the Precambrian erosion surface severely constrain arguments for deep erosion beneath the central portion of former ice sheets.

MORPHOLOGY

Recently, attention has been focused on the glaciodynamic implications of the contrast in lake density between areas underlain by Precambrian bedrock (large number of lakes) and adjacent areas underlain by flat-lying or even folded younger rocks (Sugden, 1978). Compared to the Shield, lakes are relatively rare in the younger geologic terranes that surround it, except for major water bodies that occur at the contact between Precambrian and younger terranes (i.e., Great Bear Lake, Slave Lake, Lake Athabaska, Lake Winnipeg, Lake Superior, Lake Huron, etc.) and small lakes formed over thermokarst depressions. Based on surface morphology, Sugden (1978) proposed a model for zones of deposition and erosion concentric to a hypothetical ice center in Hudson Bay. Lake-basin density was assumed to be a measure of the efficacy of glacial erosion within a region, and thus, areas exhibiting the greatest density of lakes were considered to represent landscapes dominated by glacial erosion. Erosional and/or depositional processes were assumed to be governed by systematic changes in the basal thermal regime of the ice sheet, radially away from the spreading center. Shield terrain, in general, coincided with predicted zones of warm melting and warm-freezing basal thermal regimes, in which a water layer at the base of the ice sheet, coupled with freeze-on in the warm-freezing zone, would have enhanced basal sliding and debris entrainment, effecting extensive glacial erosion within the region.

It is the author's opinion, however, that the preferential occurrence of lakes on the Shield is largely related to structural and lithological discontinuities that were subjected to particularly deep preglacial weathering, or were composed of easily eroded rock, rather than to basal thermal regime of the ice. In Precambrian terranes north of the Arctic Circle, where cold based glaciers failed to accomplish any significant erosion, a deep preglacial regolith is still in place. This must have existed over most of the more southerly parts of the Shield before being stripped during the earliest glaciations. Remnants of deeply weathered crystalline bedrock are often focused along joints and faults, and in particularly "soft" or soluble bedrock such as marble in the Precambrian terrane between Ottawa and Georgian Bay. It is not unreasonable to assume that these structures, if they had been exposed to the full force of glacial erosion, would have been eroded preferentially and would today form lake basins. Because the younger geologic terranes surrounding the Shield lack the inclined or vertical structures associated with extensive faulting and complex folding on the Shield, the regolith in these areas was probably of a much more uniform thickness and conditions were not favorable for forming undrained depressions.

THE OCEANIC RECORD

One argument against deep erosion by continental ice sheets is centred on the paucity of terrestrial glacial deposits (Flint, 1971). Proponents of deep erosion, however, argue that glacial debris was reduced to very fine grain sizes and that ocean basins acted as the ultimate repository for glacigenic sediments, trapping 90–95% of sediment from glaciated areas (Laine, 1980; Bell and Laine, 1985). Based on the volume of Cenozoic sediment in the western Atlantic Ocean, Mathews (1975) noted a decrease in Cenozoic denudation rate from south to north, apparently contradicting White's (1972) hypothesis of deep erosion beneath the north-central part of the Laurentide Ice Sheet. More rigorous studies, however, based on the volume of mid-Pliocene and Pleistocene sediment deposited on the abyssal plains off the eastern seaboard of North America, estimate the total amount of rock eroded by both glacial and glaciofluvial processes to range between 100 m (Laine, 1980) and 200 m (Bell and Laine, 1985). These authors concluded that substantial denudation of the Canadian Shield has occurred, as well as stripping of Paleozoic and younger sediments.

In order to evaluate the contention that a significant proportion of oceanic glacial sediment is derived from crystalline Shield lithologies, it is instructive to compare the grain size characteristics of terrestrial sediments derived from crystalline shield rocks with those of glacigenic oceanic deposits. Most glacially derived debris in the western North Atlantic is composed of fine-grained turbidites and ice-rafted debris, described variously as silty-clay, or sandy-clay, with an average clay content of approximately 40–50% (Laughton and Berggren, 1972). In contrast, the average grain size composition of Canadian Shield till is approximately 70% sand, 28% silt, and 2% clay (Scott, 1976). Haldorsen (1981), studying the effects of glacial comminution, has documented the relative influence of glacial crushing versus abrasion on the grain size characteristics of monolithic till from an area of sandstone in Norway. She concluded: (1) mineral grains largely retained their primary size (sand) during crushing; (2) abrasion produced cracks within individual minerals resulting in silt-sized rock flour; and

TABLE 1. AVERAGE AMOUNT OF GLACIAL EROSION CALCULATED FOR VARIOUS LITHOLOGIES USING ANALYSIS OF GLACIAL DISPERSAL DATA

Lithology	Topographic Expression	Calculated Amount of Erosion
Volcanigenic rocks of the Dubawnt Group in the Thelon Basin, Keewatin	flat-surfaced	3-5 m*
Devonian granitic intrusives, Eastern Townships, Quebec	isolated hills, 500-550 m relief	1-3 m**
Ultramafic rocks, Thetford Mines, Quebec	flat-surfaced	4.3 m**

*Kaszycki and Shilts (1979, 1980).
**Kaszycki and Shilts (unpublished data).

(3) comminution to smaller grain sizes (clay) does not occur, and, consequently, clay-enriched till can be produced only from a clay-rich substrate. These results indicate that the amount of fine, clay-sized material that could be produced by glacial erosion of crystalline shield lithologies is minimal.

In order to reconstruct a volume of Shield rock eroded from a volume of fine-grained, glacially derived oceanic sediment, these disparities in grain size must be accounted for. If all oceanic sediment was derived from crystalline source rocks, a thickness of 100 m of clay-rich oceanic sediment (50% clay) would be equal to approximately 25 times more till or approximately 2,500 m of shield till with a clay content of approximately 2%. Laine (1980) estimated the volume of sediment eroded from the central part of the Laurentide Ice Sheet (comprising Hudson Bay and the Canadian Shield) to be a minimum of 118×10^4 km^3. Accounting for preferential concentration of fines in oceanic sediment, this sediment volume corresponds to approximately 2.89×10^7 km^3 of shield till. Using an area of 7.1×10^6 km^2 for the Canadian Shield, including the Hudson Bay basin, and accounting for a decrease in porosity of approximately 50% in converting from sediment to bedrock (Laine, 1980), the depth of erosion of Shield rocks and Paleozoic rocks beneath Hudson Bay is estimated to be approximately 2.04 km, a figure far in excess of any previous estimates.

Alternatively, it may be assumed that most fine-grained detritus in oceanic glacial sediment is derived primarily from fine-grained sedimentary rocks (i.e., Paleozoic and Mesozoic limestones and shales). Tills containing a large proportion of clay-sized material are a natural product of glacial abrasion and disaggregation of these lithologies. This may suggest that most sediment deposited in the oceans is derived from Paleozoic and younger rocks, which indicates that most erosion occurred in areas underlain by comparatively soft and easily eroded sedimentary lithologies, located near the periphery of the ice sheet and also beneath Hudson Bay. These areas also coincide with areas of maximum terrestrial accumulation of glacial sediment, attesting to the erodible nature of the substrate and consequent production of large volumes of glacial debris. Shield areas, on the other hand, are composed of resistant crystalline lithologies, not easily comminuted by glacial processes. As a result, erosion in these areas was minimal and the landscape is sediment poor. A considerable amount of fine sediment also may have been contributed to the ocean basins from the presumed thick preglacial regolith, which would have been stripped by the first glaciers to traverse the Shield.

THE TERRESTRIAL RECORD

It is the authors' opinion that the disposition of glacigenic sediment over the continent is a direct reflection of the erodibility of underlying substrate. Thick sequences of glacial sediment (>200 m) occur on the Paleozoic and Mesozoic terranes rimming the periphery of the Shield, and forming the Hudson Bay and James Bay lowlands. Multiple till sequences extend a short distance onto the Shield terrane south of Hudson Bay, reflecting southward transport of easily eroded Paleozoic lithologies. In general, however, Shield terrane is sediment poor. Large areas of thin (<5 m) continuous drift cover occur within and down-ice from Proterozoic sedimentary basins occupied by comparatively erodible fine-grained clastic and volcanic rocks (e.g., the Thelon Basin in the District of Keewatin). In areas dominated by coarse crystalline lithologies, however, sediment cover is discontinuous to absent, reflecting a resistant substrate and lack of glacial erosion.

Estimates of the amounts of glacial erosion that took place over a specific area during a single glacial event have been made by mapping glacial dispersal of various rock types from well-defined sources, calculating the total volume of sediment displaced, and dividing by the area of source outcrops (Kaszycki and Shilts, 1979, 1980). Using this methodology, average amounts of erosion (Table 1) have been calculated for three areas of contrasting lithology and topography: (1) relatively soft, flat-lying Proterozoic Dubawnt volcanic and sedimentary rocks; (2) relatively resistant granitic stocks of the New Hampshire Plutonic Series, forming topographic highs in the Quebec Appalachians; and (3) non-resistant ultramafic outcrops of the Thetford Mines area of Quebec. Granitic rocks, similar to crystalline Shield lithologies, appear to have undergone the least amount of erosion, whereas larger volumes of fine-grained volcanic and sedimentary rocks have been removed. The actual amount of rock removed may have been substantially less than calculated (by as much as 50%), as all assumptions were deliberately chosen to maximize estimates of erosion. In addition, the difference in porosity between till and Shield rock was not accounted for, which may, in some cases, reduce the estimate of erosion by approximately 50% (Laine, 1980). Based on these considerations, we estimate the average amount of resistant Shield rock removed during the last glacial event to be a maximum of approximately 1–2 m, an amount that would have caused minimum modification to preexisting Shield topography.

In order to estimate the total amount of glacial erosion throughout the Quaternary, it is necessary to consider not only the number of glacial events that occurred, but also variations in the rate of erosion over this time period. Andrews (1979) suggested the possibility that the rate of glacial erosion throughout the Quaternary may have been asymmetrical, with most erosion accomplished during the first one or two glacial cycles. It is probable that the deep preglacial regolith, preserved today not only in the Precambrian terranes north of the Arctic Circle, was stripped from the southerly parts of the Shield during the earliest glacial events. Subsequent modification by later glacial cycles was probably limited to localized overdeepening of preexisting structurally controlled basins, with an average amount of erosion similar to that calculated for the last glacial event. A hypothetical erosion curve is presented in Figure 37. The total amount of glacial erosion is dependent upon the amount of erosion during

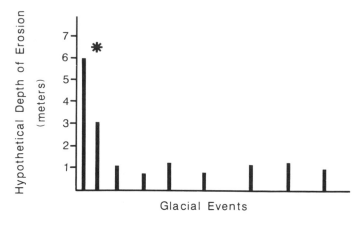

✱ maximum erosion during initial glacial cycles

Figure 37. Hypothetical erosion curve illustrating concept of asymmetric glacial erosion throughout the Pleistocene. Oldest glacial event on left.

the first one or two glacial cycles, and the total number of glacial events throughout the Quaternary. It is the authors' opinion that this value is substantially less than the 100 to 200 m of erosion of Shield terrane estimated by Laine (1980), Laine and Bell (1982), and Bell and Laine (1985), and is likely less than 20 m, a figure in agreement with Flint (1971), Sugden (1976), Gravenor (1975), Andrews (1982). Thus the present bedrock morphology of the Shield is not substantially different from that which existed before glaciation.

REFERENCES

Ambrose, J. W., 1964, Exhumed paleoplains of the Precambrian Shield of North America: American Journal of Science, v. 262, p. 817–857.

Andrews, J. T., 1979, The present ice age; Cenozoic, *in* John, B. J., ed., The winters of the world: Newton Abbott, David and Charles, p. 173–218.

——— , 1982, Comment *on* 'New evidence from beneath the western North Atlantic for the depth of glacial erosion in Greenland and North America': Quaternary Research, v. 17, p. 123–124.

Andrews, J. T., Guennell, G. K., Wray, J. L., and Ives, J. D., 1972, An early Tertiary outcrop in north-central Baffin Island, Northwest Territories, Canada; Environment and significance: Canadian Journal of Earth Sciences, v. 9, p. 233–238.

Andrews, J. T., Shilts, W. W., and Miller, G. H., 1983, Multiple deglaciation of the Hudson Bay Lowlands since deposition of the Missinaibi (last interglacial?) Formation: Quaternary Research, v. 19, p. 18–37.

Arsenault, L., Aylsworth, J. M., Cunningham, C. M., Kettles, I. M., and Shilts, W. W., 1982, Surficial geology, Eskimo Point, District of Keewatin: Geological Survey of Canada, Map 8-1980, scale 1:125,000.

Arsenault, L., Aylsworth, J. M., Kettles, I. M., and Shilts, W. W., 1981, Surficial geology, Kaminak Lake, District of Keewatin: Geological Survey of Canada, Map 7-1979, scale 1:125,000.

Aylsworth, J. M., 1986a, Surficial geology, Kamilukuak Lake, District of Keewatin: Geological Survey of Canada, Map 4-1985, scale 1:125,000.

——— , 1986b, Surficial geology, Ennadai Lake, District of Keewatin: Geological Survey of Canada, Map 5-1985, scale 1:125,000.

——— , 1986c, Surficial geology, Nueltin Lake, District of Keewatin: Geological Survey of Canada, Map 6-1985, scale 1:125,000.

——— , 1987, Surficial geology, Watterson Lake, District of Keewatin: Geological Survey of Canada, scale 1:125,000 (in press).

Aylsworth, J. M., and Shilts, W. W., 1985, Glacial features of the west-central Canadian Shield, *in* Current Research, Part B: Geological Survey of Canada Paper 85-1B, p. 375–381.

——— , 1986, Surficial geology of Coats Island, Northwest Territories: Geological Survey of Canada, Map 1633A, scale 1:125,000.

Aylsworth, J. M., Boydell, A. N., and Shilts, W. W., 1981a, Surficial geology, Tavani, District of Keewatin: Geological Survey of Canada, Map 9-1980, scale 1:125,000.

——— , 1981b, Surficial geology, Marble Island, District of Keewatin, Geological Survey of Canada, Map 10-1980, scale 1:125,000.

Aylsworth, J. M., Boydell, A. N., Cunningham, C. M., and Shilts, W. W., 1981, Surficial geology, MacQuoid Lake, District of Keewatin: Geological Survey of Canada, Map 11-1980, scale 1:125,000.

Aylsworth, J. M., Cunningham, C. M., and Shilts, W. W., 1981a, Surficial geology, Ferguson Lake, District of Keewatin: Geological Survey of Canada, Map 1-1979, scale 1:125,000.

——— , 1981b, Surficial geology, Hyde Lake, District of Keewatin: Geological Survey of Canada, Map 8-1979, scale 1:125,000.

Aylsworth, J. M., Kettles, I. M., and Shilts, W. W., 1981, Surficial geology, Dawson Inlet, District of Keewatin: Geological Survey of Canada, Map

9-1979, scale 1:125,000.

Aylsworth, J. M., Boydell, A. N., and Shilts, W. W., 1985, Surficial geology, Gibson Lake, District of Keewatin: Geological Survey of Canada, Map 1-1984, scale 1:125,000.

——, 1966, Surficial geology, Chesterfield Inlet, District of Keewatin: Geological Survey of Canada, Map 1-1985, scale 1:125,000.

Aylsworth, J. M., Cunningham, C. M., Kettles, I. M., and Shilts, W. W., 1986, Surficial geology, Henik Lakes, District of Keewatin: Geological Survey of Canada, Map 2-1985, scale 1:125,000.

Aylsworth, J. M., Cunningham, C. M., and Shilts, W. W., 1986, Surficial geology, Baker Lake, District of Keewatin: Geological Survey of Canada, Map 3-1985, scale 1:125,000.

Batterson, M. J., Taylor, D. M., and Vatcher, S. V., 1985, Quaternary mapping and drift exploration in the Strange Lake area, Labrador, *in* Current research: Newfoundland Department of Mines and Energy, Quaternary Geology Section, Report 89-1, p. 4–10.

Bell, M., and Laine, E. P., 1985, Erosion of the Laurentide region of North America by glacial and glaciofluvial processes: Quaternary Research, v. 23, p. 154–174.

Boissonneau, A. N., 1965a, Surficial geology, Algoma-Cochrane: Ontario Department of Lands and Forests, Map 5365, scale 1:506,880.

——, 1965b, Surficial geology, Algoma, Sudbury, Timiskaming, and Nipissing: Ontario Department of Lands and Forests, Map 5465, scale 1:506,880.

——, 1966, Glacial geology of northeastern Ontario; I. The Cochrane-Hearst area: Canadian Journal of Earth Sciences, v. 3, p. 559–578.

——, 1968, Glacial geology of northeastern Ontario; II. The Timiskaming-Algoma area: Canadian Journal of Earth Sciences, v. 5, p. 97–109.

Bostock, H. S., 1970, Physiographic subdivisions of Canada, *in* Douglas, R.J.W., ed., Geology and economic minerals of Canada: Geological Survey of Canada, Economic Geology Report, no. 1, p. 9–30.

Bouchard, M. A., 1980, Late Quaternary geology of the Témiscamie area, central Québec [Ph.D. thesis]: Montreal, McGill University, 284 p.

Bouchard, M. A., Cadieux, B., and Gautier, F., 1984, L'origine et les caractéristiques des lithofaciès du till dans le secteur nord du lac Albanel, Québec: Une étude de la dispersion glaciaire clastique, *in* Guha, J., and Chown, E. H., eds., Chibougamau—Stratigraphy and mineralization: Canadian Institute of Mining and Metallurgy, Special Volume 34, p. 244–260.

Boulton, G. S., Smith, G. D., Jones, A. S., and Newsome, J., 1985, Glacial geology and glaciology of the last mid-latitude ice sheet: Journal of the Geological Society of London, v. 142, p. 447–474.

Charbonneau, B. W., 1980, The Fort Smith radioactive belt, Northwest Territories, *in* Current Research, Part C: Geological Survey of Canada, Paper 80-1C, p. 45–47.

Clarke, M. D., 1983, Surficial Geology of Pelly Lake Map Sheet, District of Keewatin, Northwest Territories [B.A. thesis]: Ottawa, Carleton University.

Clayton, L., Teller, J. T., and Attig, J. W., 1985, Surging of the southwestern part of the Laurentide Ice Sheet: Boreas, v. 14, p. 235–241.

Cowan, W. R., 1968, Ribbed moraine; Till-fabric and origin: Canadian Journal of Earth Science, v. 5, p. 1145–1159.

Craig, B. G., 1964, Surficial geology of east-central District of Mackenzie: Geological Survey of Canada, Bulletin 99, 41 p.

——, 1965, Glacial Lake McConnell, and the surficial geology of parts of Slave River and Redstone River map-areas, District of Mackenzie: Geological Survey of Canada, Bulletin 122, 33 p.

——, 1968, Late-glacial and postglacial history of the Hudson Bay region, *in* Hood, P. J., ed., Earth science symposium in Hudson Bay: Geological Survey of Canada, Paper 68-53, p. 63–77.

De Geer, G., 1940, Geochronologica Suecica Principles: Kungliga Svenska/Vetenskapsakademiens Handlingat, Series 3, 18, no. 6, p. 96–130.

Derbyshire, E., 1959, Glacial drainage channels of the Schefferville-Goodwood area: McGill Sub-Arctic Research Papers, no. 6, p. 34–38.

DiLabio, R.N.W., 1978, Occurrences of disrupted bedrock on Goulburn Group: Eastern District of Mackenzie, *in* Current research, Part A: Geological Survey of Canada, Paper 78-1A, p. 499–500.

Douglas, R.J.W., ed., 1970, Geology and economic minerals of Canada: Geological Survey of Canada, Economic Geology Report, no. 1, 838 p.

Douglas, M.C.V., and Drummond, R. N., 1955, Map of the physiographic regions of Ungava–Labrador: Canadian Geographic, v. 5, p. 9–16.

Dredge, L. A., and Cowan, W. R., 1987, Quaternary geology of the Southwestern Canadian Shield, *in* Fulton, R. J., Heginbottom, J. A., and Funder, S., eds., Quaternary geology of Canada and Greenland: Geological Survey of Canada, Geology of Canada Series (in press).

Dredge, L. A., and Nixon, F. M., 1986, Surficial geology, northeastern Manitoba: Geological Survey of Canada, Map 1617A, scale 1:500,000.

Dredge, L. A., Nixon, F. M., and Richardson, R. J., 1985, Surficial geology, northwestern Manitoba: Geological Survey of Canada, Map 1608A, scale 1:500,000.

Dyke, A. S., Dredge, L. A., and Vincent, J.-S., 1982, Configuration and dynamics of the Laurentide Ice Sheet during the Late Wisconsinan maximum: Géographie physique et Quaternaire, Vol. XXXVI, p. 5–14.

Elson, J. A., 1967, Geology of glacial Lake Agassiz, *in* Mayer–Oakes, W. J., ed., Life, land, and water: Winnipeg, University of Manitoba Press, p. 37–95.

——, 1969, Washboard moraines and other minor moraine types, *in* Fairbridge, R. W., ed., Encyclopedia of geomorphology; Encyclopedia of Earth Sciences Series, Volume 3: New York, Reinhold Book Corporation, p. 1213–1219.

Eyles, N., Sladen, J. A., and Gilroy, S., 1982, A depositional model for stratigraphic complexes and facies superimposition of lodgement tills: Boreas, v. 11, p. 317–333.

Fenneman, N. M., 1938, Physiography of eastern United States: New York, McGraw Hill, 714 p.

Fisher, D. A., Reeh, N., and Langley, K., 1985, Objective reconstructions of the late Wisconsinan Laurentide Ice Sheet and the significance of deformable beds: Géographie physique et Quaternaire, vol. XXXIX, no. 3, p. 229–238.

Flint, R. F., 1943, Growth of North American ice sheet during Wisconsinan age: Geological Society of America Bulletin, v. 54, p. 325–362.

——, 1971, Glacial and Quaternary geology: New York, Wiley, 892 p.

Fulton, R. J., and Hodgson, D. A., 1979, Wisconsin glacial retreat, southern Labrador, *in* Current Research, Part C: Geological Survey of Canada, Paper 79-1C, p. 17–21.

——, 1980, Surficial geology, Goose Bay: Geological Survey of Canada, Map 22-1979, scale 1:250,000.

Fulton, R. J., Hodgson, D. A., and Minning, G. V., 1975, Inventory of Quaternary geology, southern Labrador; An example of Quaternary geology—terrain studies in undeveloped areas: Geological Survey of Canada, Paper 74-46, p. 1–14.

——, 1979, Surficial geology, Lac Brûlé: Geological Survey of Canada, Map 1-1978, scale 1:250,000.

——, 1980a, Surficial geology, Lac Melville: Geological Survey of Canada, Map 23-1979, scale 1:250,000.

——, 1980b, Surficial geology, Cartwright: Geological Survey of Canada, Map 24-1979, scale 1:250,000.

——, 1980c, Surficial geology, Groswater Bay: Geological Survey of Canada, Map 25-1979, scale 1:250,000.

——, 1980d, Surficial geology, Rigolet: Geological Survey of Canada, Map 26-1979, scale 1:250,000.

——, 1980e, Surficial geology, Snegamook Lake: Geological Survey of Canada, Map 27-1979, scale 1:250,000.

——, 1980f, Surficial geology, Kasheshibaw Lake: Geological Survey of Canada, Map 28-1979, scale 1:250,000.

——, 1981a, Surficial geology, Battle Harbour: Geological Survey of Canada, Map 19-1979, scale 1:250,000.

——, 1981b, Surficial geology, Upper Eagle River: Geological Survey of Canada, Map 20-1979, scale 1:250,000.

——, 1981c, Surficial geology, Ossokmanvan Lake: Geological Survey of Canada, Map 29-1979, scale 1:250,000.

Fulton, R. J., Hodgson, D. A., Minning, G. V., and Thomas, R. D., 1981d, Surficial geology, Winokapau Lake: Geological Survey of Canada, Map 21-1979, scale 1:250,000.

Fyles, J. G., 1955, Pleistocene features, *in* Wright, G. M., ed., Geological notes on central District of Keewatin, Northwest Territories: Geological Survey of Canada, Paper 55-17, p. 3–4.

Gravenor, C. P., 1975, Erosion by continental ice sheets: American Journal of Science, v. 275, p. 594–604.

Greene, B. A., 1974, An outline of the geology of Labrador: Newfoundland, Department of Mines and Energy, Information Circular no. 15, 64 p.

Haldorsen, S., 1981, Grain-size distribution of subglacial till and its relation to glacial crushing and abrasion: Boreas, v. 10, p. 91–105.

Hardy, L., 1977, La déglaciation et les épisodes lacustres et marins sur le versant québécois des basses terres de la baie de James: Géographie physique et Quaternaire, vol. XXXI, p. 261–273.

——, 1982, La moraine frontale de Sakami Québec subarctique: Géographie physique et Quaternaire, vol. XXXVI, p. 51–61.

Helie, R. G., 1984, Surficial geology, King William Island and Adelaide Peninsula, districts of Keewatin and Franklin, Northwest Territories: Geological Survey of Canada, Map 1618A, scale 1:250,000.

Henderson, E. P., 1959, A glacial study of central Québec–Labrador: Geological Survey of Canada, Bulletin 50, 94 p.

Higgs, R., 1978, Provenance of Mesozoic and Cenozoic sediments from the Labrador and west Greenland continental margins: Canadian Journal of Earth Sciences, v. 15, p. 1850–1860.

Hillaire-Marcel, C., Ochietti, S., and Vincent, J.-S., 1981, Sakami moraine, Quebec: A 500-km-long moraine without climatic control: Geology, v. 9, p. 210–214.

Horton, R. E., 1945, Erosion development of streams and their drainage basins: Geological Society of America Bulletin, v. 56, p. 275–370.

Hughes, O. L., 1956, Surficial geology of Smooth Rock Cochrane District, Ontario: Geological Survey of Canada, Paper 55-41, 9 p.

——, 1964, Surficial geology, Nichicun–Kaniapiskau map-area, Quebec: Geological Survey of Canada, Bulletin 106, 20 p.

——, 1965, Surficial geology of part of the Cochrane District, Ontario, Canada, *in* Wright, H. E., and Frey, D. G., eds., International studies on the Quaternary: Geological Society of America Special Paper 84, p. 535–565.

Hughes, T., Borns, H. W., Jr., Fastook, J. L., Hyland, M. R., Kite, J. S., and Lowell, T. W., 1985, Models of glacial reconstruction and deglaciation applied to maritime Canada and New England: Geological Society of America Special Paper 197, p. 139–150.

Ives, J. D., 1956, Till patterns in central Labrador: The Canadian Geographer, no. 8, p. 25–33.

——, 1960, The glaciation of Labrador–Ungava; An outline: Cahiers de Géographie de Québec, no. 8, p. 323–343.

——, 1978, The maximum extent of the Laurentide Ice Sheet along the east coast of North America during the last glaciation: Arctic, v. 31, no. 1, p. 24–53.

Ives, J. D., Andrews, J. T., and Barry, R. B., 1975, Growth and decay of the Laurentide Ice Sheet and comparisons with Fenno–Scandinavia: Naturwissenschaften, no. 62, p. 118–125.

Kaszycki, C. A., 1985, A model for glacial and proglacial sedimentation in the shield terrane of southern Ontario: Geological Society of America Abstracts with Programs, v. 17, p. 294.

Kaszycki, C. A., and DiLabio, R.N.W., 1986, Surficial geology and till geochemistry, Lynn Lake–Leaf Rapids region, Manitoba, *in* Current Research, Part B: Geological Survey of Canada, Paper 86-1B, p. 245–256.

Kaszycki, C. A., and Shilts, W. W., 1979, Average depth of glacial erosion, Canadian Shield, *in* Current Research, Part B: Geological Survey of Canada, Paper 79-1B, p. 395–396.

——, 1980, Glacial erosion of the Canadian Shield; Calculation of average depths: Atomic Energy of Canada Limited, Technical Report TR-106, 37 p.

King, G. A., 1985, A standard method for evaluating radiocarbon dates of local deglaciation: Géographie physique et Quaternaire, v. 39, no. 2, p. 163–182.

Kirby, R. P., 1961, Movements of ice in central Labrador–Ungava: Cahiers de Géographie de Québec, v. 15, no. 10, p. 205–218.

Klassen, R. A., and Bolduc, A., 1984, Ice flow directions and drift composition, Churchill Falls, Labrador, *in* Current research, Part A: Geological Survey of Canada, Paper 84-1A, p. 255–258.

Laine, E. P., 1980, New evidence from beneath the western North Atlantic for the depth of glacial erosion in Greenland and North America: Quaternary Research, v. 14, p. 125–127.

Laine, E. P., and Bell, M., 1982, Reply *to* comment *on* 'New evidence from beneath the western North Atlantic for the depth of glacial erosion in Greenland and North America': Quaternary Research, v. 17, p. 125–127.

Laughton, A. S., and Berggren, W. A., eds., 1972, Initial Reports of the Deep Sea Drilling Project, Volume 12: Washington, D.C., U.S. Government Printing Office, 1243 p.

Lee, H. A., 1959, Surficial geology of southern District of Keewatin and the Keewatin Ice Divide, Northwest Territories: Geological Survey of Canada, Bulletin 51, 42 p.

Lee, H. A., Craig, B. G., and Fyles, J. G., 1957, Keewatin Ice Divide [abs.]: Geological Society of America Bulletin, v. 68, no. 12, pt. 2, p. 1760–1761.

Low, A. P., 1896, Report on exploration in the Labrador Peninsula, Glacial geology: Geological Survey of Canada, Annual Report, p. 289–311 L.

Lundqvist, J., 1969, Problems of the so-called Rogen moraine: Sveriges Geologiska Undersokning, Ser C NR 648 (Arsbok 64 NR 5), 32 p.

Mathews, W. H., 1975, Cenozoic erosion and erosion surfaces of eastern North America: American Journal of Science, v. 275, p. 818–824.

Mayewski, P. A., Denton, G. H., and Hughes, T. J., 1981, Late Wisconsin ice sheets of North America, *in* Denton, G. H., and Hughes, T. J., eds., The last great ice sheets: New York, John Wiley and Sons, p. 67–178.

McDonald, B. C., 1969, Glacial and interglacial stratigraphy, Hudson Bay Lowland, *in* Hood, P. J., ed., Earth science symposium on Hudson Bay: Geological Survey of Canada, Paper 68-53, p. 78–99.

Ochietti, S., 1980, Le Quaternaire de la région of Trois-Rivières–Shawinigan, Québec; Contribution à la paléogéographie de la vallée moyenne du St-Laurent et corrélations stratigraphiques: Université du Québec à Montréal, 223 p.

Poole, W. H., Sanford, B. V., Williams, H., and Kelley, D. G., 1970, Geology of southeastern Canada, *in* Douglas, R.J.W., ed., Geology and economic minerals of Canada: Geological Survey of Canada, Economic Geology Report, no. 1, p. 229–304.

Prest, V. K., 1963, Red Lake–Lansdowne House area, northwestern Ontario: Geological Survey of Canada, Paper 63-6, p. 23.

——, 1970, Quaternary Geology of Canada, *in* Douglas, R.J.W., eds., Geology and economic minerals of Canada: Geological Survey of Canada, Economic Geology Report, no. 1, p. 675–764.

——, 1983, Canada's heritage of glacial features: Geological Survey of Canada, Miscellaneous Report 28, 119 p.

——, 1984, The late Wisconsinan glacier complex, *in* Fulton, R. J., ed., Quaternary stratigraphy of Canada; A Canadian contribution to IGCP Project 24: Geological Survey of Canada Paper 84-10, p. 21–36.

Prest, V. K., Grant, D. G., and Rampton, V. N., 1968, Glacial map of Canada: Geological Survey of Canada, Map 1253A, scale 1:5,000,000.

Repo, R., 1954, Om forhallandet mellan raflor och asar (on the relationship between striae and eskers): Geologi, v. 6, no. 5, 45 p.

Saarnisto, M., 1974, The deglaciation history of the Lake Superior region and its climatic implications: Quaternary Research, v. 4, p. 316–339.

Scheidegger, A. E., 1969, Stream orders, *in* Fairbridge, R. W., ed., Encyclopedia of geomorphology; Encyclopedia of Earth Sciences Series, Volume 3: New York, Reinhold Book Corporation, p. 1064–1066.

Schreiner, B. T., 1984, Quaternary geology of the Precambrian Shield, Saskatchewan: Saskatchewan Geological Survey, Report 221, 106 p.

Scott, J. S., 1976, Geology of Canadian tills, *in* Legget, R. F., ed., Glacial till: Royal Society of Canada Special Publication, no. 12, p. 50–66.

Shaw, J., 1979, Genesis of the Sveg tills and Rogen moraines of central Sweden; A model of basal melt out: Boreas, v. 8, p. 409–426.

Shaw, J., 1985, Subglacial and ice marginal environments, *in* Ashley, G. M., Shaw, J., and Smith, N. D., eds., Glacial sedimentary environments: Society of Economic Paleontologists and Mineralogists, Short Course no. 16,

p. 7–84.

Shilts, W. W., 1977, Geochemistry of till in perennially frozen terrain of the Canadian Shield; Application to prospecting: Boreas, v. 6, p. 203–212.

——, 1978, Nature and genesis of mudboils, central Keewatin, Canada: Canadian Journal of Earth Sciences, v. 15, p. 1053–1068.

——, 1980, Flow patterns in the central North American ice sheet: Nature, v. 286, no. 5770, p. 213–218.

——, 1982, Quaternary evolution of the Hudson/James Bay Region: Naturaliste Canadien, v. 109, p. 309–332.

——, 1983, Sonar studies of late and postglacial sediments in Canadian lakes: Geological Society of America Abstracts with Programs, v. 15, p. 686.

——, 1984a, Esker sedimentation models, Deep Rose Lake map-area, District of Keewatin, *in* Current research, Part B: Geological Survey of Canada, Paper 84-1B, p. 217–222.

——, 1984b, Sonar evidence for postglacial tectonic instability of the Canadian Shield and Appalachians, *in* Current research, Part A: Geological Survey of Canada, Paper 84-1A, p. 567–579.

——, 1985, Geological models for the configuration, history, and style of disintegration of the Laurentide Ice Sheet, *in* Woldenberg, M. J., ed., Models in geomorphology; The 'Binghampton' symposia in geomorphology, International series, Volume 14: London, George Allen and Unwin, p. 73–91.

Shilts, W. W., and Dean, W. E., 1975, Permafrost features under Arctic lakes, District of Keewatin, Northwest Territories: Canadian Journal of Earth Sciences, v. 12, no. 4, p. 649–662.

Skinner, R. G., 1973, Quaternary stratigraphy of the Moose River basin, Ontario: Geological Survey of Canada, Bulletin 225, 77 p.

Stockwell, C. H., McGlynn, J. C., Emslie, R. F., Sanford, B. V., Norris, A. W., Donaldson, J. A., Fahrig, W. F., and Currie, K. L., 1970, Geology of the Canadian Shield, *in* Douglas, R.J.W., ed., Geology and economic minerals of Canada: Geological Survey of Canada, Economic Geology Report, no. 1, p. 45–150.

St.-Onge, D. A., 1984, Surficial deposits of the Redrock Lake area, District of Mackenzie, *in* Current research, Part A: Geological Survey of Canada, Paper 84-1A, p. 271–277.

——, Surficial geology, Coppermine River, Northwest Territories: Geological Survey of Canada, Map 1645A, 1:500,000 scale, (in press).

Sugden, D. E., 1976, A case against deep erosion of shields by ice sheets: Geology, v. 4, p. 580–582.

——, 1977, Reconstruction of the morphology, dynamics, and thermal characteristics of the Laurentide Ice Sheet at its maximum: Arctic and Alpine Research, v. 9, p. 21–47.

——, 1978, Glacial erosion by the Laurentide Ice Sheet: Journal of Glaciology, v. 20, p. 367–391.

Sugden, D. E., and John, B. S., 1976, Glaciers and landscape, a geomorphological approach: New York, John Wiley and Sons, 376 p.

Teller, J. T., 1976, Lake Agassiz deposits in the main offshore basin of southern Manitoba: Canadian Journal of Earth Sciences, v. 13, p. 27–43.

Teller, J. T., and Clayton, L., eds., 1983, Glacial Lake Agassiz: Geological Association of Canada, Special Paper 26, 451 p.

Thomas, R. D., 1981a, Surficial geology, Montressor River, District of Keewatin: Geological Survey of Canada, Map 10-1981, scale 1:250,000.

——, 1981b, Surficial geology, Amer Lake, District of Keewatin: Geological Survey of Canada, Map 9-1981, scale 1:250,000.

Thomas, R. D., and Dyke, A. S., 1981a, Surficial geology, Lower Hayes River, District of Keewatin: Geological Survey of Canada, Map 7-1981, scale 1:250,000.

——, 1981b, Surficial geology, Darby Lake, District of Keewatin: Geological Survey of Canada, Map 8-1981, scale 1:250,000.

——, 1981c, Surficial geology, Mistake River, District of Keewatin: Geological Survey of Canada, Map 6-1981, scale 1:250,000.

——, 1981d, Surficial geology, Laughland River, District of Keewatin: Geological Survey of Canada, Map 5-1981, scale 1:250,000.

——, 1981e, Surficial geology, Woodburn Lake, District of Keewatin: Geological Survey of Canada, Map 3-1981, scale 1:250,000.

——, 1981f, Surficial geology, Pennington Lake, District of Keewatin: Geological Survey of Canada, Map 4-1981, scale 1:250,000.

Tremblay, G., 1974, Géologie du Quaternaire, régions de Rouyn–Noranda et d'Abitibi, Comtés d'Abitibi-est et d'Abitibi-ouest: Québec, Ministères des Richesses Naturelles, Dossier Public 236, 100 p.

Tyrrell, J. B., 1913, The Patrician Glacier south of Hudson Bay: Twelfth International Geological Congress, Compte Rendu, p. 523–534, [Map published 1914.]

Veillette, J. J., 1982, Écoulements glaciaires à l'ouest de la moraine d'Harricana au Québec: Commission géologique du Canada, dossier public no. 841.

——, 1983a, Les polis glaciaires au Témiscamingue; Une chronologie relative, *dans* Recherches en cours, partie A: Commission géologique du Canada, Etude 83-1A, p. 187–196.

——, 1983b, Deglaciation de la vallée supérieure de l'Outaouais, le Lac Barlow et le sud du Lac Ojibway, Québec: Géographie physique et Quaternaire, vol. XXXVII, p. 67–84.

Vincent, J. S., 1977, Le Quaternaire récent de la région du cours inférieur de La Grande Rivière, Québec: Commission géologique du Canada, Etude 76-19, 20 p.

Vincent, J. S., and Hardy, L., 1979, The evolution of glacial Lake Barlow and Ojibway, Quebec and Ontario: Geological Survey of Canada, Bulletin 316, 18 p.

Virkalla, K., 1952, On the bed structure of till in eastern Finland: Commission Geologique de Finland, Bulletin, v. 157, p. 97–109.

Washburn, A. L., 1956, Classification of patterned ground and review of suggested origins: Geological Society of America Bulletin, v. 67, p. 823–65.

White, W. A., 1972, Deep erosion by continental ice sheets: Geological Society of America Bulletin, v. 83, p. 1037–1056.

Wilson, J. T., 1938, Glacial geology of part of northwestern Quebec: Royal Society of Canada, Transactions, Section 4, v. 32, p. 49–59.

Zoltai, S. C., 1965a, Glacial features of the Quetico-Nipigon area, Ontario: Canadian Journal of Earth Science, v. 2, p. 247–269.

——, 1965b, Surficial geology, Thunder Bay district: Ontario Department of Lands and Forests, Map 5265, scale 1:506,880.

——, 1965c, Surficial geology, Kenora–Rainy River: Ontario Department of Lands and Forests, Map 5165, scale 1:506,880.

——, 1967, Glacial features of the north-central Lake Superior region, Ontario: Canadian Journal of Earth Sciences, v. 4, p. 515–528.

Manuscript Accepted by the Society September 27, 1986

ACKNOWLEDGMENTS

The authors wish to acknowledge John Scott and colleagues in the Terrain Sciences Division, Geological Survey of Canada, for moral and financial support and other help in compiling this chapter. In particular, Jean-Serge Vincent and Alan Heginbottom provided us with useful maps and information that would have been very time consuming to compile ourselves. We also thank Serge Paradis, Harvey Thorleifson, and Philip Wyatt for access to unpublished information about the Hudson Bay Lowlands and adjacent areas of the western Shield. Much original airphoto interpretation of the Shield, both west and east of Hudson Bay, was carried out by Serge Paradis under a contract awarded to Terrain Analysis and Mapping Services, Limited, of Stittsville, Ontario, and additional work was done in Labrador by Andrée Bolduc. Finally, the manuscript benefitted from careful and thorough critical reading by N. W. Rutter and R. J. Fulton. Although many of their suggestions were followed, the authors assume full responsibility for the contents of the chapter.

Geological Society of America
Centennial Special Volume 2
1987

Chapter 6

Great Plains

Waite R. Osterkamp
U.S. Geological Survey, 431 National Center, Reston, Virginia 22092
Mark M. Fenton
Alberta Research Council, 11315 87th Avenue, Edmonton, Alberta T6G 2C2, Canada
Thomas C. Gustavson
Bureau of Economic Geology, University of Texas, University Station, Box X, Austin, Texas 78712
Richard F. Hadley
Department of Geography and Geology, University of Denver, Denver, Colorado 80208
Vance T. Holliday
Department of Geography, University of Wisconsin, Madison, Wisconsin 53706
Roger B. Morrison
Morrison and Associates, 13150 West Ninth Avenue, Golden, Colorado 80401
Terrence J. Toy
Department of Geography and Geology, University of Denver, Denver, Colorado 80208

INTRODUCTION

W. R. Osterkamp and Thomas C. Gustavson

The Great Plains Physiographic Province (Fig. 1) is a generally north-south band of the western interior of North America, subcontinental in scale, that averages about 600 km in width and extends from northwestern Canada through the United States to the border with Mexico. With tongue-in-cheek, Thornbury (1965, p. 287) claimed that "in the public mind the Great Plains are thought of as a vast monotonous plain that lacks scenic interest, but has to be crossed to reach the scenic Rocky Mountains to the west." It is generally true that the Great Plains, and especially the High Plains, have low relief. But there are many scenic exceptions, largely areas of negative relief along the eroded margins of the plains and in river valleys. Areas of positive relief include structural highs such as the Black Hills (Fig. 1B) and other outliers of the Rocky Mountains. These areas of positive and negative relief are especially important, because they contain the sections where the geologic and geomorphic history of the Great Plains is best exposed.

REGIONAL SETTING AND STUDIES

The northern and eastern limits of the glaciated northernmost Great Plains are in places arbitrary, but are selected adjacent to (1) the Great Bear Plain, (2) the Kazan Region southward past Lake Athabasca, and (3) the Saskatchewan Plain (which, in Canada, is generally regarded as part of the Great Plains) into south-

ern Saskatchewan (Bostock, 1970a; Fig. 1A). Most of the western edge of the Great Plains is defined by the Rocky Mountains; the southwestern limit is bounded by the Basin and Range Province. The eastern extreme of the southern glaciated Great Plains is defined by the Missouri Coteau Escarpment, which extends from the Saskatchewan River southeastward nearly to the Missouri River in South Dakota. From the Missouri River southward into central Texas, the Great Plains is bordered by the Central Lowlands to the east. Southwestward from central Texas to the Rio Grande, the Balcones Escarpment separates the Great Plains from the West Gulf Coastal Plain (Fenneman, 1931).

Numerous publications on the geology and geomorphology of the Great Plains have appeared since the first studies of this province were reported nearly a century ago (noteworthy are Johnson, 1901; Darton, 1903, 1905). Of the more recent papers, useful reviews or regional syntheses of the surficial geology of the Great Plains were given by Hunt (1967) and Fenneman (1931) for physiography; by Thornbury (1965) for regional geomorphology; and by Wright and Frey (1965), Prest (1968), Prest and others (1968), Wright (1983), and Fenton (1984b) for Quaternary geology. A fine nontechnical summary of the evolution of the Great Plains was presented by Trimble (1980). Although information on the surficial geology of the Great Plains has continued to expand since publication of these works, they nevertheless provide valuable summaries of relatively recent knowledge.

Osterkamp, W. R., Fenton, M. M., Gustavson, T. C., Hadley, R. F., Holliday, V. T., Morrison, R. B., and Toy, T. J., 1987, Great Plains, *in* Graf, W. L., ed., Geomorphic systems of North America: Boulder, Colorado, Geological Society of America, Centennial Special Volume 2.

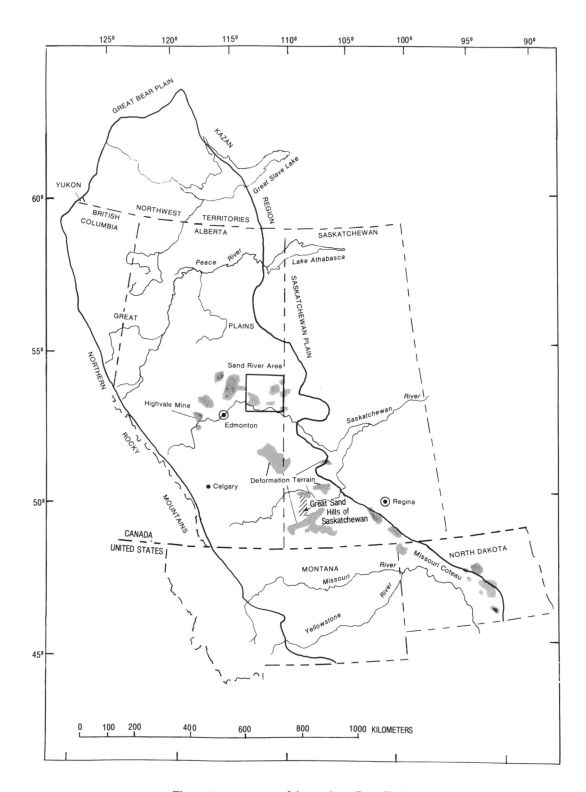

Figure 1A. Index map of the northern Great Plains.

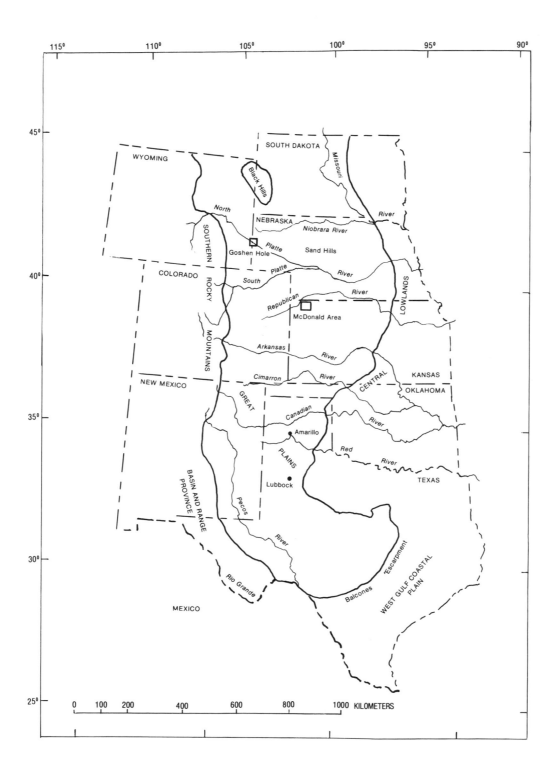

Figure 1B. Index map of the southern Great Plains.

Several basic discussions of geomorphic processes and regional geomorphology represent important steps in understanding the development of landforms of the Great Plains. Included are descriptions of the Quaternary geology and geomorphology of the Southern High Plains (Evans and Meade, 1945; Gustavson and Finley, 1985), the geomorphology of the Pecos River valley and the development of caliche as a pedogenic feature (Bretz and Horberg, 1949), the geomorphic evolution of the Canadian River valley (Gustavson, 1986), the geomorphology of southwestern Kansas (Smith, 1940), the origins and sources of loess in Nebraska (Lugn, 1962), the evidence for stagnating glacial ice on the northern Great Plains (Gravenor and Kupsch, 1959; Clayton, 1967; Parizek, 1972), and evidence for ice thrusting beneath the Laurentide Ice Sheet (Kupsch, 1962; Moran, 1971).

DEVELOPMENT AND DEGRADATION OF THE GREAT PLAINS

The Laramide orogeny in the Late Cretaceous and early Tertiary resulted in uplift of the Rocky Mountains, including the most eastern and northern structural elements of the Cordilleran Region. Tectonic and epeirogenic movement resulted in regional uplift and tilting throughout much of the Great Plains area (Gable and Hatton, 1983). From Colorado northward into Canada, Paleocene and younger fluvial, lacustrine, and paludal sediment was derived from the eroding Rocky Mountains to the west. From Texas into Kansas, regional uplift during early Tertiary time resulted in extensive incision of Cretaceous and older rocks. By middle Miocene time, however, continuing movement of sediment from the mountains eastward caused filling of the incisements and deposition of the Ogallala Formation (Schultz, 1977). Fluvial and eolian deposition continued to middle Pliocene time, followed by an extended period of landscape stability during which the well-indurated pedogenic caliche that presently mantles parts of the Ogallala Formation was formed.

During the Pleistocene the Great Plains from South Dakota northward was glaciated repeatedly, removing substantial thicknesses of pre-Pleistocene regolith. Remaining surface sediment consists mostly of Wisconsin-age diamicts, outwash, and lacustrine and paludal sediment; glacial debris older than Wisconsin age is poorly exposed. These processes ceased following retreat of the Laurentide Ice Sheet from the Great Plains between about 17,000 and 9,000 years B.P. (Prest, 1970).

In a large portion of central Nebraska, the Great Plains is underlain by extensive deposits of late-Wisconsin to late-Holocene eolian sand known as the Sand Hills (Fig. 1B). Deposits that presumably are older underlie this dune field, but exposures are very limited due to a lack of stream incision. Associated Pleistocene loess deposits may be the downwind fine-grained facies of the Sand Hills, but as noted by Holliday later in this chapter, recent evidence suggests that the two eolian facies of this area may be genetically different. Regardless of specific origins, the eolian deposits that are in large part responsible for giving much of the Great Plains its relatively flat topography were

mostly derived from valley alluvium of streams such as the Niobrara, Platte, and Pecos rivers, and from outwash carried by these streams and others headed by glaciers during Pleistocene time. In parts of the central and southern Great Plains, Quaternary sediment is predominantly loess, locally interbedded with lacustrine deposits. Where these deposits mantle the High Plains, those parts of the Great Plains where minimal fluvial incisement has occurred since the most recent depositional episode, datable organic-carbon and volcanic-ash samples indicate ages ranging from at least 1.9 Ma to the Holocene.

Major erosional episodes during the late Tertiary and Quaternary are primarily responsible for the drainage system cut into and along the margins of the Great Plains. Subsidence caused by dissolution of Permian-bedded salt in parts of New Mexico, Texas, Oklahoma, and Kansas resulted in various changes to the fluvial network of the region, including diversion of the Pecos and Canadian rivers to their present courses (Gustavson and Finley, 1985). These effects of groundwater movement helped determine the present edges of the north, west, and east sides of the Southern High Plains. Locally, drainage along the Cimarron River in southwestern Kansas has also been affected by dissolution and subsidence (Smith, 1940). In eastern Wyoming and western Kansas, Quaternary fluvial incision, also probably accelerated by groundwater erosion, has formed a complex of pediment surfaces and strath and alluvial terraces (Scott, 1965, 1982).

Quaternary erosion of the central Great Plains, which largely is drained by the Missouri River, has been mostly by fluvial processes. As first noted by Warren (1869, p. 311), however, the channel network in much of the Missouri River basin was the result of drainage rearrangement by glaciation. Preglacial drainage of the Great Plains Province in eastern Montana and North Dakota was largely to the northeast into Hudson Bay via the Missouri River and various tributaries (Colton, 1962). Flint (1955) suggested that the preglacial divide between Hudson Bay and the Gulf of Mexico was in South Dakota between the Cheyenne and Bad rivers, and that the diversion took place during the Illinoian Glaciation. Recognition of the diversion of the Missouri River is based on the observations that (1) several major tributaries feed the Missouri River from the west, but none do so from the east, (2) the Missouri River valley is trenchlike and more youthful (possibly of early Wisconsin age) in appearance than many of its westerly tributaries, and (3) in parts of North and South Dakota the river flows south, or normal to the regional west-to-east slope.

GEOMORPHIC PROCESSES AFFECTING THE GREAT PLAINS

Implicit in the foregoing discussion is the recognition that the Great Plains has been sculptured by a variety of geomorphic processes including, but not limited to, fluvial, eolian, and glacial deposition; and by erosion resulting from surface- and groundwater movement, wind transport, and glaciation. Some processes, such as eolian sorting and glaciotectonism, affected the landscape of the Great Plains by redistribution of sediment. The extent to which most of the processes have acted in an area of the Great Plains has been dependent on regional climate and geology, but as is generally true elsewhere, none of the landforms are the result of a single process.

The several contributions to this chapter consider processes currently regarded as important to the geomorphic development of the Great Plains. Perhaps as much as for any physiographic province of North America, the morphology of the Great Plains is a product of the dramatic climatic changes that have characterized Quaternary time. An overview of possible relations between these climatic variations and the resulting geomorphic processes leads to the suggestion by Morrison that landforms on the Great Plains have developed in a complex cyclic fashion. The cyclicity of these interactions is characterized by sequences of incision, lateral erosion, deposition, and landform stability. All stages of this sequence are observable in different parts of the Great Plains, and thus, this perspective provides a framework into which consideration of other geomorphic processes can be applied.

Prominent examples of climatic extremes of Pleistocene time are the various flow features that resulted from the multiple ice-sheet advances on the northern Great Plains. Most of these features, however, have been described extensively in numerous papers, and further discussion here seems unwarranted. An exception, however, is glaciotectonism, the landforms of which were identified nearly a century ago. Only recently, however, has glaciotectonism been studied in depth as a process. The resulting landforms, generalized as deformation terrain, are shown by Fenton to be common in much of the northern Great Plains (Fig. 1A). Described in relative detail are the glaciotectonic features of the Sand River area, Alberta, and highwall-stability problems at the Highvale Mine, a surface mine cutting glacially deformed sediment (Fig. 1A).

Late-Quaternary fluvial erosion on the Great Plains has been most pronounced in those areas extending roughly from the Cimarron and Arkansas river basins of Oklahoma and Kansas northward past the glacial limit of the Missouri River basin. A contribution by Hadley and Toy summarizes a series of studies, made mostly on streams of the Great Plains, that relates alluvial-channel morphologies and processes of channel adjustment to governing variables including the sediment loads and particle sizes conveyed through the channels. Approaches, mostly as power relations, describing changes in the geometries of Great Plains channels with variations in water and sediment discharges have led to recent analyses of historical channel changes in the Platte and Cimarron river basins (Fig. 1B).

The role of groundwater in affecting some geomorphic features has been recognized a surprisingly long time (see, for example, Warren, 1858, p. 23–24), but only within the last few years has the extent of the influence of groundwater been appreciated. Geomorphic features of the Great Plains that are largely caused by processes of groundwater flow and discharge are discussed by Osterkamp. In places such as Goshen Hole, in Wyoming and Nebraska, and the McDonald area of northwest Kansas (Fig. 1B),

groundwater erosion, aided by fluvial sediment transport, may have been the dominant Quaternary process shaping the present landscape. In the Southern High Plains of Texas and New Mexico, fluvial erosion is very limited, and evidence indicates a strong groundwater influence, with substantial modification by eolian processes, on the geomorphic features.

In a final contribution on geomorphic processes, Holliday discusses causes for the widespread occurrences of eolian deposition and erosion on the Great Plains. Possible source areas for the loess and dune sand that veneer much of the Great Plains as far north as southern Canada are suggested. In particular, recently published information on ages, wind directions, and particle-size ranges of the Sand Hills in Nebraska and dune fields in northeastern Colorado (Fig. 1B) is used to estimate paleoclimates at the times of deposition and possible relations to adjacent loess deposits.

The several contributions are not intended to provide an exhaustive treatment of geomorphic processes shaping the Great Plains. They do, however, highlight some of the dominant processes under current research and serve to emphasize that most landforms are the integration of a variety of processes; few features can be related to a single process. Lastly, the contributions illustrate the diversity of geomorphic features and processes that may only be possible in a physiographic province extending through a substantial latitude range and exhibiting a broad range of climates. In this respect, the Great Plains seems to be truly unique.

LONG-TERM PERSPECTIVE: CHANGING RATES AND TYPES OF QUATERNARY SURFICIAL PROCESSES: EROSION-DEPOSITION-STABILITY CYCLES

Roger B. Morrison

Charles Lyell was extraordinarily perceptive when, much ahead of his time (1830, 1833), he endorsed the principle of Uniformitarianism in geologic processes (an idea first proposed by Hutton in 1795). Subsequent scientific history has established that Uniformitarianism is approximately correct when it is applied over the whole earth and also over Phanerozoic time scales above 10^6 years in magnitude. However, Uniformitarianism becomes invalid when it is applied over shorter time spans, where perturbations were induced by changing climate and by localized tectonic episodes.

Thus, Uniformitarianism is generally invalid as a guiding principle with respect to the operation of surficial processes within the relatively short 1.65 m.y. time span of the Quaternary period. This is because surficial processes are the most affected by climatic change, and the Quaternary was marked by frequent, unusually large changes in climate on a global scale. As a result, the rates of various surficial processes (fluvial, eolian, pedogenic, etc.) changed repeatedly in cyclic fashion, so much so that at various episodes one or another type of surficial process became

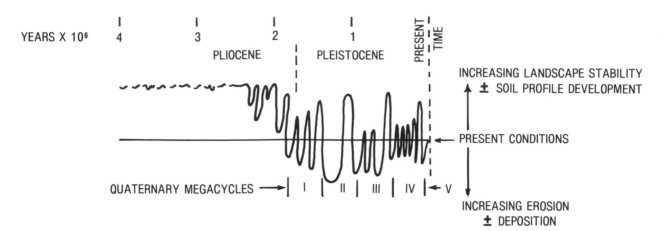

Figure 2. Diagram showing the marked change in frequency and amplitude of the erosion-deposition-landscape stability (EDS) cycles from near the end of the Pliocene to the present based on stratigraphic and geomorphic records in the Great Plains, with chronology extrapolated from oxygen-isotope and percent carbonate graphs from deep-sea core records (Shackleton and others, 1984; Zimmerman and others, 1984; Ruddiman and others, 1987).

dominant. Nowhere on earth, including the Great Plains, have surficial processes acted at a steady-state rate throughout the Quaternary.

This contribution outlines several speculations concerning the Quaternary geomorphic history of the Great Plains. (1) Four classes of geomorphic cycles operated in the unglaciated Great Plains during Quaternary time—microcycles of 10^1 to 10^2 years, mesocycles of 10^3 to 10^4 years, macrocycles of 95 ± 25 ka, and megacycles of 400 to 500 ka duration. (2) Extrinsic geomorphic thresholds, controlled by climatic change, bounded the various cycles. (3) Each cycle tended to progress through four successive stages: downcutting, lateral erosion, deposition, and landscape stability. In this discussion, therefore, the sequences are termed erosion-deposition-stability cycles.

GLOBAL CLIMATIC HISTORY DURING THE QUATERNARY

Repeated climatic changes of a global scale, and the effects of these changes, characterize the Quaternary. Late-Cenozoic stratigraphic records, both terrestrial and marine, demonstrate that beginning with the Pleistocene, climatic changes increased in amplitude and frequency by several orders of magnitude relative to those of the late Miocene and Pliocene. The increased amplitude of climatic change induced repeated glaciations during the Pleistocene as well as repeated changes in rate and type of other surficial processes extending far beyond the glacial limit, indeed worldwide.

The marine oxygen-isotope ($^{18}O/^{16}O$) record from deep-sea cores provides an important perspective. This record provides the most complete Quaternary chronology and history of climatic changes, globally averaged. It represents chiefly volume changes of ice stored on the continents during glaciations, and subordinately, temperature changes of the ocean-surface layer, both of which enhance correlation with the terrestrial glacial record.

Summary conclusions from the marine oxygen-isotope record, based on the work of Shackleton (1969), Shackleton and Opdyke (1973), and Johnson (1982), corroborated by the better loess records in Europe and China (Kukla, 1975, 1977; Fink and Kukla, 1977; Liu and others, 1985), include the following. (1) Seventeen complete interglacial-glacial (IG-G) cycles occurred during the Quaternary, not the four (United States) or five (Europe) IG-G cycles of older interpretations. Individual IG-G cycles lasted 95 ± 25 ka; thus, the IG-G cycles were similar but not identical in their duration. Also, they commonly differed in their amplitude of climatic change. Some of the IG-G cycles were cooler than normal during glacial and/or interglacial phases; others were warmer than normal during either or both phases (Fig. 2). (2) Between 12% and 13% of the last 500 ka had climates as warm or warmer than now. About the same percentage of the youngest completed IG-G cycle (Sangamon through Wisconsin) also was as warm or warmer than today (Emiliani, 1967, 1970, 1972; Johnson, 1982). (3) The Holocene is an interglacial episode that started about 10,000 years ago, following about 105,000 years of glacial conditions that were punctuated with several brief interstadials. The previous interglacial (Sangamon, in the strict sense) lasted about 15,000 years. From this record and astronomical earth-insolation data (projected into the future), it is likely that earth soon will commence upon another glacial phase, lasting at least several tens of millenia, an environment that civilized man has never experienced! Obviously, we live in an exceptional time by paleoclimatic and therefore geomorphic standards.

GEOMORPHIC RESULTS OF CLIMATIC CHANGES DURING THE QUATERNARY: EROSION DEPOSITION-STABILITY CYCLES

The climatically induced fluctuations in rate and dominant type of surficial processes through Quaternary time produced cyclic changes in landscape evolution, as demonstrated by geomorphic and stratigraphic records from throughout the United States, and apparently the world. These cycles are manifested in alluvial, eolian, lacustrine, and mixed terrains. Each cycle consists of successive episodes of (1) downcutting, (2) lateral erosion, (3) deposition, and finally (4) landscape stability, commonly with significant soil-profile development. In this paper these sequences are called "erosion-deposition-stability cycles," abbreviated to "eds cycles." In terms of duration and frequency, they range over several orders of magnitude. They appear to constitute four classes: (a) microcycles, lasting in the 10^1- to 10^2-year magnitude; (b) mesocycles of a 10^3- to 10^4-year magnitude; (c) macrocycles, lasting 95 ± 25 ka; and (d) megacycles that lasted 400 to 500 ka. The mesocycles are comparable to, but not necessarily correlative with, stadial-interstadial cycles of the marine oxygen-isotope record, and the macrocycles are comparable to glacial-interglacial cycles of this record; the megacycles correspond to groups of four to perhaps as many as six glacial-interglacial cycles.

This hierarchy of cycles implies the crossing of geomorphic-process thresholds that were at levels appropriate for each class of eds cycle in the alluvial, eolian, and pedogenic systems. Schumm (1973) originated threshold terminology in geomorphology, and he (1975, 1979, 1980), Bull (1975a, 1975b, 1979, 1980), and Coates and Vitek (1980) provided useful elaboration and examples of this concept. The thresholds of this paper are Schumm's (1979) "extrinsic thresholds," those whose crossing is due to factors that are external to the landform system, such as climatic change and tectonism. Bull (1979, 1980) points out the importance of feedbacks that complicate the flow of the process systems. Nearly all students of geomorphic thresholds have focused on those associated with microcycles; the climatic and geomorphic-process systems that induced the crossing of megacycle thresholds can only be guessed at, and certainly are far from the modern environment in the Great Plains.

Eds cycles occurred also during the late Tertiary, but their amplitudes were mostly lower than during the Quaternary and major geomorphic thresholds were crossed less frequently. As a consequence, the late-Tertiary eds megacycles lasted much longer than did those of the Quaternary, in the magnitude of 10^6 years (Fig. 2). The best evidence that this conclusion is not an artifact of fewer data for the late Miocene and Pliocene than for the Quaternary comes from deep-sea cores that extend in age into the late Tertiary.

This discussion addresses the record of Quaternary eds megacycles in the Great Plains south of the glacial limit. Here their geomorphic/stratigraphic record is mediocre as to age range and detail. More detailed and longer stratigraphic records are available from long-lasting deep-lake basins such as those of Lakes Lahontan, Searles, and Bonneville. These lacustrine records, however, give little evidence on the extent, character, and chronology of erosion episodes above the lake surfaces. To obtain these data, the subaerial geomorphology and stratigraphy must be studied along with the lacustrine stratigraphy, and this rarely has been done.

SALIENT FEATURES OF THE GREAT PLAINS THAT PERTAIN TO THEIR QUATERNARY GEOMORPHIC HISTORY

Tectonism

The whole Great Plains was affected by late-Cenozoic epeirogenic uplift, probably associated with the East Pacific Rise. The uplift during the last 10 m.y. was generally 1.5 to 2 km along the western margin of the province, decreasing to 100 to 500 m along its eastern margin (Gable and Hatton, 1983). Plio-Pleistocene uplift probably accounted for a half to a quarter of these amounts (G. R. Scott and J. W. Hawley, personal communication, 1985). However, highlands to the west, the Rocky Mountains and southeastern Basin and Range Province, rose even more than the Great Plains during the late Cenozoic (Gable and Hatton, 1983), which augmented the sediment supply to the Great Plains. As a result of these changes, the Great Plains during the Quaternary remained an intermediate high region, and was subject mostly to erosion and only local fluvial deposition.

Glacial isostasy/eustasy. Undoubtedly, various advances of the Laurentide Ice Sheet, which came as far south as northeastern Kansas in Kansan time, had isostatic depression, rebound, and forebulge effects, but these have not been well documented in the Great Plains and therefore are overlooked here. Likewise, most of the Great Plains is too far from the Gulf of Mexico to have been significantly affected by base-level changes due to glacioeustatic changes in sea level during the Pleistocene.

Tectonic stability. Apart from the regional influence of epeirogenic uplift, the Great Plains Province has been one of the most structurally stable areas of North America during the Quaternary. In most parts of the Great Plains, Quaternary tectonism can be ruled out as an important geomorphic factor. Local exceptions, however, some of which are mentioned elsewhere in this volume and its companion, "Quaternary non-glacial geology of the conterminous United States," include areas believed to have undergone tectonic uplift or depression, or collapse by solution of underlying evaporite strata.

Stratigraphic Evidence of EDS Cycles: Process Benchmarks

The several kinds of stratigraphic evidence of eds cycles are benchmarks in the record of surficial processes, providing important supplement to the landform record. They are evinced by discontinuities in stratigraphic sequences that mark significant changes in the rates, and commonly in the types, of surficial processes. They comprise the following categories.

1. Unconformities. In Quaternary successions in the Great Plains, these breaks are nearly all disconformities (with parallel strata above and below the break) caused by water or wind erosion, and diastems, representing episodes of nondeposition. Both types of unconformity were poorly recognized in older studies in the plains, partly because sophisticated methods commonly are required for their identification. Thus, recent work, induced by better tephrochronology, had led to drastic revision in interpretation at several key sites. Both types of unconformity represent "lost" geologic time in their local records. The time represented by Quaternary diastems commonly is in magnitudes of 10^1 to 10^3 years, but for the disconformities that record widespread lateral erosion or pedimentation in Quaternary sequences in the Great Plains, it commonly is 10^4 to 10^5 years.

2. Paleosols. Paleosols represent episodes of landscape stability, when little or no erosion or deposition occurred at a site. However, significant widespread soil development may occur during only part of the stability episode. Some paleosols in the Great Plains appear to be widespread and to represent episodes of regional landscape stability; probably they deserve to be designated as Geosols (North American Stratigraphic Commission, 1983). At many Great Plains sites, certain paleosols show pedologic evidence of more than one episode of soil development, commonly representing both an interglacial and one or more interstadials, and even another interglacial.

3. Marked lithologic change in the vertical stratigraphic sequence. This feature is most important as a criterion of significant change in depositional environment where the strata differ in lithofacies, such as alluvium and eolian sand. Its significance is questionable where the change is merely in particle size, such as between stream gravel and overlying finer alluvium, which might be almost coeval channel and overbank deposits of a stream whose channel migrated across its flood plain.

Deposition of alluvial and eolian units generally is very episodic; disconformities or diastems commonly separate them, and the duration of deposition and thickness of units can vary greatly. Unfortunately, these units commonly also look much alike, whether they be alluvium, eolian sand, loess, till, or lacustrine gravel or clay, making lithostratigraphy commonly an insecure means of correlation, even locally.

EROSION-DEPOSITION-STABILITY MEGACYCLES IN THE GREAT PLAINS

In alluvial sequences of the Great Plains, Quaternary EDS megacycles are displayed by alterations of a series of episodes: (1) stream downcutting, (2) chiefly lateral planation and pedimentation, (3) local alluviation, and (4) landscape stability, commonly with soil development during some or much of this episode. Thus, following a long episode of landscape stability, each megacycle was initiated by a major climatic change, greater than those that triggered lesser cycles within each megacycle, and

big enough to trigger an episode of major downcutting by streams. More details of the nature and history of these mega changes in climate, and the resulting changes in the erosional-depositional systems, await future study.

Sets of stream terraces display alluvial eds cycles, but invariably only in fragments, with long hiatuses between the various alluvial units and buried or relict soil profiles. At some localities loess and loessial-colluvial mantles on the stream terraces provide an important supplementary record that helps to fill gaps in the stratigraphic record.

EXAMPLES OF QUATERNARY EDS MEGACYCLES IN THE GREAT PLAINS

Many fine Quaternary records occur in the Great Plains, but within the available space, only the barest documentation can support the interpretations given here. Principal support is from two areas that are especially representative of this huge region and have well-studied records.

Alluvial Megacycle Sequence in the Colorado Piedmont

The Colorado Piedmont provides probably the best documentation of Quaternary EDS megacycles. It is the section of the Great Plains between the Southern Rocky Mountains and the High Plains that is drained by the South Platte and Arkansas rivers (Fig. 3). It is eroded below the High Plains surface as much as 450 m along the western border to less than 75 m along the eastern border in a steplike series of Pliocene and Quaternary erosional and depositional surfaces that are graded to former levels of both rivers. These levels comprise four pediment surfaces that represent episodes of broad stream planation, and a younger set of five strath or fill terraces (Scott, 1960, 1963, 1965, 1982; Trimble and Machette, 1979). Each surface bears a distinctive alluvial deposit, which in turn bears a relict paleosol (Fig. 4). The Nussbaum Alluvium, the oldest, mantles a pediment lying 365 to 425 m below the late-Miocene to early-Pliocene erosion surface at the east edge of the mountains, and an average of 140 m above modern streams. Glenn R. Scott (personal communication, 1985) correlates the Nussbaum with the type Blanco Formation of the Southern High Plains and with the Broadwater Formation of Nebraska, and on the basis of Blancan mammalian remains believes the Nussbaum Alluvium to be of Pliocene age. The type Blanco Formation and the calcrete it bears underlie the 1.46 Ma Guaje Ash and a strongly developed paleosol (Izett and others, 1981; V. T. Holliday, written communication, 1986). Consequently, both the Blanco Formation and the Nussbaum Alluvium probably extend into the early Pleistocene.

The Rocky Flats Alluvium rests on an extensive pediment, and near the mountain front it averages 110 m above modern streams. This alluvium has reversed polarity (Matuyama), contains mammalian remains of latest Blancan age, and correlates with the early Pleistocene. It bears a very strongly developed polygenetic paleosol that probably records at least two interglacials.

Figure 3. Map of the High Plains and other selected sections of the Great Plains Physiographic Province, showing occurrence of eolian deposits.

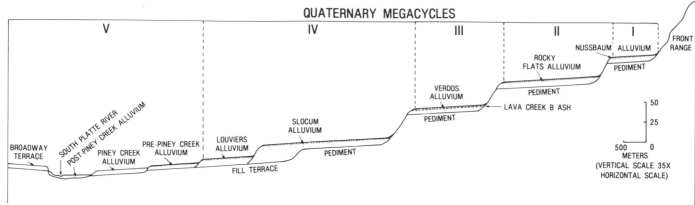

Figure 4. Morphostratigraphic relations of pediments, stream terraces, and their alluvial covers and paleosols in the Colorado Piedmont between the Front Range and the South Platte River a few kilometers north of Denver. This sequence ranges from late Pliocene to Holocene, albeit with many gaps in the record, represented by disconformities and diastems.

The Verdos Alluvium mantles another widespread pediment an average of 75 m above modern streams. It contains the Lava Creek B volcanic ash, about 620,000 years old, locally in its upper part; therefore the Verdos is early-middle Pleistocene (Kansas) in age. It bears a very strongly developed paleosol correlated with the Yarmouth Geosol.

The Slocum Alluvium mantles a moderately extensive pediment/fill-terrace system about 30 m above modern streams, and bears a strongly developed paleosol correlated with the Sangamon Geosol; therefore, this alluvium-paleosol complex is of late-middle Pleistocene age. The Louviers Alluvium, on a strath terrace 15 m above modern streams, also has a well-developed paleosol, likely also correlative with the Sangamon Geosol. Younger alluvial units, such as those of the Broadway and Piney Creek surfaces (Fig. 4), cap terraces of late-Pleistocene and Holocene age.

The sedimentary units represented in Figure 4 record about 25% tro 30%, and the paleosols perhaps another 20% to 25%, of Quaternary time in the Colorado Piedmont sequence. Thus, about half of Quaternary time is lost from this sequence. Most of the unrecorded time is represented by unconformities that underlie the various alluvial units. The unconformities beneath the Nussbaum, Rocky Flats, Verdos, and Slocum alluviums represent composites of erosion episodes that apparently completely removed sedimentary records of the early and middle parts of their megacycles. Perhaps the existing record of megacycle V, in which we live and which is in its infancy, is a partial analog of the missing meso- and macrocycle records, sedimentary and erosional, that are characteristic of the early part of a megacycle.

Correlatives of this pediment-terrace sequence occur in other parts of the Colorado Piedmont (e.g., Scott, 1982). They also occur in the "Gangplank" area near Cheyenne, Wyoming (M. E. Cooley, personal communication, 1985), along the Arkansas River, and along many other rivers that cross the High Plains section of the Great Plains.

Megacycle Sequence in the Loess Belt of Nebraska

The loess record in the Great Plains is generally superior to the alluvial record, but it remains poorly studied. It formed under a different climatic-depositional system, eolian instead of fluvial. At the sites most favorable for lodgement and preservation, it appears to have fewer and smaller unconformities than the alluvial sequences. The chief loess deposition was during times of glaciation and early interglaciation.

The most complete loessial sequences (loess, loessial colluvium and alluvium, and paleosols) are in central Nebraska, southeast of the Sand Hills region (Fig. 3), and subordinately in western Kansas (see the section on eolian processes by V. T. Holliday, this chapter). In this "loess belt," loessial sediment younger than the 0.62-Ma Lava Creek B Ash is commonly 15 to 30 m thick and locally exceeds 60 m in thickness. The sites in central Nebraska with the thickest pre-Wisconsin deposits appear to be lodgement zones on the south side of an ancient valley of the Platte River that antedated the Lava Creek Ash.

The Eustis Ash Pit–Buzzard's Roost area of south-central Nebraska has several disjunct, well-studied exposures in a line running 32 km northwest from Eustis, Nebraska (Reed and Dreeszen, 1965; Schultz and Stout, 1961, 1980; R. B. Morrison, field notes, 1966, 1972; Fredlund and others, 1985). Figure 5 summarizes data from the three principal localities: (1) Buzzard's Roost, 32 km northwest of Eustis; (2) the Gilman Canyon section (Reed and Dreeszen, 1965), 30.5 km northwest of Eustis; and (3) the Eustis Ash Pit (Fig. 6), 1.5 km southwest of Eustis.

A key marker, exposed at sites 2 and 3, is the Lava Creek B Ash, which thickens to at least 15 m in local pockets. At site 2, disconformably underlying this ash is swampy alluvium bearing a gleyed paleosol. Disconformably overlying the ash bed is the 18-to 25-m-thick Loveland loess-paleosol complex (Figs. 5, 6), comprising five main "loess" (chiefly loessial colluvium/reworked loess) units separated by five paleosols or compound

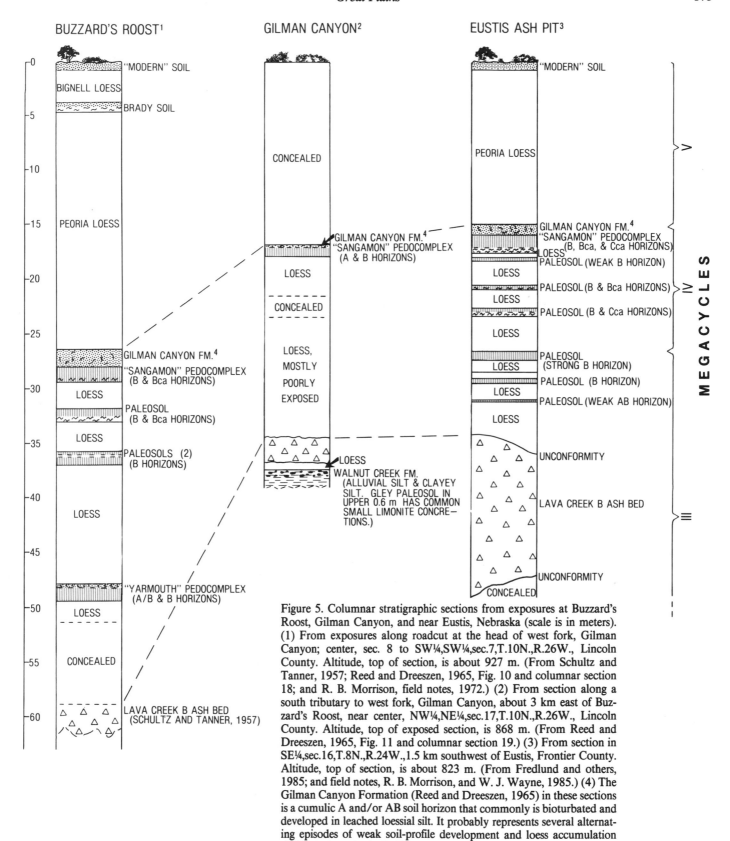

Figure 5. Columnar stratigraphic sections from exposures at Buzzard's Roost, Gilman Canyon, and near Eustis, Nebraska (scale is in meters). (1) From exposures along roadcut at the head of west fork, Gilman Canyon; center, sec. 8 to SW¼,SW¼,sec.7,T.10N.,R.26W., Lincoln County. Altitude, top of section, is about 927 m. (From Schultz and Tanner, 1957; Reed and Dreeszen, 1965, Fig. 10 and columnar section 18; and R. B. Morrison, field notes, 1972.) (2) From section along a south tributary to west fork, Gilman Canyon, about 3 km east of Buzzard's Roost, near center, NW¼,NE¼,sec.17,T.10N.,R.26W., Lincoln County. Altitude, top of exposed section, is 868 m. (From Reed and Dreeszen, 1965, Fig. 11 and columnar section 19.) (3) From section in SE¼,sec.16,T.8N.,R.24W.,1.5 km southwest of Eustis, Frontier County. Altitude, top of section, is about 823 m. (From Fredlund and others, 1985; and field notes, R. B. Morrison, and W. J. Wayne, 1985.) (4) The Gilman Canyon Formation (Reed and Dreeszen, 1965) in these sections is a cumulic A and/or AB soil horizon that commonly is bioturbated and developed in leached loessial silt. It probably represents several alternating episodes of weak soil-profile development and loess accumulation during early and middle Wisconsin time, correlative with much or all of oxygen-isotope stages 3, 4, and 5a–5d. Some workers regard it as part of the Sangamon pedocomplex *sensu lato*.

Peoria Loess

}Sangamon pedocomplex

Loveland Loess

Lava Creek B Ash

Figure 6. Exposures of loessial sequences overlying the Lava Creek B Ash, Eustis Ash Pit, Eustis, Nebraska. (Photograph by J. C. Knox.)

pedocomplexes. At its top is the Sangamon pedocomplex, which separates the Loveland loess-paleosol complex from about 15 m of overlying Peoria Loess. If the Sangamon pedocomplex can be correlated with oxygen-isotope stages 5 to 3 (Fig. 7), the Loveland represents almost 0.5 m.y. of alternating episodes of greater and lesser loess deposition and pedogenesis. The Peoria Loess, in this area nearly equal in thickness to the Loveland complex, was deposited between about 26,000 and 12,000 years ago, about 3% of Loveland time.

CONCLUSIONS: QUATERNARY GEOMORPHIC HISTORY OF THE GREAT PLAINS

1. Quaternary stratigraphic/geomorphic sequences in the Great Plains have many small to large discontinuities that record repeated change in rate of surficial processes (fluvial, eolian, mass wasting, pedogenic, etc.). Commonly the change in rate of one process versus another became large enough to change the type of surficial process that was dominant during a given episode.

2. The dominant-type episodes tended to progress in cycles comprised of four successive episode stages: (a) stream downcut-

ting, (b) lateral erosion and pedimentation, (c) deposition (alluviation, etc.), and (d) landscape stability, perhaps with significant soil development. These cycles are here called "erosion-deposition-stability cycles," or "EDS cycles."

3. In the Great Plains, the EDS cycles and their four integral stages were induced by climatic changes, for in this relatively stable region tectonism had but a trivial, very local effect.

4. Four time (duration) classes of EDS cycles appear to have operated: microcycles, lasting 10^1 to 10^2 years; mesocycles, 10^3 to 10^4 years long; macrocycles, 95 ± 25 ka in length; and megacycles, 400 to 500 ka in duration. Microcycles are minor perturbations; mesocycles are comparable to, but perhaps not correlative with, stadial-interstadial cycles of the glacial-marine record. Macrocycles resemble glacial-interglacial cycles, and megacycles correspond to 3 to perhaps 6 glacial-interglacial cycles and exhibit the greatest expression of the four dominant-process stage episodes, from downcutting to soil development.

5. At least four megacycles, and probably the start of a fifth, are evident during the Quaternary (Figs. 4, 5, and 7). Cycle I likely began about 1.8 Ma, shortly before the end of the Pliocene, and lasted 0.4 to 0.5 m.y., ending about 1.35 Ma. Cycle II lasted

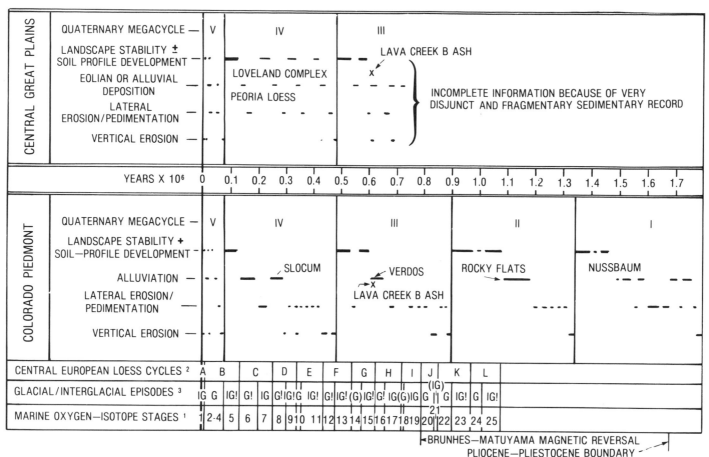

Figure 7. Changes in dominant Quaternary surficial processes in the Colorado Piedmont and central Great Plains, based on geologic, geomorphic, and pedologic records. Each of the five erosion-deposition-stability megacycles represented started with strong downcutting and ended with strong soil-profile development; cycle V remains unfinished. The sedimentary-pedologic-geomorphic record for the Colorado Piedmont covers the entire Quaternary, albeit with extensive erosional unconformities that represent removal of the earlier parts of all but the current (V) cycle. The loess record in the central Great Plains provides more detail on meso- and macrocycle events within megacycles III, IV, and V, but the exposed record is shorter than that of the Colorado Piedmont. (1) Oxygen-isotope stages and age of the Brunhes-Matuyama geomagnetic reversal from Johnson (1982). (2) Kukla (1975). (3) Episodes derived mainly from pedologic, paleontologic, and sedimentologic evidence from the Great Plains; deficient records supplemented by deep-sea oxygen-isotope and central European loess data. (Symbols: IG!, interglacial warmer than the Holocene; IG, interglacial like the Holocene; (IG), interglacial weaker than the Holocene; G!, glacial stronger than the Wisconsin glaciation; G, glacial like the Wisconsin; (G), glacial weaker than the Wisconsin.)

0.44 m.y., ending about 0.91 Ma. Cycle III (Kansas-Yarmouth) lasted about 0.42 m.y. and ended about 0.49 Ma. Cycle IV (Illinoian-Sangamon) probably lasted about 0.42 m.y., but its upper boundary is uncertain, being so close to the present. Cycle V, which we are presently in, probably began in oxygen-isotope stage 4, between 80 and 60 ka.

Megacycles I, II, and III were marked by deep downcutting, followed by extensive pedimentation. Unfortunately, these erosion events removed essentially all the geomorphic-sedimentary record of the early and middle parts of these megacycles, as also occurred for these parts of megacycle IV. The latter parts of

megacycles II and III had especially strong soil development during two or more interglacials, and megacycle IV ended with the strong Sangamon-equivalent paleosol. Thus, although the Quaternary megacycles were of nearly identical duration, they were dissimilar in terms of process history.

6. The fundamental climatic- and surficial-process controls of the EDS cycles, including various feedback mechanisms, are poorly understood, especially the extrinsic threshold controls for the longer cycles. The low thresholds of the smaller cycles were crossed at short intervals by relatively small changes in process intensity. The high thresholds for the megacycles required large

changes in process intensity, and were accompanied by major geomorphic-stratigraphic changes. Only occasionally did even the glacial-interglacial macrocycles have a climatic change large enough to induce the first stage of a megacycle—downcutting severe enough to cross a megacycle threshold.

DEFORMATION TERRAIN ON THE NORTHERN GREAT PLAINS

Mark M. Fenton

The geomorphic results of the pronounced Quaternary climatic changes noted by Morrison (Fig. 2) may be as prominent on the northern Great Plains as they are anywhere. Ice sheets that repeatedly advanced and melted profoundly affected the landscape. The earliest of the major glacial advances moved across the prairies about 2 Ma (Fenton, 1984b; Stalker, 1987), and each subsequent advance contributed to the nearly continuous cover of glacial sediment. In addition, each glaciation caused shear stresses between the advancing ice sheet and the underlying sediment or bedrock, resulting in glaciotectonism, the process whereby the overridden material is deformed in a variety of manners. The geomorphic manifestation of this process is deformation terrain, the subject of this section.

After a brief discussion of deformation-terrain studies, examples of specific landforms identified at the Sand River area, Alberta (Fig. 1A) are described. These examples are assumed representative of similar features found through much of the glaciated northern Plains. A following portion of the section details observations from a surface mine in a glacially deformed area where excavations have exposed near-surface structures and led to slope failures that provide insight into the processes of glaciotectonism and landscape modification following glacial retreat.

The glaciated northern Great Plains, and the Sand River area in particular (Bostock, 1970a, 1970b; Fenton, 1984b), are characterized by a flat to gently rolling surface traversed by broad shallow depressions, incised by narrow valleys, and punctuated by scattered groups of low hills. Regional slope is northeast, away from the Cordillera. The plains are underlain by generally flat-lying sedimentary rocks, largely Cretaceous shales in the plains of Saskatchewan and eastern Alberta, and poorly consolidated sandstones of Late Cretaceous and early Tertiary age along the east flank of the Rocky Mountains.

Overlying the bedrock is a mantle of predominantly glacial sediment. This mantle locally thickens to more than 300 m where glaciotectonism has deeply deformed the bedrock. Glaciotectonism was particularly active near the eastern margin of the Saskatchewan and Alberta plains (Bostock, 1970a, 1970b), where generally westward-flowing glaciers impinged on the east-facing scarps bordering these plains.

DEFORMATION TERRAIN

Glaciotectonic features and deformation terrain are widespread on the northern Great Plains and have been recognized from Alberta, Saskatchewan, and Manitoba southward into North Dakota, Minnesota, and South Dakota (Moran and others, 1980; Fig. 1A). Among the earliest descriptions of these features on the prairies were those by Sardeson (1898, 1905, 1906), who interpreted several large anomalous outliers of Cretaceous rock in Minnesota as being blocks transported by glacier ice. Earlier studies described the features as generally large (>1 km long) hills or ridges composed mainly of bedrock (Hopkins, 1923; Slater, 1926, 1927; Byers, 1960; Kupsch, 1962; Moran, 1971; Christiansen and Whitaker, 1976), but recent studies have indicated a wide variety of compositions and structures at individual thrust areas (Fenton and Andriashek, 1978; Moran and others, 1980; Fenton, 1983a, 1983b; Bluemle and Clayton, 1984; Fenton and others, 1985a, 1985b; Aber, 1987). An impetus for these studies and observations has been highwall stability problems in open-pit coal mines at sites of glacially thrust sediment.

GLACIOTECTONIC FEATURES IN THE SAND RIVER AREA

Deformation terrain of the Sand River area provides prominent examples of certain glaciotectonic features, perhaps better developed and of greater areal density than is typical elsewhere, but illustrative of similar features common in much of the northern Plains. The Sand River area covers about 12,000 km^2 in east-central Alberta (Fig. 1A), about 30% of which is deformation terrain (Fenton and Andriashek, 1984; Fig. 8). The features of the Sand River area differ from the "classical" thrust terrain of earlier studies by their range of morphology, size, and composition. Similar terrains have been described by Moran and others (1980) and Bluemle and Clayton (1984).

Morphology

Three types of terrain are recognized in the Sand River area: (1) hill-hole pairs, (2) hills with a fault-bounded depression, and (3) rubble terrain. These terrain types generally are easily identified on air photographs or topographic maps.

A hill-hole pair consists of a hill immediately down-glacier of a depression. These features, the depressions of which are generally occupied by lakes, form a major portion of the deformation terrain in the Sand River area. The hill may be a simple mound, or a complex of ridges perpendicular to the ice-flow direction, such as the hill south of Saddle Lake (Figs. 8 and 9). Although variable in size, the hills are generally between 15 and 45 m high and cover an area of 10 to 40 km^2; smaller features, less than 15 m in height and 2 km^2 in area, are also present. Examples include the Frenchman Lake hill, about 8 km^2 in area and 30 m high, and the Marie Lake hill, about 33 km^2 in area and 60 m in height. The largest thrust hills are south of Muriel, Sinking, and Cold lakes (Fig. 8). The Muriel–Sinking lakes thrust hill covers about 140 km^2 and exceeds 200 m in height. The Cold Lake thrust originally formed a ridge across the southern end of the lake, but a southwestward ice-sheet readvance subsequently

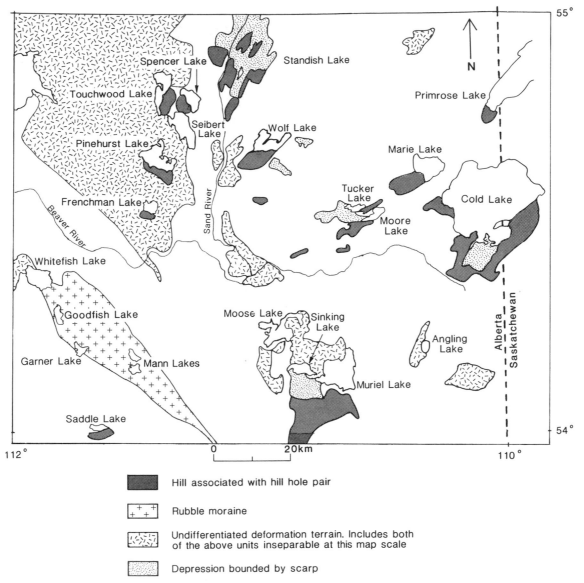

Figure 8. Map of the Sand River area, southeastern Alberta, showing the glaciotectonic features.

removed much of the southwestern part of the ridge. The remaining deposit suggests an original thrust-hill area of about 50 km^2 and a height of more than 120 m.

A hill with a fault-bounded depression is a type of hill-hole pair, the depression of which is bounded on at least one side by a straight margin that may be the trace of a vertical tear fault created during glacial excavation of the sediment. This relatively uncommon terrain is present southwest of Wolf, Primrose, Marie, and Standish lakes (Fig. 8). The Wolf Lake thrust, for example, forms a hill of arcuate ridges about 42 km^2 in area and 135 m high (Figs. 8 and 10). The hill is asymmetric with the steeper face on the down-glacier side. The fault scarps form east and west margins of the lake and parallel the glacier-flow direction. All surrounding terrain is essentially flat and undisturbed.

A variation of this terrain type results from en echelon tear faulting to produce a series of en echelon source depressions. An example of this terrain is the Standish Lake area, where en echelon tear faults form the eastern boundary of a large composite source depression (Fig. 11). A hill of arcuate ridges, about 100 m high and 12 km^2 in area, is on the otherwise flat terrain immediately down-glacier of the easternmost source depression. The depression is about 16 km^2 in area and about 2.5 km wide. Three associated en echelon depressions of similar width lie to the northwest. The thrust hills originally formed from these three depressions have been partially eroded and streamlined by glacial flow subsequent to thrusting.

A second variation of this terrain type occurs when thrust sediment is removed rather than deposited down-ice of the de-

Figure 9. Stereophotographs of hills of thrust moraine (MT) south of Saddle Lake, Sand River area.

pression. A depression results having tear-fault sides that parallel the ice-flow direction and an upstream margin perpendicular to the flow, thereby producing a roughly rectangular plan view (Fig. 8). In the Sand River area these depressions are generally occupied by bogs, indicating they are shallower than depressions up-glacier of hills, and that the ice may have removed only a thin layer of sediment during thrusting.

Rubble terrain, a group or series of glacially thrust hills, is formed of glacially thrust sediment or rubble moraine. The hills may decrease in size down-glacier (Fig. 12); an up-glacier source depression is usually recognizable.

The Whitefish Lake rubble train (Fig. 8) is an example of a glacial-rubble terrain. The rubble train extends southeastward about 60 km from the north end of Whitefish Lake, which occupies the up-glacier source depression. The western boundary, a scarp, is 50 m high in the Whitefish Lake–Garner Lake area and 30 m high in the Mann Lakes area. The eastern boundary is gradational with a change, eastward, from unmodified thrust terrain to slightly smoothed hills of fluted thrust terrain to classical fluted terrain composed of long, streamlined, southeast-trending ridges, the latter two formed by overriding of the thrust sediment. Down-glacier, or southeast of Whitefish Lake, hills of the rubble train decrease from more than 1 km^2 in area and 30 m in height to hummocks southeast of Mann Lakes with areas less than 0.12 km^2 and heights no more than 7 m.

Internal Composition and Structure

Although exposures are insufficient to describe individual features in detail, data suggest that thrust hills are formed of

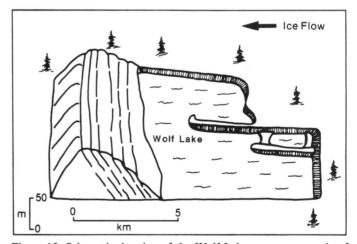

Figure 10. Schematic drawing of the Wolf Lake area, an example of hills, formed of arcuate ridges, down-glacier from a fault-bounded depression.

complexes of bedrock blocks and fragments and unconsolidated deposits of various diamictons, sand, and finer sediment. Much of the thrust material is Quaternary sediment, but it locally includes bedrock, the proportion of which was controlled mainly by the thickness of the preexisting drift. The drift is generally over 50 m thick in the Sand River area (Gold and others, 1983); thus, many of the large thrust hills are likely composed entirely of Quaternary sediment. Exceptions are (1) thrust hills south of Cold Lake containing a high proportion of folded and faulted shale and ironstone, (2) the thrust terrain northwest of Touchwood Lake,

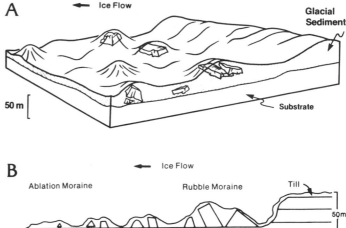

Figure 12. Schematic block diagram and section showing deposited rubble moraine (A) and rubble in transit (B).

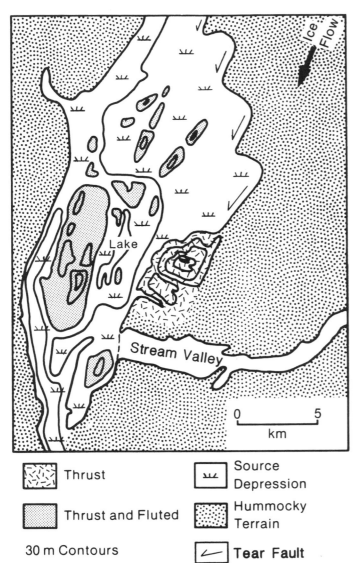

Thrust

Thrust and Fluted

30 m Contours

Source Depression

Hummocky Terrain

Tear Fault

Figure 11. Map of the Standish Lake area, showing en echelon tear faulting on east margin of the depression to form a southeast-trending series of subdepressions. Note that the hill of thrust sediment is intact south of the easternmost depression, but that other hills have been partially eroded to form fluted terrain.

where shale crops out in roadcuts, and (3) an area northeast of Whitefish Lake, where both shale and sandstone are exposed in roadcuts.

The lithologic composition of rubble-terrain features is, in general, more variable than is that of hill-hole pairs, and the blocks of preexisting sediment are generally smaller in size. Roadcuts show masses of drift and bedrock that have been folded, faulted, attenuated, and may be surrounded or mixed with syngenetic till. Whether similar variability is present in other areas of rubble terrain in the northern Great Plains depends on the availability of different sediment types.

COAL-MINE HIGHWALL STABILITY IN DEFORMATION TERRAIN: A CASE STUDY— THE HIGHVALE MINE, CENTRAL ALBERTA

Application of geomorphic research on deformation terrain is helping to solve instability problems in open-pit coal mines on the Great Plains. Glacially thrust bedrock is weakened (Figs. 13 and 14). As a result, highwalls cut into the disturbed material have a tendency to fail, posing risk to men and equipment. Failure studies, however, have also increased knowledge of the composition, structure, and genesis of deformation terrain, and have aided an understanding of the groundwater and mass-movement processes that modified the terrain following glacial retreat. As has been the case for the Sand River area, the study of specific features at the Highvale Mine, about 80 km west of Edmonton, Alberta (Fig. 1A), has led to the recognition of glacially deformed features elsewhere. In a practical manner, this knowledge of location, structure, dimensions, and hydrology of deformation terrain is essential to the safe and efficient operation of mining activity in these areas.

Geology

Bedrock at the Highvale Mine consists of Upper Cretaceous and Paleocene non-marine coal-bearing rocks (Lerbekmo and others, 1979; Moell and others, 1985; Maslowski Schutze, 1986). The coal, about 15 m thick, is in the Ardley Coal Zone of the Paskapoo Formation (Carrigy, 1970; Irish, 1970; Holter and others, 1975). Surficial deposits include extensive glacially deformed bedrock, up to 50 m thick (Fenton, 1983c; Fenton and others, 1983), overlain by a thin, discontinuous cover of till, glaciolacustrine clay, and glaciofluvial sediment (Andriashek and others,

Figure 13. Photograph of east wall of Pit 03, Highvale Mine, showing folded sandstone, mudstone, and dark-banded calcareous ironstone. Coal and a basal shear plane are below the base of the wall.

Figure 14. Photograph of thrust slabs of shale at a coal mine, central Alberta.

1979; Fig. 15). Deformation in the Highvale Mine area appears gradational between rubble terrain and hill-hole pairs.

Failures have occurred frequently near ramps leading into Pit 03 (Fenton and others, 1985b; Fig. 15), where up to 3 m of till is underlain, in order, by 12 m of sandstone, 4 to 7 m of mudstone, 1 m of interlaminated mudstone, coal, and bentonite, and a 15-m six-seam coal unit. A mass of thrust bedrock overlies the till in the western third of the area.

Glacial advances were mostly from the north, causing folding and faulting of the bedrock by compression and thrusting; the thrust planes dip generally northward. Large-scale folds (Fig. 13) and faults at Pit 03 are exposed along ramps and in smaller pits, whereas small-scale deformations are common in many exposures and cores (Fenton and others, 1983, 1985b). The base of the disturbance is a shear plane that rises southward from within the coal seams on the north to a few meters above seam 1 near the haul road (Fig. 15). The shear plane probably dies out south of the haul road. Where the shear plane immediately overlies seam 1, movement occurred along a 1-m layer of mudstone, coal, and bentonite, and is well illustrated by small folds in the coal laminae.

Groundwater Conditions

The glacially deformed overburden is nearly saturated with groundwater, and stratigraphic relations with the overlying till and underlying bedrock act to inhibit drainage. Head distributions show that groundwater flow is northward toward the highwall; hydraulic gradients are low (0.003 to 0.009 in seam 6 and 0.009 to 0.10 in the overburden), suggesting minimal drainage (Moell and others, 1985).

In the central part of Pit 03, for example, the overburden has remained poorly drained since before September 1983 (Fenton

and others, 1985b). The hydrogeologic cross section (Fig. 16) shows that by November 1984, there had been only 0.6 m of water-level decline in the overburden at site HV83-4, 110 m south of the highwall at that time. Sites farther from the highwall showed unchanged or slightly higher water levels; small increases were detected at site HV83-1.

Failure Types: Observations

Surface and subsurface data indicate the failures at the Highvale Mine are composites of (1) exfoliation or spalling, (2) block rotation, (3) block slide, and (4) subsequent in situ disintegration of the large failed blocks. The most prominent surface manifestations of the failures (Fig. 17) are series of vertical fractures caused by exfoliation and block sliding. The failures parallel the highwall face, and the failure zone extends more than 50 m back from the face. Depths of the fractures exceed 4 m and widths increase with time. Exfoliation occurs as sheets of sandstone 1 to 3 m thick and generally 3 to 10 m long.

Rotational slump blocks (Fig. 17) generally are 50 to 100 m long, semicircular in plan, and show northward movement. One slump, observed within 12 hours of failure, appeared to have moved along an existing glacial shear plane. Evidence included slickensides, indicating diagonal movement across the failure plane, and the passing of the failure plane beneath undisturbed masses of hanging-wall sediment.

In-situ disintegration of the debris that collects below the highwall follows the first three failure types. A relatively low-angle ramp of disaggregated sandstone is produced that extends onto the coal seam. In each case, removal of the ramp revealed groundwater discharge from the freshly exposed mudstone-sandstone contact and incipient sliding of the sandstone over the mudstone.

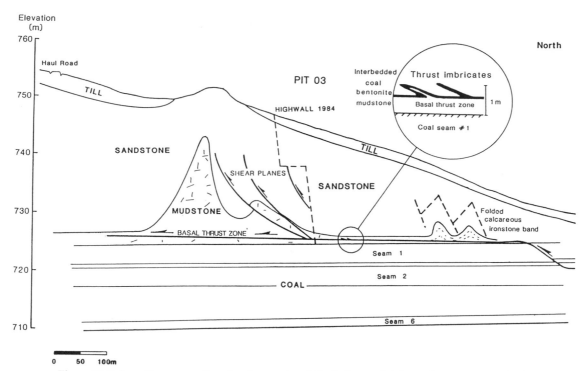

Figure 15. Schematic cross section, showing stratigraphic relations and deformation structures at Pit 03, Highvale Mine, Alberta.

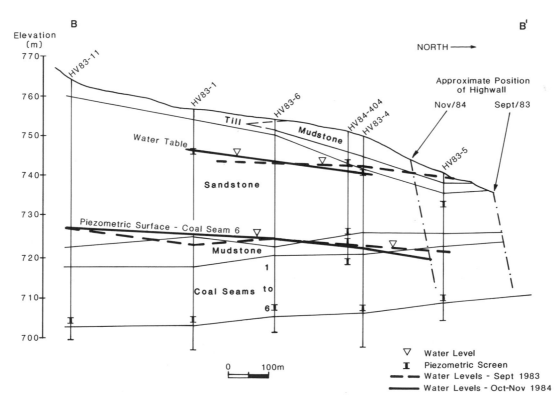

Figure 16. Hydrogeologic cross section, showing water-level changes from September 1983 to November 1984 at Pit 03, Highvale Mine, Alberta.

Failure Model

A failure model accounts for the above observations (Fig. 17), made from March through October, 1984. Intergranular movement within the sandstone during glacial deformation likely decreased rock strength, contributing to each failure type. Groundwater saturation and elevated pore-water pressures contribute to both slump and block slide, elevated pore pressures at the mudstone-sandstone contact being particularly important to the latter. The dip of the contact toward the pit also contributes to block sliding (Fig. 15). Vertical fractures may form spontaneously from the sliding, but are more likely due to opening of existing joints or a combination of joints and shear planes. Rotational slumps, particularly the large ones, are likely the result of renewed movement along glacial shear planes. Some small ones, however, likely form spontaneously due to low strength of the glacially deformed sandstone.

Deformation terrain, owing to the processes by which it forms, is inherently unstable following glacial retreat. Application of the failure model suggests that the groundwater, mass movement, and disintegration processes leading to the failures at the Highvale Mine may be indicative also of the processes that both extended small-scale structural movements of glaciotectonism and caused smoothing of the landscape through Holocene time, possibly with decreasing intensity. These studies have shown that although some deformation terrains presently exhibit a subdued topography, the geomorphic features of the landscapes evolved from specific conditions of glacially induced shear and all of the deformation terrains continue to bear signatures of those conditions. Some of the deformation features at the Sand River area were identified by their aerial-photography signatures, and it is anticipated that others will be identified in other parts of the glaciated northern Plains.

FLUVIAL PROCESSES AND RIVER ADJUSTMENTS ON THE GREAT PLAINS

R. F. Hadley and T. J. Toy

To varying degrees, the entire Great Plains has been subject to late-Cenozoic episodes of fluvial erosion and deposition. Whereas glacial and periglacial processes have dominated Quaternary landform development in the northern Plains, and groundwater and eolian processes have dominated in the southern Plains, fluvial processes have been dominant in central portions of the Great Plains. It is in this part of the region that runoff currently is sufficient and drainage networks are well-enough developed that fluvial processes remain important.

Presently, nearly all of the Great Plains is water deficient, receiving less than 600 mm of precipitation annually; most of the region receives less than 400 mm. The annual rate of potential evapotranspiration (loss of water into the atmosphere through the combined processes of surface evaporation and plant transpiration) exceeds the annual rate of precipitation in this region (Toy

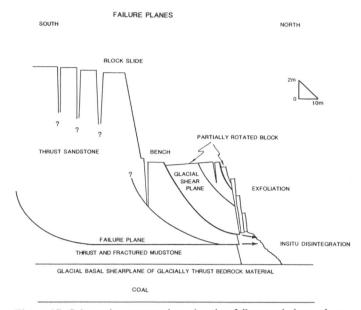

Figure 17. Schematic cross section, showing failure and shear planes, vertical fractures, and block movements at Pit 03, Highvale Mine, Alberta.

and Grim, 1984). These climatic characteristics create the semiarid environment that typifies the Great Plains (Trimble, 1980).

FLUVIAL PROCESSES

Largely as a result of variations in climate and topography, there are two types of rivers that flow generally eastward across the Plains: (1) perennial rivers that originate in the Rocky Mountains and generally carry high flows in the spring, and (2) rivers that originate on the plains and flow from a semiarid to a subhumid environment. The second group of rivers may be ephemeral, intermittent, or perennial in character.

Sediment Sources

The rivers of the Great Plains transport out of the region the erosional debris produced on the upland source areas. As pointed out by Schumm (1981), the nature and quantity of sediment produced in these source areas determine the morphologic character of the rivers. To understand the differences we find among rivers of the Great Plains, the erosional processes of upland areas are discussed briefly.

Hillslopes are the primary source of sediment transported by rivers, with flood plains and channel banks providing a secondary source. Available data (U.S. Water Resources Council, 1977) on sheet- and rill-erosion rates for tributary drainage basins in the Great Plains Physiographic Province provide average values for various land types. Average soil loss is 2.31 Mg/ha/yr (megagrams, or metric tons, per hectare per year) for forest and range-

TABLE 1. SOME UPLAND EROSION RATES IN THE GREAT PLAINS

Location	Erosion/Sediment yield	Measurement Technique	Reference
South-central Wyoming	0.25 Mg/ha/yr	Universal soil loss equation	Frickel and others, 1981 (undisturbed sites)
Southeastern Montana	3.56 Mg/ha/yr	Universal soil loss equation	Hadley and others, 1981 (undisturbed sites)
North-central Wyoming	16 m^3/km^2/yr	Survey of small reservoir	Ringen and others, 1979 (0.39 km2 contributing area)
Eastern Nebraska	7.81 Mg/ha/yr	Survey of small reservoirs	Missouri River Basin Interagency Committee, 1969 (average size of contributing area = 1.2 km^2)
Western North and South Dakota	3.52 Mg/ha/yr	Survey of small reservoirs	Missouri River Basin Interagency Committee, 1968 (average size of contributing area = 1.3 km^2)
Southwestern South Dakota	610 m^3/km^2/yr	Survey of small reservoirs	Hadley and Schumm, 1961 (average size of contributing area = 0.5 km^2)
East-central Wyoming	0.51 mm/yr	Surface lowering (LEMI)	Toy, unpublished data (technique from Toy, 1983)
Southwestern South Dakota	13.2 mm/yr	Surface lowering (erosion pins)	Schumm, 1956 (badland hillslopes)
Western North Dakota	3.18 mm/yr	Surface lowering (erosion pins)	Clayton and Tinker, 1971 (badland hillslopes, Tongue River Fm.)

land, and 9.26 Mg/ha/yr for croplands from the entire region. The National Research Council (1981) stated that wind erosion from rangelands and croplands of the Great Plains generally averages between 4.0 and 9.0 Mg/ha/yr, but may exceed 200 Mg/ha/yr from croplands on highly susceptible soils. Some of the sediment detached by wind erosion may enter stream channels directly and be transported downstream, but most eolian sediment, along with fluvial sediment, is probably transported to streams by overland flow. Eolian processes on the Great Plains are discussed in greater detail later in this chapter.

Average sheet- and rill-erosion rates mask local variabilities in land use, topography, soil types, and vegetation cover. Erosion rates for the Great Plains generally range from less than 1.1 Mg/ha/yr for forests and rangelands to as much as 20 Mg/ha/yr for croplands. These data reveal the influence of land use on erosion rates. Hadley and Schumm (1961) found more than a tenfold variation in sediment yields for areas underlain by different rock types. They also developed a direct relation between basin relief (relief ratio) and sediment yield in the Great Plains of eastern Wyoming.

Table 1 lists upland erosion rates for several Great Plains locations that illustrate the differences from place to place. Unfortunately, there are few field measurements available, and they are concentrated in the northern Plains. Data from badlands sites likely represent atypically high rates from relatively small, isolated areas. Reservoir-survey data are included for small drainage basins with the assumption that erosion rates are nearly equal to sediment-yield rates (sediment-delivery ratio = 1.0). In these small basins the probability of intermediate deposition is low.

Rainfall simulation has been used to compile additional erosion values for small plots, generally less than 0.04 ha, but due to variations in equipment and research design these data are probably only useful for comparison of land-use practices. Those interested in results of rainfall-simulation studies are referred to Lusby and Toy (1976), Gilley and others (1977), and Hofmann and others (1983).

The inverse relation between drainage area and mean annual sediment yield (Hadley and Schumm, 1961) shows that a part of upland soil erosion becomes temporarily stored at the base of hillslopes, in channels, and on flood plains. These deposits constitute secondary sediment sources during high-intensity storms or streamflow of relatively high magnitude and energy. Hadley and Schumm (1961) and Schumm and Lichty (1963), among others, showed that channels in the Great Plains experience epicycles of aggradation and degradation—possibly synonymous with the microcycles as defined by Morrison earlier in this chapter. During periods of degradation, sediment stored in the valley floor and in channels is remobilized and transported farther through the fluvial system. Although channels and flood plains are known to be secondary sediment sources, their actual roles in the overall sediment-transport process are not yet precisely defined; they indeed may be as variable as upland erosion processes.

Channel Processes

The type and quantity of sediment derived from upland erosion and delivered to the Great Plains rivers depend on characteristics of the rocks, climate, vegetation cover, and average basin slope. Because of the great variety of these characteristics in drainage basins of the Great Plains, the type and quantity of sediment transported by the rivers reflects this diversity, and determines, to some degree, stream-channel characteristics.

Several investigators have observed that the shape of alluvial channels in the Great Plains, and elsewhere, is determined by the character of the sediment transported by the stream. Schumm (1960) showed that the shapes of ephemeral and perennial stream channels are related to the type of sediment forming the channel perimeter. The effect of discharge on channel morphology was not considered. Schumm's results demonstrated the influence of silt and clay on channel shape, expressed as a width-depth ratio. He recognized the effects of mean annual discharge and mean annual flood on absolute width and depth of the channel, but concluded that neither of these changes the power-function relation between width-depth ratio and a weighted mean percent of silt and clay in the bed and banks of stable alluvial channels. This technique demonstrates the influence of sediment properties on channel geometry but cannot be used to estimate discharge from ungauged basins (Osterkamp, 1977).

A study by Osterkamp and Hedman (1982) expanded on the earlier work of Schumm (1960), Hedman and others (1974), and Osterkamp (1977), and considered data on channel geometry, channel gradient, sediment type in the channel, and discharge compiled from 252 streamflow-gauging stations in the Missouri River basin, many of which are in the Great Plains. The data were collected primarily at perennial streams; some of the smaller streams have intermittent flow. The data were analyzed by computer and produced simple and multiple power functions that relate discharge characteristics to variables of channel geometry and bed and bank sediment.

The methodology used by Osterkamp and Hedman (1982) was based on channel geometry rather than hydraulic geometry. The channel-geometry method relies on measurements taken from an identifiable geomorphic reference point or level in the channel cross section. Hydraulic-geometry measurements are related to the water surface. There are various in-channel reference levels that have been used in discharge-geometry correlations. The "active channel," defined at the upper limit as the break in the steep bank slope of the active channel and the lower limit of permanent vegetation, was used in the study by Osterkamp and Hedman (1982).

Results of the study show that channel width is best related to variables of discharge, but reduction of the standard errors of estimate can be achieved by introducing data on channel sediment, channel gradient, and discharge variability. The practical result of the study by Osterkamp and Hedman (1982) was the development of sediment-dependent equations for the purpose of general estimates of discharge characteristics. The study demonstrated that sediment variables of the channel bed and banks have a quantitative, statistically significant correspondence with the active-channel width. Figure 18 illustrates the relation between active-channel width and mean discharge for seven channel types

defined in the study. The widest channels, relative to discharge characteristics, occur in highly sandy channels. For channels of similar width, the greatest discharges occur for the high silt-clay bed channels. Figure 18 also shows that discharges decrease for channels of increasing sandiness, and increase again (nos. 6 and 7) as median particle sizes and channel armoring provide channel stability.

In addition to the effect of channel-sediment properties on width-discharge relations, Osterkamp and Hedman (1982) found that discharge variability can have a measurable effect on channel geometry. Sand channels without sufficient fine or coarse sediment to form resistant banks are most susceptible to changes in geometry due to discharge variability. Therefore, channel size and geometry result from the water and sediment transported by the channel.

Hydraulic-Geometry Relations

Hydraulic geometry, in contrast to channel geometry, was introduced by Leopold and Maddock (1953) to describe the way that flow characteristics vary with discharge. The hydraulic-geometry model expresses the changes in channel and hydraulic variables as simple power functions of discharge resulting in the following equations:

$$w = aQ^b \qquad (1)$$
$$d = cQ^f \qquad (2)$$
$$v = kQ^m \qquad (3)$$

where w = width, d = depth, v = velocity, and Q = instantaneous discharge; b, f, and m are exponents and a, c, and k are coefficients. The continuity equation expresses the product of width, depth, and velocity as equaling discharge:

$$Q = wdv \qquad (4)$$

and

$$Q = ackQ^{b+f+m}. \qquad (5)$$

Hydraulic-geometry relations for rivers have been determined for both individual cross sections, termed at-a-station hydraulic geometry, and for several cross sections in a downstream direction at a particular frequency of discharge, termed downstream hydraulic geometry. The most commonly cited average exponent values for at-a-station data are those of Leopold and Maddock (1953); these are $b = 0.26$, $f = 0.40$, and $m = 0.34$. Leopold and Maddock determined average exponent values for downstream hydraulic geometry of $b = 0.5$, $f = 0.4$, and $m = 0.1$. It is apparent from studies of hydraulic-geometry relations in various physiographic and climatic regions that average exponent values may not be useful. As pointed out by Eschner (1983), the large range of all three exponent values suggests that mean hydraulic-geometry values may be a meaningless, albeit attractive concept.

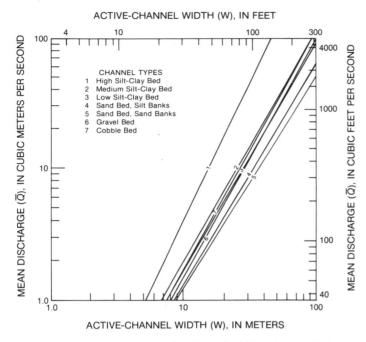

Figure 18. Relations between active-channel width and mean discharge for stream channels of specified sediment characteristics. (From Osterkamp and Hedman, 1982.)

Eschner (1983) conducted a study of hydraulic-geometry relations on the Platte River in south-central Nebraska, related to the effects of hydrology and channel-morphology changes on wildlife habitat. The purpose of the study was "to compute at-a-station hydraulic geometry at selected sites, to examine how the relations have changed with time, and to determine the significance of the relations for channel maintenance." The hydrology and morphology of the Platte River have changed considerably since settlement of the river basin in the mid-nineteenth century. Peak discharges have decreased and the channel has narrowed because of flow regulation by both on-stream and off-stream storage.

Eschner (1983) concluded that the rates of change of width, depth, and velocity with increasing discharge have varied with time. The width exponent has decreased, but similar trends are not apparent in the depth and velocity exponents. Also, Eschner found that with increasing stage and discharge, the shapes of channel cross sections, together with increased roughness generally attributable to riparian vegetation, are responsible for the complex hydraulic-geometry relations. When the flow reaches the vertical channel banks, width does not continue to increase but depth can increase rapidly (Eschner, 1983).

RIVER ADJUSTMENTS

Channel morphology adjusts to changes in discharge and sediment load if these variables are of sufficient magnitude. Alluvial channels are an integral part of a complex fluvial system;

width, depth, gradient, and sediment properties are channel variables that may change as external inputs of flow and sediment load vary with time.

When flow increases, the channel width, depth, and meander wavelength (1) increase, and the channel gradient (s) decreases (Schumm, 1968) as follows:

$$Q \propto \frac{w, d, 1}{s}. \tag{6}$$

In addition, Schumm (1968) stated that for a given discharge the dimensions of a channel are influenced by the type of sediment load that is being transported in the stream and that composes the bed and banks. An increase in the ratio of bedload to total sediment load (Q_r) will cause increases in channel width, meander wavelength, and gradient, but decreases in depth and sinuosity, P (the ratio of channel length to valley length; Schumm, 1968):

$$Q_r \propto \frac{w, 1, s}{d, P}. \tag{7}$$

With width in the numerator and depth in the denominator, equation (7) shows that a change in "type of sediment load will significantly influence channel shape" (Schumm, 1968). Osterkamp and Harrold (1981) described dynamics of alluvial stream channels, and also used the basic premise that the characteristics of the channel adjust to water and sediment discharges. They related this adjustment to implications of grade (equilibrium) or quasi-equilibrium, or that no net erosion or deposition (steady state) occurs in a given time interval. The shear stress (τ_0) along the wetted perimeter of the channel under these conditions is a function of water (Q) and sediment (Q_s) discharge:

$$\tau_0 = F (Q, Q_s). \tag{8}$$

Any change in Q or Q_s produces a change in τ_0. Osterkamp and Harrold (1981) pointed out that because "no natural stream has constant discharges of water and sediment," τ_0 in equation (8) represents time-integrated values resulting from the range of values for Q and Q_s. Equations (6), (7), and (8) show that a significant change of the hydrologic regime can cause major adjustments in river channels. There are many examples of changes in channel morphology in Great Plains rivers resulting from changes in hydrology. Among these examples of channel dynamics are changes in the Platte River system, largely due to water development, and in the Cimarron River system, due to natural variations in precipitation and streamflow.

The Platte River and its major tributaries, the North Platte and South Platte rivers, have a drainage area of about 223,000 km^2 in Colorado, Wyoming, and Nebraska. These rivers are typical of many Great Plains streams that originate in the Rocky Mountains. The North Platte River begins in the mountains of northern Colorado, flows north into central Wyoming, then bends southeastward into Nebraska. The South Platte River originates in the mountains of central Colorado and flows across the plains of northeastern Colorado into Nebraska. The North Platte

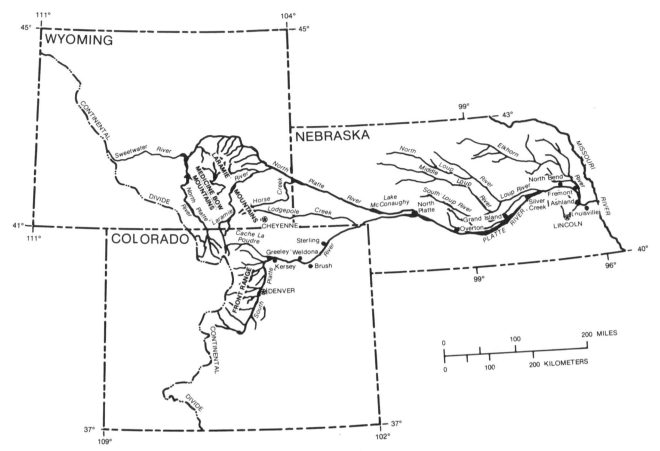

Figure 19. The Platte River basin in Colorado, Wyoming, and Nebraska. (From Eschner and others, 1983.)

and South Platte rivers join near North Platte, Nebraska to form the Platte River (Fig. 19).

Most flow in the Platte River and its tributaries upstream from the Loup River in Nebraska (Fig. 19) is derived from spring snowmelt in the Rocky Mountains (Eschner and others, 1983). Annual precipitation in the Great Plains part of the drainage basin ranges from 330 to 635 mm, and contributes limited additional flow to the channels. The major water use in the basin is for irrigation of croplands. Surface water stored in reservoirs is used for irrigation, municipal use, and hydroelectric power generation.

Water development for a variety of uses in the Platte River basin since the middle of the nineteenth century has had significant effects on the hydrology and morphology of major channels. Recent studies of hydrologic and morphologic changes in the Platte River have been conducted by Williams (1978a), Nadler and Schumm (1981), Crowley (1981), Eschner and others (1983), and Kircher and Karlinger (1983).

Kircher and Karlinger (1983) investigated the hydrologic changes that have occurred in the Platte River basin since the early part of the twentieth century. When the Platte River system was unregulated, streamflow became very low after snowmelt

subsided, and many streams became ephemeral by late summer. Today many of these streams are perennial because of flow regulation by storage and diversion, transmountain imports of water, groundwater pumpage, and return flow from irrigation. All of these factors have markedly changed the hydrology, primarily since 1935, about the time when construction began on Kingsley Dam and Lake McConaughy on the North Platte River near Ogallala, Nebraska (Fig. 19).

Changes in surface-water hydrology and channel size occurred at different times throughout the basin. The hydrologic changes are identified by shifts in levels of low flows and high flows and by the flattening of flow-duration curves, which indicates a reduction of flow variability in a downstream direction. Changes in channel morphology through time are exhibited in the plots of width and cross-sectional area of Platte River channels (Kircher and Karlinger, 1983). An example of channel narrowing that occurred in the Platte River system during the last century, using a bridge for perspective, is provided by photographs of the North Platte River in western Nebraska (Fig. 20). The narrowing represented by the photographs suggests reductions through time of both water discharge and discharge of sand-size sediment.

Changes in channel morphology of the North Platte, South Platte, and Platte rivers have been similar despite significant differences in hydrology for these rivers (Eschner and others, 1983). Construction of reservoirs and diversion of flow on the North Platte River have caused reductions of annual peak flows and mean flows on both the North Platte and Platte rivers. As pointed out by Eschner and others (1983), however, there has not been a similar reduction of peak flows on the South Platte River upstream of Julesburg, Colorado, primarily because there was very little reservoir construction. There has been diversion of flows from the South Platte River for irrigation, but this has been offset by transmountain diversions of water into the basin. The result has been virtually no net change in mean annual flows during the period of streamflow records (1903 to 1979, with some years missing; Kircher and Karlinger, 1983). Similarly, Kircher and Karlinger (1983) stated that there was no decrease in the annual peak discharge on the South Platte River during the period of record. Morphologic changes apparently occurred in response to irrigation development prior to the period of record (Eschner and others, 1983). Nadler and Schumm (1981) concluded that irrigation development changed the river from intermittent to perennial. Drought during the 1930s reduced discharge, which allowed riparian vegetation to become established and stabilize the channel. Subsequent floods have not widened the channel because of the stabilizing effect of the vegetation. Nadler and Schumm (1981) discussed the similarity of hydrologic and morphologic changes that have occurred on both the South Platte and Arkansas rivers on the plains of eastern Colorado, and attributed these changes to the effects of riparian vegetation and flow regime.

Measurements of channel width at six reaches on the Platte River in Nebraska were taken from General Land Office maps surveyed during the approximate period 1859 to 1867, and from aerial photography taken between 1938 and 1979. Changes in the intervening years between the 1860s and 1938 must be inferred (Eschner and others, 1983). Comparisons of width changes at the six sites (Fig. 21) are expressed as percentages of the "1860 width." They indicate dramatic channel narrowing that occurred in the Platte River drainage net since about 1880, as is also shown for the North Platte River at the Camp Creek bridge site by Figure 20. In general, widths of the Platte channels have decreased steadily since 1860, although near Ashland, Nebraska, below the mouth of the Loup River, channel width increased slightly between 1941 and 1971 because of the relatively unregulated flow from the Loup. Williams (1978a) measured river widths on the North Platte and Platte rivers at 35 locations from the Wyoming-Nebraska boundary to Grand Island, and found that channel widths were about 10% to 20% as wide in 1965 as they were in 1860 for most of the reach. Nadler and Schumm (1981) found that channel widths of the South Platte River in 1952 averaged about 15% of the 1867 channel widths.

Morphologic changes in Great Plains rivers have occurred in the past 100 years that result from hydrologic phenomena other than regulation of flow or groundwater pumpage for irrigation and other water-resources development uses. Schumm and

Figure 20. A: Old Camp Clark bridge spanning the North Platte River about 70 km east of the Wyoming-Nebraska State Line. The bridge was in use about 1876 to 1903. View may be from an island in the river. B: Site of old Camp Creek bridge in 1968. Present bank of the river is just below the lower edge of the photograph and is indicated by tree branches in the bottom right corner; arrows point to remaining piers of the bridge. (Photographs from Williams, 1978a.)

Lichty (1963) described channel changes on the Cimarron River in southwestern Kansas that occurred during historic time. They reported that the Cimarron River in Kansas was a narrow, meandering, perennial river with a well-developed flood plain during the period 1874 to 1913 when some record existed. The average width of the river was 15.2 m in 1874. In 1914 a major flood occurred, initiating lateral erosion that continued to widen the channel until an average width of nearly 370 m was reached in 1942. During the period 1943 to 1954 flood-plain construction occurred and the river narrowed to an average width of 168 m in 1954; from 1954 to 1960 the channel experienced periods of both widening and narrowing (Schumm and Lichty, 1963).

The sequence of channel changes for the Cimarron River appears to conform closely to the microcycles previously described in this chapter by Morrison, particularly because Schumm and Lichty (1963) were able to relate the changes directly to climatic fluctuations. Initial channel widening was attributed to the flood of 1914. The flood was followed by a long period of below normal precipitation until 1942, when another major flood occurred. Flood-plain construction after the 1942 flood occurred when precipitation was above normal but floods were only of low to moderate magnitude. These hydrologic conditions produced morphologic changes that are comparable to the Platte River channels. Riparian vegetation that became established along the channel banks trapped sediment and allowed channel narrowing to occur. Schumm and Lichty (1963) reported similar channel changes that have occurred on the Smoky Hill and Republican rivers in Kansas (Fig. 1B).

GROUNDWATER—AN AGENT OF GEOMORPHIC CHANGE

W. R. Osterkamp

Like fluvial processes, landform changes due to groundwater movement are apparent through the entire length of the Great Plains. Unlike fluvial processes, which have been most active in the central latitudes of the Plains during Quaternary time, the effects of groundwater have been most striking in the south. Only in recent years, perhaps in large part due to images of Mars sent back by the Mariner 6 and 9 spacecrafts, has the importance of groundwater systems to landforms been appreciated (Higgins, 1982, 1984). Piping, sapping, and seepage erosion are processes that long have been recognized on earth, but the significance of these processes was largely neglected. The more visible geomorphic results of solution and collapse in carbonate and evaporite terrains are well known, but even these features that result from groundwater movement have sometimes been overlooked in areas of dominantly clastic sedimentation. Recent landform studies of the Great Plains suggest that some large-scale features, such as prominent escarpments and the positions of several river valleys, as well as many smaller features, such as ephemeral-stream draws and the playa-lake basins of the High Plains section (Fig. 3), may be closely related to processes of groundwater

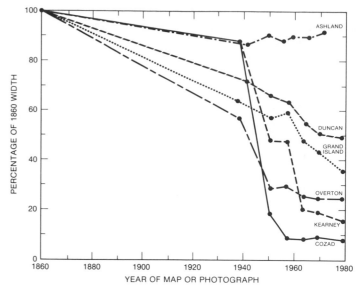

Figure 21. At-a-station changes of channel width of the Platte River, Nebraska, with time. (From Eschner and others, 1983.)

recharge, movement through the vadose and phreatic zones, and discharge.

Sapping is erosion by relatively concentrated discharge of groundwater, whereas seepage erosion is the result of diffuse discharge of groundwater, generally at lithologic contacts or other geologic boundaries. Where discharge is from poorly cemented porous beds overlain by relatively resistant rocks, these processes commonly cause undermining at the base of a cliff, subsequent cliff-face failure, and thus a steadily retreating escarpment. In areas of well-developed fractures, spring sapping may cause headward growth of drainage systems closely following the fracture systems. The related process of piping—erosion by flow of water in or from the soil or subsoil—commonly causes headward gully erosion in the unsaturated zone. Piping may be aided by chemical erosion of carbonate or other soluble cementing material, and any combination of mass-wasting, fluvial, or eolian processes may transport the erosion products from an escarpment or gully system.

Landforms caused by processes of groundwater erosion on the Great Plains appear best developed where soft Tertiary rocks contain or underlie resistant layers. Prominent examples include (1) buttes, mesas, and escarpments of Saskatchewan, Montana, and North Dakota, where sandstones and siltstones of the Paleocene Fort Union Formation are capped by clinker beds; (2) the Pine Ridge Escarpment and Goshen Hole areas of Nebraska, formed by sandstone of the Miocene Arikaree Formation overlying soft siltstone of the Oligocene White River Formation; (3) ephemeral-stream incisions encroaching the High Plains of Kansas, where headward erosion by sapping along possible fractures is extending the drainage network; and (4) the Southern High Plains of Texas and New Mexico, where well-developed caliches, particularly capping the Ogallala Formation, and thick sandstones of the Triassic Dockum Group and Permian Dewey Lake Formation have led to a plateau topography. The following discussion briefly considers the evidence for a groundwater influence on the geomorphic development of two areas—one in Nebraska (Goshen Hole) and one in northwestern Kansas. Finally, groundwater-related geomorphic processes of the Southern High Plains are described in relative detail.

GOSHEN HOLE

Goshen Hole is a roughly rectangular-shaped, escarpment-rimmed topographic depression in eastern Wyoming and western Nebraska (Fig. 22). The depression, about 5,000 km^2 in area, is an erosional feature associated with incisement of the Arikaree and underlying White River formations by the southeast-flowing North Platte River. Sandstone-capped buttes within Goshen Hole show that lateral movement of the river has not exceeded about 15 km within the 50- to 60-km-wide depression. Ephemeral stream channels of the area show a marked tendency for southeast and northeast orientation, roughly normal to the northeast- and southeast-trending escarpments bordering Goshen Hole. Alined buttes within Goshen Hole also exhibit a southeast trend, whereas the northwestern (upslope) escarpment parallels a set of faults less than 10 km to the northwest.

The upland surface surrounding most of Goshen Hole has virtually no definable drainage network, but is interrupted by numerous shallow playa depressions, some of which have irregular northeast or southeast orientations. In some cases the sandstone surface immediately above an ephemeral channel heading near the Goshen Hole rim has interior drainage to a playa depression and shows no evidence of runoff to the rill system (Fig. 22). In brief, the following features combine to indicate that escarpment retreat around Goshen Hole occurs principally by spring sapping, seepage erosion, and related processes: (1) the impossibility of fluvial erosion by the North Platte River or other perennial streams in some parts of Goshen Hole, (2) the linearity and trends of topographic features, (3) the occurrence of numerous springs and seeps immediately below escarpments, (4) high groundwater transmissivities due to secondary permeability in the White River Formation near and below escarpments, and (5) an upland surface with many closed depressions but lacking a well-defined drainage net.

In a study of erosion processes on the Colorado Plateau, Laity and Malin (1985, p. 203) identified seven characteristics suggestive of sapping and similar processes: (1) theater-shaped heads of first-order channels, (2) relatively constant valley widths, (3) high, steep sidewalls, (4) pervasive structural control, (5) hanging valleys, (6) large scale, and (7) characteristic drainage patterns that differ from fluvial networks in evolution and spacing. Although these characteristics are applied specifically to groundwater conditions in Mesozoic rocks of the Colorado Plateau, most are also applicable to the generally shallower flow systems and younger beds of the Great Plains. For example, the

northwestern escarpment of Goshen Hole (Fig. 22) is formed by a series of amphitheater-shaped cliffs. Incisions by sapping that result in nearly constant valley widths between source and confluence are common in other parts of the Great Plains, but are too short to be diagnostic around Goshen Hole. Lack of a drainage network on the Arikaree Formation largely precludes hanging valleys, but the tendency for alined gullies that head immediately below an escarpment indicates a drainage pattern caused by sapping and seepage erosion. The remaining characteristics—steep sidewalls, apparent structural control, and large scale—are all demonstrated by the geomorphic features at the margins of Goshen Hole.

Other processes influencing the geomorphology of Goshen Hole are solution and piping. Primary permeability of the upper siltstones of the White River Formation is very low, but secondary porosity and permeability are high (Crist and Borchert, 1972). Conduits appear to result from piping caused partly by dissolution of calcium carbonate cement averaging 20% to 30% by weight of the White River Formation (Denson and Bergendahl, 1961). Initial water movement may follow joints and fractures (Lowry, 1966, p. 220). These fractures are enlarged by unloading near escarpments and by turbulent flow where groundwater gradients are high. Figure 23 shows pipes developed in the White River Formation immediately below its contact with the Arikaree Formation. Collapse resulting from undermining by piping no doubt contributes to escarpment retreat in this part of the Great Plains.

NORTHWEST KANSAS

Western Kansas lies in the High Plains physiographic section (Fig. 3), and, as in other parts of the High Plains, the area is one of recently active headward dissection from the east. Sinkholes, collapse structures, and related features resulting from groundwater dissolution of Permian evaporite deposits occur in central Kansas, and have had an influence on the Quaternary evolution of the drainage network of the east-flowing Kansas River system. The long-term influence on the geomorphology of western Kansas by the evaporite dissolution is poorly understood, but the presently active local effects are well known (for example, see Merriam, 1963; Fader, 1975; Gogel, 1981).

The possible effects of seepage erosion and sapping in northwest Kansas are not as well described as the processes of chemical erosion by groundwater, but are no less conspicuous. The area is drained by the Republican River, which like other parts of the Kansas River system, does not convey runoff from the Rocky Mountains but is fed principally by effluent from the Ogallala Formation and deeper aquifers. In some areas, the headwaters of tributaries to the Republican River have almost no source of overland runoff, and headward extension of the drainage network may occur largely by processes of groundwater erosion.

The map portion simplified in Figure 24 is from the McDonald, Kansas, U.S. Geological Survey 15-minute Quadrangle.

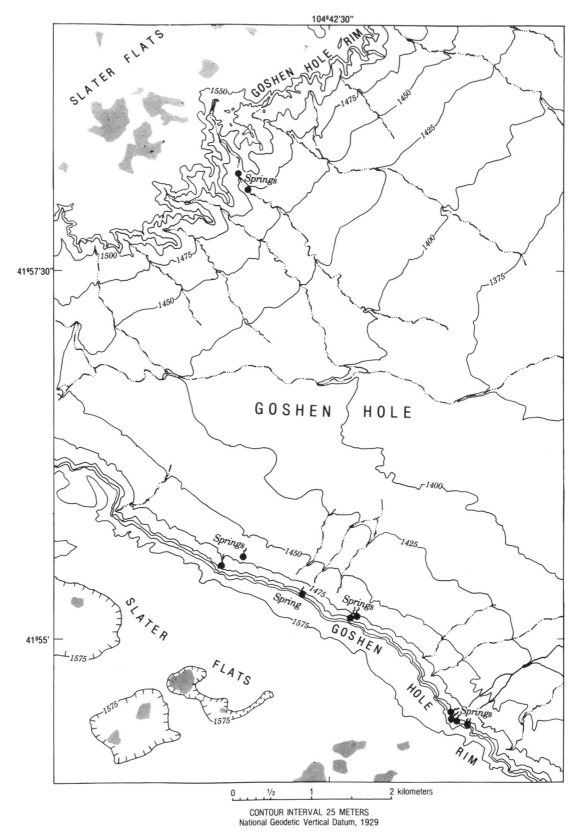

Figure 22. Map showing western end of Goshen Hole area, Wyoming. Black areas accentuate closed depressions. (Modified from Dickinson Hill Quadrangle, U.S. Geological Survey 7½-minute topographic map.)

Figure 23. Natural pipes in siltstone of the White River Formation immediately below its contact with the Arikaree Formation. (Photograph by M. E. Lowry.)

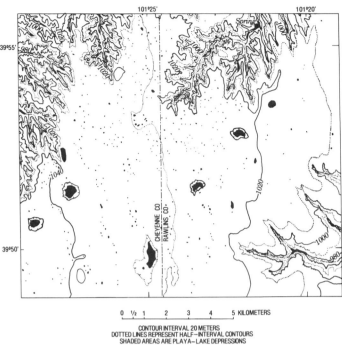

CONTOUR INTERVAL 20 METERS
DOTTED LINES REPRESENT HALF–INTERVAL CONTOURS
SHADED AREAS ARE PLAYA–LAKE DEPRESSIONS

Figure 24. Map showing area of dissected plains in northwest Kansas. Black areas accentuate lowest closed contour of the relatively large playa depressions; contours of incised areas suggest pronounced northwest trend of fracture patterns. (Modified from McDonald Quadrangle, U.S. Geological Survey 15-minute topographic map.)

Inspection shows somewhat similar geomorphic conditions as at Goshen Hole: a gentle sloping upland surface with playa depressions, a drainage network encroaching into the upland surface but not receiving direct runoff from it, and lineations suggestive of structural control (Cooley, 1984a). Steep valley sidewalls have not formed owing to soft sediments; however, most of the characteristics indicative of groundwater erosion (Laity and Malin, 1985), including relatively constant valley widths, are apparent. Headward development of the stream network by groundwater rather than by surficial-runoff processes is suggested by trends of uppermost channel reaches relative to regional slope. Many channels of the McDonald area parallel a north to north-northwest direction, nearly normal to the generally eastward slope of the High Plains surface (Fig. 24). This drainage pattern, as well as incision by some tributary channels from the southeast, appears more indicative of groundwater movement and discharge along fractures than of erosion by fluvial runoff.

Morphologies of the playa basins also exhibit possible fracture control (Fig. 24), suggesting that the depressions are recharge sites for part of the groundwater that may cause spring sapping and seepage in nearby draws. Present topographic and climatic conditions in northwestern Kansas favor low rates of runoff (about 1×10^{-4} m^3/s/km^2). Therefore, present erosional processes may be inactive relative to wetter, cooler episodes of Pleistocene time. Thus, the landforms may be largely relic.

SOUTHERN HIGH PLAINS

The recently recognized effects of groundwater processes on the development and present morphology of the Southern High Plains may be as important as for any large landform of North America. Large- and small-scale features of the area both appear to be closely related to the groundwater systems, and a variety of influences by groundwater movement and discharge is represented.

Setting

The Southern High Plains of Texas and New Mexico (Fig. 25) is an 80,000-km^2 remnant plateau formed from the southernmost remaining deposits of the Ogallala Formation. The plateau, which grades southward into the Edwards Plateau, is bounded on the north and west by escarpments rimming the Canadian and Pecos river valleys. The eastern escarpment is encroached upon and partly eroded by tributaries of the eastward-flowing Red, Brazos, and Colorado rivers, the upper reaches of which provide poorly developed drainage from the Southern High Plains. All streamflow is ephemeral, mostly in narrow, elongate draws trending southeast. As in other parts of the High Plains, much of the plateau surface lacks integrated drainage, and most adjacent through-flowing channels do not share common drainage divides but may be separated by 50 km or more of undissected plains (Gustavson and Finley, 1985).

The generally featureless surface of the Southern High Plains slopes southeastward about 1.7 m/km and is dotted by more than

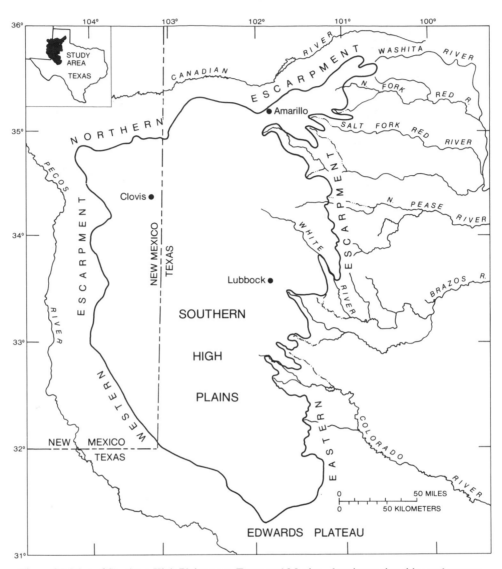

Figure 25. Map of Southern High Plains area, Texas and Mexico, showing major cities and streams.

30,000 playas, the flat-floored ephemeral lakes that are restricted to the High Plains surface with decreasing areal density northward into Wyoming and Nebraska. The playas, which are mostly less than a kilometer in diameter, occur above the zone of saturation and generally do not concentrate salts by evaporation. Many have nearly circular floors nestled in subtle depressions of the gently rolling landscape, but others that are spaced along ephemeral stream channels or define lineaments on the upland surface may be irregular or elongate in shape. Roughly 40 larger saline-lake basins also occur on the Southern High Plains. The saline lakes intersect the zone of saturation, are therefore intermittent to perennial, and are generally deeper and of more irregular shape than are most of the playas. Many of the saline lakes are bounded by one or more northeast-trending lineaments on the southeast sides where erosion has removed the Ogallala Formation, thereby exposing Cretaceous rocks (Reeves, 1970).

The Ogallala Formation of the Southern High Plains consists of up to 150 m (Reeves, 1972, p. 109) of Miocene deposits of fluvial sand and gravel grading upwards to eolian sand and silt. Numerous pedogenic caliche layers are in the Ogallala Formation and overlying Pleistocene eolian deposits; the most conspicuous of these horizons is the resistant "caprock" caliche, which forms the upper surface of the Ogallala Formation. This horizon typically is several meters in thickness and largely preserves the plateau topography of the area. Development of the caprock in part occurred because incisement by the Pecos River terminated Ogallala deposition and created a stable, relatively dry surface where pedogenic carbonate could accumulate (Osterkamp and Wood, 1984). Unlike the saline-lake deposits, which rest on Mesozoic rocks, almost all playas of the Southern High Plains occur at or above the Ogallala caprock.

The Southern High Plains is affected by two principal

groundwater systems. The upper system, or High Plains aquifer, of relatively fresh water in the Ogallala Formation and some of the underlying Mesozoic rocks, is extensively developed for irrigation, and, excepting areas of upward leakage, is recharged only by infiltration of precipitation and local runoff. Recent modeling results suggest that the average recharge rate for the Southern High Plains is low, about 0.3 cm per year (Knowles and others, 1984; Luckey and others, 1987). The lower system, the deep-brine aquifer, receives recharge from widespread outcrops of Paleozoic rocks in New Mexico. The aquifer transmits water through carbonate beds underlying the Permian evaporite deposits and discharges part of it as brine seeps in the Pecos and Canadian river valleys and in areas of Permian outcrops east of the Southern High Plains (Gustavson and Budnik, 1985; Gustavson and Finley, 1985; Orr and Kreitler, 1985; Gustavson, 1986).

Groundwater Processes of the Southern High Plains

The large-scale effects of groundwater on the geomorphology of the Southern High Plains are primarily the result of circulation through the lower of the two groundwater systems—the deep-brine aquifer. Evidence for salt dissolution and structural collapse from the Pliocene to the present beneath areas adjacent to the plateau includes (1) solute loads in streams; (2) geophysical logs from oil and gas wells; (3) cores showing brecciated collapse zones in beds overlying the evaporites; (4) folds, extension fractures, breccia chimneys and beds, and caverns in Permian outcrops adjacent to the Southern High Plains; and (5) solute contents in water from deep wells (Gustavson and Finley, 1985).

In late Miocene and early Pliocene time, continuing groundwater dissolution of Permian-age evaporite beds contributed to subsidence and collapse of overlying beds and to the development of the upper Pecos River valley in eastern New Mexico (Fig. 25). Entrenchment by the Pecos River intercepted streamflow carrying Ogallala sediment from the mountains to the west. This stream piracy, combined with formation of similar subsidence basins along the present Canadian River valley and elsewhere in the Texas Panhandle, isolated the present Southern High Plains as a topographic and hydrologic feature (Gustavson, 1982, 1986; Gustavson and Finley, 1985).

Post-Miocene dissolution of evaporite deposits has not been limited to the margins of the Southern High Plains, but has also been active beneath various parts of the interior of the plateau (Gustavson and Finley, 1985). As a result, discharge of brine may also affect small-scale geomorphic features on the surface of the Southern High Plains. In some areas of the southern part of the plateau, where the potentiometric surface of the deep-brine aquifer is higher than that of the High Plains aquifer, brines at times may have been released by upward seepage to channels developed along suspected fracture systems (Cooley, 1984b; Collins and Luneau, 1987). Elsewhere, brine seepage may be related to the expansion and deepening of the saline lake basins.

Dissolution of evaporites and formation of subsidence basins by groundwater flow through the deep-brine aquifer had three

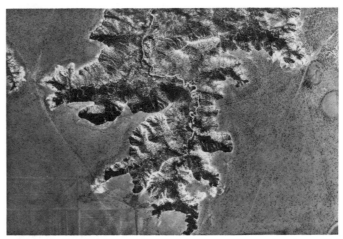

Figure 26. Aerial photograph of canyons cut into the Southern High Plains southwest of Amarillo, Texas. (Photograph by U.S. Department of Agriculture, 1967.) Note that erosion is occurring headward as theater-shaped canyons, in places exhibiting apparent fracture control; playa depressions visible along the right side are about 0.3 km in diameter.

principal effects on the geomorphic development of the Southern High Plains. It (1) led to incision by the Pecos and Canadian rivers and the termination of water and sediment supplies from the west, (2) led to accelerated rates of fluvial and groundwater erosion of rocks overlying Permian beds east of the present Southern High Plains (Osterkamp and Wood, 1984), and (3) thereby helped create on the plains topographic and hydrologic conditions that led to bounding escarpments of high relief, a poorly developed drainage network, numerous lake basins, and partial drainage, largely to the east, of groundwater in the High Plains aquifer. The topographic and hydrologic conditions initiated by chemical erosion in the deep-brine aquifer accordingly have induced the relatively recent Quaternary processes and landforms resulting from water moving through the higher aquifer system.

The most obvious small-scale geomorphic effect of groundwater movement in the High Plains aquifer is the maintenance of steep escarpments, generally by undermining the caprock caliche as sapping and seepage erode softer beds near the base of the Ogallala Formation and in the underlying sandstones of the Dockum Group (Gustavson, 1983, p. 128). Approximately 100 m of relief separate the High Plains surface from the canyon bottom shown in Figure 26. Erosion occurs headward into the nearly flat plateau and in places may be controlled by fractures; there is virtually no plateau surface drainage network contribution to the canyons. Instead, runoff collects in shallow playa depressions (right, or eastern side, Fig. 26) and infiltrates to provide recharge to the upper aquifer system (Wood and Osterkamp, 1984a).

The eastern escarpment of the Southern High Plains is a feature originated by solution and collapse in the deep-brine

Figure 27. Map summarizing topographic features of upper Frio Draw area of Southern High Plains. (Modified from Pleasant Hill and Hammond Ranch Quadrangles, U.S. Geological Survey 15-minute topographic maps; horizontal map width 14.6 km).

aquifer and modified by groundwater erosion—sapping and seepage—of the High Plains aquifer. Retreat of the eastern escarpment and cutting of nearby canyons along flow paths of increased permeability or along fractures is indicated by the irregularities of the rim. Because of limited groundwater discharge, the western escarpment has retreated very little by sapping, and therefore is relatively straight (Fig. 25); groundwater discharge along the northern escarpment adjacent to the Canadian River (Fig. 26) has caused irregularity intermediate between the two extremes (Osterkamp and Wood, 1984).

Escarpment retreat by erosion from groundwater discharge from the High Plains aquifer is still active, although recent groundwater development has resulted in reduced rates of retreat (Gustavson, 1983). The levels of saturation in the High Plains aquifer during Holocene time, however, generally have been too low to account for development by sapping of some channels on the Southern High Plains (Luckey and others, 1987). Drainages such as Frio Draw in New Mexico (Fig. 27), therefore, which by

the criteria of Laity and Malin (1985) are clearly groundwater features, are presumed to be relics of a time, possibly Pleistocene, when groundwater levels were higher than at present.

Several geomorphic processes contribute to the development of playa basins from depressions on the plains surface, but the principal processes appear to be erosion and transport related to flow in the unsaturated zone of the Ogallala Formation (Wood and Osterkamp, 1984a). Sites where water may accumulate to initiate a playa include pooled reaches along a stream channel and local depressions caused by eolian or biologic activity. Infiltration of water from playas causes dissolution of lithologic carbonates, largely as caliche, and the transport of clastic and organic particulate matter into the unsaturated zone (Fig. 28). Both processes favor local eluviation and piping and the development of secondary voids. Near-surface sandstones beneath some playas exhibit extensive leaching and microfractures, suggesting loss of strength and gentle subsidence due to gradual development of voids.

Playa floors appear to expand outward from a central position. Water ponded in playa basins concentrates organic material that moves downward during recharge events. Gas samples (Wood and Petraitis, 1984) indicate that oxidation of the organic matter in the unsaturated zone forms carbonic acid and causes continuing carbonate dissolution, increases in porosity, and expansion of the playa depression. Because organic-rich deposits of a playa lake (A, Fig. 28) are relatively impermeable, recharge is dominantly from the annular area surrounding the playa floor, thereby resulting in outward growth through time. Playa expansion may cease if the floor, relative to catchment area, becomes too great to maintain water depths sufficient to cause recharge around the playa-floor perimeter.

Processes of playa expansion by recharging groundwater are of basic importance to the present morphology of the Southern High Plains. Infiltrating water maintains the playa topography while minimizing development of stream networks. Water reaching the zone of saturation flows eastward, where, previous to extensive irrigation, discharge occurred as numerous seeps and springs (White and others, 1946; Gustavson, 1983).

Groundwater seepage to the surface from the deep-brine aquifer provides an alternative to the deflation and evaporation explanations (Price, 1940; Evans and Meade, 1945; Reeves, 1966) for some of the saline lakes that occur at the level of saturation of the plateau. At least some of the lakes overlie dissolution zones of the Permian evaporite deposits, and structural collapse has aided in lake-basin development (C. C. Reeves, Jr., Texas Tech University, personal communication, 1985). However, these lakes commonly have straight shorelines on their southeast sides that appear to be determined by fractures, but most also are fed by springs and seeps issuing from Cretaceous carbonate rocks or from the Ogallala Formation along the irregular western shorelines (Reeves, 1970).

Some of the lakes may have originated by upward leakage along fractures of water from the deep-brine aquifer. Support for this speculation is provided by a study showing that faults affect

EOLIAN PROCESSES AND SEDIMENTS OF THE GREAT PLAINS

Vance T. Holliday

Figure 28. Photograph of: A, playa deposits; B, underlying eolian deposits (Blackwater Draw Formation); and C, eroded caprock caliche of the Ogallala Formation. Exposures are at a barrow pit about 15 km north of Lubbock, Texas. A large pipe (D) formed by dissolution of carbonates and filled with eolian sand and silt exemplifies playa floors where groundwater processes have totally removed the caprock caliche.

Wind is probably the most conspicuous natural phenomenon on the Great Plains and is certainly the most prevalent meteorological feature of the region. The Great Plains is easily the most persistently windy inland area of North America and also has some of the highest average annual wind speeds of any nonmaritime region. The enduring and capricious nature of the wind is well conveyed in the following discussion related by Webb: "Does the wind blow this way here all the time?" asked the ranch visitor . . . "No, mister," answered the cowboy; "it will maybe blow this way for a week or ten days, and then it'll take a change and blow like hell for a while" (1931, p. 22).

The effect of wind on the geologic evolution of the Great Plains is significant. Much of the late Cenozoic substrate is eolian, resulting in the vast low-relief topography of the area. This openness, in turn, contributes to the present windiness of the region. The wind continues to be important geologically, often scouring the surface and producing clouds of choking dust. Indeed, the Southern High Plains is one of the dustiest regions in North America. Few areas of comparable size owe so much of their geomorphic expression to wind. Knowledge of the distribution of and processes responsible for eolian sediments on the Great Plains, therefore, is essential to understand the geomorphic evolution of the region.

A vivid example of wind erosion on the Great Plains, in this case due to drought, is the Dust Bowl of the 1930s (Fig. 29). This event most severely affected parts of Texas, New Mexico, Oklahoma, Colorado, and Kansas, but occurred throughout the Great Plains including the Prairie Provinces of Canada (Lockeretz, 1978; Gray, 1978). During the dry seasons of 1935 to 1939, approximately 93,000 km^2 of the Great Plains in the United States were damaged by wind erosion[1] (Kimberlin and others, 1977, Fig. 1). Over 12,000 km^2 of land were damaged on the Canadian Prairies (Gray, 1978).

Wind erosion has continued to be a problem on the Great Plains, with drought in the early 1950s subjecting the region to more wind erosion than during the "Dirty Thirties." During the dry seasons of 1954 to 1957, over 146,000 km^2 of land were damaged by wind erosion; in the 1975–76 dry season, over 32,000 km^2 of land were damaged (Kimberlin and others, 1977). Local wind erosion to depths greater than 1 m was noted by McCauley and others (1981) in a study of the February 23, 1977 dust storm near Clovis, New Mexico (Fig. 25). At least 4,000 km^2 have been damaged in the United States each year since the 1936–37 dry season (Kimberlin and others, 1977), and the problem also plagues the Canadian Prairies (Wheaton, 1984a, 1984b; LaDouchy and Annett, 1982).

hydraulic pressures in lower-Paleozoic rocks of the southern half of the Southern High Plains (McNeal, 1965). Although the brine is saturated with respect to calcium carbonate, mixing with the less saline water seeping from Cretaceous rocks or the overlying Ogallala Formation may cause ionic activities below saturation and the potential to dissolve the Cretaceous carbonate rocks that typically crop out in the saline basins; the process may be similar to that described by Back and others (1984). Early lake-basin expansion by chemical erosion of carbonates, aided by eolian removal of insoluble residues, therefore, may have proceeded northwestward, away from the discharge zones of brine seepage and toward the source of southeastward-moving groundwater of the Ogallala Formation. Where pressures in the deep-brine aquifer were insufficient to permit movement to the plateau surface, as in the northern part of the Southern High Plains (Orr and Kreitler, 1985), erosion of Cretaceous carbonate rocks may have occurred by the same mixing of groundwater in the Ogallala Formation with lake brines concentrated by evaporation. Isotope data from various lakes of the Southern High Plains suggest that, if these proposals are accurate, recent expansion of most saline lake basins has resulted from evaporation of water from the upper aquifer system rather than by upward seepage of brine (W. W. Wood, U.S. Geological Survey, personal communication, 1985).

Although the extent to which groundwater processes affect the geomorphic development of the Southern High Plains may be atypical, the examples summarized here suggest that these processes may be widespread on the Great Plains. As advocated by Higgins (1984), it is anticipated that awareness of groundwater processes in areas such as the Great Plains will lead to awareness of geomorphic effects in other areas as well.

---

[1]Damaged lands are those eroding at rates >3,340 Mg/km^2/yr, the rate at which damage is visible; one centimeter of erosion is equal to 4,380 to 14,640 Mg/km^2 (Kimberlin and others, 1977).

Figure 29. Photograph of the leading edge of a dust storm ("black roller") during the Dust Bowl near Lubbock, Texas. (Photograph from Texas Tech University.)

WIND AND DROUGHT ON THE GREAT PLAINS

Movement of air across the Great Plains is dominated by westerly flow from the Pacific Ocean. The air is generally dry as it reaches the Great Plains due to topographic lifting by mountains and plateaus of the Cordillera followed by subsidence over the Plains (Bryson and Hare, 1974). Exceptions to this zonal flow are a strong northerly component of airflow over the Canadian Prairies and a southerly flow over the southern Great Plains (Borchert, 1950; Bryson and Hare, 1974).

The direction of windflow varies seasonally on the Great Plains. The westerly flow of air is strongest in winter, causing relatively low precipitation, and it weakens through spring and into summer, particularly in the south, where southerly airflow begins to dominate the pattern. This influx of warm, moist Pacific air results in a summer precipitation maximum on the Great Plains. In the fall the westerlies strengthen as the southerly flow weakens. There are, of course, many perturbations to this generalized flow pattern due to frontal passages and local topographic effects (Borchert, 1950; Bryson and Hare, 1974).

Seasonality of precipitation and extreme winds has a considerable influence on the degree and amount of wind erosion (Fig. 30) and on the direction of sediment movement. Highest average wind speeds and highest winds generally occur from December through April (Johnson, 1965; Orgill and Sehmel, 1976; National Oceanic and Atmospheric Administration, 1982a; Wigner, 1984), which is the driest part of the year, when soils are most susceptible to wind deflation (Fryrear and Randel, 1972). In addition, Odynsky (1958) has shown that dune orientations on the Canadian Prairies are coincident with strong winds related to seasonal storm tracks. Seasonality of extreme winds, therefore, is considerably more important to eolian processes than are average annual prevailing winds (Carlisle and Marrs, 1982). Freeze-thaw cycles early in the windy season also help to disaggregate the sediment (Gillette, 1981).

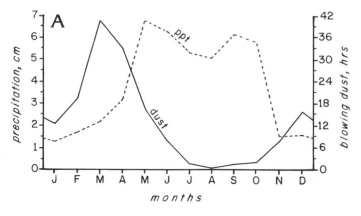

Figure 30. Graphs showing variations in precipitation and persistence of blowing dust for a typical year (A) and through the period 1947 to 1982 (B), Lubbock, Texas. Compiled from Wigner (1984).

Two topographic effects of the Great Plains are important to the distribution of eolian sediment. Escarpments along the western (upwind) side of the Plains locally increase wind speed. The "scarp effect" (McCauley and others, 1981) causes winds normal to a cliff to attain substantially higher velocities than winds blowing across more level terrain upwind. This effect may extend to several times the cliff height and, for a distance downwind, as much as ten times the cliff height. Winds are also accelerated by the "venturi effect" as they are funneled into reentrant drainages on the upwind escarpments (Carlisle and Marrs, 1982); examples occur along the southwest side of the Southern High Plains (Fig. 25).

Annual and longer climatic variations, particularly those resulting in extreme variation in mean annual precipitation across the region, are of considerable importance to eolian processes on the Great Plains. Adequate precipitation is dependent upon frequent interaction between the dry westerlies and a southerly flow of moist air. Drought accompanies a persistence in either meridional or zonal flow (Borchert, 1971; Barry, 1983), leading to higher than average temperatures and winds, loss of soil moisture, destruction of vegetative cover, and surface sediment susceptible to eolian movement. The most severe wind erosion on the Great Plains occurs after two or more years of drought (Choun, 1936; Johnson, 1965; Smith and others, 1970; Fryrear and Randel, 1972; Hagen and Woodruff, 1973; McCauley and others, 1981).

TABLE 2. AVERAGE ANNUAL WIND VELOCITIES AND MAXIMUM VELOCITIES FOR SELECTED
SITES ON THE GREAT PLAINS*

City	Period of Record	Average annual wind velocity (km/hr)	Maximum wind velocity, period of record (km/hr)
Lubbock, TX	1949-81	20.0	112.0
Amarillo, TX	1941-81	21.9	134.4
Dodge City, KS	1942-81	22.4	124.8
North Platte, NE	1952-81	16.3	115.2
Rapid City, SD	1950-81	18.1	105.6

*Data compiled from National Oceanic and Atmospheric Administration (1982a, 1982b)

DUST STORMS

Dust storms are relatively common events on the Great Plains, based on considerable historical and geologic data (Fryrear and Randel, 1972; Bark, 1978; Idso, 1976; McCauley and others, 1981; Holliday, 1984, 1985). As noted previously by Hadley and Toy (this volume), human activity may affect the intensity of wind erosion in a given area (e.g., Gray, 1978; Worster, 1979; Bonnifield, 1979), but is not responsible for eolian processes.

The relation between the atmospheric circulation producing drought and the winds causing erosion and dust storms has not been fully established, but patterns are evident. There are different types of dust storms on the Great Plains of varying size and duration (Warn and Cox, 1951; Henz and Woiceshyn, 1980; Wigner, 1984). The longest lasting, most widespread, and most spectacular are cyclogenic storms. Individual cyclogenic storms result in more erosion than do any other types of blowing dust events, but cyclogenic storms are also the most infrequent. Over time, trough-induced dust storms, a second major cause of eolian erosion, probably result in much more sediment movement than do cyclogenic storms. The general weather patterns preceding the cyclogenic and trough-induced storms differ, but the meteorological conditions during the two storms are similar. As examples representing both types, meteorological data for certain cyclogenic dust storms of the Southern High Plains in 1935 (Parkinson, 1936; Sidwell, 1938), 1950 (Warn and Cox, 1951), 1975 (Henz and Woiceshyn, 1980), and 1977 (McCauley and others, 1981) show that these storms were related to deep low-pressure systems over eastern Colorado and western Kansas. The westerly winds on the south sides of these lows brought hot, dry air from the Desert Southwest across the Southern High Plains. Jackson and others (1973) suggested that these intense cyclones and the strong westerlies across the Southern High Plains are related to surface heating and the position of a strong upper-level jet over this portion of the Plains. A similar dust storm that occurred over the Dakotas in 1934 is described by Sidwell (1938); this storm

was related to development of low pressure over northwestern Minnesota and eastern North Dakota.

EOLIAN PROCESSES AND SOILS

From studies of eolian sand to fine clay deposits near Big Spring, Texas (Fig. 3), Gillette and his colleagues (Gillette and Goodwin, 1974; Gillette and Walker, 1977; Gillette and others, 1974, 1978; Gillette, 1981), whose works are summarized below, indicate that among the important factors in soil erosion and dust suspension are content of nonerodible materials (bushes, pebbles, and boulders) and soil aggregation. An increase in either of these factors increases the threshold velocities required for soil erosion. Particles with the lowest threshold velocities are those around 0.1 mm diameter. Larger particles weight more and smaller particles are more cohesive. Therefore, soil types most susceptible to wind erosion are those dominated by fine sand followed by clayey soils that form fine sand-sized aggregates. Silty and loamy soils have the lowest erodibility because they tend to produce aggregates larger than about 1-mm diameter that have high threshold velocities.

Eolian Sorting and Sediment Characteristics

Eroded material in transport moves within a few tens of centimeters of the surface. A speed of about 16 km/hr is the threshold for significant wind erosion. Much of the Great Plains has average velocities above this level, and many areas have seasonal extremes far in excess (Table 2). In these areas, coarse sand tends to move by creep, whereas medium and finer sand moves by saltation. Silt and clay move in suspension and are entrained either as wind speed increases or by "sandblasting," whereby saltating sand causes a splash of finer material into the air or fine coatings on the sand are abraded. Sand-sized aggregates of clay also saltate and collide, causing breakage and the

Figure 31. Sheets of locally derived sand covering fields north of Lubbock, Texas, following severe dust storms of April 1 to 3, 1983. A small playa is visible in the middle distance to the right (arrow). (Photograph by V. T. Holliday.)

release of fragments into the air. Due to disaggregation and sandblasting as wind speed increases, fine-textured soils yield larger proportions of fines than do coarse soils.

The silt and clay fractions of an eolian load increase dramatically with height above the surface. A meter above the surface, however, mass transport of eolian sediment decreases significantly relative to that moved closer to the surface.

Considerable variability is apparent in sedimentological characteristics of fine material over the Great Plains, particularly over the Southern High Plains (Swineford and Frye, 1945; Warn and Cox, 1951; LaPrade, 1957; Smith and others, 1970; Gillette and others, 1978; Hobbs and others, 1985). At heights up to 9 km and distances up to 800 km downwind from source areas, suspended particles are dominantly medium-silt to clay size. In February, 1977, eolian material reaching Savannah, Georgia, from Lubbock, Texas, 1,800 km upwind, contained about 90% clay (Windom and Chamberlain, 1978). Some variation is due to differences in textures, moisture contents, and wetting-drying and freeze-thaw histories of the source materials. Despite differences in collection, analysis, and presentation of data, however, many investigators have noted a similarity among the size ranges of the particles comprising loess deposits (e.g., Swineford and Frye, 1945; Goudie, 1978).

Winds tend to remove fine material from source areas, while leaving coarser, saltating and creeping material behind (Fig. 31). With time and over large areas, this sorting process produces sheets of eolian material that are progressively finer-grained downwind, a phenomenon that is of particular interest to those working with loess (e.g., Goudie, 1978). However, data from Holliday (1982) indicated that within areas of dust production,

the sediment resulting from dust storms is similar in texture to the source sediment. The sand, silt, and clay are sorted during transport, but as the wind speed slackens, the suspended material falls back to the surface and mixes with the saltated particles. The quantities can be considerable. Lisitzin (1972) estimated that 272 × 10[6] Mg of soil, equivalent to nearly half the annual sediment load of the Mississippi River, were removed from the Great Plains by wind erosion in 1934. However, it is unclear whether this figure represented material removed locally or removed completely from the region. Warn and Cox (1951), studying dust storms in the Lubbock, Texas, area (Fig. 25) in 1950, also a period of relatively low wind erosion and dust activity, calculated that a single storm can carry an average of nearly 1,600 Mg of dust along an 8-km front (about 110 Mg/hr).

Wind erosion, of course, leads to eolian sedimentation elsewhere. Depending on sedimentological variables, material can be scattered over areas ranging from locales adjacent to the source (Fig. 31) to landscapes thousands of kilometers downwind. During the Dust Bowl period, sediment derived from the Great Plains routinely fell on much of the eastern United States; one storm reduced visibility in Baton Rouge, Louisiana, to about one kilometer (Hurt, 1981). McCauley and others (1981, Fig. 3) graphically illustrated this effect with satellite photographs of a large dust storm that originated over the Southern High Plains and moved far out over the Atlantic Ocean. Windom and Chamberlain (1978) reported dust concentrations from this cloud exceeding 0.2 mg/m^3 in Savannah, Georgia. In comparison, concentrations averaged 14.4 mg/m^3 for seven High Plains sites during dust storms of the 1950s (Hagen and Woodruff, 1973).

On the Great Plains considerable material is deposited as a

TABLE 3. COMPARISON OF DUST ACCUMULATION IN KANSAS AND NEBRASKA
(FIG. 3) DURING DRIER (1930s) AND WETTER (1960s) PERIODS

Location	Years	Dust deposition (Mg/km^2)	Comments
Kansas[1]	1933	35	individual dust fall
Nebraska[1]	1935	272	individual dust fall
St. John, KS[2]	1963-67	8.6	mean annual dust fall
Hays, KS[2]	1963-67	4.5	mean annual dust fall
Manhattan, KS[2]	1963-67	4.4	mean annual dust fall
North Platte, NE[2]	1963-67	6.4	mean annual dust fall

[1]Goudie (1978)
[2]Twiss (1983)

result of wind erosion in other areas of the region. Smith and others (1970) and Twiss (1983) presented data (Table 3) on dust accumulation at selected sites on the Great Plains from 1963 to 1967. This relatively wet period resulted in low to moderate wind erosion and dust-storm activity, yet dust accumulation was still significant. These data contrast markedly, however, with figures on dust deposition in Kansas and Nebraska during the Dust Bowl gathered by Goudie (1978; Table 3).

Soil Development

The suspension of dust and the influx of fine-grained material in areas downwind from source areas is of pedologic significance, in particular because dust appears to be the source of much of the clay that accumulates in the illuvial (Bt) horizons of soils on the High Plains (Holliday, 1982; Ruhe, 1984). Besides being important pedogenically, illuvial clay is significant as a post-depositional alteration of sediment. Some textural variation in eolian sediment may be due to this pedogenic process rather than variation in primary textures.

Carbonate dust is probably contributed to the soil as well. Warn and Cox (1951) reported that dust storms in the Lubbock area may carry up to 20% carbonate, although Holliday (1982) measured 2% to 5% carbonate in dust samples collected for a year from the same area. Carbonate is, therefore, available for translocation into the soil. Rabenhorst and others (1984), studying dust on the Edwards Plateau at the southern end of the Great Plains (Fig. 3), reported virtually no carbonate. The region is, however, one of relatively low dust-storm frequency (Orgill and Sehmel, 1976), and the period of collection (1981) was a year of low dust-storm activity (Wigner, 1984).

Several investigators (e.g., Gardner, 1972; Reeves, 1976; Gile and others, 1981; Machette, 1985) have proposed that dust is the source of the calcium carbonate that produced many calcic horizons and indurated "caliches" common in arid and semiarid

regions, such as the central and western United States. As noted earlier in this chapter, one of the best developed of such calcretes is the caprock caliche at the top of the Ogallala Formation, a strongly indurated and resistant horizon that is largely the cause of the prominent escarpments of the High Plains. Dust, therefore, appears to be of considerable geomorphic significance in addition to erosion and sedimentation on the Plains surface.

EOLIAN SEDIMENT AND RELATED LANDFORMS

Wind has deposited vast sheets of eolian sediment and has created a variety of erosional and depositional landforms on the Great Plains. The distribution of eolian sediment is well illustrated by a map of Thorpe and Smith (1952) that remains the most comprehensive statement on North American eolian deposits and provides the basis, augmented with recent data, for the following discussion.

The most extensive eolian deposits of North America are on the Great Plains of Nebraska and Kansas. The best known of these is probably loess, which is locally over 50 m thick and nearly continuous far into the Central Lowlands to the east, westward into the High Plains and piedmont of Colorado, and southward into the High Plains region of northwestern Texas and eastern New Mexico (Fig. 3). Dune fields fringe the loess area on the north and west, and most seem to be related to rivers. The Nebraska Sand Hills, forming the northern border of this area and bounded by the Niobrara and Platte rivers, is the largest sand sea in the western hemisphere (Smith, 1965). Smaller dune fields associated with the South Platte, Republican, and Arkansas rivers are in western Nebraska, eastern Colorado, and Kansas. In Colorado, Oklahoma, Texas, and New Mexico, dunes are adjacent to the Arkansas, Cimarron, Canadian, and Pecos rivers, and extend eastward through the Great Plains into the Central Lowlands (Fig. 3).

Eolian sediment is also common in other portions of the

Great Plains, but generally is thinner and less often recognized. Scattered loess deposits and dune fields associated with major drainages from South Dakota into southern Alberta and Saskatchewan were identified by Thorpe and Smith (1952). More recently, loess has been identified in scattered portions of southwestern South Dakota (Harksen, 1967), over much of the Great Plains of North Dakota (Clayton and others, 1976), and on portions of the southeastern Alberta Plains of Canada (David, 1972; Souster and others, 1977). These data suggest that similar deposits may extend into other parts of South Dakota, eastern Montana, and southern Saskatchewan and Alberta. Odynsky (1958), Prest and others (1968), and David (1977) identified scattered dune fields on the Alberta Plains (Fig. 1A). In the south, the Llano Estacado of Texas and New Mexico (Southern High Plains) is nearly covered by eolian sand and silt up to 30 m in thickness.

On the Great Plains the coarsest, sandiest deposits are generally upwind, to the west and northwest; dune fields often occur immediately downwind of rivers in these areas (Fig. 3). Eolian material generally thins and becomes finer downwind to the east and southeast. Owing to complexities in sedimentology, age, and origin, however, other relations regarding eolian deposits of the Great Plains are difficult to generalize. This is well illustrated by some of the recent research in the region.

The Great Sand Hills of Saskatchewan

The Great Sand Hills of Saskatchewan is the largest of a number of dune fields on the Canadian Prairies (David, 1972, 1977; Epp and Townley-Smith, 1980; Epp, 1986). The source of the sand is underlying glaciofluvial and glaciolacustrine sediment (David, 1977). Loess up to 80 cm thick and draped over glacial sediment is also present (David, 1972; Souster and others, 1977). Loess deposition and dune formation began after deglaciation, about 15,000 years ago. Loess deposition was more or less continuous, although a number of brief intervals of nondeposition and soil formation occurred (David, 1982). Most dune activity occurred between 15,000 and 5,000 years B.P., with only periodic activity in the last 5,000 years (David, 1982). The eolian activity appears to be related to relatively minor climatic fluctuations (greater or lesser aridity), with wetter periods favoring vegetation and stabilization of surfaces.

Loess

Clayton and others (1976) have identified the Oahe Formation, covering much of North Dakota, which is mostly loess draped over gently sloping surfaces. The loess, as generalized earlier by Morrison (this volume), accreted episodically since the late Wisconsin, Jorstad and others (1986) having demonstrated that locally there were at least nine episodes of eolian sedimentation in the Oahe Formation in the late Holocene. In places the Oahe Formation is capped by several centimeters of material deposited during the Dust Bowl years. The oldest sediment in the

Oahe Formation probably was derived in part from detritus of glacial meltwater streams. Most of the rest of the sediment was derived from hillslopes during periods of drought. Reduced effective precipitation would result in reduced vegetative cover and erosion of hillslopes. Valleys would then alluviate and provide a source of sediment to be carried by wind.

Locally, loess of Nebraska and Kansas is as old as early Pleistocene (Dreeszen, 1970; Fredlund and others, 1985), but most available information concerns the late Quaternary surficial loess (e.g., Lugn, 1935, 1968; Schultz and Stout, 1945; Condra and others, 1950; Reed and Dreeszen, 1965; Reed, 1968. Schultz, 1968). The late Quaternary loess is composed of the Gilman Canyon Formation (early Wisconsin), Peoria Loess (late Wisconsin), and Bignell Loess (early Holocene), the two younger having been studied most thoroughly (Ruhe, 1983, 1984). The greatest thicknesses are in central and southwestern Nebraska where late-Wisconsin loess is up to 10 m; thinning as well as textural fining occurs generally downwind or southeast of major rivers. Contacts between the loess units, as indicated by buried soils, are also time-transgressive downwind. Thickness and texture maps of the loess (Thorpe and Smith, 1952; Ruhe, 1983) suggest sources from the Niobrara, Platte, and Republican river valleys.

The Nebraska Sand Hills

The Nebraska Sand Hills (Fig. 3) have also generated much interest (e.g., Lugn, 1935; Smith, 1965, 1968), although the age and origin of this spectacular eolian feature are still uncertain. Recent studies of the Sand Hills (Ahlbrandt and Fryberger, 1980; Ahlbrandt and others, 1983) suggest that the dunes are late-Holocene features, possibly derived from older, unconsolidated sediment that mantled the Plains, and therefore may be genetically unrelated to the loess. In contrast, many early workers, and more recently Wells (1983), regarded the Sand Hills as a coarse, upwind facies of a single late-Pleistocene sand-silt unit. Wright and others (1985) supported a late-Pleistocene age for some dunes and a possible relation to the loess.

From analysis of modern wind regimes, Ahlbrandt and Fryberger (1980) postulated wind regimes that produced the dunes. Winds from the northwest apparently constructed the dunes, although winds on the east side of the dune field were more northerly. This analysis suggests, therefore, that the dunes formed prior to development of the present strong southerly wind component.

Colorado Deposits

There are three dune fields in northeastern Colorado (Muhs, 1985; Fig. 3). The Greeley dune field, immediately north of the South Platte River, is the smallest; the Fort Morgan dune field lies south of the river; and the Wray dune field, the largest, is on the High Plains to the east and southeast of the other two. The Fort Morgan and Wray sands were probably derived from sediment of the South Platte River. Of the three fields, the Wray sands are the finest and best sorted, apparently due to reworking during movement from the river onto the High Plains. Northeast winds produced the Fort Morgan and Wray dune fields, probably in late-Holocene time. The source and direction of movement of the Greeley sand hills is uncertain (Muhs, 1985).

Tertiary sediment is exposed in a broad area between the Fort Morgan and Wray dune fields, where all Quaternary sediment has been stripped away (Scott, 1978; Sharps, 1980). This stripping may be the result of the scarp effect of McCauley and others (1981), suggesting that northwesterly winds accelerate over the High Plains escarpment south and southeast of the South Platte River. The winds may deflate sediment on the Tertiary beds and help move sand from the Fort Morgan dune field over the escarpment and downwind into the Wray dune field.

Southern High Plains Deposits

Recent research suggests that a thick blanket of eolian sediment, the Blackwater Draw Formation, was deposited episodically through the Quaternary on the Southern High Plains of Texas and New Mexico (Gustavson and Holliday, 1985). The unit (B, Fig. 28) still may be accumulating, although the last significant period of deposition ended at least 40,000 years B.P., based on pedogenesis in the top of the deposit and radiocarbon ages from overlying dunes (Holliday, 1985). There is a marked textural change in the unit from sandy in the southwest to loess-like silty clay in the northeast, suggesting downwind fining (Seitlheko, 1975). Locally, the silty clay overlies sandier material, but in other sections in the northeastern part of the region the sediment is loesslike throughout. For example, near Amarillo almost 7 m of clay loam containing six buried soils has been identified, the lowermost soil being magnetically reversed (Lotspeich and others, 1971; Allen and Goss, 1974; Machenberg and Holliday, 1987). The Blackwater Draw Formation was probably derived from the Pecos River valley, based on textural evidence from the unit.

Below the Blackwater Draw Formation is the Ogallala Formation of Miocene-Pliocene age. The Ogallala (C, Fig. 28) was classically envisaged as a huge coalescent fan (Sellards and others, 1932; Seni, 1980), but it is now apparent that the unit contains considerable windblown material (Bachman, 1976; Hawley, 1984; Reeves, 1984; Gustavson and Holliday, 1985) and documents eolian processes in the region throughout late-Cenozoic time.

There are several dune fields along the western edge of the Southern High Plains (Fig. 3), various aspects of which have been investigated by Melton (1940), Huffington and Albritton (1941), Hefley and Sidwell (1945), and Green (1961). This work and recent studies in the Sandhills of New Mexico and Texas by Gile (1979, 1985) and in the Monahans dunes by Machenberg (1984) showed that they are in two physiographic positions: (1) along the eastern Pecos River valley, parallel to the west-facing High Plains escarpment, and (2) trending west-to-east from the Pecos River valley onto the High Plains, following breaches in the

escarpment. The first dunes have simply collected against the escarpment; the other dunes have moved around steep faces of the escarpment and are following the more gradual inclines onto the plains.

Deflation in the geologic past considerably influenced the Southern High Plains, much as it does today. Tertiary-age sediment is exposed along the western side of this region. Younger material, if ever deposited, may have been stripped off (Reeves, 1983) by the scarp effect (McCauley and others, 1981).

Playas and Associated Dunes

Probably the most conspicuous geomorphic features on the High Plains are the thousands of fresh-water, ephemeral-lake basins, or playas (Figs. 28 and 31), that dot the surface from Texas and New Mexico northward into Nebraska and Wyoming. They are most apparent and have been most extensively investigated in Texas. These basins are clearly modified by deflation (Reeves, 1966), although Wood and Osterkamp (1984a, 1984b), and Osterkamp, in the preceding portion of this chapter, suggest that hydrologic processes are important to their development.

Associated with many playas, as well as with the fewer, larger, saline lakes of Texas and New Mexico, are lee-side dunes on the east and southeast sides of the basins. The dunes are due to deflation of lacustrine sediment from the floors of the basins (Reeves, 1965; Holliday, 1985). Reeves (1965) studied these dunes to reconstruct prevailing wind directions of the last 15,000 years. The dunes probably signify seasonal winds, however, because they are active only during the dry season. Thus, changes in dune position over time may indicate either change in seasonal wind directions or change in the seasonal distribution of precipitation. Along the saline basins the dunes are often tens of meters high and occur in distinct sets (Reeves, 1965, 1966). The dunes along playas are usually less than 10 m high and occur simply as stacked sediment (Holliday, 1985). Both types of dunes developed episodically during late-Pleistocene and Holocene time (Reeves, 1965; Holliday, 1985).

Differences between Northern and Southern High Plains Deposits

In comparing eolian sediment of the Northern and Southern High Plains, several similarities mentioned above are apparent, but there are also striking differences. Sediment in the north has an essentially bimodal distribution: sand of the extensive dune fields, and loess. In contrast, sediments in the south tend to be unimodal. In the south there is a gradual west-to-east change from sand to silt and clay sizes. The north, on the other hand, does not show this textural gradation. The material in the north is generally thickest on the west, upwind, side of the region, whereas in the south the Blackwater Draw Formation is typically thin or absent on the southwest and thickens to the east and northeast. An unfortunate lack of age control for individual layers of the Blackwater Draw Formation precludes speculation on climate during deposition, but several observations suggest explanations for differences between deposits in the two regions. Streams of the northern river valleys, which were the apparent source areas, were fed by glacial meltwater, whereas the source area in the south, the Pecos River valley, received mostly runoff from precipitation. This difference may have significantly affected particle sizes of sediment deposited by the rivers and which was subsequently blown onto the plains. The top of the Blackwater Draw Formation, and thus the surface of much of the Southern High Plains, appears to be at least several tens of thousands of years older than the surface of the Northern High Plains. The surface in the south, therefore, has been subjected to more wind erosion and pedogenesis that could subdue or obscure primary lithologic variations and distribution.

SUMMARY REMARKS

The sedimentological aspects of dust storms on the Northern High Plains are not analogs of loess sedimentation because the modern winds simply rework loess already in place; the wind does not bring in significant quantities of material from sources such as stream valleys. This is demonstrated by data from Twiss (1983) showing little variation in the medial and maximum diameters of dust collected from eastern Colorado to Missouri. Also, the magnitudes, frequencies, directions, and seasonal variations of extreme winds were probably different under different climatic conditions. The possibility that modern wind-regime studies are useful for interpretation of eolian deposits on the Southern High Plains seems better than it is to the north.

Wind, dust, and geologic studies aid an understanding of other eolian processes and events. Included are wind erosion, the nature of eolian sorting and translocation processes, the effects of dust on pedogenic processes, the significance of seasonality on wind directions and extremes, and the relation between drought and eolian activity. Much of this information, particularly the latter, has been useful in attempts at reconstructing Holocene environments, especially periods of drought. The persistence of particular flow patterns reduces precipitation, and drought quickly reduces vegetative cover, leaving the surface open to wind erosion.

An understanding of wind and wind-deposited material seems to be essential to understanding the history and predicting the future of the Great Plains. Future investigations should attempt to integrate more fully the meteorological, climatological, sedimentological, pedological, and geomorphological aspects of eolian sediment and processes.

CONCLUDING REMARKS

W. R. Osterkamp

Many vacationers will continue to regard the Great Plains as a flatland that "has to be crossed to reach the scenic Rocky Mountains to the west" (Thornbury, 1965, p. 287). Many earth

scientists, however, will continue to regard the Great Plains as a province of immense interest: a region of diverse topography, landforms, climate, and geomorphic processes; and a region containing abundant information of the Cenozoic history of the North American interior.

The Great Plains is basically a depositional environment formed mostly of detritus removed from the eastern Cordillera along the western margin of the province. However, the processes of erosion, deposition, reworking of sediment, and thus—on a geologic scale—destruction of the Great Plains as a depositional landform, are the general topics of this chapter. Several of the preceding contributions have explored processes that fashion the remarkable landscapes that are often overlooked by tourists speeding toward the mountains, whereas the "Introduction" provides a basis for these considerations and contends that this long-term process of degradation is not constant, but, responding to climatic fluctuations, is episodic.

WHAT IS KNOWN OF THE GREAT PLAINS?

Contributors to this chapter have summarized the state of knowledge concerning several processes considered to be of particular importance in the geomorphic evolution of the Great Plains. The intent has been as much to emphasize recent developments and to provide speculations as it has been to outline the fund of knowledge presently accepted by those working in the Great Plains Province. Details of this outline are available from the massive literature accumulated for the Great Plains, selections of which have been cited.

Discussions in this chapter are restricted to selected topics that may focus limited attention unduly on individual geomorphic processes. Illustrating these processes, however, inescapably leads to the realization that all geomorphic features are products of a medley of stresses, a prime example being the Southern High Plains, where eolian, groundwater, and fluvial processes have combined to produce the present topography. Equally implicit in the contributions is the supposition that the tenure of one set of processes may lead to the primacy of a succeeding set, a premise consistent with Morrison's erosion-deposition-stability model but one not totally climate-dependent. In other words, the geomorphic evolution of large-scale landscapes of the Great Plains appears neither haphazard nor completely dependent on external stresses. Again citing the Southern High Plains, probable results of subsurface evaporite-bed dissolution, subsidence, and development of the Pecos and Canadian river systems were (1) escarpments defining the area as a plateau; (2) diversion of water and sediment moving from the west, hence the curtailing of significant fluvial activity on the plateau; (3) surficial stability, which enhanced sustained pedogenic and eolian processes; and (4) shallow groundwater flow advantageous to the expansion of the playa-lake basins but detrimental to the development of fluvial drainage from the plateau. The sequence of processes and feedbacks may have differed had the climatic history been different, but it seems reasonable that the evolution of landforms in the area, nonetheless, would have permitted cause-and-effect explanations regardless of the temporal and spatial distribution of external stresses.

Geomorphic processes either briefly treated or not discussed in the foregoing chapter include pedogenesis, volcanism, tectonism, lacustrine deposition, and stream diversion by glaciation. All of these processes and others have been pertinent to the geomorphic development of the Great Plains, but appear to have been of less consequence than those discussed in greater detail. Especially in the southern half of the High Plains (Fig. 3), formation of calcretes and other less indurated calcic soils has strongly influenced other geomorphic processes and thus landforms. Little large-scale tectonic movement, as noted by Morrison (this volume), has been recognized in the Great Plains for Cenozoic time, but small-scale tectonic forms such as collapse features and fracture sets may have greatly affected both fluvial and groundwater erosion. Modification of the drainage net by late-Pleistocene glaciation was extensive, but the degree to which the pre-Wisconsin ice sheets contributed to the drainage-net modifications of the northern Great Plains is not well understood. Volcanic activity, including bedded flows east of the mountains through much of Colorado and extensive ash deposition, particularly of Oligocene age as represented by the White River Formation, was of less geomorphic importance in the Great Plains than it was in areas to the west. The significance of relatively recent ash and lacustrine deposition is discussed more fully in "Quaternary non-glacial geology of the conterminous United States" (this volume).

WHAT IS NOT KNOWN OF THE GREAT PLAINS?

Directions that future geomorphic studies on the Great Plains will take, topics that will be emphasized, and certainly results and implications of those studies that will unfold, remain to be defined. Themes expressed in this chapter, however, suggest present weaknesses in the understanding of the geomorphic evolution of the Great Plains—that is, what is not known—and thus indicate possible focal points of future research.

Specific pursuits may include efforts to achieve a more precise chronology of late Cenozoic events on the Great Plains. Dated ash and soil horizons provide relatively detailed information at Eustis Ash Pit, but similar sequences have not been discovered in most other parts of the Great Plains. Inevitably, other sites will be identified that will provide dates of deposits, events, and climates; sites that may permit intercorrelation; and sites that may further test the erosion-deposition-stability model.

The first systematic studies of glacially deformed bedrock and other glaciotectonic features were made in recent years, and these studies hold promise of providing much more than increased mine safety where deformed tills overlie mineral deposits. The recent attention given to deformation terrains in the northern Great Plains is likely to result in recognition of similar features in areas where glaciotectonism has not been identified. In these areas, the deformation features may yield detailed information on

directions of ice movement. The studies at the Sand River area, for instance, show at least three distinct directions of glacial movement, each associated with a different lobe of the retreating late Wisconsin ice sheet (Andriashek and Fenton, 1987). Possible extensions of these studies may be applicable to ice-flow dynamics, mechanisms of glacial scouring and thrusting (including the positions of some large glacial lakes), and interpretations of the depths of glacial erosion by pre-Wisconsin advances, a topic heretofore virtually not considered. Continuing studies include comparisons between glaciotectonism and larger-scale deformations of the North American Cordillera and the Alps, processes showing apparent similarities.

Several of the studies discussed by Hadley and Toy (this volume) were based on data collected from streams of the Great Plains and adjacent areas to define empirical relations between discharges of water and sediment and the resulting channel morphologies and dynamics. Those studies were designed to increase the understanding of natural fluvial systems, and therefore to provide a basis for predicting channel changes resulting from either natural or imposed alterations in the streamflow and sediment loads. Hadley and Toy cite channel changes of the Platte and Cimarron river networks to exemplify how the empirical relations might be applied to disturbed fluvial systems. It seems likely that river engineers increasingly will employ the relations of these and similar studies to anticipate the effects of channel and basin modifications, a practice infrequently followed in the past. Another possible extention of the Great Plains data-based studies may be the development of theoretical descriptions of fluvial processes. Previous, perhaps relatively simplistic, attempts to formulate theoretical channel relations are summarized by Williams (1978b) and Osterkamp and others (1983).

Future research of landforms related to groundwater movement seems likely to be directed by economic and other practical considerations. In a water-deficient region such as the Great Plains, the recharge, movement, and discharge of water from the subsurface is a principal concern of people dependent on groundwater supplies. Likewise, the interactions of water with both the unsaturated and saturated zones through which it moves are pertinent to an understanding of water availability. Throughout irrigated areas of the Great Plains, but especially in the south where presently groundwater withdrawal rates greatly exceed recharge rates, geomorphic studies may prove indispensible for an understanding of regional groundwater flow systems. Studies relating landforms and groundwater of the Great Plains may also prove essential for prudent decisions concerning hazardous-waste storage, fossil-fuel development, and, of course, the long-range use of water for urban, industrial, and other nonagricultural purposes.

Considerable meteorological, climatological, and geological data are available regarding wind and eolian sediment transport on the Great Plains, but integrating these diverse data into an understanding of geomorphic processes and Quaternary history is difficult. Holliday's discussion of these processes in this chapter is a brief beginning, but much additional work will be required for the interpretation of Pleistocene events and long-term processes in order to refine the climatic-geomorphic cycles proposal of Morrison (this volume). A major reason for the difficulty in interpreting available data is that climatic and geologic conditions of today appear significantly different from those of late Pleistocene and early Holocene time. For example, winds crossing the Northern High Plains of Colorado, Kansas, and Nebraska were influenced by the continental ice sheets to the north and northeast and by ice in the Rocky Mountains. Glacial meltwater was also carrying considerable fine-grained sediment through the major drainages of the region, the probable source for the vast loess sheets of the area. As modeling efforts offer increased understanding of the effects of glacial conditions on atmospheric circulation, the signif-

icance of spatial distribution of the loess and dune records over the Great Plains likewise will provide increased understanding of Quaternary eolian processes.

As more-detailed knowledge of the glacial, fluvial, groundwater, and eolian records of the Great Plains are unraveled, the applicability of a unifying approach such as the erosion-deposition-stability model will become more apparent. Whether this model, or a similar one will prove generally useful in areas beyond a type area such as the central Great Plains remains in question. What does not seem in question, however, is that data from the Great Plains toward this end will continue to accumulate—even if tourists on their ways to the mountains occasionally do get in the way!

REFERENCES

Aber, J. S., 1987, Geomorphic and structural genesis of the Dirt and Cactus hills, Saskatchewan (in press).

Ahlbrandt, T. S., and Fryberger, S. G., 1980, Eolian deposits in the Nebraska Sand Hills, *in* Geologic and paleoecologic studies of the Nebraska Sand Hills: U.S. Geological Survey Professional Paper 1120-A, 24 p.

Ahlbrandt, T. S., Swinehart, J. B., and Maroney, D. G., 1983, The dynamic Holocene dune fields of the Great Plains and Rocky Mountain basins, U.S.A., *in* Brookfield, M. E., and Ahlbrandt, T. S., eds., Eolian sediments and processes: Amsterdam, The Netherlands, Elsevier Science Publishers, p. 379–406.

Allen, B. L., and Goss, D. W., 1974, Micromorphology of paleosols from the semiarid Southern High Plains, *in* Proceedings, International Working Meeting on Soil Micromorphology, 4th: Kingston, Ontario, Queens University, p. 511–525.

Andriashek, L. D., and Fenton, M. M., 1987, Quaternary stratigraphy and surficial geology Sand River Map Sheet 73L: Alberta Research Council Bulletin (in press).

Andriashek, L. D., Kathol, C. P., Fenton, M. M., and Root, J. D., 1979, Surficial geology: Wabamun Lake NTS 83G, Alberta Research Council Map.

Bachman, G. O., 1976, Cenozoic deposits of southeastern New Mexico and an outline of the history of evaporite dissolution: U.S. Geological Survey Journal of Research, v. 4, no. 2, p. 135–149.

Back, W., Hanshaw, B. B., and Van Driel, J. N., 1984, Role of groundwater in shaping the eastern coastline of the Yucatan Peninsula, Mexico, *in* La Fleur, R. G., ed., Ground water as a geomorphic agent: Boston, Allen and Unwin, Incorporated, p. 281–293.

Bark, L. D., 1978, History of American droughts, *in* Rosenberg, N. J., ed., North American droughts: American Association for the Advancement of Science, Selected Symposium 15, p. 9–23.

Barry, R. G., 1983, Climatic environments of the Great Plains, past and present: Transactions of the Nebraska Academy of Sciences, v. 11, p. 45–55.

Bluemle, J. P., and Clayton, L., 1984, Large scale glacial thrusting and related processes in North Dakota: Boreas, v. 13, p. 279–299.

Bonnifield, P., 1979, The Dust bowl; Men, dirt, and depression: Albuquerque, University of New Mexico Press, 232 p.

Borchert, J. R., 1950, The climate of the central North American grasslands: Annals of the Association of American Geographers, v. 40, p. 1–39.

—— , 1971, The Dust Bowl in the 1970s: Annals of the Association of American Geographers, v. 61, p. 1–22.

Bostock, H. J., 1970a, Physiographic subdivisions of Canada: Geological Survey of Canada Map 1254A, scale 1:5,000,000.

—— , 1970b, Physiographic subdivisions of Canada, *in* Douglas, R.J.W., ed., Geology and economic minerals of Canada (fifth edition) Canada Geological Survey, Economic Geology Series 1, p. 11–30.

Bretz, J. H., and Horberg, L., 1949, Caliche in southeastern New Mexico: Journal

of Geology, v. 57, p. 491–511.

Bryson, R. A., and Hare, F. K., eds., 1974, Climates of North America: Amsterdam, The Netherlands, Elsevier Science Publishers, 40 p.

Bull, W. B., 1975a, Allometric change of landforms: Geological Society of America Bulletin, v. 86, p. 1489–1498.

—— , 1975b, Landforms that do not tend toward a steady state, *in* Melhorn, W. N., and Flemal, R. C., eds., Theories of landform development: London, George Allen and Unwin, p. 111–128.

—— , 1979, The threshold of critical power in streams: Geological Society of America Bulletin, v. 90, p. 453–464.

—— , 1980, Geomorphic thresholds as defined by ratios, *in* Coates, D. R., and Vitek, J. D., eds., Thresholds in geomorphology: London, George Allen and Unwin, p. 259–263.

Byers, A. R., 1960, Deformation of the Whitemud and Eastend formations near Claybank, Saskatchewan: Royal Society of Canada, Transactions, Sec. 4, v. 53, Series 3, p. 1–16.

Carlisle, W. J., and Marrs, R. W., 1982, Eolian features of the Southern High Plains and their relationship to windflow patterns, *in* Marrs, R. W., and Kolm, K. E., eds., Interpretation of windflow characteristics from eolian landforms: Geological Society of America Special Paper 192, p. 89–105.

Carrigy, M. A., 1970, Proposed revision of the boundaries of the Paskapoo Formation in the Alberta Plains: Bulletin of Canadian Petroleum Geology, v. 18, p. 156–165.

Choun, H. F., 1936, Duststorms in the southwestern plains area: Monthly Weather Review, v. 64, p. 195–199.

Christiansen, E. A., and Whitaker, S. H., 1976, Glacial thrusting of drift and bedrock, *in* Legget, R. F., ed., Glacial till; An inter-disciplinary study: Royal Society of Canada in cooperation with the National Research Council of Canada, Royal Society of Canada Special Publication 12, p. 121–130.

Clayton, L., 1967, Stagnant-glacier features of the Missouri Coteau in North Dakota, *in* Glacial geology of the Missouri Coteau and adjacent areas (Midwest Friends of the Pleistocene in south-central North Dakota, 18th Annual Field Conference, Guidebook and miscellaneous short papers): North Dakota Geological Survey Miscellaneous Series 30, p. 25–46.

Clayton, L., Moran, S. R., and Bickley, W. B., Jr., 1976, Stratigraphy, origin, and climatic implications of late Quaternary upland silt in North Dakota: North Dakota Geological Survey Miscellaneous Series 54, 15 p.

Clayton, L., and Tinker, J. R., 1971, Rates of hillslope lowering in the badlands of North Dakota: Springfield, Virginia, National Technical Information Service W73.09121, 36 p.

Coates, D. R., and Vitek, J. D., 1980, Perspectives on geomorphic thresholds, *in* Coates, D. R., and Vitek, J. D., eds., Thresholds in geomorphology: Boston, Allen and Unwin, Incorporated, p. 3–24.

Collins, E. W., and Luneau, B. A., 1987, Fracture analyses of the Palo Duro basin area, Texas Panhandle and eastern New Mexico: The University of Texas at

Austin, Bureau of Economic Geology Report of Investigations (in press).

Colton, R. B., 1962, Geology of the Otter Creek Quadrangle, Montana: U.S. Geological Survey Bulletin 1111-G, p. 237–288.

Condra, G. E., Reed, E. C., and Gordon, E. C., 1950, Correlation of the Pleistocene deposits of Nebraska: Nebraska Geological Survey Bulletin 15A, 74 p.

Cooley, M. E., 1984a, Linear features determined from Landsat imagery in western Kansas: U.S. Geological Survey Open-File Report 84-241, 1 sheet.

—— , 1984b, Linear features determined from Landsat imagery in the Texas and Oklahoma panhandles: U.S. Geological Survey Open-File Report 84-589, 1 sheet.

Crist, M. A., and Borchert, W. B., 1972, The ground-water system in southeastern Laramie County, Wyoming: U.S. Geological Survey Open-File Report, 53 p.

Crowley, K. D., 1981, Large-scale bedforms in the Platte River downstream from Grand Island, Nebraska; Structure, process, and relationship to channel narrowing: U.S. Geological Survey Open-file Report 81-1059, 31 p.

Darton, N. H., 1903, Preliminary report on the geology and water resources of Nebraska west of the one hundred third meridian: U.S. Geological Survey Professional Paper 17, 69 p.

—— , 1905, Preliminary report on the geology and underground water resources of the central Great Plains: U.S. Geological Survey Professional Paper 32, 433 p., 72 pl.

David, P. P., 1972, Great Sand Hills, Saskatchewan, *in* Rutter, N. W., and Christiansen, E. A., eds., Quaternary geology and geomorphology between Winnepeg and the Rocky Mountains: International Geological Congress, 24th, Excursion C-22, Guidebook, p. 37–50.

—— , 1977, Sand dune occurrences of Canada: Indian and Northern Affairs, National Parks Branch, Contract 74-230, 183 p.

—— , 1982, Late Pleistocene and Holocene climatic changes based on the eolian stratigraphic record of the Canadian Prairies, An update: Geological Association of Canada Annual Meetings, Program with Abstracts, v. 7, p. 44.

Denson, N. M., and Bergendahl, M. H., 1961, Middle and upper Tertiary rocks of southeastern Wyoming and adjoining areas, *in* Short papers in the geologic and hydrologic sciences: U.S. Geological Survey Professional Paper 424-C, p. 168–172.

Dreeszen, V. H., 1970, The stratigraphic framework of Pleistocene glacial and periglacial deposits in the Central Plains, *in* Dort, W. D., Jr., and Jones, J. K., eds., Pleistocene and Recent environments of the central Great Plains: Lawrence, Kansas, University of Kansas Press, p. 9–22.

Emiliani, C., 1967, The Pleistocene record of the Atlantic and Pacific ocean sediments; Correlations with the Alaskan stages by absolute dating, and the age of the last reversal of the geomagnetic field: Progress in Oceanography, v. 4, p. 219–224.

—— , 1970, Pleistocene paleotemperatures: Science, v. 168, p. 822–825.

—— , 1972, Quaternary hypsithermals: Quaternary Research, v. 2, p. 270–273.

Epp, H. T., 1986, Prehistoric settlement response to the Harris Sand Hills, Saskatchewan, Canada: Plains Anthropologist, p. 31, 51–63.

Epp, H. T., and Townley-Smith, L., 1980, The Great Sand Hills of Saskatchewan: Regina, Saskatchewan, Saskatchewan, 156 p.

Eschner, T. R., 1983, Hydraulic geometry of the Platte River near Overton, south-central Nebraska: U.S. Geological Survey Professional Paper 1277-C, 32 p.

Eschner, T. R., Hadley, R. F., and Crowley, K. D., 1983, Hydrologic and morphologic changes in channels of the Platte River basin in Colorado, Wyoming, and Nebraska; An historical perspective: U.S. Geological Survey Professional Paper 1277-A, 39 p.

Evans, G. L., and Meade, G. E., 1945, Quaternary of the Texas High Plains: University of Texas Publications 4401, p. 485–507.

Fader, S. W., 1975, Land subsidence caused by dissolution of salt near four oil and gas wells in central Kansas: U.S. Geological Survey Water-Resources Investigations 27-75, 28 p.

Fenneman, N. M., 1931, Physiography of western United States: New York, McGraw-Hill Book Company, Incorporated, 534 p.

Fenton, M. M., 1983a, Composition of deformation terrain (Glacial thrust terrain); A clue to glacial entrainment and transport: Geological Association of Canada, Program with Abstracts, v. 8, p. A22.

—— , 1983b, Deformation terrain, mid-continent region; Properties, subdivision, recognition: Geological Society of America Program with Abstracts, v. 15, no. 4, p. 250.

—— , 1983c, Preliminary glaciotectonic map, Highvale Mine: Unpublished map prepared for TransAlta Utilities by Alberta Geological Survey, Alberta Research Council.

—— , 1984a, Deformation terrain, Canadian prairies; Morphology and sediment facies: INQUA Commission on Genesis and Lithology of Quaternary Deposits, Symposium on the Relationship between Glacial Terrain and Glacial Sediment Facies, Alberta Research Council publication, Abstracts and Program, p. 7–8.

—— , 1984b, Quaternary stratigraphy, Canadian Prairies, *in* International Geological Correlation Program Project 73/1/24; Quaternary glaciations in northern hemisphere, Canadian summary volume: Geological Survey of Canada Paper, p.57–68.

Fenton, M. M., and Andriashek, L. D., 1978, Glaciotectonic features in the Sand River area, northeastern Alberta, Canada: American Quaternary Association, 5th Biennial Meeting, Edmonton, Alberta, Abstracts, p. 199.

—— , 1984, Surficial geology, Sand River map sheet 73L: Alberta Geological Survey, Alberta Research Council Map.

Fenton, M. M., Moell, C. E., Pawlowicz, J. G., Sterneberg, G. J., Trudell, M. R., and Moran, S. R., 1983, Highwall stability project, Highvale Mine study report, December 1983: Unpublished report prepared for TransAlta Utilities by Alberta Geological Survey, Alberta Research Council, 70 p.

Fenton, M. M., Moell, C. E., Pawlowicz, J. G., and Langenbert, G., 1985a, Deformation terrain, Canadian Prairies; Landforms, lithofacies, and processes: Geological Society of America Program with Abstracts, v. 17, no. 5, p. 287.

Fenton, M. M., Langenbert, C. E., Jones, C. E., Trudell, M. R., Pawlowicz, J. G., Tapics, J. A., and Nikols, D. J., 1985b, Tour of the Highvale open pit coal mine: Guidebook prepared for the Petroleum Society of Canadian Institute of Mining and Canadian Society of Petrolem Geologists joint conference, Edmonton, Alberta, Alberta Research Council open-file report 1985-7, 55 p.

Fink, J., and Kula, G. J., 1977, Pleistocene climates in central Europe; At least 17 interglacials after the Olduavai Event: Quaternary Research, v. 7, p. 363–371.

Flint, R. F., 1955, Pleistocene geology of eastern South Dakota: U.S. Geological Survey Professional Paper 262, 173 p.

Fredlund, G. G., Johnson, W. C., and Dort, W. D., Jr., 1985, A preliminary analysis of opal phytoliths from the Eustis Ash Pit, Frontier County, Nebraska, *in* Dort, W. D., Jr., ed., Institute of Tertiary–Quaternary Studies, Tertiary-Quaternary Symposium Series: Lincoln, Nebraska Academy of Sciences, v. 1, p. 147–162.

Frickel, D. G., Shown, L. M., Hadley, R. F., and Miller, R. F., 1981, Methodology for hydrologic evaluation of a potential surface mine; The Red Rim Site, Carbon and Sweetwater counties, Wyoming: U.S. Geological Survey Water Resources Investigations 81-75, 59 p.

Fryrear, D. W., and Randel, G. L., 1972, Predicting dust in the southern Plains: Texas A&M University, Texas Agricultural Experiment Station, MP-1025.

Gable, D. J., and Hatton, T., 1983, Maps of vertical crustal movements in the conterminous United States over the last 10 million years: U.S. Geological Survey Miscellaneous Investigations Map I-1315.

Gardner, L. R., 1972, Origin of the Mormon Mesa caliche, Clark County, Nevada: Geological Society of America Bulletin, v. 83, p. 143–156.

Gile, L. H., 1979, Holocene soils in eolian sediments of Bailey County, Texas: Soil Science Society of America Journal, v. 43, p. 994–1003.

—— , 1985, The Sandhills project soil monograph: Rio Grande Historical Collections, Las Cruces, New Mexico State University, 331 p.

Gile, L. H., Hawley, J. W., and Grossman, R. B., 1981, Soils and geomorphology in the Basin and Range area of southern New Mexico; Guidebook to the Desert Project: New Mexico Bureau of Mines and Mineral Resources Memoir 39, 222 p.

Gillette, D. A., 1981, Production of dust that may be carried great distances, *in* Pewe, T. L., ed., Desert dust; Origin, characteristics, and effect on man: Geological Society of America Special Paper 186, p. 11–26.

Gillette, D. A., and Goodwin, P. A., 1974, Microscale transport of sand-sized soil aggregates eroded by wind: Journal of Geophysical Research, v. 79, p. 4080–4084.

Gillette, D. A., and Walker, T. R., 1977, Characteristics of airborne particles produced by wind erosion of sandy soil, High Plains of west Texas: Soil Science, v. 123, p. 97–110.

Gillette, D. A., Blifford, I. H., Jr., and Fryrear, D. W., 1974, The influence of wind velocity on the size distributions of aerosols generated by the wind erosion of soils: Journal of Geophysical Research, v. 79, p. 4068–4075.

Gillette, D. A., Clayton, R. N., Mayeda, T. K., Jackson, M. L., and Sridhar, K., 1978, Tropospheric aerosols from some major dust storms of the southwestern United States: Journal of Applied Meteorology, v. 17, p. 852–845.

Gilley, J. E., Bauer, M., Willis, W. O., and Young, R. A., 1977, Runoff and erosion characteristics of surface-mined sites in western North Dakota: Transactions of the American Association of Agricultural Engineers, p. 697–700, 704.

Gogel, T., 1981, Discharge of saltwater from Permian rocks to major stream-aquifer systems in central Kansas: Kansas Geological Survey Chemical Quality Series 9, 60 p.

Gold, C. M., Andriashek, L. D., and Fenton, M. M., 1983, Bedrock topography, Sand River map sheet 73L: Alberta Geological Survey, Alberta Research Council Map.

Goudie, A. S., 1978, Dust storms and their geomorphological implications: Journal of Arid Environments, v. 1, p. 291–310.

Gravenor, C. P., and Kupsch, W. O., 1959, Ice disintegration features in western Canada: Journal of Geology, v. 67, no. 1, p. 48–64.

Gray, J. H., 1978, Men against the desert: Saskatoon, Saskatchewan, Western Producer Prairie Books, 260 p.

Green, F. E., 1961, The Monahans Dunes area, *in* Wendorf, F., assembler, Paleoecology of the Llano Estacado, Fort Burgwin Research Center Publication 1: Santa Fe, The Museum of New Mexico Press, p. 22–47.

Gustavson, T. C., 1982, Structural control of major drainage elements surrounding the Southern High Plains, *in* Geology and geohydrology of the Palo Duro basin, Texas Panhandle: Texas Bureau of Economic Geology, Geological Circular 82-7, p. 176–182.

——, 1983, Diminished spring discharge; Its effect on erosion rates in the Texas Panhandle, *in* Geology and geohydrology of the Palo Duro basin, Texas Panhandle: Texas Bureau of Economic Geology, Geological Circular 83-4, p. 128–132.

——, 1986, Geomorphic development of the Canadian River valley, Texas Panhandle; An example of regional salt dissolution and subsidence: Geological Society of America Bulletin, v. 97, p. 459–472.

Gustavson, T. C., and Budnik, R. T., 1985, Structural influences on geomorphic processes and physiographic features, Texas Panhandle; Technical issues in siting a nuclear-waste repository: Geology, v. 13, p. 173–176.

Gustavson, T. C., and Finley, R. J., 1985, Late Cenozoic geomorphic evolution of the Texas Panhandle and northeastern New Mexico; Case studies of structural controls of regional drainage development: The University of Texas at Austin, Bureau of Economic Geology Report of Investigations, 42 p.

Gustavson, T. C., and Holliday, V. T., 1985, Depositional architecture of the Quaternary Blackwater Draw and Tertiary Ogallala formations, Texas Panhandle and eastern New Mexico: The University of Texas at Austin, Bureau of Economic Geology Report OF-WTWI-1985-23, 60 p.

Hadley, R. F., Frickel, D. G., Shown, L. M., and Miller, R. F., 1981, Methodology for hydrologic evaluation of a potential surface mine; East Trail Creek basin, Bighorn County, Montana: U.S. Geological Survey Water-Resources Investigations 81-58, 73 p.

Hadley, R. F., and Schumm, S. A., 1961, Sediment sources and drainage basin characteristics in upper Cheyenne River basin: U.S. Geological Survey Water-Supply Paper 1531-B, p. 133–198.

Hagen, L. J., and Woodruff, N. P., 1973, Air pollution from duststorms in the Great Plains: Atmospheric Environment, v. 7, p. 323–332.

Harksen, J. C., 1967, Quaternary loess in southwestern South Dakota: Proceedings of the South Dakota Academy of Science, v. 46, p. 32–40.

Hawley, J. W., 1984, The Ogallala Formation in eastern new Mexico, *in* Whetstone, G. A., ed., Proceedings, Ogallala Aquifer Symposium II: Texas Tech University Water Resources Center, p. 157–176.

Hedman, E. R., Kastner, W. M., and Hejl, H. R., 1974, Kansas streamflow characteristics; Part 10, Selected streamflow characteristics as related to active-channel geometry of streams in Kansas: Kansas Water Resources Board Technical Report no. 10, 21 p.

Hefley, H. M., and Sidwell, R., 1945, Geological and ecological observations of some High Plains dunes: American Journal of Science, v. 243, p. 361–376.

Henz, J. F., and Woiceshyn, P. M., 1980, Climatological relationships of severe duststorms in the Great Plains to synoptic weather patterns; Potential for predictability: National Aeronautics and Space Administration, Jet Propulsion Laboratory Publication 79-97, 27 p.

Higgins, C. G., 1982, Drainage systems developed by sapping on Earth and Mars: Geology, v. 10, p. 147–152.

——, 1984, Piping and sapping; Development of landforms by groundwater outflow, *in* LaFleur, R. G., ed., Groundwater as a geomorphic agent: Boston, Allen and Unwin, Incorporated, p. 18–58.

Hobbs, P. V., Bowdle, D. A., and Radke, L. F., 1985, Particles in the lower troposphere over the High Plains of the United States, Part I; Size distributions, elemental compositions, and morphologies: Journal of Climate and Applied Meteorology, v. 24, no. 12, p. 1344–1356.

Hofmann, L., Ries, R. E., and Gilley, J. E., 1983, Relationship of runoff and soil loss to ground cover of native and reclaimed grazing land: Agronomy Journal, v. 75, p. 599–602.

Holliday, V. T., 1982, Morphological and chemical trends in Holocene soils at the Lubbock Lake archeological site, Texas [Ph.D. thesis]: Boulder, The University of Colorado, 285 p.

——, 1984, Climatic implications of mid-Holocene eolian deposits on the Southern High Plains: Geological Society of America Abstracts with Programs, v. 16, p. 542.

——, 1985, Holocene soil-geomorphological relations in a semi-arid environment; The Southern High Plains of Texas, *in* Boardman, J., ed., Soils and Quaternary landscape evolution: Chichester, United Kingdom, John Wiley and Sons, p. 325–357.

Holter, M. E., Yurko, J. R., and Chu, M., 1975, Geology and coal reserves of the Arkley coal zone of central Alberta: Alberta Research Council Report 75-7, 41 p.

Hopkins, O. B., 1923, Some structural features of the plains area of Alberta caused by Pleistocene glaciation: Geological Society of America Bulletin, v. 34, p. 419–430.

Huffington, R. M., and Albritton, C. C., Jr., 1941, Quaternary sands on the Southern High Plains of eastern Texas: American Journal of Science, v. 239, p. 325–338.

Hunt, C. B., 1967, Physiography of the United States: San Francisco, W. H. Freeman and Company, 480 p.

Hurt, R. D., 1981, The Dust Bowl; An agricultural and social history: Chicago, Nelson-Hall, 214 p.

Hutton, J., 1795, Theory of the Earth (2 volumes): Edinburgh.

Idso, S. B., 1976, Dust storms: Scientific American, v. 235, p. 108–114.

Irish, E.J.W., 1970, The Edmonton Group of south-central Alberta: Bulletin of Canadian Petroleum Geology, v. 18, p. 125–155.

Izett, G. A., Obradovich, J. D., Naerser, C. W., and Cebula, G. T., 1981, Potassium-argon and fission-track zircon ages of Cerro Toledo Rhyolite tephra in the Jemez Mountains, New Mexico, *in* Shorter contributions to isotope research: U.S. Geological Survey Professional Paper 1199, p. 37–43.

Jackson, M. L., Gillette, D. A., Danielson, E. F., Blifford, I. H., Bryson, R. A., and Syers, J. K., 1973, Global dustfall during the Quaternary as related to environments: Soil Science, v. 116, p. 135–145.

Johnson, R. G., 1982, Brunhes–Matuyama magnetic reversal dated at 790,000 yr BP by marine-astrological correlations: Quaternary Research, v. 17,

p. 135–147.

Johnson, W. C., 1965, Wind in the southwestern Great Plains: U.S. Department of Agriculture, Conservation Research Report 6, 65 p.

Johnson, W. D., 1901, The High Plains and their utilization: Extract, 21st Annual Report of the U.S. Geological Survey, Part IV, p. 601–768.

Jorstad, T., East, T., Adovasio, J. M., Donahue, J., and Stuckenrath, R., 1986, Paleosols and prehistoric populations in the High Plains: Geoarchaeology, v. 1, p. 163–181.

Kimberlin, L. W., Hidlebaugh, A. L., and Grunewald, A. R., 1977, The potential wind erosion problem in the United States: Transactions of the American Society for Agricultural Engineering, v. 20, p. 873–879.

Kircher, J. E., and Karlinger, M. R., 1983, Effects of water development on surface-water hydrology, Platte River basin in Colorado, Wyoming, and Nebraska, upstream from Duncan, Nebraska: U.S. Geological Survey Professional Paper 1277-B, 49 p.

Knowles, T., Nordstrom, P., and Klemt, W. B., 1984, Evaluating the groundwater resources of the High Plains of Texas, Volume I: Texas Department of Water Resources, Report 288, 113 p.

Kukla, G. J., 1975, Loess stratigraphy of central Europe, *in* Butzer, K., and Isaac, G. L., eds., After the Australopithecines: The Hague, Mouton, p. 99–188.

—— , 1977, Pleistocene land-sea correlations, I, Europe: Earth–Science Review, v. 13, p. 307–374.

Kupsch, W. O., 1962, Ice-thrust ridges in western Canada: Journal of Geology, v. 70, p. 582–594.

LaDouchy, S., and Annett, C. H., 1982, Drought and dust; A study in Canada's Prairie Provinces: Atmospheric Environment, v. 16, p. 1535–1541.

Laity, J. E., and Malin, M. C., 1985, Sapping processes and the development of theater-headed valley networks on the Colorado Plateau: Geological Society of America Bulletin, v. 96, p. 203–217.

LaPrade, K. E., 1957, Dust-storm sediments of Lubbock area, Texas: Bulletin of the American Association of Petroleum Geologists, v. 41, p. 709–726.

Leopold, L. B., and Maddock, T., Jr., 1953, The hydraulic geometry of stream channels and some physiographic implications: U.S. Geological Survey Professional Paper 252, 57 p.

Lerbekmo, J. E., Singh, C., Jarzen, D. M., and Russel, D. A., 1979, The Cretaceous–Tertiary boundary in south-central Alberta; A revision based on additional dinosaurian and microfloral evidence: Canadian Journal of Earth Sciences, v. 16, no. 9, p. 1866–1869.

Lisitzin, A. P., 1972, Sedimentation in the world ocean: Society of Economic Paleontologists and Mineralogists, Special Publication 17, 218 p.

Liu Tung-sheng, An Zhisheng, Yuan Baoyin, and Han Jiamao, 1985, The loess-paleosol sequence in China and climatic history: Episodes, v. 8, p. 21–28.

Lockeretz, W., 1978, The lessons of the Dust Bowl: American Scientist, v. 66, p. 560–570.

Lotspeich, F. B., Lehman, O. R., Hauser, V. L., and Stewart, B. A., 1971, Hydrogeology of a playa near Amarillo, Texas: Texas Agricultural Experiment Station, Technical Report no. 10, 19 p.

Lowry, M. E., 1966, The White River Formation as an aquifer in southeastern Wyoming and adjacent parts of Nebraska and Colorado: U.S. Geological Survey Professional Paper 550-D, p. 217–222.

Luckey, R. R., Gutentag, E. D., Heimes, F. J., and Weeks, J. B., 1987, Digital simulation of ground-water flow in the High Plains aquifer in parts of Colorado, Kansas, Nebraska, New Mexico, Oklahoma, South Dakota, Texas, and Wyoming: U.S. Geological Survey Professional Paper 1400-D (in press).

Lugn, A. L., 1935, The Pleistocene geology of Nebraska: Nebraska Geological Survey Bulletin 10, series 2, 223 p.

—— , 1962, The origin and sources of loess in the central Great Plains and adjoining areas of the Central Lowland: University of Nebraska Studies, new series no. 26, 104 p.

—— , 1968, The origin of loesses and their relation to the Great Plains in North America, *in* Schultz, C. B., and Frye, J. D., eds., Loess and related eolian deposits of the world: Lincoln, University of Nebraska Press, p. 139–182.

Lusby, G. C., and Toy, T. J., 1976, An evaluation of surface-mine spoils area restoration in Wyoming using rainfall simulation: Earth Surface Processes, v. 2, p. 375–386.

Lyell, C., 1830, Principles of geology, being an attempt to explain the former changes of the earth's surface, by reference to causes now in operation, Volume 1: London, John Murray, 511 p.

—— , 1833, Principles of geology, Volume 3 (first edition), Murray, John, ed., London.

Machenberg, M. D., 1984, Geology of Monahans Sandhills State Park, Texas: The University of Texas, Bureau of Economic Geology, Guidebook 21, 39 p.

Machenberg, M. D., and Holliday, V. T., 1987, The Blackwwater Draw Formation type locality and its implications for Southern High Plains Quaternary stratigraphy: The University of Texas, Bureau of Economic Geology, Annual Report (in press).

Machette, M. N., 1985, Calcic soils of the southwestern United States, *in* Weide, D. L., ed., Soils and Quaternary geology of the southwestern United States: Geological Society of America Special Paper 203, p. 1–21.

Maslowski Schutze, A., 1987, Geology of the Highvale study site: Plains Hydrology and Reclamation Project: Alberta Research Council internal report (in press).

McCauley, J. F., Breed, C. S., Grolier, M. J., and Mackinnon, D. J., 1981, The U.S. dust storm of February, 1977, *in* Pewe, T. L., ed., Desert dust; Origin, characteristics, and effect on man: Geological Society of America Special Paper 186, p. 123–147.

McNeal, R. P., 1965, Hydrodynamics of the Permian Basin, *in* Young, A., and Galley, J. E., eds., Fluids in subsurface environments: American Association of Petroleum Geologists Memoir 4, p. 308–326.

Melton, F. A., 1940, A tentative classification of sand dunes; Its application to dune history in the Southern High Plains: Journal of Geology, v. 48, p. 113–174.

Merriam, D. F., 1963, The geologic history of Kansas: Kansas Geological Survey Bulletin 162, 317 p.

Missouri River Basin Interagency Committee, 1968, Missouri River basin comprehensive framework study; Sedimentation, western Dakota tributaries: Task force on sedimentation, Work Group on hydrologic analysis and projections, 6 p., 2 tables, 4 pl.

—— , 1969, Missouri River basin comprehensive framework study; Sedimentation, Platte River: Task force on sedimentation, Work Group on hydrologic analysis and projections, 5 p., 2 tables, 5 pl.

Moell, C. E., Pawlowicz, J. G., Fenton, M. M., Trudell, M. R., Jones, E. C., and Sterenberg, G., 1985, Highwall stability project, Highvale Mine study report for 1984: Unpublished report prepared for TransAlta Utilities by Terrain Sciences Department, Alberta Research Council, 140 p.

Moran, S. R., 1971, Glaciotectonic structures in drift, *in* Goldthwait, R. P., ed., Till; A symposium: Columbus, Ohio State University Press, p. 127–148.

Moran, S. R., Clayton, L., Hooke, R. L., Fenton, M. M., and Andriashek, L. D., 1980, Glacier-bed landforms of the prairie region of North America: Journal of Glaciology, v. 25, no. 93, p. 457–476.

Muhs, D. R., 1985, Age and paleoclimatic significance of Holocene sand dunes in northeastern Colorado: Annals of the Association of American Geographers, v. 75, p. 566–582.

Nadler, C. T., and Schumm, S. A., 1981, Metamorphosis of South Platte and Arkansas rivers, eastern Colorado: Physical Geography, v. 2, p. 95–115.

National Oceanic and Atmospheric Administration, 1982a, Climatic Atlas of the United States: Asheville, North Carolina, National Climatic Center, p. 73–78.

—— , 1982b, Climate of Texas: Asheville, North Carolina National Climatic Center, Climatography of the United States, no. 60.

National Research Council, 1981, Surface mining; Soil, coal, and society: Washington, D.C., National Academy Press, 233 p.

North American Stratigraphic Commission, 1983, North American stratigraphic code: American Association of Petroleum Geologists Bulletin, v. 67, p. 841–875.

Odynsky, W., 1958, U-shaped dunes and effective wind directions in Alberta: Canadian Journal of Soil Science, v. 38, p. 56–62.

Orgill, M. M., and Sehmel, G. A., 1976, Frequency and diurnal variation of dust storms in the contiguous U.S.A.: Atmospheric Environment, v. 10, p. 813–825.

Orr, E. D., and Kreitler, C. W., 1985, Interpretation of pressure-depth data from confined underpressured aquifers exemplified by the deep-basin brine aquifer, Palo Duro basin, Texas: Water Resources Research, v. 21, p. 533–544.

Osterkamp, W. R., 1977, Effect of channel sediment on width-discharge relations, with emphasis on streams in Kansas: Kansas Water Resources Board Bulletin 21, 25 p.

Osterkamp, W. R., and Harrold, P. E., 1981, Dynamics of alluvial channels; A process model, *in* Modeling components of hydrologic cycle: Littleton, Colorado, Water Resources Publications, p. 283–296.

Osterkamp, W. R., and Hedman, E. R., 1982, Perennial-streamflow characteristics related to channel geometry and sediment in Missouri River Basin: U.S. Geological Survey Professional Paper 1242, 37 p.

Osterkamp, W. R., and Wood, W. W., 1984, Development and escarpment retreat of the Southern High Plains, *in* Whetstone, G. A., ed., Proceedings of the Ogallala Aquifer Symposium II: Texas Tech University Water Resources Center, p. 177–193.

Osterkamp, W. R., Lane, L. J., and Foster, G. R., 1983, An analytical treatment of channel-morphology relations: U.S. Geological Survey Professional Paper 1288, 21 p.

Parizek, R. R., 1972, Glacial ice-contact rings and ridges: Geological Society of America Special Paper 123, p. 49–102.

Parkinson, G. R., 1936, Dust storms over the Great Plains; Their causes and forecasting: Bulletin of the American Meteorological Society, v. 17, p. 127–135.

Prest, V. K., 1968, Nomenclature of moraines and ice-flow features as applied to the Glacial Map of Canada: Geological Survey of Canada, Department of Energy, Mines, and Resources Paper 67-57, 32 p.

——— , 1970, Quaternary geology of Canada, *in* Douglas, R.J.W., ed., Geology and economic minerals of Canada (fifth edition): Canada Geological Survey, Economic Geology Series 1, p. 675–764.

Prest, V. K., Grant, D. R., and Rampton, V. N., 1968, Glacial Map of Canada: Geological Survey of Canada, Department of Energy, Mines, and Resources, 1 sheet.

Price, W. A., 1940, Caliche karst: Geological Society of America Bulletin, v. 51, p. 1938–1939.

Rabenhorst, M. C., Wilding, L. P., and Girdner, C. L., 1984, Airborne dusts in the Edwards Plateau region of Texas: Soil Science Society of America Journal, v. 48, p. 621–627.

Reed, E. C., 1968, Loess deposition in Nebraska, *in* Schultz, C. B., and Frye, J. C., eds., Loess and related eolian deposits of the world: Lincoln, University of Nebraska Press, p. 23–28.

Reed, E. C., and Dreeszen, V. H., 1965, Revision of the classification of the Pleistocene deposits of Nebraska: Nebraska Geological Survey Bulletin (new series), v. 23, 65 p.

Reeves, C. C., Jr., 1965, Chronology of west Texas pluvial lake dunes: Journal of Geology, v. 73, p. 504–508.

——— , 1966, Pluvial lake basins of west Texas: Journal of Geology, v. 74, no. 3, p. 269–291.

——— , 1970, Location, flow, and water quality of some west Texas playa lake springs: Texas Tech University Water Resources Center, 21 p.

——— , 1972, Tertiary–Quaternary stratigraphy and geomorphology of west Texas and southeastern New Mexico: East-Central New Mexico Guidebook, New Mexico Geological Society, p. 108–117.

——— , 1976, Caliche; Origin, classification, morphology, and uses: Lubbock, Texas, Estacado Books, 233 p.

——— , 1983, Pliocene channel calcrete and suspenparallel drainage in west Texas and New Mexico, *in* Wilson, R.C.L., ed., Residual deposits; Surface related weathering processes and materials: The Geological Society of London, p. 179–183.

——— , 1984, The Ogallala depositional mystery, *in* Whetstone, G. A., ed., Proceedings of the Ogallala Aquifer Symposium II: Texas Tech University, Water Resources Center, p. 129–156.

Ringen, B. H., Shown, L. M., Hadley, R. F., and Hinkley, T. K., 1979, Effect on sediment yield and water quality of a non-rehabilitated surface mine in north-central Wyoming: U.S. Geological Survey Water Resources Investigations 79-47, 23 p.

Ruddiman, W. F., Raymo, M., and McIntyre, A., 1987, Variations of the North Atlantic Ocean and Northern Hemisphere ice sheets at 41,000 years during the Matuyama: Nature, v. 320 (in press).

Ruhe, R. V., 1983, Depositional environment of late Wisconsin loess in the midcontinental United States, *in* Wright, H. E., Jr., and Porter, S. C., eds., Late–Quaternary environments of the United States, Volume 1, The late Pleistocene: Minneapolis, University of Minnesota Press, p. 130–137.

——— , 1984, Soil-climate system across the prairies in midwestern U.S.A.: Geoderma, v. 34, p. 201–219.

Sardeson, F. W., 1898, The so-called Cretaceous deposits in southeastern Minnesota: Journal of Geology, v. 6, p. 679–691.

——— , 1905, A peculiar case of glacial erosion: Journal of Geology, v. 13, p. 351–357.

——— , 1906, The folding of subjacent strata by glacial action: Journal of Geology, v. 14, p. 226–232.

Schultz, C. B., 1968, The stratigraphic distribution of vertebrate fossils in Quaternary eolian deposits in the midcontinent region of North America, *in* Schultz, C. B., and Frye, J. C., eds., Loess and related eolian deposits of the world: Lincoln, University of Nebraska Press, p. 115–138.

——— , 1977, The Ogallala Formation and its vertibrate faunas in the Texas and Oklahoma panhandles, *in* Schultz, G. E., ed., Field conference on late Cenozoic biostratigraphy of the Texas Panhandle and adjacent Oklahoma, August 4–6, 1977, Guidebook: West Texas State University, Killgore Research Center Special Publication 1, p. 5–104.

Schultz, C. B., and Martin, L. D., 1970, Quaternary mammalian sequence in the central Great Plains; in Dort, W. D., Jr., and Jones, J. D., eds., Pleistocene and Recent environments of the central Great Plains: Lawrence, University of Kansas, Special Publication 3, p. 341–353.

Schultz, C. B., and Stout, T. M., 1945, The Pleistocene loess deposits of Nebraska: American Journal of Science, v. 243, p. 231–244.

——— , 1961, Field conference on the Tertiary and Pleistocene of western Nebraska, Ninth field conference of the Society of Vertibrate Paleontology, Guidebook: Special Publication of The University of Nebraska State Museum, no. 2, 55 p.

——— , 1980, Ancient soils and climatic changes in the central Great Plains: Nebraska Academy of Sciences Transactions, v. 8, p. 184–205.

Schultz, C. B., and Tanner, L. G., 1957, Medial Pleistocene fossil vertebrate localities in Nebraska: The University of Nebraska State Museum Bulletin, v. 4, p. 59–81.

Schumm, S. A., 1956, The role of creep and rainwash on the retreat of badland slopes: American Journal of Science, v. 254, p. 693–706.

——— , 1960, The shape of alluvial channels in relation to sediment type: U.S. Geological Survey Professional Paper 352-B, p. 17–30.

——— , 1968, River adjustment to altered hydrologic regimen, Murrumbidgee River and paleochannels, Australia: U.S. Geological Survey Professional Paper 598, 65 p.

——— , 1973, Geomorphic thresholds and complex response of drainage systems, *in* Morisawa, M., ed., Fluvial geomorphology: Binghamton, New York, State University of New York, Publications in Geomorphology, p. 299–310.

——— , 1975, Episodic erosion; A modification of the geomorphic cycle, *in* Flemal, R., and Melhorn, W., eds., Theories of landform development: Binghamton, New York, State University of New York, Publications in Geomorphology, p. 69–85.

——— , 1979, Geomorphic thresholds; The concept and its applications: Institute of British Geographers, Transactions, n.s., v. 4, p. 485–515.

——— , 1980, Some applications of the concept of geomorphic threshold, *in* Coates, D. R., and Vitek, J. D., eds., Thresholds in geomorphology: Boston, Allen and Unwin, Incorporated, p. 473–485.

——, 1981, Evolution and response of the fluvial system, sedimentologic implications: Society of Economic Paleontologists and Mineralogists Special Publication no. 31, p. 19–29.

Schumm, S. A., and Lichty, R. W., 1963, Channel widening and flood-plain construction along Cimarron River in southwestern Kansas: U.S. Geological Survey Professional Paper 352-D, p. 71–88.

Scott, G. R., 1960, Subdivision of the Quaternary alluvium east of the Front Range near Denver, Colorado: Geological Society of America Bulletin, v. 71, p. 1541–1543.

——, 1963, Quaternary geology and geomorphic history of the Kassler Quadrangle, Colorado: U.S. Geological Survey Professional Paper 421-A, 70 p.

——, 1965, Nonglacial Quaternary geology of the Southern and Middle Rocky Mountains, *in* Wright, H. E., Jr., and Frey, D. G., eds., The Quaternary of the United States: Princeton, Princeton University Press, p. 243–254.

——, 1978, Map showing geology, structure, and oil and gas fields in the Sterling 1° × 2° Quadrangle, Colorado, Nebraska, and Kansas: U.S. Geological Survey Miscellaneous Investigations Map I-1092.

——, 1982, Paleovalley and geologic map of northeastern Colorado: U.S. Geological Survey Miscellaneous Investigations Map I-1378.

Seitlheko, E. M., 1975, Studies of mean particle size and mineralogy of sands along selected transects on the Llano Estacado [M.S. thesis]: Lubbock, Texas, Tech University, 69 p.

Sellards, E. H., Adkins, W. S., and Plummer, F. B., 1932, The Geology of Texas, Volume 1, Stratigraphy: The University of Texas Bulletin 3232, 1007 p.

Seni, S. J., 1980, Sand-body and depositional systems, Ogallala Formation, Texas: The University of Texas at Austin, Bureau of Economic Geology, Report of Investigations no. 105, 36 p.

Shackleton, N. J., 1969, The last interglacial in the marine and terrestrial records: Royal Society of London Proceedings, Series B. 174, p. 135–154.

Shackleton, N. J., and Opdyke, N. D., 1973, Oxygen isotope and paleomagnetic stratigraphy of equatorial Pacific core V28-238; Oxygen isotope temperatures and ice volumes on a 10/5 and 10/6-year scale: Quaternary Research, v. 3, p. 39–55.

——, 1976, Oxygen isotopic and paleomagnetic stratigraphy of Pacific core V28-239; Late Pliocene to latest Pleistocene, *in* Cline, R. M., and Hays, J. D., eds., Investigation of late Quaternary paleoceanography and paleoclimatology: Geological Society of America Memoir 145, p. 449–464.

Shackleton, N. J., and 16 others, 1984, Oxygen-isotope calibration on the onset of ice-rafting and history of glaciation in the North Atlantic region: Nature, v. 307, p. 620–623.

Sharps, J. A., 1980, Geologic map of the Limon 1° × 2° Quadrangle, Colorado and Kansas: U.S. Geological Survey Miscellaneous Investigations Map I-1250.

Sidwell, R., 1938, Sand and dust storms in vicinity of Lubbock, Texas: Economic Geography, v. 14, p. 98–102.

Slater, G., 1926, Glacial tectonics as reflected in disturbed drift deposits: Proceedings, Geologists' Association, v. 37, p. 392–400.

——, 1927, Structure of the Mud Buttes and Tit Hills in Alberta: Geological Society of America Bulletin, v. 38, p. 721–730.

Smith, H.T.U., 1940, Geologic studies in southwestern Kansas: Kansas Geological Survey Bulletin 34, p. 159–168.

——, 1965, Dune morphology and chronology in central and western Nebraska: Journal of Geology, v. 73, no. 4, p. 557–578.

——, 1968, Nebraska dunes compared with those of North Africa and other regions, *in* Schultz, C. B., and Frye, J. C., ed., Loess and related eolian deposits of the world: Lincoln, University of Nebraska Press, p. 29–47.

Smith, R. M., Twiss, P. C., Krauss, R. K., and Brown, M. J., 1970, Dust deposition in relation to site, season, and climatic variables: Soil Science Society of America Proceedings, v. 34, p. 112–117.

Souster, W. E., St. Arnaud, R. J., and Huang, P. M., 1977, Variation in physical properties and mineral composition of thin loess deposits in the Swift Current area of Saskatchewan: Soil Science Society of America Journal, v. 41, p. 594–601.

Stalker, A. M., 1987, Great Plains and Beaufort regions, *in* Fulton, R. J., and Heginbottom, J. A., eds., Quaternary geology of Canada and Greenland, Geology and economic minerals of Canada: Geological Survey of Canada, Chapter II, Part 2 (in press).

Swineford, A., and Frye, J. C., 1945, A mechanical analysis of windblown dust compared with analyses of loess: American Journal of Science, v. 243, p. 249–255.

Thornbury, W. D., 1965, Regional geomorphology of the United States: New York, John Wiley and Sons, Incorporated, 609 p.

Thorpe, J., and Smith, H.T.U., 1952, Pleistocene eolian deposits of the United States, Alaska, and parts of Canada: Geological Society of America map.

Toy, T. J., 1983, A linear erosion/elevation measuring instrument (LEMI): Earth Surface Processes and Landforms, v. 8, p. 313–322.

Toy, T. J., and Grim, D. S., 1984, Climate appraisal maps of the rehabilitation potential of strippable coal lands in the Williston Basin, Montana, North Dakota, and South Dakota: U.S. Geological Survey Miscellaneous Field Studies Map MF-1675, 3 sheets.

Trimble, D. E., 1980, The geologic story of the Great Plains: U.S. Geological Survey Bulletin 1493, 55 p.

Trimble, D. E., and Machette, M. N., 1979, Geologic map of the greater Denver area, Front Range urban corridor, Colorado: U.S. Geological Survey Miscellaneous Investigations Map I-856-H.

Twiss, P. C., 1983, Dust deposition and opal phytoliths in the Great Plains: Transactions of the Nebraska Academy of Sciences, v. 11, p. 73–82.

U.S. Water Resources Council, 1977, The nation's water resources; Appendix, Erosion and sedimentation and related resource considerations: U.S. Water Resources Council, 111 p.

Warn, G. F., and Cox, W. H., 1951, A sedimentary study of dust storms in the vicinity of Lubbock, Texas: American Journal of Science, v. 249, p. 553–568.

Warren, G. K., 1858, Preliminary report of explorations in Nebraska and Dakota, in the years 1855-'56-'57: Washington, D.C., Engineer Department, United States Army, Government Printing Office (Reprint, 1875), 125 p.

——, 1869, General considerations regarding the physical features of these rivers: U.S. Army, Corps of Engineers, Report of Chief Engineers, 1868, p. 307–314.

Webb, W. P., 1931, The Great Plains: Boston, Ginn and Company, 525 p.

Wells, G. L., 1983, Late-glacial circulation over central North America revealed by aeolian features, *in* Street-Perrott, A., Beran, M., and Ratcliffe, R., eds., Variations in the global water budget: Dordrecht, D. Reidel Publication Company, p. 317–330.

Wheaton, E. E., 1984a, Wind erosion—impacts, causes, models, and controls; A literature review and bibliography: Saskatchewan Research Council Publication E-906-48-E-84, 19 p.

——, 1984b, Climatic change impacts on wind erosion in Saskatchewan, Canada: Saskatchewan Research Council Publication E-906-16-B-84, 27 p.

White, W. N., Broadhurst, W. L., and Lang, J. W., 1946, Ground water in the High Plains of Texas: U.S. Geological Survey Water-Supply Paper 889-F, p. 381–421.

Wigner, K. A., 1984, Dust storms and blowing dust on the Texas South Plains [M.S. thesis]: Lubbock, Texas Tech University, 151 p.

Williams, G. P., 1978a, The case of the shrinking channels; The North Platte and Platte rivers in Nebraska: U.S. Geological Survey Circular 781, 48 p.

——, 1978b, Hydraulic geometry of river cross sections; Theory of minimum variance: U.S. Geological Survey Professional Paper 1029, 47 p.

Windom, H. L., and Chamberlain, C. F., 1978, Duststorm transport of sediments to the North Atlantic Ocean: Journal of Sedimentary Petrology, v. 48, p. 385–388.

Wood, W. W., and Osterkamp, W. R., 1984a, Recharge to the Ogallala Aquifer from playa lake basins on the Llano Estacado (an outrageous proposal?), *in* Whetstone, G. A., ed., Proceedings, Ogallala Aquifer Symposium II: Texas Tech University Water Resources Center, p. 337–349.

——, 1984b, Playa lake basins on the Southern High Plains of Texas, U.S.A.; A hypothesis for their development, *in* Whetstone, G. A., ed., Proceedings of the Ogallala Aquifer Symposium II: Texas Tech University Water Resources

Center, p. 304–311.

Wood, W. W., and Petraitis, M. J., 1984, Origin and distribution of carbon dioxide in the unsaturated zone of the Southerh High Plains of Texas: Water Resources Research, v. 20, p. 1193–1208.

Worster, D., 1979, Dust bowl; The Southern Plains in the 1930's: New York, Oxford University Press, 277 p.

Wright, H. E., Jr., 1983, Late-Quaternary environments of the United States: Minneapolis, University of Minnesota Press, v. 1, 407 p, v. 2, 277 p.

Wright, H. E., Jr., Almendinger, J. C., and Gruger, J., 1985, Pollen diagram from the Nebraska Sandhills and the age of the dunes: Quaternary Research, v. 24, p. 115–120.

Wright, H. E., Jr., and Frey, D. G., eds., 1965, The Quaternary of the United States: Princeton, New Jersey, Princeton University Press, 922 p.

Zimmerman, H. B., Shackleton, N. J., Backman, J., Kent, D. V., Bauldauf, J. G., Kaltenback, A. J., and Morton, A. C., 1984, History of Plio–Pleistocene climate in the northeastern Atlantic, Deep Sea Drilling Project Hole 552 A, *in* Roberts, D. G., and Schnitker, D., eds., Initial reports of the Deep Sea Drilling Project, Volume 81, p. 861–875, Washington, D.C., U.S. Government Printing Office.

MANUSCRIPT ACCEPTED BY THE SOCIETY SEPTEMBER 29, 1986

ACKNOWLEDGMENTS

TransAlta Utilities Limited, and particularly D. Nikols, gave permission to use data collected at the Highvale Mine, in the section of this chapter by M. M. Fenton. Glen Fredlund (University of Kansas), Elaine Wheaton (Saskatchewan Research Council), Henry Epp (Saskatchewan Environment), Peter David (University of Montreal), and Kenneth Wigner (National Weather Service, Lubbock) contributed valuable assistance in the preparation of the section of this chapter by V. T. Holliday.

The contributors—and reviewers—of the preceding chapter responded enthusiastically, and with admirable flexibility, despite rather severe time constraints; the sharing of their experience and insight is deeply appreciated. I think we who conspired in this geomorphic perspective of the Great Plains would agree, however, that the most important contributors are the numerous exceptional investigators who provided the wealth of information upon which these pages are based. To these people, especially, including the many whose works it has not been feasible to cite, a sincere word of thanks is extended—WRO.

Geological Society of America
Centennial Special Volume 2
1987

Chapter 7

Rocky Mountains

Richard F. Madole
U.S. Geological Survey, Box 25046, MS 966, Golden, Colorado 80225
William C. Bradley
Department of Geological Sciences, University of Colorado, Boulder, Colorado 80309
Deborah S. Loewenherz
Department of Geography, University of Illinois, Urbana, Illinois 61801
Dale F. Ritter
Department of Geology, Southern Illinois University, Carbondale, Illinois 62901
Nathaniel W. Rutter
Department of Geology, University of Alberta, Edmonton, Alberta T6G 2E3, Canada
Colin E. Thorn
Department of Geography, University of Illinois, Urbana, Illinois 61801

INTRODUCTION

Richard F. Madole

The Rocky Mountain region is one of the most topographically distinct and impressive parts of North America. The Rocky Mountains rise abruptly above the bordering regions, particularly on the east and northeast where they are flanked by plains, less so on the west and southwest where they are bounded by high plateaus. The Rocky Mountains comprise more than 100 individually named ranges that form a belt extending for slightly more than 5,000 km, from near Santa Fe, New Mexico, on the south to the Bering Sea on the north (Fig. 1). The belt varies in width from less than 100 km in the Canadian Rockies to nearly 600 km in the Middle Rockies of Wyoming and northeast Utah. The summits of the ranges rise 1,500 to 2,100 m above adjacent lowlands, to heights 1,800 to 4,400 m above sea level. The Southern Rockies of Colorado have the greatest amount of area, between 3,300 and 4,400 m, and the highest peak, Mount Elbert (4,400 m). The largest area of low mountains is in the Northern Rockies of Idaho and Montana, where summits are commonly only 2,100 to 2,400 m above sea level.

A substantial part of the Rocky Mountain region consists of lowlands, in the form of basins and fault-bounded troughs and trenches that lie between ranges. The Rocky Mountain Trench is perhaps the most spectacular fault-bounded lowland, even if it is not the most representative. It extends north from Flathead Lake, Montana, more than 1,500 km, and forms the western boundary of the Canadian Rockies. At the south end of the region, the Rio Grande rift extends northward and includes the broad, intermontane San Luis Valley and the narrower trough in which the upper Arkansas River flows. Farther north in Colorado, troughs give way to intermontane basins such as South Park and North Park, which are characteristic of the Southern and Middle Rockies and the Wyoming Basin. Basins resembling those of the Basin and Range Province, commonly referred to as Tertiary basins, occupy a substantial part of the area of the Northern Rockies of southwest Montana and part of Idaho.

The continuity of the Rockies is broken in only a few places: the Wyoming Basin, the Liard Plateau, and the basins of the Porcupine and Peel rivers (Fig. 1). These, and partial breaks such as the Snake River Plain, are the principal bases for dividing the region into several physiographic provinces. These provinces are referred to in this chapter mostly by traditional names; they include the Southern, Middle, and Northern Rockies of the United States, the Canadian Rockies, and the Arctic Rockies (Fig. 1).

A more detailed description of this vast and complex region is neither possible nor desired here; although a summary of the geologic framework and history has been included as necessary background for the sections that follow. Our purpose in this chapter is to consider the geomorphic processes that have shaped the surface morphology of these mountains, in particular to review the concepts that have evolved and the advances that have been made in geomorphology as they apply to this region. Of necessity, no attempt is made at an exhaustive survey of the region or its literature. Instead, we focus on those topics that have received the most study and on selected areas where important data, evidence, and insights have been developed. It should be emphasized, however, that the areas discussed are not the only ones that have contributed significantly to understanding the

Madole, R. F., Bradley, W. C., Loewenherz, D. S., Ritter, D. F., Rutter, N. W., and Thorn, C. E., 1987, Rocky Mountains, *in* Graf, W. L., ed., Geomorphic systems of North America: Boulder, Colorado, Geological Society of America, Centennial Special Volume 2.

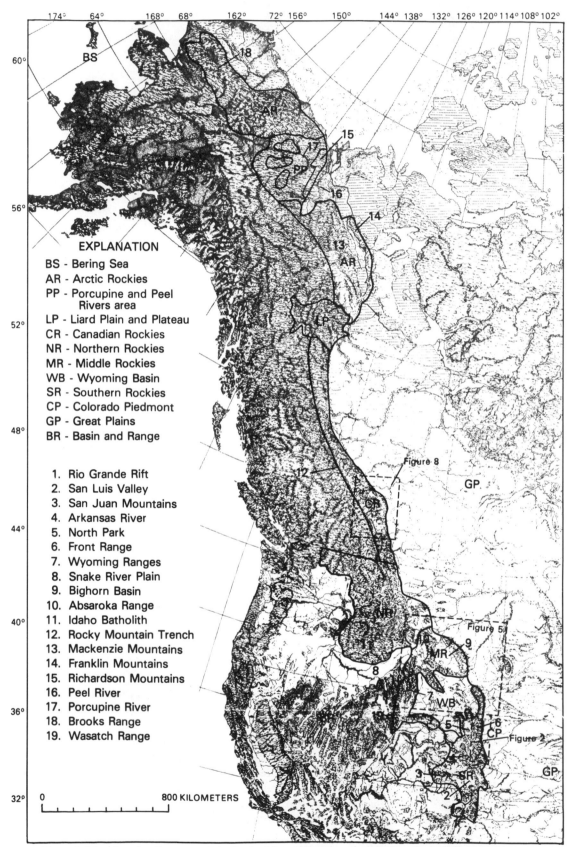

Figure 1. Map showing the extent of the Rocky Mountain region, the location of principal subdivisions and key areas mentioned in the text.

geomorphology of the Rocky Mountains. The topics and areas discussed are (1) erosion surfaces in the Front Range of the Southern Rockies, (2) fluvial processes in the Bighorn Basin of the Middle Rockies, (3) glacial processes in the central Canadian Rockies and the Jasper and Banff areas, and (4) alpine mass wasting in the Indian Peaks section of the Front Range.

The first section reviews the work done on the erosion surfaces of the Front Range in Colorado. These surfaces have attracted the attention of geomorphologists from the earliest days of research in the Rockies to the present. Whereas interest in peneplains has waned, interest in the erosion surfaces of the Rockies has continued, largely because they are fundamental to understanding the tectonic history of the region. The second section is devoted to a review of fluvial processes. Regional uplift that culminated in latest Miocene and Pliocene time initiated widespread fluvial degradation and canyon cutting that has continued to the present. Periodically during Quaternary time, degradation has been interrupted by intervals of aggradation, the record of which is preserved best in the lowlands between ranges. The Bighorn Basin is the largest and best studied of the intermontane lowlands and, as such, serves as a model in this chapter. Glacial processes are considered in the third section. Much of the scenery of the Rockies, from its southern end to the Brooks Range of Alaska, is the product of glaciation, although most of the ice that sculpted this landscape is gone. The imprint of vanished glaciers is greatest in the Canadian Rockies; hence, our consideration of glacial processes is centered on this part of the region. Finally, in the fourth section, we return to the Front Range of the Southern Rockies for a review of alpine mass wasting and the evolution of alpine landscapes.

The sections on alpine mass wasting and erosion surfaces, in a sense, represent end members in a continuum of research. The latter includes some of the earliest work in the region, done at a time when attention was on erosion and the long-term evolution of landforms. These earlier studies commonly were done on a grand scale, spatially and temporally. On the other hand, the discussion of alpine mass wasting is based chiefly on research done in the past two decades, a time when emphasis had shifted to detailed studies of process. These studies are commonly done over small areas, some no larger than an individual slope, and are concerned with time spans measured in decades or less. The other two sections contain elements of both end members and reflect the shift in emphasis from ancient forms and landscapes to modern processes and the interactions of energy and matter.

Mass movements, other than those forms associated with alpine environments, will not be included in this chapter, because the topic is treated more completely in another volume in this series. Weathering and pedogenesis also will not be discussed, because data pertaining to these topics are too sparse and widely scattered. Also, the many canyons and drainage anomalies of the region are not dealt with separately, but are described briefly here because of their historical importance. In numerous places, especially in the Southern and Middle Rockies, rivers cut across uplifts cored by resistant rocks in preference to what appear to be more logical courses on softer rocks around the uplifts. Early on, these drainage anomalies and associated canyons influenced thought on the interaction of erosion and tectonism and the concepts of antecedence and superposition (Powell, 1875; Gilbert, 1877), base level (Powell, 1875), and stream piracy (Gilbert, 1877). Their formation is an important part of the late Cenozoic history of the region.

GEOLOGIC FRAMEWORK AND HISTORY

Different parts of the region have different geologic histories, which account for fundamental differences in rocks, structures, and scenery from one area to another. Some of the more important aspects of regional geology and geologic history are described here, although it should be noted that brief summaries like this inevitably and necessarily overgeneralize. Therefore, the reader is cautioned that differences in detail between areas within the region are considerable and that conclusions drawn from the various sections of this chapter do not necessarily apply equally to all parts of the region.

As a region, the Rocky Mountains owe their identity chiefly to similarities of topography. The one thing that the various subdivisions have in common is high relief; otherwise they are characterized by diversity. There are marked differences in geology, and the differences in climate and vegetation from one end of the region to the other are profound. Compared to most other regions, even the differences from foothills to summits are exceptional. At the south end of the Rocky Mountains, forests span an altitudinal range of as much as 2,000 m, giving way to alpine tundra at their upper limit and to vegetation characteristic of desert and semiarid areas at their lower limit. However, in the Brooks Range, at the northern end of the region, slopes are largely treeless, and tundra extends from foothills to summits. The imprint of glaciation exists throughout the region, but from Glacier National Park in Montana northward, it dominates the landscape.

The Rocky Mountains include fault-bounded uplifts, folded mountains, and highlands formed by volcanism. The ages and kinds of rocks composing these ranges are equally varied. Effects of the Laramide orogeny, however, are common to the region as a whole, and the effects of a post-Laramide episode of block faulting and uplift are also widespread. Although the exact timing and character of Laramide deformation differed from place to place, the various subdivisions of the region were deformed sometime between middle Cretaceous and late Eocene; the most intense faulting and folding generally occurred in Paleocene time. Laramide deformation involved horizontal compression that shifted from east-west to northeast-southwest to north-south as the orogeny progressed (Gries, 1983). The timing and shift in direction of compression accounts for differences in the orientation of individual uplifts. A period of quiescence followed the Laramide orogeny. Then, in Neogene time (Miocene and Pliocene), widespread block faulting and uplift of major proportions produced much of the topographic relief seen in the region today.

In contrast to Laramide and post-Laramide history, differences in the pre-Laramide history between the Southern and Middle Rockies and the rest of the region are great, and give rise to much of the geologic and scenic diversity in the region. The Canadian Rockies and Northern Rockies of the United States formed along the eastern edge of an area that had received a great thickness of sediment, more than 30,000 m, from late Precambrian to early Mesozoic time. Similarly, the Brooks Range is in an area that was the locus of sedimentation during this time. On the other hand, from late Precambrian to early Mesozoic time, the area of the Southern and Middle Rockies was part of a stable platform and received only about 1,500 m of sediment. Later, especially during Cretaceous time, the area of the Southern and Middle Rockies received much more sediment, but nothing comparable to that received earlier in the northern part of the region. Moreover, most of this sediment was clastic, whereas in the Northern and Canadian Rockies it was predominantly carbonate.

Most of the ranges of the Southern and Middle Rockies are elongate anticlinal uplifts separated by broad intermontane basins. An exception to this pattern is the relatively narrow band of ranges along the western edge of the region that includes the Wasatch Range and the Wyoming Ranges (Fig. 1). These ranges are composed of complexly folded and thrust-faulted Paleozoic rocks and, in this respect, are similar to the ranges of the Basin and Range Province. As uplift proceeded during Laramide time, the relatively thin cover of sedimentary rock was stripped from the uplifts of the Southern and Middle Rockies long before orogeny had ended, and cores of Precambrian crystalline rock were exposed over broad areas. Later, volcanic activity played a role in forming mountains and high plateaus in many places in the Southern and Middle Rockies, the most notable examples being the San Juan Mountains in Colorado and the Absaroka Range in Wyoming (Fig. 1). The Southern and Middle Rockies are two provinces because the Wyoming Basin separates them topographically, even though the rocks and structures of the two provinces connect beneath the basin.

The ranges of the southern part of the Northern Rockies are characterized by block faulting similar to that of the Basin and Range Province. Farther north, however, and continuing throughout the Canadian Rockies, folding and complex, imbricate-thrust faulting of thick sequences of sedimentary rock eastward created a series of parallel sharp-crested ranges separated by narrow, linear valleys. In contrast to the broad uplifts of the Southern and Middle Rockies, Precambrian basement rock was not brought to the surface, and most of the Precambrian rock present is sedimentary. The jointing and stratification of the sedimentary rocks and the pronounced linearity produced by imbricate thrusting combine to provide much of the distinctive scenery for which the Canadian Rockies are known. The Brooks Range was also strongly folded and complexly thrust faulted as well as intruded by molten masses and metamorphosed.

Much of central Idaho is made up of separate mountain groups (Salmon River, Clearwater, Sawtooth, Coeur D'Alene)

that were formed by dissection of the plateaulike surface of the Idaho batholith. These mountains lack the north-south orientation characteristic of most ranges in the region, and they are also generally lower. The Idaho batholith is one of the largest in North America and is of Cretaceous age; the Boulder batholith of Montana and other smaller intrusives are related to the Idaho batholith. Batholiths of Mesozoic age are not characteristic of the Rocky Mountains, but they are common in regions to the north and west of the Idaho batholith.

Most of the relief seen today in the Rockies is the product of post-orogenic uplift that occurred in late Tertiary time, the Neogene block faulting and uplift described above. This uplift set the stage for the geomorphic evolution of the Rocky Mountains. It fragmented and displaced the post-Laramide erosion surface, initiated a long interval of fluid degradation that continues today, produced ridges and plateaus high enough to be glaciated, and in part established the regional pattern of major slope failure. It also created the many superposed streams and canyons of the region.

The southern part of the region has provided the most detailed documentation of the history of late Tertiary uplift. Prior to the late Oligocene uplift, the late Eocene erosion surface, which was widespread in parts of the Southern Rockies, sloped gently eastward across what is today the southern Front Range and adjoining Colorado Piedmont. Its altitude is estimated to have been about 900 m in the vicinity of Florissant, Colorado—site of Oligocene lake deposits that are critical to reconstructing the environment and altitude of the time—and slightly lower over what was to become the west edge of the Colorado Piedmont (Trimble, 1980). Today the altitude of the Piedmont in this area is about 1,800 m.

Uplift began in the Rocky Mountain region in late Oligocene time and continued into Pliocene time. Sediment derived from the rising ranges was deposited in the many intermontane basins and troughs, as well as being carried eastward and deposited on the Great Plains; hence, Miocene rocks are widespread in these areas. In late Miocene and Pliocene time, uplift accelerated. Miocene rocks deposited during earlier uplift were themselves displaced by normal faults, commonly by hundreds of meters. Most of the late Tertiary uplift occurred after the end of deposition of the Ogallala Formation on the Great Plains about 5 to 6 Ma (Trimble, 1980).

The post-Ogallala uplift was accompanied by major canyon cutting and denudation in the uplands and by the excavation of the generally soft Miocene rocks in the adjoining basins. During basin excavation, streams that had established courses on the Miocene fills were let down in many places onto older structures composed of more resistant rocks. Consequently, canyons were cut in seemingly anomalous locations, many completely across the axes of exhumed mountain ranges. In places, faulting has continued into Quaternary time and, possibly, regional uplift may still be in progress. The endogenic processes that culminated in Pliocene time set the stage for the landforms and processes discussed in the following sections.

EROSION SURFACES OF THE COLORADO FRONT RANGE: A REVIEW

William C. Bradley

INTRODUCTION

From the Appalachians to the Coast Ranges, many American mountain ranges possess high-level topography that is flat or gently rolling, out of harmony with the steep-sided canyons and glacial valleys that indent it. All who have noted this topography have wondered about it: how was it produced, what is its age, what has happened to it since it formed? These questions have been addressed by a sizable literature, much of it contributed by members of the Geological Society of America (GSA). Hence, what could be more appropriate at the time of the GSA centennial than to review the century of thinking concerning these common topographic features?

In a general way, the thinking evolved in three developmental phases. Phase One was conservative in nature. The high-level topography was interpreted as the remains of a single ancient erosion surface (most likely a peneplain), developed by stable rivers, and abandoned when uplift caused rivers to carve canyons into it. Estimates of age were limited. Phase Two was far more exuberant. Multiple erosion surfaces became the accepted interpretation; indeed, the challenge seemed to be: who could recognize the greatest number of peneplains? The activity was too frenetic to last. Phase Three was one of interpretive exhaustion, sparse literature, and retrenchment back to a single surface (commonly a pediment).

The phases followed broadly similar timetables in different parts of the country because of the popularity of certain ideas at certain times. Timetables were not identical, however. New ideas were tested in the Appalachians before they were exported to western mountains.

Perhaps because it provided the first view of mountains for eyes weary from travel across the Great Plains, or perhaps because it was so clearly visible in the rarefied atmosphere, the Colorado Front Range captured the attention of those interested in mountain topography. It stands foremost among the western ranges in the volume of literature dealing with its erosion surfaces. A review of this literature follows.

COLORADO FRONT RANGE

The Front Range (Rampart Range included) begins at the Arkansas River and extends northward 300 km to the Wyoming state line where it becomes the Laramie Range (Fig. 2). Its drainage divide, in places coinciding with the Continental Divide, is asymmetrically positioned toward the western side. Indeed, the Arkansas and South Platte rivers flow through the Front Range from sources farther west (Fig. 2). The highest topography is also asymmetrically disposed toward the west, with the exception of outliers such as Pikes Peak.

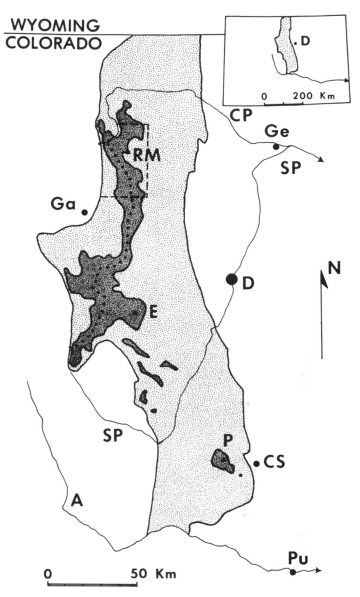

Figure 2. Generalized outline of the Colorado Front Range (light pattern) and its area above treeline (dark pattern; treeline elevation is 3,450 m), plus three of its largest rivers: A, Arkansas; SP, South Platte; and CP, Cache la Poudre. Communities: D, Denver; Ge, Greeley; CS, Colorado Springs; Pu, Pueblo; and Ga, Granby. Mountains: E, Evans; P, Pikes Peak; and RM, Rocky Mountain National Park. Line of dots: Continental Divide.

Like other elements of the Southern Rocky Mountains, the Front Range is a structural uplift involving both folding and faulting. Deformation occurred during the Laramide orogeny of Late Cretaceous to early Tertiary age. Structural basins were formed at the same time on both flanks of the uplift. Approximately 3 km of Paleozoic and Mesozoic sedimentary rocks overlay Precambrian crystalline rocks prior to uplift. Erosion during and after the Laramide has removed the sedimentary and some of

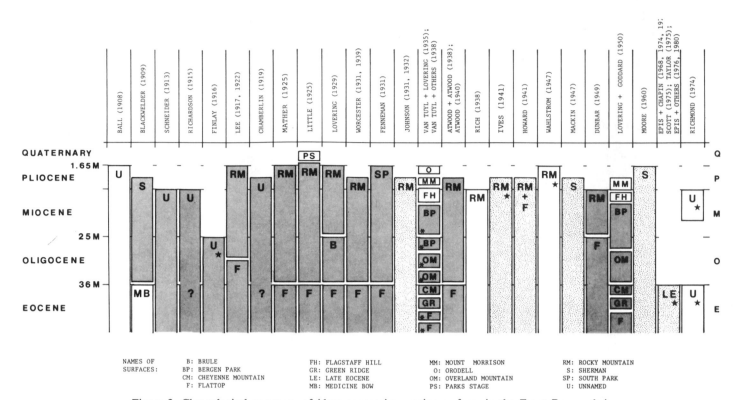

Figure 3. Chronological summary of ideas concerning erosion surfaces in the Front Range: their number, names, ages, and genesis. See text for explanation of boxes. Genetic types of surfaces are: peneplain (dark pattern), pediment or pediplain (light pattern), and "surface" (non-genetic or multi-genetic: no pattern). Stars, single surfaces that were faulted or warped so as to occur at multiple levels; Asterisks, those surfaces that carried "upper" and "lower" designations; Question marks, where a vague suggestion was made about an older, higher surface. No age is given for the Miocene-Pliocene boundary because most of the literature dates from a time when the Pliocene was considered longer than it is at present.

the crystalline rocks from the range. The sedimentary rocks survive, however, in the flanking basins where they are commonly protected by a cover of Cenozoic rocks.

Degradational processes have prevailed in the Front Range throughout most of Cenozoic time. Degradational rates have varied, however. Broad, regional uplift in late Tertiary time, accompanied by extensive block faulting in the southern part of the range, rejuvenated rivers to create canyons within the range and widespread, but gentle, erosional topography in the basins. Thus, the strong topographic contrast between the range and flanking lowlands is partly because the Precambrian crystalline rocks are more resistant to erosion than most of the Phanerozoic sedimentary rocks, and partly (in the southern area) because of the late Tertiary block faulting. (The above history is a distillation from many sources, especially Curtis [1975] and Trimble [1980].)

From the foregoing, we can make three points pertinent to erosion surfaces in the Front Range. (1) They are cut chiefly on Precambrian granitic and metamorphic rocks. (Erosional features in the flanking lowlands are not treated here; for these, see the paper that follows by Ritter [this chapter].) (2) They can be no

older than Tertiary because they are carved into rocks deformed during the Laramide orogeny. At the southern and northeastern margins of the range, there are local surfaces produced by Cenozoic exhumation of pre-Tertiary unconformities (Cross, 1894; Scott, 1975; Scott and Taylor, 1986). These surfaces are minor in extent, however, and are not discussed herein. (3) They show a broad relationship to crystalline lithology. The surfaces are flattest and most extensive in those areas where granitic rocks are most abundant: in a southern area around Pikes Peak (Pikes Peak Granite, approximately 1.0 billion years old), and in a northern area that extends onto the Laramie Range (Sherman Granite and associated intrusives, all approximately 1.4 billion years old). Between these two areas, metamorphic rocks are relatively more abundant and the erosion surfaces have greater local relief.

EROSION SURFACES

Phase One

Marvine (1874, p. 89), assistant geologist with the 1873

Hayden Survey, described the Front Range topography as follows:

. . . the mountain-zone lying between the main divide and the plains certainly impresses one as being, with few exceptions, a region of very uniform or gently undulating general elevation, carved by the powers of erosion, perhaps partly glacial but mostly by streams . . . while standing on one [of the ridges] the general level, which seems indicated in their tops, is very striking.

Marvine was looking at the forested part of the Front Range that lies east of the high tundra country (Fig. 2), occurring generally between elevations of 2,400 and 3,000 m. Is there any doubt that he recognized the existence of an ancient erosion surface? Fill up the canyons and you have an open, gently rolling landscape. Phase One had not waited for GSA to be incorporated.

Cross (1894), Emmons and others (1896), Fenneman (1905), and Davis (1911) supported Marvine. Note the cautious way in which all these authors treated the erosion surface. They recognized but a single surface; they felt no compulsion to give it a name; they avoided estimating its age; and they ignored the matter of genesis, other than relating it to streams—an exception was Davis (1911) who called it a peneplain.

Phase One then gradually quickened in pace; articles appeared more frequently and caution lessened. From Ball (1908) on, authors estimated ages for their surfaces, making it easy to compare ideas (Figs. 3 and 4). Each box in Figure 3 represents a discrete interval of time during which erosion produced the surface indicated. A typical interval consisted of certain events that produced certain topographic features. Uplift was the trigger; it was concentrated early in the interval, providing rivers with the vigor to carve canyons. As the interval progressed, downcutting slowed and canyons gave way to shallow, open valleys separated by gently rolling divides: the erosion surface. Renewed uplift then terminated that interval and its topographic development, and inaugurated a new one. Thus, it was recognized early that erosion surfaces must bear some relationship to the deformational history of a range.

Note in Figure 3 that the boxes in the lowest tier lack bottoms. This is because the early Cenozoic is not shown. Understand, too, that the sharp line at the top of each box is deceptive. It suggests the surface was precisely dated, whereas, in truth, most authors equivocated greatly about age. At the extreme, Johnson (1931, 1932) and Howard (1941) did not even discuss age; the ages cited for them are middle-of-the-road estimates based on the thinking of the day.

Ideas concerning the erosion surface had begun to crystallize. Writers late in Phase One generally supported the established thinking, although a few suggested some departures. Schneider (1913), Richardson (1915), and Finlay (1916) followed Davis (1911) in calling the surface an unnamed peneplain, but they went a step beyond in suggesting ages. Finlay (1916) also added an important new idea: that a single surface might later be faulted into multiple levels, which could be confused with surfaces of multiple ages. Blackwelder (1909) previewed Phase Two by rec-

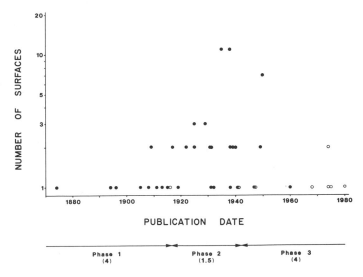

Figure 4. Semi-log plot of number of erosion surfaces versus date of publication. Data are from Figure 3 plus Marvine (1874), Cross (1894), Emmons and others (1896), Fenneman (1905), and Davis (1911). Open circles, surfaces given a multi-level appearance by later deformation. Number in parentheses below each phase shows the average number of years to produce a publication during that phase. Phase boundaries are arbitrary.

ognizing and naming two surfaces on the southern Laramie Range (state boundaries are no obstacle to erosion surfaces; his Sherman surface extends smoothly onto the northern Front Range).

Phase Two

Phase Two exploded out of its predecessor with papers by Lee (1917, 1922). He was the single most influential writer on Front Range erosion surfaces, as evidenced by the following: (1) He recognized two peneplains of different ages, an interpretation that remained popular for a quarter of a century. (2) He gave the surfaces names that were adopted by most subsequent workers—Flattop, for the smoothly rolling tundra topography in Rocky Mountain National Park (Fig. 2), and Rocky Mountain, for the broader surface below treeline first described by Marvine (1874). (3) He estimated ages for the surfaces that were broadly supported by his followers. Lee was a major contributor to what became the conventional wisdom of the 1920s and 1930s regarding erosion surfaces in the southern and middle Rocky Mountains: a late Eocene peneplain at high elevation and a late Tertiary peneplain at lower elevation.

With the exception of Chamberlin (1919), a throwback to Phase One, Lee's followers are perfectly evident on Figure 3—Mather (1925), Worcester (1931, 1939), Fenneman (1931), and the Atwood family (1938, 1940)—plus those whose greater courage or improved visual acuity enabled them to recognize three erosion surfaces: Little (1925) and Lovering (1929). There were nonconformists, as well: Johnson (1931, 1932), who was the first to suggest that the Rocky Mountain surface was a pediment whose individual segments could have been formed at different levels.

Phase Two reached its climax with papers by Van Tuyl and Lovering (1935) and Van Tuyl and others (1938), wherein 11 erosion surfaces were recognized: eight peneplains and three lesser surfaces. Reactions varied; the ideas were vigorously disputed by Rich (1935a, 1938), ignored by Atwood and Atwood (1938), and accepted, with modifications in dating, by Stark and others (1949). Alternatives were quickly suggested. Atwood and Atwood (1938), Worcester (1939), and Atwood (1940), plus Dunbar (1949) in Phase Three, appealed for a return to the conventional wisdom of Lee and others. The old paradigm, however, was already fading in popularity. Under way was a retrenchment to a single surface that was something other than a peneplain. Rich (1935a, 1938) argued that only one surface existed, that multiple genetic factors endowed it with multiple levels, and that the Flattop was not an old, cyclical erosion surface at all. Ives (1941), and Wahlstrom (1947) in Phase Three, proposed faulting or warping as a way to convert a single surface into one with multiple levels, recycling an idea that had lain dormant for 25 years (Finlay, 1916). Howard (1941) made contemporary pediments out of both the Flattop and Rocky Mountain surfaces (he used the terms "summit" and "subsummit" for these surfaces).

Phase Three

Publishing withered but did not die. Wahlstrom (1947) used volcanic rocks to show that the Flattop surface was late Tertiary in age, not early Tertiary as had previously been claimed; this dating was a key argument in his plea for a single, deformed surface. Mackin (1947) and Moore (1960) referred to the Sherman surface of Colorado-Wyoming as a pediment. Mackin (1947) made an additional point of importance: that the Flattop surface was not an ancient, cyclical erosion surface at all (agreeing with Rich [1938]), but instead was a modern surface created by periglacial processes operating above treeline during Quaternary time. Lee's (1917, 1922) paradigm was halved and its surviving member genetically altered.

Following Mackin (1947) came an echo from Phase Two in its prime: Lovering and Goddard (1950). Fifteen years of attrition had taken its toll: 11 surfaces had been consolidated to seven, and the term "peneplain" was scarcely mentioned. Except for Moore (1960), little was said for almost two decades. It was as if people despaired of ever solving the erosion-surface problems.

Resolution came in the 1960s through a U.S. Geological Survey project in the area surrounding Pikes Peak (Epis and Chapin, 1968, 1974, 1975; Scott, 1975; Taylor, 1975; Epis and others, 1976, 1980; Scott and Taylor, 1986). These men used Tertiary volcanic rocks to (1) prove the existence of a single surface, (2) date the surface as late Eocene and correlate its remnants, and (3) document the kind and amount of post-surface deformation (thereby confirming Finlay's [1916] early belief). A

single surface of late Eocene age had already been proposed by Knight (1953) for the Medicine Bow Mountains of Colorado-Wyoming, just west of the Front Range.

Finally, Richmond (1974) reiterated a two-surface model, but one that differed from all previous two-surface models. His lower surface was the older, assumed to be the late Eocene surface of farther south; his higher surface (Flattop of Lee, 1917, 1922) was dated late Tertiary by volcanic rocks, consistent with Wahlstrom (1947). Both surfaces were believed to have been displaced by faulting.

Summary

Figures 3 and 4 illustrate the evolution of ideas concerning erosion surfaces in the Front Range. Phase One favored a single surface, either one with unspecified genesis or a peneplain. Publishing was slow; it took an average of four years to produce a paper. Phase Two was characterized by multi-age peneplains and a rush to the publishers: 1.5 years to produce a paper. Phase Three brought retrenchment back toward a single surface (a pediment or one of complex origin), and again four years per paper. Also shown is a recycling of Finlay's (1916) early notion that faulting can give a single surface a multilevel appearance.

All observers of Front Range topography have accepted the validity of at least one ancient erosion surface, even though they have disagreed on age, genesis, and whether or not there is more than one. This consensus surface is widespread, and has been called Rocky Mountain, late Eocene, Sherman, South Park, and subsummit, plus some subdivision names.

The Flattop (summit) surface of the tundra country is the one whose significance has been most disputed. Many authors accepted it as a surface separate from the Rocky Mountain (Fig. 3); others considered it an original part of the Rocky Mountain surface (Ball, 1908; Davis, 1911; Finlay, 1916; Ives, 1941; Wahlstrom, 1947; Scott, 1975); still others claimed it was not even a legitimate surface (Rich, 1938; Mackin, 1947); and some ignored it altogether. Volcanic dating has clarified a part of the controversy. If high shoulders on Pikes Peak are indeed upfaulted remnants of the late Eocene surface (Epis and Chapin, 1975, p. 63; Scott, 1975, p. 233—a belief subsequently questioned by Scott and Taylor, 1986), then they cannot be correlated with the Flattop surface in Rocky Mountain National Park because the latter is late Tertiary in age (Wahlstrom, 1947; Richmond, 1974). The one point of agreement concerning the Flattop surface is that it has been strongly affected by Quaternary periglacial activities (for more on this topic, see the paper by Thorn and Loewenherz later in this chapter).

EROSION-SURFACE PROBLEMS

Erosion surfaces are difficult features to work with. This is evident from Figures 3 and 4, where the same evidence has produced more interpretive discord than harmony. Two groups of problems are identifiable.

1. Dating is at the heart of erosion-surface studies. If dating is good, other problems can be solved: the number of surfaces, their correlation, and their subsequent deformation. If dating is poor, the other problems have solutions of equal quality. This is the single factor most responsible for the disparate views shown in Figures 3 and 4.

Being erosional, surfaces have little to date. They must, of course, be younger than the youngest rocks and structures they truncate. This makes the erosion surfaces post-Laramide in age, that is, Eocene or younger. All Front Range workers have agreed on this point (Fig. 3).

The best dating of surfaces in the Front Range has come from studies of volcanic rocks that overlie or underlie the surfaces. This approach has been most successful in the southern part of the range (Epis and Chapin, 1968, 1974, 1975; Scott, 1975; Taylor, 1975; Epis and others, 1976, 1980), but it has also been used in and to the north of Rocky Mountain National Park (Wahlstrom, 1947; Richmond, 1974; Scott and Taylor, 1986).

Before radiometric dating was available, correlation of surfaces was a shaky endeavor. Established practice was to accept a correlation if flat areas, spurs, ridge crests, and hilltops had an accordance in elevation. We know better now, and only use elevation in conjunction with other evidence.

Recognition of multiple surfaces was a direct by-product of correlation by elevation. If one group of flattish features, all lying at nearly the same elevation, proved the existence of one surface, a second such group at a different elevation demonstrated a second surface. Such an interpretation may, of course, be correct. However, there are two other ways to create the same result: (a) a single surface may have multiple levels at the time of formation (Johnson, 1931, 1932; Rich, 1938; Howard, 1941; Mackin, 1947); or (b) a single surface may later be deformed into multiple levels (shown so well by Epis and his colleagues). Unless the latter two situations can be dismissed, use of multiple accordances to prove multiple surfaces is unacceptable.

2. A second group of major problems has to do with genesis. Figure 3 shows the evolution in fashionable thinking of the time: the peneplain era followed by the pediment era. This change was more than a trivial dispute in semantics. It had a profound bearing on ideas concerning Cenozoic deformation, as will be explained in the section "Structural Implications."

The problems of genesis remain largely unresolved. Most workers have been content to describe and date surfaces; few have considered the formative mechanisms. To simply call a surface a "pediment" gives little genetic insight, because there have been as many different suggestions about how pediments are formed as was true earlier in the century with peneplains.

There is more to subaerial denudation than just the processes involved in pedimentation or peneplanation. Rich (1935a, 1938) suggested that a single surface could have multi-genetic controls, one of which might be lithology. A number of workers have noted a relationship between topographic form and crystalline lithology (Ball, 1908, p. 35–36; Davis, 1911, p. 32; Eggler and others, 1969; Scott and Taylor, 1986). Rich's multi-genetic

approach has not been given the attention it deserves. In addition, other denudational processes that ought to be considered in the Front Range are those producing stepped topography (Wahrhaftig, 1965) and etchplains (see Adams, 1975, and references therein).

STRUCTURAL IMPLICATIONS

For decades, geologists used erosion surfaces as a clue to the post-Laramide deformational history of the middle and southern Rocky Mountains. Because peneplains were believed to form at low elevations and with low river gradients, substantial uplift was required to bring them to their present elevations, as well as local tilting to create their steep gradients (Davis, 1911, p. 41–42; Chamberlin, 1919, p. 162, 238). Using this kind of evidence, late Cenozoic uplift was estimated to be in the range of 1,500 to 2,700 m (Davis, 1911; Chamberlin, 1919; Dunbar, 1949, p. 406–407). Reclassifying peneplains as pediments eliminated the need for tilting and greatly reduced the amount of uplift required (Johnson, 1931, 1932; Mackin, 1947). Thereafter, uplift was estimated from displaced flora and fauna, for example, using the Florissant Lake Beds (MacGinitie, 1953; and Leopold and MacGinitie, 1972—summarized in Epis and Chapin, 1975, p. 50, 57). The curious irony is that the magnitude of uplift based on paleontology is approximately the same as it was when based on peneplains.

REMAINING PROBLEMS

Although volcanic dating has reduced much of the controversy shown in Figures 3 and 4, it has not eliminated it altogether. Three problems remain. Two have already been identified: the detailed genesis of the surfaces, and the full significance of the tundra (Flattop) surface. The third concerns the age of the consensus surface.

Workers in the southern part of the range have named the surface for its age: late Eocene (Epis and his colleagues). Those in the north have dated the surface (Sherman) as late Tertiary (Blackwelder, 1909; Mackin, 1947; Moore, 1960). In the south, volcanic rocks in earliest Oligocene time buried a surface cut on Precambrian rocks that was as widespread and flat as any surface presently seen in that area. Epis and his colleagues have argued that all the surfaces now seen there are the late Eocene surface—exhumed, dissected, and faulted, but otherwise little changed from its original shape.

In the north, the Sherman surface has been equated in age with the youngest sediments derived from it that accumulated immediately to the east. (These sediments were once called Pliocene in age; they are now called late Miocene. I sidestep this change by using the term late Tertiary.) Volcanic rocks have not been successful in defining or dating the Sherman surface because, although they exist in northern Colorado (Scott and Taylor, 1986), they are scarce, and the topography beneath them appears to have considerable local relief (W. A. Braddock, University of Colorado, oral communication, 1986).

The southern workers have rejected a late Tertiary age for the Sherman surface; they have opted instead for late Eocene (Scott and Taylor, 1986; R. C. Epis, Colorado School of Mines, oral communication, 1986; Colman, 1985, has agreed with this view). Their conclusion was based on a belief that the surface, so well dated in the south, can be traced over a broad region, including onto the adjacent Medicine Bow Mountains (Knight, 1953). They have decided, then, that the present Sherman surface is a little-changed remnant of a surface that existed in late Eocene time.

This whole matter raises two interesting questions. (1) How does one date a surface? (Not the dating technique; rather, how does one decide when a surface was formed?) Most authors in Figure 3 concluded that each surface was produced at the end of the associated erosional interval. For example, Lee (1917, 1922) dated his Rocky Mountain peneplain as end of the Pliocene. On the other hand, Epis and his colleagues, and Finlay (1916) before them, have suggested that a surface can be produced early in an erosion interval, that it can exist with insignificant change to the end of that interval, and that its age, therefore, is when it was first produced (R. C. Epis, Colorado School of Mines, oral communication, 1986). To emphasize this point, in the southern area the late Eocene surface had achieved maximum development by the time it was buried by Oligocene volcanic rocks. In nearby areas where it was not so protected, it existed as a subaerial surface throughout Oligocene time, after which faulting and uplift caused dissection. In the northern area, topographic stability evidently lasted for a longer time, probably because late Cenozoic block faulting did not affect that area to the extent that it did in the south (Colman, 1985). If Epis and his colleagues are correct, the Sherman surface, already cut by the end of Eocene time, survived without much change, and probably without burial, until late Tertiary time.

(2) A natural second question is: how much *has* the late Eocene surface been changed? Scott and Taylor (1986) reported that "close to 90 percent of the surface has been modified or destroyed." Let us ignore the areas where dissection has destroyed the surface and concentrate on the in-between areas; how much has the surface been lowered in these areas? Scott (1975, p. 233) concluded ". . . less than a few tens of meters . . .". Lovering and Goddard (1950, p. 15) believed ". . . not more than a few hundred feet. . . ."

Tertiary sediments east of the mountains contain detritus from Precambrian source rocks (Smith, 1940; Moore, 1960; Denson and Bergendahl, 1961; Denson and Chisholm, 1971; Pearl, 1971; Stanley, 1971, 1976; Izett, 1975). This indicates Tertiary erosion of the Front and Laramie ranges in post-Eocene time. Smith (1940, p. 92–94) used such evidence to estimate 100 m of surface lowering in late Tertiary time; he recognized, however, that the detritus might have come from canyon cutting rather than surface lowering, and Scott (1975, p. 232) concurred.

The area of the Sherman surface may differ from the region

to the south. It has evidently been exposed and topographically stable for more of Tertiary time, it is less dissected, and one of its granites weathers totally to grus (Eggler and others, 1969). These factors would favor lowering of the whole Sherman surface during much of Tertiary time. How much it might have been lowered between the late Eocene and late Tertiary, is a question without an answer. If the lowering has been "a few hundred feet," should we still call the Sherman a late Eocene surface? What if the lowering has been 1,000 ft? The real question being asked here is: How much lowering of the late Eocene surface is permissible for it still to be called the late Eocene surface, and beyond which it would be called a younger surface? Agreement is lacking on this fundamental point.

CONCLUDING REMARKS

Most people now believe that a single, widespread erosion surface exists in the Colorado Front Range. The names most often given to it have been: Rocky Mountain, late Eocene, and Sherman. The only reliable dating has come from associated volcanic rocks in the southern area, where a late Eocene age has been demonstrated by Epis and his colleagues. Questions remain about how the surface was produced and how much it has been lowered in post-Eocene time. A question also remains concerning the full significance of small surfaces in the tundra country that have been called Flattop. One presumes that all these questions will be answered by the time GSA has its second centennial.

This consensus surface is believed to have regional significance far beyond the Front Range (Mackin, 1947; Epis and Chapin, 1975; Scott, 1975; Colman, 1985; Scott and Taylor, 1986). Its regional extent may be even greater than these men thought (Potochnik and Damon, 1986). It may also transgress time (Potochnik and Damon, 1986); if so, the name late Eocene may be inappropriate in some areas.

FLUVIAL PROCESSES IN THE MOUNTAINS AND INTERMONTANE BASINS

Dale F. Ritter

INTRODUCTION

Fluvial processes in the Rocky Mountains and intermontane basins are intimately related to the fundamental aspects of their geological settings, and to the controlling imprints exerted by tectonics and alternating climates attendant to glacial/interglacial cycles. The largest and most intensely studied intermontane basins are found in the physiographic region known as the Middle Rocky Mountains (also Central Rocky Mountains, Wyoming Mountains). In this area, the basins are fringed by a series of high mountains, first delineated and uplifted during the Laramide orogeny. Although the Colorado Piedmont, adjacent to the Southern Rocky Mountains, has been studied in detail (Scott, 1965) and is characterized by the same fluvially developed geo-

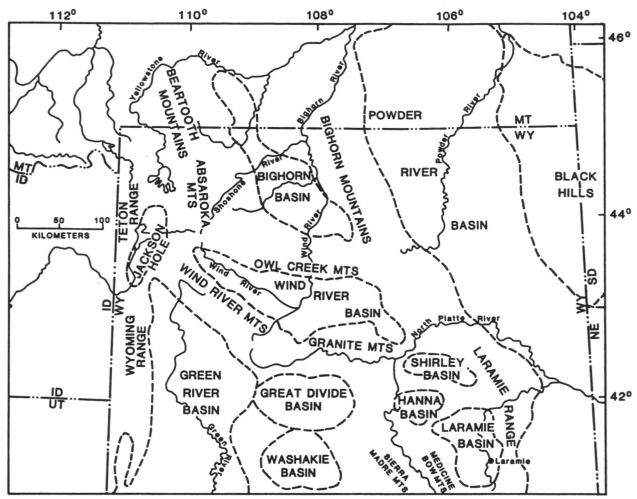

Figure 5. Major mountains and intermontane basins of Wyoming and southern Montana. (After Scott, 1965.)

morphic features, most attention here will be devoted to the basins and mountains of Wyoming and southern Montana (Fig. 5). Even in an area such as this, with similar geologic, climatic, and tectonic histories, considerable local variation exists in factors that control fluvial processes. Therefore, our intent is to make gross generalizations about processes and features, knowing full well that these will not hold in every case. Most attention will be given to basin processes, and special emphasis will be placed on details of fluvial action in the Bighorn Basin, as its geomorphology is best known and is considered by some as the model for basins of this type.

PHYSIOGRAPHIC AND GEOLOGIC CONTROLS

Mountain and Basin Settings

The mountains shown in Figure 5 are lithologically distinct from the intervening basins, being composed of resistant rocks

compared to the extremely nonresistant rocks underlying the basins. Most of the mountains (Beartooth, Wind River, Bighorn, Granite, Black Hills, and Teton ranges) are cored by highly metamorphosed granite gneisses and a suite of igneous rocks such as granites, quartz monzonites, and amphibolites. Some ranges (for example, the Absaroka Mountains) owe their resistance to an enormous accumulation of volcanic rocks. Others (e.g., Owl Creek Mountains) are etched from deformed rocks that have not been stripped of their sedimentary cover; as such, they are less pronounced topographically. Summits of peaks in the crystalline-cored ranges generally stand at elevations ranging from 2,800 m to 3,300 m.

In contrast, the basins are underlain by a sequence of sedimentary rocks that reflects the Paleozoic, Mesozoic, and Tertiary depositional history in this region. Although the sequence is complex in detail, its significance here is that rocks underlying the basins are usually easily eroded shales, mudstones, and sandstones. As a result, prolonged denudation has differentially lowered the basin floors, which commonly stand at altitudes between 1,700 m and 2,300 m.

In the mountain uplands, river action is less important in producing the landscape than other geomorphic processes such as glaciation, periglacial processes, and mass wasting (including debris flows in ephemeral channels). In addition, most topography in large mountain valleys is relict in the sense that geometry and local form reflect the work of glaciers that either no longer exist or exist only as shrunken remnants in cirques. Thus, most major valleys in the mountains, especially in areas of granitic rocks, typically have deeply eroded, U-shaped, glaciated forms. In the valley bottoms, present rivers are burdened by debris of the most recent glaciations and mass-wasted material shed from the steep valley sides. This does not mean that fluvial action played no role in developing the mountain topography. Rivers originally formed the deep valleys by vertical entrenchment, and rivers draining the glaciers have deposited thick fills of outwash that now underlie numerous terrace surfaces. In addition, alluvial fans were commonly developed by tributaries to the trunk valleys. Both fans and terraces are preserved at various levels within the valleys. However, their greatest geomorphic imprint resides in the lowest parts of the valleys beyond the limit of the most recent (Pinedale), major glacial advance. This is because advancing ice has the ability to rework older deposits and thereby destroy the depositional configuration. Also, old deposits resting high on the steep valley sides are susceptible to destruction by downslope movements.

The easily eroded basin and piedmont rocks are much more conducive to preservation of fluvial deposits and features, especially where basin excavation has continued for extended periods. Where the excavation phenomenon prevails, basin rivers are continually downcutting toward new, regionally controlled base levels. This leaves remnants of older fluvial features and deposits preserved at higher levels within the basin.

Prolonged basin excavation is totally dependent on external factors such as tectonics and climate because these determine the critical relationship between river energy and the load introduced to the fluvial system. In absence of transporting power, rivers burdened with large or coarse-grained loads will continuously deposit material. Long-term tectonic or climatic quiescence will, therefore, lead to basin filling; thus, deciphering the character of past fluvial processes becomes a problem of stratigraphy rather than geomorphology. The importance of basin degradation versus basin aggradation cannot be overstated. It represents one of the primary differences in geomorphology between different regions of the western United States. In regions experiencing continuous aggradation, the surface configuration is dominated by young alluvial plains and erosion surfaces that expand mountainward under stable or slowly rising base levels (pediments). In contrast, prolonged degradation allows preservation of features that reveal much of the Quaternary history of fluvial events.

Tectonic and Glacial Effects

Tectonism influences the regional geomorphology in two

ways: (1) it creates mountains and basins, and (2) it provides energy to degrade the basins. In both cases, the tectonic style is characterized by vertical movements caused by displacement along nearly vertical faults by upthrusting and underthrusting, and by regional epeirogenic uplift. (See Lowell, 1983, for review.) In general, deformation during the Laramide orogeny created the regional framework by producing simultaneous sinking of the basins and elevation of the mountains. Elevation differences between the Precambrian rocks of the basins and the mountains (structural relief) can be as much as 15 km, and the axes of the structural basins do not always coincide with the axes of the modern topographic basin floors.

Movement along mountain-front faults commonly has produced broad anticlinal folds in the overlying sedimentary sequence. In some cases, folding has resulted from passive draping of the sedimentary pile over the edge of the rising basement, which produced nearly vertical or slightly overturned Paleozoic and Mesozoic rocks. Resistant units in the piedmont zone, such as the Madison Limestone, have been etched into hogbacks or cuestas by differential erosion. Basinward, the vertical orientation dies out into normal, low-angle stratigraphic sequences or minor folds. Thus, the piedmont zone differs from the basin because it generally has a more humid climate, and it is underlain by highly deformed rocks.

Degradation of the basins was prompted by regional Tertiary uplift, which increased both mountain and basin elevations and changed the basin climates from subtropical to semiarid and arid. Almost every basin in the Middle Rocky Mountains was subjected to degradation following late Tertiary epeirogenic uplift (for reviews, see Eardley, 1951; Love, 1960; Scott, 1965). This final epeirogeny, combined with the onset of Quaternary climatic deterioration, rejuvenated the basin rivers and initiated a general excavation phase, which has persisted to the present.

Superimposed on the tectonic influence is the fact that most mountain ranges in the region have experienced repeated glaciation, which resulted in several notable effects. First, climatic zones in the piedmont shifted basinward during glacial periods and mountainward during interglacials. In some cases, the changes were severe enough to engender periglacial activity in areas that are now arid to semiarid (Mears, 1981). This had a direct effect on the type of soils developed (Reheis, 1984). Second, and more important to our task here, coarse gravel was periodically transferred from the mountains into the basins, filling the trunk valleys that traversed the basin floors or spreading as sheets across the basin rocks. These deposits are decidedly more resistant to erosion than the underlying Mesozoic and Tertiary basin rocks. As a result, in a prevailing degradational environment, remnants of terraces underlain by coarse, mountain-derived gravels are preserved at various levels in most basins. Many workers agree that a large portion of the sand-and-gravel deposits are actually outwash, which is additional evidence of the direct influence that glaciation has had on basin geomorphology.

The glacial history of the Middle Rocky Mountains is beyond the scope of this paper. It should be noted, however, that early analyses of Quaternary history of intermontane basins labored under the constraint that only two (Gilbert, 1890; Atwood, 1909; Capps, 1909) or three (Alden, 1912, 1932; Atwood and Mather, 1912, 1932; Blackwelder, 1915; Fryxell, 1930) major glaciations occurred in the Rocky Mountains. Although more glaciations were long suspected, it was not until Richmond (1957, 1962, 1964, 1965) presented evidence for a greatly expanded pre-Wisconsin glacial history that this constraint on interpretations of basin history could be removed. In addition, since Richmond's initial work, it has become clear that major glacial periods included multiple substages (Mears, 1974).

SPATIAL AND TEMPORAL CONSIDERATIONS: THE GENERAL MODEL

The general evolutionary model of basin geomorphology that existed prior to 1950 is best exemplified by Mackin's classic analysis of the Bighorn Basin (Mackin, 1937, 1948). In addition, several basic concepts pertaining to fluvial processes and landforms were developed in geomorphic studies of this basin (Mackin, 1936, 1937, 1948). Mackin's model integrates the three physiographic elements that are found throughout the Middle Rocky Mountain area: (1) a relatively flat surface in the mountain ranges, known as the subsummit surface, (2) integration of drainages that pass through large canyons cutting across the trends of mountain ranges, and (3) river terraces found in each basin. In brief, Mackin argued that the subsummit surface was cut as a late Miocene or early Pliocene pediment at the culmination of a long episode of basin aggradation. During this filling interval, the floor of the Bighorn Basin progressively rose until sediment was being deposited high along the flanks of the surrounding mountains, completely covering the Owl Creek and Pryor mountains and the north end of the Bighorn Mountains (Fig. 5). Streams rising in the mountains were graded to the level of the basin floor and formed the subsummit surface by pedimentation. At this time, drainage was integrated by stream capture, placing the Wind River/Bighorn River across the position of the buried mountains.

The high-level pedimentation ended when Pliocene regional uplift initiated a cycle of basin excavation, which still persists. The uplift allowed streams to be superimposed across the resistant mountain rocks, creating the deep canyons. Most Tertiary sediments deposited during the earlier filling episode were removed from the basin. Terraces formed during the degradational phase were interpreted by Mackin as evidence for periodic pauses in the general downcutting trend. During these episodes, rivers eroded laterally rather than vertically, creating broad rock-cut terraces that, in a larger sense, are partially completed pediments.

Although the above model is a masterpiece of synthesis, it may not precisely fit the geomorphic history found in other basins, especially with regard to timing of particular events. For example, some authors believe that the high-level surfaces may be relicts from Eocene erosion accomplished before major epeirogeny occurred (Knight, 1953; Scott, 1973). The model has also

been questioned for the Bighorn Basin itself. For example, Love (1960) suggested that integration of the Wind River and Bighorn River drainages occurred during the Eocene rather than the late Tertiary, and Ritter (1975) questioned whether the northern part of the Bighorn Basin actually filled to the level proposed by Mackin. In addition, Love (1960) pointed out that pediment formation cannot occur during a period of tectonic instability. In northwestern Wyoming, therefore, pedimentation probably operated in middle Pliocene, and uplift with associated basin degradation probably did not begin until late Pliocene. Nonetheless, the model generally fits the stratigraphic and tectonic history of the region, and, as indicated earlier, it is most important to recognize that late Tertiary uplift in all basins has prompted pronounced basin excavation throughout the Quaternary. In light of this, it is important to examine the fluvial processes and features that would dominate such a degradational environment.

PROCESSES AND FEATURES

The most important control of fluvial processes and features in basins of the Middle Rocky Mountains is the transfer of coarse-grained sediment from the surrounding mountains into and across the adjacent basins. This alluvium is more resistant to erosion than the underlying bedrock and, therefore, tends to armor the basin rocks. This simple control in an environment of pervasive degradation led to three fluvially related concepts that have been fully developed in geomorphic studies of these basins: (1) the creation and preservation of terraces, (2) graded rivers, and (3) stream piracy.

River Terraces

By any standard, the most important fluvial features in the basins are river terraces; they indicate the drainage lines at various times during excavation of any basin, and they are capped by fluvial deposits. As indicated earlier, terrace gravels are preserved in almost all basins of the region (Scott, 1965). The degree of preservation of any terrace depends on its age; old terraces consist of widely separated remnants, and younger terraces, in general, are continuously intact from the basins into the mountain domain. Historically, the mode of origin for terraces in this region has been a subject of considerable controversy. The differing opinions revolved around whether the terrace gravels are fills produced by the influx of outwash during episodes of mountain glaciation, or whether they represent load that was in transit during the formation of rock-cut benches or partially completed pediments. As indicated above, Mackin (1937) believed that the terrace gravels in the Bighorn Basin were of the latter type. In his scheme, terraces were erosional in origin, formed under widespread lateral cutting by rivers during episodes of fluvial stability when vertical erosion had ceased. As each stream meandered across the valley bottom, it truncated the underlying bedrock while simultaneously depositing a layer of point-bar gravel on top of the eroded surface (Fig. 6). By this process, the gravel was

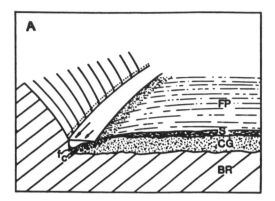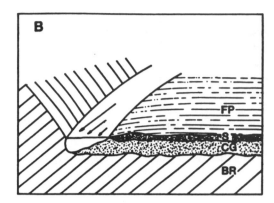

Figure 6. Formation of rock-cut terrace by lateral planation. At low flow (A), channel bottom is occupied by channel gravel and fine alluvium. These represent fill (f) deposited on top of eroded surface (c) during waning phase of high-flow event. During subsequent high-flow event (B) all channel material is entrained; underlying bedrock is truncated and channel position shifts laterally. FP, flood plain; S, silt; CG, channel gravel; BR, bedrock. (After Mackin, 1937.)

spread as a sheet of laterally accreted debris covering an underlying bedrock surface of low relief. Each episode of terrace formation ended when the system reverted to its dominant excavation trend; rivers subsequently downcut to a lower level where the process was repeated.

Rock-cut terraces, as envisioned by Mackin, must have the following two characteristics. First, the surface eroded across the bedrock must be essentially flat and be a virtual mirror image of the surface on top of the gravel. Second, the gravel layer overlying the truncated bedrock must be thin and must maintain a constant thickness across the bench (transverse to the valley axis). This follows because, during floods, all the bed material must be entrained in order to expose the underlying bedrock to erosion. During waning phases of any flood, new point-bar material is deposited in the channel bottom on top of the eroded surface (Fig. 6). Thus, the depth to which a river can scour becomes the controlling factor on the gravel thickness; any thickness greater than the scour depth insures that the bedrock cannot be exposed for erosion. In most rivers, the depth of scour is 1.75 to 2 times the maximum depth attained during a flood (Wolman and Leopold, 1957; Palmquist, 1975). Therefore, for most rivers in basins of the Middle Rocky Mountains, gravel thickness deposited by lateral accretion should not be greater than 6 to 9 m (Mackin, 1937, 1953).

Mackin's interpretation of the terraces was presumably influenced by other early studies, which did not entertain the possibility that the gravels were fills having a glaciofluvial origin (Alden, 1932; Rouse, 1934; Pierce and Andrews, 1941). In fact, terrace gravels were thought to predate and underlie moraines; consequently, there was thought to be no genetic relationship between the two. However, studies after Mackin's work provided evidence that some of the terraces in the Bighorn Basin were the result of valley filling. Moss and Bonini (1961), using seismic profiles across the Cody terrace near Cody, Wyoming, demon-

strated clearly that the bedrock surface beneath the terrace gravel is extremely irregular. The gravel, therefore, does not maintain a constant thickness, nor is the eroded bedrock surface a mirror image of the terrace tread. In addition, gravel thickness (as much as 30 m) far exceeds any reasonable depth of scour for the Shoshone River. They concluded that the gravel of the Cody terrace represents a fill which buried a bedrock topography that existed prior to the influx of the coarse-grained alluvium. Ritter (1967), using drill-hole data, found the same situation in several terraces along the front of the Beartooth Mountains. In that setting, some of the terrace gravels examined were high above the modern rivers and, definitely, are pre-Wisconsin in age.

Thus, it seems that a number of terraces examined by Mackin are not rock-cut in origin but, instead, represent episodes of filling that interrupted the general trend of excavation in the Bighorn Basin. The key question, however, is whether the gravel fills are glaciofluvial in origin and, therefore, related to mountain glaciations.

Terrace formation by glaciofluvial filling has been postulated for many years (Holmes and Moss, 1955; Eschman, 1955; Nelson, 1954). However, proof of a relationship between glaciation and basin terraces is complicated because the gravel-terrace surfaces are difficult to trace through mountain gorges to the moraines, and because isolated terrace remnants near a mountain front cannot always be distinguished from pediments (Bryan and Ray, 1940; Ray, 1940). Nonetheless, Moss (1974) was able to present reasonable arguments that Mackin's Powell and Cody terrace gravels are genetically related to the Bull Lake and Pinedale moraines in the Absaroka Mountains. Moss noted a distinct increase in terrace gradient, and in gravel thickness and particle size, when tracing the gravels from the Bighorn Basin to the morainal positions. In addition, the terrace gravels physically interfinger with the morainal deposits.

In light of this, it is interesting to note that many early

commentaries of basin geomorphology accepted a glaciofluvial origin for younger terrace gravels but denied that origin for gravels older than Bull Lake (Scott, 1965). Older landforms and alluvium were almost invariably considered to be pediments that formed in interglacial periods. It is equally interesting to note that no explanation was presented for this remarkable change in fluvial dynamics during Quaternary time. Most likely the interpretations stemmed from several philosophic barriers existing at the time. First, older gravels are often isolated from the mountains and few, if any, moraines of pre–Bull Lake age remain intact. Therefore, proof of a genetic relationship, such as that provided by Moss for the younger gravels, is almost impossible. Second, the highest gravels in many basins were often considered to be Tertiary in age and, accordingly, were considered to have predated any known glaciation. For example, the highest and oldest terrace gravel in the Bighorn Basin underlies Tatman Bench, the surface of which stands 360 m above the present Greybull River and is underlain by 14 m of andesitic sand and gravel. Because of the enormous local relief, the alluvium was assigned various Tertiary ages. Recently, however, Rohrer and Leopold (1963) dated pollen in this deposit and established its age as possibly late Pliocene, but probably early Pleistocene. This date places all gravels in the Bighorn Basin within the realm of glacial history and removes the stigma attached to an interpretation that gravels presumed to be Tertiary in age are glaciofluvial. Third, as implied earlier, all works prior to Richmond's (1957, 1962b, 1964, 1965) development of an expanded Rocky Mountain glacial sequence were done under the restriction that only three major glacial episodes were believed to have occurred in the area. A geologist faced with numerous terrace levels standing high above the modern rivers would be hesitant to postulate a glaciofluvial origin for their alluvial caps if it meant trying to force a correlation between multiple terrace levels and an obviously insufficient number of glaciations.

The philosophical approach to the interpretation of gravel origins has shifted during the last several decades toward a more detailed understanding of the process of formation. For our purposes here, the new analytical thrust involves (1) studying modern outwash deposits and their relationships with existing glaciers, (2) estimating fluvial conditions necessary to entrain coarse-grained material, and (3) determining how fluvial systems respond to climatic change.

It now seems clear that glaciofluvial filling should be expected to have occurred during every glacial episode. Church and Ryder (1972) presented cogent arguments that the deglacial phase of a glacial cycle is characterized by a profound increase in sediment yield, often by an order of magnitude over normal yield rates. This large influx of sediment occurs at the beginning of deglaciation and rapidly subsides after a short interval. As a result of this influx, the glaciofluvial systems are overwhelmed, and large outwash trains are produced quickly. In addition, Ryder (1971) has shown that the episode is also characterized by the formation of tributary alluvial fans, which often interfinger with or overlie the outwash bodies.

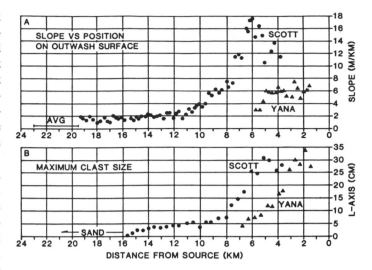

Figure 7. Change in slope of outwash surface (A) and in outwash particle size (B) away from the terminus of modern glaciers, southeast Alaska. (Adapted from Boothroyd and Ashley, 1975.)

The characteristics of outwash are now better known because numerous studies have been conducted in modern glaciofluvial environments to determine the relationship between outwash sedimentation and fluvial mechanics (Church, 1972; Boothroyd and Ashley, 1975; Church and Gilbert, 1975; Fahnestock, 1963, 1969; Bradley and others, 1972). In almost every study, maximum particle size and gradient of the outwash surface decreased dramatically away from the glacier in a zone that is usually located within 12 km of the glacier front (Fig. 7). Downstream from this zone, particle size and slope decrease more slowly. These are precisely the characteristics noted by Moss (1974) while tracing the Shoshone River terraces to the morainal zones.

Meltwater floods and floods caused by destruction of ice-dammed lakes occur in the ice-margin environment (Bradley and others, 1972) and transport enormous boulders for short distances. Transport of very coarse debris is limited because the flow parameters that control particle transport (velocity, depth, shear stress, and stream power) all seem to be drastically reduced within 10 km of the ice front (Fahnestock, 1963; Boothroyd and Ashley, 1975). In addition, ice-front hydrology is characterized by extreme and frequent variations in discharge (Fahnestock, 1963). The flow conditions needed to reentrain a large clast, once it has come to rest after a flow event, are greater than those previously needed to transport the same particle. Therefore, if slope, shear stress, and velocity decrease downstream, an even larger flow event is required to reinitiate movement. Eventually, clasts are moved to a position where any flow that is likely to be generated at the ice margin cannot cause entrainment, and the particle remains as a lag boulder. This does not mean that extremely large boulders are never moved far in a glacially related environment. Rare, catastrophic flow events associated with ice damming have transported enormous boulders for extreme distances (Birkeland, 1968; Baker, 1973).

The apparent relationship between maximum particle size and flow conditions has prompted another line of research designed to reconstruct paleohydrology from the properties of gravel deposits (Baker, 1973, 1974; Baker and Ritter, 1975; Bradley and Mears, 1980; Costa, 1983). In essence, this technique is applied to old terrace deposits in an attempt to show that the gravels were transported and deposited under flow regimes (tractive forces or critical velocities) that are inconsistent with the modern, interglacial conditions. For example, Baker (1974) analyzed terrace gravels on the Colorado Piedmont by using the dimensionless relationship

$$\frac{DS}{1.65d} = 0.06,$$

where D is depth in meters, S is slope, and d is the average intermediate axis in meters of the 10 largest boulders. By this approach, the paleodepth could be calculated assuming that the terrace slope was equal to the energy gradeline for the Pleistocene rivers. Furthermore, slope and cross-sectional area measured in a paleochannel preserved in the Broadway terrace surface allowed a rough estimate of the discharge during gravel deposition. Comparison of that discharge to modern records showed that the Pleistocene flood discharge was probably an order of magnitude greater than the major flow events of the present Clear Creek. Similar approaches in British Columbia have led to the same conclusion (Clague, 1975).

The analysis of paleohydrology using clast size and flow equations is fraught with difficulties and requires numerous assumptions (Baker, 1974; Church, 1978). Nonetheless, even though it is unrealistic to expect that ancient flow magnitudes can be determined in specific terms, estimates of the largest flow can be compared with modern rivers within broad ranges of error (Church, 1978; Desloges and Gardner, 1984). In addition, we now have available a rather detailed understanding of proglacial deposits and bedform properties. Bedforms, such as ripples and dunes, have distinct hydrologic controls, and bar types change according to their positions relative to the ice margin. These analyses should provide the basis for sedimentological comparisons of ancient gravel deposits and, also, for reconstruction of channel morphology and hydrology.

Another approach developed recently centers on how fluvial systems might respond to major climatic changes such as those associated with the glacial/interglacial cycle. The basic premise of this research is that climate change will produce tangible modifications in river load and discharge. These, in turn, will require a threshold response in the river, leading to alteration of channel morphology or river pattern. This type of response, called river metamorphosis (Schumm, 1969, 1977), has been aptly demonstrated in Australia (Schumm, 1968) and in central Texas (Baker and Penteado-Orellana, 1977). River metamorphosis is based on a large number of empirical relationships for calculating flow

characteristics of old rivers from their preserved morphology and sediment properties (Dury, 1965, 1976; Schumm, 1969). For our purposes, the presumed increase in discharge and coarse-grained bedload attendant to glaciofluvial transport should lead to greater width, meander wavelength, and width-depth ratios, and to lower sinuosity compared to modern rivers (streams become wider, straighter, and shallower).

It seems clear, from studies of modern proglacial systems, that channel-pattern thresholds may be present when proceeding downstream from the ice margin. Boothroyd and Ashley (1975) showed transitions from straight to braided and, finally, to meandering channels away from the ice margins. This sequence is consistent with other studies showing that threshold conditions controlled by slope and discharge, or slope and sediment load, exist in river systems (Lane, 1957; Leopold and Wolman, 1957; Ackers and Charlton, 1971; Schumm and Khan, 1972). Therefore, with downstream changes in discharge, slope, and bedload characteristics in glaciofluvial systems, it is conceivable that different patterns and fluvial mechanics prevailed along the longitudinal extent of an outwash terrace. In fact, Moss and Whitney (1971) and Moss (1974) entertained this possibility for the Shoshone River terraces, suggesting that downstream changes in fluvial controls would allow lateral spreading of outwash gravel if conditions initiate a meandering, rather than braided, pattern. The implication here is that on any given terrace the dominant process of origin might change from filling near the mountains to Mackin-type lateral planation in the distal reaches. Such a scenario is quite possible where the proper load-discharge conditions exist.

The above discussion is intended to support the conclusion that most terrace gravels in the Middle Rocky Mountains represent glaciofluvial deposition. Indeed, many workers may readily accept this premise. It should be pointed out, however, that terraces may result from processes other than those discussed above. For example, Schumm (1973) suggested that the response to a threshold-crossing event may be very complex. Changing from glacial to interglacial conditions, therefore, will probably engender the downcutting needed to create a terrace, but the entrenchment is not necessarily continuous. Instead, the long-term adjustment to new fluvial controls may include minor episodes of filling and cutting, which produce subsidiary terraces unrelated to a major climatic reversal (Womack and Schumm, 1977). In some cases, a series of purely erosional terraces may result during the adjustment phase. Moss (1974), for example, identified five distinct erosional terrace levels, all cut in the alluvium of the Cody terrace, which were formed before the major post-Pleistocene entrenchment of the Shoshone River. Ritter (1982) found a similar series of erosional terraces along the Nenana River in Alaska, which must have formed during the general downcutting trend of the river between two glacial episodes.

Valley filling can also occur as the result of stream piracy (Mackin, 1936; Ritter, 1967, 1972). The process of piracy will be discussed later, but it is important to recognize that diversion of a mountain river (carrying coarse-grained bedload) into the valley of a pirating river (adjusted to the transport of fine-grained load) provides a geomorphic setting conducive to filling in the valley of the pirating river.

Deposition of gravel represents only half of the story involved in terrace origin because the creation of any terrace requires an episode of downcutting to produce its topographic form. An interesting aspect of terraces in the Wyoming basins is that they are commonly unpaired except close to the mountain fronts; in other words, a terrace does not have a topographic counterpart standing at the same elevation on the other side of the valley. Mackin believed that this phenomenon occurred because a moderate component of downcutting occurred while the river laterally eroded across the valley bottoms. However, the unpaired terrace distribution can also be explained by preferential downcutting of rivers at the valley margins. Placement of the rivers at the lateral margins of the gravel deposits allows them to entrench the erodible basin rocks rather than cutting through the more resistant coarse-grained debris (Ritter, 1967).

The greater resistance of terrace gravels versus basin rocks results in inverted topography, another characteristic of basin geomorphology. Because gravels are fluvial in origin, they must have occupied the valley bottoms during their deposition, even though many gravels now stand on drainage divides, the highest topographic entities. This phenomenon obviously requires unequal downwasting of the surface and high rates of incision and denudation. Incision rates in the Bighorn Basin have been calculated to be approximately 0.15 m/1,000 yrs using radiometrically dated ashbeds as controls on the absolute values (Palmquist, 1979, 1983; Reheis, 1984). Incision rates are, of course, not denudation rates because the latter assume equal lowering over all parts of the basin. Nonetheless, incision rates may estimate the maximum rate of denudation. Thus, at the maximum denudation rates for the Bighorn Basin, approximately 6 to 8 m.y. is needed to remove the 900 to 1,200 m of post-Eocene sediment postulated in the Mackin (1937) model of filling. A denudation rate of this magnitude over such a large area would be extremely high (Schumm, 1977).

Graded Rivers

The concept of grade in fluvial mechanics stems from early geologic and engineering studies, which stressed the point that rivers attain an equilibrium condition between channel morphology, prevailing discharge, and load characteristics (e.g., Gilbert, 1877). Historically, controversy arose about the utility of the concept, because the general perception existed that a graded condition represented a balance between erosion and deposition. In addition, Davis (1899, 1902) introduced an evolutionary connotation to the concept by linking it to a particular stage in his geographical cycle. In an attempt to clarify the meaning of grade, Mackin (1948, p. 471) defined the graded river as " . . . one in which, over a period of years, slope is delicately adjusted to provide, with available discharge and with prevailing channel

characteristics, just the velocity required for the transportation of the load applied from the drainage basin."

It is now clear that river morphology adjusts rapidly to fundamental controls, and that every river displays an inherent balance, called quasi-equilibrium, which is identified by hydraulic geometry (Leopold and Maddock, 1953; Wolman, 1955). This perception not only negated Davis' evolutionary connotation of grade (Hack, 1960), but also prompted the thought that all rivers must be graded, thereby eliminating the uniqueness of a graded river. In an excellent review, Knox (1976) pointed out that Mackin's words "over a period of years" imply that grade requires morphologic stability over time, a condition that is not necessary or expected in quasi-equilibirum. Therefore, Knox (1976, p. 179) suggested that a graded river is better defined as "one in which the relationship between process and form is stationary and the morphology of the system remains relatively constant over time." Knox also suggested that ungraded streams are those undergoing rapid change in properties such as longitudinal profiles and (or) cross-channel morphology. Thus, episodic adjustments occur when thresholds of stability are passed, and rivers are temporarily ungraded until they develop a new, stable channel controlled by new hydrologic and load conditions. The conceptual separation of quasi-equilibrium and grade was presupposed by the recognition of the importance of different time scales in geomorphic analyses, and by the understanding that a single fluvial parameter may operate as a dependent or independent variable on different time scales (Schumm and Lichty, 1965; Chorley and Kennedy, 1971).

The significance of the above discussion to the geomorphology of Wyoming basins is simply that Mackin's examples of landforms resulting from rivers in a graded condition are the Bighorn Basin terraces developed by the Shoshone River, especially the Cody and Powell terraces. However, in the perception of grade expressed by Knox, and accepting the glaciofluvial origin of the gravels of the Cody and Powell terraces (Moss, 1974), it is doubtful that rivers experiencing rapid aggradation of outwash (Church and Ryder, 1972), frequent high-magnitude flow (Fahnestock, 1963), and changing gradient and pattern (Boothroyd and Ashley, 1975; Moss, 1974) could be graded. Thus, the Shoshone terraces probably result from alternating episodes of ungraded conditions representing spasms of filling and cutting, and, by doing so, the terraces become questionable examples of the graded condition.

Stream Piracy

Stream piracy is a prominent process operating in basin and piedmont zones. It relates primarily to the different load characteristics and flow regimes developed in rivers heading in the mountains as compared to those rising in the basins. Minor tributary rivers heading in the basins usually have low gradients adjusted to the transport of fine-grained sediment derived from the basin rocks. In contrast, major rivers heading in the mountains develop steep gradients needed to transport coarse-grained gravel loads. This fundamental control means that, at equal distance upstream from the confluence of a tributary and a master river, the tributary valley will stand at an elevation well below that of the trunk valley. The process is indirectly related to the erosional resistance of different materials because, as indicated earlier, rivers have difficulty downcutting through a gravel substrate. Therefore, trunk rivers are often poised on lag concentrates, while tributaries are free to entrench the basin rocks. Captures occur when drainage divides are breached or sapping extends the tributary into the poised trunk valleys. Eventually, the master-river flow is diverted into the valley of the tributary. Diversions of this type have long been recognized and have been employed to explain the distribution of terrace gravels (Mackin, 1936, 1937; Ritter, 1967, 1972; Ritter and Kauffman, 1983), alluvial fan deposits (Denny, 1965), and the origin of pediments (Rich, 1935b; Hunt and others, 1953; Schumm, 1977).

The importance of piracy to basin geomorphology is twofold. First, the diverted river may not be able to transport its coarse bedload on the low gradients of the tributary channel or valley bottom. As a result, deposition of gravel may occur until a gradient is developed that is capable of transporting the coarse load. In the sense of Knox (1976), the river is temporarily ungraded until it adjusts its slope to the character of the load. Second, if filling does result, it is critical to recognize that the topography underlying the gravel fill was cut by the tributary river prior to capture. Therefore, the surface beneath the gravel fill was developed by one river (the tributary), and deposition was accomplished by a different river (the diverted trunk).

The geomorphic implication of the above is that all gravel fills are not necessarily glacially, tectonically, or climatically controlled; some gravel fills may be the result of stream piracy. Piracy is most likely random in spatial and temporal terms, and the process probably functions continuously in piedmont regions, even though the precise location of the capture changes. This obviously complicates the analysis of fluvial geomorphology in the Wyoming basins, but it is a factor that cannot be ignored in regional and local geomorphic studies.

SUMMARY

In summary, the early ideas concerning fluvial processes in the Wyoming basins are subject to modification. There is little doubt that the perception of dominant excavation throughout the Quaternary still holds. However, the meaning of preserved gravel-covered surfaces has been questioned enough to warrant a reevaluation. An expanded glacial sequence and our increased understanding of process mechanics in high-energy fluvial systems support an interpretation that much of the basinward transport of coarse gravel was accomplished by glaciofluvial action. Therefore, landforms such as terraces may not be examples of graded river mechanics. In addition, ancillary fluvial actions such as complex responses to threshold-crossing events, river metamorphosis, and stream piracies are complicating factors, which demand that each local situation be examined in detail.

GLACIAL PROCESSES IN THE CENTRAL CANADIAN ROCKY MOUNTIANS

Nathaniel W. Rutter

INTRODUCTION

The central Canadian Rocky Mountain area is ideal for discussing and illustrating glacial processes. Here, we have outstanding examples of glacial alpine morphology, an abundance of alpine glaciers, and wide-spread montane glacier deposits displaying a variety of landforms. Specifically, the area includes the Banff and Jasper national parks east of the Continental Divide and the Rocky Mountains Forest Reserve that terminates near the edge of the mountains and the prairies (Fig. 8). The specific style of glaciation in this area may not be totally applicable to the style of glaciation in other areas of the Rocky Mountains, but certainly the gross effect of glaciation, be it erosional or depositional, probably isn't much different, and should be applicable.

The objective of this section is to review the research concepts that have been applied to the Rocky Mountains in general, and the central Canadian Rocky Mountains in particular, in order to elucidate the style of alpine glaciation and the processes involved in producing the present-day landscape. To achieve this objective, a certain amount of landscape description is necessary as well as a brief account of the glacial history. An example, in some detail, will be given on glacial process studies recently carried out in the Banff area involving the analyses of diamicton facies to determine the mode of deposition by the glacier and the origin of the landforms that were produced.

GEOMORPHIC AND GLACIER SETTING

There is little doubt that the central Canadian Rocky Mountains developed their "basic" geomorphic expression prior to Pleistocene glaciation. The Laramide orogeny was responsible for the northwest structural trend of the mountains, which consists mainly of faulted and folded Paleozoic carbonate and clastic rocks thrusted over Mesozoic shale, siltstone, coal, and sandstone. In the Main Ranges, peaks reach heights of over 3,500 m, such as Temple Mountain and the highest peak in Alberta—Mount Columbia (Fig. 8). However, most of the higher mountains average between about 2,400 m and 3,000 m in elevation. The basic drainage pattern was no doubt developed during mountain building. The major rivers and streams follow the topographic lows, parallel to the northwest-trending mountains—some flowing southeastward and others flowing northwestward, depending upon the position of divides within the valleys. More difficult to explain are the positions of outlet valleys that cut across major mountain ranges, and eventually make their way to the east. Although there are no definitive answers, there have been suggestions that their origin may center around exhumation, downcutting during mountain building, and glacial stream diversions. As will be explained later, glacial stream diversion is a plausible

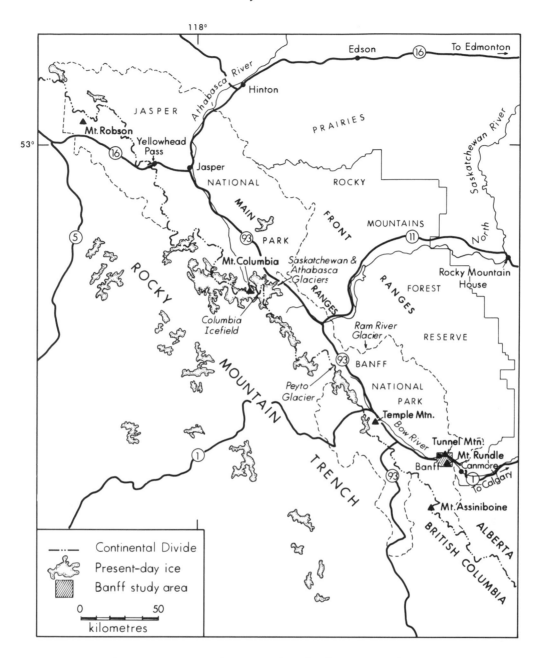

Figure 8. General map of the central Canadian Rocky Mountains area.

explanation for some stream patterns. Although preglacial gravels and old erosional surfaces are evident, the major modifications since the Rockies were established in their present form have been caused by periglacial and glacial processes.

STYLE AND HISTORY OF GLACIATION

The present-day icefields, the cirque and valley glaciers, and the abundance of alpine glacial erosional and constructional landforms give ample opportunity for study of the style and extent of Holocene and Pleistocene glaciations. Unfortunately, there has been no quantitative reconstruction of ice masses during discrete time intervals, such as has been so expertly accomplished by Pierce (1979) in the Yellowstone Park area of Wyoming.

Erosional Style

This area contains the most outstanding examples of glacial alpine morphology in the central Canadian Rockies, consisting of cirques, arêtes, cols, horns, U-shaped and hanging valleys, and rock steps, leaving little doubt that widespread glaciation took place in the past. U-shaped valleys extend from cirques into glaciated tributary valleys, which, in turn, enter major outlet valleys such as the Bow, North Saskatchewan, and Athabasca. Classical glacial sculpturing is displayed to the eastern edge of the mountain front, extending over distances of at least 150 km from ice sources near the Continental Divide. Hanging valleys, rock drumlins, roche moutonnées, and glacially induced valley steps are widespread.

Postglacial erosion and modification have masked the effects of glaciation in many areas so that it is difficult to precisely define the upper limit of maximum ice. The distribution and lithology of erratics show that most ice originated in the area of the Continental Divide, although, during times of extensive glaciation, ice from British Columbia passed over low parts of the Continental Divide, such as Yellowhead Pass (1,133 m) and into the Jasper area. In the Banff townsite area, erratics indicate that only mountains above about 2,400 m were ice free during maximum glaciation. This would make ice thicknesses about 1,000 m (Rutter, 1972). Cols, "smoothed-over" summit areas, and erratics indicate that, at times, ice passed over divides of lower mountain ranges. Toward the eastern edge of the mountains, summits about 2,000 m high were covered with ice at least during maximum glaciation. What emerges then, from observing glacial erosional features, is that during maximum glaciation a fingerlike network of mostly topographically controlled valley glaciers emanated from the cirque areas down alpine valleys, feeding into the major outlet valleys that flowed out onto the prairies to the east.

The number of glaciations is difficult to discern from erosional features. Two types of features observed in the Banff area aid in demonstrating that more than one event took place. First, breaks are present in valley-side slopes that mark the upper limit of a "U-shaped" valley within a glaciated valley. On the northeast flank of Mount Rundle near Canmore (Fig. 9), for example, the slope of the valley wall decreases above the limit of a less extensive glaciation. Second, there are glacier stream diversions across mountain ranges whose canyons were later glaciated. A good example is the diversion of the Bow River, through the north end of Mount Rundle, which now separates Tunnel Mountain from Mount Rundle (Figs. 8 and 9). A glacially controlled, subglacial, or side stream cut about 500 m vertically through the bedrock. Ice would have covered the southern part of Mount Rundle in the case of a subglacial stream, or kept pace with ablation in the case of a side-glacial stream. The canyon has since been glacially smoothed, indicating that ice subsequently passed through the stream-cut canyon. Although these erosional features aid in demonstrating that multiple glaciation took place in the central Canadian Rocky Mountains, it is the glacial deposits that provide the main evidence.

Depositional Style

Postglacial alluviation, colluviation, and erosion have hindered the preservation of glacial and interglacial deposits. Only on the floor of the major outlet valleys and some of the tributary valleys are remnants of relatively complete sections of Pleistocene deposits present. Most sections form bluffs 50 to 100 m in height along valley margins or protected areas, and consist of till overlying outwash gravel. There are few sections that consist of two or more distinct tills. Within or near cirques and within some of the higher alpine valleys, Holocene or late Pleistocene till deposits are found.

The depositional landforms of the central Canadian Rockies display a variety of styles. Within or near the lips of cirques, there are commonly multiple, arcuate end moraines (including push or recessional moraines) a few meters high and up to tens of meters wide. Downvalley, up to a few kilometers from the cirque lips, lateral moraines, ground moraine, and indistinct till mounds, which may be end moraines, are present locally. Farther downvalley, in the major tributary and outlet valleys, there is a departure from this trend. Here, many of the valleys are floored by streamlined surfaces, which are oriented parallel to the valley walls. These surfaces extend beyond the mountain front to the east and into the Rocky Mountain Trench to the west. They commonly are streamlined, constructional (drumlinoid, fluted) surfaces consisting of glacial drift in the lower parts of the valleys, and crag and tail deposits and streamlined bedrock surfaces on the slopes. The streamlined features are from 1 to 2 km long, a few hundred meters wide, and as much as 30 m high, although most that consist of glacial drift are usually less than a meter or two high. There are essentially no lateral moraines, hummocky or ground moraine, or arcuate-end moraines present in the major tributary and outlet valleys. This is in direct contrast to many of the valleys of the United States Rockies, such as in the Front Range of Colorado and the Wind River Range of Wyoming (Richmond, 1965; Porter and others, 1983) where "classical" moraines are the rule, although streamlined surfaces are observed in some places, such as in the Yellowstone National Park area of Wyoming (Pierce, 1979).

Other prominent features associated with glaciation of the central Canadian Rocky Mountains are high-level outwash terraces along the major valleys, a few eskers, and relatively expressionless surfaces underlain by ground moraine.

Glacial History

Most of the effort in Quaternary research in the central Canadian Rocky Mountains has been directed toward deciphering the glacial history. Comments about glacial features were made before the turn of the century, mostly in bedrock geology reports by such luminaries as Dawson (1888a, 1888b, 1890, 1891) and McConnell (1896).

It wasn't until the 1970s that detailed glacial geological reports began to appear, mostly dealing with stratigraphy, limits of glaciation, and relative and absolute dating of events, and in some cases summarizing the glacial history throughout the region. Evidence of glaciation has been mainly from multiple till sections,

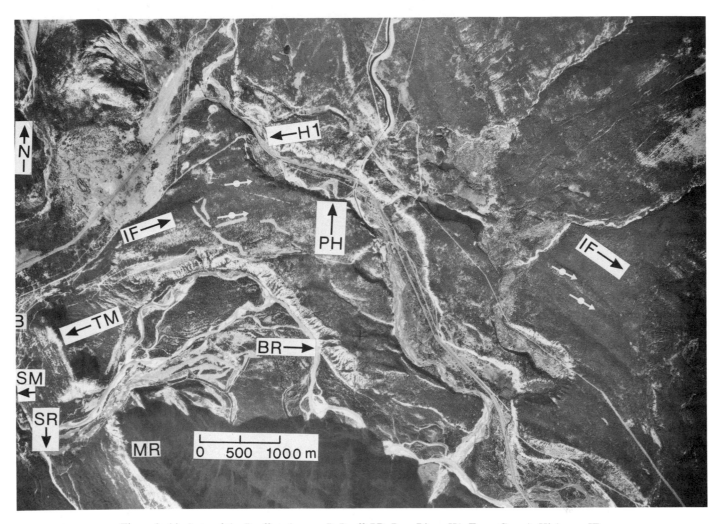

Figure 9. Airphoto of the Banff study area. B, Banff; BR, Bow River; H1, Trans-Canada Highway; IF, Former ice-flow direction; PH, Powerhouse Section; MR, Mount Rundle; SM, Sulphur Mountain; SR, Spray River; TM, Tunnel Mountain. Drumlinoid feature indicated by arrow with dot.

till distribution, geomorphic features, stratigraphic markers such as tephras, and a few absolute dates. These include reports by McPherson (1970), Shaw (1972), Rutter (1972, 1984), Roed (1975), Boydell (1978), Luckman and Osborn (1979), Jackson (1980), Levson (1986), and Mandryk (1986).

As many as four Pleistocene advances considered Illinoian or younger, and several Holocene advances have been recorded, although absolute dating control is generally lacking. The best evidence is recorded for the last three Pleistocene events. The oldest terminates beyond the mountain front, the next younger event in the area of the mountain front, and the youngest terminates well up the major valleys. Some ^{14}C dating control suggests that the two youngest events are late Wisconsin in age. However, the debate continues on the absolute age and the extent of the late Wisconsin ice and the number of advances involved. It is becoming clear that "finger matching" or correlation of events with

similar areal distribution of glacial deposits in various drainage systems may not be appropriate (Gibbons and others, 1984). The final Pleistocene deglaciation probably took place 10,000 to 12,000 years ago (Rutter, 1984). There is evidence for limited renewed glacial activity in the late Wisconsinan or early Holocene through parts of the Canadian Rocky Mountains. Mazama tephra (6,600 yrs B.P.) overlies deglacial deposits and, therefore, places a minimum age on this activity. After this, wide areas of the alpine region remained ice free until a few hundred years ago, as indicated by the preservation of Mazama, St. Helens "Y" (ca. 3,400 yrs B.P.), and Bridge River (ca. 2,600 yrs B.P.) tephras within many nonglacial deposits that overlie glacial deposits (Luckman and Osborn, 1979; Osborn, 1985; M. A. Reasoner, University of Alberta, oral communication, 1985). The advances of the last few hundred years were minor and are recorded by moraines in ice-free cirques and in front of present-day glaciers.

PRESENT-DAY GLACIERS

By the late 1800s, when the first observations were made on several alpine glaciers, the theory that glaciers had advanced and retreated beyond their present positions, at least in mountainous regions, had been accepted by most. Many naturalists and scientists photographically recorded ice-front positions, to be compared with later photographs; made comments on the motions of various glaciers; and synthesized what was known about the observed glaciers in the central Canadian Rockies (Vaux and Vaux, 1907; Vaux, 1910; Vaux, 1911; Wheeler, eight papers between 1907 and 1920a, and 1920b, 1932, 1933; Sherzer, 1907; Palmer, 1924; Thorington, four papers between 1927 and 1945; McCoubrey, 1937; McFarlane, 1946; Meek, 1948; McCauley, 1958; Meier, 1960; Paterson, 1962; Gardner and others, 1964). Heusser (1956) made a valuable contribution on the behavior of glaciers in the Canadian Rocky Mountains by summarizing available data and combining it with botanical information to reconstruct postglacial environments. The data presented by various workers indicates that ice fronts of alpine glaciers have fluctuated several times throughout the past few hundred years and that an overall recession is presently taking place.

There are literally thousands of glaciers in the central Canadian Rocky Mountains. By far the most abundant are cirque glaciers or "slab-type" glaciers (classified by Ommanney [1972] as mountain glaciers and glacierates), a few are icefields, and many are outlet and valley glaciers. In addition, there are hundreds of rock glaciers, many presumably cored with remnant glacier ice. Glaciers are distributed along a northwest-southeast belt in the area of the Continental Divide (Fig. 8). Although local variations in aspect, insolation, topography, and temperature control the elevation of glaciers to a large extent, it is the increase in precipitation over the higher parts of the mountains that is a major factor in their distribution (see section on Past and Present Snowlines). For example, from Calgary to the Continental Divide, a distance of about 120 km, the total precipitation increases from about 450 mm to nearly 1,000 mm per year.

The dominant ice mass in the region is the Columbia Icefield, which straddles the Continental Divide in the area of the boundary between Banff and Jasper national parks (Fig. 8). Meltwater from the icefield finds its way to the Pacific Ocean, Hudson Bay, and the Arctic Ocean. The average elevation of the icefield is roughly 3,000 m; its area, including outlet glaciers, is no more than 100 m (Paterson, 1972).

Several glaciers such as the Saskatchewan and Athabasca (outlet glaciers of the Columbia Icefield) and the Peyto and Ram River glaciers have been investigated in regard to flow, mass balance, and other physical aspects (Paterson and Savage, 1963; Savage and Paterson, 1963; Ostrem, 1966; Reid and Charbonneau, 1972, 1975, 1979, 1981). Still others have been observed to determine the annual rate of recession (Henock, 1972). Probably the most classic work is that by Meier (1960) on the mode of flow of the Saskatchewan Glacier; this was one of the earliest works to test the theory of alpine glacier flow.

Using the well-studied Athabasca and Saskatchewan glaciers as examples of typical central Canadian Rocky Mountain outlet glaciers, we find that they are warm based, have velocity components of both internal deformation and basal sliding, ELAs (equilibrium line altitudes; the lines dividing the glaciers or former glaciers into accumulation and ablation zones; Andrews, 1975) of about 2,600 m, slightly negative mass balances (at least for the last 20 years or so), ice thickness from about 100 m to 450 m, average velocities on the order of tens of meters per year near their snouts, and basal shear stresses between 0.5 and 1.5 bars (Meier, 1960; Paterson, 1969, 1972). Among other things, these investigations have supplied data and provided a valuable check on the reconstruction of past glaciers, such as Pierce's (1979) in the Yellowstone Park area.

With the exception of a few ELA determinations, cirque glaciers have not been investigated in the Canadian Rockies. It is assumed that their movement is essentially rotational, like the movement of a Norwegian cirque glacier, as determined by McCall (1960). Rotational movement is suggested by the presence or arcuate-end moraines near the snouts of many cirque glaciers. One of the better studies on a cirque glacier elsewhere in the Rocky Mountains is by Reheis (1975) on the source, transportation, and deposition of the Arapaho Glacier, Front Range, Colorado.

ROCK GLACIERS

Although rock glaciers are peripheral to the subject of glaciation, some mention should be made because hundreds have been observed in the Canadian Rocky Mountains, many of which have formed as a result of glaciation. In the Jasper National Park area, rock glaciers are best developed below massive quartzite outcrops of the Main Ranges. Most are small, less than 0.25 km^2 in area, and are divided fairly equally between those associated with present or former glaciers within or close to cirques and those developed below talus slopes (Luckman, 1981). The morphology, lichen cover, surface weathering, and stratigraphic relationships of the rock glaciers examined by Luckman and Crockett (1978) and Luckman and Osborn (1979) suggest that they have been inactive for a long period of time, probably developing and subsequently ceasing to be active more than 6,600 yrs B.P. (age of Mazama ash). The reader is referred to the many reviews of rock glaciers, some including rock glaciers observed in other parts of the Rocky Moutains (Ives, 1940; Wahrhaftig and Cox, 1959; Johnson, 1967; Potter, 1972; Benedict, 1973b; Whalley, 1974a; White, 1976a; White, 1979).

PRESENT AND PAST SNOWLINES

Determination of snowlines is an integral part of our interest in the distribution and behaviour of glaciers, be it for short-term practical goals such as water supply or for an understanding of the broad controls over present-day glacier distribution, necessary for a correct interpretation of evidence for lower snowlines and more extensive glaciation in the past.

Glacier equilibrium line altitude (ELA) has been widely used to infer present and Pleistocene climatic conditions. Regional trends of present-day and Pleistocene ELAs relate to modern precipitation patterns in many areas (Meierding, 1982). ELAs delineate the height distribution of the glacier snowline and not the regional or climatic snowline (lower limit of snow surviving at the end of the summer season). ELAs of past glaciers cannot be directly measured, so a number of indices have been developed as surrogates for ELAs, particularly those that existed in alpine regions (Charlesworth, 1957; Ostrem, 1966; Flint, 1971; Andrews, 1975; Leonard, 1984). Meierding (1982) compared various indices for the Front Range, Colorado, and evaluated each index based upon the assumption that regional ELA trends in a relatively small mountain range can be approximated by simple trend surfaces. Indices included cirque-floor altitudes, toe-to-headwall altitude ratios (THAR), glaciation threshold (see glaciation limit below), maximum altitude of lateral moraines, and accumulation area ratios (AAR). All methods have merit, the degree of which is determined mainly by the amount of data, the ease of accumulating data, and the availability of stratigraphic information on past glaciations.

In the Canadian Rocky Mountains, ELAs of present-day glaciers vary a great deal depending on physical and climatic factors. Ostrem (1966) applied an indirect method of estimating climatic snowlines called the glaciation limit (GL). It is defined as the average difference in elevation between unglaciated and glaciated summits. Ostrem has suggested that the climatic snowline is about 100 m below the GL and that glacier ELAs lie about 100 to 400 m below the GL. Smoothed-out iso-glacihypses (GL elevations) form roughly northwest-trending lines varying in elevation from 2,600 m in the west to about 3,100 m toward the east (Fig. 10). The increase in altitude toward the east is a result of a decrease in precipitation, with increased continentality (Ostrem, 1966). Therefore, if ELAs lie about 100 to 400 m below the GL, present-day glaciers should be nestled in cirques and valleys a few hundred meters below these lines, locally controlled by such factors as exposure and insolation. When considering measured ELAs from various glaciers, such as the Athabasca, this concept does indeed seem to be valid.

Evidence of climatic change in the past is provided by glacier-free cirques, best developed on the north and northeast flanks of the central Canadian Rocky Mountains. In the Banff area there are as many as three steplike cirques within the same valley, with the highest and lowest separated by about 600 m of elevation (Rutter, 1966). The complexity of multiple cirques in the same valley and the incomplete stratigraphic record have made any estimates of former climatic snowlines difficult to assess. Trenhaile (1977) points out that factors such as preglacial stream-cut hollows that fall within the altitudinal range but on either side of the ELA, may provide a focus for cirque formation not related directly to the ELA. Also, the influence of geological and topographical settings may control where nivation is most

effective and, therefore, locally control the elevation of cirques and cirque glaciers. All that can be said at present is that, until there is a better understanding of the stratigraphic record and of cirque formation, the lowest Pleistocene ELAs were a few hundred meters below present-day ELAs.

In the United States Rockies, present-day ELAs stand at about 3,700 m in Colorado, about 2,500 m in northwestern Montana, and about 2,100 m in northern Idaho (Porter and others, 1983; Leonard, 1984). During the maximum ice advance of the last glaciation, ELAs were some 1,000 m lower than they are at present in the United States Rocky Mountains and probably were depressed no more than half that amount in the Brooks Range, Alaska (Porter and others, 1983). Although increased precipitation may have been an important controlling factor, the lowering of the mean annual temperature may have been on the order of 8°C to 12°C in the United States Rockies during the last glaciation (Porter and others, 1983; Nelson and others, 1984).

GLACIAL EROSION

The abundance of glacial erosional features observed in the central Canadian Rocky Mountains leaves little doubt that confined, warm-based glaciers formerly occupied most of the mountain valleys. Although glacial-erosion studies per se have not been carried out in the Canadian Rockies, there are many conclusions that can be made on the nature of glacial erosion by observations of the landscape, comparisons with other areas, and applying the principles and theory of glacial erosion.

There are at least two fundamental processes of glacial erosion—abrasion and plucking (quarrying). Abrasion requires basal debris, sliding of basal ice, and transport of debris within the ice down toward the bedrock. Factors affecting rate and type of abrasion include ice thickness, basal water pressure, relative hardness of rock particles and bedrock, particle characteristics, and the efficient removal of rock flour. Because of the low yield stress of ice, plucking requires the presence of loosened, highly jointed material to be an effective erosional agent.

The very nature of the terrain, consisting largely of folded, faulted, and jointed, poorly resistant shale and carbonate rocks, makes the central Canadian Rockies particularly susceptible to glacial abrasion and plucking. Striations, grooves, chattermarks, roche moutonnées, and rock drumlins are commonly observed on the glaciated surfaces of the carbonate rocks but are much less distinct or absent on the more resistant quartzites and other siliceous rocks. Carbonate clasts are commonly striated and faceted within the tills. The till matrix (<2 mm) and glacial lacustrine deposits consist of a high proportion of clay-sized rock flour. In addition, angular blocks, up to 2 m in diameter, of local bedrock are found in basal till suggesting that plucking may be a factor in their origin.

The rates of glacial erosion that produced the spectacular U-shaped valleys and cirques in the central Canadian Rockies depend upon a number of variables. These include the glacier system (basal ice temperature, velocity, ice thickness), the bed-

234   *R. F. Madole and Others*

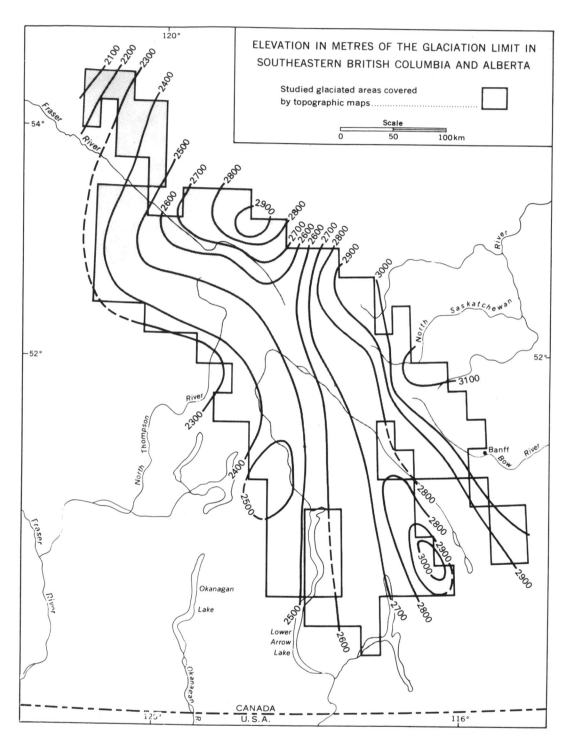

Figure 10. Iso-glacihypses (glaciation limit) for the central Canadian Rocky Mountains. Iso-glacihypses are dashed where inferred (from Ostrem, 1966).

rock system (lithology, jointing, bedding altitude), and the geomorphic system (cross section, roughness, altitudinal fall) (Sugden and John, 1976). Although it would be impossible to quantify the variables involved, an estimate on the amount of glacial erosion can be obtained by two approaches (Andrews, 1975). One deals with short-term (1 to 1,000 years) measurements of material carried in suspension and solution, and as bottom load in glacier meltwater streams; the other deals with gross change of the landscape morphology, which results from long-term processes (10,000 to 1 million years) such as the change in cross-sectional areas between a V-shaped river valley and a U-shaped glaciated valley. In the long-term approach, one considers the potential volume of material either deposited or eroded by glaciers and converts the volume to a rate of erosion by estimating the total age of the features (Andrews, 1975). Rates of erosion calculated by both approaches vary a great deal, from about 50 to 1,000 mm/1,000 years. These methods have not been applied to the central Canadian Rockies. Until we know more on the total age of the features, as well as have more data on other factors, the exercise would be futile. However, we can speculate on the total amount of glacial erosion. The present-day drainage pattern within the mountains and across the prairies to the east is not much different than in preglacial times. All ancient channel deposits, stream courses, and Tertiary gravels can be explained by drainage patterns similar to those of today. This leads to the conclusion that there has not been the "wholesale" change of morphology and drainage that might be expected had a great amount of glacial erosion taken place. In fact, there is reason to believe that valley glaciers merely smoothed out protuberances and rough surfaces while eroding no more than tens of meters of material during the modification of valleys from V-shaped to U-shaped (Rutter, 1980).

Cirques are major erosional features in glaciated alpine regions. Other glacial alpine features such as cols, arêtes, and horns form as a result of cirque formation. In the central Canadian Rocky Mountains, well-developed, amphitheater-shaped, valley-head cirques are found throughout the northwest-trending Main Ranges, parallel to the Continental Divide. Some are better developed with more "classical" shapes than others; most have steep headwalls, some have a rock threshold enclosing a tarn, and most have a ratio of height to length of about 1 to 3. Adjacent, and back-to-back valley-head cirques commonly form horns in the area of the Continental Divide; Mount Assiniboine, on the Alberta–British Columbia border, is probably the finest example of a horn in North America (Fig. 8). As mentioned earlier, some valleys have as many as three successive "valley-head" cirques. In addition, cirques that form discrete basins but do not have the "classical" shape of valley-head cirques occur along the flanks of many valleys.

Although many writers have described the general morphology of cirques, there is a paucity of quantitative data on length, width, and height, and how they relate or compare to other cirques. Exceptions to this are discussed by Andrews (1975). The length-to-height ratio appears to be particularly useful in defining shape and comparing the similarities and differences of cirques in various regions. In the central Canadian Rockies, studies concerning cirques are rare except for studies related to present and past ELAs (Ostrem, 1966; Trenhaile, 1975). The only two detailed studies are by McLaren and Hills (1973) and Trenhaile (1976). In McLaren and Hills' study (1973), 15 variables were measured in 53 cirques, to predict the maximum extent of a specific late Wisconsinan advance. Results indicated that maximum ice extent can be predicted within certain limits with 68% accuracy. In Trenhaile's (1976) study, four parameters (elevation, orientation, area, elongation) were used to express cirque morphology. Among other things, results showed that cirques are most frequent, best developed, largest, and at lowest elevations on the northeastern flanks of mountains.

Most cirques develop by glacial action from previously water-eroded features or from nivation hollows (Embleton and King, 1975). Any theory of cirque erosion has to account for lengthening, widening, and deepening as a function of time (Andrews, 1975). Although there are many unanswered questions, we can say that headwall shattering by freeze-thaw action is one of the key factors in cirque development. It appears that the most favorable environment for freeze-thaw action is the exposed wall rock above the glacier. This action will lengthen and widen the cirque but will have little effect in deepening it (Embleton and King, 1975). For deepening to occur, some sort of erosion at the base of the glacier is needed, which precludes a warm- (wet-) based glacier. One proposal by Weertman (1971) stressed the importance of long-term changes of the 0°C isotherm, producing a prolonged series of freeze-thaw cycles that could result in considerable erosion. Another proposal involves glacial abrasion and plucking, and entails rotational movement of the glacier acting as a rigid body sliding over its bed, as discussed by McCall (1960). This, in fact, may be the main cause of overdeepening of cirques.

GLACIAL CONSTRUCTIONAL LANDFORMS

The central Canadian Rocky Mountains are distinguished by constructional, streamlined (drumlinoid, fluted) landforms, mostly in glacial drift in many of the major valleys. Our understanding of the origin of these landforms is still incomplete, although in the past few years there has been a dramatic increase in our understanding of glacial processes involved in the development of depositional landforms. This understanding has resulted from the observation of modern glaciers, the detail analyses of glacial diamictons, the development of theoretical concepts, and the organization of data into useful classifications (Harrison, 1957; Hartshorn, 1958; Boulton, 1968, 1970, 1972, 1978; Boulton and Eyles, 1979; Embleton and King, 1975; Goldthwait, 1971; Shaw, 1971, 1977a, 1977b, 1979; Andrews, 1975; Stankowski, 1976; Eyles, 1979; Eyles and others, 1982; Lawson, 1979, 1981, 1982; Schlüchter, 1979; Proudfoot, 1985; Levson, 1986; Mandryk, 1986).

It is well established that a warm- (wet-) based valley glacier deforms and flows internally as well as sliding along its base.

Debris is initially added to the glacier surface from rock falls and, at its base, from abrasion and plucking of bedrock or of preexisting glacial deposits or other material. Depending upon flow characteristics, this material can be continually shifted from one position to another during transportation—carried either supraglacially, englacially, or subglacially. However, supraglacial drift probably never gets into a subglacial position. What is not as well established is identifying the process involved from the character of the final depositional product. Not only is the process of release from the glacier unclear, its position of deposition and possible reworking are difficult to ascertain. In addition, the glacial dynamics involved in producing the resultant landforms are often speculative.

The final products of glacial-debris transport have been organized into a number of genetic till (diamicton) classifications (Boulton, 1976a; Dreimanis, 1976; Lawson, 1981). Among many categories, most classifications at least subdivide till into supraglacial (ablation) and basal. Basal till is further subdivided into basal meltout and lodgment types. Although this is as far as we have been able to subdivide till in the central Canadian Rocky Mountains at the present time, this has helped in our understanding of till deposition and landform formation.

As the streamlined (drumlinoid, fluted), constructional landforms are the most widespread glacial landform in the region, an explanation of their origins is needed. The voluminous literature on the descriptions and theories of streamlined landforms includes good reviews by Smalley and Unwin (1968), Menzies (1979), and Smalley (1981). One approach that has been used in the Banff and Jasper areas is the identification of a variety of diamicton facies; the interpretations of their origin have been helpful in understanding the processes of sedimentation and the associated landforms. Although such studies have also been carried out in the Jasper area (Levson, 1986), the example explained in detail below is from a section in the Banff area studied by Mandryk (1986).

The Powerhouse section is located along the Trans-Canada Highway 5.5 km northeast of the town of Banff, Alberta. The section is oriented northwest-southeast and is situated in the middle of a 7-km-wide, broad, U-shaped portion of the Bow River valley (Figs. 8 and 9). Mountains comprised of carbonate rocks bound the Bow River valley area; valley bottom elevation is about 1,375 m, with surrounding mountains about 2,740 m. The ground surface above the Powerhouse section has drumlinoid topography, the trend of which is parallel to the former ice-flow direction in the valley.

Descriptions of the various sedimentary units making up the Powerhouse section are given in Figure 11. Using these data, interpretations were made regarding the genesis of each sedimentary unit and glacial history.

The sediment comprising the Powerhouse section is interpreted to have been deposited subglacially as lodgment till. Evidence that it was deposited subglacially is based upon the following points. First, the drumlinoid features present on the surface indicate that the underlying till may have formed beneath

a glacier (Muller, 1974; Boulton, 1976a; Menzies, 1979). Second, most fabric orientations are parallel to the former ice-flow direction and, therefore, indicate subglacial till (Kruger, 1979; Haldorsen, 1982; Fig. 11). Third, the homogeneous roundness (Fig. 11) throughout the section with an absence of frost-shattered, less-rounded material toward the surface indicates that all the material was transported subglacially (Dreimanis, 1976). Fourth, the vertically homogeneous till-matrix graphic statistics (Fig. 11) do not show the large increases in mean grain size (Mz) or improved sorting (S.D.I.) upward that would be expected if a supraglacial-till component was present (Boulton and Eyles, 1979).

The following three characteristics indicate that the till is lodgment and not basal meltout. First, the gravel-rich layers (Fig. 11) are interpreted to be horizons of lodged clast, a common property of lodgment tills (Boulton, 1976b; Kruger, 1979). Second, the gravel layers are contained in a 24-m thickness of till. This is within the observed thickness range of lodgment tills (Boulton, 1976b) and beyond the thickness range of basal meltout till (Boulton, 1976b; Lawson, 1979). Third, the petrological composition of a series of samples collected along a vertical column is homogeneous (Fig. 11). This is expected in lodgment till and is interpreted to be the result of the plastering process during basal-debris deposition, which destroys any stratification inherited from within the glacial ice by smearing it out (Hyvaerinen and others, 1973; Haldorsen, 1977).

Once sedimentary units at the Powerhouse section were assigned genetic interpretations, the geologic history and the evolution of the geomorphology of the deposits could be established. Initial glaciation of the Banff area was marked by the advance of a valley glacier in which englacial clasts were randomly oriented. There was compressive flow of the glacier in the vicinity of the Powerhouse section, depositing diamicton with a few sand and gravel lenses and a poorly oriented clast fabric. As glaciation progressed, the valley glacier increased in thickness and achieved a wider valley profile, which caused extending flow. Lodgment of till continued, now depositing clasts with a strong orientation. Eventually, deglaciation occurred, to be followed by the subsequent readvance of the ice during the next glaciation.

During the second advance of ice into this area, a subaerial, proglacial stream deposited the upper gravel bed. The second glacial advance progressed in the same manner as the first. The initial compressive ice flow during the second advance locally produced more free water, resulting in more sand and gravel lenses within the diamicton than during the first advance. The valley glacier had very little supraglacial debris; evidence indicates that most of the debris transported by the glacier was in a basal position. Debris was melting out of the base onto a surface that reflected the shape of the glacier's base. This former surface is reflected in the modern-day, valley-floor drumlinoid topography (Fig. 9). During deglaciation, the glacier thinned but still flowed, maintaining the drumlinoid topography. When flowage stopped, the glacier must have been debris-poor because, as melting continued, the drumlinoid surface was not masked. The preservation

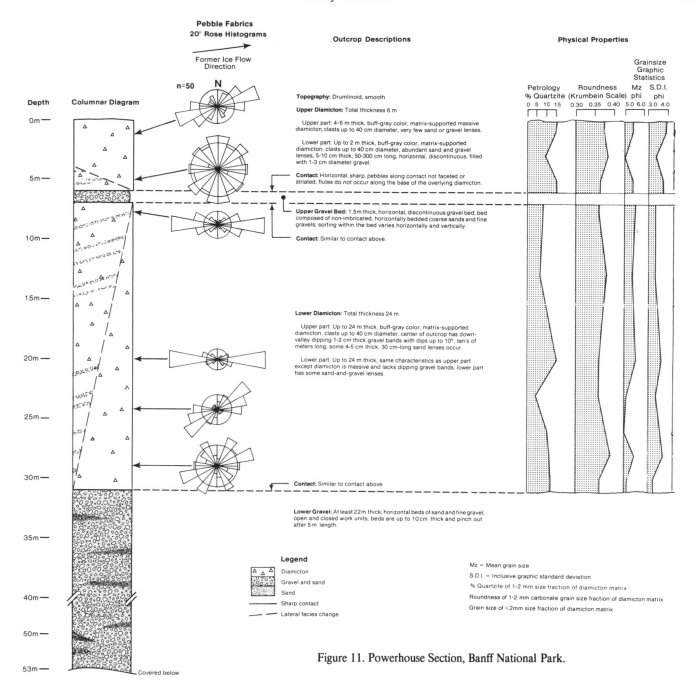

Figure 11. Powerhouse Section, Banff National Park.

of a nonhummocky, drumlinoid topography implies that most debris had melted out of the glacier's base while clean ice still filled the valley. This clean ice was essentially free of an insulating cover of supraglacial debris and, therefore, would have melted rapidly. Although this study may not explain all streamlined, constructional landforms found in the region, it goes a long way toward explaining the processes involved.

The relatively young, arcuate-end (including push and recessional) and lateral moraines, and ground moraine found in the higher alpine areas of the central Canadian Rocky Mountains have not been described in detail, nor their origin explained in

this paper. There is an extensive literature on these types of landforms, including reviews by Prest (1968), Flint (1971), Embleton and King (1975), Andrews (1975), and Sugden and John (1976).

End, push, and recessional moraines mark the snout of a glacier, whereas lateral moraines mark the former side-margin limits of a glacier within a valley. All these forms may contain a variety of till types and may be deposited by a variety of processes. For example, end moraines can be formed by (1) bulldozing and overriding of loose material as the glacier advances; (2) the accumulation of till at the glacier margin composed of debris that was dumped and subsequently slid into position after being carried from subglacial, englacial, and supragla-cial sources; (3) the formation of ice-cored moraines and the gradual letting down of debris as the ice core melts; and (4) the fracturing and thrusting of frozen consolidated sediments (Andrews, 1975). Further, more prominent end moraines may require relatively longer periods to form, implying a steady-state equilibrium for the glacier. There is evidence that such conditions existed in the central Canadian Rockies, only during the latter part of the Pleistocene period or within the Holocene period.

OTHER GLACIAL OR ASSOCIATED LANDFORMS

The streamlined landforms, commonly difficult to discern within the mountains, are better developed to the east beyond the mountain front. Here, swarms of well-formed drumlins spread eastward out of the valleys and extend mainly southeastward along the mountain front. In one area at least, south of the Athabasca River, there is a transition from drumlins into rogen moraine. Why these forms are better developed beyond the mountain front has not been investigated.

Within the mountains, there are surprisingly few ice-contact fluvial deposits, such as eskers and kames, although there are a number of subsurface deposits that have been identified as being of ice-contact fluvial origin (Rutter, 1972). As suggested earlier, within the mountains, stagnant ice containing supraglacial and englacial debris seems to have been a relatively minor component of this glacial environment during most of the Pleistocene. Outside the mountain front, however, there are wide areas of ice-contact glacial landforms.

Outwash forms discontinuous, high-level terraces about 20 to 40 m above the present drainage in major outlet valleys such as the Bow, North Saskatchewan, and Athabasca. It commonly abuts till deposits a few meters above the terrace level and forms steps with younger terraces below.

Eolian deposits related to glaciation are a minor part of the landscape. Sand dunes have been formed from material derived from exposed, sandy lacustrine plains east of Canmore and north of Jasper. Discontinuous patches of loess, up to about 0.5 m thick, locally overlie glacial and nonglacial deposits throughout the region.

CONCLUSIONS

There is no doubt that a great deal of information on glaciation has been derived from studies in the central Canadian Rocky Mountains. Most of the effort has been placed on Quaternary stratigraphy and glacial history. Unfortunately, the story is incomplete due to the lack of complete sections, datable materials, and such diagnostic landforms as lateral and end moraines. Even though the mountains display outstanding examples of glacial erosional and constructional landforms, there have been very few studies related directly to them—either here or in other parts of the Rocky Mountains. The fragmented information that we now have only gives us cause to speculate, often erroneously, on the origins of the landforms. More basic, systematically acquired, quantitative data from a population of landforms are needed in order to better interpret their origins. It is becoming apparent that an explanation of constructional landform origins cannot be separated from the sedimentological processes involved in their formation. The approach now being undertaken in the Banff and Jasper area can aid in explaining why the depositional style varies between different parts of the Rockies, that is, why there are end and lateral moraines in many Pleistocene deposits of the United States Rockies but not in many parts of the central Canadian Rocky Mountains.

ALPINE MASS WASTING IN THE INDIAN PEAKS AREA, FRONT RANGE, COLORADO

Colin E. Thorn and Deborah S. Loewenherz

INTRODUCTION

Alpine geomorphology is, by definition, restricted to the zone above timberline (Löve, 1970) in mountainous (Barsch and Caine, 1984) areas. Despite these limitations, it is a topic of concern in both mountain and periglacial geomorphology. These subdisciplines include diverse research themes, which, in the case of mountain geomorphology, are illustrated in Figure 12. The great range of research scales that have been investigated in the Rocky Mountains is exemplified in this chapter by Bradley's discussion of erosion surfaces and Rutter's illustration of glacial processes. Alpine mass wasting has been studied over a similar range of scales.

At the regional scale, alpine and (or) periglacial mass wasting has been directly associated with the erosion surfaces examined by Bradley (this chapter). Although the mass-wasting process was a central concept in the explanations of these surfaces in the studies by Russell (1933), Mackin (1947), and Thompson (1968), its actual role was seen as being distinctly different by each study. In the last two decades, however, research at this scale has fallen into disfavor and, as Bradley points out, publication of it "withered but did not die."

The move away from regional studies (which may be assigned to the "relief generation and history" category in Fig. 12) has simultaneously been a move toward "process dynamics and activity" (Fig. 12). The core components of the latter approach have been field measurement of processes and construction of process-response models. This methodological approach, with its inherent logistical restrictions, has resulted in studies of small spatial scale and short temporal duration. The process-geomorphology approach has also heightened awareness of local variability and resulted in a reluctance on the part of researchers to extrapolate results over the regional scales previously employed. This trait is typified by the current trend of using local names to identify alpine glacial sequences.

Any attempt to summarize alpine mass-wasting research in the Rockies is confronted by an entirely dichotomous situation.

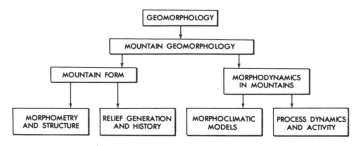

Figure 12. A conceptual outline of traditional themes in mountain geomorphology (after Barsch and Caine, 1984; published with permission of J. D. Ives, University of Colorado).

On the one hand, there is a limited number of dated, regional studies; by contemporary standards, these lack adequate verification. On the other hand, there is a burgeoning number of contemporary process studies, which by their very design contribute to our understanding of landform evolution only at microscales or mesoscales. The difficulty is compounded by the absence of an adequate body of geomorphic theory with which to link the two.

In our opinion, scale linkage is the greatest intellectual challenge confronting geomorphologists. Consequently, our intention is to summarize the alpine mass-wasting–process research of the last two decades within a methodological framework that lies between those frameworks of traditional regional studies and those of process studies. At the present time, the Indian Peaks, in the Front Range of Colorado, is the only intensively studied area to which this perspective has been applied and, for this reason, is used here as a case study. It is important to emphasize that the Indian Peaks are merely the illustrative material for what is intended as a methodological contribution. The area is not being presented as a type site for the Rocky Mountains; indeed, we believe it is a profound misconception to view the Rocky Mountains as a geomorphic entity that could be typified by a single area.

THE INDIAN PEAKS

Using the paradigm most frequently accepted by contemporary process geomorphologists, the complex structural history of the study area is taken as a set of given constraints. This history has been examined by Wahlstrom (1940, 1947), and described by Richmond (1974). The area has extensive exposure of Precambrian gneisses, granodiorite, and monzonite, plus some early Cenozoic stock materials. The Eocene erosion surfaces discussed by Bradley (this chapter), although warped and dissected, create unusually large upland surfaces throughout the area. The Indian Peaks (Fig. 13) themselves have been uplifted some 300 m above summits to the south and north. Niwot Ridge, which is partly mantled by a diamicton of uncertain age and origin (Madole, 1982), is one of the large erosional remnants.

Most peaks in this area are 3,900 to 4,100 m in elevation;

North Arapaho, at 4,118 m, is the highest. The north–south–trending Continental Divide and abutting valley bottoms exhibit classical glacial features with relative relief as great as 700 m from the summits to the cirque bottoms. Many of the west-east ridges that run eastward from the Divide are arêtes at their western ends, but become broad, gently rolling, and regolith covered at their eastern ends; Niwot Ridge is a prime example of this phenomenon.

Alpine glaciation, while recognized as commonly operating at a smaller scale than regional structural activity, is also normally accepted by process geomorphologists as another large-scale process that serves to precondition the environment within which mass wasting functions. This viewpoint leads to the assumption that present-day mass-wasting forms postdate the "last glaciation." However, it must be realized that at this scale the "last glaciation" is recognized as having occurred at significantly different times on various parts of the landscape. This leads to an important distinction between those mass-wasting forms that appear to require glacial preconditioning of the landscape (e.g., forms derived from glacially eroded free faces) and those that appear not to require such preconditioning (e.g., turf-banked lobes).

The glacial chronology of the Indian Peaks has been studied extensively (Madole, 1976, 1986; Meierding and Birkeland, 1980). Absolute-age dates are scarce due to a paucity of material suitable for isotope dating, although local studies have contributed considerably to development of relative age-dating techniques (Carroll, 1974; Dowdeswell, 1982). Type names from Wyoming (pre-Bull Lake, Bull Lake, and Pinedale) are generally used for the older glaciations, all of which produced large-scale valley-glacier systems reaching 10 to 15 km eastward from the divide.

Between about 5,000 and 300 B.P., small glaciers formed at three different times (Benedict, 1973a), but were confined almost exclusively to established cirques or to very short distances downvalley. At present, the immediate Indian Peaks area contains seven cirques to the west of the divide and eleven to the east. Two of those to the west are presently occupied by small drift glaciers or Ural-type glaciers (Outcalt and MacPhail, 1965); four cirques to the east have similar glaciers. White (1982) discussed both the structural geology and glacial history of the region in considerably greater detail.

The climate of the region is known reasonably well by mountain standards, thanks to the pioneering work of Marr (1967; Marr and others, 1968a, 1968b), to the subsequent extension of Marr's work and analysis by Barry (1973), and to the compilation of current climatic summaries (M. V. Losleben, University of Colorado, written communication, 1983). Marr organized the original network of stations by major vegetation types; D1 (Fig. 13), the original alpine-tundra–type site at 3,750 m elevation on the western end of Niwot Ridge, has one of the longest continuous high-elevation records in North America, dating back to 1952.

From a geomorphologist's point of view, it is particularly

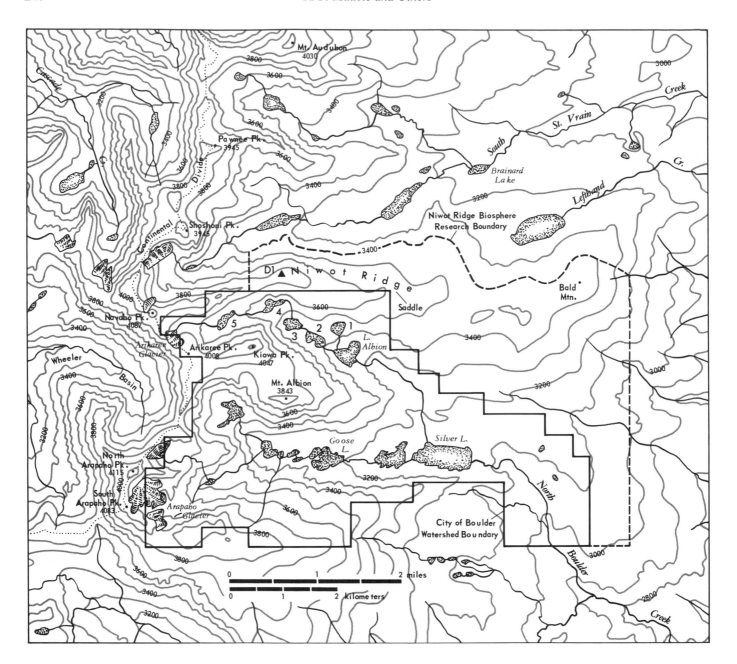

Figure 13. The Indian Peaks region, Front Range, Colorado (after Ives and Dow, 1981; published with permission of J. D. Ives, University of Colorado).

important to note Barry's (1973) comments concerning the difficulty of measuring precipitation accurately in alpine environments. This problem is exacerbated by high windspeeds; D1 has an annual mean windspeed of 10.3 m/sec and mean monthly windspeeds >11.8 m/sec from October through March. Since much of the precipitation falls as snow, it is quickly redistributed by the wind, resulting in a decoupling of the meteorological screen and ground climates that has profound geomorphic implications (Fahey, 1973; Thorn, 1979a). While snowpack accumulation and ablation patterns, as well as glacier-mass balances, are critical geomorphic inputs, they are only presently being examined in this light as part of the University of Colorado's Long-Term Ecological Research Program (T. N. Caine, University of Colorado, written communication, 1985). Existing snowpack research includes snow-course records (Washicheck, 1972), studies of alpine snowpack management (Martinelli, 1959a, 1959b, 1975), and microscale accumulation modeling (Berg, 1986). A number of mass-balance studies (Alford, 1980; Johnson, 1979) have revealed annual inputs of as much as 3-m water equivalent for the local drift glaciers.

Carroll (1976), Caine (1982), and Greenland and others (1984) provide the only hydrologic data that may be applied directly to geomorphology. Carroll's 1-year study of the upper Green Lakes valley revealed a very high watershed efficiency with runoff exceeding 90% of inputs (excluding wintertime sublimation). Partial contributing-area concepts were applied to both hydrology and geomorphology by Caine; to date, the primary result is revelation, but not resolution, of extremely complex patterns in both contexts.

Niwot Ridge is, and has been, the focus of an enormous amount of botanical research on alpine tundra. From a geomorphic perspective, two papers are of particular interest. Webber and May (1977) found that substrate moisture, snow cover, and substrate disturbance (in order of importance) are the primary controls on plant growth. In 1978, Komárková and Webber published a vegetation map of the alpine zone of Niwot Ridge that has provided an interesting stratification criterion for geomorphic studies.

This summary provides a glimpse into the depth and diversity of the research that has already been completed in the Indian Peaks; the data from this research constitutes an enormous resource with which to support studies of geomorphic processes. For those who wish to pursue these topics further, Ives (1980), the journal *Arctic and Alpine Research,* and Halfpenny and others (1986) provide the most appropriate starting points; in addition, many of the topics are developed on a global scale in Ives and Barry (1974).

A METHODOLOGICAL FRAMEWORK

The intermediate methodological framework presented here was developed by Caine (1971, 1974, 1976a, 1979, 1982, 1984, 1986). A portion of this framework is derived from work in the San Juan Mountains of southwestern Colorado (Caine, 1976a, 1976b, 1979), but the upper Green Lakes valley (above the outlet of Green Lake 4, Fig. 13) has been the primary research area. The 1971 paper is a preliminary one in which the frailties of our grasp of process-form links were exposed and a scale-linkage scheme spanning plot, slope, and basin scales was presented.

Beginning with the 1974 paper, Caine developed a model focused upon sediment fluxes within a drainage basin. The basic landform elements considered in this model are depicted in Figure 14. These individual elements are grouped into two morphological subsystems: the hillslopes and the drainage network; the latter will be considered here only briefly. Within and across these morphological subsystems, three cascading systems (sediment fluxes) move material; these cascading subsystems are coarse-debris (~>8 mm), fine-sediment (~<8 mm), and geochemical. The attraction of this framework is that it is centered on a larger scale than the standard process-response model, thus providing process geomorphology with necessary generalization. On the other hand, it is at a scale considerably smaller than that employed in basin-wide or regional studies and, therefore, provides these studies with necessary detail. The addition of a glacial

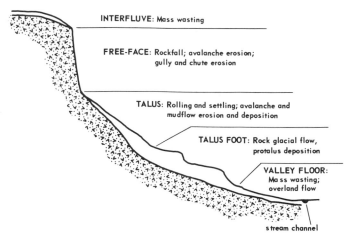

Figure 14. Landform elements commonly present in the Indian Peaks region, Colorado (from Caine, 1974; published with permission of N. T. Caine, J. D. Ives and R. G. Barry, eds., and Methuen and Company, Limited).

subsystem to the three cascading subsystems (Barsch and Caine, 1984) is rejected here because it appears to create a mixed set of transportation-classification criteria.

Caine's (1974) framework, as modified by his later papers, will be used as the organizational format hereafter. Other methodological contributions will also be considered within appropriate sections of the framework. Only skeletal components of available process measurements will be presented, although a comprehensive set of citations will be made. Where there are sufficient data, each sediment flux will be considered separately within each of Caine's hillslope subsystem components. Thereafter, the fluxes will be reconsidered from a basin-wide perspective.

THE INTERFLUVE

Research on interfluves within the Indian Peaks has been centered almost exclusively on Niwot Ridge (Fig. 13). This massive ridge starts at the divide as Navajo Peak, a horned peak, and extends eastward for about 2 km as an arête, which terminates a little to the west of the D1 weather station. This western end of the ridge falls into Caine's (1984) "high alpine" category, as it is composed of largely unvegetated bedrock and coarse debris. The ridge then extends some 8 km farther eastward beyond D1 as a broad, rolling, alpine tundra surface, recognized by Caine (1984) as the "alpine tundra" geomorphic zone. It is this much larger alpine-tundra portion of the ridge that has been the focus of geomorphic, as well as botanic, research. The alpine-tundra segment of Niwot Ridge has probably never been subject to direct glaciation (Madole, 1982), although it has experienced considerable variation in the intensity of its periglacial regime (Benedict, 1970a). The geomorphic forms found on the alpine tundra are mesoscale to microscale spatially, and they are probably exclusively periglacial in origin and development.

Coarse-Debris Flux

Very little research has been conducted in the region on coarse-debris processes within interfluves. Block slopes, defined by White (1981) as "a sheet or blanket of rocks on a mountain slope at an angle of >10 degrees, extending parallel to the contour, over solid or weathered rock, with no visible rock source or cliff above," occur within upper alpine zones. However, aside from brief descriptions (Wallace, 1967; White, 1976a, 1981), they have received no attention. Although White (1981) recorded low rates of movement (1 to 2 cm/yr) on such features, he believed that they are relict. The genesis of block slopes, block fields, and block streams remains obscure. It is commonly assumed that the blocks were produced by freeze-thaw weathering and that movement originally occurred with the blocks included within a matrix of fines. However, neither of these situations has ever been observed directly, either in the Indian Peaks or elsewhere.

Coarse debris within the alpine-tundra zone has mostly been studied by virtue of its incorporation in processes that are dominated by fines; therefore, the bulk of the discussion will be in the following section, as this provides greater continuity. However, one or two salient points are raised here.

Frost creep (Benedict, 1970a) has emerged as an important mechanism for the local redistribution of coarse debris, which is the only apparent fate of coarse debris within the alpine-tundra zone. Benedict demonstrated that frost creep is dependent upon soil saturation during the fall freeze. Nevertheless, both present and past distributions of frost creep indicate that it reflects a drier regime than solifluction (gelifluction of Washburn, 1980). Morphological evidence of frost-creep–dominated microenvironments is restricted to unvegetated areas with generous supplies of coarse debris; here stone stripes produce stone-banked lobes and terraces where gradients decrease. All such forms also indicate that coarse material is moving downslope faster than abutting fines are.

The sorting that produces large-scale patterned ground creates highly localized movement of coarse debris. At present, this process is restricted to wet sites on high cirque floors, but it has been widespread in the past on the crests of windswept knolls (Benedict, 1970a). Highly localized movement of cobble-size material may also be produced by the burrowing of pocket gophers (*Thomomys talpoides* as noted in Thorn, 1978a); this is of significance primarily because it represents subsurface-to-surface movement. Another type of local redistribution of coarse material occurs within the unvegetated portions of late-lying snow patches where the washing out of fine material may leave lag surfaces (Thorn, 1979a), or even create local concentration in some instances.

Fine-Sediment Flux

One of the most significant conceptual changes to emerge from research in the Indian Peaks is recognition of the widespread importance of eolian infall to the alpine zone (Nash, written communication, 1968; Burns, 1980; Thorn and Darmody, 1980, 1985a, 1985b). This is a pervasive influence because the influx occurs across the entire alpine surface without regard to local variations in topography or surface type. The work by Thorn and Darmody (1985a) revealed that the influx falling onto Niwot Ridge and the Green Lakes valley is of uniform character; it has a mean grain size in the 7.38 to 3.21φ (0.0060 to 0.1081 mm) range, is not highly weathered, and has a heterogeneous mineralogy that makes determination of its provenance all but impossible. However, long-distance eolian transport into the region from the south is known to occur (Willard, 1979), and fallout carried in from the northwest from the Mount St. Helens eruptions was identified by Thorn and Darmody in their 1980 samples.

Caine (1974) estimated annual eolian input to represent a layer 0.005 mm thick. On the upper alpine-zone interfluves, this influx will be quickly blown and washed across bedrock surfaces and, thereby, concentrated in sheltered areas. It is presumably moved beneath the block-slope surfaces as well, although our knowledge of such processes is negligible. On alpine-tundra surfaces, its fate varies considerably depending upon surface texture and the availability of overland flow. The significance of this variation falls mainly into the geochemical domain, but there are some mechanical by-products. Probably the zones of greatest concentration occur along the downslope margins of large seasonal snowpatches where unvegetated core areas and abundant meltwaters combine to produce efficient redistribution.

Although it predates the emphasis upon eolian infall, Benedict's (1966, 1969, 1970a, 1970b) seminal, 7-year study of periglacial mass-wasting processes and forms did much to establish the Indian Peaks as a research center and still stands as an important milestone in alpine, periglacial mass-wasting research.

Benedict (1970a) undertook a detailed appraisal of frost creep and solifluction, including present and past rates of movement, and spatial and temporal interaction of processes and the ensuing forms. The result is a comprehensive summary, complete with context, of processes that have usually been studied only for much shorter periods and in much less detail. Methodologically, the significance of the study lies in the hierarchical placement of contemporary processes within a Holocene scale and careful separation of process and form. The distinction between process and form is critical in geomorphology. Benedict (1970a) used the process terms "solifluction" and "frost creep," and the form terms "turf-banked, stone-banked lobe and terrace;" and thereby overcame the misleading but inherent restrictions associated with morphogenetic terms such as "solifluction lobe." His field measurements clearly demonstrated the validity of this approach. A useful extension of the separation principle would be replacement of the term "solifluction" by "gelifluction," thereby incorporating Washburn's (1980) subdivision of solifluction.

Another methodological contribution was Fahey's (1973, 1974) work on diurnal and seasonal frost heaving, conducted along a transect running from D1 down into vegetation zones below the alpine tundra. This work served to establish the decoupling that exists between meteorological-screen temperatures and those at shallow depths within the soil, especially in the presence of a snowcover. The low frequency of even shallow diurnal freeze-thaw cycles within the alpine tundra, and an even greater scarcity of ground-heave cycles, led Fahey to discount their geomorphic significance. As with research elsewhere, Fahey's work shifted emphasis onto the role of seasonal freezing in the development of patterned ground.

Further reexamination of traditional concepts is reflected in Thorn's (1976; 1987) work on nivation, a concept that involves the intensification and (or) acceleration of widespread weathering and transport processes beneath and (or) around the margins of late-lying snow patches. A large snow-patch site on Niwot Ridge with an unvegetated core area revealed only localized redistribution of sediment within the confines of the snow patch, but at rates 20 to 30 times greater than those recorded on snowfree slopes; at another snow-patch site that was entirely vegetated, such transfer was negligible.

The standard process-response approach was avoided by Bovis (1978, 1982), who undertook plot studies designed to contrast soil-loss rates between major plant communities above and below timberline. Bovis found substantially greater losses beneath a Ponderosa-pine–bunch-grass cover than on either dry tundra or tundra meadow. Integration of Bovis and Thorn's studies (Bovis and Thorn, 1981), using Komárková and Webber's (1978) vegetation map to stratify results, revealed order-of-magnitude differences in sediment yields from different portions of the alpine tundra. Across the alpine section of Niwot Ridge, snow patches occupy only 3% of the area, but yield about 50% of the sediment total; tundra meadow occupies about 50% of the area, but yields only about 5% of the sediment total. Such results support Schumm's (1979) suggestion of the presence of step-function patterns in geomorphology.

In recent years, Thorn (1982a; Frank and Thorn, 1985; C. E. Thorn, J. E. Burt, D. E. May, unpublished data, 1986) has continued to pursue the use of plant communities as a means of stratifying surficial geomorphic processes on the alpine tundra. Order-of-magnitude differences in the burrowing of pocket gophers *(Thomomys talpoides)*, as well as in the frequencies and magnitudes of shallow freeze-thaw cycles, have been demonstrated between some alpine-plant communities. One important attribute of this approach is the feasibility of extrapolating geomorphic results using the remote sensing of plant communities as a surrogate measure (Frank and Thorn, 1985).

All the processes discussed here represent redistribution, which logically includes the topic of patterned-ground development. This topic is not developed here as it necessitates reference to a large, but separate literature. Benedict (1969, 1979), Fahey (1975), Ray and others (1983), and Rissing and Thorn (1985) have all worked on aspects of patterned ground in the region. The paper by Ray and others (1983) is especially noteworthy as it proposes a general hypothesis derived from linear-stability theory for the regularity observed in patterned ground. In ending this

section, it is important to reemphasize that all the surficial processes discussed in this section also mobilize material at the fine end of the coarse-debris category; in virtually all instances, it is the movement of fines that promotes the movement of coarse material as a secondary phenomenon.

Geochemical Flux

Investigation of chemical processes was developed late in studies of the Indian Peaks. This neglect was a reflection of the prevailing perspective in alpine-periglacial studies, which has traditionally assigned low rates and little importance to chemical processes in cold environments. Reappraisal is now taking place and work undertaken in the Indian Peaks has contributed to this trend.

Thorn (1976) found that beneath a large snow patch resting on colluvium, surface lowering was divided approximately equally between chemical and mechanical processes, and also that chemical processes within the snow patch area were 2 to 5 times greater than on nearby snow free areas (Thorn, 1975). However, such sites represent only a small fraction of the alpine-tundra surface. Dixon (1983, 1986) has pursued detailed geochemical work on a limited number of catenary sequences. His results suggest that alpine geochemical processes are similar to those found in temperate regions and occur quite rapidly. However, work is only just beginning (Caine, 1982) on geochemical processes at a basin scale, and the complex pattern of partial contributing areas created by a melting snowpack overlying a highly varied surface presents a research problem that will require much more attention before it can be resolved.

Burns' (1980; Burns and Tonkin, 1982) study of pedogenesis throughout the Indian Peaks region provides an areal context for some of the geochemical work underway. Strong relationships emerged between soil types and vegetation communities across the alpine tundra and, in turn, these may be tied directly to snowcover. Burns did not undertake direct catenary studies, but his work did reveal such patterns when his sites were synthesized into slope sequences. One extremely important suggestion made by Burns (1980) is that chemical weathering of fines (primarily derived from eolian influx) dominates pedogenesis and that chemical weathering of coarse debris contributes very little.

In summary, Thorn's (1976) and Caine's (1982) determinations of the relative importance of chemical solution, Dixon's (1986) identification of processes akin to temperate environments, Dixon's (1986) and Caine's (1982) identifications of complex transfer patterns (at the slope and basin scales, respectively), and Burns' (1980) suggestions concerning domination by eolian input would seem to suggest that the geochemical situation on the interfluves merits serious reappraisal unfettered by traditional preconceptions.

THE FREE FACE

The free face, although widespread, has received no direct attention in the Indian Peaks. However, structural controls, with associated hydrothermal activity, and glaciation appear to dominate free-face formation in the area. At a much smaller scale, individual weathering mechanisms must be involved. Thorn (1979b, 1980, 1982b) examined the role of bedrock freeze-thaw weathering as part of his evaluation of nivation, and White (1976b) discussed this same topic. Both workers are among those who currently question the widespread, but unverified, assumption that freeze-thaw weathering is the dominant bedrock-disruption mechanism in alpine environments. Most local investigations of free-face evolution have been pursued by indirect means, principally by studying slide-rock accumulation, and it is to talus that we now turn.

THE TALUS

Talus in the Indian Peaks is derived largely from glacially eroded free faces and must, therefore, postdate the most recent glaciation at each location. The movement of fines and water through talus is poorly understood because it most commonly occurs beneath a surficial layer of coarse slide rock. In contrast, knowledge of the behavior of coarse debris is quite well known, and this is the only flux for which there is sufficient local data to merit discussion.

Coarse-Debris Flux

White's (1968) work in the Indian Peaks led him to subdivide talus into rockfall, alluvial, and avalanche types—a now widely accepted categorization. Each type exhibits characteristic sorting (and fabric in the case of rockfall; McSaveney, 1971), occurs in predictable topoclimatic contexts, and exhibits distinctive downslope profiles (White, 1981). White (1981), Caine (1974, 1984), and Wallace (1967) have all noted the extremely erratic nature of talus shift, while the long-term measurements of the first two authors have revealed very low rates of movement in the region, generally less than 5 cm/yr.

Virtually all of this type of research in the Indian Peaks has been on rockfall talus. This may be an important limitation because Caine (1974) suggested that one of the important distinctions of alluvial talus is its ability to deliver material all the way to the valley bottom. Although small-scale snow avalanching is common in the area, large-scale avalanches are all but absent. In general, Caine (1974) suggested that talus processes fail to deliver material to the stream/lake system and that movement either ceases at the foot of the talus or is taken up by other transfer mechanisms.

THE TALUS FOOT

Valley-wall or lobate rock glaciers, a talus-foot phenomenon, are abundant throughout the Front Range of Colorado. As a discrete type of landform, they have received considerable attention (Outcalt and Benedict, 1965; Wallace, 1967; White, 1976a, 1981), and they have not generally been a contentious topic. However, research has been focused upon the surficial layer of coarse debris that mantles them. The amalgam of ice, fines, and coarse debris that forms the bulk of their mass is largely unknown. Accordingly, discussion here must be confined exclusively to their role in the coarse-debris flux.

Coarse-Debris Flux

Lobate rock glaciers, which apparently require a couloir or similar form to concentrate debris and snow delivery, are ice-cemented rather than ice-cored, and have surface features indicative of flow. Rates of movement are broadly related to size (White, 1981); large rock glaciers (or those on particularly steep slopes) flow at annual rates of 50 to 160 cm/yr, while smaller ones flow at much lower rates of 5 to 10 cm/yr. If estimates of their discharge are accurate, these features are second only to talus itself as a transfer mechanism of cliff debris.

Other characteristic forms in the talus-foot zone include aprons of very large boulders and protalus ramparts (White, 1981). The position of the very large boulders is a direct product of freefall sorting. The protalus ramparts are assumed to have been formed by the classical method of coarse material running across seasonal snow patches at the base of a talus; none in the region is presently active.

THE VALLEY FLOOR

Examination of mass wasting within a valley bottom is an inherently difficult task because of the interaction with fluvial processes. In the Indian Peaks, the impact of recent glaciation dominates valley-bottom surfaces, which are extremely variable, being mixtures of bedrock outcrops, large boulders, and dense vegetation. With the exception of tongue-shaped rock glaciers, mass-wasting forms in the valley bottom have received virtually no attention in the Indian Peaks, although Caine (1974) reported that none of these surfaces exhibits any signs of being particularly active. There are some special contexts, however, that produce distinctive forms, which represent extremely localized transfer. Ephemeral ponds may promote well-developed sorted polygons, for example, and a complex of only poorly understood factors may create subnival boulder pavements (White, 1972; Hara and Thorn, 1982).

A number of cirque-floor, valley-bottom, or tongue-shaped rock glaciers occur within the Indian Peaks (Outcalt and Benedict, 1965), although attention has fallen disproportionately on the valley-bottom Arapaho rock glacier (White, 1971a, 1971b, 1975, 1976a, 1981; Benedict, 1973b; Whalley, 1974a). This is very large by local standards and, according to Wallace (1967), measures about 250 m wide and 610 m long, is 21 to 27 m thick in its lower section, and has surface-flow forms about 4 to 6 m high. Unlike lobate rock glaciers, the valley-bottom rock glaciers have been a major source of controversy in the literature. Debate has centered on whether these features represent the product of several rockfalls, single catastrophic rockfalls, degenerate glaciers,

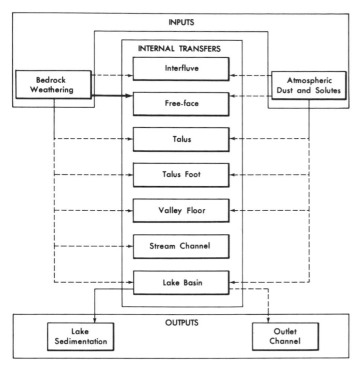

Figure 15. Outline of sediment fluxes in the Indian Peaks region, Colorado (from Caine, 1974; published with permission of N. T. Caine, J. D. Ives and R. G. Barry, eds., and Methuen and Company, Limited). Shafts of arrows (dashed, solid, and thick) show increasing relative magnitude of waste transfers in the alpine system.

THE DRAINAGE BASIN

Caine (1974, 1982, 1984, 1986) has created a new scale perspective on geomorphic processes within the Indian Peaks by amalgamating processes into fluxes and examining their significance at the basin scale. This change has been achieved by standardizing original process-study data as surface-lowering rates or as work (Caine, 1976c), adding his own long-term measurements (Caine, 1984), and placing everything within the framework described earlier. This may be envisaged as a flow diagram (Fig. 15).

The crest of Niwot Ridge has a surface-lowering rate of less than 0.01 mm/yr (Caine, 1974), an extremely low one by high-elevation standards. Similar calculations for cliff retreat reveal rates of about 0.76 mm/yr. Therefore, removal of material from cliff faces appears to be 10 to 100 times faster than from the interfluves. While cliff-retreat rates are in good agreement with debris-discharge estimates for talus and lobate rock glaciers, available evidence suggests that the discharged coarse debris does not enter the fluvial/lacustrine network. This means that, to a very large degree, it is valid to regard the interfluves, valley walls, and streams and lakes as three separate systems.

The coarse-debris flux on the valley sides is the dominant alpine process today with respect to mass moved over distance (Caine, 1984). It is a transfer that is internal to the cliff-talus and lobate-rock-glacier complexes. This makes it a closed system in the contemporary scene and one that is presumably "opened" only by glaciation. Such opening must be complicated both spatially and temporally, as valley glaciers have had variable down-valley extent and may not always occupy the entire width of the valley. Furthermore, they have exhibited variable erosive power during the Holocene (Reheis, 1975).

Sediment of grain sizes comparable to that in the eolian influx dominate local alpine-lake sediments (Caine, 1974; Andrews and others, 1982, 1985). This implies that only eolian material is capable of passing through the entire mass-wasting system into the fluvial/lacustrine one; in passing, it should be noted that it also indicates the high efficiency of alpine lakes as sediment traps. Thus, any calculations derived from lake sediments must include adjustment for this input. Holocene lake-sedimentation rates imply surface-lowering rates of 4 to 10 mm/1,000 yr without adjustment (Caine, 1984). However, as Caine noted, it is not possible to compare contemporary process rates with values averaged over the entire Holocene, particularly when it is known that both glacial erosion (Reheis, 1975) and interfluve mass wasting (Benedict, 1970a) have fluctuated markedly over this time. In addition, the absolute rate of eolian influx, its relative importance, and the efficiency with which eolian material is transferred are all subject to variation during the same time.

Direct estimates of contemporary rock weathering in the alpine zone are extremely sketchy and crude (Caine, 1984), but suggest surface lowering of about 10 mm/1,000 yr for the entire Green Lakes valley and half that for the upper Green Lakes

or some combination thereof. The uncertainty has also been heightened by the fact that, on the east side of the divide, the valley-bottom rock glaciers are generally ice cored (including the Arapaho one; Benedict, 1973b), whereas those on the west side of the divide are all ice cemented.

White (1971a, 1971b, 1975) measured long-term movement and debris discharge on the Arapaho rock glacier. The movement was recorded by resurveying marked boulders; the emergent pattern was one of extremely erratic motion, averaging about 5 cm/yr. Movement at the snout was dominated by boulders falling from high up on the face. White (1975) considered the genesis of rock glaciers in detail; they appear to be degenerate, debris-rich ice glaciers fed by many rockfalls, as opposed to being the result of a single catastrophic event.

As with most rock glaciers, the coarse surface gives way to finer material at depth within the Arapaho rock glacier; however, there are no data concerning the behavior of these fines. Although White (1971b) has calculated the debris discharge of this rock glacier, neither this discharge value, nor that of any other valley bottom–rock glacier has been incorporated into the sediment transfer model developed by Caine.

valley (T. N. Caine, University of Colorado, written communication, 1985). Estimates from stream solutes are only about 4 mm/1,000 yr for the alpine zone, but are much higher in the subalpine. Stream-solute loads may be influenced by water reactions with eolian infall, but, as the contributing partial areas are so poorly understood, this issue is unresolved.

In summary, Caine's findings have revealed that the Indian Peaks region is a relatively inactive alpine zone with a poorly articulated geomorphic regime at the basin scale. The surfaces exhibit distinct step-function behavior at a number of scales with respect to both water and sediment transfer (Caine, 1982; Thorn, 1978b). Zones of high activity are surrounded by an inactive matrix that makes localized redistribution the dominant contemporary mode for coarse debris. In the clastic realm, only the eolian input appears capable of traversing the entire system. The geochemical scene appears to require profound reevaluation, particularly with respect to the relationship between bedrock weathering and water solutes. This entire complex is subject to important seasonal fluctuations in contributing area as a response to the seasonal melting of highly variable snowpack accumulation. The interaction between various surface types and available meltwaters is sufficiently complex in and of itself that, while patterns of water and sediment yields are usually coincident, they are occasionally disjunct (Caine, 1982).

DISCUSSION

Prior to discussing the issues in a broader vein, it is appropriate to summarize the salient characteristics of the Indian Peaks. Most importantly, the region must be recognized as a rather inactive alpine environment with surface-lowering rates as much as two orders of magnitude lower than some other alpine landscapes. The pedestrian rate of change is produced by low-energy inputs (the area lacks glacial energy) acting upon very resistant surfaces. Next in significance are the poorly developed linkages within basins, which leave interfluves, valley sides (hillslopes), and the stream/lake network as largely independent entities with respect to coarse-debris transfers. Eolian infall appears to dominate the fine-clastic flux and is transferred throughout the alpine zone, apparently accumulating in the lakes. The pattern of geochemical transfers is complex and, although still poorly understood, of much greater importance than has been generally recognized. In short, the strength of the Indian Peaks' geomorphic record lies in knowledge of what may be called the pattern of landscape organization.

In attempting to extend this knowledge, it is important to consider whether the Indian Peaks have experienced or are currently experiencing the full range of alpine mass-wasting processes. The landform array suggests that the majority of commonly reported processes are present, or have been in the recent past, with several notable exceptions.

Among the exceptions appears to be the low frequency of rockfalls. In a 20-year period, Caine (1986) reported that only three events in the Indian Peaks involved more than 5 m^3. These results may be contrasted with rockfall data for the Kananaskis and Lake Louise regions of Alberta where Gardner (1980, 1982; Gardner and others, 1983) not only reported high rockfall frequencies, but also presented data for very large-magnitude/low-frequency events; this type of event was termed a bergsturtz by Whalley (1974b). Large rockfalls and bergsturtz must not only be considered for their role in free-face development, but also for their capacity to deliver material to the valley bottom and, thereby, link the hillslope and drainage network subsystems.

Flood-debris flows (Mears, 1979; Gardner, 1982) also constitute an intermediate process combining both fluvial and mass-wasting attributes. Whatever classification difficulty they present, debris flows are enormously effective in transporting material. It seems reasonable to assert that, when they are more fully integrated into the spectrum of mass-wasting mechanisms, they may well be recognized as one of the preeminent transfer processes in some alpine regions. Yet another process that may effectively link hillslopes and drainage-network subsystems, as well as create distinctive landforms, is snow avalanching. This process is poorly represented within the Indian Peaks but is very important in many parts of the Rockies (Luckman, 1972, 1981).

A final geomorphic factor that is only spasmodically present in the Indian Peaks is permafrost (Ives and Fahey, 1971; Ives, 1973). There are no well-developed ideas that establish the geomorphic significance of the difference between deep seasonal freezing and permafrost. Indeed, the variability of soil-moisture content in either context is likely to be at least as important a consideration as the precise distinction between deep seasonal and annual freezing. Nevertheless, any consideration of the Rockies at large must include an appreciation of the superimposition of an increasingly arctic regime upon the alpine zone as one moves northward.

While the Indian Peaks lack some important process components, the primary weakness in the record, that is, the small spatial scales and short temporal scales studied, is shared with all other alpine regions in the Rockies. As noted at the outset of this paper, and also by Gardner (1982), geomorphology in general, as well as alpine geomorphology specifically, is bipolar. Therefore, in considering the most appropriate way to extend the Indian Peaks' record, we are examining but one example of the general question of how to make process geomorphology more relevant to landscape development.

It is quite apparent that sediment fluxes represent landform maintenance rather than landform initiation. Landform initiation has not been considered in any comprehensive way in alpine mass wasting. Occasional examples may be found, but these generally consider high-magnitude/low-frequency events whose by-products have only been superficially modified (e.g., a bergsturtz may subsequently exhibit the behavior of a rock glacier). One period that would appear to be extremely important for landform initiation is the paraglacial one that follows deglaciation of a landscape (Church and Ryder, 1972). Clearly, this presumes that alpine mass wasting progresses in a previously glaciated environment and, while this may be generally true, it raises interesting questions about alpine areas where this may not be the case; the Yukon Territory is an area of potential examples.

Another topic of considerable importance is the stability or potential decay of alpine mass-wasting forms in changing environments. Evidence from the Indian Peaks suggests that some forms have exhibited movement rates that range over several orders of magnitude. This suggests great stability of forms once initiated, but also means that present rates of movement may be irrelevant to much of the history of the forms. Form stability in a changing periglacial regime has been considered by Thorn and Loewenherz (1987) using examples from the Indian Peaks. A number of the spatial and temporal complexities associated with inputs versus responses are well illustrated by Olyphant's theoretical modeling of rock-glacier movement (1983; written communication, 1986) and by his consideration of topoclimate (Olyphant, 1985).

Clearly, the present temporal scale of observation in process geomorphology cannot be extrapolated over time spans appropriate to regional-landscape development. This problem will not be rectified by merely accumulating data from diverse regions (which will only create a more comprehensive, but still contemporary spatial picture); nor will another 20 or 30 years of the same type of records produce a significant improvement, as the sample will still be too small and will remain just as unrepresentative of some paleoenvironments.

The enormous contrast in geomorphic scale, represented by Bradley's contribution to this chapter versus Rutter's and our own, and which Ritter discussed in the context of past and present concepts in fluvial evolution, highlights an intellectual gulf. It may be drawn even more sharply by pointing out that the wealth of mass-wasting data summarized in this section would seem to be applicable to at least some of the erosion-surface explanations described by Bradley (Mackin, 1947), but it is not. This merely serves to illustrate that data are worthless without a theoretical framework. At present, our knowledge of alpine mass wasting in the Rocky Mountains is highly localized and confined to microscales and mesoscales. Extension to macroscale or regional perspectives will require not only additional data, but, much more importantly, a concerted effort to develop new methodological schemes of which the one presented here can only be considered a first step.

SUMMARY AND CONCLUSIONS

Richard F. Madole

The Rocky Mountain region was one of the last parts of North America to be explored geologically. The region was difficult to reach from both the east and west, and its mountain barriers were shunned as much as possible in the search for railroad routes. It was not until the two decades prior to the founding of the Geological Society of America that scientists, principally those of the Hayden and Powell surveys, began to systematically investigate the Rockies. Since that time, however,

the region has attracted more than its share of geologic and geomorphic attention, which is understandable considering that (1) the range of altitude and latitude spanned encompasses an exceptionally large number of environments, processes, and landforms of interest to geomorphologists, (2) the interaction between tectonism and erosion is more readily observed here than in most other regions, and (3) the natural beauty of the region alone attracts researchers.

The geomorphic research done in the Rocky Mountains over the past century has paralleled the general development of geomorphic thought elsewhere. Early work was done on a grand scale, spatially and temporally, and focused on the long-term evolution of landforms and landscapes. Even as exploration was in progress, Powell and Gilbert recognized that the region offered special insights into the broad relationships between uplift and erosion; these insights subsequently contributed to the development of concepts such as antecedent streams, superimposition, and base level. A tendency that continued for decades was to study large areas from a historical perspective, which generally encompassed the period from the end of Laramide time to the present. Blackwelder's (1915) post-Cretaceous history of the mountains of central western Wyoming is one of many examples of this tendency. Gradually, emphasis shifted from landforms and landscapes to the processes that formed them and from regional studies to detailed studies of comparatively small areas, a trend that has accelerated markedly in the last few decades.

Consistent with the early-day tendency to focus on large features, among the first topics to be investigated were the erosion surfaces of the Rocky Mountains. Although erosion surfaces have been identified in most ranges of the Southern and Middle Rockies and in some ranges of the Northern Rockies; they have been studied the most in the Front Range of Colorado, and findings in this area provide an explanation for their distribution elsewhere in the region. Erosion surfaces are cut on Precambrian crystalline rock; they are best developed on granitic rocks, and are less well developed on metamorphic rocks. The uplifts of the Southern and Middle Rocky Mountains are cored with crystalline rock, much of which is granitic, and are more favorable to the development of erosion surfaces than other parts of the region. In contrast, the complexly thrust faulted sedimentary rocks that compose the northern part of the Northern Rockies and the Canadian Rockies are not favorable to the development and preservation of erosion surfaces. Interest in erosion surfaces is likely to continue in the Southern and Middle Rocky Mountains, because the surfaces here bear on interpretations of regional tectonic history, especially as it affects adjoining basins where there are various mineral deposits of economic value.

Work on the erosion surfaces played a key role in the evolution of study of fluvial processes and landforms in this region. Mackin (1937, 1947), in his classic papers on the Bighorn Basin, cited most frequently for thoughts on the concept of the graded stream and the origin of stream terraces, was integrating work on (1) the age and origin of the subsummit erosion surface, (2) an explanation of drainages that pass through canyons cut across the trends of mountain ranges, and (3) river terraces. In addition, Mackin (1947) also challenged the peneplain concept for the sub-summit surface in favor of a pediplain origin, and suggested that the summit surface, equivalent to the Flattop surface of the Front Range, was a noncyclic surface produced during Pleistocene time by alpine mass wasting.

Mackin's work focused attention on several aspects of Rocky Mountain geomorphology and laid important foundations for work that followed, including the challenges to his interpretation of the origin of river terraces in the Bighorn Basin and to his definition of a graded stream. The former engendered considerable controversy. Mackin's thoughts on the origin of terraces are coupled to views on the broader subject of erosion surfaces and cycles, topics that were "center stage" in the geomorphology of the Rockies at the time. Mackin recognized that the river terraces could be traced upstream to moraines, but believed that they passed beneath the moraines and, therefore, existed prior to glaciation. Mackin's attention was on the erosion of the rock-cut surface beneath the gravel terrace deposits. The gravel was an incidental by-product of lateral planation under conditions of stability by a graded stream. Hence, the splendid river terraces of the Bighorn Basin were viewed as products of graded streams, stability, and lateral planation. Eventually, the validity of all aspects of this view was questioned.

Mackin's model requires that the terraces consist of a uniformly thin cover of gravel over an evenly beveled bedrock surface. In time, these attributes were found to be lacking in places by others, most notably by Moss, who saw the river terraces of the Bighorn Basin and the moraines to which they grade as contemporaneous and advocated a glaciofluvial origin for the terraces. This origin had been widely accepted elsewhere in the Rockies without debate.

The key to understanding the fluvial processes and landforms in the basins of the Middle Rockies (and it applies equally to the Southern Rockies as well as to other parts of the Rocky Mountain region) is the transfer of coarse-grained sediment from the surrounding mountains into and across the adjacent basins (Ritter, this chapter). The rocks of the basin are mostly fine grained and much less resistant to erosion than the coarse-grained sediment (gravel) derived from the mountains. Once deposited, the gravel resists erosion and the older basin-fill sediment is removed preferentially. This simple control is responsible for much of the landscape of Rocky Mountain basins, including the many prominent gravel-capped benches and mesas, pirated streams, graded rivers, the terraces and their tendency to be unpaired, and the sculpting of the basin rocks generally.

In the past few decades, attention has shifted from the erosion surface beneath the terrace deposits to the deposits themselves, and to questions about how the sediment composing these deposits is entrained, transported, and deposited. Detailed studies of modern environments and processes are leading to a clearer understanding of more-ancient landforms. Such studies indicate that valley filling is associated with glaciation, chiefly during a short interval at the beginning of deglaciation when the amount of sediment entering the system is considerably greater than at other times. Studies of outwash, fluid mechanics, and glaciofluvial landforms in modern environments explain characteristics observed in Pleistocene terrace deposits, such as the dramatic increase in maximum particle size and terrace-tread gradient in a zone near the glacier front. Consideration of the relationships between flow conditions and particle transport has led to paleohydrologic reconstructions based on the properties of gravels. The particle size of clasts has been used to obtain measures of paleodepth that, when combined with other field data, allow rough estimates to be made of paleodischarges. Reconstruction of paleodepths and discharges enables comparisons to be made between modern and Pleistocene fluvial systems, which in turn permits consideration of how fluvial systems might respond to climatic change and to river metamorphosis.

Although most terrace gravels in the region are now widely accepted as being glaciofluvial, each terrace deposit and level does not necessarily relate to a glacial advance, and some are of nonglacial origin. Several erosional terrace levels can be cut in a single terrace deposit in postglacial time or between glaciations (Moss, 1974; Ritter, 1982), and stream piracy is capable of producing gravel terrace deposits quite independently of either tectonic or climatic controls.

Glacial landforms and deposits exist throughout the Rocky Mountain region. Most, however, were formed by glaciers that have long since disappeared. The cirque glaciers common to many parts of the region today formed in Holocene time, most just a few centuries ago, and are not remnants of the much larger valley glaciers of the Pleistocene. The Pleistocene glaciers were more like the present-day glaciers of the Columbia icefield, studies of which have provided important insights into the glacial processes that were once much more extensive in the Rocky Mountains. Even so, to fully understand erosional and depositional glacial landforms requires some reconstruction of the former glaciers and their environments.

Studies of existing glaciers have, in recent decades, led to an understanding of glacier mechanics. Knowledge of glacier mechanics, in turn, has been applied to studies of glacial landforms and deposits (Pierce, 1979). Now, the reverse is being attempted, and glacier dynamics are being reconstructed on the basis of the attributes of glacial deposits. Detailed studies of the sediment composing glacial deposits and the vertical sequence of facies present are used to reconstruct glacier dynamics and associated erosional and depositional processes. This approach utilizes information obtained from studies of sedimentology and processes in modern glacial environments, of glaciology, and of landforms. Sediment properties and associated landforms are used to identify tills of different origins and to distinguish between till and diamictons of nonglacial origin. Sediment and landform data are also used to determine whether glaciers were thick or thin, active or stationary, and were undergoing extensional or compressive flow at the study site.

In contrast to studies of glacial features, many studies of alpine mass wasting in the Rockies focus more on contemporary processes than on ancient landforms and deposits. This can be attributed to the fact that the summits in much of the region have been shaped in the recent past and are still being shaped today to some degree by intense frost action, solifluction, running water, ice, and other agents. The ongoing mass wasting tends to obscure, if not remove, the evidence of past mass wasting. Consequently, the imprint of alpine mass wasting on the landscape is not discernible as far back in time as the imprints of other processes discussed, such as fluvial or glacial erosion and deposition.

Most alpine mass-wasting studies have been conducted in the past three decades, during the time when emphasis shifted from regional studies of landscapes and landforms to detailed studies of process in present-day environments. Investigations of present-day processes commonly must reduce the spatial scale of the study in order to observe change in the landscape within what is geologically an extremely short time, generally at the most a matter of decades. Hence, the discussion of alpine mass wasting is at a much different scale than that of fluvial processes in the Bighorn Basin, although the principal landform elements—interfluve, slope, and basin—are analogous, and the transfer of material fluxes within and across these landscape elements is equally important. The discussion in these two sections exemplify the polarity in geomorphology described by Gardner (1982) between considerations of long-term landform evolution and present-day processes, and they illustrate the shift in scale and emphasis from the time of Blackwelder and Mackin to the present.

Landform elements—interfluves, slopes, and basin floors—are considered in terms of material fluxes. Rates of transfer vary in space, and obviously in time. Many processes were more vigorous in Pleistocene time than at present and some landforms and deposits—block fields and large patterned ground, for example—are relicts of that time. In general, interfluves have been lowered and flattened, free faces have retreated, talus (including lobate rock glaciers) has accreted and moved by creep, and basin (valley) floors are still dominated by the effects of the last glaciation.

Frost creep and solifluction (gelifluction) are major processes by which materials are transferred on interfluves. Frost creep requires moisture and soil saturation during autumn freeze, but operates in a drier regime than solifluction. Both vegetation and geomorphic processes are controlled by soil moisture and overland flow. Vegetation influences processes, including pedogenesis, on the interfluves. Frost creep, frost heave, overland flow, and soil loss are all greater where vegetation is sparse or absent.

Early studies of the alpine landscape concentrated on landforms and their distribution; most of the landforms that were studied, including rock glaciers, talus, and block fields, were composed of coarse-grained material. Within the past decade, attention has shifted to fine-grained materials and the geochemical flux. It has become apparent that a major amount of eolian sediment is being transported into the alpine area, although this condition may be significant only in the alpine areas of the Southern and Middle Rockies because of their proximity to arid and semiarid lowlands, which serve as major source areas for eolian material. The chemical weathering of this eolian influx is thought to be the dominant control on pedogenesis in the alpine zone of the Front Range.

Geochemical studies indicate that chemical weathering and solution are important processes in the alpine zone, and that weathering processes there are similar to those in temperate regions. Chemical weathering rates are controlled by a complex set of factors that includes the abundance of meltwater and the residence time and availability of organics to promote reactions. Patterns of accumulation and loss due to chemical weathering on slopes and within the basin are also complex, and soil geochemistry and water geochemistry differ in the Indian Peaks area of the Front Range.

The transfer of coarse debris on valley sides is the dominant alpine process operating at present and talus and the related landforms, rock glaciers and protalus ramparts, have been the most-studied features of alpine mass wasting. Talus formation is the most active of these processes and movement of lobate rock glaciers is the second most active; protalus ramparts do not appear to be forming at present. The lobate rock glaciers of the Front Range are ice-cemented, rather than ice-cored, and have rates of movement broadly proportional to size and slope.

The forms on the basin or valley floor mostly reflect the effects of the last glaciation. Tongue-shaped rock glaciers are among the few features present on the valley floor; they are present at valley heads and, like talus and lobate rock glaciers, postdate Pinedale glaciation. Their origin has been the subject of controversy both with respect to the origin of the debris and of the landform itself. Opinion has been divided as to whether the debris was supplied by ongoing rockfall or individual, catastrophic rockfalls, and also whether they are of nonglacial or glacial origin. The latter view holds that these forms are simply stagnant, debris-covered glaciers. Studies in the Front Range have fueled the controversy by disclosing that both ice-cored (glacial) and ice-cemented (nonglacial) tongue-shaped rock glaciers exist in the area.

The Indian Peaks region is a relatively inactive, low-energy alpine environment with surface-lowering rates as much as two orders of magnitude less than in some alpine regions (Caine, 1974). Landform elements are largely independent systems with respect to the transfer of coarse material; such transfer is localized within each system. In contrast, fine material, chiefly from the eolian sediment that is received in all parts of the alpine zone, is transferred between all landform elements. The geochemical system is probably dominated by interaction between water and eolian sediment, rather than between water and residuum derived from bedrock. Mass wasting in an alpine zone of the Front Range may not be representative of areas in the northern part of the Rocky Mountain region where the dominant rock types and geologic structures are markedly different.

REFERENCES

Ackers, P., and Charlton, F. G., 1971, The slope and resistance of small meandering channels: Proceedings of the Institution of the Civil Engineers (London), 1970, Suppl. 15, Paper 7362 S, p. 349–370.

Adams, G. F., ed., 1975, Planation surfaces; Peneplains, pediplains, and etchplains, in Benchmark papers in geology, Volume 22: Stroudsburg, Pennsylvania, Dowden, Hutchinson, and Ross, Incorporated, 476 p.

Alden, W. C., 1912, Pre-Wisconsin glacial drift in the region of Glacier National Park, Montana: Geological Society of America Bulletin, v. 23, p. 687–708.

—— , 1932, Physiography and glacial geology of eastern Montana and adjacent areas: U.S. Geological Survey Professional Paper 174, 133 p.

Alford, D. L., 1980, The orientation gradient; Regional variations of accumulation and ablation in alpine basins, in Ives, J. D., ed., Geoecology of the Colorado Front Range; A study of alpine and subalpine environments: Boulder, Colorado, Westview Press, p. 214–223.

Andrews, J. T., 1975, Glacial systems; An approach to glaciers and their environments: North Scituate, Massachusetts, Duxbury Press, 191 p.

Andrews, J. T., Birkeland, P. W., Kihl, R., Litaor, M., Nichols, H., and Stuckenrath, R., 1982, Holocene lake sediments from subalpine and alpine lakes, Colorado Front Range; Baseline studies: Geological Society of America Abstracts with Programs, v. 14, p. 433.

Andrews, J. T., Birkeland, P. W., Harbor, J., Dellamonte, N., Litaor, M., and Kihl, R., 1985, Holocene sediment record, Blue Lake, Colorado Front Range: Zeitschrift für Gletscherkunde und Glazialgeologie, v. 21, p. 25–34.

Atwood, W. W., 1940, The physiographic provinces of North America: Boston, Ginn and Company, 536 p.

Atwood, W. W., and Atwood, W. W., Jr., 1938, Working hypothesis for the physiographic history of the Rocky Mountain region: Geological Society of America Bulletin, v. 49, p. 957–980.

Atwood, W. W., Jr., 1909, Glaciation of the Uinta and Wasatch mountains: U.S. Geological Survey Professional Paper 61, 96 p.

Atwood, W. W., Jr., and Mather, K. F., 1912, The evidence of three distinct glacial epochs in the Pleistocene history of the San Juan Mountains, Colorado: Journal of Geology, v. 20, p. 385–409.

—— , 1932, Physiography and Quaternary geology of the San Juan Mountains, Colorado: U.S. Geological Survey Professional Paper 166, 176 p.

Baker, V. R., 1973, Paleohydrology and sedimentology of Lake Missoula flooding in eastern Washington: Geological Society of America Special Paper 144, 79 p.

—— , 1974, Paleohydraulic interpretation of Quaternary alluvium near Golden, Colorado: Quaternary Research, v. 4, p. 94–112.

Baker, V. R., and Penteado-Orellana, M. M., 1977, Adjustment to Quaternary climatic change by the Colorado River in central Texas: Journal of Geology, v. 85, p. 395–422.

Baker, V. R., and Ritter, D. F., 1975, Competence of rivers to transport coarse bedload material: Geological Society of America Bulletin, v. 86, p. 975–978.

Ball, S. H., 1908, Part I, General geology, in Spurr, J. E., and Garrey, G. H., eds., Economic geology of the Georgetown Quadrangle (together with the Empire District), Colorado: U.S. Geological Survey Professional Paper 63, p. 29–96.

Barry, R. G., 1973, A climatological transect on the east slope of the Front Range, Colorado: Arctic and Alpine Research, v. 5, p. 89–110.

Barsch, D., and Caine, N., 1984, The nature of mountain geomorphology: Mountain Research and Development, v. 4, p. 287–298.

Benedict, J. B., 1966, Radiocarbon dates from a stone-banked terrace in the Colorado Rocky Mountains, U.S.A.: Geografiska Annaler, v. 48A, p. 24–31.

—— , 1969, Microfabric of patterned ground: Arctic and Alpine Research, v. 1, p. 45–48.

—— , 1970a, Downslope soil movement in a Colorado alpine region; Rates, processes, and climatic significance: Arctic and Alpine Research, v. 2, p. 165–226.

—— , 1970b, Frost cracking in the Colorado Front Range: Geografiska Annaler, v. 52A, p. 87–93.

—— , 1973a, Chronology of cirque glaciation, Colorado Front Range: Quaternary Research, v. 3, p. 584–599.

—— , 1973b, Origin of rock glaciers: Journal of Glaciology, v. 12, p. 520–522.

—— , 1979, Fossil ice-wedge polygons in the Colorado Front Range; Origin and significance: Bulletin of the Geological Society of America, v. 90, p. 173–180.

Berg, N. H., 1986, Blowing snow at a Colorado alpine site; Measurements and implications: Arctic and Alpine Research, v. 18, p. 147–161.

Birkeland, P. W., 1968, Mean velocities and boulder transport during Tahoe-age floods of the Truckee River, California-Nevada: Geological Society of America Bulletin, v. 79, p. 137–142.

Blackwelder, E., 1909, Cenozoic history of the Laramie region, Wyoming: Journal of Geology, v. 17, p. 429–444.

—— , 1915, Post-Cretaceous history of the mountains of central western Wyoming: Journal of Geology, v. 23, p. 97–117, 193–217, 307–340.

Boothroyd, J. C., and Ashley, G. M., 1975, Processes, bar morphology, and sedimentary structures on braided outwash fans, northeastern Gulf of Alaska, in Jopling, A., and McDonald, B., eds., Glaciofluvial and glaciolacustrine sedimentation: Society of Economic Paleontologists and Mineralogists Special Publication 23, p. 193–222.

Boulton, G. S., 1968, Flow tills and related deposits on some Vestspitsbergen glaciers: Journal of Glaciology, v. 7, p. 391–412.

—— , 1970, On the deposition of subglacial and melt-out tills at the margins of certain Svalbard glaciers: Journal of Glaciology, v. 9, p. 231–245.

—— , 1972, Modern arctic glaciers as depositional models for former ice sheets: Journal of the Geological Society of London, v. 128, p. 361–393.

—— , 1976a, The origin of glacially fluted surfaces; Observations and theory: Journal of Glaciology, v. 17, p. 287–310.

—— , 1976b, A genetic classification of tills and criteria for distinguishing tills of different origin, in Stankowski, M., ed., Till; Its genesis and diagenesis: Warsaw, Poland, University im A. Mickiewicza w Poznaniu, Seria Geografia, v. 12, p. 65–80.

—— , 1978, Boulder shapes and grain-size distributions of debris as indicators of transport paths through a glacier and till genesis: Sedimentology, v. 25, p. 773–799.

Boulton, G. S., and Eyles, N., 1979, Sedimentation by valley glaciers; A model and genetic classification, in Schlüchter, C., ed., Moraines and varves; Origin, genesis, classification, in Proceedings, International Quaternary Association Symposium on Genesis and Lithology of Quaternary Deposits, Zurich, 1978: Rotterdam, Netherlands, A. A. Balkema, p. 11–23.

Bovis, M. J., 1978, Soil loss in the Colorado Front Range: Zeitschrift für Geomorphologie, Supplementband, v. 29, p. 10–21.

—— , 1982, Spatial variation of soil loss controls, in Thorn, C. E., ed., Space and time in geomorphology, The Binghamton Symposia in Geomorphology Series, no. 12: London, George Allen and Unwin, p. 1–24.

Bovis, M. J., and Thorn, C. E., 1981, Soil loss variation within a Colorado alpine area: Earth Surface Processes and Landforms, v. 6, p. 151–163.

Boydell, A. N., 1978, Multiple glaciations in the Foothills, Rocky Mountain House area, Alberta: Canada, Alberta Research Council Bulletin 36, 35 p.

Bradley, W. C., and Mears, A. I., 1980, Calculations of flows needed to transport coarse fraction of Boulder Creek alluvium at Boulder, Colorado: Geological Society of America Bulletin, v. 91, no. 3, pt. 2, p. 1057–1090.

Bradley, W. C., Fahnestock, R. K., and Rowekamp, T. T., 1972, Coarse sediment transport by flood flows on Knik River, Alaska: Geological Society of America Bulletin, v. 83, p. 1261–1284.

Bryan, K., and Ray, L. L., 1940, Geologic antiquity of the Lindenmeier site in Colorado: Washington, D.C., Smithsonian Institution Miscellaneous Collections Publication 3554, 76 p.

Burns, S. F., 1980, Alpine soil distribution and development, Indian Peaks, Colorado Front Range [Ph.D. thesis]: Boulder, University of Colorado, 360 p.

Burns, S. F., and Tonkin, P. J., 1982, Soil geomorphic models and the spatial distribution and development of alpine soils, in Thorn, C. E., ed., Space and time in geomorphology, The Binghamton Symposia in Geomorphology Series, no. 12: London, George Allen and Unwin, p. 25–43.

Caine, T. N., 1971, A conceptual model for alpine slope process study: Arctic and Alpine Research, v. 3, p. 319–329.

—— , 1974, The geomorphic processes of the alpine environment, *in* Ives, J. D., and Barry, R. G., eds., Arctic and alpine environments: London, Methuen, p. 721–748.

—— , 1976a, The influence of snow and increased snowfall on contemporary geomorphic processes in alpine areas, *in* Steinhoff, H. W., and Ives, J. D., eds., Ecological impacts of snowpack augumentation in the San Juan Mountains of Colorado: Fort Collins, Colorado State University, p. 145–200.

—— , 1976b, Summer rainstorms in an alpine environment and their influence on soil erosion, San Juan Mountains, Colorado: Arctic and Alpine Research, v. 8, p. 183–196.

—— , 1976c, A uniform measure of subaerial erosion: Geological Society of America Bulletin, v. 87, p. 137–140.

—— , 1979, The problem of spatial scale in the study of contemporary geomorphic activity on mountain slopes *with special reference to* the San Juan Mountains: Studia Geomorphologica Carpatho-Balcanica, v. 13, p. 5–22.

—— , 1982, Water and sediment fluxes in the Green Lakes Valley, Colorado Front Range, *in* Halfpenny, J. C., ed., Ecological studies in the Colorado alpine; A festschrift for John W. Marr: Boulder, University of Colorado, Institute of Arctic and Alpine Research Occasional Paper no. 37, p. 1–12.

—— , 1984, Elevational contrasts in contemporary geomorphic activity in the Colorado Front Range: Studia Geomorphologica Carpatho-Balcanica, v. 18, p. 5–31.

—— , 1986, Sediment movement and storage on alpine slopes in the Colorado Rocky Mountains, *in* Abrahams, A. D., ed., Hillslope processes, The Binghamton Symposia in Geomorphology Series, no. 16: Boston, Massachusetts, George Allen and Unwin, p. 115–137.

Capps, S. R., 1909, Pleistocene geology of the Leadville Quadrangle, Colorado: U.S. Geological Survey Bulletin 386, 99 p.

Carroll, T., 1974, Relative age dating techniques and a late Quaternary chronology, Arikaree Cirque, Colorado: Geology, v. 2, p. 321–325.

—— , 1976, An estimate of watershed efficiency for a Colorado alpine basin, *in* Proceedings of the 44th Western Snow Conference, Calgary, Alberta, Canada, Fort Collins, Colorado, Colorado State University, p. 69–77.

Chamberlin, R. T., 1919, The building of the Colorado Rockies: Journal of Geology, v. 27, p. 145–164, 225–251.

Charlesworth, J. K., 1957, The Quaternary era—with special reference to its glaciation: London, Edward Arnold, 2 vols., 1700 p.

Chorley, R. J., and Kennedy, B. A., 1971, Physical geography; A systems approach: London, Prentice-Hall International, 370 p.

Church, M., 1972, Baffin Island sandurs; A study of Arctic fluvial processes: Ottawa, Ontario, Canada, Geological Survey of Canada Bulletin 216, 208 p.

—— , 1978, Paleohydrological reconstructions from a Holocene valley fill, *in* Miall, A., ed., Fluvial sedimentology: Canada Society of Petroleum Geologists Memoir 5, p. 743–772.

Church, M., and Gilbert, R., 1975, Proglacial fluvial and lacustrine environments, *in* Jopling, A., and McDonald, B., eds., Glaciofluvial and glaciolacustrine sedimentation: Society of Economic Paleontologists and Mineralogists Special Publication 23, p. 22–100.

Church, M., and Ryder, J. M., 1972, Paraglacial sedimentation; A consideration of fluvial processes conditioned by glaciation: Geological Society of America Bulletin, v. 83, p. 3059–3072.

Clague, J. J., 1975, Sedimentology and paleohydrology of late Wisconsinan outwash, Rocky Mountain Trench, southeastern British Columbia, *in* Jopling, A., and McDonald, B., eds., Glaciofluvial and glaciolacustrine sedimentation: Society of Economic Paleontologists and Mineralogists Special Publication 23, p. 223–237.

Colman, S. M., 1985, Map showing tectonic features of late Cenozoic origin in Colorado: U.S. Geological Survey, Miscellaneous Investigation Series, Map I-1566, scale 1:1,000,000.

Costa, J. E., 1983, Paleohydraulic reconstruction of flash-flood peaks from boulder deposits in the Colorado Front Range: Geological Society of America Bulletin, v. 94, p. 986–1004.

Cross, W., 1894, Pikes Peak folio, Colorado: U.S. Geological Survey Geologic Atlas, Folio 7, 5 p.

Curtis, B. F., ed., 1975, Cenozoic history of the Southern Rocky Mountains: Geological Society of America Memoir 144, 279 p.

Davis, W. M., 1899, The geographical cycle: Geographical Journal, v. 14, p. 481–504.

—— , 1902, Base-level, grade, and peneplain: Journal of Geology, v. 10, p. 77–111.

—— , 1911, The Colorado Front Range, a study in physiographic presentation: Association of American Geographers Annals, v. 11, p. 21–83.

Dawson, G. M., 1888a, Recent observations on the glaciation of British Columbia and adjacent regions: Geological Magazine, v. 5, p. 347–350.

—— , 1888b, Recent observations on the glaciation of British Columbia and adjacent regions: American Geologist, v. 3, p. 249–253.

—— , 1890, On the glaciation of the northern part of the Cordillera: American Geologist, v. 6, p. 153–162.

—— , 1891, On the later physiographical geology of the Rocky Mountain region in Canada with special reference to change in elevation, and to the history of the glacial period: Transactions of the Royal Society of Canada in the year 1890, v. 8, sec. 4, p. 3–74.

Denny, C. S., 1965, Alluvial fans in the Death Valley region, California and Nevada: U.S. Geological Survey Professional Paper 466, 62 p.

Denson, N. M., and Bergendahl, M. H., 1961, Middle and upper Tertiary rocks in southeastern Wyoming and adjoining areas: U.S. Geological Survey Professional Paper 424-C, p. C168–C172.

Denson, N. M., and Chisholm, W. A., 1971, Summary of mineralogic and lithologic characteristics of Tertiary sedimentary rocks in the Middle Rocky Mountains and the northern Great Plains: U.S. Geological Survey Professional Paper 750-C, p. C117–C126.

Desloges, J. R., and Gardner, J. S., 1984, Process and discharge estimation in ephemeral channels, Canadian Rocky Mountains: Canadian Journal of Earth Sciences, v. 21, p. 1050–1060.

Dixon, J. C., 1983, Chemical weathering of late Quaternary cirque deposits in the Colorado Front Range [Ph.D. thesis]: Boulder, University of Colorado, 174 p.

—— , 1986, Solute movement on hillslopes in the alpine environment of the Colorado Front Range, *in* Abrahams, A. D., ed., Hillslope processes, The Binghamton Symposia in Geomorphology Series, no. 16: Boston, Massachusetts, George Allen and Unwin, p. 139–159.

Dowdeswell, J. A., 1982, Relative dating of late Quaternary deposits using cluster and discriminant analysis, Audubon Cirque, Mount Audubon, Colorado Front Range: Boreas, v. 11, p. 151–161.

Dreimanis, A., 1976, Tills, their origin and properties, *in* Legget, R. F., ed., Glacial till: The Royal Society of Canada Special Publication 12, p. 11–49.

Dunbar, C. O., 1949, Historical geology: New York, J. Wiley and Sons, Incorporated, 567 p.

Dury, G. H., 1965, Theoretical implications of underfit streams: U.S. Geological Survey Professional Paper 452-C, p. C1–C43.

—— , 1976, Discharge prediction, present and former, from channel dimensions: Journal of Hydrology, v. 30, p. 219–245.

Eardley, A. J., 1951, Structural geology of North America: New York, Harper, 624 p.

Eggler, D. H., Larson, E. E., and Bradley, W. C., 1969, Granites, grusses, and the Sherman erosion surface, southern Laramie Range, Colorado-Wyoming: American Journal of Science, v. 267, p. 510–522.

Embleton, C., and King, C.A.M., 1975, Glacial geomorphology: London, Edward Arnold, 573 p.

Emmons, S. F., Cross, W., and Eldridge, G. H., 1896, Geology of the Denver Basin: U.S. Geological Survey Monograph 27, 556 p.

Epis, R. C., and Chapin, C. E., 1968, Geologic history of the Thirtynine Mile volcanic field, central Colorado: Colorado School of Mines Quarterly, v. 63, no. 3, p. 51–85.

———, 1974, Stratigraphic nomenclature of the Thirtynine Mile volcanic field, central Colorado: U.S. Geological Survey Bulletin 1395-C, 23 p.

———, 1975, Geomorphic and tectonic implications of the post-Laramide, late Eocene erosion surface in the Southern Rocky Mountains, *in* Curtis, B. F., ed, Cenozoic history of the Southern Rocky Mountains: Geological Society of America Memoir 144, p. 45–74.

Epis, R. C., Scott, G. R., Taylor, R. B., and Chapin, C. E., 1976, Cenozoic volcanic, tectonic, and geomorphic features of central Colorado, *in* Epis, R. C., and Weimer, R. J., eds., Studies in Colorado field geology: Professional Contributions of Colorado School of Mines no. 8, p. 323–338.

———, 1980, Summary of Cenozoic geomorphic, volcanic, and tectonic features of central Colorado and adjoining areas, *in* Kent, H. C., and Porter, K. W., eds., Colorado geology: Denver, Rocky Mountain Association of Geologists, p. 135–156.

Eschman, D. F., 1955, Glaciation of the Michigan River basin, North Park, Colorado: Journal of Geology, v. 63, p. 197–213.

Eyles, N., 1979, Facies of supraglacial sedimentation on Icelandic and alpine temperate glaciers: Canadian Journal of Earth Sciences, v. 16, p. 1341–1361.

Eyles, N., Sladen, J. A., and Gilroy, S., 1982, A depositional model for stratigraphic complexes and facies superimposition in lodgement tills: Boreas, v. 11, p. 317–333.

Fahey, B. D., 1973, An analysis of diurnal freeze-thaw and frost heave cycles in the Indian Peaks region of the Colorado Front Range: Arctic and Alpine Research, v. 5, p. 269–281.

———, 1974, Seasonal frost heave and frost penetration measurements in the Indian Peaks region of the Colorado Front Range: Arctic and Alpine Research, v. 6, p. 63–70.

———, 1975, Non-sorted circle development in a Colorado alpine location: Geografiska Annaler, v. 57A, p. 153–164.

Fahnestock, R. K., 1963, Morphology and hydrology of a glacial stream—White River, Mount Rainier, Washington: U.S. Geological Survey Professional Paper 422-A, p. A1–A70.

———, 1969, Morphology of the Slims River, *in* Bushnell, V. C., and Ragle, R. H., eds., Icefield Ranges Research Project; Scientific results: American Geographical Society and Arctic Institute of North America, v. 1, p. 161–172.

Fenneman, N. M., 1905, Geology of the Boulder District, Colorado: U.S. Geological Survey Bulletin 265, 101 p.

———, 1931, Physiography of western United States: New York, McGraw-Hill Book Company, 534 p.

Finlay, G. I., 1916, Colorado Springs folio, Colorado: U.S. Geological Survey Geologic Atlas, Folio 203, 17 p.

Flint, R. F., 1971, Glacial and Quaternary geology: New York, John Wiley and Sons, 892 p.

Frank, T. D., and Thorn, C. E., 1985, Stratifying alpine tundra for geomorphic studies using digitized aerial imagery: Arctic and Alpine Research, v. 17, p. 179–188.

Fryxell, F. M., 1930, Glacial features of Jackson Hole, Wyoming: Augustana Library Publication 13, 129 p.

Gardner, J. S., 1980, Frequency, magnitude, and spatial distribution of mountain rockfalls and rockslides in the Highwood Pass area, Alberta, Canada, *in* Coates, D. R., and Vitek, J. D., eds., Thresholds in geomorphology, The Binghamton Symposia in Geomorphology Series, no. 9: London, George Allen and Unwin, p. 267–295.

———, 1982, Alpine mass wasting in contemporary time; Some examples from the Canadian Rocky Mountains, *in* Thorn, C. E., ed., Space and time in geomorphology, The Binghamton Symposia in Geomorphology Series, no. 12: London, George Allen and Unwin, p. 171–192.

Gardner, J. S., Nelson, J. G., and Ashwell, I. Y., 1964, Alpine studies in the upper Red Deer River valley: Canadian Alpine Journal, v. 47, p. 137–147.

Gardner, J. S., Smith, D. J., and Deslodges, J. R., 1983, The dynamic geomorphology of the Mount Rae area; A high mountain region in southwestern Alberta: Waterloo, Department of Geography, University of Waterloo, 237 p.

Gibbons, A. B., Megeath, J. D., and Pierce, K. L., 1984, Probability of moraine survival in a succession of glacial advances: Geology, v. 12, p. 327–330.

Gilbert, G. K., 1877, Report on the geology of the Henry Mountains: Washington, D.C., U.S. Geographical and Geological Survey, Rocky Mountain Region, 160 p.

———, 1890, Lake Bonneville: U.S. Geological Survey Monograph 1, 438 p.

Goldthwait, R. P., ed., 1971, Till; A symposium: Columbus, Ohio State University Press, 402 p.

Greenland, D., Caine, N., and Pollak, O., 1984, The summer water budget and its importance in the alpine tundra of Colorado: Physical Geography, v. 5, p. 221–239.

Gries, R., 1983, North-south compression of Rocky Mountain foreland structures, *in* Lowell, J. D., and Gries, R., eds., Rocky Mountain foreland basins and uplifts: Denver, Colorado, Rocky Mountain Association of Geologists, p. 9–32.

Hack, J. T., 1960, Interpretation of erosional topography in humid temperate regions: American Journal of Science, Bradley volume, v. 258-A, p. 80–96.

Haldorsen, S., 1977, The petrography of tills; A study from Ringsalen, southeastern Norway: Norges Geologiske Undersokelse, v. 336, 36 p.

———, 1982, The genesis of tills from Astadalen, southeastern Norway: Norsk Geologisk Tidsskrift, v. 62, p. 17–38.

Halfpenny, J. C., Ingraham, K. P., Mattysse, J., and Lehr, P. J., 1986, Bibliography of alpine and subalpine areas of the Front Range, Colorado: University of Colorado, Institute of Arctic and Alpine Research Occasional Paper 43, 114 p.

Hara, Y., and Thorn, C. E., 1982, Preliminary quantitative study of alpine subnival boulder pavements, Colorado Front Range, U.S.A.: Arctic and Alpine Research, v. 14, p. 361–367.

Harrison, P. W., 1957, A clay-till fabric; Its character and origin: Journal of Geology, v. 65, p. 275–308.

Hartshorn, J. H., 1958, Flow till in southeastern Massachusetts: Geological Society of America Bulletin, v. 69, p. 477–482.

Henock, W.E.S., 1972, Glacier variations: International Symposia on the Role of Snow and Ice in Hydrology, Guidebook, Sept. 1972, p. 91–96.

Heusser, C. J., 1956, Postglacial environments in the Canadian Rocky Mountains [Alberta–British Columbia]: Ecological Monographs, v. 26, p. 263–302.

Holmes, G. W., and Moss, J. H., 1955, Pleistocene geology of the southwestern Wind River Mountains, Wyoming: Geological Society of America Bulletin, v. 66, p. 629–653.

Howard, A. D., 1941, Rocky Mountain peneplanes or pediments: Journal of Geomorphology, v. 4, p. 138–141.

Hunt, C. B., Averitt, P., and Miller, R. L., 1953, Geology and geography of the Henry Mountains region, Utah: U.S. Geological Survey Professional Paper 228, 234 p.

Hyvaerinen, L., Kauranne, K., Yltetyinen, V., 1973, Modern boulder tracing in prospecting, *in* Jones, M. J., ed., Prospecting in areas of glacial terrain: London, Institute of Minerals and Metallurgy, p. 87–95.

Ives, J. D., 1973, Permafrost and its relationship to other environmental parameters in a midlatitude, high-altitude setting, Front Range, Colorado Rocky Mountains, *in* North American Contribution; Permafrost; Second International Conference, Yakutsk, U.S.S.R., 1973: Washington, D.C., National Academy of Sciences, p. 121–125.

———, ed., 1980, Geoecology of the Colorado Front Range; A study of alpine and subalpine environments: Boulder, Westview Press, 484 p.

Ives, J. D., and Barry, R. G., eds., 1974, Arctic and alpine environments: London, Methuen, 999 p.

Ives, J. D., and Dow, V., 1981, The Indian Peaks Wilderness Area shaded relief map: Boulder, Colorado, International Mountain Society, scale 1:50,000.

Ives, J. D., and Fahey, B. D., 1971, Permafrost occurrence in the Front Range Colorado Rocky Mountains, U.S.A.: Journal of Glaciology, v. 10, p. 105–111.

Ives, R. L., 1940, Rock glaciers in the Colorado Front Range: Geological Society of America Bulletin, v. 51, p. 1271–1294.

———, 1941, Thrust faulting around Crater Lake, Colorado: Pan-American Geologist, v. 76, p. 259–274.

Izett, G. A., 1975, Late Cenozoic sedimentation and deformation in northern Colorado and adjoining areas, *in* Curtis, B. F., ed., Cenozoic history of the Southern Rocky Mountains: Geological Society of America Memoir 144, p. 179–209.

Jackson, L. E., Jr., 1980, Quaternary stratigraphy and history of the Alberta portion of the Kananaskis Lakes map area (82-J) and its implications for the existence of an ice-free corridor during Wisconsin time: Canadian Journal of Anthropology, v. 1, p. 9–10.

Johnson, D., 1931, Planes of lateral corrasion: Science, v. 73, p. 174–177.

———, 1932, Rock fans of arid regions: American Journal of Science, v. 23, p. 389–416.

Johnson, J. B., 1979, Mass balance studies on the Arikarce Glacier [Ph.D. thesis]: Boulder, University of Colorado, 287 p.

Johnson, R. B., 1967, Rock streams on Mount Mestas, Sangre de Cristo Mountains, southern Colorado: U.S. Geological Survey Professional Paper 575-D, p. D217–D220.

Knight, S. H., 1953, Summary of the Cenozoic history of the Medicine Bow Mountains, Wyoming: Wyoming Geological Association, 8th Annual Field Conference, Guidebook, p. 65–76.

Knox, J. C., 1976, Concept of the graded stream, *in* Melhorn, W. N., and Flemal, R. C., eds., Theories of landform development, Sixth Annual Geomorphology Symposia Series: Binghamton, State University of New York, p. 169–198.

Komárková, V., and Webber, P. J., 1978, An alpine vegetation map of Niwot Ridge, Colorado: Arctic and Alpine Research, v. 10, p. 1–29.

Kruger, J., 1979, Structures and textures in till indicating subglacial deposition: Boreas, v. 8, p. 323–340.

Lane, E. W., 1957, A study of the shape of channels formed by natural streams in erodible material: Omaha, Nebraska, U.S. Army Corps of Engineers, Engineering Division, Missouri River, M.R.D. Sediments Series, no. 9, 106 p.

Lawson, D. E., 1979, Sedimentological analysis of the western terminus region of the Matanuska Glacier, Alaska: U.S. Army Corps of Engineers, Cold Regions Research and Engineering Laboratory Report 79-9, 112 p.

———, 1981, Sedimentological characteristics and classification of depositional process and deposits in the glacial environment: U.S. Army Corps of Engineers, Cold Regions Research and Engineering Laboratory Report 81-27, 16 p.

———, 1982, Mobilization, movement, and deposition of active subaerial sediment flows, Matanuska Glacier, Alaska: Journal of Geology, v. 90, p. 279–300.

Lee, W. T., 1917, The geologic story of the Rocky Mountain National Park, Colorado: U.S. National Park Service Publication, 89 p.

———, 1922, Peneplains of the Front Range and Rocky Mountain National Park, Colorado: U.S. Geological Survey Bulletin 730-A, 17 p.

Leonard, E. M., 1984, Late Pleistocene equilibrium-line altitudes and modern snow accumulation patterns, San Juan Mountains, Colorado, U.S.A.: Arctic and Alpine Research, v. 16, p. 65–70.

Leopold, E. B., and MacGinitie, H. D., 1972, Development and affinities of Tertiary floras in the Rocky Mountains, *in* Aham, A. G., ed., Floristics and paleofloristics of Asia and eastern Northern America: Amsterdam, Elsevier Publishing Company, Chap. 12, p. 147–200.

Leopold, L. B., and Maddock, T., 1953, The hydraulic geometry of stream channels and some physiographic implications: U.S. Geological Survey Professional Paper 252, 57 p.

Leopold, L. B., and Wolman, M. G., 1957, River channel patterns; Braided, meandering, and straight: U.S. Geological Survey Professional Paper 282-B, p. 39–85.

Levson, V. M., 1986, Wisconsinan glacial sediments and depositional models in the mountain environment, Jasper National Park, Canada [M.Sc. thesis]: Edmonton, University of Alberta, 187 p.

Little, H. P., 1925, Erosional cycles in the Front Range of Colorado and their correlation: Geological Society of America Bulletin, v. 36, p. 495–512.

Löve, D., 1970, Subarctic and subalpine; Where and what?: Arctic and Alpine Research, v. 2, p. 63–73.

Love, J. D., 1960, Cenozoic sedimentation and crustal movement in Wyoming: American Journal of Science, Bradley volume, v. 258-A, p. 204–214.

Lovering, T. S., 1929, Geologic history of the Front Range, Colorado: Colorado Scientific Society Proceedings, v. 12, p. 59–111.

Lovering, T. S., and Goddard, E. N., 1950, Geology and ore deposits of the Front Range, Colorado: U.S. Geological Survey Professional Paper 223, 319 p.

Lowell, J. D., 1983, Foreland deformation, *in* Lowell, J. D., and Gries, R., eds., Rocky Mountain foreland basins and uplifts: Denver, Colorado, Rocky Mountain Association of Geologists, p. 1–8.

Luckman, B. H., 1972, Some observations on the erosion of talus slopes by snow avalanches in Surprise Valley, Jasper National Park, *in* Slaymaker, H. O., and McPherson, H. J., eds., Mountain geomorphology: Vancouver, Tantalus Research Limited, p. 85–92.

———, 1981, The geomorphology of the Alberta Rocky Mountains; A review and commentary: Zeitschrift für Geomorphologie, v. 37, p. 91–119.

Luckman, B. H., and Crockett, K. J., 1978, Distribution and characteristics of rock glaciers in the southern part of Jasper National Park: Canadian Journal of Earth Sciences, v. 15, p. 540–550.

Luckman, B. H., and Osborn, G. D., 1979, Holocene glacier fluctuations in the middle Canadian Rocky Mountains: Quaternary Research, v. 11, p. 52–77.

MacGinitie, H. D., 1953, Fossil plants of the Florissant beds, Colorado: Carnegie Institute of Washington Publication 599, 198 p.

Mackin, J. H., 1936, The capture of the Greybull River: American Journal of Science, v. 31, p. 373–385.

———, 1937, Erosional history of the Big Horn Basin, Wyoming: Geological Society of America Bulletin, v. 48, p. 813–894.

———, 1947, Altitude and local relief of the Bighorn area during the Cenozoic: Wyoming Geological Association, 2nd Annual Field Conference, Guidebook, p. 103–120.

———, 1948, Concept of the graded river: Geological Society of America Bulletin, v. 59, p. 463–512.

———, 1953, Stream planation near Colorado Springs, Colorado: Geological Society of America Bulletin, v. 64, p. 705–710.

Madole, R. F., 1976, Glacial geology of the Front Range, Colorado, *in* Mahaney, W. C., ed., Quaternary stratigraphy of North America: Stroudsburg, Pennsylvania, Dowden, Hutchinson and Ross, p. 297–318.

———, 1982, Possible origins of till-like deposits near the summit of the Front Range in north-central Colorado: U.S. Geological Survey Professional Paper 1243, p. 1–31.

———, 1986, Lake Devlin and Pinedale glacial history, Front Range, Colorado: Quaternary Research, v. 25, p. 43–54.

Mandryk, G. B., 1986, Pleistocene valley glacier sedimentology and stratigraphy in the Banff-Canmore area of Alberta, Canada [M.Sc. thesis]: Edmonton, University of Alberta, 290 p.

Marr, J. W., 1967, Ecosystems of the east slope of the Front Range in Colorado: Boulder, University of Colorado Press, Series in Biology, no. 8, 134 p.

Marr, J. W., Clark, J. M., Osburn, W. S., and Paddock, M. W., 1968a, Data on mountain environments, Front Range, Colorado; Four climax regions; Part 3, 1959–1964: Boulder, University of Colorado Press, Series in Biology, no. 29, 181 p.

Marr, J. W., Johnson, A. W., Osburn, W. S., and Knorr, O. A., 1968b, Data on mountain environments. Front Range, Colorado; Four climax regions; Part 2, 1953–1958: Boulder, University of Colorado Press, Series in Biology, no. 28, 171 p.

Martinelli, M., Jr., 1959a, Some hydrologic aspects of alpine snowfields under summer conditions: Journal of Geophysical Research, v. 64, p. 451–455.

———, 1959b, Alpine snowfields; Their characteristics and management possibilities: International Association of Scientific Hydrology Publication, v. 48, p. 120–127.

———, 1975, Water-yield improvement from alpine areas: U.S. Department of Agriculture, Forest Service Research Paper RM-138, p. 1–16.

Marvine, A. R., 1874, Report for the year 1873, *in* 7th Annual Report of U.S. Geological and Geographical Survey of the Territories (Hayden's Survey):

Washington, D.C., p. 83–192.

Mather, K. F., 1925, Physiographic surfaces in the Front Range of northern Colorado and their equivalents on the Great Plains [abs.]: Geological Society of America Bulletin, v. 36, p. 134–135.

McCall, J. G., 1960, The flow characteristics of a cirque glacier and their effect on glacial structure and cirque formation, *in* Lewis, W. V., ed., Norwegian cirque glaciers: Royal Geographical Society Research Series, no. 4, p 39–62.

McCauley, C., 1958, Geographic study of mountain glaciation in the Northern Hemisphere—Part 2B, Western Canada and Arctic Canada: American Geographical Society, 108 p.

McConnell, R. G., 1896, Report on an exploration of the Finlay and Omineca rivers: Geological Survey of Canada Annual Report, v. 7, p. 6c–40c.

McCoubrey, A. A., 1937, Glacier observations, 1936 and 1937: Canadian Alpine Journal, v. 25, p. 113–116.

McFarlane, W. T., 1946, Glacier investigation in Banff, Yoho, and Jasper national parks: Canadian Alpine Journal, v. 29, p. 265–273.

McLaren, P., and Hills, L. V., 1973, Cirque analysis as a method of predicting the extent of a Pleistocene ice advance: Canadian Journal of Earth Sciences, v. 10, p. 1211–1225.

McPherson, H. J., 1970, Landforms and glacial history of the upper North Saskatchewan valley, Alberta, Canada: Canadian Geographical Journal, v. 14, p. 10–26.

McSaveney, E. R., 1971, The surficial fabric of rockfall talus, *in* Morisawa, M., ed., Quantitative Geomorphology, Second Annual Geomorphology Symposia Series: Binghamton, State University of New York, p. 181–197.

Mears, A. I., 1979, Flooding and sediment transport in a small alpine drainage basin in Colorado: Geology, v. 7, p. 53–57.

Mears, B., Jr., 1974, The evolution of the Rocky Mountain glacial model, *in* Coates, D. R., ed., Glacial Geomorphology, Fifth Annual Geomorphology Symposia Series: Binghamton, State University of New York, p. 11–40.

—— , 1981, Periglacial wedges and the late Pleistocene environment of Wyoming's intermontane basins: Quaternary Research, v. 15, p. 171–198.

Meek, V., 1948, Glacier observations in the Canadian Cordillera: Canadian Geographical Journal, v. 37, p. 190–209.

Meier, M. F., 1960, Mode of flow of the Saskatchewan Glacier, Alberta, Canada: U.S. Geological Survey Professional Paper 351, 70 p.

Meierding, T. C., 1982, Late Pleistocene glacial equilibrium-line altitudes in the Colorado Front Range; A comparison of methods: Quaternary Research, v. 18, p. 289–310.

Meierding, T. C., and Birkeland, P. W., 1980, Quaternary glaciation of Colorado, *in* Kent, H. C., and Porter, K. W., eds., Colorado geology: Rocky Mountain Association of Geologists, p. 165–173.

Menzies, J., 1979, A review of the literature on the formation and location of drumlins: Earth Science Reviews, v. 14, p. 315–359.

Moore, F. E., 1960, Summary of Cenozoic history, southern Laramie Range, Wyoming and Colorado, *in* Weimer, R. J., and Haun, J. D., eds., Guide to the geology of Colorado: Geological Society of America–Rocky Mountain Association of Geologists–Colorado Scientific Society publication, p. 217–222.

Moss, J. H., 1974, The relationship of river terrace formation to glaciation in the Shoshone River basin, western Wyoming, *in* Coates, D. R., ed., Glacial Geomorphology, Fifth Annual Geomorphology Symposia Series: Binghamton, State University of New York, p. 293–314.

Moss, J. H., and Bonini, W. E., 1961, Seismic evidence supporting a new interpretation of the Cody Terrace near Cody, Wyoming: Geological Society of America Bulletin, v. 72, p. 547–556.

Moss, J. H., and Whitney, J. H., 1971, Diversity of origin of the Cody and Powell terraces along the Shoshone River, Bighorn Basin, Wyoming: Geological Society of America Abstracts with Programs, v. 3, p. 652–653.

Muller, E. H., 1974, Origin of drumlins, *in* Coates, D. R., ed., Glacial Geomorphology, Fifth Annual Geomorphology Symposia Series: Binghamton, State University of New York, p. 187–204.

Nelson, A. R., Madole, R. F., Evanoff, E., Scott, G. R., and Piety, L. A., 1984, Quaternary paleotemperature estimates using amino-acid ratios measured on terrestrial gastropods from fluvial sequences in Colorado [abs.]: American Quaternary Association, Program and Abstracts, 8th Biennial Conference, Boulder, Colorado, p. 92.

Nelson, R. L., 1954, Glacial geology of the Frying Pan River drainage, Colorado: Journal of Geology, v. 62, p. 325–343.

Olyphant, G. A., 1983, Computer simulation of rock-glacier development under viscous and pseudoplastic flow: Geological Society of America Bulletin, v. 94, p. 499–505.

—— , 1985, Topoclimate and the distribution of Neoglacial facies in the Indian Peaks section of the Front Range, Colorado, U.S.A.: Arctic and Alpine Research, v. 17, p. 69–78.

Ommanney, C.S.L., 1972, Glacial inventory: International Symposia on the Role of Snow and Ice in Hydrology, Guidebook, Sept. 1972, p. 84–87.

Osborn, G. D., 1985, Holocene tephrostratigraphy and glacial fluctuations in Waterton Lakes and Glacier national parks, Alberta and Montana: Canadian Journal of Earth Sciences, v. 22, no. 7, p. 1093–1101.

Ostrem, G., 1966, The height of the glaciation limit in southern British Columbia and Alberta: Geografiska Annaler, v. 48A, no. 3, p. 126–138.

Outcalt, S. I., and Benedict, J. B., 1965, Photointerpretation of two types of rock glaciers in the Colorado Front Range, U.S.A.: Journal of Glaciology, v. 5, p. 849–856.

Outcalt, S. I., and MacPhail, D. D., 1965, A survey of neoglaciation in the Front Range of Colorado: Boulder, University of Colorado Studies, Series in Earth Sciences, no. 4, 124 p.

Palmer, H., 1924, Observations on the Freshfield Glacier, Canadian Rockies: Journal of Geology, v. 32, p. 432–441.

Palmquist, R. C., 1975, Preferred position model and subsurface symmetry of valleys: Geological Society of America Bulletin, v. 86, p. 1391–1398.

—— , 1979, Estimated ages of Quaternary terraces, northwestern Wyoming: Geological Society of America Abstracts with Programs, v. 11, no. 7, p. 491.

—— , 1983, Terrace chronologies in the Bighorn Basin, Wyoming: Wyoming Geological Association, 34th Annual Field Conference, Guidebook, p. 217–231.

Paterson, S. T., 1962, Glaciological research on Athabaska Glacier in 1961: Canadian Alpine Journal, v. 45, p. 149.

Paterson, W.S.B., 1969, The physics of glaciers: New York, Pergamon Press, 250 p.

—— , 1972, Athabasca and Saskatchewan glaciers: International Symposia on the Role of Snow and Ice in Hydrology, Guidebook, Sept. 1972, p. 88–91.

Paterson, W.S.B., and Savage, J. C., 1963, Geometry and movement of the Athabasca Glacier: Journal of Geophysical Research, v. 68, p. 4513–4520.

Pearl, R. H., 1971, Pliocene drainage of eastern Colorado and northwestern Kansas: Mountain Geologist, v. 8, p. 25–30.

Pierce, K. L., 1979, History and dynamics of glaciation in the northern Yellowstone National Park area: U.S. Geological Survey Professional Paper 729-F, 90 p.

Pierce, W. G., and Andrews, D. A., 1941, Geology and coal resources of the region south of Cody, Park County, Wyoming: U.S. Geological Survey Bulletin 921-B, p. 99–180.

Porter, S. C., Pierce, K. L., and Hamilton, T. D., 1983, Late Wisconsin mountain glaciation in the western United States, *in* Porter, S. C., ed., Late Quaternary environments of the United States, Volume 1, The late Pleistocene: Minneapolis, University of Minnesota Press, p. 71–111.

Potochnik, A. R., and Damon, P. E., 1986, Tectonic and geomorphic implications of post-Laramide erosion: Geological Society of America Abstracts with Programs, v. 18, p. 403.

Potter, N., Jr., 1972, Ice-cored rock glacier, Galena Creek, northern Absaroka Mountains, Wyoming: Geological Society of America Bulletin, v. 83, p. 3025–3058.

Powell, J. W., 1875, Exploration of the Colorado River of the West and its tributaries: Washington, D.C., U.S. Government Printing Office, 291 p.

Prest, V. K., 1968, Nomenclature of moraines and ice-flow features as applied to the glacial map of Canada: Geological Survey of Canada Paper 67-57, 32 p.

Proudfoot, D. N., 1985, Lithostratigraphic and genetic study of Quaternary sediments in the vicinity of Medicine Hat, Alberta [Ph.D. thesis]: Edmonton, University of Alberta, 248 p.

Ray, L. L., 1940, Glacial chronology of the Southern Rocky Mountains: Geological Society of America Bulletin, v. 51, p. 1851–1917.

Ray, R. J., Krantz, W. B., Caine, T. N., and Gunn, R. D., 1983, A model for sorted patterned ground regularity: Journal of Glaciology, v. 29, p. 317–337.

Reheis, M. C., 1975, Source, transportation, and deposition of debris on Arapaho Glacier, Front Range, Colorado, U.S.A.: Journal of Glaciology, v. 14, p. 407–420.

—— , 1984, Climatic and chronologic controls on soil development, northern Bighorn Basin, Wyoming and Montana [Ph.D. thesis]: Boulder, Colorado, University of Colorado, 346 p.

Reid, I. A., and Charbonneau, J.O.G., 1972, Glacier surveys in Alberta: Department of the Environment, Inland Waters Directorate, Report Series 22, 17 p.

—— , 1975, Glacier surveys in Alberta—1971: Department of the Environment, Inland Waters Directorate Report Series 43, 18 p.

—— , 1979, Glacier surveys in Alberta—1977: Department of the Environment, Inland Waters Directorate, Report Series 65, 17 p.

—— , 1981, Glacier surveys in Alberta—1979: Department of the Environment, Inland Waters Directorate, Report Series 69, 19 p.

Rich, J. L., 1935a, Comment *on* "Physiographic development of the Front Range": Geological Society of America Bulletin, v. 46, p. 2046–2051.

—— , 1935b, Origin and evolution of rock fans and pediments: Geological Society of America Bulletin, v. 46, p. 999–1024.

—— , 1938, Recognition and significance of multiple erosion surfaces: Geological Society of America Bulletin, v. 49, p. 1695–1722.

Richardson, G. B., 1915, Castle Rock folio, Colorado: U.S. Geological Survey Geologic Atlas, Folio 198, 14 p.

Richmond, G. M., 1957, Three pre-Wisconsin glacial stages in the Rocky Mountain region: Geological Society of America Bulletin, v. 68, p. 239–262.

—— , 1962, Quaternary stratigraphy of the La Sal Mountains, Utah: U.S. Geological Survey Professional Paper 324, 135 p.

—— , 1964, Three pre-Bull Lake tills in the Wind River Mountains, Wyoming; A reinterpretation: U.S. Geological Survey Professional Paper 501-D, p. D104–D109.

—— , 1965, Glaciation of the Rocky Mountains, *in* Wright, H. E., Jr., and Frey, D. G., eds., The Quaternary of the United States: Princeton, New Jersey, Princeton University Press, p. 217–230.

—— , 1974, Raising the roof of the Rockies: Rocky Mountain Nature Association, Incorporated, 81 p.

Rissing, J. M., and Thorn, C. E., 1985, Particle size and clay mineral distribution within sorted and nonsorted circles and the surrounding parent material, Niwot Ridge, Front Range, Colorado, U.S.A.: Arctic and Alpine Research, v. 17, p. 153–163.

Ritter, D. F., 1967, Terrace development along the front of the Beartooth Mountains, southern Montana: Geological Society of America Bulletin, v. 78, p. 467–484.

—— , 1972, The significance of stream capture in the evolution of a piedmont region, southern Montana: Zeitschrift für Geomorphologie, v. 16, p. 83–92.

—— , 1975, New information concerning the geomorphic evolution of the Bighorn Basin: Wyoming Geological Association, 27th Annual Field Conference, Guidebook, p. 37–44.

—— , 1982, Complex river terrace development in the Nenana Valley near Healy, Alaska: Geological Society of America Bulletin, v. 93, p. 346–356.

Ritter, D. F., and Kauffman, M. E., 1983, Terrace development in the Shoshone River valley near Powell, Wyoming, and speculations concerning the sub-Powell terrace: Wyoming Geological Association, 34th Annual Field Conference, Guidebook, p. 197–203.

Roed, M. A., 1975, Cordilleran and Laurentide multiple glaciation, west-central Alberta, Canada: Canadian Journal of Earth Sciences, v. 12, p. 1493–1515.

Rohrer, W. L., and Leopold, E. B., 1963, Fenton Pass Formation (Pleistocene?), Bighorn Basin, Wyoming: U.S. Geological Survey Professional Paper 475-C, p. 45–48.

Rouse, J. T., 1934, The physiography and glacial geology of the valley region, Park County, Wyoming: Journal of Geology, v. 42, p. 738–752.

Russell, R. J., 1933, Alpine land forms of western United States: Geological Society of America Bulletin, v. 44, p. 927–950.

Rutter, N. W., 1966, Multiple glaciation in the Banff area, Alberta: Bulletin of Canadian Petroleum Geology, v. 14, p. 613–626.

—— , 1972, Geomorphology and multiple glaciation in the area of Banff, Alberta: Geological Survey of Canada Bulletin 206, 31 p.

—— , 1980, Erosion by Pleistocene continental ice sheets in the area of the Canadian Shield: Atomic Energy Control Board, 112 p.

—— , 1984, Pleistocene history of the western Canadian ice-free corridor, *in* Fulton, R. J., ed., Quaternary stratigraphy of Canada; A Canadian contribution to IGCP Project 24: Geological Survey of Canada Paper 84-10, p. 50–56.

Ryder, J. M., 1971, The stratigraphy and morphology of paraglacial alluvial fans in south-central British Columbia: Canadian Journal of Earth Sciences, v. 8, p. 279–298.

Savage, J. C., and Paterson, W.S.B., 1963, Borehole measurements in the Athabasca Glacier: Journal of Geophysical Research, v. 68, p. 4521–4536.

Schlüchter, C., 1979, Moraines and varves; Origin, genesis, classification: Proceedings, International Quaternary Association Symposium on Genesis and Lithology of Quaternary Deposits, Zurich, 1978, Rotterdam, Netherlands, A. A., Balkema, 441 p.

Schneider, H., 1913, Physiography of Golden and vicinity and its relation to the geologic structure: Colorado School of Mines Quarterly, v. 8, p. 3–10.

Schumm, S. A., 1968, River adjustment to altered hydrologic regimen, Murrumbidgee River and paleochannels, Australia: U.S. Geological Survey Professional Paper 598, 65 p.

—— , 1969, River metamorphosis: Proceedings of the American Society of Civil Engineers, Journal of the Hydraulics Division, v. 95, p. 255–273.

—— , 1973, Geomorphic thresholds and complex response of drainage systems, *in* Morisawa, M., ed., Fluvial Geomorphology, Fourth Annual Geomorphology Symposia Series: Binghamton, State University of New York, p. 299–310.

—— , 1977, The fluvial system: New York, John Wiley and Sons, 338 p.

—— , 1979, Geomorphic thresholds; The concept and its applications: Transactions of the Institute of British Geographers, v. 4, p. 485–515.

Schumm, S. A., and Khan, H. R., 1972, Experimental study of channel patterns: Geological Society of America Bulletin, v. 83, p. 1755–1770.

Schumm, S. A., and Lichty, R. W., 1965, Time, space, and causality in geomorphology: American Journal of Science, v. 263, p. 110–119.

Scott, G. R., 1965, Nonglacial quaternary geology of the Southern and Middle Rocky Mountains, *in* Wright, H. E., Jr., and Frey, D. G., eds., Quaternary of the United States: Princeton, New Jersey, Princeton University Press, p. 243–254.

—— , 1973, Tertiary surfaces and deposits of the Southern Rocky Mountains and their recognition: Geological Society of America Abstracts with Programs, v. 5, no. 6, p. 510–511.

—— , 1975, Cenozoic surfaces and deposits in the Southern Rocky Mountains, *in* Curtis, B. F., ed., Cenozoic history of the Southern Rocky Mountains: Geological Society of America Memoir 144, p. 227–248.

Scott, G. R., and Taylor, R. B., 1986, Map showing late Eocene erosion surface, Oligocene–Miocene paleovalleys, and Tertiary deposits in the Pueblo, Denver, and Greeley 1° × 2° quadrangles, Colorado: U.S. Geological Survey Miscellaneous Investigation Series Map I-1626, scale 1:250,000.

Shaw, J., 1971, Mechanism of till deposition related to thermal conditions in a Pleistocene glacier: Journal of Glaciology, v. 10, p. 363–373.

—— , 1972, Pleistocene chronology and geomorphology in the Rocky Mountains in south and central Alberta, *in* Slaymaker, H. O., and McPherson, H. J., eds., Mountain geomorphology; Geomorphological processes in the Canadian Cordillera: Vancouver, Canada, British Columbia Geographical Series no. 14, Tantalus Research Limited, p. 37–45.

—— , 1977a, Tills deposited in arid polar environments: Canadian Journal of Earth Sciences, v. 14, p. 1239–1245.

—— , 1977b, Till body morphology and structure related to glacier flow: Boreas,

v. 6, p. 189–201.

——, 1979, Genesis of the Sveg tills and Rogen moraines of central Sweden; A model of basal meltout: Boreas, v. 8, p. 409–426.

Sherzer, W. H., 1907, Glaciers of the Canadian Rockies: Smithsonian Institution, Smithsonian Contributions to Knowledge, v. 34, p. 135.

Smalley, I. J., 1981, Conjectures, hypotheses, and theories of drumlin formation: Journal of Glaciology, v. 27, p. 503–505.

Smalley, I. J., and Unwin, D. J., 1968, The formation and shape of drumlins and their distribution and orientation in drumlin fields: Journal of Glaciology, v. 7, p. 377–390.

Smith, H.T.U., 1940, Geologic studies in southwestern Kansas: State Geological Survey of Kansas Bulletin 34, 212 p.

Stankowski, M., ed., 1976, Till; Its genesis and diagenesis: Warsaw, Poland, Univ. im A. Mickiewicza w Poznaniu, Seria Geografia, v. 12, 263 p.

Stanley, K. O., 1971, Tectonic implications of Tertiary sediment dispersal on the Great Plains east of the Laramie Range: Wyoming Geological Association, 23rd Annual Field Conference, Guidebook, p. 65–70.

——, 1976, Sandstone petrofacies in the Cenozoic High Plains sequence, eastern Wyoming and Nebraska: Geological Society of America Bulletin, v. 87, p. 297–309.

Stark, J. T., Johnson, J. H., Behre, C. H., Jr., Powers, W. E., Howland, A. L., Gould, D. B., and others, 1949, Geology and origin of South Park: Geological Society of America Memoir 33, 188 p.

Sugden, D. E., and John, B. S., 1976, Glaciers and landscape; a geomorphological approach: London, Edward Arnold, 376 p.

Taylor, R. B., 1975, Neogene tectonism in south-central Colorado, in Curtis, B. F., ed., Cenozoic history of the Southern Rocky Mountains: Geological Society of America Memoir 144, p. 211–226.

Thompson, W. F., 1968, New observations on alpine accordances in the western United States: Annals of the Association of American Geographers, v. 58, p. 650–669.

Thorington, J. M., 1927, The Lyell and Freshfield glaciers, Canadian Rocky Mountains, 1926: Smithsonian Institution, Smithsonian Miscellaneous Collection, v. 78, p. 1–8.

——, 1945, Notes on Freshfield and Lyell glaciers: American Alpine Journal, v. 5, p. 435–436.

Thorn, C. E., 1975, Influence of late-lying snow on rock-weathering rinds: Arctic and Alpine Research, v. 7, p. 373–378.

——, 1976, Quantitative evaluation of nivation in the Colorado Front Range: Geological Society of America Bulletin, v. 87, p. 1169–1178.

——, 1978a, A preliminary assessment of the geomorphic role of pocket gophers in the alpine zone of the Colorado Front Range: Geografiska Annaler, v. 60A, p. 181–187.

——, 1978b, The geomorphic role of snow: Annals of the Association of American Geographers, v. 68, p. 414–425.

——, 1979a, Ground temperatures and surficial transport in colluvium during snowpatch meltout, Colorado Front Range: Arctic and Alpine Research, v. 11, p. 41–52.

——, 1979b, Bedrock freeze-thaw weathering regime in an alpine environment: Earth Surface Processes, v. 4, p. 211–228.

——, 1980, Alpine bedrock temperatures; An empirical study: Arctic and Alpine Research, v. 12, p. 73–86.

——, 1982a, Gopher disturbance; Its variability by Braun–Blanquet vegetation units in the Niwot Ridge alpine tundra zone, Colorado Front Range, U.S.A.: Arctic and Alpine Research, v. 14, p. 45–51.

——, 1982b, Bedrock microclimatology and the freeze-thaw cycle; A brief illustration: Annals of the Association of American Geographers, v. 72, p. 131–137.

——, 1987, Nivation; A geomorphic chimera, in Clark, M. J., ed., International perspectives in periglacial research: Chichester, United Kingdom, John Wiley and Sons (in press).

Thorn, C. E., and Darmody, R. G., 1980, Contemporary eolian sediments in the alpine zone, Colorado Front Range: Physical Geography, v. 1, p. 162–171.

——, 1985a, Grain-size distribution of the insoluble component of contempor-

ary eolian deposits in the alpine zone, Front Range, Colorado: Arctic and Alpine Research, v. 17, p. 433–442.

——, 1985b, Grain-size sampling and characterization of eolian lag surfaces within alpine tundra, Niwot Ridge, Front Range, Colorado, U.S.A.: Arctic and Alpine Research, v. 17, p. 443–450.

Thorn, C. E., and Loewenherz, D. S., 1987, Interaction between spatial and temporal trends in alpine periglacial landform evolution; Implications for paleo-character, in Boardman, J., ed., Periglacial processes and landforms in Britain and Ireland: Cambridge, Cambridge University Press (in press).

Trenhaile, A. S., 1975, Cirque elevation in the Canadian Cordillera: Annals of the Association of American Geographers, v. 65, p. 517–529.

——, 1976, Cirque morphometry in the Canadian Cordillera: Annals of the Association of American Geographers, v. 66, p. 451–462.

——, 1977, Cirque elevation and Pleistocene snowlines: Zeitschrift für Geomorphologie, v. 21, p. 445–459.

Trimble, D. E., 1980, Cenozoic tectonic history of the Great Plains contrasted with that of the Southern Rocky Mountains; A synthesis: Mountain Geologist, v. 17, p. 59–69.

Van Tuyl, F. M., and Lovering, T. S., 1935, Physiographic development of the Front Range: Geological Society of America Bulletin, v. 46, p. 1291–1350.

Van Tuyl, F. M., Johnson, J. H., Waldschmidt, W. A., Boyd, J., and Parker, B. H., 1938, Guide to the geology of the Golden area: Colorado School of Mines Quarterly, v. 33, 32 p.

Vaux, G., Jr., 1910, Observations on glaciers in 1909: Canadian Alpine Journal, v. 2, p. 126–130.

Vaux, G., Jr., and Vaux, W. S., 1907, Glacier observations: Canadian Alpine Journal, v. 1, p. 138–148.

Vaux, M. M., 1911, Observations on glaciers in 1910: Canadian Alpine Journal, v. 3, p. 127–130.

Wahlstrom, E. E., 1940, Audubon-Albion stock, Boulder County, Colorado: Geological Society of America Bulletin, v. 51, p. 1789–1820.

——, 1947, Cenozoic physiographic history of the Front Range, Colorado: Geological Society of America Bulletin, v. 58, p. 551–572.

Wahrhaftig, C. A., 1965, Stepped topography of the southern Sierra Nevada, California: Geological Society of America Bulletin, v. 76, p. 1165–1190.

Wahrhaftig, C. A., and Cox, A., 1959, Rock glaciers in the Alaska Range: Geological Society of America Bulletin, v. 70, p. 383–436.

Wallace, R. G., 1967, Types and rates of alpine mass movement, west edge of Boulder County, Colorado, Front Range [Ph.D. thesis]: Columbus, The Ohio State University, 199 p.

Washburn, A. L., 1980, Geocryology; A survey of periglacial processes and environments: New York, Halsted Press, 406 p.

Washicheck, J. N., 1972, Summary of snow survey measurements for Colorado and New Mexico: Denver, Colorado, Soil Conservation Service, 208 p.

Webber, P. J., and May, D. E., 1977, The magnitude and distribution of below-ground plant structures in the alpine tundra of Niwot Ridge, Colorado: Arctic and Alpine Research, v. 9, p. 157–174.

Weertman, J., 1971, Shear stress at the base of a rigidly rotating cirque glacier: Journal of Glaciology, v. 10, p. 31–38.

Whalley, B., 1974a, Rock glaciers and their formation as part of a glacier debris-transport system, in the collection Reading geographical papers: Whiteknights, Reading, England, United Kingdom, University of Reading, Department of Geography, Paper no. 24, 60 p.

——, 1974b, The mechanics of high-magnitude, low-frequency rock failure and its importance in a mountainous area, in the collection Reading geographical papers: Whiteknights, Reading, England, United Kingdom, University of Reading, Department of Geography, no. 27, 48 p.

Wheeler, A. O., 1907, Observations on the Yoho Glacier: Canadian Alpine Journal, v. 1, p. 149–159.

——, 1920a, Motion of the Yoho Glacier 1917, 1918, and 1919: Canadian Alpine Journal, v. 11, p. 121–146.

——, 1920b, Notes on the glaciers of the Main and Selkirk ranges of the Canadian Rocky Mountains: Canadian Alpine Journal, v. 11, p. 121–146.

——, 1932, Glacial change in the Canadian Cordillera; The 1931 expedition:

Canadian Alpine Journal, v. 20, p. 120–142.

——, 1933, Records of glacial observations in the Canadian Cordillera, 1933 and 1934: Canadian Alpine Journal, v. 22, p. 172–185.

White, P. G., 1979, Rock glacier morphometry, San Juan Mountains, Colorado: Geological Society of America Bulletin, v. 90, no. 6, p. 515–518.

White, S. E., 1968, Rockfall, alluvial, and avalanche talus in the Colorado Front Range: Abstracts for 1967, Geological Society of America Special Paper 115, p. 237.

——, 1971a, Debris falls at the front of Arapaho rock glacier, Colorado Front Range, U.S.A.: Geografiska Annaler, v. 53A, p. 86–91.

——, 1971b, Rock glacier studies in the Colorado Front Range, 1961 to 1968: Arctic and Alpine Research, v. 3, p. 43–64.

——, 1972, Alpine subnival boulder pavements in the Colorado Front Range: Geological Society of America Bulletin, v. 83, p. 195–200.

——, 1975, Additional data on Arapaho rock glacier in the Colorado Front Range, U.S.A.: Journal of Glaciology, v. 14, p. 529–530.

——, 1976a, Rock glaciers and block fields, review and new data: Quaternary Research, v. 6, p. 77–97.

——, 1976b, Is frost action really only hydration shattering? A review: Arctic and Alpine Research, v. 8, p. 1–6.

——, 1981, Alpine mass movement forms (non-catastrophic); classification description and significance: Arctic and Alpine Research, v. 13, p. 127–137.

——, 1982, Physical and geological nature of the Indian Peaks, Colorado Front Range, *in* Halfpenny, J. C., ed., Ecological studies in the Colorado alpine; A festschrift for John W. Marr: Boulder, University of Colorado, Institute of Arctic and Alpine Research Occasional Paper no. 37, p. 1–12.

Willard, B. E., 1979, Plant sociology of alpine tundra, Trail Ridge, Rocky Mountain National Park, Colorado: Colorado School of Mines Quarterly, v. 64,

p. 1–119.

Wolman, M. G., 1955, The natural channel of Brandywine Creek, Pennsylvania: U.S. Geological Survey Professional Paper 271, 56 p.

Wolman, M. G., and Leopold, L. B., 1957, River flood plains; Some observations on their formation: U.S. Geological Survey Professional Paper 282-C, p. 87–109.

Womack, W. R., and Schumm, S. A., 1977, Terraces of Douglas Creek, northwestern Colorado; An example of episodic erosion: Geology, v. 5, p. 72–76.

Worcester, P. G., 1931, Peneplanation and base-leveling in the Rocky Mountains [abs.]: Colorado–Wyoming Academy of Science Journal, v. 1, no. 3, p. 27–28.

——, 1939, A textbook of geomorphology: New York, D. Van Nostrand Company, 565 p.

MANUSCRIPT ACCEPTED BY THE SOCIETY OCTOBER 9, 1986

ACKNOWLEDGMENTS

We are indebted to Arthur L. Bloom (Cornell University), James S. Gardner (University of Waterloo), Robert L. Schuster and Eugene S. Ellis (U.S. Geological Survey) for reviewing drafts of the entire chapter. Bradley also thanks Rudy C. Epis (Colorado School of Mines) and Glenn R. Scott (U.S. Geological Survey) for thoughtful reviews of the section on the erosion surfaces of the Front Range. Thorn and Loewenherz thank Nel Caine (University of Colorado) and Sidney E. White (Ohio State University) for reviews of the section on alpine mass wasting, and also acknowledge the many researchers who have worked in the Indian Peaks, but accept full responsibility for any errors made in discussing their work.

Geological Society of America
Centennial Special Volume 2
1987

Chapter 8

Colorado Plateau

William L. Graf
Department of Geography, Arizona State University, Tempe, Arizona 85287
Richard Hereford
U.S. Geological Survey, 2255 North Gemini Drive, Flagstaff, Arizona 85601
Julie Laity
Department of Geography, California State University at Northridge, Northridge, California 91330
Richard A. Young
Department of Geological Sciences, State University College of Arts and Science, Geneseo, New York 14454

GEOMORPHOLOGIC RESEARCH IN THE COLORADO PLATEAU

William L. Graf

INTRODUCTION

Field studies in the Colorado Plateau occupy an honored place in the development of geomorphic theory. The purpose of this chapter is to briefly review the foundational, regional, and process-oriented studies in the region, and to provide a review of promising threads of inquiry set in a context of more than a century of geomorphologic research in the region.

The Colorado Plateau has sharply defined boundaries that separate it from neighboring geomorphic provinces (Fig. 1; for details see Thornbury, 1965). On the west, faults and the perimeters of volcanic plateaus mark the boundary between the Colorado Plateau and the Basin and Range Province. The boundary extends across the southern edge of the plateau where it is less radically defined, but is nonetheless visible on the surface in the form of an uplifted edge of sedimentary rocks known as the Mogollon Rim, which extends from northwest Arizona diagonally into north-central New Mexico. The eastern and northern boundaries are delineated by the contact between sedimentary rocks and upthrust or folded crystalline rocks of the Rocky Mountains.

The plateau is a definable tectonic unit relatively easily separated from other provinces, but it shows considerable internal variation (Fig. 1; for details see Hunt, 1974a). The interior of the kidney-shaped Colorado Plateau Province reveals a series of subsections that depend on geologic and geomorphologic definition. The centrally located Canyon Lands Section is dominated by gently folded sedimentary rocks, while the western High Plateaus Section reveals widespread accumulations of volcanic materials.

The Uinta Basin Section in the north and the Grand Canyon–Datil sections in the south have mountains, cliffs, and dissected terrain. The Navajo Section is largely a sedimentary platform with isolated buttes, mesas, folded mountains, and volcanic plugs.

The exploration and investigation of this spectacular landscape have given rise to so many contributions to geomorphology that it is not possible to even mention all of them, let alone analyze them in detail. Instead, in this chapter we briefly review three groups of contributions and investigate three more series of studies in greater detail. The following subsections of this introduction contain (1) brief reviews of the products of public land surveys of the Colorado Plateau that formed the foundation field studies of much of modern geomorphology, (2) regional reconnaissance studies that contributed to the understanding of arid and semiarid surficial conditions, and (3) more recent process-oriented investigations. Major sections of the remaining chapter are devoted to recent developments in the understanding of the Tertiary history of the Grand Canyon Section, twentieth-century fluvial processes in the Canyon Lands and Navajo sections, and the analogs for other planets provided by the plateau province.

FOUNDATION STUDIES

In the period from 1847 to 1880, surveys of the public lands of the Colorado Plateau region provided geologists and geographers the opportunity to study previously unknown landscapes. Their writings containing observations and deductions laid the first paving stones on the road to the development of the science of American geomorphology. These foundation works were contained in the reports of exploration and survey expeditions in association with interpretations of the long-term geologic history of the region. Among the most important authors were Newberry, Marvine, Powell, Dutton, and Gilbert.

Newberry accompanied two exploratory efforts by the U.S.

Graf, W. L., Hereford, R., Laity, J., and Young, R. A., 1987, Colorado Plateau, *in* Graf, W. L., ed., Geomorphic systems of North America: Boulder, Colorado, Geological Society of America, Centennial Special Volume 2.

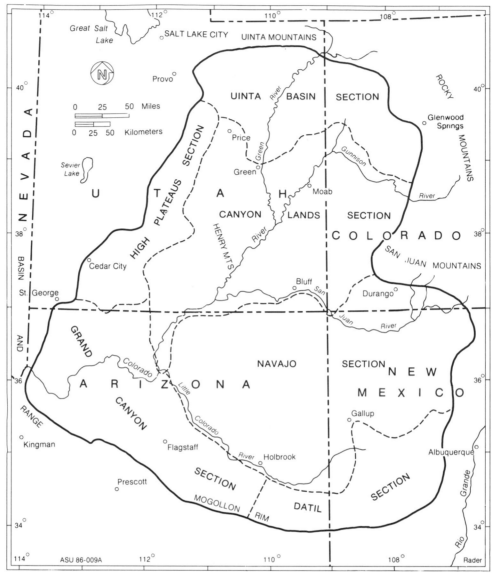

Figure 1. Index map of the Colorado Plateau Province. Redrawn from a base map by C. B. Hunt.

Army Topographic Engineers, one by Ives up the Colorado River in 1847 (Newberry, 1861), and one by Macomb to the junction of the Green and Colorado rivers in 1849 (Newberry, 1876). Newberry's great contribution was the recognition of the importance of fluvial processes in the sculpture of the arid and semiarid landscape. Subsequently, the work of running water has dominated geomorphic research in the Colorado Plateau. Newberry's work was influenced by Marvine (1874), who also had some influence over his better-known friend G. K. Gilbert (Davis, 1922).

Of the four great geological surveys of the western United States (King's, Wheeler's, Powell's, and Hayden's), the most important for the development of geomorphology was John Wesley Powell's Geographical and Geological Survey of the Rocky Mountain Region (Bartlett, 1962). Despite its name, the Powell Survey focused on the Colorado Plateau region and provided an intellectual stage upon which Powell and his two major assistants, Gilbert and Clarence E. Dutton, were able to develop their theories of landscape development. Powell was clearly the early conceptual leader of this threesome who developed a series of fundamental principles concerning fluvial processes (Chorley and others, 1964). Dutton provided an artistic flair for description that brought to reality for his scientific readers a landscape that challenged the imagination (Diller, 1911). Gilbert was the scientist who translated observations into generalizations based on concepts of applied physics and general systems thinking (Pyne, 1980).

Powell's intellectual contributions were mainly at the regional scale where he proposed alternative explanations for the courses of the rivers of the Colorado Plateau (Fig. 2). He clearly articulated the opposing explanations for the deeply eroded canyons as either superposition or antecedence. In the superposition hypothesis, Powell envisioned the major rivers flowing on an ancient surface of low gradients above the present landscape, which developed as the streams were rejuvenated by tectonic uplift (Powell, 1895). In the antecedent hypothesis, he envisioned the stream courses being developed on relatively horizontal geologic structures that later were slowly folded and dissected by streams that cut across the relatively newer structures. Subsequent investigations indicated that a combination of these explanations was most likely to be correct (e.g., Hunt, 1969).

Dutton's expansive descriptions of the Colorado Plateau landscape were the beginnings of several important contributions. Working mostly in the Grand Canyon and High Plateau areas, he provided clear statements about the limiting role of base level in fluvial erosion systems (Dutton, 1880, 1881). He accepted Powell's view of antecedent rivers for the interpretation of modern landscapes, and used as examples geologic discontinuities exposed in the walls of the Grand Canyon. Dutton (1882) also provided early comprehensive field studies of alluvial fans (which he termed alluvial cones) by showing direct causal linkages among the characteristics of fan planimetric shape, profile form, particle size distribution, and fluvial transport processes.

Gilbert's only monograph-length publication pertaining to the Colorado Plateau region has been his most lasting contribution, and is the cornerstone of process geomorphology. In his discussion of the Henry Mountains region, his description and explanation of "land sculpture" introduced a significant philosophical approach that was different from that of Powell (Gilbert, 1877). Gilbert sought explanation of the development of slope and stream profiles by outlining a series of lawlike statements which, taken together, formed a coherent body of geomorphic theory. The most fundamental proposition of this body of theory was the concept of dynamic equilibrium, wherein the interaction of the resistance of various rock units with the erosive action of running water produced characteristic varying slope angles.

The foundation studies by scientists with the early public land and resource surveys of the Colorado Plateau region provided the basic data for the sweeping generalizations of William Morris Davis (Davis, 1899). Davis relied heavily on the results of the western surveys in formulating his theory of evolution for landscapes, which was to dominate American geomorphology for four decades. Concepts related to antecedent, consequent, and superimposed streams, base level, peneplains, equilibrium, energy gradients, and the development of general erosion surfaces are frequently associated with Davis and his wide-ranging theories, but their foundations are in part in field studies of the Colorado Plateau.

REGIONAL STUDIES

During the first third of the twentieth century, while the Davisian intellectual juggernaut rolled onward, the central Colo-

Figure 2. A legacy of photographic data. Views of the neck of Bowknot Bend looking east on the Green River about 100 km upstream from the confluence with the Colorado River. Top, 1871 photo by J. K. Hillers of the 1871 Powell expedition (Hillers' photo number 476); Middle, 1921 photo by E. C. LaRue while surveying for dam sites (LaRue photo number 9); Bottom 1968 photo by E. M. Shoemaker while reoccupying some of Hillers' photo stations (Shoemaker photo number 13). All photos from U.S. Geological Survey Photographic Library, Denver, Colorado. Note the stabilization of the sediment on the point bar on the left side of the photo between the 1921 and 1968 views coincidental with the invasion of the phreatophyte plant tamarisk. Talus appears to be stable.

rado Plateau became the site of a series of regional geologic and geographic studies (Fig. 3). Although they were largely descriptive, these studies also provided interpretations of the plateau landscape and its processes, as well as the logical starting point for most modern research efforts. Three workers, Herbert E. Gregory, Howard D. Miser, and Charles B. Hunt, provided especially informative literature, while a number of other authors concerned mostly with economic geology provided supporting geomorphologic evidence for explanation of the region.

Gregory worked on plateau-related research for almost half a century, ultimately producing U.S. Geological Survey Professional Papers covering the Navajo Section and most of the southern Canyon Lands and High Plateaus sections (Gregory, 1917, 1938, 1950; Gregory and Moore, 1931). Without his comprehensive regional works with their necessary background, subsequent topical investigations would have been impossible. Gregory's perceptive interpretations in geomorphic matters addressed weathering, eolian processes, and landscape evolution, but his most significant geomorphic contributions were in the investigation of streams that had become entrenched in alluvial fills. Gregory (1917) laid the basis for the debate about the causes of stream entrenchment by clearly defining the competing hypotheses: climatic change, over-grazing, and catastrophic floods. Generally he believed that climate was the major causal mechanism.

Hunt and Miser investigated more limited areas of the Colorado Plateau, but their contributions to geomorphic theory were substantial. Hunt led a general geologic and geographic reconnaissance of the Henry Mountains region, visiting the study area of Gilbert half a century after the first investigation. Much of Hunt's work was directed to problems of structure and lithology, but he also devoted attention to geomorphic issues of river channel change (Hunt and others, 1953). Through extensive analysis of historical accounts, Hunt showed that recent stream channel entrenchment and catastrophic widening of major streams occurred as the result of a single flood event, which triggered a series of basin-wide erosion responses beginning in the 1890s. On a longer time scale, he showed how stream capture on pediments was an important component of the fluvial history of the area.

Miser's work was focused on the San Juan River and its canyon through the southeastern Colorado Plateau. Fueled by the search for sites for potential high dams in the region, Miser mapped the canyon accurately and provided the first analyses of fluvial processes there (Miser, 1924a, 1924b). His description of sand waves on channel floors during floods stimulated interest in bed processes. He also deduced that in many parts of the canyon floods mobilized all of the alluvium, so that erosion of the canyon floor was probably occurring even though it usually appeared to be covered with unconsolidated material. In other canyon reaches he detected deposits that he explained as the products of upstream erosion in the plateau tributaries.

Most of the remaining areas of the central plateau were subjects of detailed reconnaissance studies by the U.S. Geological Survey that provided firm footing for specialized research that

was to follow (Fig. 4). Made at a time when precious metals, coal, and oil were suspected to occur in the region, most of these studies emphasized economic geology. However, they also touched on geomorphic problems and illustrated a landscape that was unique, but that also offered a setting for field investigations of general principles.

The regional studies of the Colorado Plateau are significant for modern geomorphology in two ways. First, taken together they provided one of the first truly comprehensive descriptions of a geomorphic region. From them emerges a picture of ordered complexity, a region with a coherent geologic structure that is occasionally interrupted by volcanic intrusions, resulting in a distinctive landscape. Second, the regional studies were dominated by description, but the authors provided interpretations that led directly to important generalizations in the mid–twentieth century. Gregory's speculations on channel entrenchment foresaw later general conclusions by scores of researchers (reviewed by Graf, 1983a), his work on weathering led to larger theories (Koons, 1955), and his early comments on eolian processes were the earliest efforts at research now pursued by planetary geologists (reviewed later in this chapter). Hunt's investigation in the Henry Mountains provided indispensible background for later more specialized efforts in the same area (e.g., Graf, 1983b). Miser's work on the San Juan also presaged more detailed investigations into modern processes by Leopold and others (1964, p. 227–233) and into longer term landscape development by Leopold and Bull (1979).

PROCESS-ORIENTED STUDIES

Geomorphology in the 1980s seems preoccupied with process-oriented studies (Gregory, 1985), and recent work in the Colorado Plateau appears to be no exception. Three groups of studies illustrate the general trends: spatial variation of fluvial processes in the Henry Mountains region, environmental change and its impacts in the Chaco River Basin, and flood stratigraphy in the Escalante River Basin.

Because of the early studies by Powell, Gilbert, and Hunt, the Henry Mountains area and the nearby canyons of the Colorado River have a rich background of geomorphic observations upon which the modern researcher can build process-oriented studies. This evidence, supplemented by additional field measurements and survey records, permitted the analysis of two general issues: changes in the processes of the master stream of the region with environmental quality consequences, and the construction of a model of the spatial variation of fluvial processes in regional stream systems. Along the Colorado River, vegetation changes in the last century have radically altered the riparian environment of the Colorado and Green rivers, with concomitant changes in the near-channel hydraulic properties of the river (Graf, 1978). The development of high dams on the major rivers had tended to stabilize rapids in the canyons downstream by eliminating the highest flood flows, so that the rivers now behave differently than in pre-dam periods (Graf, 1979, 1980).

Figure 3. Early research in the central Colorado Plateau. Upper left, The Powell Survey begins its second voyage down the Green and Colorado rivers, embarking from the railroad town of Green River, Wyoming, in May, 1871 (Hillers photo number 462, U.S. Geological Survey Photographic Library, Denver); Upper right, Almon Harris Thompson, topographer with the Powell Survey who prepared the first comprehensive topographic map of the region, poses on "Old Paint" in 1875 (Hillers photo number 661, U.S. Geological Survey Photographic Library, Denver); Middle left, digging out Charles B. Hunt's car from a sand drift near Avery Springs in the Henry Mountains in the middle 1930s (Hunt photo number 34, U.S. Geological Survey Photographic Library Denver; Lower left, H. E. Gregory uses his Ford Model A for scale in a 1929 view of the sediments exposed in an arroyo wall along Comb Wash near Bluff, Utah (Gregory photo number P2000, 9-2122, G(833), Special Collection of the Marriott Library, University of Utah, Salt Lake City); Lower right, dam-site surveyors take a lunch break and a swim on the Lower Green River, September 19, 1921 (La Rue photo number 28, U.S. Geological Survey Photographic Library, Denver).

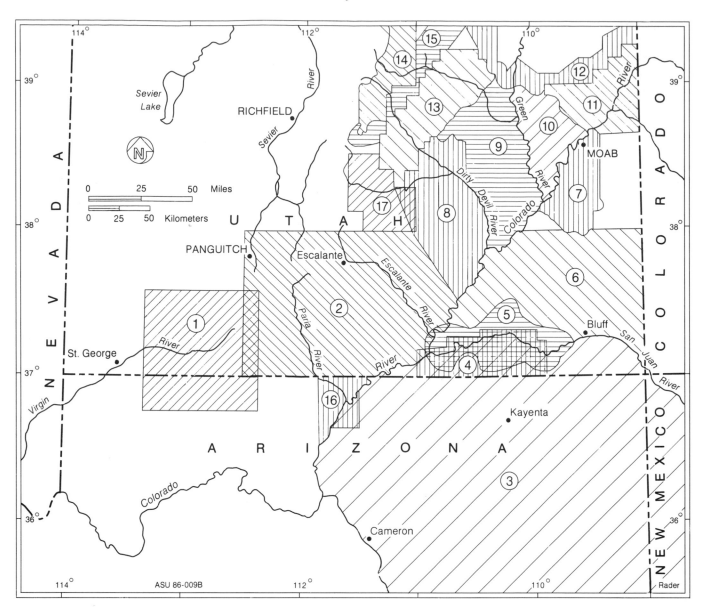

Figure 4. Reconnaissance study of areas of the central Colorado Plateau. 1, Gregory (1950); 2, Gregory and Moore (1931); 3, Gregory (1917); 4, Baker (1936); 5, Miser (1924a, 1924b); 6, Gregory (1938); 7, Baker (1933); 8, Hunt and others (1953); 9, Baker (1946); 10, McKnight (1940); 11, Dane (1935); 12, Fisher (1936); 13, Gilluly (1929); 14, Spieker (1931); 15, Lupton (1916); 16, Phoenix (1963); 17, Smith and others (1963). Modified from an original figure by Gregory (1950, p. 3).

The spatial variation of erosion and sedimentation in the regional streams of the plateau province is relatively consistent and apparently is derived from the distribution of stream power (Graf, 1982a). In small channels, there is a distinct distance decay in erosional responses to system perturbations, so that if base level of a small stream is lowered through downcutting by a flood on the master stream, the tributary responds by downcutting to lesser degrees as distance increases away from the master stream (Graf, 1982b). More generally, there is a typical spatial arrangement of stream power throughout the drainage systems in the Henry

Mountains region, whereby stream power is at a maximum along those streams with about 50 to 100 km^2 drainage area upstream (Graf, 1983b). Other areas upstream and downstream tend to have lower values of stream power. Thus, changes in erosion and sedimentation are usually most radical in the intermediate-sized basins (Graf, 1985).

Investigations of the sedimentology of Chaco Canyon have also led to fruitful generalizations. From its source area in the eastern Colorado Plateau of northwest New Mexico, the Chaco River drains arid and semiarid table lands and empties into the

San Juan River. In its middle reaches the river flows through a wide shallow canyon partially filled with alluvium that was the site of pre-Columbian Indian cities and agricultural enterprises. The archeological evidence left by these early societies and the abrupt demise of the culture several hundred years ago have led to intensive investigations of geomorphological and sedimentological records in search of explanations for possible environmental changes. In the late 1970s and early 1980s a clear picture of the hydrogeomorphic events of the past several hundred years became apparent. Work by Hall (1977, 1980) and Betancourt and Van Devender (1981), for example, showed that the climate and vegetation in the area of Chaco Canyon had radically changed during the past several hundred years, and that the changes from dry to relatively moist conditions are matched in time by channel entrenchment and erosion (Hall, 1983).

The sedimentary facies of Quaternary deposits in Chaco Canyon, irrespective of the linkages with vegetation evidence, have been instructive for interpretation of processes. Love's (1979, 1983) extensive analysis of the sedimentology of the alluvial deposits in the canyon showed that recent declines in precipitation coincided with the deposition of new facies, a result consistent with the findings of Hall. Love also found that other episodes more distant in time must have been climatically controlled, and that the connective mechanism was the ability of the stream to transport its sediment as indicated by stream power. Buried channels throughout the alluvial deposits of the canyon bear testimony to the repetitive nature of the fluvial processes, and the various depositional units provide an interpretive key to climatic and geomorphic changes that are probably typical of the southern Colorado Plateau. The connections between climate and surface processes are probably more widely applicable. In a recent compendium (unfortunately of limited distribution), Wells and others (1983) drew together the diverse geomorphic and archeologic research pertaining to Chaco Canyon and demonstrated the utility of the Chaco Canyon as a representative research area.

Another group of workers has also recently focused on process-oriented studies as revealed by sedimentological evidence. The analysis of the flood history of various tributaries of the Escalante River in the western Colorado Plateau has relied in part on the analysis of slack water flood deposits, a technique developed in other areas (Kochel and Baker, 1982). The flood deposits of the Escalante tributaries occur as discontinuous bodies of terrace and flood-plain material along stream courses deeply entrenched in sandstone. They have been dated using radiocarbon and dendrochonologic methods.

Although much of this work has yet to reach formal print, it promises important contributions to geomorphic theory. Preliminary conclusions (Boison and Patton, 1983, 1985; Patton and Boison, 1986) indicate two important process-related findings. First, individual basins appear to exhibit substantial variation from general trends, and the onset of erosion or deposition may not be coincidental over large areas (unlike the findings of Hereford as reported later in this chapter). Thus, depositional units

may have a variety of ages depending on location. Second, there appears to be an important link between landslide activities on surfaces of the upper watershed and depositional processes along stream courses of the lower watershed. Bursts of sediment derived from active landslides and carried by floods may explain many of the alluvial deposits.

The recent process-related research in the Henry Mountains, Chaco Canyon, and Escalante Basin represent contributions of field studies in the plateau to the development of geomorphic theory. Each of these groups of studies reveals specific information about a specific area, but the studies also lead to wider generalizations and development of fundamental geomorphic principles. The following sections explore in greater detail three lines of research in other areas of the Colorado Plateau that are likely to improve our geomorphologic understanding and appreciation.

LANDSCAPE DEVELOPMENT DURING THE TERTIARY

Richard A. Young

OVERVIEW AND PERSPECTIVE

The spectacular landscapes of the Colorado Plateau not only attracted the scientific interest of the early explorer-geologists and natural scientists, but numerous specific geologic structures and relationships were so dramatically exposed that they appear to have virtually compelled the geologists on the early expeditions to speculate on their origins. Many of the early geologic reports are excessive in their use of superlatives to describe the plateaus and canyons. A statement by Dutton (1882, p. 92) makes the impact clear. "It would be difficult to find anywhere else in the world a spot yielding so much subject matter for the contemplation of the geologist; certainly there is none situated in the midst of such dramatic and inspiring surroundings." Powell said of the Canyon region (1895), "The resources of the graphic arts are taxed beyond their powers in attempting to portray its features. Language and illustration combined must fail."

Most of the features are not unique to the plateau, but their stark forms and excellent exposure encouraged study and lively debate. In addition, many of the simple structural features are enormous, dominating the traveler's view for many miles, and they had obviously influenced the evolution of the adjacent drainage systems. It is ironic that some of the very features that were first used by Powell to demonstrate or formulate the theories of antecedence and superposition were misinterpreted; such as the antecedence of the Lodore Canyon along the Green River and the Colorado River across the Kaibab Upwarp. The theories have occasionally survived, but to explain the relationships in other places. Hunt (1969, p. 61–66) has summarized and tabulated the many ideas and contributions of 34 geologists from 1861 through 1968 relating to the Grand Canyon and its relationship to the Colorado Plateau.

Most of the early geologists viewed the plateau as having been a broad surface of marine and lacustrine deposition until early Eocene time. An initial period of uplift accompanying the emptying of the Eocene Lakes was envisioned to have caused a great period of regional denudation, still near sea level, between early Eocene and late Miocene time. The major canyon cutting and regional incision by tributaries were relegated to a Miocene-Quaternary episode associated with several thousand feet of differential Pliocene-Pleistocene uplift and the availability of more runoff as air masses were intercepted by the uplifted Rocky Mountains. It should be remembered throughout this discussion that the Pliocene-Miocene boundary was shifted from 11 Ma to 5 Ma in the early 1970s. This creates some confusion in the literature when the terms late Miocene or early Pliocene are used, especially when radiometric and paleontological ages are indiscriminately combined from references written both prior to and following this change in the boundary position.

Early geologists recognized that the early Tertiary regional denudation extended into the Great Basin (Lee, 1908; Davis, 1930), and they also realized that the Great Basin had been uplifted relative to sea level along with the plateau. The early geologists also assumed, lacking evidence to the contrary, that the Colorado River had basically maintained its present course once it became established, following the emptying of the Eocene lakes. The river was thought to have cut downward more rapidly as the uplift supposedly increased after early Miocene time. Thus, it was variously presumed to be antecedent or superposed across the Tertiary structures.

After the early reports of Powell and Dutton, as more localized studies of numerous subregions of the plateau accumulated in the literature, many of the broader generalizations about the timing of major events and the accompanying development of landforms eventually began to be questioned, especially the age of the Grand Canyon in northern Arizona. The most specific challenge came from the work of Longwell (1946) who suggested that the Colorado River could not have existed at the west edge of the plateau until late in the Miocene or early Pliocene epoch, because it was demonstrably younger than the Miocene(?) Muddy Creek Formation in the Grand Wash Trough (Lake Mead). This observation required either an alternative outlet for the Colorado River between late Eocene and early Pliocene time, or some different drainage system to explain the "great denudation" cycle of earlier workers. Longwell's observations to the west of the plateau were gradually confirmed during the 1960s and 1970s as more specific chronologic information near Lake Mead (Lucchitta, 1972, 1979; Blair and others, 1977) and more local studies elsewhere provided evidence incompatible with the simple scenarios of earlier workers (McKee and others, 1964; McKee and McKee, 1972; Price, 1950).

The increasing numbers of radiometric ages throughout the plateau, while helping to put local chronologies on sounder footing, created regional inconsistencies in proposed theories. Unfortunately, there is a lack of volcanic rocks with ages falling between Eocene and early Miocene time in key areas across the

plateau. The infusion of detail often led to greater differences of opinion rather than agreement on the timing of regional Tertiary events (McKee and McKee, 1972; McKee and others, 1964; Hunt, 1969). Because a number of the basic ideas concerning the early Tertiary evolution of the plateau were challenged by specific, detailed studies, the broader community of geologists not directly involved in plateau studies probably came to doubt the validity of most of the earlier works, although they contained many sound observations within the confused chronologies.

The origin of the plateau landscape had been firmly associated with the gradual evolution of the Colorado River and its tributaries by the writings of Powell and other respected geologists. However, it became increasingly difficult to explain the presumed Eocene to Miocene landscape changes without an obvious drainage outlet to account for the relatively sudden appearance of the Colorado River at the Hurricane fault, and, presumably, across the western plateaus in the last 5 million years or less (for locations see Fig. 5). A variety of different theories eventually emerged that did not completely resolve the dilemma but dealt with specific aspects of the problem.

Generally, these theories fell into two categories. Some theories allowed the Colorado River to take another course from the vicinity of the Kaibab Upwarp (McKee and others, 1964; Lucchitta, 1975), while others required the early denudation of the plateau to have continued longer, thus compressing the evolution of the Colorado River into the late Miocene and Pliocene epochs (Hunt, 1956, 1969, 1974b). Some hypotheses combined aspects of both approaches. However, no one was able to compile a convincing geological scenario encompassing the evolution of the entire plateau and explaining all the conflicting data. New field studies in crucial areas were frustrated by the absence of early Tertiary volcanic rocks or fossiliferous sediments (Young, 1966).

Hunt (1956) reviewed the available evidence, added observations from studies of his own, and devised a synthesis that required the Colorado River to exit from the plateau until Miocene or Pliocene time at the Peach Springs Canyon–Truxton Valley reentrant in the plateau edge along the southern extension of the Hurricane fault. He suggested that the river later became "anteposed" west of the Hurricane fault in its present course during the late-stage differential uplift and tilting (Hunt, 1956, Figs. 61, 62).

Subsequently, Young (1966) showed that the large Peach Springs Canyon–Truxton Valley reentrant was filled with several hundred feet of pre–middle Miocene, deeply-weathered gravels of westerly or southerly provenance. This very old valley contained no Colorado River sediments above or beneath the extensive 17- to 18-Ma Peach Springs Tuff and related volcanics that underlie most of the valley.

Further refinements in explanation followed. McKee and others (1964) edited a symposium volume that attempted to separate the factual field evidence from the unsupported hypotheses, and they developed a two-stage capture hypothesis for the youthful western Grand Canyon. Hunt (1969) devised an alternative explanation that would still permit an ancestral Colorado River

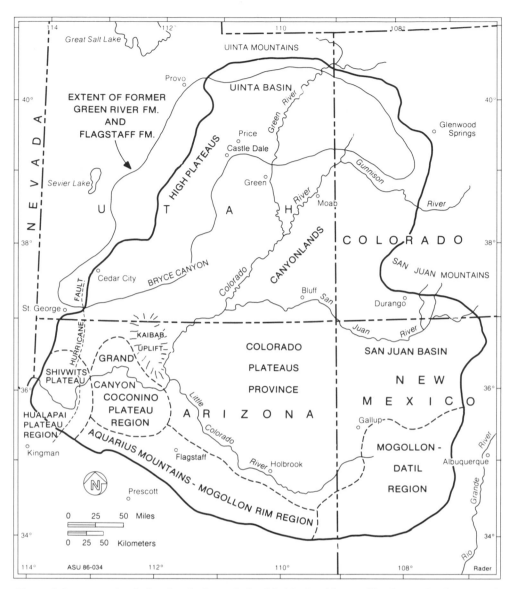

Figure 5. Important detailed regions in the analysis of the Tertiary history of landscape development of the Colorado Plateau during the Tertiary.

to exit southward at Peach Springs Canyon prior to the middle of the Miocene. However, he did not resolve the conflicting data of Young (1966) and failed to explain specifically how the course of the river might have "switched" from the proposed Peach Springs Canyon channel to its present position across the Hualapai Plateau, without leaving evidence of such a presumably recent change in the extensive postvolcanic sediments. Subsequently Hunt (1974b) attempted to explain the late shift in course from the Peach Springs outlet by a complex sequence of events involving a near drying up of the evolving ancestral Colorado River drainage and massive underground piping in the Lake Mead region.

By this time the revised geologic history of the adjacent Great Basin (Lucchitta, 1972; Anderson, 1971; Eberly and Stan-

ley, 1978; Frost and Martin, 1982) and progress on further studies in the Peach Springs–Hualapai Plateau region (Young and Brennan, 1974; Young and McKee, 1978; Young, 1979) were beginning to clearly demonstrate that the old northeast-trending valley at Peach Springs, eroded along the Hurricane fault zone, was part of a more complex older Laramide drainage system that demanded a north to northeasterly outlet onto the plateau. Furthermore, this old erosion surface, related to the Rim gravels of north-central Arizona (Price, 1950), bore no obvious relation to the much younger Colorado River. This has recently been confirmed by the discovery of probably early Eocene (or older) viviparid gastropods on the Coconino Plateau in sediments stratigraphically equivalent to the oldest Peach Springs Canyon deposits (Young and Hartman, 1984; Young, 1982; Hartman, 1984).

Smiley and others (1984) published a volume on the *Land-scapes of Arizona* that was an outgrowth of a 1973 symposium. This volume, published ten years later than intended, contained a more plausible synthesis, which explains the late Tertiary evolution of the Colorado River by stream capture west of the Kaibab Upwarp (Lucchitta, in Smiley and others, 1984, Ch. 13). This approach retained some of the ideas of McKee and others (1964) and suggestions by Young in the same volume. It is similar to the version of the capture hypothesis appearing in a little-circulated NASA report by Goetz and others (1975), also contributed by Lucchitta. In spite of the recent publication date of the Smiley and others (1984) symposium volume, the long delay in printing unfortunately excluded the more recent data bearing on the age of the Coconino Plateau drainage (Young, 1979, 1982; Young and Hartman, 1984). However, all of these more recent hypotheses involve some degree of headward erosion by a younger drainage system from west of the Hurricane fault across the Hualapai and western Coconino plateaus to capture the ancestral Colorado River near the Kaibab Upwarp.

Although this preoccupation with the origin of the Colorado River on the western plateau may seem to ignore the Tertiary geomorphology of much of the rest of the plateau, the close historical association of the Colorado River's origins with the geomorphic evolution of the entire plateau has caused the debate to center around the chronology of the Grand Canyon's origin and its presumed relationship to a plateau uplift; a logical but unfortunate assumption. If the Colorado Plateau did not owe much of its present geomorphic diversity to the gradual incision by the Colorado River system, then what unknown events might have been responsible for an important portion of the post-Eocene landscape evolution? It has become increasingly difficult to account for a late Tertiary differential uplift of the plateau proper coincident with the apparent Pliocene incision of the western Grand Canyon. Lucchitta (1979) summarized the evidence for less than 900 m of differential uplift between Yuma and the mouth of the Grand Canyon, essentially during Pliocene time. Thus, the dilemma has become more specifically defined. If the plateau had been uplifted beginning in Eocene time and extensively eroded, how could the Colorado River not have developed by middle Miocene time (an interval of 35 to 40 m.y.)? If a well-integrated ancestral Colorado River drainage did not account for much of the erosion and the geomorphic features of the plateau, what erosional event(s) did shape the surface and when? If another drainage system or another Colorado River outlet existed for most of the Tertiary, where is the evidence? If the post-Miocene uplift of the plateau was not in excess of 1,800 m, as Powell (1880) suggested, what accounted for the dramatic late Tertiary incision of the Colorado River?

Increasing aridity during the Miocene had originally been proposed to account for a middle Tertiary slowing of erosion rates. The increased Pliocene and Quaternary moisture in the uplifted Rocky Mountains was presumed to explain some of the recently increased downcutting. However, the topographic relationships of the 1.2-Ma inner canyon lavas to the Grand Canyon

gorge near Toroweap Canyon (McKee and others, 1968) indicates that little vertical downcutting of the channel occurred in concert with the increased Pleistocene moisture supply.

All of these contradictions and new data led to the understandable distrust of the simple plateau geomorphic histories that had been proposed. It is crucial for geomorphologists studying the landforms and the processes on the plateau to be able to accurately distinguish between the effects of any period of early Tertiary denudation and the magnitude of influences of subsequent mid-Tertiary events that appear to predate the rapid Pliocene incision of the modern Grand Canyon. The older interval would include the regional responses of the plateau block to the Laramide compression and marginal uplift, which occurred in Arizona mainly between 75 to 50 Ma (Shafiqullah and others, 1980). A period of middle Tertiary structural deformation must reflect some of the stresses associated with the main Basin and Range disturbance from late Oligocene through middle Miocene time, approximately 25 to 12 Ma.

The geomorphic elements of portions of the plateau landscapes, especially in Arizona, have evolved during a long interval of climatic variation (semitropical to arid) throughout much of the Tertiary, and must have been accompanied by at least two or three episodes of uplift and deformation. Obviously, geomorphological studies of the resulting complex landscapes must attempt to deal with all these tectonic, spatial, temporal, and climatic variables.

STRATIGRAPHIC EVIDENCE

It has recently been documented that the southern margin of the Colorado Plateau in Arizona was uplifted and eroded down to a Kaibab-Moenkopi surface before middle Eocene time (Young and Hartman, 1984; Young, 1985). The specific evidence takes the form of viviparid gastropods (extinct in North America by middle Eocene time) found in lacustrine limestones interbedded with the thick crystalline-pebble Rim gravels on the Coconino Plateau south of Grand Canyon (Goetz and others, 1975; Young, 1982). "Rim gravel" is a term applied to arkosic sediments and lag gravels containing conspicuous Precambrian clasts derived from ranges south and west of the plateau during late Cretaceous (?) and early Tertiary time. Although now presumed to have a direct relation to the obvious marginal Laramide uplifts in central Arizona, the widespread gravels have previously been assigned ages ranging from the Triassic through the Pliocene. They were first interpreted as Tertiary (Miocene) by Price (1950) in north-central Arizona, and their general distribution and significance were recently summarized by Peirce and others (1979) for the Mogollon Rim region.

The lacustrine Flagstaff Formation (late Paleocene to early Eocene) in southern Utah may be most nearly equivalent in age to the dated Rim gravels on the Colorado Plateau. The basal Flagstaff beds presently range in elevation from 2,000 to 3,000 m in elevation between Richfield and Castle Dale, Utah, whereas the deeply weathered gravels containing fossiliferous limestones south of the Grand Canyon occur near elevations of 2,000 m.

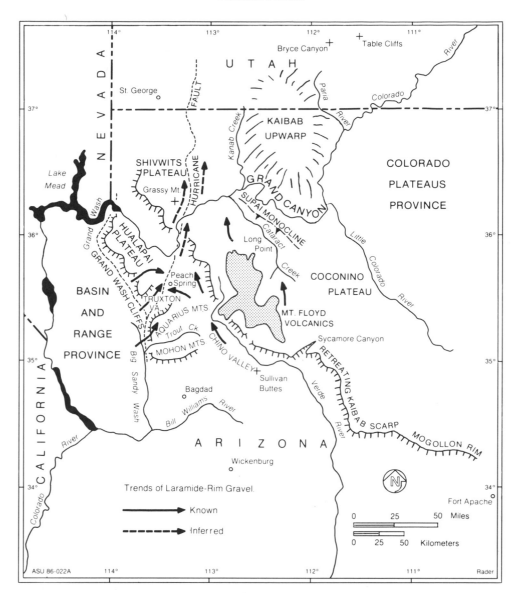

Figure 6. The western Colorado Plateau showing the known (solid arrows) and inferred (dashed arrows) Laramide–Rim gravel drainage trends.

Young and McKee (1978) previously speculated on a connection between the Rim gravel equivalents on the western plateau and the early Tertiary basins in southern Utah. The late Cretaceous-Paleocene paleogeographic reconstructions in Mallory (1972) are consistent with such a setting for the southwestern plateau in Laramide time. Any subsequent ancestral Colorado River immediately following this episode of northerly drainage could have continued to enter the remnants of such basins from the east or south until much later in the Tertiary. This would be especially true for all the Colorado River tributaries upstream from the Little Colorado River.

In addition to the accumulating evidence for extensive Laramide erosion around the plateau margin, there is also convincing evidence for an extended period of deep weathering and tectonic quiescence (late Eocene to early Oligocene) following the end of the Laramide orogeny (Gresens, 1981). The stratigraphic evidence bearing on the early to middle Tertiary paleogeographic reconstructions for critical areas, with inferred connections to the southern Utah basins, are discussed briefly in this section with emphasis on those regions where deposits of pre-Miocene or pre-Oligocene age have best been documented (Fig. 6).

MOGOLLON RIM–AQUARIUS MOUNTAINS

Peirce (1984) has recently presented arguments for extending the Mogollon Rim (retreating Kaibab Formation scarp) westward to the Shivwits Plateau to mark the southwestern margin of the plateau. Although this has an appealing physiographic simplicity, it has the disadvantage of placing a relatively undeformed portion of the plateau, as well as the western Grand Canyon, in the so-called transition zone. It also creates a somewhat arbitrary boundary that truncates the Hurricane, Toroweap, and associated north-south faults obliquely. The resulting line does not appear to conform to the probable crustal thickening at the plateau margin that is defined by geophysical and seismic data. In fact, the southwest margin of the plateau is not readily defined by any geologic contact on existing maps. The Laramide-initiated erosion beveled the uplifted plateau margin, exposing diverse Precambrian rocks over a broad region near the Aquarius Mountains (Goff and others, 1983). Younger volcanic rocks and restricted outcrops of Paleozoic formations obscure the basic field relations on the state geologic map such that the area more closely resembles the adjacent Basin and Range Province. The magnitude of offsets along normal faults west of the Chino Valley is poorly known, but the style of faulting and monoclinal flexing appears similar to those exposed further to the north between the Coconino Plateau and the Grand Wash fault (Goff and others, 1983).

For the purposes of this discussion, the Tertiary rocks on or adjacent to the Mogollon Rim from the Fort Apache region westward to the vicinity of Trout Creek (Mohon Mountains) are grouped together because they occur along the narrow zone marking the contact between relatively undeformed Paleozoic rocks and Precambrian rocks near the plateau margin. A resolution of the best definition of the "true" margin of the plateau is not critical to this discussion. This transitional zone includes most of the exposures of Rim gravels described in the earlier literature from Arizona and the only stratigraphic sections in Arizona where pre-Miocene volcanic rocks have been found capping the early Tertiary sediments on the plateau or close to its margin. Peirce and others (1979) have summarized most of the data, but several key localities with dated volcanics need to be mentioned. Volcanic flows with indicated ages (approximate) rest on the early Tertiary erosion surface or on early Tertiary Rim gravel deposits at the following places along the southern margin of the plateau: Trout Creek, 24 Ma (Young and McKee, 1978); Aquarius Mountains, 24± Ma (Goff and others, 1983); Fort Apache region, 28 Ma (Peirce and others, 1979); Sullivan Buttes, 23–26 Ma (Roden and others, 1979); and the Mogollon-Datil region in eastern New Mexico, 28–38 Ma (Elston and others, 1973).

Elsewhere in southern Arizona, rocks in the 29 to 39 Ma age range rest on this regional early Tertiary surface (Shafiqullah and others, 1980). The bulk of these rocks appear to be of Oligocene age. K-Ar ages from Rim gravel clasts in the Fort Apache region indicate that the northeast-flowing Laramide drainage must have still been active after early Eocene time (54 Ma) but probably had ended long before the middle of the Oligocene (Peirce and others, 1979). The scattered fluvial gravels representing events from late Cretaceous through early Eocene time have not been found to be fossiliferous and have not been of much use in

deciphering the details of the early Tertiary events. Only the gradual availability of significant numbers of radiometric ages has finally produced the very basic framework outlined here.

Along this southern margin of the plateau and in the adjacent "transition zone", younger Tertiary faulting accompanying Basin and Range extension led to reworking of some of the older sediments, thus creating younger fluvial deposits that have been confused with the sediments from which they were derived (McKee and McKee, 1972; Peirce and others, 1979). This led to some premature estimates of the timing of the main plateau uplift (McKee and McKee, 1972).

GRAND CANYON–COCONINO PLATEAU

The stripped Kaibab erosion surface in the Grand Canyon region, from the Echo Cliffs to the Hurricane fault, includes remnants of younger Mesozoic and Cenozoic sedimentary rocks, generally capped by Miocene or Pliocene lavas. The Tertiary sediments near Long Point, Arizona, south of Grand Canyon beneath the north edge of the Mount Floyd volcanic field, contain fossiliferous limestones with viviparid gastropods of probable early Eocene age (or older). These gastropods (Young and Hartman, 1984; Young, 1985) are most similar to forms such as *Viviparus trochiformis,* described from the lower Flagstaff limestones (La Rocque, 1960; Kitzmiller, 1981) of late Paleocene-early Eocene age. Hartman (1984) has recently completed an exhaustive revision of the systematics of these and related forms that occur throughout North America from Utah to Alberta and from the Dakotas westward into Montana. He has revised the confusing terminology, inaccurate correlations, and nomenclature errors in the numerous museum and U.S. Geological Survey collections, and has demonstrated that, regionally, these gastropods became extinct by middle Eocene time. Overall, the existing paleontologic and radiometric age data between Grand Canyon and the Mogollon Rim indicate that the regional surface (Powell surface of Davis, 1930; Tertiary Peneplain of Robinson, 1907) in north-central Arizona had certainly developed long before middle Eocene time. The drainages coming off the Laramide uplifts were transporting volcanic clasts erupted from late Cretaceous to early Eocene time (Peirce and others, 1979; Young, 1985) toward the margins of shallow lakes south of Grand Canyon, presently preserved as sediments near elevations of 2,000 m. All the paleontologic and radiometric age constraints on fossils, capping lavas, and clast ages indicate that the Rim gravel drainages had to still be active in early Eocene time, but may have been ponded or otherwise disrupted shortly thereafter.

Any regional drainage system capable of eroding the plateau down to and below the Kaibab Formation over such a broad region had to be a well-integrated system graded to a lower base level north of the present Grand Canyon. This is clearly contrary to some present-day topographic relationships and logically points to a Laramide tectonic framework when the adjacent mountains were higher, the plateau was more strongly tilted toward the north-northeast, and there were extensive lake basins of Paleocene and Eocene age in southern and central Utah.

WESTERN GRAND CANYON–HUALAPAI PLATEAU

Longwell's observations concerning the late Miocene-Pliocene age of the Muddy Creek Formation and its restriction on the age of the Colorado River in the Grand Wash Trough have been further verified by the work of Lucchitta (1966, 1972) and paleontological evidence summarized by Blair and others (1977), as well as radiometric age determinations by Damon and others (1978). The adjacent Hualapai Plateau has been the focus of ongoing studies by Young (1966, 1979, 1982, 1985) and Young and Brennan (1974) that have gradually elaborated on the antiquity of the pre–Colorado River drainage incised through the upturned edge of the plateau and converging on the Hurricane fault zone in lower Peach Springs Canyon (Young and McKee, 1978). The oldest Tertiary rocks in the ancient, northeast-trending canyon system are reddish-hued, deeply weathered arkosic sediments derived from the Precambrian ranges to the west and southwest of the plateau. This drainage was disrupted by movements along compressional monoclinal folds of undoubted Laramide age (Young, 1979, 1982; Huntoon, 1981).

Ponding in the local Tertiary canyons upstream from the monoclines produced thick, fresh-water limestones and fine red claystones on the central Hualapai Plateau and near Peach Springs (Young, 1979). The basal gravels on the Hualapai Plateau are stratigraphically equivalent to the Rim gravel sediments, which extend in a nearly unbroken sheet (discontinuously) from the western margin of the plateau to the San Francisco volcanic field. The only minor elevation discontinuity in the outcrops is over a short distance (4 km) across the Hurricane fault at Peach Springs Canyon where all Tertiary sediments were removed by headward eroding tributaries as a result of the differential movements on the fault.

Many of the outcrops do not show on the state geologic map, but appear on maps by Young (1966), Koons (1948, 1964), Blissenbach (1952), and Agrams and Squires (Goetz and others, 1975). Although they exhibit slight regional variations in clast lithology, reflecting the slight variations in source terraces, all these Rim gravel equivalents can be traced discontinuously to the south where they were originally described along the Mogollon Rim. When seen in relatively undisturbed exposures, the "gravels" consist of reddish to pinkish, weathered arkosic sands, silts, and clays with gravel lenses or boulder horizons. The pebbles and boulders may comprise less than 20 percent of the section, but extensive lag gravels on surfaces of low relief have concentrated the resistant lithologies. The most obvious lithologic difference from west to east is in the variation of the ratio of quartzites to other crystalline rocks. East of the Hurricane fault it is common for 20 to 50 percent of the pebbles in a collection to be quartzites, reflecting their abundance to the south in north-central Arizona. Pebble types on the western Hualapai Plateau derived from Precambrian exposures to the west average 30 to 60 percent igneous and metamorphic clasts with only 5 to 15 percent quartzite clasts (Young, 1966).

The remaining clasts reflect Paleozoic rock types extending in the direction of the inferred source or from nearby scarps. The middle Miocene Peach Springs Tuff (Young and Brennan, 1974) occurs in most of the key stratigraphic sections west of the Hurricane fault, with younger, more locally derived sediments between the Rim gravels and the volcanic rocks. An unconformity marked by deep weathering separates the older reddish-hued Rim gravel equivalents and lake beds from the younger, locally derived gravels, which had filled most of the older canyons prior to the onset of local volcanism. The magnitude of the buried weathered zone is especially apparent in samples from a 475 m drill hole in the Truston Valley sampled at 3-m intervals and carefully examined by Young (1982). This uranium exploration hole penetrated 238 m of Rim gravel (Music Mountain Conglomerate of Young, 1966) in the lowest portion of the hole of which the upper 40 m were extensively weathered.

The erosion surface on the Hualapai Plateau can be extended southward through the Truxton Valley to the edge of the Mohon Mountains–Aquarius Mountains region where the oldest gravels are covered by Oligocene volcanics (Young and McKee, 1978). This broad surface, contiguous with the regions previously described to the east, preserves a deeply incised, partially exhumed, pre-Oligocene drainage system and erosion surface. The convergence of the individual drainages on the Hurricane fault from the west, southwest, and southeast (Young, 1982) in the vicinity of Peach Springs, the distribution of lacustrine limestone facies, the consistency of lacustrine sedimentation on the upstream sides of the monoclines, the deep weathering of the basal gravels, and the reddish coloration of the lower part of the section all point to an early Tertiary age for the original drainage and the rocks that first filled the buried canyons (Young, 1979, 1982).

The only obstacles to a more complete reconstruction of a drainage outlet northward along the Hurricane fault (converging with the Coconino Plateau drainage to the east) are the structural complexities of the Hurricane fault movements and the amount by which the northeasterly regional tilting might have been diminished since the Laramide. The 24-km-wide, 1,000-m-deep erosional reentrant in the plateau margin along the southern extension of the Hurricane fault (Truxton Valley) establishes the obvious antiquity of the fault zone, as does the discordant Precambrian geology seen where the fault emerges at the edge of the plateau. Monoclinal flexing of the present downthrown block at the Colorado River suggests that the Hurricane fault also may have a complex deformational history that included reversed Laramide dislocation prior to late Tertiary normal movement, as has been demonstrated for other plateau faults to the east and the west (Huntoon, 1974, 1981).

An increase in the regional dip of less than one degree, and restoration of a minimum of 460 m Miocene or younger displacement at the Colorado River, are sufficient to open a northerly drainage outlet for the Paleocene-Eocene drainage into Utah between the Shivwits Plateau and the Kaibab Upwarp. While some of the details are still speculative, and limited epeirogenic warping must be invoked, the regional setting and corroborative evidence are compelling. There were uplifted Laramide highlands to the south and west, Cretaceous-Eocene basins to the north, and a broad regional pre-Eocene erosion surface in between, covered with 60 to 90 m of sedimentary cover traceable to the marginal uplifts. There are many geologic settings where more fragmentary evidence is accepted as proof of a connection between source-lands and basins. Of all the areas available for study of the early Tertiary history, it appears that the Hualapai-Coconino Plateaus may contain the most useful, most diverse, and oldest record of the early Tertiary history of the plateau drainage.

HIGH PLATEAUS—SOUTHERN UTAH TO UINTA BASIN

The chronology and stratigraphy of the Tertiary basins of the high plateaus in southern and central Utah has been refined by the studies of Bowers (1972), Anderson and others (1975), and Ryder and others (1976). The Tertiary stratigraphy relevant to this discussion begins with the late Cretaceous–Paleocene, fluvio-lacustrine North Horn Formation, extending from the Uinta Basin to the latitude of Bryce Canyon. Without interruption, deposition of the Flagstaff Limestone began during late Paleocene time in the southwestern portion of the basin (Ryder and others, 1976). The Flagstaff Limestone is the lowest member of the Green River Formation, and the top of the limestone is close to the Paleocene-Eocene boundary. Feldspathic sands of "southerly provenance" were supplied to the south flank of the basin during late Paleocene time, in contrast to the quartz-rich sediment coming off the Uinta Uplift and the Sevier orogenic belt (Ryder and others, 1976). Deposition of the Green River Formation completed the early Tertiary sequence on both sides of the Uinta Uplift during the Eocene. By late Eocene or early Oligocene time the remaining traces of the large lakes were covered by the coarse alluvial Uinta and Duchesne river formations eroded from the flanks of the Uinta Uplift.

The lower member of the Flagstaff Formation near Castle Dale, Utah, contains abundant gastropods, including species of *Viviparis, Lioplacodes,* and *Physa* (Kitzmiller, 1981). The same genera collected by Young (1982) and Young and Hartman (1984) have demonstrated that the viviparids all became extinct by middle Eocene time in North America.

The southernmost outcrops of the Flagstaff Limestone in Utah are near Richfield at elevations from 1,600 to 2,700 m. The paleogeographic reconstructions of the Green River and Flagstaff lakes by Hintze (1973) indicate that the edge of the Flagstaff Limestone basin extended to the Table Cliffs–Bryce Canyon area, which is only about 160 km north of the postulated lake basins reconstructed by Young (1982) on the Coconino Plateau. The Utah basins were over 480 km in length from southwest to northeast.

Bowers (1972) equates the Pine Hollow and Wasatch formations near the Table Cliffs–Bryce Canyon region with the North Horn and Flagstaff formations, respectively. The basal limestone members of the Wasatch Formation are correlated with the lower to middle Eocene Flagstaff Limestone on the basis of tentative gastropod identifications. Thus, except for the Grand Canyon to Bryce Canyon section along the drainage basins of Kanab Creek and the Paria River, it is possible to propose stratigraphic correlations from northern Arizona into southern Utah for early Paleocene to early Eocene rocks. The highest Rim gravel remnants crop out at elevations between 1,600 and 2,100 m on the plateau in Arizona, but the Table Cliffs–Bryce Canyon strata are presently between 2,300 and 3,000 m in elevation. As noted for the Hualapai Plateau region, these elevation differences do not take into account post-Laramide deformation, epeirogenic warping, or erosion of an unknown thickness of the Rim gravel deposits in northern Arizona. Both the late Eocene and early Oligocene regional erosion and the middle to late Tertiary Basin and Range extension have changed relative surface elevations by unknown amounts. Several thousands of meters of potential epeirogenic uplift of the western plateau in Pliocene time are probable (Lucchitta, 1979). Over the distances involved, less than one-half of one degree of regional tilting could reverse the present relative elevations between Utah and Arizona.

SOUTH CENTRAL COLORADO–SAN LUIS UPLIFT

Dickinson and others (1986) have recently described the distribution of Paleogene feldspathic petrofacies related to basement source regions now buried beneath the Oligocene volcanic cover of the San Juan Mountains. Their general discussion of early Tertiary facies relationships in the Cordilleran foreland region of central Utah documents how Laramide tectonics influenced sediment dispersal throughout the north-central portion of the Colorado Plateau, especially the provenance of the Paleocene-Eocene sediments filling the Uinta Basin (Fig. 7). There appears to be good evidence for a major, northwest-flowing Paleogene paleodrainage system between the Monument and Uncompahgre uplifts that fed a large delta complex bordering Lake Uinta. According to Dickinson and others (1986), this drainage system would have been essentially contemporaneous with the northwest-flowing drainage crossing the Grand Canyon region (Young and McKee, 1978). This regional provenance study (Dickinson and others, 1986) fills an important gap in our understanding of the early Tertiary paleogeography in a critical region and is consistent with the emerging picture of the Paleogene drainages broadly converging on the depositional basins of the northern plateau from all directions. The existence of this regional framework for sedimentation through early to middle Eocene time provides important constraints for the evolution of the post-Laramide landscape and the origin of the Colorado River system.

TECTONIC FRAMEWORK

Shafiqullah and others (1980) and Haxel and others (1984) bracket the Laramide orogeny in Arizona as a period of late Cretaceous (75 to 80 Ma) to early Eocene (50 to 58 Ma) volcanism, plutonism, thrust faulting, and metamorphism accompanying crustal compression and shortening under northeast-south-

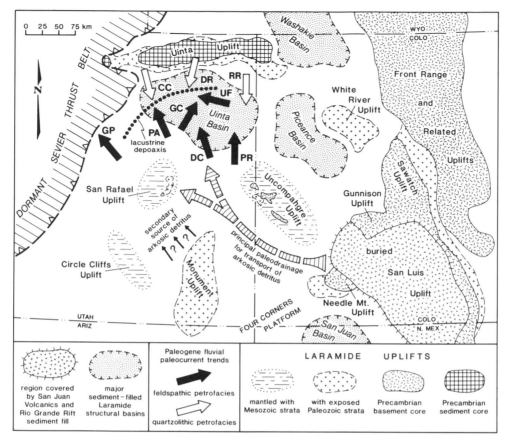

Figure 7. Inferred mid-Paleocene to mid-Eocene dispersal of arkosic detritus from Laramide uplifts of southern Rocky Mountains and Colorado Plateau (from Dickinson and others, 1986). CC, Currant Creek Formation; DC, Desolation Canyon; DR, Duchesne River Formation; GC, Gate Canyon and Sunnyside area; GP, Gunnison Plateau; PA, Price area; PR, P.R. Spring area; RR, Raven Ridge area; UF, Uinta Formation. See discussion in Dickinson and others (1986, Fig. 10) for complete explanation.

west directed stresses. In the northwest corner of Arizona near Lake Mead the stresses appear to become more east-west directed (Drewes, 1978). Drewes (1978) proposed there were two peaks of activity occurring near 75 and 55 Ma. On the plateau this episodic nature of the Laramide deformation is reflected in the fact that some of the compressional monoclines disrupted rocks of early Eocene (?) age during a reactivation that clearly postdates most of the Laramide initiated erosion (Young, 1979, 1982). The primary phase of late Cretaceous–Paleocene Laramide uplift had been essentially completed in order to explain the regional erosion beneath the Rim gravels. As much as 600 m of Mesozoic and Paleozoic sediments may have been removed from the region. The later episode of deformation is necessary to explain the ponding of late Rim gravel age, especially on the relatively featureless Coconino Plateau (Young, 1979, 1982). Miller (1962) indicates that a period of slightly older Cretaceous uplift in central Arizona may have accompanied the waning of the Sevier orogeny in Utah, but the magnitude of this event seems to be much less than the Laramide deformation as Cretaceous marine rocks are present above and below the hiatus.

Young (1985) has proposed that the main phase of the Laramide orogeny resulted in rapid northeastward scarp recession (more than 1 km/10^6 yr), in direct contrast with the order of magnitude slower recession rates accompanying later Tertiary events, including the incision of the Grand Canyon drainage. Such rapid scarp migration accompanying uplift would be the result of the limited drainage incision on the plateau proper while the Cretaceous-Eocene lake basins in southern Utah constituted the local base level. Such a geomorphic setting could accelerate strike valley tributary planation parallel to cuesta scarps (discussed under "Geomorphic Framework").

The prolongation of the Laramide activity into the Eocene is substantiated by the fact that igneous rocks emplaced in the early Eocene were nonetheless eroded and incorporated in the Rim gravels (Peirce and others, 1979). The only direct evidence of local intrusive activity on or near the plateau is a small, 65-Ma stock south of Lake Mead that intrudes Cambrian sedimentary rocks on the Grand Wash Cliffs and is unconformably overlain by Miocene lava flows (Young, 1979). Thus the Rim gravel episode may be inferred to have continued from late Cretaceous or early Paleocene time until the early Eocene when a final (?) significant pulse of deformation ponded the drainages at or near monoclines and uplifts on the plateau. The Kaibab Upwarp is the only significant Laramide structure between the Coconino Plateau and the early Tertiary basins of southern Utah. An absence of significant tectonism from middle Eocene to middle Oligocene time is suggested by the late Eocene to early Oligocene lack of drainage incision and the evidence for deep weathering in soils throughout western North America (Telluride erosion surface of San Juan region, summarized by Gresens, 1981).

The next recognized tectonic event that could have produced regional drainage readjustments is the Basin and Range episode and the contemporaneous plateau faulting that is best dated in Colorado (Epis and Chapin, 1975). Basin and Range extension occurred between 35 and 12 Ma all around the southern plateau margin, and Damon (Shafiqullah and others, 1980) proposed peaks of activity in the mid-Tertiary at about 32 Ma (New Mexico), 26 Ma (eastern Arizona), and 21 Ma (Sonoran Desert). In northwestern Arizona most of the dated volcanic rocks fall in the 24 to 12 Ma range, with an apparent clustering between 15 and 18 Ma (Young and others, 1985). Volcanism on or near the plateau margin buried many of the older valleys filled with early Tertiary Rim gravels. Some reworking of the Rim gravel sediments must have occurred during Basin and Range faulting at the plateau margin and more or less continuously during the rest of the late Tertiary. This may have produced several generations of younger lag gravels with different apparent ages in local settings (Peirce and others, 1979; McKee and McKee, 1972).

The lack of diversity of ages for volcanic units on the western plateau makes the documentation of late Miocene and Pliocene faulting difficult. However, the Hurricane and Grand Wash faults can be shown to have experienced pre- and post-middle Miocene movement where they are crossed by the Peach Springs Tuff (Young and Brennan, 1974). Most Miocene volcanic units are involved in faulting to some degree, and Pliocene volcanics are displaced along the Hurricane fault in Utah and in northern Arizona. Generally, however, it is not possible to distinguish between the Basin and Range related faulting and that which appears to be associated with Pliocene-Pleistocene volcanism along particular faults north and south of the Grand Canyon.

The major regional effect of the Basin and Range extension and tectonism may have been a decrease in the stronger (?) northeast regional tilting induced by the northeast-directed Laramide stresses. Backtilting or collapse around the plateau margin is indicated by the regional distribution of the Peach Springs Tuff across the transition zone (Young and Brennan, 1974), and such a regional "relaxation" would explain why the Laramide canyons incised in the southwest plateau margin appear to end in topographic cul-de-sacs at the present time. A 0.5° to 1° increase in the regional dip from Peach Springs to the Utah border would create a 1,500 to 3,000 m elevation reversal, obviously much more than is required to tectonically account for the existing discrepancies.

Tectonic warping and faulting during the late Oligocene to middle Miocene Basin and Range disturbance appears to be responsible for, or at least coincident with, the emergence of better integrated, gradually incising drainage systems across the plateau (Hunt, 1956, 1969). An increase in structural relief at the western plateau margin itself would create or rejuvenate any headward-eroding streams on the Hualapai Plateau drainage in the manner postulated by McKee and others (1964) or Lucchitta (1975). This tectonically induced headward erosion of the main Colorado River canyon from Lake Mead, has been duplicated by smaller drainages near Peach Springs and the Truxton Valley. In this region the main scarp east of the Hurricane fault zone has been breached by headwardly eroding streams that have reversed former northeast-flowing drainages for 13 to 16 km farther onto the plateau (Young, 1985). Much of the region south of Peach Springs now drains to the west in contrast to the conditions prevailing when the Truxton Valley filled (Twenter, 1962; Young, 1979, 1985). The topographic and structural differentiation of the Colorado Plateau Basin and Range boundary has obviously had the effect of accelerating the headward erosion locally and, probably enhanced channel downcutting along the western Grand Canyon gorge after the inferred capture.

CLIMATIC SETTING

Paleobotanical and paleontological studies of the Paleocene and Eocene lake deposits in Utah and Colorado have provided ample proof of the mild, humid-temperate to subtropical conditions accompanying the waxing and waning of Lake Uinta (Ryder and others, 1976). Such conditions prevail today at latitudes between 30° and 35° with massive atmospheric subsidence (Smiley and others, 1984). The slightly more southerly position of the North American continent during Paleocene to Eocene time would have created a warmer climate with possible monsoonal rainfall in portions of Arizona. However, outside the Paleocene and Eocene basins of Utah, Colorado, and New Mexico, there are relatively few detailed regional paleoclimatic records from which to reconstruct precise climatic histories.

Following the early Tertiary and subsequent northward shift of the continent, the gradual appearance of mountain barriers to the west and epeirogenic warping of the entire Rocky Mountain-Great Basin region accompanied a transition to cooler, dryer conditions on the plateau in middle to late Tertiary time. At the same time greater runoff may have gradually been available to the through-flowing Colorado tributaries that head in the Colorado Rockies.

Regionally, during late Eocene to early Oligocene time, there is convincing evidence of the widespread development of low relief and deep soils composed of residual quartz, clay, and sesquioxides like those presently forming in the southeastern United States (Gresens, 1981). These soils are associated with a continental-scale erosion surface that has been equated with the Telluride Erosion Surface of the San Juan Mountain region (Gresens, 1981). Gresens correlated the deep weathering profiles

and evidence for subdued topography with a global slowing of
plate motions during the Eocene following the Laramide orogeny
(about 40 Ma). Such an event is hypothesized to cause a short-
term sea level rise due to rapid erosion of the remaining tectonic
relief and filling of subduction trenches with sediments. Subsequently, sea level falls as the ocean-spreading centers cool and
sink (Gresens, 1981). This tectonically driven interplay between
convergence, continental erosion, and sea level change has been
hypothesized to correlate with the weathering and erosion interval in late Eocene–early Oligocene time from the Caribbean
through Mexico and the southwestern U.S. into Washington
State and British Columbia. Gresens (1981) noted that this setting
would explain why Oligocene rocks formed following renewed
tectonism were often composed of the fine clay and quartz sands
from the intensely weathered residual overburden.

Such a mid-Tertiary tectonically quiet interval would help
explain who no period of marked drainage incision is recognized
in the late Eocene–early Oligocene record on the plateau. As long
as the Uinta Basin and its adjacent counterparts remained relatively low, and moisture levels remained below those of the
Paleocene-Eocene epochs, drainage incision would be impeded
until an external base level became connected to the plateau
interior. The Rim gravels on the Hualapai Plateau certainly show
evidence of this interval of deep chemical weathering as a zone of
completely disintegrated clasts in the sediments some distance
below the volcanic interval (Young, 1979, 1982). Throughout
the Hualapai Plateau, the sediments above the weathered horizon
but below the Miocene volcanics are relatively unweathered, less
red in color than those above, more carbonate cemented, and
show an abrupt increase in coarseness. The contrast is especially
obvious where rocks above and below the zone of weathering are
lithologically similar, but the uppermost, pre–middle Miocene
clasts appear no more weathered than alluvial fan deposits of
Pliocene age (Young, 1979).

Toward the middle of the Miocene, only the plateau margin
and the broad Laramide Plateau uplifts may have projected
through the Tertiary sedimentary cover as low uplands. Even
now, rocks of Paleocene to early Eocene age from Arizona (Coconino Plateau), New Mexico (San Juan Basin), Utah (Wasatch
Plateau), and Colorado (San Juan Basin) all still lie near common
elevations from 1,800 to 2,100 m, in spite of an unknown
amount of epeirogenic warping that must have postdated Laramide time. This further attests to the apparent lack of relief in
mid-Tertiary time when gently sloping alluvial plains may have
covered much of northern Arizona.

All of the broad, general aspects of the early to middle
Tertiary climates appear to be consistent with a slow or limited
rate of Eocene-Oligocene landscape development prior to the
existence of an external drainage outlet connected to a lower base
level. At this level of generalization, with such broad, limited
climatic evidence, it is difficult to completely separate the tectonic
and climatic influences. However, all the facts are consistent with
a lack of regional drainage integration or incision until middle
Miocene time or later.

GEOMORPHIC EVIDENCE

Geomorphological studies fall into a broad spectrum of
methods ranging from the historical sequence approach to
process-oriented studies. A study of the Colorado Plateau, focused on the origin of the Colorado River, must necessarily be of
the former type. Of equal importance, however, are the many
process-oriented studies that have made valuable contributions to
the development of geomorphic theory. Aside from the classical
studies of plateau landforms, too numerous to list, the study of the
origin of segmented cliffs in sandstones (Oberlander, 1977) and
the study of sapping in the development of theater-headed valley
networks (Laity and Malin, 1985) serve to illustrate the continued potential of the plateau as a valuable geomorphic laboratory for terrestrial and planetary studies. Nonetheless, the thrust of
this discussion is focused on what can be deduced from the long-
term geomorphic development of the plateau landscape. The
foundations laid in the preceding sections provide the time constraints for the emergence of a plateau-wide, integrated drainage
system.

The critical region, the northern Arizona–southern Utah section, demonstrates the following. (1) The uplift of the plateau
margin accompanying Laramide compression created a broad
erosion surface much like the surface presently observed over
much of northern Arizona. (2) The early erosion interval must
have been associated with cliff recession rates from 1.5 to 3.8
km/10^6 yr (Young, 1985), depending upon what date one assumes for the beginning of cliff recession (uplift). (3) The Laramide surface was buried by as much as 60 to 90 m of Rim gravel
derived from the present Basin and Range Province from the
western Grand Canyon to the edge of the San Francisco volcanic
field and for some undetermined distance north of the Coconino
Plateau near Grand Canyon. Incised canyons around the plateau
margin preserve sections of Rim-gravel–equivalent sediments in
excess of 225 m (750 ft) thick (Young, 1982). (4) The Laramide
was followed by late Eocene–early Oligocene tectonic quiescence
and regional weathering as the climate became less humid.
(5) The Colorado River began to emerge as an integrated
drainage system in late Miocene to Pliocene time and was completed by headward erosion and capture from the west.

It must be emphasized that the southern margin of the plateau had already been strongly uplifted and was undergoing extensive planation *during* the interval of Paleocene-Eocene
sedimentation in the Uinta Basin and surroundings. The 2,400 m
or so of structural relief still recorded in the Kaibab structural
datum is probably residual from that period, reduced by an unknown amount of mid-Tertiary "back tilting." Late Laramide
disturbances ponded drainages during the final phase of Rim
gravel deposition, accounting for the transition to lacustrine clays
and limestones.

Young (1985) has compared Eocene scarp positions associated with the Laramide drainage of the Hualapai Plateau with
the amount of subsequent scarp retreat accompanying the development of the Colorado River at the base of the same scarp.

Laramide scarp recession was at least an order of magnitude more rapid than post-Laramide scarp recession. Such rapid lateral erosion (2.5 to 3.8 km/10^6 yr) accompanying uplift could have been the result of the limited drainage incision on the plateau proper while the drainages were graded to the Utah basins. Lateral strike tributaries, if not incising, could efficiently undercut southwest-facing cuesta scarps and accelerate lateral planation. In other words, the paleogeography, climate, and tectonics favored lateral planation by strike valley tributaries over the process of drainage incision. Scarp migration since that time, back from the Colorado River along some of the same scarps, has only resulted in an average rate of recession of 160 m/10^6 yr, an order of magnitude slower. Headward erosion rates by ephemeral streams through scarps near Peach Springs are similar (170 m/10^6 yr). These figures closely match the broad range of cliff recession rates calculated by Cole and Mayer (1982) from studies in Utah, Arizona, and the Sinai (100 to 2,000 m/10^6 yr). This dramatic decrease in scarp recession rates on the plateau coincides with the significant regional late Miocene–Pliocene drainage incision. The slower rates may indicate that rapid cliff recession in cuestaform landscapes is favored only by stable or rising base levels, whereas drainage incision causes irregular scarp dissection by random headward growth of scarp-face tributaries.

This early Laramide lateral planation and deposition, combined with the mid-Tertiary weathering, must represent the "great denudation" of Dutton and Powell. However, these early geologists had no way of determining that the main phase of denudation on the southern plateaus *accompanied* rather than followed the filling of the Uinta Basin.

SUMMARY

This regional analysis suggests that the broader aspect of landscape evolution by uniform scarp retreat can dominate only when drainage incision is absent. If these basic observations are correct they provide an important perspective on ancient landscapes where such surfaces have formed on gently dipping sedimentary rocks. It could also help explain why fragments of multiple erosion surfaces, terraces, and remnants of easily eroded sediments are often preserved in regions of considerable relief like the Grand Canyon. Vertical incision, while creating spectacular local relief, tends to haphazardly destroy surrounding evidence of older erosion surfaces by slow, uneven headward erosion along randomly spaced tributaries, rather than by efficient planation. This may be reflected in the observation that fragmentary relict erosion surfaces have generally been identified in areas of severe erosion by characteristics such as accordant summits, straths, pediments, and "peneplains." Such broad features may form relatively quickly in geologic time, whereas significant uplift accompanied by drainage incision may only destroy them at rates that are an order of magnitude slower (Young, 1985).

Finally, the broad geomorphic framework of the Colorado Plateau now appears to have been shaped largely by Laramide cliff recession under climatic and base level controls very different from those associated with Pliocene canyon incision. The Colorado River canyons should be viewed as less significant, albeit spectacular, etchings on the broad cuestaform landscape created during the Laramide.

One of the questions that remains unanswered as a result of this analysis is the relative importance of the Rim gravel deposits in the preservation of the Laramide landscape. It seems likely that much of the plateau was buried beneath 60 m or more of sediment by middle Eocene time. The effect this cover had on reducing the rate or amount of Oligocene through Pliocene modification of the presently preserved surface and landforms remains uncertain.

THE SHORT TERM: FLUVIAL PROCESSES SINCE 1940

Richard Hereford

INTRODUCTION

Analyses of geomorphic and sedimentologic processes operating on a human time scale complement the previous investigation into the long-term history of landscape development. In this section of the chapter we discuss the post-1940 history of geomorphic processes in sand-bed streams of the Colorado plateau. Many channels on the plateau were wide and entrenched before 1940, a situation that developed since about 1880 during the widely recognized episode of arroyo cutting in the Southwest (Cooke and Reeves, 1976). Beginning shortly after 1940, however, aggradation became the main process in many streams. This aggradation reduced channel width and resulted in widespread development of flood plains.

These young deposits and landforms have not been widely studied. Most of the research on streams in the Southwest was done before 1960, according to a recent overview of the subject (Graf, 1983a), which was before flood plains were conspicuous landforms. Studies done since about 1960 have recognized this young alluvium, but have attributed its origin to different mechanisms, even within the same drainage basin (e.g., Leopold, 1976; Patton and Schumm, 1981).

Modern alluviation on the Colorado Plateau has been, for the most part, in channels having continuous or discontinuous longitudinal profiles. Discontinuous channel deposition is reasonably well known from field and experimental studies (Pickup, 1975; Schumm, 1977; Malde and Scott, 1977; Patton and Schumm, 1981). Such streams alternate between deposition and erosion independently of external controls such as climate.

Deposition in continuous-channel systems, however, has not been widely studied (Graf, 1983b), although this type of channel is probably typical of many streams on the plateau (Graf, 1985). These streams produce sediment in hillslope areas and transport it to downstream depositional sites (Graf, 1982a). Continuous-channel streams respond to changes in the hillslope system, changes brought about by variations in precipitation and climate.

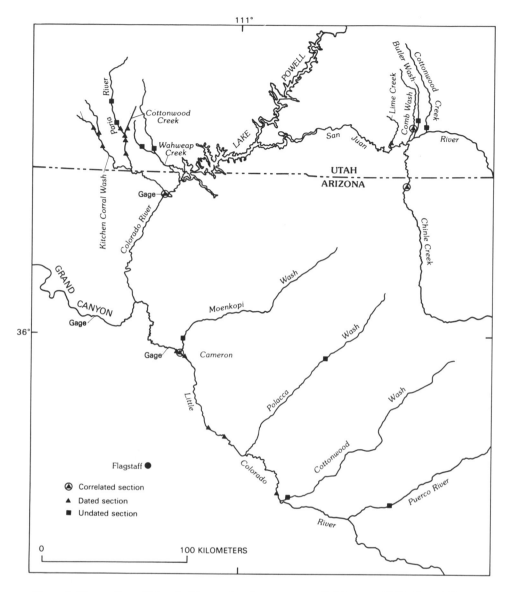

Figure 8. The southern Colorado Plateau showing streams studied for post-1940 alluvial history.

The following subsections review the post-1880 alluvial history of continuous-channel streams, describe the sedimentology and stratigraphy of the flood-plain alluvium, and discuss some possible causes of the shift from erosion to aggradation in the early 1940s. Flood-plain deposits discussed here are mainly in the Navajo and Canyon Lands sections (Hunt, 1956) of the southern Colorado Plateau (Fig. 8), but observations there are probably representative of the larger area. The flood-plain deposits were measured, sampled, described, and dated in the Paria River basin, the Little Colorado River valley (Hereford, 1979, 1983, 1984, 1986), Chinle Creek, Comb Wash, and Lime Creek (Fig. 8). In addition, reconnaissance studies of several other streams were made.

The typical Colorado Plateau stream is ephemeral; runoff

generally occurs in direct response to precipitation other than snowfall. Precipitation is biseasonal with winter and summer precipitation maxima. The region is noted for the intensity of its summer rainfall, which occurs mainly through convective storms. Large frontal storms are typical of the winter months and provide snow or low intensity rainfall.

HISTORY OF STREAM PROCESSES SINCE 1880 ON THE COLORADO PLATEAU

Alluvial history

Three events have dominated the alluvial history of the Colorado Plateau since 1880: (1) arroyo development and stream

Figure 9. Photographs of the Little Colorado River channel in the vicinity of Cameron, Arizona. Top, View downstream at the Cameron bridge about 1911 to 1920. Right, Aerial photograph of the channel in the Cameron area in November 1936. North is at top of the photograph, river flows northwest. Note shadow of U.S. 89 bridge in upper center. Figure 9A was taken on south side of the river immediately east of the bridge.

Figure 10. Upstream panoramic view of the flood plain and channel of Cottonwood Creek, a tributary of the Paria River in southern Utah, showing the channel (C), flood plain (FP), and cottonwood terrace (CT), which is probably equivalent to the Naha Formation of Arizona.

entrenchment beginning about 1880, (2) aggradation through flood-plain alluviation beginning in the early 1940s, and (3) incision of the flood plains since about 1980. The cause of the post-1880 entrenchment has been the topic of a large and complex literature (reviewed by Graf, 1983a). Alluviation since 1940, the topic of this section of the chapter, is generally thought to be the result of land use, climatic change during the twentieth century, or channel filling unrelated to climate or other external factors.

Although the causes of entrenchment are poorly understood, the chronology and effects of erosion are well known from several studies (Gregory, 1917; Bryan, 1925; Gregory and Moore, 1931; Bailey, 1935; Colton, 1937; Thornthwaite and others, 1942). Streams began to erode within 10 or 20 years of 1880, shortly after settlement of the area. Channels were rapidly deepened and widened, causing problems for early settlers. Valuable farmland was destroyed as the channels dissected alluvial valleys; the water table lowered, making it difficult to obtain water for farming and domestic use.

Channels that were eroded in the late 1800s remained wide and deep through the early 1900s, as shown by photographs from this period (Fig. 9). These photographs typically show barren to lightly vegetated, wide, and sandy channels with a braided pattern. Flood plains were not present; the only depositional forms were channel bars. These deposits, named the "older channel alluvium" (Hereford, 1984), are discontinuous, thin, and not well preserved in the stratigraphic record. This older alluvium was probably a transient deposit that was frequently reworked. Thus, the alluvium records erosion or net-sediment transport during the early 1900s.

Flood plains began to form in the channels in the early 1940s and are presently distinctive landforms in many alluvial valleys (Fig. 10). This recent aggradation has been reported by several workers in large and small streams distributed over a substantial portion of the Colorado Plateau and adjoining regions (Emmett, 1974; Leopold, 1976; Dunne and Leopold, 1978; Love, 1979, 1983; Hereford, 1984). During aggradation, rates of channel enlargement and headcut migration apparently declined (Leopold and others, 1966; Wells and others, 1983, Fig. 8.10).

Photographs and stratigraphic data record the development of flood plains in the early 1940s. Photographs, probably the most detailed information source, clearly show a change in channel configuration in the mid-1900s. Aerial photographs taken in the late 1930s over much of the Navajo Section of the plateau show streams with wide channels that were essentially free of vegetation (Fig. 9B; Hereford, 1984, Fig. 6). In contrast, aerial photographs of the same region taken in the early 1950s by the Army Map Service show narrow channels with vegetated flood plains; channels in the 1950s were typically half of their earlier width.

Ground-based photographs (Graf, 1983b, Figs. 5, 6; Hawley and Love, 1983, Fig. 1) confirm the aerial-photograph interpretation. An interesting example is Comb Wash in southeast Utah, which was photographed in the early 1900s by Gregory (1938, pl. 19A). This photograph (Fig. 11A) shows a wide, sandy channel, which contrasts with the narrow channel present now in Comb Wash (Fig. 11B). Two to three meters of sediment have accumulated in the channel since about 1939, the germination date of the oldest tamarisk partially buried in the flood-plain alluvium (Fig. 11B).

Figure 11. Photographs showing changes in the channel of Comb Wash. Top, Comb Wash in the early 1900s (Gregory, 1938, pl. 19A). Bottom, Similar view in June 1985. Flood-plain sediment has built up above the bedrock ledges behind the third horseman. Note the elevated flood plain with cutbanks and tamarisk trees at far side of channel. Flood plain was incised in 1980.

Several photographs taken at the Paria River gaging station record about 30 years of flood-plain development. The channel at the station was photographed by U.S. Geological Survey personnel in 1931, 1939, 1951, and the mid-1960s (available from Water Resources Division, Tucson, Arizona). The 1931 and 1939 photographs show the characteristic wide channel of the early 1900s. The 1951 photograph, in contrast, shows a narrow channel that was partially filled with sand. Photographs taken in the mid-1960s show a well-developed flood plain similar to the present one.

Recent stratigraphic studies using tree-ring dating techniques (Hereford, 1984, 1986) have established the chronology of ag-

gradation. These studies show that deposition of the Little Colorado and Paria rivers flood-plain alluvium began in the early 1940s. Aggradation probably began on these streams after a shift to lower flood discharges in 1942. Thereafter, until about 1980, flood plains aggraded intermittently during overbank floods.

Incision of the flood plains beginning about 1980 has been the latest development. Hereford (1984) found that the flood plain of the Little Colorado River was incised during a single flood in 1980. Another study (Hereford, 1986) showed that the Paria River flood plain was also incised in 1980 during floods in February and September (Fig. 12). This recent incision is apparently widespread because many streams show evidence of flood

Figure 12. Erosional relationship between the channel and flood-plain deposits of the Paria River. Scale is 1.4 m long. Contact between flood-plain and channel alluvium is the vertical line to right of center. FPA, flood-plain alluvium; CA, channel alluvium; C, colluvium.

plain erosion dating from the late 1970s to early 1980s (Graf, 1985). Its immediate cause was probably large floods, but these floods were no larger than floods that in previous years deposited sediment on the flood plains. This erosional event is too recent to determine whether it is a major reversal in process or a brief erosional interval within a continuing aggradational period.

Deposition in discontinuous and continuous channels

Several studies have shown that flood plains or flood plain-like landforms have developed in semiarid regions by different processes (Patton and Schumm, 1981; Graf, 1982a, 1983b). The type of longitudinal channel profile, whether discontinuous or continuous, distinguishes the two processes. These channel profiles are probably the two principal types in Colorado Plateau streams (Graf, 1985).

In discontinuous channels, deposition follows erosion localized at a migrating knickpoint or headcut (Schumm and Hadley, 1957; Patton and Schumm, 1975, 1981; Malde and Scott, 1977). Most of the stream power in discontinuous channels is dissipated at the migrating knickpoint; consequently, sediment is deposited a short distance below the headcut (Pickup, 1975). Sediment is derived from erosion of an alluvial-valley fill; the stream essentially reworks earlier deposits with little new sediment supplied to the channel system.

In continuous channel systems, sediment is deposited without knickpoint migration or significant channel enlargement. Sediment in these channels is derived principally from erosion of lightly weathered bedrock at the head of the stream. Thus, new sediment is added to the channel system as it becomes available through weathering in the hillslope system. A substantial portion of the stream power in continuous channels is available to transport this sediment into the channel system (Graf, 1982a, Fig. 9.5).

Depositional processes in continuous channels will be discussed in a following section. We will emphasize continuous-channel deposition because such channels are directly linked to the hillslopes. Thus, they are likely to respond to variations in energy input to the hillslope system. These variations affect the transport capacity of the channels resulting in either sediment storage, transport, or erosion.

Discontinuous channels appear to be typical of the shortest streams on the Colorado Plateau. Discontinuous channels, therefore, probably contribute only a small portion of total sediment yield and runoff. This observation is consistent with the quantitative studies of Leopold and others (1966) in an adjoining region; they reported that knickpoint migration and channel enlargement did not produce significant quantities of sediment. The distribution of continuous and discontinuous channels, however, has not been mapped on a regional basis. Such mapping would improve our understanding of sediment yield and runoff.

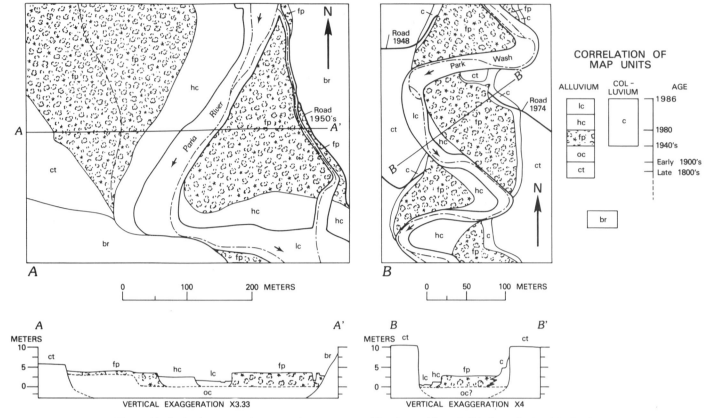

Figure 13. Geologic maps and cross sections constructed by plane table and alidade methods of the Paria River and Park Wash. Park Wash is an arroyo similar to Cottonwood Creek. lc, lower channel; hc, high channel; fp, flood plain; oc, older channel alluvium; ct, cottonwood terrace; br, bedrock. Cottonwood terrace is equivalent to Naha Formation described in text.

GEOMORPHOLOGY OF THE FLOOD PLAINS

Flood plains on the Colorado Plateau have several physical characteristics in common with most flood plains. Vegetation is abundant, consisting of a variety of riparian plants, saltcedar, cottonwood, and willow, as well as nonriparian plants such as juniper, russian olive, and big sage. Many of these plants are partially buried in the alluvium. Several of them, including saltceder, cottonwood, and big sage, produce annual growth rings (Ferguson, 1964; Fritts, 1976; Hereford, 1984). Ring counts from living plants of these species have been used to establish the chronology of flood plain deposition (Hereford, 1984, 1986).

The flood-plain surface has little relief except for levees and overbank channels (Love, 1979, Fig. 1; Hereford, 1984, Fig. 4). Levees, rising less than 1 m above the surface, border the channels of many flood plains. Overbank channels, which are shallow, poorly formed linear depressions, are common features on larger streams such as the Little Colorado River.

Geomorphic and geologic setting of the flood plains

The geomorphic and geologic setting of the flood plains mapped in Figure 13 are typical of the area (Fig. 8). Flood plains are inset beneath a terrace of late Holocene age, which is recognized by its height above the channel, by the presence of senescent cottonwood trees (Hereford, 1984) that date from the late 1800s, and locally by structures built during settlement of the region in the 1860s to 1880s.

In most places, the alluvium forming the terrace is probably equivalent to the Naha Formation (Hack, 1942; Dunne and Leopold, 1978, p. 691). This formation, which was deposited from about A.D. 1450 to 1880, forms an extensive valley-fill deposit. Aggradation of the Naha was terminated during arroyo cutting and general stream entrenchment that began about 1880.

Spatial Distribution of continuous-channel flood plains in the drainage network

Flood plains are not present throughout the drainage networks; rather, they occur principally in the high-order channels having large drainage areas. This spatial distribution probably results from the nonlinear downstream distribution of stream power (Graf, 1982a) and sediment-transport capacity in the channel system: transport capacity and stream power do not increase monotonically downstream. In low-order channels having small drainage areas, stream power and sediment-transport capacity are high (Fig. 14) because of steep slopes and narrow channels. Sediment is eroded and transported through these stream segments.

High-order channels, in contrast, have reduced transport capacity and are potential sediment-storage sites. Thus, sediment eroded from low-order basins is transported to high-order channels downstream from the zone of maximum transport capacity (Fig. 14). In this zone, sediment is stored on flood plains during overbank floods.

This spatial distribution of transport capacity provides a physical explanation for the stratigraphy of flood plain alluvium. A study of the Little Colorado River showed that the flood-plain deposits are correlative over a 170-km reach of the river (Hereford, 1984). (In the Paria River basin, dated flood plain deposits are correlative along the Paria River and principal tributaries [Hereford, 1986].) Thus, the deposits are not time-transgressive. Correlation of deposits would be expected if the zone of maximum transport capacity remained stationary through time in the drainage network. Discontinuous-channel deposition, in contrast, probably results in time-transgressive stratigraphy: the deposits become progressively younger upstream due to knickpoint migration.

FLOOD PLAIN SEDIMENTOLOGY

Source of the sediment

The sedimentary rocks on the Colorado Plateau, particularly the widespread Mesozoic-age sandstone, siltstone, and shale, weather and decompose rapidly (Colbert, 1956; Schumm and Chorley, 1966), providing the abundant sediment yield for which the region is noted. The continuous-channel streams trace directly to lightly vegetated hillslopes and cliffs developed on these rocks (Fig. 15). Sediment is supplied to the channel system through erosion of disaggregated bedrock on the hillslopes and cliffs.

Sedimentology of the flood-plain alluvium

For the most part, flood-plain alluvium ranges in size from sand to clay and minor gravel. Median grain size of the sand is typically very fine to fine. The sand is very poor to poorly sorted. Flood-plain sediment is finer grained then sediment in the channels (Hereford, 1984), which is typically fine to coarse gravel with minor clay. The thickness of the alluvium is variable, but 1.5 to 5 m is characteristic of the streams in the southern Colorado Plateau.

Three units, recognized on the basis of grain size, bedding type and thickness, and stratigraphic position, are apparently widespread in Colorado Plateau streams (Fig. 16). The units are (1) a basal gravel, (2) an intermediate thin-bedded unit, and (3) an upper thick-bedded unit. These units overlie the previously mentioned older channel alluvium, predominantly of fine to medium sand and local gravel, which is slightly coarser than the

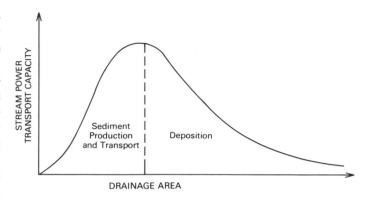

Figure 14. Generalized distribution of stream power/sediment-transport capacity as a function of location in the drainage network (modified from Graf, 1982a, Fig. 9.5).

flood-plain alluvium. Stratification in the older alluvium is in parallel sets 10–15 cm thick. This older alluvium is probably equivalent to the channel bars in the early photographs of the channel (Figs. 9, 11A). The basal gravel has a sandy matrix with clast size ranging from granule to small cobble; it is thin, discontinuous, and interfingers locally with the intermediate unit.

The intermediate unit typically consists of several thin, lenticular sand beds with poorly defined contacts. It is drab colored, sparingly carbonaceous, and has clay streaks and laminations as much as 1 cm thick. Bedding thickness varies but averages between 5 and 10 cm, which is thinner than bedding in the upper unit. In the Paria River basin and probably elsewhere, the intermediate unit is more poorly sorted and is about 1 phi size finer than the upper unit. Large, mature saltcedar trees that germinated in the 1940s to mid-1950s are rooted on bedding surfaces in this unit.

The upper unit, the thickest of the three units, consists of several beds that have sharp nongradational contacts. The entire unit is generally lighter colored than the intermediate unit, although individual beds vary in color from light gray to moderate reddish orange. Bed thickness is variable but averages between 10 and 20 cm. The contact between beds is irregular and marked by duff layers several centimeters thick, composed of leaves from saltcedar, pinyon, juniper, and big sage. Sedimentary structures are not common; however, ripple-cross lamination, climbing ripple structure, and centriclinal cross-stratification (Underwood and Lambert, 1974) are locally present.

A marker bed, or interval of several beds, is present near the top of the upper unit. The interval is distinctive in the stratigraphic column because of its finer grain size as compared with beds above and below it. At the Paria River basin and Little Colorado River sections, the interval is a single bed. At the Chinle Creek, Comb Wash, and Lime Creek sections, the marker consists of several thin beds of silty, clayey sand interbedded with silty sand.

Figure 15. Photograph showing sediment source area of a tributary to Kitchen Corral Wash. Scale is 1.4 m in length. Light-colored alluvium in channels was derived from hillslope formed on Petrified Forest Member of the Chinle Formation (Triassic). Older alluvial deposits (vegetated area in foreground) do not contribute significant amounts of sediment.

Depositional processes

The basal gravel and intermediate unit apparently represent the initial channel-fill deposits of the flood plain. Thin bedding, fine grain size, and poor sorting of the intermediate unit suggest that it was deposited during a period of low flow and infrequent large floods, a discharge regimen that was typical of the 1940s and 1950s (Gatewood and others, 1963; Hereford, 1984). Gradational contacts in the intermediate unit, which suggest low current velocities and lack of intense scour, also indicate low flood discharge. In contrast, thicker bedding, coarser grain size, and better sorting indicate that the upper unit was deposited during a period of higher flows. Sharp erosional contacts between beds also suggest high-velocity currents. Each bed in the unit probably records a single overbank flood.

Vertical accretion and sediment storage

Flood plains are generally thought to form through lateral (Wolman and Leopold, 1970) or vertical accretion (Schumm and Lichty, 1963), or through some combination of the two processes (Stene, 1980). On the Colorado Plateau, vertical accretion has probably been the dominant mechanism of flood-plain construc-

tion. This is indicated by the vertical arrangement of beds (Figs. 13, 16), the lateral continuity of stratification in transverse flood-plain sections, and the absence of epsilon surfaces (Allen, 1963), which are distinctive crosscutting contacts suggestive of lateral accretion. Moreover, flood-plain alluvium overlaps the older channel deposits (Fig. 16), which indicates vertical accretion.

Vertical accretion implies that sediment was stored rather than redistributed across the channel. Furthermore, because the channel remains approximately fixed during vertical accretion, alluvium deposited on the flood plains is preserved. Channel stability and sediment storage were not typical of the period 1880 through the mid-1940s. Therefore, the poorly preserved, discontinuous deposits of the older channel alluvium probably record erosion or sediment transport, whereas the flood-plain deposits record progressive sediment accumulation.

REGIONAL CORRELATION OF FLOOD-PLAIN ALLUVIUM

As previously discussed, the shift from erosion to aggradation has been noted by several workers. Most studies have re-

Figure 16. Photograph showing stratification in flood-plain alluvium at the Paria River gaging station in northern Arizona. Scale is 1.4 m in length. OCA, older channel alluvium; BI, basal and intermediate units; UU, upper unit; MB, marker bed; CA, post-1980 channel alluvium.

ported that aggradation began in the 1940s. A large area was affected, and aggradation was probably typical of most streams on the Colorado Plateau.

In addition, stratigraphic studies using tree-ring dating and marker beds permit regional correlation of the flood-plain units. Figure 17 illustrates the stratigraphic correlation of the units at four widely spaced sections. The intermediate unit was not recognized in an earlier study of the Little Colorado River (Hereford, 1984), but has been recognized at Chinle Creek, Comb Wash, Lime Creek (not shown in Fig. 17), and throughout the Paria River basin. Dates from at or near the base of the unit range from 1939, a maximum date inferred from the previously described photographs at the Paria River gage, to 1946 at an upstream Paria River section. Thus, deposition of the basal and intermediate units probably began after 1939 and before 1946. The top of the intermediate unit dates between 1954 and 1956 (Fig. 17); therefore, deposition of the upper unit began after 1954 to 1956 and lasted until 1980. The marker bed of the upper unit was deposited in the period 1972 to 1974.

A high level of depositional synchroneity between drainage basins is shown by the widespread distribution of the marker interval. This quite likely resulted from distinctive weather conditions effecting runoff in a large area. Indeed, 1972 to 1973 was a time of anomalous global weather patterns (Kukla and Kukla,

1974) that effected the Colorado Plateau. October 1972, for example, was the wettest October in 92 years at Flagstaff, Arizona (Hereford, 1984).

Flood-plain alluviation appears to have been broadly synchronous throughout a substantial portion of the Colorado Plateau (Fig. 17). This implies that alluviation was controlled by external factors—probably climate—rather than internal geomorphic controls. Furthermore, alluviation was progressive: erosion and deposition evidently did not alternate in a random manner. Thus, the streams evidently responded in a direct and simple manner to sediment input from the source area; otherwise correlation among drainage basins would not be possible.

CAUSES OF THE SHIFT FROM EROSION TO AGGRADATION IN THE 1940s

Land use and human activity

Conservation measures, such as construction of dams and stock tanks, were implemented on the Colorado Plateau after enactment of the Taylor Grazing Act of 1934. These structures were designed to reduce sediment load and peak-flood discharge. Their effect on the larger streams, however, is probably limited as suggested by the discharge histories of the Little Colorado and

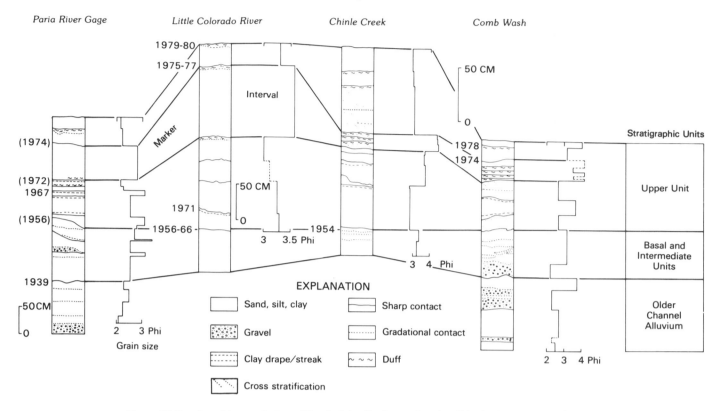

Figure 17. Stratigraphic correlation of flood-plain alluvium at four localities in the southwest Colorado Plateau. Dated horizons are indicated on the left side of the columns. Dates in parentheses at the Paria River gage are based on correlation with other sections in the Paria River basin. Estimated grain size shown by dashed lines.

Paria rivers, which do not show a gradual decline in discharge as might result from increased water use and impoundment. These streams and many others on the plateau are free flowing through most of their lengths and lack significant impoundments. Therefore, land use alone probably does not account for the shift from erosion to aggradation.

The introduction of nonnative riparian plants, specifically tamarisk (or saltcedar), has impacted the channel systems (Graf, 1978). Tamarisk was brought into the Colorado Plateau in the late 1800s; it spread through the channel systems in the 1930s and 1940s (Christensen, 1962; Robinson, 1965) and it presently is the dominant vegetation on the flood plains of many streams. These plants have undoubtedly affected sedimentation because they trap sediment (Hadley, 1961), but it seems unlikely that they were the single cause of aggradation. First, flood plains are present in the absence of tamarisk or where tamarisk is not abundant. Second, the plant was under cultivation in several riverside settlements, but photographs of streams taken near the settlements in the early 1900s do not have tamarisk in them (Figs. 9, 11A; Hereford, 1984). Finally, although tamarisk spread rapidly through the master streams (Colorado and Green rivers; Graf, 1978), it did not appear in the smaller streams until the late 1930s and early 1940s after a decline in flood discharge—the same decline associated with aggradation beginning in the early 1940s.

Geomorphic controls inherent to the fluvial system

Erosion and aggradation can occur simultaneously in the channel system, as shown by a number of experiments and field observations (Schumm, 1977; referred to as the semiarid cycle of erosion by Schumm and Hadley, 1957). Basically, sediment accumulates in a widened channel downstream from a migrating knickpoint. This phase of the cycle begins to operate after sediment accumulates on the channel floor to a threshold slope. Above this slope the stream becomes unstable and the system reverses operation, leading to knickpoint migration and erosion. A characteristic of the semiarid cycle of erosion is that erosional and depositional events do not correlate within or between drainage basins (Patton and Schumm, 1981); each basin has a different alluvial history. The simultaneous initiation of aggradation and regional correlation of flood-plain alluvium in southern Colorado Plateau streams apparently do not support the semiarid erosional cycle.

Flood discharge and climatic variation

A change in flood discharge emerges as the most likely and immediate cause of the shift from erosion to aggradation. Although discharge records are not available for many streams, the

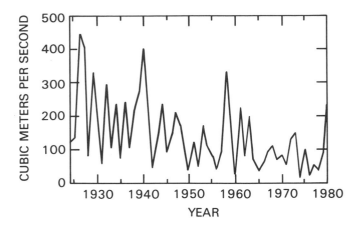

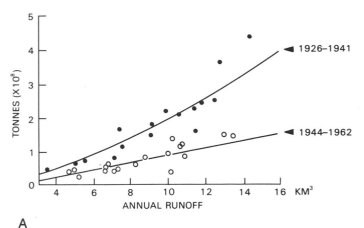

A

Figure 18. Annual flood series of the Paria River, 1924 to 1980. Note the 55 percent decline in flood discharge after 1940.

few that are available show a decline during the early 1940s (Fig. 18). That a major hydrologic change occurred is suggested by the sediment yield of the Colorado River in the Grand Canyon (Fig. 19). Sediment yield at this station declined after 1942 or 1943; however, the runoff did not change (Fig. 19A). This decline resulted in part from improved land use and conservation measures initiated in the 1930s (Hadley, 1974, 1977), but the abrupt decline shown by the sediment-yield time series (Fig. 19B) suggests that other factors also contributed to the decline. First, sediment began to accumulate on flood plains in tributary channels at about this time (Hereford, 1984, 1986). Second, this accumulation was concurrent with a decline in peak-flood discharge in the tributaries (Fig. 18). It seems likely that the decline in flood discharge was linked to a change in precipitation such as a decrease in the frequency of intense precipitation events.

A change in the climate of the Northern Hemisphere was approximately concurrent with the change in flood discharge and in fluvial process from erosion to aggradation. The hemispheric mean surface air temperature (MSAT) rose by slightly more than 0.50° C between 1880 and 1940; thereafter, MSAT declined by a similar amount (Bryson, 1974; Kukla and others, 1977; Douglas and others, 1982; Jones and others, 1982, Fig. 2). These changes were probably caused by a shift in atmospheric circulation (Kalnicky, 1974) that may have altered rainfall patterns (Bryson, 1974; Reitan, 1980; Zishka and Smith, 1980). Hence, the fluctuation in MSAT is linked through atmospheric circulation to surface runoff and fluvial processes.

IMPLICATIONS FOR INTERPRETATION OF FLUVIAL PROCESSES IN SEMIARID REGIONS

The modern alluvial history of continuous-channel streams in the Colorado Plateau underscores the importance of flood regimen in semiarid-region fluvial processes, at least on time scales of a century or less. A succession of large floods will

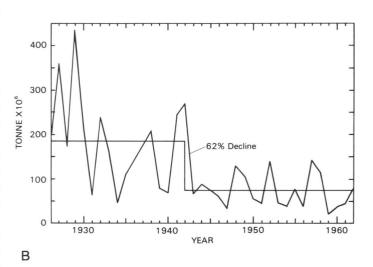

B

Figure 19. (A) Annual runoff and sediment yield of the Colorado River in Grand Canyon National Park. Solid circles, 1926 to 1941; open circles, 1944 to 1962. Sediment-yield records are complete except for 1942 water year (October to September accounting period). Sediment yield for 1942 was estimated from measured annual runoff using a least squares regression from the 1926 to 1941 data. (B) Time series of sediment yield of the Colorado River at Grand Canyon, Arizona, 1926 to 1962, showing the 62 percent decline in sediment yield beginning about 1942. Horizontal lines are the median sediment yield for the two periods, 1926 to 1941 and 1942 to 1962.

probably result in channel instability and high sediment yield. In contrast, a succession of small floods probably will reduce sediment yield and favor aggradation through flood-plain alluviation. Valley morphology, water discharge, and sediment yield are sensitive to changes in the frequency of large floods, which in turn are probably controlled by changes in the frequency of high intensity precipitation.

Arroyo cutting and filling have been explained by three climatic hypotheses (Graf, 1983a), in which channel filling follows either a shift to (1) drier conditions, (2) wetter conditions, or (3) a change in the frequency, intensity, or seasonality of precipitation. The dominant effect of floods in the alluvial history of continuous-channel streams favors the third hypothesis. The terms "wet" and "dry," which are generally based on annual precipitation totals, do not necessarily reflect flood regimen. A change in the frequency of flood-producing precipitation or precipitation frequency probably is not directly related to total annual precipitation (Leopold, 1951; Englehart and Douglas, 1985). Large floods, for example, might occur during dry years; conversely, small floods might occur during relatively wet years.

THE COLORADO PLATEAU IN PLANETARY GEOLOGY STUDIES

Julie Laity

INTRODUCTION

The study of landforms on extraterrestrial surfaces is accomplished largely by the study of photo images acquired by orbiting spacecraft. A geomorphologist studying these images must rely on reasoning by analogy to interpret complex landforms, many of which may have formed by processes that are of little significance on Earth. The interpretation of such images is constrained therefore, by familiarity with terrestrial environments. An additional interpretation problem is that there is no unique correlation between form and process in the evolution of landscapes: similar morphologies may result from a number of different erosional or depositional activities. For this reason, auxiliary data on planetary climate, materials, and surface ages become critical in the final interpretation. The problems are exemplified by the diverse origins proposed for Martian channels and valleys, where wind, water, and glacial ice have all been invoked as explanations (Baker, 1982).

Uncertainty as to the effectiveness of certain geomorphic processes on other planets stems in part from the lack of knowledge of their operation or importance in terrestrial environments. It follows that analog studies have proven to be of considerable value in planetary studies. The Colorado Plateau has been a particularly fruitful area of investigation, owing to its limited rainfall, diversity of landforms, and accessibility to scientists involved in planetary research. The purpose of this section is to review the role of the Colorado Plateau in planetary geology studies. Topics include (1) sapping processes of valley and slope development; (2) tectonic structures, including grabens and fracture patterns; and (3) eolian processes, incorporating depositional features such as crescentic, topographically-controlled, and linear dunes as well as erosional forms such as yardangs.

SAPPING PROCESSES

Valleys on Mars

The Mars Channel Working Group (1983) published a consensus paper arguing that water in some form was an essential ingredient in the various channel- and valley-forming processes on Mars. Martian "valleys" are distinguished from "channels" by

an apparent absence of bedforms that indicate direct fluid flow. In addition, many of these valleys have network properties and overall morphologies very different from those of terrestrial "rainfall" drainage systems. Recent research has invoked sapping mechanisms for the primary genesis of theater-headed valleys on Mars (Milton, 1973; Sharp, 1973; Sharp and Malin, 1975; Baker, 1980; Pieri, 1980; Higgins, 1982; and others). The main argument for sapping processes is based on morphology. The most telling characteristic of sapping on Earth is the presence of cuspate valley terminations. Other features in many Martian and terrestrial systems include straight valleys that are relatively constant in width, high and often steep valley sidewalls, and hanging valleys. Martian valleys differ considerably in network pattern from the dendritic form common to terrestrial streams (Pieri, 1980), and they show a lack of tributary competition for undissected intervalley terrain. Terrestrial dendritic networks show a statistically significant tendency for tributary junction angles to increase with increased recipient stream size, whereas Martian valleys do not.

Sapping is a process leading to the undermining and collapse of valley head and side walls by weakening or removal of basal support as the result of enhanced weathering or erosion by concentrated fluid flow at a site of seepage. Groundwater sapping on Earth results from discharge of a permanent groundwater system. It is a sustained process that responds to long-term climatic shifts. Erosion can continue even through drier periods, albeit at reduced rates corresponding to diminished groundwater outflow.

Except for morphologic features, there is no direct evidence that Martian valleys are formed by sapping, nor is there any comprehensive or universally accepted theory of how sapping would operate in the Martian environment. If the valleys evolved from a nonflow process such as ground-ice sapping, there remains the question of how such a process operates to create a network of valleys. Numerous problems are posed by the necessary reliance on photogeologic interpretation for assessing the role of sapping. These include (1) an inability to determine the limiting boundaries of a subsurface groundwater basin, (2) lack of good topographic information, and (3) masking of original processes by subsequent activity induced by climatic change, such as mass movements on valley side slopes.

Topographic criteria that may prove useful are the constant valley widths, the form of the longitudinal profile and of valley terminations, structural controls, and the network space-filled characteristics in a basin (Laity and Malin, 1985). Viewed at an orbital scale, a close affinity in morphology exists between many of the Martian valleys, such as Nirgal Vallis, and many theater-headed valleys on Earth. The role of terrestrial fieldwork, including that on the Colorado Plateau, is to isolate the detailed criteria that will allow further evaluation of possible sapping processes on Mars.

Valley development by sapping on Earth

Large-scale valley networks formed predominantly by sapping are relatively uncommon on earth because the processes involve a unique interplay of lithologic, stratigraphic, structural, and climate controls. Characteristic requirements include (1) a permeable aquifer underlain by an impermeable boundary, (2) a rechargeable groundwater system, (3) a free face at which subsurface water can emerge, and (4) some form of structural or lithological inhomogeneity that locally increases the hydraulic conductivity of the material and along which valleys grow. The frequency and magnitude of precipitation events are also factors. Low intensity, long-duration rainfall favors groundwater recharge and enhances sapping; high rates of rainfall of short duration favor runoff. Excessive runoff destroys sapping features; nevertheless, some runoff appears essential to remove the products of mass wasting.

Overland flow is the principal process of drainage development on Earth and plays a major role in the evolution of landscapes on the Colorado Plateau (Powell, 1875; Gilbert, 1877; Dutton, 1882; Davis, 1901; Gregory, 1917, 1938; Miser, 1925; Gregory and Moore, 1931; Hunt, 1956; Hunt and others, 1953; Cooley, 1958). Throughflow processes on the Colorado Plateau are limited by thin and poorly developed soils and by pervasive jointing of the bedrock that permits infiltrated water to move rapidly toward the groundwater system. In terms of the larger canyons, throughflow processes are probably insignificant relative to surface and groundwater contributions. Groundwater processes result from the deep percolation of water that enters the permanent groundwater flow system and emerges at a zone of seepage. Baseflow resulting from seepage into a channel sustains a number of intermittent and perennial streams on the Colorado Plateau.

The geomorphic significance of sapping in the maintenance of sandstone cliffs in the American Southwest has long been recognized. The role of groundwater erosion in the headward retreat of canyons has received much less attention. Bryan (1928) described the development of niches and alcoves by sapping in the walls of the Cliff House Sandstone, Chaco Canyon, New Mexico. Water seeps out of the rock, often above thin shaly lenses, and breaks up the sandstone by solution of the cement, by frost action, and by crystallization of salts on alcove walls. Bryan (1954) attributed the retreat of the north wall of Chaco Canyon to basal sapping at the contact of the porous Cliff House Sandstone and the underlying dark shale of the Menefee Formation. Joint blocks undermined by seepage become detached from the wall and slowly move away from the cliff face.

Slab movement was further documented by Schumm and Chorley (1964), who studied the movement and collapse of a huge monolithic block in the canyon. They noted that the rate of movement was seasonal, increasing during the winter owing to frost action and wetting of the shale by snow melt. A similar process occurs in the erosion of a sandstone caprock in the Chuska Mountains, New Mexico. Joint blocks of Chuska Sandstone move laterally as block glides are aided by piping of loose sand beneath the caprock and above the underlying shale (Watson and Wright, 1963). Vertical jointing increases percolation of surface water and breaks the caprock into blocks.

Ahnert (1960) examined several sandstones in his study of cliff retreat in the southwest. He identified the importance of basal sapping in the maintenance of cliffs and the development of valleys with theater heads. He also attributed the vertical nature of Entrada Sandstone cliffs in New Mexico to the emergence of groundwater at the cliff base, immediately above the less permeable Carmel Formation. In the Dakota and Point Lookout sandstones (New Mexico) the distribution of valleys with semicircular heads is related to the dip direction of the beds. Groundwater moves down the back scarp, so that valleys grow in an updip direction. Canyons continue to enlarge until the rock is completely removed, often leading to the formation of wind gaps along the front scarp. Ahnert considered sapping to be more significant during pluvial times than at present.

Of the regions on the Colorado Plateau displaying theater-headed valleys, the best and most areally extensive are developed in rocks of the Glen Canyon Group. These consist, in ascending order, of the Wingate Sandstone, the Kayenta Formation of mud-, silt-, and sandstone, and the Navajo Sandstone. Theater-headed valley development appears to be related principally to the nature, occurrence, and attitude of the Navajo Sandstone. The Navajo Sandstone (Triassic/Jurassic) is considered by most workers to represent an extensive dune deposit of an ancient interior desert. Cliffs of this material appear massive, with horizontal bedding planes commonly more than 15 m apart. The porosity of the sandstone ranges from 15 to 35 percent.

Valley growth in the Navajo Sandstone has been attributed to eolian action, to waterfall erosion, and to groundwater seepage and associated mass movement. Stokes (1964) argued that the most significant process is wind erosion, indicated by the alignment of canyons parallel to the prevailing wind direction. Waterfall erosion was cited by Hunt and others (1953) for the creation of box-headed canyons surrounding the Henry Mountains. Although many geologists have noted that the Navajo Sandstone/Kayenta Formation contact is an important spring zone (McKnight, 1940; Baker, 1946; and others), only the early work of Gregory (1917; Gregory and Moore, 1931) extended such observations to the role of seepage in the erosional development of the canyons. He believed the steep walls forming box canyons in the formation resulted in "disintegration that was aided by groundwater and guided by the texture and structure of rocks," concluding that "stream erosion has played little part in the production of these features" (Gregory and Moore, 1931).

The role of jointing in network pattern and morphology has received some attention. Campbell (1973) grouped canyon morphologies into classes based on the degree of joint control, and argued that the meanderlike forms developed in the Navajo Sandstone on the Shonto Plateau, Arizona, result from the collapse of alcoves that have enlarged by groundwater seepage emerging from joints in the bedrock. The widening of valleys by the collapse of massive slabs developed by exfoliation jointing and undermined by seepage is discussed by Bradley (1963) and by Robinson (1970).

A detailed examination of the role of groundwater in the

formation of the canyons in the Navajo Sandstone was conducted by Laity (1980, 1983) in the Glen Canyon region of south-central Utah, near the confluence of the Colorado and San Juan rivers. Air photographs show deeply entrenched theater-headed valleys with a morphology analogous to Martian valleys. The canyons show clear indications of groundwater seepage and of related mass movement. In order to assess the role of groundwater, the study investigated canyon morphology, drainage pattern, space-filling evolution of networks in drainage basins, the form of the longitudinal profile, observed seepage related to theoretical sapping models, alcove distribution, small-scale weathering processes, and structural, lithologic, and hydrologic relationships. This examination indicates that groundwater sapping is the predominant mechanism of valley growth in theater-headed valleys. The term "sapping" is used to encompass several processes by which the emergence of groundwater reduces the support of steep cliffs and contributes to their collapse. Other processes are essential to break down the debris produced by wall collapse and to remove the disintegrated materials.

Several geomorphic considerations that support a mechanism of sapping emerged from the foregoing study. These include the high transmissivity of the aquifer, observed alcove formation and/or undermining of cliff faces at sites of seepage, a spatial distribution of seepage outflow along canyon walls that supports theoretical groundwater sapping models, a strong degree of structural control of canyon growth, and drainage networks that differ considerably from those formed by surface erosion. The significance of some of these factors is detailed below.

1. Transmissivity of the aquifer. Studies show that the highly transmissive quality of Navajo Sandstone is a function of its permeability and its geometrical configuration and continuity (Jobin, 1956, 1962). The recharge potential of the aquifer is high because of its widespread surface exposure at low dip angles and pervasive fracturing. Movement of water is generally slow owing to the fine-grained nature of the material. Sandstone yields water in part from intergranular openings only partially filled with cement, and in part from fractures. Sandstone samples show porosity to be excellent in the lower Navajo Sandstone. Carbonates and clays are either minor constituents or when present, are distributed in patches.

Towards the base of the formation, the Navajo Sandstone becomes mineralogically more complex. Carbonate cementation is widespread and intergranular spaces are filled with clays. One explanation for this may be that water circulating through the formation was more saturated with respect to the various solids to be precipitated, and the presence of the underlying Kayenta Formation slowed water flow and promoted deposition. The result is that maximum groundwater outflow occurs in a zone a meter or two above the actual Navajo Sandstone/Kayenta Formation contact.

The Kayenta Formation is a complex, fluvially derived material that acts as an aquiclude for downward-moving groundwater in the overlying sandstone. Samples obtained from the uppermost strata show a significant porosity reduction resulting from an abundance of clay filling the pores, the development of quartz overgrowths, and extensive carbonate cementation (Laity, 1983).

2. Erosion at the Seepage Face. Groundwater emerging at seeps in the cliff face results in slow surface grain release or sloughing. The seepage face is frequently a surface deposit of efflorescent salts 100 to 200 μm in thickness. Beneath this is a sandy calcareous material 1 to 2 cm in thickness that adheres weakly to the wall rock. SEM micrographs show that tunnels 200 to 600 μm in diameter are formed within the evaporite deposit. These maintain flow to the surface and accommodate the volume of converging groundwater at the seepage face. As the weathering surface grows in thickness, the weakened material sloughs off in flakes, aided in winter by freeze-thaw cycles and the formation of icicles.

The geomorphic consequence of slow surface grain release and of sloughing is the enlargement of cavities and alcoves. These undermine basal support for the cliff and cause eventual slab failure at the canyon head and sidewalls. This process is facilitated by exfoliation joints that develop parallel to the canyon. Maintenance of the sidewalls and continued growth of the canyons requires removal of talus. Mechanically weak sandstone shatters readily upon impact and is further comminuted by active weathering processes.

3. Distribution of Seepage Outflow. The distribution of springs and the relative discharge of groundwater along canyon walls are compatible with a general model of drainage extension by sapping processes. Dunne (1980) and others propose that groundwater discharge should be greatest at the canyon head and decrease away from it. Seepage discharge measurements at theater heads make up 20 to 40 percent of total flow, with secondary contributions derived from sidewall springs and lateral inflow (Laity and Malin, 1985). The significant percentage of discharge emanating from the head spring, and the decreasing rate of inflow in a downstream direction, support a model of growth by sapping processes. Also, seepage alcoves are more common in the headward reaches of the valleys.

In the initial stages of valley development, growth of the main canyon proceeds more rapidly than that of tributary canyons owing to a larger subsurface drainage area. In time, a network of canyons is formed by mass-movement–induced headward retreat. Given a constant climate, headward growth rates by sapping probably decline as the system enlarges and the drainage area of the spring heads lessen. In an advanced stage of valley development, lateral retreat by sidewall seepage approaches that of headward retreat, and valleys widen. They may continue to grow until adjacent tributaries merge, leaving only isolated buttes and remnants of the original surface. A new drainage system of surface erosion develops on the exposed bedrock.

4. Network Pattern and Morphology. Within the Navajo Sandstone, two populations of valleys are developed with markedly different geomorphic features. The first group exhibits theater heads, longitudinal profiles with high steplike discontinuities, and sparse, commonly asymmetric, structurally controlled

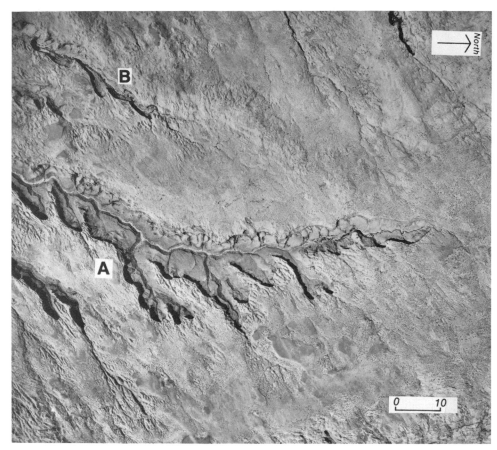

Figure 20. Bowns and Long canyons. Long Canyon (A; 11 km in total length) and the smaller Bowns Canyon (B; 7 km in length) are tributaries in the Colorado River in the Glen Canyon region. They lie adjacent to one another on the gently dipping limb of the Waterpocket Fold. The canyon networks are asymmetrically branched, with tributaries heading in an up-dip direction and aligned with the regional jointing. Theater-headed valley terminations are indicative of groundwater sapping processes.

network patterns. The second group is characterized by tapered terminations of first-order tributaries, a relatively smooth concave-up profile, and a more dendritic network pattern. The differences in form of these valleys are attributed primarily to structural constraints that determine the relative effectiveness of surface erosion or groundwater erosion. Of particular importance is the dip direction of the beds relative to the orientation of the valleys. Where groundwater converges toward the canyon head and runoff values are low, headcuts develop; where groundwater flows downdip away from valley heads and seepage is minimal, and where runoff is high, profiles are concave up and theater heads do not occur.

The actual network pattern of valleys in the Navajo Sandstone strongly reflects the structural controls of dip direction and of regional jointing. Valleys that lie transverse to the slope of a monocline develop an asymmetrically branched pattern, having either right- or left-handed tributaries (Fig. 20). Valleys that grow directly up a monocline have short branches or are unbranched. The development of a more elongate and symmetrically branched network may occur within a gently plunging syncline. Finally, where groundwater flows away from valley heads and overland flow processes predominate, the pattern is commonly more dendritic with tapered terminations.

Implications for Mars

To address the question of drainage formation by sapping on Mars, it is first necessary to understand the erosional mechanism on Earth and thereby establish reasonable photogeologic criteria of network pattern and valley morphology. The canyons of the Colorado Plateau are large in scale and morphologically similar to Martian valleys in a number of features. The network spatial evolution, rate of erosion, and degree of structural control differ from fluvially eroded networks. Although the constituent materials, scale, climate, structure, and groundwater/ice conditions of Mars cannot be replicated at any terrestrial site, it is hoped that detailed study of the unique landscapes produced by sapping on Earth will allow for a careful reconsideration of this process on Mars.

TECTONIC GEOMORPHOLOGY

Tectonic studies on the Colorado Plateau have addressed problems of graben formation, joint inheritance, and astrobleme formation. The geometry, kinematics, and mechanics of a system of simple grabens in Canyonlands National Park, Utah, has received considerable attention (McGill and Stromquist, 1974; Stromquist, 1976; Golombeck, 1979; McGill and Stromquist, 1979; Golombeck and McGill, 1983). Many of the characteristics of these grabens have proved useful in the study of simple grabens on the Moon, Mars, and Ganymede, even though the specific mechanisms of origin differ. The most direct application of the Utah graben studies was by Golombeck (1979), whose study of Lunar grabens was used to constrain the thickness of the megaregolith. A study of jointing in the Grand Canyon (McGill, 1980) dealt with the question of whether prominent joint patterns in surface rocks are inherited from underlying older rocks, and therefore tell little or nothing about younger stress fields; or, whether they are independent of older stress patterns. This is of interest because of the need to interpret fracture patterns that are common in the crusts of other planets. Finally, Shoemaker and Herkenhoff (1984) reexamined Upheaval Dome in Canyonlands National Park, Utah, and argued that this feature represents a deeply eroded astrobleme of impact origin. Previously, McKnight (1940) had suggested that the structure owed its form to plastic flowage of salt and other rocks in the underlying Paradox Formation.

Grabens

Grabens and graben systems are important tectonic structures on terrestrial planetary bodies. Although the Moon is tectonically inactive in relation to the Earth, the presence of wrinkle ridges and grabens indicates that for at least part of lunar history there have been stresses sufficient to deform or fault crustal rocks. Almost all lunar grabens show definite areal or geometric relations with large impact basins (Quaide, 1965). Tectonic information provided by the grabens includes clues to the thickness of the surface layer based on models of the three-dimensional geometry of grabens, constraints on the total expansion of the Moon, and constraints on the tectonic development of basins (Golombeck and McGill, 1983).

Most terrestrial grabens have bounding faults with convergent dips of about 60°, but other dip angles are possible. The floor of the graben is usually horizontal, but may be tilted or involved in secondary faulting if one or both of the bounding faults is listric. Photographs of lunar grabens indicate that tilting or secondary faulting of the floor is rare and that faults bounding grabens apparently have convergent 60° dips. The symmetry and simplicity of lunar grabens also imply a simple geometry in the third dimension (Golomech and McGill, 1983). Most grabens on Earth are more complex than lunar grabens, but simple ones do exist, such as those of the Needles District, Canyonlands National Park.

The grabens of Canyonlands National Park, southeastern Utah, occur in an area near the confluence of the Colorado and Green rivers and extend for 25 km south and southwest along the east side of Cataract Canyon of the Colorado River (Fig. 21). The grabens occur in a 460-m-thick plate of bedded brittle rocks comprising the upper Hermosa, Rico, and Cutler formations (Pennsylvanian and Permian) overlying ductile evaporites of the Paradox Member, Hermose Formation (Pennsylvanian). The rocks dip 4° northwest toward Cataract Canyon of the Colorado River in a region where the river has eroded through the brittle plate down into the evaporites. The free face of the canyon permits the evaporites to flow down the slope, causing extension and faulting in the brittle plate. The result is a spectacular series of more-or-less systematically spaced grabens, with obvious regularities in graben size and spacing. Graben widths are characteristically 150 to 200 m, and average depths range from 25 to 75 m.

Stresses in the brittle plate cause graben curvature to be concave in the direction of the dip. The uniformity of graben spacing results from an essentially uniform strength and thickness of the brittle plate. Field and experimental evidence suggests that the bounding graben faults are initiated at or close to the contact between the brittle and ductile rocks, and are propagated both upward and laterally (McGill and Stromquist, 1979).

The grabens of Canyonlands were chosen for study because they are essentially uneroded, and because the mechanism responsible for the graben-forming stresses is simple and well understood. The primary relief of the grabens provides an excellent opportunity to study the interactions of structure and geomorphology during graben growth. This is important because the results need to be applied to planets that are themselves characterized by exceedingly slow rates of surface change.

Planetary fracture patterns

Studies of fracture patterns on the Colorado Plateau have been undertaken in the Grand Canyon (McGill, 1980) and at the Chesler Canyon lineament, Canyonlands National Park, Utah (Potter and McGill, 1981). The purpose is to understand the fractures seen on planetary surfaces and to determine the stress histories responsible for these patterns. In principle, joint and fault arrays can be used to interpret the orientation of stress fields that cause brittle failure. Difficulties arise, however, because most materials are not simple, homogeneous, and isotropic, and because the stress fields can change with time. In addition, there is concern about the inheritance of fracture patterns in surface rocks from underlying older rocks. A reconnaissance study of fracture patterns at various stratigraphic levels in the Grand Canyon by Hodgson (1961) suggested a significant correlation between basement fracture trends and overlying Paleozoic sedimentary units. The limited extent of this initial study led McGill (1980) to attempt a more systematic comparison of Precambrian and Paleozoic fractures in a variety of geographic localities in the Grand Canyon. Sites were chosen both close to and far from faults, and at points of direct contact between the Tapeats Sandstone (Cam-

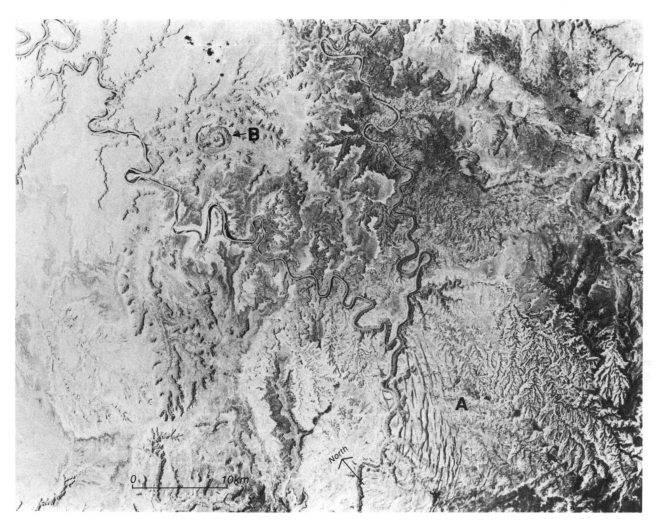

Figure 21. Upheaval Dome and the grabens of the Needles District. These features are located in southeastern Utah in Canyonlands National Park. The grabens (A) occur near the confluence of the Colorado and Green rivers and extend for 25 km along the east side of Cataract Canyon of the Colorado River. They result from the flow of ductile evaporites down a slope towards Cataract Canyon that causes extension and faulting in an overlying brittle plate. Graben curvature is concave in the direction of the dip. Upheaval Dome (B) conforms to the structure expected from a deeply eroded astrobleme.

brian) and basement to find direct evidence for or against the propagation of individual joints. The results of this research suggest that the dominant joint trends in the Tapeats Sandstone do not correspond with trends in the basement structures.

Impact origin of Upheaval Dome

A recent interpretation of the origin of Upheaval Dome by Shoemaker and Herkenhoff (1984) presents evidence that it conforms with the structure expected for a deeply eroded astrobleme (Fig. 21). The convergent displacement of the rocks forming the dome is thought to correspond with the deformation that results from collapse of a transient cavity produced by high-speed impact. The presence at depth of plastic beds in the Paradox Forma-

tion may have further facilitated growth in the dome. Erosion appears to have removed as much as 2 km of rock since the impact structure was formed, giving an approximate date of impact in latest Cretaceous or Paleogene time.

EOLIAN PROCESSES

Introduction

Earth-based and spacecraft observations of dust-storm activity obscuring the surface of Mars reveal that eolian processes are active at the present time. Also, evidence of various landforms—including dunes, yardangs, and mantling windblown sediments—suggests that eolian processes have been important in the history

Figure 22. Transverse dunes (A) and isolated barchan dunes (B). This field represents part of the north circumpolar erg on Mars and is located at latitude 77°N, longitude 51° (Viking Orbiter frame 514B23).

of Mars. Knowledge of the physical and, to some extent, of the chemical conditions that govern eolian processes is paramount because of concerns about the present-day effectiveness of wind erosion on Mars and the nature of activity in the past. Some important information necessary to interpret Martian environments includes (1) the frequency and magnitude of surface winds, (2) the range of wind speeds at which particles are entrained and transported, (3) characteristics of grain saltations, (4) the nature of rock weathering, and (5) the effects of electrostatically charged particles (Greeley and others, 1981).

Efforts to understand the Martian eolian environment have concentrated on laboratory simulations (primarily wind tunnel work), analyses of spacecraft data, and field studies of eolian processes on Earth. Unfortunately, theoretical work and wind tunnel experiments have not provided a global picture or suffi-

ciently comprehensive solutions to observations of apparent wind-pattern trends, surface color variations, and the distribution of dunes on Mars. Analog studies are important because uncertainty as to the effectiveness of wind on Mars stems in part from a lack of knowledge of its importance on Earth.

The global distribution of dunes on Earth and on Mars differs greatly. The most extensive terrestrial sand seas, or ergs, are found in the mid- to low-latitude deserts. On Mars, dunes are dominant in the high latitudes, forming an almost continuous collar around the north polar cap (Fig. 22). In the south, by contrast, dunes occur mainly as numerous discrete crater floor dune fields. Within areas of dune accumulation on Mars, most forms appear to be related to crescentic dunes. The lack of parabolic dunes on Mars is attributed to the absence of vegetation, whereas dome and star dunes are thought to occur in a few localities, such as marginal areas of crater floor dune fields (Cutts and Smith, 1973; Breed, 1977). The apparent absence of longitudinal dunes, coupled with a general lack of dunes over much of the surface of Mars, suggests that Martian eolian sediments are not presently migrating and are topographically restricted to a small number of sites.

Wind-derived landscapes of the Colorado Plateau include sand sheets, dunes, lag gravel plains, yardangs, deflation hollows, and fluted cliffs. All major types of dunes are represented, generally by small, simple forms (Breed and others, 1984). Deflation and abrasion are major shaping processes where bedrock is exposed to strong, unidirectional winds. The landscape also records relict eolian features that are representative of earlier climatic conditions.

The primary role of analog studies on the Colorado Plateau has been to address the effectiveness of wind as a geomorphological agent and to quantitatively analyze topographic controls on sand accumulation (McCauley and Breed, 1980). Additionally, the study of dune forms that are abundant on Earth but absent on Mars (e.g., linear dunes), will help to further clarify sediment transport history, climate change, and wind regimes on Mars.

Wind deposition

The largest dune field on Mars occurs in the north polar region and covers an area of at least 3,500 km². In some areas, the dunes form a nearly continuous sheet of ridges with arcuate segments (barchanoid and transverse dunes); these ridges almost completely mask the underlying topography. In other areas, particularly around large topographic features, the sheet is discontinuous and breaks up into strings of crescentic dunes or isolated barchans. Most of the dunes experience several wind directions, resulting in elongated horns, secondary slip faces, or changes in the dune symmetry (Tsoar and others, 1979).

The massed crescentic ridges of Mars are similar in plan to dunes typical of desert basin ergs and dune fields on Earth. Crescentic dunes, which represent the largest percentage of active dunes on the Colorado Plateau, include barchans and barchanoid or transverse ridges. These dunes, oriented perpendicular to the effective wind direction, are characterized by gentle, convex upwind slopes and steep (32° to 34°) concave downwind or lee slopes. Along the flood plain of the Little Colorado River, barchanoid ridges often form in low areas confined between valley walls (Fig. 23). The sediment is derived primarily from the dry streambed. In northeastern Arizona inactive vegetated barchanoid dunes cover large parts of the Hopi Reservation (Breed and others, 1984).

As analogies for Martian dunes, crescentic dunes have received little attention. The dunes tend to be less massive, and dune fields cover smaller areas than those of Asia and Africa. Nevertheless, these dunes merit further study because the ample climatic and geomorphic information available for the Colorado Plateau allows for more detailed analysis than for many other regions.

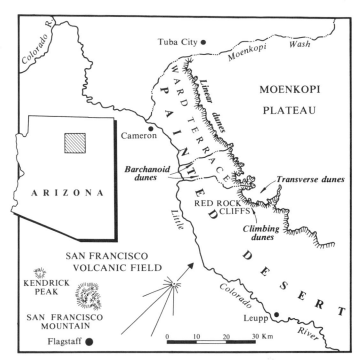

Figure 23. The Painted Desert region. The "sand rose" represents the relative frequency per year of winds of sand moving strength (from Breed and others, 1984). Barchanoid dunes develop in the flood plain of ephemeral tributaries to the Little Colorado River. Climbing dunes ascend 60 m from Ward Terrace to the top of the Red Rock Cliffs, where the sand recombines to form transverse dunes. The Moenkopi Plateau has the largest array of linear dunes in North America.

Most terrestrial dunes occur within large flat basins or playas. Except for the northern plains and some large impact basins (e.g., Hellas), Mars lacks such large smooth basins. Hence, many of the Martian dunes are topographically controlled and occur within craters or on the slopes of craters.

Two common dune forms are climbing dunes and falling dunes. Tsoar and Greeley (1980) suggest that climbing dunes occur on the windward side of an obstacle where the slope angle is less than 35°; owing to this shallow angle, the work of secondary eddies in the lee of the dune becomes insignificant and thus a blanket of sand develops, covering the crater lip. On Mars, dunes often approach a crater as barchans, surmount the crater rim as climbing dunes, and then descend to the crater floor as falling dunes.

On the Colorado Plateau, climbing and falling dunes are well represented in a variety of topographically controlled environments along the western edge of the Painted Desert in northern Arizona. This region is dominated by persistent southwesterly winds with speeds in the range of 30 to 40 knots. Sand grains, which originate in the flood plain of the Little Colorado River, travel northeast across the painted Desert toward the Moenkopi Plateau, forming numerous fields of barchan dunes. As the actively moving dunes approach the base of the plateau cliffs, they lose their distinct barchanoid shape and assume a complex morphology. Elsewhere they form climbing dunes that funnel up abandoned watercourses. The dunes climb 60 m from the Ward Terrace to the top of the Red Rock Cliffs at Paiute Trail Point (Fig. 23). At the head of the escarpment the sand recombines to form dome dunes and transverse dunes. The sand grains that comprise these dunes have travelled 15 km from their source region along the Little Colorado River and have climbed through 165 m of elevation (McCauley and Breed, 1980). No climbing dunes presently reach the top of Moenkopi Plateau, which lies 158 m in altitude above the Paiute Trail Point.

In order to understand the topographic control of wind and the associated modification of dune morphology, an instrumented field study is underway in the Painted Desert (McCauley and Breed, 1980). It includes in situ measurement of annual and seasonal wind regimes both below and on top of the Red Rock Cliffs. Sand traps installed within 10 m of the wind recorder measure the sand flux in relation to wind velocity and direction, and height above the sand surface. Sand samples have been collected from more than 100 sites in order to determine the relationship between sand movement and grain size parameters.

Breed and Breed (1979) believed that the Painted Desert was involved in multiple periods of dune formation, subject to episodic destruction by periods of heightened fluvial activity. They postulated that, during the Pleistocene, climbing dunes fronted the Moenkopi Plateau, thus providing a ramp for saltating grains to reach the top of the plateau. Erosional retreat of the plateau cliffs, diminished winds, and a lessened sand supply have apparently abated this process. At present, the linear dunes of the Moenkopi Plateau are being worn away by deflation and fluvial erosion (Breed and others, 1984).

On Earth, linear dunes account for at least half of all dunes in global sand seas. However, large, regionally distributed longitudinal dunes appear to be virtually absent on Mars (Breed and others, 1979; Tsoar and others, 1979). A study of the wind regimes in which linear dunes develop on Earth and the role of these dunes as distributional systems for sand may help to explain the reasons for their absence on Mars. Such answers may lie in a lack of sediment (available dune sand has already migrated to sites of accumulation), climatic change, or wind regimes that are unfavorable for their development.

Sand ridges are a form of linear dune that are straight in plan view, have side slopes of approximately equal steepness, and have length-to-width ratios frequently in excess of 75:1. Sand ridges in northeast Arizona have average lengths of about 3.65 km, widths of 43 m, and heights of 2 to 10 m (Breed and others, 1984). They extend over 65,000 km² of the Colorado Plateau (Hack, 1941; Stokes, 1964).

Very few of the linear dunes on the Colorado Plateau appear to be active today. Most have structureless cores that consist of slightly reddened sand and are partially hardened by caliche deposits. Overlying the core is a veneer of white sand derived from deflation of the Navajo Sandstone. Volcanic ash collected from

the base of a dune near Gold Spring is at least 100,000 years old, and possibly as old as 300,000 years (Breed and others, 1984). Active dune segments are found along the southern edge of the Moenkopi Plateau in a zone of very low rainfall (less than 200 mm/yr).

Field studies in northern Arizona by Breed and Breed (1978) suggest that linear dunes form where the wind regime includes both seasonally opposing winds (that pile sand first on one slope of the dune, then on the other); and strong prevailing winds that coincide with, and reinforce, the seasonally bimodal component. The strong prevailing west-southwest winds of the Colorado Plateau are responsible for the longitudinal growth of the dunes and for their relatively straight shape. Linear dune growth seems to require a long wind fetch, and, thus, these dunes are characteristic of areas with few topographic obstructions.

Wind erosion

Over most of the surface of Mars, wind erosion features are comparatively scarce despite billions of years of exposure. Extremely slow erosion rates (on the order of 10^{-2} to 10^{-3} μm/yr) have been estimated, based on the preservation of myriads of small craters in Chryse Planitia and other areas. Wind erosion may have been more efficient in other areas of Mars. Many craters in high latitudes are surrounded by low platforms or pedestals that may have formed by wind erosion of intercrater areas (McCauley, 1973; Arvidson and others, 1976). Estimates of erosion for these areas are around 10^{-1} μm/yr. The pattern of erosion in etched terrain and fretted terrain is also suggestive of wind action on exposed vertical faces.

Wind erosion features of the Colorado Plateau range from such large-scale forms as deflation hollows, yardangs, and wind-fluted cliffs, to smaller features including blowouts and scour grooves. Wind is an effective agent of erosion where vegetation is lacking, the wind regime is strong and persistent, the materials are physically susceptible, and where a lag material does not armor the surface. Sand abrasion is not essential, owing to erosion by deflation, whereby fine grains are carried away in atmospheric suspension. Deflation is considered by Breed and others (1984) to be more significant than abrasion in the development of yardangs on the Colorado Plateau. The streamlining of bedrock outcrops develops by winnowing of loose grains from bedrock surfaces. Dust storms occurring east of the Little Colorado River in the spring are evidence of the role of deflation in modifying the surface.

Clear evidence for wind erosion on Mars is provided by arrays of yardangs up to 50 km in length and 1 km in width. The largest field of yardangs is found in southern Amazonis, where they occur along with pedestal craters and fluted escarpments (Ward, 1979). Martian yardangs are probably of fairly recent origin because they are abundant in areas of low crater densities. The relative freshness of the wind-stripped surface implies that the rock must thus be easily erodible. Regional streaks may parallel the yardangs.

On Earth, yardangs are strong indicators of unidirectional winds, although they may be structurally controlled in part and may, in turn, funnel diverse winds. Though rare on the Colorado Plateau as a whole, yardangs occur in the Painted Desert. They are developed in uniformly bedded, reddish to grayish claystone and siltstone of Mesozoic age. The swelling clays that form the slopes are resistant to wind erosion when wet, but dry to a crumbly surface that releases loose fine silt and clay particles that are entrained by the wind. The largest individual yardangs formed in this material are on the Ward Terrace and are about 50 m long (Breed and others, 1984). The southern part of the Kaibito Plateau northwest of Tuba City includes yardangs (10 to 20 m long) cut into the consolidated Navajo Sandstone. Imperfections in their shape result from crossbedding in the ancient lithified dunes, and from joint patterns. These yardangs are well preserved because the highly permeable nature of the sandstone limits the amount of erosion by surface runoff.

Geomorphic investigations on the Colorado Plateau can isolate factors that are important in determining morphologic variation between yardangs, thereby aiding in the interpretation of Martian yardangs. Some of the factors that are important are (1) rock lithology and structure, (2) wind strength, frequency, and directionality, (3) amounts of locally available abrasive sediments and, for the Earth only, (4) rainfall and vegetation.

Summary

The morphology of the Martian surface suggests that wind erosion of consolidated rocks is extremely slow (about 10^{-3} mm/yr), despite repeated exposure to violent dust storms. Higher erosion rates are found on poorly consolidated materials, and vertical faces are probably more susceptible to wind erosion than are horizontal surfaces. Yardangs are the primary indicators of wind erosion. Evidence for the accumulation and erosion of windblown debris on Mars is provided by dunes, streaks, and splotches. Most of the dunes are crescentic in form and occur in a zone around the north polar cap. Linear dunes, the second most common type on Earth, appear to be absent on Mars.

The Colorado Plateau includes a number of wind-derived landforms that are relatively accessible and have proven to be useful Earth-analog sites, providing indicators on how eolian processes may have modified the surface of Mars. To date, research into yardang formation and into the topographic control of dunes appears to be the most fruitful line of inquiry. Some of the conclusions from these field studies are that: (1) topography may locally intensify wind erosion and affect the redistribution of surface materials, (2) that structural influence must be carefully evaluated if yardang orientations are used to interpret wind directions on Mars, (3) that deflation is an important agent of yardang formation, even in the absence of sand abrasion, and (4) that the absence of linear dunes on Mars suggests that dune sediments have moved to permanent areas of accumulation.

SUMMARY

Owing to the limitations of photogeologic interpretation, terrestrial and planetary analogs must be exhaustively pursued for their value as explanations. Nevertheless, where Earth analog sites are carefully chosen and a problem is clearly defined, such as the Utah grabens, valuable insight has been provided on planetary processes. Conversely, the quest to solve planetary problems has helped develop new ideas about terrestrial geology. This is true, for example, in the increasing recognition of the role of groundwater in valley development and a revived interest in the role of wind erosion. Eolian studies have provided many interesting ideas on sand transport and deflation, but await further supportive process-oriented research. Finally, planetary geologists have occasionally used their knowledge to develop new explanations for unique features, such as the astrobleme origin of Upheaval Dome.

CONCLUSIONS
William L. Graf

Field studies in the Colorado Plateau have contributed to the development of geomorphic theory from the earliest days of the science. The efforts of Newberry, Marvine, Powell, Dutton, and Gilbert led directly to early generalizations. Regional studies outlined the characteristics of the plateau and yielded insights into surficial processes. Recent process-oriented studies in the Henry Mountains region, Chaco Canyon, and the Escalante Basin, along with analyses of Tertiary landscape evolution, recent alluvial histories, and planetary geologic analogs promise additional theoretical contributions.

Future geomorphologic undertakings in the Colorado Plateau are likely to address nonfluvial processes, tectonic geomorphology, and questions related to environmental quality. Mass-movement processes are common in the region, especially those of Pleistocene and early Holocene age, but they remain little studied. Recent catastrophic mass movements may also receive attention, as in the rockfall studies of Schumm and Chorley (1964, 1966). Tectonic geomorphology as a subdiscipline offers direct connections to the most recent trends in geologic research, and it is reasonable to expect that the connection between surface evidence and subsurface crustal processes will be the subject of further work in the plateau province. Finally, as economic development of the Colorado Plateau continues, the role of geomorphic processes in influencing environmental quality is likely to become increasingly important. The fluvial systems of the plateau feed water into massive management projects, so that sediment and the materials it contains (heavy metals, herbicides, and pesticides) will probably become a major issue in terms of science as well as public policy. The original scientific explorations of the Colorado Plateau were driven by the need to assess the resources of the public lands of the region. The management of those public lands is now likely to drive new and hopefully equally productive scientific endeavors.

REFERENCES CITED

Ahnert, F., 1960, The influence of Pleistocene climates upon the morphology of cuesta scarps on the Colorado Plateau: Annals of the Association of American Geographers, v. 50, p. 139–156.

Allen, J.R.L., 1963, The classification of cross-stratified units with notes on their origin: Sedimentology, v. 2, p. 93–114.

Anderson, J. J., Rowley, P. D., Fleck, R. J., and Nairn, A.E.M., 1975, Cenozoic geology of southwestern high plateaus of Utah: Geological Society of America Special Paper 160, 88 p.

Anderson, R. E., 1971, Thin skin distension in Tertiary rocks of southeastern Nevada: Geological Society of America Bulletin, v. 82, p. 43–58.

Arvidson, R. E., Carusi, A., Coradini, M., Pulchignoni, M., Federico, C., Funicello, R., and Salomone, M., 1976, Latitudinal variation of wind erosion of crater ejecta on Mars: Icarus, v. 27, p. 503–516.

Bailey, R. W., 1935, Epicycles of erosion in the valleys of the Colorado Plateau Province: Journal of Geology, v. 43, p. 337–355.

Baker, A. A., 1933, Geology and oil possibilities of the Moab District, Grand and San Juan counties, Utah: U.S. Geological Survey Bulletin 841, 95 p.

—— , 1936, Geology of the Monument Valley–Navajo Mountain region, San Juan County, Utah: U.S. Geological Survey Bulletin 865, 106 p.

—— , 1946, Geology of the Green River Desert–Cataract Canyon region, Emery, Wayne, and Garfield counties, Utah: U.S. Geological Survey Bulletin 951, 122 p.

Baker, V. R., 1980, Some terrestrial analogs to dry valley systems on Mars, in Reports of Planetary Geology Program, 1979–1980: NASA Technical Memorandum 81776, p. 286–288.

—— , 1982, The channels of Mars: Austin, University of Texas Press, 198 p.

Bartlett, R. A., 1962, Great surveys of the American West: Norman, University of Oklahoma Press, 362 p.

Betancourt, J. L., and Van Devender, T. R., 1981, Holocene vegetation in Chaco Canyon, New Mexico: Science, v. 214, p. 656–658.

Blair, W. H., McKee, E. H., and Armstrong, A. K., 1977, Age and environment of deposition; Hualapai Limestone member of the Muddy Creek Formation: Geological Society of America Abstracts with Program, v. 9, no. 4, p. 390.

Blissenback, E., 1952, Geology of the Aubrey Valley south of the Hualapai Indian Reservation, northwest Arizona: Plateau, v. 24, no. 4, p. 119–127.

Boison, P. J., and Patton, P. C., 1983, Late Pleistocene and Holocene alluvial stratigraphy of three tributaries of the Escalante River, south-central Utah [abs.], in Wells, S. G., Love, D. W., and Gardner, T. W., eds., Chaco Canyon Country, Albuquerque, New Mexico: Field Trip Guidebook, 1983 Field Conference, American Geomorphological Field Group, p. 243–244.

—— , 1985, Sediment storage and terrace formation in the Coyote Gulch basin, south-central Utah: Geology, v. 13, p. 31–34.

Bowers, W. E., 1972, The Canaan Peak, Pine Hollow, and Wasatch formations in the Table Cliff region, Garfield County, Utah: U.S. Geological Survey Bulletin 1331-B, 39 p.

Bradley, W. C., 1963, Large-scale exfoliation in massive sandstones of the Colorado Plateau: Geological Society of America Bulletin, v. 74, no. 5, p. 519–527.

Breed, C. S., 1977, Terrestrial analogs of the Hellespontus dunes, Mars: Icarus, v. 30, p. 326–340.

Breed, C. S., and Breed, W. J., 1978, Field studies of sand-ridge dunes in central Australia and northern Arizona, in Reports of Planetary Geology Program, 1977 to 1978: NASA Technical Memorandum 79729, p. 216–218.

—— , 1979, Windforms of central Australia and a comparison with some linear dunes on the Moenkopi Plateau, Arizona, in El-Baz, F., ed., Scientific results of the Apollo–Soyuz Missions: NASA Special Paper 412, p. 319–358.

Breed, C. S., Grolier, M. J., and McCauley, J. F., 1979, Morphology and distribution of common "sand" dunes on Mars; Comparison with Earth: Journal of Geophysical Research, v. 84, no. B14, p. 8183–8204.

Breed, C. S., McCauley, J. F., Breed, W. J., McCauley, C. K., and Cotera, A., 1984, Eolian (wind-formed) landscapes, in Smiley, T. L., Nations, J. D., Péwé, T. L., and Schafer, J. P., eds., Landscapes of Arizona: University Press

of America, p. 359–413.

Bryan, K., 1925, Date of channel trenching (arroyo cutting) in the arid Southwest: Science, v. 62, p. 338–344.

—— , 1928, Niches and other cavities in sandstone at Chaco Canyon, New Mexico: Zeitschrift für Geomorphologie, v. 3, p. 128–140.

—— , 1954, The geology of Chaco Canyon, New Mexico, in relation to the life and remains of the historic peoples of Pueblo Bonito: Smithsonian Miscellaneous Collection, v. 122, no. 7, 65 p.

Bryson, R. A., 1974, A perspective on climatic change: Science, v. 194, p. 753–760.

Campbell, I. A., 1973, Controls of canyon and meander form by jointing: Area, v. 5, no. 4, p. 291–296.

Chorley, R. J., Dunn, A. J., and Beckinsale, R. P., 1964, The history of the study of landforms or the development of geomorphology, vol. 1, Geomorphology before Davis: London, Methuen and Company and John Wiley and Sons, 678 p.

Christensen, E. M., 1962, The rate of naturalization of Tamarix in Utah: American Midland Naturalist, v. 68, p. 51–57.

Colbert, E. H., 1956, Rates of erosion in the Chinle formation: Plateau, v. 28, p. 73–76.

Cole, K. L., and Mayer, L., 1982, Use of packrat middens to determine rates of cliff retreat in the eastern Grand Canyon, Arizona: Geology, v. 10, p. 597–599.

Colton, H. S., 1937, Some notes on the original condition of the Little Colorado River; A side light on the problems of erosion: Museum Notes, Museum of Northern Arizona, v. 10, p. 17–20.

Cooke, R. U., and Reeves, R. W., 1976, Arroyos and environmental change in the American Southwest: Oxford, Oxford University Press, 213 p.

Cooley, M. E., 1958, Physiography of the Glen-San Juan Canyon area, Part I: Plateau, v. 31, no. 2, p. 21–23.

Cutts, J. A., and Smith, R.S.U., 1973, Aeolian deposits and dunes on Mars: Journal of Geophysical Research, v. 78, p. 4139–4154.

Damon, P. E., Shafiqullah, M., and Scarborough, R. B., 1978, Revised chronology for critical stages in the evolution of the lower Colorado River: Geological Society of America Abstracts with Program, v. 10, no. 3, p. 101–102.

Dane, C. H., 1935, Geology of the Salt Valley anticline and adjacent areas, Grand County, Utah: U.S. Geological Survey Bulletin 863, 184 p.

Davis, W. M., 1899, The geographical cycle: Geographical Journal, v. 14, p. 481–504.

—— , 1901, An excursion to the Grand Canyon of the Colorado: Harvard College Museum Zoology Bulletin 38, p. 107–201.

—— , 1922, Biographical memoir of Grove Karl Gilbert (1843–1918): National Academy of Science, v. 21, mem. 5, 303 p.

—— , 1930, The Peacock Range, Arizona: Geological Society of America Bulletin, v. 41, p. 293–313.

Dickinson, W. R., Lawton, T. F., and Inman, T. F., 1986, Sandstone detrital modes, Central Utah foreland region; Stratigraphic record of Cretaceous-Paleogene tectonic evolution: Journal of Sedimentary Petrology, v. 56, p. 276–293.

Diller, J. S., 1911, Major Clarence Edward Dutton: Bulletin of the Seismological Society of America, v. 1, p. 137–142.

Douglas, A. V., Cayan, D. R., and Namia, J., 1982, Large-scale changes in North Pacific and North American weather patterns in recent decades: Monthly Weather Review, v. 110, p. 1851–1862.

Drewes, H., 1978, The Cordilleran orogenic belt between Nevada and Chihuahua: Geological Society of America Bulletin, v. 89, p. 641–657.

Dunne, T., 1980, Formation and controls of channel networks: Progress in Physical Geography, v. 4, no. 2, p. 211–239.

Dunne, T., and Leopold, L. B., 1978, Water in environmental planning: New York, W. H. Freeman, 818 p.

Dutton, C. E., 1880, Report on the geology of the High Plateaus of Utah: Washington, D.C., U.S. Geographical and Geological Survey of the Rocky

Mountain Region, 307 p.

——, 1881, The physical geology of the Grand Canyon district, U.S. Geological Survey, 2nd Annual Report, p. 47–166.

——, 1882, Tertiary history of the Grand Canyon region: U.S. Geological Survey Monograph 2, 264 p.

Eberly, L. D., and Stanley, T. B., Jr., 1978, Cenozoic stratigraphy and geologic history of southwestern Arizona: Geological Society of America Bulletin, v. 89, p. 921–940.

Elston, W. E., Damon, P. E., Coney, P. J., Rhodes, R. C., Smith, E. I., and Bikerman, M., 1973, Tertiary volcanic rocks, Mogollon-Datil Province, New Mexico, and surrounding region; K-Ar dates, patterns of eruption, and periods of mineralization: Geological Society of America Bulletin, v. 84, p. 2259–2274.

Emmett, W. W., 1974, Channel aggradation in western United States as indicated by observations at Vigil Network sites: Zeitschrift für Geomorphology, Supplement Band 21, v. 2, p. 52–62.

Englehart, P. J., and Douglas, A. V., 1985, A statistical analysis of precipitation frequency in the conterminous United States, including comparisons with precipitation totals: Journal of Climate and Applied Meteorology, v. 24, p. 350–362.

Epis, R. C., and Chapin, E. E., 1975, Geomorphic and tectonic implications of the post-Laramide, Late Eocene erosion surface in the southern Rocky Mountains: Geological Society of America Memoir 144, p. 45–74.

Ferguson, C. W., 1964, Annual rings in big sagebrush: Tucson, University of Arizona Press, 95 p.

Fisher, D. J., 1936, The Book Cliffs coal field in Emery and Grand Counties, Utah: U.S. Geological Survey Bulletin 852, 104 p.

Fritts, H. D., 1976, Tree rings and climate: London, Academic Press, 567 p.

Frost, E. G., and Martin, D. L., 1982, Mesozoic-Cenozoic tectonic evolution of the Colorado River region, California, Arizona, and Nevada: San Diego, California, Cordilleran Publishers, 608 p.

Gatewood, J. S., Wilson, A., Thomas, H. E., and Kister, L. R., 1963, General effects of drought on water resources of the Southwest; 1942–56: U.S. Geological Survey Professional Paper 372-B, p. B1–B55.

Gilbert, G. K., 1877, Report on the geology of the Henry Mountains: Washington, D.C., Geographical and Geological Survey of the Rocky Mountain Region, 160 p.

Gilluly, J., 1929, Geology and oil and gas prospects of part of the San Rafael Swell, Utah: U.S. Geological Survey Bulletin 806-C, p. 19–130.

Goetz, A.F.H., Billingsley, F. C., Gillespie, A. R., Abrams, M. J., Squires, R. L., Shoemaker, E. M., Lucchitta, I., and Elston, D. P., 1975, Application of ERTS images and image processing to regional geologic problems and geologic mapping in northern Arizona: NASA Technical Report 32-1597, Jet Propulsion Laboratory, 188 p.

Goff, F. E., Eddy, A. C., and Arney, B. H., 1983, Reconnaissance geologic strip map from Kingman to south of Bill Williams Mountain, Arizona: Los Alamos National Laboratory Map LA-9202-MAP, scale 1:48,000, 4 sheets.

Golombeck, M. P., 1979, Structural analysis of Lunar grabens and the shallow crustal structure of the Moon: Journal of Geophysical Research, v. 84, no. B9, p. 4657–4666.

Golombeck, M. P., and McGill, G. E., 1983, Grabens, basin tectonics, and the maximum total expansion of the Moon: Journal of Geophysical Research, v. 88, p. 3563–3578.

Graf, W. L., 1978, Fluvial adjustments to the spread of tamarisk in the Colorado Plateau region: Geological Society of America Bulletin, v. 86, p. 1491–1501.

——, 1979, Rapids in canyon rivers: Journal of Geology, v. 87, p. 533–551.

——, 1980, The effect of dam closure on downstream rapids: Water Resources Research, v. 16, p. 129–136.

——, 1982a, Spatial variation of fluvial processes in semiarid lands, *in* C. E. Thorne, ed., Space and time in geomorphology: London, George Allen and Unwin, p. 193–217.

——, 1982b, Distance decay and arroyo development in the Henry Mountains, Utah: American Journal of Science, v. 282, p. 1541–1554.

——, 1983a, The arroyo problem; Paleohydrology and paleohydraulics in the short term, *in* Gregory, K. J., ed., Background to paleohydrology: London, John Wiley and Sons, p. 279–302.

——, 1983b, Downstream changes in stream power in the Henry Mountains, Utah: Annals of the Association of American Geographers, v. 73, p. 373–387.

——, 1985, The Colorado River; Instability and basin management: Washington, D.C., Association of American Geographers, 88 p.

Greeley, R., White, B. R., Pollack, J. B., Iverson, J. D., and Leach, R. N., 1981, Dust storms on Mars; Considerations and simulations, *in* Péwé, T., ed., Desert dust; Origin, characteristics, and effect on man: Geological Society of America Special Paper 186, p. 101–121.

Gregory, H. E., 1917, Geology of the Navajo Country: U.S. Geological Survey Professional Paper 93, 161 p.

——, 1938, The San Juan Country; A geographic and geologic reconnaissance of southeastern Utah: U.S. Geological Survey Professional Paper 188, 123 p.

——, 1950, Geology and geography of the Zion Park region, Utah and Arizona: U.S. Geological Survey Professional Paper 220, 200 p.

Gregory, H. E., and Moore, R. C., 1931, The Kaiparowits region; A geographic and geologic reconnaissance of parts of Utah and Arizona: U.S. Geological Survey Professional Paper 164, 161 p.

Gregory, K. J., 1985, The nature of physical geography: London, Edward Arnold, 262 p.

Gresens, R. L., 1981, Extension of the Telluride erosion surface to Washington State, and its regional and tectonic significance: Tectonophysics, v. 79, p. 145–164.

Hack, J. T., 1941, Dunes of the western Navajo Country: Geographical Review, v. 31, p. 240–263.

——, 1942, The changing physical environment of the Hopi Indians of Arizona: Papers of the Peabody Museum of American Archaeology and Ethnology, v. 35, 85 p., 12 plts.

Hadley, R. F., 1961, Influence of riparian vegetation on channel shape, northeastern Arizona: U.S. Geological Survey Professional Paper 424-C, p. C30–C31.

——, 1974, Sediment yield and land use in the southwest United States: International Association of Scientific Hydrologists Publication 113, p. 996–998.

——, 1977, Evaluation of land-use and land-treatment practices in semiarid western United States: Philosophical Transactions of the Royal Society of London, v. 278, p. 543–554.

Hall, S. A., 1977, Late Quaternary sedimentation and paleoecologic history of Chaco Canyon, New Mexico: Geological Society of America Bulletin, v. 88, p. 1593–1618.

——, 1980, Geology of archeologic sites and associated sand dunes, San Juan County, New Mexico: unpublished report to Navajo Cultural Resource Management Program, 42 p.

——, 1983, Holocene stratigraphy and paleoecology of Chaco Canyon, *in* Wells, S. G., Love, D. W., and Gardner, T. W., eds., Chaco Canyon Country, Albuquerque, New Mexico: Field Trip Guidebook, 1983 Field Conference, American Geomorphological Field Group, p. 219–236.

Hartman, J. H., 1984, Systematics, biostratigraphy, and biogeography of latest Cretaceous and early Tertiary Viviparidae (Mollusca, Gastropoda) of southern Saskatchewan, western North Dakota, eastern Montana, and northern Wyoming [Ph.D. thesis]: Minneapolis, University of Minnesota, 919 p.

Haxel, G. B., Tosdal, R. M., May, D. J., and Wright, J. E., 1984, Latest Cretaceous and early Tertiary orogenesis in south-central Arizona; Thrust faulting, regional metamorphism, and granitic plutonism: Geological Society of America Bulletin, v. 95, no. 6, p. 631–653.

Hawley, J. W., and Love, D. W., 1983, Summary of the hydrology, sedimentology, and stratigraphy of the Rio Puerco Valley, *in* Wells, S. G., Love, D. W., and Gardner, T. W., eds., Chaco Canyon Country, Albuquerque, New Mexico: Field Trip Guidebook, 1983 Conference, American Geomorphological Field Group, p. 33–36.

Hereford, R., 1979, Preliminary geologic map of the Little Colorado River Valley between Cameron and Winslow, Arizona: U.S. Geological Survey Open-

File Report 79-1574, scale 1:62,500.

——, 1983, Effect of climate and a geomorphic threshold on the historic geo-morphology and alluvial stratigraphy of the Paria and Little Colorado rivers, southwest Colorado Plateaus [abs.], *in* Wells, S. G., Love, D. W., and Gardner, T. W., eds., Chaco Canyon Country, Albuquerque, New Mexico: Field Trip Guidebook, 1983 Conference, American Geomorphological Field Group, p. 267.

——, 1984, Climate and ephemeral-stream processes; Twentieth-century geo-morphology and alluvial stratigraphy of the Little Colorado River, Arizona: Geological Society of America Bulletin, v. 95, p. 654–668.

——, 1986, Modern alluvial history of the Paria River drainage basin, southern Utah: Quaternary Research, v. 25, p. 293–311.

Higgins, C. G., 1982, Drainage systems developed by sapping on Earth and Mars: Geology, v. 10, p. 147–152.

Hintze, L. F., 1973, Geologic history of Utah: Brigham Young University Geology Studies for Students no. 8, 181 p.

Hodgson, R. A., 1961, Reconnaissance of jointing in Bright Angel Area, Grand Canyon, Arizona: American Association of Petroleum Geologists Bulletin, v. 45, p. 95–97.

Hunt, C. B., 1956, Cenozoic geology of the Colorado Plateau: U.S. Geological Survey Professional Paper 279, 99 p.

——, 1969, Geologic history of the Colorado River: U.S. Geological Survey Professional Paper 669-C, p. 59–130.

——, 1974a, Natural regions of the United States and Canada: San Francisco, W. H. Freeman, 726 p.

——, 1974b, Geology of the Grand Canyon: Flagstaff, Arizona, Museum of Northern Arizona, p. 129–141.

Hunt, C. B., Averitt, P., and Miller, R. L., 1953, Geology and geography of the Henry Mountains region, Utah: U.S. Geological Survey Professional Paper 228, 234 p.

Huntoon, P. W., 1974, Geology of the Grand Canyon; Post-Paleozoic structural geology of the eastern Grand Canyon, Arizona: Flagstaff, Museum of Northern Arizona, p. 82–115.

——, 1981, Grand Canyon monoclines; Vertical uplift or horizontal compression: University of Wyoming Contributions to Geology, v. 19, no. 2, p. 127–134.

Jobin, D. A., 1956, Regional transmissivity of the exposed sediments of the Colorado Plateau as related to uranium deposits: U.S. Geological Survey Professional Paper 300, p. 207–211.

——, 1962, Relation of the transmissive character of the sedimentary rocks of the Colorado Plateau to the distribution of uranium deposits: U.S. Geological Survey Bulletin 1125, 151 p.

Jones, P. D., Wigley, T.M.L., and Kelly, P. M., 1982, Variations in surface air temperatures; Part 1, Northern Hemisphere, 1881–1980: Monthly Weather Review, v. 110, p. 59–70.

Kalnicky, R. A., 1974, Climatic change since 1950: Annals of the Association American Geographers, v. 64, p. 100–112.

Kitzmiller, J. M., II, 1981, Geology of Joe's Valley Reservoir Quadrangle, Sanpete and Emery counties, Utah [M.S. thesis]: Provo, Utah, Brigham Young University, 80 p.

Koons, E. D., 1948, Geology of the eastern Hualapai Reservation: Plateau, v. 20, p. 53–60.

——, 1955, Cliff retreat in the southwestern United States: American Journal of Science, v. 253, p. 44–52.

——, 1964, Structure of the eastern Hualapai Indian Reservation, Arizona: Arizona Geological Society Digest, v. VII, p. 97–114.

Kukla, G. J., and Kukla, H. J., 1974, Increased surface albedo in the Northern Hemisphere: Science, v. 183, p. 709–714.

Kukla, G. J., Angell, J. K., Korshover, J., Dronia, H., Hoshiai, M., Namias, J., Rodewald, M., Yamamoto, R., and Iwashima, T., 1977, New data on climatic trends: Nature, v. 270, p. 573–580.

Laity, J. E., 1980, Groundwater sapping on the Colorado Plateau, *in* Reports of Planetary Geology Program, 1980: NASA Technical Memorandum 82385, p. 358–360.

——, 1983, Diagenetic controls on groundwater sapping and valley formation, Colorado Plateau, revealed by optical and electron microscopy: Physical Geography, v. 4, p. 103–125.

Laity, J. C., and Malin, M. C., 1985, Sapping processes and the development of theater-headed valley networks on the Colorado Plateau: Geological Society of America Bulletin, v. 96, p. 203–217.

La Rocque, A., 1960, Molluscan faunas of the Flatstaff Formation of central Utah: Geological Society of America Memoir 78, 100 p.

Lee, W. T., 1908, Geological reconnaissance of a part of western Arizona: U.S. Geological Survey Bulletin 352, 96 p.

Leopold, L. B., 1951, Rainfall frequency; An aspect of climatic variation: Transactions American Geophysical Union, v. 32, p. 347–357.

——, 1976, Reversal of erosion cycle and climatic change: Quaternary Research, v. 6, p. 557–562.

Leopold, L. B., and Bull, W. B., 1979, Base level, aggradation, and grade: Proceedings of the American Philosophical Society, v. 123, p. 168–202.

Leopold, L. B., Wolman, M. G., and Miller, J. P., 1964, Fluvial processes in geomorphology: San Francisco, W. H. Freeman, 522 p.

Leopold, L. B., Emmett, W. W., and Myrick, R. M., 1966, Channel and hillslope processes in a semiarid area, New Mexico: U.S. Geological Survey Professional Paper 352-G, p. G193–G253.

Longwell, C. R., 1946, How old is the Colorado River? American Journal of Science, v. 244, p. 871–935.

Love, D. W., 1979, Quaternary fluvial geomorphic adjustments in Chaco Canyon, New Mexico, *in* Rhodes, D. D., and Williams, G. P., eds., Adjustments of the fluvial system: Dubuque, Iowa, Kendall/Hunt, p. 277–308.

——, 1983, Quaternary facies in Chaco Canyon and their implications for geomorphic-sedimentologic models, *in* Wells, S. G., Love, D. W., and Gardner, T. W., eds., Chaco Canyon Country, Albuquerque, New Mexico: Field Trip Guidebook, 1983 Conference, American Geomorphological Field Group, p. 195–206.

Lucchitta, I., 1966, Cenozoic geology of the upper Lake Mead area adjacent to the Grand Wash Cliffs, Arizona [Ph.D. thesis]: State College, Pennsylvania State University, 218 p.

——, 1972, Early history of the Colorado River in the Basin and Range Province: Geological Society of America Bulletin, v. 83, p. 1933–1948.

——, 1975, The Shivwits Plateau, *in* Goetz, A.F.H., Billingsly, F. C., Gillespie, A. R., Abrams, M. J., Squires, R. L., Shoemaker, E. M., Lucchitta, I., and Elston, D. P., eds., Application of ERTS images and image processing to regional geologic problems and geologic mapping in northern Arizona: NASA Technical Report 32-1597, p. 41–72.

——, 1979, Late Cenozoic uplift of the southwestern Colorado Plateau and adjacent lower Colorado River region: Tectonophysics, v. 61, p. 63–95.

Lupton, C. T., 1916, Geology and coal resources of Castle Valley in Carbon, Emery, and Sevier counties, Utah: U.S. Geological Survey Bulletin 628, 88 p.

Malde, H. E., and Scott, A. G., 1977, Observations of contemporary arroyo cutting near Santa Fe, New Mexico, U.S.A.: Earth Surface Processes, v. 2, p. 39–54.

Mallory, W. W., ed., 1972, Geologic Atlas of the Rocky Mountain Region: Denver, Rocky Mountain Association of Geologists, 331 p.

Mars Channel Working Group, 1983, Channels and valleys on Mars: Geological Society of America Bulletin, v. 94, p. 1035–1054.

Marvine, A. R., 1874, The stratigraphy of the east slope of the Front Range: Washington, D.C., U.S. Geographical and Geological Survey, 120 p.

McCauley, J. F., 1973, Mariner 9 evidence for wind erosion in the equatorial and mid-latitude regions of Mars: Journal of Geophysical Research, v. 78, p. 4123–4137.

McCauley, C. K., and Breed, W. J., 1980, Topographically controlled dune systems on Earth and Mars, *in* Reports of Planetary Geology Program, 1979–1980: NASA Technical Memorandum 81776, p. 255–256.

McGill, G. E., 1980, Planetary fracture patterns; Influence of inheritance on stress analysis, *in* Reports of Planetary Geology Program, 1979–1980: NASA Technical Memorandum 81776, p. 80–82.

McGill, G. E., and Stromquist, A. W., 1974, A model for graben formation by subsurface flow; Canyonlands National Park, Utah: Amherst, University of Massachusetts, Department of Geology and Geography Contribution no. 15, 79 p.

—— , 1979, The grabens of Canyonlands National Park, Utah; Geometry, mechanics, and kinematics: Journal of Geophysical Research, v. 84, no. B9, p. 4547–4563.

McKee, E. D., and McKee, E. H., 1972, Pliocene uplift of the Grand Canyon region; Time of drainage adjustment: Geological Society of America Bulletin, v. 83, p. 1923–1932.

McKee, E. D., Wilson, R. F., Breed, W. J., and Breed, C. S., 1964, Evolution of the Colorado River in Arizona: Flagstaff, Museum of Northern Arizona, 67 p.

McKee, E. D., Hamblin, W. K., and Damon, P. E., 1968, K-Ar age of the lava dams in Grand Canyon: Geological Society of America Bulletin, v. 79, p. 133–136.

McKnight, E. T., 1940, Geology of the area between the Green and Colorado rivers, Grand and San Juan counties, Utah: U.S. Geological Survey Bulletin 908, 147 p.

Miller, H. W., 1962, Cretaceous rocks of the Mogollon Rim area in Arizona, *in* Weber, R. H., and Pierce, H. W., eds., Mogollon Rim region: New Mexico Geological Society Field Conference Guidebook 13, 93 p.

Milton, D. J., 1973, Water and processes of degradation in the Martian landscape: Journal of Geophysical Research, v. 78, p. 4037–4047.

Miser, H. D., 1924a, Geological structure of San Juan Canyon and adjacent country, Utah: U.S. Geological Survey Bulletin 751-D, p. 115–155.

—— , 1924b, The San Juan Canyon, southeastern Utah; A geographic and hydrographic reconnaissance: U.S. Geological Survey Water-Supply Paper 538, 80 p.

—— , 1925, Erosion in San Juan Canyon, Utah: Geological Society of America Bulletin, v. 36, p. 365–377.

Newberry, J. S., 1861, Colorado River of the West: American Journal of Science, v. 33, p. 387–403.

—— , 1876, Geological report of the exploring expedition from Santa Fe, New Mexico, to the junction of the Grand and Green rivers of the great Colorado of the West in 1859: Washington, D.C., U.S. Government Printing Office, 159 p.

Oberlander, T. M., 1977, Origin of segmented cliffs in massive sandstones of southeastern Utah, *in* Doehring, D. O., ed., Geomorphology in arid regions: Boston, Allen and Unwin, p. 79–114.

Patton, P. C., and Boison, P. J., 1986, Process and rates of Holocene terrace formation in Harris Wash, Escalante River Basin, south-central Utah: Geological Society of America Bulletin, v. 97, p. 369–378.

Patton, P. C., and Schumm, S. A., 1975, Gully erosion, northwestern Colorado; A threshold phenomenon: Geology, v. 3, p. 88–90.

—— , 1981, Ephemeral-stream processes; Implications for studies of Quaternary valley fills: Quaternary Research, v. 15, p. 24–43.

Peirce, H. W., 1984, The Mogollon escarpment: Fieldnotes, Arizona Bureau of Geology and Mineral Technology, v. 14, p. 8–11.

Pierce, H. W., Damon, P. E., and Shafiqullah, M., 1979, An Oligocene (?) Colorado Plateau edge in Arizona: Tectonophysics, v. 61, p. 1–24.

Phoenix, D. A., 1963, Geology of the Lees Ferry area, Coconino County, Arizona: U.S. Geological Survey Bulletin 1137, 86 p.

Pickup, G., 1975, Downstream variations in morphology, flow conditions and sediment transport in an eroding channel: Zeitschrift für Geomorphologie, v. 19, p. 443–459.

Pieri, D. C., 1980, Martian valleys; Morphology, distribution, age, and origin: Science, v. 210, p. 895–897.

Powell, J. W., 1875, Exploration of the Colorado River of the West and its tributaries, 1869–1872: Washington, D.C., U.S. Government Printing Office, Smithsonian Institute Publication, 291 p.

—— , 1880, U.S. Geological Survey Second Annual Report, 588 p.

—— , 1895, Canyons of the Colorado: Meadville, Pennsylvania, Flood and Vincent, 400 p.

Potter, D. B., and McGill, G. E., 1981, Field analysis of a pronounced topographic lineament, Canyonlands National Park, Utah, *in* O'Leary, D. W., and Earle, J. L., eds., Proceedings of the Third International Conference on Basement Tectonics: Basement Tectonics Committee Publication no. 3, p. 169–176.

Price, W. E., 1950, Cenozoic gravels on the rim of Sycamore Canyon, Arizona: Geological Society of America Bulletin, v. 61, p. 501–508.

Pyne, S. J., 1980, Grove Karl Gilbert; A great engine of science: Austin, University of Texas Press, 280 p.

Quaide, W., 1965, Rilles, ridges, and domes; Clues to maria history: Icarus, v. 4, p. 374–389.

Reitan, C. H., 1980, Trends in the frequencies of cyclone activity over North America: Monthly Weather Review, v. 107, p. 1684–1688.

Robinson, E. S., 1970, Mechanical disintegration of the Navajo Sandstone in Zion National Park, Utah: Geological Society of America Bulletin, v. 81, p. 2799–2806.

Robinson, H. H., 1907, The Tertiary peneplain of the plateau district and adjacent country in Arizona and New Mexico: American Journal of Science, v. 24, p. 109–120.

Robinson, T. W., 1965, Introduction, spread, and areal extent of saltcedar (*Tamarix*) in the western states: U.S. Geological Survey Professional Paper 491-A, p. A1–A12.

Roden, M. F., Smith, D., and McDowell, F. W., 1979, Age and extent of potassic volcanism on the Colorado Plateau: Earth and Planetary Science Letters, v. 34, p. 279–284.

Ryder, R. T., Fouch, T. D., and Elison, J. H., 1976, Early Tertiary sedimentation in the western Uinta Basin, Utah: Geological Society of America Bulletin, v. 87, p. 496–512.

Schumm, S. A., 1977, The fluvial system: New York, John Wiley and Sons, 338 p.

Schumm, S. A., and Chorley, R. J., 1964, The fall of Threatening Rock: American Journal of Science, v. 262, p. 1041–1054.

—— , 1966, Talus weathering and scarp recession in the Colorado Plateaus: Zeitschrift für Geomorphologie, v. 10, p. 11–36.

Schumm, S. A., and Hadley, R. F., 1957, Arroyos and the semiarid cycle of erosion: American Journal of Science, v. 225, p. 164–174.

Schumm, S. A., and Lichty, R. W., 1963, Channel widening and floodplain construction along Cimarron River in southwestern Kansas: U.S. Geological Survey Professional Paper 352-D, p. 71–88.

Shafiqullah, M., Damon, P. E., Lynch, D. J., Reynolds, S. J., Rehrig, W. A., and Raymond, R. H., 1980, K-Ar geochronology and geologic history of southwestern Arizona and adjacent areas: Arizona Geological Society Digest, v. XII, p. 201–260.

Sharp, R. P., 1973, Mars; Troughed terrain: Journal of Geophysical Research, v. 78, no. 20, p. 4063–4072.

Sharp, R. P., and Malin, M. C., 1975, Channels on Mars: Geological Society of America Bulletin, v. 86, p. 593–609.

Shoemaker, E. M., and Herkenhoff, K. E., 1984, Impact origin of Upheaval Dome, Utah, *in* Reports of Planetary Geology Program, 1983: NASA Technical Memorandum 86246, p. 93.

Smiley, T. L., Nations, J. D., Péwé, T. L., Schafer, J. P., eds., 1984, Landscapes of Arizona: Lanham, Maryland, University Press of America, 505 p.

Smith, J. F., Jr., Huff, L. C., Hinrichs, E. N., and Luedke, R. G., 1963, Geology of the Capital Reef area, Wayne and Garfield counties, Utah: U.S. Geological Survey Professional Paper 363, 102 p.

Spieker, E. M., 1931, The Wasatch Plateau coal field, Utah: U.S. Geological Survey Bulletin 819, 210 p.

Stene, L. P., 1980, Observations on lateral and overbank deposition; Evidence from Holocene terraces, southwestern Alberta: Geology, v. 6, p. 314–317.

Stokes, W. L., 1964, Incised, wind-aligned stream patterns of the Colorado Plateau: American Journal of Science, v. 262, p. 808–816.

Stromquist, A. W., 1976, Geometry and growth of grabens, Lower Red Lake Canyon area, Canyonlands National Park, Utah: Amherst, University of Massachusetts, Department of Geology and Geography Contribution no. 28,

118 p.

Thornbury, W. D., 1965, Regional geomorphology of the United States: New York, John Wiley and Sons, 609 p.

Thornthwaite, C. W., Sharpe, C.T.S., and Dosch, E. F., 1942, Climate and accelerated erosion in the arid and semiarid southwest with special reference to the Polacca Wash drainage basin, Arizona: U.S. Department of Agriculture Technical Bulletin 808, 134 p.

Tsoar, H., and Greeley, R., 1980, Dunes related to obstacles on Earth and Mars; Observations and simulation, *in* Reports of Planetary Geology Program, 1979–1980: NASA Technical Memorandum 81776, p. 257–258.

Tsoar, H., Greeley, R., and Peterfreund, A., 1979, Mars; The north polar sand sea and related wind patterns: Journal of Geophysical Research, v. 84, no. B14, p. 8167–8180.

Twenter, F. R., 1962, Geology and promising areas for groundwater development in the Hualapai Indian Reservation, Arizona: U.S. Geological Survey Water-Supply Paper 1576-A, 38 p.

Underwood, J. R., Jr., and Lambert, W., 1974, Centroclinal cross strata; A distinctive sedimentary structure: Journal of Sedimentary Petrology, v. 44, p. 1111–1113.

Ward, A. W., 1979, Yardangs on Mars; Evidence of recent wind erosion: Journal of Geophysical Research, v. 84, p. 8147–8166.

Watson, R. A., and Wright, H. E., 1963, Landslides on the east flank of the Chuska Mountains, northwestern New Mexico: American Journal of Science, v. 261, p. 525–548.

Wells, S. G., Love, D. W., and Gardner, T. W., eds., 1983, Geomorphic processes on the alluvial floor of the Rio Puerco, *in* Chaco Canyon Country, Albuquerque, New Mexico: Field Trip Guidebook, 1983 Field Conference, American Geomorphological Field Group, p. 37–39.

Wolman, M. G., and Leopold, L. B., 1970, Flood plains, *in* Dury, G. H., ed., Rivers and river terraces: London, MacMillan, p. 166–196.

Young, R. A., 1966, Cenozoic geology along the edge of the Colorado Plateau in northwestern Arizona [Ph.D. thesis]: St. Louis, Washington University, 167 p.

—— , 1979, Laramide deformation, erosion, and plutonism along the southwest-ern margin of the Colorado Plateau: Tectonophysics, v. 61, p. 25–47.

—— , 1982, Paleogeomorphologic evidence for the structural history of the Colorado Plateau margin in western Arizona, *in* Frost, E. G., and Martin, D. L., eds., Mesozoic-Cenozoic tectonic evolution of the Colorado River region, California, Arizona, and Nevada: San Diego, Cordilleran Publishers, p. 29–39.

—— , 1985, Geomorphic evolution of the Colorado Plateau margin in west-central Arizona; A tectonic model to distinguish between the causes of rapid, symmetrical scarp retreat and scarp dissection, *in* Hack, J. T., and Morisawa, M., eds., Tectonic geomorphology, Binghamton Symposia in Geomorphology, International Series 15: London, Allen and Unwin, p. 261–278.

Young, R. A., and Brennan, W. J., 1974, The Peach Springs Tuff; Its bearing on structural evolution of the Colorado Plateau and development of Cenozoic drainage in Mohave County, Arizona: Geological Society of America Bulletin, v. 85, p. 83–90.

Young, R. A., and McKee, E. H., 1978, Early and middle Cenozoic drainage and erosion in west-central Arizona: Geological Society of America Bulletin, v. 89, p. 1745–1750.

Young, R. A., and Hartman, J. H., 1984, Early Eocene fluviolacustrine sediments near Grand Canyon, Arizona; Evidence for Laramide drainage across northern Arizona into southern Utah: Geological Society of America Abstracts with Program, v. 16, p. 703.

Young, R. A., McKee, E. H., Hartman, J. H., and Simmons, A. M., 1985, Similarities and contrasts in tectonic and volcanic style and history along the Colorado Plateau-to-Basin and Range transition zone in western Arizona; Geologic framework for Tertiary extensional tectonics: Conference on Heat and Detachment in crustal extension on continents and planets, Houston, Lunar and Planetary Institute, p. 152–155.

Zishka, K. M., and Smith, P. J., 1980, The climatology of cyclones and anticyclones over North America and surrounding ocean environs for January and July, 1950–77: Monthly Weather Review, v. 108, p. 387–401.

MANUSCRIPT ACCEPTED BY THE SOCIETY NOVEMBER 21, 1986

Geological Society of America
Centennial Special Volume 2
1987

Chapter 9

Basin and Range

John C. Dohrenwend
U.S. Geological Survey, 345 Middlefield Road, MS 901, Menlo Park, California 94025

INTRODUCTION

The Basin and Range physiographic province is an ideal natural laboratory for geomorphologic research. The more than 400 mountain ranges and intervening basins within the province provide similar topographic settings with similar tectonic origins for large populations of landforms (e.g., alluvial fans, pediments, fault scarps, faceted spurs, range front embayments, pluvial lake shorelines). Samples from these landform populations can be selected and studied in a manner somewhat analogous to the design of a controlled laboratory experiment, that is, by keeping most geomorphic factors (e.g., lithology, structure, tectonic history, climatic history, physiographic setting, etc.) closely constrained, and then studying the effects of variation in the remaining factors on specific aspects of process or form. However, this approach has not been widely utilized. Recent research emphasis has encouraged the intensive study of local areas, and regional studies of specific processes or landforms have not been widely pursued. Moreover, the province is relatively remote and largely undeveloped so that access to many areas is difficult. Consequently, the geomorphology of large areas of the Basin and Range remains, for the most part, unstudied and unknown.

General Physiography

Generally defined, the Basin and Range physiographic province is that area of southwestern North America characterized by relatively evenly spaced, subparallel mountain ranges and intervening alluviated basins formed by high-angle extensional faulting. This distinctive combination of physiography and structure is more extensively developed in this region than in any other part of the world. The region extends from southern Idaho and Oregon through most of Nevada and parts of western Utah, eastern California, western and southern Arizona, southwestern New Mexico, and northern Mexico (Fig. 1). It forms an irregularly shaped region more than 1,500 km long and 500 to 1,000 km wide (with an area of 0.8 million to more than one million km² depending on whether the region is defined physiographically or tectonically). Boundaries with the adjacent physiographic provinces are generally gradational, and at least in part,

somewhat arbitrary. Within the United States, the province describes an irregular arc that curves around the western and southern margins of the Colorado Plateau. Other boundaries include the Sierra Nevada batholith and southern Cascade Range on the west and northwest, the Columbia Plateaus on the north, and the Wasatch Range of the middle Rocky Mountains on the northeast. In Mexico, Basin and Range structure extends southward from Arizona and New Mexico along both flanks of the Sierra Madre Occidental.

The physiography of the Basin and Range Province is distinguished by a conspicuous pattern of alternating, more or less regularly spaced, sublinear uplands and lowlands. The uplands are abrupt, steeply sloping, and deeply dissected; the lowlands are typically broad, gently sloping, and largely undissected. Relief within the uplands generally ranges between 500 and 1,500 m; relief within the intervening basins is generally less than 500 m. However, relief between range crests and adjacent basin axes locally exceeds 3,000 m. Indeed, the 3,454 m separating Badwater (−86 m) in Death Valley from Telescope Peak (3,368 m) on the crest of the adjacent Panamint Range and the 3,362 m between the Salton Sea (about −70 m) and San Jacinto Peak (3,292 m) represent the greatest amount of local relief within the southwestern United States.

Within a region as large and structurally complex as the Basin and Range Province, physiographic variability is to be expected. In central Nevada, imposing mountain ranges as much as 180 km long and 25 km wide locally rise to altitudes above 3,500 m. Altitudes along the axes of intervening basins generally range between 1,500 and 2,000 m, and mean range relief usually exceeds 700 m. The ranges are consistently aligned along generally north to northwest trends and comprise more than 40% of the total surface area. In contrast, the ranges of the Mojave Desert in southeastern California are significantly smaller and lower. Range lengths are commonly less than 25 km and widths less than 6 km. Altitudes of range crests and basin axes are approximately 1,000 m lower than in central Nevada, and mean range relief is on the order of 500 m. Range trends are variable, and range areas account for generally less than 30% of the total surface area. Be-

Dohrenwend, J. C., 1987, Basin and Range, *in* Graf, W. L., ed., Geomorphic systems of North America: Boulder, Colorado, Geological Society of America, Centennial Special Volume 2.

304 J. C. Dohrenwend

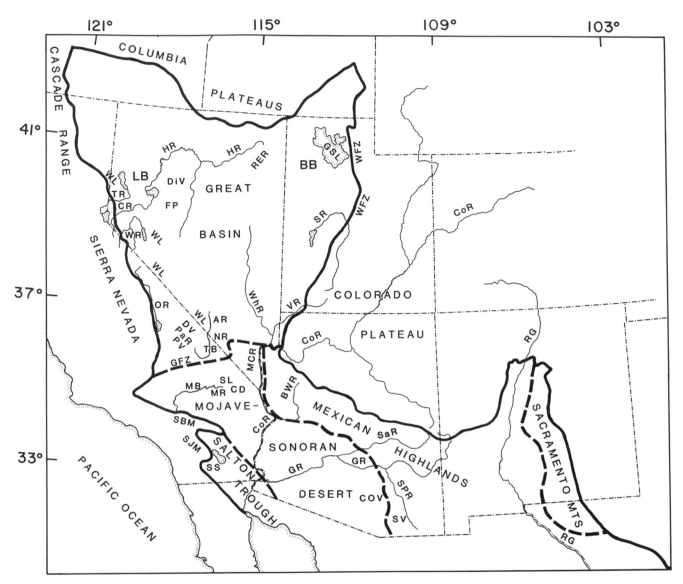

Figure 1. Generalized map of the Basin and Range Province showing approximate locations of the geographic features referenced in this chapter: Amargosa River–AR; Bill Williams River–BWR; Bonneville Basin–BB; Carson River–CR; Canada del Oro Valley–CoV; Cima Dome–CD; Colorado River–CoR; Death Valley–DV; Dixie Valley–DiV; Fairview Peak–FP; Garlock Fault Zone–GFZ; Gila River–GR; Great Salt Lake–GSL; Lahontan Basin–LB; Humboldt River–HR; Manix Basin–MB; McCullough Range–MCR; Mojave River–MR; Nopah Range–NR; Owens River–OR; Panamint Range–PaR; Panamint Valley–PV; Rio Grande–RG; Ruby-East Humboldt Range–RER; Salt River–SaR; Salton Sea–SS; San Bernardino Mtns–SBM; San Jacinto Mtns–SJM; San Pedro River–SPR; Siever River–SR; Silver Lake playa–SL; Sonoita Valley–SV; Tecopa Basin–TB; Truckee River–TR; Virgin River–VR; Walker Lane–WL; Walker River–WR; Wasatch Fault Zone–WFZ; White River–WhR.

cause of such contrasts, the Basin and Range Province has been subdivided into five sections (Fig. 1): the Great Basin, the Mojave-Sonoran Desert, the Salton Trough, the Mexican Highlands, and the Sacramento Mountains (Fenneman, 1931).

The Great Basin, largest of these sections, includes all of the Basin and Range Province north of the Garlock fault zone and the eastward extension of its trend through southernmost Nevada.

The Great Basin is not, as its name suggests, a large regional depression. Rather, it could be more accurately described as a great bulge consisting of a central area of elevated basins and ranges flanked on three sides by large areas of significantly lower terrain: the 140,000 km^2 Bonneville basin on the east, the 115,000 km^2 Lahontan basin on the west, and a less well defined lowland comprising much of southern Nevada on the south. With the exception of limited areas along its northern, northwestern, and southeastern margins, all drainage within the Great Basin is internal. More than 200 distinct drainage areas can be defined, and more than 100 of these are isolated closed basins, many of which supported perennial lakes during late Wisconsin time (Snyder and others, 1964; Mifflin and Wheat, 1979). Significant perennial fluvial systems within the Great Basin are limited to the Truckee, Carson, Walker, and Owens rivers along the east flank of the Sierra Nevada, the Sevier and Jordan rivers along the west flank of the Wasatch Mountains, and the Humboldt River, which feeds an annual flow of approximately 0.5 million acre feet northward and westward from the elevated center of Nevada into the Lahontan basin (Hunt, 1974).

South of the Great Basin, the Mojave-Sonoran Desert section includes three physiographically distinct areas: the Mojave Desert, the lower Colorado River valley; and the Sonoran Desert. Compared to the Great Basin, the topography of this region is subdued. Large areas are essentially lacking in morphologic evidence of active tectonism (Bull and McFadden, 1977; Bull, 1978; Morrison, 1985). Although morphometrically similar, the Mojave and Sonoran deserts display dramatic contrasts in drainage integration and basin dissection. Drainage within the Mojave is predominantly internal and most of the closed basins of the region are undissected except in proximal piedmont areas. Drainage within the Sonoran Desert has been largely integrated into the Gila, Salt, or Bill Williams river systems; and basin dissection, although relatively superficial, is more continuous and widespread than in the Mojave. The valley of the lower Colorado River separates these two areas, and within this valley, piedmonts have been deeply and extensively dissected in response to base-level lowering along the river. Prior to construction of large reservoirs and diversions, the average annual discharge of this stream was approximately 14.9 million acre feet (18.4 billion m^3; Hely, 1967).

The Salton Trough bounds the southwest margin of the Mojave Desert. This basin, as much as 135 km wide and 225 km long, is a fault-bounded depression containing more than 6,000 m of late Cenozoic sedimentary deposits (Loeltz and others, 1975). As the northern continuation of the Gulf of California and a locus of spreading between the Pacific and North American plates, this tectonic depression is an area of high topographic relief. It is bounded on three sides by high mountain ranges; San Jacinto Peak, the highest point in these ranges, rises to an altitude of 3,292 m. In contrast, more than 5,400 km^2 of the basin surface lies below mean sea level and the present surface of the Salton Sea lies at an altitude of about –70 m.

The Mexican Highlands is a transitional terrain of varied topography and structure between the Sonoran Desert to the southwest and the Colorado Plateau to the north, and its physiographic boundaries with these areas are generally gradational and indistinct. The ranges of the Mexican Highlands trend north to northwest; crestal altitudes generally range between 1,500 and 2,700 m, 900 to 1,300 m above the axes of intervening basins (Morrison, 1985). Drainage within the highlands is more com-

pletely integrated than in any other area of the Basin and Range. Most of the Arizona basins drain via the Gila River system to the Sonoran Desert, and most of the New Mexico basins drain to the Rio Grande Rift. Within several of the eastern Arizona basins, deep dissection induced by this drainage integration has exposed thick sections of syntectonic and post-tectonic basin fill (Menges and McFadden, 1981; Morrison, 1985).

Situated along the eastern margin of the Mexican Highlands, the Sacramento Mountains section forms a transitional zone, approximately 80 to 150 km wide and 500 km long, between the Rio Grande Rift and the Great Plains. Basins within this narrow zone are the product of downwarping as well as faulting; the larger ranges are gently eastward dipping fault blocks of Paleozoic sedimentary strata. These transitional structures bear little resemblance to the majority of the basins and ranges in other areas of the province (Hunt, 1974).

General Geologic and Tectonic Setting

The pre-Quaternary geology of the Basin and Range Province can be summarized as a late Precambrian and Paleozoic continental margin assemblage, grading from miogeoclinal strata on the east to eugeoclinal strata on the west. This assemblage was complexly deformed by late Paleozoic to early Mesozoic orogenies, intruded in western areas by Mesozoic granitic rocks, broadly overlain by Cenozoic volcanic rocks, and extensively deformed by at least two phases of middle to late Cenozoic extensional faulting (Stewart, 1978, 1980). As Stewart points out, the orientation and distribution of the extensional faulting clearly follows structural patterns established well before late Cenozoic time. The eastern margin of the Basin and Range Province corresponds closely with the eastern margin of the late Precambrian and Paleozoic miogeocline, and the pattern of late Cenozoic extensional faulting is closely associated with the earlier patterns of Mesozoic and earliest Cenozoic tectonic deformation and early to middle Cenozoic igneous activity (Stewart, 1978).

The earlier phase of extension within the Basin and Range was marked by the initiation and maintenance of widespread, shallow detachment faulting. This deformation began between approximately 30 and 35 Ma (at the approximate time of the intersection of the East Pacific Rise and the subduction zone of the Farallon plate along the western margin of North America) and locally continued into the time of the later extensional phase (Stewart, 1978; Eaton, 1982). The later phase of extension, which is responsible for the classic Basin and Range structure of high-angle block faulting, began at approximately 17 Ma (Christiansen and McKee, 1978; Eaton, 1982). In the Great Basin and in the rift valley of the Rio Grande, this style of deformation has continued episodically to the present day; but, since approximately 14 Ma, tectonic activity has tended to become progressively more concentrated towards the margins of the Great Basin. Extensional deformation generally ended in the Sonoran Desert between 10 and 6 Ma and in the Mexican Highlands between 6 and 4 Ma (Shafiqullah and others, 1980; Menges and McFadden,

1981; Morrison, 1985). However, a discontinuous, northwest-trending band of late Pliocene and Pleistocene normal faulting across central Arizona delineates the remnant of a Quaternary continuation of extensional deformation in the western part of the Mexican Highlands (Morrison and others, 1981).

In addition to these temporal variations, large domains of consistently tilted basin-range fault blocks form distinct transverse zones that subparallel the extension direction of the province. Major antiformal and synformal axes, approximately orthogonal to the extension direction, form the domain boundaries within each zone. This regional tilt pattern may be related to stress relief extending outward from initial sites of rupture during the later phase of late Cenozoic extension (Stewart, 1980).

With the possible exception of the northern part of the African rift system, the Basin and Range Province is unique for its combination of thin continental crust, anomalously low seismic velocities in the underlying upper mantle, a pronounced layer of low seismic velocity and high electrical conductivity in midcrust, high heat flow, a long history of episodic magmatism, regional uplift, and pronounced geophysical and physiographic bilateral symmetries (Stewart, 1978; Eaton, 1982). These characteristics indicate broadly distributed rifting within a thin, brittle, rheologically-layered carapace of continental lithosphere as the driving mechanism behind the formation of classic Basin and Range fault-block structures. Development of these structures has apparently proceeded at relatively low strain rates and has been strongly influenced by lateral traction from the northward-migrating Pacific plate (Eaton, 1982). It has been suggested that the Basin and Range is time transgressive, growing northward as the Mendocino triple junction migrates northward with the Pacific plate (Dickinson and Snyder, 1979). Generally, the Great Basin has been more recently active than the Mojave-Sonoran Desert. However, range and basin morphometry and range front morphology suggest that tectonic activity within the Great Basin is highly variable and does not follow a clearly defined south-to-north trend. Also, Quaternary quiescence within the Sonoran Desert may be associated with the opening of the Gulf of California (Coney, 1983). Whatever the case, development of the "classic" Basin and Range fault-block structures has varied significantly in time and space, and this variation is a fundamental determinant of the present-day Basin and Range landscape.

Purpose and Scope

This chapter is not intended as a general review of Basin and Range geomorphology. Rather, it is focused on a discussion of two of the more actively investigated areas of process-related geomorphic research in the Basin and Range; the impact of Quaternary climatic change on arid geomorphic systems, and the degradational evolution of landforms in arid environments. As with many aspects of the earth sciences, few of the processes or combinations of processes related to these subjects can be studied directly. Either they are presently inactive or they operate much too slowly to be productively studied by direct observation. Con-sequently, most existing knowledge regarding these processes has been gained indirectly from systematic analyses of landforms and surficial deposits. One of the more productive approaches to the indirect study of geomorphic processes involves the identification of systematic associations between landforms and the climatic, tectonic, geologic, physiographic, and evolutionary factors that have influenced their development. Information about the processes that have operated to produce the landforms is then inferred from these associations. For example: (1) morphometric relations between alluvial fans and their source areas indicate that discharge and sediment yield are primary determinants of alluvial fan development (Hooke, 1968); (2) systematic relations between fault-scarp height and slope can be accurately modelled by the diffusion equation suggesting that one or more of the diffuse processes of hillslope degradation (creep, sheet wash, rill wash, or raindrop impact) are responsible for fault-scarp degradation (Hanks and others, 1984); and (3) comparison of drainage networks on a series of accurately dated lava-flow surfaces can be used to study the processes of drainage evolution (Dohrenwend and others, 1987). Process-related research within the Basin and Range Province has proceeded largely via such indirect approaches.

INFLUENCE OF CLIMATIC CHANGE ON EROSIONAL AND DEPOSITIONAL PROCESSES

Geomorphic and sedimentologic processes in the Basin and Range are particularly sensitive to climatic change. Moderate to high relief, pronounced seasonality, low precipitation, and sparse vegetation combine to produce a variety of climatically sensitive geomorphic and sedimentologic thresholds and a corresponding variety of climatically indicative landforms and sedimentary deposits. Relict landforms formed in response to late Pleistocene and early Holocene climatic conditions include: cirques and niva-tion hollows; talus-armored flatirons and colluvial wedges; talus and pavement-covered sand ramps and stabilized dunes; beach ridges, tufa mounds, wave-cut scarps, and dissected sublacustrine erosion surfaces; zones of fluvial stripping and dissection along mountain fronts; and planar piedmont surfaces veneered by well-sorted stone pavements. Relict sedimentary deposits include: glacial moraines, colluvial aprons, sand ramps and sand sheets, accretionary eolian mantles, tufa-cemented beach sands and gravels, and the relatively well sorted alluvial sands and gravels of late Pleistocene fan surfaces. Such relicts are abundant and widely distributed, and they provide the means to determine the influence of climate and climatic change on geomorphic processes in arid environments. Indeed, over the past two decades, the study of these relicts has fostered many significant contributions to the general understanding of the timing and magnitude of Quaternary climatic changes and the influence of climate on surficial geologic processes. Although the signal of climatic change, as expressed in the response of erosional and depositional systems, is not easily filtered from a noisy background of tectonic, lithologic, and evolutionary influences, it is apparent that climatic change has significantly affected the geomorphology and surficial stratigraphy of the region, even in areas of rapid vertical tectonism.

Recent estimates of late Pleistocene climatic conditions in the American Southwest have been attempted by, among others, Brackenridge (1978), Mifflin and Wheat (1979), Van Devender and Spaulding (1979), McCoy (1981), Benson (1981), Barry (1983), Galloway (1983), Spaulding and others (1983), and Dohrenwend (1984). However, paleoclimatic reconstruction is a difficult game, and the conclusions of these several studies are, at best, only in partial agreement and, at worst, directly contradictory. Alternative interpretations for full-glacial climatic conditions range from cold and dry (Galloway, 1970, 1983; Brakenridge, 1978; McCoy, 1981) to mild and wet (Mifflin and Wheat, 1979; Wells, 1979). Estimates of full-glacial mean annual temperatures range from 2.8° to 16°C colder than at present and estimates of full-glacial mean annual precipitation range from 50% to approximately 200% of present levels. Part of this variation is undoubtedly due to differences in the methods and assumptions used to generate these estimates; however, a substantial part may reflect real differences in the magnitude and character of the late Pleistocene-to-Holocene climatic transition in different areas of the Basin and Range Province (Spaulding and Graumlich, 1986). Whatever the case, several trends appear to be generally consistent throughout the larger part of the province: (1) pluvial lake levels were high to intermediate between approximately 25 and 12 ka, declined rapidly after approximately 12 ka, and were generally low or dry between approximately 10 and 5 ka (Smith and Street-Perrott, 1983); (2) during times approximately synchronous with the occurrence of high lake levels (approximately 30 to 11 ka), pinyon-juniper woodlands were present in many areas of the province where desert scrub communities are now stable (Van Devender, 1973, 1977; Van Devender and Spaulding, 1979, Spaulding and others, 1983); and (3) the effective moisture of latest Pleistocene and earliest Holocene climatic regimes was significantly greater than the effective moisture of the modern climate (Barry, 1983; Spaulding and others, 1983). Moreover, documentation of short-lived early Holocene lakes in several basins fed by nonglaciated source areas in regions as widely separated as central New Mexico (Bachhuber, 1982; Fleischouer and Stone, 1982) and eastern California (Wells and others, 1984b, 1987) indicate significant fluctuation in effective moisture during the time of the latest Pleistocene-to-Holocene climatic transition.

These changes in effective moisture and in the distribution and density of vegetation are inferred to have induced a variety of interrelated changes in the operation and relative importance of the various geomorphic processes that are or have been active within the Basin and Range. The general trend from a cooler environment with greater effective moisture and a relatively dense vegetation cover to a warmer environment with less effective moisture and a relatively sparse vegetation cover brought about the general desiccation and deflation of pluvial lake beds and distal piedmont areas. Concomitant deposition of sand sheets and silt- and salt-rich eolian veneers on the surrounding piedmonts and uplands and degeneration of the regional vegetation cover initiated a variety of changes in geomorphic process. Eolian silts form the primary constituent of the nearly ubiquitous vesicular A horizons that characterize the uppermost parts of most desert piedmont soils, and eolian-borne salts (primarily calcium carbonate and calcium sulfate) commonly engulf subjacent soil horizons (McFadden and others, 1984; McFadden and Bull, 1987). Moreover, eolian silts and salts are inferred to have substantially accelerated processes of physical weathering and colluviation on hillslopes, and in concert with the impact of decreased vegetation densities on hillslope runoff, these processes are credited with a general destabilization and stripping of upland hillslopes during the early Holocene (Bull, 1979; Mayer and Bull, 1981; Mayer and others, 1984; Wells and others, 1982, 1984b, 1987). This relatively short-lived increase in hillslope sediment yield caused a major pulse of early Holocene alluvial fan aggradation. Subsequent complex responses of hillslope-piedmont systems included the stripping and dissection or proximal piedmont areas and the progradation of distal piedmont areas to form the "telescoping" fan surfaces that characterize most piedmont areas in the southwestern Basin and Range today. For example, increased specific runoff yields and decreased sediment yields from stripped hillslopes have increased the power of piedmont streams to transport coarser materials to more distal piedmont areas (Mayer and others, 1984). Thus, in the Basin and Range, many geomorphic processes were profoundly affected by the Pleistocene-to-Holocene climatic transition.

Pulses of Eolian Activity

Eolian processes are particularly sensitive to conditions of climate and climatic change. Within the southwestern Basin and Range, several diverse geomorphic phenomena preserve a record of significant fluctuations in eolian activity during middle and late Quaternary time. (1) Climbing and falling dunes, stabilized by veneers of talus gravel and deeply trenched by ephemeral streams are common throughout the central and eastern Mojave Desert and the southern Great Basin (Smith, 1967; Smith, 1984; Whitney and others, 1985). Vertical sequences of buried soils within these sand ramps indicate alternating periods of eolian activity and relative inactivity. For example, climbing dunes along the southwest flank of the Sheep Hole Mountains in the south-central Mojave contain as many as seven laterally continuous buried soils (Smith, 1967), and a 0.24 to 0.73 Ma sand ramp on the northeast margin of the Amargosa desert in the south-central Great Basin contains a similar vertical sequence of buried soils (Whitney and others, 1985). (2) Buried soils also occur in accretionary mantles of eolian silt on dated lava flows of the Cima volcanic field in the eastern Mojave Desert. These soils indicate at least three periods of major eolian activity separated by periods of relatively inactivity during the past 0.5 to 0.6 m.y. (McFadden and others, 1984, 1986). (3) Evidence of fluctuations in eolian activity is also preserved by microchemical laminations in rock varnish on K-Ar–dated lava flows of the Cima and Coso volcanic fields.

Systematic oscillations in the Mn:Fe ratios of these varnish laminations are inferred to reflect past fluctuations in eolian alkalinity and suggest as many as 13 significant oscillations in eolian activity over the past 1.1 m.y. (Dorn, 1984b). Thus, pulses of eolian deposition have repeatedly affected the piedmonts and upland areas of the southwestern Basin and Range throughout most of Quaternary time.

The association of these pulses of eolian activity with times of significant climatic change is suggested by the coincidence of the late Pleistocene-to-Holocene climatic transition with what may have been the last major pulse of eolian deposition in the eastern Mojave Desert. Stratigraphic relations between eolian deposits, alluvial fan deposits, and ^{14}C dated shorelines along the west margin of Silver Lake playa north of Baker, California, document a brief but significant pulse of eolian deposition between approximately 10.5 and 9.5 ka (Wells and others, 1984b; 1987). In addition, deposits of eolian silt, as much as 1 m thick, occur on the surface of a latest Pleistocene lava flow in the nearby Cima volcanic field (Wells and others, 1984a, 1985); and Pleistocene soils within accretionary mantles on the older Cima lava flows are commonly buried by deposits of eolian silt, as much as 30 cm thick, that support a weakly developed Holocene soil (McFadden and others, 1984; 1986).

Hillslope Processes

Hillslopes are complex landforms, produced primarily by degradational processes, that form the dominant landscape element in most upland areas. In arid and semiarid environments, a typical hillslope profile consists of four main components: a relatively narrow upper convexity, a free face, a straight segment, and a broad, well developed lower concavity (Carson and Kirkby, 1972). Either the free face or the straight segment may be absent, but it is unusual for both of these components to be missing. The free face is generally considered to be gravity-controlled and loosening-limited, the straight segment either gravity-controlled or wash-controlled and transport-limited, and the basal concavity wash-controlled and transport-limited. For example, strong positive correlations between hillslope gradient and mean particle size on debris-covered straight segments and basal concavities in the Mojave Desert (Abrahams and others, 1985) indicates that these slope components are at least partly adjusted to hydraulic process (i.e., are wash-controlled). Comprehensive reviews of arid and semiarid hillslope process and morphology have been prepared by Wilson (1968), Carson and Kirkby (1972), Young (1972), and Cooke and Warren (1973).

As the primary degradational elements of upland areas, hillslopes are the ultimate source of nearly all erosional debris in arid and semiarid geomorphic systems. Consequently, the impacts of tectonic movement or climatic change on these systems are largely effected through changes in water discharge and sediment yield from hillslopes. Of particular relevance to this discussion are the general responses of hillslope systems to climatic change.

Decreasing effective moisture during the Pleistocene-to-Holocene transition, produced by a combination of temperature increase and (or) precipitation decrease, affected hillslopes by reducing the density of vegetation cover. This primary effect induced several self-enhancing responses within hillslope systems: (1) decreased vegetation densities increased surface runoff, which acted to increase soil erosion, reduce average soil thickness, and thus accelerate the decrease in vegetation densities; (2) decreased vegetation densities also decreased water infiltration, which acted to further reduce vegetation densities; and (3) increased soil erosion decreased average soil thickness and increased the area of impermeable surfaces, which acted to further increase both surface runoff and soil erosion (Bull, 1979, Fig. 7). All of these responses combined to rapidly strip upper hillslopes of their late Pleistocene mantle of soil and loose debris. Hillslope sediment yields increased dramatically during this stripping; however, sediment yields decreased just as dramatically upon completion of the stripping. The net result was a rapid change from vegetation- and soil-covered, transport-limited upper hillslopes with relatively low water discharges and low sediment yields to largely denuded, loosening-limited upper hillslopes with high water discharges and low sediment yields.

It is likely that this multiple response was enhanced by a concurrent increase in the flux of eolian silt and salts into hillslope systems. An increased salt flux would have accelerated the mechanical weathering of hillslope clasts and increased sediment availability on hillslopes (Wells and others, 1984b, 1987). Salt weathering is a highly effective physical weathering process on distal alluvial fan surfaces adjacent to the Death Valley salt pan (Goudie and Day, 1980); and salts in aerosolic dust are inferred to be an important factor in the physical weathering of basaltic boulders and cobbles in the Cima volcanic field in the eastern Mojave Desert (Wells and others, 1984a; 1987). Moreover, the accumulation of significant amounts of eolian silt on talus-mantled hillslopes may have promoted debris-flow activity (Wells and others, 1982). The high surface roughness of talus slopes provides an effective trap for eolian fines and, at the same time, promotes rapid infiltration of rainfall and slope wash. If rapidly saturated, a thick matrix of eolian silt within a talus mantle on a steep hillslope would provide the critical strength needed for a viscous slurry flow. Both of these eolian-initiated processes would have increased the effectiveness of hillslope stripping and exacerbated the response of hillslope systems to the Pleistocene-to-Holocene climatic transition.

Alluvial Fan Processes

Alluvial fans are ubiquitous landforms in piedmont areas of the Basin and Range. Coalescing fans form the piedmont slopes along most tectonically active range fronts and veneer the fringing pediments of most stable range fronts. Individual fans form generally fan-shaped cone segments that radiate downslope from the point where the fan-building stream crosses the mountain front. In most areas, the lateral extent of any individual fan is restricted by adjacent fans so that the typical planform of individ-

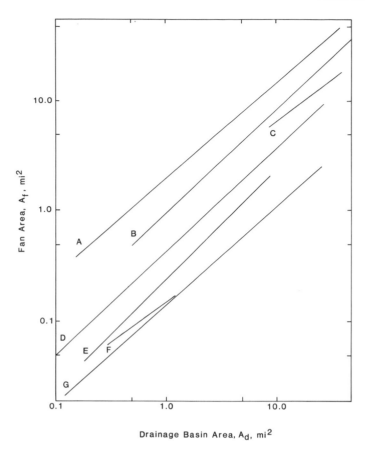

Figure 2. Fan area versus drainage basin area relations for various geographic locations and drainage basin lithologies (redrawn from Hooke, 1968, Fig. 3):
(A) $A_f = 2.1 A_d^{0.91}$; shale source, San Joaquin Valley (Bull, 1964).
(B) $A_f = 0.96 A_d^{0.98}$; sandstone source, San Joaquin Valley (Bull, 1964).
(C) $A_f = 1.05 A_d^{0.76}$; Death Valley, west side (Hooke, 1968).
(D) $A_f = 0.42 A_d^{0.94}$; Owens Valley (Hooke, 1968).
(E) $A_f = 0.24 A_d^{1.01}$; Cactus Flats (Hooke, 1968).
(F) $A_f = 0.16 A_d^{0.75}$; quartzite source, Deep Springs Valley (Hooke, 1968).
(G) $A_f = 0.15 A_d^{0.90}$; Death Valley, east side (Hooke, 1968).

ual fans is markedly elongate. Cross-fan profiles are typically convex; radial profiles are generally concave. Fan gradients in the western United States typically range between 0.5° and 7.5° (Anstey, 1965); however, gradients of as much as 25° occur along some tectonically active range fronts (Denny, 1965). Fan radii usually range between 0.5 and 10 km (Anstey, 1965), although fans with radii of as much as 20 km occur in some areas.

General Depositional Processes. The depositional processes operating on alluvial fans have been ably and comprehensively discussed by Hooke (1967, 1968), Bull (1972, 1977), and Nilsen (1982). Fans are formed by alluvial deposition resulting from changes in the hydraulic geometry of streamflow from confined mountain valleys to open piedmont areas. When a mountain stream leaves the confines of its valley to spread out on the adjoining piedmont surface, flow width increases, flow depth and

velocity decrease, flow discharge decreases (as surface water infiltrates into the permeable fan deposits), and deposition occurs (Bull, 1972, 1977). Two general processes deliver sediment to the fan surface: streamflow and debris flow. The relative importance of these processes varies greatly from one fan to another and may vary significantly through time on any individual fan. Some fans appear to be dominated by debris-flow processes (Blackwelder, 1928; Beaty, 1963). Sediments transported by streamflow can be deposited as (1) sheetflood deposits, where channelized flow spreads out at the downstream ends of fan channels; (2) channel deposits, where low-magnitude flow events backfill part of the distributary channel network; or (3) sieve deposits, where the fan surface is sufficiently permeable to allow the complete infiltration of stream discharge (Bull, 1972, 1977; Hooke, 1967). Sediments transported by debris flow are deposited wherever the basal shear stress exceeds the yield strength of the flow. This condition may be induced by water loss from the flow, a decrease in hydraulic gradient as the fan slope decreases, or a decrease in flow depth as the flow spreads laterally (Hooke, 1967). Thus, on fans where both streamflow and debris-flow processes are operative, the proportion of debris-flow deposits generally decreases downfan.

Fan Morphometry. Significant morphometric relations between alluvial fans and their source areas indicate that source area characteristics strongly influence fan morphology and process (Bull, 1968, 1977; Denny, 1965; Melton, 1965; Hooke, 1967, 1968; Hooke and Rohner, 1977). Fan area, A_f, is related to source basin area, A_s, by the equation:

$$A_f = cA_s^n, \qquad (1)$$

where c and n are regression constants that vary with a variety of source area characteristics, including lithology, fracture density, basin relief, tectonic stability, and climate (Bull, 1977; Hooke, 1968; Hooke and Rohner, 1977). Values of c typically range between 0.1 and 2.0 with lower values generally characteristic of resistant, blocky-weathering rock types and tectonically active range fronts, and higher values generally characteristic of readily erodible rock types, back-tilted piedmonts, and stable range fronts (Fig. 2). Values of n are typically less than unity, indicating an inverse relation between source area size and sediment yield (Hooke, 1968). Fan slope, S_f, is related to source basin relief, H, and source basin area, A_s, by the equation:

$$S_f = y (H/A_s^{0.5})^z; \qquad (2)$$

where the regression constants y and z also depend on source area characteristics (Melton, 1965). Values of y generally range between 5 and 10, and values of z generally range between 0.6 and 1.0. The parameter, $H/A_s^{0.5}$, is a dimensionless measure of basin ruggedness.

These general morphometric relations between fan area, fan gradient, source basin area, and source basin ruggedness indicate a strong link between overall fan morphometry and source basin discharge and sediment production. Larger source areas generate larger storm discharges, and the higher flow velocities and bed shear stresses associated with these larger discharges are able to transport on a lower slope the same caliber of material that smaller discharges can transport on a higher slope. Therefore, the greater quantities of sediments derived from larger source areas are transported farther from range fronts and form larger, more gently sloping alluvial fans (Hooke, 1968).

As source area discharge and sediment production characteristics change because of climatic, tectonic, evolutionary, or other influences, the general morphology of the associated alluvial fan will also change. Long-term changes, such as a progressive evolutionary change in source basin morphometry will likely affect source basin and fan morphometry as well as fan morphology; however, short term changes, such as an abrupt climatically induced change in source area discharge or sediment yield, will probably only superficially affect fan morphology. Therefore, although the morphometric relations between alluvial fans and their source basins provide an elegant conceptual basis for explaining overall fan morphology, this approach is not well suited for the analysis of short-term change in surficial morphology, particularly those induced by late Quaternary climatic changes.

Influence of Climatic Change on Fan Processes. Alluvial fans in stable environments of the Basin and Range (where rates of differential uplift and (or) base level change are very slow) typically exhibit a number of common surficial characteristics: (1) surface segmentation where the ages and slopes of the fan segments generally decrease in a downfan direction (e.g., Bull, 1977; Wells, 1978; Dohrenwend, 1982a); (2) a marked contrast in the surface morphology of the various ages of fan segments: younger surfaces usually exhibit a constructional bar and swale microtopography, intermediate surfaces are typically planar with well-developed stone pavements, and older surfaces are commonly degraded to ballenas (e.g., Bull, 1974; Wells, 1978; Hoover and others, 1981; Dohrenwend, 1982a); (3) a tendency for the dissection and partial stripping of proximal piedmont areas (e.g., Bull, 1977; Wells, 1978; Wells and others, 1984b, 1987); (4) thin veneers (generally less than 2 m thick) of Holocene and late Pleistocene alluvium in intermediate piedmont areas and the presence of laterally continuous buried soils beneath this veneer (Sharp, 1984; Wells and others, 1984b, 1987); (5) two distinct types of piedmont drainage, distributary drainage radiating from fan apices and subparallel tributary drainage arising on abandoned fan segments (e.g., Cooke and Warren, 1973; Wells, 1977; Dohrenwend, 1982a; Wells and others, 1984b, 1987); and (6) a general tendency for the tributary drainage to be more deeply incised than the distributary drainage (Denny, 1965; Wells, 1977). Similar surficial characteristics are generally shared by alluvial fans in active tectonic environments (e.g., Lustig, 1965; Denny, 1965, 1967; Hooke, 1967; Dohrenwend, 1982a; Ponti, 1985) and in areas of rapid baselevel change (e.g., Bull, 1974; Wilshire and others, 1983; Dohrenwend, 1985). It would appear, therefore, that either (1) different combinations of external influences elicit similar responses and effect conver-

gent morphologic trends within alluvial fan systems, or (2) the effects of climatic fluctuations have dominated alluvial fan systems, at least during the late Quaternary (e.g., Lustig, 1965; Christenson and Purcell, 1985; Ponti, 1985).

Effects of the Pleistocene-to-Holocene climatic transition on alluvial fans in the southwest Basin and Range are well documented by geomorphic, sedimentologic, and soils stratigraphic studies. A distinctive and ubiquitous bar and swale microtopography characterizes most Holocene fan surfaces but is absent or extremely subdued on nearly all late Pleistocene fan surfaces (Fig. 3). This conspicuous difference in surface morphology is most likely the combined result of two general influences: (1) the profound degradation of bar and swale microtopography on late Pleistocene fan surfaces by sheet flooding, surface creep, lateral flowage of saturated silt, soil shrink-swell, and salt splitting (Denny, 1965; Cooke, 1970a; Bull, 1974; Wells and others, 1984b, 1987; McFadden and Bull, 1987); and (2) the deposition of generally finer materials during the late Pleistocene than during the Holocene (Mayer and Bull, 1981; Mayer and others, 1984). Both of these influences are either directly or indirectly associated with the Pleistocene-to-Holocene transition.

General desiccation and deflation of pluvial lake beds and distal piedmont areas resulted in the deposition of silt- and salt-rich eolian veneers on late Pleistocene and early Holocene piedmont surfaces. The salts accelerated the splitting and general disintegration of larger surface clasts. The silts, a primary constituent of the vesicular Avk horizons that underlie most late Pleistocene and early Holocene stone pavements, comprise the medium for the surface creep, lateral flow, and shrink-swell processes that have acted to smooth these surfaces (Wells and others, 1984b, 1987; McFadden and others, 1984; McFadden and Bull, 1987).

In southwestern Arizona, the mean particle sizes of late Pleistocene and Holocene alluvial fan deposits are significantly different (Mayer and others, 1984). The mean particle sizes of late Pleistocene deposits are finer near source areas and decrease less rapidly with increasing distance from source areas than the mean particle sizes of Holocene deposits. Latest Holocene deposits are coarsest and exhibit the most rapid downstream decrease in mean particle size. These relations indicate a relatively wet late Pleistocene environment, where thick soils and extensive vegetation limited hillslope runoff and stream power, followed by an arid Holocene environment, where sparse vegetation and soils cause rapid hillslope runoff and high stream power. As a result, more competent Holocene streamflows have been able to transport coarser materials farther from source areas (Mayer and Bull, 1981; Mayer and others, 1984). The presence of relict sheetflood bedforms on late Pleistocene and early Holocene fan surfaces (Fig. 3B) supports this inference of greater stream power during early and middle Holocene flood events (Wells and Dohrenwend, 1985). Such flood events, which mobilized sediment with mean grain sizes of 2 to 8 mm on inactive fan surfaces, may also be at least partly responsible for the early and middle Holocene dissection of proximal piedmont areas.

Reconstruction of the late Quaternary history of the lower

Figure 3. Alluvial fans of the southwestern Basin and Range. A, At least four ages of Holocene alluvial fan surfaces showing conspicuous bar and swale morphology and progressive rock varnish development. Aerial view northeast across the lower eastern piedmont of the Soda Mountains, eastern Mojave Desert. B, Late Pleistocene fan surfaces with relict sheetflood bedforms on darkly varnished, interlocking stone pavements. Aerial view south across the lower eastern piedmont of the Soda Mountains, eastern Mojave Desert.

TABLE 1. ABBREVIATED GLOSSARY OF SOIL TERMINOLOGY*

Master Soil Horizons

A horizon	Formed at the surface and characterized by accumulation of organic matter intermixed with the mineral fraction; not dominated by properties characteristic of B horizons; in desert areas, pale horizons with little alteration or organic matter are considered A horizons.
B horizon	Formed below the A horizon; characterized by (1) destruction of most or all original rock and sedimentary structure, (2) illuvial concentration of silicate clay, iron, aluminum, carbonates, gypsum, or silica, (3) removal of carbonates, (4) residual concentration of sesquioxides, and (5) formation of pedogenic structure.
C horizon	Materials little affected by pedogenic processes and lacking in the properties of A or B horizons.

Subordinate Designations within Master Horizons

k	accumulation of carbonates
t	accumulation of silicate clay
v	presence of vesicular pores
w	cambic or color B horizon

*Modified from McFadden and Bull, in press.

Soda Mountains piedmont in the eastern Mojave Desert supports these interpretations (Wells and others, 1984b, 1987). Geomorphic, sedimentologic, and soils stratigraphic analyses of dated fan surfaces and related eolian and colluvial deposits documents the complex response of this relatively small hillslope-piedmont system to the Pleistocene-to-Holocene climatic transition. Increased sediment yields from destabilized hillslopes initiated the deposition of an early Holocene alluvial veneer across most of the piedmont surface. The preexisting late Pleistocene piedmont surface only locally escaped burial in a few widely scattered areas. Hillslope sediment yields subsequently decreased but hillslope discharges remained high. In addition, piedmont discharges increased, partly induced by a likely intensification of monsoonal conditions and related convection storm activity. Consequently, proximal piedmont areas were incised and partly stripped, and thin middle and late Holocene veneers (derived primarily from dissection of proximal piedmont areas) were deposited in intermediate and distal piedmont areas by both hillslope and piedmont drainage systems.

In summary, although the morphometry of alluvial fans in semiarid and arid environments is largely controlled by drainage basin characteristics, the surface morphology and stratigraphy of these fans has also been strongly influenced by climatic change. Climatically induced changes in hillslope-piedmont systems associated with the Pleistocene-to-Holocene transition were profound and widespread, and most fan surfaces today bear the conspicu-

ous overprint of this climatic change regardless of regional variations in tectonic activity or local conditions of baselevel change (Bull, 1978). From central Nevada to southern Arizona, the distinctive and nearly ubiquitous bar and swale topography of Holocene fans contrasts sharply with the smooth, pavement-covered surfaces of late Pleistocene fans.

INFLUENCE OF CLIMATIC CHANGE ON STABLE PIEDMONT SURFACES

The effects of climatic change can profoundly affect the formation, modification, and eventual destruction of stable piedmont surfaces. Once removed from the influence of active fluvial processes by channel incision, abandoned piedmont surfaces experience a period of relative stability characterized by the progressive development of soils, stone pavements, and rock varnish. The processes involved in the formation of these stable surface features are profoundly affected by changes in effective moisture, vegetation density, and eolian activity; all manifestations of climatic change. Thus, development of soils, stone pavements, and rock varnish on stable surfaces not only provides useful indicators of relative age but also preserves a record of the influence of climatic change on geomorphic processes in arid regions. Eventually, however, impermeable soil horizons form on these abandoned piedmont surfaces increasing surface runoff and enhancing their dissection, stripping, and, finally, destruction.

Soil Development

Soil development exerts a profound influence on the preservation and degradation of piedmont surfaces in the Basin and Range. During at least the middle and late Quaternary, climatic changes have strongly affected both the processes and rates of soil development in this region. Many soil constituents are derived from rainwater and eolian dust; and climatically controlled fluctuations in precipitation and dust influx have resulted in significant variations in the development of vesicular, argillic, and calcic soil horizons. In addition, most soils are polygenetic having experienced major alterations between arid and semiarid climatic conditions. The following brief discussion of soils development on stable piedmont surfaces in the southwestern Basin and Range is synthesized from McFadden (1982), McFadden and others (1984, 1986) and McFadden and Bull (1987).

Soil Profile Development. Soils on well-drained piedmont surfaces in the southwest Basin and Range reflect the dominant influence of an arid-to-semiarid climatic history. High temperatures and low precipitation limit soil moisture and produce shallow depths of leaching; sparse vegetation and high soil temperatures minimize organic-matter accumulation. The resulting soils possess: (1) thin, organic-poor, commonly silt rich and vesicular A horizons (Av); (2) strongly reddened argillic B horizons (in Pleistocene soils); and (3) secondary carbonate accumulations (McFadden, 1982; McFadden and Bull, 1987). An abbreviated glossary of soil terminology is provided in Table 1.

Holocene soils are weakly developed. Older Holocene soils typically possess well-developed Av horizons, thin discontinuous to continuous Bw horizons, and thin discontinuous carbonate coatings on clast sides and bottoms. The Av horizon is characterized by abundant vesicles lined with thin calcareous clay films. This horizon typically possesses a loamy texture and a coarse to very coarse prismatic structure (Fig. 4). The prominent polygonal network of vertical cracks that defines this prismatic structure indicates significant shrink-swell within the horizon. Ped tops and sides are commonly less red and less effervescent than ped interiors, and segregated secondary carbonate often forms micropendants or lamallae on ped bottoms. These features suggest intense leaching along vertical ped boundaries. Discontinuous Bw horizons occur beneath larger surface clasts on middle Holocene surfaces and very thin but continuous Bw horizons are usually present in soils on early Holocene surfaces. The presence of the Bw horizon indicates a moist, slightly alkaline microenvironment conducive to localized hydrolysis, oxidation, and iron oxyhydroxide accumulation. These processes accompany the progressive alteration of the Av ped interiors and represent the first stages of genetic B horizon development (McFadden, 1982; McFadden and others, 1984, 1986; McFadden and Bull, 1987).

Pleistocene soils are moderately to strongly developed. In addition to the almost ubiquitous Av horizon, these soils possess clay-rich, reddish brown argillic B horizons and Stage II to Stage III calcic horizons (calcic horizon classification is summarized in Table 2). On uneroded piedmont surfaces, the thickness, redness, and clay content of the argillic B horizon increases with increasing surface age. Pedogenic iron oxyhydroxides progressively accumulate in this horizon, partly as a result of the weathering of minerals such as biotite and hornblende. The strong red color indicates that hematite is the principal iron oxyhydroxide. The rate of iron oxyhydroxide accumulation relative to clay accumulation is greater in semiarid soils than in arid soils largely because of more intense weathering and less accumulation of aerosolic dust in semiarid soils. Moreover, the more strongly leached soils of semiarid thermic regions contain only weakly to moderately calcareous Bk horizons; whereas secondary carbonate may completely plug the Bk horizons in soils of arid hyperthermic regions, and Stage IV petroclacic horizons are locally preserved on partly stripped piedmont surfaces of early Pleistocene and (or) Pliocene age (McFadden, 1982; McFadden and others, 1984, 1986; McFadden and Bull, 1987).

Influence of Climatic Change on Soil Development. It is clear that climatic change has significantly affected soil development in the southwestern Basin and Range. Climatic changes have altered soil water balance, thereby changing the intensity and depth of leaching and the depths of carbonate and clay translocation. For example, older Pleistocene soils in semiarid areas of the Mojave Desert lack thick, well-developed calcic horizons; however, early Holocene soils in these same areas typically possess Stage II calcic horizons, indicating that calcic horizons form relatively rapidly under present climatic conditions. If the older Pleistocene soils had been exposed to a climatic

A

B

Figure 4. Vesicular A horizon (Av). A, Side view showing the coarse prismatic structure of the Av peds; B, Top view (with the stone pavement removed) showing the polygonal geometry of the peds.

regime similar to the Holocene throughout the history of their development, they should possess the strongly developed calcic or petrocalcic horizons that occur in soils of similar age elsewhere in the southwest (Machette, 1985). Therefore, past climatic conditions (relatively moist with relatively little eolian activity) are clearly reflected in the development of these soils (McFadden and Bull, 1987). Soils in accretionary mantles on dated lava flows of the Cima volcanic field indicate similar climatic influence. The nature of the latest Pleistocene and (or) Holocene soils in this area indicates conditions of shallow leaching and relatively high influx of eolian silt, silicate clay, carbonate, and soluble salts. In contrast, the nature of the older Pleistocene soils indicates a long period of surface stability with limited eolian influx and a leaching regime that was sufficiently intense to induce significant chemical alteration and deep translocation of carbonate and soluble salts (McFadden and others, 1984, 1986).

In addition, climatic changes have induced substantial variations in eolian activity that have caused major fluctuations in the influx of eolian fines and salts to soils systems. For example, three major stages of soil development related to at least three major pulses of eolian activity can be recognized in multiple loess mantles on dated lava flows of the Cima volcanic field (McFadden and others, 1984, 1986). Weakly developed soils occur on flows as young as 0.015 Ma and as old as 0.14 Ma, indicating that the youngest loess mantle is less than 15,000 yrs old and represents the only loess event of sufficient magnitude to bury large areas of the Cima field during the past 0.14 m.y. Well-developed soils, buried beneath the youngest loess mantle on 0.25-Ma and older flows, record a stable period of soil development with relatively little eolian influx on a 0.14- to 0.25-Ma loess mantle. Similar stratigraphic relations indicate the presence of at least one additional major loess event between 0.58 and 0.70 Ma (McFadden and others, 1984, 1986). During these periods of major eolian activity, eolian fines and salts would have quickly infiltrated into the permeable alluvial deposits and gravelly soils of piedmont areas thus accelerating soil development in these areas.

Secondary Carbonate Accumulation

The following brief discussion of secondary carbonate accumulation on stable piedmont surfaces in the Basin and Range is generalized from Lattman (1973), Bachman and Machette (1977), Gile and others (1979), and Machette (1985).

Accumulations of secondary carbonate are common on stable piedmont surfaces in those areas where predominantly warm-dry climates have prevailed during most of the Quaternary. Within the Basin and Range Province, rates of carbonate accumulation vary widely and are strongly influenced by climatic, lithologic, and topographic factors. In most cases, the carbonate originates from sources external to the piedmont, primarily solid carbonate in eolian dust and silt and Ca^{2+} dissolved in rainwater (Gile and others, 1979). (Obvious exceptions to this generalization are accumulations formed on calcareous parent materials dominated by limestone, dolomite, or carbonate-cemented sandstone clasts.) Two general types of carbonate accumulations can occur, pedogenic and nonpedogenic. Pedogenic accumulations, or calcic-soil horizons, form within soil profiles by the precipitation of carbonates originally present in rainwater or leached by the rainwater from eolian fines deposited in upper soil horizons. Continued accumulation generally proceeds from coatings on pebbles, to segregated filaments and nodules, to massive accumulations that gradually plug the calcic horizon (Table 2). Eventu-

TABLE 2. STAGES OF CALCIUM CARBONATE MORPHOLOGY IN CALCIC SOILS AND PEDOGENIC CALCRETES DEVELOPED IN NONCALCAREOUS PARENT MATERIALS UNDER ARID AND SEMIARID CLIMATES OF THE AMERICAN SOUTHWEST[*]

Stage	Diagnostic Morphologic Characteristics
I	Thin discontinuous coatings on clasts, usually on undersides; faint coatings on ped faces and a few filaments in soil; coatings and filaments are sparse to common.
II	Continuous, thin to thick coatings on tops and bottoms of clasts; soft, 0.5 to 4 cm nodules; some carbonate in matrix; coatings and nodules are common.
III	Massive accumulations between clasts; many coalesced nodules; continuous dispersion of carbonate in matrix (K fabric); matrix is moderately to firmly cemented.
IV	Indurated, 0.2- to 1.0-cm-thick laminae in upper part of Km horizon; cemented platy to weak tabular structure; laminae may drape over fractured surfaces; Km horizon is 0.5 to 1 m thick.
V	Dense, indurated laminae >1 cm thick; thin to thick pisolites; laminar carbonate coats vertical faces and fractures; strong platy to tabular structure; Km horizon is 1 to 2 m thick.
VI	Multiple generations of laminae, pisolites, and breccia; recemented, indurated and dense; many case hardened surfaces; thick, strong tabular structure; Km horizon is commonly >2 m thick.

[*]Modified from Machette, 1985, Table 1.

ally, over hundreds of thousands to millions of years, dense pedogenic calcretes containing more than 50% $CaCO_3$ can develop (Machette, 1985). Nonpedogenic accumulations, including gully-bed cementation, laminar layers on interfluves and hillslopes, and case-hardened surfaces are usually associated with lateral surface or subsurface water flow (Lattman, 1973). This process involves the evaporative concentration of Ca^{2+} to levels of supersaturation and the subsequent precipitation of $CaCO_3$ in areas of intermittent surface flow (e.g., sheet wash, rill wash, gully flow, etc.), temporary surface ponding, or where subsurface flow either discharges or reaches a near-surface position (Lattman, 1973; Machette, 1985).

Controls on Carbonate Accumulation. In stable piedmont areas, pedogenic carbonate accumulation is primarily controlled by (1) the supply of Ca^{2+} ions to the piedmont surface and (2) the amount of moisture available to translocate the Ca^{2+} ions into the subjacent soil (Machette, 1985). The combined effect of these two controls can be summarized in a general descriptive model that portrays the rate of carbonate accumulation as a two-dimensional space defined in terms of Ca^{2+} influx and moisture availability (Fig. 5; Machette, 1985). This space is divided into moisture-limited and influx-limited areas: moisture-limited conditions occur where the Ca^{2+} influx exceeds the leaching capacity of the available moisture, influx-limited conditions occur

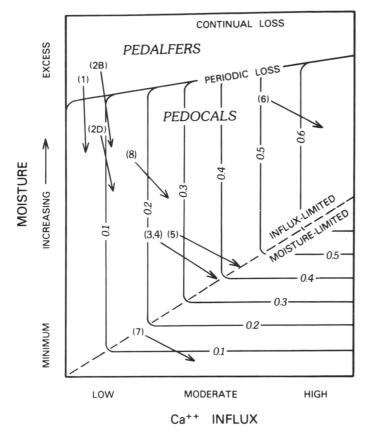

1. *Buena Vista, CO*   5. *Las Cruces, NM*

2. *Boulder-Denver, CO*   6. *Roswell-Carlsbad, NM*

3. *Albuquerque, NM*   7. *Vidal Junction, CA*

4. *San Acacia, NM*   8. *Beaver, UT*

Figure 5. Model of carbonate accumulation rates under varying conditions of moisture and Ca^{2+} influx (rates in g of $CaCO_3/cm^2/10^3$yrs shown by vertical and horizontal lines). Soil data for eight chronosequences are plotted under the conditions and rates of accumulation inferred for pluvial episodes; arrows show inferred conditions during interpluvial episodes (2B = Boulder, CO; 2D = Denver, CO). Most areas have higher accumulation rates during interpluvials; an exception, Vidal Junction in the eastern Mojave Desert (7), has lower accumulation rates during interpluvials (from Machette, 1985, Fig. 5).

where the leaching capacity is greater than the Ca^{2+} influx. The boundary between these two conditions represents a balance between Ca^{2+} supply and moisture availability. Changes under moisture-limited conditions are largely independent of variations in Ca^{2+} influx; conversely, changes under influx-limited conditions are essentially independent of fluctuations in moisture availability. In any given area, changes that decrease moisture and (or) increase Ca^{2+} influx, such as occurred during the Pleistocene-to-Holocene climatic transition, may induce either an increase or a

decrease, or relatively little change, in the local rate of carbonate accumulation (depending on the magnitude of the imposed climatic change and the original position within the moisture/influx space of the model). In central and southern New Mexico, for example, increased accumulation rates, caused by increased Ca^{2+} influx under influx-limited conditions, were probably little affected by decreased rainfall. However, in low elevation areas of the eastern Mojave Desert, rates of accumulation probably decreased in spite of an increase in Ca^{2+} influx, because generally influx-limited conditions during the late Pleistocene changed to the strongly moisture-limited conditions of the Holocene (Machette, 1985).

In addition to climatic influences, the accumulation of secondary carbonate is also strongly controlled by lithologic, geographic, and topographic factors. Carbonate accumulations on the piedmonts of south-central and southern Nevada are very strongly controlled by the composition of the piedmont detritus: (1) essentially all fans dominated by carbonate detritus contain significant secondary carbonate accumulations, and older fan surfaces exhibit strongly developed calcretes; (2) fans of noncarbonate sedimentary rock detritus generally show less well-developed accumulations; (3) fans of basic igneous rock detritus exhibit a wide variation in carbonate accumulation from extremely well-developed to very weakly developed; and (4) fans of acidic igneous material are, at most, almost always very weakly cemented (Lattman, 1973). These relations are generally consistent throughout most of the Mojave Desert as well.

Carbonate accumulation on noncalcareous fans is strongly influenced by the proximity of external sources for eolian-borne carbonate dust and silt. In southern Nevada, the most strongly developed carbonate accumulations on noncalcareous fans occur downwind of playas that are flanked on their upwind sides by ranges composed of carbonate bedrock (Lattman, 1973). In the Mojave Desert, a similar association occurs in areas downwind of the Manix basin (Hale, 1985); and the well-developed calcic horizons of cumulic soils in eolian silt on lava flows of the Cima volcanic field are situated immediately downwind of the extensive playas of Soda and Silver lakes (McFadden and others, 1984, 1986).

The wide variability of carbonate accumulation, noted by Lattman (1973), on piedmonts composed of basaltic and (or) andesitic debritus is partly explained by the effectiveness of this material as an eolian carbonate trap. The characteristic surface roughness of basalt and andesite clasts promotes eolian deposition and the higher maximum surface temperatures occurring on these clasts enhance $CaCO_3$ precipitation (Wells and Schultz, 1979). Thus, under influx-limited conditions, the presence of cobble- to boulder-sized clasts of basic igneous rock can induce the development of dense laminated calcretes in areas of high eolian carbonate flux, but only relatively weakly developed accumulations in areas of low eolian carbonate flux. The strong contrast in carbonate cementation on the windward versus leeward piedmonts of the McCullough Range south of Las Vegas (Lattman, 1973) serves as a compelling example of this effect.

Local topographic factors exert a strong influence on the type of secondary carbonate accumulation, particularly where accumulation rates are high (Lattman, 1973). Gully-bed cementation is generally limited to the floors and lower channel walls of piedmont drainageways. Other laminar layer accumulations may form on gently to steeply sloping interfluve areas; however, these accumulations are usually best developed on broad, gently sloping interfluves and in broad interfluve depressions. In these areas, the laminar layers commonly overlie the relatively impermeable Avk horizon. Case hardening forms on surface exposures of carbonate dominated detritus and of calcic soil horizons. Case-hardened surfaces range in slope from approximately horizontal to vertical but are most conspicuous on the steeply sloping walls of gullies and arroyos (Lattman, 1973).

Weathering of Secondary Carbonate Accumulations. In arid and semiarid environments, secondary carbonate accumulations weather primarily by mechanical breakup (Lattman, 1983). Large angular fragments of calcrete and cemented gravel (the caliche rubble of Lattman, 1973) form distinctive mantles on older piedmont surfaces capped by well-indurated carbonate accumulations (Lattman, 1973, 1977) and on the degraded surfaces of Pliocene basalt flows (Wells and others, 1984a; McFadden and others, 1984). Locally, these rubble mantles have been recemented into layers of calcrete breccia. Freezing and thawing of water in fractures within surface exposures or near-surface accumulations of calcrete is considered to be the dominant mechanism for this breakup (Lattman, 1973, 1977). Long-term weather data for basin-floor elevations in southern Nevada document an average of 35 freeze-thaw cycles from December through February, and widespread precipitation occurs most frequently during these same winter months. Moreover, well-indurated calcretes are relatively impermeable and, therefore, only superficially affected by solution. Softer, less well indurated accumulations may be more susceptible to solution weathering; however, these poorly-indurated carbonate layers typically become case hardened (to as much as 0.5 m) upon exposure and subsequently respond to weathering in the same manner as other well-indurated calcretes (Lattman, 1977). Thus, mechanical breakup by water freeze and thaw probably dominates the degradation of exposed secondary carbonate accumulations in arid and semiarid environments of the Basin and Range.

Stone Pavement Development

Stone pavements are armored surfaces of pebble, cobble, and boulder veneers, one to two clasts thick, that overlie deposits of sand, silt, and clay. They range from flat, well-sorted, tightly interlocking cobble mosaics (Fig. 6) to irregular, poorly sorted, noninterlocking pebble to boulder veneers. Pavements occur in a variety of relatively stable geomorphic environments where vegetation is sparse, but they are particularly prominent and abundant in hot desert regions. In the Basin and Range, they are developed in such diverse geomorphic settings as abandoned alluvial surfaces, stabilized sand sheets and sand ramps, eolian mantles on lava flow surfaces, and residual mantles on gentle hillslopes.

Until recently, stone pavements have been viewed primarily as the residual lags of eolian deflation. Indeed, as Cooke (1970a, p. 561) points out, "the deflation hypothesis has become enshrined as a general explanation of pavement formation in almost every geomorphological text published in this century." However, an increasing body of evidence indicates that most stone pavements, in the southwestern United States at least, have not been formed by deflation.

According to Cooke (1970a), pavement formation involves a diverse variety of geomorphic processes that operate at different levels of relative importance as the pavements develop through time (Fig. 7). These processes include upward migration of particles under the influence of freezing and thawing, alternate solution and recrystallization of salts, cycles of saturation and desiccation, raindrop impact, soil wash, creep, and, in areas where loose fine material is concentrated at the surface, deflation. Most stone pavements become established in relatively stable areas (such as abandoned alluvial fan surfaces) where the processes of stone sorting, surface creep, and clast disintegration are dominant over concentrated wash processes. Weathering, creep, and sheetwash processes effectively smooth the original bar and swale microtopography of the fan surface (McFadden and Bull, 1987). These processes remain active after pavement formation but effect little change on well-established pavements. Disintegration of surface clasts by physical weathering, particularly salt weathering, is sensitive to climatically induced variations in the flux of eolian fines. However, the importance of this process probably decreases with time as pavement clasts are progressively reduced in size by earlier disintegration.

Salt Weathering. Mechanical disintegration of surface clasts by salt splitting is an important process in the smoothing of younger pavement surfaces. Aerosolic salts are a significant component of the soils beneath most stone pavements, particularly in lower piedmont areas where their abundance is indicated by high soil electroconductivities and the presence of secondary gypsum (McFadden, 1982; Wells and others, 1987). Intense salt weathering of surface clasts is especially conspicuous near the margins of saline playas (Goudie and Day, 1980) and along the late Pleistocene shorelines of pluvial lakes (Wells and others, 1984b, 1987). Medium- to coarse-grained plutonic and metamorphic rocks are particularly susceptible to this process, whereas fine-grained lithologies are relatively resistant. Consequently, older pavements are typically depleted in coarse-grained clasts and enriched in fine-grained clasts (McFadden and Bull, 1987; Wells and others, 1987).

Vertical Sorting of Pavement Clasts. The importance of vertical stone sorting as a primary process of pavement formation is suggested by an associative relation between stone pavements and the vesicular A horizons of desert soils (Springer, 1958; Cooke, 1970a; Howard and others, 1977; McFadden and others, 1984, 1985, 1986). In the Basin and Range, vesicular A horizons, composed of essentially clast-free silt and clay, are present beneath most stone pavements. Moreover, the better-sorted, more tightly interlocking pavements are usually underlain by the

Figure 6. Stone pavements in the southwestern Basin and Range. A, Well-developed interlocking pavement of basaltic clasts overlying a 1- to 3-m-thick eolian mantle on the surface of a 0.85-Ma lava flow, Cima volcanic field, Mojave Desert. The absence of topographic highs on the flat surface of the eolian mantle and the absence of a buried pavement at the top of a well-developed buried soil within this eolian mantle indicate that the pavement was formed by vertical lifting. B, Tightly interlocking stone pavement dominated by limestone and dolomite clasts on a late(?) Pleistocene fan surface, western piedmont of the Nopah Range, southwestern Great Basin.

thicker, more strongly developed vesicular horizons. The concentration of coarse clasts in surface pavements and the absence of coarse clasts in the subjacent vesicular horizons suggests that the pavement clasts have moved upward through the vesicular horizon to the surface. The efficacy of this process has been demonstrated by laboratory experiments in which pebbles have been induced to migrate upward through mixtures of silt and clay in response to repeated cycles of wetting and drying (Springer, 1958; Cooke, 1970a). Upward migration of pavement clasts is also suggested by field situations where pavements have reformed in areas that were artificially stripped of their original surface pavements. For example, a stone pavement on an abandoned road south of Barstow in the central Mojave Desert is sufficiently well developed that "the margin of the road is difficult to discern" because the pavement "continued all but unbroken from the road onto the surrounding" undisturbed surface (Elvidge, 1982, p. 347). Vertical lifting is also necessary to explain the presence of interlocking cobble to boulder pavements on eolian deposits of silt and fine sand that bury the surfaces of Quaternary lava flows in the Mojave Desert (Wells and others, 1984a, 1985). Pavement-covered eolian deposits as much as 1 m thick mantle the crests of topographic highs on these flows (Fig. 6A), a situation where vertical lifting of clasts would appear to be the only possible mechanism for pavement formation.

The primary mechanism for upward migration of pavement clasts involves the repeated expansion and contraction of the Avk horizon (Springer, 1958). "Through swelling of the soil, the stone is lifted slightly. As the soil shrinks, cracks are produced around the stone and within the soil. Because of its large size the stone cannot fall into the cracks, but finer particles may either fall or be washed into the cracks. The net effect is an upward displacement of the stone. If this process is repeated many times by wetting and drying or freezing and thawing, the stone may eventually be brought to the surface" (Springer, 1958, p. 65). The shrink-swell potential of vesicular A horizons is indicated by (1) the characteristic vertical polygonal cracking of these horizons and associated doming of the ped surfaces (McFadden and others, 1984, 1986; McFadden and Bull, 1987); and (2) x-ray diffraction analyses of vesicular A horizons collected from several localities in southeastern California that reveal the presence of a variety of expanding lattice clays, particularly of the mortmorillonite group (Howard and others, 1977).

Syndepositional Lifting of Surface Clasts. All of these findings indicate that upward migration of subsurface clasts through existing soil horizons is a viable process for pavement development. However, studies of soil formation in eolian mantles on lava flows of the Cima volcanic field show that this process is not necessarily required for the formation and maintenance of pavements. Soil-stratigraphic relations demonstrate that entire preexisting Pleistocene stone pavements on these eolian mantles have been gradually lifted above the Pleistocene soil by Holocene accumulation of eolian fines beneath pavement clasts (McFadden and others, 1985, 1986). Subsurface incorporation of eolian fines via the well-developed vertical crack systems within

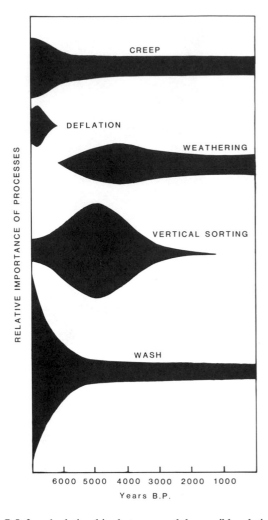

Figure 7. Inferred relationships between, and the possible relative importance of, pavement-forming processes through time at pavement sites in the Panamint Valley and Calico Mountain areas of the California desert (based on the general relations shown in Cooke, 1970a, Fig. 14).

the Avk horizon during dry periods alternates with crack closure and compressional doming of the Avk peds during wet periods. This process has enabled the progressive development of cumulic soil horizons by gradual incorporation of eolian material, while at the same time continually maintaining preexisting stone pavements at the land surface (McFadden and others, 1985, 1986).

In either case, whether produced by vertical stone sorting through a preexisting soil horizon or by continual vertical lifting at the land surface, it is apparent that most stone pavements are not the product of eolian deflation. Rather, they are produced by the *accumulation* of eolian fines and the concomitant lifting of surface clasts by the expansion and contraction of these fines during repeated cycles of wetting and drying.

Because of the apparent genetic association between Avk soil horizons and stone pavements, at least in the southwestern Basin and Range, it would appear that the rate of pavement development on natural unpaved surfaces is largely a function of Avk horizon development. Well-developed pavements have reformed in less than 25 years on surfaces where the original pavements were artificially stripped but the underlying soil horizons were left largely undisturbed (Elvidge, 1982). However, both pavements and Avk horizons are generally absent or only very weakly developed on late Holocene alluvial fan surfaces and become progressively better developed with age on middle Holocene to late Pleistocene surfaces (Wells and others, 1984b, 1987; McFadden and Bull, 1987). Thus, because pavement development is closely associated with Avk horizon development and because Avk horizon development is at least partly controlled by eolian activity, which is, in turn, closely linked to climatic change, it would appear that stone pavement development is also linked to the effects of climatic change.

Rock Varnish Formation

The phenomenon of rock varnish has been a subject of intense geomorphic study and controversy for more than 100 years. As a result, an imposing body of literature concerning rock varnish has been developed. This literature has been ably and comprehensively summarized by Dorn and Oberlander (1981b, 1982) and the following brief discussion has been abstracted from their work.

Rock varnish is a chemical coating, commonly 10 to 30 microns thick, that accretes on rock surfaces in many diverse terrestrial environments, ranging from arctic and alpine settings to a variety of humid environments. However, it is usually best developed and is certainly most conspicuous in hot deserts; consequently, it is often referred to as desert varnish. Rock varnish is comprised mainly of clay minerals and oxides of manganese and iron with lesser amounts of detrital SiO_2 and $CaCO_3$ and trace elements including Mg, Ca, K, Na, Ti, and Cu. These constituents are derived primarily, if not entirely, from sources external to the underlying rock. The "classic" rock varnish, the type that has received the most attention in the literature, is a dark manganese-rich coating that commonly displays a conspicuous lustrous sheen; however, orange to dusky brown varnishes are also widespread, particularly in hyper-arid environments.

Several attributes of rock varnish are particularly relevant to a discussion of the influence of climatic change on arid geomorphic processes. These include the external source of varnish constituents, the process of varnish formation, varnish micromorphology, and chemical microlaminations within varnish coatings.

External Origin of Varnish Constituents. The external origin of the constituents of rock varnish is now well established. (1) Clay minerals make up as much as 70% of the dark manganese-rich varnishes and as much as 90% of the orange to brown manganese-poor varnishes (Potter and Rossman, 1977, 1979). (2) Varnish may coat a weathering rind or may form in direct contact with unaltered host rock (Allen, 1978; Elvidge, 1979). (3) Moreover, many varnish substrates are almost completely devoid of manganese (Elvidge, 1979; Dorn and Ober-

lander, 1982); and even when present, manganese fails to show any evidence of a concentration gradient that would suggest translocation from the substrate into the varnish coating (Allen, 1978). (4) Indeed, alternating Mn-rich (20 wt% MnO_2) and Mn-poor (3 wt% MnO_2) laminations are common in many varnishes (Perry and Adams, 1978; Dorn and Oberlander, 1981b, 1982; Dorn, 1984b). This evidence indicates that airborne dust is the principal source of most rock varnish constituents (Allen, 1978; Elvidge, 1979; Elvidge and Moore, 1979; Dorn and Oberlander, 1981a, 1981b, 1982; Dorn, 1984b).

Microbial Origin of Varnish. The microbial origin of rock varnish is also well documented (Perry and Adams, 1978; Dorn and Oberlander, 1981a, 1981b, 1982). Manganese-rich rock varnishes have been produced by laboratory cultures of manganese-fixing bacteria collected from varnish-covered rocks (Dorn and Oberlander, 1981a). In addition, both field and laboratory observations indicate that varnish develops by the nucleation of microscopic patches that grow both laterally and vertically until they coalesce into a uniform dark coating (Perry and Adams, 1978; Dorn and Oberlander, 1981a). The resulting varnish micromorphology is often characterized by botryoidal microlaminated structures reminiscent of algal stromatolites. Moreover, microbial fixing of rock varnish is consistent with the general occurrence of rock varnish in desert areas. (1) Rock surfaces in deserts are usually low in organic nutrients and thus are ideal for slow-growing Mn-oxidizing bacteria that would be outcompeted in richer environments. (2) Mn-rich varnish is often very well developed where water intermittently flows over rock surfaces; sites favorable to microbial colonization. (3) Sites of Mn-rich varnish have near-neutral pH levels that are favorable for Mn-oxidizing microbes; whereas sites of Mn-poor orange varnish have high pH levels that are not favorable for these bacteria (Dorn and Oberlander, 1981a).

The microbial origin of rock varnish also helps to explain the widely varying rates of varnish formation in different environments. In laboratory experiments where limiting environmental factors were absent or closely controlled, varnishes have formed in less than six months (Dorn and Oberlander, 1981a). In periglacial and arctic environments, varnishes have been observed to form in 40 years, and in riverine environments, varnishes have formed in less than 100 years (Dorn and Oberlander, 1982). In arid environments, however, varnishes form much more slowly, and progressive varnish development with increasing age is conspicuous on Holocene to late Pleistocene alluvial surfaces (Bull, 1974; Wells and others, 1984b, 1987).

Varnish Micromorphology. That the general process of varnish formation, microbial fixing of eolian dust, is sensitive to conditions of climate and climatic change should hardly be surprising. Indeed, this sensitivity is indicated in many varnishes by the preservation of what appears to be a faithful record of climatic change (Dorn and Oberlander, 1982; Dorn, 1984a, 1984b). The surface micromorphology of most rock varnishes in desert areas can be fitted into a continuum that varies from lamellate to botryoidal (Fig. 8). Clay-rich varnish with a lamellate

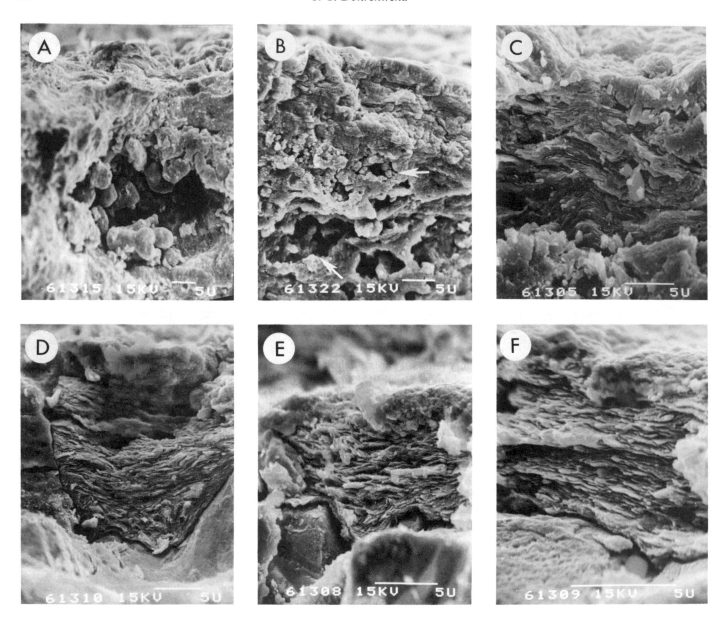

Figure 8. Varnish micromorphological stratigraphy from localities along the west side of Silver Lake, eastern Mojave Desert. A, Well-developed botryoids under lamellate structures in varnish from a late Pleistocene pavement surface (estimated at 36.6 ± 4.5 ka by cation-ratio dating). Sample was collected from an area well above the 10 to 15.5 ka highstand of Lake Mojave. B, Varnish from the same late Pleistocene surface but collected from an area closer to the high Lake Mojave shoreline. The subsurface botryoids (upper arrow) are less well developed, suggesting a slightly more dusty paleo-eolian microenvironment. Lower arrow identifies the varnish-rock contact. C, and D, Varnish with an entirely lamellate micromorphology on two cobbles from the high Lake Mojave shoreline. E, and F, Varnish from a Holocene alluvial surface (estimated at 6.6 ± 0.4 ka by cation-ratio dating) also has an entirely lamallate micromorphology (from Dorn, 1984a, Fig. 7).

micromorphology develops when the rate of oxide accumulation at varnish nucleation points is less than the rate of clay accumulation. Conversely, clay-poor varnish with a botryoidal micromorphology develops when the rate of oxide accumulation is greater than the rate of clay accumulation. Thus, lamellate varnish indicates a relatively high eolian dust flux and botryoidal varnish indicates a relatively low dust flux; and variations between lamellate and botryoidal micromorphology appear to record long-term temporal variations in eolian activity. For example, varnishes collected from constructional rock surfaces on dated late Pleistocene lava flows in the Cima volcanic field usually display a microstratigraphy of lamellate over botryoidal morphology, thus indicating a change from less dusty to more dusty (and presumably more arid) conditions (Dorn, 1984a). A similar microstratigraphy occurs on varnished pavement clasts of late Pleistocene fan surfaces that are truncated by the 15.5- to 10-ka shoreline of

Silver Lake; however, varnish collected from this shoreline and from adjacent Holocene surfaces cut across this shoreline are entirely lamellate (Dorn, 1984a; Wells and others, 1984, 1987). Thus, it would appear that varnish micromorphology preserves a record of the Pleistocene-to-Holocene climatic transition and, therefore, is potentially useful for the differentiation of late Pleistocene and Holocene surfaces.

Varnish Microchemical Laminations. Alternating Mn-rich and Mn-poor microlaminae in rock varnish were first described by Perry and Adams (1978). The Mn-oxidizing bacteria that form varnish are inhibited by alkalinity, and contemporary Mn-rich varnishes possess near-neutral pH levels, whereas Mn-poor varnishes possess alkaline pH levels. Therefore, these microlaminations probably reflect changes in the alkalinity of the eolian environment. Microprobe profiles of Mn:Fe ratios in varnish collected from exposed constructional surfaces on K-Ar–dated lava flows of the Coso and Cima volcanic fields demonstrate a systematic relation between the number of microchemical laminations and the age of the underlying volcanic rock (Dorn, 1984b, Figs. 1 and 2). Five to six laminations occur in varnish on 0.135- to 0.14-Ma-rocks and as many as 25 laminations occur in varnish on 1.1-Ma-rocks (Fig. 9). Presumably, at least 13 oscillations in eolian alkalinity have occurred during the last 1.1 m.y. These oscillations have been tentatively interpreted as cycles of relatively low effective moisture with low vegetation densities and dry playas followed by relatively high effective moisture with higher vegetation densities and shallow lakes (Dorn, 1984b).

In summary, the process of rock-varnish formation appears to be strongly linked to climatic conditions and, therefore, to climatic change. Variations in effective moisture may affect varnish formation directly through changes in the activity of the varnish-forming bacteria and indirectly through changes in the

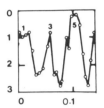

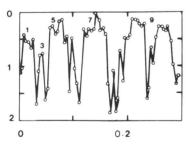

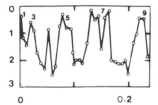

Figure 9. Microchemical laminations in rock varnish on volcanic deposits in the Coso Range as determined by electron microprobe analyses. The distance between each microprobe data point is 2 μm. However, the horizontal scales of each section have been normalized to the K-Ar age of the underlying volcanic rock. The peaks (low Mn:Fe ratios) have been assigned an arbitrary notation of sequential odd numbers indicating probable periods of more alkaline eolian conditions. The troughs (high Mn:Fe ratios) reflect less-alkaline conditions (modified from Dorn, 1984b, Fig. 1).

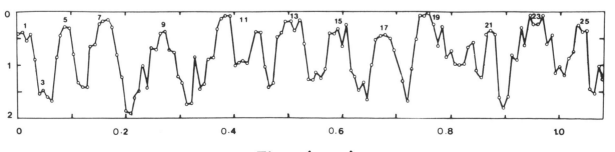

Time (m.y.)

eolian flux that supplies the varnish constituents. The net result is a climatically sensitive process that appears to preserve a general record of climatic change in the southwestern Basin and Range and provide a potentially powerful means for dating stable Quaternary surfaces in this arid region.

DEGRADATIONAL EVOLUTION OF TECTONIC AND VOLCANIC LANDFORMS

One of the most fruitful areas of current geomorphological research in the Basin and Range Province concerns the degradation of tectonic and volcanic landforms. Within the Basin and Range, a variety of these landforms of widely varying scale comprise a substantial proportion of the landscape (e.g., fault scarps, cinder cones, lava-flow surfaces, pediments, range fronts, and at the regional scale, the very mountain ranges and intermontane basins themselves). These landforms range widely in age and are situated within a variety of physiographic, geologic, tectonic, and climatic settings. Analysis of carefully selected samples of these landforms (where most geomorphically significant influences are constrained within narrow limits) allows the systematic study of landform degradation through tens of thousands to millions of years and provides significant insight into the types, rates, and magnitudes of processes responsible for that degradation. To date, such studies have focused on the degradation of (1) fault scarps in unconsolidated alluvium, (2) fault-bounded range fronts, (3) cinder cones and lava flow surfaces, and (4) pediments.

Fault-Scarp Degradation

Fault scarps in unconsolidated surficial deposits comprise a large and readily definable population of relatively simple landforms that are particularly amenable to quantitative geomorphic analysis. The first descriptions of fault scarps in unconsolidated alluvium within the Basin and Range Province were presented by G. K. Gilbert in his classic monograph on Lake Bonneville (1890). However, detailed analysis of these landforms did not begin until nearly 90 years later.

Descriptive Model of Scarp Degradation. Using qualitative field observations and longitudinal profile measurements along a large number of Quaternary fault scarps across 10,000 km^2 of the north-central Great Basin and photographs of original scarp morphology of surface faulting associated with the 1954 Fairview Peak–Dixie Valley earthquakes, Wallace (1977) developed the first detailed qualitative model of fault-scarp degradation in the Basin and Range. A schematic summary of the salient characteristics of this model is presented in Figure 10. Fault scarps described by this model are limited to those developed in cohesionless frictional materials such as unconsolidated alluvial deposits or colluvium. The initial fault scarp is a loosening-limited free face. This free face is rapidly modified by crestal failure and upper scarp recline. Individual clasts and large blocks of partly indurated material fall from the face to accumulate as a basal-debris apron along the foot of the scarp. Thus, the free face is

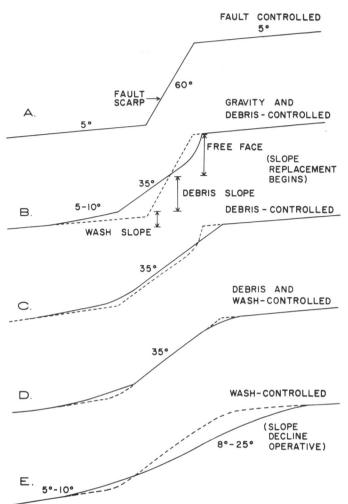

Figure 10. Sequence of fault-scarp degradation. The dotted line in each profile indicates the incremental change from the previous profile (from Wallace, 1977, Fig. 3).

rapidly replaced by a transport-limited debris slope with an initial slope angle that approximates the angle of internal friction of the loose debris. (Nash [1984] has termed this process of free-face retreat and debris-apron formation "raveling.") Even as the debris slope is forming, slope-wash processes begin modifying its form via alluvial deposition near the base and rilling and gullying along the main slope. As long as the loosening-limited free face survives, the crest of the scarp maintains a sharp break in slope with the original piedmont surface. However, after the free face is eliminated, this sharp crestal break is quickly rounded by creep and slope-wash processes. The result is a slightly rounded uniformly sloping step in the original piedmont surface (Fig. 10D). For fault scarps cutting poorly indurated alluvium and colluvium in north-central Nevada, free faces are eliminated within a few hundred to, at most, a few thousand years, and significant rounding of the scarp crest occurs within 10,000 yrs (Wallace, 1977).

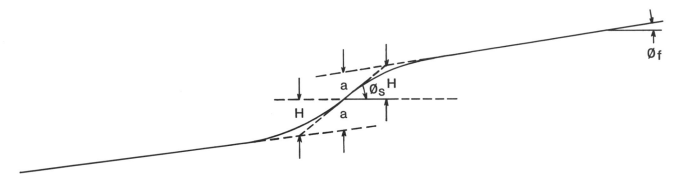

Figure 11. The geometry of a fault scarp. 2a is the scarp offset, 2H is the scarp height, ϕs is the scarp slope angle, ϕf is the far-field slope angle. Where $\phi f > 0$, 2a and 2H are not the same (redrawn from Hanks and others, 1984, Fig. 1).

The Diffusion Equation Model. After elimination of the free face by raveling, slope-wash processes begin to progressively lower the smooth, uniformly sloping debris slope that has taken its place. The downslope profile of this smooth, transport-limited hillslope segment resembles the graphic form of the error function (Fig. 11), a relatively simple mathematical expression that forms the fundamental solution to the one-dimensional diffusion equation for step-like initial conditions (Nash, 1980; Hanks and others, 1984). The following discussion of the application of the diffusion equation to the problem of fault-scarp degradation is synthesized from Nash (1980, 1984), Mayer (1982), Coleman and Watson (1983), and Hanks and others (1984).

To a first approximation, the degradational evolution of a two-dimensional downslope profile of a fault scarp in unconsolidated alluvium can be quantitatively described by the expression:

$$\frac{\delta u}{\delta t} - k\frac{\delta^2 u}{\delta x^2} = 0 \qquad (3)$$

where x is the horizontal dimension, u is the relative elevation, t is time, and k is the mass diffusivity (an empirically determined constant that depends on the physical properties of the scarp-forming materials and the climatic environment of the scarp). In words, this relation states that convex-upward topography erodes and concave-upward topography aggrades and that the rate of change of a hillslope segment is a function of its curvature (i.e., sharp breaks in slope degrade more rapidly than smooth slopes).

Equation (3) assumes: (1) downhill movement of mass at a rate proportional to the local topographic gradient; (2) conservation of mass, at least at a local scale; (3) no general landscape lowering; (4) an unchanging local baselevel below the scarp; (5) scarp degradation under a condition of constant process rates (as averaged over periods of years or decades); (6) scarp formation by a single event; (7) orthogonality between the scarp and the downslope profile; and (8) negligible contour curvature along the scarp in the vicinity of the profile. It is important to note that assumptions 1 and 2 are probably only approximately satisfied on relatively planar, uniformly sloping, undissected piedmont segments where the various processes of erosion and transport are more or less uniformly distributed. Scarps modified by channelized fluvial processes do not satisfy these basic assumptions and, therefore, cannot be adequately modeled by this relation. Assumptions 3 and 4 are reasonable for most scarps offsetting unconsolidated alluvium on abandoned piedmont segments. Both field relations throughout the Great Basin and geomorphic theory suggest that in such situations scarp degradation occurs much more rapidly than general piedmont erosion. Assumption 5 is also a workable approximation on abandoned piedmont segments because these areas are *relatively* insensitive, morphologically, to Quaternary climatic fluctuations. Of course, climatic fluctuations and concomitant fluctuations in the rates of degradational processes acting on piedmont areas do impose an undetermined amount of error upon estimates of absolute scarp age determined by diffusion equation analysis. However, estimates of relative age should be largely unaffected. Assumptions 6, 7, and 8 merely constrain the selection of fault-scarp segments that are suitable for diffusion equation analysis.

The solution to equation (3) for a scarp at x = 0 with an offset of 2a imposed upon a surface with slope angle ϕf (see Fig. 11) is:

$$u(x,t) = a \, \text{erf}\,[x/2(kt)^{1/2}] + x \tan \phi f \qquad (4)$$

where $\text{erf}\,[x/2(kt)^{1/2}]$ is the error function of argument $x/2(kt)^{1/2}$. The maximum slope in this situation occurs at x = 0 (the center of the scarp) and is defined by:

$$\tan \phi s = \frac{\delta u}{\delta x}\bigg|_{x=0} = \frac{a}{(\pi kt)^{1/2}} + \tan \phi f. \qquad (5)$$

Equation (5) predicts that the maximum scarp slope (tan ϕs) for any given k, 2a, and ϕf should decrease with time, and this prediction is supported by abundant field evidence. Field measurements of maximum scarp-slope angle for fault scarp systems in north-central Nevada (Wallace, 1977), western Utah (Bucknam and Anderson, 1979), northwestern Arizona (Mayer, 1982, 1984), and central New Mexico (Machette, 1982) document a general decrease in maximum scarp-slope angle with increasing scarp age. Moreover, if the value of k is known or can be closely approximated, equation (3) can be used to estimate scarp age from field measurements of 2a, ϕs, and ϕf (Nash, 1980, 1984; Coleman and Watson, 1983; Hanks and others, 1984). However, to determine the appropriate value for k, at least one scarp of known age with the same aspect, within the same climatic environment, and underlain by the same geologic materials must be measured and analyzed (Nash, 1984).

Equation (3) also predicts that the maximum scarp slope (tan ϕs) for any given k, ϕf and t should increase with increasing scarp offset, 2a. This prediction explains the strong positive correlations between maximum scarp-slope angle and scarp height reported by Bucknam and Anderson (1979) for fault scarps offsetting unconsolidated deposits in western Utah (Fig. 12).

Limitations and Implications of the Diffusion Equation Model. A major limitation of the application of the diffusion equation to the problem of scarp degradation is the mass diffusivity term, k. The many and varied influences of the physical properties of the scarp material, the present climate and the climatic history of the scarp site, and the resulting geomorphic processes that have acted to degrade the scarp reside in this empirical term. In other words, the diffusion equation itself embodies little, if any, specific information about the actual processes of hillslope transport and erosion. However, the fact that this relation successfully describes the morphological evolution of degrading scarps provides some insight into the processes of degradation on relatively gently sloping planar piedmont surfaces in the Basin and Range. The fundamental assumptions of the diffusion equation, namely downslope movement at a rate proportional to the slope and local conservation of mass, would appear to represent at least approximate conditions for the actual processes acting on these planar piedmont surfaces. This, in turn, suggests that one or more of the diffuse processes of hillslope creep, raindrop impact, sheet wash, and rill wash, are sufficiently active to cause profound smoothing and eventual elimination of irregular segments on these surfaces within periods of thousands to hundreds of thousands of years depending on local climatic and geologic conditions.

The dependence of the diffusion equation model on accurate, empirically based estimates of the mass diffusivity term, k, indicates the need for an infusion of process-oriented research into the problem of scarp degradation and the morphologic dating of fault scarps. Detailed, quantitative knowledge of the actual processes responsible for scarp degradation, the relationships between these processes, the range of variation in the physical characteristics of the materials underlying the scarps, the variability of microclimate, and the local impact of this variability on hillslope processes are representative of the many problems that need to be addressed if diffusion equation analysis of scarp degradation is to proceed beyond its present empirical basis. Geomorphic analyses of marine, lacustrine, or fluvial terrace scarps of known age should help to resolve these questions. For example, Pierce and Coleman (1986) show that fluvial scarp degradation in central Idaho varies systematically with scarp height and scarp orienta-

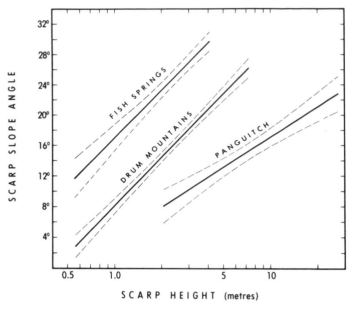

Figure 12. Maximum scarp-slope angle versus scarp height for three locations in west-central Utah:
Fish Springs, $\phi s = 17.0 + 23.6 \log H$, $R^2 = 0.82$, $n = 38$;
Drum Mountains, $\phi s = 8.07 + 21.1 \log H$, $R^2 = 0.88$, $n = 49$;
Panguitch, $\phi s = 3.90 + 13.2 \log H$, $R^2 = 0.88$, $n = 11$;
(from Bucknam and Anderson, 1979, Fig. 5 and Table 1).

tion (microclimate). For 2-m-high scarps, the degradation rate on south-facing scarps is twice that of north-facing scarps; and for 10-m-high scarps, degradation is ten times greater on south-facing scarps. Many similar studies are needed to augment and refine present knowledge concerning fault-scarp degradation.

Range Front Evolution

More than 400 separate mountain ranges form the corrugated landscape of the Basin and Range. These ranges provide another population of landforms that is potentially useful for the study of landform degradation. They vary from small isolated uplands a few hundred meters high and a few kilometers wide and long to imposing uplands as much as 3 km high, 25 km wide, and 180 km long. Although the majority share similar tectonic origins and histories, marked regional contrasts in their morphometric and morphologic characteristics are associated with regional differences in the timing and style of classic Basin and Range structural deformation. K-Ar dating of volcanic rocks associated with Basin and Range deformation indicate that the later extensional phase of general block faulting (initiated as early as 17 Ma in some areas) generally ended between 6 and 10 Ma in the Mojave and Sonoran deserts and between 4 and 6 Ma in the Mexican Highlands, but is still continuing along the margins of the Great Basin (Christiansen and McKee, 1978; Shafiqullah and others, 1980). Trend-surface analyses of the general range morphometry within the Basin and Range demonstrate general in-

TABLE 3. MOUNTAIN FRONT TECTONIC ACTIVITY CLASSES*

Tectonic activity class	Local base level processes**	Mountain front sinuosity	Valley floor width/valley depth ratios	Typical Landforms
Highly Active				
Class 1	$\Delta u/\Delta t \geq \Delta cd/\Delta t + \Delta pa/\Delta t$	1.2 to 1.6	0.5 to 0.9	Unentrenched fans; steep hillslopes on all materials; elongate drainage basins; narrow valley floors.
Moderately Active				
Class 2 Class 3 Class 4	$\Delta u/\Delta t < \Delta cd/\Delta t > \Delta pd/\Delta t$	1.5 to 3.4	0.5 to 3.6	Entrenched fans with old soils on fanheads; large more equant drainage basins with wider valley floors. Class 2: V-shaped mountain valleys. Class 3: U-shaped mountain valleys. Class 4: Embayed mountain front.
Inactive				
Class 5	$\Delta u/\Delta t \ll \Delta cd/\Delta t \pm \Delta pd/\Delta t$	1.8 to 7.2	2.1 to 47.0	Fringing pediment, dissected or undissected; mountain front embayments; large integrated stream systems; steep hillslopes limited to resistant materials.

*Modified from Bull and McFadden (1977) and Bull (1984).

**u = uplift, cd = channel downcutting, pa = piedmont aggradation, pd = piedmont degradation, t = time.

creases in the range length, width, relative area, and average relief from the Mojave and Sonoran deserts to the Mexican Highlands to the Great Basin (Lustig, 1969). This general correspondence between range morphometry and tectonic history across broad ranges of climatic history, lithologic composition, and geologic structure suggests the profound influence of tectonic environment and neotectonic history on landscape development.

Impact of Vertical Tectonics on Alluvial Systems. The impact of vertical tectonic movements on the processes and morphologies of alluvial systems along fault-bounded range fronts can be conceptualized by relatively simple inequalities that relate uplift, erosion, and deposition. The following discussion of this idea is synthesized from the work of Bull (1973, 1978, 1984) and Bull and McFadden (1977). General processes affecting the local base level of a stream draining a fault-bounded range front include the relative uplift of the mountain mass, channel downcutting within the mountains, and erosion or deposition on the piedmont immediately adjacent to the range front. The relative rates of these general processes affect the loci of hillslope, valley, and piedmont erosion and deposition, and therefore, determine the general morphology of the mountain front, its drainage basins, and adjoining piedmont. For example, continued rapid uplift of a fault-bounded mountain mass relative to an adjacent basin results in the formation of narrow, steep-sided, high-gradient valleys within the mountain mass and a thick accumulation of alluvial fan deposits adjacent to the mountain front. Within an individual drainage system, channel downcutting in the mountain basin would tend to cause alluvial deposition on the adjacent piedmont and entrenchment of the fan apex. However, this tendency for entrenchment is counteracted by uplift of the mountain

basin relative to the adjacent piedmont; and as long as the rate of relative uplift equals or exceeds the combined rates of channel downcutting and alluvial fan deposition, fanhead entrenchment will not occur. This relation, conceptualizing a tectonically induced threshold for fanhead deposition, can be symbolically expressed as: $\Delta u/\Delta t \geq \Delta cd/\Delta t + \Delta pa/\Delta t$; where u = magnitude of relative uplift, cd = magnitude of channel downcutting, pa = magnitude of piedmont aggradation and t = time. Five relations relating tectonic activity to fluvial process thresholds and resultant piedmont landforms are summarized in Table 3. Additional relations could probably be devised for diagnostic landforms within the mountain basins or along the mountain front. Such relations represent the first step towards quantification of the influence of tectonic activity on range-front processes and range-front degradation through time. Definition of these tectonically related process thresholds provides a theoretical foundation for the use of diagnostic landforms as quantitative indicators of tectonic activity.

Bull and McFadden (1977) and Bull (1978) applied these concepts to a semiquantitative geomorphic analysis of tectonic activity along approximately 80 mountain ranges in the Mojave Desert and southwestern Great Basin. Using the presence or absence of diagnostic landforms (entrenched alluvial fans, range front pediments), mountain front sinuosities, ratios of valley-floor–width to valley-depth, general frequencies and distributions of steep mountain hillslopes, and approximate ages of the oldest piedmont surfaces adjacent to range fronts, each range-front segment was assigned to one of the categories of tectonic activity defined in Table 3. Marked differences in both local and regional landscapes were identified by this procedure. Range fronts along the flanks of Death Valley and Panamint Valley and along the several Quaternary strike-slip fault zones of the central Mojave Desert were all found to possess morphologic characteristics diagnostic of moderate to high vertical tectonic activity. Elsewhere within the Mojave Desert, and within a 20 to 30 km wide zone along the southern boundary of the Great Basin (immediately north of the Garlock fault), nearly all range fronts display morphologic characteristics indicating a lack of vertical tectonic activity.

As Bull and McFadden (1977) point out, a fundamental limitation of inferring tectonic activity from range-front morphology arises from the fact that differential vertical uplift is, of course, only one of many factors that can significantly affect the behavior of fluvial systems. Distinguishing between the possible effects of these various influences is, at best, a nontrivial problem that usually requires general knowledge of the regional context and specific field-based knowledge of local lithologic, morphometric, and base-level influences. Another significant limitation is that range-front morphology may not be a sensitive indicator of temporal variations in uplift rates. For example, rapid differential uplift at a relatively constant rate followed by stability, *or* rapid uplift that gradually decreases with time, *or* slow continuing uplift might all produce similar range-front morphologies at various stages during the degradational history of a mountain range. Resolution of this problem requires more precise knowledge of the temporal and spatial patterns of late Tertiary tectonism within the Basin and Range against which regional patterns of range-front morphology can be compared and analyzed. Thus, inferences of tectonic activity based on casual observations of range-front morphology should be considered tentative.

A computer simulation based on a random-walk probability algorithm, in which stream discharge is approximated by an inverse relation between stream gradient and stream length, provides some additional insight into the fluvial degradation of relatively stable upland areas in the Basin and Range (Mayer and others, 1981). When applied to a three-dimensional array of topographic data and assuming a condition of base level stability, the model predicts an increase in mountain-front sinuosity and the development of range-front pediments and inselbergs. The model also predicts that the rate of channel downcutting will be greatest in headwater areas and will progressively decrease downstream toward the stable base level. This prediction is consistent with the theoretical deduction that stream power generally exceeds critical power in headwater areas but generally falls short of critical power in lower piedmont areas constrained by a stable base level (Bull, 1979). It is also consistent with the general pattern of long-term upland erosion in the Reveille Range of central Nevada where reconstruction of latest Miocene and early Pliocene paleotopography (based on the distribution of extensive erosion surface remnants capped by dated late Cenozoic lava flows) documents a pattern of maximum erosion in crestal and upper flank areas and approximate equilibrium in midpiedmont areas of that range (Dohrenwend and others, 1985).

Development of Faceted Spurs. The evolution of faceted spurs is another aspect of the morphology of fault-bounded range fronts that is potentially useful for analysis of the response of degradational processes to relative vertical tectonic movement. Hamblin (1976) interpreted the notched morphology of faceted spurs along the Wasatch Front in central and northern Utah as evidence of long-term discontinuous movement along the Wasatch fault. According to Hamblin, these notches are the remnants of small range-front pediments preserved at the apices of successive generations of faceted spurs. This interpretation implies movement along the range-bounding fault in an active-quiet pattern of recurrent periods of quasi-continuous vertical movement (long series of discrete displacement events of a few meters each with a recurrence interval ranging from hundreds to a few thousands of years) separated by periods of essentially no movement and pediment formation. It also implies the relatively rapid development of range-front pediments during periods of tectonic stability.

This model is supported, at least in part, by a similar interpretation of faceted-spur development along fault-bounded range fronts in north-central Nevada (Wallace, 1977). If repeated faulting in steps of a few meters occurs every 10,000 years or so, the base of the faceted spur should be progressively extended so that 'the lower part of the facet tends at all times to be a younger surface than the upper part'; and erosion on the flanks of the spur and on the face of the facet would tend to lessen facet slope with time (Wallace, 1977, p. 642). This conceptual description corresponds well with Hamblin's model for intervals of essentially continuous vertical movement. Wallace also noted that sharp-crested spurs appear to be relatively stable landforms that remain as prominent landscape elements for relatively long periods of time because repeated uplift induces rejuvenation, which in turn would perpetuate the sharp-crested form. However, Wallace did not mention the presence of notched facets.

The correspondence between specific characteristics of range-front morphology and general aspects of vertical tectonic activity suggests that systematic regional analyses of specific landforms is a potentially powerful approach for determining the influence of differential vertical uplift on hillslope, fluvial, and piedmont processes within the Basin and Range. Unfortunately, few investigations of this sort have been attempted.

Geomorphic Characteristics of Metamorphic Core Complexes. Not all mountain ranges in the Basin and Range are the product of high-angle normal faulting. In some areas, metamorphic core complexes, the exhumed remnants of low-angle detachment faulting, survive as deeply dissected domal mountain masses. These antiformal ranges apparently formed in response to tectonic denudation caused by middle Tertiary detachment faulting (Spencer, 1984). Generally defined, metamorphic core complexes are asymmetric antiformal uplifts of metamorphic and plutonic rocks that are overlain by unmetamorphosed sedimentary and volcanic rocks and separated from these cover rocks by one or more gently to moderately dipping zones of detachment faulting (Coney, 1980; Spencer, 1984). In most cases, the unmetamorphosed upper plate rocks have been highly attenuated and

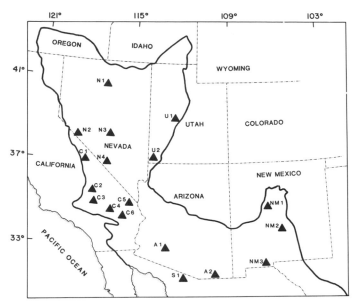

Figure 13. Map showing the locations of latest Tertiary and Quaternary basaltic volcanic fields within the Basin and Range Province. **California,** C1, Big Pine; C2, Coso; C3, Black Mountain; C4, Pisgah–Lavic Lake; C5, Cima; C6, Amboy. **Nevada,** N1, Battle Mountain; N2, Aurora; N3, Lunar Crater–Reveille Range; N4, Crater Flat. **Utah,** U1, Black Rock Desert; U2, Caliente–Saint George Basin. **Arizona,** A1, Sentinel; A2, San Bernardino. **Sonora,** S1, Pinacate. **New Mexico,** MN1, Los Lunas; MN2, Carrizozo; NM3, Potrillo.

pervasively sliced by numerous listric normal faults associated with the detachment faulting; and the lower plate rocks, metamorphosed at mid-crustal depths, have been largely exhumed by this faulting. These exhumed tectonic landforms occur along the axis of the North American Cordillera from northern Mexico to southern Canada (Coney, 1980), and they are particularly abundant within an arcuate 200-km-wide zone, which extends across the eastern part of the Mojave-Sonoran Desert section of the Basin and Range Province from northern Mexico to southern Nevada (Coney, 1980; Spencer, 1984; Reynolds and Spencer, 1985).

Several structurally controlled landforms are commonly associated with the metamorphic core complexes of the Mojave-Sonoran Desert (Pain, 1985, Fig. 1). Adjacent to the metamorphic core, the pervasively faulted upper plate rocks typically form a blocky landscape with a rectangular drainage pattern. Within the core, the brecciated upper part of the lower plate commonly forms a suite of distinctive landforms: (1) a small ledge underlain by erosionally resistant microbreccia often marks the upper surface of the lower plate; (2) a belt of gently rounded hills underlain by erosionally weak chloritic breccia typically lies between this small ledge and the range front; (3) slightly-dissected, triangular-shaped facets along the range front coincide with the dome-shaped surface of the mylonitic front beneath the zone of chloritic breccia; and (4) near the range crest, areas of low local relief

probably represent highly degraded continuations of these dome facets (Pain, 1985).

These structurally-controlled landforms are the result of complex interactions between tectonic and erosional processes. Although they were ultimately produced by erosional exhumation, the controlling structures were formed in response to middle Tertiary detachment faulting and reached surficial levels during or immediately after this faulting, largely as a result of tectonic denudation (Spencer, 1984). Additional research is needed to more comprehensively document the geomorphic characteristics of metamorphic-core complexes and to more precisely define the relative roles of the tectonic and surficial processes responsible for their exhumation.

Degradation of Volcanic Landforms

Volcanic landforms form rapidly, can be related to original constructional forms with some degree of certainty, and are datable by K-Ar analyses. Thus, the long-term geomorphic processes that modify these landforms can be accurately quantified by selecting a series of landforms, similar in origin and original form but significantly different in age, and comparing the morphologic differences between these landforms for different time intervals. At least 18 latest Tertiary and Quaternary basaltic volcanic fields are located within the Basin and Range Province (Fig. 13). At least three of these fields have been sufficiently active and long-lived to provide ideal locations for investigations of volcanic landform evolution (Duffield and others, 1980; Dohrenwend and others, 1984a; Turrin and Dohrenwend, 1984; Turrin and others, 1984), and one of these fields, the Cima volcanic field, has been intensively studied in this regard.

The Cima volcanic field contains more than 40 basaltic cones and associated lava flows that cover an aggregate area of approximately 150 km^2 on the crests and flanks of several large, actively downwasting pediment domes in the eastern Mojave Desert. Both the cones and flows of this field have been geomorphologically analyzed to determine the rates and trends of their degradational evolution and to determine the processes responsible for that degradation (Dohrenwend and others, 1984a, 1984b, 1986a; Wells and others, 1984a, 1987; McFadden and others, 1984, 1986).

Degradation of Cinder Cones. The cinder cones of the Cima field display several morphologic trends that are closely related to cone age (Dohrenwend and others, 1986a): average and maximum side-slope angles decrease with increasing age; fringing debris aprons expand, stabilize, and then decrease in size with progressive cone degradation; and cone drainage evolves from irregularly spaced rills and gullies to integrated drainage networks of large gullies and small valleys. These trends form the basis of an empirical model that documents: (1) rapid initial stripping of a loose cinder mantle from upper cone slopes accompanied by rapid debris-apron formation; and (2) a gradual transition, between 0.25 and 0.6 m.y., from uniform stripping of upper slopes to localized dissection of both cone slopes and debris apron (Fig. 14). This transition is apparently controlled by a concomitant change from widely distributed subsurface drainage within the pervious cinder mantle to concentrated surface flow across the heterogeneous assemblage of agglutinate masses, flows, and shallow intrusives that form the cone interior. This model also indicates that, for the past million years at least, the primary mode of degradation of the Cima cones has been general downwasting accompanied by a progressive decline in cone height and maximum slope angle. Although the degradational pattern of the Cima cones (decline in cone height and decrease in maximum slope induced by erosional lowering of crestal slopes and aggradational raising of basal slopes) is superficially similar to the pattern of degradation of "raveled" fault scarps in unconsolidated alluvium, fundamental differences between cone geometry and fault-scarp geometry prevent direct application of diffusion equation analysis to the problem of cone degradation.

Degradation of Lava Flow Surfaces. The lava flows in the Cima field also display a systematic pattern of surface degradation (Fig. 15; Wells and others, 1984a, 1985; McFadden and others, 1984). Flows younger than 0.25 Ma support patchy eolian mantles and weakly developed soils. Constructional relief on these flow surfaces is being reduced by bedrock rubbling, colluviation, and eolian deposition. 0.25- to 0.60-Ma flows are almost completely covered by accretionary eolian mantles supporting interlocking stone pavements and thick soils with well-developed argillic horizons. These flows are undergoing relatively little erosional modification. Flows older than 0.60 Ma are characterized by partly stripped soils with Stage III carbonate accumulations and are dominated by surface runoff and fluvial dissection. These temporal variations in surface morphology and stability are the result of a temporal sequence of interrelated surficial processes (eolian deposition, pedogenesis within the eolian deposits, and eventually fluvial erosion). Early stages of surface modification are dominated by burial with eolian silt. Formation of a well-developed soil with a clay-rich B-horizon within this silt mantle and carbonate engulfment of this soil progressively reduce the infiltration capacity of the flow surface to the point where surface runoff becomes dominant. Subsequent fluvial erosion strips the accretionary mantle and associated soils, exposing both the underlying carbonate horizons and the massive flow interiors (Wells and others, 1984a, 1985).

Drainage Development on Lava-Flow Surfaces. Drainage network development on the Cima lava flows and on flows of the Lunar Crater volcanic field in central Nevada is consistent with this process-response model of flow-surface degradation (Dohrenwend and others, 1987). Drainage networks on representative lava flows of various ages are shown in Figure 16; and average values of stream frequency and drainage density for drainage networks on 18 flows are plotted in Figure 17. Drainage on 0.01-to 0.15-Ma flows is severely limited and is largely controlled by original flow forms. On 0.15- to 0.4-Ma flows, simple drainage networks are developed on 1- to 4-m-thick mantles of flow rubble and eolian silt. Relatively complex networks have

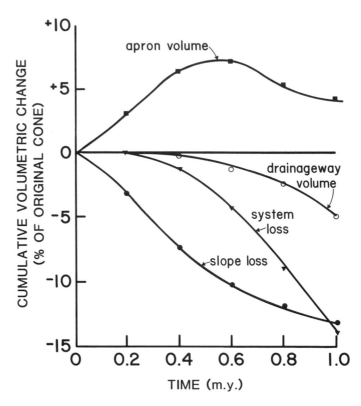

Figure 14. Model of cone degradation over the first million years. Cumulative volumetric change (expressed as % of original cone volume) versus time (from Dohrenwend and others, 1986a, Fig. 7.)

developed on 0.4- to 1.1-Ma flows and have partly stripped the accretionary mantles on most flows older than 0.6 Ma.

Substituting time for space, several generalizations can be made regarding the development of these drainage networks. Drainage development proceeds in a manner that is similar to the qualitative model proposed by Glock (1931). A brief period of rapid elongation extends master drainages to the full length of most flows within about 0.3 m.y. Elaboration then replaces elongation as the principal mode of network extension and proceeds by the addition of short tributaries to master drainages and by the formation of small, low-order drainages along flow margins. Drainage densities increase first rapidly, then more slowly, and apparently reach maximum extension sometime after 1.1 m.y. During elaboration, the relation between stream frequency (F) and drainage density (D) approximates $F = 0.65D^2$, which is essentially the same as the empirical relation for mature drainage basins (Melton, 1958) and the theoretical relation for large topologically random drainage networks (Shreve, 1969). These results are consistent with experimental analyses of drainage-network evolution (Parker, 1977; Parker and Schumm, 1982) and provide general confirmation of this work in field situations where lithologic, structural, and tectonic influences are essentially constant.

Long-Term Erosion Rates

Analysis of the general processes responsible for the degradational evolution of mountain ranges or range fronts through time requires some knowledge of the rates at which these processes have operated. Estimating average erosion rates from modern sediment yields or long-term sedimentation rates is not well suited to the Basin and Range. The region's complex history of late Cenozoic tectonism and climatic change precludes the use of modern sediment yields as reliable indicators of long-term erosion rates, even assuming that existing data on modern fluvial processes were adequate to generate reliable estimates of modern sediment yield. Moreover, available late Cenozoic sedimentary sections are neither continuous enough nor sufficiently well dated to provide accurate estimates of range erosion.

Probably a more reliable approach for estimating average erosion rates in the Basin and Range involves the geomorphic analysis of erosion surfaces capped by late Tertiary and Quaternary basaltic lava flows. Average downwasting rates can be calculated by measuring the average vertical distance between a flow-capped erosion surface and the level of an adjacent modern erosion surface, then dividing this height difference by the K-Ar age of the lava flow. Estimates based on this approach for several widely separated upland areas within the Great Basin and Mojave Desert are compiled in Table 4. These estimates, ranging between 0.8 and 4.7 cm/10^3yr, are average values for periods of 0.67 to 10.8 m.y. Average rates of backwearing or slope retreat can be estimated in a similar manner by determining the average horizontal distance between the maximum possible extent of a capping lava flow and the present eroded margin of that flow, then dividing this horizontal distance by the K-Ar age of the lava flow. Estimated rates of slope retreat determined by this method range between 6 and 35 cm/10^3yr (Table 4). Of course, the uncertainties inherent in determining the original extent of most lava flows are substantial; therefore, these estimates should be considered only general approximations. Moreover, the armoring effect of blocky basalt talus inhibits the backwearing of basalt-capped hillslopes, further limiting the general applicability of these estimates. However, these difficulties notwithstanding, the rates of slope retreat summarized in Table 4 are comparable to the 10 to 100 cm/10^3yr rates of slope retreat and pediment formation that have been previously estimated or assumed for diverse areas of the Basin and Range (Wallace, 1978; Menges and McFadden, 1981; Mayer and others, 1981; Saunders and Young, 1983).

Formation and Development of Pediments

Pediments are prominent landscape elements in many areas of the Basin and Range, and in the Mojave and Sonoran deserts they have long been the subject of geomorphological scrutiny. Indeed, more papers have been published on this subject than on any other aspect of geomorphology in this region. Unfortunately, the net result of this long history of investigation is not altogether

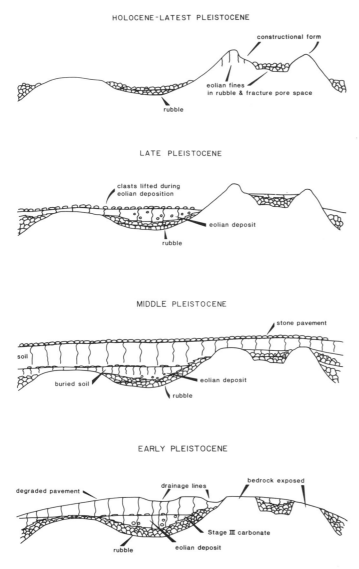

Figure 15. Late Tertiary and Quaternary degradation of basaltic lava flow surfaces, Cima volcanic field, eastern Mojave Desert. Holocene and latest Pleistocene surfaces (<0.002 Ma)—unstable surfaces with rubbled constructional forms and local accumulations of eolian silt and fine sand. Late Pleistocene surfaces (0.06 to 0.14 Ma)—partly stable surfaces with accumulations of eolian silt and fine sand in topographic lows; stone pavements underlain by weakly developed soils occur on the eolian deposits. Middle Pleistocene surfaces (0.2 to 0.6 Ma)—stable surfaces almost completely mantled by thick deposits of eolian fines with interlocking stone pavements underlain by complex, moderately to strongly developed soils. Early Pleistocene surfaces (>0.6 Ma)—slowly degrading, partly stripped surfaces with strongly developed soils; characterized by well-developed drainage networks, degraded stone pavements with pedogenic carbonate rubble, and local areas of exposed bedrock (modified from Wells and others, 1984b, Fig. 9).

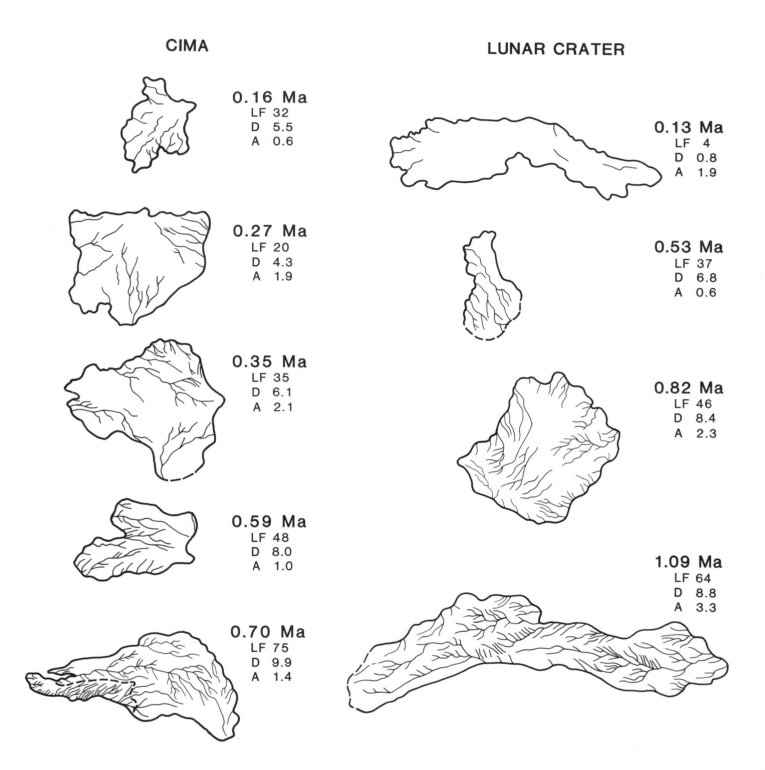

Figure 16. Drainage network evolution on basaltic flow surfaces of the Cima and Lunar Crater volcanic fields. LF, link frequency (#/km^2); D, drainage density (km/km^2); A, flow area (km^2); K-Ar ages (Ma) from Dohrenwend and others (1984a) and Turrin and others (1984).

TABLE 4. EROSION RATES IN UPLAND AREAS OF THE SOUTHWESTERN BASIN AND RANGE INFERRED FROM COMPARISON OF
ACTIVE AND RELICT BASALT-CAPPED EROSION SURFACES

Location	Minimum age of relict erosion surface (m.y.)	Maximum average downwasting rate (cm/10^3 yr)	Maximum possible backwearing rate (cm/10^3 yr)	Reference
Buckboard Mesa, southern Great Basin	2.82	4.7	–	Carr, 1984
Cima volcanic field, Mojave Desert	4.48	1.1	22–28	Dohrenwend and others, 1986 and unpublished data
Cima volcanic field, Mojave Desert	3.88	2.8	6–35	Dohrenwend and others, 1986 and unpublished data
Cima volcanic field, Mojave Desert	3.64	2.5	8	Dohrenwend and others, 1986 and unpublished data
Cima volcanic field, Mojave Desert	0.85	2.8	30	Dohrenwend and others, 1986 and unpublished data
Cima volcanic field, Mojave Desert	0.67	3.0	–	Dohrenwend and others, 1986
Lunar Crater volcanic field, central Great Basin	2.86	0.8	–	Turrin and Dohrenwend, 1984 and unpublished data
Lunar Crater volcanic field, central Great Basin	1.08	1.1	–	Turrin and Dohrenwend, 1984 and unpublished data
Reveille Range, central Great Basin	5.70	3.1	22	Dohrenwend and others, 1986 and unpublished data
Reveille Range, central Great Basin	5.58	–	18	Dohrenwend and others, 1986 and unpublished data
Reveille Range, central Great Basin	3.79	1.6	18	Dohrenwend and others, 1986 and unpublished data
White Mountains east flank, southwest Great Basin	10.8	2.4	–	Marchand, 1971
White Mountains east flank, southwest Great Basin	10.8	2.0	–	Marchand, 1971

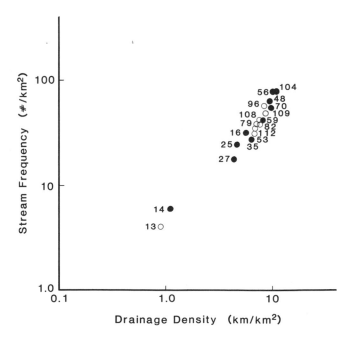

Figure 17. Stream frequency versus drainage density for basaltic flow surfaces. Solid circles designate flows in the Cima field; open circles designate flows in the Lunar Crater field. Numbers indicate K-Ar ages in 10^4 yrs.

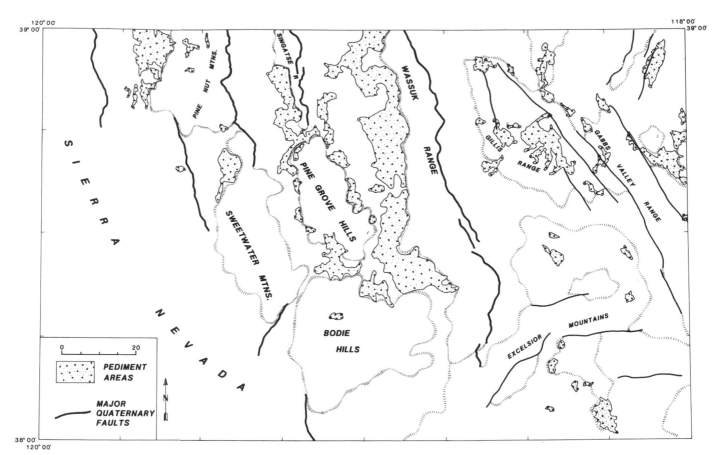

Figure 18 (continued on facing page). Distribution of pediments in the west-central Basin and Range. A, Walker Lake 1° × 2° Quadrangle; B, Tonopah 1° × 2° Quadrangle. Scale: 0-20 km.

clear. Individual conclusions are varied and frequently contradictory; and the "pediment problem" has thus far successfully eluded a general solution. Pediments are complex landforms that are difficult to define and correspondingly difficult to study. "Disagreement often begins with the problem of definition . . . and continues through all subsequent phases of inquiry. Here is a subject dominated by almost unbridled imagination; a subject in need of organization and direction" (Cooke and Warren, 1973, p. 188). Despite such problems, a significant body of factual knowledge concerning pediments has been developed.

Lithologic Associations. Pediments in the Basin and Range have formed on a wide variety of lithologies; however, most of these pediments (particularly the more extensive ones) bevel easily erodible rock types. Within the Mojave and Sonoran deserts, bedrock pediments are preferentially developed on deeply weathered, coarse-crystalline granitic rocks (Gilluly, 1937; Mammerickx, 1964; Warnke, 1969; Cooke, 1970b; Moss, 1977; Oberlander, 1972, 1974; Dohrenwend and others, 1984b; 1986b). Locally, however, extensive pediments also cut discordantly across volcanic and metamorphic rocks (Gilluly, 1937; Mammerickx, 1964; Cooke and Reeves, 1972; Dohrenwend, unpublished data). In the west-central Great Basin, pediments are

preferentially developed on middle to late Cenozoic terrigenous sedimentary rocks, predominantly poorly indurated fluvial and lacustrine strata, and on middle to late Tertiary volcanic rocks, predominantly lahars and nonwelded ash flow tuffs (Gilbert and Reynolds, 1973; Dohrenwend, 1982a, 1982b). Pediments also bevel late Cenozoic terrigenous sediments along the Ruby and East Humboldt ranges of northeast Nevada (Sharp, 1940), in the Furnace Creek area of Death Valley (Denny, 1967), in the San Pedro, Sonoita, and Canada del Oro basins of southeast Arizona (Melton, 1965; Menges and McFadden, 1981), and along the valley of the Rio Grande (Denny, 1967); and extensive pediment surfaces cut across Pleistocene lacustrine deposits in the Lahontan basin (R. Morrison, personal communication, 1985) and the Tecopa basin (Hillhouse, 1987; Dohrenwend, 1985). Moreover, many alluvial piedmonts in the Basin and Range are in fact complex surfaces of erosion and deposition cut across older Quaternary basin fill and veneered by thin mantles of late Pleistocene and Holocene alluvium (Sharp, 1984; Dohrenwend, 1982a; Wells and others, 1984b; Wilshire and others, 1983).

Tectonic Environment. The bedrock pediments of the Basin and Range are mostly located in stable or quasi-stable environments where erosional and depositional processes have

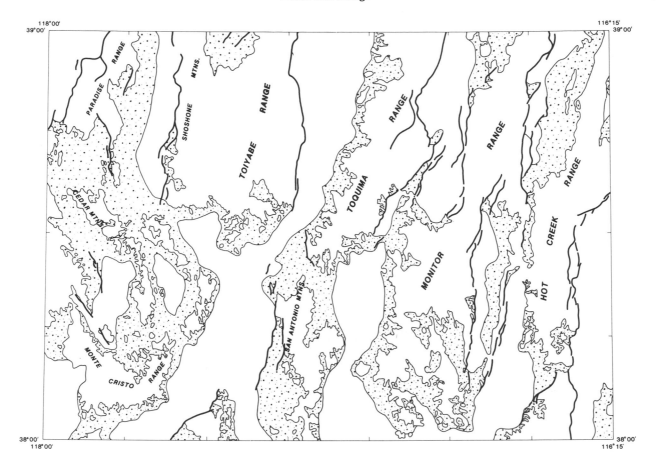

been approximately balanced for relatively long periods of time. Although pediments have not been systematically mapped or correlated with tectonic environment province wide, regional trends suggest a general correspondence between tectonic stability and pediment development. A general morphometric comparison between the Sonoran and Mojave deserts suggests that pediments are generally larger in some areas of the Sonoran Desert than in the western Mojave Desert (Cooke, 1970b). Also, regional surficial geologic mapping of a 30,000 km^2 area in western and central Nevada indicates that pediments in the more stable central part of the Great Basin are generally larger and more continuous than pediments near the more tectonically active western margin of that region (Dohrenwend, 1982a, unpublished data). In addition, the pediments of west-central Nevada are preferentially developed in local settings of relative stability within an otherwise tectonically active region (Fig. 18). Large pediments bevel broad, gently sloping piedmonts near the nodal lines on the backtilted flanks of the larger range blocks. Smaller pediments occur around the peripheries of upwarped structural highs, on backtilted flanks of small blocks along major strike-slip faults within the Walker Lane, on upfaulted piedmont segments, and within small embayments adjacent to active fault bounded range fronts (Dohrenwend, 1982a, 1982c). However, the most extensive pediments within this region occupy an area of relatively subdued topog-

raphy along the eastern margin of the Walker Lane where shallow middle Miocene detachments occur at or close to the present land surface (G. F. Brem, D. A. John, personal communications, 1985). Similar spatial associations between areas of shallow Miocene detachment faulting and extensive pediments occur across large areas of the Sonoran Desert and along a northwest-trending zone of extensive pediments west of the valley of the lower Colorado River in the eastern Mojave Desert (Dohrenwend, unpublished data). Range and basin morphometry and range-front morphology within these areas suggests significantly less high-angle normal faulting during the late Tertiary and Quaternary than in other areas of the province. Consequently, the generally low local topographic relief (ca. 100 to 200 m?) that has been inferred for active areas of shallow deformation along closely spaced listric normal faults (Zoback and others, 1981) has probably remained low in these areas since Miocene time.

Pediment Age. The pediments of the southwestern Basin and Range are, at least in part, relict forms of considerable age. The ubiquitous presence of deep-weathering profiles beneath extensive pediments on granitic rocks suggest at least preQuaternary ages for the original surfaces of these pediments (Oberlander, 1972, 1974; Moss, 1977). Moreover, the local burial of Tertiary pediment remnants by late Miocene and Pliocene basalt flows (Fig. 19) provides convincing evidence for a long

Figure 19. Granite Springs pediment dome (foreground) and a late Tertiary pediment buried by 3.6- to 4.7-Ma lava flows of the Cima volcanic field (on the skyline). The Granite Springs pediment cuts across deeply weathered quartz monzonite; the buried late Tertiary pediment truncates quartz monzonite and terrigenous sedimentary strata. Aerial view east towards the northern part of the Cima volcanic field.

history of pediment evolution (Oberlander, 1972, 1974; Dohrenwend and others, 1984b, 1985, 1986b). North of the San Bernardino Mountains in the southwestern Mojave, 8- to 9-Ma lava flows bury remnant pediment surfaces cut across saprolitic-weathering profiles. These pediment remnants now stand as much as 150 m above adjacent pediment surfaces (Oberlander, 1972, 1974). In the vicinity of Cima dome in the northeastern Mojave, 3.5- to 5.0-Ma lava flows bury remnant pediments cut across less intensely weathered materials; and flows younger than 1.0 Ma bury pediment remnants locally capped by soils similar to those developed on adjacent Quaternary surfaces. Differences in height between buried pediment remnants and adjacent actively downwasting pediment surfaces in the Cima area are systematically related to the ages of the lava flows overlying the pediment remnants (Dohrenwend and others, 1984b, 1986b). These relations indicate that, since late Miocene time, pediment surfaces cut across deeply weathered granitic rocks have evolved more or less continuously by progressive stripping of a thick late Tertiary weathering mantle (Oberlander, 1972, 1974; Moss, 1977; Dohrenwend and others, 1984b, 1986b).

Morphometric Analysis. Combined with an incomplete knowledge of neotectonic history and a lack of adequate age control for most pediments, this long history of pediment evolution has confounded numerous attempts to define the process of

pediment formation in the southwestern Basin and Range. Neither the present morphologic characteristics of these pediments nor the processes involved in their continuing evolution are clearly related to the processes of original pediment formation. It is hardly surprising, therefore, that morphometric analyses of these pediments have met with indifferent success. Mammerickx (1964) concluded that differences in pediment slope could not be satisfactorily explained by lithologic differences in pediment substrate or by variations in the size of drainage basins tributary to the pediments. Cooke (1970b) determined a very weak inverse relation between average pediment slope and pediment length and a weak positive correlation between average pediment slope and pediment association relief/length ratio. However, at least some correlation between these parameters is to be expected as a consequence of their interdependent geometry. Conventional morphometric analysis assumes that "relations among morphometric variables for a group of forms distributed in space indicate the changing relations among those variables that an individual form may experience through time" (Cooke, 1970b, p. 36). This assumption is probably not valid for the granitic pediments of the southwestern Basin and Range because of the relict nature of these landforms. "Conventional morphometric analysis of a relict landscape that has been stripped to a fossil weathering front can at best reveal anomalies that indicate lack of adjustment between form and process in the landscape" (Oberlander, 1974, p. 866). The late Tertiary and Quaternary evolution of these landforms does not necessarily reflect their middle Tertiary origin, and present processes of pediment modification may be very different from past processes of pediment formation.

Processes of Pediment Formation and Modification. As Cooke and Warren (1973) aptly point out, attempts to deduce the formation of an existing landform from observations of the processes presently operating on that landform are often unsuccessful becuase this approach commonly confuses cause and effect. For example, the observation that pediment surfaces are typically dissected along mountain fronts has been cited to support the idea that pediments are "born dissected" (Gilluly, 1937; Rahn, 1967). However, a similar spatial association is characteristic of alluvial piedmonts, yet "congenital dissection" has not been proposed for alluvial fans. Clearly, mere spatial association does not establish a cause-and-effect relation between form and process, particularly in the case of long-lived landforms that have evolved through complex tectonic and climatic histories (where the relative dominance of the various geomorphic processes associated with these landforms may have changed dramatically through time). Moreover, it should be equally apparent that, even if a cause-and-effect relation between form and process does exist, it may be unidirectional. For example," . . . sheetflooding cannot produce a planar surface because a planar surface is necessary for sheetflooding to occur," and, " . . . widespread weathering and occasional removal of weathered debris is unlikely to produce a pediment surface; it is more likely . . ." that these processes would " . . . maintain a pre-existing form . . ." (Cooke and Warren, 1973, p. 189). Thus, the assumption that pediment-forming and pediment-modifying processes are one and the same is not justified.

Pediment-Modifying Processes. Processes presently operating on the granitic pediment surfaces of the Mojave Desert are probably dominated by lateral migration of surface flow concentrated in shallow gullies and anastomosing washes (Rahn, 1967; Cooke and Mason, 1973; Dohrenwend and others, 1984b, 1986b). Most active pediment surfaces are integral parts of the piedmont plain; under stable baselevel conditions, their surfaces are likely divisible into three general process zones: an upper zone of erosion, an intermediate zone of transportation, and a lower zone of aggradation (Dohrenwend and others, 1984b, 1986b). "The precise form of the plain at any one time, and the distribution of the erosional and depositional forms within it, are reflections of the relationships among variables in an open system . . .", specifically, relations between rates of debris supply and removal (Cooke and Mason, 1973, p. 59). If the rate of debris supply is greater than the rate of removal, the boundaries between these process zones will migrate upslope; conversely, if the rate of debris removal is greater than the rate of supply, the boundaries will migrate downslope. That is, the precise extent of each zone on any given piedmont is determined by the ratio of stream power to critical power for the drainages on that piedmont. Stream power exceeds critical power in the zone of erosion,

approximately equals critical power in the zone of transportation, and falls short of critical power in the zone of aggradation (Bull, 1979). These concepts are supported by morphostratigraphic relations in the Apple Valley area of the southwestern Mojave (Cooke and Mason, 1973) and by patterns of pediment dome erosion in the area of the Cima volcanic field in the northeastern Mojave (Dohrenwend and others, 1984b, 1986b). However, because the pediments of both areas are granitic pediments that are not backed by mountain fronts, even this simplistic model does not necessarily apply to all active pediment surfaces.

Other processes known to be actively modifying present pediment surfaces in the Basin and Range include sheetwash, rill wash, surface and subsurface weathering, and soil formation. These processes have been documented by morphologic and stratigraphic analysis and by anecdotal description in studies that have been comprehensively summarized by Hadley (1967) and Cooke and Warren (1973). However, the relative importance of these various processes has not been convincingly demonstrated, either observationally or theoretically.

The possibility that the relative importance of various pediment-modifying processes may change through time tends to obscure relationships between these processes and pediment form. An instructive example of significant process change through time is provided by the evolutionary history of the granitic pediments of the Mojave and Sonoran deserts. The granitic pediments of the Mojave region were apparently fully developed before 9 Ma and have been subsequently modified only by removal of a thick saprolitic regolith formed during Miocene time (Oberlander, 1972, 1974). According to Oberlander, the late Miocene Mojave was probably a semiarid landscape characterized by extensive cut and fill surfaces of low relief surmounted by steep-sided hills and supporting an open woodland interspersed with grassy plains. Active pedimentation within this landscape appears to have involved an approximate balance between regolith erosion (probably by slope wash, rill wash, and channelized flow processes) and regolith renewal (by chemical breakdown of granitic rock along a subsurface weathering front). Increasing aridity and an accompanying deterioration in vegetative cover induced by uplift of the Transverse ranges apparently upset this balance and triggered the stripping of these regolith-covered pediments and concomitant formation of the relatively smooth surfaces of the present pediments. This general scenario is supported by a variety of evidence, including remnants of saprolitic-weathering profiles beneath late Tertiary lava flows, continuity between these relict profiles and boulder mantles on granitic hillslopes, the elevated positions of these relict surfaces (as much as 150 m) above present erosion surfaces, and general grusification to depths of as much as 40 m below present pediment surfaces. However, Oberlander's further assertion that the regolith stripping was accomplished by parallel rectilinear backwearing of regolith-covered slopes is contradicted by relations between K-Ar–dated remnants of Tertiary and Quaternary pediment surfaces and modern pediment surfaces in the eastern Mojave (Dohrenwend and others, 1984b, 1986b). These relations

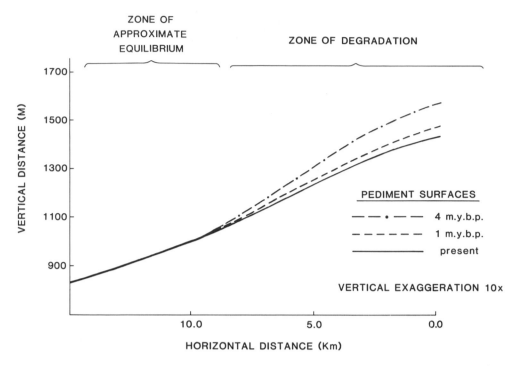

Figure 20. Empirical model of pediment dome evolution in the vicinity of the Cima volcanic field. The Pliocene and Pleistocene surfaces were reconstructed using a smoothed plot of average downwasting rates versus distance from dome summits. Crestal and upper flank areas have downwasted, midflank areas have remained in a state of approximate equilibrium (i.e., little or no downwasting or aggradation over time), and lower flank areas (not shown) have probably aggraded. Horizontal distance is measured from the dome summit (from Dohrenwend and others, 1986b).

document a Pliocene and Pleistocene history of essentially continuous downwearing on the crests and upper flanks of several large pediment domes in the area of the Cima volcanic field (Fig. 20). In either case, whether backwearing or downwearing, this scenario only relates to pediment modification; the process of pediment formation (i.e., development of the extensive regolith-covered surfaces prior to 9 Ma) remains unresolved.

Pediment-Forming Processes. Deducing the processes of pediment formation is a much more difficult task than documenting the processes of pediment modification; the relative difficulty is perhaps best illustrated by the fact that more than 80 years of careful investigation has failed to generate a generally accepted model of pediment formation. Factors contributing to this difficulty include: the length of time generally required for pediment formation, the complex tectonic and climatic history of the Basin and Range; the tendency to confuse pediment-modifying and pediment-forming processes; the general lack of reliable time control on most pediments; and the practical difficulties inherent in mapping the subsurface limits of pediment surfaces.

Perhaps one of the most durable statements on the formation of fringing pediments was made by Sharp (1940) when summarizing his conclusions regarding the pediments along the Ruby-Humboldt Range in northeastern Nevada. "(1) Pediments are formed by lateral planation, weathering, rill wash, and rain wash. The relative efficacy of these various processes is different under different geologic, topographic, and climatic conditions. (2) Lateral planation is most effective along large permanent streams and in areas of soft rocks. (3) Weathering, rill wash, and rain wash are most effective in areas of ephemeral streams, hard rocks, and low mountain mass. (4) All variations from pediments cut entirely by lateral planation to those formed entirely by other processes are theoretically possible, although in this area the end members of the series were not observed and perhaps do not actually exist in nature" (Sharp, 1940, p. 368). Although specifically concerned with fringing pediments along the Ruby–East Humboldt Range, this summation serves as a general descriptive model for pediment development along range fronts. Also, it emphasizes the probability that different pediments have most likely been formed by different combinations of processes. Indeed, the extensive pediments on deeply weathered granite in the Mojave and Sonoran deserts may well have formed by a combination of processes that is quite different from those responsible for the formation of fringing pediments in the Great Basin.

Historically, most investigations of pediment formation in the Basin and Range have assumed parallel slope retreat as the general mode of pediment formation and have attempted to determine the specific processes responsible for this backwearing. Thus, most investigations have focused on erosion of interfluvial areas along mountain fronts (i.e., formation of the piedmont junction or piedmont angle). Piedmont-angle formation and interfluve retreat are usually attributed to: lateral trimming by mountain-sourced drainage, differences between degradational processes operating on the mountain front and on the pediment, or some combination of these two general explanations (Cooke and Warren, 1973; Hadley, 1967).

It seems clear from qualitative morphological comparisons of active fault-bounded range fronts (where slickensides are locally preserved on steep-faceted spurs and the trace of the principal range-bounding fault is coincident with the base of the range front) and inactive fault-bounded range fronts (where steep-rounded spur ends form an abrupt angle with the encroaching piedmont and the trace of the principal range-bounding fault lies as much as several kilometers basinward of the present mountain front) that general backwearing along range fronts is a dominant mode of range-front degradation in the Basin and Range. In those few instances where field relations permit estimates of average long-term erosion rates (Table 4), rates of range-front retreat range up to as much as 35 cm/10^3 yr (0.35 km/10^6yr). Assuming these maximum rates are a reliable indicator of the general efficacy of backwearing in the late Cenozoic Basin and Range, then the formation of the relatively narrow pediments fringing the fault-bounded range fronts of the Great Basin (e.g., the 1- to 2-km-wide pediment along the east flank of the Ruby Mountains [Sharp, 1940]; or the 1-km-wide pediment along the west flank of the East Range [Wallace, 1978]) would require at least several million years, not an unreasonable estimate. Formation of the much more extensive pediments of the Mojave and Sonoran deserts would require periods in excess of 10 million years, a somewhat less reasonable estimate.

Remembering that (1) many of the larger granitic pediments of the Mojave and Sonoran deserts were fully developed before 9 Ma (Oberlander, 1974; Dohrenwend and others, 1984b, 1986b), (2) extreme extension on shallow detachment surfaces continued in this region until as late as 15 to 13 Ma (Zoback and others, 1981), and (3) "classic" basin and range deformation in this area occurred primarily between 13 and 10 Ma, but may have continued until sometime between 10 and 6 Ma (Shafiqullah and others, 1980); it would appear that sufficient time was not available for the formation of the larger pediments by backwearing or slope retreat in an environment of general tectonic stability. Either these pediments were not fully developed by 9 Ma (which appears unlikely in view of the evidence to the contrary), or general range-front backwearing in this region proceeded at a significantly faster rate than available evidence would suggest, or some other process or combination of processes was responsible for the formation of these truncated bedrock surfaces.

One possible alternative mechanism involves the extension and enlargement of embayments along range fronts. Embayments make up a significant proportion of the overall lengths of degraded mountain fronts (Bull and McFadden, 1977; Bull, 1978; Parsons and Abrahams, 1984). Extension and enlargement of embayments necessarily results in the removal of the intervening divides and, therefore, in the retreat of the range front (Gilluly, 1937; Sharp, 1940; Parsons and Abrahams, 1984). Thus, range-front retreat is the net result of spur retreat and divide removal. In addition, concentrated fluvial action is significantly more important within embayments than along outward-facing segments of the range front (Parsons and Abrahams, 1984), and embayment slopes are more likely to be eroded by vigorous lateral planation. Consequently, embayment enlargement is likely to proceed more rapidly than spur retreat, and divide removal should significantly increase the rate of range-front retreat over that which could be effected by spur retreat alone.

Another possible model involves the downwasting and regrading of deeply weathered terrains with initially low or moderately low relief. (This model differs from that proposed by Oberlander [1972, 1974] only in the specification of this initial condition.) Areas of active shallow detachment style extensional formation are thought to be characterized by generally low (less than 200 m?) local relief (Zoback and others, 1981). In such areas, erosional processes would likely proceed at significantly lower rates than in areas of rapid high-angle normal faulting; and if such areas were subjected to a somewhat moister climate than at present (as argued by Oberlander), extensive deeply weathered areas of low relief would likely develop. Downwasting at rates comparable to those documented for areas of the central and southern Basin and Range over the past 1 to 10 m.y. (Table 4) could, within about 4 to 8 m.y., reduce local highs in these terrains to produce the extensive late Miocene pediment domes of the Mojave Desert. This speculative hypothesis is untested, but it does serve to illustrate the possibility of a genetic relation between pediment domes and detachment deformation. Additional research regarding the age and tectonic associations of pediment domes is needed before the question of the origin of these relict landforms can be fully resolved.

CONCLUSIONS

The landscape of the Basin and Range Province is a complex mosaic of juxtaposed relict and active geomorphic elements. Tectonic elements include constructional landforms of all scales, the most significant being the basins and ranges themselves. Under the influence of a generally semiarid to arid late Tertiary and Quaternary climate, degradation of this tectonic framework has been slow; and partly degraded tectonic landforms millions of years old still dominate even the more tectonically quiescent landscapes of the province. However, the impact of time is conspicuous in these tectonically inactive areas, providing numerous examples of degradational evolution including progressive range-front degradation and pediment development. Superimposed upon this partly degraded tectonic framework are the subordinate effects of base level change and climatic change. Locally, base level change has been an important factor; but even in areas of significant incision (e.g., the valley of the lower Colorado River, or the Manix and Tecopa basins of the Mojave Desert), the effects of such change have been limited both in magnitude and in space. In contrast, climatic change has conspicuously affected the development of landforms and the distribution and character of surficial deposits throughout the province. Quaternary climate has varied significantly, producing a variety of climatically relict landforms and deposits (including "telescoped" fans; partly stripped colluvial wedges; sand dunes, sheets, and ramps; and lacustrine plains and shorelines). Relicts of the Pleistocene-to-Holocene transition are particularly widespread and conspicuous; however in most areas, even these effects constitute little more than a superficial overprint on the underlying tectonic framework.

Although the effects of tectonism generally dominate the overall landscape of the Basin and Range, the influences of climate and climatic change have profoundly affected the operation and distribution of geomorphic processes throughout the province. Climatically induced transgressions of process thresholds (e.g., eolian erosion and deposition, soil development, pavement formation, hillslope erosion, and piedmont erosion and aggradation) have been common and widespread within this semiarid to arid region. Climatic change has affected geomorphic processes directly through changes in insolation, precipitation, evapotranspiration, and wind velocity, and indirectly through changes in vegetation, soils, surficial deposits, and landforms. Indirect influences are more subtle and more difficult to predict, but can be equally, if not more, significant in changing the distribution, character, and rates of geomorphic processes. These process perturbations have left a rich and complex record of climatic change in the surficial deposits, soils, rock varnish, and landforms of the region.

The preponderance of evidence suggests that Quaternary climatic changes have been approximately synchronous over large areas of the Basin and Range. Therefore, the effects of these changes are potentially useful for regional correlation of the Quaternary record. Such correlations enable the approximate and (or) relative dating of various surficial deposits and landscape elements, and thus provide the principal means for regional evaluation of neotectonic activity as well as documenting regional histories of process change. Of course, not all of the geomorphic responses to a climatic change are necessarily synchronous with that change; complex interactions between processes, landforms, and deposits affect the timing and distribution of process operation. Moreover, the timing, magnitude, and even the direction of specific geomorphic responses to a climatic change depends, at least in part, on antecedent climatic and geomorphic conditions, particularly in the semiarid to arid climate of the Basin and Range. Such complications render climatic interpretation of the geomorphic and stratigraphic record difficult and often frustrate attempts to relate form and process.

Because many elements of the surficial geologic record

of the Basin and Range are relict, knowledge of presently active geomorphic processes is not sufficient by itself to fully interpret this record. Knowledge of past processes and the history of process change is also necessary. The interrelations between geomorphic processes, landforms, and surficial deposits vary temporally as well as spatially and are affected by sequence in time as well as by distribution in space. For example, degradation of fault scarps and range fronts is complicated by renewed and (or) continuing tectonic activity. In the case of most fault scarps in alluvium, the rate of scarp degradation is rapid relative to the recurrence of surface rupture, so that the resulting compound scarps can be recognized and analyzed by a variety of means. In the case of most range fronts, however, separation of the effects of tectonic activity and time is generally much more difficult. Range-front morphometry, age and distribution of piedmont deposits, and diagnostic landforms can be used to define general classes of differential vertical movement along upfaulted range fronts. However, several problems remain: distinguishing range fronts with different tectonic histories as well as different levels of tectonic activity; classifying areas of horizontal tectonic movement; and recognizing range fronts where assumptions of similar initial forms and similar initiating processes are not valid.

Such complications notwithstanding, analysis of progressive degradational changes within time series of relict landforms appears to be one of the better ways to resolve contradictions between theoretical models of landscape evolution. Such studies can be particularly productive where landforms with similar composition, structure, tectonic history, and initial form can be reliably dated; and these situations are relatively common in the Basin and Range. Landforms that have been carefully studied in this regard include: normal fault scarps, fluvial scarps, and lacustrine scarps in alluvium; fault-bounded range fronts; abandoned piedmont surfaces; and cinder cones and lava flows. Each of these various types of landforms degrades in response to a unique combination of processes, which is determined by the scale, lithology and structure, physiographic setting, and geometry of the degrading landforms. Degradation operates progressively and cumulatively (i.e., each stage is affected by all preceeding stages and affects all subsequent stages). Moreover, although the processes involved have undoubtedly been subjected to accelerations and decelerations associated with climatic change, these climatically induced variations generally operate within intervals of graded time that are averaged out over the substantially longer periods of cyclic time typically required for the degradational evolution of most larger landforms. Thus, the progressive degradational changes recorded by these landforms, being relatively insensitive to the effects of climatic change, are directly relevant to the general problem of landscape evolution.

REFERENCES

Abrahams, A. D., Parsons, A. J., and Hirsh, P. J., 1985, Hillslope gradient–particle size relations; Evidence for the formation of debris slopes by hydraulic processes in the Mojave Desert: Journal of Geology, v. 93, p. 347–357.

Allen, C. C., 1978, Desert varnish of the Sonoran Desert: Optical and electron microanalysis: Journal of Geology, v. 86, p. 743–752.

Anstey, R. L., 1965, Physical characteristics of alluvial fans: U.S. Army Material Command, U.S. Army Natick Laboratories, Technical Report ES-20, 109 p.

Barry, R. G., 1983, Late-Pleistocene climatology, *in* Porter, S. C., ed., Late Quaternary environments of the United States, Volume 1, The late Pleistocene: Minneapolis, University of Minnesota Press, p. 390–407.

Bachhuber, F. W., 1982, Quaternary history of the Estancia Valley, central New Mexico: New Mexico Geological Society Guidebook 33, p. 343–346.

Bachman, G. O., and Machette, M. N., 1977, Calcic soils and calcretes in the southwestern United States: U.S. Geological Survey Open-File Report 77-794, 163 p.

Beaty, C. B., 1963, Origin of alluvial fans, White Mountains, California and Nevada: Association of American Geographers Annals, v. 53, p. 516–535.

Benson, L. V., 1981, Paleoclimatic significance of lake level fluctuations in the Lahontan Basin, Nevada: Quaternary Research, v. 16, p. 390–403.

Blackwelder, E., 1928, Mudflow as a geologic agent in semiarid mountains: Geological Society of America Bulletin, v. 39, p. 465–484.

Brackenridge, G. R., 1978, Evidence for a cold, dry full-glacial climate in the American Southwest: Quaternary Research, v. 9, p. 22–40.

Bucknam, R. C., and Anderson, R. E., 1979, Estimation of fault-scarp ages from a scarp-height–slope-angle relationship: Geology, v. 7, p. 11–14.

Bull, W. B., 1968, Alluvial fans: Journal of Geological Education, v. 16, p. 101–106.

——— , 1972, Recognition of alluvial-fan deposits in the stratigraphic record: Society of Economic Paleontologists and Mineralogists Special Publication 16, p. 63–83.

——— , 1973, Local base-level processes in arid fluvial systems: Geological Society of America Abstracts with Programs, v. 5, p. 562.

——— , 1974, Geomorphic tectonic analysis of the Vidal region, *in* Woodward, McNeil and Associates, Information concerning site characteristics, Vidal Nuclear Generating Station Units 1 and 2, Appendix 2.5B, amendment 1 (Geology and Seismology): Los Angeles, Southern California Edison Company, 66 p.

——— , 1977, The alluvial fan environment: Progress in Physical Geography, v. 1, p. 222–270.

——— , 1978, Tectonic geomorphology of the Mojave Desert: unpublished technical report, U.S. Geological Survey Project no. 14-08-001-G-394, 188 p.

——— , 1979, Threshold of critical power in streams: Geological Society of America Bulletin, Part I, v. 90, p. 453–464.

——— , 1984, Tectonic geomorphology: Journal of Geological Education, v. 32, p. 310–324.

Bull, W. B., and McFadden, L. D., 1977, Tectonic geomorphology north and south of the Garlock Fault, California, *in* Doering, D. O., ed., Geomorphology in arid regions: State University of New York at Binghamton, Publications in Geomorphology, p. 115–138.

Carr, W. J., 1984, Regional structural setting of Yucca Mountain, southwestern Nevada, and late Cenozoic rates of tectonic activity in part of the southwestern Great Basin, Nevada and California: U.S. Geological Survey Open-File Report, OFR-84-854, 109 p.

Carson, M., and Kirkby, M., 1972, Hillslope form and process: London, Cambridge University Press, 475 p.

Christenson, G. E., and Purcell, C., 1985, Correlation and age of Quaternary alluvial-fan sequences, Basin and Range Province, southwestern United States, *in* Weide, D. L., Soils and Quaternary geology of the southwestern United States: Geological Society of America Special Paper 203, p. 115–122.

Christiansen, R. L., and McKee, E. H., 1978, Late Cenozoic volcanic and tectonic evolution of the Great Basin and Columbia intermontane regions: Geological

Society of America Memoir 152, p. 283–311.

Coney, P. J., 1980, Cordilleran metamorphic core complexes; An overview: Geological Society of America Memoir 153, p. 7–31.

——— , 1983, The plate tectonic setting of Cordilleran deserts, *in* Wells, S. G., and Haragan, D. R., eds., Origin and evolution of deserts: Albuquerque, University of New Mexico Press, p. 81–97.

Coleman, S. M., and Watson, K., 1983, Ages estimated from a diffusion equation model for scarp degradation: Science, v. 221, p. 263–265.

Cooke, R. U., 1970a, Stone pavements in deserts: Annals of the Association of American Geographers, v. 60, p. 560–577.

——— , 1970b, Morphometric analysis of pediments and associated landforms in the western Mojave Desert, California: American Journal of Science, v. 269, p. 26–38.

Cooke, R. U., and Mason, P., 1973, Desert Knolls pediment and assoicated landforms in the Mojave Desert, California: Revue de Geomorphologie Dynamique, v. 20, p. 71–78.

Cooke, R. U., and Reeves, R. W., 1972, Relations between debris size and the slope of mountain fronts and pediments in the Mojave Desert, California: Zeitschrift für Geomorphologie, v. 16, p. 76–82.

Cooke, R. U., and Warren, A., 1973, Geomorphology in deserts: Berkeley and Los Angeles, University of California Press, 394 p.

Denny, C. S., 1965, Alluvial fans in the Death Valley region, California and Nevada: U.S. Geological Survey Professional Paper 446, 62 p.

——— , 1967, Fans and pediments: American Journal of Science, v. 265, p. 81–105.

Dickinson, W. R., and Snyder, W. S., 1979, Geometry of subducted slabs related to San Andreas transform: Journal of Geology, v. 87, p. 609–627.

Dohrenwend, J. C., 1982a, Surficial geology, Walker Lake 1° by 2° Quadrangle, Nevada-California: U.S. Geological Survey Miscellaneous Field Studies Map, MF-1382-C.

——— , 1982b, Late Cenozoic faults, Walker Lake 1° by 2° Quadrangle, Nevada-California: U.S. Geological Survey Miscellaneous Field Studies Map, MF-1382-D.

——— , 1982c, Tectonic control of pediment distribution in the western Great Basin: Geological Society of America Abstracts with Programs, v. 14, p. 161.

——— , 1984, Nivation landforms in the western Great Basin and their paleoclimatic significance: Quaternary Research, v. 22, p. 275–288.

——— , 1985, Patterns and processes of middle and late Quaternary dissection in the Tecopa Basin, California, *in* Hale, G. R., ed., Quaternary lakes of the eastern Mojave Desert, California: Friends of the Pleistocene, Pacific Cell 1985 Annual Meeting Guidebook, p. 113–144.

Dohrenwend, J. C., Abrahams, A. D., and Turrin, B. D., 1987, Drainage development on basaltic lava flows, Cima volcanic field, southeast California, and Lunar Crater volcanic field, south-central Nevada: Geological Society of America Bulletin, v. 98, (in press).

Dohrenwend, J. C., McFadden, L. D., Turrin, B. D., and Wells, S. G., 1984a, K-Ar dating of the Cima volcanic field, eastern Mojave Desert, California: Geology, v. 12, p. 163–167.

Dohrenwend, J. C., Wells, S. G., Turrin, B. D., and McFadden, L. D., 1984b, Rates and trends of late Cenozoic landscape degradation in the area of the Cima volcanic field, Mojave Desert, California, *in* Dohrenwend, J. C., ed., Surficial geology of the eastern Mojave Desert, California: Geological Society of America 1984 Annual Meeting Guidebook, p. 101–115.

Dohrenwend, J. C., Turrin, B. D., and Diggles, M. F., 1985, Topographic distribution of dated basaltic lava flows in the Reveille Range, Nye County, Nevada; Implications for late Cenozoic erosion of upland areas in the Great Basin: Geological Society of America Abstract with Programs, v. 17, p. 351.

Dohrenwend, J. C., Wells, S. G., and Turrin, B. D., 1986a, Degradation of Quaternary cinder cones in the Cima volcanic field, Mojave Desert, California: Geological Society of America Bulletin, v. 97, p. 421–427.

Dohrenwend, J. C., Wells, S. G., McFadden, L. D., and Turrin, B. D., 1986b, Pediment dome evolution in the eastern Mojave Desert, California, *in* Gardiner, V., ed., Proceedings of the First International Conference on Geomorphology: London, Wiley-Interscience (in press).

Dorn, R. I., 1984a, Geomorphological interpretation of rock varnish in the Mojave Desert, *in* Dohrenwend, J. C., ed., Surficial geology of the eastern Mojave Desert, California: Geological Society of America 1984 Annual Meeting Guidebook, p. 150–161.

—— , 1984b, Cause and implications of rock varnish microchemical laminations: Nature, v. 310, p. 767–770.

Dorn, R. I., and Oberlander, T. M., 1981a, Microbial origin of desert varnish: Science, v. 213, p. 1245–1247.

—— , 1981b, Rock varnish origin, characteristics, and usage: Zeitschrift für Geomorphologie, v. 25, p. 420–436.

—— , 1982, Rock varnish: Progress in Physical Geography, v. 6, p. 317–367.

Duffield, W. A., Bacon, C. R., and Dalrymple, G. B., 1980, Late Cenozoic volcanism, geochronology, and structure of the Coso Range, Inyo County, California: Journal of Geophysical Research, v. 85, p. 2381–2404.

Eaton, G. P., 1982, The Basin and Range Province; Origin and tectonic significance: Annual Reviews Earth and Planetary Science, v. 10, p. 409–440.

Elvidge, C. D., 1979, Distribution and formation of desert varnish in Arizona [M.S. thesis]: Tempe, Arizona State University, 109 p.

—— , 1982, Reexamination of the rate of desert varnish formation reported south of Barstow, California: Earth Surface Processes and Landforms, v. 7, p. 345–348.

Elvidge, C. D., and Moore, C. B., 1979, A model for desert varnish formation: Geological Society of America Abstracts with Programs, v. 11, p. 271.

Fenneman, N. M., 1931, Physiography of the western United States: New York, McGraw-Hill, 534 p.

Fleishouer, A. L., Jr., and Stone, W. J., 1982, Quaternary geology of Lake Animas, Hidalgo County, New Mexico: Socorro, New Mexico Bureau of Mines and Mineral Resources Circular 174, 25 p.

Galloway, R. W., 1970, The full-glacial climate in the southwestern United States: Annals of the Association of American Geographers, v. 60, p. 245–256.

—— , 1983, Full-glacial southwestern United States; Mild and wet or cold and dry?: Quaternary Research, v. 19, p. 236–248.

Gilbert, C. M., and Reynolds, M. W., 1973, Character and chronology of basin development, western margin of the Basin and Range Province: Geological Society of America Bulletin, v. 84, p. 2489–2510.

Gilbert, G. K., 1890, Lake Bonneville: U.S. Geological Survey Monograph 1, 438 p.

Gile, L. H., Peterson, F. F., and Grossman, R. B., 1979, The Desert Soil Project monograph: Washington, D.C., U.S. Soil Conservation Service, 984 p.

Gilluly, J., 1937, Physiography of the Ajo region, Arizona: Geological Society of America Bulletin, v. 48, p. 323–348.

Glock, W. S., 1931, The development of drainage systems; A synoptic view: Geographical Reviews, v. 21, p. 474–482.

Goudie, A. S., and Day, M. J., 1980, Disintegration of fan sediments in Death Valley, California, by salt weathering: Physical Geography, v. 1, p. 126–137.

Hadley, R. F., 1967, Pediments and pediment-forming processes: Journal of Geological Education, v. 15, p. 83–89.

Hale, G. R., 1985, Mid-Pleistocene overflow of Death Valley toward the Colorado River, *in* Hale, G. R., ed., Quaternary lakes of the eastern Mojave Desert, California: Friends of the Pleistocene, Pacific Cell 1985 Annual Meeting Guidebook, p. 113–144.

Hamblin, W. K., 1976, Patterns of displacement along the Wasatch fault: Geology, v. 4, p. 619–622.

Hanks, T. C., Bucknam, R. C., Lajoie, K. R., and Wallace, R. E., 1984, Modification of wave-cut and faulting-controlled landforms: Journal of Geophysical Research, v. 89, p. 5771–5790.

Hely, A. G., 1967, Lower Colorado River water supply; its magnitude and distribution: U.S. Geological Survey Professional Paper 486-D, 54 p.

Hillhouse, J. W., 1987, Late Tertiary and Quaternary geology of the Tecopa basin, southeastern California: U.S. Geological Survey Miscellaneous Geologic Investigations Map, scale 1:48,000, (in press).

Hooke, R. LeB., 1967, Processes on arid-region alluvial fans: Journal of Geology, v. 75, p. 438–460.

Hooke, R. LeB., 1968, Steady-state relationships on arid-region alluvial fans in closed basins: American Journal of Science, v. 266, p. 609–629.

Hooke, R. LeB., and Rohner, W. L., 1977, Relative erodibility of source-area rock types, as determined from second-order variations in alluvial fan size: Geological Society of America Bulletin, v. 88, p. 1177–1182.

Hoover, D. L., Swadley, W. C., and Gordon, A. J., 1981, Correlation characteristics of surficial deposits with a description of surficial stratigraphy in the Nevada Test Site region: U.S. Geological Survey Open-File Report 81-512, 27 p.

Howard, R. B., Cowen, B., and Inouye, D., 1977, Reappraisal of desert varnish formation: Geological Society of America Abstracts with Programs, v. 9, p. 438–439.

Hunt, C. B., 1974, Natural regions of the United States and Canada: San Francisco, W. H. Freeman and Company, 725 p.

Lattman, L. H., 1973, Calcium carbonate cementation of alluvial fans in southern Nevada: Geological Society of America Bulletin, v. 84, p. 3013–3028.

—— , 1977, Weathering of caliche in southern Nevada, *in* Doering, D. O., ed., Geomorphology in arid regions: State University of New York at Binghamton, Publications in Geomorphology, p. 27–50.

—— , 1983, Effect of caliche on desert processes, *in* Wells, S. G., and Haragan, D. R., eds., Origin and evolution of deserts: Albuquerque, University of New Mexico Press, p. 101–109.

Loeltz, O. J., Irelan, B., Robison, J. H., and Olmstead, F. H., 1975, Geohydrologic reconnaissance of the Imperial Valley, California: U.S. Geological Survey Professional Paper 486-K, 54 p.

Lustig, L. K., 1965, Clastic sedimentation in Deep Springs Valley, California: U.S. Geological Survey Professional Paper, 352-F, p. 131–192.

—— , 1969, Trend surface analysis of the Basin and Range Province and some geomorphic implications: U.S. Geological Survey Professional Paper, 500-D, 70 p.

Machette, M. N., 1982, Quaternary and Pliocene faults in the La Jencia and southern part of the Albuquerque-Belen basins, New Mexico; Evidence of fault history from fault-scarp morphology and Quaternary geology, *in* Grambling, J. A., and Wells, S.G., eds., Albuquerque country II: New Mexico Geological Society, 33rd Annual Field Conference Guidebook, p. 161–170.

—— , 1985, Calcic soils of the southwestern United States; *in* Weide, D. L., ed., Soils and Quaternary geology of the southwestern United States: Geological Society of America Special Paper 203, p. 1–22.

Mammerickx, J., 1964, Quantitative observations on pediments in the Mojave and Sonoran deserts (southwestern United States): American Journal of Science, v. 262, p. 417–435.

Marchand, D. E., 1971, Rates and modes of denudation, White Mountains, eastern California: American Journal of Science, v. 270, p. 109–135.

Mayer, L., 1982, Quantitative tectonic geomorphology with applications to neotectonics of northwestern Arizona [Ph.D. thesis]: Tucson, University of Arizona, 512 p.

—— , 1984, Quaternary faulting along the west side of the Virgin Mountains, Arizona and Nevada, *in* Dohrenwend, J. C., ed., Surficial geology of the eastern Mojave Desert, California: Geological Society of America 1984 Annual Meeting Guidebook, p. 175–183.

Mayer, L., and Bull, W. B., 1981, Impact of Pleistocene-Holocene climatic change on particle size distribution of fan deposits in southwestern Arizona: Geological Society of America Abstracts with Programs, v. 13, p. 95.

Mayer, L., Mergner-Keefer, M., and Wentworth, C. M., 1981, Probability models and computer simulation of landscape evolution: U.S. Geological Survey Open-File Report 81-656, 31 p.

Mayer, L., Gerson, R., and Bull, W. B., 1984, Alluvial gravel production and deposition; Useful indicator of Quaternary climatic changes in deserts, *in* Schick, A. P., ed., Channel processes; Water, sediment, catchment controls: Catena Supplement 5, p. 137–151.

McCoy, W. D., 1981, Quaternary aminostratigraphy of the Bonneville and Lahontan basins, western United States, with paleoclimatic implications [Ph.D. thesis]: Boulder, University of Colorado.

McFadden, L. D., 1982, The impacts of temporal and spatial climatic changes on alluvial soils genesis in southern California [Ph.D. thesis]: Tucson, University of Arizona, 430 p.

McFadden, L. D., and Bull, W. B., 1987, Quaternary soil development in the Mojave Desert, California, *in* Whitley, D. S., ed., Late Pleistocene archeology and environments in California: Salt Lake City, University of Utah Press (in press).

McFadden, L. D., Wells, S. G., Dohrenwend, J. C., Turrin, B. D., 1984, Cumulic soils formed in eolian parent materials on flows of the Cima volcanic field, Mojave Desert California, *in* Dohrenwend, J. C., ed., Surficial geology of the eastern Mojave Desert, California: Geological Society of America 1984 Annual Meeting Guidebook, p. 134–149.

McFadden, L. D., Wells, S. G., and Dohrenwend, J. C., 1985, The influence of eolian flux rates and climatic change on the development of stone pavements and associated soils in the Cima volcanic field, Mojave Desert, California: Geological Society of America Abstracts with Programs, v. 17, p. 368.

McFadden, L. D., Wells, S.G., and Dohrenwend, J. C., 1986, Cumulic soils formed in eolian parent materials on flows of the Cima volcanic field, Mojave Desert, California: Catena, v. 13, p. 361–389.

Melton, M. A., 1958, Geomorphic properties of mature drainage systems and their representation in an E4 phase space: Journal of Geology, v. 66, p. 35–54.

——, 1965, The geomorphic and paleoclimatic significance of alluvial deposits in southern Arizona: Journal of Geology, v. 73, p. 1–38.

Menges, C. M., and McFadden, L. D., 1981, Evidence for a latest Miocene to Pliocene transition from Basin-Range tectonic to post-tectonic landscape evolution in southeastern Arizona: Arizona Geological Society Digest, v. 13, p. 151–160.

Mifflin, M. D., and Wheat, M. M., 1979, Pluvial lakes and estimated pluvial climates of Nevada: Nevada Bureau of Mines and Geology Bulletin 94, 57 p.

Morrison, R. B., 1985, Pliocene/Quaternary geology, geomorphology, and tectonics of Arizona, *in* Weide, D. L., ed., Soils and Quaternary geology of the southwestern United States: Geological Society of America Special Paper 203, p. 123–146.

Morrison, R. B., Menges, C. M., and Lepley, L. K., 1981, Neotectonic maps of Arizona: Arizona Geological Society Digest, v. 13, p. 179–183.

Moss, J. H., 1977, Formation of pediments: scarp backwearing or surface downwasting? *in* Doering, D. O., ed., Geomorphology in arid regions: State University of New York at Binghamton, Publications in Geomorphology, p. 51–78.

Nash, D. B., 1980, Morphologic dating of degraded normal fault scarps: Journal of Geology, v. 88, p. 353–360.

——, 1984, Morphologic dating of fluvial terrace scarps and fault scarps near West Yellowstone, Montana: Geological Society of America Bulletin, v. 95, p. 1413–1425.

Nilsen, T. H., 1982, Alluvial fan deposits: American Association of Petroleum Geologists Memoir 31, p. 49–86.

Oberlander, T. M., 1972, Morphogenesis of granitic boulder slopes in the Mojave desert, California: Journal of Geology, v. 80, p. 1–20.

——, 1974, Landscape inheritance and the pediment problem in the Mojave desert of southern California: American Journal of Science, v. 274, p. 849–875.

Pain, C. F., 1985, Cordilleran metamorphic core complexes in Arizona; A contribution from geomorphology: Geology, v. 13, p. 871–874.

Parker, R. S., 1977, Experimental study of drainage basin evolution and its hydrologic implications: Fort Collins, Colorado State University Hydrology Paper 90, 51 p.

Parker, R. S., and Schumm, S. A., 1982, Experimental study of drainage networks, *in* Bryan, R., and Yair, A., eds., Badland geomorphology and piping: Norwich, Geo Books, Cambridge University Press, p. 153–168.

Parsons, A. J., and Abrahams, A. D., 1984, Mountain mass denudation and piedmont formation in the Mojave and Sonoran deserts: American Journal

of Science, v. 284, p. 255–271.

Perry, R. S., and Adams, J., 1978, Desert varnish; Evidence of cyclic deposition of manganese: Nature, v. 276, p. 489–491.

Pierce, K. L., and Colman, S. L., 1986, Effect of height and orientation (microclimate) on geomorphic degradation rates and processes, late-glacial terrace scarps in central Idaho: Geological Society of America Bulletin, v. 97, p. 869–885.

Ponti, D. J., 1985, The Quaternary alluvial sequence of the Antelope Valley, California, *in* Weide, D. L., ed., Soils and Quaternary geology of the southwestern United States: Geological Society of America Special Paper 203, p. 79–96.

Potter, R. M., and Rossman, G. R., 1977, Desert varnish; The importance of clay minerals: Science, v. 196, p. 1446–1448.

——, 1979, The manganese- and iron-oxide mineralogy of desert varnish: Chemical Geology, v. 25, p. 79–94.

Rahn, P., 1967, Inselbergs and nickpoints in southwestern Arizona: Zeitschrift für Geomorphologie, v. 10, p. 217–225.

Reynolds, S. J., and Spencer, J. E., 1985, Cenozoic extension and magmatism in Arizona, *in* Papers presented to the conference on heat and detachment in crustal extension on continents and planets: Houston, Texas, Lunar and Planetary Institute, Contribution no. 575, p. 128–132.

Saunders, I., and Young, A., 1983, Rates of surface processes on slopes, slope retreat, and denudation: Earth Surface Processes and Landforms, v. 8, p. 473–501.

Shafiqullah, M., Damon, P. E., Lynch, D. J., Reynolds, S. J., Rehrig, W. A., and Raymond, R. H., 1980, K-Ar geochronology and geologic history of southwestern Arizona and adjacent areas: Arizona Geological Society Digest, v. 12, p. 201–260.

Sharp, R. P., 1940, Geomorphology of the Ruby–East Humboldt Range, Nevada: Geological Society of America Bulletin, v. 51, p. 337–372.

——, 1984, Alluvial microstratigraphy, Mojave Desert, San Bernardino County, California: California Geology, v. 12, p. 139–145.

Shreve, R. L., 1969, Stream lengths and basin areas in topologically random channel networks: Journal of Geology, v. 77, p. 397–414.

Smith, H.S.U., 1967, Past versus present wind action in the Mojave Desert region, California: Bedford, Massachusetts, U.S. Air Force Cambridge Research Laboratories, AFCRL-67-0683, 26 p.

Smith, R.S.U., 1984, Eolian geomorphology of the Devils Playground, Kelso Dunes, and Silurian Valley, California, *in* Dohrenwend, J. C., ed., Surficial geology of the eastern Mojave Desert, California: Geological Society of America 1984 Annual Meeting Guidebook, p. 162–174.

Smith, G. I., and Street-Perrott, F. A., 1983, Pluvial lakes of the western United States, *in* Porter, S. C., ed., Late Quaternary environments of the United States, Volume 1, The late Pleistocene: Minneapolis, University of Minnesota Press, p. 190–212.

Snyder, C. T., Hardman, G., and Zdenek, F. F., 1964, Pleistocene lakes in the Great Basin: U.S. Geological Survey Miscellaneous Geologic Investigations Map I-416, scale 1:1,000,000.

Spaulding, W. G., and Graumlick, L. J., 1986, The last pluvial climatic episodes in the deserts of southwestern North America: Nature, v. 320, p. 441–444.

Spaulding, W. G., Leopold, E. B., and Van Devender, T. R., 1983, Late Wisconsin paleoecology of the American Southwest, *in* Porter, S. C., ed., Late Quaternary environments of the United States, Volume 1, The late Pleistocene: Minneapolis, University of Minnesota Press, p. 259–293.

Spencer, J. E., 1984, Role of tectonic denudation in warping and uplift of low-angle normal faults: Geology, v. 12, p. 95–98.

Springer, M. E., 1958, Desert pavement and vesicular layer of some desert soils in the desert of the Lahontan Basin, Nevada: Proceedings Soil Science Society of America, v. 22, p. 63–66.

Stewart, J. H., 1978, Basin-Range structure in western North America; A review: Geological Society of America Memoir 152, p. 1–31.

——, 1980, Regional tilt patterns of late Cenozoic Basin-Range fault blocks, western United States: Geological Society of America Bulletin, v. 91, p. 460–464.

Turrin, B. D., and Dohrenwend, J. C., 1984, K-Ar ages of basaltic volcanism in the Lunar Crater volcanic field, northern Nye County, Nevada; Implications for Quaternary tectonism in the central Great Basin: Geological Society of America Abstracts with Programs, v. 16, p. 679.

Turrin, B. D., Dohrenwend, J. C., Wells, S. G., and McFadden, L. D., 1984, Geochronology and eruptive history of the Cima volcanic field, eastern Mojave Desert, California, *in* Dohrenwend, J. C., ed., Surficial geology of the eastern Mojave Desert, California: Geological Society of America 1984 Annual Meeting Guidebook, p. 88–100.

Van Devender, T. R., 1973, Late Pleistocene plants and animals of the Sonoran Desert; A survey of ancient packrat middens in southwestern Arizona [Ph.D. thesis]: Tuscon, University of Arizona, 179 p.

—— , 1977, Holocene woodlands in southwestern deserts: Science, v. 198, p. 189–192.

Van Devender, T. R., and Spaulding, W. G., 1979, The development of vegetation and climate in the southwestern United States: Science, v. 204, p. 701–710.

Wallace, R. E., 1977, Profiles and ages of young fault scarps, north-central Nevada: Geological Society of America Bulletin, v. 88, p. 1267–1281.

—— , 1978, Geometry and rates of change of fault-generated range fronts, north-central Nevada: U.S. Geological Survey Journal of Research, v. 6, p. 637–650.

Warnke, D. A., 1969, Pediment evolution in the Halloran Hills, central Mojave Desert, California: Zeitschrift für Geomorphologie, v. 13, p. 357–389.

Wells, P. V., 1979, An equable glaciopluvial in the West; Peniglacial evidence of increased precipitation on a gradient from the Great Basin to the Sonoran and Chihuahuan deserts: Quaternary Research, v. 12, p. 311–325.

Wells, S. G., 1977, Geomorphic controls of alluvial fan deposition in the Sonoran Desert, southwestern Arizona, *in* Doering, D. O., ed., Geomorphology in arid regions: State University of New York at Binghamton, Publications in Geomorphology, v. 27–50.

—— , 1978, Geomorphic framework of an open drainage basin in the Basin and Range Province of southwestern Arizona: Geological Society of America Abstracts with Programs, v. 10, p. 153.

Wells, S. G., and Dohrenwend, J. C., 1985, Relict sheetflood bed forms on late Quaternary alluvial-fan surfaces in the southwestern United States: Geology, v. 13, p. 512–516.

Wells, S. G., and Schultz, J. D., 1979, Some factors influencing Quaternary calcrete formation and distribution in alluvial fill of arid basins: Geological Society of America Abstracts with Programs, v. 11, p. 538.

Wells, S. G., Ford, R. L., Grimm, J. P., Martinez, G. F., Pickle, J. D., Sars, S. W.,

and Weadock, G. L., 1982, Development of debris mantled hillslopes; An example of feedback mechanisms in desert hillslope processes: American Geomorphological Field Group Field Trip Guidebook, 1982 Conference, p. 141.

Wells, S. G., Dohrenwend, J. C., McFadden, L. D., Turrin, B. D., and Mahrer, K. D., 1984a, Types and rates of late Cenozoic geomorphic processes on lava flows of the Cima volcanic field, Mojave Desert, California, *in* Dohrenwend, J. C., ed., Surficial geology of the eastern Mojave Desert, California: Geological Society of America 1984 Annual Meeting Guidebook, p. 116–133.

Wells, S. G., McFadden, L. D., Dohrenwend, J. C., Bullard, T. F., Feilberg, B. F., Ford, R. L., Grimm, J. P., Miller, J. R., Orbock, S. M., and Pickle, J. D., 1984b, Late Quaternary geomorphic history of Silver Lake, eastern Mojave Desert, California; An example of the influence of climatic change on desert piedmonts, *in* Dohrenwend, J. C., ed., Surficial geology of the eastern Mojave Desert, California: Geological Society of America 1984 Annual Meeting Guidebook, p. 69–87.

Wells, S. G., Dohrenwend, J. C., McFadden, L. D., Turrin, B. D., and Mahrer, K. D., 1985, Late Cenozoic landscape evolution on lava flow surfaces of the Cima volcanic field, Mojave Desert, California: Geological Society of America Bulletin, v. 96, p. 1518–1529.

Wells, S. G., McFadden, L. D., Dohrenwend, J. C., 1987, Influence of late Quaternary climatic changes on geomorphic and pedogenic processes on desert piedmonts, Mojave Desert, California: Quaternary Research (in press).

Whitney, J. W., Swadley, W. C., and Shroba, R. R., 1985, Middle Quaternary sand ramps in the southern Great Basin, California and Nevada: Geological Society of America Abstracts with Programs, v. 17, p. 750.

Wilshire, H. G., Reneau, S. L., Schellentrager, G. W., and Barmore, R. L., 1983, Quaternary geomorphology of the Mohave Mountains, western Arizona: Geological Society of America Abstracts with Programs, v. 15, p. 387.

Wilson, L., 1968, Slopes, *in* Fairbridge, R. W., ed., Encyclopedia of geomorphology: New York, Reinhold Book Corporation, p. 1002–1020.

Young, A., 1972, Slopes: Edinburgh, Oliver and Boyd, 288 p.

Zoback, M. L., Anderson, R. E., and Thompson, G. A., 1981, Cenozoic evolution of the state of stress and style of tectonism of the Basin and Range Province of western United States: Philosophical Transactions of the Royal Society of London, Series A, v. 300, p. 407–434.

MANUSCRIPT ACCEPTED BY THE SOCIETY SEPTEMBER 29, 1986

Geological Society of America
Centennial Special Volume 2
1987

Chapter 10

Central America and the Caribbean

Thomas W. Gardner
Department of Geosciences, The Pennsylvania State University, University Park, Pennsylvania 16802
William Back
U.S. Geological Survey, 431 National Center, Reston, Virginia 22092
Thomas F. Bullard
Department of Geology, University of New Mexico, Albuquerque, New Mexico 87131
Paul W. Hare
Department of Geosciences, The Pennsylvania State University, University Park, Pennsylvania 16802
Richard H. Kesel
Department of Geography and Anthropology, Louisiana State University, Baton Rouge, Louisiana 70803
Donald R. Lowe
Department of Geology, Louisiana State University, Baton Rouge, Louisiana 70803
Chris M. Menges
Department of Geology, University of New Mexico, Albuquerque, New Mexico 87131
Sergio C. Mora
Department of Geology-ICE, Universidad de Costa Rica, Apdo. 10032-1000, San Jose, Costa Rica
Frank J. Pazzaglia
Department of Geosciences, The Pennsylvania State University, University Park, Pennsylvania 16802
Ira D. Sasowsky
Department of Geosciences, The Pennsylvania State University, University Park, Pennsylvania 16802
Joseph W. Troester
Department of Geology, University of Puerto Rico, Mayaguez, Puerto Rico 00708
Stephen G. Wells
Department of Geology, University of New Mexico, Albuquerque, New Mexico 87131

OVERVIEW OF CARIBBEAN GEOMORPHOLOGY

Thomas W. Gardner

Central America and the Caribbean islands encompass an extraordinarily diverse suite of geomorphic processes and landforms created by the interaction of distinctly different climatic, tectonic, and lithologic domains. The purpose of this overview is to review some of the significant types of geomorphic contributions from the region; to provide a summary of the climatic and tectonic setting of the region; and to introduce the four sections that comprise this chapter.

Climatic extremes within the region are significant for their impact on process and form diversity. For example, during the Pleistocene a climate gradient created by topographic relief produced glacial deposits in the Cordillera de Talamanca, Costa Rica (Hastenrath, 1973), that are only several kilometers from eco-

nomic laterite and bauxite oxisols in the adjoining Valle de El General (Castillo and others, 1970). Modern climatic zones range from wet tropical with annual rainfall in excess of 4.0 m in parts of Costa Rica, Guatemala, and Panama to dry tropical steppe with annual rainfall not exceeding 1.0 m in parts of Nicaragua, Honduras, and the Caribbean. Some of these climatic zones at lower elevations that did not experience the dramatic Pleistocene climatic oscillations of North America provide superb locations for the study of climate, process, and form without the overprint of relict climatic landforms.

Superimposed on the climatic diversity is an active plate setting that juxtaposes areas of rapid crustal movement and complex geology with relatively more stable carbonate platforms, for example, in the area of the Yucatan and Motagua fault zone. This ultimately causes the close spatial association of inceptisols on recent volcanic flows and oxisols tens of meters thick in parts of Central America. It is the plate tectonic setting that provides the

Gardner, T. W., Back, W., Bullard, T. F., Hare, P. W., Kesel, R. H., Lowe, D. R., Menges, C. M., Mora, S. C., Pazzaglia, F. J., Sasowsky, I. D., Troester, J. W., and Wells, S. G., 1987, Central America and the Caribbean, *in* Graf, W. L., ed., Geomorphic systems of North America: Boulder, Colorado, Geological Society of America, Centennial Special Volume 2.

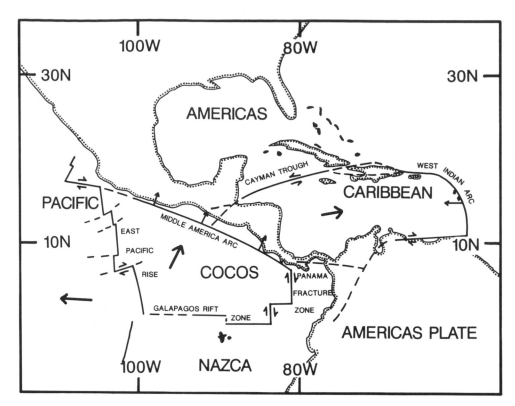

Figure 1. Lithospheric plates (names in bold type) of the Central American region. Plate boundaries are illustrated in heavy lines. Geologic structures that define plate boundaries are denoted by small letters. Heavy arrows indicate direction of plate motion with respect to Americas plate. Light arrows indicate direction of relative plate motion at the boundaries (after Molnar and Sykes, 1969).

region with its most notable and/or widespread attributes, namely, volcanism (McBirney, 1958; Weyl, 1980, Ch. 5), earthquakes (Nelson and Ganse, 1980), karsts (Back, this chapter), and general tectonic instability. Numerous destructive earthquakes (Ms > 7.0) have occurred throughout the region within historic time from Hispaniola in the Caribbean (Sykes and Ewing, 1965) to Costa Rica and Panama in Central America (Morales and Montero, 1984; Nelson and Ganse, 1980). Many have generated volumetrically significant landslides (Mora, unpublished data) and classic disruptions of the fluvial system (Winslow and McCann, 1985).

The Caribbean and Central American region is contained within part or all of four plates (Fig. 1), which have created the major physiographic features (Fig. 2). Present-day tectonic activity in Central America and the Caribbean is caused primarily by the interaction of two relatively small plates, the Caribbean and Cocos plates, with surrounding plates. The northern Caribbean plate boundary extends from Guatemala on the west to the Lesser Antilles on the east (Fig. 1). Eastward into Guatemala from its triple junction in the Middle America Trench, the boundary is expressed as the Motagua, Polochic, and Chamelecon fault zone, where up to 140 m of offset on fluvial terraces of the Rio Motagua indicate a maximum left lateral slip rate along the Motagua

fault of 6.0 mm/yr over the past 10,000 years (Schwartz and others, 1979). Eastward, the boundary becomes the tectonically active Cayman Trough, passing either north of Hispaniola (Molnar and Sykes, 1969) or through the northern portion of the island (Goreau, 1980) as the Septentrional fault system (Winslow and McCann, 1985), where geomorphic features such as truncated meanders, beheaded tributaries, and offset streams (Winslow and McCann, 1985) suggest recent left-lateral slip. Calculated slip rates of 20–40 mm/yr are suggested from offset lithologic units (Winslow, personal communication, 1985). Raised marine terraces on northern Hispaniola suggest vertical uplift rates of 0.3–0.4 mm/yr (Dodge and others, 1983; Winslow, personal communication, 1985). The Caribbean plate boundary continues eastward and eventually southward as the Puerto Rico Trench. Vertical uplift rates along the eastern boundary have been estimated at 0.25– 0.44 mm/yr from dated coral terraces on Barbados (Bender and others, 1979).

Another major source of tectonic and volcanic activity is created along the western boundary of the Caribbean plate where the Cocos plate is subducting at the Middle America Trench (Figs. 1 and 2). The Cocos plate, bounded on the east by the Panama Fracture Zone, and on the south by the Galapagos Rift Zone, is moving northeastward relative to the Caribbean plate.

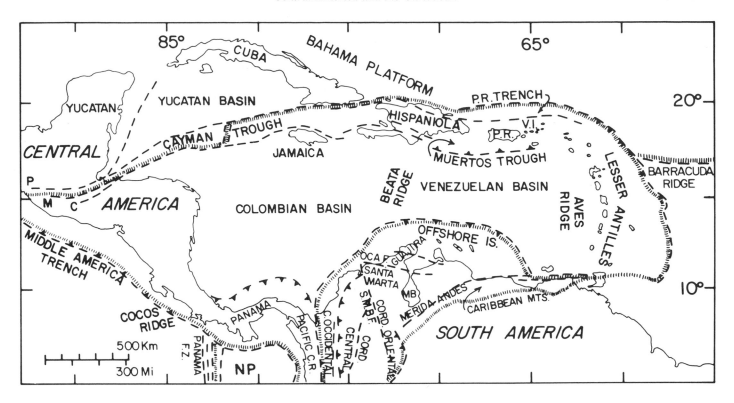

Figure 2. Physiography and modern plate boundaries in Middle America. Short hatching marks plate boundaries; dashed lines are faults. Abbreviations are: E., fault; F.Z., fracture zone; C.R., coast range; Cord., Cordillera; S.M.B.F., Santa Marta–Bucaramanga fault; P.R., Puerto Rico; V.I., Virgin Islands; MB, Maracaibo Basin; NP, Nazca plate. Central American faults are M, Motagua; P, Polochic; C, Chamelecon (after Mattson, 1984).

Convergence (normal to the trench) is estimated at 100 mm/yr along the western Pacific coast of Costa Rica where subduction of the aseismic Cocos Ridge (Fig. 2) has led to rapid vertical uplift an order of magnitude (1.0–4.0 mm/yr) greater than on the northern or eastern Caribbean plate boundary (Gardner and others, this chapter). Thus, the Caribbean and Central America are important areas for evaluating theories of large-scale tectonic activity and plate motion. Geomorphology will play a significant role in constraining these theories by providing data on the rates and timing of tectonic movements.

The active tectonic configuration has created a complex geologic setting in Central America (Fig. 3). The region is divided into two fundamental units, which differ in their geologic history and structure (Dengo, 1969; deBoer, 1979). The Chortis block in Guatemala, Honduras, El Salvador, and northern Nicaragua includes continental crust of Paleozoic metamorphic and igneous rocks (Figs. 3 and 4) overlain by moderately deformed Mesozoic redbeds, carbonates, and locally volcanic rocks, and Cenozoic marine, continental, and volcanic rocks with total aggregate thicknesses in excess of 4 km (Case and others, 1984).

The southern unit from southern Nicaragua to Panama includes late Mesozoic (Cretaceous) oceanic crust and ophiolites (Santa Elena, Costa Rica) overlain by Cretaceous, Tertiary, and Quaternary marine and volcanic rocks (Figs. 3 and 4). This sequence was complexly deformed during the late Tertiary and Quaternary and intruded by Neogene granodiorites (Cordillera de Talamanca). Superimposed on the entire sequence are Neogene and Quaternary volcanics (Fig. 5) of the Middle America Volcanic Province (Case and others, 1984). Volcanic rocks are dominantly calc-alkaline, but locally contain tholeiitic and alkaline components.

The Caribbean and Central America are regions of active, dynamic landscapes adjusting to rapid tectonic movement, eustatic sea-level change, youthful volcanism, and driven, in general, by warm, wet tropical climates. Given the geologic diversity, geomorphologic research has encompassed numerous subdisciplines that have provided significant contributions to Pleistocene sea-level reconstructions (Bender and others, 1979; Dodge and others, 1983); karst geomorphology (this chapter); fluvial system response to seasonal, extreme floods (Gupta, 1975); humid tropical, alluvial fan sedimentation (Schramm and Nummedal, 1982); remote sensing for tectonic mapping, geomorphic/geologic mapping, and soil/landscape evaluation (Rebillard and others, 1982; Erb, 1982; Segovia and others, 1980); various aspects of tectonic geomorphology (Schwartz and others, 1979; Hare and Gardner, 1985; Winslow and McCann, 1985;

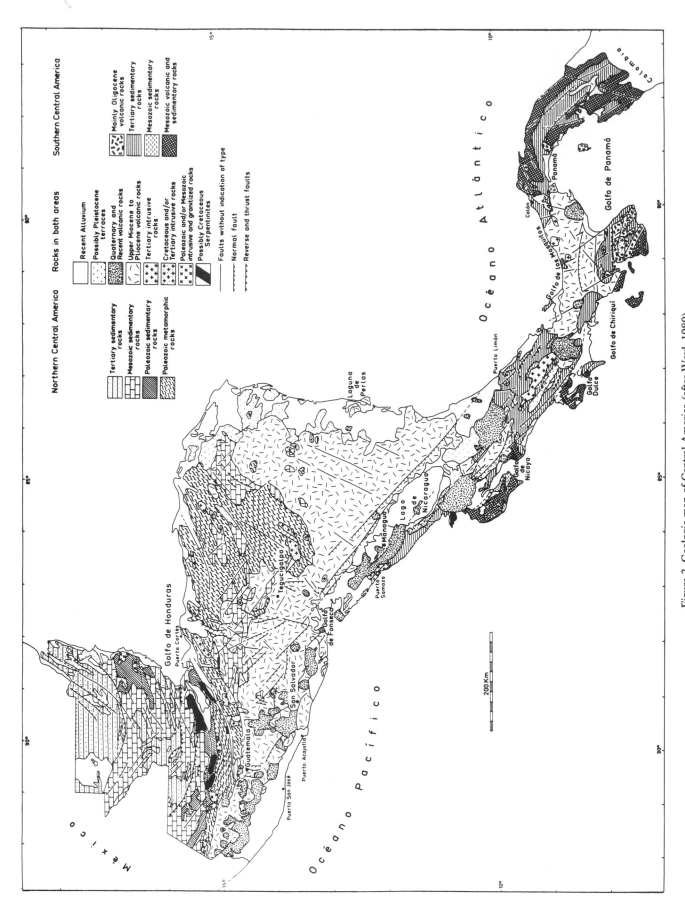

Figure 3. Geologic map of Central America (after Weyl, 1980).

Northern Central America

Tertiary sedimentary rocks

Mesozoic sedimentary rocks

Paleozoic sedimentary rocks

Paleozoic metamorphic rocks

Rocks in both areas

Recent Alluvium

Possibly Pleistocene terraces

Quaternary and Recent volcanic rocks

Upper Miocene to Pliocene volcanic rocks

Tertiary intrusive rocks

Cretaceous and/or Tertiary intrusive rocks

Paleozoic and/or Mesozoic intrusive and granitized rocks

Possibly Cretaceous Serpentinites

Faults without indication of type

Normal fault

Reverse and thrust faults

Southern Central America

Mainly Oligocene volcanic rocks

Tertiary sedimentary rocks

Mesozoic sedimentary rocks

Mesozoic volcanic and sedimentary rocks

México

Puerto San José
Puerto Acajutla
San Salvador
Guatemala
Puerto Cortés
Golfo de Honduras
Tegucigalpa
Golfo de Fonseca
Puerto Somoza
Managua
Lago de Nicaragua
Laguna de Perlas
Puerto Limón
Golfo de Nicoya
Golfo Dulce
Golfo de Chiriquí
Golfo de los Mosquitos
Colón
Panamá
Golfo de Panamá
Colombia

Océano Pacífico

Océano Atlántico

200 Km

Gardner and others, this chapter; Menges and others, this chapter); pedogenesis; and volcanic landforms (Murata and others, 1966; Ulate and Corrales, 1966; Waldron, 1967; Kesel, 1973).

It was a difficult task to solicit manuscripts for this chapter from the wide range of active and existing research topics. My choices, given the limited space, were based on two criteria: (1) that the research be current, and (2) that it represent the most important type of research that the region has to offer. In making these choices I have certainly neglected significant research in many subdisciplines. In reading this chapter you will find that three of the four contributions focus on Costa Rica. This was not by design, but resulted from the withdrawal or missed deadlines of several articles.

The chapter is divided into four sections, excluding this overview. The first section addresses karst landforms and processes of the Yucatan, Puerto Rico, Jamaica, and Costa Rica in relation to the tectonic, depositional, and diagenetic histories. The second and third sections explore the relationships between tectonic geomorphology and aseismic ridge subduction at the Middle America Trench along the Pacific coast of Costa Rica. Those contributions examine the relationship between morphometry of mountain fronts and drainage basins and active, differential uplift. The fourth section describes alluvial fan formation and its Quaternary history within forearc and back-arc settings in Costa Rica.

KARST OF THE CARIBBEAN

Joseph W. Troester, William Back and Sergio C. Mora

GENERAL PROCESSES

The spectacular, varied, and discontinuous karst of the Caribbean is the result of a combination of factors including migration of the Caribbean Plate island arc and associated subduction, overthrusting and fracturing, tectonic uplift, carbonate deposition and diagenesis, groundwater flow and discharge, thermodynamic and kinetic controls of mineral dissolution, and sea-level fluctuations. Important controlling factors responsible for development of karst features include: (1) environments of deposition of the soluble rocks, (2) characteristics of the bedding and lithification of these rocks, (3) tectonic history, (4) nature of the flowpaths of the groundwater, (5) altitude of the limestone above base level, and (6) climate, vegetation, and soil cover. Regions of heavy rainfall and dense vegetation supply the two agents responsible for dissolution of carbonate rocks, that is, water and carbon dioxide gas.

Although karst is widespread throughout the Caribbean, the extremely active tectonism of the region causes the karst to be discontinuous. The greatest area of karst extends from the Yucatan Peninsula southward through Belize and Guatemala. The extent of this karst area is, in part, due to the relative stability of this region of North and Central America, which was not subjected to the migration of the Caribbean Plate as were the Greater Antilles. This region has been in essentially its same location since

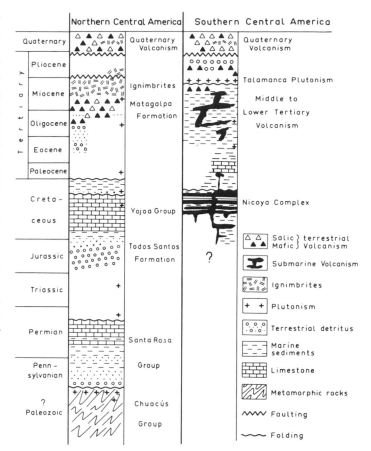

Figure 4. Generalized geologic column through northern and southern Central America (after Weyl, 1980).

pre-Mesozoic time, when the Gulf of Mexico was opened by separation of North and South America and rotation of the Yucatan Peninsula. This separation permitted the Caribbean Plate to migrate through the area that is now Costa Rica and part of Nicaragua from the Pacific Ocean up through the proto-Caribbean Sea. The islands of the Greater Antilles formed in response to the subduction associated with the northeastward migration of the Caribbean Plate into the proto-Caribbean. The Caribbean Plate collided with the Bahama Platform in Eocene time, and since then the movement has been more eastward along major strike-slip faults. During the migration, largely in Cretaceous time, carbonate platforms were forming around the margins of the sea. The karst of the Caribbean is developed on these Cretaceous rocks, and on the Tertiary rocks, which were deposited largely after active volcanism ceased in the Greater Antilles. The exception to this is the karst formed on the Jurassic rocks that are overthrust onto the Cretaceous and Eocene rocks in western Cuba.

The basic control on the development of karst features is the distribution of the primary and secondary permeability, which in

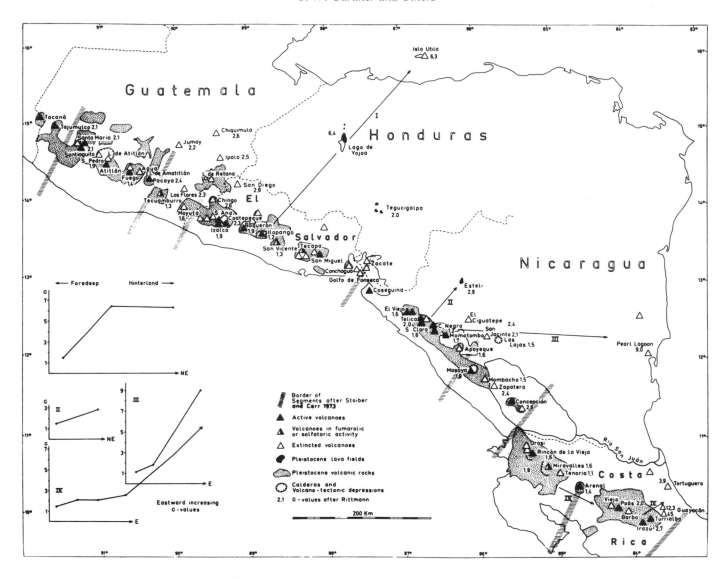

Figure 5. Quaternary volcanism in Central America (after Weyl, 1980).

turn controls the occurrence and movement of water that is geo-chemically undersaturated with respect to the carbonate minerals. Distribution of primary permeability is influenced by the environment of deposition, which determines the nature of the carbonate sediments, and to some degree the diagenesis and lithification processes. Distribution of secondary permeability is controlled largely by dissolution along fractures, joints, and faults. Secondary permeability is developed by the processes of late diagenesis and groundwater alteration, during which not only solution occurs, but also a redistribution of porosity and permeability takes place by precipitation and transformation of minerals, such as neomorphism of aragonite to calcite and dolomitization of aragonite and calcite.

Karst features form after lithification and during the development of a fully functioning aquifer system. Where the carbon-

ate rocks are approximately horizontal or gently dipping and have a relatively permeable, less soluble cover, karstification may begin over an entire region. Streams cut through the cover and provide discharge points for circulation systems that develop in the limestone as water infiltrates vertically downward, through the cover into the carbonate rocks. Sinkholes will develop throughout the area after the circulation system has developed enough to form lateral solution passages in the upper part of the zone of saturation.

The equilibrium geochemistry of the dissolution of calcite and other carbonate minerals is well known; essential points are summarized here using calcite as an example. Carbon dioxide dissolves in water to form carbonic acid, which disassociates and dissolves calcite. At equilibrium the following reactions must be satisfied simultaneously:

$$CO_2 \text{ (gas)} \rightleftharpoons CO_2 \text{ (aqueous)} \quad\quad K_{CO_2}$$
$$H_2O + CO_2 \text{ (aqueous)} \rightleftharpoons H_2CO_3$$
$$H_2CO_3 \rightleftharpoons HCO_3^- + H^+ \quad\quad K_1$$
$$HCO_3^- \rightleftharpoons CO_3^{2-} + H^+ \quad\quad K_2$$
$$CaCO_3 \rightleftharpoons Ca^{2+} + CO_3^{2-} \quad\quad K_c$$

where the K's are the equilibrium or disassociation constants for each reaction.

The saturation index (SI_c) measures the departure of the water from equilibrium with calcite, and is defined as

$$SI_c = \log(IAP/K_c)$$

where IAP is the ion activity product of the calcium and carbonate ions. However, because bicarbonate is the dominant carbonate species instead of carbonate, the equation is usually given as:

$$SI_c = (K_2[HCO_3^-][Ca^{2+}])/(K_c[H^+])$$

where the quantities in brackets are the activities of the various ions. The saturation index equals zero at equilibrium. Negative values of SI_c indicate subsaturation, while positive values indicate supersaturation.

The partial pressure of carbon dioxide, P_{CO_2}, that is in equilibrium with the solution is also calculated using the above equilibrium constants, as follows:

$$P_{CO_2} = ([HCO_3^-][H^+])/(K_1K_{CO_2})$$

The partial pressure of carbon dioxide in the normal atmosphere is $10^{-3.5}$ (0.03%). It can be much higher, up to about 10^{-1} (10%), in the soil atmosphere because of plant root respiration and decaying organic material.

However, equilibrium alone does not explain all of carbonate dissolution chemistry. White (1977) has shown that it takes a few days for groundwater to reach equilibrium with the carbonate rocks that surround it. This is approximately the same length of time it takes water to flow through a well-developed conduit karst system. Therefore, much of the groundwater in a karst system will be out of equilibrium with respect to the surrounding rock, and carbonate dissolution kinetics must be included to understand the rates of the processes that are occurring.

Calcite dissolution kinetics has been studied by several researchers (see Berner and Morse, 1974; Plummer and others, 1979; Herman, 1982; White, 1984). Berner and Morse (1974) found three different kinetic regimes in their study of the dissolution of calcite, two of which are important in the formation of karst, according to White (1984). Most water entering a karst region is undersaturated with respect to calcite and is in Berner and Morse's region 2 ($-4 < SI_c < -0.6$). In this region the dissolution rate of calcite is proportional to the partial pressure of carbon dioxide. As the water dissolves calcite and comes closer to saturation, it enters Berner and Morse's region 3, which extends from near saturation to equilibrium ($-0.6 < SI_c < 0$). In this region the calcite dissolution rate decreases rapidly to nearly zero.

The characteristic karst landforms, from karren to tower karst, form mainly by differential solution. In arctic and alpine regions, the relief of karst landforms is often about 1 m, whereas the average sinkhole depth in temperate regions ranges from 5 to 10 m (White and White, 1979). The relief in humid tropical karst regions is even larger. Troester and others (1984) found that the average sinkhole depth in the northern karst belt of Puerto Rico is 19 m, and that the average depth for the Cervicos karst region of the Dominion Republic is 23 m.

White (1984) gave three main reasons for differential solution. First, increased discharge: once depressions are established, more runoff is funneled into them, and they grow larger. Second, increased P_{CO_2}: sinkholes collect thicker soils and greater amounts of vegetation and organic debris, and consequently higher concentrations of carbon dioxide, causing greater dissolution. And third, faster dissolution rates: during intense rainfall the subsaturated water often flows on or near the surface to the lowest point, where it is funneled into the subsurface; in general, the greater the subsaturation, the faster the dissolution rate. While each of these may be a reason for differential solution, they do not explain why tropical karst has greater relief than temperate karst because all three of these processes operate in all climates.

White (1984) suggests that differential solution is highest in the tropics not only because the rainfall is greater, but also much more intense. The best cone karst in Puerto Rico is developed in an area that receives more than 2,500 mm of rain annually, and, according to the U.S. Weather Bureau (1961), the 2-year 30-minute rainstorm has an intensity of 100 mm/hour; this is double the intensity found in many karst areas in temperate climates. The high amount of precipitation and the flashy storms lead to increased soil erosion by running water, and slumping on steep slopes. These processes remove the soil from the upper surfaces and deposit it in the bottoms of the sinks. The rapid infiltration of subsaturated water that occurs in these sinks would be in Berner and Morse's (1974) region 2 kinetic regime, where dissolution of calcite is proportional to P_{CO_2}. On the bare limestone surfaces on the upper portions of the cone and tower karst, the P_{CO_2} is atmospheric, $10^{-3.5}$. However, according to Miotke (1973) the P_{CO_2} in the bottoms of the sinks is about $10^{-1.5}$. Because P_{CO_2} is greater by a factor of 100, the dissolution rate at the bottom of the sink is 100 times greater than on the bare rock exposed on the higher surfaces in the cone and tower karst. This accelerated dissolution, caused in part by the more intense rainfall, results in the greater amount of differential solution found in humid tropical karst.

KARST FEATURES

The Caribbean is known internationally for its variety of karst landforms, which include sinkholes, positive relief features, river caves, and deep canyons. Many different names have been proposed for karst in the humid tropics. However, there appear to be two distinct humid tropical karst landform end members: cone karst (also known as Kegelkarst or cockpit karst) and tower karst

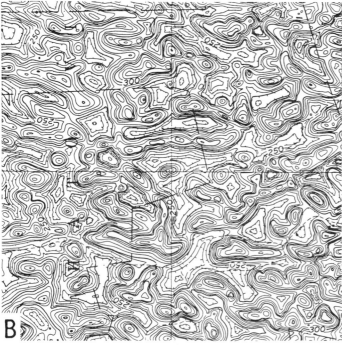

Figure 6. (A) An oblique aerial photograph of an area of cone karst near Arecibo, Puerto Rico. (B) A typical 4-km² section of cone karst from the Utuado Quadrangle, Puerto Rico. The contour interval is 10 m. Grid lines are one km apart.

(also known as Turmkarst or mogote karst). Cone karst is composed of many closely spaced cone- or tower-shaped hills separated by deep sinkholes, usually with little level ground. Tower karst is characterized by isolated limestone hills separated by areas of alluvial deposition.

The cone karst of Puerto Rico, described by Monroe (1976; Figs. 6A and 6B), the cockpit karst of Jamaica, described by Versey (1972), and the Cervicos karst region of the Dominican Republic described by Palmer (1983; Fig. 7) are all similar. In each of these areas, the karst is formed on nearly horizontal, uniform, middle Tertiary limestones with high rainfall (2,000 to 2,500 mm/yr), warm tropical temperatures (averaging 25° to 26° C), and significant relief, with little or no soil. The depressions are round or oval, with steep, concave, and sometimes vertical sides, with a depth of tens of meters, and separated by cone-shaped hills connected by sharp ridges.

There are, however, some small differences in the various cone karst areas. Most of the depressions in Jamaica drain toward a deep vertical shaft. Baker and others (1986) explored several of these to depths of 80 m, without intersecting the water table. However, in Puerto Rico, the water in the depressions appears to drain through numerous small openings, and large caves are seldom found in the bottoms of the sinks. The vertical shafts in Jamaica may have developed because of the greater relief available, due to more recent tectonics, or because the limestones are more dense. Another possible explanation is that the shafts are present in Puerto Rico, but obscured by landslide debris and the slightly larger amount of soil found in the bottoms of the depressions there.

Among caves found in the Caribbean region, the larger and longer ones are formed by undersaturated water flowing into the carbonates from an area of less soluble rocks (Troester and White, 1986). These cave systems are composed of active river passages, and often include older, abandoned, upper-level passages formed by the same river. Several rivers that now cross the karst belts in narrow gorges or canyons may at one time have crossed the limestone in caves. Examples of caves include the Rio Camuy and the Rio Encantado cave systems in Puerto Rico, River Head Cave in Jamaica (Fincham, 1977), and Rio Grande Cave in Belize (Dougherty, 1985). Examples of narrow canyons include the Rio Tanama and Rio Guajataca in Puerto Rico.

Puerto Rico

Puerto Rico has a central mountainous area composed of Late Jurassic to Eocene volcanics and volcanoclastic rocks. Carbonate deposition began in the Early Cretaceous and continued sporadically into the early Tertiary. All of these rocks were folded, faulted, and intruded by large granodiorite Late Cretaceous plutons. The small karst areas that have developed on the scattered limestone outcrops of this region are grouped together and usually called the Cretaceous karst region of Puerto Rico. After active volcanism and tectonism ceased, a section of middle Oligocene to Pliocene limestones and terrigenous sediments were deposited unconformably over the older rocks along the north and south coasts (Briggs and Akers, 1965; Monroe, 1973, 1975, and 1980; Moussa and Seiglie, 1975; Meyerhoff, 1975b; and Frost and others, 1983). These rocks make up the northern and southern karst belts of Puerto Rico.

The northern karst belt (Monroe, 1976) is the largest and best-known karst area in Puerto Rico. It is 10 to 20 km wide and extends for over 100 km along the north-central and northwestern coasts. It contains good examples of both cone and tower karst.

Cone karst is characterized by many closely spaced cone- or tower-shaped hills separated by deep sinkholes (Figs. 6A, 6B, and 7). As mentioned above, it is formed by the combination of intense rainfall, soil erosion, slumping, and solution. One of the largest sinkholes in Puerto Rico is under the Arecibo Observatory. It is nearly circular, with a diameter of 500 m. It is 70 m from the bottom of the sink to the lowest point on the divide that separates it from the adjacent sinkholes. The highest point on this divide is 155 m from the bottom. This sinkhole, like most of the others in the area, has a concave slope, with the upper parts approaching vertical. Before the construction of the Arecibo Observatory, the drainage in this sink was internal through the porous limestone at the bottom. This drainage was disrupted by construction, and the water must now be pumped out. No large caves or evidence of them has been found in this sinkhole, even though the Rio Ta-

Figure 7. A typical 9-km² section of the cone karst in the Cervicos karst area, Anton Sanchez Quadrangle, Dominican Republic. The contour interval is 20 m. The grid lines are 1 km apart.

nama flows through a very large cave just south of the observatory.

The other major type of humid tropical karst is tower karst. It is characterized by isolated limestone hills separated by areas of alluvium or other detrital sand (Monroe, 1976). Tower karst forms by the same combination of processes mentioned above. Erosion caused by the torrential rain storms of the tropics, as well as small landslides, removes the rock and soil from the sides of sinks and deposits it in the bottom. The water flowing through the soils in the bottom becomes enriched in carbon dioxide and dissolves the limestone there faster than on the bare limestone sides. The difference between cone karst and tower karst is the presence or absence of sediment between the hills. If the material in the bottom of the sink is not removed by the drainage system, the sinkhole will acquire a flat bottom as it slowly fills with sediment. If the sediment supply to the sink is great enough, the sink will fill entirely, and the area becomes one of tower karst, with only the hills that separate the original sinks exposed.

The best examples of tower karst in Puerto Rico are along the northern edge of the northern karst belt, where the towers usually rise about 30 m above the blanket sands (Figs. 8A and 8B). Because the towers in Puerto Rico do not usually have the classical tower shape, but instead are more often rounded, they are normally called mogotes. The classic mogote shape is discussed at length in Monroe (1976), where he shows the effect of the prevailing winds in making them asymmetrical. The sedi-

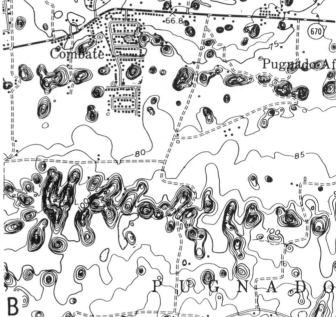

Figure 8. (A) A photograph of mogotes rising out of a pineapple field on the blanket sands, near Cruce Davila, Barceloneta, Puerto Rico. (B) A typical 4-km^2 section of mogote karst from the Manati Quadrangle, Puerto Rico. The contour interval is 5 m.

ments that separate the mogotes were called blanket sands by Briggs (1966). They are mostly river sediment that was distributed by longshore currents and were probably deposited during the Pliocene into depressions that had already begun to form. The blanket sands accumulated in these depressions and began to weather to their current composition of quartz sand and clay residium. This clayey soil provided a place for plant growth and a good reservoir for carbon dioxide. Miotke (1973) found that the soil atmosphere in the wet, clayey blanket sand ranged from 1.5 to 7 percent. This is mugh higher than is possible on the almost bare, steep limestone slopes of the mogotes. Water that percolates down through the blanket sands becomes enriched in carbon dioxide and is able to dissolve much more limestone than the water that soaks into the mogotes. The resulting differential solution continues to lower the limestone below the blanket sands faster than the surrounding mogotes. As the limestone below the blanket sands is dissolved, the sand and clay are moved down into the underlying cavities by piping, occasionally causing catastrophic sinkhole collapses in the overlying blanket sands (Miotke, 1973).

Another karst landform found in Puerto Rico is the zanjon, which was first defined and described by Monroe (1964). He gave the name zanjones to sets of parallel depressions that occur in thin-bedded outcrops of the Lares Limestone of Oligocene age. According to Monroe, they range from a few centimeters to about 3 m wide, from 1 to 4 m deep, and to more than 100 m long. A zanjon is similar to a bogaz or a solution corridor as defined by Monroe (1964) and Sweeting (1973). These features appear to form by the widening and deepening of one joint set.

Many of the zanjones in Puerto Rico are parallel or sub-parallel to the great southern Puerto Rico fault zone and the great northern Puerto Rico fault zone. These fault zones displace all of the pre-Oligocene rocks, with the Lares Limestone being the oldest carbonate not cut by these faults. One of the best examples of zanjones is near the Rio Guajataca, where the zanjones in the Lares trend west or west-northwest, paralleling the great southern Puerto Rico fault zone.

Other significant features of the northern karst belt of Puerto Rico are the rivers that begin in the mountainous interior of the island and flow north across the limestones to the Atlantic Ocean. The larger rivers, such as the Rio Grande de Arecibo and the Rio Grande de Manati, cut across the karst belt in wide alluviated valleys. These rivers have high discharges and sediment loads.

The carbonate chemistry of the Rio Grande de Arecibo was studied by Troester and White (1986), who found that the water which flowed from the volcanic interior of the island was subsaturated with respect to calcite when it reached the karst region. As it flowed across the karst belt its calcium and bicarbonate content increased, and the water obtained saturation. The partial pressure of carbon dioxide remained slightly above atmospheric.

Some of the smaller rivers that cross the karst belt flow through caves or deep narrow canyons. For example, the Rio Encantado Cave System near Florida, which has recently been explored and mapped (Ganter, 1986) to a length of almost 20 km, appears to have only 4 km^2 of catchment area on the insoluble rocks of the interior of the island (Troester and White, 1983). The water from this small catchment area sinks into the limestone and has been traced north through a series of short disconnected caves to the Rio Encantado Cave System, which flows into the Rio Grande de Manati.

The Rio Camuy is another river that begins in the interior of the island and then flows through the northern karst belt. Part of its course is through the 8-km-long Rio Camuy Cave System described by Beck (1977), Gurnee (1967), Gurnee and Gurnee (1974), and Monroe (1976). Most of the rest of its course is through a narrow canyon.

The Rio Tanama also begins in the interior of the island. It flows north through the karst in a narrow canyon on its way to the Rio Grande de Arecibo and the ocean. Along its way, the river flows through nine tunnels or arches, four of which are entirely travertine (Gurnee, 1972, 1977). The longest tunnel passes almost under the Arecibo Observatory. Monroe (1976) suggests that much of the Rio Tanama's deep, narrow canyon is a collapsed river cave like the Rio Camuy.

Troester and White (1986) studied the carbonate chemistry of the Rio Camuy and the Rio Tanama. In both rivers, the water from the interior of the island is subsaturated. As the rivers flow through the karst, their discharge is augmented by groundwater from springs and seeps, which has higher concentrations of Ca and HCO_3^- and a higher partial pressure of carbon dioxide. Under normal flow conditions, the water reaches saturation in the water-filled portions of the Rio Camuy Cave System. However, in the large air-filled passages and in the canyons, the water looses carbon dioxide and becomes supersaturated. Normally, the Rio Tanama and the Rio Camuy become supersaturated, while the Rio Grande de Arecibo remains near equilibrium because (1) a smaller portion of its flow is derived from the karst and (2) it has much less contact with the limestone because it is flowing on a bed of alluvium. Under high flow conditions, the large amount of runoff from the interior of the island causes subsaturated water to flow through the entire river system.

Interestingly, the Ca^{2+} and HCO_3^- content and partial pressure of carbon dioxide in these rivers is not noticeably greater than rivers in carbonate areas of temperate climates. The major difference in carbonate chemistry is the total lack of seasonal variation such as that reported by Shuster and White (1971, 1972).

Jamaica

The karst of Jamaica provides a good example of both structural and lithologic controls in its development. The following several paragraphs are summarized from Versey (1972) and Sweeting (1958).

The most extensive karst development in Jamaica occurs in the White Limestone Formation that was deposited during middle Eocene to lower Miocene time. The formation is more than 1,500 m thick where it is fully developed, but discontinuities of deposition and erosion in the higher elevation leave a maximum thickness of about 500 m. The core of Jamaica is composed primarily of Cretaceous pyroclastic rocks with some conglomer-

ates, shales, and minor amounts of limestone. At the end of the Cretaceous, these rocks were folded, uplifted, and intruded by several granodiorite bodies.

Along the north coast, and through two extensions, one to the southeast and one to the southwest, the White Limestone Formation is represented by well-bedded chalk. This facies represents sedimentation of micritic reef-derived material in an outer neritic environment. The bulk of the sediment is made up of planktonic microfauna. The chalks have reacted to stress by folding rather than by faulting, and large-scale fracturing is not characteristic as in other parts of the White Limestone Formation. The chalks are characterized by primary rather than secondary diagenetic porosity. These two features, the primary porosity and the general absence of fractures, result in a primary permeability that is greater than secondary permeability. The groundwater moves through the body of rock rather than through specific solution channels. Definite lines of groundwater flow have developed only along the largest fractures. In the absence of these fractures, the transmissivity values are low. Along the north coast, this low permeability acts as an effective barrier to the northward flow of groundwater.

Over a large part of central Jamaica, the lowest 100 to 150 m of the White Limestone Formation is composed of recrystallized carbonates; part of which is composed of diagenetic dolomite, and part of which is recrystallized limestone. There is no primary porosity or permeability in these rocks, and the water flows entirely through the fissures. Throughout the deposition of the White Limestone, the central part of Jamaica constituted a fault-bounded shoal on which deposition of biogenic limestone was taking place. This shoal was then uplifted and its sediments lithified. Subsequently it has been extensively block faulted and its surface deeply dissected by karst processes.

The White Limestone did not have imposed on it a drainage pattern inherited from overlying strata. The karst features had their origin solely in the character and attitude of the limestones and the manner of the uplift and tectonism. The major karst features are related to the great block-faulting episode that occurred in the late Tertiary, which has been mainly responsible for the variety of the present karst forms. The complexities introduced by this faulting, such as generating drainage barriers and creating distinct catchments within the karst, together with the erosion to base level of some streams, have resulted in major modification of the earlier flow that was entirely subterranean. A major event was erosional breaking of the Tertiary limestone cover and exposure of the underlying Cretaceous pyroclastic rocks in several inliers that occur in Jamaica. Drainage from these inliers was onto and almost immediately beneath the karst plateaus surrounding them. Whereas previously the erosional force had been the rainwater that fell on the carbonates, now large quantities of subsaturated water were funneled into the carbonates. In addition, sediment carried with the water eroded the limestone by abrasion. This sediment is still being transported by the karst streams flowing beneath the plateaus and is either being carried to the sea or is trapped in the interior valleys.

Figure 9. Aerial photograph of Chakalal Lagoon showing collapsed cave-roof blocks of limestone now submerged.

The main control on the physiography of the limestone outcrop is block faulting, which is more pronounced and younger in the south than it is in the north. The karst features are superimposed on the structural elements. In some areas, the karst erosional processes have accentuated the structures, and in others, the sediments carried by the streams have alluviated some of the interior valleys, masking much of the structure. The interior valleys have been structurally depressed rather than being formed primarily by erosion. They are generally downfaulted on at least one side, and many of these faults act as underground barriers to the groundwater flow. The groundwater is therefore forced to the surface during high water stages, causing solution at the foot of the surrounding hills and frequently causing widespread alluviation by the sediments carried in the conduit water, resulting in the flat floor characteristic of poljes. Impedance of the drainage may be caused by a reversal of the dip, by the upfaulting of the karst base, or by change in the limestone facies from a permeable to an impermeable nature. These poljes are generally linear, with the long axis perpendicular to the overall direction of groundwater flow.

Many of the valleys are drained by sinkholes in the downgradient side; thus rapid inflow can exceed the karst system's drainage capacity, causing flooding. The underground barriers formed by the faulting and the occurrence of more impermeable facies of the limestone also have direct control on the occurrence of springs in that they both act as impediments to the subterranean flow, forcing the water to the surface. In some areas the springs are heavily laden with sediments eroded from the Cretaceous inliers alluviating the valleys.

Yucatan

In the Yucatan, structure, dissolution, and groundwater discharge have combined to develop another process that produces karst features. The occurrence of sinkholes (cenotes) and the configuration of the shoreline along the east coast of the Yucatan are controlled largely by dissolution of limestone resulting from mixing of the discharging fresh groundwater and the encroachment of Caribbean Sea water into the carbonate aquifer. The salt water both in the Caribbean Sea and in the aquifer is supersaturated with respect to calcite, whereas the fresh water is either at equilibrium or supersaturated. Mixing of these two solutions produces a blend that can be subsaturated with respect to calcite and therefore capable of effective dissolution of limestone in these coastal areas (Back and others, 1984, 1986).

The groundwater flow pattern is controlled largely by the initial permeability and porosity of the calcareous sediments composing this aquifer and the fracturing that has produced secondary permeability along which groundwater flows and further enhances the permeability by dissolution. Incipient coves occur in those areas where fractures intersect the coastline and produce points of significant groundwater discharge. As the groundwater continues to flow in the subterranean fracture-solution channels, greater dissolution occurs, and the self-perpetuating process gradually forms an underground solution network developing into a labyrinth of caves. Near the shore, the roofs of these caves collapse as a result of dissolution of supporting material (Fig. 9). The enlargement of the lagoons by this process permits the additional erosional activity of wave action and biogenetic erosion by the

Figure 10. Carbonate coast being modified by groundwater discharge and wave action.

grazing and browsing marine invertebrates. The continuing groundwater discharge and dissolution form large coastal cave systems; thus the lagoons continue to enlarge. The headlands separating the lagoons are eroded primarily by wave action. Coalescence of lagoons forms a series of crescent-shaped beaches (Fig. 10) that extend along much of the east coast of the Yucatan (Back and others, 1984).

Many of the cenotes in this region are aligned along the fractures and were formed by dissolution of mixed groundwater. A body of salt water extends under the northern part of the peninsula of the Yucatan at a depth no greater than about 70 m. It is possible that some of the cenotes formed in the interior, northern part of the peninsula have been formed by dissolution occurring at the interface between the regional fresh groundwater lens and the extensive body of salt water in the carbonate aquifers on which it floats. Some submarine sinkholes and the "blue-holes" of the Bahamas may originate by this process.

Costa Rica

All reef and carbonate platforms in Costa Rica showing evidence of karstification were formed in shallow marine environments that were raised above sea level by accelerated uplift during the early to middle Tertiary. Karst structures in Costa Rica have been classified as mature to incipient depending on the occurrence and evolution of caves, avens, sinkholes, aquifers, and karren (lapies).

The best-known karsts are from Corredor, Cajon, and Damas (Mora, 1979b; Mora and Valdes, 1983; AEC, 1986). The Corredor site near Ciudad Neily contains a closed blind valley draining through a cave that, in turn, feeds an aquifer. Springs, cropping out near the town, are intensively exploited for human use. Damas karst exhibits an intricate system of galleries, halls, and several peripheral springs permanently yielding considerable volumes of water (Mora and Valdes, 1983). At the Cajon site, many sinkholes, karren areas, caves, and springs have been detected, but no underground exploration has yet been attempted, although this site has been investigated as a possible site for a large hydropower dam (Mora, 1979a; Grant, 1973).

Karstification is well developed at two locations in the Barra Honda Formation of Oligocene age at the Colorado and Barra Honda sites. The maximum thickness for the formation does not exceed 250 m. The formation usually "crowns" the top of the group of hills, which are remnants of the original carbonate platform. Peripheral outcrops around these hills do not show outcrops thicker than 90 m. However, caves exist that are as deep as 240 m in interior areas. This morphology suggests developments of synsedimentary cuvettes. The Barra Honda site, a national park since 1974, is by far the most mature (Fig. 11) and best-known karst in the country.

Little is known about the Colorado karst even though its geologic setting is well understood (Rivier, 1983). However, even though it appears to be a young karst, well-developed karst features other than lapies (karren), small caves, and sinkholes could exist in the unexplored areas. No big springs have been reported from this site, which is exploited as a source of calcium carbonate.

Barra Honda, Rosario, Corralillo, and Quebreda Honda Cerros have extensive networks of karst structures, but only Barra

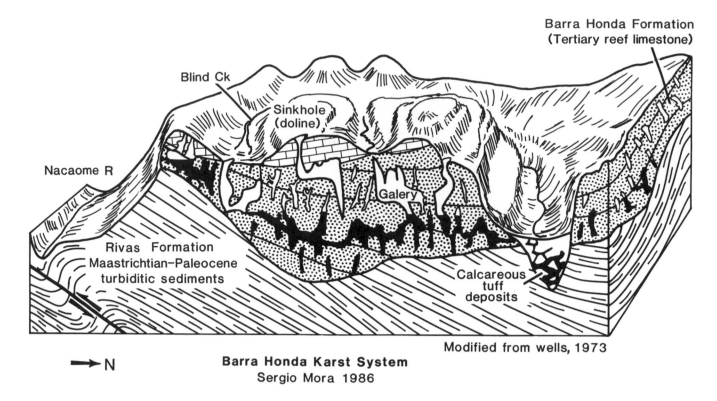

Figure 11. Cave map of Barra Honda karst, Costa Rica.

Honda's have been explored underground. The Cerro Barra Honda karst has reached a moderately mature stage. It has karren (lapies), sinkholes, avens, protodolines, and caves of mostly vertical development with a maximum depth of 240 m. Peripheral springs, which are extensively exploited as a water supply for nearby communities, yield important volumes of water even during long dry seasons (Mora, 1978, 1981).

SHORELINE RETREAT

Although the necessary field and laboratory work have not been done, it is probable that much of the shoreline retreat observed on islands in the Caribbean results from mixing zone phenomena. For example, the shoreline of Mona Island off the west coast of Puerto Rico is characterized by steep cliffs with huge blocks that have collapsed onto the narrow beach. This collapse may be due to the undercutting of the limestone by the dissolution processes occurring in the mixing zone at the interface between the fresh and salt water within the aquifers of the island. Also, the large carbonate plain along the south coast of westernmost Cuba is characterized by numerous sinkholes that may be formed by this process. We further speculate that Cienga Bay on the peninsula of Zapata in the same region of Cuba may be an enormous collapsed cave system.

Increased permeability that has been noted in coastal carbonate aquifers such as in Cayman (Chidley and Lloyd, 1977)

may result from the subterranean dissolution occurring in the mixing zone.

Another coastal karst feature of a microscale that may be in part formed by mixing of ocean water and rainwater is the coastal lapies, also known as fakir beds, that characterize many of the Caribbean shore areas. These pits and pinnacles are generally attributed to the biogenetic erosion of algae and browsing invertebrates that destroy the limestone in the pits and leave the pinnacles as microerosional remnants. Many of the pits are covered by a deposit of sodium chloride and it may be that the brackish water formed by dissolution of salt by rainwater in these pits is a contributing influence in the pitting of the limestone.

BEACH ROCK

The formation of beachrock can be considered a karst process and is widespread throughout many of the shorelines of the Caribbean islands. The beachrock is calcareously cemented beach deposits occurring in the surf zone. The deposition of the calcareous cement is caused by the outgassing of carbon dioxide as established by Hanor (1978) in his classic paper on St. Croix. Based on nonstatistical observations of the writers of this section, it appears that the most consolidated and extensive beachrock deposits are formed along beaches where a reasonable amount of water infiltrates in a back-beach area of storm deposits covered with abundant vegetation. The vegetation generates carbon dioxide gas, which is dissolved in the infiltrating groundwater that becomes supersaturated with respect to calcite. As the water migrates seaward and discharges along the beach, outgassing causes cementation by the precipitation of calcite. The decrease in permeability resulting from the cementation at the discharge points forces the water to migrate farther seaward and the discharge points and consequent cementation extend farther seaward. As the area of groundwater discharge is forced to expand downgradient from the formed beachrock, the dispersed calcium carbonate no longer occurs in concentrated amounts sufficient to form the coherent cement. Therefore, beachrock occurs in linear bands in the surf zone parallel to the beach. Beachrock deposits may form the solid platform necessary for sensile marine fauna and permit the growth of coral reefs.

SUMMARY

Although karst is widely distributed throughout much of the Caribbean region, we have selected only a few of the areas that best exemplify the processes that develop tropical karst. The best understood areas are Puerto Rico and Jamaica, followed by the Yucatan, Dominican Republic, and Costa Rica. Excellent work is being done on the karst of Cuba, but the restrictions imposed on American scientists for communication with Cuban scientists make their work largely unknown to us.

The major controls on the development of tropical karst include the limestone, the climate, and the relief. The formation of karst features is dependent on the purity of the limestone, distribution of its initial permeability, and its association with rocks of other lithologies. The tropical climate—high rainfall and warm temperatures—produces the two most important constituents for the dissolution of limestone: water and carbon dioxide gas, produced by the abundant vegetation. The elevation and relief cause a gravity flow system to develop, which transports this water and carbon dioxide through the soluble limestone. Superimposed on these basic controls are the modifying processes of diagenesis, tectonism, and sea-level fluctuations. The consequences of these processes are (a) continuous modification of the landscape to produce karst that is a combination of solution features and erosional remnants, (b) alteration of stream channels and drainage patterns, (c) modifications of coastline configuration, and (d) development of permeable aquifers and host rock for ore deposits.

The many processes required for development of karst have a complex, but not chaotic, interrelationship, and our understanding is progressively improving. However, before full understanding of these processes and the mutual effects can be achieved, some aspects of geochemistry, hydrology, and geology involved in the formation of karst must be studied. Laboratory research on the kinetics of carbonate dissolution needs to be applied to field situations. To do this, special attention must be paid to the effects of intense rainfall events on carbonate geochemistry in cave streams and in the unsaturated zone. Quantification of the fluid and sediment transport on the sides of sinks and through karst drainage systems is needed to understand the relative importance of the different processes. Karst research has and will continue to have input in the understanding of the complex geologic history and tectonics of the Caribbean region.

EVOLUTION OF DRAINAGE SYSTEMS ALONG A CONVERGENT PLATE MARGIN, PACIFIC COAST, COSTA RICA

Thomas W. Gardner, Paul W. Hare, Frank J. Pazzaglia, and Ira D. Sasowsky

INTRODUCTION

The use of landscape features as indicators of vertical tectonism began in earnest over 50 years ago with the classic debates between Penck (1925) and Davis (1932). Initiated by uplift, Davis' (1899) "ideal cycle" allows for systematic changes in the landscape through geologic time with the development of the penultimate landscape feature, a peneplain. Davis realized, however, that "old age" may never be achieved due to tectonic interruptions. Though such interruptions mark the beginning of a new cycle in some respects, Davis (1899, p. 499) suggested that the landscape "can only be understood by considering what had been accomplished in the preceeding cycle previous to its interruption." Though other variables such as climatic change must be considered, the landscape can record its tectonic history.

Studies of large-scale geomorphic surfaces that invoke erosional processes ranging from Appalachian peneplanation (Davis, 1889; Johnson, 1931; Campbell, 1933; Ashley, 1935), to the multiple upland surfaces—treppen—of New England (Meyerhoff and Hubbell, 1929; Meyerhoff, 1975a), to African pediplanation (Dixey, 1942; King, 1947, 1951; Fair and King, 1954) offer some useful and occasionally controversial examples of regional geomorphic surfaces as tectonic indicators. Furthermore, systematic variation in surface elevation and drainage adjustments are used as indicators of tectonic deformation. Notable examples come from the Dead Sea rift zone (Quennell, 1956), the East African rift zone (Doornkamp and Temple, 1966), and the Mesozoic rift along the Appalachians (Judson, 1975).

The Pacific coast of Costa Rica, an area of rapid and pronounced vertical tectonism, provides an excellent location to examine the interaction between landscape evolution and vertical tectonism. The goal of this paper is to explore the relationships among large-scale geomorphic surfaces (20-100 km^2 in extent), drainage basin morphometry, longitudinal river profiles, drainage adjustment, and variable rates of vertical tectonism astride and peripheral to a subducting aseismic ridge.

The Pacific coast of Costa Rica is part of a forearc and magmatic arc region created by northeastward subduction of the Cocos plate beneath the Caribbean plate at the Middle America Trench (Figs. 2 and 12). Subduction of oceanic lithosphere is often cited as the driving force of deformation in arc and forearc regions. Recent studies have investigated various aspects of *re-*

gional forearc deformation in response to specific styles of subduction behavior. Molnar and Atwater (1978) attributed variations in tectonic style (simple, cordilleran, extensional) in the overriding plate to variations in age (and buoyancy) of the subducting plate. Jordan and others (1983) argued that significant variations in uplift patterns, volcanism, and thin-skinned deformation seen in the Andean region can be correlated with the change from subhorizontal to more normally dipping subduction of the Nazca plate. McNutt (1983), using gravity/topography response-function modeling, demonstrated that the elevation of the Klamath Mountains of northern California is supported by the rigidity of the subducting Gorda plate. Thus, much of the *regional* uplift and deformation in the forearc and arc regions has been attributed to the behavior of the subducting plate. In general, the nature of the processes is known, although the specifics are unresolved.

The *local* response of a forearc region to subduction also has been investigated, with primary emphasis placed on unraveling the tectonic history of accretionary wedge sedimentary rocks. Two recent studies in this vein are the work of Wang and Shi (1984) and Stockmal (1983). Wang and Shi investigated the thermal history of material involved in deformational flow in the accretionary prism using a finite element model. Stockmal utilized slip-line field theory to investigate deformation in the forearc driven by the basal shear of a "normally" subducting slab.

Both groups of studies have tried to explain deformation and uplift associated with typical subduction regimes. Unfortunately, the data available do not fully allow the models to focus on the specific correlations between variations in the subduction process and the resulting tectonic signature of a region. The subduction process is further complicated by subduction of buoyant ocean lithosphere in the form of oceanic plateaus, very young lithosphere, or an aseismic ridge, which may be responsible for many interesting geologic responses, the more noticeable being a decrease in volcanism, arc polarity reversals, low-angle or shallow subduction, a decrease in seismic potential, an atypical trench morphology, and anomalous vertical crustal motions (Van Andel and others, 1971; Vogt and others, 1976; Kelleher and McCann, 1976; Pilger, 1981; Wadge and Burke, 1983; McCann and Sykes, 1984). These phenomena are broadly understood, but their specific causes and quantification remain largely unaddressed.

The goal of this paper is to better understand one specific case among these phenomena—the vertical motions of the crust in response to subduction of oceanic lithosphere containing the aseismic Cocos Ridge (Fig. 12)—by using the directions, magnitudes, rates, and locations of vertical crustal deformations obtained from such geomorphic indicators as marine and fluvial terraces, geomorphic surfaces, stream patterns, and drainage basin morphometries. The Pacific coast of Costa Rica is an unsurpassed site for such a study. First, both "normal" (Bourgois and others, 1984) and aseismic ridge subduction (Van Andel and others, 1971) occur there, and a coastal transect from the northern Nicoya Peninsula to the Panama border (Fig. 12) can cover each, thereby providing a comparison of effects. By "normal"

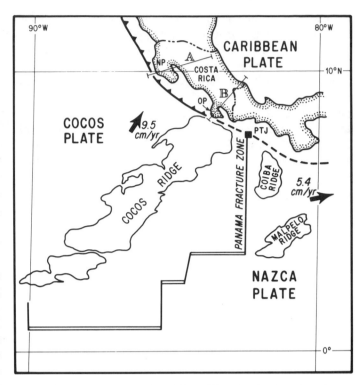

Figure 12. Plate tectonic setting of Costa Rica showing location of cross sections in Figure 14. PTJ, Panama Triple Junction; NP, Nicoya Peninsula; OP, Osa Peninsula.

here is meant typical intraoceanic subduction with little or no accretion of sediments and few complex deformation zones (Lundberg, 1983; Bourgois and others, 1984). Second, the history of Cocos Ridge subduction (8 Ma to present) is reasonably well constrained by known plate motions.

TECTONIC HISTORY

At approximately 8 Ma (constrained by the timing of separation of the Cocos and Malpelo ridges), the Panama Fracture Zone grew northward to intersect the Middle America Trench in the vicinity of the Nicoya Peninsula (Fig. 13). With the development of the Panama Triple Junction at this time, subduction beneath Costa Rica east of the Panama Triple Junction (controlled now by Nazca/Caribbean relative plate motions) became much more oblique with a decrease in relative velocity (Fig. 13). West of the Panama Triple Junction (vicinity of Nicoya Peninsula), subduction continued at a relatively high convergence rate (Cocos/Caribbean relative motion). In crossing the Panama Triple Junction from northwest to southeast, one would have observed a decrease in convergence (normal to the trench) from 10 cm/yr to approximately 3.5 cm/yr. A decrease in volcanic activity (perhaps cessation?) would be expected with this slowdown in convergence, consistent with geochronological data (detailed later).

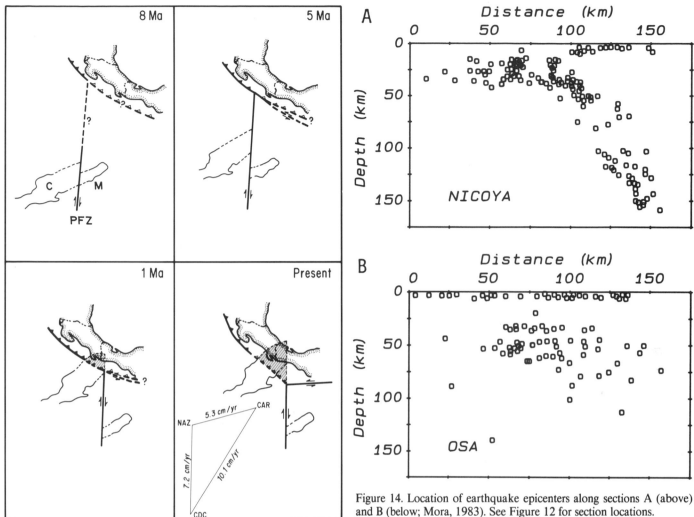

Figure 13. Plate motions in the eastern equatorial Pacific at four times beginning with 8 Ma, based on the RM2 plate motion model of Minster and Jordan (1978) utilizing Cocos (COC)/Caribbean (CAR), Cocos/ Nazca (NAZ), and Nazca/Caribbean relative plate motions. A relative velocity triangle for the present location of the Panama Triple Junction is also shown. C, Cocos Ridge; M, Malpelo Ridge; PFZ, Panama Fracture Zone.

Figure 14. Location of earthquake epicenters along sections A (above) and B (below; Mora, 1983). See Figure 12 for section locations.

The Panama Triple Junction migrated southeasterly (at approximately 3.5 cm/yr) reestablishing Cocos/Caribbean mode convergence for Costa Rica until about 1 Ma. At 1 Ma (soon after passage of the Panama Triple Junction through the region), the Cocos Ridge impinged on the Middle America Trench in the vicinity of the Osa Peninsula. The interaction between this buoyant ridge and the subduction zone has been primarily limited to the region of the Osa Peninsula, with approximately 100 km of relative convergence occurring up to the present (Fig. 13). Over this 1-m.y. period, subduction beneath Nicoya Peninsula (based on these reconstructions) should have been "normal" (i.e., unaffected by Cocos Ridge interaction). The concept of "normal"

subduction the vicinity of the Nicoya Peninsula and differences between the Nicoya and Osa peninsulas are supported by seismicity patterns (Fig. 14). Both a well-defined Benioff zone observed across the Nicoya Peninsula and current volcanic activity inland of the Nicoya Peninsula support "normal" subduction in the region. Clearly, seismic (Fig. 14), volcanic (Figs. 5 and 15), and ground acceleration patterns (Fig. 16) are quite different near the Osa Peninsula. The occurrence of earthquakes to depths of 100 km is evidence for some form of subduction or underthrusting, but the lack of a well-defined Benioff zone (and diffuse seismicity) argues for a significant degree of intraplate shortening.

Because convergence rates and age of subducting lithosphere do not change dramatically across the Costa Rican portion of the Middle America Trench, the primary differences in tectonic response between the Nicoya Peninsula and Osa Peninsula are most likely a consequence of the existence (Osa Peninsula) or lack (Nicoya Peninsula) of a subducting aseismic ridge. As a result, the Nicoya Peninsula can serve as the "control," allowing

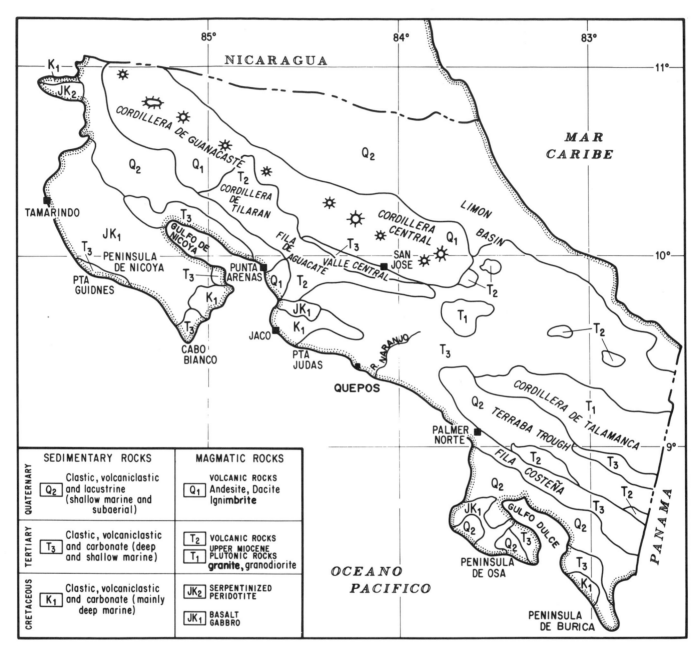

Figure 15. General geologic map of Costa Rica (modified from Weyl, 1980).

us to distinguish between the tectonic response to aseismic ridge subduction and the tectonic response to "normal" subduction along the Middle America Trench.

The spatial and temporal relationships among the Neogene and Quaternary plutonic, volcanic, and sedimentary rocks in Costa Rica (Fig. 15) support this plate tectonic reconstruction. Plutonism with dominantly quartz monzonite affinities occurred along the axis of the Cordillera de Talamanca in late Miocene time. Radiometric ages for the plutons range from 11.4 to 8.5 Ma (Bellon and Tournon, 1978; Bergoeing, 1983). As suggested by Weyl (1980), this plutonism resulted from ongoing convergence

and subduction of the Cocos plate at the Middle America Trench. This is consistent with our model (Fig. 13) at 8 Ma just prior to the intersection of the Panama Fracture Zone and the Middle America Trench.

Plate reconstruction predicts that volcanic activity should have decreased in that part of Costa Rica southeast of the intersection of the Panama Fracture Zone with the Middle America Trench after about 8 Ma. Volcanic rocks make up a very small portion of the stratigraphic column in southwestern Costa Rica (Fig. 17). Several andesitic volcanic units (Paso Real and Rio Pey formations in Fig. 17) of limited areal extent, with radiometric

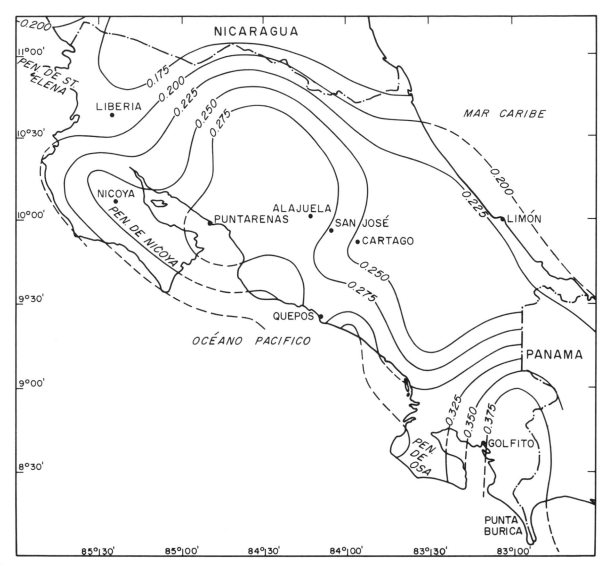

Figure 16. Preliminary map of isoacceleration lines for Costa Rica (after Shah and others, 1976).

ages of 5.0 Ma (Kesel, 1983), outcrop in the Limon Basin and along the flanks of the Terraba Trough (Fig. 15). Volcanic activity should have continued in the northwestern portion of Costa Rica where subduction of the Cocos plate beneath the Caribbean plate continued. Basaltic to andesitic volcanism in the Valle Central and Cordillera de Tilaran yield radiometric ages of 5 to 2–2.3 Ma (Aguacate and Monte Verde formations in Fig. 17; Weyl 1980; Bergoeing, 1983). The volumetrically most significant post-Pliocene volcanism occurred in the Cordillera de Guanacaste in northwestern Costa Rica. Dacitic to rhyodacitic ignimbrites and tuffs with radiometric ages ranging from 1.6 to 0.14 Ma (Bagaces and Liberia formations in Fig. 17) blanket much of the mainland of northwestern Costa Rica (Weyl, 1980; Bergoeing and others, 1982).

At about 1 Ma in our model (Fig. 13), the aseismic Cocos

Ridge impinged on the Middle America Trench in the vicinity of Peninsula de Osa. Rapid uplift should have resulted. Post 1 Ma uplift of southwestern Costa Rica, particularly in the Valles Central, Cordillera de Talamanca, Fila Costena, and Peninsula de Osa, is well constrained stratigraphically and geomorphologically. The youngest marine beds in the Terraba Trough are reported as mid-Pliocene from microfauna in an unnamed claystone stratigraphically above the Gatun Formation (Fig. 18; Kesel, 1983). The oldest terrestrial units (Fig. 18) are alluvial volcaniclastic sandstones and conglomerates of the late Pliocene to early Pleistocene Paso Real Formation. Kesel (1983) suggested that uplift of the Fila Costena postdated the middle Pliocene marine clay and was probably middle Pleistocene to Holocene in age. Alluvial fans along the southern flank of the Cordillera de Talamanca with granitic clasts from the Cordillera yield radio-

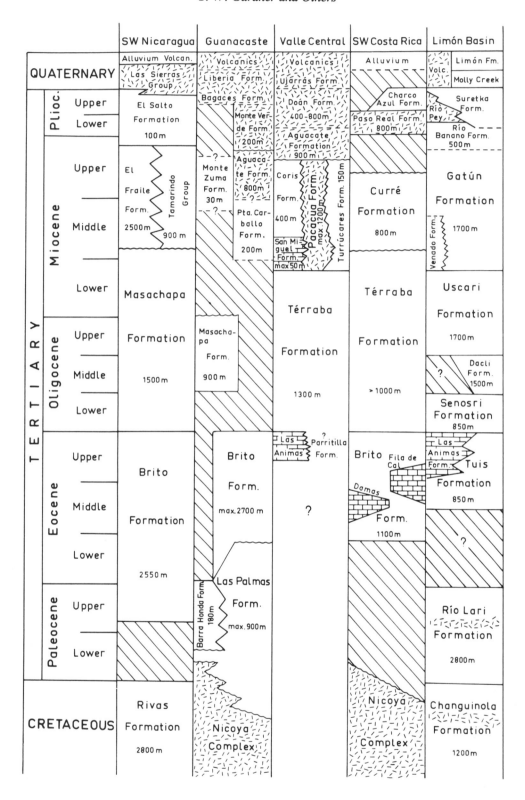

Figure 17. General stratigraphic correlation chart for the Cenozoic of Costa Rica (from Weyl, 1980).

metric ages ranging from 34,500 to 8,800 yr B.P., suggesting that uplift and unroofing of the Talamanca plutons were well established by late Pleistocene time (Kesel and Lowe, this chapter).

Late Pleistocene and Holocene uplift astride the Cocos Ridge is suggested by fluvial terraces along the Rio Terraba near Palmer Norte (Fig. 15; Kruckow, 1974). A 190-m terrace has been radiometrically dated at 7,500 yr B.P., suggesting an unbelievably high uplift rate of 25 mm/yr (Alt and others, 1980). Quaternary continental deposits (31,200 yr B.P. radiometric date) have a near vertical attitude, and Holocene beach sands show evidence of normal faulting (Alt and others, 1980). More reasonable uplift rates ranging from 1.0 to 4.0 mm/yr are estimated from radiometric dates (this chapter) on marine terraces along the Peninsula de Osa (Madrigal, 1977).

Northwest of Peninsula de Osa along the Pacific coast and away from the aseismic Cocos Ridge, uplift is less pronounced (Menges and others, this chapter) as might be expected from the plate reconstructions. Numerous marine and fluvial terraces are preserved along Peninsula de Nicoya (Alt and others, 1980; Hare and Gardner, 1985), but few have been dated. Bergoeing (1983) reported a 6,620 yr B.P. radiometric age for a 7 m terrace near Cabo Blanco (Fig. 15), indicating markedly less uplift than for terraces of the Rio Terraba. No faulting of Quaternary units has been observed on the Peninsula de Nicoya. Recent releveling surveys (Miyamura, 1975) and the elevation of bioerosional benches (Fischer, 1980) along the coast near Puntarenas and the Golfo Dulce indicate general subsidence.

Given this plate tectonic setting and the supporting geology, it should be possible to discern differences in the morphology of the landscapes that result from differential Quaternary uplift immediately astride (Osa Peninsula) and laterally adjacent to (Nicoya Peninsula) the subducting, aseismic Cocos Ridge. Deformation of large-scale erosional surfaces, basin morphometries, fluvial patterns, and mountain-front morphometries (Menges and others, this chapter) will be used to elucidate this uplift history.

NICOYA PENINSULA

Physiographically, the Nicoya Peninsula is part of the Middle America Forearc Ridge (Case and others, 1984). The bedrock geology of the Nicoya Peninsula is largely composed of the Nicoya Complex, a Jurassic-Cretaceous basic igneous complex of pillow basalts, diabase dikes, small gabbroic intrusions, and intercalated pelagic sediments. Cretaceous to Tertiary sedimentary rocks unconformably overlie the Nicoya Complex locally along the margin of the peninsula (Fig. 19).

Four distinct geomorphic surfaces have been delineated on the Nicoya Peninsula: the Santa Cruz, Cobano, La Mansion, and Cerro Azul surfaces (Figs. 20 and 21). The Santa Cruz Surface with an average elevation of 20 m is an aggradation surface of low relief constructed by active flood plain sedimentation of Rio Canas and its tributaries (Fig. 20). Ponded drainage, marshes, and seasonal lagoons with occasional hills rising abruptly from the flood plains characterize this surface. The drainage from this

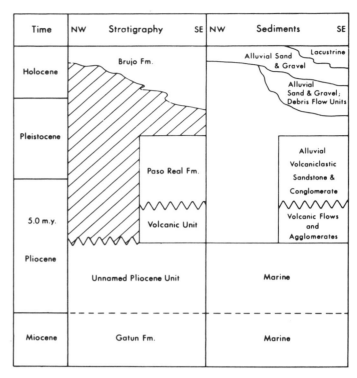

Figure 18. Stratigraphic sequence for a portion of the Terraba Trough, southwest Costa Rica. Note refinement of the Paso Real Formation over that of Figure 17 (from Kesel, 1983).

large, basinlike portion of the northern part of the Nicoya Peninsula is funnelled through a relatively small gap to the aggradational plains of the Rio Tempisque that separates the peninsula from the mainland.

The Cobana Surface (Fig. 21), marked by accordant summits, is largely coincident with and truncates gently dipping exposures of the Miocene to Pliocene Montezuma Formation (Fig. 19; Dengo, 1962; deBoer, 1979), a shallow water, transgressive marine sequence (Lundberg, 1981). The Cobano Surface (Fig. 21) is presently being dissected by joint-controlled rectangular drainage. Flat-topped summits indicate that erosion has been insufficient to appreciably lower the elevation of the Cobano Surface. The flat-topped residuals, at elevations of 160 to 200 m, suggest similar amounts of tectonic uplift since formation of the surface.

The La Mansion Surface is developed along major river valleys in the mountains of the central and southern portion of the Nicoya Peninsula. The surface is characterized by wide, open valleys, and a distinct break in slope at the base of flanking hills. The La Mansion Surface is largely composed of stratified fluvial sands and gravels. Clasts are dominantly basalt and chert derived from the Nicoya Complex. Incision by the modern drainage has created a depositional terrace typically 4–10 m above stream level. Rivers in the central part of the peninsula have incised to bedrock and do not as yet possess active flood plains. In these valleys, the surface exists as a nearly continuous terrace. The La

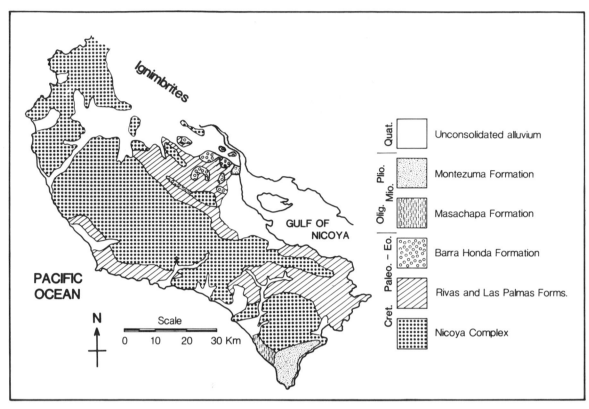

Figure 19. Generalized geologic map of the Nicoya Peninsula (modified from the Mapa Geologico de Costa Rica, 1982).

Mansion Surface is more discontinuous to the south where active flood plains are developing.

Developed within the mountainous backbone of the peninsula, the Cerro Azul Surface is characterized by the presence of a prominent topographic scarp, fluvial knickpoints, and a oxisol 5–10 m thick on the Nicoya Complex (Hare and Gardner, 1985). The total area of the Cerro Azul Surface is approximately 265 km². The surface is not continuous but exists as numerous fragments (Fig. 21), which decrease on average from 650 m in the northwest to 390 m in the southeast. Topographic relief within individual blocks of the Cerro Azul Surface may average up to 200 m, indicating incomplete planation. The Cerro Azul Surface, as yet undated, is postulated to have formed by fluvial erosion while the Nicoya landmass was lower with respect to a steady oceanic base level. It was fragmented following uplift by headward elongation and elaboration of the extending fluvial system (Hare and Gardner, 1985) following models proposed by Glock (1931) and Schumm and Parker (1973), to produce a stepped landscape—treppen—similar in form and origin to those described by Penck (1925) and Meyerhoff (1975a).

Assuming that the Cerro Azul Surface represents a time line (that it was at one time everywhere "graded" to a single sea-level stand) and allowing for initial topographic irregularities, the present elevation of the surface may reflect tectonic deformation. Trend surface analyses (Fig. 22) of elevation data collected on an

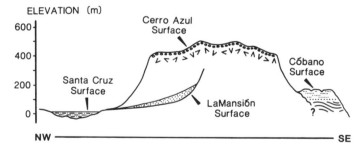

Figure 20. General topographic relationships between the four geomorphic surfaces of the Nicoya Peninsula. Illustration is a composite section drawn along a northwest-southeast line through the peninsula (after Hare and Gardner, 1985).

0.5 km² grid were used to model the irregular Cerro Azul Surface (Hare and Gardner, 1985).

First- through fourth-order trend surfaces were fit to the entire data set, ignoring possible structural discontinuities. A fourth-order surface (Fig. 22a), which best fits the data, shows an elongate dome with a crest elevation of over 650 m located at the Esperanza block. The fourth-order trend surface suggests that cross-trend arching is present only from the Esperanza to Cerro

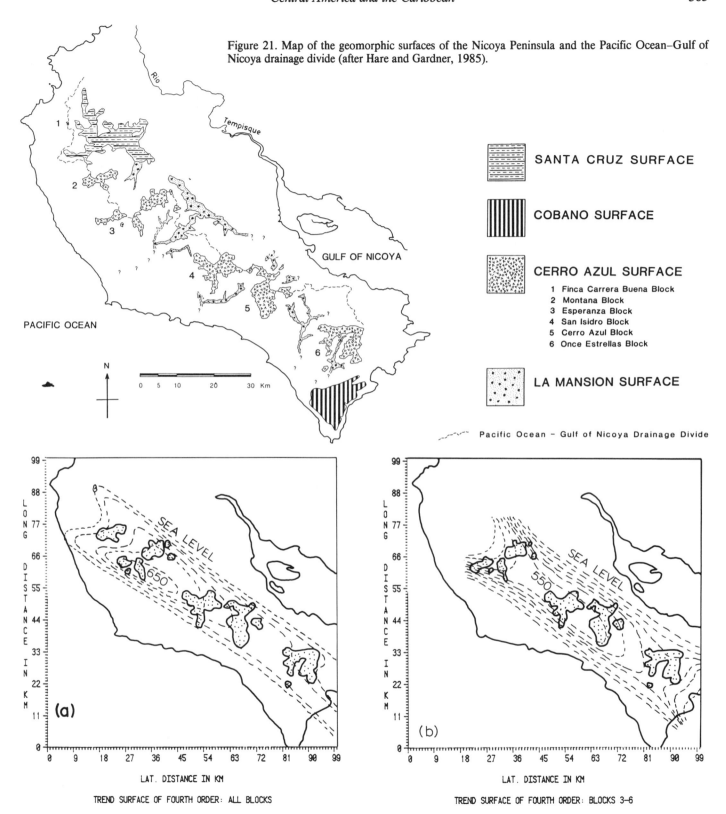

Figure 21. Map of the geomorphic surfaces of the Nicoya Peninsula and the Pacific Ocean–Gulf of Nicoya drainage divide (after Hare and Gardner, 1985).

Figure 22. (a) Trend surfaces for all blocks of the Cerro Azul Surface (stippled) with fourth-order trend (n = 1,017, R^2 = 0.419, F–statistical probability = 0.0001), and (b) trend surfaces for the Esperanza, San Isidro, Cerro Azul, and Once Estrellas blocks of the Cerro Azul Surface with fourth-order trend (n = 921, R^2 = 0.456, F–statistical probability = 0.0001). Block outlines are located on Figure 21; contour interval is 50 m. Highest contour is labeled in a (after Hare and Gardner, 1985).

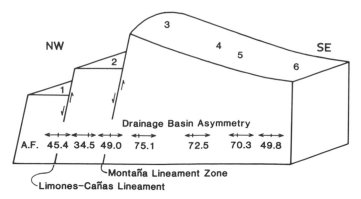

Figure 23. Faulted half-dome model developed from statistical analyses of the Cerro Azul Surface and Landsat lineament analyses showing two structural discontinuities northwest of the Esperanza block (3). Drainage asymmetry factors (A.F.) are also shown in their relative spatial positions for major rivers draining southwestward into the Pacific. Block numbers are located on Figure 21 (after Hare and Gardner, 1985).

Azul blocks; the Once Estrellas block is not part of an actual elongate dome.

Trend surfaces were also calculated for the Esperanza to Once Estrellas blocks (Fig. 22b) to allow incorporation of the structural discontinuities suggested by LANDSAT and regression analyses (Hare and Gardner, 1985). The fourth-order surface models a southeast-plunging antiform, which dies out at the Once Estrellas block. Maximum elevation of the antiform is 725 m at the Esperanza block.

Statistical analysis of the Cerro Azul Surface suggests that neotectonic deformation of the surface can be modeled as a faulted-domal uplift (Fig. 23). This deformational model is supported by additional geomorphic evidence. Morphometric analysis of major fluvial systems reveals important variation in drainage basin asymmetry. Drainage basin asymmetry is calculated from an asymmetry factor (A.F.) based on drainage area due to the inaccuracies in stream delineation on topographic maps:

$$A.F. = 100 \ (Ar/At)$$

where Ar is the drainage area on the downstream right of the main drainage line and At is the total drainage area. As an example, if a block of homogeneous crustal material is uplifted on its western edge, an easterly flowing stream with equal tributary contribution from the north and south will develop. However, if the above occurs and the blocks are tilted to the north, the easterly flowing drainage network will display "left-sided" asymmetry, with the trunk stream shifted northward and most of the tributaries entering the trunk stream from the south. A persistent occurrence of this feature over several adjacent basins provides a strong indication of centers of uplift as well as tilt directions of

tectonic blocks. This measure can only be employed with reasonable assurance for homogeneous lithologies where climatic-induced slope asymmetry can be ruled out. Any structural or lithologic control of heterogeneous strata (i.e., tilted sedimentary rocks) can produce similar asymmetries during basin development without tectonic activity. For this area of Costa Rica, the Nicoya Complex underlies the Cerro Azul Surface and can be considered homogeneous at the basin scale. Furthermore, the tropical vegetation shows no strong aspect asymmetry at this latitude. Though the asymmetry factor varies in a downstream direction for any given river on the Nicoya Peninsula, southward asymmetry occurs only for rivers between the Esperanza (3) and Once Estellas (6) block, such as Rios Quiriman, Ora, and Bonbo (Figs. 21 and 23). To the south, the asymmetry factor suggests that Rio Ario is an aerially balanced system; to the north of the Esperanza block the asymmetry factor indicates balanced systems or systems asymmetric to the north, in agreement with the trend surface model.

Extensive development of the Cerro Azul Surface coupled with the differential uplift, suggests that a hypsometric analysis (Strahler, 1952) of Nicoya Peninsula drainage basins might provide a useful technique for delineation of that surface, and a quantitative measure of stage of basin development. Hypsometric analysis provides a method for comparison of basins of unequal size and relief with respect to stage of geomorphic evolution and interruptions to the "normal" erosion cycle (Strahler, 1952). Hypsometric curves were constructed following the methods of Haan and Johnson (1966). Eight areally extensive basins on the Nicoya Peninsula that drain in a southwesterly direction to the Pacific Ocean were selected for analysis. Elevations were sampled on a grid with a density of 16 points per square kilometer.

Strahler (1952) presented examples of curves for landscapes in youthful, mature, and monadnock phases. The progression from youthful to monadnock phase shows a curve that begins as strongly convex upward, becomes sigmoidal, and ends strongly concave upward. The curves for the study basins (Fig. 24) show the mature to monadnock forms that are somewhat unexpected for the tectonically active landscape.

Results of the hypsometric analysis (Table 1, column 5) show similar values for the hypsometric intergral for all basins, indicating a similar stage of basin evolution. The expected relationship between the elevation of the Cerro Azul Surface (Table 1, column 3) and a change in slope on the hypsometric curve (column 6) is not apparent, however. An attempt was made to correlate the average elevation, minimum elevation, and extent of the Cerro Azul Surface (columns 2, 3, and 9) with standard parameters derived from the curves (column 4, 5, 6, 7, 8, 10, and 11), but no significant trends emerged, although Strahler (1952) had demonstrated sensitivity of some of those parameters to stage of basin development. The apparent insensitivity of the hypsometric method is probably due to the relative relief within the Cerro Azul Surface and the small percentages of the Cerro Azul Surface present in the basins (0 to 25.6 percent).

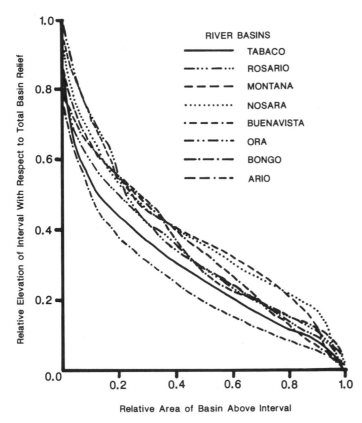

RIVER BASINS

————————	TABACO
—··—·—···	ROSARIO
— — — —	MONTANA
············	NOSARA
·—·—·—·—	BUENAVISTA
—··—·—···	ORA
—·—·—·—	BONGO
— —·— —	ARIO

Relative Elevation of Interval With Respect to Total Basin Relief

Relative Area of Basin Above Interval

Figure 24. Hypsometry curves for 8 drainage basins on the Nicoya Peninsula, Costa Rica. Horizontal axis indicates proportion of total basin area accounted for by land mass above a given elevation. Vertical axis is a normalized measure of elevation calculated by dividing the mean of the class interval by the total basin relief for each basin.

OSA AND BURICA PENINSULAS

Physiographically, the Osa and Burica peninsulas are part of the Middle America Forearc Ridge (Case and others, 1984) and are underlain by relatively homogeneous Cretaceous sea-floor basalts of the Nicoya Complex (Fig. 15). Lower-lying areas along the margin of the peninsulas have a thin veneer of Cretaceous, Tertiary, and Quaternary sediments. However, only minor lithologic control of the fluvial systems is attributed to these units.

Normal subduction is disrupted along this portion of the Costa Rican coast because the aseismic Cocos Ridge is impinging upon the Caribbean Plate at this location (Figs. 12, 13, and 14). Rapid tectonic uplift of portions of the Osa area are attributed to this aseismic ridge subduction. A series of six marine terraces and beach ridges (Madrigal, 1977) extends up to 10 m above sea level along the southeast coast of the Osa Peninsula south of Puerto Jimenez (see Fig. 25 for general location). The lowest marine terrace, 2.5 m above sea level, has an age of 590 ± 40 yr B.P. (Dicarb Radioisotope Co., DIC-3154), yielding an uplift rate of 4.2 mm/yr. An intermediate terrace approximately 8 m above

sea level contains shell material with a radiometric age of 22,780 +940/−1060 yr B.P. (DIC-3362). Assuming a low sea-level stand at 120 m below present sea level would yield an approximate uplift rate of 5.6 mm/yr. Fluviomarine deposits 10 km west of Puerto Jimenez previously mapped as Pliocene Charco Azul Formation, which occur 20–30 m above sea level, have been radiometrically dated at 34,530 ± 1210/1420 yr B.P. (DIC-3153). Assuming a sea level stand approximately 40 m below modern sea level would yield an approximate uplift rate of 1.7 mm/yr. This rate is similar to the 1.25 mm/yr rate calculated for uplifted deep-sea Plio-Pleistocene sediments from the adjacent Burica Peninsula along the Panamanian border (Corrigan and Mann, 1986).

Analyses of fluvial geomorphic features on the Osa and Burica peninsulas in southwestern Costa Rica are used to describe neotectonism associated with subduction of the aseismic Cocos Ridge. Incidents of drainage basin asymmetry, slope patterns, drainage divide migration, stream capture, and stream antecedence/superimposition are used to divide the area into distinct tectonic blocks.

Fluvial systems on an uplifted landmass respond to lowered base level by downcutting and divide migration. This basic observation was documented by Gilbert (1880) in his famous Law of Divides. Tectonic regimes characterized by nonuniform warping, doming, faulting, and tilting complicate Gilbert's simple law. Such processes like those concerning rifting and block faulting, described in Gerson and others (1985), Doornkamp and Temple (1966), and Van de Graaff and others (1977), produce many characteristic geomorphic features in the landscape. Three of the more easily observed of those geomorphic features are drainage basin asymmetry, stream capture, and stream superimposition/ antecedence.

To determine the magnitude of those geographic features on the Osa and Burica peninsulas, topographic data were collected from 1:50,000 topographic base maps. Data include drainage divides, basin outlines, longitudinal profiles of trunk, streams, and topographic linear features.

The most persistent and well-defined topographic linear features were used to create boundaries of distinct topographic blocks. The following analyses were made independent of these blocks with the intent that the ancillary data would substantiate, nullify, or modify their existence.

In each basin, the stream with the greatest length was designated as the trunk stream and subsequently subdivided into 3,000-m-long intervals, beginning at the headwaters. Starting at the stream's headwaters and moving downstream, the drainage area to the right and to the left was calculated using a polar planimeter for each 3,000 m interval. A basin is said to be "left-" or "right-" sided if, when viewing the basin from the headwaters to the mouth, the trunk stream is shifted to the side with the lesser areal extent (see inset, Fig. 25).

A simple slope analysis, using an opisometer to measure stream length, compiled the average stream slope of each drainage basin. The study area was then partitioned into nine sections

TABLE 1. PARAMETERS DERIVED FROM HYPSOMETRIC CURVES FOR DRAINAGE BASINS ON THE NICOYA PENINSULA, COSTA RICA

Basin	(1) Block	(2) % Cerro Azul	(3) Average Elevation (m) Cerro Azul	(4) r	(4) z	(5) Hypsometric Integral (%)	(6) Elevation of Inflection Points (m)	(7) Slope @ Inflection	(8) Integral above Inflection (%)
Tobaco	2	1.5	392		0.4	28.9	155	.349	26.2
	3	3.0	650						
	T	4.5							
Rosario	3	14.4	650		1.0	35.2	680	1.333	9.3
							558	2.100	14.6
							184	0.426	31.5
Montana	3	9.2	650	0.05	0.5	38.8	347	0.415	27.8
Nosara	3	6.6	650		0.5	38.1	246	0.539	31.6
Buenavista		0.0	---		0.33	34.4	277	0.460	25.2
Ora	4	15.7	552		0.4	31.6	190	0.432	28.5
	5	4.0	552						
	T	19.7							
Bongo	5	11.2	552		0.5	24.5	110	0.317	23.2
	6	2.3	415						
	T	13.5							
Ario	6	25.6	415		0.5	34.3	391	0.757	14.1
							292	1.040	21.8
							132	0.320	31.0

Basin	(9) Minimum elevation of Cerro Azul Surface (m)	(10) Integral above Minimum Elevation of Cerro Azul (m)	(11) Integral above % Cerro Azul
Tobaco	300	19.0	3.6
	480	9.1	
Rosario	500	16.1	11.3
Montana	600	11.7	7.5
Nosara	420	17.2	5.3
Buenavista	---	---	---
Ora	400	17.8	12.0
	400	17.8	
Bongo	400	10.2	8.0
	300	14.8	
Ario	300	21.2	16.3

Footnotes:
(1) Fault blocks containing drainage basin; see Figure 21. T indicates total % Cerro Azul Surface in the drainage basin.
(2) Percentage of the total basin area in which the Cerro Azul Surface is present.
(3) Average elevation of the Cerro Azul Surface as reported by Hare and Gardner (1985).
(4) Hypsometric curves can be described by a function of the general form:

$$y = \frac{d-x}{x} \cdot \frac{a}{d-a} \; z$$

The ratio $\frac{a}{d} = r$ is determined by the slope of the curve at its inflection point. The exponent z controls the general location of the curve (Strahler, 1952).
(5) Percentage of area underneath the hypsometric curve; that is, percent volume of landmass remaining from original surface.
(6) Elevation in basin corresponding to inflection point on the curve.
(7) Slope of the curve at its inflection point(s).
(8) Area underneath the portion of the curve above the inflection point. This represents the volume of landmass above a certain elevation.
(9) Elevation taken from topographic maps.
(10) Represents volume of landmass at or above Cerro Azul Surface.
(11) Area under the curve which accounts for the given % of Cerro Azul Surface observed in the basin.

based on drainage divide orientation. The average slope of each section was obtained by averaging the slopes of every basin contained in that particular section. Finally, the region was then analyzed in a qualitative manner by noting and recording instances of apparent stream capture, superimposition, barbed tributaries, and drainage divide migration. Inferences about directions of neotectonic movement were based on regional consistencies in basin asymmetry, regional slopes, and fluvial pattern adjustments. Figure 25 shows that groups of adjacent basins have the same magnitude and pattern of asymmetry. The best example of this occurs on the east coast of the Burica Peninsula, where

basins 40 to 49 show left-sided asymmetry with about 40 percent of basin area to the left, indicating tilt to the north.

This simple areas analysis clearly shows different sections or blocks of the region tilting in response to neotectonic deformation. In many cases, boundaries between these blocks are well defined by a basin that does not show appreciable asymmetry, such as the basin of the Rio Tigre (Fig. 25, basin no. 18) in the east-central section of the Osa Peninsula.

Stream-slope analysis substantiated two hypotheses; one, that uplift is generally occurring on the southwestern margins of the peninsulas, and the other, that the drainage divide is slowly

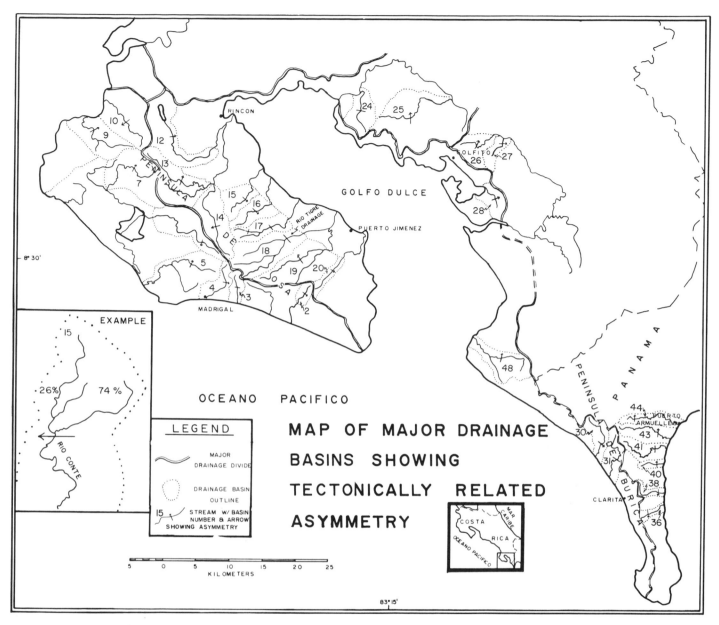

Figure 25. Map of drainage basin asymmetry. Inset shows 'right sided' basin. Arrows always indicate the basin side with the lesser areal extent. Note that groups of adjacent basins show similar directions of asymmetry.

migrating eastward. In every case, the western segments, or the segments near the coast, show slopes two or three times as great as eastern or inland drainage (Fig. 26). If slope magnitude and drainage divide position can be used as a surrogate for tectonic activity, the Osa and Burica peninsulas would be less active than the coastal area along the eastern Golfo Dulce near Golfito, a region of recent large earthquakes (Ms >6.5; Miyamura, 1980) and high ground acceleration potential (Fig. 16).

The eastern edge of the Golfo Dulce also displays well-developed examples of stream capture, drainage divide migration, and stream superimposition or antecedence (Fig. 27). Among the

most interesting features present are the marshy sag ponds and disrupted drainage. Currently, these areas drain westwardly, antecedent across uplifting areas adjacent to the coast, avoiding the obvious structural spillway. However, some of these streams have barbed tributaries, indicating that the entire drainage network may have originally flowed southeastward toward Puerto Armuelles.

Before uplift began, the major drainage direction was predominantly to the southeast. Volcanism to the north and east produced wedges of alluvial fan material, which, coupled with eruption events, succeeded in damming the relatively sluggish

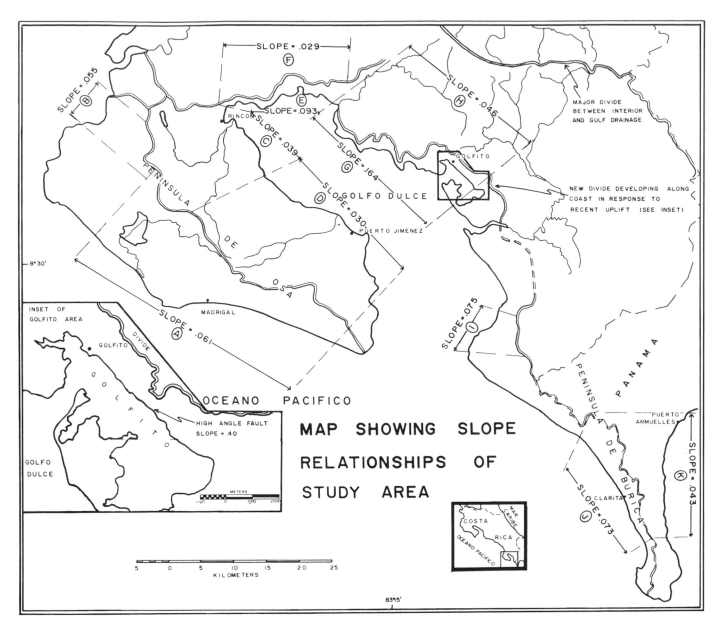

Figure 26. Slope relationships in the region. Note that all eastern slopes are consistently two to three times as steep as western slopes. Also note the extremely steep slopes in areas of rapid uplift, such as the area around Golfito (see inset).

southeast-flowing drainage and creating a drainage divide along the current position of the Costa Rican–Panamanian border. All of the drainage north and west of this divide was then shunted westward. This newly formed westward drainage is antecedent to recent coastal uplift.

High rates of uplift in the Golfito area are responsible for bisecting the older southeast-flowing drainage and for shunting all of the drainage northwest of Golfito to the northwest. The reversal of this drainage is clearly substantiated by the occurrence of barbed tributaries northwest of Golfito. This drainage avoids the main structural spillway to the northwest, instead turning to the

southwest where it is antecedent across the coastal uplift (Fig. 27).

The cumulative geomorphic data provide the basis for partitioning the region into 15 tectonic blocks (Fig. 28). Boundaries for these blocks coincide with many of the linear features first identified from the topographic base. In addition, many of them are found to coincide with mapped high-angle faults. The arrows indicate the direction of tilt for each block. There are areas of both convergence and divergence. Arrow directions clearly show discrete block tilting, generally to the north. A regional interpretation of neotectonism based on these findings indicates that at least

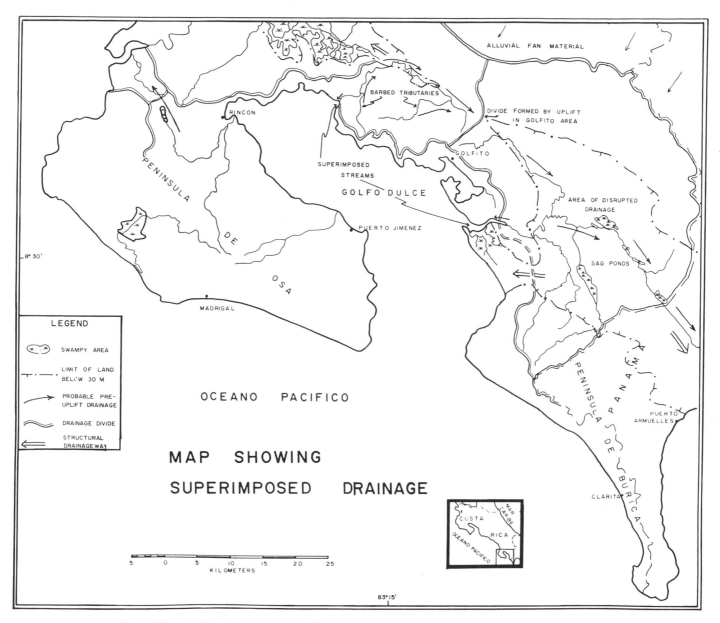

Figure 27. Disturbed and superimposed drainage of the study area. Barbed tributaries are more abundant but could not be shown due to scale. Note the well-developed structural spillways.

five areas seem to act as central uplifts. These areas also contain some of the steepest slopes found in the study area, a further indication that they are currently an uplift center.

A traverse of the study area from west to east would result in crossing a steep, west-dipping slope, then a gentle east-dipping slope, a deep valley (Golfo Dulce), and another steep west-dipping slope, a gentle east slope, and a deep valley. It is postulated here that the Golfo Dulce and the linear valley east of Golfito are actually the same feature, differing only in magnitude and representing low-angle thrust faults (Ministerio de Industria, Energia, y Minas, 1982; see Fig. 29) with uplift occurring on their western edges (northward tilt). Figure 28 clearly shows that the

blocks tilt predominantly to the north. However, several depressions in the area could also be high-angle reverse faults or graben features bounded by normal faults (Castillo, 1984).

SUMMARY

The Pacific Coast of Costa Rica, paralleling the Middle America Trench and astride the subducting aseismic Cocos Ridge, is ideally situated for tectonic geomorphic studies. Of the fundamental variables that control surface morphology and geomorphic system evolution, geology, climate, and tectonism, only tectonism varies significantly in time and space. Geologically, significant portions of the Osa, Burica, and Nicoya peninsulas are

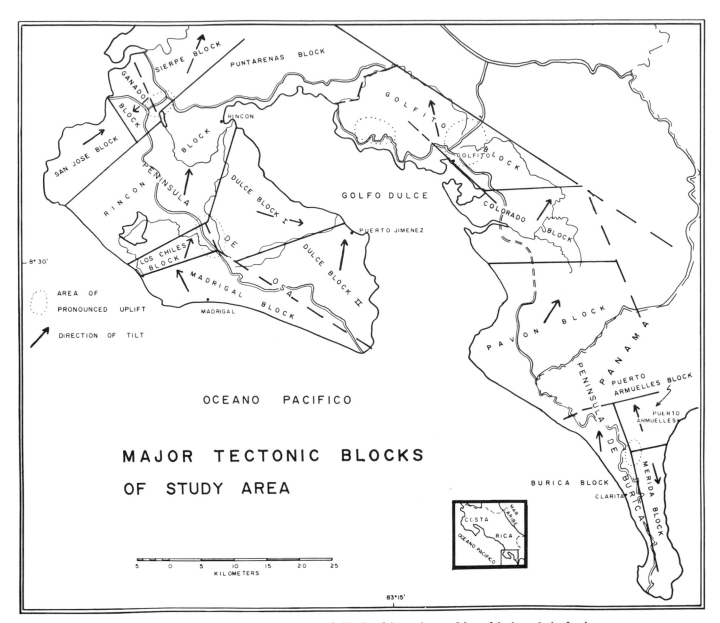

Figure 28. Map showing the 15 major tectonic blocks of the study area. Most of the boundaries for these blocks were first established during an early recognition of major linear features on a topographic base. Subsequent geomorphic analyses modified and refined these boundaries into the present 15 blocks.

underlain by ocean floor basalt, which at the watershed-scale is reasonable homogeneous. Furthermore, late Quaternary climates at the low latitude (9°N) and low elevation (average of 200 m above sea level) have been very stable.

It is apparent from the stratigraphic record, uplifted marine terraces, modern seismic activity, and present distribution of surface tectonic features (i.e., faults, sag ponds) that the Osa Peninsula region immediately astride the Cocos Ridge is tectonically more active than the Nicoya Peninsula region, some 200 km northwest of the Cocos Ridge. Drainage basins along the coastal zone display distinct morphological characteristics that can be

directly related to local tectonic environment and overall proximity to the Cocos Ridge. Old geomorphic surfaces record uplift and tilting of the oceanic crust. Ancillary morphologic data, like drainage basin asymmetry, are consistent with tilt directions estimated from elevation data. Basin and divide asymmetry, as well as recent stream diversions, suggest that in the Osa area crustal fragmentation during uplift is more extensive, and tilting less systematic, than in the Nicoya area.

In the next section, Menges and others will document the effect of proximity to the Cocos Ridge on mountain-front morphometry.

MORPHOMETRY OF TECTONIC LANDSCAPES ALONG A CONVERGENT PLATE BOUNDARY, PACIFIC COAST OF COSTA RICA

Chris M. Menges, Stephen G. Wells, and Thomas F. Bullard

The Pacific coastal region of Costa Rica extends across the tectonic boundary between the forearc and magmatic arc region of an active convergent margin. This plate boundary has evolved in a complex manner since the mid-Miocene (8 Ma), characterized by the southeastward migration of the Panama Triple Junction until 1 Ma, when the aseismic Cocos Ridge impinged on the Middle America Trench offshore along the southernmost coastal area of Costa Rica (Figs. 12 and 13; Minster and Jordan, 1978; Cross and Pilger, 1982). Partial subduction (100 km of relative convergence) of the aseismic ridge during the past 1 m.y. correlates with arc segmentation, which, in turn, is characterized by flattening in the angles of the associated Benioff-Wadati seismic zone and absence of Quaternary volcanism within the magmatic arc (Figs. 14 and 15; Stoiber and Carr, 1973; Cross and Pilger, 1982). The southwesternmost coastal region of Costa Rica has experienced significant amounts of Quaternary deformation (Miyamura, 1975; Fisher, 1980; Harpster and others, 1981) in response to this plate-ridge interaction. Thus, this region provides a unique opportunity to study how spatial variations in the plate tectonic framework of this convergent margin have affected the style and rates of Quaternary deformation and uplift in the onshore forearc region.

Specifically, the goal of this paper is to compare the tectonic geomorphology of the coastal mountain chain and associated piedmont in (1) the region of the aseismic ridge-plate interaction, and (2) a region directly to the north where subduction of more typical oceanic lithosphere occurs. Several workers have documented Quaternary regional uplift, active surface faulting, and historical seismicity in this part of Costa Rica (Miyamura, 1975; Fischer, 1980; Alt and others, 1980; Harpster and others, 1981). However, none of these studies used systematic analyses of tectonic landforms to define regional patterns of tectonic activity along this section of the plate boundary. In this study, regional morphometric analyses were applied to selected landforms in the coastal mountains and piedmonts of Costa Rica to define the relative amounts of tectonic activity on known or inferred faults that have generated topographic escarpments or mountain fronts. Morphometric analyses in tectonic geomorphology studies refer to the measurement on topographic maps of quantitative landform parameters (e.g., mountain-front sinuosity, valley heights and widths, and longitudinal stream profiles) that have been defined so as to identify adjustments of fluvial and hillslope systems to the degree of relative Quaternary uplift localized along mountain-front structures. Most of the analyses used in this study were developed by Bull (1973), Bull and McFadden (1977), and Bull (1978) in studies of the tectonic geomorphology of active extensional and compressional terrains in arid to semiarid regions

of the southwestern United States. This type of regional morphometric analysis has not previously been applied to forearc thrust fault systems in the convergent plate boundaries in a tropical region.

GEOLOGIC AND GEOMORPHIC SETTINGS OF STUDY AREAS

Two major study areas were selected along the coastal region of southwestern Costa Rica between latitude 9°30′N and 8°N (Fig. 29), within warm wet tropical climate zones (mean annual temperature = 26°C; mean annual precipitation = 4,000 mm). Region I includes the coastal ranges and highlands near Puerto Quepos and Dominical, where oceanic lithosphere of the Cocos plate is being subducted under the Caribbean plate. Region II lies to the south and includes the coastal ranges south of Palmar Sur and the highlands of the Osa Peninsula, which are situated onshore from the zone where an aseismic ridge has been partially subducted (Figs. 12 and 13). The major lithologic units of these regions are summarized in Figure 15, and the major structural elements of the study areas are illustrated in Figure 29.

The coastal ranges extend from north of Quepos to south of Palmar Sur (Fig. 29), and typically reach elevations in excess of 1,000 m, with some range block elevations in excess of 2,500 m. A coastal piedmont, up to 10 km wide and extending from the coastal range foothills to the shoreline, discontinuously occurs along the Pacific coast. The structural geology of the coastal mountain chain is characterized by a complex anastomosing series of northwest-trending, moderately northeast-dipping imbricate thrusts and high-angle reverse faults (Mora, 1984). These faults typically define the fronts of subparallel mountain ranges that step upward and inland from the coast (Figs. 29 and 30). Locally, antithetic structures (normal and reverse faults dipping southwestward) and tear faults occur perpendicular to the predominant northwest structural grain. Thirty to forty kilometers inland, folding occurs along axes that parallel the thrust faults (Fig. 29; Mora, 1984). The major lithologic units of the coastal range include deep and shallow marine clastic rocks with interbedded carbonate, volcanic, and igneous units of Cretaceous to Tertiary age (Fig. 15; Castillo, 1984; Ministerio de Industria, Energia, y Minas, 1982). The coastal piedmont section is characterized by shallow marine and fluvial clastic deposits of Quaternary age.

Major drainage lines, such as Rio Naranjo (Fig. 31a), head in the coastal ranges, drain perpendicular to the coastal piedmont, and form a broad series of stepped terrace surfaces and valley fills tens of meters thick that extend up to 60 m above modern base level (Table 2; Mora and Valdes, 1983). Ages of the terrace deposits have not been determined for most river systems; however, relative terrace ages can be established by soil chronosequences for terraces on rivers such as Rio Naranjo in region I (Table 2). To the south, along the Rio Terraba, a 190-m terrace has a questionable radiometric date of 7,500 yr B.P., suggesting an exceptionally high uplift rate in region II of 25 mm/yr (Alt

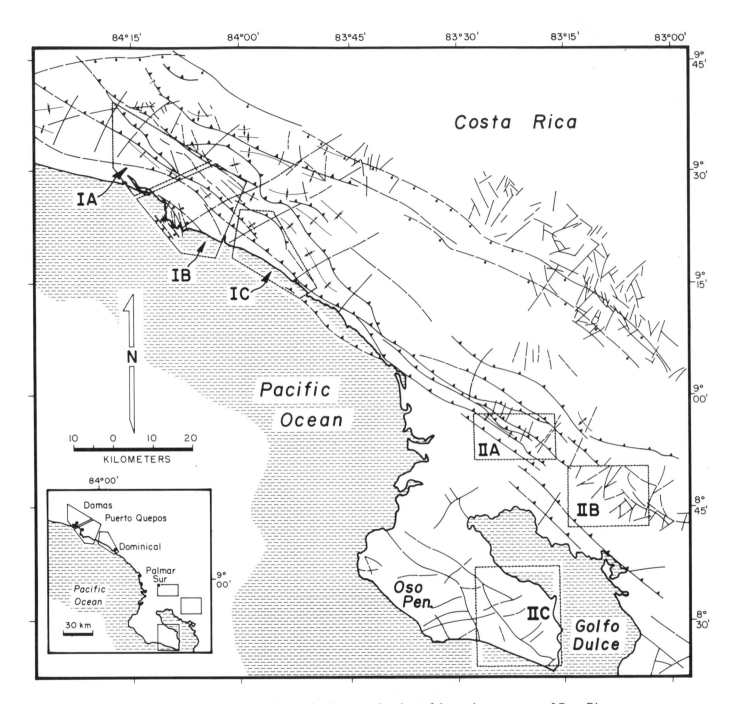

Figure 29. Generalized map of the regional structural geology of the southwestern coast of Costa Rica. Fault traces (dashed where inferred) are shown with heavier lines than fold axis traces. Types of structures include: thrust faults (barbs on upper plate), normal faults (bar and ball on downthrown side), unclassified faults (traces with no symbols), and folds (arrows indicating dip directions of limbs). The fine dotted lines indicate the approximate boundaries of the six subareas studied, labeled by region (Roman numerals) and subarea (letters). Source maps for this compilation are Mora (1984) in the northern and central portions, and Ministerio de Industria, Energia, y Minas (1982). The specific geologic map for subarea IIB did not show any northwest-trending thrust faults, although such structures probably do exist in this subarea, based on the detailed mapping of thrust fault systems in similar geologic and physiographic settings in areas to the north.

and others, 1980). In addition, Quaternary continental deposits (31,000 yr B.P.) along the piedmont in region II display nearly vertical dips, and Holocene beach sands are displaced by normal faulting (Alt and others, 1980). Two prominent marine wave-cut terraces occur along headland regions at 1.5 and 3.0 m above sea level along the entire coastline area, and beach ridges occur discontinuously along embayments developed in the coastal piedmont deposits.

Region II differs from region I (Fig. 31a) in that it contains portions of the Osa Peninsula that are characterized by low-relief upland blocks surrounded by a discontinuous coastal piedmont. The tectonic setting is characterized by structural blocks bounded by normal faults that apparently downstep toward the eastern and southern coasts. The uplifted blocks are composed primarily of Cretaceous Nicoya Complex units (ophiolitic suite of graywackes, cherts, shales, carbonates, pillow basalts, and mafic intrusives; Ministerio de Industria, Energia, y Minas, 1982). The normal faults have been observed to offset fluviomarine deposits radiometrically dated at 34,530 ± 1,210/1,420 yr B.P. (DIC-3153), which were previously mapped as the Pliocene Charco Azul Formation. Likewise, a series of beach ridges occur at 8 to 10 m in elevation along the coastal piedmont area; the lowest ridge–marine terrace complex has been radiometrically dated at 590 ± 40 yr B.P. (DIC-3154). Although the Osa Peninsula differs in tectonic style from the coastal ranges, it has undergone profound Quaternary uplift and deformation, which has impacted the fluvial systems (Gardner and others, this chapter).

METHODS

This study utilized landform analyses that focus on the responses of the fluvial system to discrete tectonic perturbations created by localized relative uplift along fault zones intersecting stream channels. The theory underlying these morphometric analyses defines from three to five relative classes of tectonic activity. These classes are associated with descriptive equations that interrelate local baselevel processes (tectonic uplift, stream downcutting, basin sedimentation, and erosion) to fluvial systems crossing structurally controlled topographic mountain fronts (Bull, 1973; Bull and McFadden, 1977; Bull, 1978). Each tectonic activity class is characterized by a distinct assemblage of landforms that may be quantitatively identified by specific ranges in the values of morphometric parameters measured from topographic maps and/or aerial photographs.

This study defines the relative degree of tectonic activity along mountain fronts, located within six subareas of regions I and II (Fig. 31a, b), on the basis of five main types of landform parameters measured from 1:50,000 scale topographic maps with 20-m contour intervals. These parameters are mountain-front sinuosity, the degree of facet development and dissection, valley height-width ratios, longitudinal stream profiles, and an index of the concavity of these stream profiles (Tables 3 and 4). Theoretical and operational definitions of these parameters are summarized in Table 3. The operational procedures for the definition and

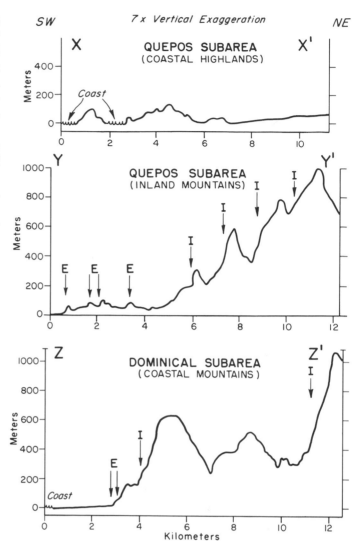

Figure 30. Topographic cross sections of the coastal plains and mountains of region I (profile locations indicated on Fig. 31a). All profiles with the same vertical exaggeration are oriented southwest to northeast, approximately perpendicular to the coastline. The positions of distinct external (E) and internal (I) topographic mountain fronts, as defined in text, are indicated along each profile.

measurement of all parameters must be carefully standardized prior to analyses to best accommodate the scale of the landforms and topographic maps.

The sinuosity and facet-dissection indices defined the morphologies of mountain fronts, which were then used to assign the fronts to four general classes of relative tectonic activity. These classes were not directly correlated with those previously established by Bull and McFadden (1977) or Bull (1978) because of the possible complications introduced by the large climate differences existing between humid tropical Costa Rica and the more arid regions of those studies. The valley height-width ratios, longitudinal stream profiles, and concavity indices were used to

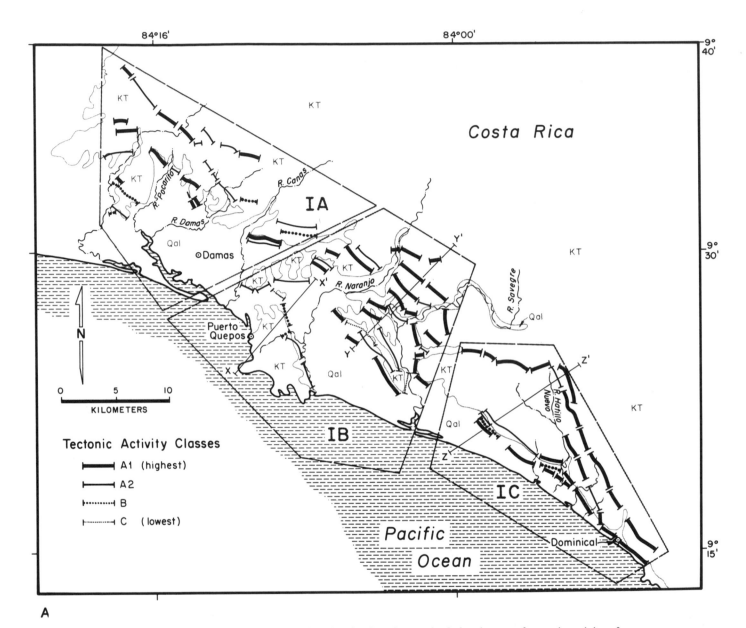

A

Figure 31 (this and facing page). Map showing locations and relative degrees of tectonic activity of individual mountain fronts in region I (Fig. 31a) and region II (Fig. 31b) that were analyzed for sinuosity and facet characteristics. Refer to Figure 29 for relative positions of these two regions. The boundaries of each subarea are outlined and labeled within boxes. The relative tectonic activity classes correspond to designations in histograms of Figure 32 (refer to text for discussion of mountain fronts and tectonic activity classes). Note the increasing frequency of class A1 (most active) from north to south in regions I and II, culminating in a very high proportion of Al fronts in subarea IIB. Also shown are: (a) the geologic contacts between Cretaceous and Tertiary age bedrock (KT) and Quaternary alluvium (Qal), compiled from Ministerios de Industria, Energia, y Minas (1982) maps; and (b) the location of selected cities and rivers. The transects labeled X–X′, Y–Y′, and Z–Z′ in Figure 31a indicate the positions of the topographic profiles in Figure 30.

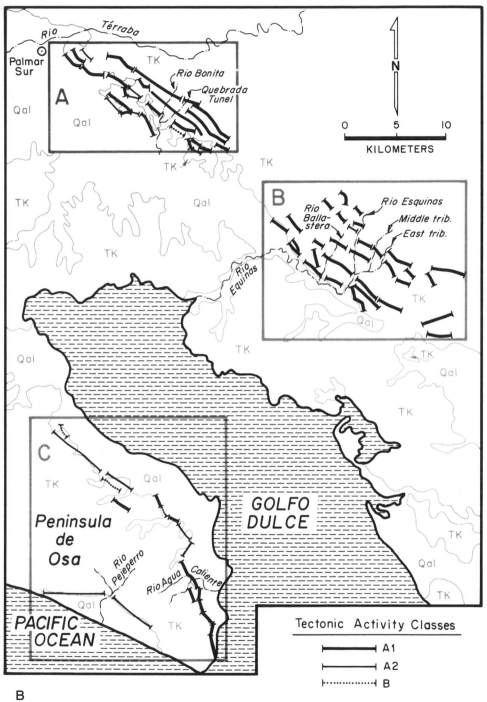

TABLE 2. SOIL PROFILE DESCRIPTIONS AND CHRONOSEQUENCE FOR TERRACES OF THE RIO NARANJO, PROVINCIA PUNTARENAS, COSTA RICA*

Terrace Number (1 = oldest)	Maximum Height above Base Level (m)	Characteristic Horizon	Total B Horizon Thickness (cm)	Color matrix mottle	Texture[1]	Structure[2]	Consistency[3]	Clay films[4]
Qt1	60	B3	460+	2.5YR 4/8 10YR 4/8	cl	mcsbkmpn	s,p	3,pf
Qt2	20	B4	106	10YR 6/4 5YR 5/8	sicl	cvcsbk	ss,p	2,pf,po
Qt3	9	B2	83	7.5YR 4/6 7.5YR 5/6	sic	fmsbk	ss,p	2,pf
Qt4	5	B2	36	10YR 5/3 7.5YR 5/8	sil	cabkcpn	ss,p	----
Qt5	3	Cn	0	10YR 3/3 7.5YR 5/6	fs-c	mcgnsbk	n.a.	n.o.

*Profile descriptions by S.G. Wells, J.B. Ritter, and T.F. Bullard. Soil conditions were moist during the descriptions.

[1] c = clay; si = silt; l = loam.

[2] f = fine; m = medium; c = coarse; vc = very coarse; sbk = subangular blocky; pr = prismatic; abk = angular blocky; gr = granular.

[3] 2 = common; 3 = many; pf = on ped faces; po = pore linings; n.o. = not observed.

collectively isolate perturbations in channel and valley morphologies possibly related to the presence and relative magnitudes of tectonic activity on one or more mountain fronts crossed by a given stream. Parametric and nonparametric statistical tests were applied to the morphometric data (Table 5) to identify statistically significant differences in data groups indicative of variations in tectonic activity among the subareas.

MOUNTAIN FRONTS

Definition of mountain fronts

Mountain fronts were defined as topographic escarpments with measurable relief equal to or exceeding one contour interval (Figs. 30 and 31). Larger escarpments were subdivided for analysis into discrete segments with generally similar geologic and physiographic characteristics, based on criteria including: (1) intersection with crosscutting drainages large in scale relative to the fronts, (2) abrupt deflections in orientation, (3) changes in lithology, where known, and (4) abrupt changes in the general physiography of a mountain front (e.g., relief, dissection, steepness), relative to adjoining front segments. Mountain fronts were classified as internal or external, depending upon whether they occur within the mountainous terrain itself or form part of the physiographic boundary between the main mountain chain and the adjoining coastal plain (Fig. 30), following the scheme of Bull (1978). This classification accommodates the commonplace occurrence within the bedrock range block of prominent linear topographic escarpments that equal or exceed the lateral continuity and relief characteristics of fronts defining the boundary

between the mountain chain and coastal piedmont. Both external and internal mountain front escarpments commonly coincide with, or are subparallel to the mapped traces of thrust faults (Figs. 29 and 31).

Mountain-front sinuosity

Mountain-front sinuosity is an index of the degree of irregularity or sinuosity along the base of a topographic escarpment. The utility of this parameter is based on the tendency of active structures to maintain straight profiles in plan view, relative to the more irregular profiles produced by erosional processes, along the base of associated topographic escarpments.

Mountain-front sinuosity (S) is defined as the ratio of the observed length along the margin of the topographic mountain-piedmont junction (Lmf) to the overall length of the mountain front (Ls), or S = Lmf/Ls (Table 3). Values of S tend toward 1.0 on the most tectonically active fronts and increase along fronts with decreasing amounts of tectonic uplift relative to basal erosion or pedimentation.

Facets

A facet is a triangular- to polyhedral-shaped hillslope that lies between two adjacent drainages within a given mountain-front escarpment. In theory, a high percentage of the length of very active fronts will display prominent, large facets that are generated and/or maintained by recurrent faulting along the base of the escarpments (Bull, 1978). Less tectonically active fronts will contain fewer, smaller, and/or more internally dissected

Table # 3 Summary of Morphometric Parameters used in Tectonic Landform Analysis of Individual Mountain Fronts (MF)

Morphometric Parameter	Definition	Mathematical Derivation	Measurement [1] Procedure	Purpose	Potential Difficulties	Source
S	Sinuosity of topographic mountain fronts.	$\frac{[2]\ [3]}{Lmf/Ls}$	LmF / Ls (plan view)	Define the degree of topographic modification (embayment and/or pedimentation) of mountain front from the position of possible controlling tectonically active structures.	-Define actual topographic junction. -Definition of discrete mountain front segments.	Bull, 1978 Bull and McFadden, 1977
Facet (%)	Percent faceting along mountain fronts.	$\frac{[4]}{Lf/Ls}$	valley / Lmf / LF / LS / LF	Define the percentage of given mountain front with well defined triangular facets, using ratio of cumulative lengths of facets to length of overall mountain front.	-Systematic definition of individual facet. -Discrimination between faceted and unfaceted (more highly dissected) sections of a mountain front.	this study
Fd	Percent dissected Mountain Fronts (in given region)	$\frac{\#\ of\ dissected\ Mf}{total\ \#\ MF}$	Lmf (dissected MF) / Ls (undissected MF) / Ls Lmf	Define the percentage of total mountain fronts that have been dissected by large drainage basins into distinct facets.	-Systematic definition of mountain front dissection from topographic maps.	this study
Ffd	Percent dissected facets (on mountain fronts in a given region)	$\frac{\#\ of\ Mf,\ with\ dissected\ facets}{total\ \#Mf}$ (in given region)	(MF, with dissected facets) Lmf / Ls / (MF, with undissected facets) Lmf / Ls	Define the percentage of mountain fronts that contain facets with significant internal dissection (as indicated by degree of contour crenulation along facet base).	-Systematic identification of dissected vs undissected facets, especially on highly dissected MF with many narrow facets.	this study
Vf	Valley floor - valley height ratio	$\frac{[5]\ Vfw}{[\frac{[6](Eld-Esc)+[7](Erd-Esc)}{2}][9]}$ [8]	Eld / Erd / Esc / Vfw (cross-section)	Define the ratio of the width of the valley floor to the mean height of two adjacent divides, measured at given locations along a stream channel within the range block.	-Resolution of topographic maps in defining Vfw and divide elevations. -Need to minimize variations in stream size (length area). -Effect of change in lithology.	Bull, 1978 Bull and McFadden, 1977
K	Stream profile concavity	∫ stream long profile, ie., area under % height - % distance longitudinal profile curve.	1.0 / O / O 1.0 (plot derived from longitudinal profile measured from topographic maps)	Define the degree of concavity displayed on a longitudinal stream profile, with lower K values associated with more concave-up profiles. Each profile is normalized by converting measured heights and lengths to percentage of total height and length, respectively, of profile.	-Resolution of stream profiles on topographic maps. -Interpretational problems in discriminating between tectonic and lithologic inputs to profile irregularities.	Adapted from Shepard, 1979 and Kelson, 1986

FOOTNOTES TO TABLE 3

[1] Sketch illustrating measurement techniques used to derive each morphometric parameter. Sketches for all parameters except Vf and K illustrate the base of a mountain front in plan view as defined by contour patterns on a topographic map (mountainous area in stipple).

[2] mf - length of topographic mountain front (ie. topographic junction between mountain front and piedmont).

[3] S - Shortest length that parallels or circumscribes overall mountain front, drawn with minimum curvature tangential to outer (basinward) margin of front.

[4] F - Cumulative length of all facets developed along a given mountain front, the length of each facet is measured tangential to its basal contour.

[5] Vfw - Width of valley floor, including floodplain and/or low terraces, as defined by topographic contours.

[6] Eld - Altitude of left divide, immediately above Vfw measurement.

[7] Erd - Altitude of right divide, measured immediately above Vfw measurement.

[8] Esc - Altitude of stream channel at location of Vfw measurement.

[9] The valley floor width and divide height are measured from topographic map contours along a cross-valley transect oriented perpendicular to stream channel.

TABLE 4. SUMMARY OF MOUNTAIN FRONT SINUOSITY, FACETING, AND CONCAVITY DATA FOR REGIONS I AND II

Subarea	Number of Fronts	Mountain Front Sinuosity					Facets			
		range	mean	std. dev.	median	mode	mean (%)	std. dev. (%)	Fd[a] (%)	Ffd[b] (%)
Region I										
A	32	2.0-1.0	1.4	0.26	1.3	1.2	46	34	92	89
B	39	3.1-1.0	1.5	0.52	1.2	1.1	34	34	91	82
C	29	2.1-1.0	1.2	0.27	1.1	1.0	69	33	63	50
Region II										
A	38	1.9-1.0	1.2	0.24	1.1	1.0	16	26	----	----
B	37	1.5-1.0	1.1	0.10	1.0	1.0	41	33	----	----
C	26	2.2-1.0	1.5	0.35	1.3	1.1	18	22	----	----

	Concavity			
Subarea	n	mean	median	std. dev.
Region I				
A	--	----	----	----
B	--	----	----	----
C	--	----	----	----
Region II				
A	3	0.199	0.172	0.076
B	4	0.334	0.319	0.040
C	5	0.201	0.213	0.039

Footnotes:

[a]Percent of dissected mountain fronts.

[b]Percent of mountain fronts with dissected facets.

---- data not available.

TABLE 5. SUMMARY OF STATISTICAL TESTS OF MORPHOMETRIC DATA AMONG SUBREGIONS IN THE STUDY AREA

A. T-tests for Statistical Differences in the Mean Mountain Front Sinuosities ($\underline{S}$) of Subregions in Regions I and II

Ho: There is no statistical difference between the means of the mountain front sinuosities of the two areas tested.

Test	t	t_o	Decision	Significance Level
A vs B	0.977	2.660	Accept Ho	$\alpha = 0.005$
A vs C	2.900	2.660	Reject Ho	$\alpha = 0.005$
B vs C	2.793	2.660	Reject Ho	$\alpha = 0.005$
A vs B	3.047	2.660	Reject Ho	$\alpha = 0.005$
A vs C	3.387	2.660	Reject Ho	$\alpha = 0.005$
B vs C	6.233	2.660	Reject Ho	$\alpha = 0.005$

B. Wilcoxon Rank Sum Test for Statistical Differences in the Concavity Indices (K) of Stream Profiles Within the Three Subareas of Region II.

Ho: There is no statistical difference between the means of K of the two areas tested.

Test	Ru	Tu	Decision	Significance Level
A vs B	22	18	Reject Ho	$\alpha = 0.005$
A vs C	12.5	21	Accept Ho	$\alpha = 0.005$
B vs C	30	28	Reject Ho	$\alpha = 0.005$

facets due to the combined effects of the establishment of more drainages across the front, the internal dissection of existing facets, and the lateral migration of larger streams crossing the front (Table 3). Dissection of the overall escarpments can be semiquantitatively assessed to indicate relative tectonic activities of the fronts (Table 3). A given mountain front, as defined earlier, is considered to be dissected if it contains one or more large drainages that have produced distinct facets and/or escarpment hillslopes with highly crenulated contours. Individual facets are themselves considered to be internally dissected if they are defined by crenulated sets of contours at lower to midslope heights. The more tectonically active fronts tend to be less dissected, ranging from an essentially laterally continuous undissected escarpment (defined as 100 percent faceting, Table 3) to a nearly continuous front with only a few large distinct facets that tend to have minimal internal dissection.

VALLEY SYSTEMS

Valley cross sections and longitudinal profiles

Channel and valley morphologies of small- to medium-sized drainage lines (5 to 20 km length) that cross one or more transversely oriented mountain fronts may reflect the impact of local baselevel perturbations imposed by relative uplift along tectonically active structures associated with the frontal escarpments. Thus, detailed longitudinal profiles and valley height-width ratios were measured from 20 m contours to isolate anomalous patterns in channel or valley forms attributable to tectonically active mountain fronts.

Longitudinal stream profiles were plotted at 7.5 to 9.5x vertical exaggeration in order to highlight any irregularities in channel slope. Another means of quantitatively comparing the form of longitudinal stream profiles is the dimensionless concavity index (C) of Shepherd (1979), which measures the degree to which a normalized profile deviates from an ideal 45° line. Each profile is normalized by converting the measured heights and lengths to percentages of the total height and length of the profile (Table 3). This study uses a similar concavity index (K) of Kelson (1986) that is defined as the area under the normalized stream profile described above. Thus, K, analogous to the hypsometric integral of Strahler (1952) and Schumm (1956), compares stream relief to stream length. Concave-up profiles have lower K values, which may suggest prolonged basin and channel degradation associated with longer time periods since baselevel lowering, whereas high K values suggest less channel downcutting, continued baselevel lowering, and/or less time since a single baselevel fall.

Transverse valley profiles were defined using the valley floor–valley height ratio parameter of Bull and McFadden (1977) and Bull (1978). This index is defined as the ratio of width of the valley floor (Vfw) to the mean height of the right and left divides (Erd and Eld, respectively) measured at one site along the stream with respect to the elevation of the stream channel (Esc; Table 3). Thus, the Vf ratio discriminates those reaches of the stream with low Vf values dominated by downcutting in response to local baselevel fall from those reaches with high Vf values characterized by more lateral stream planation.

However, Vf ratios may also vary widely among streams with different drainage basin areas, discharges, and underlying bedrock lithologies. Consequently, Vf ratios were not directly used in this study to estimate the relative classes of tectonic activity of specific fronts, as this would require comparison of Vf values among streams of variable size and lithology. Instead, Vf values were combined with the longitudinal profile and/or concavity data to indicate changes in valley and profile morphologies, suggesting localized uplift in channel reaches upstream from mountain fronts crossed by a given stream.

TECTONIC INFLUENCE ON MOUNTAIN FRONTS

Region I

The sinuosity and facet-dissection characteristics of 30 to 40 mountain fronts were analyzed in each of the three subareas of region I (Figs. 29 and 31a; Table 4). These subareas were defined on the basis of (1) the number, distribution, and general form of discrete topographic mountains; (2) the position of the external range boundary relative to the coast (Fig. 30); (3) variations in the size and patterns of river networks draining these fronts; (4) the presence/absence of bedrock coastal highlands (Figs. 30

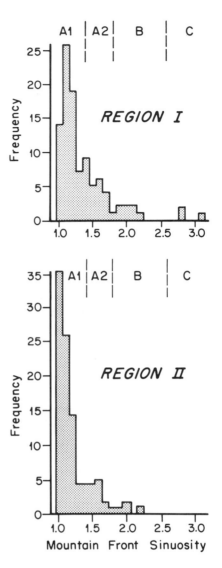

Figure 32. Composite histograms of mountain-front sinuosity values measured on individual mountain fronts shown on Figure 31 in all subareas of regions I and II. The relative tectonic activity classes estimated from the data distributions are labeled (Al to C, from highest to lowest rates of tectonic activity) above histograms; these classes correspond to coding of mountain fronts on Figure 31.

and 31); and (5) the quality and distribution of available geological and structural control. Sinuosity is low in all subareas, with 50 percent of the mountain fronts displaying S values less than 1.3, and 89 percent less than 1.8 (Table 4; Fig. 32). Values of sinuosity of fronts range from 3.1 to 1.0, with means of 1.4, 1.5, and 1.2 for subareas A, B, and C, respectively. These data indicate that higher sinuosity values are contributed mainly from subareas A and B. Statistical tests indicate a significant difference between the mean sinuosity of fronts in subarea C versus subareas A or B, although the sinuosities of the latter two subareas are statistically indistinguishable (Table 5). The mountain fronts, in general, tend to be more sinuous and dissected in coastal areas than in the more interior mountainous regions.

Mountain fronts in subareas A and B exhibit less faceting and a greater amount of facet dissection than do fronts in subarea C (Table 4). For example, more than a third of the mountain fronts in subareas A and B are so dissected that no identifiable facet segments exist (% facet = 0). In contrast, a third of the fronts in subarea C are so little dissected that 80 to 100 percent of their lengths are contained in at most a few large, internally undissected facets.

Mountain fronts were then assigned to four classes of relative tectonic activity, Al (highest) through C (lowest; Figs. 31a and 32). These classes, specifically established for the Costa Rica study, were based primarily on groupings within the mountain-front sinuosity data (Fig. 32), supplemented by facet characteristics. The low sinuosities and percent faceting associated with the majority of mountain fronts in all subareas suggests a generally high degree of tectonic activity along most of the topographic boundary of the coast ranges of region I. Furthermore, the combined morphometric data suggest that the mountain fronts of subarea C are more tectonically active than those of either subareas B or A. The latter northernmost subarea appears to contain the lowest relative levels of tectonism within sampled parts of the coastal ranges.

Region II

Mountain-front sinuosities in region II (Figs. 31b and 32) range from 1.0 to 2.2; the means of S for this region are 1.2, 1.1, and 1.5 for subareas A, B, and C, respectively (Table 4). Differences between the means of all three subareas are statistically significant at the 99 percent confidence level (Table 5).

The degree of facet development present along individual mountain fronts (% faceting, Table 3) ranges from 0 to 100 percent. Means of percent faceting for subareas A and C are low (16 and 18%, respectively), relative to the higher mean percent of subarea B (41%; Table 4). Subareas A and C also contain a higher proportion of dissected fronts with no facet development (68 and 42%, respectively) compared to subarea B (19%; Table 4).

These morphometric data suggest generally high levels of tectonic activity along mountain fronts throughout region II. Differences in mountain-front sinuosities and faceting are inferred to represent variations in tectonic activity among the three subareas (Table 4). The very low mean sinuosity (1.1) and low standard deviation ($\alpha = 0.10$) in subarea B imply especially active faulting along these mountain fronts (Table 4).

TECTONIC INFLUENCES ON VALLEY SYSTEMS

Region I

Most of the 14 streams analyzed in region I that cross mountain fronts display steeply sloping and irregular profiles characterized by a series of concave and convex segments, with steeper

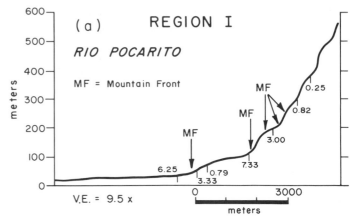

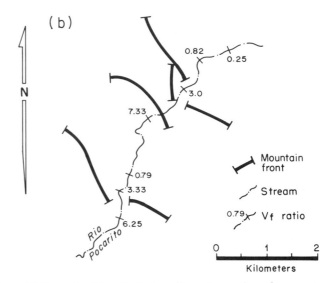

Figure 33. Example of longitudinal profile and map view of representative stream (Rio Pocarito) in region I (Fig. 31a) that crosses a subparallel series of transversely oriented mountain fronts. Vf ratio values (defined in text and Table 3) are labeled (small numbers) on both the longitudinal profile and plan views. Note the convex profile anomalies and general decrease in Vf values that occur in stream reaches above intersections with individual mountain fronts.

gradient reaches located at or small distances upstream from profile intersections with the topographic escarpments oriented transverse to stream channels (e.g., Fig. 33). Streams cutting across the coastal highlands of subarea B, or in valleys that parallel mountain fronts near the coastal plain, have more regular and gently sloping longitudinal profiles. Streams in valleys that parallel mountain fronts in the mountainous interior have profiles resembling the steeper, benched profile segments of streams that transect interior fronts.

Vf values along the same 14 streams range from 0.11 to 32, with typical values of 0.2 to 7. Individual Vf ratios fluctuate

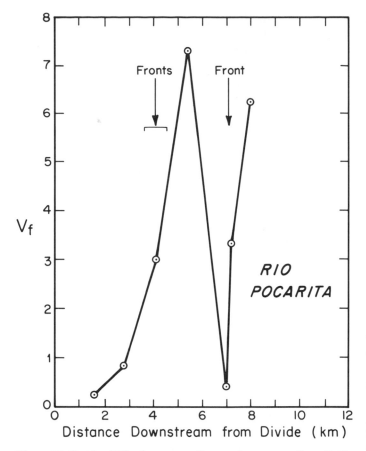

Figure 34. Graph of Vf values versus distance downstream from divide (increasing to right) along the Rio Pocarito (shown in Fig. 31a). Note the abrupt large decrease and increase in Vf values measured in stream reaches above and below (respectively) the intersection of the stream channel with transversely oriented topographic mountain fronts (indicated on graph). Note also the generally low Vf values observed in the headwater areas near the divide.

stream from the front (Figs. 33 and 34). This pattern is repeated several times downstream along river profiles that cross multiple subparallel transverse mountain fronts with sufficient spacing to allow development of the Vf fluctuations described above. Bull (1978) observed similar decreases and increases in the Vf ratios measured along stream reaches located upstream from a series of transversely oriented mountain fronts and associated thrust faults, along the tectonically active southern border of the San Gabriel Mountains of southern California.

The combined Vf ratios and longitudinal profiles of many streams in all three subareas suggest that the profile of a stream reach undergoes abrupt steepening of gradient at or immediately upstream from individual mountain fronts. These gradient changes resemble large composite knickpoints imparted by local tectonic uplift along known or inferred faults along these fronts (Figs. 29, 31a, and 33). These reaches commonly coincide with abrupt decreases in Vf ratios indicative of downcutting within narrow V-shaped canyons; this is interpreted as a stream response to local baselevel changes imparted by relative local uplift at the mountain front downstream. Farther upstream, more gently inclined gradients and increasing Vf ratios reflect decreased stream downcutting and/or increased lateral planation. These stream reaches commonly are located within embayments that form in the interior of uplifted blocks between adjacent subparallel mountain fronts. The presence and relative magnitude of these profile and Vf anomalies observed in streams crossing a given front suggest localized uplift along the causative structure at rates exceeding the ability of the fluvial system to incise and remove the tectonically induced gradient anomaly. Variations in gradient and Vf values are evident to varying degrees in most moderate- to small-sized rivers that cross low-sinuosity mountain fronts in the region, regardless of the underlying lithologies. Some individual profile anomalies may reflect changes in rock resistance not evident at the scale of available geologic maps.

Region II

Longitudinal profiles of seven streams in region II exhibit several alternating convex and concave reaches similar to the profile gradient changes described above on streams in region I (Figs. 33 and 35). Distinct steepening of gradients in region II likewise occurs upstream from points where streams cross mapped or inferred faults and associated mountain fronts (Fig. 35). These convex segments are similarly interpreted to reflect localized uplift along the mountain fronts at rates and/or amounts exceeding the capacity of streams to alter or remove imposed channel slope discontinuity. Abrupt changes in rock types in both regions I and II may control some of the gradient change, but the effect is difficult to evaluate precisely with available geologic mapping and limited field access. A further complication is introduced by repetition of some rock types by thrust faults associated with many of the fronts in regions A and B. Still, the composite convex-concave profile anomalies occur to some degree on most streams crossing transverse mountain fronts of

significantly along the profile on a given stream; however, several general patterns are identifiable on most streams in the region that cross transversely oriented mountain fronts (Figs. 33 and 34). Vf ratios generally decrease in value downstream in the headwaters region, but remain somewhat low (<1 to 2), probably reflecting persistent channel downcutting in these reaches. Farther downstream, the general ranges of Vf ratios increase, presumably in response to the increasing amounts of lateral stream migration and valley-floor widening, but on many stream profiles this general trend is obscured by abrupt, pronounced Vf fluctuations near channel intersections with the topographic mountain fronts. The most commonly observed pattern comprises: (1) large Vf values (>2–3) at distances of one or more kilometers upstream from the escarpment; (2) a sharp decrease in the Vf ratio to <1.0, associated with development of narrow, deep canyons, along channel reaches at, or adjacent to, the intersection with the front; and (3) a return to large Vf values within broader valleys down-

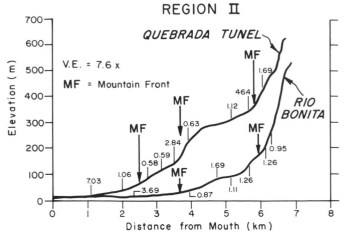

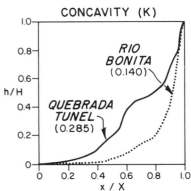

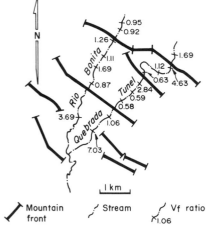

Figure 35. Example of longitudinal profiles, map views, and normalized concavity profile of two representative streams (Quebrada Tunel and Rio Bonita) in region II (Fig. 31b) that cross a subparallel series of transversely oriented mountain fronts. Vf ratio values (defined in text and in Table 3) are labeled (small numbers) on both profiles and map views. Note a correspondence among Vf ratio fluctuations, convex profile anomalies, and stream–mountain-front intersections that resembles patterns observed in Figure 33. The concavity index (K) is derived from the normalized longitudinal profile graph, which consists of plotting measured horizontal distances (x)/total length of profile (X) (as abscissa) versus measured heights (h)/cumulative height of profile (H) (as ordinate). The index (K) is defined as the area under this curve and is given in parenthesis next to each profile.

both regions I and II; the prevalence and position of these gradient changes relative to the escarpments suggest that the profile anomalies reflect tectonic activity on the front-bounding faults.

Concavity index measures of longitudinal stream profiles in region II suggest additional differences in the relative tectonic activity among the three subareas (Fig. 35; Table 4). The concavity indices for the three subareas range from 0.140 to 0.393 (Table 4), and statistical tests on the means from the three subareas indicate three distinct populations (Table 5). Subarea B has the highest mean, suggesting a greater degree of tectonic activity along the mountain fronts crossed by these streams relative to those in subareas A and C.

Vf ratios in the streams studied in region II range from 0.47 to 8.40. Individual Vf ratios measured along a given stream often display fluctuating patterns related to transecting mountain fronts that resemble those described earlier in region I (Figs. 33 and 35). This relationship is best expressed across the interior mountain fronts of subareas A and B. The Vf ratios measured upstream from interior fronts in these areas are at least several times lower than the Vf ratios in subarea C or the more exterior fronts in subareas A and B. Also, Vf variations tend to be more erratic along streams in the Osa Peninsula of subarea C, which lacks the high relief, mapped thrust faults, and multiple subparallel mountain fronts of the other subregions (Figs. 29 and 31). These relationships suggest that the relative influence of tectonic uplift is greater along interior mountain fronts of subareas A and B; the effects of composite relief, which is greater in the interior mountain fronts of the subareas, plus the decreased stream power in these headwater areas, may influence the fluvial response to the mountain-front tectonic activity.

MOUNTAIN-FRONT MORPHOMETRY AND PLATE TECTONIC SETTING

Within the two study regions (I and II), differences occur in the morphologic and morphometric expression of exterior and interior mountain fronts associated with an active tectonic environment. Low sinuosity and a high percentage of undissected facets characteristically occur within the interior of the coastal range, suggesting a relatively higher degree of tectonic uplift in these areas. The mountain fronts nearer the coastline are characterized by higher sinuosities and a lower percentage of undissected facets, indicative of decreased levels of tectonic activity. However, some of these morphometric differences may reflect increased erosional embayment of lower-relief coastal mountains by larger streams, such as the Rio Savegre and Damas. Divergence of river terraces toward the mountain fronts and into the range blocks supports increased rates of uplift during the Quaternary (Table 2). Vf ratios, longitudinal stream profiles, and concavity indices provide additional morphometric data supporting relative variations in the tectonic activity among the subareas of the two study regions. The combined data suggests a general southward trend of increasing tectonic activity among mountain fronts in three subareas of region I, with subarea C the most

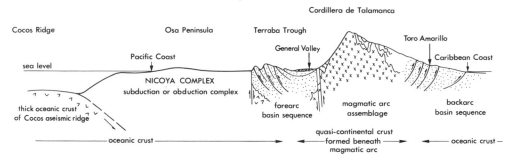

Figure 36. Cross section through southeastern Costa Rica showing inferred relationships of principal structural and geomorphic elements.

active (Fig. 31). This southward increase in the relative levels of tectonic activity continues through region II, where the southernmost Rio Terraba subarea (B) contains the lowest values in mountain-front sinuosity and the highest values of percent faceting and concavity; these features are all associated with very high relative degrees of tectonic uplift.

Field evidence supports the morphometric trends and indicates a relative increase in tectonic activity from north to south. Fault scarps greater than 1.0 m in height, which rupture surface soils, have been been observed in region IC and in region II. In region IIA, terrestrial deposits 31,000 years old are dipping vertically, and in region IIC, fluviomarine sediments 34,500 years old have been uplifted between 20 and 30 m above mean sea level. In addition, six marine terraces extend up to 10 m above mean sea level along the southeastern coast of the Osa Peninsula (Madrigal, 1977); one of the youngest, 2.5 m above high-tide level, is dated at 590 ± 40 yr B.P. (DIC-3254). Increased uplift and deformation of latest Quaternary deposits and terraces in region II is consistent with the morphometric measurements on the mountain fronts.

Morphometric data reflecting a southward increase in vertical tectonic activity is consistent with both the available geologic data and changes in the plate tectonic setting along the adjacent convergent boundary, which are characterized by the occurrence of a partially subducted aseismic ridge offshore from region II (Figs. 12 and 13). Thus, this study provides a regional reconnaissance highlighting the location and regional variation in neotectonic activity on boundary thrust fault systems developed along convergent plate margins. These results suggest that variations in the plate tectonic setting along convergent boundaries can be detected by landscape morphometric analyses.

ALLUVIAL FAN FORMATION IN A MAGMATIC ARC SETTING, COSTA RICA

Richard H. Kesel and Donald R. Lowe

INTRODUCTION

Magmatic arcs provide two, often closely juxtaposed, but quite different, settings for the formation of alluvial fans: (1) rap-

idly constructed, steep-sided strato-volcanoes that serve as sources of juvenile volcanic and volcaniclastic debris for the growth of surrounding alluvial aprons, and (2) tectonically uplifted arc segments that yield deep-level volcanic, sedimentary, metamorphic, and plutonic detritus to fans constructed on adjacent down-dropped blocks.

Southern Costa Rica is a magmatically and tectonically active portion of the Middle American Magmatic Arc developed along the convergent boundary between the Cocos and Caribbean plates (Fig. 1, 12, and 36). The central and northern parts of the country include the presently active chain of Holocene and Pleistocene volcanoes, the Cordillera Central. The southeast part includes a high, tectonically uplifted but magmatically inactive segment of the arc, the Cordillera de Talamanca.

This paper reports on the results of studies of humid-climate alluvial fans developed in both of these settings (Fig. 37). The Rio Toro Amarillio fan is one of a series of active, coalescing alluvial fans developed in the back-arc along the eastern Caribbean side of the Cordillera Central. The Rio General valley, a structural forearc depression developed along the southwestern Pacific margin of the Cordillera de Talamanca, is partially filled by a series of late Pleistocene and early Holocene alluvial fans heading in the adjacent uplifted Talamancan block. In this paper, the sedimentology and morphology of these fans are compared with regard to their climatic, tectonic, and magmatic settings.

RIO TORO AMARILLO FAN

Physical Setting

The Rio Toro Amarillo is one of several rivers that form an active, coalescing alluvial fan deposited between the Cordillera Central and the Caribbean coastal plain (Fig. 38). The fans (Fig. 36) represent transverse basin-type fill (Miall, 1981). The Rio Toro Amarillo drains the northern flank of Volcan Irazu and Volcan Turrialba, the southernmost volcanoes of the Cordillera Central (Figs. 5 and 37). These volcanoes are made up entirely of Quaternary rocks, including andesitic lava, tuff, ash, and volcaniclastic sediments (Weyl, 1980). The headwaters of the Rio Toro Amarillo extend to within 1 km of the Irazu crater at an elevation

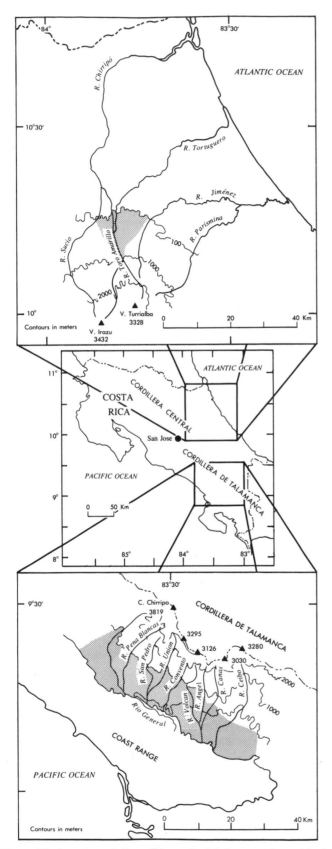

Figure 37. Location of Costa Rican alluvial fan complexes (shaded areas).

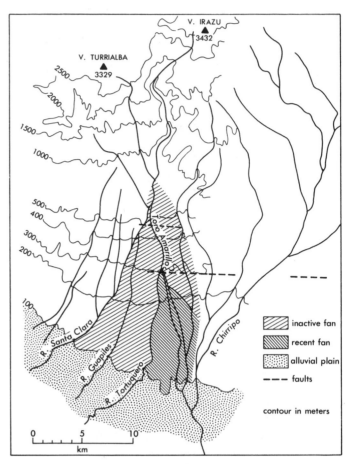

Figure 38. Morphology and topography of the Rio Toro Amarillo fan.

of 3,364 m. On the flanks of the volcanoes, the Rio Toro Amarillo and its tributaries are confined to steep-walled, V-shaped valleys cut into the volcanic rocks. Downstream, heading at an elevation of 800 m, the Toro Amarillo alluvial fan extends for 20 km, eventually merging at its distal end with a broad, low-gradient alluvial plain crossed by meandering rivers (Fig. 38).

The Rio Toro Amarillo area has an average annual precipitation of 450 cm; every month is wet except for March, when rainfall may drop to less than 2 cm. The vegetation cover is mainly wet tropical forest except on the older fan surfaces where forest has been replaced by grasslands used for pasture.

Fan Morphology

Across the proximal portion of the fan, the river is incised below the fan surface from depths that decrease from 100 m at the fan apex to 10 m approximately 10 km downfan. Along this incised portion of the Rio Toro Amarillo, the adjacent fan surface is inactive and not subject to flooding and sediment accumulation. The eastern half of the fan is inactive throughout its entire

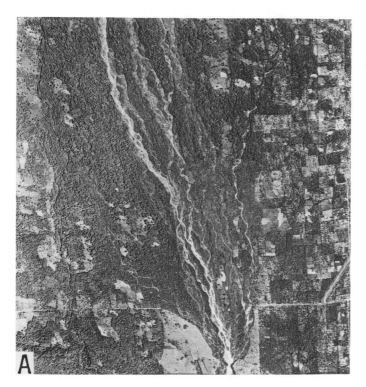

Figure 39. Aerial photographs of the proximal part of the modern, active Rio Toro Amarillo fan showing progressive westward shift in channel location and effects of increased sediment discharge due to 1963 to 1965 eruptions of Volcan Irazu. (A) 1960, prior to eruptions active channel is narrow and located along eastern margin of fan, remainder of fan vegetated; (B) 1972, seven years after eruption active channel is divided between central and western parts of fan, forming side zones of sediment aggradation; (C) 1981, 16 years after eruption, channel is narrow and located along western margin of fan. Fault separating inactive and active fan segments is located near lower edge of photos. North is toward the top. Airplane runway in lower right is approximately 1.1 km long (after Kesel, 1985).

length. The lower end of the incised western portion of the fan is terminated by a northeast-trending, high-angle fault (Fig. 38 and 39). The fault is marked by a southeast-facing scarp up to 10 m high, which serves as an intersection point (Hooke, 1967). Downstream, the channel is no longer incised and spreads out to form the modern, active braided alluvial fan surface, which covers only the western half of the lower part of the entire Toro Amarillo fan (Fig. 38).

Most of the inactive portion of the fan, both in proximal areas and flanking the active depositional fan surface on the east (Fig. 38), is a heavily vegetated, rolling, boulder-covered surface. The mean gradient of this surface is about 0.044 m/m. The principal morphological elements are relict boulder-dominated longitudinal bars, up to 100 m wide and 500 m long, standing 0.5 to 2 m above intervening, inactive braid channels.

The modern portion of the fan surface can be subdivided into three zones based on topography, vegetation, and recent fluvial activity: (1) an eastern, well-vegetated, topographically high zone, active prior to 1960; (2) a central partially vegetated zone, approximately 10 m lower than the eastern zone, active from 1960 to 1981; and (3) a western zone 5 m lower than the central zone, and the site of the present Toro Amarillo Channel. The development of these zones has been discussed by Kesel and Lowe (1987).

For a period of unknown length prior to 1960, The Rio Toro Amarillo occupied the eastern zone. By 1960 (Fig. 39A), the locus of fluvial activity had shifted to the central zone. Aerial photographs taken in 1960 (Fig. 39A) show that the active zone includes several anastomosing 10- to 50-m-wide channels separated by vegetated islands and longitudinal gravel bars. During high water stages, flow spreads outside of the central zone, occupying some of the older, recently abandoned channels on the eastern part of the fan.

From 1963 to 1965, Volcan Irazu experienced a series of major eruptions and accompanying earthquakes. Although most of the ash produced during these eruptions fell on the western slopes of the Cordillera Central (Murata and others, 1966), a considerable amount of debris was introduced to the Rio Toro Amarillo drainage basin. This high sediment influx had a profound effect on fan activity, which was still apparent in 1972 (Fig. 39B). The large sediment supply apparently resulted in rapid aggradation in the then-active central zone, which triggered a partial shift of fan activity to the western margin of the fan surface. By 1981 (Fig. 39C), fluvial activity was limited to a single active braided channel system located along the western part of the fan. Today much of the western fan surface that was active during the period of high sediment discharge following the 1963 to 1965 eruptions has been incised by the present river and has been partially revegetated (Fig. 39C).

Fan Stratigraphy

Only the uppermost 100 m of the Toro Amarillo fan sediments are exposed, even in the most deeply incised portions of the proximal river channel. Exposed sediments consist largely of moderately sorted, well-rounded to subangular cobble gravels representing facies Gm of Miall (1977). These gravels show a crude horizontal stratification reflecting the dominance of large longitudinal braid bars during fan build-up. The uppermost gravels in the incised proximal fan include several massive, mud-matrix debris flow deposits (Gm of Miall, 1977) that suggest fan incision may have been immediately preceded and perhaps initiated by a period of unusual tectonic or volcanic activity. This activity may have been related to that which formed the transverse fault lower on the fan.

On the presently active fan, there are nowhere more than 6 m of fan sediments exposed. These consist largely of coarse cobbles and gravels deposited as longitudinal bars (Kesel and Lowe, 1987). Gravels deposited during the post 1963 to 1965 period of activity are relatively clean, well-sorted sediments, in contrast to older, poorly sorted, sand-matrix-rich underlying units depsited during earlier active periods, suggesting that sediment character is related to varying conditions of sediment influx onto the fan surface.

The active fan surface also shows finer gravel and sand caps developed on vegetated bar tops (Kesel and Lowe, 1987). Overall, the deposits on the active fan fine rapidly downfan to fine gravel and sand where the fan merges into the low-gradient coastal plain (Kesel, 1985).

RIO GENERAL VALLEY

Physical Setting

The Rio General valley is a southeast-northwest–trending topographic trough situated between the Cordillera de Talamanca to the northwest, reaching an elevation of 3,819 m, and the Coast Range to the southwest, having a maximum elevation slightly above 1,300 m (Fig. 37). The Cordillera de Talamanca is a volcanically inactive portion of the Middle American Magmatic Arc (Fig. 36) composed mainly of Eocene to Miocene marine strata, basaltic and andesitic volcanic rocks, and upper Miocene quartz dioritic and granodioritic plutonic rocks (Weyl, 1957, 1971). The Coast Range and the Rio General valley collectively represent an uplifted, synclinally deformed, Tertiary forearc basin. The Coast Range is made up of middle Eocene to Pliocene marine sedimentary rocks and Pliocene volcanic and volcaniclastic rocks. The sequence strikes northwest, parallel to the range, and dips northeast, toward and beneath the Rio General valley (Henningsen, 1966).

The structural and topographic depression represented by the Rio General valley was formed by differential uplift of the bounding mountain ranges. Uplift of the Cordillera de Talamanca probably began during late Pliocene or early Pleistocene time and has continued into the Holocene (Dengo, 1962; Henningsen, 1966). The range is strongly asymmetric with a steep, fault-bounded western scarp and a broad, more gently sloping eastern flank. Uplift along the western scarp has raised upper Miocene rocks from elevations of 400 to 900 m along the edge of the Rio General valley to elevations in excess of 3,000 m along the range crest. Uplift of the Coast Range also began during the Quaternary and has continued into the Holocene (Kesel, 1983).

The floor of the Rio General valley is covered by a series of coalescing alluvial fans constructed by streams flowing from the Cordillera de Talamanca (Fig. 40), exemplifying the transverse basin-fill pattern associated with forearc basins (Miall, 1981). Fan morphology, clast composition, and paleocurrent patterns indicate that fan sediments were derived by erosion of uplifted magmatic arc rocks exposed in the Cordillera de Talamanca. At their distal extremities, the fans lap onto strongly folded and faulted Tertiary forearc basin sediments of the Coast Range (Fig. 36).

The average annual precipitation in the Rio General valley is 330 cm, with a four-month dry season from December to March. A dry month usually has less than 1 cm of rainfall, whereas the wettest month, October, has more than 55 cm. The natural vegetation is tropical moist forest changing to montane wet forest in the Cordillera de Talamanca (Tosi, 1969). Vegetation on the fan surfaces has been altered to grassland by burning and deforestation associated with agricultural activity (Talbot and Kesel, 1975).

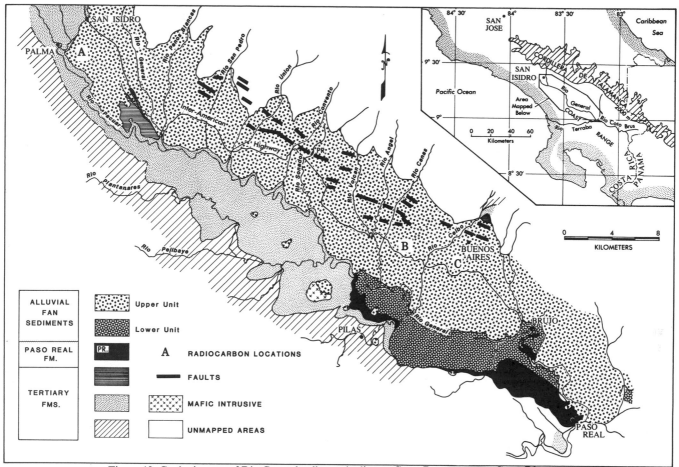

Figure 40. Geologic map of Rio General valley and adjacent Coast Range, eastern Costa Rica.

Fan Morphology

The Rio General valley fans head in streams draining the steep, fault-bounded western slope of the Cordillera de Talamanca. The fans extend 10 to 16 km from the base of the mountain front and have marked concave upward longitudinal profiles. Major morphological characteristics of the fans are summarized in Table 6.

The fan-forming streams are now deeply incised below the adjacent alluvial fan surfaces throughout their length. Because sediment discharge from the drainage basins is now confined to the floor of the incised valleys, the upper fan surfaces no longer receive sediment and are inactive. The presence of vegetated channel banks and channel bars suggests that little lateral sediment accumulation is occurring within the channels and that the channels presently are receiving minimal amounts of sediment from their drainage basins.

The depth of channel incision and channel geometry vary with position on the fans. In proximal areas, the streams have deep, narrow, steep-walled valleys (Fig. 41) incised from 20 to 80 m below the fan surfaces. Gradients of the proximal fan surfaces range from 0.033 to 0.109 m/m, whereas the gradients of the incised proximal channels are 0.019 to 0.101 m/m (Table 6). Because the proximal channel gradient is lower than the gradient of the fan surface, the amount of channel incision decreases toward the midfan region. As the amount of incision decreases, the width of the incised channels increases. The midfan channels are flat floored, up to 1 km wide, and incised 5 to 15 m below the fan surfaces (Fig. 42).

On the distal parts of the fans, streams are incised from 90 m in the northwestern part of the valley to 420 m in the southeast (Table 6). The downfan increase in incision reflects differential uplift of the distal portions of the fans, which have reverse gradients of up to 0.032 m/m. During this uplift, major streams were able to downcut and maintain their courses within steep, narrow, incised valleys. Smaller streams, however, were unable to downcut rapidly and many have reversed their flow directions. In many cases, these streams were ponded to form lakes, and the largest, Laguna Chonta on the Rio Ceibo fan at the southeastern end of the valley, was approximately 6 km across (Fig. 43). Since formation of this lake, deeply entrenched tributaries of the major incised streams have eroded headward, integrating the drainage

TABLE 6. ALLUVIAL FAN CHARACTERISTICS*

	General-Penas Blancas	Cajon-San Pedro	Union	Convento-Sonador	Volcan	Guapinol-Canas	Ceibo	Toro Amarillo
Fan area (km^2)	32	63	55	18	78	67	159	162
Basin area (km^2)	43	44	92	26	50	66	179	138
Basin elevation (m.a.s.l.)	3345	3280	3440	2840	3126	3030	3280	3364
Fan gradient (m/m)	0.089	0.082	0.058	0.109	0.084	0.63	0.033	0.044
Channel gradient (m/m)	0.079	0.074	0.042	0.101	0.075	0.055	0.019	0.33[1] (0.026)[2]
Apex elevation (m.a.s.l.)	1050	1000	800	1050	860	740	460	780
Apex incision (m)	80	80	80	20	60	22	75	120
Distal elevation (m.a.s.l.)	620	654	600	585	510	---	535	---
Distal incision (m)	90	90	130	140	240	---	230	---

*For location, see Figure 37.

[1]Active fan.

[2]Incised portion.

except for small ponds and swamps that remain without external drainage. Numerous terraces formed along the valley margins during channel incision. Longitudinal profiles of these terraces diverge from the modern valley floors and the upper fan surfaces in the midfan area.

Based on variations in the thickness of soil developed on the inactive fan surfaces (Kesel and Spicer, 1985; Fig. 44), Rio General valley fans appear to be segmented similar to the segmented alluvial fans described by Bull (1964). Segmented alluvial fans typically show slope breaks reflecting shifts in the loci of fan deposition (Bull, 1964; Hooke, 1972; Heward, 1978). Slope breaks have not been documented on the General valley fans, but soil profiles increase in thickness downfan, suggesting that fan surfaces are progressively older downfan. Bull (1964) and Heward (1978) suggest that this pattern develops where the rate of uplift at the source exceeds the rate of channel incision on the fan, causing active fan sedimentation to become localized immediately adjacent to the fan apex (Fig. 45A). In contrast, the Rio Ceibo fan complex underwent channel entrenchment in the proximal fan area causing a shift in the locus of deposition to the midfan region (Fig. 45B). Fans such as these form where the rate of channel incision exceeds the rate of uplift (Bull, 1964; Heward, 1978). On this fan, soils in the midfan area are thinner than in proximal and distal areas.

Alluvial fan stratigraphy

The vertical sequence of sediments in the Rio General valley fans is exposed in incised and uptilted beds along their distal margins. Fan sediments in the Rio General valley unconformably overlie Tertiary marine and volcanic rocks and reach a maximum thickness of 900 m at the distal end of the Rio Ceibo fan (Fig. 46). In general, the fan sediments form a lenticular body that thickens away from the base of the Cordillera de Talamanca. The 900 m thick sequence of sediments in the Rio Ceibo fan can be divided into two main units, here informally designated the upper and lower members (Fig. 46). Each member is composed of a lower unit of coarse pebble and cobble gravel and an upper unit of sand, clay, and fine gravel.

The gravel unit of the lower member includes about 320 m of moderately to poorly sorted, poorly lithified pebble to cobble gravel, muddy debris-flow deposits, and lenticular sand beds. The unit has a mean dip of 19° to the northeast, toward the valley axis. The gravels are predominantly well rounded and massive to crudely stratified. They represent the Gm facies of Miall (1977) and reflect the buildup of a succession of longitudinal braid bars. The clasts are composed mainly of volcanic and metavolcanic rocks. Imbrication indicates that the debris was derived from the Cordillera de Talamanca. Soil profiles within the gravel sequence mark hiatuses in local fan activity.

The gravel unit of the lower member represents an older series of alluvial fans developed along the Talamancan front. The distal extent of these fans is unknown. It is possible that the Coast Range had not formed yet and that the fans graded downslope into a coastal plain system.

The lower member gravel is overlain by approximately 80 m of poorly sorted, thick-bedded to massive volcaniclastic sand containing sparse fine gravel beds and debris flow depsits. Vegetal material is common throughout the unit. Debris flow deposits at Brujo (Fig. 40) contain large chunks of tree limbs, several of

Figure 41. Fanhead entrenchment of the Rio San Pedro–Cajon fan.

Figure 42. Braided valley-bottom surface with elongate bars and boulder clasts along the mid-fan area. Upper fan surface is visible in background.

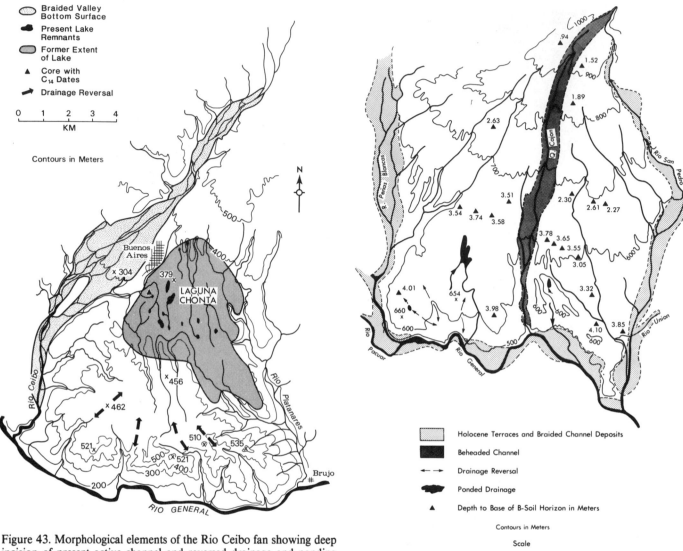

Figure 43. Morphological elements of the Rio Ceibo fan showing deep incision of present active channel and reversed drainage and ponding resulting from uplift of distal fan.

Figure 44. Thickness of soil development on Rio San Pedro–Cajon fan indicating progressively older soils downfan (from Kesel and Spicer, 1985).

which have yielded radiocarbon dates exceeding 37,000 yr B.P. (GX4976, GX4977, Geochron laboratory). Fossilized wood from trees in growth position within the sand unit at the top of the lower member also produced a radiocarbon date greater than 37,000 yr B.P. (GX4978). A K-Ar sample from a boulder that represents the most common rock type in the debris flows was dated at 3.8×10^6 years (Baksi, personal communication, 1985), indicating that some debris was derived from Tertiary volcanic rocks, probably the underlying Paso Real Formation (Fig. 40 and 46).

Gravel and sand of the lower member lie with angular unconformity beneath sediments of the upper member (Fig. 40 and 46). This unconformity apparently reflects a period of Coast Range uplift and erosion prior to initial deposition of the overlying gravels. The upper gravel member marks a second phase of

alluvial fan formation along the Talamancan front. This phase is correlated with the construction of the still preserved but incised fans. In distal parts of the fans, these younger gravels dip up to 10° toward the northeast. Upper member gravels are composed of well-rounded, moderately sorted pebble to cobble gravel. Near the top, the clasts reach a maximum diameter of 2 m, but most beds are composed of debris 10 to 20 cm in diameter. The gravels represent facies Gm of Miall (1977) deposited as longitudinal braid bars. The upper gravel differs from the lower gravel in that it contains up to 25 percent clasts of plutonic rock in addition to clasts of volcanic rock. The presence of plutonic debris and imbrication measurements indicate that the gravels were derived

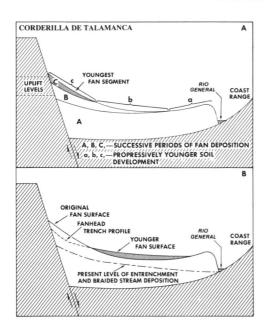

Figure 45. Alluvial fan profiles in Rio General valley with uptilted distal margins. (A) segmented type of fan where relative uplift exceeds the rate of stream incision, (B) fan in which stream incision exceeds the rate of uplift.

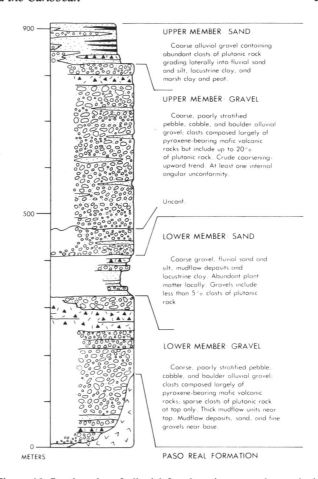

Figure 46. Stratigraphy of alluvial fan deposits exposed near incised, distal end of Rio Ceibo fan.

from the Cordillera de Talamanca. The clasts of plutonic rock increase in abundance upsection, indicating progressive uncovering by erosion of the igneous core of the Cordillera de Talamanca. There are a number of small outliers of alluvial gravels containing clasts of plutonic rock unconformably overlying Tertiary sediments southwest of the Rio General within the edge of the Coast Range (Fig. 40). These remanants indicate that the younger generation of alluvial fans extended farther to the southwest than their present limits would suggest, but have largely been removed by erosion during later uplift of the Coast Range.

The upper sandy unit of the upper member forms a lenticular body that is present only in the topographically lower midfan region of the present fans. Where present, these sediments show very thin weathering profiles. They reach a thickness of 70 m on the Rio Volcan-Canas and Ceibo fans and at their distal ends dip 3° toward the valley axis. Uplift of the Coast Range was probably recurrent during deposition of the upper member as indicated by the greater northeastward dip (up to 10°) on the underlying coarse gravel unit. The upper sand unit shows considerable lateral variability. On the Rio Ceibo fan it is composed of interstratified gravel, volcanic sand, silt, and lacustrine clay. On the Rio Volcan-Canas fan complex, the unit consists predominantly of bedded gravels with clasts up to 30 cm in diameter and only a minor amount of sand, silt, and clay.

Radiocarbon age determinations have been obtained at three locations in this uppermost sand and fine gravel unit (Fig. 47). A radiocarbon date of 10,290 ±275 yr B.P. (GX8490)

was obtained from the base of a 2-m-thick accumulation of organic-rich, lacustrine clay overlying the weathered gravels (Fig. 47A) of the Rio Genral–Rio Penas Blancas fan complex at the northwest end of the valley (Fig. 40). The clay was deposited in a lake formed by drainage reversal and ponding that resulted from uptilting of the distal end of the fan. This sample provides only a minimum age for lake formation because it was collected from the lake margin. Three dates were obtained from a 50-m-thick exposure in an incised channel downfan from the midpoint of the Rio Volcan-Canas fan complex (Fig. 47B). The dates ranged from 34,480 +3,940/–1,400 yr B.P. (GX10704) near the base of the sandy section to 17,050 ±550 yr B.P. (GX3720) near the midpoint of the sequence. A thin organic-rich layer at the base of a 2-m-thick clay unit at the top of the Laguna Chonta section (Fig. 47C) yielded a radiocarbon date of 12,830 ±395 yr B.P. (GX4114). A similar bed of peaty clay 0.5 m higher in the section yielded a date of 8,810 ±245 yr B.P. (GX 4113). These dates indicate that the upper sand and gravel unit accumulated over the last 35,000 yr B.P., and that much of this time is probably represented by the uppermost few meters of fine lacustrine sediments. They also date the approximate time at which the fan surface

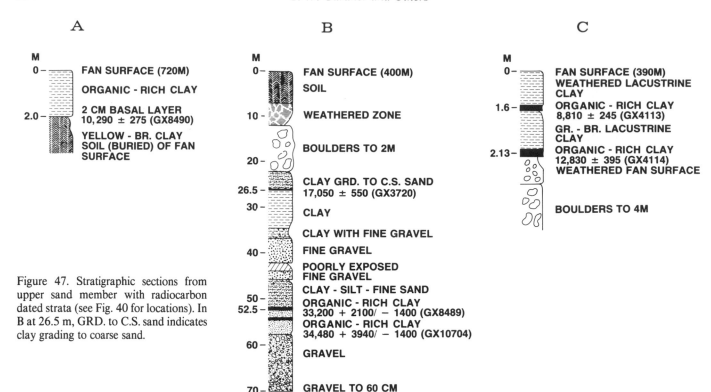

Figure 47. Stratigraphic sections from upper sand member with radiocarbon dated strata (see Fig. 40 for locations). In B at 26.5 m, GRD. to C.S. sand indicates clay grading to coarse sand.

became inactive in distal and midfan regions at about 35,000 yr B.P. whereas deposition in proximal areas may have continued to about 12,000 yr B.P.

Controls on alluvuial fan formation

Alluvial fans in the Rio General valley clearly formed in response to the Pleistocene uplift of the Cordillera de Talamanca. Their subsequent evolution was influenced both by uplift of the Cordillera de Talamanca and the Coast Range. The sedimentary sequence in the southeastern part of the valley indicates two phases of alluvial fan building. Each was characterized by deposition of a basal unit of upward coarsening and thickening pebble to cobble gravel overlain by a thinner, finer-grained unit of fluvial and lacustrine sediments. Similar fan sequences have been interpreted as resulting from short-lived, but pronounced uplift (Steel and Wilson, 1975; Steel and others, 1977), or continuous uplift accompanied by dissection of the source area (Heward, 1978). The uplift of the Talamancan block relative to the forearc basin has occurred along a series of frontal normal faults now largely buried beneath Quaternary fan gravels. In contrast, the Coast Range includes a series of major, northeast-dipping reverse faults (Mora, 1979a, b). Present surfaces of the Rio General valley fans are cut by a number of faults running more-or-less parallel to the Talamancan front (Fig. 47), and displacing parts of the uppermost sandy unit, which is at least as young as 13,000 to 17,000 yrs old. However, the morphologic expression of these faults indicates relative uplift of the southwestern block, the opposite

sense of movement as that associated with uplift of the Cordillera de Talamanca. Quite possibly, the style of deformation within the valley has changed during the latest Quaternary.

Although tectonism has clearly played the major role in the development and evolution of the Rio General valley fans, the possibility of some climatic influence cannot be eliminated. There is clear evidence of Pleistocene glaciation in the Cordillera de Talamanca, where a Wisconsin-age snow line has been calculated at 3,500 m and moraine deposits extend down to 3,320 m (Hastenrath, 1973). Although the source areas of debris for the Rio General valley fans, with one exception, have maximum elevation of less than 3,300 m and, therefore, were not directly influenced by glaciation (Table 6), it is possible that periglacial effects extended to considerably lower elevations. The possible role of glaciation, meltwater runoff form glaciers, and reduction in protective forest cover due to cooling in periglacial zones cannot be evaluated at this time, but could have contributed to the generation of coarse detritus that was moved downslope and deposited on the Rio General valley fans.

RATES OF UPLIFT, INCISION, AND SEDIMENTATION

The rate of uplift plays a major role in determining the rates of sedimentation and channel incision on alluvial fans. There have been few direct measurements of uplift rates in Central America. Based on geodetic resurveys over the last 20 years,

Miyamura (1975) estimated that the Cordillera Central is being uplifted at a rate of at least 2 m/1,000 yr. Estimates of the maximum uplift rates of the Cordillera Central and the Cordillera de Talamanca can be made indirectly, if Bull (1977) and Heward (1978) are correct in suggesting that fanhead entrenchment occurs when the rate of incision exceeds the rate of uplift.

Within the incised inactive portion of the Rio Toro Amarillo fan, radiocarbon dates were obtained from two wood samples located 8 m below the uppermost inactive surface, which is 40 m above the present channel. The samples, taken from the same sedimentary unit about 20 m apart, yielded dates of 10,725 ±380 yr B.P. (GX12202) and 10,285 ± 115 yr B.P. (GX12201) with an average of 10,500 yr B.P. This maximum age for upper fan surface indicates the Rio Toro Amarillo had a mean incision rate of about 3.8 m/1,000 yr at this point, which is about 6 km below the apex. At the apex, where incision is about 120 m, the suggested incision rate is about 12 m/1,000 yr.

The ages of the proximal, upper surfaces of the Rio Ceibo and Rio Canas-Guapinol fans can also be estimated from radiocarbon dates. The Ceibo fan surface was actively receiving boulder-sized gravels until approximately 13,000 years ago, when deposition of the overlying lacustrine clay began (Fig. 47C) with apparently no subsequent contribution of sediment from a Talamancan source. Assuming fanhead entrenchment began immediately prior to clay deposition, an incision rate of 5.8 m/1,000 yr is necessary to produce a channel incised 75 m (Table 6) below the fan surface.

The channel of the Rio Canas-Guapinol fan is entrenched 22 m (Table 6) below the fan surface at its apex. The age of the fan surface was estimated by using the radiocarbon dates from organic clays in a sediment sequence exposed by channel incision near midfan (Fig. 47B). An average rate of 1.6 m/1,000 yr was calculated from this 26-m-thick sequence that is bracketed by a basal date of 34,480 +3,940 yr B.P. and an upper date of 17,050 ±550 yr B.P. The youngest dated stratum is overlain by 26.5 m of sediment, including a 10-m-thick unit of massive boulders. Assuming that the boulder unit represents a single depositional event and can be excluded from the long-term sedimentation rate, the extrapolated age of the fan surface is estimated to be between 7,300 and 6,200 years old. Even if this age is considerably in error, the fan surface is no older than the youngest-dated stratum. Using this estimated age for the fan surface, the rate of fanhead entrenchment would be approximately 3 m/1,000 yr. The age estimate for the fan surface at this site also indicates that the incision rate for this 70-m-deep channel may have been in excess of 10 m/1,000 yr. The age of the distal fan surface is uncertain. The rate of channel incision along the outer fan margin may be in the neighborhood of 12 m/1,000 yr, a very rough approximation based on an extrapolation of the rate from the fan apexes.

Channel incision reaches a maximum along the distal fan margins which have been uptilted by uplift of the Coast Range (Table 6). The apparent increase in the rate of channel incision away from the Cordillera de Talamanca is similar to findings by Kulm and others (1981) that the seaward edge of the Peruvian forearc basin is uplifting faster than the landward portion. The distal end of the Ceibo fan, uplifted to a higher elevation than the apex (Fig. 43), provides unequivocal evidence that the Coast Range has been uplifting at a faster rate than the Cordillera de Talamanca, at least during late Pleistocene and Holocene times. This difference in uplift rates also appears to be reflected by denudation rates for river basins that drain the two areas. The average annual suspended sediment discharge from the rivers of the Cordillera de Talamanca and the Coast Range is 167 tons/km^2/yr and 451 tons/km^2/yr, respectively (Instituto Costarricense de Electricidad, 1981). This translates into denudation rates (using a specific gravity of 2.64; Selby, 1974) of 0.06 m/1,000 yr and 0.17 m/1,000 yr for the two areas.

These data provide the basis for speculation concerning the interrelationship between uplift rates, seismic and volcanic activity, and total erosion, which can include both channel incision and slope denudation (Kennedy, 1962). During the Holocene, rates of fanhead entrenchment appear to have exceeded the measured uplift rate of Miyamura (1975) and suggest that this estimate of 2 m/1,000 yr may be a reasonable rate for this entire time period. The present denudation rate for the Cordillera de Talamanca, even if increased by 20 percent to accommodate for bedload, is at least an order of magnitude lower than the incision rates. By comparison, areas in Japan with uplift rates of 1 to 2 m/1,000 yr have a denudation rate that is estimated at 6 m/1,000 yr (Yoshikawa, 1974; Yoshikawa and others, 1981, p. 104). This difference in denudation rates may reflect differences between the two areas in the numbers and frequency of seismic activity. Since 1890, Central America has experienced 24 major earthquakes of magnitude 7 or greater (Carr and Stoiber, 1977), whereas Japan has had 96 during the same period (Yonekura, 1972). Yoshikawa and others (1981) indicated that periods of seismic activity in Japan are episodic and have recurrence intervals that range from 1,000 (p. 63) to 8,600 (p. 169) yr. In the Cordillera de Talamanca, channel erosion is rapid and appears to exceed the rate of uplift, but slope denudation is much less effective in lowering interfluve areas as indicated by the denudation rate data. This may represent conditions during low sediment yield periods characterized by low seismic activity. It is interesting to note also that the sedimentation rate during the low yield period and prior to fanhead incision is only slightly lower than the uplift rate of Miyamura (1975). Sedimentation rates for the coarse gravel divisions could not be determined, but must be at least several times greater than during the low sediment yield period.

DISCUSSION

The Quaternary Rio Toro Amarillo and Rio General valley fans represent the main types of alluvial fans deposited within and adjacent to magmatic arcs. Fans like the Toro Amarillo develop in response to the rapid construction of the active arc through volcanism. The Rio Toro Amarillo fan occurs in a back-arc setting, but similar fans are equally likely to form in forearc regions. Fans like those in the Rio General valley form as a result of the rapid tectonic uplift of magmatic arc blocks. The Rio General valley fans have grown in response to tectonic uplift of the magmatically inactive Tertiary arc, represented by the Cordillera de Talamanca. Although the Rio General valley fans were deposited in a forearc structural depression, similar fans deriving debris from the Cordillera de Talamanca exist on the Caribbean coast in a back-arc setting.

Do these fan types exhibit characteristic differences that can be related to their differing tectonic/magmatic settings? If we were to consider ideal, end-member systems, the answer to this question is clearly yes. Although the rates of construction and overall morphologies of volcanically constructed versus tectonically uplifted blocks may be similar, faulting, and sometimes folding, associated with tectonic uplift have profound effects on adjacent fans that may not characterize fans formed in response to volcanic buildup.

The Rio General valley fans show transverse faults and related uplifted fan blocks that are now being reduced by erosion. An older generation of fans, represented by the lower member of the Quaternary fan gravels, lies unconformably beneath those of the modern fans. Progressive reverse dip within the distal gravel of the modern fans records progressive uplift of the Coast Range. Extensive lacustrine and peat deposits on the fan surfaces reflect ponding due largely to uplift of the distal part of the fans, but small ponds have also formed behind blocks uplifted along transverse frontal faults on proximal areas of the fan surfaces (Fig. 40). Formation, incision, and deactivation of major alluvial fans has occurred at least twice in the evolution of the Rio General valley, probably in response to uplift of both the Talamancan and Coast Range blocks.

Although fans developed adjacent to passively constructed volcanic edifices might lack evidence of concurrent tectonism, arcs are tectonically as well as magmatically active areas. The Rio Toro Amarillo fan appears to lack many of the structural complexities found in the Rio General valley fans, in part, perhaps, because the shallow depths of incision do not reveal deeper, more deformed units, if they exist. The Rio Toro Amarillo fan is, however, cut by at least one major transverse fault that has profoundly altered depositional patterns on the fan surface. However, the absence of a direct structural control on the buildup of the arc itself, as reflected by the absence of bounding frontal faults along the Cordillera Central, suggests that adjacent fans may be less directly affected by faulting during their formation, and their deposits may lack many of the breaks, unconformities, stratigraphic complexities, and lacustrine facies found on the Rio General valley fans.

Both fan types are constructed largely of coarse pebble, cobble, and boulder gravels of Gm facies (Miall, 1977). The fan surfaces and deposits show preserved surface morphologies and internal structure, respectively, indicating gravel deposition as large longitudinal braid bars. The gravels in both fan systems are extremely coarse, with individual particles commonly up to 2 m in diameter and mean diameters generally in the range of 10 to 15 cm. Our preliminary observations suggest that gravels deposited during and following the recent 1963 to 1965 eruption of Volcan Irazu are better sorted than those deposited during proceeding periods of fan activity. The extremely episodic character of deposition on and around volcanoes has been described by Vessell and Davies (1981) and it is possible that the volume and style of eruption may influence the rates and, hence, styles of associated fan sedimentation. Similar sedimentation bursts may characterize tectonically influenced fans as a result of episodic tectonism and earthquakes.

The only clear difference we have identified between the Rio Toro Amarillo and Rio General valley fan deposits is clast composition. Clasts making up the Toro Amarillo fan gravels are composed exclusively of fresh or hydrothermally altered andesitic volcanic rocks, breccias, or volcaniclastic sediments (Kesel and Lowe, 1987). Rio General valley fan gravels vary in composition, but most include abundant, heterogenous volcanic and metavolcanic rocks representing a wide variety of individual flow types, a variety of metasedimentary rocks, and plutonic rocks. Overall, the geometries and deposits of both fan types attest to the efficiency of coarse debris production under humid-climate regimes and to the competence of the fan streams in moving that debris.

Previous studies have shown that fan area can be related to source basin area and that this relationship can be influenced by lithology, slope, and tectonic activity within the source basin (Hooke, 1968; Bull, 1977). Costa Rican fans (Fig. 48) are larger than many arid and semiarid fans having source basins of comparable size, most of which are also located in tectonically active areas (Bull, 1964; Beaumont, 1972; Rockwell and others, 1985). They are also larger than fans in Honduras (Schramm, 1981), Puerto Rico (McClymonds and Ward, 1966), and Jamaica (Wescott and Ethride, 1980). The ratio of the size of the Himalayan-derived Kosi River fan to the area of its source basin is the same as that for the Costa Rican fans. The Rio General valley fans may actually be somewhat larger relative to their source basin areas because they have been reduced in size by erosion of the distal portions during Coast Range uplift. These relationships emphasize the extremely rapid uplift rates characterizing the Quaternary evolution of Costa Rica.

CONCLUSIONS

Costa Rican alluvial fans represent fluvial systems that have developed within a tectonic and volcanic framework associated with the Middle American Magmatic Arc. They are typical of fan complexes that develop in arc settings. Fan construction occurs mainly during periods of high sediment production in adjacent source basins resulting from tectonic and/or volcanic activity. These periods are separated by longer, inactive phases of low sediment yield from source basins, reworking of fan sediments, and possible channel incision. The importance of tectonism in formation of Costa Rican fans is suggested by the presence of observable cross-fan faults and by the divergence of valley-side terraces toward the proximal and distal portions of the fans,

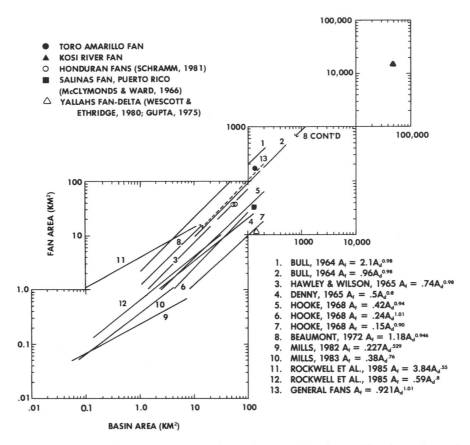

- ● TORO AMARILLO FAN
- ▲ KOSI RIVER FAN
- ○ HONDURAN FANS (SCHRAMM, 1981)
- ■ SALINAS FAN, PUERTO RICO
  (McCLYMONDS & WARD, 1966)
- △ YALLAHS FAN-DELTA (WESCOTT &
  ETHRIDGE, 1980; GUPTA, 1975)

1. BULL, 1964 $A_f = 2.1A_d^{0.98}$
2. BULL, 1964 $A_f = .96A_d^{0.98}$
3. HAWLEY & WILSON, 1965 $A_f = .74A_d^{0.98}$
4. DENNY, 1965 $A_f = .5A_d^{0.8}$
5. HOOKE, 1968 $A_f = .42A_d^{0.94}$
6. HOOKE, 1968 $A_f = .24A_d^{1.01}$
7. HOOKE, 1968 $A_f = .15A_d^{0.90}$
8. BEAUMONT, 1972 $A_f = 1.18A_d^{0.946}$
9. MILLS, 1982 $A_f = .227A_d^{.529}$
10. MILLS, 1983 $A_f = .38A_d^{.76}$
11. ROCKWELL ET AL., 1985 $A_f = 3.84A_d^{.55}$
12. ROCKWELL ET AL., 1985 $A_f = .59A_d^{.8}$
13. GENERAL FANS $A_f = .921A_d^{1.01}$

Figure 48. Relationship of fan area to source basin area for Costa Rican fans and fans from other regions (modified from Kesel, 1985).

segmented fan surfaces, abnormal and variable depths of fanhead trenching, and upward coarsening megasequences of fan sediments.

The results of this study indicate that fans formed in most arc settings are likely to be influenced by tectonic activity. However, different parts of the arc are likely to experience significantly different rates, periods, and styles of uplift. Rio General valley fans clearly formed in response to Pleistocene uplift of the Cordillera de Talamanca, but much of their later history has been controlled by the subsequent, relatively rapid uplift of the Coast Range. Present tectonic styles within the Rio General valley may be unlike previous regimes under which the fans developed and could lead to complete removal of fan sediments by erosion. Adjacent to the magmatically constructed modern arc, and perhaps particularly in the back-arc setting, continued volcanism may provide a more hospitable environment for long-term preservation of fan deposits in the geologic record.

SUMMARY

Thomas W. Gardner

This chapter on the geomorphology of the Caribbean and Central America has focused on three topics from among the diversity of geomorphic processes and landforms of the region. These topics include karsts, alluvial fans, and tectonic landscapes along convergent plate margins. They were included in the chapter because they represent a subset of geomorphic processes that characterize the region.

Karstification is a significant and areally extensive geomorphic component of the Caribbean landscape, from the relatively stable Yucatan Platform to the active plate boundaries of the Caribbean islands and southwestern Costa Rica. Karst landforms and the processes of karstification are related to the tectonic environment, depositional setting, subsequent diagenesis, and groundwater movement.

Geomorphic analyses of active, wet alluvial fans are scarce and difficult to acquire. Rates and types of processes on wet alluvial fans in forearc and back-arc settings in Costa Rica provide new insights into the Quaternary history, stratigraphy, and origins of those types of deposits.

Systematic analyses of tectonic landscapes along convergent plate margins with rapid and differential rates of Quaternary tectonic deformation provide critical data on the timing, rates, and magnitudes of that deformation. Analyses of mountain-front morphometry, drainage-basin morphometry, and deformed or uplifted marine and fluvial surfaces not only contribute to our understanding of the dynamics of landscape processes in active tectonic settings but also help constrain rates and timing of deformation; data that is critically needed to evaluate various aspects of plate-tectonic models.

While providing a review of current research in the region, it is hoped that the chapter will stimulate new and continued research efforts.

REFERENCES CITED

Alt, J. N., Harpster, R. E., and Schwartz, D. P., 1980, Late Quaternary deformation and differential uplift along the Pacific coast of Costa Rica: Geological Society of America Abstracts with Programs, v. 12, p. 378–379.

Ashley, G. H., 1935, Studies in Appalachian mountain sculpture: Geological Society of America Bulletin, v. 46, p. 1395–1436.

Asociacion Espeleologica Costarricense (AEC), 1986, Descubren primero lago subterraneo en el pais: La Republica, v. 27, no. 1, p. 82.

Back, W., Hanshaw, B. B., and Van Driel, J. N., 1984, Role of groundwater in shaping the eastern coastline of the Yucatan Peninsula, Mexico, *in* LaFleur, R. G., ed., Ground water as a geomorphic agent: Boston, Allen and Unwin, Incorporated, p. 281–293.

Back, W., Hanshaw, B. B., Herman, J. S., and Van Driel, J. N., 1986, Differential dissolution of a Pleistocene reef in the groundwater mixing zone of coastal Yucatan, Mexico: Geology, v. 14, p. 137–140.

Baker, L. L., Devine, E. A., and DiTonto, M. A., 1986, Jamaica, The 1985 expedition of the NSS Jamaica Cockpits Project: National Speleological Society News, v. 44, no. 1, p. 4–15.

Beaumont, P., 1972, Alluvial fans along the foothills of the Elburz Mountains, Iran: Paleogeography, Paleoclimatology, Paleoecology, v. 12, p. 251–273.

Beck, B. F., 1977, Hydrogeology of a cave utilized for public water supply; Rio Caumy Cave, Puerto Rico, U.S.A., *in* Dilamarter, R. R., and Csallany, S. C., eds., Hydrologic problems in karst regions: Bowling Green, Western Kentucky University, p. 249–261.

Bellon, H., and Tournon, J., 1978, Contribution de la geochronometrie K-Ar a l'etude du magmatisme de Costa Rica, Amerique Centrale: Bulletin Societe Geologie France, v. 20, p. 955–959.

Bender, M. L., Fairbanks, R. G., Taylor, F. W., Matthews, R. K., Goodard, J. G., and Broecker, W. S., 1979, Uranium-series dating of Pleistocene reef tracts of Barbados, West Indies: Geological Society of America Bulletin, v. 90, p. 577–594.

Bergoeing, J. P., 1983, Dataciones Radiometricas en algunas Muestras de Costa Rica: Instituto Geografico Nacional Informe Semestral, Enero a Junio, p. 71–86.

Bergoeing, J. P., Brenes, L. G., Malavassi, I. E., and Jimenez, B. R., 1982, Guaital Mapa Geomorfologico del Pacifico Norte de Costa Rica: Instituto Geografico Nacional, escala 1:100,000.

Berner, R. A., and Morse, J. W., 1974, Dissolution kinetics of calcium carbonate in sea water; IV, Theory of calcite dissolution: American Journal of Science, v. 274, p. 108–134.

Bourgois, J., Azema, J., Baumgartner, P. O., Tourmon, J., Desmet, A., and Ouboin, J., 1984, The geologic history of the Caribbean-Cocos plate boundary with special reference to the Nicoya Ophiolite complex (Costa Rica) and D.S.D.P. results (Legs 67 and 84 off Guatemala; A synthesis: Tectonophysics, v. 108, p. 1–32.

Briggs, R. P., 1966, The blanket sands of northern Puerto Rico, *in* Caribbean Geological Conference, 3rd, Kingston, Jamaica, 1962, Transactions: Jamaica Geological Survey Publication 95, p. 60–69.

Briggs, R. P., and Akers, J. P., 1965, Hydrogeologic map of Puerto Rico and adjacent islands: U.S. Geological Survey Hydrologic Inventory Atlas HA-197, scale 1:240,000.

Bull, W. B., 1964, Geomorphology of segmented alluvial fans in Fresno County, California: U.S. Geological Survey Professional Paper 352-E, p. 89–129.

——, 1973, Local base level processes in arid fluvial systems: Geological Society of America Abstracts with Programs, v. 5, p. 562.

——, 1977, The alluvial-fan environment: Progress in Physical Geography, v. 1, p. 221–270.

——, 1978, Geomorphic tectonic activity classes of the south front of the San Gabriel Mountains, California: Unpublished Final Report, U.S. Geological Survey Contract no. 14-08-0001-G-394, 59 p.

Bull, W. B., and McFadden, L. D., 1977, Tectonic geomorphology north and south of the Garlock fault, California, *in* Doehring, D. O., ed., Geomorphology in arid regions: Proceedings, Eighth Annual Geomorphology Symposium, State University of New York, Binghamton, p. 115–138.

Campbell, M. R., 1933, Chambersburg (Harrisburg) peneplain in the piedmont of Maryland and Pennsylvania: Geological Society of America Bulletin, v. 44, p. 553–573.

Carr, M. J., and Stoiber, R. E., 1977, Geologic setting of some destructive earthquakes in Central America: Geological Society of America Bulletin, v. 88, p. 151–156.

Case, J. E., Holcombe, T. L., and Martin, R. G., 1984, Map of geologic provinces in the Caribbean region, *in* Bonini, W. E., Hargraves, R. B., and Shagam, R., eds., The Caribbean–South American plate boundary and regional tectonics: Geological Society of America Memoir 162, p. 1–31.

Castillo, M. R., 1984, Geologia de Costa Rica; Una sinopsis: San Jose, Editorial de la Universidad de Costa Rica, 182 p.

Castillo, M. R., Madrigal, G. R., and Sandoval, M. F., 1970, Nota geotecnica sobre el Yacimiento de laterita bauxitica del Valle de El General: Ciudad Universitaria Direccion de Geologia, Minas y Petroleo, Informes Tecnicos y Notas Geologicas, v. 34, p. 1–22.

Chidley, T.R.E., and Lloyd, J. W., 1977, The hydrogeological assessment of freshwater lenses in oceanic islands, *in* Morel-Seytoux, H. J., ed., Surface and subsurface hydrology: Fort Collins, Colorado, Water Resources Publication, p. 232–245.

Corrigan, J. C., and Mann, P., 1986, Rapid, outer-forearc uplift related to attempted subduction of an aseismic ridge; The Burica Peninsula, Panama–Costa Rica: Geological Society of America Abstracts with Programs, v. 18, p. 571–572.

Cross, T. A., and Pilger, R. H., 1982, Controls of subduction geometry, location of magmatic arcs, and tectonics of arc and back arc regions: Geological Society of America Bulletin, v. 93, p. 545–562.

Davis, W. M., 1889, The rivers and valleys of Pennsylvania: National Geographic Magazine, v. 2, p. 183–253.

——, 1899, The geographic cycles: Geographical Journal, v. 14, p. 481–504.

——, 1932, Piedmont benchlands and primarrumpfe: Geological Society of America Bulletin, v. 43, p. 399–440.

deBoer, J., 1979, The outer arc of the Costa Rican orogen (oceanic basement complexes of the Nicoya and Santa Elena peninsulas): Tectonophysics, v. 56, p. 221–259.

Dengo, G., 1962, Tectonic-igneous sequences in Costa Rica: Geological Society of America, Petrographic Studies, A volume to honor A. F. Buddington, p. 133–161.

——, 1969, Problems of tectonic relations between Central America and the Caribbean: Gulf Coast Association Geological Society Transactions, v. 19, p. 311–320.

Denny, C. S., 1965, Alluvial fans in the Death Valley region of California and Nevada: U.S. Geological Survey Professional Paper 466, 62 p.

Dixey, F., 1942, Erosion cycles in central and southern Africa: Geological Society of South Africa Transactions, v. 45, p. 151–181.

Dodge, R. E., Fairbanks, R. G., Benninger, L. K., and Maunasse, F., 1983, Pleistocene sea levels from raised coral reefs of Haiti: Science, v. 219, p. 1423–1425.

Doornkamp, J. C., and Temple, P. H., 1966, Surface drainage and tectonic instability in part of southern Uganda: Geographical Journal, v. 132, p. 238–252.

Dougherty, P. H., 1985, Belize; The Rio Grande project: National Speleological Society News, v. 43, p. 329–334.

Erb, D. K., 1982, Geologic remote sensing in "difficult terrain," photogeomorphology in the humid tropics, *in* Proceedings, International Symposium on Remote Sensing of Environment, second thematic conference, Ann Arbor, Michigan, Environmental Resources Institute, p. 365–374.

Fair, T.J.D., and King, L. C., 1954, Erosional land-surfaces in the eastern marginal areas of South Africa: Geological Society of South Africa Transactions, v. 57, p. 19–26.

Fincham, A. G., 1977, Jamaica underground: Geological Society of Jamaica, 247 p.

Fischer, R., 1980, Recent tectonic movements of the Costa Rican Pacific coast: Tectonophysics, v. 70, p. T25–T33.

Frosft, S. H., Harbour, J. L., Beach, D. K., Reulini, M. J., and Harris, P. M., 1983, Oligocene reef tract development southwestern Puerto Rico: Miami Beach, Florida, University of Miami, Sediments IX, 144 p.

Ganter, J., 1986, U.S. Exploration '86: National Speleological Society News, v. 44, p. 328–329.

Gerson, R., Grossman, S., and Bowman, D., 1985, Stages in the creation of a rift valley; Geomorphic evolution along the southern Dead Sea Rift, *in* Hack, J., and Morisawa, M., eds., Tectonic geomorphology: Proceedings of the 15th Geomorphology Symposia Series, Binghamton, p. 53–75.

Gilbert, G. K., 1880, Report on the geology of the Henry Mountains [Utah]: U.S. Geographical and Geological Survey of the Rocky Mountain Region, Washington, D.C., U.S. Government Printing Office, p. 18–98.

Glock, W. S., 1931, The development of drainage systems; A synoptic view: Geographical Review, v. 21, p. 475–482.

Goreau, P. D., 1980, The geophysics and tectonic evolution of the eastern Cayman Trough and western Hispaniola [abs.]: Caribbean Geological Conference, 9th Santo Domingo Resumenes, p. 27–28.

Grant, L., 1973, Dye test for Boruca's limestone: Comm-Alcoa-Hensley, Schmide-Instituto Costarricense de Electicidad, 9 p.

Gupta, A., 1975, Stream characteristics in eastern Jamaica; An environment of seasonal flow and large floods: American Journal of Science, v. 275, p. 825–842.

Gurnee, J., ed., 1967, Conservation through commercialization; Rio Camuy development proposal: National Speleological Society Bulletin, v. 29, no. 2, p. 27–72.

Gurnee, R. H., 1972, Exploration of the Tanama: Explorers Journal, v. 51, no. 3, p. 159–171.

—— , 1977, Exploration of the Tanama, *in* Sloane, B., ed., Cavers, caves, and caving: Rutgers University Press, p. 369–392.

Gurnee, R. H., and Gurnee, J., 1974, Discovery at the Rio Camuy: New York, Crown Publishers, 183 p.

Haan, C. T., and Johnson, H. P., 1966, Rapid determination of hypsometric curves: Geological Society of America Bulletin, v. 77, p. 123–126.

Hanor, J. S., 1978, Precipitation of beachrock cements; Mixing of marine and meteoric waters versus CO_2-degassing, *in* Back, W., and Freeze, R. A., eds., Chemical hydrogeology; Benchmark papers in geology, Stroudsburg, Pennsylvania, Hutchinson Ross Publishing Company.

Hare, P. W., and Gardner, T. W., 1985, Geomorphic indicators of vertical tectonism along converging plate margins, Nicoya Peninsula, Costa Rica, *in* Hack, J., and Morisawa, M., eds., Tectonic geomorphology: Proceedings of the 15th Geomorphology Symposia Series, Binghamton, p. 76–104.

Harpster, R. E., Alt, J. N., and Schwartz, D. P., 1981, Evidence for Quaternary deformation along the western edge of the Coast Range, southwestern Costa Rica: Geological Society of America Abstracts with Programs, v. 13, p. 59.

Hastenrath, S., 1973, On the Pleistocene glaciation of the Cordillera de Talamanca: Zeitschrift für Gletscherkunde und Glazialgeologie, v. 9, p. 105–121.

Hawley, J. W., and Wilson, W. E., 1965, Quaternary geology of the Winnemucca area, Nevada: Las Vegas, Nevada, Desert Research Institute, University of Nevada Technical Report No. 5, 94 p.

Henningsen, D., 1966, Notes on stratigraphy and paleontology of Upper Cretaceous and Tertiary sediments in southern Costa Rica: American Association Petroleum Geologists Bulletin, v. 50, p. 562–580.

Herman, J. S., 1982, The dissolution kinetics of calcite, dolomite, and dolomitic rocks in the CO_2–water system [Ph.D. thesis]: The Pennsylvania State University, 218 p.

Heward, A. P., 1978, Alluvial fan sequence and megasequence models, with examples from Westphalian D–Stephanian B coalfields, northern Spain, *in* Miall, A. D., ed., Fluvial sedimentology: Canadian Society Petroleum Geology Memoir 5, p. 669–702.

Hooke, R. LeB., 1967, Processes on arid-region alluvial fans: Journal of Geology, v. 75, p. 438–460.

—— , 1968, Steady-state relationships on arid-region alluvial fans in closed basins: American Journal Science, v. 266, p. 609–629.

—— , 1972, Geomorphic evidence for Late-Wisconsin and Holocene tectonic deformation, Death Valley, California: Geological Society America Bulletin, v. 83, p. 2073–2098.

Instituto Costarricense de Electricidad, 1981, Boletin de sedimento en suspension no. 1, San Jose, Costa Rica, 92 p.

Johnson, D., 1931, Stream sculpture on the Atlantic slope: New York, Columbia University Press, 142 p.

Jordan, T. E., Isacks, B. I., Allmendinger, R. W., Brewer, J. A., Ramos, V. A., and Ando, C. J., 1983, Andean tectonics related to the geometry of the Nazca plate: Geological Society of America Bulletin, v. 94, p. 341–361.

Judson, S., 1975, Evolution of Appalachian topography, *in* Melhorn, W. N., and Flemal, R. G., eds., Theories of landform development: Binghamton, New York, 6th Annual Geomorphology Symposia Series, p. 29–45.

Kelleher, J., and McCann, W., 1976, Buoyant zones, great earthquakes, and unstable boundaries of subduction: Journal Geophysical Research, v. 81, p. 4885–4896.

Kelson, K., 1986, Long-term tributary adjustments to base-level lowering, northern Rio Grande rift, New Mexico [M.S. thesis]: Albuquerque, University of New Mexico, 212 p.

Kennedy, W. Q., 1962, Some theoretical factors in geomorphological analysis: Geology Magazine, v. 99, p. 304–312.

Kesel, R. H., 1973, Notes on the lahar landforms of Costa Rica: Zeitschrift für Geomorphologie Neve Folge, Supplement Band 18, p. 78–91.

—— , 1983, Quaternary history of the Rio General valley, Costa Rica: National Geographic Society Research Reports, v. 15, p. 339–358.

—— , 1985, Alluvial fan systems in a wet-tropical environment, Costa Rica: National Geographic Research, v. 1, p. 450–469.

Kesel, R. H., and Lowe, D. R., 1987, Geomorphology and sedimentology of the Toro Amarillo alluvial fan in a humid tropical environment, Costa Rica: Geografiska Annaler (in press).

Kesel, R. H., and Spicer, B. E., 1985, Geomorphological relationships and ages of soils on alluvial fans in the Rio General valley, Costa Rica: Catena, v. 12, p. 149–166.

King, L. C., 1947, Landscape study in southern Africa: Geological Society of South Africa Proceedings, v. 50, p. 23–52.

—— , 1951, The geomorphology of the eastern and southern districts of southern Rhodesia: Geological Society of South Africa Transactions, v. 54, p. 33–64.

Kruckow, T., 1974, Landhebung im Valle Central und Wachstum der Kustenebenen in Costa Rica (Mittelamerika): Jahrbuch witthert zu Bremen, v. 18, p. 247–263.

Kulm, L. D., Prince, R. A., French, W., Johnson, S., and Masias, A., 1981, Crustal structure and tectonics of the central Peru continental margin and trench, *in* Kulm, L. D., Dymond, J., Dasch, E. J., and Hussong, D. M., eds., Nazca plate; Crustal formation and Andean convergence: Geological Society of America Memoir 154, p. 445–468.

Lundberg, N., 1981, Evolution of the slope landward of the Middle America Trench, Nicoya Peninsula, Costa Rica, *in* Leggett, J. K., ed., Forearc geology: Geological Society of London Special Publication, p. 431–447.

—— , 1983 Development of fore-arcs of intra-oceanic subduction zones: Tectonics, v. 2, p. 51–61.

Madrigal, R. G., 1977, Terrazas marinas y tectonismo en Peninsula de Osa, Costa Rica: Revista Geografica, v. 86–87, p. 161–166.

Mapa Geologico de Costa Rica (preliminary edition), 1982, Nicoya, Costa Rica: Direction de Geologia, Minas, y Petroleo, escala 1:200,000.

Mattson, P. H., 1984, Caribbean structural breaks and plate movements, *in* Bonini, W. F., Hargraves, R. B., and Shagam, R., eds., The Caribbean-South American plate boundary and regional tectonics: Geological Society of America Memoir 162, p. 131–152.

McBirney, A. R., 1958, Active volcanoes of Nicaragua and Costa Rica, *in* Catalogue of active volcanoes of the world including solfatava fields, Part 6, Central America: Naples, International Association of Volcanology, p. 107–146.

McCann, W. R., and Sykes, L. R., 1984, Subduction of aseismic ridges beneath the Caribbean plate: Implications for the tectonics and seismic potential of the northeastern Caribbean; Journal of Geophysical Research, v. 89, p. 4493–4519.

McClymonds, N. E., and Ward, P. E., 1966, Hydrologic characteristics of the alluvial fan near Salinas, Puerto Rico: U.S. Geological Survey Professional Paper 550-C, p. C231–C234.

McNutt, M. K., 1983, Influence of plate subduction on isostatic compensation in northern California: Tectonics, v. 2, p. 399–415.

Meyerhoff, H. A., 1975a, The Penckian model—with modifications, *in* Melhorn, W. N., and Flemal, R. C., eds., Theories of landform development: Proceedings Sixth Annual Geomorphology Symposium, State University of New York, Binghamton, p. 45–68.

—— , 1975b, Stratigraphic and petroleum possibilities of middle Tertiary Rocks in Puerto Rico (discussion): American Association Petroleum Geologists Bulletin, v. 59, p. 169–172.

Meyerhoff, H. A., and Hubbell, M., 1929, The erosional landforms of eastern and central Vermont: Vermont State Geologist Biennial Report, p. 315–381.

Miall, A. D., 1977, A review of the braided river depositional environment: Earth Science Review, v. 13, p. 1–62.

—— , 1981, Alluvial sedimentary basins; Tectonic setting and basin architecture, *in* Miall, A. D., ed., Sedimentation and tectonics in alluvial basins: Geological Association of Canada Special Paper 23, p. 1–33.

Mills, H. H., 1982, Piedmont-Cove deposits of the Dellwood Quadrangle, Great Smokey Mountains, North Carolina, U.S.A.; Morphometry: Zeitschift für Geomorphologie, Neue Folge, Band 26, p. 163–178.

—— , 1983, Piedmont evolution at Roan Mountain, North Carolina, U.S.A.: Geografiska Annaler, v. 65A, p. 111–126.

Ministerio de Industria, Energia, y Minas, 1982: Mapas Geologicos de Costa Rica, Golfito, Quepos, y Talamanca Quadrangles, scale 1:200,000.

Minster, J. B., and Jordan, T. H., 1978, Present day plate motions: Journal of Geophysical Research, v. 83B, p. 5331–5354.

Miotke, F. D., 1973, Subsidence of the surface between mogotes and Puerto Rico east of Arecibo: Caves and Karst, v. 15, p. 1–12.

Miyamura, S., 1975, Recent crustal movements in Costa Rica disclosed by relevelling surveys: Tectonophysics, v. 29, p. T191–T198.

—— , 1980, Sismisidad de Costa Rica: San Jose, Editorial Universidad de Costa Rica, 190 p.

Molnar, P., and Atwater, T., 1978, Interarc spreading and Cordilleran tectonics related to the age of the subducted oceanic lithosphere: Earth and Planetary Sciences Letters, v. 41, p. 827–857.

Molnar, P., and Sykes, L. R., 1969, Tectonics of the Caribbean and Middle America regions from focal mechanisms and seismicity: Geological Society of America Bulletin, v. 80, p. 1639–1684.

Monroe, W. H., 1964, The zanjon, a solution feature of the karst topography in Puerto Rico: U.S. Geological Survey Professional Paper 501-B, p. 126–129.

—— , 1973, Stratigraphic and petroleum possibilities of middle Tertiary rocks in Puerto Rico: American Association of Petroleum Geologists Bulletin, v. 57, p. 1086–1099.

—— , 1975, Stratigraphic and petroleum possibilities of middle Tertiary rocks in Puerto Rico (reply): American Association of Petroleum Geologists Bulletin, v. 59, p. 172–175.

—— , 1976, The karst landforms of Puerto Rico: U.S. Geological Survey Professional Paper 899, 69 p.

—— , 1980, Geology of the middle Tertiary formations of Puerto Rico: U.S. Geological Survey Professional Paper 953, 93 p.

Mora, C. S., 1978, Estudio Geologico de los Cerros Barra Honda y alrededores, Tesis de Bachillerato, Escuela de Geologia, Universidad de Costa Rica, 199 p.

—— , 1979a, Geologicos y geotecnicos para el Proyecto Hidroelectrico Boruca: San Jose, Costa Rica Departamento de Geologia Instituto Costarricense de Electricidad, 2 v., 275 p.

—— , 1979b, Estudio geologico de una parte de la region sureste del valle de General, provincia de Puntarenas, Costa Rica [thesis]: Universidad de Costa Rica, 321 p.

—— , 1981, Barra Honda: San Jose, Costa Rica, Editorial Universidad Estatal a Distancia, 115 p.

—— , 1983, Una Revision y actualizacion de la clasificacion morfotectonica de Costa Rica, segun la teoria de la techonica de placas: Universidad Nacional, Boletin de Vulcanologia, no. 13, p. 18–36.

—— , 1984, Faults and folds of the south and southeast region of Costa Rica: scale 1:200,000. Instituto Costarricense de Electricidad, San Jose, Costa Rica, Departmento de Geologio.

Mora, S., and Valdes, R., 1983, Estudio geologico-goetecnico para el Proyecto Hidroelectrico Savegre: San Jose, Costa Rica, Departmento de Geologia Instituto Costarricense de Electricidad, 3 v., 420 p.

Morales, L. D., and Montero, W., 1984, Los tremblores senlidos en Costa Rica durante. 1973–1983 y surelacion con la sismicidad del pais: Revisita Geologica de America Central, v. 1, p. 29–56.

Moussa, M. T., and Seiglie, G. A., 1975, Stratigraphic and petroleum possibilities of middle Tertiary Rocks in Puerto Rico (discussion): American Association Petroleum Geologists Bulletin, v. 59, p. 163–168.

Murata, K. J., Dondoli, C., and Saenz, R., 1966, The 1963–65 eruptions of Irazu Volcano, Costa Rica (The period of March 1963 to October 1964): Bulletin Volcanologique, v. 29, p. 765–793.

Nelson, J. B., and Ganse, R. A., 1980, Map of significant earthquakes 1900–1979: Riverdale, Maryland, National Oceanic and Atmospheric Administration, scale 1:1,000,000.

Palmer, R., 1983, The Dominican Republic, A brief study of the caving potential: Caves and Caving, no. 21, p. 25–27.

Penck, W. O., 1925, Die piedmontflachen des sudlichen schwarzwaldes: Berlin, Gesellschaft für Erdkunde, v. 1, p. 81–108.

Pilger, R. H., 1981, Plate reconstructions, aseismic ridges, and low-angle subduction beneath the Andes: Geological Society of America Bulletin, v. 92, p. 448–456.

Plummer, L. N., Parkhurst, D. L., and Wigley, T.M.L., 1979, Critical review of the kinetics of calcite dissolution and precipitation, *in* Jenne, E. A., ed., Chemical modeling in aqueous systems: American Chemical Society Symposium series 93, p. 537–573.

Quennell, A. M., 1956, The structure and geomorphic evolution of the Dead Sea Rift: Quarterly Journal of the Geological Society of London, v. 114, p. 2–18.

Rebillard, P., Dixon, T., and Farr, T., 1982, Geologic observations of the northern boundary of the Caribbean Plate across Central America as seen by Seasat and SIR-A, *in* Actes du symposium international de la commission VII d104e la societe internationale de photogrammetrie et teledetection, Toulouse, p. 593–599.

Rivier, F., 1983, Sintesis geologica y mapa del area del Bajo Tempisque, Costa Rica: Informe Semestral, enero-junio, Instituto Geografico Nacional, 275 p.

Rockwell, T. K., Keller, E. A., and Johnson, D. L., 1985, Tectonic geomorphology of alluvial fans and mountain fronts near Ventura, California, *in* Morisawa, M., and Hack, J. T., eds., Tectonic geomorphology: Boston, Allen and Unwin, p. 183–207.

Schramm, W. E., 1981, Humid tropical alluvial fans, northwest Honduras [M.S. thesis]: Louisiana State University, 184 p.

Schramm, W. E., and Nummedal, D., 1982, Braided stream sedimentation in a humid tropical environment: Geological Society of America Abstracts with Programs, v. 14, p. 80.

Schumm, S. A., 1956, Evolution of drainage systems and slopes in badlands at Perth Amboy, New Jersey: Geological Society of America Bulletin, v. 67, p. 597–646.

Schumm, S. A., and Parker, R. S., 1973, Implications of complex response of drainage systems for Quaternary alluvial stratigraphy: Nature Physical Science, v. 243, p. 99–100.

Schwartz, D. P., Cluff, L. S., and Donnelly, T. W., 1979, Quaternary faulting along the Caribbean–North American plate boundary in Central America: Tectonophysics, v. 52, p. 431–445.

Segovia, A. V., Foss, J. E., and Sole, E. A., 1980, Evaluation of soils and landscapes of the seasonal tropics by means of remote sensing images: an interdisciplinary study, *in* Proceedings, International Symposium of Remote Sensing, no. 14, San Jose, p. 915–927.

Selby, M. J., 1974, Rates of denudation: New Zealand Journal of Geography, no. 56, p. 1–13.

Shah, H., Mortgat, Ch. P., and Lubetking, L., 1976, A progress report: Seismic risk analysis of Costa Rica: Stanford University, John A. Blume Earthquake Engineering Center, Department of Civil Engineering, 29 p.

Shepherd, R. G., 1979, River channel and sediment responses to bedrock lithology and stream capture, Sandy Creek drainage, central Texas, *in* Rhodes, D. D., and Williams, G. P., eds., Adjustments of the fluvial system: Proceedings of the 10th Annual Geomorphology Symposium, State University of New York, Binghamton, p. 255–276.

Shuster, E. T., and White, W. B., 1971, Seasonal fluctuations in the chemistry of limestone springs; A possible means for characterizing carbonate aquifers: Journal of Hydrogeology, v. 14, p. 93–128.

—— , 1972, Source areas and climatic effects in carbonate groundwaters determined by saturation indices and carbon dioxide pressures: Water Resources Research, v. 8, p. 1067–1073.

Steel, R. J., and Wilson, A. C., 1975, Sedimentation and tectonism (Permo-Triassic) on the margins of the North Minch Basin: Journal of the Geological Society of London, v. 131, p. 183–202.

Steel, R. J., Maehle, S., Nilsen, H., Roe, S. L., Spinnangr, A., 1977, Coarsening upward cycles in the alluvium of Hornelen Basin (Devonian) Norway; Sedimentary response to tectonic events: Geological Society of America Bulletin, v. 88, p. 1124–1134.

Stockmal, G. S., 1983, Modeling of large scale accretionary wedge deformation: Journal of Geophysical Research, v. 88, p. 8271–8287.

Stoiber, R. E., and Carr, M. J., 1973, Quaternary volcanic and tectonic segmentation of Central America: Bulletin Volcanologique, v. 37, p. 304–325.

Strahler, A. N., 1952, Hypsometric (area-altitude) analysis of erosional topography: Geological Society of America Bulletin, v. 69, p. 279–300.

Sweeting, M. M., 1958, The Karstlands of Jamaica: Geographical Journal, v. 124, p. 184–199.

—— , 1973, Karst landforms: New York, Columbia University Press, 362 p.

Sykes, L., and Ewing, M., 1965, The seismicity of the Caribbean: Journal of Geophysical Research, v. 70, p. 5065–5074.

Talbot, L. M., and Kesel, R. H., 1975, The tropical savanna ecosystem, *in* Kesel, R. H., ed., Grassland ecology; A symposium: Geoscience and Man, v. 10, Louisiana State University, p. 15–26.

Tosi, J. A., Jr., 1969, Mapa Ecologico Republica de Costa Rica: San Jose, Costa Rica, Central Cientifico Tropical, scale 1:500,000.

Troester, J. W., and White, W. B., 1983, Criteria for development of master trunk drainage systems in tropical climates: National Speleological Society Bulletin, v. 45, p. 34.

—— , 1986, Geochemical investigations of three tropical karst drainage basins in Puerto Rico: Ground Water, v. 24, p. 475–482.

Troester, J. W., White, E. L., and White, W. B., 1984, A comparison of sinkhole depth frequency distributions in temperate and tropical karst regions, *in* Beck, B. F., ed., Sinkholes; Their geology, engineering and environmental impact: Proceedings, First Multidisciplinary Conference on Sinkholes, A. A. Balkema, Rotterdam, p. 65–73.

Ulate, C. A., and Corrales, M. F., 1966, Mud floods related to the Irazu Volcano eruptions: American Society of Civil Engineers, Hydrology Division Journal, v. 92, p. 117–129.

U.S. Weather Bureau, 1961, Generalized estimates of probable maximum precipitation and rainfall-frequency data for Puerto Rico and Virgin Islands: Technical Paper no. 42, 94 p.

Van Andel, T. H., Heath, G. R., Malfait, B. T., Heinrichs, D. F., and Ewing, J. I., 1971, Tectonics of the Panama Basin, eastern equatorial Pacific: Geological Society of America Bulletin, v. 82, p. 1489–1508.

Van de Graaff, W.J.E., Crowe, R.W.A., Bunting, J. A., Jackson, P., and Jackson, M. J., 1977, Relict early Cainozoic drainage in arid western Australia: Zeitschrift für Geomorphologie, v. 21, p. 379–400.

Versey, H. R., 1972, Karst of Jamaica, *in* Herak, M., and Stringfield, V. T., eds., Karst: New York, Elsevier Publishing Company, p. 445–466.

Vessell, R. K., and Davies, D. D., 1981, Nonmarine sedimentation in an active forearc basin, *in* Ethridge, F. G., ed., Recent and ancient nonmarine depositional environments; Models for exploration: Society of Economic Paleontologists and Mineralogists Special Publication no. 31, p. 31–45.

Vogt, P. R., Lovine, A., Bracey, D. R., Hey, R. N., 1976, Subduction of aseismic ocean ridges; Effects on shape, seismicity, and other characteristics of consuming plate boundaries: Geological Society of America Special Paper 172, 59 p.

Wadge, G., and Burke, K., 1983, Neogene Caribbean plate rotation and associated Central American tectonic evaluation: Tectonics, v. 2, p. 633–643.

Waldron, H. H., 1967, Debris flows and erosion control problems caused by ash eruptions of Irazu Volcano, Costa Rica: U.S. Geological Survey Bulletin no. 1241-I, 37 p.

Wang, C. Y., and Shi, Y. L., 1984, On the thermal structure of subduction complexes; A preliminary study: Journal of Geophysical Research, v. 89, p. 7709–7718.

Wells, S. G., 1973, Geology of the Cerros Barra Hondo region in the Nicoya Peninsula, Costa Rica: National Park Service of Costa Rica, 28 p.

Wescott, W. A., and Ethridge, F. G., 1980, Fan delta sedimentology and tectonic setting Yallahs fan delta, southeast Jamaica: American Association of Petroleum Geologists, v. 64, p. 374–399.

Weyl, R., 1957, Beitrage zur Geologie der Cordillera de Talamanca Costa Rica (Mittelamerika): Neues Jahrbuch Geologie und Palaontologie Abhandlung, v. 105, p. 123–204.

—— , 1971, Die Morphologisch-tektonische Gliederung Costa Rica (Mittelamerika): Erdkunde, v. 25, p. 221–230.

—— , 1980, Geology of Central America: Gerlin-Stuttgart, Gerbruder Borntraeger, 371 p.

White, E. L., and White, W. B., 1979, Quantitative morphology of landforms in carbonate rock basins in the Appalachian highlands: Geological Society of America Bulletin, v. 90, p. 385–396.

White, W. B., 1977, Role of solution kinetics in the development of karst aquifers: International Association of Hydrogeologists, Memoir 12, p. 203–517.

White, W. B., 1984, Rate processes; Chemical kinetics and karst landform development, *in* LaFleur, R. G., ed., Groundwater as a geomorphic agent: Boston, Allen and Unwin, p. 227–248.

Winslow, M. A., and McCann, W. R., 1985, Neotectonics of a subduction/strike slip transition zone; the northeastern Dominican Republic: Geological Society of America Abstracts with Programs, v. 17, p. 753.

Yonekura, N., 1972, A review on seismic crustal deformation in and near Japan: Bulletin, Department of Geography, University of Tokyo, v. 4, p. 17–50.

Yoshikawa, T., 1974, Denudation and tectonic movement in contemporary Japan: Bulletin, Department of Geography, University of Tokyo, v. 6, p. 1–14.

Yoshikawa, T., Kaizuka, S., and Ota, Y., 1981, The Landforms of Japan: University of Tokyo Press, 222 p.

ACKNOWLEDGMENTS

Gardner and others gratefully acknowledge Sr. F. M. Rudin (Director, Instituto Geografico Nacional, Costa Rica) and Dr. Sergio Mora (Instituto Costariccense de Electricidad, Costa Rica) for logistical and field support, aerial photographs, and topographic maps. Partial funding was provided to P. W. Hare by the Standard Oil Company of California and to T. W. Gardner by the

Department of Geosciences, Pennsylvania State University. Figure 8 was provided by Kevin Furlong. Data for Figure 18 and Table 1 were provided by M. Guebert, C. Lemieux, S. O'Hare, G. Veni, and D. White.

Menges and others would like to thank Sr. F. M. Rudin (Director, Instituto Geografico Nacional) and Dr. Sergio Mora (Instituto Costariccense de Electricidad, Costa Rica) for assistance in field support and transportation. Sr. Luis Chaves Montes (Instituto Geografico Nacionalo, Costa Rica) was a constant source of information concerning field logistics. Geologic maps were provided by Sr. Jose Fco. Castro Muñoz (Ministerior de Industria, Energia, y Minas, Costa Rica). Partial funding was provided by the Mellon Inter-American Field Research grant awarded by the Latin American Institute and by the Research Allocations Committee of the University of New Mexico; this support for the field studies is gratefully acknowledged. We would also like to thank the following people for contributing to the collection and analysis of portions of the morphometric data: J. Appel, P. Karas, K. Nelson, J. Persico, J. Ritter, L. Wandsnider, and J. Wesling.

Research for the section on alluvial fan formation was supported by grants to R. H. Kesel from the National Geographic Society, the Zemurray Foundation, and the Association of American Geographers, and by National Science Foundation Grant number EAR 7822754 to D. R. Lowe. Kesel and Lowe would like to thank Professor Manuel Arguello in Heredia, Costa Rica for his help and logistic support.

Research for the section on tectonic geomorphology in Costa Rica was partially supported by National Science Foundation Grant number EAR-8615277 to T. W. Gardner and S. G. Wells.

MANUSCRIPT ACCEPTED BY THE SOCIETY DECEMBER 15, 1986

Geological Society of America
Centennial Special Volume 2
1987

Chapter 11

Columbia and Snake River Plains

Victor R. Baker
Department of Geosciences, University of Arizona, Tucson, Arizona 85721
Ronald Greeley
Department of Geology, Arizona State University, Tempe, Arizona 85287
Paul D. Komar
School of Oceanography, Oregon State University, Corvallis, Oregon 98331
Donald A. Swanson and Richard B. Waitt, Jr.
U.S. Geological Survey, David A. Johnston Cascades Volcano Observatory, 5400 MacArthur Boulevard, Vancouver, Washington 98661

INTRODUCTION

Victor R. Baker

This chapter treats two areas of the northwestern United States characterized by great late Cenozoic outpourings of basaltic lava. The western part of this region is underlain by flood basalt of the Columbia River Basalt Group. As discussed below by Waitt and Swanson, this area constitutes the Columbia Plain. The laterally extensive, thick cooling units of the Columbia River Basalt Group contrast with the thinner, less extensive flows of the Snake River Plain, which lies to the southeast. Lavas of the Snake River Plain were emplaced coincident with Pliocene-Quaternary rifting, from about 5 Ma to present.

Much of the Columbia Plain is overlain by tens of meters of Pleistocene loess, which is extensively dissected to form the rolling topography of the Palouse Hills. In the northeast, broad channels were carved into the Palouse loess and underlying basalt by cataclysmic Pleistocene floods. The characteristic flood erosion of the basalt impressed early settlers as a scaring of the earth by removing its protective soil, and the term "scabland" was applied to it.

The best known geomorphic studies in the Columbia Plateau centered on the role of cataclysmic flooding in the origin of its landscape. The central figure in this research for over half a century was J Harlen Bretz (Fig. 1), a glacial geologist at the University of Chicago. Bretz (1923a) used the name "Channeled Scabland" to describe the area of loess-mantled northeastern Columbia Plain that was scoured by flood channels. In 20 major articles and monographs, mostly published between 1923 and 1932, Bretz shocked the geologic community with his hypothesis that the plexus of proglacial channels that forms the Channeled Scabland was derived from a relatively brief, immense flood. This

flood, which Bretz (1925) named the Spokane Flood, spilled across preglacial stream divides, and eroded the loess topography to form linear scarps, hanging valleys, and high-level bars of flood gravel. Bretz ascribed phenomenal power to the flood, including its ability to deeply scour the basalt and to transport immense boulders (Fig. 2). The controversy that erupted over the Spokane Flood hypothesis became one of the most fascinating episodes in the history of geology (Baker, 1981). It was not until the 1960s that the cataclysmic flood origin of the Channeled Scabland was largely accepted in the geologic community.

In recent years a major interest in the Columbia and Snake River Plains has derived from discoveries in the exploration of the Moon and Mars, where extensive plains of basaltic lavas form the planetary surfaces. The basaltic flows of the Columbia and Snake River Plains are excellent terrestrial analogs for these features. Moreover, the discovery of great channels on Mars, morphologically similar to the Channeled Scabland, has opened a new debate concerning the geomorphic role of cataclysmic flooding (Baker, 1985).

GEOMORPHIC EVOLUTION OF THE COLUMBIA PLAIN AND RIVER

R. B. Waitt, Jr., and D. B. Swanson

INTRODUCTION AND REGIONAL SETTING

Since early geologic reconnaissances in central and eastern Washington, geologists have tried to relate drainage and topography to widespread volcanic accumulations and conspicuous geologic structures (Russell, 1900, 1901; Smith, 1901; Willis, 1903; Calkins, 1905). Several hypotheses of drainage evolution were later debated for the topographically complicated western

Baker, V. R., Greeley, R., Komar, P. D., Swanson, D. A., and Waitt, R.B., Jr., 1987, Columbia and Snake River Plains, *in* Graf, W. L., ed., Geomorphic systems of North America: Boulder, Colorado, Geological Society of America, Centennial Special Volume 2.

Figure 1. Professor J Harlen Bretz in the study of his home near Chicago, Illinois. This photograph was taken in 1977 when "Doc" Bretz was 95 years old.

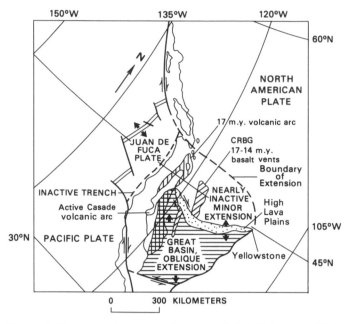

Figure 3. Plate-tectonic context of western United States from Christiansen and McKee (1978, Fig. 13.8) outlining tectonic features and volcanic rocks in relation to interaction between Pacific, Juan de Fuca, and North American plates. Arrows indicate directions of relative motion. CRBG, Columbia River Basalt Group.

Figure 2. Boulder of Columbia Plateau basalt eroded from the Grand Coulee region and transported to the Ephrata Fan by cataclysmic outburst flooding. The boulder measures 18 m by 11 m by 8 m. A scour hole developed around this boulder is shown in Figure 34B.

part of the Columbia basin (Flint, 1938a; Warren, 1941; Waters, 1955; Mackin, 1961). More recent geologic mapping, geochronology, and structural and process analyses invite a fresh assessment of regional landscape and drainage evolution of the Columbia basin. Unlike the northern Appalachian Mountains, where classic theories of landscape and drainage evolution were developed, the Columbia drainage basin contains a rich assemblage of dated Cenozoic volcanic and clastic material of known emplacement mechanisms that constrain any theory of drainage and landscape evolution. This subchapter summarizes a fuller report (Waitt and Swanson, in review).

Regional topography and drainage in the Pacific Northwest are broadly controlled by volcanism and tectonism related to plate-tectonic convergence along the margin of western North America. During the late Cenozoic, the Juan de Fuca plate has been converging obliquely northeastward against the North American plate and subducting beneath it (Fig. 3; Atwater, 1970, Fig. 10; Christiansen and McKee, 1978, Fig. 13.8). A dominant feature of the landscape since the middle Tertiary has been a Cascade highland with calc-alkalic volcanoes. East of the Cascade arc, backarc spreading perhaps was responsible for the Miocene eruption of the Columbia River Basalt Group along extensional fissures. Continuing stresses, no doubt related to plate interactions, resulted in younger Miocene northwest-trending right-lateral megashears, east-west extension manifested by north-trending dikes of the Columbia River Basalt Group, and east-trending compressional folds and faults (Fig. 3; Wise, 1963).

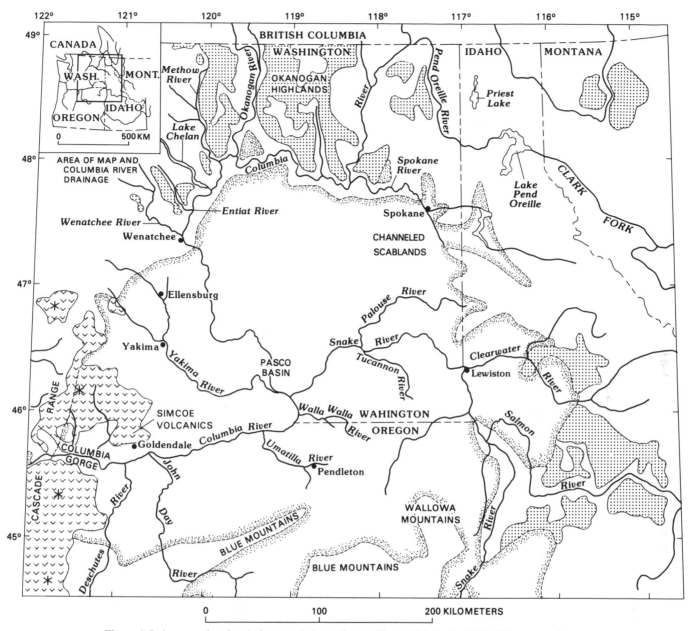

Figure 4. Index map showing drainage and places discussed in text. Light stipple highlights approximate limits of Columbia River Basalt Group delineated by dashed line. Heavy dot pattern indicates generalized distribution of incised erosion surface peripheral to Columbia River Basalt. "V" pattern indicates areas of extensive volcanism younger than Columbia River basalt in and near Cascade Range, including stratovolcanoes (*). Insert shows extent of Columbia River drainage basin.

In central Washington and adjacent areas, more than 200,000 km^3 of Miocene tholeiitic flood basalt of the Columbia River Basalt Group underlies a 160,000-km^2 area (Fig. 4). The region as a whole is appropriate termed the 'Columbia Intermontane Province' (Freeman and others, 1945; Thornbury, 1965). The popular designation 'Columbia Plateau' is a geomorphic misnomer: relative to rugged, older highlands on all sides, the area is a plain whose center seems to be as low now as in any previous age (Fig. 5). We call it the 'Columbia Plain' or 'basalt plain', analogous to 'Snake River Plain' for the subprovince of similar character to the southeast. The surface of the Columbia Plain is intricately warped but generally saucer shaped—nearly at sea level at Pasco basin but 600–2,000 m higher at its rims. In the southwest and south the basalt is folded into a valley-and-ridge topography with as much as 1,800 m of relief. An arm of the basalt extends westward across the Cascade Range.

Figure 5. Photograph of raised relief of map of Columbia Plain and peripheral uplands in low-angle illumination from west (left). Yakima folds show in southwest part of plain; the connected structural high of Naneum Ridge (N) and Hog Ranch axis (H) is delineated by dash-dot line. River valleys and transient channels: CR, Columbia River; YRv, Yakima River; SpR, Spokane River; ClR, Clearwater River; SR, Salmon River; SnR, Snake River; SCSn, lower Salmon-Clearwater-Snake River; B, Box-canyon segment of Columbia; CS, Channeled Scabland; GC, Grand Coulee. Ridges and lows: PB, Pasco basin; N, Naneum Ridge anticline; H, Hog Ranch axis; HH, Horse Heaven Hills anticline; RH, Rattle-snake Hills anticline; SaM, Saddle Mountains anticline; SM, Snipes Mountain anticline; U, Umtanum anticline; YRg, Yakima Ridge anticline; BM, Blue Mountains uplift; KV, Kittitas Valley; D-U, Dalles-Umatilla syncline. Cities: C, Chelan; E, Ellensburg; Y, Yakima; L, Lewiston; S, Spokane; W, Wenatchee.

FORMATION OF COLUMBIA PLAIN AND DRAINAGE DERANGEMENT BY BASALT FLOWS

Great lava floods of the Columbia River Basalt Group (hereafter informally abbreviated as 'Columbia River basalt') erupted from linear vents in southeastern Washington and adjacent Oregon and Idaho and descended, spreading north and west (Waters, 1939, 1955; Mackin, 1961; Swanson, 1967; Swanson and others, 1980). The basalt flooded a basin whose depth initially was perhaps 400 m or more, judged from the 200–500 m of local relief on the subbasalt surface exposed around the basin perimeter (Waters, 1939; Swanson, 1967; Camp, 1981). From 17.5 to 14.5 Ma, basalt flows buried prebasalt topography and created a plain across eastern Washington and northeastern Oregon (Fig. 6). From 14.5 to 6 Ma, the flows were progressively more confined within broad lows. Several flows escaped westward across the ancestral Cascade Range along a trough 75 km wide.

The surface of a rapidly moving, low-viscosity basalt flow sloped north and west toward its distal margin probably less than 0.1°. Between eruptions the whole section probably sagged and deformed the surface into a gentle saucer shape with relief of 10 m and more over distances of 100 km. Offlap of successive basalt flows from the margins of the plain after 15 Ma (Mackin, 1961; Swanson and others, 1979; Hooper and Camp, 1981) shows that basining occurred during, as well as after, basalt accumulation. The total accumulation of basalt forms a lens 400 km in diameter and 4 km or more thick at its center.

The saucer shape of the plain and the 4-km-thick lens of basalt can be explained largely, though not entirely, by sagging due to loading by the basalt and to transfer of magma to the surface. Isostatic lowering of the basin as it was loaded can account for most of the subsidence of the bottom of the lens. If the new load of basalt came from beyond the basin and added crustal thickness, then after isostatic compensation the top of the basalt lens is calculated to be 200–300 m higher in the middle than at the rim (Waitt and Swanson, in review). Most known feeder dikes are on the fringe of the plain, but a more central position of the emptied magma chambers under the basin could account for some of the net basining. If tapped from directly below the plain, the cooled surface basalt might be denser and less voluminous than the deep magma it replaced. The result would be a net lowering of the surface at the center by about 100 m (Waitt and Swanson, in review). The much greater 600–2,000 m actual sag of the basalt surface and the existence of a basin before eruption indicate the operation of other processes, such as backarc crustal thinning, perhaps by subcrustal erosion.

Sedimentary deposits interfinger with the Columbia River Basalt Group around the margin of the basalt plain. Along the west margin, silicic volcaniclastic debris was shed from active volcanoes in the ancestral Cascade Range, and a subarkose facies was derived from the north. Basaltic hyaloclastic debris, as well as interbasalt sandstone and mudstone near the margin of the Columbia Plain, are evidence that basalt flows dammed pre-existing

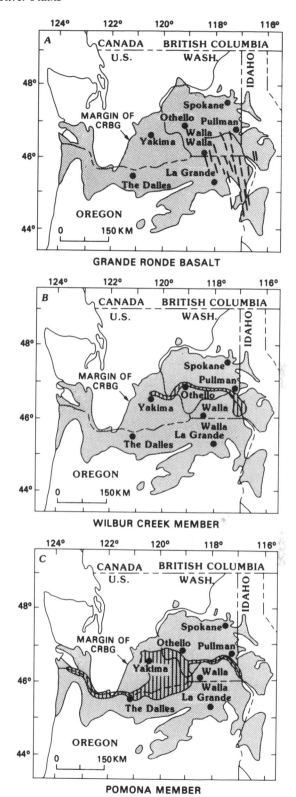

Figure 6. Maps showing distribution of selected units of the Columbia River Basalt Group (CRBG; after Swanson and others, 1979). A, 15 Ma Grande Ronde Basalt; B and C, 13 to 12 Ma Ponoma and Wilbur Creek Members of Saddle Mountains Basalt (lined areas). NNW lines on A are feeder dikes.

streams (Waters, 1955; Mackin, 1961; Schmincke, 1967; Griggs, 1976; Swanson and others, 1979, 1982; Camp, 1981). Between eruptions, shallow lakes and fluvial plains migrated as far as 50 km inside the basalt margin, aided by basinward tilting between extrusions.

Between Spokane and Wenatchee, the Spokane and Columbia Rivers flow in a great counterclockwise arc around the margin of the basalt field. This drainage originated from highland streams that, dammed by the basalt margin, flowed along the natural gutter where the margin of the basalt, initially sloping gently north and west, met the peripheral highlands (Figs. 5, 7; Waters, 1939; Mackin and Cary, 1965).

DRAINAGE REVISIONS BY TECTONISM AND BASALT FLOWS

Prior theories of structure-transverse drainage

Through-flowing rivers transverse to orogenic structures, long considered anomalous, are the subject of an extensive literature. W. M. Davis's (1889) theory of successive capture of longitudinal consequent streams by transverse subsequent streams accounted for only some of the perceived anomalies in northern Appalachian drainage. His theory was later challenged; alternatives were promoted (Johnson, 1931; Strahler, 1945). Four main theories have emerged to account for structure-transverse drainage in different regions: (1) predeformation streams successfully maintain courses across rising structures (antecedent streams); (2) streams originating on an unconformable cover are superposed onto buried transverse structures (superposed consequent streams); (3) streams develop by headward growth in weak rocks exposed during erosion of folded strata (subsequent streams); (4) streams form in lows between the main positive structures (consequent streams).

In most regions, transverse drainage has been inferred to be antecedent or superposed, thus independent of structures it crosses. Oberlander (1965) found that structure-transverse drainage anomalies in the Zagros Mountains of Iran resulted in part from longitudinal and transverse subsequent streams that had grown headward in weak conformable rocks that extended across the structures at higher erosional levels. In contrast to drainage in the northern Appalachians or the Zagros Mountains, the Columbia River drainage has a fairly consistent and simple relation to transverse structures. In central Washington, most trunk streams flow along synclines—picking their way from syncline to syncline, dodging around the noses of plunging anticlines, and crossing at structural saddles those anticlines and mountain ranges that cannot be altogether avoided. Most of the trunk rivers originated as basalt-consequent or structure-consequent streams.

Columbia and Yakima rivers in south-central Washington

The western part of the basalt plain is crenulated into anticlinal ridges and synclinal valleys trending east and southeast

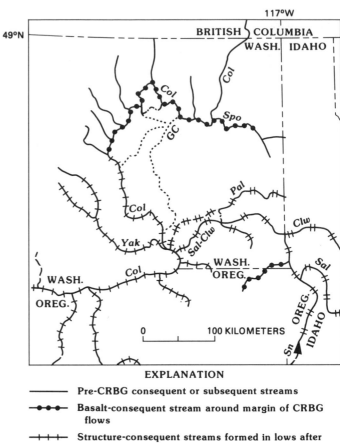

EXPLANATION

—————— Pre-CRBG consequent or subsequent streams

●–●–●–● Basalt-consequent stream around margin of CRBG flows

+–+–+–+ Structure-consequent streams formed in lows after CRBG flows

+H–H–H Basalt-consequent and structure-consequent streams formed by and after CRBG flows

–H– –H– CRBG-consequent, structure-consequent stream, course abandoned because of subsequent capture

— — — Stream captured by post-CRBG damming (arrow, site of overflow)

—·—·— Stream permanently deranged by Pleistocene glaciers or floods

········· Temporary glacier-deranged stream course

Figure 7. Inferred origin of major rivers in Pacific Northwest. Rivers: Col, Columbia; Sal, Salmon; Clw, Clearwater; Sn, Snake; Yak, Yakima; Spo, Spokane; Pal, Palouse; GC, Grand Coulee. CRBG, Columbia River Basalt Group.

(Fig. 5). Naneum Ridge anticline, with more than 1,800 m of relief, continues southward as the Hog Ranch anticlinal axis (Mackin, 1961) and forms the divide between Columbia and Yakima tributaries. Several hypotheses have tried to explain the courses of the Yakima River across anticlines, and the Columbia River which avoids most of them. (1) Both rivers predate the structure; they maintained their courses across rising anticlines (Russell, 1900; Smith, 1901; Calkins, 1905). (2) A great fan of silicic volcaniclastic debris that shed eastward from the Cascades pushed the Columbia River eastward to Pasco basin (Waters,

1955). (3) The Yakima River developed in a structural low between the Cascade Range to the west and the Hog Ranch axis to the east (Schmincke, 1964); both the Columbia and Yakima Rivers were defeated by the rising Horse Heaven anticline and diverted eastward to Pasco basin (Flint, 1938a; Warren, 1941).

Volcaniclastic-fan hypothesis. Provenance of ancient stream deposits of the Columbia Plain can be inferred from lithology of constituents: streams draining southern Washington and northern Oregon carry diverse volcanic clasts; those draining northern Washington and Idaho carry gneissic and quartzitic clasts; those draining central Idaho carry granitic, metavolcanic, and quartzitic clasts. Interbasalt and suprabasalt sedimentary deposits along the southwest basalt-plain margin are largely silicic volcaniclastic debris from the Cascade Range (Waters, 1955; Schmincke, 1967; Swanson, 1967; Swanson and others, 1979, 1982), where plutons dated from 15.5 to 13 Ma (Engels and others, 1976) may be intrusive equivalents of volcanoes that shed the debris. Hypothesized eastward diversion by a great fan of volcaniclastic debris shed from the Cascade Range (Waters, 1955; Mackin and Gary, 1965), which was proposed to explain the general course of the Columbia River between Wenatchee and Wallula Gap, was based on an inference that Cascadian silicic-volcaniclastic debris overlies the youngest basalt flow near Yakima. But in fact, quartzite-bearing conglomerate (northern provenance) overlies the youngest (10.5 Ma) basalt flow at Snipes Mountain southeast of Yakima (Schmincke, 1964, 1967). Therefore the Columbia River must have continued to flow southwestward through the Yakima area *after* the last of the silicic volcaniclastic debris had accumulated from the west (Fig. 8).

Tectonism and topography. Quartzite-bearing gravel on and south of Horse Heaven anticline shows that the Columbia River formerly flowed south-southwest from Yakima to the present Columbia Gorge, probably after 10.5 Ma (Fig. 8). Having stratigraphically disproved Waters' (1955) fan-diversion idea, Schmincke (1967) revived an earlier idea (Calkins, 1905; Mackin, 1961) that the rising Umtanum anticline diverted the Columbia eastward at Priest Rapids. But diversion of the Columbia there by Umtanum anticline was just one of many steps in the evolution of the Columbia and its tributaries.

Rivers in the Columbia Plain follow structural lows. Developing structures must have shifted the drainage into topographically low courses (Fig. 7). The growing Naneum Ridge anticline diverted the ancestral Columbia drainage eastward at Wenatchee; continued southward growth of the Naneum-Hog anticline shifted the Columbia east from Ellensburg and kept the upper Yakima River west, thus dividing the two rivers. Growth of anticlines located to the south shifted the ancestral Columbia River still farther east and the Yakima River farther west to synclinal lows. Growing Horse Heaven anticline diverted the lower Yakima River east to the lower Yakima syncline and Pasco basin. Both rivers had to cross the longest anticlines somewhere; they do so where the rising structures elevated the basalt surface least or last.

South of Kittitas Valley the Yakima River crosses several east-trending anticlinal ridges in spectacular water gaps (Fig. 5). Because there is no evidence in the region of any unconformable cover through which superposition could have occurred, these gaps have long invited the idea of antecedence (Russell, 1900; Smith, 1901; Waters, 1955). Antecedence implies that the stream course is independent of the structures. But as Schmincke (1964, p. 366) noted, the river course roughly corresponds to a cross-structural sag toward which anticlines plunge eastward from the Cascade Range and westward from the Hog Ranch axis—thus the structurally lowest route from the Kittitas Valley syncline. The present course of the Yakima preferentially along a structural low must have originated in topographic lows between the ends of anticlines at early stages of growth. Continued growth of the anticlines eventually caused the river to incise them.

The hypothesis that major rivers developed in topographic lows between growing structural highs explains most features of the principal drainage pattern of the Columbia Plain: (1) from the structurally high west, north, and east margins of the basalt plain, all major rivers flow to the structural low, Pasco basin; (2) the Columbia abandons its basalt-margin course and turns sharply across the basalt at Wenatchee just upstream of the largest anticline; (3) below this turn the Columbia follows structural lows to Pasco basin, dodging around some plunging anticlines and crossing others at structural saddles; (4) the Yakima River flows mostly along synclines—Kittitas Valley and lower Yakima Valley; (5) even the famous structure-transverse segment of the Yakima River below Ellensburg generally coincides with a structural sag across the belt of anticlinal ridges; and (6) from Pasco basin the united drainage (lower Columbia River) crosses Horse Heaven anticline at a structural saddle and follows synclines and structural saddles to and across the Cascade Range.

The regional correspondence of major rivers with structural lows indicates that most rivers are consequent streams that formed in topographic lows determined by tectonism. After the rivers acquired the lowest courses available to them, further uplift caused some river segments to incise. Thus even spectacularly incised strucure-crossing rivers originated as consequent streams.

Some basalt flows and intercalated sedimentary beds of the Columbia River Basalt Group, about 14.5 Ma and younger in age, become thinner toward the crests of anticlines and away from the present Columbia River southeast of Wenatchee (Schmincke, 1967; Swanson and others, 1982; Reidel, 1984)— evidence that rising anticlines had diverted paleodrainage east from the basalt margin by 14.5 Ma. Yet the vast extent of individual basalt flows and sedimentary interbeds indicates that structural relief then was only a few meters. The distribution of some late Columbia River basalt flows as broad lobes that come within 25–40 km of the west basalt margin (Fig. 6C) shows that before 10.5 Ma, structural relief there was slight. In the western part of the plain, some late basalt flows advanced westward down narrow valleys locally floored with metavolcanic-clast (Salmon River) gravel (Fig. 6B). Quartzite-rich (northern provenance) cobble conglomerate overlying a 10.5-Ma basalt in the Horse

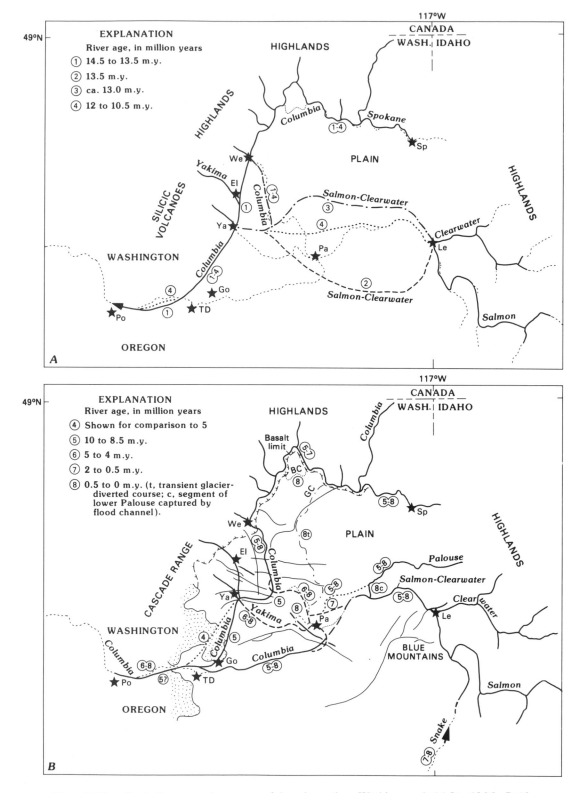

Figure 8. Maps illustrating successive courses of rivers in southern Washington: A, 14.5 to 10 Ma; B, 10 Ma to present. BC, Box Canyon; GC, Grand Coulee. Present-day cities: El, Ellensburg; Go, Goldendale; Le, Lewiston; Pa, Pasco; Po, Portland; Sp, Spokane; TD, The Dalles; WE, Wenatchee; Ya, Yakima. Fine dotted lines, present drainage lines; hatchured line, approximate limit of Columbia River Basalt Group; thin lines, anticlinal axes; stipple, fields of late Cenozoic volcanic rocks in eastern Cascade Range.

Heaven area shows that the ancestral Columbia River was diverted east to Pasco basin later (Fig. 6). Mild folding began as early as 15 Ma, but tectonism rearranged much of the Columbia drainage after 10.5 Ma (Fig. 8).

Salmon-Clearwater River

Emerging from crystalline-rock highlands, the modern Salmon and Clearwater rivers unite at a sharp downwarp in the basalt at Lewiston. Their union long predates the entry of the upper Snake into that area (Fig. 8; Wheeler and Cook, 1954). Uplift of flows dated 13.5–12 Ma (Camp, 1981, Fig. 4) formed lows where the ancestral Salmon and Clearwater Rivers formed. Most flows in the Lewiston basin younger than 14 Ma are intracanyon flows, and cobble gravel below the youngest flow shows that a large river existed before 6 Ma (Camp, 1981; Webster and others, 1982). At Lewiston, the Salmon-Clearwater (Snake) River turns west at the south base of the enormous Lewiston structural upwarp (Fig. 8) and follows a broad structural and topographic low westward. The lower Salmon-Clearwater (Snake) River and tributary Tucannon River (Fig. 4) flow down the structure contours of a depositional slope steepened by Pliocene tilting toward Pasco basin (Swanson and others, 1980; Camp and Hooper, 1981).

A 14–13 Ma flow delineates a former topographic low that curved far south of the present lower Salmon-Clearwater (Snake) River; locally beneath this flow is gravel containing metavolcanic and other clasts from central Idaho. The ancestral Salmon-Clearwater River thus flowed far south but later shifted north to its present course (Fig. 8). Between 12.5 and 6 Ma, at least five intracanyon flows, some underlain by gravel rich in metamorphic and metavolcanic rock types (eastern provenance), followed a west-draining canyon eroded in the basalt by the ancestral Salmon-Clearwater River 10 to 85 km north of the modern river (Swanson and others, 1980). The younger intracanyon flows are nearly coincident with the eastern segment of the modern canyon, though farther west the valley at times branched to the north (Fig. 8). In a special review on Pasco basin, Fecht and others (1986) show the sequence of stream courses in greater detail.

Pasco basin and lower Columbia River

Major rivers that drain much of the Pacific Northwest unite at Pasco basin, the structural and topographic low of the region. A general offlap, and thickening toward the basin of Columbia River basalt flows dating 14.5 to 10.5 Ma, show that topographic basining began at least by 14 Ma and perhaps earlier (Reidel, 1984); younger flows became more restricted to the Pasco area, which shows that the Pasco basin subsided during as well as after basalt extrusion.

The major rivers were diverted into Pasco basin successively (Fig 8). Intracanyon flows show that the lower Salmon-Clearwater (Snake) River flowed that way by 13 Ma; a conglomerate rich in metavolcanic clasts beneath an 8.5-Ma intracanyon basalt at Wallula Gap shows that the Salmon-Clearwater River left Pasco basin near the present gap. The ancestral Columbia River delivered quartzite-rich conglomerate to the Yakima-Goldendale area farther west until after 10.5 Ma, missing Pasco basin. The lack of northern rock types in teh conglomerate at Wallula Gap shows that the Columbia River did enter Pasco basin until after 8.5 Ma.

Diversion of the Columbia and lower Yakima Rivers to Pasco basin after 8.5 Ma probably resulted from the rise of the Horse Heaven anticline (Fig. 5), a formidable and continuous structure between the eastern Cascade Range and the Blue Mountains uplift. The Salmon-Clearwater and Columbia Rivers had to cross the anticline somewhere; they do so at the structurally lowest sag. This coincidence of modern drainage to structural lows again shows that tectonism was the main influence on drainage.

Below Wallula Gap the Columbia River bends westward into a structural and topographic low, the Dalles-Umatilla syncline. The river crosses the Cascade Range in a complex structural saddle, where a 12-Ma intracanyon basalt flow is underlain and overlain by quartzite-bearing gravel, evidence of a major river across the Cascade Range (Tolan and Beeson, 1984). After 12 Ma the position of the Columbia River across the Cascade Range was partly maintained by volcanic highs to the north and south.

Late basin fills

In several structural and topographic basins, the Columbia River Basalt Group and interfingering sedimentary rocks are overlain by younger clastic accumulations that delineate areas where postbasalt downwarping outpaced stream downcutting. Northern and central Pasco basin is floored by 300 m of siltstone, sandstone, conglomerate, and rare claystone beds (Newcomb and others, 1972), much of it consisting of upward-fining sequences typical of a meandering-river floodplain. Vertebrate fossils indicate that the deposit dates from 4 to 5 Ma (Gustafson, 1978). The logical cause of the aggradation is slackening of through-flowing drainage because Horse Heaven anticline continued to rise.

The basin deposit commonly has been inferred to be related to the Columbia River (Flint, 1938a; Gustafson, 1978). But a conglomerate bed in the middle of the section consists of metavolcanic and quartzite clasts of eastern provenance, and imbrication also reveals westward paleocurrents: the Salmon-Clearwater River, not the Columbia, deposited the conglomerate. Therefore, at about 5 Ma the lower Salmon-Clearwater River was flowing westward some 20 to 30 km north of its present course (Fig. 8), while the Columbia River lay somewhere west of its present course.

The basin deposit is capped by a thick caliche layer (Newcomb and others, 1972; Gustafson, 1978) indicating that the upper surface was exposed to weathering for a long time after the rivers began to incise the surface. Since about 4 Ma the Salmon-Clearwater and Columbia Rivers have incised 170 m below this surface and through at least that thickness of basalt at Wallula

Gap. The deposit dips south beneath Pleistocene flood deposits and below the grade of the rivers, so downwarping continued in southern Pasco basin until after 4 Ma.

Partial or complete ponding of drainage may occur if some structures rise while others subside across the courses of through-flowing rivers. The basin deposits accumulated in response to the development of Pasco basin and other synclines. Downwarping of these deposits below river grade implies that the basining that had begun in Miocene time continued into the Pliocene or later.

HIGH PERIPHERY OF COLUMBIA PLAIN

Upland erosion surface

In parts of mountainous areas surrounding the Columbia Plain, an erosion surface of low to moderate relief forms plateau-like divides below which canyons are sharply cut 300–1,500 m (heavy shading on Fig. 4). Discussion of upland erosion surfaces grew unpopular after the 1930s because many so-called "pene-plains" in the Rocky Mountains were reinterpreted as pediments (Bradley, 1936; Mackin, 1947; Hunt and others, 1953), and the concept of cyclicity in landscape evolution was itself challenged (King, 1953; Hack, 1960; Melhorn and Flemal, 1975). Nonetheless, a deeply and sharply incised upland erosion surface anomalous to any noncyclic fluvial landscape—an 'inequilibrium landscape' of Strahler (1952a)—has been independently noted in Washington and Idaho (Waters, 1939; Ross, 1947; Mackin and Gary, 1965).

In the eastern Cascade Range we reexamined Waters' (1939) idea that the erosion surface (Fig. 4) was resurrected from beneath Columbia River basalt (Tabor and others, 1982, 1987). Nonconformable contacts of some outliers of basalt are steep against older gneiss that rises above the basalt. The west margin of the basalt only partly filled narrow, steep-walled valleys. No evidence exists that basalt ever extended much farther, though the erosion surface extends an additional 50 km (Fig. 4).

A discontinuous low-relief surface of the Okanogan Range (Fig. 4), extending north from the Columbia River to the international boundary, lies 650–1,700 m above sharply incised tributaries of the Methow and Okanogan rivers. This surface must have developed beyond the basalt margin, for a basalt carapace could not have been so neatly stripped from the surface and leave no outliers. Part of central Idaho also has an extensive upland surface with accordant summits, conspicuous where the Salmon drainage is cut as much as 1,500 m below it (Fig. 4). The erosion surface in Idaho is younger than the upper Oligocene to lower Miocene volcanic rocks that it bevels (Ross, 1947).

Thus a discontinuous upland erosion surface is expressed in crystalline-rock tracts west, north, and east of the Columbia Plain. Most of this surface could not have lain beneath basalt and seems instead contemporaneous with and younger than the basalt. The basalt drowned pre-existing trunk streams and kept local base levels high for 10 m.y.—a situation in which surrounding highlands could have eroded to gentle relief. The Cascade Range

and other highlands then rose in late Miocene to Pliocene time. An incised upland erosion surface is an expected result of such a history.

Incision of upper Columbia River

The Spokane-Columbia River between Spokane and Wenatchee roughly follows the contact between ancestral highlands and the originally gently north- and west-sloping depositional surface of basalt (Figs. 4, 5, 9). After 14.5 Ma, the surface of Pasco basin subsided relative to the basalt-plain margins. Locked in elevating courses, the Spokane and upper Columbia rivers incised. From the Spokane River confluence to Wenatchee, the Columbia River has downcut as much as 750 m into crystalline rocks beneath the basalt (Figs. 9, 10).

LATE TECTONIC AND VOLCANIC MODIFICATIONS

Stratigraphy, together with a general absence of geomorphic evidence of capture, suggest that the tectonism responsible for drainage changes occurred mostly in Miocene and Pliocene time. But modern seismicity and fault scarps in Quaternary material show that movement has continued to the present (Waitt, 1979). The lower Yakima River shows geomorphic evidence of a drainage shift that illustrates a process involved in earlier drainage diversions. The Yakima flows northeastward, between anticlines, past the head of Badger Coulee, a broad southeast-trending valley perched about 30 m higher which is clearly an abandoned river course (Fig. 11). This drainage shift probably occurred because growing adjacent anticlines raised Badger Coulee above the level of the alternate course on the north. The Yakima River then overflowed northward, or the Columbia overflowed between migrating meanders.

Young basalt flows overlapping the southwest margin of the Columbia River basalt are outflows from the high Cascades south of Mount Rainier (Luedke and others, 1983). The many volcanic outpourings from the Cascades deranged many small drainages at the edge of the Columbia Plain. The Simcoe volcanic field (5–2 Ma) in southern Washington may even have aided eastward diversion of the Columbia River (Figs. 4, 8B). The most radical Pliocene or Pleistocene drainage change was northward overflow of the Snake River, which drains most of southern Idaho, into Salmon (Columbia) drainage (Figs. 4, 8B). The idea for such a capture, initially based on geomorphic criteria such as barbed tributaries upstream and a gorge downstream of the capture site (Wheeler and Cook, 1954), is supported by studies of fossil fish of the lake ponded in the western Snake River Plain (Smith and others, 1982). The capture must have occurred after 2.5 Ma when the lake still existed (Kimmell, 1982). The cause of damming of a now-vanished course of the Snake River, probably to the southwest, must have been Pliocene volcanism and tectonism.

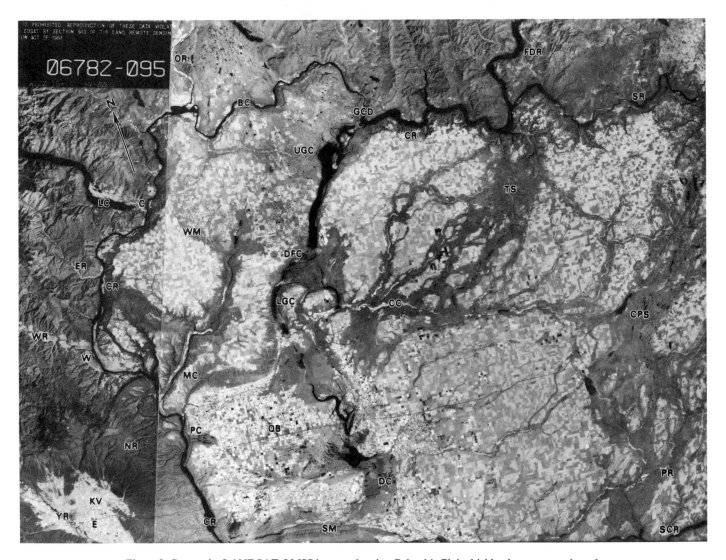

Figure 9. Composite LANDSAT-5 MSS images, showing Columbia Plain, highlands on west and north, scabland channels carved by great Lake Missoula floods, and limit of last-glacial Okanogan lobe of Cordilleran ice sheet. Left, #6782-095, 16 Aug., 1985; right, #6802-039, 25 Aug., 1985. Cities: C, Chelan; E, Ellensburg; W, Wenatchee; S, Spokane. Rivers and lakes: OR, Okanogan River; LC, Lake Chelan; CR, Columbia River; BC, Box Canyon; ER, Entiat River; WR, Wenatachee River; YR, Yakima River; PR, Palouse River; SR, Spokane River; SCR, Salmon-Clearwater River; FDR, Lake Franklin D. Roosevelt dammed behind GCD, Grand Coulee Dam. Ridges and valleys: SM, Saddle Mountain anticline; NR, Naneum Ridge anticline; KV, Kittitas Valley; QB, Quincy Basin. Glacial and scabland features: WM, Withrow moraine; CPS, Cheney-Palouse scabland tract; TS, Telford scabland tract; CC, upper Crab Creek; UGC, upper Grand Coulee; LGC, lower Grand Coulee; MC, Moses Coulee; DC, Drumheller scabland channels; DFC, Dry Falls cataract; PC, Potholes cataract.

Pleistocene glacial and catastrophic-flood modifications

Late Miocene to Pleistocene uplift of mountains of the Pacific Northwest and growth of stratovolcanoes created extensive high-altitude areas, which eventually became glaciated because of global cooling from late Miocene onward (Donnelly, 1982). Several episodes of Pleistocene glaciation, the most recent 20,000 to 13,000 yr ago, modified the northern fringe of the Columbia Plain.

The Cordilleran ice sheet advanced southward as lobes along trunk valleys (Fig. 12; Waitt and Thorson, 1983). Only the Okanogan lobe advanced far out onto the Columbia Plain, in the northwest. An early Pleistocene advance of the Okanogan lobe may have shifted the Columbia River southward from its former basalt-margin position to ice-marginal Box Canyon across basalt (Figs. 8B, 9; Flint, 1935). The last-glacial Okanogan lobe scoured the basalt surface, built a large terminal moraine during its maximum stand, and formed many moraines, eskers, and other depositional forms during ice recession. Geomorphic details of surface landscape in the northwest part of the Columbia Plain were

Figure 10. Oblique aerial photograph showing view northwest across northwestern part of Columbia Plain bounded by incised Columbia River valley. Limit of last-glacial Okanogan lobe of Cordilleran ice sheet is delineated by outer (toward left) limit of morainal topography and large boulders, beyond which is Columbia River basalt covered by undisturbed loess developed into unbroken wheat fields. Square outlined by roads and fencelines in center of view is 1 mi (1.6 km) on a side.

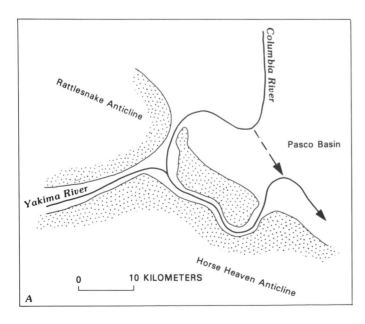

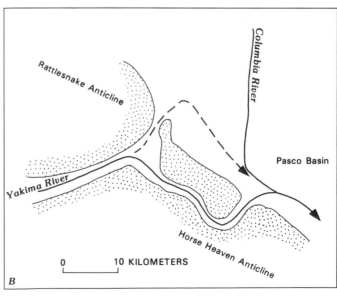

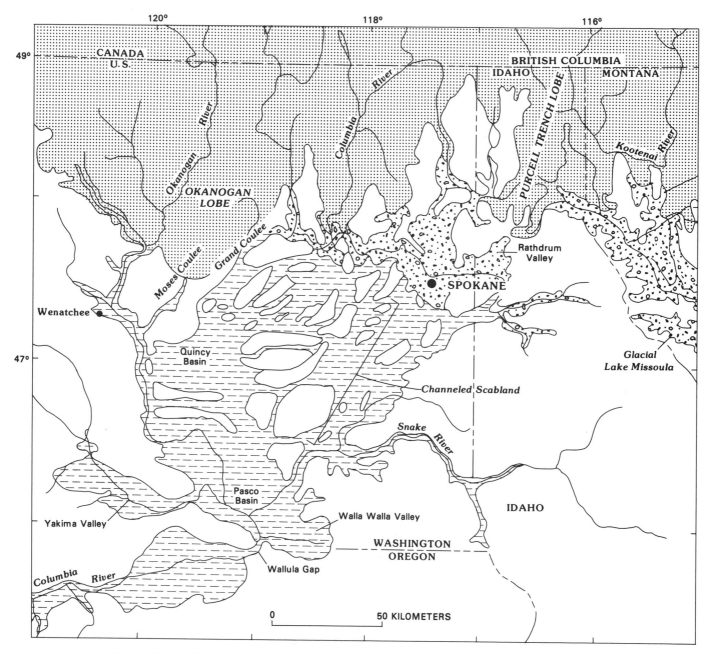

Figure 12. Late Pleistocene Cordilleran ice sheet (dot pattern), ice-dammed lakes (coarse pattern), and courses of catastrophic Lake Missoula floods (dash pattern). Modified from Waitt (1985, Fig. 1).

Figure 11. Drainage revision at Badger Coulee, showing two possible *pre*capture drainage patterns and alternative overflow paths to effect capture. A, capture by meander cutoff of Columbia River; B, capture by overflow of Yakima River; stipple indicates highlands; dashed line shows present course of Yakima River.

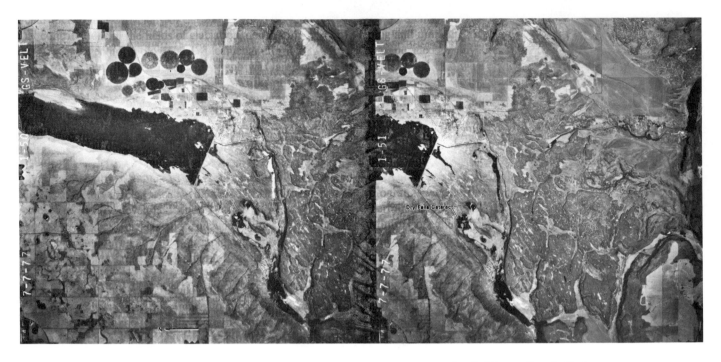

Figure 13. Stereopair of aerial photographs of Dry Falls cataract at head of lower Grande Coulee, showing details of flood-eroded scabland into the Columbia River basalt.

utterly changed (Figs. 9, 10). During the last two glaciations, the Okanogan lobe temporarily diverted the Columbia River across the Columbia Plain through the Grand Coulee and channels farther south, shown by a dot-dash line on Figure 8B.

The Purcell Trench lobe of the Cordilleran ice sheet dammed the Clark Fork River to form glacial Lake Missoula of maximum volume 2,500 km^3 in mountain valleys well east of the Columbia Plain. The lake discharged through the ice dam as stupendous floods that swept south across the northern part of the Columbia Plain to Pasco basin (Fig. 12) and down the Columbia River valley (Bretz and others, 1956; Baker, 1973a; Waitt and Thorson, 1983). The scoured floodpaths on the Columbia Plain constitute the Channeled Scabland, a great anastomosis of flood channels and recessional gorges replete with a bizarre assemblage of erosional and depositional landforms such as rock basins, giant cataract alcoves (Fig. 13), large residual 'islands,' great bars of gravel, and giant current dunes. From its preflood valley, the lower Palouse River was diverted southward through a flood-formed coulee to the Snake River valley (Figs. 8, 9; Bretz and others, 1956). The extraordinary landscape of the Channeled Scabland is now thought to have been carved by dozens of separate floods (Waitt, 1985). Although the floods cumulatively lasted only months, some scabland channels are 100 to 400 m deep. These briefly cut subfluvial channels are more voluminous than many valleys carved by gradually incising stream systems over millions of years.

THE SPOKANE FLOOD DEBATE AND ITS LEGACY

Victor R. Baker

THE SPOKANE FLOOD CONTROVERSY

In 1926 the presidential address of the American Association for the Advancement of Science was delivered by a geomorphologist. For his topic William Morris Davis chose "The Value of Outrageous Geological Hypotheses." He observed, "We shall be indeed fortunate if geology is so marvelously enlarged in the next thirty years as physics has been in the last thirty. But to make such progress violence must be done to many of our accepted principles" (Davis, 1926).

The chance to see geology's reception to this challenge came but a year later. A 44-year old glacial geologist named J Harlen Bretz was invited to give a lecture to the Geological Society of Washington, D.C. The topic was "Channeled Scabland and the Spokane Flood." Bretz (1927) described work he had begun several years earlier (Bretz, 1923a, 1923b, 1925) on anomalous physiographic features in eastern Washington. His detailed field evidence included great cataracts (Fig. 14), flood bars (Fig. 15), and streamlined hills of loess (Fig. 16) that he argued could not be explained by any hypothesis other than a great flood of water. However, his audience, mostly members of the U.S. Geological

Figure 14. West Potholes cataract on the western rim of the Quincy Basin. Bretz (1923b) showed that this and three other cataracts operated simultaneously as spillways for cataclysmic flood water entering the Quincy basin. The scale of the cataract also indicates the magnitude of this flow. The water spilled from a 3-km-wide lip, over 100-m cliffs into 40-m-deep plunge basins of two great alcoves. Boulders as large as 30 m in diameter were carried from the cataract to a 50-m-high gravel bar, which is not cut by the modern Columbia River. Bretz (personal communication, 1978) recalls that an early topographic map of this cataract inspired his scabland studies.

Survey, was decidedly hostile to this view. A series of prepared discussions followed in which alternative views, mostly involving normal stream action and possible glacial diversions, were held as superior to the cataclysmic flood hypothesis (Baker, 1978a).

Bretz (1928a, 1928b) subsequently answered his critics, and he initiated new studies showing diverse effects of the cataclysmic flooding (Bretz, 1928c, 1929, 1930, 1932). However, others began work in the Channeled Scabland for the expressed purpose of disproving his theory. The most detailed of various revisionist studies was done by Richard Foster Flint (1938b), who made numerous errors of interpretation (Bretz and others, 1956; Baker, 1978a). Flint seems to have systematically ignored incontrovertible evidence of cataclysmic flood flows. Flint's colleague at Yale, Aaron Waters (written communication, 1978) recalls informing him that a study of West Bar (Fig. 17) would provide a phenomenal example of flood processes. Flint ignored the suggestion, but he observed other examples of flood bars and giant current ripples without realizing their significance (Baker, 1978a).

Bretz argued that the field evidence, if properly studied, would eventually vindicate his theory. During the 1930s and 1940s a growing minority came to accept the flood hypothesis with minor modifications (e.g., Allison, 1933, 1941). Moreover, a dramatic twist was added to the story in 1940 at a meeting of the American Association for the Advancement of Science in Seattle, Washington. Howard Meyerhoff (written communication, 1978), who attended that meeting, recalls a key moment in a session organized to debate various proposed origins of the Channeled Scabland. Attention was focused on the well-prepared presentation by Flint in contrast to the view of Bretz, who was not present.

Near the end of the session Joseph T. Pardee rose to speak on the topic "Ripple Marks (?) in Glacial Lake Missoula." Par-

Figure 15. Flood bar, described as "Bar 2" by Bretz and others (1956), located near Wilson Creek, Washington. Bretz (1928b) described numerous examples of flood-emplaced bars in the Channeled Scabland. This bar is 1 km long and 25 m high.

dee, who had worked in western Montana and in the Channeled Scabland for the U.S. Geological Survey, had been present at the infamous 1927 meeting in Washington, D.C. Although he had published on the scablands (Pardee, 1922), he had not spoken at that earlier meeting. Indeed, there are indications that in 1922 Pardee possessed key evidence supporting Bretz's hypothesis, but that he was dissuaded from revealing this by his superior, W. C. Alden, and by other colleagues of the U.S. Geological Survey (Bretz, written communication, 1974). It is also probable that in 1922, before Bretz's work on the topic, Pardee had independently realized the cataclysmic origin of the Channeled Scabland, but had been prevented from publishing the idea by his superiors. Until 1940 he had remained silent on the topic.

In Pardee's talk and subsequent publication (Pardee, 1942), he described evidence for an immense glacial lake in western Montana (Fig. 18). This lake, which he had earlier named Lake Missoula (Pardee, 1910), impounded approximately 2500 km^3 of water behind a lobe of the late Pleistocene Cordilleran ice sheet in northern Idaho (Fig. 19). Moreover, Pardee described evidence that the dam had failed suddenly, thereby rapidly draining the lake, resulting in what he termed "unusual currents." The lake

drainage created immense flood bars (Fig. 20) as well as great ripple marks composed of gravel and boulders, rising to heights of up to 15 m and spaced as much as 200 m apart. Pardee described the outflow of this lake to the west into the Columbia Plateau region. He never stated a connection to the Channeled Scabland, perhaps generously leaving that point to Bretz.

Encouraged by a growing group of supporters, Bretz, then nearly seventy years old, returned in 1952 for his last field work on the Channeled Scabland problem. With the aid of considerable new data provided by George E. Neff of the U.S. Bureau of Reclamation, it was possible to refute all alternatives to the cataclysmic flood hypothesis (Bretz and others, 1956; Bretz, 1959). The new data revealed numerous examples of giant current ripples throughout the Channeled Scabland, which showed the forcefull passage of the Missoula Flood.

BRETZ AND UNIFORMATIARIANISM

Why was scientific sentiment so strongly against Bretz? William Morris Davis, true to his championing of the outrageous hypothesis, was one of the few prominent earth scientists to sup-

Figure 16. Streamlined loess hill, 60 m high, in Cheney-Palouse scabland tract near Hooper, Washington. Bretz (1923a) first recognized that such hills were shaped by the action of cataclysmic flood water that eroded their upstream ends to convergent prows pointing up the scabland tracts.

port Bretz's work (Bretz, personal communication, 1978). Even Kirk Bryan, who had worked extensively in the field area (Bryan, 1927) sided with his colleagues of the U.S. Geological Survey. As described by Hubbert (1980, p. 1092), who had himself been unjustly rebuked by the Survey hierarchy, "The might and majesty of the United States Geological Survey were marshalled in opposition to the flood hypothesis." How could members of the organization that had nurtured the innovative research of Powell, Gilbert, and Dutton have been so vehemently opposed to this creative idea?

In the experimental sciences, laboratory experiments almost invariably refine principles that are presupposed by the prevailing consensus of "normal science" (Kuhn, 1962, 1978). This consensus, or paradigm, ensures that the quantitative regularity of nature that is sought is conditioned by the regularity that is expected. In the earth sciences, the "laboratory" is Earth itself, but the operation of paradigms remains the same. The paradigm of Huttonian-Lyellian tradition was that valleys form very slowly by the prolonged action of processes that we can observe today in the landscape. Fluvial landform development, like science itself, pro-

ceeds progressively and with order. Bretz's hypothesis was anathema to this tradition, and his critics believed that they were defending science itself by revealing supposed inadequacies of the Spokane Flood hypothesis.

The Spokane Flood Debate is often described as a conflict between catastrophism and uniformitarianism. Unfortunately, both concepts have been much maligned through overuse, misuse, and misinterpretation. Uniformitarianism is generally associated with the work of James Hutton, who advocated that river valleys are produced by the rivers flowing in them. This concept was amplified by Playfair (1802, p. 102–115), who categorically denied a role for cataclysmic events in the origin of the valleys. A profound irony of the Hutton-Playfair "fluvialist" school is that it was derived for Scottish valleys, which are now known to have been shaped by glaciers (Davies, 1969). Indeed, the competing "diluvialist" school recognized through excellent field work that numerous anomalies could not be explained by fluvialist theory: erratic boulders, alpine lakes, hanging valleys, cross-axial drainage, misfit streams, and dry valleys. These anomalies were eventually resolved by Louis Agassiz, whose glacial theory was

Figure 17. West Bar on the Columbia River showing its spectacular cover of giant current ripples (gravel waves). The latter have an average height of 8 m and a chord (spacing) of 160 m.

embraced by the fluvialist school as the resolution of an embarrassing dilemma. Continental glaciation, which lacked modern analogs of comparable scale, was thereby added to the list of orderly processes that had gradually shaped the surface of the planet.

We now know that the landscapes studied by the founders of uniformitarianist ideas in geology developed at nonuniform rates by past processes that differ from those seen to be operating today. This is only ironic if one holds to several fallacious views of uniformitarianism (Shea, 1982). Gould (1965) elaborated the misunderstanding by distinguishing two types of uniformitarianism: (1) methodologic uniformitarianism—the principle that the basic laws of nature are invariant with time; and (2) substantive uniformitarianism—the statement that similar processes and rates prevailed in the geologic past as at present. Hubbert (1967) showed that substantive uniformitarianism is an anachronism. It evolved early in the nineteenth century as an alternative geological explanation to assumptions of special creation and interference by divine providence.

Extricated from its obsolete substantive components, modern uniformitarianism is essentially equivalent to the principle of scientific parsimony, or Occam's razor (Albritton, 1967). Among competing hypotheses for explaining the natural world, the simplest hypothesis, that the basic laws of nature (if we can discern them) are constant, will generally prevail. This is not merely a fundamental principle of geology but one for all science. It is, as stated by Hubbert (1967, p. 29), "... a succinct summation of the totality of all experimental and observational evidence." Methodologic uniformitarianism follows as a direct consequence of the inductive scientific method.

The most profound irony of the Spokane Flood debate is that Bretz, not his critics, remained true to the essence of modern uniformitarianism. His critics were defending an anachronism. Substantive uniformitarianism became unnecessary after the flowering of scientific research in the late 1800s. Careful reading of Bretz's papers will show that they were argued in the best tradition of methodologic uniformitarianism. As he reflected on his scabland studies, Bretz (written communication, 1978) wrote: "I have always used Chamberlin's method of multiple working hypotheses. I applied Occam's razor to select the most appropriate hypothesis, but always with due regard for possible dull places in the tool."

Figure 18. Strandlines of glacial Lake Missoula at Missoula, Montana. Pardee (1910) recognized these as evidence of a great glacial lake that occupied intermontane basins in western Montana. The lake was impounded by a lobe of the Cordilleran ice sheet that dammed the Clark Fork River drainage at Pend Oreille Lake. Lake level reached a maximum altitude of 1,265 m, or about 625 m deep at the ice dam (Pardee, 1942).

Problems with the definition of uniformitarianism have led to proposals to abandon the term (Gould, 1984). Schumm (1985) favors its retention in geomorphology, but limited to the following concepts: (1) the permanence of natural laws (methodologic uniformitarianism), (2) continuity (principle of accountability), (3) simplicity (principle of scientific parsimony), and (4) the tendency of open systems to adjust toward a steady state of optimum efficiency (principle of optimality or equilibrium). Schumm (1985) argues that these specific principles of philosophic uniformity, combined with analogic reasoning, provide the basis for extrapolation in geomorphology. Uncertainty in such extrapolation arises from problems of scale, location, convergence, divergence, singularity, sensitivity, and complexity of landforms (Schumm, 1985).

It is interesting to speculate on the history of geology if the fluvial-diluvial debates of the early nineteenth century had occurred in different settings. In the semiarid Colorado Plateau, fluvial processes obviously dominate the landscape, as elucidated by Gilbert (1877). There are no anomalies better explained by diluvial concepts than by fluvial ones. Substantive uniformitarianism may never have risen to confuse later generations.

Had the fluvial-diluvial debates occurred in the Channeled Scabland, diluvial concepts and catastrophism would have been found compatible with methodologic uniformitarianism. Had Bretz done his studies a century earlier or half a century later, his ideas would have been given a greatly different reception.

LEGACY OF THE GREAT SCABLANDS DEBATE

The "Great Scablands Debate," as it was dubbed by Gould (1980) was unusual in scientific tradition. Because of Bretz's stubborn and prolonged defense of his outrageous hypothesis, he defied what might be called the "Principle of Lost Discoveries." As stated by Beveridge (1950), this principle can be described as follows: "In nearly all matters the human mind has a strong tendency to judge in the light of its own experience, knowledge, and prejudices, rather than on the evidence presented. Thus, new ideas are judged in the light of prevailing beliefs. If the ideas are

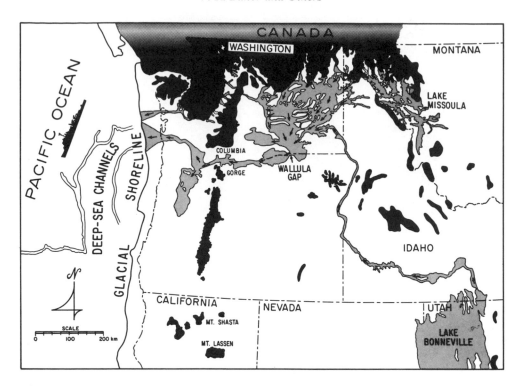

Figure 19. Map showing regions of the northwestern United States affected by cataclysmic flooding in the late Pleistocene. Note the relationship of glacial Lake Missoula to flooded areas in eastern Washington. Flooding from Lake Missoula occurred between approximately 17,000 and 12,000 yrs B.P. (Baker and Bunker, 1985). Flooding from Lake Bonneville occurred approximately 15,000 years ago (Scott and others, 1983; Currey and Oviatt, 1985).

too revolutionary, that is to say, if they depart too far from reigning theories and cannot be fitted into the current body of knowledge, they will not be acceptable. When discoveries are made before their time they are almost certain to be ignored or meet with opposition that is too strong to overcome, so in most instances they might as well not have been made."

Even after the general acceptance of the flood origin of the Channeled Scabland, sentiment persisted that the phenomenon was a unique curiosity. Bretz himself had argued that he was studying a special case, different from prevailing geomorphic experience. He hoped this approach would persuade his critics to examine his data and field areas rather than to dismiss his hypothesis out of hand (Bretz, 1928b). Standard textbooks continued to treat the topic as controversial (Thornbury, 1969). Flint's (1971) widely used textbook, *Glacial and Quaternary Geology,* made only the following oblique reference to the Missoula floods: "Similar features, collectively known as channeled scabland, were widely created east of the Grand Coulee by overflow of an ice-margin lake upstream (Bretz, 1969)" (Flint, 1971, p. 232).

As geomorphologists and sedimentologists began to actively seek evidence of jökulhlaups and related cataclysmic floods, evidence of such phenomena has mounted. Malde (1968) described the Bonneville Flood, which produced scabland features on the Snake River Plain of Idaho. The concept of cataclysmic flooding also has great stratigraphic relevance (Ager, 1980). Today there is a renaissance of interest in the role of episodic sedimentation and rare events in the stratigraphic record (e.g., Sadler, 1981; Dott, 1983).

Recent work has shown that numerous proglacial floods attended the late Wisconsin retreat of Laurentide ice from central North America. The Kankakee Flood affected the Illinois River system (Willman and Frye, 1970). Another large late Pleistocene jökulhlaup affected the Wabash Valley of Indiana (Fraser and others, 1983). The upper Mississippi Valley contains immense slack-water deposits emplaced up tributary mouths by late-glacial flooding (J. C. Knox and R. C. Kochel, personal communication, 1985). The complex of overspills associated with glacial Lake Agassiz (Teller and Clayton, 1983) led to numerous cataclysmic floods. Kehew and Lord (1986) describe typical scabland-like landforms in proglacial lake spillways of the northern Great Plains, including longitudinal grooves, inner channels, streamlined hills, anastomosis, and gravel bars.

Thus, a variety of new research is providing mounting evidence that large-scale proglacial flooding was an important phenomenon during recession of the great late Quaternary ice sheets

Figure 20. Eddy bar formed by rapid, cataclysmic drainage of Lake Missoula in its constricted reach near Perma, Montana. The bar rises over 250 m above the modern Flathead River.

(Baker, 1983). In some areas, such as the northwestern United States, cataclysmic flooding played the dominant role in landscape development. However, the most spectacular twist in the scabland story came with the discovery in 1972 that the surface of Mars had been eroded by great channels (McCauley and others, 1972; Masursky, 1973; Milton, 1973). Because of Bretz's work, numerous similarities between certain Mars channels and the Channeled Scabland could be used to demonstrate an origin by cataclysmic flooding (Baker and Milton, 1974). A new debate on channel origins ensued with pronounced parallels to the Spokane Flood Debate (Baker, 1978a, 1985).

J Harlan Bretz had the tremendous good fortune of living long enough to see the progress of geomorphology catch up to the innovation of his research. In 1979 he was awarded the Geological Society of America's highest honor, the Penrose Medal. At age 97, over a half-century after the infamous meeting of the Washington Academy of Science that denounced his hypothesis, Bretz's observations of cataclysmic flood processes were being used to explain other geological phenomena on both Earth and Mars.

CATACLYSMIC FLOOD PROCESSES AND LANDFORMS

Victor R. Baker and Paul D. Komar

INTRODUCTION

The Spokane Flood controversy preceded the swing to the quantitative studies that characterized geomorphology in the 1950s and 1960s (Mackin, 1963). The view that geomorphology is mainly concerned with forces acting on resistant materials at the Earth's surface (Strahler, 1952b) did not play a significant role in the controversy. Gilluly (1927) did employ some simple calculations in his effort to show the unreasonable nature of Bretz's cataclysmic flooding. However, his assumptions were flawed, as he later readily admitted (Waters, 1969).

Bretz realized that the origin of the Channeled Scabland would eventually need to be investigated in quantitative, physical terms. In questioning his own cataclysmic flood hypothesis, Bretz (1932, p. 81) stated, "Somewhere must lurk an unrecognized weakness.

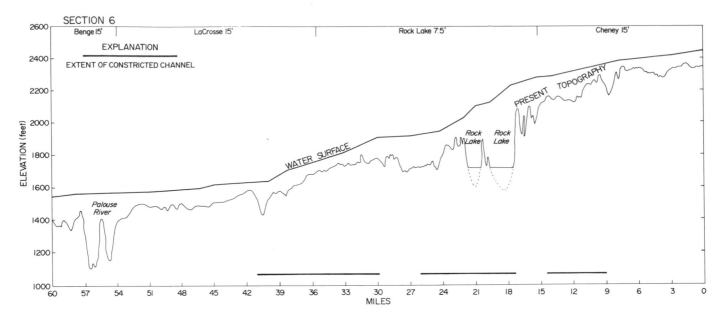

Figure 21. High water-surface profile for cataclysmic flood flows in the eastern Cheney-Palouse scabland tract. Note the marked steepening of the water-surface gradient through constricted reaches. The
deep scour basins of Rock Lake are located in a particularly constricted section with a water-surface
gradient of 0.01 or 10 m/km.

Where is it? If it exists, ir probably lies in the hydraulics of the concept." He added (Bretz, 1932, p. 83), ". . . we do not know enough about great flood mechanics to make any conclusions valid. . . . Hydraulic competency must be allowed the glacial streams, however much it may differ from that of stream floods under observation." In this section we review what has been learned about great flood mechanics.

PALEOHYDROLOGY AND PALEOHYDRAULICS

Several fortuitous circumstances combined to allow the relatively precise reconstruction of maximum flood flows that affected the region of the Channeled Scabland during the late Pleistocene. The flood paleochannels are remarkably well preserved. Post-flood geomorphic processes on the Columbia Plateau have done little to erase geomorphic evidence of the cataclysmic flows. Most important are the high-water indicators (Baker, 1973a), including small divide crossings, eroded channel margins, flood sediment, and ice-rafted erratics. These are used to plot high-water surface profiles along the paleoflood channels (Fig. 21). Such profiles define water-surface gradients that can be used in paleohydraulic calculations.

Flood Flow Calculations. The first paleohydraulic discharge estimate for flooding in the Channeled Scabland was made by Bretz (1925). Applying the Chezy formula,

$$Q = V A = C A R^{\frac{1}{2}} S^{\frac{1}{2}}, \qquad (1)$$

where Q is discharge in m³ sec⁻¹, V is mean flow velocity (m

sec⁻¹), A is cross-sectional area (m²), C is a roughness coefficient, R is hydraulic radius (m), and S is energy slope. Bretz estimated the peak discharge at Wallula Gap (Fig. 19) as 1.9×10^6 m³ sec⁻¹. This estimate probably erred on the low side, as Bretz (1925) noted, because of problems involved in applying experience on roughness in small-scale flows to the immense Missoula Flood discharges.

Another estimate of Lake Missoula outflow was made by Pardee (1942, p. 178–179). Pardee considered flow at Eddy Narrows in the lake basin itself. He considered the Chezy formula (1) and the Manning formula:

$$Q = V A = n^{-1} A R^{2/3} S^{\frac{1}{2}}, \qquad (2)$$

where n is the semi-empirical Manning roughness coefficient, the other parameters being defined as in equation (1). Pardee also observed that pebbles 76 mm in diameter were carried in the flood flow to form eddy deposits high above the channel floor at Eddy Narrows. From data provided by Hjulström (1935) he calculated a mean flow velocity associated with this particle size and multiplied by cross-sectional area to obtain a peak flow. The mean of these various procedures yielded a peak discharge estimate of 10.9×10^6 m³ sec⁻¹ and a mean flow velocity of 20 m sec⁻¹.

The most comprehensive paleohydraulic analysis of Missoula flooding to date was performed by Baker (1973a). His approach used the slope-area procedure (Dalrymple and Benson, 1967; Baker, 1973a, p. 17–19). The method accounts for nonuni-

TABLE 1. HYDRAULIC GEOMETRY FOR SOME MISSOULA FLOOD CHANNELS AT MAXIMUM STAGE

Location	Width (m)	Depth (m)	Velocity (m/sec^{-1})	Froude Number	Discharge (10^6 m^{3}sec^{-1})
Rathdrum Prairie	5000	175	20	0.5	17.5
Soap Lake	1500	100	30	1.0	4.5
Lind Coulee	2100	75	17	0.6	2.7
Wilson Creek	5000	80	15	0.5	6
Staircase Rapids	3000	40	30	1.5	3.6

form flow, thereby improving on estimates based on equations (1) and (2). However, the slope-area method is very sensitive to the high-water marks. Cross sections must be located where the high-water line can be specified.

Table 1 summarizes paleohydraulic parameters calculated by Baker (1973a) for various Missoula Flood paleochannels. Mean flow velocities ranged from 8 to 30 m sec^{-1} at peak flow stages. Constricted reaches in the western Channeled Scabland achieved the highest velocities. Such velocities would be impossible in relatively shallow river flow (less than 5 m deep) because the water would cavitate (Fig. 22). However, Missoula Flood flows were exceptionally deep (60 to 120 m). The steep water surface gradients (2 to 12 m km^{-1}) resulted in the exceptionally high mean flow velocities.

The scaling of Missoula Flood flows by depth and velocity is very important. Flow regime (Simons and others, 1965) is defined by the flow Froude Number F,

$$F^2 = V^2 (g D)^{-1}, \tag{3}$$

where g is the acceleration of gravity and D is the flow depth. Lower regime flow, common in tranquil rivers, is defined by F <1.0. Whereas upper regime flow, characterizing rapid or shooting flow in rivers, has F >1.0. The great depth of scabland channels predominantly resulted in lower regime hydraulics (Fig. 21). A few scabland constrictions achieved upper regime flow and cavitating phenomena (Fig. 22).

Flow Modeling. A relatively new approach to paleoflood reconstruction is the use of step-backwater computation of water-surface profiles (Chow, 1959, p. 274–280). This technique has been applied to studies of paleofloods in rigid-boundary bedrock canyon streams (O'Connor and others, 1986; Ely and Baker, 1985; Partridge and Baker, 1987; Webb, 1985). The major improvement achieved by this approach is a more accurate accounting of flow-energy losses associated with steady, nonuniform (gradually-varied) flow in irregular channels. Moreover, relatively sophisticated computer routines are available for the application of step-backwater analysis to geomorphic problems (Feldman, 1981; Hydrologic Engineering Center, 1982). These

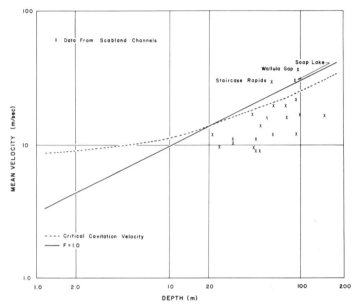

Figure 22. Velocity-depth diagram for Missoula Flood flows. The F = 1.0 line is calculated from equation (3), and the critical cavitation velocity is calculated according to the analysis by Baker (1979). At this velocity absolute pressure in the rapidly flowing water is reduced below the liquid vapor pressure according to Bernoulli's Law. Cavitation can lead to phenomenal erosive effects when vapor bubbles implode against flow boundaries. Note that scabland hydraulic data (Baker, 1973a) were mainly subcritical (F <1.0) and noncavitating. However, particularly pronounced constrictions (named data points) were supercritical (F >1.0) and cavitating.

methods deal with the unsimplified one-dimensional energy equation for flow profiles (see Fig. 23 for definitions):

$$z_1 + y_1 + \alpha_1 (V_1^2 0.5 g^{-1}) = z_2 + y_2 + \alpha_2 (V_2^2 0.5 g^{-1}) + h_e, \tag{4}$$

where α_1 and α_2 are velocity head coefficients at cross sections 1 and 2 (Chow, 1959) and h_e is the head loss between cross sections.

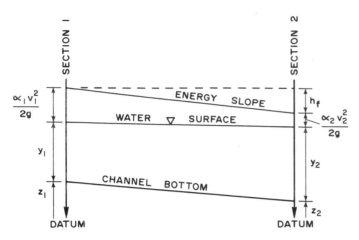

Figure 23. Schematic diagram illustrating the basic variables used in energy-balance flow modeling.

Equation (4) expresses the conservation of energy for gradually varied flow within an incremental reach of channel. The sum of a flow's potential and kinetic energy must equal that of a downstream cross section less any head loss (h_e) between sections. The computer programs generate accurate energy-balanced water-surface profiles associated with given discharges for reaches where the channel geometry is well characterized by multiple cross sections and where reasonable estimates of friction-loss coefficients (Manning's n and expansion-contraction loss coefficients) can be made. To determine paleoflood discharges, various discharges are routed through the reach until one is obtained that results in an energy-balanced water-surface profile that matches the profile defined by the geologic evidence. Uncertainties in the friction-loss coefficients have little effect on the computed profile with respect to the importance of the modeled discharge (Webb, 1985; Partridge and Baker, 1987).

Using these procedures, a modeling experiment was conducted for the Lake Missoula breakout reach, Rathdrum Prairie in northern Idaho (Fig. 24). Topographic information was obtained from the U.S. Geological Survey Greenacres, Coeur D'Alene, Rathdrum, Spirit Lake, and Athol 15 minute Topographic Quadrangles of Washington and Idaho. Following Baker (1973a), a Manning n value of 0.04 was used in the final analysis. Effective flow (flow in the downstream direction) was assumed to be confined to the main portion of the valley (Fig. 24).

Modeling various discharges through the present geometry of the Rathdrum Prairie with the U.S. Army Corps of Engineers HEC-2 Water Surface Profiles flow model (Hydrologic Engineering Center, 1982) indicates that a minimum discharge of approximately 17.5×10^6 m³ sec⁻¹ is required to exceed or equal the high-water mark indicators identified by Baker (1973a; Fig. 25). Considering reasonable uncertainties in the friction-loss coefficients and downstream flow conditions, numerous routing trials indicate that a plus-or-minus figure of about 2.5×10^6 m³ sec⁻¹ is appropriate for this discharge estimate.

The simulated flow profile is physically reasonable. At the downstream portion of the reach (cross sections 1 and 2), the water-surface profile is relatively flat and velocities are low (Fig. 25), reflecting backwater effects of hydraulic ponding in the Spokane Basin. Upstream, the flow profile is steeper and the velocities greater (Fig. 25), where water poured out of the Pend Oreille Lake basin adjacent to the breakout point. The flow is modeled as attaining critical conditions at the point where water entered the modeled reach (cross section 6), as may be expected for flows undergoing marked width contraction (Matthai, 1967).

Average velocities predicted by the flow model range up to 27 m/sec⁻¹ (Fig. 24) depending on local channel geometry. Channel bed shear stresses calculated from the modeled hydraulics vary from 4×10^4 dynes/cm² at section 1, to 3×10^5 dynes/cm² at section 6. These are preliminary calculations, but future work may relate such results to preserved flood features, providing a more precise quantification of depositional and erosional processes associated with catastrophic flooding.

Simulation Modeling. A completely independent approach to modeling Missoula Flood discharges involves consideration of its ice dam. Clarke and others (1984) developed a computer simulation model of Missoula outburst floods based on glaciological equations that govern enlargement by water flow of a tunnel penetrating through the ice dam. By analogy to observed jökulhlaups from modern ice-dammed lakes (Mathews, 1973; Nye, 1976), the model generates outflow hydrographs for a variety of failure scenarios. Predicted peak flows range from approximately 3 to 14×10^6 m³ m³ sec⁻¹, and durations range from 8 to 20 days. It is interesting that the largest figure is close to flow modeling of Rathdrum Prairie discharge derived from geomorphic evidence of high-water marks.

Table 2 summarizes the physical parameters of glacial Lake Missoula and its outburst flooding. These figures illustrate the immense size of the physical system that produced the Channeled Scabland.

Sediment Transport. The Missoula floods exhibited phenomenal sediment transport capability as evidenced by boulders entrained by the flows (Fig. 2). This is to be expected, since the peak flows had immense velocities (Tables 1 and 2). They also had extremely high bed shear stresses, τ (dynes/cm²), given by

$$\tau = \gamma \, D \, S, \qquad (5)$$

where γ is the specific weight of the fluid (10^3 dynes/cm³ for clear water), D is the flow depth (cm), and S is the energy slope. The combination of these factors yielded phenomenal power per unit area of bed, ω, calculated as

$$\omega = \frac{\gamma \, Q \, S}{W} = \tau \, V, \qquad (6)$$

where Q is discharge, V is mean flow velocity, and W is bed width.

The concept of stream power is intimately tied to sediment transport capability (Bagnold, 1966, 1977, 1980). It can also be

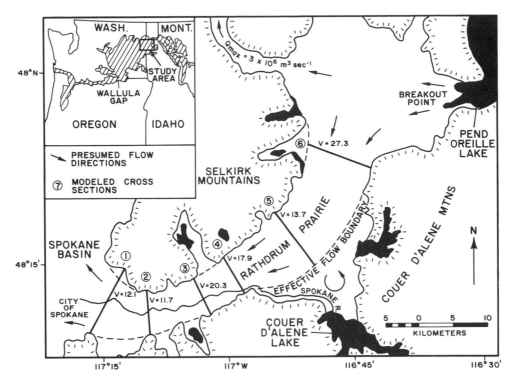

Figure 24. The modeled reach of Rathdrum Prairie and its relationship to local and regional Channeled Scabland geography. Six cross sections were utilized to characterize the channel geometry within the reach. Mean flow velocities (V) are indicated for each of the numbered cross sections. The maximum potential discharge (Q_{max}) for the flow-pathway to the north was calculated with the Manning Equation, assuming (1) an energy slope similar to that predicted for the Rathdrum Prairie reach and (2) that the entire cross-sectional area of the valley was conveying water. The inset map shows the generalized extent of glacial Lake Missoula in Montana and the generalized extent of cataclysmic flooding downstream of the lake (cross hatching).

thought of as the driving term in a threshold criterion that is balanced by resisting power (Bull, 1979). Unfortunately, it is not yet possible to express resisting power in the same physical terms as stream power. Moreover, the use of power to predict sediment transport is limited by the inability to adequately define the efficiency of power utilization and its partitioning in accomplishing work by natural streams.

Typical values of τ and ω for rivers in flood are 10^2 dynes/cm^2 and 10^4 ergs/cm^2-sec respectively. However, peak Missoula Flood flows, reconstructed by Baker (1973a), yield values of up to 10^5 dynes/cm^2 and 10^8 ergs/cm^2-sec respectively. In other words, the Missoula Flood peak flows, by these measures, were three to four orders of magnitude more intense than floods of major rivers, such as the Amazon or Mississippi. It is therefore not surprising that Bretz's critics in the 1920s and 1930s could not believe rivers could accomplish what Bretz proposed. Bretz was describing a phenomenon on an immensely different physical scale than that envisioned by his critics.

The effects of such dynamic flood conditions are especially noteworthy as to the sizes of sediment carried in the various modes of transport; that is, as bedload, suspension, and washload. As discussed by Komar (1980), it is theoretically possible to

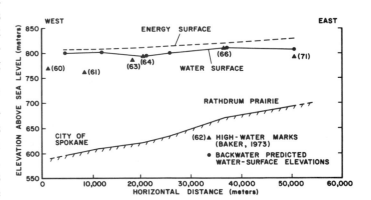

Figure 25. Longitudinal profile of the modeled reach with step-backwater-predicted water-surface and energy-surface profiles for a discharge of 17.5×10^6 m^3 sec^{-1}. This is the minimum discharge that results in a computed profile that equals or exceeds the geologic stage indicators identified by Baker (1973a) over the length of the reach.

TABLE 2. PHYSICAL PARAMETERS FOR THE LATE PLEISTOCENE MISSOULA FLOODS

Lake Missoula, Montana		
Maximum Lake Elevation	1265 m	(Pardee, 1942)
Maximum Lake Volume (approximate)	2200 km^3	(Pardee, 1942)
Maximum Water Depth at Ice Dam	635 m	(Pardee, 1942)

Lake Outburst Parameters		
Simulated Outflow discharges	2.7 x 10^6 m^{3}sec^{-1}	(Clark and others, 1984)
	14 x 10^6 m^{3}sec^{-1}	
Flow Modeled Outburst Discharges	17.5 x 10^6 m^{3}sec^{-1}	(this paper)
	($\pm$2.5 x 10^6 m^{3}sec^{-1})	
Slope-Area Discharge	21 x 10^6 m^{3}sec^{-1}	(Baker, 1973a)

Peak Flow Dynamics of the Soap Lake Constriction	
Water-Surface Slope	0.01
Flow Depth	100 m
Bed Shear Stress, τ	10^5 dynes/cm^2
Mean Flow Velocity, V	30 m/sec
Power per Unit Area, ω	3 x 10^8 ergs/cm^2-sec

extend the domains of these transport modes to evaluate intense floods. Figure 26 represents this approach. Note that typical floods having $\tau = 10^2$ dynes/cm^2 could entrain and transport grains of about 1 cm diameter, while sediment finer than 1 mm would be in suspension, and the washload would consist of sediments finer than 0.1 mm. A truly exceptional flood with $\tau = 10^4$ dynes/cm^2 could transport gravel as coarse as 10 to 30 cm in suspension and sand-sized diameters of 0.4 to 0.8 mm and finer in the washload. The increase in τ to 10^5 to 10^6 dynes/cm^2 for the Missoula Flood is seen to extend entirely off scale in Figure 26, the result indicating that even boulders could have been transported in suspension with the sand-sized grains in the washload. Estimates based on this analysis of sediment transport in clear water actually are likely to be conservative, especially for the Missoula floods, where erosion of the loess of the Palouse Formation could have resulted in high concentrations of washload. These high concentrations reduce the settling velocities of the coarser sediments so that they could be transported more readily in suspension than indicated by the curves of Figure 26. Such analyses of the modes of sediment transport confirm that the Missoula floods had a considerable capacity for eroding and transporting seidments.

Stream Power and Flood Erosion. The Missoula floods constituted the largest known discharges of fresh water on Earth. However, discharge is not the critical factor in flood erosion. The spectacular erosion of rock channels in the Channeled Scabland was largely accomplished in a matter of hours during peak flood flows (Baker, 1973a, 1981). Disbelief that fluvial flows, even such large flows, could accomplish so much in a short time has contributed to a view that multiple large cataclysmic floods must have occurred. Although multiple outbursts of Lake Missoula did occur during the late Pleistocene (Waitt, 1980, 1984, 1985; Baker

and Bunker, 1985), many of those may have been too small to accomplish significant geomorphic work. It is proposed here that the maximum flood discharges indicated by the paleoflood high-water evidence could have generated sufficient stream power to produce the indicated erosion. Much of the erosion could have been accomplished in one or a few exceptionally large floods.

Power multiplied by time gives potential work. For the Missoula Flood, with power per unit area of 10^8 ergs/cm^2-sec applied for about 3 hours (10^4 seconds), the potential work is 10^{14} ergs/cm^2. By comparison, the same potential work would require 10^{14} seconds (3×10^5 years) of continual glacial action at a power per unit area of 1 erg/cm^2-sec. Even a major river in flood, such as the Mississippi, only generates about 10^4 ergs/cm^2-sec. The river would have to be in continuous flood for 10^{10} seconds (300 years) to produce equivalent work. However, the river probably achieves this power only 1 percent or less of the time, so its required time would be at least 3×10^4 years.

Of course, accomplished work of erosion depends on the efficiency of power expenditure, a parameter that is not well known in geomorphic applications. Nevertheless, assuming a range of perhaps an order of magnitude in efficiency factors, we see that the Missoula Flood action can indeed accomplish as much physical work in hours that would require thousands of years of action by other geomorphic agents, such as glacial ice.

An additional consideration is that of thresholds of power necessary to produce erosion (Bull, 1979). Agents generating low power during most of their operation may only rarely achieve critical values necessary for significant work. Thus, rivers may only achieve necessary thresholds to erode bedrock during large, rare floods (Baker, 1977). The phenomenal power levels achieved by the Missoula Flood peak flows seem to have exceeded critical power thresholds for the erosion of intact rock.

EROSIONAL AND DEPOSITIONAL LANDFORMS

The Channeled Scabland contains a wealth of landforms characterizing deep flood flows in bedrock stream systems. Somewhat similar features can be recognized in smaller bedrock-canyon streams affected by precipitation-induced floods (Baker, 1984). Moreover, many of the scabland landforms serve as analogs to features in the outflow channels of Mars (Baker and Milton, 1974; Baker, 1978b, 1982).

Scabland Erosion Complexes. Experimental studies of fluvial erosion in simulated bedrock show that a sequence of erosional forms develops with time (Shepherd and Schumm, 1974). First to appear in the experiments are longitudinal grooves, potholes, and transverse erosional ripples. The grooved surface is subsequently incised to form a conspicuous inner channel that may migrate upstream by knickpoint recession. The inner channel may evolve to form an elongate basin, incised below base level (Schumm and Shepherd, 1973).

A roughly analogous sequence seems to characterize scabland channelways. Figure 27 shows a series of hypothetical cross sections that illustrate how Missoula floodwater eroded the Columbia Plateau basalt. The first flood water encountered a plateau capped by loess hills (top diagram in Figure 27). The phase (I) was followed by a series of changes as the underlying basalt was exposed and eroded (lower diagrams in Figure 27). First, the high velocity water quickly exposed the underlying basalt, leaving occasional remnants of streamlined loess (Phase II). The surface of the uppermost basalt flow was then encountered. This probably yelded to groove development, possibly associated with longitudinal roller vortices that manifested macroturbulent structure in the flow field (Baker, 1979). Continued incision would encounter layers in the basalt that would display very different resistance qualities to the turbulent flood water. The first exposure of well-developed columnar jointing, perhaps at the top of a flaring colonnade along the irregular cooling surface, could introduce a very different style of erosion (Phase III). Large sections of columns could now be removed at this site with the simultaneous development of vertical vortices. Such vortices, called "kolks," are thought to characterize especially intense flood flow phenomena (Matthes, 1947; Baker, 1984). With enlargement and coalescence of the resultant potholes, the surface assumed the bizarre butte and basin topography that characterizes much of the Channeled Scabland (Phase IV). The eventual topographic form was the development of a prominent inner channel (Phase V). Such inner channels may have been initiated at downstream structural steps in the basalt, which then migrated headward by cataract recession.

An excellent example of this sequence can be seen at Dry Falls (Figs. 11 and 28). Dry Falls is a great cataract 5.5 km wide and 120 m high that formed in the Grand Coulee, largest of the scabland channelways (Bretz, 1932). The falls display a series of horseshoe-shaped headcuts with plunge-pool lakes in their alcoves (Figs. 11 and 28). A prominent inner channel extends downstream, comprising a series of closed basins containing

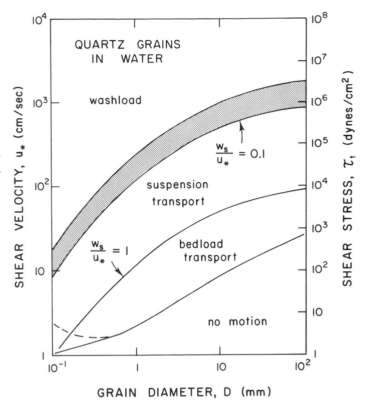

Figure 26. Curves of bed shear stress τ required for the transition of a certain grain size from bedload to suspension transport to washload. Curve derivation is discussed by Komar (1980).

lakes. The whole complex was excavated as cataclysmic flood water incised the zone of fractured basalt along the Coulee monocline (Bretz, 1932, 1969). The basalt surface north of the falls is marked by a prominent series of longitudinal grooves, aligned with the paleoflow direction of the cataclysmic flood and cut across structural trends (Baker, 1981). The grooves measure up to 3 m in height and are spaced 30 to 60 m apart (Fig. 29). These lead right up to the lips of the cataract.

Groove development is not especially common in the Channeled Scabland, probably because it was relatively easy for the flood flows to erode the basalt entablature layers to expose the underlying columnar jointed rock. Consequently, butte-and-basin topography (Fig. 30) prevails in most scabland tracts. Potholes, rock basins, and minor channels characterize such surfaces. The microtopography can be among the most bizarre of any landscape on Earth. As described by Bretz (1932, p. 26–28): "The channels run uphill and downhill, they unite and they divide, they deepen and they shallow, they cross the summit, they head on the back-slopes and cut through the summit; they could not be more erratically and impossibly designed."

Streamlined Erosional Residuals. Bretz (1923a, p. 624–626) first recognized that hundreds of isolated loess hills in the scabland tracts of the eastern Channeled Scabland pos-

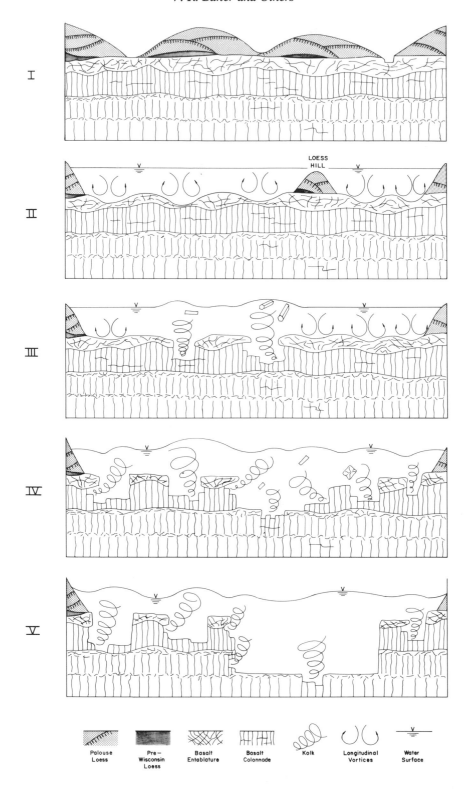

Figure 27. Hypothetical sequence of flood erosion at a typical scabland divide crossing. See text for discusion.

Figure 28. Dry Falls cataract complex.

Figure 29. Longitudinal grooves immediately upstream of Dry Falls cataract.

Figure 30. Butte-and-basin scabland development in Lenore Canyon.

sessed a remarkable shape. Their steep, ungullied bounding hill-slopes converge to prow-like upstream terminations, yielding an overall streamlined or airfoil shape (Fig. 16). Baker (1973b) compared the planimetric shapes of these hills to the lemniscate form (Fig. 31) used by Chorley (1959) in a similar comparison with drumlins. The lemniscate is a streamlined geometric shape controlled by a dimensionless parameter K,

$$K = \frac{\pi L^2}{4 A_s}, \qquad (7)$$

where L is the length of the form and A_s is its area. In a detailed analysis of the lemniscate, Komar (1984) found that it provides a close representation of symmetrical airfoils (Joukowski Sections) and can therefore serve in analyses of streamlined landforms. The Scabland Islands have a close visual resemblance to lemniscates having the same length-to-width L/W proportions (Fig. 31). Quantitative studies of their morphology by Baker and Kochel (1978) and Baker (1979) further confirm this correspondence, the analyses revealing significant relationships involving hill lengths, widths, and areas (Fig. 32) that are consistent with the mathematics of the geometric lemniscate (Komar, 1984).

Streamlining has the purpose of reducing the drag or resist-ance to a flowing fluid, many river islands (Fig. 33) as well as the Scabland hills achieving that troutlike form. Baker (1979) hypo-thesized that the streamlined shape of the erosional residuals in the Channeled Scabland and in Martian outflow channels devel-oped their characteristic shapes and length-to-width L/W pro-portions to minimize the flow resistance. Direct measurements on airfoils in both air and water demonstrate that such a minimiza-tion of the drag occurs for proportions in the range L/W = 3 to 4 (Komar, 1983). The Scabland residuals were found by Baker and Kochel (1978) and Baker (1979) to have an average L/W = 3.25, the Martian streamlined landforms having an average L/W = 3.25, confirming that these streamlined landforms had been eroded to shapes and proportions that minimized the drag. Iso-lated islands in the Columbia River and other major rivers have similar proportions, appearing much like the da Vinci drawing of Figure 33 (Komar, 1984). In a series of flume experiments, Komar (1983) examined the processes by which water erosion shapes streamlined landforms to achieve minimum drag. These several studies of the geometry and origin of streamlined land-forms demonstrates a marked consistency between those on Earth and Mars, thus supporting the hypothesis that the Martian streamlined islands, and thus the outflow channels, were eroded

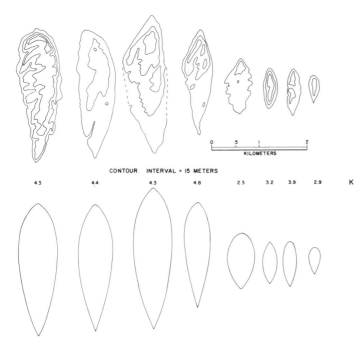

CONTOUR INTERVAL = 15 METERS

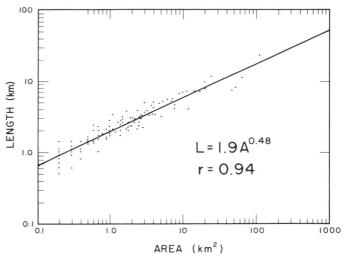

$$L = 1.9 A^{0.48}$$
$$r = 0.94$$

Figure 32. Relationship of length to area for planimetric shapes of streamlined scabland loess hills. Note that the approximate regression result, $\ell^2 = 4A$, yields $K = \pi$ when substituted into equation 7.

Figure 31. Comparison of typical streamlined scabland loess hills (top) to equivalent lemniscate loops calculated from Chorley's (1959) K-factor (equation 7). The lemniscates approximate streamlined airfoil shapes (Komar, 1984).

by large-scale floods, probably in a manner analogous to that which carved the Channeled Scabland.

Scour Marks. Obstacles to flow in scabland channels resulted in deformation of flood flow streamlines, thus producing distinctive scour marks. The hydrodynamics of such scour involved vortex formation at the obstacle front and in its wake. The process at the front of a blunt-nosed obstacle can be envisioned as separation of the flow boundary layer and a rolling of streamlines ahead of the obstacle to form a horseshoe-shaped vortex (Fig. 34A). Both the horseshoe vortex and the wake vortex may produce distinctive scour holes near large scabland boulders (Fig. 34B. Whether deposition or erosion occurs depends on the Reynolds number N, where

$$N = \frac{\rho \, Va \, T}{\mu}, \qquad (8)$$

where ρ is the fluid density, Va is the approach velocity, T is the obstacle diameter, and μ is the dynamic viscosity. At a high Reynolds number (generally indicating high flow velocity), the wake vortex system scours in the lee of obstacles. However, at a low Reynolds number (low flow velocity), the wake vortex is weaker than the surrounding flow field. Entrained sediment in the flood flow is then conveyed into the wake region by this velocity gradient. The result is deposition in the lee of the obstacle. Such features are especially common in scabland channels and are termed "pendant bars" (Fig. 35).

Bars. Most rivers that transport coarse gravel assume a braided pattern and are dominated by linguoid and transverse bars that are low in profile. Such rivers tend to have high width-to-depth ratios for their channel cross sections. Scabland channelways, in contrast, were relatively narrow and deep, a quality restricted to resistant-boundary streams, such as those occurring in bedrock (Baker, 1984). A consequence of this flow phenomenon is the development of bars that are tens of meters high, dominated internally by foreset bedding (Baker, 1973a). The bars occur in complexes associated with scabland erosion, streamlined residual hills, and giant current ripples (Fig. 36). The scabland bars reflect a variety of flow conditions related to their emplacement. Many have superimposed giant current ripples that can be related to hydraulic conditions prevailing at their time of formation.

Gravel Waves. Bretz and others (1956) applied the name "giant current ripples" to the large-scale transverse gravel waves of the Channeled Scabland (Fig. 37). These bedforms have chords (spacings) that generally range from 20 to 200 m and heights that range from 1 to 15 m. One of the most impressive examples of gravel waves (Fig. 38) covers an area of approximately 40 km² near Spirit Lake, Idaho, immediately downstream of the breakout point of glacial Lake Missoula (Fig. 24). In plan view the gravel waves have crest forms that are analogous to the smaller sand ripple forms of rivers (Fig. 39). However, the scabland bedforms are composed of pebbles, cobbles, and boulders (Fig. 40). The internal structure is dominated by foreset bedding in open-work gravel. Gravel-wave stoss slopes are armored with cobbles and boulders.

The gravel waves emplaced by Missoula flooding can be classed hydraulically as dunes (Allen, 1982, p. 330). Because of this, it has been suggested that they be named "giant current dunes" (Waitt, 1984). Like the common flow regime dunes of

V. R. Baker and Others

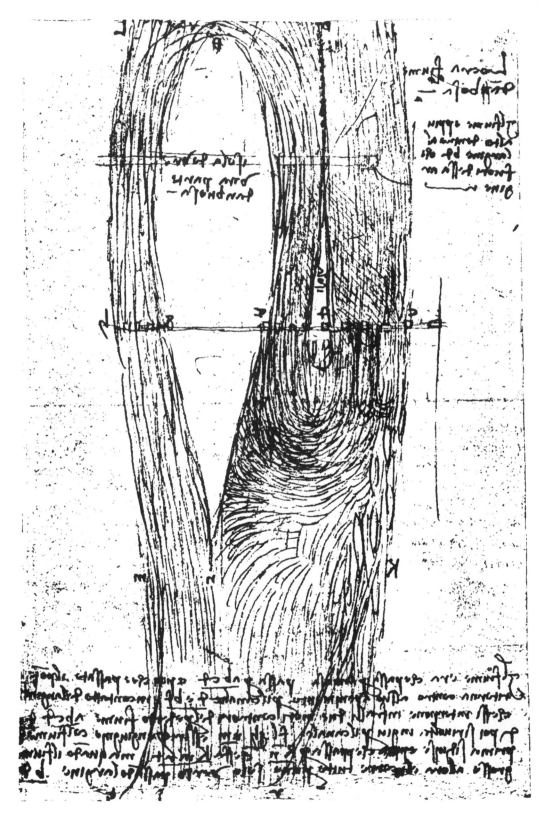

Figure 33. Drawing by Leonardo da Vinci of a streamlined island in the Loire River at Amboise, France. The ratio of island length to width is 4:1, the same as noted in streamlined islands by Baker (1979) and Komar (1983, 1984).

(A)

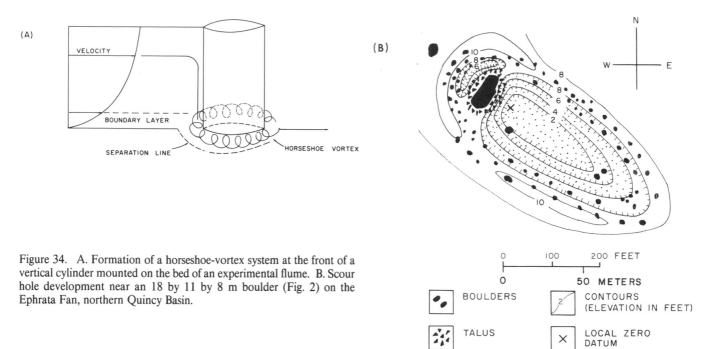

(B)

Figure 34. A. Formation of a horseshoe-vortex system at the front of a vertical cylinder mounted on the bed of an experimental flume. B. Scour hole development near an 18 by 11 by 8 m boulder (Fig. 2) on the Ephrata Fan, northern Quincy Basin.

Figure 35. Pendant bar in the Snake River Canyon immediately downstream from the Palouse River confluence. Note the giant current ripples (gravel waves) on the bar surface and the railway bridge for scale.

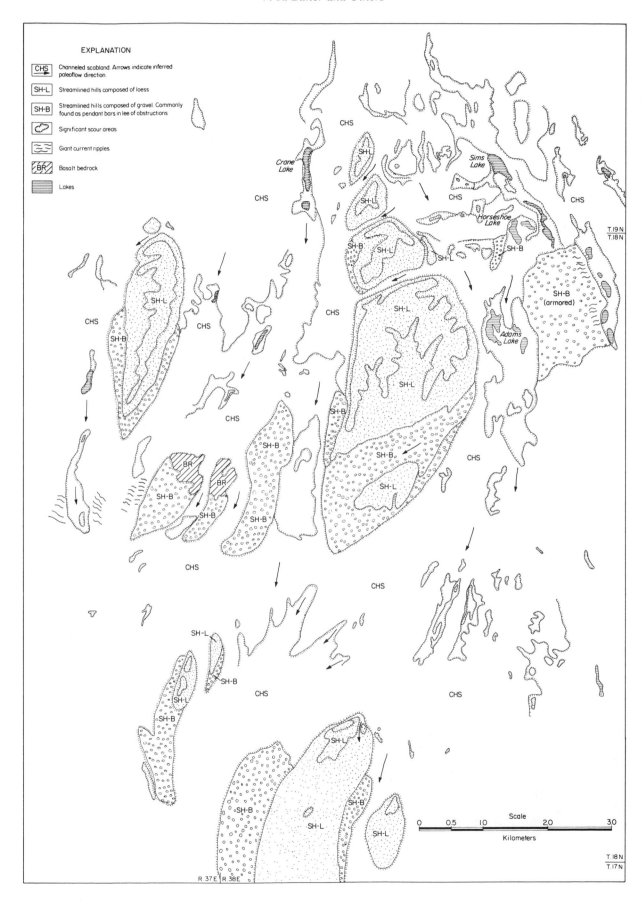

Figure 37. Giant current ripples on the great flood bar at Marlin, Washington. The bar occurs on a preflood divide between Crab Creek and Canniwai Creek, showing that Missoula Flood flows reduced these preflood valleys to channel-floor topography. The large-scale gravel waves average 60 m in spacing and 2 m in height. They developed beneath a maximum floodwater depth of 60 m and at a mean flow velocity of 12 m/s.

sand-bed streams (Simons and others, 1965; Harms and Fahnestock, 1965), the scabland gravel waves are scaled to flow depth (Baker, 1973a). However, unlike sand-bed fluvial dunes, the scabland gravel waves do not have smaller ripples superimposed upon them. The sediment is too coarse for the formation of small-scale ripples. Indeed, the complexity of bedforms in gravel-bed streams is such (Smith, 1978) that it may be useful to retain a separate category for large-scale gravel mesoforms that develop in relatively narrow deep fluvial channels. Similar bedforms have been observed following major floods on bedrock rivers in Texas and Australia (Baker, 1984). Nevertheless, because of the role of these bedforms in the Spokane Flood controversy, perhaps it is just as well to bow to historical precedent and retain the term "giant current ripple." Using this name certainly does little to worsen the already confused state of bedform terminology.

Figure 36. Geomorphic map of the Macall area in the central Cheney-Palouse scabland tract. Note the association of gravel bars, streamlined loess hills, scoured basalt, and giant current ripples (gravel waves).

The hydraulic significance of the giant current ripples is illustrated in Figure 41. By comparing ripple dimensions to maximum flood paleohydraulic parameters, Baker (1973a) showed that stream power per unit area of stream bed, equation 6, best explained ripple morphology. The remarkable consistency in scale to the hydraulics of the scabland paleofloods indicates the basic physical controls on the most spectacular flood known on Earth.

THE CHANNELS OF MARS

In 1972, the Mariner 9 spacecraft orbiting Mars returned pictures of immense channels and valleys dissecting the planet. The resulting controversy over the origin of those features (Baker, 1985) has many parallels to the controversy over the origin of the Channeled Scabland. Indeed the channels of Mars, like the Channeled Scabland, constitute an important geomorphic anomaly. The search for their origin is one of the great geomorphic problems of our time.

Figure 38. Giant current ripples near Spirit Lake, Idaho. These gravel waves average 85 m in spacing and 4 m in height.

The term "channel" is properly restricted to one class of Martian troughlike landforms that display at least some evidence for large-scale fluid flow on their floors. The principal landform of interest here is the outflow channel (Sharp and Malin, 1975), which shows evidence of flows emanating from zones of regional collapse known as "chaotic terrain" (Fig. 42). The outflow channels are immense, as much as 100 km wide and 2,000 km in length. It was recognized shortly after their discovery that they possessed a similar suite of bedforms and morphological relationships to what was exhibited in the Channeled Scabland (Baker and Milton, 1974). Included are streamlined uplands (Fig. 43), longitudinal grooves and inner-channel cataracts (Fig. 44), depositional fan complexes (Fig. 45), anastomosis (Fig. 42), scour marks, and others (Baker, 1982, 1985). Particularly important are the scour marks (Baker, 1978b) and the necessity for the responsible fluid to possess a free upper surface (Fig. 46). The entire assemblage of landforms can only be explained by the action of a fluid with physical properties similar to turbulent water. The present consensus is that water in concert with entrained debris and ice was the primary agent of outflow channel genesis (Mars Channel Working Group, 1983).

Hydraulics. The paleohydraulics of Martian outflow channels can be investigated in a manner similar to that discussed for the Channeled Scabland. For example, a reach on one outflow channel, Kasei Vallis, was analyzed by flow modeling (Fig. 47). It was assumed that canyon depths (calculated by shadow measurements on Viking Orbiter spacecraft images) correspond to maximum probable depths of paleoflood water. Channel slope was established at 2.5 m/km from the regional radar data (Lucchitta and Ferguson, 1983). Roughness (Manning's n) was taken as 0.02, or about one-half a reasonable terrestrial value. For six cross sections (Fig. 47) a discharge of 20×10^6 m^3 sec^{-1} results in a modeled water-surface profile that nearly matches the canyon depths. In general, the channel slopes, discharges, and Froude numbers are remarkably similar to the maximum Missoula Flood values calculated by slope-area (Baker, 1973a) and flow-modeling procedures, as described above.

Implications. The outflow channels of Mars raise numerous questions. How could aqueous erosion have been generated on such an immense scale? How could so much water have been generated to form the channels and not be evident on the surface today? These questions are stimulating current research. Various

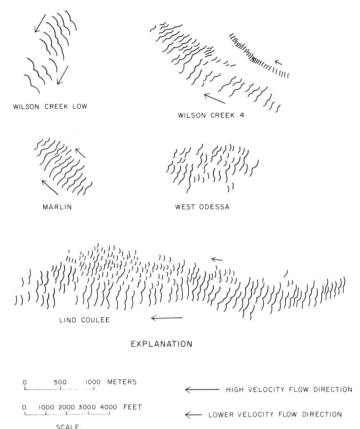

Figure 39. Plan views of giant current ripple sets. The patterns were traced from aerial photographs and show a pronounced decrease in bedform size and spacing away from the thread of maximum flood flow velocity. (From Baker, 1973a, Fig. 46.)

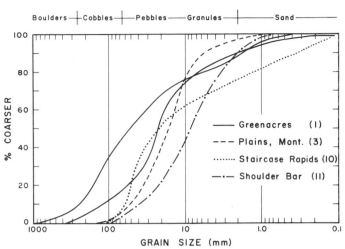

Figure 40. Grain-size distributions for sediment comprising giant current ripples in the Channeled Scabland. The numbers for the ripple sets are keyed to locations and descriptions in Baker (1973a, Appendix II).

investigators have evaluated channel formation in terms of sediment transport theory (Komar, 1979), ice-covered rivers (Wallace and Sagan, 1979; Carr, 1983), streamlining (Baker, 1979; Komar, 1983), and cavitation and macroturbulence (Baker, 1979). Theories for fluid release have been proposed by Maxwell and others (1973), Masursky and others (1977), Carr (1979), Nummedal and Prior (1981), and Lucchitta and Ferguson (1983). Despite the ingenuity of the various fluid-release proposals, none has yet been demonstrated by the available data. The lack of an identified source for the Martian floods reminds one of the scabland debate of the 1920s and 1930s when the source of the "Spokane Flood" was unknown (Baker, 1978a).

Although many aspects of channel and valley development on Mars remain unexplained, the continued study of this geomorphic puzzle has much to offer our science. The problem—the origin of valleys—is the same fundamental question that inspired James Hutton and John Playfair to establish a scientific basis for the study of landforms. We now have two planets on which to compare the development of fluvial landscapes.

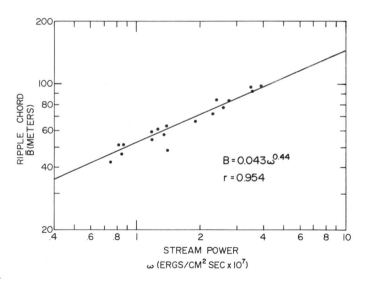

Figure 41. Mean ripple chord ($\bar{B}$) as a function of stream power (ω) for giant current ripples of the Channeled Scabland (modified from Baker, 1973a, Fig. 54).

Figure 42. Complex of outflow channels south of Chryse Planitia showing chaotic-terrain source areas (C) and flow directions (arrows). Note the overall anastomosing character of channels eroded into cratered plateau uplands. Locally, small remnants of eroded plateau materials have streamlined shapes (JPL Viking Orbiter mosaic 211-5821.)

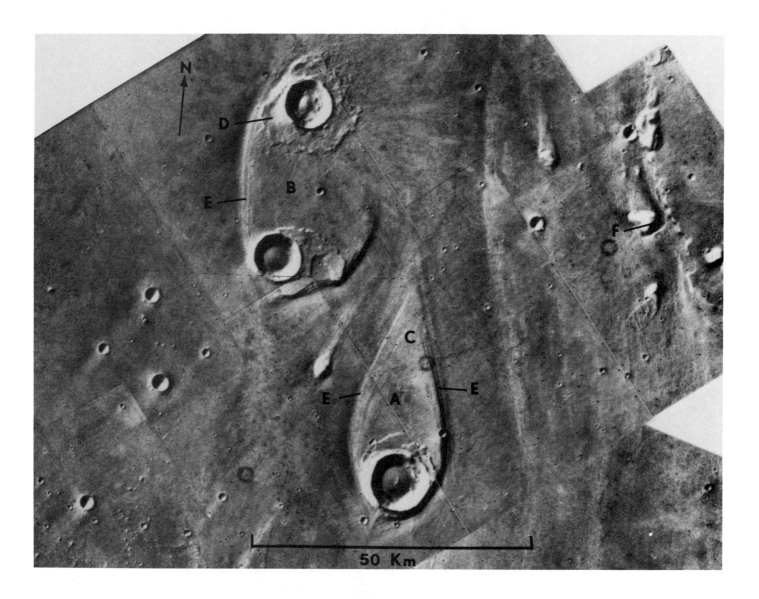

Figure 43. Streamlined uplands (A, B) at the mouth of Ares Vallis in Chryse Planitia. Note the tapered downstream tail (C) and the erosion of ejecta blankets from upstream crater (compare noneroded crater at D). The responsible fluid flowed north and scoured bedrock as shown by the terracelike benches on the upland flanks (E). A prominent scour hole surrounds the small island at F. The scarp surrounding island A is about 600 m high, and that surrounding island B is about 400 m high. (JPL Viking Orbiter Mosaic 211-4985.)

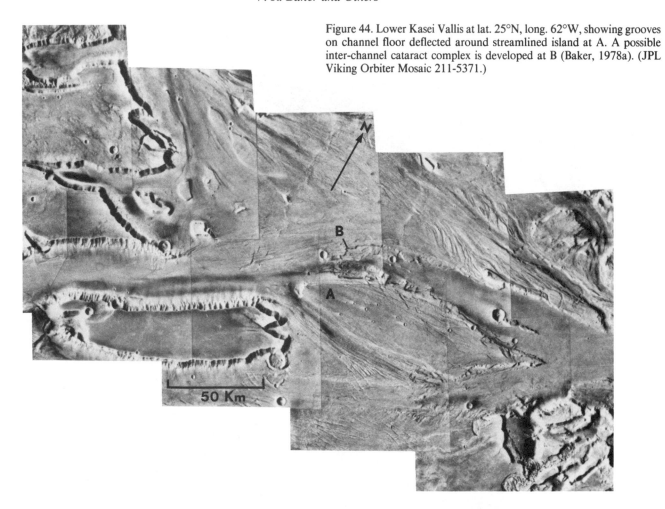

Figure 44. Lower Kasei Vallis at lat. 25°N, long. 62°W, showing grooves on channel floor deflected around streamlined island at A. A possible inter-channel cataract complex is developed at B (Baker, 1978a). (JPL Viking Orbiter Mosaic 211-5371.)

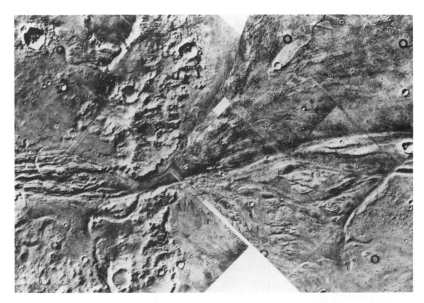

Figure 45. Fanlike complex of probable flood bars (right) at mouth of constriction in Maja Vallis (left). Flow from left to right debouched from the constriction onto the Chryse Planitia lowland. A fan complex developed that was subsequently scoured by flow in a manner similar to scouring of the Ephrata Fan of the Channeled Scabland. This Viking Orbiter mosaic depicts a scene approximately 150 km across.

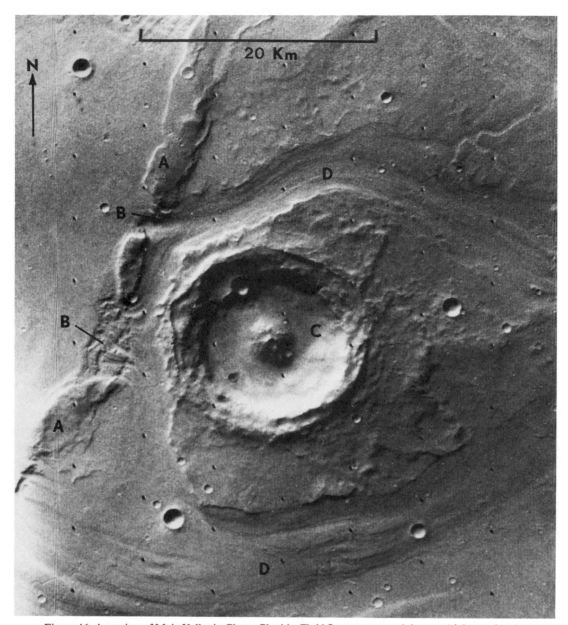

Figure 46. A portion of Maja Valles in Chryse Planitia. Fluid flow was toward the east (right on photo). Fluid ponded west of the marelike ridge (A) and spilled through gaps (B) as it overflowed low points in the ridge. It also spilled around crater (C), scouring channels (D) in its ejecta blanket. (Viking Orbiter picture 20A62.)

Figure 47. Sketch map of a portion of North Kasei Vallis (lat. 27°N, long. 70°W) prepared from JPL Viking Orbiter Mosaic 211-5882. Note locations of cross sections used in flow modeling of this outflow channel.

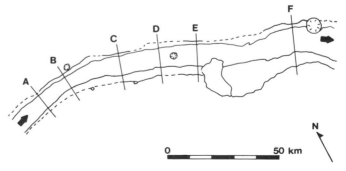

NORTH KASEI CHANNEL

THE SNAKE RIVER PLAIN

Ronald Greeley

INTRODUCTION

The geological exploration of the solar system has advanced from earth-bound telescopic views of the Moon to sophisticated data returned via spacecraft from most of the major solid-surface objects in the solar system. Although initially studied by astronomers, analysis of planetary surfaces now falls within the realm of the geologist. However, such studies are not without controversy, particularly as the traditional method for testing geological hypotheses—field checking—is presently difficult, if not impossible, for most planetary objects. Consequently, the goal of most planetary geology studies is to understand surface processes in fundamental terms that can be extrapolated to a wide variety of planetary environments. Such studies typically involve laboratory simulations, theoretical modelling, and interpretation of spacecraft data. In addition, an important element of planetary geology is the study of terrestrial analogs—features (and processes) on Earth that are similar to those on other planets.

Basaltic volcanism appears to have taken place on all the terrestrial planets. Evidence for basalts includes lava samples returned from the Moon, photogeologic interpretations of lava flows on Mars and Mercury, and analysis of data from landed spacecraft and radar images from the Soviet Venera missions to Venus. Thus, since the mid-1960s, planetary geologists have focused on study of basaltic regions on Earth as a means of understanding similar volcanic terrains on other planets (Balsaltic Volcanism Study Project, 1981).

The Snake River Plain (Fig. 48) has been particularly important in these studies (King, 1982). The plain forms a broad arch across the southern part of Idaho, extending more than 500 km eastward from the border with Oregon to the Yellowstone Plateau. The youngest flows (about 2,000 yrs B.P.) preserve key surface features and are only sparsely covered with soils and vegetation. Vent and vent-related features are well exposed, and in several places, such as canyons cut by the Snake River, flows can be seen in cross section. In addition, an overprint of aeolian features, modification by the catastrophic flood from Pleistocene Lake Bonneville, and localized ground-sapping processes make the plain a good analog for Mars where the interplay of volcanic aeolian, and fluvial events can be documented (Carr, 1981).

This paper reviews those aspects of the plain that have been important in planetary geology, including the influence that the plain has had on understanding geomorphology of basaltic terrains.

BASALTIC FEATURES ON THE SNAKE RIVER PLAIN

The factors governing the form of volcanoes have long been of interest (e.g., Cotton, 1952). Recently, this interest has been

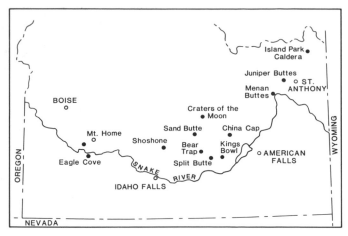

Figure 48. Map of Snake River Plain showing principal features described in text.

stimulated by the discovery of extremely long lava flows on the Moon (Fig. 49) and enormous volcanoes on Mars (Fig. 50). Consequently, several authors have analyzed the factors that appear to influence volcanic landforms, particularly in other planetary environments (Greeley, 1977; Head and others, 1981; Whitford-Stark, 1982). Because volcanism on most planets appears to be dominated by mafic magmas, the emphasis of most of these studies has been on basaltic volcanism. In compiling data on the major basaltic region on Earth, it became apparent that the two-fold classification of basaltic volcanism into Hawaiian styles of eruption and flood eruptions was inadequate for some areas, especially the Snake River Plain. Thus, an intermediate type of basaltic volcanism was proposed (Greeley, 1976, 1982a, 1982b; Table 3). Termed "plains" volcanism (typified by the Snake River Plain), this style is represented in parts of Iceland, the Modoc Plateau of California, parts of the Deccan Plateau, and on other planets.

Flood eruptions typically involve very high rates of effusion (about 1 km^3/day) from fissures that produce widespread, simple cooling units exceeding about 10 to 20 m in thickness (Fig. 51). Features such as spatter cones, lava tubes, and channels either do not form or are seldom preserved. In contrast, large shield volcanoes, formed from Hawaiian-type eruptions, involve moderate rates of effusion (about .01 km^3/day) from central vents that produce multiple flow units 3 to 5 m thick. Flows are often emplaced through lava tubes and channels; mild fire-fountaining may produce cinder cones, but these constitute a relatively small volume in comparison to the flows.

"Plains" volcanism shares characteristics of both flood and Hawaiian eruptions. Like Hawaiian volcanism, plains volcanism involves multiple, thin (3 to 5 m) flow units erupted from central vents and minor fire fountaining to produce cinder cones. Like flood eruptions, the vents are often aligned on rift zones, and some of the flows are fissure fed. The surface of flow accumula-

Figure 49. Mosaic of Apollo 17 images showing a series of complex lava flows on the Moon in the Imbrium basin. These flows, which appear to originate from fissures (out of the mosaic to the lower left), can be traced more than a thousand kilometers out of view toward the upper right. Despite their great length these flows are less than 40 m thick and evidently were extremely fluid lavas that erupted at very high rates of effusion (Apollo 17 photographs AS17-155-23714 through 23716).

Figure 50. Viking Orbiter image of Olympus Mons, Mars, one of the largest shield volcanoes (600 km in diameter) in the Solar System. This volcano is bounded by a scarp that in places rises higher than 3 km and has a central complex caldera more than 80 km across. Despite its great size, high resolution images show that the volcano is composed of individual lava flows, many of which were fed through lava tubes and channels, similar to shield volcanoes on Earth (Viking Orbiter frame 646-A-28).

TABLE 3. BASALTIC VOLCANOES

Volcanic Landform	Eruptive Style	Typical surface	Comments
1. Flood basalt plateau	Fissure eruption, high volume--highly fluid; few flow features	Pahoehoe; massive, dense, low vesicularity; vertical jointing	Common on planetary surfaces (1)
2. Basalt "plains"	Fissure and central vent, moderately high volume and rate; fluid lavas; lava tubes and channels	Pahoehoe, tube and toe-fed, aa; variable textures, from massive to vesicular; vertical joints and horizontal units.	Common on planetary surfaces (2)
3. Basaltic ash plains	Explosive	Pumiceous	Rare
4. Shield volcano	Central vent with associated fissures; high rate, moderate volume, fluid lavas, lava tubes and channels	Pahoehoe, tube and toe-fed, vesicular, platy jointing	Common on planetary surfaces (2)
5. Composite cone	Central vent, explosive alternating with flows, tubes and channels infrequent	Pyroclastics and vesicular to dense flows; pahoehoe and aa	Commonly associated with plate subduction zones (3)
6. Cinder cone	Explosive, infrequent flows	Pyroclastics	Common on planetary surfaces (2)
7. dome	Central vent, low volume, low rate, viscous lavas	Block flows	Rare

(1) Earth, Moon, Mars, Mercury, Venus(?)
(2) Earth, Moon, Mars, Venus(?)
(3) Earth, Mars, Venus(?)

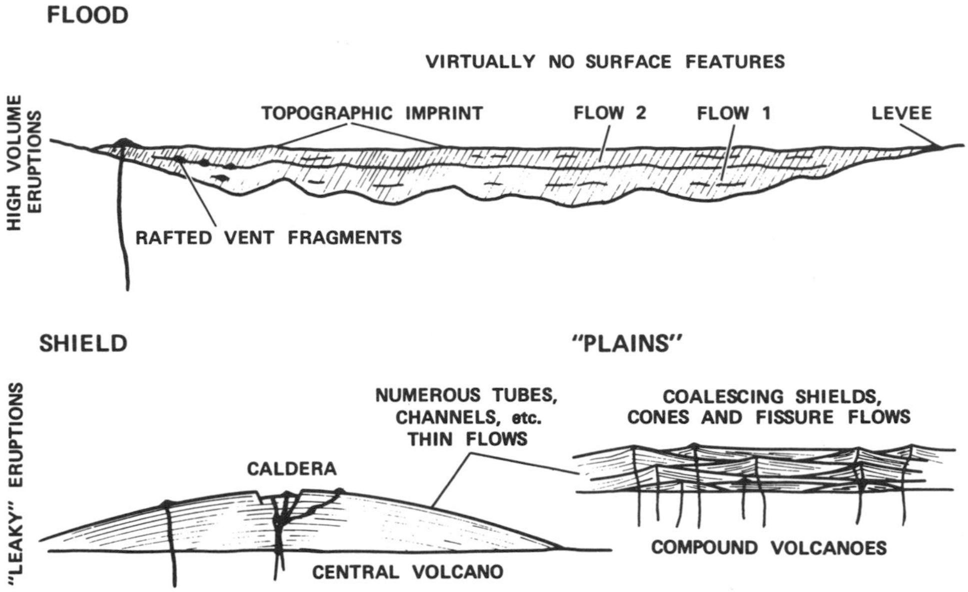

FLOOD

SHIELD "PLAINS"

Figure 51. Stylized diagrams showing the characteristics of flood basalt regions (top diagram), shield volcanoes resulting from Hawaiian-style volcanism (lower left), and "plains" volcanism (lower right), which combines characteristics of both flood lava terrains and shield volcanoes.

tion is planar because the vents are spread over a wide area, not focused in a central zone.

Flow accumulations. Most flows on the Snake River Plain accumulate as: (1) small, low shields, (2) fissure flows, and (3) major tube-fed flows (Fig. 52). All were probably emplaced relatively slowly—often advancing only a few meters per hour. They formed "toey" lava flows with hummocky surfaces of several meters relief. Pressure ridges (Fig. 53) and collapse craters are common. *Low shields* (Greeley, 1982a), described by Noe-Nygaard (1968) for features in the Faroe Islands, are low-profile volcanoes having slopes of about 0.5° and diameters of about 15 km. In the Snake River Plain, the summit area of low shields commonly steepens and has one or more irregular craters. Low shields are often aligned on rift zones, with the shields along each rift zone being roughly of the same age. The shields are composed of numerous thin flow units, some of which are fed by small lava tubes and channels. It appears that once the shields reach some maximum size, effusion ceases and/or shifts to other vent positions, leading to multiple coalescing low shields.

The Wapi lava field, a typical low shield, covers about 300 km^2 of the plain (Champion, 1973; Champion and Greeley, 1977). The field consists of multiple flows and flow units that are typically less than 10 m thick. The total thickness, however, probably exceeds 100 m near the center of the field and near Pillar Butte, one of the main vent areas. Its surface is typified by ridges, plateaus, tumuli, and collapse depressions—all of which are primary features formed during the emplacement of lava flows (Fig. 53).

Fissure flows are commonly associated with rift zones and consist of sheets of lava erupted from linear vents. Although fissure flows can cover dozens of square kilometers (as in the craters of the Moon area), their areal extent and total volume are less than a tenth as large as flood eruptions. Moreover, eruption at point-sources along the fissure may produce localized spatter and cinder cones, both of which are rare or absent in flood basalt plateaus. Figure 54 shows some fissure flows and associated spatter cones near Kings Bowl.

Major lava tube flows play an important role in maintaining the relatively flat surface of basaltic plains by filling the areas between low shields. Lava tubes develop in Hawaiian-type eruptions in which the rate of effusion is moderate (10–100 m^3/sec), of long duration (many days or weeks), and sporadic, thus producing multiple flow units. Only recently has the role of lava tubes and channels in the emplacement of basalt flows been assessed (Holcomb, 1980; Greeley, 1986). Lava tubes and, to a lesser extent, channels efficiently transport lavas great distances from their vents. Lava tube systems longer than 20 km are common in many basalt areas, including the eastern Snake River Plain, where perhaps more than a fifth of the flows contain lava tubes and channels (Greeley, 1982a). This type of flow is long, narrow, somewhat sinuous, and is emplaced by tubes typically larger than 10 m across. Although the sources for the tubes are not always obvious, the flows mainly come from low shields, pit craters, and fissures.

Flows associated with Bear Trap lava tube and Shoshone Ice Cave lava tube are examples of major tube-fed flows. The Shoshone Ice Cave flow is one of the youngest and least altered flows on the plain, and the ice cave is part of a complex lava

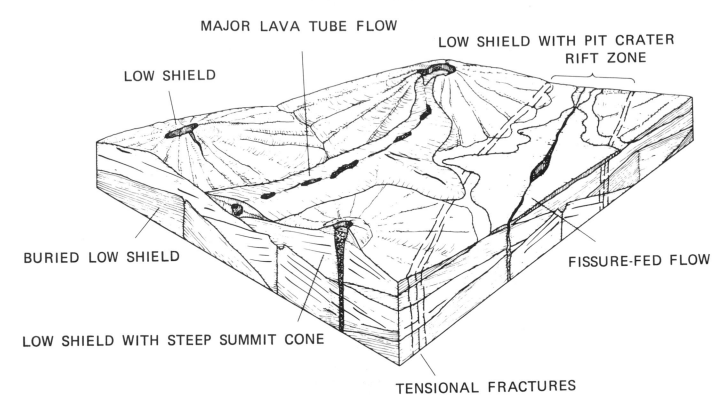

MAJOR LAVA TUBE FLOW

LOW SHIELD WITH PIT CRATER
RIFT ZONE

LOW SHIELD

BURIED LOW SHIELD

LOW SHIELD WITH STEEP SUMMIT CONE

FISSURE-FED FLOW

TENSIONAL FRACTURES

Figure 52. Block diagram illustrating the principal features of plains volcanism, typified by the Snake River Plain. These terrains are composed of multiple, coalescing low shields, fissure-fed lava flows, and major lava tube flows (after Greeley, 1982a).

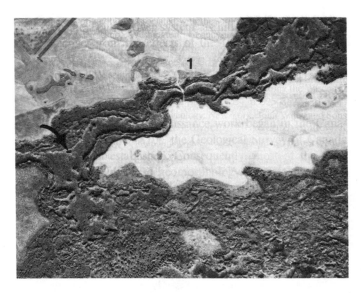

Figure 53. Vertical aerial photograph showing pressure ridges (1) and small collapse craters (arrow) on part of the Wapi lava flow (dark area).

tube/channel system. Many of the formerly roofed parts have collapsed, leaving open trenches. The compound flows associated with this system cover about 210 km^2 and consist of both aa and pahoehoe flow units. Numerous small distributary lava tubes and channels were important in carrying lava away from the main tube. This caused a subtle topographic arch to form along the axis of the main tube, a feature typical of many lava tubes.

Bear Trap lava tube and the flows associated with it make up one of several tube-fed systems south of Craters of the Moon. The tube can be traced by a series of collapsed segments more than 21 km westward, where it becomes buried by younger lava flows (Greeley and King, 1975). The axial trace of the tube is defined by a broad topographic swell, which partly controlled emplacement of subsequent lava flows. The Bear Trap lava tube flows were also controlled by topography generated by an older lava tube system to the south (Fig. 55). Although the total extent of the flows associated with Bear Trap lava tube cannot be determined because the northern flank of the tube-arch and the distal end of the flow are buried, the exposed part of the flow is estimated to cover about 60 km^2.

These three major types of eruptions—producing low shields, fissure flows, and tube-fed flows—maintain the generally flat surface of basaltic plains (Fig. 52), such as the Snake River Plain.

Figure 54. Oblique aerial view northward along the Kings Bowl rift and lava field showing spatter cones (foreground); arrow points to ash erupted from the fissure.

Figure 55. Stylized cross section showing flows associated with Bear Trap lava tube and an older, unnamed lava tube; the axes of lava tubes are frequently topographic highs, which can control the location of subsequent flows (COM, Craters of the Moon).

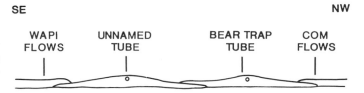

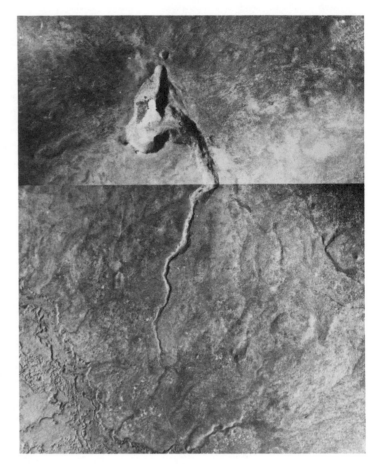

Figure 56. Comparison of a small crater (Aratus CA) in Mare Serenitatis on the Moon (left side) with Bear Crater on the Snake River Plain (right side). Area of lunar photograph is about 17 km by 25 km; terrestrial photograph is about 2.5 km by 3.3 km. Both features are vents for basaltic lavas (after Greeley, 1973).

CRATERS

Several types of craters occur in the Snake River Plain, including calderas, pit craters, maar craters, and small collapse depressions. Calderas are rare and tend to be small, except for the Island Park caldera (Christiansen, 1982) on the northeastern boundary of the plain. Pit craters on the Snake River Plain are generally less than 1 km in diameter and are commonly found at the summit of low shields. Many pit craters fed sinuous lava channels (Fig. 56).

Maar craters and tuff rings are found in several parts of the plain. The largest, Menan Buttes (Fig. 57), may have formed by interaction of magma with surface water, possibly the ancestral Snake River. Others, such as Sand Crater (Fig. 58) and Split Butte (Fig. 59), are young maar craters that apparently have involved groundwater (Womer and others, 1982). They consist of tephra deposits around central craters 500 to 800 m in diameter that developed in lava flows. They appear to have formed by an early phreatomagmatic phase and a later effusive phase that emplaced a lava lake, which then subsided to form the craters.

There are differences in eruptive processes between Sand Crater and Split Butte, however, which are reflected in their morphology and in the tephra. Sand Crater is nearly symmetrical, whereas Split Butte is strongly asymmetrical, with much thicker tephra deposits on the downwind side of the vent. Evidently, Split Butte was active during strong westerly winds, whereas Sand Crater formed during relatively calm winds. The average height-width ratio for Split Butte is about 1:25, whereas at Sand Crater, the ratio is about 1:15. Heiken (1971) has suggested that the height-width ratio is partly dependent upon the depth of water-magma interaction, with deeper interaction resulting in higher deposits.

Collapse depressions are circular to elongate rimless craters of about 3 to more than 70 m across, which are not associated with vents. Although commonly found on some basalt flows, presumably they could form on flows of other compositions so long as the rheological properties were similar. The flows are always thin (<20 m), frequently involve multiple flow units, and involve pahoehoe lavas, which often have pressure ridges, pressure plateaus, and lava "toes" and "tongues." Thus, collapse depressions are not to be expected (nor have they been found) on

Figure 57. Oblique aerial view across Menan Buttes, maar crater/cones formed in the eastern Snake River Plain.

Figure 58. Oblique aerial view of Sand Crater, a tuff-cone, which is astride a small fissure. This feature
formed by phreatomagmatic eruptions and contained a late-stage lava lake.

flood lavas. Within a given lava field, some flows have abundant collapse depressions, whereas other flows lack them. The factors that govern the size of collapse depressions are not known, or at best are poorly understood. Hatheway (1971) suggested that the upper size is limited by the flow thickness and noted that the larger collapse depressions occur in the thicker flow units on the McCartys lava flow of New Mexico.

AEOLIAN FEATURES

The Snake River Plain is not particularly well known for aeolian features. However, remnants of dunes, climbing and fall-

ing dunes, windblown ash deposits, and one of the highest single dunes in North America—the Bruneau dune—all occur on the Snake River Plain and have provided insight into aeolian activity as a planetary process.

Bruneau Dune Field. The Bruneau dune field lies near Bruneau, Idaho, on the southern margin of the Snake River Plain about 30 km south of Mountain Home (Greeley and others, 1971; Murphy and Greeley, 1972; Murphy, 1973). The main part of the field is within Eagle Cove, a semicircular depression about 5.5 km across, formed by a meander of the Snake River. The Bonneville Flood (Malde, 1968), dated at about 15,000 years B.P. (Currey and others, 1983), probably scoured the cove and

Figure 59. Oblique aerial view of Split Butte showing the prominent tephra rim and the basalt pit crater contained within the tephra rim. Prevailing winds from the west (right side of photograph) caused tephra deposits to accumulate preferentially to the east (left side) of vent area.

Figure 60. Vertical aerial photograph showing Eagle Cove (dashed line) and the Bruneau dune field; field has been subdivided into: central dune complex (1), west dune field (2), "Little Sahara" dune field (3), and north dune field (4). Prevailing winds are from the northwest (left side of photograph) toward the east, although winds from the opposite direction also occur (see Fig. 63).

can be taken as an upper limit for the age of the dunes within the cove. The floor of the cove intersects the local water table, which accounts for small lakes northwest of the main dunes. In winter, the water is often frozen.

Murphy (1973) subdivided the field into several smaller dune deposits (Fig. 60). The central dune complex is 1,800 m long and rises 160 m above the floor, equal to the rim height of the cove. It includes the most prominent dune mass, a star dune that has two prominent arms, two subsidiary arms, and a central crater (Fig. 61). The west rim field, on the outer flank of the cove, consists of small transverse dunes that feed into the cove. The "Little Sahara" dune field lies between the central dune complex

and the southern rim of the cove. The north dune field forms an arc around the north side of the central dune complex and is composed of transverse dunes. In addition to these fields, all of which are active, numerous stabilized dunes occur in the surrounding areas (Fig. 62). The dune sands are derived from the weathering of the Plio-Pleistocene Bruneau and Glenns Ferry formations (Murphy, 1973), both of which consist predominantly of poorly consolidated fine-grained sediments and are exposed in a valley west of Eagle Cove and in the Cove rim.

Surface-wind data for 18 years (1932 to 35, 1949, 1951 to 65) are available from Mountain Home Air Force Base, 17 km northeast of Eagle Cove. Wind roses compiled from these data

Figure 61. Oblique aerial view of central dune complex showing prominent star-dune arms and central crater (1), "Little Sahara" dune field is in foreground; lakes (2) and marsh area (3) represent interaction of floor of Eagle Cove and local water table.

Figure 62. Oblique aerial view of Bruneau dune field showing Snake River (foreground) and Eagle Cove, an old meander loop of the Snake River. The west dune field (1), dunes stabilized by vegetation (2), and central dune complex (3) are also shown.

WIND ROSE DIAGRAM, MOUNTAIN HOME AIR FORCE BASE, IDAHO,
FOR PERIODS 1932–33, 1943–1945, 1949, 1951–1965

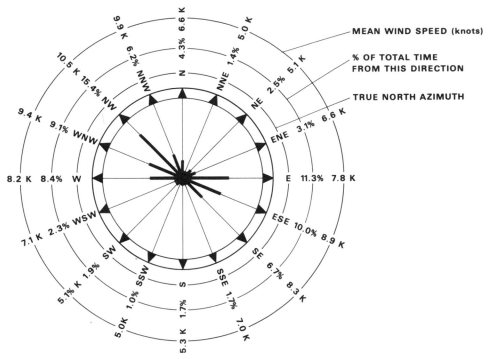

Figure 63. Wind rose diagram compiled from data obtained at Mountain Home Air Force Base, Idaho.

show two dominant wind directions (Fig. 63), southeasterly (generally occurring in the summer) and northwesterly (occurring principally in the winter).

The evolution of the Bruneau dunes can be derived from the local geology, topography, and meteorology. Both local and regional topography influence the wind patterns. Regionally, the major winds are channeled by the Snake River depression. Locally, the winds are channeled by gullies in the rim wall. Principally during the winter, the western dune field spills sand into the cove where it forms the "Little Sahara" dune field. Sand is also fed into the core over the northwest rim; additional sands are fed into the system from the gullies on the south rim. It is difficult to determine whether the Bruneau dune field is experiencing net accumulation, net loss, or is in, a steady state. Analysis of aerial photographs covering a 26-yr period and discussions with local residents suggest little recent change in the overall form (Murphy, 1973). Some sands exit the cove via climbing dunes on the east-southeast rim, although the amount leaving the alcove would appear to be less than the flux into the cove, judging from the sizes of the respective dune fields. The height of the central dune complex, 160 m, is about the same elevation as the surrounding rim, suggesting that the star dune is "capped" by wind shear across the alcove.

Once inside the cove, sands appear to be buffetted back and forth by the 180° swing of the prevailing winds, leading to the maintenance of the star dunes, the principal arms of which are transverse to the two dominant wind directions.

St. Anthony Dunes. The St. Anthony dunes cover an area in excess of 250 km[2] and are found primarily west of St. Anthony, Idaho (Russell, 1902; Stearns and others, 1938; Koscielniak, 1973). Most of the active dunes are transverse and parabolic forms; climbing and falling dunes also occur where parts of the field cross the Juniper Buttes, which are remnants of silicic volcanoes. The trailing arms of older dunes have been stabilized by vegetation (Fig. 64).

Although Stearns and others (1938) suggested that the sources for the sands were Mud Lake and Market Lake, it seems more likely that most of the sands are derived from deposits of the Snake River and Henry's Fork. Koscielniak (1973) found that some sands are also contributed from basalt flows over which the dunes pass, and from outcrops in Juniper Buttes, but these are only minor constituents of the dune sands.

The prevailing winds are southwesterlies, which drive the dunes into Juniper Buttes. The dune field splits, with a small branch of the field passing north of Juniper Buttes and a main mass curving around the southeast side of the buttes. From his analysis of the orientation of the older, stabilized dunes and the axes of the currently active dunes, Koscielniak (1973) suggested that there has been a shift in prevailing winds of about 24° between the development of the two dune sets.

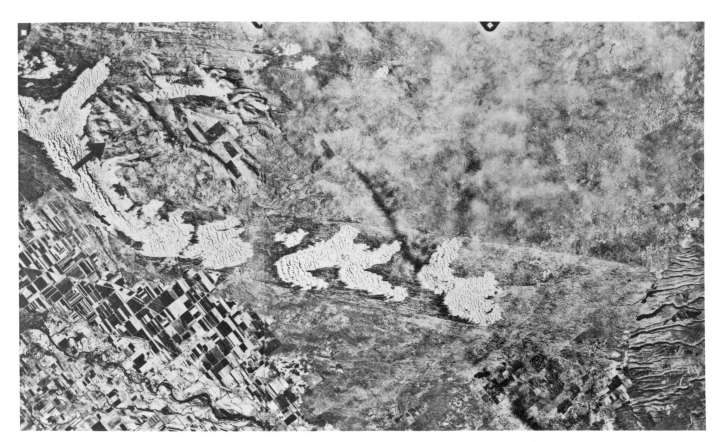

Figure 64. Mosaic of high-altitude (U-2 aircraft) photographs of a sand dune field west of the town of St. Anthony. Wind is from the left to the right; main part of sand dune field passes southeast of Juniper Buttes (Juniper Buttes indicated by arrow); small arm of field also passes to the northwest side (off picture). Small segment of dune field crosses the buttes to form climbing and falling dunes (see Fig. 65). Linear traces of trailing dunes are shown between the main dune patches (NASA-Ames photograph 172-186, frames 5694 and 5696).

In addition to the northern and southern branches, some of the sands from the field extend up and over Juniper Buttes as climbing and falling dunes, passing through a slight depression between North and South "Junipers." These dunes tend to follow the topographically lowest paths over the buttes. They approach the buttes as a series of parabolic dunes, the forms of which become distorted near the summits. As the sands tumble downslope, they develop into a series of imbricating falling dunes (Fig. 65).

Other aeolian features. Several areas on the Snake River Plain display dune relics and other features related to wind. Long sand ridges parallel the Snake River near American Falls (Fig. 66). Some of the ridges predate the younger lava flows, shown by superposition of the Wapi flow. Most of the ridges appear to be remnants of parabolic dunes, similar to those near St. Anthony, and probably are not longitudinal dunes. Most of the sand ridges are stabilized by vegetation, but blowouts have produced locally active patches of sand (Ann Lewis, personal

communication, 1982). Although these ridges have been briefly described (Greeley and King, 1977), the source(s) of the sands and the origin of the dunes have not been studied. Similar sand ridges occur in the northeastern part of the plain, near Mud Lake.

Volcanic eruptions of ash have produced deposits that were influenced by the wind. For example, ash from the phreatic eruptions at Kings Bowl is visible on aerial photographs as light-toned streaks emanating from the fissure (Fig. 54). Wind apparently was also instrumental in distributing ash at Split Butte (Fig. 59).

Niccum (1969) described high albedo streaks visible on aerial photographs as soil lineaments. On the ground, the lineaments support more dense vegetation than the adjacent surfaces and are composed of "loosely packed, brown sandy soils," whereas the surrounding soil is light gray silt. Niccum suggested that some soil lineaments are windblown volcanic deposits, perhaps similar to those formed at Kings Bowl. However, some light-toned streaks visible on aerial photographs are artifacts of range fires. Fires sweep the plain and are driven by winds. Ash deposits from the

Figure 65. Oblique aerial view of dunes crossing the Juniper Buttes. Prevailing winds are from the background toward the foreground. Climbing dunes are on the far side of the buttes and the series of falling dunes are seen in the middle of the picture.

Figure 66. High-altitude (U-2 aircraft) photograph of the Snake River Plain showing linear sand ridges parallel to the trend of the Snake River. Younger age of the Wapi lava flow (upper part of picture) is shown by superposition of the lavas on the sand dunes (NASA-Ames photograph 72-186).

burned vegetation leaves jagged ground patterns, which persist for many years until the sagebrush and other vegetation is reestablished.

POSSIBLE PLANETARY ANALOGS

In this section, planetary features are discussed as possible analogs to those seen in the Snake River Plain. Features of basaltic "plains" include low shields on the Moon and Mars, small flow features on lunar lava plains, lava tubes and channels on the Moon and Mars, and collapsed depressions on the Moon. Aeolian features, including dunes and wind streaks, are well documented on Mars.

Low shields. Images of Mars near Viking Lander I reveal a field of about 60 low-profile structures, tentatively identified as low shields (Greeley and others, 1977). They average 2.5 km across; many have summit knobs—giving a two-part profile— and summit craters. In size and morphology, the Martian features closely resemble those on the Snake River Plain, Idaho. Small shield-shaped features have also been identified on the Moon (Fig. 67) in the Orientale basin (Greeley, 1976). They are low-profile structures a few kilometers across, some of which have steeper elements near the central fissure that could be spatter and pyroclastic deposits. The areas on the Moon and Mars in which these features occur are inferred to have involved volcanism of a style comparable to that in the Snake River Plain.

Lunar flow features. The lunar maria, known to be basalt flows, generally lack well-defined lava-flow fronts, the Imbrium flows being exceptions (Schaber, 1973). However, some mare features resemble other flow features in the Snake River Plain. Although small features such as pahoehoe ropes could not survive degradation by impact cratering, larger features (>10 m) might survive on the younger lunar lavas (Greeley and Schultz, 1977). These features include: (a) tumuli, (b) collapse depressions, (c) plateaus, (d) ridges, and (e) "moats" around some pre-flow features. These forms produce distinctive hummocky surfaces on the Snake River Plain, and similar mare surfaces occur on the Moon and Mars.

Mare units in the lunar Oceanus Procellarum contain numerous interlinking depressions and patches of smooth-surfaced units. Within the Flamsteed Ring are numerous, small (50–150 m) rings characterized by a narrow depression encircling a low-relief mound, called *ring moats* (Schultz, 1976; Schultz and Greeley, 1975; Schultz and others, 1976). These features may form on the margins of thin (10 m) basalt flows surrounding preexisting relief, such as volcanic cones and tumuli, and are similar to certain features in the Snake River Plain (Fig. 68).

Ring-moat structures and the textured mare surfaces may provide important clues to the style of mare basalt emplacement where distinct flow termini are absent. Some lunar units may be analogous to relatively thin (10–20 m), compound lava flows of low viscosity, which erupted from numerous local vents on the Snake River Plain.

Sinuous rilles, lava tubes, and associated features.

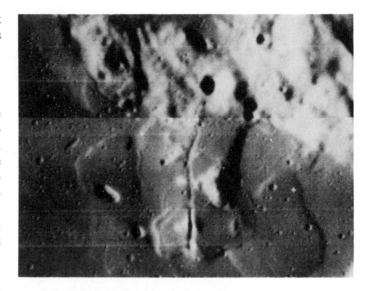

Figure 67. Enlarged Lunar Orbiter framelets showing coalescing low shields associated with mare lavas in the Orientale Basin on the Moon. Shields appear to be related to a linear fissure vent; steep features on the fissure may be pyroclastic deposits. Area of photograph is 22 km by 35 km; north is toward the bottom of photograph (Lunar Orbiter IV photograph 181 H$_2$).

Sinuous rilles have long been recognized on the lunar surface (Fig. 69), and their origin was a matter of considerable debate in the 1960s. From field studies conducted in the Snake River Plain and elsewhere, and using arguments based principally on their geomorphology, it was demonstrated that most sinuous rilles are lava channels and collapsed lava tubes (Oberbeck and others, 1969; Greeley, 1971a, 1971b). Analysis of the distribution of lunar sinuous rilles and their relation to other volcanic features show that they are not randomly located on mare units. Some areas, such as the Crisium basin, contain few sinuous rilles, whereas other basins, such as Imbrium, have abundant sinuous rilles. By terrestrial analogy, their presence implies eruptions of plains-type basalts, that is, sporadic eruptions of lava, in contrast to flood-type eruptions. Absence of obvious flow fronts and flow margins indicates the very thin character of the flow units associated with the rilles—a feature also typical of plains-type basalt flows.

Lava tubes were first recognized on Mars on the flank of Olympus Mons, the largest shield volcano in the solar system. Viking Orbiter pictures show that most of the large Martian shield volcanoes are composed of tube-fed and channel-fed flows similar to shields on Earth.

Aeolian features. Aeolian processes are well documented on Mars and are suspected on Venus (Greeley and Iversen, 1985). Viking images reveal with remarkable clarity a wide variety of wind-related features, both from orbit and from the two landing sites (Sharp and Malin, 1984). Although the largest dune field occurs in the north polar area, smaller dune fields are found in

Figure 68. Oblique aerial view of China Cap, a small scoria cone on the Snake River Plain. A shallow moat surrounds the cone and was formed by the encroachment of superposed lava flows.

nearly all parts of Mars. Of particular interest are dunes that occur within impact craters. In some respects, these are analogous to the Bruneau dune field (Fig. 62) and involve the interplay of sand supply, local and regional wind patterns, and topographical control. Figure 70 shows climbing-falling dunes on Mars, similar to those seen at Juniper Buttes.

Wind streaks (Fig. 71) constitute the most abundant aeolian features on Mars. Termed "variable features," they appear, disappear, and change in size, shape, and orientation with time, evidently in response to near-surface winds. As such, they are used as "wind vanes" to model local wind patterns. They occur in two principal forms, dark wind streaks and light streaks. In general, dark streaks change on shorter time scales (approximately a few weeks) than light streaks. In both cases, they are considered to represent mobile windblown sediment whose form and orienta-

tion is controlled by the topographic features with which it is associated. Most dark streaks are considered to represent a dark substrate exposed by erosion of surficial materials. In contrast, light streaks are thought to be deposits of dust. Although not necessarily directly analogous with the wind streaks seen in the Snake River Plain, the ash patterns near Kings Bowl (Fig. 54) and those recognized by Niccum (1969) may share attributes with the light streaks on Mars, in that both are deposits of fine-grained windblown particles.

SUMMARY

The identification of volcanic features and distinctive flow textures on some planetary surfaces that appear to be analogous to the Snake River Plain suggests similar styles of eruption and modes of emplacement. Features include lava tubes and channels,

Figure 69. Sinuous rilles near the crater Prinz on the Moon. The morphology and geological setting of these and similar features suggest that they are lava channels/collapsed lava tubes.

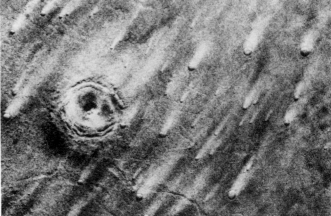

Figure 70. Viking Orbiter image showing an impact crater on Mars and series of climbing/falling dunes associated with the crater. Dunes appear to be barchan dunes moving in direction indicated by arrow. Crater shown is about 28 km in diameter (Viking Orbiter frame 571-B-53).

Figure 71. Light/dark streaks on Mars imaged by Viking Orbiter. Top. "Bright" streaks in Hesperia Planum (26°S lat., 247°W long.), considered to be dust deposits in the "wake" of craters; wind is from lower left; area shown is about 300 km by 200 km (Viking Orbiter frame 350A01). Bottom. "Dark" streaks in southern Tharsis area (28°S lat., 125°W long.), considered to be erosional areas swept free of windblown sediments in the "wake" of craters; wind is from the left; area shown is about 200 km by 140 km (Viking Orbiter frame 603A25).

low shields, cinder and spatter cones, ring-moat structures, and hummocky-textured surfaces. Many of the extraterrestrial features, however, are much larger than their possible Earth analogs. Although some differences may be attributed to differences in planetary environment, including different gravitational accelerations on the planets, the reasons are poorly understood.

Although it is often tempting to apply terrestrial experience directly to the interpretation of extraterrestrial surfaces, caution must be exercised to take into account differences in scaling and environment. Moreover, the possibility of volcanism on the planets and satellites in forms not represented on the Earth is very real.

Few studies of dune forms associated with topographic features, such as hills and craters, have been conducted. However, analyses of the Bruneau dunes and the dunes near St. Anthony provide insight into the interplay of dune sands, winds, and topography for features seen on Mars. Analysis of wind streaks on the Snake River FPlain also sheds light on the possible nature of similar features seen in many parts of Mars.

In conclusion, studies of the geomorphic features and surface evolution on the Snake River Plain have been invaluable as a means for understanding the geology of other planets and satellites. Although the primary objective of these studies has been for extraterrestrial applications, the results have also contributed toward a better geological understanding of the Snake River Plain.

REFERENCES

Ager, D., 1980, The nature of the stratigraphical record: New York, John Wiley, 122 p.

Albritton, C. C., 1967, Uniformity, the ambiguous principle, *in* Albritton, C. C., ed., Uniformity and simplicity: Geological Society of America Special Paper 89, p. 1–2.

Allen, J.R.L., 1982, Sedimentary structures, their character and physical basis: Amsterdam, Elsevier, v. 1, 594 p., v. 2, 664 p.

Allison, I. S., 1933, New version of the Spokane Flood: Geological Society of America Bulletin, v. 44, p. 675–722.

—— , 1941, Flint's fill-hypothesis for channeled scabland: Journal of Geology, v. 49, p. 54–73.

Atwater, T., 1970, Implications of plate tectonics for the Cenozoic tectonic evolution of western North America: Geological Society of America Bulletin, v. 81, p. 3513–3536.

Bagnold, R. A., 1966, An approach to the sediment transport problem from general physics: U.S. Geological Survey Professional Paper 422-I, 37 p.

—— , 1977, Bed load transport by natural rivers: Water Resources Research, v. 13, p. 303–312.

—— , 1980, An empirical correlation of bedload transport rates in flumes and natural rivers: Proceedings of the Royal Society, v. 372A, p. 453–473.

Baker, V. R., 1973a, Paleohydrology and sedimentology of Lake Missoula flooding in eastern Washington: Geological Society of America Special Paper 144, 79 p.

—— , 1973b, Erosional forms and processes for the catastrophic Pleistocene Missoula floods in eastern Washington, *in* Morisawa, M., ed., Fluvial geomorphology: London, Allen and Unwin, p. 123–148.

—— , 1977, Stream channel response to floods, with examples from central Texas: Geological Society of America Bulletin, v. 88, p. 1057–1071.

—— , 1978a, The Spokane Flood controversy and the Martian outflow channels: Science, v. 202, p. 1249–1256.

—— , 1978b, A preliminary assessment of the fluid erosional processes that shaped the Martian outflow channels: Proceedings of the 9th Lunar and Planetary Science Conference, p. 3205–3223.

—— , 1979, Erosional processes in channelized water flows on Mars: Journal of Geophysical Research, v. 84, p. 7985–7993.

—— , 1981, Catastrophic flooding; The origin of the Channeled Scabland: Stroudsburg, Pennsylvania, Dowden, Hutchinson and Ross, 360 p.

—— , 1982, The channels of Mars: Austin, University of Texas Press, 198 p.

—— , 1983, Late Pleistocene fluvial systems, *in* Porter, S. C., ed., The late Pleistocene: Minneapolis, University of Minnesota Press, p. 115–129.

—— , 1984, Flood sedimentation in bedrock fluvial systems, *in* Koster, E. H., and Steel, R. J., eds., Sedimentology of gravels and conglomerates: Canadian Society of Petroleum Geologists Memoir 10, p. 87–98.

—— , 1985, Models of fluvial activity on Mars, *in* Woldenberg, M., ed., Models in geomorphology: London, Allen and Unwin, p. 287–312.

Baker, V. R., and Bunker, R. C., 1985, Cataclysmic late Pleistocene flooding from glacial Lake Missoula; A review: Quaternary Science Reviews, v. 4, p. 1–41.

Baker, V. R., and Kochel, R. C., 1978, Morphometry of streamlined forms in terrestrial and Martian channels: National Aeronautics and Space Administration, Proceedings of the 9th Lunar and Planetary Science Conference, p. 3193–3203.

Baker, V. R., and Milton, D. J., 1974, Erosion by catastrophic floods on Mars and Earth: Icarus, v. 23, p. 27–41.

Basaltic Volcanism Study Project, 1981, Basaltic volcanism on the terrestrial planets: New York, Pergamon Press, Incorporated, 1286 p.

Beveridge, W.J.B., 1950, The art of scientific investigation: New York, Vintage Books, 239 p.

Bradley, W. H., 1936, Geomorphology of the north flank of the Uinta Mountains: U.S. Geological Survey Professional Paper 185-I, p. 163–204.

Bretz, J H., 1923a, The Channeled Scabland of the Columbia Plateau: Journal of Geology, v. 31, p. 617–649.

—— , 1923b, Glacial drainage on the Columbia Plateau: Geological Society of America Bulletin, v. 34, p. 573–608.

—— , 1925, The Spokane Flood beyond the Channeled Scabland: Journal of Geology, v. 33, p. 97–115, 232–259.

—— , 1927, Channeled Scabland and the Spokane Flood: Journal of the Washington Academy of Science, v. 17, p. 200–211.

—— , 1928a, Alternative hypotheses for Channeled Scabland: Journal of Geology, v. 36, p. 193–223, 312–341.

—— , 1928b, Bars of the Channeled Scabland: Geological Society of America Bulletin, v. 39, p. 643–702.

—— , 1928c, The Channeled Scabland of eastern Washington: Geographical Review, v. 18, p. 446–477.

—— , 1929, Valley deposits immediately east of the Channeled Scabland of Washington: Journal of Geology, v. 37, p. 393–427, 505–541.

—— , 1930, Valley deposits immediately west of the Channeled Scabland: Journal of Geology, v. 38, p. 385–422.

—— , 1932, The Grand Coulee: American Geographical Society Special Publication 15, 89 p.

—— , 1959, Washington's Channeled Scabland: Washington Department of Conservation, Division of Mines and Geology Bulletin no. 45, 57 p.

—— , 1969, The Lake Missoula floods and the Channeled Scabland: Journal of Geology, v. 77, p. 505–543.

Bretz, J H., Smith, H.T.U., and Neff, G. E., 1956, Channeled Scabland of Washington; New data and interpretations: Geological Society of America Bulletin, v. 67, p. 957–1049.

Bryan, K., 1927, The "Palouse soil" problem: U.S. Geological Survey Bulletin 790, p. 21–45.

Bull, W. B., 1979, Threshold of critical power in streams: Geological Society of America Bulletin, v. 90, p. 453–464.

Calkins, F. C., 1905, Geology and water resources of a portion of east-central Washington: U.S. Geological Survey Water-Supply Paper 18, 37 p.

Camp, V. E., 1981, Geologic studies of the Columbia Plateau; Upper Miocene basalt distribution, reflecting source locations, tectonism, and drainage history in the Clearwater embayment, Idaho: Geological Society of America Bulletin, Pt. I, v. 92, p. 669–678.

Camp, V. E., and Hooper, P. R., 1981, Geologic studies of the Columbia Plateau; Late Cenozoic evolution of the southeast part of the Columbia River basalt province: Geological Society of America Bulletin, Pt. I, v. 92, p. 659–668.

Carr, M. H., 1979, Formation of Martian flood features by release of water from confined aquifers: Journal of Geophysical Research, v. 84, p. 2955–3007.

—— , 1981, The surface of Mars: New Haven, Connecticut, Yale University Press, 226 p.

—— , 1983, The stability of streams and lakes on Mars: Icarus, v. 56, p. 476–495.

Champion, D. E., 1973, The relationship of large scale surface morphology to lava flow direction, Wapi lava field, southeast Idaho [M.S. thesis]: Buffalo, State University of New York, 44 p.

Champion, D. E., and Greeley, R., 1977, Geology of the Wapi lava field, Snake River Plain, Idaho, *in* Greeley, R., and King, J. S., eds., Volcanism of the eastern Snake River Plain, Idaho: NASA CR-154621, p. 133–151.

Chorley, R. J., 1959, The shape of drumlins: Journal of Glaciology, v. 3, p. 339–344.

Chow, V. T., 1959, Open-channel hydraulics: New York, McGraw-Hill, 680 p.

Christiansen, R. L., 1982, Late Cenozoic volcanism of the Island Park area, eastern Idaho, *in* Bonnichsen, B., and Breckenridge, R. M., eds., Cenozoic geology of Idaho: Idaho Bureau of Mines and Geology Bulletin 26, p. 345–368.

Christiansen, R. L., and McKee, E. H., 1978, Lake Cenozoic volcanic and tectonic evolution of the Great Basin and Columbia Intermontane regions: Geological Society of America Memoir 152, p. 283–311.

Clarke, G.K.C., Mathews, W. H., and Pack, R. T., 1984, Outburst floods from glacial Lake Missoula: Quaternary Research, v. 22, p. 289–299.

Cotton, C. A., 1952, Volcanoes as landscape forms: New York, Hafner, 416 p.

Currey, D. R., and Oviatt, C. G., 1985, Durations, average rates, and probable causes of Lake Bonneville expansions, stillstands, and contractions during the last deep-lake cycle, 32,000 to 10,000 years ago, *in* Kay, P. A., and Diaz, H. F., eds., Problems of and prospects for predicting Great Salt Lake levels; Papers from a conference held in Salt Lake City, March 26–28, 1985: Salt Lake City, University of Utah, Center for Public Affairs and Administration, p. 9–24.

Currey, D. R., Oviatt, C. G., and Plyler, G. B., 1983, Stansbury shoreline and Bonneville lacustrine cycle: Geological Society of America Abstracts with

Program, v. 15, p. 301.

Dalrymple, T., and Benson, M. A., 1968, Measurement of peak discharge by the slope-area method: U.S. Geological Survey, Techniques of Water-Resources Investigation, Book 3, Chapter A2, 12 p.

Davies, G. L., 1969, Earth in decay; A history of British geomorphology, 1578–1878: New York, Elsevier, 390 p.

Davis, W. M., 1889, The rivers and valleys of Pennsylvania: National Geographic Magazine, v. 1, p. 183–253.

——, 1926, The value of outrageous geological hypotheses: Science, v. 63, p. 463–468.

Donnelly, J. W., 1982, Worldwide continental denudation and climatic deterioration during the late Tertiary; Evidence from deep-sea sediments: Geology, v. 10, p. 451–454.

Dott, R. H., Jr., 1983, Episodic sedimentation; How normal is average? How rare is rare? Does it matter?: Journal of Sedimentary Petrology, v. 53, p. 5–23.

Ely, L. L., and Baker, V. R., 1985, Reconstructing paleoflood hydrology with slackwater deposits; Verde River, Arizona: Physical Geography, v. 5, p. 103–126.

Engels, J. C., Tabor, R. W., Miller, F. K., and Obradovich, J. D., 1976, Summary of K-Ar, Rb-Sr, U-Pb, and fission-track ages of rocks from Washington State prior to 1975 (exclusive of Columbia Plateau basalts): U.S. Geological Survey Miscellaneous Field Studies Map MF-710.

Fecht, K. R., Reidel, S. R., and Tallman, A. M., 1986, Paleodrainage of the Columbia River system on the Columbia Plateau of Washington State—a summary: Richland, Washington, Rockwell-Hanford Operations, document RHO-BW-SA-318p, 55 p.

Feldman, A. D., 1981, HEC models for water resources system simulation, theory, and experience: Advances in Hydroscience, v. 12, p. 297–423.

Flint, R. F., 1935, Glacial features of the southern Okanogan region: Geological Society of America Bulletin, v. 46, p. 169–194.

——, 1938a, Summary of late-Cenozoic geology of southeastern Washington: American Journal of Science, v. 35, p. 223–230.

——, 1938b, Origin of the Cheney-Palouse scabland tract: Geological Society of America Bulletin, v. 49, p. 461–524.

——, 1971, Glacial and Quaternary geology: New York, John Wiley, 892 p.

Fraser, G. S., Bleuer, N. K., and Smith, N. D., 1983, History of Pleistocene alluviation of the middle and upper Wabash Valley, in Shaver, R. H., and Sunderman, J. A., eds., Field trips in midwestern geology, Volume 1: Bloomington, Indiana Geological Survey, p. 197–224.

Freeman, O. W., Forrester, J. D., and Lupher, R. L., 1945, Physiographic divisions of the Columbia Intermontane Province: Annals of the Association of American Geographers, v. 35, no. 2, p. 53–75.

Gilbert, G. K., 1877, Report on the geology of the Henry Mountains: Washington, D.C., U.S. Geographical and Geological Survey of the Rocky Mountain Region, 160 p.

Gilluly, J., 1927, Discussion; Channeled Scabland and the Spokane Flood: Washington Academy of Science Journal, v. 17, no. 8, p. 203–205.

Gould, S. J., 1965, Is uniformitarianism necessary?: American Journal of Science, v. 263, p. 223–228.

——, 1980, The panda's thumb: New York, Norton, 343 p.

——, 1984, Toward the vindication of punctuational change, in Berggren, W. A., and Van Couvering, J. A., eds., Catastrophe and earth history: Princeton, Princeton University Press, p. 9–34.

Greeley, R., 1971a, Lunar Hadley Rille; Consideration of its origin: Science, v. 172, p. 722–725.

——, 1971b, Lava tubes and channels in the lunar Marius Hills: The Moon, v. 3, p. 289–314.

——, 1973, Comparative geology of crater Aratus CA (Mare Serenitatis) and Bear Crater (Idaho), in Apollo 17 Preliminary Science Report: National Aeronautics and Space Administration Special Paper 330, p. 30–1, 30–6.

——, 1976, Modes of emplacement of basalt terrains and an analysis of mare volcanism in the Orientale Basin: Proceedings, 7th Lunar Scientific Conference, p. 2747–2759.

——, 1977, Volcanic morphology, in Greeley, R., and King, J. S., eds., Vol-

canism of the eastern Snake River Plain, Idaho: NASA CR-154621, p. 6–22.

——, 1982a, The Snake River Plain, Idaho; Representative of a new category of volcanism: Journal of Geophysical Research, v. 87, p. 2705–2712.

——, 1982b, The style of basaltic volcanism in the eastern Snake River Plain, Idaho, in Bonnichsen, B., and Breckenridge, R. M., eds., Cenozoic geology of Idaho: Idaho Bureau of Mines and Geology Bulletin 26, p. 407–421.

——, 1986, The role of lava tubes in Hawaiian volcanoes, in U.S. Geological Survey Professional Paper 1350 (in press).

Greeley, R., and Iversen, J. D., 1985, Wind as a geological process on Earth, Mars, Venus, and Titan: Cambridge, Cambridge University Press, 333 p.

Greeley, R., and King, J. S., 1975, Geologic field guide to the Quaternary volcanics of the south-central Snake River Plain, Idaho: Idaho Bureau of Mines and Geology Pamphlet 160, 49 p.

——, eds., 1977, volcanism of the eastern Snake River Plain, Idaho: NASA CR-254621, 308 p.

Greeley, R., and Schultz, P. H., 1977, Possible planetary analogs to Snake River Plain basalt features, in Greeley, R., and King, J. S., eds., Volcanism of the eastern Snake River Plain, Idaho: NASA CR-154621, p. 233–251.

Greeley, R., Koscielniak, D. E., and Hodge, D. S., 1971, Bruneau sand dune field, Idaho, and its possible implications in Martian geology: EOS American Geophysical Union Transactions, v. 52, p. 860.

Greeley, R., Theilig, E., Guest, J. E., Carr, M. H., Masursky, H., and Cutts, J. A., 1977, Geology of Chryse Planitia: Journal of Geophysical Research, v. 82, p. 4093–4109.

Griggs, A. B., 1976, The Columbia River Basalt Group in the Spokane Quadrangle, Washington, Idaho, and Montana: U.S. Geological Survey Bulletin 1413, 39 p.

Gustafson, E. P., 1978, The vertebrate faunas of the Pliocene Ringold Formation, south-central Washington: University of Oregon, Museum of Natural History Bulletin 23, 62 p.

Hack, J. T., 1960, Interpretation of erosional topography in humid temperate regions: American Journal of Science, v. 258-A, p. 80–97.

Harms, J. C., and Fahnestock, R. K., 1965, Stratification, bed forms, and flow phenomena (with an example from the Rio Grande): Society of Economic Paleontologists and Mineralogists, Special Publication 12, p. 84–115.

Hatheway, A. W., 1971, Lava tubes and collapse depressions [Ph.D. thesis]: Tucson, University of Arizona, 353 p.

Head, J. W., and 9 others, 1981, Distribution and morphology of basalt deposits on planets, in Basaltic Volcanism Study Project, Basaltic volcanism on the terrestrial planets: New York, Pergamon Press, Incorporated, p. 700–800.

Heiken, G. H., 1971, Tuff rings; Examples from the Fort Rock–Christmas Lake Valley basin, south-central Oregon: Journal of Geophysical Research, v. 76, p. 5616–5626.

Hjulström, F., 1935, Studies in the morphological activity of rivers as illustrated by the River Fyris: Geological Institute, University of Upsala Bulletin, v. 25, p. 221–528.

Holcomb, R. T., 1980, Kilauea Volcano, Hawaii; Chronology and morphology of the surficial lava flows [Ph.D. thesis] Stanford University, 321 p.

Hooper, P. R., and Camp, V. E., 1981, Deformation of the southeast part of the Columbia Plateau: Geology, v. 9, p. 323–328.

Hubbert, M. K., 1967, Critique of the principle of uniformity, in Albritton, C. C., ed., Uniformity and simplicity: Geological Society of America Special Paper 89, p. 3–33.

——, 1980, Presentation of the Penrose Medal to J Harlan Bretz; Citation: Geological society of America Bulletin, part II, v. 91, p. 1091–1094.

Hunt, C. B., Averitt, P., and Miller, R. L., 1953, Geology and geography of the Henry Mountains region, Utah: U.S. Geological Survey Professional Paper 228, 234 p.

Hydrologic Engineering Center, 1982, HEC-2 water surface profiles, user's manual: Davis, California, U.S. Army Corps of Engineers, 173 p.

Johnson, D. W., 1931, Stream sculpture on the Atlantic slope: New York, Columbia University Press, 142 p.

Kehew, A. E., and Lord, M. L., 1986, Origin and large-scale erosional features of glacial-lake spillways in the northern Great Plains: Geological Society of

America Bulletin, v. 97, p. 162–177.

Kimmel, P. G., 1982, Stratigraphy, age, and tectonic setting of the Miocene-Pliocene lacustrine sediments of the western Snake River Plain, Oregon and Idaho, *in* Bonnichsen, B., and Breckenridge, R. M., eds., Cenozoic geology of Idaho: Idaho Bureau of Mines and Geology Bulletin 26, p. 559–578.

King, J. S., 1982, Selected volcanic features of the south-central Snake River Plain, Idaho, *in* Bonnichsen, B., and Breckenridge, R. M., eds., Cenozoic geology of Idaho: Idaho Bureau of Mines and Geology Bulletin 26, p. 439–451.

King, L. C., 1953, Canons of landscape evolution: Geological Society of America Bulletin, v. 65, p. 721–752.

Komar, P. D., 1979, Comparisons of the hydraulics of water flows in Martian outflow channels with flows of similar scale on Earth: Icarus, v. 37, p. 156–181.

——, 1980, Modes of sediment transport in channelized water flows with ramifications to the erosion of the Martian outflow channels: Icarus, v. 42, p. 317–329.

——, 1983, Shapes of streamlined islands on Earth and Mars; Experiments and analyses of the minimum-drag form: Geology, v. 11, p. 651–654.

——, 1984, The lemniscate loop-comparisons with the shapes of streamlined landforms: Journal of Geology, v. 92, p. 133–145.

Koscielniak, D. E., 1973, Eolian deposits on a volcanic terrain near Saint Anthony, Idaho [M.S. thesis]: Buffalo, State University of New York, 28 p.

Kuhn, T. S., 1962, The structure of scientific revolutions: Chicago, University of Chicago Press, 172 p.

——, 1978, The essential tension: Chicago, University of Chicago Press, 366 p.

Lucchitta, B. K., and Ferguson, H. M., 1983, Chryse Basin channels; Low-gradients and ponded flows: Journal of Geophysical Research, v. 88, p. A553–A568.

Luedke, R. G., Smith, R. L., and Russell-Robinson, S. L., 1983, Map showing distribution, composition, and age of late Cenozoic volcanoes and volcanic rocks of the Cascade Range and vicinity, northwestern United States: U.S. Geological Survey Miscellaneous Geologic Investigations Map I-1507, scale 1:500,000.

Mackin, J. H., 1947, Altitude and local relief in the Bighorn area during the Cenozoic: Guidebook for Wyoming Geological Association Field Conference in the Bighorn Basin, p. 103–120.

——, 1961, A stratigraphic section in the Yakima Basalt and the Ellensburg Formation in south-central Washington: Washington Division of Mines and Geology, Report of Investigations 19, 45 p.

——, 1963, Rational and empirical methods of investigation in geology, *in* Albritton, C. C., ed., The fabric of geology: Reading, Massachusetts, Addison-Wesley Publishing Co., p. 135–163.

Mackin, J. H., and Cary, A. S., 1965, Origin of Cascade landscapes: Washington Division of Mines and Geology Information Circular 41, 35 p.

Malde, H. E., 1968, The catastrophic late Pleistocene Bonneville flood in the Snake River Plain, Idaho: U.S. Geological Survey Professional Paper 596, 52 p.

Mars Channel Working Group, 1983, Channels and valleys on Mars: Geological Society of America Bulletin, v. 95, p. 1035–1054.

Masursky, H., 1973, An overview of geological results from Mariner 9: Journal of Geophysical Research, v. 78, p. 4009–4030.

Masursky, H., Boyce, J. M., Dial, A. L., Schaber, G. G., and Strobell, M. E., 1977, Classification and time of formation of Martian channels based on Viking data: Journal of Geophysical Research, v. 82, p. 4016–4038.

Mathews, W. H., 1973, Record of two jökulhlaups: International Association of Scientific Hydrology Publication no. 95, p. 99–110.

Matthai, H. F., 1967, Measurement of peak discharge at width contractions by indirect methods: U.S. Geological Survey Techniques of Water Resources Investigations, book 3, chapter A4, 44 p.

Matthes, G. H., 1947, Macroturbulence in natural stream flow: EOS American Geophysical Union Transactions, v. 28, p. 255–262.

Maxwell, T. A., Otto, E. P., Picard, M. D., and Wilson, R. C., 1973, Meteorite impact; A suggestion for the origin of some stream channels on Mars: Geol-

ogy, v. 1, p. 9–16.

McCauley, J. F., and 7 others, 1972, Preliminary Mariner 9 report on the geology of Mars: Icarus, v. 17, p. 289–327.

Melhorn, W. N., and Flemal, R. C., eds., 1975, Theories in landform development: Binghamton, State University of New York, Publications in Geomorphology, 306 p.

Milton, D. J., 1973, Water and processes of degradation in the Martian landscape: Journal of Geophysical Research, v. 78, p. 4037–4047.

Murphy, J. D., 1973, Bruneau dune and the geology of Eagle Cove, Idaho [M.S. thesis]: Buffalo, State University of New York, 77 p.

Murphy, J. D., and Greeley, R., 1972, Sand dunes at Eagle Cove (Bruneau), Idaho; Possible analogs to Martian eolian features: EOS America Geophysical Union Transactions, v. 53, p. 1035.

Newcomb, R. C., Strand, J. R., and Frank, F. J., 1972, Geology and groundwater characteristics of the Hanford Reservation of the U.S. Atomic Energy Commission, Washington: U.S. Geological Survey Professional Paper 717, 78 p.

Niccum, M. R., 1969, Geology and permeable structures in basalts of the east-central Snake River Plain near Atomic City, Idaho [M.S. thesis]: Pocatello, Idaho State University, 134 p.

Noe-Nygaard, A., 1968, On extrusion forms in plateau basalts: Reykjavik, Visindafelag Islendinga, anniversary volume, p. 10–13.

Nummedal, D., and Prior, D. B., 1981, Generation of Martian chaos and channels by debris flows: Icarus, v. 45, p. 77–86.

Nye, J. F., 1976, Water flow in glaciers; jökulhlaups, tunnels and veins: Journal of Glaciology, v. 17, p. 181–207.

Oberbeck, V. R., Quaide, W. L., and Greeley, R., 1969, On the origin of lunar sinuous rilles: Modern Geology, v. 1, p. 75–80.

Oberlander, T., 1965, The Zargos streams; A new interpretation of transverse drainage in an orogenic zone: Syracuse Geographical Series, no. 1, 168 p.

O'Connor, J. E., Webb, R. H., and Baker, V. R., 1986, Paleohydrology of pool and riffle pattern development, Boulder Creek, Utah: Geological Society of America Bulletin, v. 97, p. 410–420.

Pardee, J. T., 1910, The glacial Lake Missoula, Montana: Journal of Geology, v. 18, p. 376–386.

——, 1922, Glaciation in the Cordilleran region: Science, v. 56, p. 686–687.

——, 1942, Unusual currents in glacial Lake Missoula, Montana: Geological Society of America Bulletin, v. 53, p. 1569–1600.

Partridge, J. B., and Baker, V. R., 1987, Palaeoflood hydrology of the Salt River, Arizona: Earth Surface Processes and Landforms (in press).

Playfair, J., 1802, Illustrations of the Huttonian theory of the earth: Edinburgh, William Creech, 528 p.

Reidel, S. P., 1984, The Saddle Mountains; The evolution of an anticline in the Yakima fold belt: American Journal of Science, v. 284, p. 972–978.

Ross, C. P., 1947, Geology of the Borah Peak Quadrangle, Idaho: Geological Society of America Bulletin, v. 58, p. 1085–1160.

Russell, I. C., 1900, A preliminary report on the geology of the Cascade Mountains in northern Washington: U.S. Geological Survey, 20th Annual Report, Part 2, p. 83–210.

——, 1901, Geology and water resources of Nez Perces County, Idaho: U.S. Geological Survey Water-Supply Paper 53, 85 p.

——, 1902, Geology and water resources of the Snake River Plains of Idaho: U.S. Geological Survey Bulletin 199, 192 p.

Sadler, P. M., 1981, Sediment accumulation rates and the completeness of stratigraphic sections: Journal of Geology, v. 89, p. 569–584.

Schaber, G. G., 1973, Lava flows in Mare Imbrium; Geologic evaluation from Apollo orbital photography: Proceedings, 4th Lunar Science Conference, p. 73–92.

Schmincke, H.-U., 1964, Petrology, paleocurrents, and stratigraphy of the Ellensburg Formation and interbedded Yakima Basalt flows, south-central Washington [Ph.D. thesis]: Baltimore, The Johns Hopkins University, 425 p.

——, 1967, Flow directions in Columbia River basalt flows and paleocurrents of interbedded sedimentary rocks, south-central Washington: Geologische Rundschau, v. 56, p. 992–1019.

Schultz, P. H., 1976, Moon morphology: Austin, University of Texas Press, 626 p.

Schultz, P. H., and Greeley, R., 1975, Lunar ring-moat structures; Constraints on degradational models: EOS American Geophysical Union Transactions, v. 56, p. 1015.

Schultz, P. H., Greeley, R., and Gault, D. E., 1976, Degradation of small mare surface features: Proceedings, 7th Lunar Science Conference, p. 985–1003.

Schumm, S. A., 1985, Explanation and extrapolation in geomorphology; Seven reasons for geologic uncertainty: Japanese Geomorphological Union Transactions, v. 6-1, p. 1–18.

Schumm, S. A., and Shepherd, R. G., 1973, Valley floor morphology; Evidence of subglacial erosion: Area, v. 5, p. 5–9.

Scott, W. E., McCoy, W. D., Shroba, R. R., and Rubin, M., 1983, Reinterpretation of the exposed record of the last two cycles of Lake Bonneville, western United States: Quaternary Research, v. 20, p. 261–285.

Sharp, R. P., and Malin, M. C., 1975, Channels on Mars: Geological Society of America Bulletin, v. 86, p. 593–609.

—— , 1984, Surface geology from Viking landers on Mars; A second look: Geological Society of America Bulletin, v. 95, p. 1395–1412.

Shea, J. H., 1982, Twelve fallacies of uniformitarianism: Geology, v. 10, p. 455–460.

Shepherd, R. G., and Schumm, S. A., 1974, An experimental study of river incision: Geological Society of America Bulletin, v. 85, p. 257–268.

Simons, D. B., Richardson, E. V., and Nordin, C. F., Jr., 1965, Forms generated by flow in alluvial channels: Society of Economic Paleontologists and Mineralogists, Special Publication 12, p. 34–52.

Smith, G. O., 1901, Geology and water resources of a portion of Yakima County, Washington: U.S. Geological Survey Water-Supply Paper 55, 68 p.

Smith, G. R., Swirydczuk, K., Kimmell, P. G., and Wilkinson, B. H., 1982, Fish biostratigraphy of the Miocene to Pliocene sediments of the western Snake River Plain, Idaho, in Bonnichsen, B., and Breckenridge, R. M., eds., Cenozoic geology of Idaho: Idaho Bureau of Mines and Geology Bulletin 26, p. 519–541.

Smith, N. D., 1978, Some comments on terminology for bars in shallow rivers, in Miall, A. D., ed., Fluvial sedimentology: Canadian Society of Petroleum Geologists Memoir 5, p. 85–88.

Stearns, H. T., Crandall, L., and Steward, W. G., 1938, Geology and groundwater resources of the Snake River Plain in southeastern Idaho: U.S. Geological Survey Water-Supply Paper 774, 268 p.

Strahler, A. N., 1945, Hypotheses of stream development in the folded Appalachians of Pennsylvania: Geological Society of America Bulletin, v. 56, p. 45–88.

—— , 1952a, Hypsometric (area-altitude) analysis of erosional topography: Geological Society of America Bulletin, v. 63, p. 1117–1141.

—— , 1952b, Dynamic basis of geomorphology: Geological Society of America Bulletin, v. 63, p. 923–938.

Swanson, D. A., 1967, Yakima basalt of the Tieton River area, south-central Washington: Geological Society of America Bulletin, v. 78, p. 1077–1110.

Swanson, D. A., Wright, T. L., Hooper, P. R., and Bentley, R. D., 1979, Revisions in stratigraphic nomenclature of the Columbia River Basalt Group: U.S. Geological Survey Bulletin 1457-B, 59 p.

Swanson, D. A., and 7 others, 1980, Reconnaissance geologic map of the Columbia River Basalt Group, Pullman and Walla Walla Quadrangle, southeast Washington and adjacent Idaho: U.S. Geological Survey Miscellaneous Investigations Map I–1139, scale 1:250,000.

Swanson, D. A., Byerly, G. R., and Bentley, R. D., 1982, Columbia River Basalt Group, in Tabor, R. W. and 5 others, eds., Geologic map of the Wenatchee 1:100,000 Quadrangle, Washington: U.S. Geological Survey Miscellaneous Investigations Map I-1311.

Tabor, R. W., Waitt, R. B., Frizzell, V. A., Jr., Swanson, D. A., Byerly, G. R., and Bentley, R. D., 1982, Geologic map of the Wenatchee 1:100,000 Quadrangle, Washington: U.S. Geological Survey Miscellaneous Investigations Map I-1311.

Tabor, R. W., and 8 others, 1967, Geologic map of the Chelan 30-minute by 60-minute Quadrangle, Washington: U.S. Geological Survey Miscellaneous Investigations Map MI-1661 (in press), scale 1:100,000.

Teller, J. T., and Clayton, L., eds., 1983, Glacial Lake Agassiz: Geological Association of Canada Special Paper 26, 451 p.

Thornbury, W. D., 1965, Regional geomorphology of the United States: New York, John Wiley and Sons, 609 p.

—— , 1969, Principles of geomorphology: New York, John Wiley, 594 p.

Tolon, T. L., and Beeson, M. H., 1984, Intracanyon flows of the Columbia River Basalt Group in the lower Columbia River gorge and their relationship to the Troutdale Formation: Geological Society of America Bulletin, v. 95, p. 463–477.

Waitt, R. B., Jr., 1979, Late Cenozoic deposits, landforms, stratigraphy, and tectonism of Kittitas Valley, Washington: U.S. Geological Survey Professional Paper 1127, 18 p.

—— , 1980, About forty last-glacial Lake Missoula jökulhlaups through southern Washington: Journal of Geology, v. 88, p. 653–679.

—— , 1984, Periodic jökulhlaups from Pleistocene glacial Lake Missoula; New evidence from varved sediment in northern Idaho and Washington: Quaternary Research, v. 22, p. 46–58.

—— , 1985, Case for periodic, colossal jökulhlaups from glacial Lake Missoula: Geological Society of America Bulletin, v. 95, p. 1271–1286.

Waitt, R. B., and Thorson, R. M., 1983, The Cordilleran ice sheet in Washington, Idaho, and Montana, in The Late Pleistocene, Volume 1 of Late Quaternary environments of the United States: Minneapolis, University of Minnesota Press, p. 53–70.

Wallace, D., and Sagan, C., 1979, Evaporation of ice in planetary atmospheres; Ice-covered rivers on Mars: Icarus, v. 39, p. 385–400.

Warren, C. R., 1941, Course of the Columbia River in southern central Washington: American Journal of Science, v. 239, p. 209–232.

Waters, A. C., 1939, Resurrected erosion surface in central Washington: Geological Society of America Bulletin, v. 50, p. 638–659.

—— , 1955, Geomorphology of south-central Washington, illustrated by the Yakima East Quadrangle: Geological Society of America Bulletin, v. 66, p. 663–684.

—— , 1969, James Gilluly, pioneer of modern geological ideas: Earth-Science Reviews/Atlas, v. 5, p. A19–A27.

Webb, R. H., 1985, Late Holocene flooding on the Escalante River, south-central Utah [Ph.D. thesis]: Tucson, University of Arizona, 204 p.

Webster, G. D., Pankratz-Kuhns, M. J., and Waggoner, G. L., 1982, Late Cenozoic gravels in Hells Canyon and the Lewiston Basin, Washington and Idaho, in Bonnichsen, B., and Breckenridge, R. M., eds., Cenozoic geology of Idaho: Idaho Bureau of Mines and Geology Bulletin 26, p. 669–683.

Wheeler, H. E., and Cook, R. F., 1954, Structural and stratigraphic significance of the Snake River capture, Idaho-Oregon: Journal of Geology, v. 62, p. 525–536.

Whitford-Stark, J. L., 1982, Factors influencing the morphology of volcanic landforms; An Earth-Moon comparison: Earth Science Review, v. 18, p. 109–168.

Willis, R., 1903, Physiography and deformation of the Wenatchee-Chelan district, Cascade Range: U.S. Geological Survey Professional Paper 19, p. 41–101.

Willman, H. B., and Frye, J. C., 1970, Pleistocene stratigraphy of Illinois: Illinois State Geological Survey Bulletin 94, 204 p.

Wise, D. U., 1963, An outrageous hypothesis for the tectonic pattern of the North American Cordillera: Geological Society of America Bulletin, v. 74, p. 357–362.

Womar, M. B., Greeley, R., and King, J. S., 1982, Phreatic eruptions of the eastern Snake River Plain of Idaho, in Bonnichsen, B., and Breckenridge, R. M., eds., Cenozoic geology of Idaho: Idaho Bureau of Mines and Geology Bulletin 26, p. 453–464.

Manuscript Accepted by the Society November 10, 1986

ACKNOWLEDGMENTS

We thank Richard G. Craig and Harold E. Malde for their reviews of this chapter. Jim E. O'Connor assisted with the hydraulic calculations reported in Figures 24 and 25.

Geological Society of America
Centennial Special Volume 2
1987

Chapter 12

The Interior Mountains and Plateaus

Michael J. Bovis
Department of Geography, University of British Columbia, Vancouver, British Columbia V6T 1W5, Canada

INTRODUCTION

This chapter evaluates the most significant contributions to geomorphology from a region of considerable climatic, topographic, and geologic diversity (Fig. 1). Climate ranges from subtropical semidesert, through temperate montane and alpine environments, to boreal forest and arctic tundra. This wide range of conditions has prompted the appearance of significant publications in virtually every subdiscipline of geomorphology. The region is renowned for original contributions in topics as diverse as erosional surfaces, volcanic landforms, glaciation, periglacial landforms, and fluvial and hillslope processes. These form the major themes of this chapter. There are two criteria for the inclusion of a work in this review. The first, and most restrictive, is that it contain a fundamental contribution to geomorphic theory not previously stated elsewhere. If this were followed rigorously, relatively few works would be included. A second criterion is that the work be a significant contribution to the geomorphology of the region. Although this chapter is not an account of regional geomorphology, the second criterion has been adopted for the following reasons. First, it allows contributions to basic theory to be included. Second, it places great value on regional or local studies that have contributed to knowledge of the morphology, the evolution, and the geographic distribution of particular landforms. Since complexity of landform evolution and distribution is the rule, deeper theoretical understanding of landforms in this region, as elsewhere, has occurred only after a prolonged period of empirical investigation.

In this chapter, relatively little attention is paid to Quaternary paleoenvironments and topics such as glacial stratigraphy and correlation. These have been thoroughly reviewed in works such as Wright and Frey (1965), Porter (1983), and Wright (1983). Moreover, other volumes in the Decade of North American Geology (DNAG) series have addressed this wider field of "Quaternary science, " of which Quaternary geomorphology is but a part. In contrast to the large review literature on the Quaternary in general, relatively little attention has been paid to assessing contributions to the study of landforms, particularly within the Interior Mountains and Plateaus region. The most comprehensive review of American work is Thornbury (1965),

which though still a valuable reference on regional geomorphology, does not cover work completed in the last 20 years. Works on the northern Cordillera, such as Holland (1964) and Wahrhaftig (1965a) are now similarly dated, but still serve as basic reference material. In comparison with allied disciplines such as regional tectonics, Cordilleran geomorphology has lacked a coherent statement of its major accomplishments.

An important consideration in such a review is to decide on the depth of the historical perspective in each subarea. In some cases, work completed since 1960 is emphasized, as seems appropriate given the scope of earlier regional overviews. However, some research topics saw great development in the study region during the late nineteenth century and first half of the twentieth century, notable examples being volcanic landforms, Cenozoic erosional history, and glacial erosion. In those sections, therefore, full acknowledgment is made of the contributions made by these ground-breaking studies. A final word is in order concerning the relative weighting given to each part of the study region, since the reader might conclude that too much emphasis has been placed on the Cascades and Sierra Nevada. While these areas are quite small relative to the entire region, nonetheless they have produced many significant contributions to geomorphic knowledge. This in no way diminishes the intrinsic value to geomorphology of other regions, such as the Canadian Cordillera, or the intermontane belt of Alaska. This review, however, must be influenced primarily by work in print, rather than by the potential for geomorphic studies.

Regional Topography

The study area extends as a sinuous belt of rugged mountain and dissected upland topography from the subtropical, southern tip of the Sierra Nevada, through the Cascade Range of Oregon and Washington, the Columbia Mountains, the Interior Plateau, and northern interior ranges of British Columbia, and terminates in the subarctic to arctic upland and lowland zone of the Yukon and Kuskokwim river basins in Alaska (Fig. 1). The area comprises thirty physiographic regions, many of which are geologically distinct from one another.

Bovis, M. J., 1987, The Interior Mountains and Plateaus, *in* Graf, W. L., ed., Geomorphic systems of North America: Boulder, Colorado, Geological Society of America, Centennial Special Volume 2.

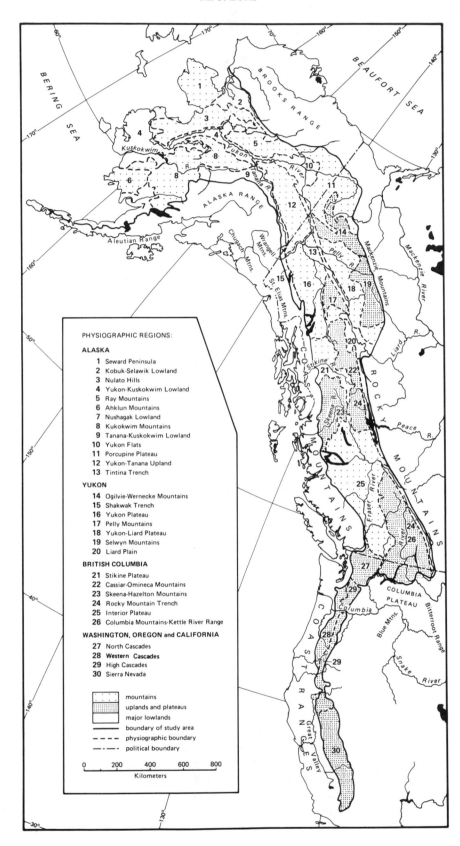

PHYSIOGRAPHIC REGIONS:

ALASKA

1 Seward Peninsula
2 Kobuk-Selawik Lowland
3 Nulato Hills
4 Yukon-Kuskokwim Lowland
5 Ray Mountains
6 Ahklun Mountains
7 Nushagak Lowland
8 Kukokwim Mountains
9 Tanana-Kuskokwim Lowland
10 Yukon Flats
11 Porcupine Plateau
12 Yukon-Tanana Upland
13 Tintina Trench

YUKON

14 Ogilvie-Wernecke Mountains
15 Shakwak Trench
16 Yukon Plateau
17 Pelly Mountains
18 Yukon-Liard Plateau
19 Selwyn Mountains
20 Liard Plain

BRITISH COLUMBIA

21 Stikine Plateau
22 Cassiar-Omineca Mountains
23 Skeena-Hazelton Mountains
24 Rocky Mountain Trench
25 Interior Plateau
26 Columbia Mountains-Kettle River Range

WASHINGTON, OREGON and CALIFORNIA

27 North Cascades
28 **Western Cascades**
29 High Cascades
30 Sierra Nevada

mountains
uplands and plateaus
major lowlands
boundary of study area
physiographic boundary
political boundary

0   200   400   600   800
Kilometers

Figure 1. Physiographic regions. (Compiled from Feneman, 1931; Bostock, 1948; Holland, 1964; and Wahrhaftig, 1965a).

Elevations are generally highest in the Sierra Nevada, especially along the elevated eastern rim of the range where many peaks, including Mount Whitney, rise to more than 3,800 m, and are flanked to the west by large areas higher than 3,000 m. The fault-defined eastern escarpment in the Mount Whitney area is one of the highest in the Americas. The western slope of this huge asymmetric fault block declines to the Great Valley of California, and is deeply incised by the southwest-flowing headwaters of the Sacramento and San Joaquin rivers. The elevation of the Sierra Nevada declines notably in a northwesterly direction toward Lassen Peak, which marks the southern limit of the volcanic Cascades Physiographic Province. Over the 1,100 km distance from Lassen Peak, in northern California, to Meager Mountain in southwest British Columbia, the Cascades mountain belt contains more than a dozen major stratovolcanic centers (Fig. 2), the highest of which (Mounts Shasta, Adams, and Rainier) approach the elevations of some of the High Sierra peaks. Such is the sustained, high elevation of the Sierra-Cascade belt that, over a distance of 1,700 km, it is traversed by only the Columbia, Klamath, and Pit rivers.

In the Canadian and Alaskan sections of the Interior Mountains and Plateaus, only the Columbia Mountains, with two peaks higher than 3,500 m, are at all comparable in elevation with the Sierra-Cascade belt. By comparison, the Interior Plateau of British Columbia is relatively subdued, averaging only 1,000 m, with isolated ranges rising to 1,800 m. Major rivers, such as the Fraser and Columbia, are incised 500 to 1,000 m below the upland surface. To the north of the Interior Plateau lies the Cassiar-Omineca mountain belt, dissected by the east-flowing Peace and Liard river systems; and the Skeena Mountains, cut by the west-flowing Skeena, Nass, and Iskut rivers. Isolated peaks attain 2,400 m, but most areas are below 2,000 m. To the north of these ranges lies the Stikine Plateau, developed between 600 and 1,200 m, and dominated on its western margin by the much higher Mount Edziza volcanic complex, which attains 2,700 m. The Yukon Plateau, drained by the Teslin and Pelly tributaries of the Yukon River system, averages 1,000 m in elevation, with the highest points attaining almost 2,000 m in the Plateau proper, and 2,500 m in the Ogilvie Mountains, which form its northeastern boundary. All of the ranges that lie between the Interior Plateau and the Alaskan border retain the dominant northwest-to-southeast strike of rocks in this part of the Cordillera.

The Alaskan portion of the study region consists of a series of upland blocks, separated by large intermontane basins, aligned parallel to the mainly east-west trend of the Cordillera in Alaska. Most of the uplands of central Alaska are drained by the Yukon River system and generally lie below 1,500 m. By contrast, the west-to-east trending Brooks Range, which forms the northern boundary of the region, has large areas above 1,500 m, and several peaks over 2,500 m in the eastern half of the range.

Regional Geology and Tectonics

Introduction. Emphasis is placed here on the geologic character and development of the physiographic regions de-

scribed above, though for completeness, brief reference is made to adjacent regions that lie just outside of the study area. Much of this section is based on the summary works dealing with Mesozoic and Cenozoic paleogeography (Howell and McDougall, 1978; Armentrout and others, 1979), as well as more recent regional syntheses (e.g., Coney and others, 1980; Ernst, 1981; Monger and others, 1982; and Cowan, 1985).

Many physiographic boundaries are broadly conformable with major tectonic and stratigraphic discontinuities, which demarcate distinctive terranes within the Cordilleran orogen. The strike of rock units and the topographic grain are, therefore, more or less coincident in many areas. It is now generally accepted that the Cordillera west of the craton boundary is a "collage" of terranes, accreted to the continental margin, primarily during the Mesozoic era. The lateral extent of each terrane is evident from the juxtapositioning of rock assemblages of contrasting lithological and paleontological characteristics.

The Canadian and Alaskan Intermontane Belts. Monger and others (1982) identify several tectonostratigraphic belts within the Canadian and Alaskan cordilleras, although two belts, the Omineca and the Intermontane, constitute much of the northern half of the study area. The northwest-trending Omineca Crystalline Belt runs through the Columbia, Cassiar, Omineca, and Pelly ranges, and comprises granitic plutons and high-grade metamorphics, which evolved from Middle Jurassic through Cretaceous time. This plutonic and metamorphic welt is lodged against the Paleozoic miogeoclinal sediments that form much of the Rocky Mountain Belt, and is separated from them by the Rocky Mountain Trench, a major tectonic boundary in this part of the Cordillera. The earlier Jurassic magmatic and deformation events in the Omineca Belt probably coincided with the suturing of the easternmost of the exotic terranes that make up most of the Intermontane Belt.

A similar pattern of terrane accretion throughout Mesozoic and early Cenozoic time is now known to have taken place in much of Alaska. Compared with the Canadian Cordillera, a very intricate collage was assembled, involving principally Paleozoic and Mesozoic marine sediments and Mesozoic volcanic arc complexes (Jones and others, 1978). The trend of sutures suggests a northerly to northeasterly direction of subduction, thereby establishing the distinctive curvature of the regional strike, which has persisted to the present. The Cretaceous and early Tertiary periods also saw the development of a roughly parallel, curved set of dextral transcurrent faults in Canada and Alaska. Principal shears include the Tintina and Rocky Mountain trenches, and the Pinchi, Denali, and Fraser River zones. Right-lateral offsets of around 450 km have been calculated for the northern part of the Cordillera.

During the first half of the Cenozoic, extension of the Queen Charlotte fault zone restricted magmatic arc development to southern British Columbia—related to subduction of the Juan de Fuca plate—and to the Aleutian-Wrangell volcanic belt—related to Pacific plate subduction. The Alaskan zone was very active in late Cenozoic time, producing a thick, andesitic volcanic se-

472 M. J. Bovis

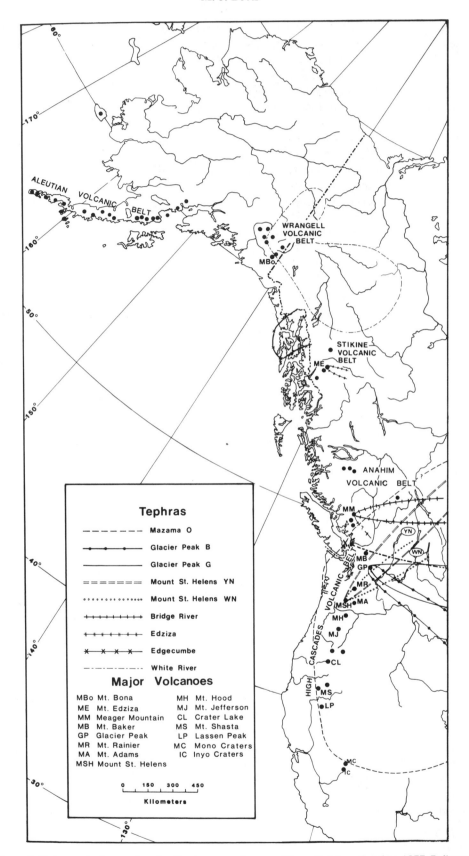

Figure 2. Holocene tephras and major Quaternary volcanoes. (Compiled from Souther, 1977; Beikman, 1980; Clague, 1981a; and Sarna-Wojcicki and others, 1983.)

quence in the Wrangell Mountains (Fig. 2). By comparison, Cenozoic volcanics in interior Alaska are discontinuous, and of relatively small extent.

The large areas of Eocene to early Miocene volcanics in the Interior Plateau of British Columbia range from basaltic to rhyolitic in composition (Souther, 1977). Late Miocene and Pliocene time saw the eruption of large volumes of plateau basalt in the Chilcotin and Fraser plateau areas of southwest British Columbia. The lavas are relatively flat lying in the Fraser River area, but are notably upwarded toward the west, indicating significant post-Miocene uplift of the Coast Mountains (Roddick, 1966). Rocks of similar age and composition erupted in the Stikine volcanic belt. Quaternary volcanism, again mainly basaltic, has been particularly active in the Stikine belt, including the massive Mount Edziza complex (Fig. 2). In the Anahim belt, the timing of volcanism is progressively later toward the east, probably related to easterly migration of a mantle hotspot (Bevier and others, 1979).

The Cascades. This mountain belt consists of two distinctive terranes with completely different geologic histories: the plutonic-metamorphic North Cascades block, which extends northward from latitude 47°N into the Cascade belt of southwest British Columbia; and the Cenozoic volcanic Cascades, which stretch from the headstreams of the Yakima River south to Lassen Peak in northern California (Fig. 1, 2). The North Cascades underwent a prolonged period of erosion, following uplift and overthrusting in late Cretaceous time. This exposed the plutonic and metamorphic core of the range. The range has been arched upward along a north-south axis in the past 10 m.y. as part of the late Cenozoic reelevation of much of the Cordillera. The stratovolcanoes of Mount Baker and Glacier Peak were built in the Quaternary period, at roughly the same time as the other major volcanic peaks of the central and southern Cascades.

The volcanic Cascades are subdivisible into an older, more subdued Western Cascade belt, comprising Eocene to Miocene accumulations; and the High Cascades, which include Plio-Quaternary basaltic shield volcanoes surmounted by andesitic to dacitic stratovolcanoes of Quaternary age (Hammond, 1979). Many of these composite cones have erupted over the Holocene, notably Mount St. Helens, Mount Mazama (Crater Lake), Mount Rainier, and Glacier Peak (Sarna-Wojcicki and others, 1983). Lassen Peak is somewhat different from the majority of Cascade volcanoes since it consists of a huge dacitic dome complex, which has breached through an earlier, deeply dissected stratovolcanic sequence.

The Sierra Nevada. The Mesozoic tectonic development of the Sierra Nevada has certain broad similarities with the Canadian and Alaskan regions described above, in that subduction and the accretion of sedimentary wedges occurred from the upper Paleozoic through the late Mesozoic. As in the northern Cordillera, an imbricate pattern of sedimentary wedges and interjected ophiolite slices was emplaced along a series of westerly verging thrusts. Large masses of sediment and arc volcanics were subsequently metamorphosed along several suture zones. Granitic intrusions associated with several magmatic arcs of Jurassic age

occurred along the length of the Klamath-Sierra belt. From late Jurassic to the Oligocene, the Franciscan subduction complex was accreted to the western continental margin. A huge granitic batholith was emplaced along the Sierra Nevada axis, representing the 'root' of the Franciscan magmatic arc. Regional uplift associated with these intrusions caused earlier Sierra cover rocks and plutons to be eroded and shed westward into the Great Valley.

From the early Miocene to the middle Pliocene, several phases of volcanism affected the Sierra, especially its northern half, where early Tertiary valleys were completely inundated by tuffs and flows. This was followed in the middle Miocene to early Pliocene by andesitic breccias, lavas and mudflows, and subsequently by latite flows, some of which are now preserved as mesas on interfluves (Dalrymple, 1963). Quaternary volcanism has been confined mainly to the Mono Lake–Mammoth Lakes area, and the Inyo Craters (Fig. 2). As in the Cascade belt, volcanism has yielded valuable tephras for the correlation of glacial and nonglacial deposits on the east flank of the Sierra (Dalrymple, 1964; Dalrymple and others, 1965). Prominent among these tephras is the Bishop tuff, dating from about 700 ka (Izett and others, 1970).

Cenozoic Development of Landforms

The physiographic history of the study area has been covered by monographs and texts dealing either wholly, or in part, with the geomorphology of particular regions. Notable among these are Fenneman (1931), Bostock (1948), Holland (1964), Thornbury (1965), Wahrhaftig (1965a), McKee (1972), and Huber (1981). In addition, passing reference to Cenozoic erosional history has been made in many articles dealing with regional tectonics.

Tertiary History. The several episodes of orogenic activity in Late Cretaceous to late Eocene time established some of the present topographic alignments within the Interior Mountains and Plateaus region. Throughout the Paleogene, prolonged subaerial denudation occurred, most notably in the more strongly uplifted crystalline areas of the Sierra Nevada, the North Cascades, and the Omineca crystalline belt of eastern British Columbia. Erosional detritus was shed into adjacent, less elevated zones such as the California Great Valley, the Puget-Willamette depression, and the Columbia embayment, as well as numerous fault-defined features within the Canadian Cordillera, such as the Fraser Valley and the Rocky Mountain Trench. Thick Eocene to Oligocene successions accumulated peripheral to eroding uplands in Alaska.

There is widespread evidence that by the Miocene epoch, much of the study area had been reduced to a lower relief than exists at present. Compared with Eocene and Pliocene times, both of which saw great orogenic events, the intervening Oligocene and Miocene epochs were periods of relative tectonic quiescence, in which erosion generally outstripped the rate of uplift. The major exception to this was the Miocene orogeny that affected

474   M. J. Bovis

much of southern Alaska. Extensive areas of middle to late Tertiary erosion surfaces remain in the Sierra Nevada, the Interior
Plateau of central British Columbia, the Stikine and Yukon plateaus, and the uplands of central Alaska. All of these surfaces
were dissected by a cycle of canyon cutting produced by renewed
uplift during the Plio-Pleistocene period.

In Alaska the intermontane belt consists of a series of elongated, structural downwarps containing thick fills of Tertiary sediments, derived from the Mesozoic to Paleozoic rocks of the
intervening uplands. Reduction of many areas to a surface of low
relief probably progressed at the same rate as the filling of adjacent, subsiding lowlands. Remnants of one or more erosion surfaces are widespread in these uplands. Anomalous drainage
patterns are common in interior Alaska, and have been discussed
previously by Wahrhaftig (1958, 1965a) and Thornbury (1965).
Although many of the intermontane lowlands are connected by
shallow troughs and passes, the trunk streams and major tributaries do not always follow them, but instead cut through the uplands in deep "interlowland" gorges. Such a pattern suggests
drainage superimposition. Plio-Quaternary uplift of the Alaska
Range, of the order of 2,000 to 3,000 m, has warped the late
Tertiary erosion surface. Despite this rapid uplift, which created
some of the highest peaks in North America, several north-
flowing tributaries of the Tanana River have maintained their
courses through the Range in deep gorges.

Extensive erosion surface remnants occur also in northern
British Columbia and central Yukon (Bostock, 1948; Tempelman-Kluit, 1980) and are particularly well preserved in the less
elevated area of the Yukon Plateau. More strongly uplifted areas
such as the Selwyn, Cassiar, Skeena, and Omineca mountains,
have been almost completely dissected, and preserve few remnants of the upland surface. Relatively little is known of the
Tertiary drainage history of this large region. Recently, Tempelman-Kluit (1980) has reconstructed events within southern
Yukon Territory. In the Miocene epoch, an ancestral southwest-
flowing system, involving most of the present Yukon River headstreams, flowed by a direct route to the Pacific via the Dezadeash
and Alsek valleys. This trunk stream alignment persisted as the St.
Elias and Coast mountains began to rise, and continuity of tributaries was maintained across the Shakwak and Tintina trenches as
these evolved by normal faulting. Final disruption of this system
to produce the present northwest-flowing Yukon River occurred
by glacial diversion during several Pleistocene glaciations.

The amount of Plio-Pleistocene uplift in the Interior Plateau
of British Columbia was apparently greatest in its southern portion, as shown by the present northward inclination of the plateau
basalt layers, and the greater incision of the south-flowing Fraser
River in this section. The south-to-north slope of the erosion
surface in much of the Fraser River basin, against the present
river gradient, is macroscale evidence of a late Tertiary reversal of
the river flow direction (Lay, 1940).

The presence of a thick Miocene sedimentary sequence in
the Rocky Mountain Trench (Clague, 1974) indicates that this
major topographic feature was already well developed as a half-

graben by that time. The present, remarkable alignment of the
Kootenay, upper Columbia, and upper Fraser drainages is probably attributable, therefore, to Cenozoic normal faulting along the
trench. The pronounced "elbow" on the Columbia River strongly
suggests capture of the trench section of the river by a more
vigorous, south-flowing proto-Columbia. This river, together
with the Kootenay and Okanagan, constituted the ancestral
drainage running roughly parallel to the regional strike within the
southern part of the uplifted Omineca Crystalline Belt.

In the North Cascades, erosion surface remnants are much
less extensive than in the Interior Plateau because of strong uplift
accompanied by fluvial dissection that have prevailed since the
early Pliocene. The Skagit River is considered to be antecedent to
this uplift. In contrast with the North Cascades, the pre-Pliocene
evolution of relief in the volcanic Cascades depended to a large
extent on volcanic outpourings, in addition to differential uplift
and erosion (Hammond, 1979). The Pit, Klamath, and Columbia
rivers maintained their antecedent courses across the rising Cascade chain. Pre-Pliocene antecedence of the Columbia, in approximately its present position, is established by the presence of
Mio-Pliocene plateau basalts within the Columbia River Gorge,
and by the existence of fluvial deposits at the western end of the
gorge, derived from sources far to the east and northeast.

The crystalline rocks of the Sierra Nevada preserve evidence
of at least two Tertiary cycles of planation (Matthes, 1930; Dalrymple, 1963). The relatively flat upland interfluves probably
date back at least to the Miocene. Broad valleys were cut into this
surface in the late Miocene, when the Sierra Nevada was notably
lower than present. The final cycle of canyon cutting began in the
Pliocene, in response to uplift along the eastern fault zone. The
past 10 m.y. have seen over 2,300 m of uplift (Huber, 1981),
causing the headstreams of the present west-flowing consequent
drainage lines to be severed close to the Owens Valley axis.

Pleistocene History. By the beginning of the Quaternary,
the main features of the modern bedrock topography had developed, although some mountain areas were probably lower than
present. Changes in the macro-topography, caused by glaciation,
were minor when compared with those wrought by fluvial erosion during the longer Mio-Pliocene interval. Many areas of the
Interior Mountains and Plateaus region preserve a record of two,
and in some cases, four separate glaciations. In major lowlands
such as the Puget, overlapping till layers occur, separated by
nonglacial deposits. In mountain areas such as the Sierra Nevada,
the Cascades, and the Brooks Range, nested morainal sequences
are usually observed (Porter and others, 1983; Waitt and Thorson, 1983). Many upland areas of British Columbia and Yukon,
though located toward the core of the Cordilleran glacier complex (Fig. 3), also preserve depositional evidence of at least two
glaciations (Hughes and others, 1972; Rutter, 1976; Clague,
1981a). Correlation between the various glaciated areas prior to
the latest Wisconsin glaciation is usually difficult because of poor
dating control on organic and volcanic materials interbedded
with glacial deposits.

In the late Wisconsin, as in earlier glaciations, the largest ice

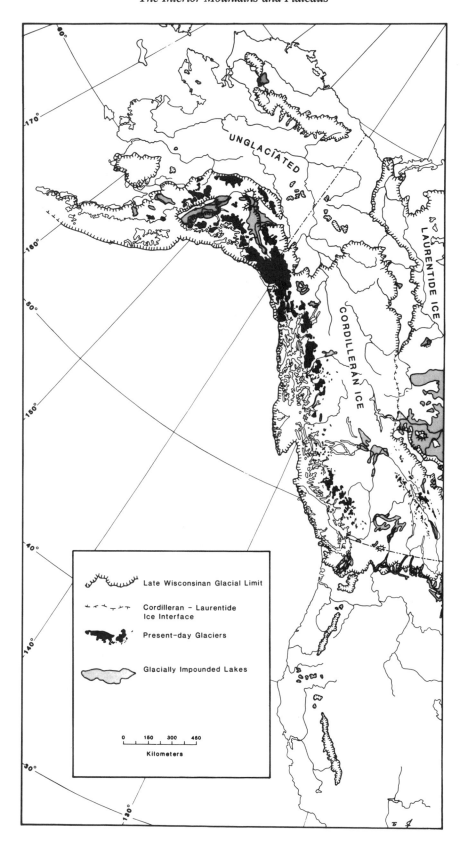

Figure 3. Late Wisconsinan glacial limits of the western Cordillera, and major areas of glaciolacustrine sediments. (Compiled from Crandell, 1965; Wahrhaftig and Birman, 1965; Prest and others, 1967; Péwé, 1975; Porter and others, 1983; and Waitt and Thorson, 1983.)

body by far was the Cordilleran glacier complex (Fig. 3). This formed primarily by the coalescence of piedmont lobes from the high mountain ranges of British Columbia and southern Yukon (Clague, 1981a). Virtually all of the intermontane belt was inundated by ice, with the exception of most of interior Alaska and northwest Yukon, which remained ice-free throughout the Quaternary. These areas were subjected instead to a harsh tundra climate, as revealed by fossil ice wedges and other paleoenvironmental indicators (Péwé, 1975, 1983). The fringes of these northern intermontane areas were covered by piedmont lobes from the Alaska, Wrangell, and St. Elias ranges, which were somewhat separate from the main Cordilleran glacier complex (Denton, 1974; Péwé, 1975; Rampton, 1981). The southern limit of the Cordilleran ice at the maximum of the last glaciation extended from the Puget Lowland, across the northern part of the North Cascades, through to the northern boundary of the Columbia Plateau. Ice pushed forward as three major lobes: the Puget, the Okanogan, and the Columbia.

Smaller icecaps and glacier complexes accumulated in the Brooks Range, the Cascades, and the Sierra Nevada, where long alpine-type glaciers extended down major river valleys for tens of kilometers through otherwise unglaciated areas. In the southern Cascades, glaciation was largely restricted to the Mount Shasta and Lassen Peak areas (Porter and others, 1983). Numerous small cirque and valley glaciers also developed in the highest parts of the Klamath Mountains.

It is now known that the pattern of ice buildup and retreat during the late Wisconsin was by no means synchronous throughout the Interior Mountains and Plateaus region. The maximum extent of the Cordilleran ice, at least in the south, was attained between 15 and 14 ka. Alpine glaciers in the Brooks, Alaska, Columbia, Rocky, and Cascade mountains, however, appear to have reached their maximum extent around 17 to 18 ka, and had retreated significantly by the time of the Cordilleran maximum (Clague, 1981a; Porter and others, 1983; Waitt and Thorson, 1983). By about 10 ka, the Cordilleran ice had retreated north of the 49th parallel, and glaciers in the Cascade-Sierra belt were not substantially larger than present. In the interior mountains and plateaus of British Columbia and Yukon, deglaciation was somewhat slower, but by 9.5 ka ice caps and glaciers were not much larger than today in the Columbia, Skeena, and Cassiar mountains.

The types of glacial erosional landforms, and their degree of development throughout the study region, depended on ice thickness and velocity, the direction of ice motion relative to valley alignments, rock lithology and structure, and the general style of glaciation in a given area. High-mountain areas such as the Sierra-Cascade Belt, the Columbia Mountains, the Skeena Mountains, and the higher parts of the Stikine and Yukon plateaus, all display many of the classic forms of alpine glaciation, since each possessed a large area above the glaciation threshold during both the advance and retreat phases. However, many of the spectacular glacial troughs, cirques, and overdeepened lake basins must be regarded as the products of multiple glaciation, in that the pat-

terns of ice accumulation, motion, and wastage were probably similar in successive glaciations. Within the Canadian intermontane area, on the other hand, ice buildup occurred primarily by the coalescence of piedmont-type glaciers, which led to considerable smoothing and fluting of bedrock, producing a more subdued glaciated landscape than in the adjacent mountain areas. Ice discharging from the intermontane area also caused deep scouring along valleys running parallel to the direction of ice motion.

Glacial depositional landforms also vary markedly according to the style of glaciation and deglaciation. Mountain areas occupied by oscillating, alpine-type glaciers during the final stages of the late Wisconsin glaciations preserve morainal ridges marking recessional pauses, or minor readvances, during deglaciation. These are well preserved in the Cascades, and strikingly so in the Sierra Nevada. In both areas, morainal topography from earlier Wisconsin and pre-Wisconsin glaciations is also well preserved, with the earliest, outermost sequences most strongly modified by weathering and erosion (Porter and others, 1983; Shroba and Birkeland, 1983).

In contrast to nested moraines in the mountain valleys, much of the intermontane belt typically preserves a blanket of glacial drift, derived from one or more Wisconsin glaciations (Clague, 1986; Fulton and Smith, 1978; Tipper, 1971). The record of sedimentation from Quaternary nonglacial intervals is generally sparse on account of limited sedimentation, and the general stream incision during these periods (Clague, 1986). Thick basal and ablation tills cover large areas of plateaus and interfluves, but are often covered by late glacial and postglacial deposits along the major valleys. Basal till is often sculpted into drumlin forms, with glacially moulded bedrock occurring where drift is thin or absent. In contrast to the surrounding mountain areas, deglaciation of large plateau and upland areas occurred by downwasting (Fulton, 1967). Blockage of the major valleys by stagnant ice, or by retreating lobes of Cordilleran ice, created large proglacial lakes in southern and central British Columbia, the middle Stikine Valley, and southwest Yukon (Fig. 3). All of these areas preserve thick sequences of glaciolacustrine silts and clays overlying till or recessional outwash. Extensive networks of glacial meltwater channels, often discordant with the present drainage, were cut across many plateau areas either by streams issuing directly from an ice front, or by proglacial lakes overtopping divides (Tipper, 1971). Large esker complexes were developed on the gently undulating plateaus of central British Columbia, and in the more rugged plateau areas of southern Yukon, often in association with major meltwater channels. Glacial lakes were also impounded by the Puget Lobe of the Cordilleran ice sheet in the southern Puget Lowland and along the west flank of the North Cascades; and by the Okanogan Lobe in the Chelan, Okanogan, and middle Columbia Valleys. The complex late-Pleistocene history of the middle Columbia River in response to the Lake Missoula floods is told in detail in Chapter 11 of this volume (Columbia Plateau).

Pleistocene conditions within the unglaciated areas of

Yukon and Alaska have been described by Hughes and others (1972), and Péwé (1975, 1983). At present, most of central Alaska and Yukon lies within the zone of discontinuous permafrost (Fig. 4). During the Pleistocene glaciations, continuous permafrost prevailed over this large unglaciated region, and produced ice-wedge polygons, solifluction deposits, and large cryoplanation terraces, the latter strikingly displayed in the Yukon-Tanana region and other upland areas of central Alaska and Yukon. During the Pleistocene in Alaska, silt and clay were deflated from broad outwash plains at far above the present rate. Loess deposits, some predating the Illinoian glaciation, are therefore important within the subarctic unglaciated regions. Deposits are thick along the major river valleys and extend as tapering wedges into the surrounding uplands (Péwé, 1975).

Holocene History. Changing conditions within the fluvial, glacial, and mass-wasting systems over the Holocene have resulted in significant landform adjustments, many of which are still in progress. Changes in all three systems have been closely interlinked in many areas due to the forcing of glacial and hydrologic fluctuations by shifts in Holocene climate. Volcanism has also exerted a strong, local influence on Holocene landform development, principally in the High Cascades, and in the Stikine volcanic belt.

Although the more subtle climatic changes over the Holocene have not had obvious landform expression, the change from glacial to nonglacial conditions, via an intervening period of 'paraglacial' sedimentation in the early Holocene, had a major impact on landform development in the major river valleys (Church and Ryder, 1972). Many trunk streams in the Canadian and Alaskan parts of the Cordillera show evidence of late Pleistocene to early Holocene valley filling, followed by one or more phases of valley cutting, in which the fill was sculpted into stream-cut terraces. Many rivers have cut through well over 100 m of fill to the underlying bedrock. Valley filling was a response to high sediment loads from steep slopes mantled by unconsolidated, unvegetated drift. This produced thick colluvial wedges and debris fans grading into, and overlying, the main valley outwash train. Progressive stabilization of drift topography by changes in slope angle, revegetation, and, in some cases, outright depletion of drift cover, led to a significant decrease in sediment supply, and a change from net filling to net cutting. Stream incision has been less pronounced in valleys adjacent to large modern icefields. Present-day fluvial conditions along the Donjek and White rivers in Yukon, for example, may provide modern counterparts of early Holocene conditions in many of the larger valleys of the intermontane belt.

Holocene glacial fluctuations are well documented in the Interior Mountains and Plateaus region. Most work has been conducted in the Sierra Nevada and Cascades and is summarized by Porter and others (1983). Significantly fewer contributions have come from the Canadian and Alaskan parts of the study area (e.g., Denton and Stuiver, 1966; Rampton, 1970; Denton and Karlen, 1977). The overall pattern was one of glacier retreat during the first 4,000 to 5,000 years of Holocene time, corresponding to the Altithermal interval. In the past 4,000 to 5,000 years, significant Neoglacial readvances have occurred. The most recent, widely referred to as the Little Ice Age, saw glaciers in much of the study area attain their maximum post-Pleistocene dimensions within the past 200 years. Rock glaciers have also readvanced within the Neoglacial period, principally in the Sierra Nevada (Burke and Birkeland, 1979), the Kluane Range (Johnson, 1978), and in the Brooks Range (Ellis and Calkin, 1984). Spectacular examples, just beyond the study area, are described by Wahrhaftig and Cox (1959).

Postglacial volcanic activity within the study area has been confined to the Cascades and, in British Columbia, to the Anahim and Stikine belts. Of these, the High Cascade belt has produced the largest volume of volcanic ejecta, and related deposits. The most notable case is Mount Mazama (Crater Lake) in Oregon, where a series of explosive eruptions, around 6,700–7,000 years B.P. (Sarna-Wojcicki and others, 1983), deposited several tens of cubic kilometres of tephra over an area of at least 1.7×10^6 km^2, thereby creating the most useful stratigraphic marker in western North America (Fig. 2). Explosive volcanism, though on a lesser scale, has occurred at least seven times over the Holocene at Mount St. Helens, making it the most active of all Cascade volcanoes. In comparison with the widespread Mazama and St. Helens tephra layers, other Cascade volcanoes have generated relatively localized tephra falls within the past 10,000 years.

Postglacial activity in the Anahim Volcanic Belt of central British Columbia has produced small pyroclastic cones and blocky basaltic flows up to 16 km in length (Hickson and Souther, 1984). In the Mount Edziza complex of northern British Columbia, Holocene volcanic accumulations cover hundreds of square kilometers to the north and south of Mount Edziza (Souther and others, 1984), and include dozens of basaltic cinder cones and extensive blocky lava fields on the flanks of the main shield volcano. One of the most recent eruptions from this area occurred at about 1,300 years ago, and distributed tephra as a narrow band more than 10 km to the east (Clague, 1981a).

Holocene explosive volcanism in areas immediately peripheral to the study region has generated several tephras (Fig. 2), the most notable examples being (from north to south): the two White River Ash beds of Yukon and eastern Alaska, derived from the Mount Bona area of the St. Elias Mountains, and dating from 1.5 to 1.8 ka and 1.2 ka (Lerbekmo and Campbell, 1969); the Bridge River Ash bed of southern British Columbia and western Alberta, derived from Meager Mountain in the southern Coast Mountains, and dating from about 2.4 ka (Nasmith and others, 1967); and two unnamed tephras, which occur widely in the central and southern Sierra Nevada, derived from the Mono-Inyo craters zone, and dating from about 1.2 ka and 0.7 ka (Wood, 1977).

Holocene volcanism has also been associated with catastrophic rockslides and debris flows, especially in the Cascades. These releases of material have had a significant effect on the morphology of both the volcanic cones as well as the adjacent valleys. For example, collapse of the north flank of Mount St.

M. J. Bovis

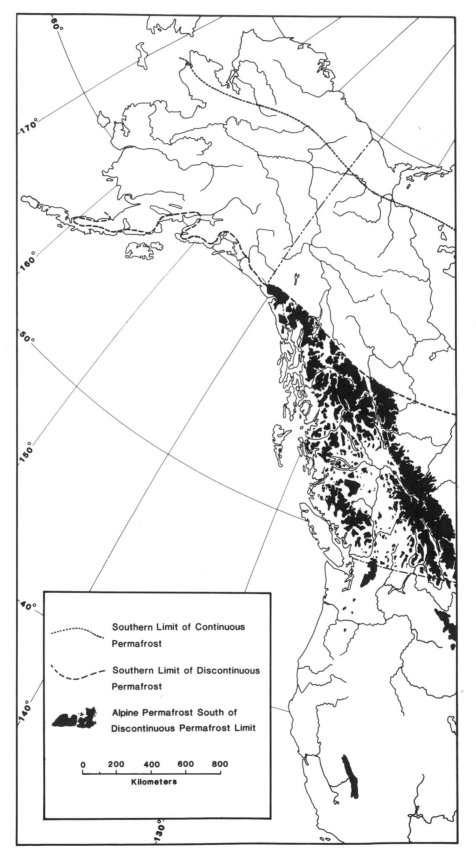

Figure 4. Permafrost regions. (Compiled from Brown, 1967; and Péwé, 1975, 1983.)

Helens during the 1980 eruption resulted in burial of about 60 km^2 by 2.8×10^9 m^3 of debris to an average depth of 45 m. Large volcanic mudflows (lahars) also occurred during or shortly after this great eruption. Large lahars have occurred also at Mount Baker, Mount Hood, Mount Shasta, and above all, at Mount Rainier, where the Osceola Mudflow dating from about 5 ka attained a volume close to 2×10^9 m^3 (Crandell, 1971). There is little doubt that the most catastrophic mass movement events within the study region have been, and will continue to be, associated with volcanoes.

Outside of the Cascades, rock avalanches have been reported from the Sierra Nevada (Keefer, 1984), the metamorphic Cascade belt of southern British Columbia (Mathews and McTaggart, 1978), the sedimentary Skeena Mountains (Eisbacher, 1971), the volcanic and sedimentary Stikine Plateau (Alley and Young, 1978), and the metamorphic and sedimentary zone of the Shakwak Trench in Yukon (Clague, 1981b). Rock slumps (see Varnes, 1978, for terminology), like rock avalanches, have caused significant postglacial morphological changes in many valleys. Large lakes are impounded by such debris, and there is a history of temporary damming of major rivers such as the Columbia (Palmer, 1977) and the Thompson tributary of the Fraser River (Evans, 1984). Slumps and earthflows are important within the deeply weathered, older volcanics of the Western Cascades (Swanson and James, 1975; Swanston and Swanson, 1976; Swanson and Swanston, 1977); in the volcanics and sediments of the Interior Plateau of British Columbia (Vandine, 1980; Bovis, 1985; Cruden, 1985); along the Columbia River in the Omineca Crystalline Belt (Piteau and others, 1978); in the volcanics and sediments of the Stikine and Yukon plateaus (Alley and Young, 1978; Eisbacher, 1979a; Rampton, 1981); and in clay-rich materials along the Nenana River in Alaska (Wahrhaftig, 1958). Large slumps have also developed in unconsolidated glacial drift, especially glaciolacustrine silts and clays. Such failures have been reported along the Columbia Valley upstream of Grand Coulee Dam (Jones and others, 1961); and within the Fraser-Thompson drainage (Ryder, 1976; Evans, 1982; Thomson and Mekechuk, 1982).

A very widespread type of slope denudation is gelifluction, particularly within the boreal and subarctic parts of the Intermontane Belt, which lie within the belt of discontinuous permafrost, and within the extensive area of alpine permafrost (Brown, 1967; Péwé, 1975, 1983; Fig. 4). Deformation of seasonally thawed ground as gelifluction lobes has affected valley side slopes and interfluves in much of the study area north of 56°N. Another landform typical of the discontinuous permafrost zone is the ice-cored mound known as the open-system pingo. These are fairly numerous within the Yukon-Tanana upland.

A notable landform in many cirques of the interior uplands and mountains is the rock glacier. Some may preserve a glacier-ice core, others may be simply ice-cemented talus. Rock glaciers are strikingly developed in the Kluane Ranges, the Brooks Range, and the Ogilvie Mountains (Vernon and Hughes, 1966; Johnson, 1978; Ellis and Calkin, 1984). Intense frost action on cirque headwalls, coupled with heavy snowfalls, also favors rock glacier development in the large alpine permafrost areas of the Columbia Mountains, the North Cascades (Thompson, 1962), and the Sierra Nevada (Péwé, 1983).

HISTORY OF GEOMORPHOLOGICAL RESEARCH

Introduction

Assignment of a precise date for the commencement of geomorphological research in the region involves arbitrary decisions concerning the distinction between scientific studies of landforms and reconnaissance descriptions of topography and drainage. For example, Waitt (1982) identifies a geomorphic component in the journals of Lewis and Clark of the early nineteenth century. Geomorphology was placed on a somewhat firmer footing with the publication of the first report of the Geological Survey of California (Whitney, 1865). This founding of a scientific research organization by government decree established a pattern for the ensuing 120 years, which have seen the domination of geologic and geomorphic research by federal and state (or provincial) surveys in both the United States and Canada. However, geomorphology has traditionally formed only a small part of all earth-science work in the region, and its level of funding by government agencies has varied appreciably over time, according to national priorities and goals (Leviton and others, 1982). The dictates of both war and economic crisis can be held largely accountable for the curtailment of much geomorphic research in the period 1914 to 1945, in favor of the more pressing needs of mineral and water-resource inventory. The past 40 years have seen a marked increase in the level of funding for geomorphic research in both government surveys and universities. Since the allocation of research funds has largely respected the international boundary, the history of research in the Canadian Cordillera is dealt with in a separate section.

Sierra Nevada and Cascades

As already noted, state-funded work began relatively early in California, motivated primarily by an interest in the age and origin of the auriferous gravels of the western Sierra Nevada. Exploration of the Sierra quickly led to the identification of spectacular glaciated landforms, stream-cut canyons, and evidence of a prolonged erosional history. The opening up of the Yosemite area in the 1870s, and its relative proximity to the populous regions of California, served as a very powerful influence on both bedrock and landform studies in western North America during the late nineteenth century. With the notable exception of Whitney's (1880) monograph on the auriferous gravels, the annual reports, bulletins, monographs, and professional papers of the U.S. Geological Survey (USGS) came to dominate the regional literature. Several of these are acknowledged classics. The other major influence on work in the Sierra Nevada was the geology department at the University of California, which produced original contributions in the fields of glaciation, denudation chronology, tectonic geomorphology, and volcanic landforms. By 1920, the golden age of the extended monograph was largely over, at least for the USGS. Many of the notable works of the interwar era emanated from the universities, a good example being the work on Cascade volcanoes.

The major research themes in the Sierra Nevada and Cascades were well defined during this first period of research, and include: glacial erosion, multiple glaciation and glacial chronology; the amount and timing of uplift of the Sierra Nevada; and the recent history of the Cascade volcanoes. Most of these themes have been greatly developed since 1945. The most recent advances in the field of glaciation include regional evidence for multiple glaciation, an absolute chronology for glacial events, and the development of relative-age criteria. The absolute chronology of uplift in the Sierra Nevada has been reasonably well established, together with more refined estimates of its amount. Most recently, attempts have been made to reconcile the history of uplift with tectonic models of the western continental margin.

In the volcanic Cascades, work has shifted over the past 40 years from volcanic morphology to petrography and the tectonic setting of volcanism. Apart from studies of glacial chronology, the most notable geomorphic research has concerned volcanically triggered mass movements such as lahars and volcanically triggered rock avalanches. Mass movement has also become an important research priority in the U.S. Forest Service over the past 25 years in the context of timber harvesting, especially in the Western Cascades. An allied theme that has emerged since the late 1970s is sediment budgeting on a drainage basin scale. Long-term studies are underway to assess the variation of sediment production and delivery rates within steep forested catchments, as well as in the Toutle River system, as hillslopes and valley fills slowly recover from the effects of the 1980 Mount St. Helens eruptions.

The Canadian Intermontane Belt

When geological reconnaissance work began in the Canadian Cordillera in the 1870s, the Geological Survey of Canada was already well established. Consequently, most of the early important contributions to geomorphology are associated with this agency. Prominent among contributors in the period 1875 to 1900 was G. M. Dawson, who is credited with originating the Cordilleran Ice Sheet concept; with recognition of the distinctive glaciolacustrine deposits of the southern interior of British Columbia; with identification of widespread erosion surfaces in the Interior Plateau; and with description of extensive postglacial fan and terrace deposits from the valleys of the same region. Dawson also conducted extensive reconnaissance work in northern British Columbia and Yukon, which led to refinement of the ice sheet concept, and the recognition of large unglaciated areas in the Yukon.

In the first half of the twentieth century, many of the research themes initiated by Dawson languished, as bedrock mapping and mineral exploration took first priority. Only glaciation was pursued in any depth, especially in central and northern British Columbia where evidence for multiple Cordilleran glaciation was assembled. Large areas of northern British Columbia and Yukon remained unexplored geologically, but the increasing availability of air photographs allowed regional physiographic overviews to be compiled.

Since 1960, many studies of Quaternary phenomena have been conducted by the Terrain Sciences Division of the Geological Survey of Canada, as part of a regional terrain mapping program, which most recently has included geologic hazards. Terrain mapping has also been carried out by the provincial Ministry of Environment in British Columbia. Studies also began on periglacial landforms and permafrost, especially in the unglaciated parts of Yukon. With the increase in federal funding for university research over the past 25 years, more detailed studies have been accomplished in the areas of glacial, fluvial, periglacial, and hillslope geomorphology. Notable advances have occurred in the fields of deglaciation processes, the effects of glaciation on fluvial sedimentation, the geomorphic effects of river diversions, and a variety of mass-movement processes. Geotechnical studies of hillslope stability have also figured prominantly over this same period, which has seen the completion of many public works projects in the intermontane belt.

Central Alaska

The first geomorphic studies of note in Alaska were the reconnaissance investigations published in the Annual Reports of the U.S. Geological Survey in the late 1890s and early 1900s. A major figure of this period was Brooks, who compiled the first comprehensive account of Alaskan physiography and climate (Brooks, 1906). He also instigated a program of regional geologic studies, many of which appeared as Bulletins of the Geological Survey and some of which include brief descriptions of landforms. In contrast with the more accessible coastal areas of the state, much of interior Alaska received relatively little attention from geomorphologists in the period 1920 to 1940. This same gap in research activities, identifiable in the Canadian Cordillera and the conterminous states, is attributable to general economic conditions. This era saw the beginnings of research on periglacial phenomena, motivated initially more by the practical needs of land clearance and construction than by geomorphic enquiry. The Second World War saw a considerable improvement in the highway system in the state, and this, combined with the military construction program, brought to light many interesting periglacial phenomena. These became the topic of detailed investigation in the postwar era, but it was not until the late 1950s that a coherent picture of ground-ice forms and gelifluction phenomena began to emerge. Attention has been focussed since on open-system pingos, fossil ice-wedge polygons, cryoplanation terraces, and rock glaciers. A somewhat separate theme has concerned Quaternary sea-level fluctuations in western Alaska, which includes the history of the Bering Land Bridge, and the evidence for former periglacial conditions on the Alaskan Continental Shelf.

CENOZOIC EROSION SURFACES

Introduction

Subaerial erosion surfaces have been documented in many parts of the study region over the past eighty to one hundred years. However, with the notable exception of the Sierra Nevada, interest in denudation chronology has waned in the past forty years or so, probably because of the difficulty of resolving geomorphic evolution from erosional evidence alone. The large majority of studies have attributed multi-level erosion surfaces to fluvial planation followed by baselevel change. Most of the Sierra Nevada studies adopt this line of reasoning. However, Wahrhaftig (1965b) has shown that much of the stepped topography of the unglaciated part of the range is explicable by differential erosion, influenced by differential weathering, that need not have any baselevel implications. In alpine and subarctic regions, some surfaces are probably due to cryoplanation (Thompson, 1962; Péwé, 1975). These alternative explanations are considered following a review of tectonically controlled erosion in the Sierra Nevada.

Erosional History of the Sierra Nevada

The relatively large number of erosion surface studies in the Sierra Nevada can be attributed in part to the availability of both erosional and depositional evidence, and to the relatively early exploration of the region for the hydraulic mining of auriferous gravels. The earliest studies, from 1865 to about 1915, were concerned almost exclusively with the age and origin of these gravels, which occur as a series of buried paleochannel segments well above the level of the present southwest-flowing trunk streams in the northern half of the Sierra. The apparently undisturbed condition of most of the gravels led Whitney (1880) to conclude that abandonment of the high channel system could not have been accomplished by downcutting caused by tectonic uplift, but rather was more likely due to climatic change. The role of tectonic uplift was first clearly stated by Gilbert (1883), who reasoned that canyon cutting had resulted from an increase in the slope of the west flank of the Sierra, in response to tectonic uplift along the eastern fault zone. Canyon-cutting was explained, therefore, by an increase in stream power (paraphrasing Gilbert). It is ironic that Gilbert was correct in this instance, yet quite wrong in his insistence that river terraces would, in the main, be an expression of tectonic uplift alone. The role of uplift in canyon cutting was further elaborated by Le Conte (1886).

The simple tilt model, introduced by Gilbert, and extended by Le Conte, was not based on any measurements of the present inclination of the Tertiary channel segments. These data were supplied by Lindgren (1893), from the Yuba and American river basins, where extensive hydraulic mining of auriferous gravel had

afforded direct access to the bedrock-gravel unconformity. Many of the steeply sloping channel segments, with gradients of 0.02 to 0.04, occurred on southwest-flowing, former trunk streams. Paleo-tributaries, running roughly orthogonal to this direction, had generally lesser slopes, as would be expected if southwesterly tilting of the range had occurred. Modern stream channels were found to slope at 0.006 to 0.008, irrespective of channel direction. Lindgren concluded that uplift at the crest of the Sierra must have amounted to 1,100 to 1,300 m, assuming rigid tilting of the mountain block according to the Gilbert model. The chronology of Le Conte was largely retained, in that the gravels were considered to have accumulated throughout the Mio-Pliocene, the overlying volcanics at the close of the Pliocene, with canyon cutting occurring post-Pliocene.

The culmination of this first period of work on erosional history was Lindgren's (1911) monograph on the Tertiary gravels. From very extensive field investigations, he concluded that the north half of the Sierra had been tilted more or less as a rigid block, a fact also noted by Ransome (1898) in the Stanislaus River basin. However, an important qualification of the earlier tilt model was that the eastern fault scarp had grown as much, if not more, by graben sinking as by absolute elevation of the Sierra block, which Lindgren deduced from the different timing of dislocations on various strands of the eastern fault system. This suggested that regional up-arching of the Sierra, and part of the western Basin-Range, must have occurred prior to downfaulting of the eastern blocks, which verified Le Conte's (1889) schematic model of the Basin-Range, and largely conforms with modern ideas on the development of the Basin-Range structures (Stewart, 1978). Lindgren's work is therefore an interesting case of geomorphological analysis clarifying regional tectonics.

Although reference to widespread erosion surfaces on the western flank of the range goes back as far as Gilbert (1883), this aspect had been neglected in early publications on the auriferous gravels. Lawson (1904) identified three major topographic levels in the Kern River area of the southern Sierra Nevada, the most extensive of which, the Chagoopa Plateau, Lawson thought had been dissected by major tectonic uplift during the latter half of the Quaternary to form Kern Canyon. This chronology is now known to be in error (Dalrymple, 1963), but Lawson's work provided a framework for later work in Yosemite Valley (Matthes, 1930), drew attention to a large, unglaciated area in the southern Sierra, and reported basaltic lavas overlying erosion surfaces and filling certain canyons. It is fitting that one of Dalrymple's K-Ar samples was actually collected by Lawson in 1903.

Knopf (1918) identified erosional bevels more than 2,000 m above Owens Valley along the eastern fault scarp, which he correlated with Lawson's Chagoopa Plateau. This suggested that planation, following the first phase of uplift, must have affected the western part of the Basin-Range as well. This reinforced Lindgren's notion of epeirogenic up-warping prior to downfaulting of the eastern block. Knopf shifted Lawson's second phase of uplift to the beginning of the Quaternary, a date " . . . more in

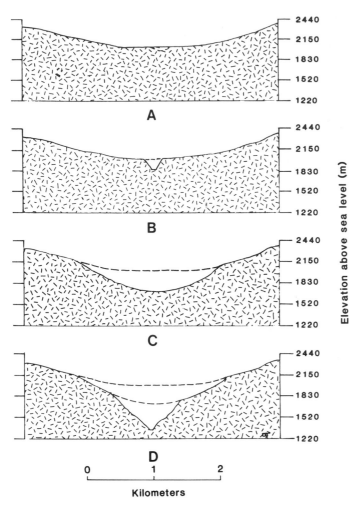

Figure 5. Erosional stages of Merced Canyon, Sierra Nevada, prior to the Quaternary glaciation. A, depicts the Broad Valley stage of development, in late Miocene time; B, the inception of canyon cutting after the first strong tilting of the Sierra block; C, shows the Mountain Valley stage of Pliocene age; and D, the Mountain Canyon incision of plio-Quaternary age. (Redrawn from Matthes, 1930.)

harmony with the length of time indicated by the great erosional work performed since the uplift . . ." (Knopf, 1918, p. 88).

The well-known monograph on the "Geologic History of Yosemite Valley" by Matthes (1930) provided a detailed account of erosion surfaces in the Merced River basin of the central Sierra Nevada. He recognized three main erosional stages—the Broad Valley, the Mountain Valley, and the Canyon stage—which corresponded approximately with those of Lawson (1904); but, whereas the Kern River stages had been deduced from widespread erosion surfaces, Matthes' stages were derived by projecting the longitudinal profiles of hanging tributary valleys to the Merced River axis (Fig. 5). The existence of hanging valleys due to fluvial dissection had been noted previously by Turner (1900, p. 271). Matthes' results suggested two major baselevel changes, the first producing the Mountain Valley incision in the

early Pliocene, the second the Mountain Canyon stage at the beginning of the Pleistocene. Matthes took the analysis a step further by reconstructing the longitudinal profile of the Merced River at each of these stages, which led to an estimate of the amount of uplift at the crest of the Sierra. This involved many assumptions about former valley gradients, and therefore was more speculative than the analysis of valley cross-profiles.

By the late 1950s, challenges were being mounted to previously accepted ideas concerning the magnitude and timing of uplift in the Sierra Nevada, notably from paleobotanists familiar with the Cenozoic floras of the Sierra Nevada and adjacent Basin-Range (Axelrod, 1957, 1962). Axelrod's data suggested a relatively late, Pleistocene uplift, but this conflicted with the physiographic evidence, which seemed to require a longer time span for canyon development. The problems involved in an unequivocal interpretation of spore and pollen data, as indicators of elevation change, were discussed later by Christensen (1966), by which time potassium-argon dates on Cenozoic volcanics were available (Dalrymple, 1963). The dates indicated that the Mountain Canyon stage of incision was well underway by 3 Ma, which would place the Mountain Valley and Broad Valley stages further back, in the early Pliocene and Miocene respectively, as Matthes (1930) had suggested. These critical dates ended a long period of speculation concerning the timing of uplift. There still exists uncertainty as to the amount of uplift, as shown by the latest interval estimates by Huber (1981). This paper, and earlier works which have sought to estimate uplift rates from a combination of stratigraphic and physiographic evidence (Hudson, 1960; Christensen, 1966), all share the same problems of having to assume some reasonable value for the preuplift, longitudinal gradient of a given river, and an appropriate model for the subsequent uplift (e.g., rigid tilting, tilting with warping; Fig. 6). There is additional uncertainty as to linear versus nonlinear extrapolation of the reconstructed profile to the present divide, or in Huber's case, beyond the divide. Despite these uncertainties, enough data have now accumulated to demonstrate, beyond a reasonable doubt, that early to mid-Pliocene uplift initiated the Mountain Canyon stage of cutting; from that time, uplift has been steady, if not accelerating (Huber, 1981). The computed rates of uplift are compatible with uplift of the Sierra to the glaciation threshold in the first quarter to one-third of the Quaternary (the McGee glaciation), rather than in the last third of Pliocene time, as previously proposed by Curry (1966).

Works on the erosional history of the Sierra span nearly a century. They demonstrate the importance of verifying first-order assessments, based on morphological evidence, with proper stratigraphic and absolute-age control. It is remarkable that earlier estimates of uplift amount and timing, by Lindgren (1911) and Matthes (1930), were tolerably correct despite the lack of absolute-age methods. The distinctive contribution of Sierran studies lies in their application of classical, geomorphological techniques to the elucidation of tectonic history. At first, geomorphological analysis provided the impetus for ideas on Cordilleran tectonics. The inverse procedure of applying modern tectonic

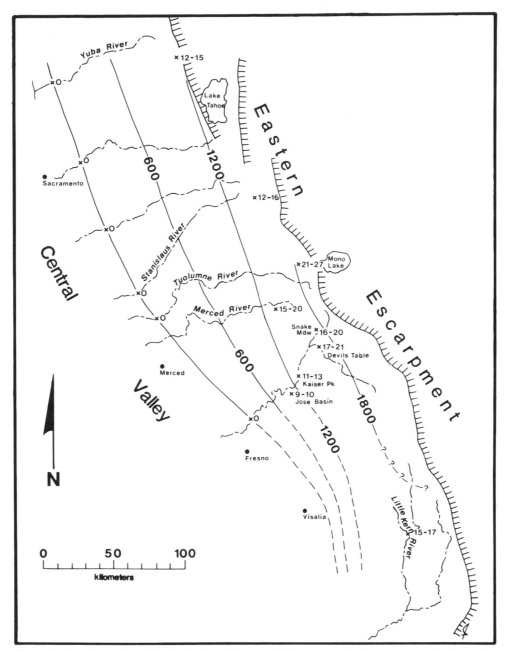

Figure 6. Late Cenozoic uplift of the Sierra Nevada. Uplift contours, in feet, are controlled by point estimates, in hundreds of feet, denoted by 'X' symbols. (Redrawn from Christensen, 1966.)

concepts to landform development in this region (e.g., Hay, 1976; Crough and Thompson, 1977) provides a new perspective on the causes of uplift, in addition to ensuring a healthy future for studies of erosional and tectonic histories.

Other Interpretations of Erosion Surfaces

Several writers have questioned whether all upland erosion surfaces can be used to infer former base levels of erosion. One of the earliest critics of the conventional erosion cycle model of Cordilleran topography was Daly (1912), who challenged the earlier interpretations of Dawson (1895) concerning the erosional history of south-central British Columbia. Dawson had argued for early- to mid-Tertiary fluvial planation, followed by late Tertiary or early Quaternary dissection—that is, two erosion cycles—the latest represented by glacially modified mountain canyons. Daly regarded Cordilleran rocks as of comparable hardness to those forming the Appalachians. There, he noted, all

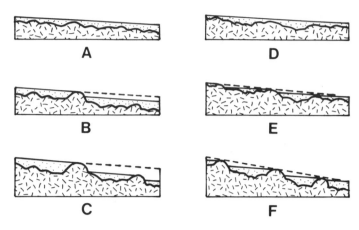

Figure 7. Formation of erosion steps in the Sierra Nevada. Sequence A to C, depicts step development under conditions of steady, vertical uplift. Sequence D to F, depicts step formation during a single episode of regional tilting. (Redrawn from Wahrhaftig, 1965b.)

of Tertiary time had been required for the excavation of the present valleys through the late-Mesozoic summit peneplain (the Schooley Surface). In the Cordillera he argued for a single cycle of cutting, in which the present mountain-canyon landscape had developed through all of post-Laramide time. He attributed summit accordance in part to uniform divide lowering, once sharp interfluves had been created by the recession of valley-side slopes; and in part to an assumed increase in the rate of denudation with altitude, which would tend to reduce the variation in peak heights over time. Daly, like Dawson before him, was also impressed with the more efficient production and removal of rock waste in the alpine tundra, and argued that this could account for erosional bevels close to timberline. This phenomenon had been noted earlier in the Alps of Europe. Daly's reservations concerning erosional capacity were voiced at a time when the greatly differing tectonic histories of the eastern and western seaboards were not known in detail. However, although minor erosional features may be due to the timberline effect, extensive erosion bevels, some tens of square kilometers in extent, are more likely to be fragments of former fluvial planation surfaces.

Thompson (1962) provided further, detailed evidence for the existence of erosional bevels closely coinciding with timberline, and largely reiterated the climatic-geomorphic control cited earlier by Daly and Dawson. Such "alp slopes" were better developed on north-facing slopes in the North Cascades. Since Thompson's work, these features have not been studied in detail, although similar features have since been observed in the southern Coast Mountains of British Columbia. Objections to climatic-geomorphic control have alluded to the substantial Holocene fluctuations of timberline, which would lead to a less-concentrated erosional effect. A counterargument might propose that alp slopes have evolved over a much longer span of the Quaternary, and therefore might reflect an average position for timberline during the many nonglacial interludes.

The final case reviewed here concerns the prominent stepped topography on the west flank of the Sierra Nevada, discussed by Wahrhaftig (1965b). He noted that retention of granitic greuss cover was prerequisite to rapid chemical disintegration of the underlying bedrock in the Sierra Nevada; where exposed, bedrock was considered to weather more slowly, at least by chemical processes. Local exhumation of bedrock during downcutting would create local base levels, below which upstream areas could not be lowered by fluvial planation. Retention of some greuss cover in areas above nickpoints would then favor continued weathering, which, coupled with lateral planation, would gradually create large steps (Fig. 7). The important implication of this is that the elevation of a given step may be largely random, and although initiated by downcutting from a fall in baselevel, may contain no information concerning the position of this former baselevel. This, in turn, casts some doubt on the interpretation of planation surfaces identified elsewhere in the granitic terrain of the Sierra Nevada by Lawson (1904) and Matthes (1930), in that some canyon cutting may have been initiated by tectonic uplift, whereas other instances may owe their initiation to progressive removal of rock steps.

VOLCANIC LANDFORMS

Introduction

Volcanic rocks of widely varying ages crop out over large areas of the study region, but only three areas preserve well-formed, primary volcanic landforms: the High Cascades and, in British Columbia, the Anahim and Stikine volcanic belts. The stratovolcanoes of the High Cascades have received the most attention by far, and an extensive literature has accumulated over the past 100 years on all aspects of volcanism in this region. However, relatively few papers have dealt exclusively with cone morphology and its development, the most notable examples being published prior to 1945. Since this classic period of volcanic studies, most work has focused on petrographic and tectonic aspects both in the Cascades and in British Columbia. It therefore lies outside of the scope of this review.

Volcanoes As Landforms

Two names figure prominently in the literature on Cascade volcanic landforms: J. S. Diller, most of whose studies appeared prior to 1905; and Howel Williams, most of whose monographs were published from the late 1920s to the early 1940s. Diller is most noted for his studies on Lassen Peak (1891), Mount Shasta (1895), and Crater Lake (Diller and Patton, 1902). One of Diller's major contributions was to bring to the attention of a wide audience the magnitude and comparative recency of Cascade cone building, and to describe in general terms the relative importance of lava flows, pyroclastic accumulations, and acidic domes in the architecture of the volcanoes. Particular attention was given to Crater Lake, the origin of whose 'caldera' had

Figure 8. Crater Lake, Oregon, with Wizard Island in the foreground. Note the large truncated valley in the east wall of the crater.

invited comment since the mid-1880s (Fig. 8). Diller and Patton (1902) established the former existence of a major volcano at Crater Lake, termed Mount Mazama, from the presence of radial dykes cutting the crater rim. Evidence of local glacial deposition and erosion, the latter clearly displayed as truncated, U-shaped valleys in the cliffs surrounding Crater Lake, suggested that the volcano had been comparable in elevation to Mount Shasta, a size now thought to be exaggerated. Two main factors led Diller and Patton to reject a purely explosive origin for the crater. The first was the apparent paucity of the encircling tephra deposits, compared with the computed volume of the former volcano. Secondly, the discrepancy between the mainly dacitic composition of the tephra and the andesitic composition of much of the remaining cone led them to expect more andesitic debris in the tephra if a cataclysmic explosion had taken place. They proposed, therefore, that the caldera had formed mainly by subsidence, but did not deny the existence of explosive activity during the final eruption.

The Crater Lake problem was taken up by Smith and Swartzlow (1936), who argued strongly in favor of explosion on the basis of very widespread tephra deposits and the general absence of faults so characteristic of Hawaiian calderas. Williams (1941, 1942) reexamined the explosive hypothesis, and concluded that much of the missing material was accounted for in voluminous glowing avalanche deposits that extend for tens of kilometers from the eruption site. Relatively little was accounted for in air-fall tephra, which he estimated to be slightly over

8 km^3. The discrepancy between the sum of these two volumes and that of the missing cone was accounted for by subsidence. At that time, the widespread occurrence of Mazama tephra throughout the Pacific Northwest was not yet appreciated (Wilcox, 1965), and this distribution, combined with the subsequent discovery of very large Mazama rock fragments at distances of up to 40 km from Crater Lake, leaves little doubt that explosive volcanism, on a catastrophic scale, played an important part in the final destruction of Mount Mazama. The volume of air-fall tephra alone is now estimated to exceed 30 km^3 (Williams and Goles, 1968). Although the computed volumes of material involved have changed over time, the combination of glowing avalanches, explosive ejection of debris, and subsidence, proposed by Williams, is still widely accepted today.

Williams also conducted detailed studies at several other centers of Quaternary volcanism in the Cascades. His study of the Lassen Peak area (1932) is notable in several respects, and a review of this monograph alone will perhaps suffice to demonstrate the importance of his work, at least in terms of the interpretation of volcanic landforms. First, it provided an account of one of the largest dacite domes on record at that time, namely Lassen Peak. Williams demonstrated that this large, steep-sided mass had been slowly thrust upward as gas-deficient, super-viscous magma, following previous eruption of more fluid dacite. Although plug domes had been reported elsewhere, notably at Mount Pelée, nothing on this scale had been reported previously in North America, and it presented a sharp contrast to the stratovolcanoes of the rest of the Cascades, whose structure was generally known at that time. In contrast to the relatively slow, intermittent growth of stratovolcanoes, Williams proposed that very large domes might be built within several years to decades. He also made comparisons between this massive dome and the much smaller feature built after the 1914 to 1917 eruptions at Lassen Peak.

The Lassen Peak study also identified a wide variety of cone forms, according to a divergent sequence of magmas. A succession was documented from subdued, basaltic shield volcanoes, through large andesitic stratovolcanoes, only the roots of which remained, followed by eruptions of relatively acidic lavas that had culminated in dome building. A similar, divergent sequence of magmas was identified later at other volcanic centers in central and southern Oregon, and subsequently led to the recognition of the southern High Cascades as a somewhat different petrographic province from the more uniform, andesitic to dacitic sequence characteristic of stratovolcanoes in the remainder of the range.

An important factor in understanding the overall form of Cascade volcanoes is the extent of glacial erosion during the Pleistocene. Most of the northern cones have been deeply eroded, especially Mount Rainier, where it is estimated that more than 50% of the original cone volume has been removed (Fiske and others, 1963). By contrast, the remarkably symmetrical shape of Mount St. Helens, prior to the 1980 eruption, was clear evidence of considerable postglacial activity, which had largely obliterated the effects of glaciation. A similar effect was noted at Mount Shasta, which, though probably suffering less intense glaciation than the northern cones, has also seen notable postglacial cone-building both at the summit, and in the flanking Shastina cone (Christiansen and Miller, 1976).

Relatively little work was done on the volcanoes of northern British Columbia prior to the mid-1960s. The paper by Mathews (1947) on the flat-topped 'tuyas' of the Kawdy Plateau is one of the few publications concerned with volcanoes as landforms. Several features led Mathews to conclude a subglacial origin for these volcanoes. First, they were flat topped, but had not been eroded into that configuration. Secondly, pillow lavas pointed to subaqueous extrusion. Thirdly, the lack of spreading of the relatively fluid lavas beyond the base of the volcanoes indicated confinement by some material. He therefore proposed that lavas had been extruded into subglacial lakes, melted into the base of the Cordilleran Ice Sheet at the end of the last glaciation. Most of Mathew's interpretations have since been confirmed by Allen and others (1982), who have compared these British Columbian examples with similar forms in Iceland.

GLACIATION

Introduction

Much of the work on glacial geomorphology accomplished in the study region in the past 130 years has been concerned with establishing the number of glaciations, their correlation between each mountain area, and the paleogeography of ice bodies and adjacent glacial lakes during each glaciation. Significantly fewer studies have dealt with the relative importance of glacial, versus fluvial, erosion in the development of mountain and upland topography, and the mechanics of glacial erosion and deposition. In this review, no attempt is made to summarize contributions in the wider field of Quaternary paleoenvironments, nor is much attention paid to questions of glacial chronology and correlation. The state of knowledge in these areas has recently been reviewed elsewhere (Porter, 1983; Wright, 1983) and has received detailed coverage in other DNAG volumes. In that this is a review of contributions to glacial geomorphology, as distinct from Quaternary science, a fairly restricted range of topics is addressed.

Glacial Erosion

At the time of the earliest work on glacial phenomena in the region, such as the Whitney survey of 1865, the glacial theory was already widely accepted, largely due to extensive field verification in the eastern and central United States and Canada, and in no small measure to the influential voice of Dana in his 1855 address to the American Association for the Advancement of Science (Merrill, 1924, p. 631–632). Many North American geologists were convinced that large areas of the continent had been glaciated, based largely on the widespread distribution of drift deposits and glacial striae. Ideas concerning the erosive power of ice were as yet poorly developed. Many argued that glacial modification had been limited to smoothing by abrasion, and even that large ice bodies had protected the underlying topography from denudation. A second viewpoint, largely originating with Ramsay (1862, 1864) in the English-speaking literature, argued that glaciation had effected profound erosional changes. The important dialogue in Europe in the latter half of the nineteenth century, concerning the general validity of these opposing viewpoints, had its parallel in contemporary American and Canadian writings, some of which, it will be shown, contributed greatly to our understanding of glacial erosion.

The Whitney survey recognized the role of glacial erosion in the Sierra Nevada (Whitney, 1865), and also emphasized that the main elements of the landscape had evolved in preglacial times. The origin of Yosemite Valley, however, presented an enigma, since the relatively small volume of morainal debris in Merced Canyon was not consistent with extensive glacial erosion. Accordingly, the famous fault hypothesis was proposed by Whitney to account for the sheer canyon walls. The antithesis of this view of the valley is seen in the early writings of John Muir (1874), who advocated that virtually the entire canyon system was due to glacial erosion. Although overstating the case for glacial erosion, Muir made the important connection between jointing and the removal of rock by ice.

In the 1870s and 1880s, proponents of glacial erosion emphasized abrasion, with great attention given to glacial striae, so much so that areas lacking striae, even though possessing thick drift mantles, were often considered to be unglaciated. In this context, the work of Le Conte (1875) is significant. He noted that

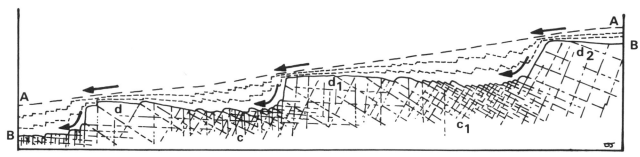

Figure 9. Development of longitudinal valley steps by glacial quarrying. Section A–A represents the preglacial canyon floor. Well-jointed rock (C, C_1) is quarried away, whereas massive outcrops at d, d_1, and d_2 are only abraded by ice. (Redrawn from Matthes, 1930.)

much of the erosional evidence in the Lake Tahoe area could not be explained by abrasion, but rather by the tearing away of large blocks by ice, followed by rounding in subglacial transport. These blocks, in his view, were to be clearly distinguished from angular morainal materials, which had been transported on top of the glacier. This paper is therefore one of the earliest North American references to glacial quarrying. Like Muir, Le Conte recognized the important role of glacial erosion in the formation of the Yosemite area, referring to some of the glacially modified domes as huge moutonée forms.

Russell's 1889 monograph on the Mono Lake area is regarded by many as a classic within the glacial literature; it documented evidence for multiple glaciation in the eastern Sierra Nevada, and provided a valuable discussion of the development of hanging valleys and glacial troughs. Notwithstanding Russell's great contribution to regional geomorphology, it is evident that earlier work by Helland (1877) on glacial quarrying, and by McGee (1883) on the origin of 'U'-shaped valleys, greatly influenced his thinking on glacial erosion processes. McGee's 1883 paper had contained one of the clearest statements to date on the transformation of 'V'-shaped to 'U'-shaped valleys. (In his 1894 paper, McGee subsequently presented a model for hanging valley development, which was based mainly on his earlier observations in the eastern Sierra Nevada.) An important point, also emphasized by Russell, was that discordance between hanging valleys and glacial troughs could be produced by valley widening, and need not have required a great erosional deepening of the trunk valley.

The seeds of later work were also sown in the eastern Sierra Nevada, since Willard Johnson served as Russell's principal topographer on the Mono Lake project. The famous bergschrund hypothesis of glacial sapping (Johnson, 1904) was based on the observation of subglacial frost-shattering of rock, and glacial entrainment of rock waste, at the base of a bergschrund on Mount Lyell. From this he concluded that the great depth of a cirque or a stepped glacial valley was due to headward recession of a rock-wall by subglacial sapping, at a rate controlled by the changing bergschrund position. Although the introduction of the somewhat restrictive bergschrund control was an original contribution,

Johnson's mechanism was, in many respects, very similar to that proposed earlier by Helland. Initially his hypothesis was widely accepted, both in Europe by geologists as renowned as A Penck, and in North America by Matthes (1900). In the study region, Gilbert (1904) used it to explain the topographic asymmetry of the High Sierra. Subsequently, objections were voiced to Johnson's proposal that cirques, or even glacial troughs, could develop more or less independently of the preglacial topography. Others noted that many cirque forms had been excavated far below the level of the deepest recorded bergschrund. Eventually the less restrictive meltwater hypothesis of Lewis (1938, 1940) came to be adopted.

The work of Matthes (1930) was conceived in this transitional period between the statement of Johnson's hypothesis and the subsequent innovations of Lewis. Matthes' monograph is still one of the finest in the geomorphic literature, and with the exception of absolute chronology, most of his interpretations have stood the test of time. It is perhaps no accident that Johnson and Matthes had both served as topographers before making major contributions to glacial geomorphology. Matthes, in fact, is almost as well known for his topographic map of Yosemite Valley as he is for his landform interpretations. Both writers clearly acknowledged the opportunity that making topographic maps had afforded them in their late interpretive work.

Matthes' greatest contributions to glacial erosion concerned the role of jointing in glacial quarrying, and an assessment of the relative contributions of glacial and preglacial erosion in canyon formation. Work in the Yosemite region after Johnson's paper had tended to advocate headward recession of rock steps to account for the great depth of glacial canyons. Matthes observed that virtually all major steps in the longitudinal profile of the Yosemite canyons corresponded to massive, joint-free zones, which could not conceivably have receded several kilometers. Instead, he proposed that the intervening, well-jointed rock between the massive outcrops had been selectively quarried away during several glaciations (Fig. 9). This interpretation has since had a very wide application (Flint, 1971).

An important section of the monograph deals with the extent of glacial modification of the canyon landscape (Matthes,

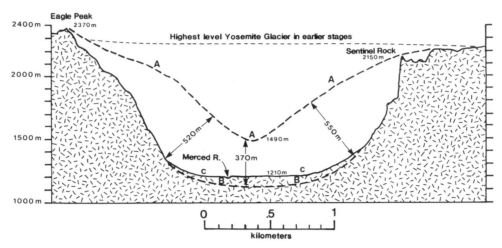

Figure 10. Glacial modification of Yosemite Valley, as deduced from the reconstructed, preglacial cross profile. (Redrawn from Matthes, 1930.)

1930, 84–103), and is probably the section most frequently quoted in geomorphology textbooks. The analysis hinged on a tolerable reconstruction of the form of the Merced and Tenaya canyons prior to glaciation: that is, on the valley form at the time of the Canyon Stage of incision in Plio-Pleistocene time. Although such an exercise involved some degree of speculation, which Matthes clearly acknowledged, comparison of the reconstructed and present-day profiles at selected points provided a clear indication of the level of glacial transformation (Fig. 10). Matthes concluded that valley widening had produced most of the form change, outstripping vertical erosion by about 2:1, except in Tenaya Canyon, where structure had favored greater vertical scour. This interpretation harks back to McGee (1894), and provided quite a different view from that of Johnson (1910), who had explained the hanging valleys of the region by vertical scour of the trunk glacier. Matthes' more refined interpretation was made possible by his recognition of hanging valleys beyond the limits of glaciation, which led to a more cautious appraisal of these forms in the glaciated canyons.

A major contribution by Matthes was the documentation of multiple glaciation in the Sierra Nevada. Although Russell (1889) had first presented clear morainal evidence of multiple glaciation, Matthes documented its erosional implications. Careful mapping of moraines and glacially polished surfaces showed that Wisconsinan glaciers had largely occupied preexisting glacial troughs, a fact that complied with the long-observed paucity of moraines in relation to the size of the canyons. This important perspective on the progressive development of glacial landforms also allowed him to deduce the contribution of glacial quarrying in the shaping of large-scale dome landforms.

The Cordilleran Ice Sheet

Introduction. In a preceding section on Pleistocene history, reference was made to two distinctive styles of glaciation. At the maximum of the last glaciation, summit icefields fed alpine-type valley glaciers in the Sierra Nevada, much of the Cascade Range, and the Brooks Range. The remaining glaciated areas were occupied by the Cordilleran glacier complex. In the Sierra-Cascade belt, the overall style of glaciation and deglaciation was known with some degree of certainty, even in the earliest studies conducted before 1890. Subsequent work has been concerned with establishing the number of glacial and interglacial episodes, their time of occurrence, and, therefore, their correlation with events elsewhere on the continent. Considerable effort has also been spent on establishing the environmental conditions associated with each major glaciation. Studies in British Columbia and Yukon, of course, have also been concerned with these aspects of the Quaternary, but differ from studies in the Sierra-Cascade belt in that considerable discussion has centered on the overall style of glaciation and deglaciation. In this section, we trace the history of ideas concerning the growth and decay of the Cordilleran glacier complex.

Ice Sheet Growth and Decay. Early European work on glaciation in the Alps was built around the concept of a summit icefield, or 'mer de glace', nourishing tributary valley glaciers. As studies of glacial erratics progressed, it became apparent that large-scale coalescence of piedmont glaciers had occurred to form major ice sheets. The Scandinavian Ice Sheet had been proposed prior to 1880, and evidence for the Laurentide and Cordilleran ice sheets published prior to 1890. One of the main architects of the Cordilleran Ice Sheet concept was G. M. Dawson (1888), who is not to be confused with J. W. Dawson, the die-hard critic of glaciation (Merrill, 1924, p. 639). In the southern interior of British Columbia, Dawson observed evidence of glaciation in places far removed from the known ice accumulation zones. Evidence for glacial inundation of the highest summits of the Interior Plateau (2,000 m above sea level) came to light in the early 1870s. In the report of the 1877 field season (Dawson, 1879,

WEST   EAST

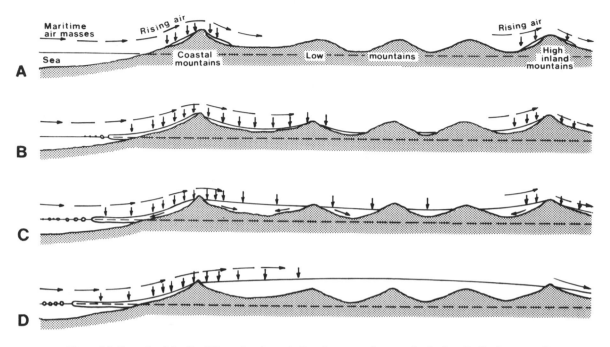

Figure 11. Growth of the Cordilleran ice sheet. A. Development of mountain glaciers. B. Coalescence of valley glaciers to form trunk glaciers. C. Development of mountain ice sheet. D. Maximum ice sheet phase. (Redrawn from *Glacial and Quaternary Geology,* by R. F. Flint. Copyright 1971, John Wiley & Sons, Inc. Reprinted by permission of J. Wiley & Sons, Inc.)

p. 135B), an extensive south-moving ice sheet was proposed, and later a major intermontane flow center around latitudes 55° to 59°N (Dawson, 1888). According to Dawson, glaciation had occurred when the Cordillera was notably higher than present, with deglaciation taking place following, or concurrent with, regional subsidence. This scheme was based on an erroneous interpretation of the glaciolacustrine silts of the interior valleys as fjord sediments. By 1912, Daly had adopted the view that the laminated silts had accumulated in ice-dammed lakes, an interpretation subsequently enlarged upon by Mathews (1944) and Fulton (1965).

By 1920, challenges had been mounted to Dawson's ice-sheet concept, principally by Tyrrell (1919), who advocated that valley glaciation had dominated much of the intermontane belt. Kerr (1934) was one of the first writers to examine the growth of the Cordilleran Ice Sheet in any detail, proposing four stages: the alpine, intense alpine, mountain ice sheet, and Cordilleran Ice Sheet. Kerr proposed that the ice divide was located to the east of the Coast Mountains during the Cordilleran Ice Sheet stage. This idea was taken up by Flint (1943), who postulated that eastward shift of the divide had occurred, similar to that proposed earlier for the Scandinavian Ice Sheet by Hansen (1894). Westward ice movement through the Coast Mountains, against the topographic

gradient, was proposed, but no specific sites were mentioned (Fig. 11). Ice buildup to an intermontane dome was explained by topographic confinement, which favored a low mass turnover in comparison with windward slopes, where rapid flow to a calving ice shelf had occurred. Flint also proposed, but did not prove, that westward migration of the Rocky Mountain ice divide had occurred for similar reasons.

The next significant commentary on the ice-cap hypothesis was that of Tipper (1971). Although accepting Dawson's original notion of an ice dome in northern British Columbia, he questioned its extension to the south-central area of the province where, as far as can be judged, Dawson had never proposed westward ice movement across the Coast Mountains. Tipper made a strong case for dominant eastward to northeastward movement from the Coast Mountains, reinforcing a view expressed earlier in Armstrong and Tipper (1948). He therefore proposed ice buildup along the lines of Kerr's model, but stopped short of an ice dome centered to the east of the Coast Mountains. Some glacial erratics suggested westward ice movement through the southern Coast Mountains, but this was referred to an earlier Wisconsinan glaciation, in which a full ice cap was considered to have developed.

The ice-cap topic has been reviewed recently by Clague (1981a), who suggested that high-elevation striae, not valley-bottom indicators of ice motion, should be used to determine the ice-sheet configuration at the glacial maximum. Clague (1984) has since evaluated drumlin and striae evidence for possible east-to-west ice motion across Bulkley Valley in northern British Columbia. However, since his flow indicators are located at 600 to 750 m, and the Skeena Mountains rise to over 2,100 m just 30 km east of the valley, the striae and drumlins could have been produced by temporary, westward outflow from the Skeena Mountains during Kerr's 'mountain ice sheet' stage, rather than during the 'Cordilleran ice sheet' stage. It is evident that data on high-elevation striae and other ice-flow indicators are as yet insufficient to prove or disprove the existence of an interior ice dome at the maximum of the last glaciation. A major obstacle to this line of enquiry is that ice retreat in the critical Coast Mountains area involved active valley and piedmont glaciers, which followed similar flow paths to those of the advance phase. It therefore seems unlikely that positive evidence of westward, uphill flow would have survived, although some ridge-top locations would not have been covered by ice during deglaciation.

A second line of enquiry concerns the pattern of deglaciation. Tipper (1971) argued that a fully developed ice cap would imply a residual ice dome over the Interior Plateau, thick enough to leave a roughly concentric series of recessional moraines. These have yet to be recognized in the region. In fact, the deglaciation model put forward by Fulton (1965, 1967) suggests the contrary—namely, that ice thinned rather rapidly over south-central British Columbia, leaving residual masses of stagnant valley ice associated with the distinctive glaciolacustrine regime of the interior valleys. This evidence for rapid deglaciation over interior British Columbia would tend to argue against an ice dome thick enough to permit westward flow of ice against the regional topographic gradient. However, it could also be argued that a climatically 'dead' ice dome would waste away with few or no moraines forming.

In summary, the ice-cap concept, devised by Dawson, refined by Kerr, and widely publicised by Flint (1943, 1971), has to be regarded as a working hypothesis, a point frequently overlooked in many citations of the earlier literature. Given the topographic complexity of the intermontane belt, it seems unlikely that the pattern of ice buildup in the much simpler Scandinavian mountain chain can serve as a model for Cordilleran conditions. The most recent review of late Wisconsin ice sheets in North America (Prest, 1984) casts considerable doubt on long-accepted ideas concerning the configuration of the Laurentide Ice Sheet. Much less attention was paid to the Cordilleran Ice Sheet, but it is evident from this brief assessment that a thorough review of late Wisconsin conditions in the intermontane belt is also appropriate.

PERIGLACIAL LANDFORMS

Introduction

The term periglacial is used here to describe cold-climate, nonglacial phenomena that typically occur in arctic and subarctic latitudes, and in the alpine tundra of temperate mountains (Fig. 4). There has been considerable debate as to whether the term should be confined strictly to permafrost conditions (Washburn, 1980) since not all features classed as periglacial require permafrost for their formation. That discussion does not weigh heavily here because virtually all of the important periglacial work reviewed was carried out in the continuous and discontinuous permafrost zones of Alaska and Yukon. Many fundamental contributions to the theory of ground-ice phenomena have been made by studies on the Arctic coastal plain and in the Arctic Archipelago. This work is reviewed fully in Chapter 14, this volume (Arctic Lowlands). The alpine tundra parts of the study area, when compared with those of the Rocky Mountain region, have produced comparatively few contributions to the theory of periglacial landforms.

The following categories of periglacial phenomena are examined: ground-ice forms, including ice wedges, ice-wedge casts, and open-system pingos; and forms produced by slope movement, of which two, rock glaciers and cryoplanation terraces, have received particular attention. There has been much less work accomplished in the study area on sorted polygons. As stated earlier, in the introduction to 'Glaciation', a general review of work dealing with Quaternary paleoenvironments is beyond the scope of this chapter. A thorough review of this aspect of periglacial work is provided by Péwé (1983).

Ground-Ice Phenomena

Ice Wedges and Ice-Wedge Casts. One of the earliest statements in English on the theory of ice-wedge development was that of Leffingwell (1915), who proposed the now-accepted idea that polygonal fractures could be initiated by thermal contraction. However, this idea had been proposed earlier by several European workers in the nineteenth century (See Washburn, 1980, for references). From 1920 to the early 1940's, there were few studies of ice wedges in central Alaska and Yukon. A variety of ground-ice forms had come to light between 1900 and 1940, largely as a result of placer mining in the Tanana Valley and other parts of central Alaska. One of the earliest studies of ice wedges in central Alaska (Rockie, 1942) arose from an agricultural investigation of ground subsidence in the Fairbanks area. Forested land had been cleared some thirty years earlier, and had caused wastage of ice wedges within the silty soils. The resulting high-centered polygons, many with a local relief of 2 to 3 m, had rendered much of the farmland uncultivable. Although Rockie proposed that wastage of ground ice had caused the subsidence, he made no connection between the high-centered polygons of his area and the low-centered features described previously by Leffingwell (1919).

A much more extensive assessment of permafrost conditions in central Alaska was made by Taber (1943). At that time, the paper constituted one of the most detailed English language treatments of ground-ice phenomena. The three-dimensional geometry of ice-wedge polygons in silt was examined in numer-

Figure 12. Flat-topped ice wedge in silt, near Livengood, Alaska. (From Péwé, 1975; photograph by T. L. Péwé.)

ous hydraulic mining exposures. Taber concluded correctly that ice wedges had formed after, rather than concurrently with, the deposition of host sediments. Many of the wedges he examined had flat tops, and lacked any sign of wedge-to-surface ice veins, as might be expected if ice wedges were actively forming by thermal contraction (Fig. 12). Instead of inferring from the observed morphology that the ice wedges were relict forms, Taber rejected Leffingwell's thermal contraction theory, and proposed that ice wedges had grown by segregation during downward freezing. Although some of Taber's interpretations were wrong, his extensive reporting of thawed-down ice wedges, below the present permafrost table, has been used by later workers as a criterion for identifying inactive ice wedges (See Burn et al., 1986). Taber also provided one of the first accounts of regional variation in the thickness and continuity of permafrost in North America.

Considerable interest in ground-ice phenomena arose from the wartime construction program in Alaska. One of the outcomes of this was the extensive compilation by Muller (1945), based largely on Russian and European work, much of which was previously unknown to North American workers. Many new ideas were quickly applied to the interpretation of Alaskan fea-

tures. The ground-subsidence problem discussed by Rockie was reexamined by Péwé in considerably more detail. Work began in the late 1940s, with a summary of findings appearing in Péwé (1954). He observed that silt in the 'thermokarst mounds' was largely undeformed, which showed that they had formed not by frost heave but by downwasting of peripheral ice layers. The features described were classic examples of what are now called 'degraded ice-wedge polygons.' The distinction between high- and low-centered polygons was not made in this paper, but was outlined a year later in the report by Hopkins and others (1955). This provided a detailed summary of postwar research in Alaska conducted by the U.S. Geological Survey. The study was motivated primarily by the need for information on groundwater reserves in permafrost areas, but also provided a valuable, regional documentation of the role played by groundwater flow in determining the locations of permafrost layers within the discontinuous zone. This work demonstrated that residual ground ice was largely confined to materials of lower hydraulic conductivity, such as eolian silts, particularly in the relatively flat-lying areas along major river valleys, where hydraulic gradients are generally low. Climatic, topographic, and vegetative determinants of permafrost were also illustrated. A preliminary map of Alaskan permafrost zones was also included, presenting considerably more detail than that of Taber (1943).

The 1950s saw a considerable improvement in the understanding of factors responsible for active ice-wedge development and the preservation of inactive ice wedges. This in turn led to more research on the paleoclimatic significance of inactive wedges and ice-wedge casts. Several papers in the first International Permafrost Conference of 1963 (published 1966) examined ice wedges, but only one (Péwé, 1966) focussed explicitly on fossil features. Péwé made the important distinction between conditions conducive to permafrost development and those sufficient for ice-wedge formation. Using the thermal regime of modern, active ice-wedge environments, Péwé concluded that mean annual temperatures of –6°C or lower would be required. Subsequent research, summarized by Black (1976), has shown that the rate of cooling, as well as the water content and thickness of the active layer, are also important determinants of ground cracking in a given material type. Therefore, it may not be possible to apply a uniform estimate of paleotemperature to all areas preserving fossil ice wedges.

One of the most detailed studies of ice-wedges casts in the intermontane belt of Alaska is that of Péwé and others (1969), conducted at Donnelly Dome, just north of the Alaska Range. The existence of clear wedge forms at this site has since been questioned by Black (1976). However, given the low cohesive strength of the gravelly materials, well-preserved wedges would not be expected. Ice wedges apparently melted during the same Holocene climatic amelioration that had produced the "thaw unconformity" typical of ice wedges in silts of the discontinuous permafrost zone (Fig. 12). This study therefore underlined the need to take material texture into account when assessing the paleoclimatic implications of ice-wedge casts, since ice wastage at

Donnelly Dome was attributed in part to efficient groundwater flow through high-conductivity gravels.

Pingos. One of the earliest scientific studies of pingos in English (Porsild, 1938) dealt, in part, with the intermontane belt of Alaska. Porsild introduced the term 'pingo' into the scientific literature, and proposed the cryostatic mechanism for what are now termed 'closed-system' pingos, formed when permafrost steadily encroaches upon, and encapsulates, a thawed layer within the zone of continuous permafrost. Porsild also differentiated the closed-system pingo from the hydraulic type of ice-cored mound, produced by upward deformation of a relatively thin permafrost layer under groundwater pressure within the zone of discontinuous permafrost. The hydraulic type had been noted previously by Leffingwell (1919), and was subsequently termed the 'open-system' pingo by Müller (1963). Pingos received little study in North America during the 1940's, and in the next fifteen years, most work was confined to closed-system pingos on the Arctic coastal plain (See Chapter 14, this volume for a review of this literature). Some reference was made to open-system pingos in the interior, unglaciated parts of Yukon and Alaska (Mackay, 1966), but little was known of pingo distribution within this zone of discontinuous permafrost. Mackay outlined the major factors required for the growth of open-system pingos: local relief to create an adequate hydraulic gradient; groundwater recharge through discontinuous permafrost in the zone of higher elevation; hydraulic contact with conducting materials in the adjacent lowland; and a permafrost layer near the top of the lowland sediments, thick enough to produce hydraulic confinement, yet thin enough to allow upward deformation as a pingo mound. Subsequent pingo research in this part of the study area therefore has focussed on a refinement of the factors responsible for open-system pingo growth. There has also been some comment on the distribution of open-system pingos with respect to the northern limits of the Wisconsinan glaciation.

Definitive works on open-system pingos in Alaska have been written by Holmes and others (1966, 1968), based largely on work in the upper Tanana Valley and the Yukon-Tanana Upland. These studies provided the first documentation of nearly 300 pingos within forested areas, although isolated cases had come to light previously. Most pingos were estimated to be no older than 7,000 years, based on the radiocarbon age of deformed host sediments, and surface soil development suggested that many were less than 5,000 years old. Most examples were located close to the base of south- and southeast-facing slopes, since thermal conditions in these areas ensured adequate groundwater recharge, as well as lowland permafrost sufficiently thin to allow upward deformation by hydraulic pressure (Fig. 13). In some cases, it was found that hydraulic pressure in excess of 20 psi (140 kN/m^2) would be required to overcome the sum of the ice-soil overburden pressure and the bending resistance of the permafrost layer. Therefore, it was proposed that some pingos had grown by a combination of artesian and freezing pressure. No data were given on subpingo piezometric head, but the maximum required hydraulic pressure (equivalent to an excess head of about 14 m),

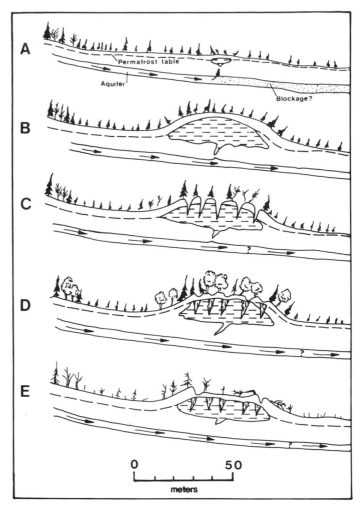

Figure 13. Hydraulic model for growth of open-system pingos. A, Winter, year 1. Groundwater under artesian pressure seeps towards surface and freezes. B, Winter, several years later. A large ice laccolith has formed. Channel to surface obstructed, water outlet shifts elsewhere. C, Winter, several tens of years later. Frost cracks develop (width exaggerated). D, Summer, shortly after stage C. Ice wedges develop in pingo and in pingo overburden. E, Summer, tens to hundreds of years later. Deep thaw follows summer fire. Central part of pingo melts uniformly. (Redrawn from Holmes and others, 1968.)

does not seem unreasonably large, given a local relief of well over 100 m in many valleys.

The distribution of open-system pingos in central Yukon was examined by Hughes (1969) in a short, though thought-provoking paper. The central question concerned the almost complete absence of pingos from that part of the discontinuous permafrost zone covered by ice during the late Wisconsin (McConnell) glaciation. The possibility was raised of a systematic change in permafrost thickness at the glacial boundary as a cause of the pingo distribution. This would imply that material within the glacial limit was not yet in thermal equilibrium, which seems unlikely. Terrain differences were also noted on either side of the glacial limit: within the McConnell limit, there were fewer of the narrow, valley-bottom sites associated with most pingos in the unglaciated region. Hughes also proposed that differences in the character of surficial materials on either side of the glacial limit might be an important causitive factor. This topic does not seem to have been followed up, although it was noted by Brown and Péwé (1973) as a future research priority.

This review of selected ground-ice forms shows that significant contributions have been made to the interpretation of fossil forms. Most ice wedges in the region are now known to be relict from the last glaciation, when continuous permafrost probably occurred over most of the unglaciated region, and temperatures were generally low enough to produce widespread ground cracking. The existence of flat-topped ice wedges below the present permafrost table points to postglacial warming, followed by climatic deterioration. Slow encroachment of permafrost into areas preserving ice-wedge casts points to a similar climatic trend, which moreover, is generally consistent with the known chronology of neoglaciation. Significant contributions have also been made in the area of open-system pingos. However, compared with the considerable recent advances in the understanding of closed-system pingos on the Arctic coastal plain, many questions remain concerning the hydrology and distribution of open-system pingos.

Slope Movement and Nivation Phenomena. A complex suite of geomorphic processes, usually combined under the headings nivation and gelifluction, produce distinctive periglacial landforms on alpine and subarctic hillslopes in the study region. Two forms that have received particular attention are rock glaciers and cryoplanation terraces. Rock glaciers are quite widespread within the alpine and subarctic mountains of the study region, but despite this relative abundance of forms, fewer than a dozen significant papers have been published on this topic during the past 75 years. This total includes the most important works conducted just beyond the study area boundary in Alaska and Yukon.

The term 'rock glacier' was introduced by Capps (1910) in a study conducted on the southern flank of the Wrangell Mountains. Capps proposed that freezing and thawing of interstitial ice was responsible for rock glacier motion, and that some forms were still moving. Earlier workers had proposed that rock glaciers were large landslides. Following Capps' ground-breaking study, only passing reference was made to rock glaciers in Alaska over the next 50 years, prior to the definitive work of Wahrhaftig and Cox (1959).

Elsewhere in the study region, only one important work on rock glaciers was carried out over this same period, that of Kesseli (1941) in the central Sierra Nevada. Kesseli doubted the general validity of the 'creep' mechanism proposed by Capps, and concluded a glacial origin from the tongue-shaped deposits and their relatively great transport distance from parent rock faces. Most forms were therefore regarded as relict, but local reactivation by small cirque glaciers during the nineteenth century was deduced from concentric sets of tranverse ridges. Localized contemporary movement was attributed to residual glacier ice, or to secondary,

postdepositional creep. The latter mechanism was not discussed, nor was buried glacier ice conclusively proven, but neither of these points disproved the hypothesis that most of the features were glacial relicts.

Wahrhaftig (1949) described multiple rubble sheets at Jumbo Dome, located some 20 km north of the Wisconsinan glacial limit in southern Alaska, that apparently never was glaciated. Wahrhaftig documented five phases of rubble deposition, the youngest of which preserved rock glacier morphology. Earlier phases were deeply dissected by gullies, within which the younger phases were stratigraphically inset. Rubble deposition was referred to glacial periods, with rubble dissection taking place in the ensuring interglacials. Although this opinion was not supported by an absolute chronology, the earliest deposits were noted to be completely severed by erosion from their source areas, suggesting a pre-Wisconsin age. The mode of rubble emplacement was not discussed in detail, but the presence of rock-glacier forms in an unglaciated area showed, contrary to Kesseli, that glacier ice need not be present for large-scale rubble movement on slopes flatter than 15°.

The paper by Wahrhaftig and Cox (1959) is one of the most detailed studies of rock glaciers ever conducted, and their morphological classification of rock glaciers as lobate, tongue-shaped, and spatulate is still widely used. Since the work was conducted just beyond the study area boundary, it is given only a brief summary here. Ice-cemented debris was attributed to meltwater freezing in a permafrost environment, and it was noted that coarse, blocky debris would favor the accretion of large amounts of interstitial ice. Rock glacier movement was then explained by the deformation of interstitial ice, probably over a considerable depth. This suggested a rheological similarity between glaciers and rock glaciers. An eight-year record of movement showed significant displacement of surface markers. Large pits at the head of some rock glaciers were shown to result from the wastage of a glacier-ice core. This showed that both 'ice-cemented' and 'ice-cored' debris could exist in the same local environment, though these precise terms were coined by Potter (1972) several years later. The paleoclimatic significance of rock glaciers was also elaborated. Two phases of growth were noted, separated by a warmer period during which dissection of the earlier deposits had occurred. The exact chronology was not known, but the proposed rock glacier fluctuations were broadly consistent with Holocene paleoclimate in the region. A similar pattern of Holocene rock glacier fluctuations has since been documented in the Brooks Range by Ellis and Calkin (1984).

In Yukon Territory, rock glaciers are well developed in the Wernecke Mountains and have been documented by Vernon and Hughes (1966). They, like Wahrhaftig and Cox (1959) before them, recognized spoon-shaped depressions at the head of several rock glaciers, but Vernon and Hughes proposed that such depressions could be used to separate debris-covered glaciers (i.e., glacier-ice-cored rock glaciers) from rock glaciers (ice-cemented rock glaciers). This stated connection between surface morphology and internal structure therefore predates Potter's (1972) dis-

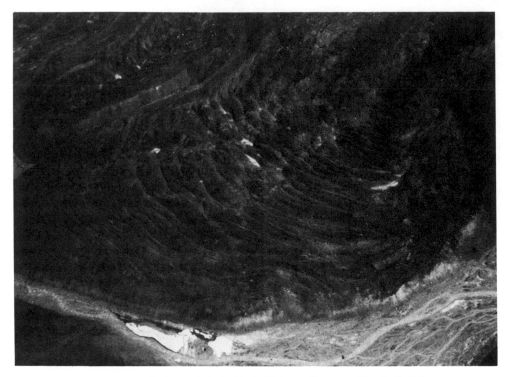

Figure 14. Talus rock glacier, Grizzly Creek, Yukon. (Photograph by P. G. Johnson.)

tinction between ice-cored and ice-cemented rock glaciers, as indicated by White (1976).

Spectacular cases of glacier-ice-cored rock glaciers have been reported from the Kluane and Donjek ranges of southwest Yukon by Johnson (1978, 1980, 1984; Fig. 14). Johnson (1980) has documented a continuum from active glaciers, through stagnant glaciers and ice-cored moraines, to glacier-ice-cored rock glaciers. Many of these show evidence of multiple phases of movement over the Holocene. Johnson (1978, 1984) also recognizes the talus-derived and avalanche types of rock glacier. Both of these have had a completely different history of movement from the ice-cored type, and may have involved catastrophic failure, rather than the slow, rheological creep supposed to occur in ice-rich talus. Catastrophic failure almost certainly explains the avalanche forms, since they generally lack the complex pressure ridging of the glacier-ice-cored and talus-derived types. However, Johnson (1978) has reported permafrost conditions in some of the talus-derived forms and this, coupled with the existence of near-surface ice-cemented debris, may imply a role for slow creep in some of the talus-derived rock glaciers.

Cryoplanation terraces show a pronounced regional concentration in the unglaciated uplands of interior Alaska and Yukon. Relatively few examples have been reported elsewhere in the study region (Washburn, 1980, Ch. 7; Péwé, 1983). One of the earliest North American references to cryoplanation terraces was made by Cairnes (1912) in a study conducted along the Alaska-Yukon border. Cairnes used the term 'equiplanation' to describe

the erosive effect of prolonged nivation activity. The role of nivation had earlier been outlined by Matthes (1900), but the scale of the erosion features described by Cairnes far surpassed the nivation hollows of Matthes. Cairnes attributed the disintegration of low bedrock cliffs at the rear of each nivation terrace to observable freeze-thaw processes, but did not specify how the frost-riven rubble could be transported away on very low slopes.

Gelifluction processes were given more explicit treatment in Eakin (1916). Many examples of what he termed 'altiplanation' terraces were described in the uplands that lie between the Yukon and Koyukuk rivers, principally the Indian River Upland. Altiplanation was described as ". . . a special phase of solifluction that . . . expresses itself in terrace-like forms and flattened summits . . ." (p. 78). Eakin noted that fine material, and therefore solifluction activity, were more notable close to the base of each altiplanation scarp. He therefore specified solifluction as the debris-transport process on the terraces, but unlike Cairnes made no reference to the retreat of terrace scarps by nivation. Instead he assumed that solifluction would gradually reduce the elevation of each terrace tread.

These early studies contributed to both periglacial terminology and to the understanding of denudation in cold climates. However, as with rock glaciers, relatively little new work was accomplished on cryoplanation terraces in the study region until the 1950s, save for passing reference in regional studies of geology and geomorphology. Péwé and others (1965b) made no reference to altiplanation terraces, and only brief mention was given in Péwé and others (1965a) and Wahrhaftig (1965a). Rather more work had been accomplished in Central Europe and the Soviet Union by this time, particularly on the separation of true cryoplanation terraces from structural benches. By the late 1960s, Péwé (1970) had demonstrated from the stratigraphy of superjacent solifluction deposits in the Fairbanks area that some terraces predated the Illinoian glaciation. This suggested that many of the larger terraces had taken tens to hundreds of thousands of years to form.

Much of the North American literature on cryoplanation terraces has recently been reviewed by Péwé (1975), Reger and Péwé (1976), Washburn (1980, 1985), and Péwé (1983). The review by Reger and Péwé (1976) reiterated that most terraces were true erosion features, discordant to the local rock structure, and had probably formed by the retreat of scarps through nivation. Virtually all terraces examined were considered inactive at present, which led to an assessment of the climatic regime necessary for reactivation. They concluded that depressed Wisconsinan temperatures would cause more effective nivation, and therefore notable scarp retreat. The permafrost table is presently less than 2 m below the surface of most terraces, and it was argued that a thinner, more easily saturated active layer during the Wisconsin glaciation would have promoted more efficient debris removal by gelifluction. In his review of this work, Washburn (1980) suggested the need for caution when estimating the temperature regime required for terrace reactivation, particularly since it had not been demonstrated that terrace debris was completely stable under the present climate at all elevations.

An outstanding problem concerns the mechanics of coarse rubble transport over distances of up to 500 m on slopes as low as 5°. The thaw-consolidation theory of McRoberts and Morgenstern (1974), which proposes high pore pressures, and therefore strength loss in thawing, saturated soils, would apply only to fine-grained materials. A more generally acceptable mechanism is the two-sided frost creep effect described by Mackay (1981), in which ice lenses can grow by downward freezing within the active layer, as well as by upward freezing from the permafrost table. However, on low-angle slopes, the potential frost creep would be small, unless a large amount of ice segregation were possible. Reger and Péwé (1976) mention that rubble forming the terraces typically contains many ice lenses, but it is not known if this ice is accumulating at present.

FLUVIAL PROCESSES

Introduction

Much of the work accomplished on fluvial geomorphology has dealt with the response of alluvial channels to high rates of sediment input from the relatively steep terrain that makes up a large part of the study region. Three types of study are recognized. First are those dealing with fluvial sedimentation conditioned by glaciation: that is, rivers responding to 'paraglacial' conditions. The second category has considered the impact of human activi-

ties on accelerated sediment production, the most notable cases being the hydraulic mining debris on the west flank of the Sierra Nevada, and major river diversions within the Canadian Intermontane Belt. The third topic deals with the impact of slope movement on alluvial channels, and includes a discussion of sediment budgeting. This work occupies the interface between hillslope and fluvial processes, but is considered here under fluvial processes since it often includes a discussion of sediment routing through watersheds.

Fluvial Sedimentation under Paraglacial Conditions

The term 'paraglacial' was introduced by Ryder (1971a) and Church and Ryder (1972) to denote nonglacial processes conditioned by glaciation. It includes accelerated wastage of drift from recently deglaciated slopes, the associated rapid aggradation of trunk streams, and fan building by tributaries. Although the term paraglacial was new, several studies of glacially conditioned sedimentation predate the work of Church and Ryder. One of the earliest references to thick accumulations of nonglacial material in Cordilleran valleys was that of Dawson (1895), who drew attention to what are now termed paraglacial alluvial fans in the valleys of central British Columbia. Dawson deduced from stratigraphic relations that many fans dated from the earliest part of the postglacial. He seems to have been influenced in no small way by the excellent paper of Drew (1873), in which the term 'fan' was introduced, and the well-known relationship between fan slope and fan size first stated.

After this important preliminary work, no further studies of note were carried out on paraglacial alluvial fans in the Interior Mountains and Plateaus region until the early 1960s. In the intervening period, a few qualitative descriptions of Quaternary fluvial deposits in Alaska had appeared in regional geologic accounts. Mertie (1937) referred to extensive alluvial aggradation in the Yukon and Tanana valleys, which he correlated with glaciation in the Alaska Range. Particular attention was drawn to the Tanana Valley, where enormous volumes of debris had displaced the Tanana River to the north side of its valley. Mertie also noted extensive river terraces in the same valley, which were interpreted essentially as a fill-and-cut sequence. More detailed, though largely descriptive, studies were made by Fernald (1960), who documented a similar northward deflection of the Kuskokwim River, and by Williams (1962), who reported valley outwash and alluvial fan formation in the Yukon Flats area, related to glaciation in the Brooks Range. In that both studies were concerned more with the chronology of the materials than with their morphology and development, they are not considered further here.

Prior to the early 1960s, few studies had examined fluvial processes in present-day paraglacial and proglacial environments. Over the ensuing 10 years, several important works were completed on braided channels close to glacier termini. Fahnestock (1963) examined a segment of the glacially fed White River that drains Mount Rainier. A large amount of bedload transport was observed in the predominantly braided channel network over a

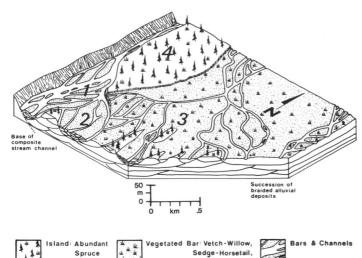

Figure 15. Schematic diagram of braided channel system in Donjek Valley (Redrawn from Williams and Rust, 1969). Level 1: Fluvial activity at all river stages. Unvegetated. Level 2: Subject to widespread fluvial action only during flood stages. Some channels active at low stages. Vetch-willow association, with minimum age 12 years. Level 3: Mainly overbank flooding. Very few channels active at low stages. Denser vegetation than level 2, mainly vetch-willow, and sedge-horsetail associations, with minimum age 27 years. Level 4: Mid-channel islands, with abandoned channels (not shown). Dense, mature spruce cover, minimum age 212 years.

two-year period, with net aggradation occurring in the first year, net degradation in the second. In addition to affirming the acknowledged importance of bedload transport in the formation of braiding, Fahnestock added the factor of highly variable discharge, driven by variations in the delivery of glacial meltwater. Under rapidly increasing discharge, scour of non-armored bed segments would occur, in addition to bank erosion, resulting in a pulse of sediment that would be locally redeposited as small bars. This, in turn, would promote anabranching and the overall maintenance of a braided pattern.

Some of the largest outwash trains occur in southwest Yukon, along the Donjek and White River tributaries of the Yukon River. Williams and Rust (1969) identified four levels in the braided outwash of Donjek Valley, the highest being a low terrace created by recent incision of the braided channel network (Fig. 15). Between this level and the perennial channel network were two intermediate levels of braided channels, occupied only during flood stages. The slow vertical shift of the channel system they attributed to increasing discharge related to recession of Donjek Glacier.

An important component of the glacially conditioned sedimentation cycle in south-central British Columbia, described by Church and Ryder (1972) is the paraglacial alluvial fan (Fig. 16). In the long period following Dawson's work, studies of alluvial fans in the American Southwest had come to dominate the literature on fans (Davis, 1905; Blackwelder, 1928; Blissenbach,

1954). These works led subsequently to the quantitative studies by Bull and others in the 1960s, most of which are summarized in Bull (1977). Ryder (1971b) noted several important differences between paraglacial and arid-region fans. One of the most significant was that paraglacial fans had been constructed relatively quickly from glacial materials eroded from catchments whose bedrock form was essentially preglacial or glacial in origin, whereas in arid regions, both fans and tributary catchments had evolved jointly over much of Quaternary time. It was argued, therefore, that fan-catchment relations formulated in arid regions might not comply with those in other environments. In particular, fan slope, for a given catchment area, was notably steeper in the paraglacial case, a difference most readily attributed to the higher sediment concentration of paraglacial fan-forming flows. It was also suggested that the paraglacial cycle was probably too short for catchment characteristics to be imprinted on fan morphology to the same degree achieved in the extra-glacial areas. Computed sedimentation rates for the early Holocene in south-central British Columbia (Church and Ryder, 1972) are far higher than those reported from contemporary paraglacial environments, the latter data admittedly based on rather short observation records. This discrepancy probably hinges on the lesser supply of glacigenic materials in these areas, when compared with the thick upland drift deposited by Cordilleran glaciations in central British Columbia.

The work of Ryder (1971a, 1971b) has provided an important bridge between studies of Pleistocene conditions, which had long dominated the regional literature, and studies of contemporary processes which, at that time, were only just beginning. This link is summarized by the schematic geomorphic work curve of Church and Ryder (1972; Fig. 17), and was subsequently emphasized by Slaymaker and McPherson (1977), in their overview of geomorphic processes in the Canadian Cordillera. The location of Mazama tephra within 2 m of the surface of many fills, coupled with the abandonment of many fan surfaces by tributary incision, suggests a phase of very rapid sedimentation in the first one-third of the Holocene.

The paraglacial concept has been extended recently to fans along the Nenana Valley in Alaska by Ritter and Ten Brink (1986). They document fans conformable with trunk-valley outwash sequences, and deduce from this stratigraphic fact that fans probably formed immediately after the outwash had attained its terminal level. The authors emphasize, however, that a paraglacial age for each fan unit identified has yet to be demonstrated by absolute dating.

The topic of late Quaternary sedimentation in the Interior Plateau of British Columbia has been addressed most recently by Clague (1986) and Ryder and Church (1986). Clague's findings basically support the earlier paraglacial model of Church and Ryder (1972). The stratigraphic evidence points to relatively short periods of very rapid sedimentation, associated with glaciations, separated by much longer nonglacial interludes, during which either limited sedimentation, or erosion of valley fills, took place.

Figure 16. Stereopair of dissected paraglacial debris fan, Fraser Valley, near Churn Creek. Formerly more extensive fan has been dissected and segmented by falling baselevel in the early? Holocene. Level 1 is the original fan surface. Levels 2 and 3 are progressively younger, stream-cut surfaces. (Province of British Columbia photos B.C. 7713: 218, 219).

Ryder and Church (1986) present a detailed discussion of river terraces in the middle Fraser Valley. They report a relatively late (about 7 ka) onset of downcutting near Lillooet, after which time several degradational and aggradational terraces have been produced. In this particular area, river sedimentation has been influenced by backwater from one or more early Holocene landslides, so it is not yet clear whether their chronology is generally applicable to the surrounding region. They raise the possibility that incision may have commenced as the cooler, wetter conditions around 6 to 7 ka produced an increase in Fraser River discharge.

Human Modification of the Fluvial Environment

Over the past century, the natural regime of many rivers within the study area has been disturbed dramatically by human activities. The works reviewed here illustrate the role of both increased sediment supply and increased stream discharge on the destabilization of alluvial river channels. A classic study of the influence of hydraulic mining debris was conducted by Gilbert (1917) in the Sierra Nevada. The study was a milestone in fluvial geomorphology; it contained one of the first field-scale tests of sediment transport concepts, some of which Gilbert had formu-

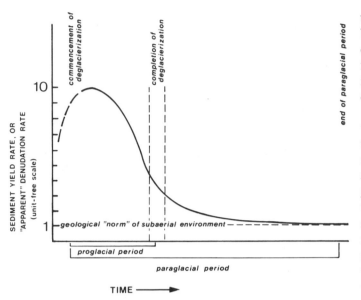

Figure 17. Model of Holocene sedimentation. (Redrawn from Church and Ryder, 1972.)

lated in an adjunct laboratory study (Gilbert, 1914). It also contained one of the earliest attempts at what is now termed 'sediment budgeting'; estimates were made of the volume of sediment delivered to channels by surveying abandoned hydraulic excavations.

The work commenced in 1905 in response to an appeal, made by the California Miners' Association to the Federal Government, to reconsider an earlier federal injunction on hydraulic mining activities. This had been successfully lodged some twenty years earlier by inhabitants of the Sacramento Valley, whose riparian lands had been devastated by downstream sedimentation. A large part of the report is concerned with the impact of mining debris on navigation in San Francisco Bay. This review is restricted to Gilbert's findings in the canyons and foothills of the Sierra Nevada.

One of the areas most affected by mining debris was the Yuba River basin, and, accordingly, Gilbert made his most detailed observations there. Laboratory findings concerning the relationship of stream capacity to slope, among others, were used to interpret the infilling of canyon channels as a necessary response to greatly increased load. Valley filling had occurred in a downstream direction, causing an increase in channel slope, and therefore capacity. Discharge, meanwhile, had remained fairly constant. Following the curtailment of hydraulic mining activities in the late nineteenth century, sediment loads were greatly reduced and river incision through, in some cases, tens of meters of fill had produced a second great pulse of sediment to the eastern tributaries of the Sacramento River system. By the early 1900s, some attempt had been made to contain the sediment with barrage structures, but such works were conducted in ignorance of

the mechanics of sediment transport. Scour directly below a large barrage on the Yuba River not only destabilized the channel downstream, as Gilbert had foreseen, but also led eventually to failure of the barrage itself. This released a large pulse of sediment into the Sacramento River system, further exacerbating the downstream sedimentation problem.

Channel incision through the fill behind the barrage provided a unique chance to estimate annual bedload transport, since the barrier had failed after only one year. Examination of the now-dissected sediment wedge behind the failed barrage revealed a small-scale replica of the entire sedimentation cycle in the region. Initially the barrier had caused a marked upstream reduction of channel slope by inducing deposition of a prism of sediment. The longitudinal slope of this deposit was insufficient for transport of coarse gravel to the barrier. Therefore, coarse material accumulated upstream, increasing the channel slope, which eventually allowed coarse gravel to be transported across the fill surface to the barrier. Although the barrier had caused filling in an upvalley direction, whereas the mining debris had caused downvalley filling, a delayed delivery of coarse material, similar to that noted in the barrier deposit, had been observed in all trunk streams. This provided a large-scale field demonstration of the interrelationships between capacity, load calibre, and slope, investigated during the laboratory work.

Gilbert's monograph on sediment transport was the first to document a large-scale modification of the fluvial system wrought by human action. His findings were enlarged upon by subsequent engineering work in fluvial hydraulics in the succeeding twenty years, but were not generally recognized by fluvial geomorphologists until much later.

The second type of study reviewed in this section concerns significant changes in the behavior of alluvial channels wrought by large-scale flow diversions (Kellerhals and others, 1979). These have been fairly numerous in Canada, with most designed to augment flow at hydroelectric plants. Most examples are located in thinly peopled, remote areas and so have not received much public attention. Such diversions constitute large-scale geomorphic 'experiments.' They are worthy of close attention, since a number of cases demonstrate the present predictive limitations of fluvial hydraulics. A notable example within the study region is the Nechako-Cheslatta diversion, in which the impounded upper Nechako River was discharged via a spillway through the much smaller Cheslatta River system, before rejoining the Nechako River below Kenney Dam (Fig. 18). Between 1956 and 1975, an estimated total diversion volume of 7.8×10^{10} of water passed through Cheslatta River, converting it from an initially tortuous, meandering, single-thread channel only 5 m wide at bankfull stage, to an entrenched single-thread channel with many bars, and 75 to 150 m wide at bankfull. Incision 5 to 10 m below floodplain level had converted large wetland areas into drained, low terraces.

In such large-scale diversions, future discharges can often be predicted tolerably well, or, as in the case detailed here, even controlled. In addition, as Kellerhals and others note, valley slope

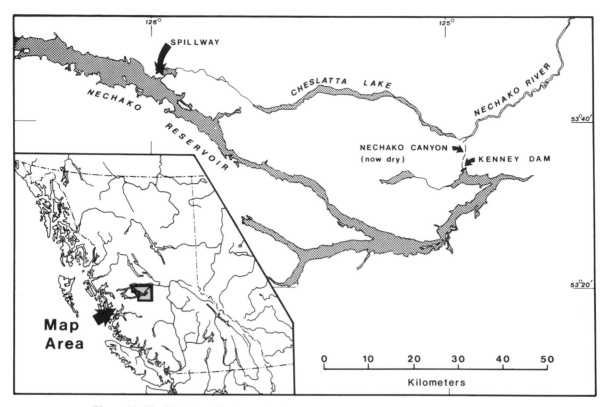

Figure 18. The Nechako-Cheslatta diversion (Redrawn after Kellerhals and others, 1979.)

can be accurately surveyed. However, prediction of the future river behavior requires a knowledge of the types of materials that might be exposed to erosion, once the original channel becomes incised in response to increased discharge. This information is difficult to obtain where the prediversion stream is not already incised, or where a complex suite of glacigenic materials comprises the valley fill. Islands of coarse debris, or bedrock ledges, may serve as local baselevel points, above which the steady-state channel form may be predictable if it traverses sand or gravel. By contrast, there is considerable uncertainty as to the vertical stability of channels cut in finer-grained materials, such as silts, or tills with low boulder content, since the deficiency of coarse clasts retards the natural self-armoring tendency of channels cut in heterogeneous materials.

Hydroelectric projects that involve flow regulation without augmentation from diversion usually drastically change the natural flow pattern. The most notable example in the Canadian Intermontane Belt is the Peace River power project, based on the W.A.C. Bennett Dam at Portage Mountain. In a preliminary study, Kellerhals and Gill (1973) noted a roughly five-fold reduction of annual peak flows, and a two- to three-fold increase in annual low flows. Within the first four years of the project's operation, measurable reductions in the dimensions of the mainstem channel were observed, associated with rapid encroachment of phreatophytes onto formerly active bar surfaces. In addition, the timing of tributary freshet peaks with regulated mainstem

flows well below natural levels led to notable tributary incisions at formerly accordant stream junctions.

Kellerhals (1982) has since provided additional documentation of morphological changes produced by flow regulations, reduction of sediment load, and reduced river competence to transport sediment. An important point made in this paper is that properly documented case histories may provide an effective means of improving predictive capabilities in situations where a 'first principles' approach to predicting channel morphology is uninviting.

Impact of Slope Processes on Alluvial Channels

One of the most catastrophic transformations of any North American fluvial system within the historical period occurred in the catchments of the Toutle and Cowlitz rivers as a result of the Mount St. Helens eruptions of 1980. This review highlights recent findings within the Toutle-Cowlitz river system by Haeni (1983), Lehre and others (1983), and Anderson and others (1985). The recent findings of Collins and Dunne (1986), concerning sediment production from hillslopes, are also summarized briefly.

Haeni (1983) has described the massive aggradation in the Cowlitz and Columbia rivers, which occurred in response to large mudflows triggered within the North Fork Toutle River catchment in May 1980. The usual 12-m navigation channel on the lower Columbia was reduced to slightly more than 4 m in depth, primarily because the slug of sediment arrived at a time of tidally reversed flow. Although the Columbia navigation channel was quickly reopened by dredging, channel capacity along the Cowlitz River remained severely reduced through aggradation, a situation that has since been alleviated by the Corps of Engineers through a large-scale program involving dredging, dike rehabilitation, and the construction of debris retention structures on the Toutle River system. This work falls more within the realm of engineering hydraulics than fluvial geomorphology, concerned as it is with the design and performance of mitigation works, and accordingly is not considered further here.

Anderson and others (1985) examined sedimentation in Coldwater Lake, impounded by the debris-avalanche deposit in adjacent North Fork Toutle River. A strong seasonal maximum of deposition coincided with the fall rains, each precipitation event creating a layer proportional to the duration and intensity of rainfall. Turbidite deposition during the wet season, followed by slow clay accumulation in the dry season produced a varve-like layered structure to the deposit. Sediment yield was estimated to be at least 10 times greater than in comparable, steep-forested catchments subjected to the same precipitation regime, since sediment traps had probably significantly underestimated the sedimentation rate during the winter periods. An interesting observation made by Anderson and others concerns the possible parallels between the graded varves, found in glaciolacustrine deposits and Precambrian sedimentary sequences, and those found in the Mount St. Helens area. They suggest that the transient hillslope instability within the blast zone may serve as a model for erosional conditions in rapidly deglaciated terrain, or in ancient landscapes prior to the evolution of soil-binding vegetation.

Within the North Fork Toutle River basin, Lehre and others (1983) constructed a posteruption sediment budget. Sediment budgeting involves a specification of sediment transport processes and their rates of operation within a catchment, as well as a cataloging of major sediment storage sites. Most data were derived from photogrammetry and measured cross sections, supplemented by measurements of hillslope erosion on the tephra-covered slopes, as described by Collins and others (1983). Debris deposition from the eruption was estimated at 68×10^6 m^3 from tephra, $2,500 \times 10^6 m^3$ from debris avalanche, and $12 \times 10^6 m^3$ from mudflow material. Over the ensuing year, export of sediment from the North Fork to the main Toutle River amounted to about $35 \times 10^6 m^3$, of which some 75% was derived from dissection of the debris avalanche deposit, principally by mudflows released shortly after the eruption (Fig. 19). This new channel network showed greatest headward extension and widening during the 1980 fall rains, with much lesser development taking place during the second six months of the 1980–81 observation period. The remaining 25% of sediment exported was tephra, stripped from hillslopes by sheetwash, gullying, and, on the steeper slopes, by shallow-seated sliding. With the removal of the relatively impermeable, upper part of the tephra deposit, hillslope erosion has declined rapidly. Additional data on this topic

have recently been provided by Collins and Dunne (1986). Figure 20 shows the marked decline in the rate of tephra erosion since the experiment began. This is attributed mainly to an increase in infiltration rate in the rilled areas consequent upon the removal of fine-grained tephra and the exposure of coarser-grained pyroclastic surge deposits beneath.

The work of Lehre and others (1983) is based in principle on earlier studies conducted in the Coast Ranges of Oregon by Dietrich and Dunne (1978). The extensive Coast Range literature on sediment budgets is reviewed by Kelsey in Chapter 13 of this volume. Although posteruption conditions within the Toutle River system were completely different from those in this earlier case-study region, an interesting point common to both studies is the relatively large, valley-floor reservoir of stored sediments. In the Oregon example, this storage volume was equivalent to 10^2 to 10^3 times the mean sediment discharge of the basin. It was noted that mobilization of relatively small quantities of this material as secondary denudation could significantly change the rank ordering of primary denudation processes, such as solution and creep, in the overall sediment budget of the catchment. In the Toutle River example, the residence time of stored sediment appears to be hundreds to thousands of times shorter, principally because of continued instability of the channel systems cut into the huge debris avalanche deposit, the main sediment storage element within the present catchment. It is estimated that this will continue to be the major sediment source for the foreseeable future, far outstripping the contributions from primary erosion of tephra and colluvium on the surrounding hillslopes. It will be interesting to observe, over the next few decades, how the relative contributions of primary and secondary erosion change within the Toutle River catchment.

Sediment budgeting and sediment routing have also been important research topics in the forested experimental watersheds of the Western Cascades (Swanson and others, 1982; Swanson and Fredricksen, 1982). This work is a logical outgrowth of earlier work on the sediment contributions from processes such as creep, stream-side slumping, and earthflow in both logged and unlogged areas (Swanson and Swanston, 1977). The geotechnical aspects of this work are reviewed later under 'Hillslope Processes.' The major components and linkages within the forested catchment system are shown in Figure 21. Quantifying linkages among these elements would be a challenge, even in undisturbed watersheds. Timber harvesting adds significant complications. A crucial point emphasized by Swanson and Fredriksen (1982) is the role played by large organic debris in augmenting sediment storage in relatively steep, torrent-prone channels. If such debris is disturbed or removed as a result of logging operations, sediment yield may increase markedly, and may be erroneously attributed to accelerated hillslope erosion if the watershed is treated as a 'black box.' Careful sediment budgeting therefore allows the relative contributions to sediment yield to be apportioned among: (a) stream sediments stored prior to timber harvesting; (b) sediment delivered to channels from streamside sources after harvesting; and (c) sediment contributions from accelerated debris

Figure 19. Mount St. Helens and the debris avalanche of May 1980, showing extent of subsequent dissection in July 1986. Blast-zone slope in the foreground.

avalanche and slump-earthflow activity on the surrounding logged terrain. Although the management implications of this type of study are beyond the scope of this review, it is clear that rational land management in the Cascades, and elsewhere, requires that the full complexity of geomorphic processes be considered within a watershed framework.

HILLSLOPE PROCESSES

Introduction

A wide spectrum of readjustments to hillslope form has occurred in the study region over the Holocene in response to a diversity of geologic, topographic, and climatic conditions. Mass-movement processes are emphasized in this review since, from the standpoint of morphologic changes, they have clearly dominated hillslope development over the Holocene within the Interior Mountains and Plateaus region. Three types of mass movement are reviewed in detail: volcanic debris flows, rock avalanches (sturzstroms), and slumps and earthflows. Because much of the work reviewed deals with catastrophic events, several important works have focussed on hillslope hazards. These

contributions are reviewed in the sections on volcanic debris flows and rock avalanches.

Volcanic Debris Flows (Lahars)

Considerable public attention has been drawn to this type of mass movement as a result of the 1980 eruption of Mount St. Helens. However, the potential for destructive lahar events in the Cascades had been outlined in considerable detail over the 25 years prior to this event (Crandell and Waldron, 1956; Mullineaux and Crandell, 1962; Crandell, 1971; Crandell and others, 1979). Prior to the mid-1950s, relatively little was known of the recent eruptive history of the Cascade volcanoes. Volcanic mudflows had been documented during the 1914 to 1917 eruptions at Lassen Peak, and also by Verhoogen (1937) at Mount St. Helens. However, the paper by Crandell and Waldron marks the beginning of modern research on the geomorphic effects of volcanic eruptions in the region, since it provided the first detailed documentation of lahar deposits, some of which had been mapped previously as glacial till in the Puget Lowland (Willis, 1898).

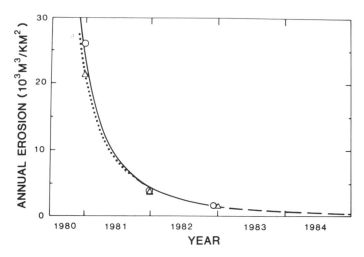

Figure 20. Decrease in tephra erosion from hillslopes in the Mount St. Helens area. Measured annual erosion of tephra and colluvium (circles and solid curve), and of tephra alone (triangles and dotted curve), for the Toutle River basin during 1980-83. Projected erosion for 1983-84 shown as a dashed line (From Collins and Dunne, 1986).

Although much innovative work has been accomplished on lahars within the Cascades, it should be noted that the earlier work in this region benefitted greatly from the literature describing historical lahar events in Indonesia and Japan. Detailed stratigraphic work was conducted on lahar deposits in the 1960s, especially at Mount Rainier (Fig. 22) and Mount St. Helens. A Holocene chronology of activity was gradually built up from radiometric dating of lahar deposits, and from the known age of interstratified tephras, principally those originating from Mount Mazama and Mount St. Helens (Mullineaux and Crandell, 1962; Crandell, 1971). By 1980 a considerable amount was known of the frequency, magnitude, and causes of lahars in the Cascades, and detailed mapping of lahar deposits had been used to produce hazard maps. The recent eruptive history at Mount St. Helens had suggested that a major eruption was likely to occur before the end of the twentieth century. Premonitory seismic activity during March 1980 prompted a detailed assessment of the probable slope stability hazards, and also led to preparations for the monitoring of events as they occurred. Although the magnitude of the north-flank rock avalanches, and the area devastated by the blast from the initial eruption, both far exceeded previous expectations, the trajectories followed by lahars, and their overall magnitudes, were generally within the limits anticipated from previous stratigraphic mapping.

A detailed assessment of the aftermath of the eruptions is given in Lipman and Mullineaux (1981). Articles in this volume dealing with lahars are Janda and others (1981), Cummans (1981), and Miller and others (1981). The paper by Janda and others demonstrated the complexity of the lahar generation process. Although lahar initiation by large-scale rock avalanching had been inferred at Mount Rainier by Crandell (1971), the most

probably cause of volcanic mudflows was previously considered to be hot, pyroclastic flows. These could be transformed into lahars after melting large volumes of summit ice and snow, and subsequently entraining large masses of unstable stratovolcanic material. This process was observed during the May eruptions, but the largest mudflows were generated several hours after the climactic explosion by failure of the saturated debris avalanche, which had slid and flowed into the upper North Fork Toutle Valley during the first few minutes of the eruption. The flow behaviour of this large 'unexpected' event was less erratic than that displayed by the smaller, more mobile lahars produced by pyroclastic events. Several weeks after the eruption, flow parameters were estimated from the morphology, inclination, and thickness of lahar deposits by Fink and others (1981). Using techniques previously developed by Johnson (1970), estimates were obtained for yield strength, Bingham viscosity, and average velocity. Velocity estimates were derived from the super-elevation of debris at selected cross sections, and assume that no 'fluid' energy was sapped by internal resistance as the flows passed through the channel.

Allied to these preliminary studies of lahars generated by the 1980 eruption is the detailed work on debris-flow dynamics being conducted at Muddy Creek, on the southeast flank of Mount St. Helens. The important recent findings have been summarized by Pierson (1986). Of particular interest has been the documentation of the transition from hyperconcentrated streamflows to true debris flows. This occurs typically at sediment concentrations of about 75% by weight. Pierson's results demonstrate the dramatic change in rheological properties of sediment-laden flows in the concentration range of 75% to 80%. Pierson has also suggested that the Manning flow equation may be calibrated to the calculation of debris-flow velocity, provided that a restricted range of flow-sediment concentrations is used in the analysis. Although this analysis is preliminary, it suggests an alternative to velocity computations that not only require knowledge of Bingham viscosity and strength (Johnson, 1970), but also require specification of channel shape at the time of the debris flow. Finally, the research at Muddy Creek is building a real-time data base on debris flow dynamics which, though probably not exceeding that collected by Japanese geomorphologists in the 1970s (e.g., Okuda and others, 1980), is certainly long overdue in North American research on debris flows.

Rock Avalanches

Several catastrophic rockslope failures, involving complete disintegration of the failed mass as a high-velocity rock avalanche (sturzstrom), have occurred in the study region over the past 25 years. Principal among these are the Little Tahoma Peak rock avalanches at Mount Rainier in 1963, probably triggered by a steam eruption (Crandell and Fahnestock, 1965; Fahnestock, 1978); the Hope-Princeton rockslide of 1965 in southwest British Columbia, apparently triggered by a small earthquake (Mathews and McTaggart, 1978); several small rock avalanches in the

HILLSLOPE   CHANNEL

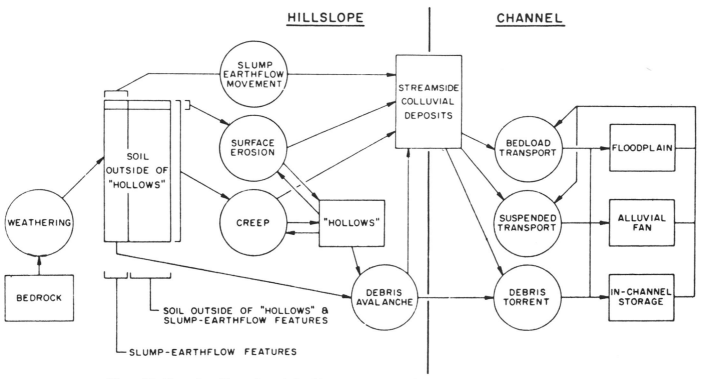

Figure 21. Flow chart illustrating relationships among storage sites and transport processes in steep, volcanic terrain in the Western Cascades, Oregon. (From Swanson and Fredriksen, 1982.)

Sierra Nevada in 1980, associated with moderate- to high-magnitude earthquakes (Keefer, 1984); and the huge rock avalanche at Mount St. Helens, generated by the 1980 eruption (Voight and others, 1983). In addition, prehistoric rock avalanches have been reported from other parts of the study region by Palmer (1977), Eisbacher (1971, 1979b), Clague (1981b), Cruden (1985), and Crandell and others (1984). Many of these failures appear to have been triggered by either seismic or volcanic events.

Several authors have considered the dynamics of rock avalanches. Bruce and Cruden (1977) analyzed the Hope-Princeton rockslide as a plane friction event, and obtained good agreement between the ultimate angle of shearing resistance of the slide material (about 20°) and the gradient of the 'farboschung' line—defined as the ratio of total vertical drop to total horizontal distance travelled, measured from the crown of the failure surface to the toe of the slide mass. Although no strong inferences were drawn from this close numerical agreement, it is evident that the 'fahrboschung' line describes the average rate of kinetic energy loss, but does not specify the details of this loss. However, without detailed observations of rock avalanches in motion, it would be difficult to justify a more complex model. For this reason, work on rock avalanche dynamics has involved considerably more uncertainty than comparable work on snow avalanches, for which a large film record of flow events exists.

The massive rock slope failure at Mount St. Helens provided

a unique opportunity to observe the transition from initial rockslide to fully developed debris avalanche (Voight and others, 1983). The motion of separate failed masses was studied from a series of photographs that commenced some 25 seconds after the release of the first and largest rockmass, which by that time had moved approximately 700 m downslope. Rock avalanche motion was described by a general expression that had been applied earlier by Mellor (1968) to snow avalanche motion:

$$\tau = a_0 + a_1 v + a_2 v^2 \tag{1}$$

where τ is motion resistance per unit area of flow, a_0 is Coulomb strength, depicted as velocity independent; $a_1 v$ represents viscous resistance, proportional to the first power of velocity; and $a_2 v^2$ describes resistance generated during high velocity turbulent flow, and includes aerodynamic drag. Relative to snow avalanches, however, flowing debris possesses very large 'viscous' resistance, which led the authors to propose a reduced model involving Coulomb and viscous terms alone. Motion of the first block during the first 40 seconds was described adequately by plane friction. This conclusion was reached by substituting rockmass acceleration into the equation of motion and solving for friction coefficient. At first sight this appears to support previous plane-frictional analyses of rock avalanches, but the precise nature of the forces resisting the motion remains uncertain. Rapid acceleration to over 50 m/sec. indicates a considerable strength drop after

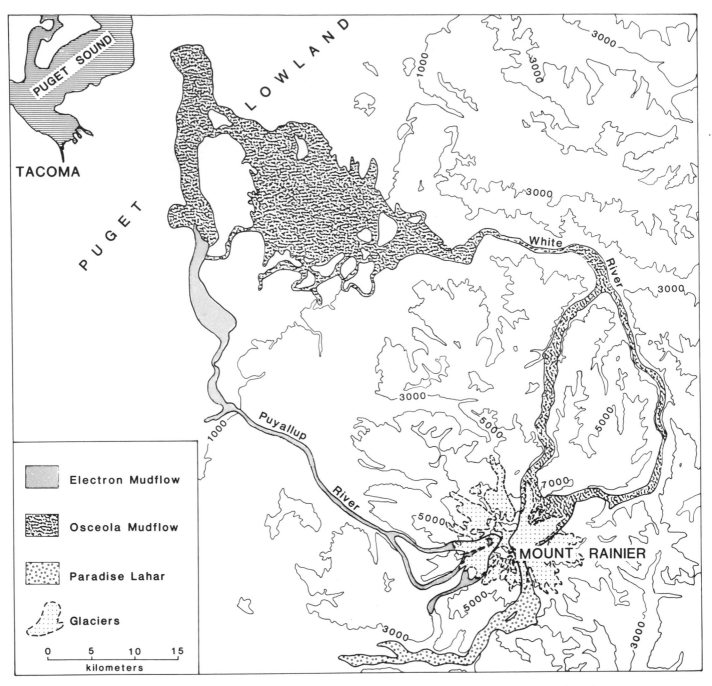

Figure 22. Major volcanic mudflows from Mount Rainier. For clarity, shading depicting mudflows is not extended into areas currently occupied by glaciers. (Redrawn from Crandell, 1971.)

the initial failure, probably from a combination of dilatation and pore pressure rise, the latter produced in part by gas pressure.

Strength calculations were made from the thickness of deposits in the final flow phase, and velocities were obtained from superelevation data. From the calculated strength, texture, water content, and final inclination of this debris, it is evident that motion toward the distal margin was better described as a debris flow, rather than a rock avalanche. Therefore, although unique

scientific data were obtained from the Mount St. Helens film sequence, it is unclear whether they can be applied readily to true rock avalanches. Many of these appear to have had very little matrix fines or positive fluid pressures, and their motion is thought to be controlled primarily by a complex, velocity-dependent frictional term that bears little mechanical resemblance to viscous resistance.

Fahnestock (1978) also considered rock avalanche dynamics in his analysis of the Little Tahoma Peak events on Mount Rainier, and concluded that the rock avalanche may have ridden on a cushion of compressed air, after becoming air launched. From the evidence presented, there is little doubt that one of the rock avalanches was air launched during part of its motion. However, an appreciable volume of snow was incorporated in the debris, and it is possible that the resulting snow-rock mixture could have reached its terminus without the aid of an air cushion.

The prehistoric Mount Shasta event recently described by Crandell and others (1984) is apparently the largest known Quaternary landslide. An estimated 26 km^3 of volcanic debris and Quaternary sediments cover an area of 450 km^2 in the Shasta River Valley, northwest of the volcano. Movement probably originated as a series of seismically triggered rock avalanches, since there is no clear stratigraphic evidence of a large volcanic event accompanying the failure. Large-scale fluidization is thought to have occurred from entrainment of saturated Quaternary sediments along the valley floor. The 43 km runout of the debris on an average slope of less than 1° was unexplained. However, a conventional 'fahrboschung' analysis, using the most probable basal friction coefficient of about 0.1, obtained from the Mount St. Helens event, suggests that the summit of Mount Shasta may have stood more than 4,000 m above the valley floor at the time of the massive slope failure. Any reduction of the assumed friction coefficient would reduce this computed elevation.

One of the most dense concentrations of rockslope failures in the study area occurs in the Skeena Mountains of north-central British Columbia (Eisbacher, 1971). Most of the 25 examples he identified within a 750 km^2 area of sedimentary rocks are deep-seated failures, best classed as rock slumps according the Varnes' (1978) classification. Several rock avalanches were also reported, produced by more shallow-seated, dip-slope failures. Eisbacher provided no more than a reconnaissance account of these features, but his observations suggest that the dipslope model for rockslope failure, developed by Cruden (1976) in the carbonate zone of the Canadian Rockies, may be applicable to a wider region. In a later paper, based on work conducted in the Mackenzie Mountains, Eisbacher (1979b) documented many large rock avalanches, some with runouts as long as 7 km. Many of his examples are anomalous, in that releases have occurred on slopes as flat as 14°, considerably lower than the ultimate friction angle of the carbonate rock material. To overcome this mechanical difficulty, Eisbacher proposed a 'roller bearing' model for avalanche release. This requires strong seismic shaking and the disintegration of a basal-rock layer into roughly equant blocks, the rotation of which allows downslope translation, and subsequent disintegration, of the overlying rock mass.

This brief review of some of the largest slope failures in the study region has shown that many problems remain unsolved concerning the mechanics of large-scale movement. The opportunities for real-time data collection, afforded by the Mount St. Helens event, are few and far between. If questions still remain concerning the motion of this event, the prospects for resolving the mechanical details of prehistoric rock avalanches seem remote. The examples discussed here cover a very wide range of material types and flow boundary conditions, and it is therefore improbable that a single model will account for all cases. Preliminary attempts to predict rock-avalanche runout from the gradient and height of slope from which the release occurs have been reported by Scheidegger (1973) and Davies (1982). Some of the larger avalanches discussed here (Hope-Princeton, Mount St. Helens, Mount Shasta) were rendered more mobile by the entrainment of large volumes of water and saturated sediments from the valley floor directly below the failure zone. This introduces a serious complication into runout models build primarily on slope geometry and assumed friction coefficients.

Slumps, Earthflows, and Rock Creep

Slumps and Earthflows. Weathered and altered volcanic or sedimentary rocks crop out over large areas of the Interior Plateau of central British Columbia and the Western Cascades of Oregon. In both regions, slumping is widespread along the margins of valleys incised deeply into these relatively weak rock units (Palmer, 1977; Eisbacher, 1979a; Evans and Cruden, 1981), and retrogressive slumping has created huge, slow-moving earthflows (Swanston and Swanson, 1976; Swanson and Swanston, 1977; Vandine, 1980; Bovis, 1985; Rawlings, 1986; Fig. 23). Although many of these slumps and earthflows were noted long ago during the earliest geological surveys, they have only received detailed attention within the past 20 years or so. Geotechnical studies associated with major public works projects or with timber harvesting, account for a large part of this recent work.

Many landslides occur along the Columbia River Gorge, and were identified in the 1930s during reconnaissance for the Lower Columbia hydroelectric projects. Many of the original reports are unpublished, but are summarized by Palmer (1977). Most of the failures along the gorge can be classed as large, slow-moving slumps, some of which have become elongated into forms resembling earthflows. Most can be attributed to softening of clay units, and to the development of significant pore pressures generated by vigorous groundwater recharge through the overlying, more permeable volcanics. One of the largest earthflows on the Washington shore is close to 5 km in length, and has moved more than 10 m in years of high precipitation, causing disruption of trunk power lines.

Large slow-moving earthflows, juxtaposed with slumps and zones of deep soil creep, have also been reported in the experimental forest watersheds of the Western Cascades in Oregon (Swanson and Swanston, 1977). Most of the slope movement occurs in areas of altered and deeply weathered volcaniclastic materials, and shows a well-developed seasonal variation related to groundwater recharge. It is common to find active slump or earthflow masses surrounded by large areas of 'inactive' landslide terrain. This association is found in other regions of the Cordillera, notably the Coast Range of California (Kelsey, 1978) and parts of southwest British Columbia (Bovis, 1985), but to date most research has been directed to the most actively moving parts

Figure 23. Large, slow-moving earthflow near Big Bar Creek, middle Fraser Valley. Earthflow has breached through early? Holocene stream-cut terraces.

of a given mass-movement complex. Work in the Western Cascades is important, therefore, not only for its documentation of factors influencing the movement regime of active earthflows, but also for its recognition of zones of deep-creep deformation. These relatively small, distributed shear displacements are scarcely within the domain of conventional limit equilibrium analysis, yet they may produce well-developed failure planes in the future as the clay-rich material weakens by strain softening. Widespread creep deformation therefore presents a challenge to forest engineers who must assess the probable impacts of timber harvesting within such potentially unstable watersheds.

Many slumps and earthflows occur within the predominantly volcanic and sedimentary terrain of the Interior Plateau of British Columbia. Most are found at, or close to, the steep margins of major valleys incised up to 1,000 m below the plateau level. Prior to the 1970s, the landslide literature of this large region consisted of a few detailed engineering reports. No regional studies of landslide types had been carried out. Since 1970 many landslides have been identified through regional terrain mapping (Ryder, 1976, 1981), and recent regional overviews of landslides in the Canadian Cordillera (Mollard, 1977; Eisbacher,

1979a; Cruden, 1985) are able to cite a growing literature. However, given the diversity of slope-movement types, and the size of the region, detailed studies have yet to be conducted over much of the Plateau area. Large-scale rock slumps are particularly widespread in the Tertiary volcanics of the central Interior Plateau region (Evans and Cruden, 1981). Some large block failures occur where joints or faults diverge downslope in plan view, whereas slopes with convergent discontinuities are comparatively stable. This is an important observation, since most assessments of rockslope stability refer to a vertical section oriented parallel to the direction of major principal stress, and ignore the effects of joints on lateral yielding normal to this conventional plane of analysis.

Many cases of retrogressive slumping in Quaternary sediments have been documented along the deeply incised Fraser and Thompson rivers of the Interior Plateau (Ryder, 1976, 1981; Evans, 1982), many apparently related to disturbance from irrigation or excavation. The most detailed study to date is that of Thomson and Mekechuk (1982). They conclude that recent reactivation of a large landslide in glaciolacustrine materials is due primarily to the slow buildup of artesian pore pressures, which they attribute to the recent increase in mean annual precipitation in the Interior Plateau. This conclusion is congruent with the significant upward trend in Fraser River discharge over the past 30 years. The important contribution of this study is therefore the suggestion that secular climatic shifts may cause a reactivation of 'inactive' landslides if the local stratigraphy favors pore-pressure buildup.

Some of the largest, noncatastrophic slope failures in British Columbia are elongated earthflows, many of which exceed $10^7 m^3$ (Fig. 23). Since the publication of the first detailed study by Vandine (1980), more than 20 additional cases have been identified along the middle Fraser River and its tributaries (Bovis, 1985). Some of the larger cases probably were initiated shortly after deglaciation; however, many appear to postdate the Mazama ashfall of some 7,000 years ago. It is thought, therefore, that the wetter, cooler climatic phases of the post-Hypsithermal period may have initiated, or reactivated, failure in areas where groundwater discharge is impeded by a confining layer of clay-rich material.

Rock Creep. Within the past 20 years or so, a variety of anomalous landforms, such as uphill-facing (antislope) scarps, and ridge-top grabens or half-grabens, have come to light in the steep mountain topography of the western Cordillera. A recent review of this growing literature on so-called 'gravitational' deformation features is given by Radbruch-Hall (1978). Most cases seem to involve a partial collapse of an entire alpine mountain ridge, caused perhaps by stress relief brought about by deglaciation. As far as can be judged, antislope scarps were reported within the Interior Mountains and Plateaus region prior to their recognition in any other part of North America. Mylrea (1969) identified anomalous 'linears' above the Mica Dam on the upper Columbia River, but concluded that they were probably ice-sculpted features. Mollard (1977) considered these same features to be the product of differential isostatic rebound, an interpretation discussed subsequently by Bovis (1982). Elsewhere in the study region, Beget (1985) has since produced one of the most detailed studies of antislope scarps in North America, based on work near Glacier Peak in the North Cascades. Beget's work is significant for its documentation of multiple phases of scarp development, the earliest probably related to glacial downwasting, the most recent probably related to nonglacial debuttressing of slopes, apparently by landsliding. Antislope scarps and similar features related to gravitational deformation of high mountain ridges appear to have been confused with seismic fault scarps, both in the North Cascades and elsewhere. Their proper interpretation is therefore of more than academic interest. At present, the number of cases on record is simply too few to make any firm pronouncements concerning their origin and chronology.

CONTRIBUTIONS TO GEOMORPHOLOGIC THEORY AND THOUGHT

Erosion Surfaces and Long-term Landscape Evolution

Studies of the erosional history of the Sierra Nevada have

contributed materially to knowledge of the pace of landscape development on a timescale of millions to tens of millions of years. This is due in large measure to the availability of both erosional and depositional evidence, including datable volcanics. There are few other areas of North America in which geomorphology and regional tectonics have been so closely meshed, and in which denudation chronology has been able to fultill a role beyond landform description. Reconstructions of the longitudinal profiles of Tertiary valleys on the west slope have demonstrated the reality of rigid tilting of a large fault-defined block (Lindgren, 1911; Matthes, 1930; Christensen, 1966; Huber, 1981). This, in turn, has led to inquiry as to the regional tectonic framework for such a style of uplift (Hay, 1976; Crough and Thompson, 1977). In the past 20 years, absolute dating has provided a time frame for canyon cutting in hard crystalline rock under conditions of steady to accelerating uplift. The ambiguity of erosional evidence has been underscored by the work of Wahrhaftig (1965b), particularly in granitic terrain that displays such a variety of forms according to the local weathering environment.

Volcanic Landforms

Major contributions to volcanic landforms development have come from studies in the High Cascades. The work of Diller (1891, 1895) and Williams (1932) identified the range of cone forms and their correlation with magma type, and contributed to the theory of volcanic dome formation. Papers by Diller and Patton (1902) and Williams (1941, 1942) greatly enlarged understanding of explosive volcanism and caldera formation. The importance of glacial erosion, combined with postglacial activity, in the development of the present form of many stratovolcanoes in the Cascades was subsequently demonstrated by Fiske and others (1963) and Christiansen and Miller (1976).

Mathews (1947) first identified the distinctive 'tuya' volcanic forms in northern British Columbia, and enlarged upon previously held theories developed in Iceland concerning their subglacial origin. Most of Mathew's original interpretations have since been confirmed in the more comprehensive paper by Allen and others (1982).

Glaciation

Glacial studies in the Sierra Nevada have contributed greatly to the theory of glacial erosion, and to the documentation of multiple glaciation in the Cordillera. The latter has not been emphasized in this chapter, but major original works include Matthes (1930) and Blackwelder (1931). More recent contributions to glacial chronology are reviewed in Porter (1983). Fundamental contributions to the theory of glacial erosion span a wide range of landform scale, from small-scale crescentic gouges (Gilbert, 1906), through meso-scale forms such as riegel steps (Russell, 1889; Johnson, 1904; Matthes, 1930) and ridge asymmetry (Gilbert, 1904), to glacial canyons and hanging valleys (McGee, 1894; Matthes, 1930). The work of Matthes has been particularly influential in other glaciated areas, and forms much of the basis of presently accepted ideas on glacial erosion. Matthes greatly enlarged on earlier work by Le Conte (1875) and Russell (1889) concerning the magnitude of glacial erosion and the role of jointing in differential glacial erosion. Matthes' Yosemite studies also clarified the relative importance of glacial and preglacial erosion in canyon formation, and thereby contributed to the understanding of long-term landscape evolution.

The major original contribution from the Canadian Cordillera was the documentation of the Cordilleran Ice Sheet by Dawson (1879, 1888), a concept subsequently extended by Kerr (1934), Flint (1943), and Tipper (1971). Dawson also pointed to the existence of a large unglaciated area in Yukon Territory, which was subsequently traced into central Alaska.

Periglacial Landforms

Virtually all of the original contributions stem from work completed in the unglaciated areas of Alaska and Yukon. Although the theory of ice-wedge development was largely worked out on the Arctic coastal plain, the recognition of inactive forms in central Alaska by Péwé (1954, 1966) has contributed greatly to the reconstruction of Quaternary environmental conditions in this region. The preservation of fossil ground ice forms has been closely tied to the hydraulic conductivity of host sediments. This indicates the need for caution in the interpretation of both ice wedges and ice-wedge casts.

The original work on pingos by Porsild (1938) was conceived partly in west-central Alaska, and identified the two distinctive types of ice-cored mounds. Subsequent work has enlarged understanding of open-system pingos (Mackay, 1966; Holmes and others, 1968; Hughes, 1969), and has emphasized the importance of local hydrologic conditions, permafrost thickness, and possibly glacial history in their development.

Much of the original work on rock glaciers was conducted just beyond the boundary of the study region (Capps, 1910; Wahrhaftig and Cox, 1959). The work of Wahrhaftig (1949) was the first to identify multiple phases of rubble accumulation and to suggest a correlation with glacial episodes. Vernon and Hughes (1966) first made explicit the distinction between glacier-ice-cored and ice-cemented rock glaciers, though this distinction was implicit in the paper by Wahrhaftig and Cox (1959). More recently, Johnson (1978, 1980, 1984) has documented the morphological distinctions between glacier-ice-cored rock glaciers, talus rock glaciers, and avalanche rock glaciers, and has noted the different chronology of Holocene movement associated with each type of feature.

Most of the important North American work on cryoplanation terraces has been conducted in the intermontane belt of Alaska, the notable papers being Cairnes (1912), Eakin (1916), Péwé (1970, 1975, 1983), and Reger and Péwé (1976). It is now realized that many of these erosional benches may have taken tens to hundreds of thousands of years to form, with the most intense phases of development taking place during Quaternary glaciations.

Fluvial Geomorphology

The earliest fluvial work in the study region postdates that of the Powell survey in the Colorado Plateau, and therefore was able to apply the fundamental concepts of baselevels, grade, and fluvial planation developed in that region. However, an important new perspective on fluvial phenomena was provided by Lindgren (1893), who compared the present inclination of auriferous gravels in Tertiary channels on the west slope of the Sierra Nevada with their maximum probable inclination at the time of deposition. He argued that gradients steeper than 0.05 were unlikely in an aggrading gravel river, and thereby used a hydraulic concept to obtain a minimum figure for postdepositional tilting of the range. This then translated to uplift at the crest of the Sierra Nevada; probably one of the earliest uses of paleohydraulic analysis. Certainly the use of a hydraulic argument to calculate tectonic uplift was new at the time.

Some fifteen years after the publication of Lindgren's first work on the auriferous gravels, Gilbert began formulating his ideas on the dynamics of contemporary channels in the same field area. Field observations in the Sierra Nevada allowed a verification of laboratory-determined relationships between channel slope, channel shape, discharge, sediment load, and sediment calibre in a way not previously accomplished. New concepts were used to account for present river behavior, as well as to predict likely future trends in erosion and sedimentation. The performance of engineering structures under conditions of high sediment load was also assessed. The work of Gilbert (1914, 1917), therefore, served as the basis for field and experimental work in hydraulics over the ensuing 20 years, and also laid the foundations of hydraulic geometry. The full significance of this work was not generally appreciated by geomorphologists until the great quantitative revival of fluvial geomorphology in the early 1950s.

The more recent contributions to fluvial work have been concerned more with 'river behavior' than with fundamental hydraulic research. However, given the difficulty of predicting fluvial responses from what are invariably transient inputs of water and sediment, these studies are important in illustrating the general behavior of channels in regions of high sediment supply. Fahnestock's (1963) study of a glacially fed, braided stream provided important documentation of very rapid changes in both channel form and position. A significant finding was the role of wide fluctuations in discharge within a meltwater environment in the maintenance of a braided network under conditions of abundant sediment supply. More recent work, exemplified by Williams and Rust (1969) and Miall (1977) has considered the behavior of much larger braided systems. These studies reaffirm that a braided pattern may exist under conditions of aggradation, degradation, or equilibrium of a valley fill, and that the vertical position of the channel network may be susceptible to influence from discharge fluctuations induced by climatic change over the past 100 years. On a longer time scale, massive fluctuations in sediment supply, wrought by deglaciation, have been shown by Church and Ryder (1972) to produce major transformations of entire valley topography, on a time scale of probably one thousand years or less. This provides an important conceptual yardstick by which we may judge the adequacy of present-day 'process' studies, and suggests that detailed work on the relatively long Holocene record of fluvial sedimentation may be more rewarding than studies of contemporary river behavior if our task is to comprehend landscape evolution, a point subsequently emphasized in Church (1980). Kellerhals (1982) has emphasized the fact that large-scale engineering manipulations of rivers can be regarded as geomorphic 'experiments,' which provide a valuable data base for predicting future river behavior.

Important contributions have been made in the area of sediment budgeting within steep forested catchments. Some of the most recent findings are summarized in Swanson and others, (1982). As yet, this approach to catchment dynamics has not had widespread application within the study area outside of the Western Cascades. The main contribution of sediment budgeting research, quite apart from its practical application to land-management problems involving sediment transfer, is its provision of a methodology that allows detailed 'process' studies, both on hillslopes and in river channels, to be applied to the study of landscape evolution.

Hillslope Processes

Important contributions to the theory of catastrophic mass movement phenomena have been made in the region. Prominent among these are the large lahars of the High Cascades. The work of Crandell and Waldron (1956) and Crandell (1971) has added greatly to knowledge of the chronology and dimensions of these events. Observations during and after the Mount St. Helens eruptions by Janda and others (1981), Cummans (1981), and Fink and others (1981) provide important data on the initiation, flow velocity, and material properties of large lahars. More recently, Pierson (1986) has provided valuable, real-time data on the flow regime of debris flows at Mount St. Helens. Of particular interest has been the transition between so-called hyperconcentrated fluvial flows and true debris flows.

The role of seismic and volcanic events in the triggering of rock avalanches has been demonstrated, or strongly implied, by Mathews and McTaggart (1978), Voight and others (1983), Keefer (1984), and Crandell and others (1984). The real-time analysis of the rock avalanche by Voight and others (1983) is unique within the mass-movement literature, though the general application of their findings to the relatively dry, rubbly materials found in most other rock avalanches is less certain. Other work on dynamics (Bruce and Cruden, 1977), has shown that simple plane friction, based on the ultimate angle of shearing resistance, produces results compatible with the present configuration of the failure surface and the topography of the rock-avalanche debris.

The most important findings on slow mass movement, not emphasized in other regions, concern the transition from slow, distributed creep deformation to large-scale translational sliding (Swanson and Swanston, 1977). Work in the Western Cascades has focussed also on the mechanics of slump-earthflow movements in relation to surface and subsurface hydrology, in addition to documenting the impacts of forest harvesting on slope movement processes. In view of the large translational sliding component in many features hitherto described as 'earthflows,' Bovis (1986) has recently suggested that this term be abandoned in favor of 'earthslide,' and that the term 'earthflow' be restricted to failures in sensitive soils.

The role of recent climatic shifts in the reactivation of landslides has been identified by Thomson and Mekechuk (1982). On a longer time scale, Bovis (1985) has proposed that postglacial fluctuations in the dimensions of large earthslides in the Canadian Cordillera may be driven by piezometric fluctuations influenced by Holocene climatic changes.

FUTURE WORK

Discussion of the most profitable directions for future research is subdivided here into three categories, based on the time frame of the observations. First are studies of long-term landscape development, on a timescale of millions to hundreds of thousands of years. As noted in the section on 'Cenozoic Erosion Surfaces,' this end of the geomorphic timescale has been largely neglected in the past 40 years, with the notable exception of the Sierra Nevada, where it has been a long-standing research theme. The story of the Tertiary evolution of the rest of the study area, in particular the Canadian and Alaskan parts of the Cordillera, has hardly been told, with the notable exception of the southern Yukon (Tempelman-Kluit, 1980). Ironically, the value of, and need for, such studies of long-term landscape development is more readily seen in recent overviews of regional tectonics, rather than in recent geomorphic writings from the Interior Mountains and Plateaus. The statement in Monger and Price (1979, p. 776) is particularly appropriate:

It seems to us that a profitable, long-term scientific endeavor, attempted locally but never to date for the whole Cordillera, would be to trace backwards in time the evolution of regional physiographic features. Such a study would involve integrating evidence from peneplanation, . . . drainage-pattern changes, changes of floras, . . . studies of rates of uplift on terranes like the St. Elias Mountains, and various aspects of volcanology, geochronometry, and geophysics. Ideally the results would begin to bridge the gap between 'present-day' and historical geodynamics.

There are, of course, a number of obstacles to the pursuit of this topic in all areas, not the least of which is the obliterating effect of glaciation. However, the nonpursuit of long-term studies is probably related more to the current emphasis on present-day processes and short-term changes, rather than to a dearth of long-term evidence. The recent studies by Clague (1974), Huber (1981), Tempelman-Kluit (1980), and Waitt (1979) are representative of what hopefully will be a growing number of studies within this longer time frame.

The second timescale concerns changes on the order of thousands to tens of thousands of years. Most areas have seen only moderate changes in the macro-topography over this time interval, but profound changes to the topography within individual valleys have occurred from the combined effects of glacial erosion and deposition, and postglacial fluvial and mass-movement processes. Since the late-glacial and postglacial sedimentary record is often less ambiguous than the preglacial record, and is most easily tied to present-day landforms, it seems profitable to focus on the Holocene as the first step in making a connection between studies of present-day processes and the evolution of the surrounding macro-topography. In contrast to the large amount of Holocene stratigraphic work, relatively few studies have addressed landform evolution on the timescale of 10,000 years, at least within the study region.

Although the sedimentary history of many Cordilleran valleys is known in general terms, detailed studies of the pace of landform development are few, particularly with regard to fluvial and mass-movement processes. By comparison, the chronology and extent of Neoglaciation, and the attendant landform changes, are known in considerable detail. The explanation for this discrepancy seems to be that greater continuity of effort has occurred in the realm of glacial stratigraphy over the past 40 years, in comparison with studies of nonglacial landform development. The latter have seen an overriding emphasis in recent years on contemporary landform development, at the expense of the longer-term picture. The important themes requiring further work are the history of filling and cutting within the major valleys, principally over Holocene time; the chronology of large-scale mass-movement events, again mainly over the Holocene; and the development history of large periglacial features, such as rock glaciers and cryoplanation terraces.

An important realization of the past 20 years concerns the impact of catastrophic events such as rock avalanches and volcanically triggered mass movements in landform development over the Holocene. These not only transform hillslopes, but radically affect the fluvial system, and may even create new lakes. Although the effects of such failures may be localized, there are probably thousands of major slope movement events within the Interior Mountains and Plateaus region, and although work on the mechanics and causes of such failures is well advanced, a documentation of their wider geomorphic impact remains to be carried out.

Studies of geomorphic change over periods of decades to centuries provide vital information concerning process mechanisms, especially within the fluvial, hillslope, and periglacial domains. Geomorphic and engineering work clearly overlap on this timescale, though the motives for maintenance of instrumented networks are usually different between these two disciplines. In contrast to the relatively long hydrologic records in the Cordillera, data records on geomorphic change are generally shorter than 10 years. While a decade may be adequate to characterize the more rapid slope movement processes, it is often inadequate to measure the more subtle rates of change within fluvial and periglacial environments. Various methods are being used to extend this meager data record, including stratigraphy, supported by radiocarbon dating, and dendrochronology—the full potential of the latter being far from fully developed. Outstanding problems related to this shortest time scale fall partly in the domain of 'present-day' processes, and include the hydrology of open-system pingos, resolution of which may lead to a better understanding of their present distribution; the role of secular climatic changes in the initiation, or reactivation of large-scale slope movement phenomena; and sediment sources and routing within Cordilleran drainage basins. The latter topic is a formidable undertaking, given the size and complexity of many watersheds. Sediment budgeting, coupled with detailed studies of hillslope and stream-channel processes, is probably the most fruitful approach.

REFERENCES

Allen, C. A., Jercinovic, M. J., and Allen, J.S.B., 1982, Subglacial volcanism in north-central British Columbia and Iceland: Journal of Geology, v. 90, p. 699–715.

Alley, N. F., and Young, G. K., 1978, Environmental significance of geomorphic processes in the northern Skeena Mountains and southern Stikine Plateau: Victoria, British Columbia Ministry of Environment, Resource Analysis Branch Bulletin 3, 83 p.

Anderson, R. Y., Nuhfer, E. B., and Dean, W. E., 1985, Sedimentation in a blast-zone lake at Mount St. Helens, Washington; Implications for varve formation: Geology, v. 13, p. 348–352.

Armentrout, J. M., Cole, M. R., and Terbest, H., Jr., eds., 1979, Cenozoic paleogeography of the western United States: Pacific Coast Paleogeography Symposium, 3, Society of Economic Paleontologists and Mineralogists, Pacific Section, 335 p.

Armstrong, J. E., and Tipper, H. W., 1948, Glaciation in north-central British Columbia: American Journal of Science, v. 246, p. 283–310.

Axelrod, D. I., 1957, Late Tertiary floras and the Sierra Nevada uplift: Geological Society of America Bulletin, v. 68, p. 19–46.

——, 1962, Post-Pliocene uplift of the Sierra Nevada, California: Geological Society of America Bulletin, v. 73, p. 183–198.

Beget, J. E., 1985, Tephrochronology of antislope scarps on an alpine ridge near Glacier Peak, Washington, U.S.A.: Arctic and Alpine Research, v. 17, p. 143–152.

Beikman, H. M., 1980, Geologic map of Alaska: U.S. Geological Survey Mono 80-1, scale 1:2,500,000.

Bevier, M. L., Armstrong, R. L., and Souther, J. G., 1979, Miocene peralkaline volcanism in west-central British Columbia; Its temporal and platetectonics setting: Geology, v. 7, p. 389–392.

Black, R. F., 1976, Periglacial features indicative of permafrost; Ice and soil wedges: Quaternary Research, v. 6, p. 3–26.

Blackwelder, E., 1928, Mudflow as a geologic agent in semiarid mountains: Geological Society of America Bulletin, v. 39, p. 465–484.

——, 1931, Pleistocene glaciation in the Sierra Nevada and Basin ranges: Geological Society of America Bulletin, v. 42, p. 865–922.

Blissenbach, E., 1954, Geology of alluvial fans in semi-arid regions: Geological Society of America Bulletin, v. 65, p. 175–190.

Bostock, M. S., 1948, Physiography of the Canadian Cordillera with special reference to the area north of the fifty-fifth parallel: Geological Survey of Canada Memoir 247, 106 p.

Bovis, M. J., 1982, Uphill-facing (antislope) scarps in the Coast Mountains, southwest British Columbia: Geological Society of America Bulletin, v. 93, p. 804–812.

——, 1985, Earthflows in the Interior Plateau, southwest British Columbia: Canadian Geotechnical Journal, v. 22, p. 313–334.

——, 1986, The morphology and mechanics of large-scale mass movement, with particular reference to southwest British Columbia, *in* Abrahams, A., ed., Hillslope processes, 16th Geomorphology Symposium, State University of New York, Buffalo: Boston, Allen and Unwin, p. 319–341.

Brooks, A. H., 1906, The geology of Alaska, a summary of existing knowledge: U.S. Geological Survey Professional Paper 45, 327 p.

Brown, R.J.E., 1967, Permafrost in Canada: Geological Survey of Canada, Map 1246A, scale 1:7,603,200.

Brown, R.J.E., and Péwé, T. L., 1973, Distribution of permafrost in North America and its relationship to the environment; A review, 1963–1973: Proceedings, Permafrost, Second International Conference, p. 71–100.

Bruce, I., and Cruden, D. M., 1977, The dynamics of the Hope Slide: International Association of Engineering Geology, v. 16, p. 94–98.

Bull, W. B., 1977, The alluvial-fan environment: Progress in Physical Geography, v. 1, p. 222–270.

Burke, R. M., and Birkeland, P. W., 1979, Reevaluation of multiparameter relative dating techniques and their application to the glacial sequence along the eastern escarpment of the Sierra Nevada, California: Quaternary Research, v. 11, p. 21–51.

Burn, C. R., Michel, F. A., and Smith, M. W., 1986, Stratigraphic, isotopic, and mineralogical evidence for an early Holocene thaw unconformity at Mayo, Yukon Territory: Canadian Journal of Earth Sciences, v. 23, p. 794–803.

Cairnes, D. D., 1912, Differential erosion and equiplanation in portions of Yukon and Alaska: Geological Society of America Bulletin, v. 23, p. 338–348.

Capps, S. R., 1910, Rock glaciers in Alaska: Journal of Geology, v. 18, p. 359–375.

Christensen, M. N., 1966, Late Cenozoic crustal movements in the Sierra Nevada of California: Geological Society of America Bulletin, v. 77, p. 163–182.

Christiansen, R. L., and Miller, C. D., 1976, Volcanic evolution of Mount Shasta, California: Geological Society of America Abstracts with Programs, v. 8, p. 360–361.

Church, M., 1980, Records of recent geomorphological events, *in* Cullingford, R. A., Davidson, D. A., and Lewin, J., eds., Timescales in geomorphology: Chichester, John Wiley and Sons, p. 13–29.

Church, M., and Ryder, J. M., 1972, Paraglacial sedimentation; A consideration of fluvial processes conditioned by glaciation: Geological Society of America Bulletin, v. 83, p. 3059–3072.

Clague, J. J., 1974, The St. Eugene Formation and the development of the Southern Rocky Mountain Trench: Canadian Journal of Earth Sciences, v. 11, p. 916–938.

——, 1981a, Late Quaternary geology and geochronology of British Columbia: Geological Survey of Canada Paper 80-35, 41 p.

——, 1981b, Landslides at the south end of Kluane Lake, Yukon Territory: Canadian Journal of Earth Sciences, v. 18, p. 959–971.

——, 1984, Quaternary geology and geomorphology, Smithers-Terrace-Prince Rupert area, British Columbia: Geological Survey of Canada Memoir 413, 71 p.

——, 1986, The Quaternary stratigraphic record of British Columbia; Evidence for periodic sedimentation and erosion controlled by glaciation: Canadian Journal of Earth Sciences, v. 23, p. 885–894.

Collins, B. D., and Dunne, T., 1986, Erosion of tephra from the 1980 eruption of Mount St. Helens: Geological Society of America Bulletin, v. 97, p. 896–905.

Collins, B. D., Dunne, T., and Lehr, A. K., 1983, Erosion of tephra-covered hillslopes north of Mount St. Helens, Washington; May 1980–May 1981: Zeitschrift für Geomorphologie, Supplemendband 46, p. 103–121.

Coney, P. J., Jones, D. L., and Monger, J.W.H., 1980, Cordilleran suspect terranes: Nature, v. 288, p. 329–333.

Cowan, D. S., 1985, Structural styles in Mesozoic and Cenozoic mélanges in the western Cordillera of North America: Geological Society of America Bulletin, v. 96, p. 451–462.

Crandell, D. R., 1965, The glacial history of western Washington and Oregon, *in* Wright, H. E., Jr., and Frey, D. G., eds., The Quaternary of the United States: Princeton, New Jersey, Princeton University Press, p. 341–353.

——, 1971, Postglacial lahars from Mount Rainer volcano: U.S. Geological Survey Professional Paper 677, 75 p.

Crandell, D. R., and Fahnestock, R. K., 1965, Rockfalls and avalanches from Little Tahoma Peak on Mount Rainier, Washington: U.S. Geological Survey, Bulletin 1221-A, 30 p.

Crandell, D. R., and Waldron, H. H., 1956, A recent volcanic mudflow of exceptional dimensions from Mount Rainier, Washington: American Journal of Science, v. 254, p. 349–362.

Crandell, D. R., Mullineaux, D. R., and Miller, C. D., 1979, Volcanic hazards studies in the Cascade Range of the western United States, *in* Sheets, P. D., and Grayson, D. K., eds., Volcanic activity and human ecology: New York, Academic Press, p. 195–219.

Crandell, D. R., Miller, C. D., Glicken, M. X., Christiansen, R. L., and Newhall, C. G., 1984, Catastrophic debris avalanche from ancestral Mount Shasta volcano, California: Geology, v. 12, p. 143–146.

Crough, S. T., and Thompson, G. A., 1977, Upper mantle origin of Sierra Nevada uplift: Geology, v. 5, p. 396–399.

Cruden, D. M., 1976, Major rock slides in the Rockies: Canadian Geotechnical Journal, v. 13, p. 8–20.

Cruden, D. M., 1985, Rockslope movements in the Canadian Cordillera: Canadian Geotechnical Journal, v. 22, p. 528–540.

Cummans, J., 1981, Chronology of mudflows in the South Fork and North Fork Toutle River following the May 18 eruption, *in* Lipman, P. W., and Mullineaux, D. R., eds., The 1980 eruptions of Mount St. Helens, Washington: U.S. Geological Survey Professional Paper 1250, p. 479–486.

Curry, R. R., 1966, Glaciation about 3,000,000 years ago in the Sierra Nevada: Science, v. 154, p. 770–771.

Dalrymple, G. B., 1963, Potassium-argon ages of some Cenozoic volcanic rocks of the Sierra Nevada, California: Geological Society of America Bulletin, v. 74, p. 379–390.

——, 1964, Cenozoic chronology of the Sierra Nevada, California: University of California Publication in Geological Sciences, v. 47, 41 p.

Dalrymple, G. B., Cox, A., and Doell, R. R., 1965, Potassium-argon age and paleomagnetism of the Bishop Tuff, California: Geological Society of America Bulletin, v. 76, p. 665–674.

Daly, R. A., 1912, Geology of the North American Cordillera at the forty-ninth parallel: Geological Survey of Canada Memoir 38, 857 p.

Davies, T.R.H., 1982, Spreading of rock avalanche debris by mechanical fluidization: Rock Mechanics, v. 15, p. 9–24.

Davis, W. M., 1905, The geographical cycle in an arid climate: Journal of Geology, v. 13, p. 381–407.

Dawson, G. M., 1879, Report on exploration in the southern part of British Columbia: Geological Society of Canada, Report of Progress 1877-78.

——, 1888, Recent observations on the glaciation of British Columbia and adjacent regions: Geological Magazine, v. 5, p. 347–350.

——, 1895, Report on the area of the Kamloops map sheet: Geological Society of Canada, Annual Report for 1894, v. 7, pt. B, p. 1–427.

Denton, G. H., 1974, Quaternary glaciations of the White River Valley, Alaska and Yukon Territory: Geological Society of America, v. 85, p. 871–892.

Denton, G. H., and Karlen, W., 1977, Holocene glacial and tree-line variations in the White River valley and Skolai Pass, Alaska and Yukon Territory: Quaternary Research, v. 7, p. 63–111.

Denton, G. H., and Stuiver, M., 1966, Neoglacial chronology, northeastern St. Elias Mountains, Canada: American Journal of Science, v. 264, p. 577–599.

Dietrich, W. E., and Dunne, T., 1978, Sediment budget for a small catchment in mountainous terrain: Zeitschrift für Geomorphologie, Supplement Band 29, p. 191–206.

Diller, J. S., 1891, A late volcanic eruption in northern California and its peculiar lava: U.S. Geological Survey, Bulletin, v. 79, 33 p.

—— , 1895, Mount Shasta, a typical volcano: National Geographic Society, Monograph, v. 1, p. 237–268.

Diller, J. S., and Patton, H. B., 1902, The geology and petrography of Crater Lake National Park: U.S. Geological Survey Professional Paper 3, 167 p.

Drew, F., 1873, Alluvial and lacustrine deposits and glacial records of the upper Indus basin: Geological Society of London Quarterly Journal, v. 29, p. 441–471.

Eakin, H. M., 1916, The Yukon-Koyukuk region, Alaska: U.S. Geological Survey Bulletin, v. 631, 88 p.

Eisbacher, G. H., 1971, Natural slope failure, northeastern Skeena Mountains: Canadian Geotechnical Journal, v. 8, p. 384–390.

—— , 1979a, First-order regionalization of landslide characteristics in the Canadian Cordillera: Geoscience Canada, v. 6, p. 69–79.

—— , 1979b, Cliff collapse and rock avalanches (sturzstroms) in the Mackenzie Mountains, northwestern Canada: Canadian Geotechnical Journal, v. 18, p. 309–334.

Ellis, J. M., and Calkin, P. E., 1984, Chronology of Holocene glaciation, central Brooks Range, Alaska: Geological Society of America Bulletin, v. 95, p. 897–912.

Ernst, W. G., ed., 1981, The geotectonic development of California, Rubey Memorial Volume 1: Englewood Cliffs, New Jersey, Prentice–Hall, 706 p.

Evans, S. G., 1982, Landslides and surficial deposits in urban areas of British Columbia, A review: Canadian Geotechnical Journal, v. 19, p. 269–288.

—— , 1984, The landslide dam on Thompson River, near Ashcroft, British Columbia: Geological Survey of Canada Paper 84-1A, p. 655–658.

Evans, S. G., and Cruden, D. M., 1981, Landslides in the Kamloops Group in south-central British Columbia, a progress report: Geological Survey of Canada Paper 81-1B, p. 170–177.

Fahnestock, R. K., 1963, Morphology and hydrology of a glacial stream: U.S. Geological Survey Professional Paper 422A, 70 p.

—— , 1978, Little Tahoma Peak rockfalls and avalanches, Mount Rainier, Washington, U.S.A., in Voight, B., ed., Rockslides and avalanches, Volume 1: Amsterdam, Elsevier Scientific, p. 181–196.

Fenneman, N. M., 1931, Physiography of western United States: New York, McGraw-Hill, 534 p.

Fernald, A. T., 1960, Geomorphology of the upper Kuskokwim region, Alaska: U.S. Geological Survey Bulletin 1071-G, p. 191–279.

Fink, J. H., Malin, M. C., D'Alli, R. E., and Greeley, R., 1981, Rheological properties of mudflows associated with the spring 1980 eruptions of Mount St. Helens volcano, Washington: Geophysical Research Letters, v. 8, p. 43–46.

Fiske, R. S., Hopson, C. A., and Waters, A. C., 1963, Geology of Mount Rainier National Park, Washington: U.S. Geological Survey Professional Paper 444.

Flint, R. F., 1943, Growth of the North American ice sheet during the Wisconsin Age: Geological Society of America Bulletin, v. 54, p. 325–362.

—— , 1971, Glacial and Quaternary geology: New York, John Wiley and Sons, p. 128–129.

Fulton, R. J., 1965, Silt deposition in late-glacial lakes of southern British Columbia: American Journal of Science, v. 263, p. 553–570.

—— , 1967, Deglaciation studies in Kamloops region, an area of moderate relief, British Columbia: Geological Survey of Canada Bulletin 154, 36 p.

Fulton, R. J., and Smith, G. W., 1978, Late Pleistocene stratigraphy of south-central British Columbia: Canadian Journal of Earth Sciences, v. 15, p. 971–980.

Gilbert, G. K., 1883, Whitney's climatic changes: Science, v. 1, p. 141–142, 169–173, 192–195.

—— , 1904, Systematic asymmetry of crest lines in the High Sierra of California: Journal of Geology, v. 12, p. 579–588.

—— , 1906, Crescentic gouges on glaciated surfaces: Geological Society of America Bulletin, v. 17, p. 303–316.

—— , 1914, The transportation of debris by running water: U.S. Geological Survey Professional Paper 86, 263 p.

—— , 1917, Hydraulic mining debris in the Sierra Nevada: U.S. Geological Survey Professional Paper 105, 154 p.

Haeni, F. P., 1983, Sediment deposition in the Columbia and lower Cowlitz rivers, Washington-Oregon, caused by the May 18, 1980, eruption of Mount St. Helens: U.S. Geological Survey Circular 850-K, 21 p.

Hammond, P. E., 1979, A tectonic model for evolution of the Cascade Range, in Cenozoic paleogeography of the western United States: Pacific Coast Paleogeography Symposium, 3rd, Pacific Section, Society of Economic Paleontologists and Mineralogists, Los Angeles, p. 219–237.

Hansen, A. M., 1894, The glacial succession in Norway: Journal of Geology, v. 2, p. 123–144.

Hay, E. A., 1976, Cenozoic uplifting of the Sierra Nevada in response to North American and Pacific plate interactions: Geology, v. 4, p. 763–766.

Helland, A., 1877, On the ice-fjords of North Greenland, and on the formation of fjords, lakes, and cirques in Norway and Greenland: Geological Society of London Quarterly Journal, v. 33, p. 142–176.

Hickson, C. J., and Souther, J. G., 1984, Late Cenozoic volcanic rocks of the Clearwater–Wells Gray area, British Columbia: Canadian Journal of Earth Sciences, v. 21, p. 267–272.

Holland, S. S., 1964 (reprinted 1976), Landforms of British Columbia, A physiographic outline: British Columbia Department of Mines and Petroleum Resources Bulletin 48, 138 p.

Holmes, W. G., Foster, H. L., and Hopkins, D. M., 1966, Distribution and age of pingos in central Alaska; 1st Permafrost International Conference, Lafayette, Indiana: Washington, D.C., National Research Council, p. 88–93.

Holmes, W. G., Hopkins, D. M., and Foster, H. L., 1968, Pingos in central Alaska: U.S. Geological Survey Bulletin 1241-H.

Hopkins, D. M., Karlstrom, T.N.V., Black, R. F., Williams, J. R., Péwé, T. L., Fernald, A. T., and Muller, E. H., 1955, Permafrost and groundwater in Alaska: U.S. Geological Survey Professional Paper, v. 264-F, p. 113–146.

Howell, D. G., and McDougall, K. A., 1978, Mesozoic paleogeography of the western United States: Pacific Coast Paleogeography Symposium 2, Society of Economic Paleontologists and Mineralogists, Pacific Section, 573 p.

Huber, N. K., 1981, Amount and timing of late Cenozoic uplift and tilt of the central Sierra Nevada, California: U.S. Geological Survey Professional Paper 1197, 28 p.

Hudson, F. S., 1960, Post-Pliocene uplift of the Sierra Nevada: Geological Society of America Bulletin, v. 71, p. 1547–1574.

Hughes, O. L., 1969, Distribution of open-system pingos in central Yukon Territory with respect to glacial limits: Geological Survey of Canada Paper 69-34, 8 p.

Hughes, O. L., Rampton, V. N., and Rutter, N. W., 1972, Quaternary geology and geomorphology, southern and central Yukon (northern Canada): 24th International Geological Congress, Montreal, P. Q., Field Excursion A-11, Guidebook, 59 p.

Izett, G. A., Wilcox, R. E., Powers, H. A., and Desborough, G. A., 1970, The Bishop ash bed, a Pleistocene marker bed in the western United States: Quaternary Research, v. 1, p. 121–132.

Janda, R. J., Scott, K. M., Nolan, K. M., and Martinson, M. A., 1981, Lahar movement, effects, and deposits, in Lipman, P. W., and Mullineaux, D. R., eds., The 1980 eruptions of Mount St. Helens, Washington: U.S. Geological Survey Professional Paper 1250, p. 461–478.

Johnson, A. M., 1970, Physical processes in geology: San Francisco, Freeman, Cooper and Company, 577 p.

Johnson, D. W., 1910, The origin of Yosemite Valley: Appalachia, v. 12, p. 138–146.

Johnson, P. G., 1978, Rock glacier types and their drainage systems: Canadian Journal of Earth Sciences, v. 15, p. 1496–1507.

—— , 1980, Glacier–rock glacier transition in the southwest Yukon Territory, Canada: Arctic and Alpine Research, v. 12, p. 195–204.

—— , 1984, Paraglacial conditions of instability and mass movement; A discussion: Zeitschrift für Geomorphologie, v. 28, p. 235–250.

Johnson, W. D., 1904, The profile of maturity in Alpine glacial erosion: Journal of Geology, v. 2, p. 569–578.

Jones, D. L., Silberling, N. J., and Hillhouse, J. W., 1978, Microplate tectonics of Alaska; Significance for the Mesozoic history of the Pacific Coast of North

America, *in* Howell, D. G., and McDougall, K. A., eds., Mesozoic paleo-geography of the western United States: Los Angeles, California, Pacific Section, Society of Economic Paleontologists and Mineralogists, p. 71–74.

Keefer, D. K., 1984, Landslides caused by earthquakes: Geological Society of America Bulletin, v. 95, p. 406–421.

Kellerhals, R., 1982, Effect of river regulation on channel stability, *in* Hey, R. W., Bathurst, J. C., and Thorne, C. R., eds., Gravel-bed rivers: Chichester, England, John Wiley and Sons Limited, p. 685–715.

Kellerhals, R., and Gill, D., 1973, Observed and potential downstream effects of large storage projects in northern Canada: 11th International Congress on large dams, Transactions, Madrid, Spain, p. 731–754.

Kellerhals, R., Church, M., and Davies, L. B., 1979, Morphological effects of interbasin river diversions: Canadian Journal of Civil Engineering, v. 6, p. 18–31.

Kelsey, H. M., 1978, Earthflows in Franciscan mélange, Van Duzen River basin, California: Geology, v. 6, p. 361–364.

Kerr, F. A., 1934, Glaciation in northern British Columbia: Royal Society of Canada Transactions, Section IV, v. 28, p. 17–31.

Kesseli, J. E., 1941, Rock streams in the Sierra Nevada, California: Geographical Review, v. 31, p. 203–227.

Knopf, A., 1918, A geologic reconnaissance of the Inyo Range and the eastern slope of the southern Sierra Nevada: U.S. Geological Survey Professional Paper 110, 130 p.

Lawson, A. C., 1904, Geomorphogeny of the upper Kern basin: University of California Publications in Geology, v. 3, p. 291–376.

Lay, D., 1940, Fraser River Tertiary drainage-history in relation to placer-gold deposits: British Columbia Department of Mines Bulletin, v. 3, 32 p.

Le Conte, J., 1875, On some of the ancient glaciers of the Sierra Nevada: American Journal of Science, v. 110, p. 126–139.

———, 1886, A post-Tertiary elevation of the Sierra Nevada as shown by the river beds: American Journal of Science, v. 32, p. 167–181.

———, 1889, On the origin of normal faults and of the structure of the Basin region: American Journal of Science, v. 38, p. 257–263.

Leffingwell, E. de K., 1915, Ground-ice wedges, the dominant form of ground-ice in the north coast of Alaska: Journal of Geology, v. 23, p. 635–652.

———, 1919, The Canning River region, northern Alaska: U.S. Geological Survey Professional Paper 109, 251 p.

Lehre, A. K., Collins, B., and Dunne, T., 1983, Post-eruption sediment budget for the North Fork Toutle River drainage, June 1980–May 1981: Zeitschrift für Geomorphologie, Supplement Band 46, p. 143–163.

Lerbekmo, J. F., and Campbell, F. A., 1969, Distribution, composition, and source of the White River ash, Yukon Territory: Canadian Journal of Earth Sciences, v. 6, p. 109–116.

Leviton, A. E., Rodda, P. U., Yochelson, E., and Aldrich, M. L., eds., 1982, Frontiers of geological exploration of western North America: San Francisco, Pacific Division, American Association for the Advancement of Science, 248 p.

Lewis, W. V., 1938, A meltwater hypothesis of cirque formation: Geological Magazine, v. 75, p. 249–265.

———, 1940, The function of meltwater in cirque formation: Geographical Review, v. 30, p. 64–83.

Lindgren, W., 1893, Two Neocene rivers of California: Geological Society of America Bulletin, v. 4, p. 257–298.

———, 1911, The Tertiary gravels of the Sierra Nevada: U.S. Geological Survey Professional Paper 73, p. 9–81.

Lipman, P. W., and Mullineaux, D. R., 1981, The 1980 eruptions of Mount St. Helens, Washington: U.S. Geological Survey Professional Paper 1250, 844 p.

Mackay, J. R., 1966, Pingos in Canada, *in* Proceedings, 1st Permafrost International Conference, Lafayette, Indiana: National Academy of Science Publication 1287, p. 71–76.

———, 1981, Active layer slope movement in a continuous permafrost environment, Garry Island, Northwest Territories, Canada: Canadian Journal of Earth Sciences, v. 18, p. 1666–1680.

Mathews, W. H., 1944, Glacial lakes and ice retreat in south-central British Columbia: Royal Society of Canada Transactions, Series 2, v. 38, p. 39–57.

———, 1947, "Tuyas," flat-topped volcanoes in northern British Columbia: American Journal of Science, v. 245, p. 560–570.

Mathews, W. H., and McTaggart, K. C., 1978, Hope rockslides, British Columbia, Canada, *in* Voight, B., ed., Rockslides and avalanches, Volume 1: Amsterdam, Elsevier Scientific, p. 259–275.

Matthes, F. E., 1900, Glacial sculpture of the Bighorn Mountains, Wyoming: U.S. Geological Survey 21st Annual Report, v. II, p. 173–190.

———, 1930, Geologic history of the Yosemite Valley: U.S. Geological Survey Professional Paper 160, 137 p.

McGee, W. J., 1883, Glacial cañons: American Association for the Advancement of Science, Summarized Proceedings for 1883, p. 238.

———, 1894, Glacial cañons: Journal of Geology, v. 2, p. 350–364.

McKee, B., 1972, Cascadia; The geologic evolution of the Pacific Northwest: New York, McGraw-Hill, 394 p.

McRoberts, E. C., and Morgenstern, N. R., 1974, The stability of thawing slopes: Canadian Geotechnical Journal, v. 11, p. 447–469.

Mellor, M., 1968, Avalanches: Hanover, New Hampshire, U.S. Army, Cold Regions Research and Engineering Laboratory, Monograph III A3d, 215 p.

Merrill, G. P., 1924, The first one hundred years of American geology: New Haven, Yale University Press, 773 p.

Mertie, J. B., Jr., 1937, The Yukon-Tanana region, Alaska: U.S. Geological Survey Bulletin 872, 276 p.

Miall, A. D., 1977, A review of the braided river depositional environment: Earth Science Reviews, v. 13, p. 1–62.

Miller, C. D., Mullineaux, D. R., and Crandell, D. R., 1981, Hazards assessments at Mount St. Helens, *in* Lipman, P. W., and Mullineaux, D. R., eds., The 1980 eruptions of Mount St. Helens, Washington: U.S. Geological Survey Professional Paper 1250, p. 789–802.

Mollard, J. D., 1977, Regional landslide types in Canada, *in* Coates, D. R., ed., Landslides: Boulder, Colorado, Geological Society of America, Reviews in Engineering Geology, v. 3, p. 29–56.

Monger, J.W.H., and Price, R. A., 1979, Geodynamic evolution of the Canadian Cordillera; Progress and problems: Canadian Journal of Earth Sciences, v. 16, p. 770–791.

Monger, J.W.H., Price, R. A., and Tempelman-Kluit, D. J., 1982, Tectonic accretion and the origin of the two major metamorphic and plutonic welts in the Canadian Cordillera: Geology, v. 10, p. 70–75.

Muir, J., 1874, Studies in the Sierra, II, Mountain sculpture, Origin of Yosemite Valleys: reprinted in Sierra Club Bulletin, v. 10, p. 62–77.

Müller, F., 1963, Observations on pingos: National Research Council Canada, Technical Transactions, v. 1073, 117 p.

Muller, S. W., 1945, Permafrost or permanently frozen ground and related engineering problems (second edition): U.S. Geological Survey Special Report, Strategic Engineering Study 62, 231 p.

Mullineaux, D. R., and Crandell, D. R., 1962, Recent lahars from Mount St. Helens, Washington: Geological Society of America Bulletin 73, p. 855–869.

Mylrea, F. H., 1969, Geology of Mica damsite, Columbia River, British Columbia: Geological Association of Canada Proceedings, v. 20, p. 57–64.

Nasmith, H., Mathews, W. H., and Rouse, G. E., 1967, Bridge River ash and some other Recent ash beds in British Columbia: Canadian Journal of Earth Sciences, v. 4, p. 163–170.

Okuda, S., Suwa, H., Okunishi, K., Yokoyama, K., and Nakano, M., 1980, Observations on the motion of a debris flow and its geomorphological effects: Zeitschrift für Geomorphologie Supplement Band, v. 35, p. 142–163.

Palmer, L., 1977, Large landslides of the Columbia River Gorge, Oregon and Washington, *in* Coates, D. R., ed., Landslides: Boulder, Colorado, Geological Society of America, Reviews in Engineering Geology, v. 3, p. 69–83.

Péwé, T. L., 1954, Effect of permafrost on cultivated fields, Fairbanks area, Alaska: U.S. Geological Survey Bulletin 989-F, p. 315–351.

———, 1966, Ice-wedges in Alaska; Classification, distribution, and climatic significance, *in* Proceedings, Permafrost International Conference, Lafayette, Indiana: Washington, D.C., Natinal Academy of Sciences Publication 1287, p. 76–81.

——, 1970, Altiplanation terraces of early Quaternary age near Fairbanks, Alaska: Acta Geographic Lodziensia, v. 14, p. 357–363.

——, 1975, Quaternary geology of Alaska: U.S. Geological Survey Professional Paper 835, 145 p.

——, 1983, The periglacial environment in North America during Wisconsin time, *in* Porter, S. C., ed., Late-Quaternary environments of the United States, Volume 1, The late-Pleistocene: Minneapolis, University of Minnesota Press, p. 157–189.

Péwé, T. L., Ferrians, O. J., Jr., Nichols, D. R., and Karlstrom, T.N.V., 1965a, Guidebook for Field Conference F–Central and South-central Alaska, International Association Quaternary Research, 7th Congress, U.S.A.: Lincoln, Nebraska, Nebraska Academy of Science, 141 p.

Péwé, T. L., Hopkins, D. M., and Giddings, J. L., 1965b, The Quaternary geology and archaeology of Alaska, *in* Wright, H. E., Jr., and Frey, D. G., eds., The Quaternary of the United States: Princeton, New Jersey, Princeton University Press, p. 355–374.

Péwé, T. L., Church, R. E., and Andresen, M. J., 1969, Origin and paleoclimatic significance of large-scale patterned ground in the Donnelly Dome area, Alaska: Geological Society of America Special Paper 103, 87 p.

Pierson, T. C., 1986, Flow behavior of channelized debris flows, Mount St. Helens, Washington, *in* Abrahams, A. D., ed., Hillslope processes, 16th Geomorphology Symposium, State University of New York, Buffalo: Boston, Allen and Unwin, p. 269–296.

Porsild, A. E., 1938, Earth mounds in unglaciated Arctic northwestern America: Geographical Review, v. 28, p. 46–58.

Porter, S. C., ed., 1983, Late-Quaternary environments of the United States, Volume 1, The late Pleistocene: Minneapolis, University of Minnesota Press, 407 p.

Porter, S. C., Pierce, K. L., and Hamilton, T. D., 1983, Late Wisconsin mountain glaciation in the western United States, *in* Porter, S. C., ed., Late-Quaternary environments of the United States, Volume 1, The late Pleistocene: Minneapolis, University of Minnesota Press, p. 71–111.

Potter, N., Jr., 1972, Ice-cored rock glacier, Galena Creek, northern Absaroka Mountains, Wyoming: Geological Society of America Bulletin, v. 83, p. 3025–3058.

Prest, V. K., 1984, The late Wisconsin glacier complex, *in* Fulton, R. J., ed., Quaternary stratigraphy of Canada: Geological Survey of Canada Paper 84-10, p. 22–36.

Prest, V. K., Grant, D. R., and Rampton, V. N., 1967, Glacial map of Canada: Geological Survey of Canada, Map 1253 A, scale 1:5,000,000.

Radbruch-Hall, D. H., 1978, Gravitational creep of rock masses on slopes, *in* Voight, B., ed., Rockslides and avalanches, Volume 1: Amsterdam, Elsevier Scientific, p. 607–657.

Rampton, V. N., 1970, Neoglacial fluctuations of the Natazhat and Klutan glaciers, Yukon Territory, Canada: Canadian Journal of Earth Sciences, v. 7, p. 1236–1263.

——, 1981, Surficial materials and landforms of Kluane National Park, Yukon Territory: Geological Survey of Canada Paper 79-24, 37 p.

Ramsay, A. C., 1862, On the glacial origin of certain lakes in Switzerland, the Black Forest, Great Britain, Sweden, North America, and elsewhere: Geological Society of London Quarterly Journal, v. 18, p. 185–204.

——, 1864, On the erosion of lakes and valleys: Philosophical Magazine, v. 4, p. 293–311.

Ransome, F. L., 1898, Some lava flows of the western slope of the Sierra Nevada: U.S. Geological Survey Bulletin 89, 71 p.

Rawlings, G., 1986, Active slide in bentonitic clays, Upper Hat Creek, British Columbia: Canadian Geotechnical Journal, v. 23, p. 164–173.

Reger, R. D., and Péwé, T. L., 1976, Cryoplanation terraces; Indicators of a permafrost environment: Quaternary Research, v. 6, p. 99–109.

Ritter, D. F., and Ten Brink, N. W., 1986, Alluvial fan development and the glacial-glaciofluvial cycle, Nenana Valley, Alaska: Journal of Geology, v. 94, p. 613–625.

Rockie, W. A., 1942, Pitting on Alaskan farm lands; A new erosion problem: Geographical Review, v. 32, p. 128–134.

Roddick, J. A., 1966, Coast crystalline belt of British Columbia, *in* Tectonic history and mineral deposits of the western Cordillera: Canadian Institute of Mining and Metallurgy Special Volume 8, p. 73–82.

Russell, I. C., 1889, Quaternary history of Mono Valley, California: U.S. Geological Survey 8th Annual Report, p. 261–394.

Rutter, N. W., 1976, Multiple glaciation in the Canadian Rocky Mountains with special emphasis on northeastern British Columbia, *in* Mahaney, W. C., ed., Quaternary stratigraphy of North America: Dowden, Hutchinson and Ross, Incorporated, p. 409–440.

Ryder, J. M., 1971a, The stratigraphy and morphology of paraglacial alluvial fans in south-central British Columbia: Canadian Journal of Earth Sciences, v. 8, p. 279–298.

——, 1971b, Some aspects of the morphometry of paraglacial alluvial fans in south-central British Columbia: Canadian Journal of Earth Sciences, v. 8, p. 1252–1264.

——, 1976, Terrain inventory and Quaternary geology, Ashcroft, British Columbia: Geological Survey of Canada Paper 74-49, 17 p.

——, 1981, Terrain inventory and Quaternary geology, Lytton, British Columbia: Geological Survey of Canada Paper 79-25, 20 p.

Ryder, J. M., and Church, M., 1986, The Lillooet terraces of Fraser River; A paleoenvironmental enquiry: Canadian Journal of Earth Sciences, v. 23, p. 869–884.

Sarna-Wojcicki, A. M., Champion, D. E., and Davis, J. O., 1983, Holocene volcanism in the conterminous United States and the role of silicic volcanic ash layers in correlation of latest-Pleistocene and Holocene deposits, *in* Wright, H. E., Jr, ed., Late-Quaternary environments of the United States, Volume 2, The Holocene: Minneapolis, University of Minnesota Press, p. 52–77.

Scheidegger, A. E., 1973, On the prediction of the reach and velocity of catastrophic landslides: Rock Mechanics, v. 5, p. 231–236.

Shroba, R. R., and Birkeland, P. W., 1983, Trends in late-Quaternary soil development in the Rocky Mountains and Sierra Nevada, *in* Porter, S. C., ed., Late-Quaternary environments of the United States, Volume 1, The late Pleistocene: Minneapolis, University of Minnesota Press, p. 145–156.

Slaymaker, O., and McPherson, H. J., 1977, An overview of geomorphic processes in the Canadian Cordillera: Zeitschrift für Geomorphologie, v. 21, p. 169–186.

Smith, W. D., and Swartzlow, C. R., 1936, Mount Mazama; Explosion versus collapse: Geological Society of America Bulletin, v. 47, p. 1809–1830.

Souther, J. G., 1977, Volcanism and tectonic environments in the Canadian Cordillera; A second look, *in* Baragar, W.R.A., Coleman, L. C., and Hall, J. M., eds., Volcanic regimes in Canada: Geological Association of Canada Special Paper 16, p. 3–24.

Souther, J. G., Armstrong, R. L., and Harakal, J., 1984, Chronology of the peralkaline, late Cenozoic Mount Edziza Volcanic Complex, northern British Columbia, Canada: Geological Society of America Bulletin, v. 95, p. 337–349.

Stewart, J. M., 1978, Basin-range structure in western North America; A review, *in* Smith, R. B., and Eaton, G. P., eds., Cenozoic tectonics and regional geophysics of the western Cordillera: Geological Society of America Memoir 152, p. 1–31.

Swanson, F. J., and Fredriksen, R. L., 1982, Sediment routing and budgets; Implications for judging impacts on forestry practices: U.S. Forest Service General Technical Report PNW-141, p. 129–137.

Swanson, F. J., and James, M. E., 1975, Geology and geomorphology of the H. J. Andrews Experimental Forest, Western Cascades, Oregon: U.S. Department of Agriculture Forest Service Research Paper PNW-188, 14 p.

Swanson, F. J., and Swanston, D. N., 1977, Complex mass-movement terrains in the Western Cascade Range, Oregon, *in* Coates, D. R., ed., Landslides: Boulder, Colorado, Geological Society of America, Reviews in Engineering Geology, v. 3, p. 113–124.

Swanson, F. J., Janda, R. J., Dunne, T., and Swanston, D. N., eds., 1982,

Sediment budgets and routing in forested drainage basins: U.S. Forest Service General Technical Report PNW-141, 165 p.

Swanston, D. N., and Swanson, F. J., 1976, Timber harvesting, mass erosion, and steepland forest geomorphology in the Pacific Northwest, *in* Coates, D. R., ed., Geomorphology and engineering: Stroudsberg, Dowden, Hutchinson and Ross, p. 199–221.

Taber, S., 1943, Perennially frozen ground in Alaska; Its origin and history: Geological Society of America Bulletin, v. 54, p. 1433–1548.

Tempelman-Kluit, D., 1980, Evolution of physiography and drainage in southern Yukon: Canadian Journal of Earth Sciences, v. 17, p. 1189–1203.

Thompson, W. F., 1962, Cascade alp slopes and gipfelfluren as climageomorphic phenomena: Erdkunde, v. 16, p. 81–94.

Thomson, S., and Mekechuk, J., 1982, A landslide in glacial lake clays in central British Columbia: Canadian Geotechnical Journal, v. 19, p. 296–306.

Thornbury, W. D., 1965, Regional geomorphology of the United States: New York, John Wiley, 609 p.

Tipper, H. W., 1971, Glacial geomorphology and Pleistocene history of central British Columbia: Geological Survey of Canada Bulletin 196, 89 p.

Turner, H. W., 1900, The Pleistocene geology of the south-central Sierra Nevada, with especial reference to the origin of Yosemite Valley: Proceedings, California Academy of Sciences, v. I, Geology, 3rd Series, p. 312–314.

Tyrrell, J. B., 1919, Was there a "Cordilleran Glacier" in British Columbia?: Journal of Geology, v. 27, p. 55–60.

Vandine, D. F., 1980, Engineering geology and geotechnical study of Drynoch landslide, British Columbia: Geological Survey of Canada Paper 79-31, 34 p.

Varnes, D. J., 1978, Slope movement types and processes, *in* Schuster, R. L., and Krizek, R. J., eds., Landslides, analysis and control: Washington, D.C., National Academy of Sciences, p. 11–33.

Verhoogen, J., 1937, Mount St. Helens, A recent Cascade volcano: University of California, Department of Geological Sciences Bulletin, v. 24, p. 263–302.

Vernon, P., and Hughes, O. L., 1966, Surficial geology, Dawson, Larsen Creek, and Nash Creek map-areas, Yukon Territory: Geological Survey of Canada Bulletin 136, 25 p.

Voight, B., Janda, R. J., Glicken, H., and Douglass, P. M., 1983, Nature and mechanics of the Mount St. Helens rockslide-avalanche of 18 May 1980: Geotechnique, v. 33, p. 243–273.

Wahrhaftig, C., 1949, The frost-moved rubbles of Jumbo Dome and their significance in the Pleistocene chronology of Alaska: Journal of Geology, v. 57, p. 216–231.

—— , 1958, Quaternary geology of the Nenana River valley and adjacent parts of the Alaska Range: U.S. Geological Survey Professional Paper 293-A, p. 1–68.

—— , 1965a, Physiographic divisions of Alaska: U.S. Geological Survey Professional Paper 482, 52 p.

—— , 1965b, Stepped topography of the southern Sierra Nevada, California: Geological Society of America Bulletin, v. 76, p. 1165–1190.

Wahrhaftig, C., and Birman, J. H., 1965, The Quaternary of the Pacific mountain system in California, *in* Wright, H. E., Jr., and Frey, D. G., eds., The Quaternary of the United States: Princeton, New Jersey, Princeton University Press, p. 299–340.

Wahrhaftig, C., and Cox, A., 1959, Rock glaciers in the Alaska Range: Geological Society of America Bulletin, v. 70, p. 383–436.

Waitt, R. B., Jr., 1979, Late Cenozoic deposits, landforms, stratigraphy, and tectonism in Kittitas valley, Washington: U.S. Geological Survey Professional Paper 1127, 18 p.

—— , 1982, Quaternary research in the Northwest 1805–1979 by early government surveys and the U.S. Geological Survey, and prospects for the future, *in* Leviton, A. E., and others, eds., Frontiers of geological exploration of west-

ern North America: San Francisco, California, American Association for the Advancement of Science, Pacific Division, p. 167–192.

Waitt, R. B., Jr., and Thorson, R. M., 1983, The Cordilleran Ice Sheet in Washington, Idaho, and Montana, *in* Porter, S. C., ed., Late-Quaternary environments of the United States, Volume 1, The late Pleistocene: Minneapolis, University of Minnesota Press, p. 53–70.

Washburn, A. L., 1980, Geocryology; A survey of periglacial processes and environments: New York, John Wiley, 406 p.

—— , 1985, Periglacial problems, *in* Church, M., and Slaymaker, O., eds., Field and theory; Lectures in geocryology: Vancouver, University of British Columbia Press, p. 166–202.

White, S. E., 1976, Rock glaciers and block fields, review and new data: Quaternary Research, v. 6, p. 77–97.

Whitney, J. D., 1865, Geology I; Report of progress and synopsis of the fieldwork, from 1860–1864: Philadelphia, Sherman and Company, 498 p.

—— , 1880, The auriferous gravels of the Sierra Nevada of California: Harvard College Museum of Comparative Zoology Memoir 6, no. 1, 659 p.

Wilcox, R. E., 1965, Volcanic-ash chronology, *in* Wright, H. E., Jr., and Frey, D. G. eds., The Quaternary of the United States: Princeton, New Jersey, Princeton University Press, p. 807–816.

Williams, H., 1932, Geology of the Lassen Volcanic National Park, California: University of California Publications in Geological Sciences, v. 21, p. 195–385.

—— , 1941, Calderas and their origin: University of California Publications in Geological Sciences, v. 25, p. 239–346.

—— , 1942, The geology of Crater Lake National Park, Oregon, with a reconnaissance of the Cascade Range southward to Mount Shasta: Carnegie Institute of Washington Publication 540, 162 p.

Williams, H., and Goles, G., 1968, Volume of the Mazama ash-fall and the origin of Crater Lake caldera, *in* Dole, H. M., ed., Andesite Conference Guidebook: Oregon Department of Geology and Mineral Industries Bulletin 62, p. 37–41.

Williams, J. R., 1962, Geologic reconnaissance of the Yukon Flats district, Alaska: U.S. Geological Survey Bulletin 1111-H, p. H289–H331.

Williams, P. F., and Rust, B. R., 1969, The sedimentology of a braided river: Journal of Sedimentary Petrology, v. 39, p. 649–679.

Willis, B., 1898, Drift phenomena of Puget Sound: Geological Society of America Bulletin, v. 9, p. 111–162.

Wood, H., 1977, Distribution, correlation, and radiocarbon dating of late Holocene tephra, Mono and Inyo craters, eastern California: Geological Society of America Bulletin, v. 88, p. 89–95.

Wright, H. E., Jr., ed., 1983, Late-Quaternary environments of the United States, Volume 2, The Holocene: Minneapolis, University of Minnesota Press, 277 p.

Wright, H. E., Jr., and Frey, D. G., eds., 1965, The Quaternary of the United States: Princeton, New Jersey, Princeton University Press, 922 p.

Manuscript Accepted by the Society September 29, 1986

ACKNOWLEDGMENTS

The writer thanks Dr. J. J. Clague and Professor O. Slaymaker for critical reviews of an earlier version of this chapter. Professor P. G. Johnson kindly provided photographs of periglacial features. The writer acknowledges the assistance of Douglas Johnson in the bibliographic research for this chapter and for drafting many of the figures. Sandy Lapsky assisted in the typing of the manuscript. Trevor Gale assisted in the map compilations.

Geological Society of America
Centennial Special Volume 2
1987

Chapter 13

Pacific Coast and Mountain System

Daniel R. Muhs
U.S. Geological Survey, Box 25046, MS 963, Denver Federal Center, Denver, Colorado 80225
Robert M. Thorson
Department of Geology and Geophysics, University of Connecticut, Storrs, Connecticut 06268
John J. Clague
Geological Survey of Canada, 100 W. Pender St., Vancouver, British Columbia V6B 1R8, Canada
W. H. Mathews
Department of Geological Sciences, University of British Columbia, Vancouver, British Columbia V6T 2B4, Canada
Patricia F. McDowell
Department of Geography, University of Oregon, Eugene, Oregon 97403
Harvey M. Kelsey
Department of Geology, Western Washington University, Bellingham, Washington 98225

INTRODUCTION

Daniel R. Muhs

The Pacific Coast and Mountain System (PCMS) includes the coastal mountains and plains fronting the Pacific Ocean from the United States–Mexico border in the south to the northernmost part of the Alaska Range in Alaska, a latitudinal span of approximately 31°. In Alaska, the region encompasses the Alaska-Aleutian Range, the coastal mountain ranges to the south (the Kodiak, Kenai, Chugach, St. Elias, and Fairweather ranges), and the lowlands in between. In British Columbia, the PCMS includes, from west to east, the Insular Ranges on Vancouver Island and the Queen Charlotte Islands, the Coastal Depression, which includes the coastal lowlands and the continental shelf, and the Coast Mountains. The PCMS in Washington and Oregon comprises the Olympic Peninsula, the Coast Ranges, the Klamath Mountains, and the coastal plain bordering the Pacific Ocean. The Klamath Mountains extend into extreme northern California, but the Coast Ranges, which comprise the bulk of the PCMS in this region, continue from the Klamath Mountains and extend southward over most of the northern two-thirds of California. South of the Coast Ranges in California, the PCMS includes the Transverse and Peninsula ranges. In addition, a narrow coastal plain extends along much of the California coast.

The great latitudinal variation of the PCMS, along with elevations that extend from sea level to almost 7,000 m, make the PCMS in North America a region of extreme climatic contrasts. Conditions have been cold and moist enough to support glaciers in Alaska since the Miocene, whereas southern California proba-

bly has had a Mediterranean climate since the Miocene. Within California alone, mean annual precipitation varies from about 150 to 3,000 mm. Vegetation reflects this climatic diversity, and also plays an important role in the geomorphology of the region. In Alaska, British Columbia, and northern Washington, the geomorphology of the PCMS displays the effects of climatic change. South of these subregions, the effects of climatic change on landforms are less obvious, but global climatic changes, which resulted in sea-level fluctuations, have had important effects on the landscape.

The physical geography of the PCMS is dependent on the ongoing tectonic processes found there. Much of the PCMS of Alaska and British Columbia is accretionary fragments or terranes that have collected as a result of plate motions. Furthermore, the PCMS is the locus of interaction between the Pacific, North America, Juan de Fuca, and Explorer plates. Plate boundaries take the form of transform faults in some regions, and in others they form subduction zones where there is active volcanism. In all parts of the PCMS, tectonic activity has been and continues to be an important influence on the regional geomorphology.

This chapter is a review of recent contributions to the geomorphology of the PCMS. Most of the referenced studies have been published in the last 25 years, and the majority of these have been published in the last 15. The large number of published studies on the PCMS of North America in recent years is indicative of the tremendous enthusiasm and effort by geomorphologists working in this region. This chapter represents the most recent integrated review of the geomorphology of the PCMS, but there are also several papers, maps, and books that can give the

Muhs, D. R., Thorson, R. M., Clague, J. J., Mathews, W. H., McDowell, P. F., Kelsey, H. M., 1987, Pacific Coast and Mountain System, *in* Graf, W. L., ed., Geomorphic systems of North America: Boulder, Colorado, Geological Society of America, Centennial Special Volume 2.

reader a general background on the geology and physiography of
the region, while emphasizing regional description. These include
works by Williams (1958), Wahrhaftig and Birman (1965),
Wahrhaftig (1965), Shepard and Wanless (1971), Norris and
Webb (1976), Slaymaker and McPherson (1977), Beikman
(1980), and Baldwin (1981).

The focus of the chapter will be on geomorphic processes
rather than descriptive geomorphology. Geomorphology as a
science is well beyond the stage where landform studies are limited to regional mapping, description, and classification. Modern
geomorphologists are usually interested in understanding landform genesis and the dynamics of geomorphology, and it is in this
realm that geomorphologists working in the PCMS have made
important contributions to their discipline. An understanding of
how landforms are generated in this region requires an appreciation of Quaternary stratigraphy, paleoclimatology, sea-level
changes, and neotectonics. Therefore, much of the discussion will
focus on landscape evolution in Quaternary time.

This chapter represents the efforts of six geomorphologists.
All the contributors are individuals who are presently working on
aspects of the geomorphology of the PCMS and have field experience in various parts of the region. The chapter contributions
pertain to specific subregions rather than to individual topics or
geomorphic processes. There are two principal reasons for this
organizational scheme. First, the PCMS itself has distinctive subregions, and certain geomorphic processes are more important in
some subregions than they are in others. In a sense, therefore,
subregional division implies topical division in the PCMS. Second, division of the chapter into subregions emphasizes the rich
interaction of various geomorphic processes in particular subregions. Landforms are often not formed by a single process, and
this is particularly true of the PCMS.

As is the case with most review papers, certain topics will
receive more coverage than others, depending on the interests of
the reviewer. Constraints of space force us to limit ourselves to a
few topics in each subregion. It has been our aim, however, to
pick those topics for review and discussion that have been the
most important contributions of the PCMS to geomorphology.

GEOMORPHIC PROCESSES OF THE ALASKAN PACIFIC COAST AND MOUNTAIN SYSTEM

Robert M. Thorson

INTRODUCTION AND SETTING

Southern Alaska forms the northern and western half of the
Pacific Coast and Mountain System (PCMS) in North America
(Fig. 1). From the western tip of the Aleutian chain to the Queen
Charlotte Islands in British Columbia, it is larger than all other
parts of the system combined, spanning more than 3,000 km and
42° longitude. This region contains the highest (Mount McKin-
ley, 6,912 m), lowest (Aleutian Trench), wettest, and coldest
parts of the PCMS. Its landscape, which reflects these extremes,

results primarily from prolonged and continuing glaciation of
lithologically diverse terrain at the convergent boundary between
the Pacific plate and North America.

This section summarizes our limited understanding of the
dominant geomorphic processes in the Alaska PCMS. The dearth
of "process-oriented" studies, particularly those involving drainage basin development, denudation, and slope processes make
certain generalizations unavoidable. Reconnaissance field observations are substituted for specific studies where such gaps in our
knowledge are particularly acute, and generalizations drawn from
such observations should be considered as hypotheses for future
work. Following a sketch of the region's geologic and climatic
setting, specific processes—volcanic, glacial, fluvial, slope, and
coastal—are described separately. This traditional, but organiza-
tionally necessary breakdown is overly simplistic because process
interaction in alternating glacial-interglacial settings is largely responsible for the geomorphology. This theme is developed in a
later section, which is followed by brief descriptions of selected
problems for future research.

Much of Alaska is a mosaic of lithologically distinct tectonostratigraphic terranes that have few equivalents in western North
America. Consequently, considering this region as an extension of
the Pacific Coast and Mountain System in the lower United
States is a physiographic, and perhaps untenable, oversimplifica-
tion. Using the boundaries established by Wahrhaftig (1965), the
PCMS in Alaska includes the Aleutian Islands, the regions south
of and including the Aleutian-Alaska Range, and the panhandle
of southeast Alaska. Refer to Wahrhaftig (1965) for an excellent
descriptive summary of the region's physiography, and to Beikman (1980) for an overview of the bedrock geology. For the sake
of brevity, and in keeping with the theme of other parts of this
chapter, the southern coastal ranges will be emphasized.

The Aleutian archipelago is a series of late Cenozoic volcanoes that surmount older complexly-deformed rocks along the
crest of the submarine Aleutian Ridge. Summit ice caps and small
valley glaciers radiate from their summits (600 to 2,500 m),
which stand well above snowline. Of the 78 volcanoes located in
the arc, 66 have been active during Holocene time, and 17 contain large calderas (Coats, 1950). The Aleutian Range dominates
the Alaskan Peninsula. Glacierized Quaternary stratovolcanoes
there stand above more subdued, ice-scoured older rocks at intervals along the length of the range. Dozens of low passes through
the range were cut by large glaciers draining northward from
Pleistocene ice caps on the continental shelf to the south. The
Alaska Range to the north is a higher, more rugged, and heavily
glaciated system that continues eastward into Canada. The
Aleutian-Alaska Range forms a single arcuate mountain system
that defines the northern limit of the PCMS. The coastal mountains to the south—the Kodiak, Kenai, Chugach, St. Elias, and
Fairweather ranges, from west to east—define a second arcuate
mountain system that is subparallel to the Alaska-Aleutian Range
system. These mountains consist largely of weakly metamorphosed volcanic sedimentary assemblages with abundant plutonic
bodies. The most extensive system of glaciers in North America,

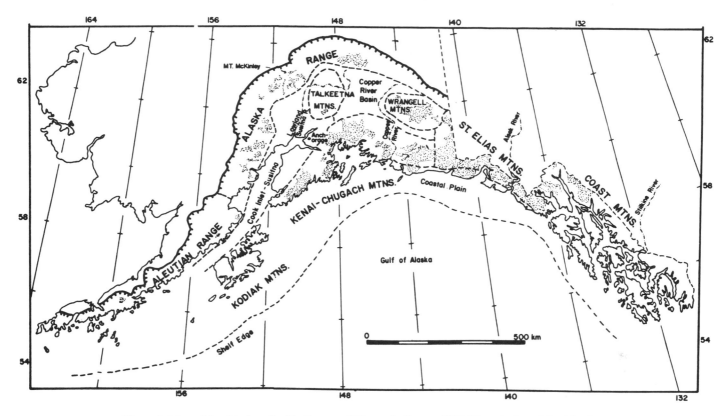

Figure 1. Map of the Alaskan Pacific Coast and Mountain System (PCMS) exclusive of the Aleutian Islands showing unranked physiographic subdivisions (after Wahrhaftig, 1965). Stippled areas indicate modern glacierized regions shown on U.S. Geological Survey Alaska Map A (scale 1:5,000,000), that served as source regions for glacial lobes in Figure 2. The modern ice cover may be overrepresented in some areas. Heavy hachured line shows northern border of PCMS, and heavy dashed lines show primary subdivisions.

covering more than 100,000 km² (Field, 1975, p. 4), cloaks these coastal ranges.

Between these two arcuate mountain systems is a complex structural lowland, which includes, from west to east, the Cook Inlet–Susitna Lowland, the Talkeetna Mountains, and the Copper River basin. The Wrangell Mountains, a massive Neogene complex of shield and composite volcanoes with extensive summit ice fields, indent the lowland province at its eastern limit. The mainland portion of southeast Alaska and its archipelago is an extensively glaciated, lithologically diverse region that is fragmented by large transform faults, which serve as loci for its largest fiords. The continental shelf off southern Alaska, which narrows toward the east, is cut by many large submarine valleys that apparently were carved by glaciers during times of lowered sea level (Molnia, 1986). A narrow strip of coastal plain lies between Cape Spencer and the Copper River Delta.

All of southern Alaska consists of allochthonous terranes, a series of fault-bounded disjunct crustal fragments that accreted to North America during subduction of the Pacific plate; some terranes may have moved as much as 9,000 km (Jones and others,

1982b). Most accreted during Mesozoic time, but some of the southern terranes may have accreted as late as early Tertiary time. Structural complexity within small areas (Cowan and Boss, 1978) and compressional tectonic styles are characteristic of the region. Underplating during subduction may be contributing to rapid uplift of the Chugach and adjacent ranges (Plafker and others, 1985). Faults are generally either landward-dipping thrusts or strike-slip faults, and nearly all exhibit postglacial displacements. A wide range of lithologies are present, but flysch and volcanics predominate (Beikman, 1980).

Southern Alaska is one of the most seismically active areas in the world, with earthquakes commonly above Richter magnitude six (Stephens and others, 1985). During the 1964 Good Friday Alaska Earthquake (Eckel, 1970), the ground was measurably displaced for more than 250,000 km², and uplift of as much as 15 m (2 m average) occurred over a broad area of emergence. Geologic evidence for uplift is apparent along much of south coastal Alaska, but its causes and rates remain obscure because so little is known about the glacio-isostatic record. Hudson and others (1982), following earlier work by Derksen (1975) and D. J. Miller (1953) on raised marine terraces bordering the northern Gulf of Alaska, indicated that much of the coast has been uplifted at a rate exceeding 0.5 cm/yr during the Holocene, and presumably during much of Quaternary time. At these rates the highest peaks bordering southeast Alaska could have been raised in as little as 4.5 m.y. Using seismic reflection data from the Gulf of Alaska, Bruns and Plafker (1976) detected 30 structural highs in sediments no older than Miocene in age. Tide gauge records in a broad area centered on Glacier Bay (Hicks and Shofnos, 1965; Hudson and others, 1982) indicate land emergence of as much as 4 cm/yr, most of which is glacio-isostatic. Tilting of salt marsh sediments in lower Cook Inlet (Thorson and others, 1980) suggest landward tilting of as much as 1 m/km/100 yrs. Stratigraphic studies of recent faulting (e.g., Detterman and others, 1974; Bruhn, 1978; Thorson, 1978) confirm episodic Holocene movement of many large border faults.

Most of south-central Alaska has a subarctic transitional maritime-continental climate characterized by heavy winter snowfall averaging 1 to 5 m (Johnson and Hartman, 1969; Hackett and Santeford, 1980). Precipitation is heaviest at the coast, where northeast-moving storms precipitate orographically as snow in the coastal mountains. Inland from the southern ranges, the climate is more continental. Southeast Alaska has a cold maritime climate with mean annual precipitation exceeding 5 m (Péwé, 1975), and a mountain snowpack averaging 5 to 7 m deep. The Aleutians have a cloudy, windy, cold maritime climate. Across the PCMS in Alaska, mean annual precipitation ranges from 80 to 550 cm/yr and mean annual temperatures range from –2° to +6°C.

The modern glaciation threshold (snowline) is as low as 800 m in areas bordering the northern Gulf of Alaska, but rises steeply inland to as much as 2,000 m within 200 km of the coast (Hamilton and Thorson, 1983). The Cook Inlet–Susitna Lowland, and to a lesser extent the Copper River basin, have lower glaciation thresholds because Pacific moisture is funneled northeastward between the coastal and inland arcuate mountain systems. Snowlines are highest in precipitation shadows north of the Wrangell Mountains and the Alaska Range. Glaciers are temperate (isothermal) for most of their lengths, but subpolar characteristics are evident on the highest, coldest summits (Benson and Motyka, 1979).

Given the available moisture, low temperatures, and frequent freezing cycles, weathering in the Alaskan PCMS is dominantly mechanical, and probably controlled largely by frost action. However, recent investigations by Dixon and others (1984) indicate that chemical weathering processes may play a significant and interdependent role in high-altitude rock disintegration. Extensive areas of bedrock throughout the region are usually mantled with outcrop rubble, a disordered veneer of lichen-covered blocks and fine fragments that are commonly sorted into weakly developed polygonal patterns (e.g., Nelson and Reed, 1978). Most soils in the region are Inceptisols with a thin O or A horizon above a cambic B horizon (Reiger and others, 1979). Entisols and Histosols are also common in high-

land and lowland terrain, respectively, and Spodosols are common in well-drained localities in southeast and south-central Alaska. Mollisols are limited to tundra areas and those areas underlain by calcareous rocks.

GEOMORPHIC PROCESSES

Most of the landscape in the Alaska PCMS results from deep selective erosion by alpine glacier complexes along linear zones of structural (faults) and/or lithologic weakness. Deposition is commonly limited to interior and coastal lowlands, and nearly all deposits are late Pleistocene or younger (<125 ka; Karlstrom and others, 1964). Erosion and deposition beneath modern glaciers are perhaps the only geomorphic processes that operate exclusively of other processes, yet even their rates and intensities are influenced by uplift rates and by periglacial slope processes. The geomorphology of areas beyond existing glaciers is polygenetic.

Volcanic Processes

Alaska's southern border is flanked by a chain of Neogene andesitic stratovolcanoes that extends more than 2,800 km from Attu Island in the west to Hayes Volcano in the northernmost Aleutian Range. Active volcanoes are also located in the Wrangell and St. Elias mountains, and near Sitka in southeast Alaska. Volcanoes in the Wrangell Mountains, one of the largest accumulations of lava on earth (Hogan and others, 1978), are similar to shields, but they are surmounted by ice-capped stratovolcanoes. Volcanism in the Aleutians began at least by mid-Miocene time and has been particularly active in Quaternary (Hogan and others, 1978; Kennett and Thunell, 1975) and historic (Simkin and others, 1981) time.

Glacier-volcano interactions are common owing to their mutual aerial extent. Volcanoes usually project well above snowline in areas with high accumulation rates, resulting in summit ice caps and small valley glaciers that commonly cloak active vents; new vents continue to be discovered beneath mountain ice sheets (Miller and Smith, 1976; Yount and others, 1985). Glaciers develop within, fill, and spill out of most calderas, many of which are Holocene in age. Recent eruptions are commonly at least partly phreatic, because the increased heat flow melts large quantities of ice, resulting in accelerated glacier flow and greatly increased stream discharge (Yount and others, 1985; Motyka, 1977). Eruptions beneath thick ice, such as in the Togiak Valley (Coonrad and Hoare, 1976), may produce flat-topped buttes known as tuyas.

Pyroclastic material, a common eruptive product, influences other processes as well. Coarse tephra is usually restricted to proximal regions near sources, but some ash flows carry well beyond the volcanic flanks (Miller and Smith, 1977). Thick accumulations of coarse proximal pyroclastic material on volcanic flanks commonly results in widespread catastrophic debris flows (lahars) that mantle the lower slopes of many volcanoes (Yehle,

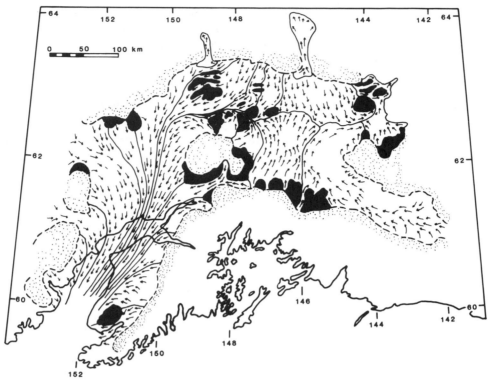

Figure 2. Map of the interior portions of the Cordilleran Ice Sheet in south-central Alaska showing flow directions and unglaciated areas (black) during the maximum extent of ice during early Wisconsin time (late Wisconsin glaciers were more restricted).

1980; Riehle and others, 1981). Thick pyroclastic accumulations also alter glacier mass balance and stream sediment load. Finer-grained tephra is widely distributed and nearly ubiquitous throughout southern Alaska, where it is commonly reworked into alluvial and lacustrine settings (Westgate and others, 1983; Thorson and Romick, 1984).

Glacial Processes

Nearly all of the Alaska PCMS was covered by the Cordilleran Ice Sheet as recently as about 15,000 years ago; the ice sheet apparently remained extensive throughout much of the Wisconsin Stage (Hamilton and Thorson, 1983). Consequently, glacial erosion is probably the single most widespread and important geomorphic process to have affected the area. The dearth of pre-Wisconsin deposits throughout much of the region, the sharpness of glacial-erosion features, and the voluminous quantities of glaciomarine drift on, and ice-rafted material beyond, the continental shelf indicates that much of southern Alaska is a geologically-young, glacially eroded landscape with limited post-glacial modification.

At its maximum late Wisconsin extent, the Cordilleran Ice Sheet consisted of a contiguous mass of independent glacier systems that can be divided into four main parts (Hamilton and Thorson, 1983; Fig. 2). North of the crest of the Alaskan-Aleutian Range (in the orographic shadow from coastal sources of moisture), the ice sheet consisted of a coalesced mass of independent valley glacier systems. Regions south of the divide along the south coastal ranges were heavily inundated by ice, which drained southward beyond the present coast, mantling much of the continental shelf with glacial deposits and carving many large valleys across it (Molnia, 1986). Glaciers between the Alaskan-Aleutian and south coastal range divides flowed peripherally into the Copper River basin to the east, and merged into the Cook Inlet–Susitna ice stream to the west, which drained southward to a tidewater terminus (Schmoll and Yehle, 1986). A large independent ice cap formed on the exposed continental shelf seaward of the Aleutian Range; a true ice shelf may have developed north of it (Black, 1983).

The Yakataga Formation, a 3,000 m thick sequence of interbedded marine tillites, gravel, and sand that is widely exposed in coastal mountains bordering the northern Gulf of Alaska (Miller, 1953), records multiple episodes of extensive tidewater glaciers during the last 6 m.y. (Armentrout, 1983). Evidence for at least 12 glacial episodes between 2.7 and about 10 Ma is also preserved beneath lava flows in the Wrangell Mountains (Denton and Armstrong, 1969). Marine cores from sediments in the Gulf of Alaska indicate repeated intervals when tidewater glaciers fringed much of the southern coast (von Huene and others, 1976). These records indicate that prolonged glaciation and rapid uplift have been the dominant geomorphic processes throughout Neogene time.

Morphologic evidence for repeated glaciation is less clear. South of the Alaska Range, drift and erosional surfaces equivalent to marine isotope stage 2 lie within valleys that show no more than three clearly distinct older glacial episodes (e.g., Karlstrom, 1964). Clearly, the major valleys have developed through repeated glaciation. It is not known whether the absence of older records indicates periodic inundation and erosion of the entire landscape by glaciers, or whether the oldest glacial landforms have been completely eliminated by weathering and slope processes. Many of the alpine glacial systems are well integrated into dendritic networks that are elongated along strike for more than 100 km; parabolic-shaped glacial valley walls commonly stand more than 1,000 m high. In the northern Talkeetna Mountains, transection passes drain southward across the crest of the range in nearly all of the major drainages, suggesting a long history of progressive valley integration. Accordant summits in the Chugach Range and Talkeetna Mountains were interpreted by Miller (1958) as ancient fluvial erosional surfaces that were subsequently incised. More detailed research is required before this interpretation can be verified; alternative possibilities include cryoplanation at levels controlled by regional climate and surfaces of aerial scouring beneath former ice caps.

The PCMS in Alaska presently has nearly every possible morphologic classification of glaciers exclusive of shelves and ice caps (Field, 1975). Cirque and valley glaciers and systems of valley glaciers are extremely common, as well as mountain ice caps, ice fields, mountain ice sheets, and tidewater glaciers. Classic alpine and coastal glacial features are ubiquitous. Depositional features commonly include thin and patchy till sheets near valley bottoms that thicken downstream, well-developed end moraines, extensive areas of hummocky dead-ice terrain, and widespread accumulations of stratified meltwater deposits. Rogen moraines and drumlins are very rare, and restricted to broad lowlands such as the lower Susitna Valley (Reger and Updike, 1983).

Glacial geology in Alaska has historically been stratigraphic in its approach, with few studies dealing directly with the origins of glacial features (Hamilton and others, 1986; Péwé, 1975; Coulter and others, 1965). Recent exceptions to this trend are the works of Mickelsen (1971), Goodwin (1984), Lawson (1979), Evenson and Clinch (1986), Gustavson and Boothroyd (1982), Updike (1982), and others on glaciogenic sedimentation, which have given clearer insight into modern processes, and therefore a guide to interpreting older deposits. A common theme of much of this research is the importance of subglacial meltwater in eroding and transporting sediment within the glacial system. Surging glaciers and their deposits have also received considerable attention (Post, 1969; Kamb and others, 1985; Driscoll, 1980a, 1980b; Watson, 1980). From these studies it is now clear that (1) surging is triggered by a buildup of basal water pressure that accelerates the sliding required to maintain high water pressure (positive feedback), (2) surge moraines are morphologically and lithologically distinct from nonsurge moraines, and (3) surge moraines

stabilize largely through surface melting, debris flowage, and topographic inversions. Continuing investigations by the U.S. Geological Survey on tidewater glaciers (Meier and others, 1980; Brown and others, 1982) have led to a predictive law for their behavior, which explains their asynchronous chronological pattern. Fundamental controls for tidewater glaciers include centerline water depth, fiord geometry, and rate of subglacial sedimentation (Mann, 1986).

My unpublished studies of the Cordilleran Ice Sheet reveal two aspects of the glacial geomorphology that have yet to be thoroughly documented. What were the dynamics of glacier flow in the Cook Inlet–Susitna Lowland, which I interpret to be a large, low-gradient, fast-moving ice stream that conveyed accumulating ice from as much as 60 percent of the interior Cordilleran Ice Sheet? Reconstructed flow patterns, based on topographically scoured lineaments, morainal patterns, and fluted terrain, suggest that at its maximum extent the ice stream drained as many as 15 separate lobes and was more than 125 km across and 400 km long. Altitudinal limits for the ice stream, defined on the basis of trimlines, faceted spurs, hanging valleys, and other features, indicate that it sloped soutward from about 1,500 to 600 m with a surface gradient about 3.2 m/km. The reconstructed basal shear stress of less than 0.4 bars suggests that basal sliding accounted for nearly all of its motion. A high water table in the glacier, perhaps in conjunction with a tidewater terminus and isostatic depression, may have facilitated such high rates of sliding.

A second observation concerns the broad glaciated plateau just leeward of the crest of the Talkeetna Mountains. Many north-draining glacial valleys in this area are significantly larger than would be expected from the limited number of cirques draining into them. In addition, the heads and walls of many tributary valleys are heavily scoured and grade continuously upward to unmodified, low-relief surfaces. I hypothesize that these features result from selective linear erosion beneath a mountain ice cap, which may have been largely frozen to its bed (Sugden and John, 1976, p. 203). This speculation suggests that the thermal characteristics of former glacier systems must be investigated before the glacial landscape can be properly interpreted.

Fluvial Processes

The fluvial geomorphology of the Alaska PCMS is difficult to summarize because the wide variety of fluvial features at various scales have generally received only reconnaissance descriptions. Process-oriented studies are rare; examples include Bradley and others (1972), Boothroyd and Ashley (1975), and Ritter (1982). All of the large rivers in southern Alaska are glaciofluvial to some extent because they originate as meltwater streams and because they drain terrain with an abundant supply of recent unconsolidated glacial sediment. Southern Alaska's largest rivers, the Susitna, Copper, Alsek, and Stilkine, carry tremendous quantities of suspended glacial flour and sand wash load. For example, the Copper River has a mean annual solid load of 100,000,000 metric tons (Hayes, 1977), and the thickness of Holocene muds beyond its mouth exceeds 300 m in some places (Molnia, 1983).

The abruptness of the rugged south coast of Alaska and the short distance northward to the drainage divide ensures that few southern Alaska rivers are more than a few hundred kilometers long. The Susitna and Copper rivers and their master tributaries traverse gravelly and coarsely braided upper reaches where they drain modern glaciers, but merge into wide vegetated flood plains with meandering sandy channels in their lower reaches. The Susitna ends in an extensive salt marsh, the Copper River in an embayed tidal delta, and the Alsek and Stilkine in partly filled fiords. Discharges are highest during late spring during peak snow melt, and lowest in winter, owing to reduced ablation and subarctic temperatures.

The headwaters of most tributary drainages originate as clearly defined V-shaped channels that descend from steep, rubbly mountain slopes toward the master streams. The gradients, shapes, and drainage areas of first-order streams vary widely in response to rock type. As they merge downslope into higher-order drainages, the tributaries commonly become deeply incised as they enter major glacial valleys, the floors of which serve as local base levels over most of the region. Large, broadly concave, gravelly alluvial fans are commonly built outward into the glacial troughs, in some cases crossing them completely. Fan apexes are commonly incised by single gravelly channels, which may become distributary in the distal reaches of the fan.

Fan size is controlled by many factors. In small basins that lack glacial troughs, fan size is apparently controlled by basin relief, area, and drainage density; lithologies most susceptible to periglacial breakdown, such as schist and bedded pelitic rocks, result in higher drainage densities and increasing fan size. In tributary drainages that contain one or more small glacial valleys, the alluvial fans are generally larger and more deeply incised. This observation suggests that fans were most actively aggrading during the short interval when the master drainage was deglaciated but when the tributaries were still occupied by glaciers. Alternatively, rapid fan accumulation following glacial recession (the paraglacial interval of Church and Ryder, 1972) may have resulted from the abundant sediment supply of recently deglaciated valleys. Large subangular boulders on fan surfaces well beyond the mountain front occur on many, but by no means a majority, of the smaller tributary fans. Two possible explanations for their occurrence are catastrophic discharges related to release of impounded water during deglaciation, or repeated slush-flow avalanches, which are common when thick granular snowpacks melt rapidly.

I am not aware of any studies of comparative drainage density for the Alaskan PCMS, but geologic and hydrologic controls of potential significance are as follows: (1) much of the precipitation arrives as snow, which melts rapidly and drains away before the ground is thawed; (2) colluvial downslope movements including frost creep, gelifluction, and talus accretion are rapid, perhaps precluding the development and maintenance

of a dense channel network; (3) most first-order drainages develop in coarse outcrop rubble with exceptionally high infiltration rates; and (4) permafrost is widely present, increasing the resistance of fine-grained unconsolidated material to rill erosion.

Process-oriented fluvial geomorphologic studies have focused most on outwash streams draining existing glaciers. Detailed studies of the Scott and Yana outwash fans, located west of the Copper River and in the Malaspina foreland, respectively, were conducted by Boothroyd and Ashley (1975). Conclusions drawn from their work are probably relevant to mid-sized (5-50 km long) outwash streams that are not characterized by outburst discharges. They recognized three outwash fan facies—upper, middle, and lower—with characteristic bed materials, channel geometry, bedforms, and slope. In the unvegetated upper fans, all but the highest flows are confined to a single channel that has a steep concave longitudinal slope increasing upstream from 7.6 to 17.6 m/km. In this reach, longitudinal bars—which grow by slipface migration—are characteristic, and clasts are strongly imbricated upstream and oriented transverse to flow. The midfan is characterized by multiple braided channels incised below vegetated bars, with slopes in the range of 2.6 m/km, common longitudinal bars, and clasts finer than 10 cm long. The lower fan consists of braided sandy channels with linguoid bars and slopes lower than 2 m/km. My observations suggest that the facies and their characteristics described by Boothroyd and Ashley (1975) are also representative of larger streams, but the midfan and lower fan facies are significantly elongated, and transitions are less abrupt. In larger rivers, long meandering reaches are present in regions of low surface gradient. Regional generalizations regarding hydraulic geometry, stream regimen (incising versus aggrading), and recent history are premature.

The two largest rivers in the Alaskan PCMS (the Susitna and Copper rivers) both cut directly across one or more mountain ridges as narrow canyons more than 300 m deep, even though much lower drainage routes are available. These rivers are but two examples of hundreds of large streams with courses transverse to rock structure, that must have complex and polygenetic drainage histories (e.g., Oberlander, 1965). Cenozoic drainage development has been greatly influenced by rapid and differential rates of uplift for various regions, varying intensities of glacial erosion, asynchronous patterns of glacial advances and retreats, transection drainage from former ice fields, orientation relative to orographic barriers and solar radiation, and drainage of large glacial lakes. It is also premature to speculate on the origin of the courses for even the best known of Alaska's larger rivers. Many smaller canyons that cut directly across lower ridges may have resulted from subglacial drainage. Asymmetrical valley cross profiles and drainage networks are abundant, and clearly result from a variety of causes.

Fluvial systems can be greatly affected by periodic release of glacially impounded meltwater (jokulhaups), which may increase the normal discharge by many orders of magnitude. Post and Mayo (1971) identified dozens of streams with potential outburst flood hazard, the best known of which is the Knik River, just northeast of Anchorage. Initial discharges in the Knik River of 142 to 170 m^3/sec increased to as much as 20,047 m^3/sec during the annual drainage of Lake George, which was impounded by the Knik Glacier each winter (the lake has not reformed since 1966; Reger and Updike, 1983, p. 242). These brief outburst floods are responsible for the morphological and alluvial characteristics of the Knik River valley train (Bradley and others, 1972). Outburst flood hazards are also being investigated on volcanoes (Benson and Motyka, 1979) because melting during historic eruptions has greatly increased stream discharge. Floods resulting from rapid drainage of nonglacial lakes also occurs (e.g., Reger and Updike, 1983, p. 215). The morphology and bed character of many streams are probably influenced by episodic flooding, but too little is known about modern stream hydrology or Holocene alluvial history to permit valid regional interpretations at this time.

Terraces are present on nearly all moderate-sized streams, but their interpretation is clouded by the many factors that could be responsible for valley incision. Ritter's (1982) reinterpretation of a part of the terrace sequence in the Nenana River Valley demonstrated the role of glacial retreat, lake drainage, and tectonism in terrace development. Pleistocene outwash terraces can commonly be traced to former ice front positions, but lower terraces can have a variety of causes, which may include: (1) variations in the supply of bedload, (2) sea-level variations, (3) outburst flooding, (4) localized uplift, and (5) disruptions of gradient based on common river icings (aufeis), marginal glacier advances, or tributary aggradation. A regional pattern of terrace development may not be present.

Slope Processes

Slope processes in southern Alaska are significantly affected by the rigorous periglacial climate and the young, oversteepened nature of modern slopes (Péwé, 1975). High altitude slopes with near-surface bedrock are greatly influenced by the rapid mechanical disintegration associated with frost riving and frost shattering, with fragment size related to the bedding and jointing characteristics of the local bedrock. On gentle slopes, extensive areas of rubble are present, with netlike and polygonal patterns frequently developing in response to frost sorting. Blockfields and streams (felsenmeer) are presently active in many areas. Broad gentle slopes at intermediate elevations are dominated by solifluction.

On steeper slopes, which usually occur along the walls of cirques and glacial valleys, talus is extremely abundant. Normally, closed spaced talus cones merge downslope into a nearly continuous mantle parallel to the valley walls. Talus at lower elevations is commonly partly stabilized and lichen covered, suggesting that much of it is now relict, perhaps from early postglacial time. However, the apparent patterns of talus growth vary not only from valley to valley but within different portions of the same valley. For example, along Tsusena Creek in the northern Talkeetna Mountains, fresh talus is now building outward and burying older lichen-covered Holocene talus near the valley

mouth, but upstream, partially stabilized talus cones are slightly entrenched, with modern cones developing at their bases.

Rock glaciers are abundant in south-central Alaska inland from the Pacific coast (e.g., Richter, 1976). Steep slopes, abundant talus, and permafrost are required for their formation, which explains their absence in the Aleutian Islands and much of southeast Alaska. Lobate, tongue-shaped, spatulate, and complex forms are all present, and their distribution crudely parallels the modern snowlines and glaciation threshold (Wahrhaftig and Cox, 1959; Nelson and Reed, 1978). They are typically restricted to cirques and upper portions of glacial valleys, and measured rates of movement up to 73 cm/yr have been demonstrated. Wahrhaftig and Cox (1959) demonstrated that the stabilization and reactivation patterns of rock glaciers in the southern Alaska Range reflect regional paleoclimatic variations during Holocene time. Multistage rubble sheets and felsenmeer on more gentle slopes also reflect temporal variations in the intensity of rubble movement (Wahrhaftig, 1949; Ferrians and others, 1983, p. 141).

Lower slopes are characterized by broad concave fans and aprons of colluvium that occur transitionally between the higher talus slopes and valley bottom alluvium. Leveed and bouldery colluvial fans are especially common; they may result from aggradation during repeated slush flows, which may be analogous to the debris flows of arid fans. Deposits of debris avalanches and small landslides are also common on lower slopes.

Deposits of large landslides are common, and result from a variety of causes. Rockfall avalanches that may have moved over compressed air (Shreve, 1966), large slumps, and translational block slides were initiated by ground shaking associated with the 1964 Alaska earthquake (Eckel, 1970), and many prehistoric slides may also have failed during ground disturbances. One of the most well-known rockfalls occurred at the head of Lituya Bay during a 1958 earthquake along the Fairweather fault (Miller, 1960). It created a wave that surged to a maximum height of 530 m before it moved rapidly along the length of the bay to the Pacific. Large debris flows, block glides, and deep-seated separation failures in bedrock are reported in quadrangle maps from south-central Alaska (Williams and Johnson, 1980; Nelson and Reed, 1978; Yehle, 1980). Volcanic debris flow deposits as much as 35 m thick have been recognized at the base of the Wrangell Mountains (Yehle, 1980). The Bluff Point Landslide near Homer (Reger, 1978) is a single slump more than 5 km long that occurred in weakly consolidated Tertiary strata.

Coastal Processes

The configuration of Alaska's south coast is constantly changing in response to the ongoing rapid retreat of tidewater glaciers, which have created isostatic uplifts associated with current deglaciation, tectonism, and many fiords tens of kilometers long within the last century. Localized uplifts of presumed isostatic origin (Hicks and Schofnos, 1965; Clark, 1978a) near Glacier Bay are causing emergence in excess of 3 cm/yr. Long-term (Holocene) uplift rates may approach 1 cm/yr for much of the Alaska coast between the Malaspina Glacier and Juneau (Hudson and others, 1976). During the 1964 Alaska earthquake, more than 1,000 km of coastline was measurably displaced by as much as 15 m, and numerous tidal deltas failed, producing tsunamis and seiche waves that greatly modified the coastline. Tremendous sediment loads from countless glacial meltwater rivers are brought into this dynamic setting and reworked by high-energy coastal processes.

Rugged fiord-indented coastal settings occur everywhere except along the narrow coastal plain between the Copper River and Icy Point. In settings facing the Gulf of Alaska, steep headlands stand above gravel storm beaches and broad abrasion platforms. The more protected heavily vegetated fiord walls typically stand above more narrow gravel beaches and scattered tideflats, with the shoreline morphology controlled by the regional slope, shoreline stability, exposure to storms, fetch, and sediment supply. The morphology of the fiord heads reflects the balance between glacial meltwater sedimentation, which can exceed rates of 2.3 m/yr, and tidal currents, which can exceed 6 knots (Carlson and others, 1979). Fiord heads without existing glaciers, many of which have tidal ranges in excess of 5 m, are commonly occupied by either prograding, gravelly deltas or by extensive salt marshes and broad tideflats.

Molnia (1986) presents a speculative model for the geomorphic origin of the Gulf of Alaska coast. He suggests that the entire coastal plain, which consists of chenierlike ridges extending as much as 7 km inland, is less than 1,000 years old and developed primarily by aperiodic progradation in response to wave reworking of glacial meltwater sediment. Hayes (1977, p. 112) classifies this area as a mesotidal collision shoreline, noting that it is unusual to find a depositional shoreline of this magnitude on such a tectonically active coast. Its anomalous occurrence results from the exceptionally high sediment input, especially from the Copper River, but wave energy is also exceptionally high. For example, recent storm surge deposits, which extend an average of 8.2 m above sea level for more than 250 km of this coastal zone, may have been associated with storm waves between 15 and 20 m high (Molnia, 1980).

East of the Copper River, the coastal zone is largely an outwash plain that is modified seaward into a series of beach ridges and backshore dunes (Hayes, 1977). Coastal sediment fines westward from gravel to sand, and eolian processes become more important in that direction. Cuspate features at a variety of scales are present along the entire length of the coastal plain. The Copper River Delta is classified by Galloway (1975, Fig. 3) as intermediate between a tide-dominated and wave-dominated delta. The broadly braided outwash plain of the Copper River merges gradually into a salt marsh plain where it becomes progressively indented by multiple tidal channels normal to the coast, and contains a seaward margin of barrier islands. Beach morphology and depositional facies are clearly illustrated by Hayes (1977). Eolian modification of the delta and its barrier islands reflect both strong onshore winds and offshore adiabatic winds.

Differential tectonic movements and isostatic effects asso-

ciated with late Pleistocene and current deglaciation render any interpretations of regional sea-level changes problematic at best. Coastal terraces occur as high as 488 m along the northern Gulf of Alaska (Miller, 1958). More recent detailed studies of these terraces (Sirkin and others, 1971; Hudson and others, 1976) have confirmed their tectonic origin. Cirques with their floors below present sea level occur at many places from the eastern Aleutians to Chatam Strait in southeast Alaska; they apparently resulted from some combination of sea-level regression, extreme snowline depression, and tectonic submergence. Much of the southern Alaskan continental shelf was exposed during times of minimum late Pleistocene sea levels. It provided a locus for the growth of large coalescing piedmont lobes and ice sheets, which were especially well developed along the Pacific margin of the Aleutian Range (Hamilton and Thorson, 1983). Molnia (1983) reported that much of the shelf morphology east of Cook Inlet is also glacial in origin, resulting largely from seaward expansion of ice originating in coastal mountains. Mann (1983) found evidence for more limited, and highly variable glaciation of the Alaska shelf during the last glaciation.

Glacioisostatic effects also have and are continuing to affect the southern coast. Clark (1978b) predicted that the Alaskan panhandle, south-central Alaska, and the Aleutian Islands should have experienced emergence and submergence followed by gradual emergence, respectively, as a result of global isostatic effects. Glacioisostatic rebound in southeast Alaska is clearly evident from late Pleistocene raised marine sediments and coastal terraces up to 150 m elevation (Ackerman and others, 1979; Miller, 1973). Glaciomarine facies in upper Cook Inlet (Schmoll and others, 1972) also suggest deglaciation as early as 15 ka and postglacial emergence.

Eustatic sea-level variations have influenced the geomorphology of southern Alaska primarily by creating more emergent land above Pleistocene snowlines. Because of nearly complete glaciation of the south coast, eustatic sea-level lowering had minimal direct effects on stream regimen and coastal morphology. In the Aleutian Islands, where isostatic and tectonic effects have been less severe, a scanty eustatic sea-level record is present (Morris and Bucknam, 1972). Holocene sea level in the western Aleutians apparently reached its maximum elevation of +2 to 3 m above 3 to 4 ka (Black, 1980) followed by late Holocene emergence. On Amchitka Island, near the western limit of the Aleutian chain, Gard (1980) documented a series of three marine terraces formed during the last (marine isotope stage 5e) and previous interglacial high sea-level stands; the lowest of these terraces is dated by U-series methods at about 127 ka (Szabo and Gard, 1975).

Other Geomorphic Processes

Eolian deposits, consisting of loess, dune sand, cover sands, and cliff-head dunes occur widely but locally throughout much of southern Alaska (Black, 1951). Process-oriented papers on eolian sediments are even less common than those on fluvial sediments

SIMPLIFIED GEOMORPHIC MODEL

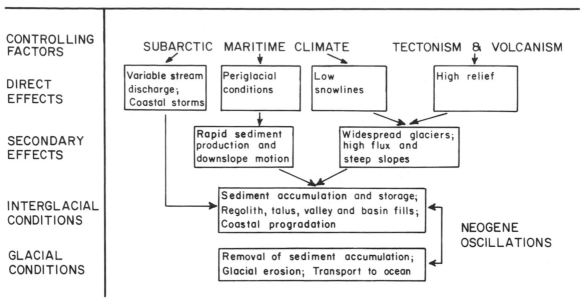

Figure 3. Schematic diagram showing important components of a simplified model for the geomorphology of the Alaska Pacific Coast and Mountain System.

(Thorson and Bender, 1985). Loess is common in lowland regions because of the large number of active valley trains with associated adiabatic and katabatic winds. Although widespread, it is usually thin (<1 m), and drapes existing landforms rather than creating true loess landscapes. Thick loess deposits, which thin and fine exponentially downwind, are restricted to blufftops leeward of actively braided meltwater channels. For example, loess exceeds 7 m in thickness immediately downwind from the Knik River flood plain near Palmer, Alaska (Reger and Updike, 1983, p. 224). Dunes are more common behind sandy beaches on the Gulf of Alaska coasts, where a variety of forms (longitudinal dominant) exist. Cliff-head dunes (largely parabolic) are commonly present above low bluffs adjacent to actively braided flood plains. Amorphous blankets of eolian sand up to 5 m thick with subhorizontal stratification are widespread in coastal lowlands with abundant sand supply. Lea (1985) interprets the absence of both dune morphology and evidence for slip-face migration in these cover sands as indicating deposition under periglacial conditions more mesic than those that would favor good dune morphology.

Permafrost strongly influences the surface character of lowlands throughout south-central Alaska. Although its distribution is poorly known, permafrost is apparently absent at low elevations in southeast Alaska, the coastal plain, and the Aleutian Islands. Permafrost is very discontinuous in lower Cook Inlet, but becomes increasingly common in areas of more continental climate to the north and east. In the Copper River basin, permafrost is widespread, close to melting temperatures, and 30-60 m thick (Péwé, 1975). Permafrost is shallowest in either fine-grained or

organic soils or on north-facing slopes. The primary geomorphic effects of lowland permafrost result from its impermeability and thermal sensitivity. Thousands of the bogs and small ponds in the Copper River basin are thaw lakes, which develop in response to the greater thermal absorption of water relative to the thick surface vegetation. Eventual drainage of the ponds results in the reestablishment of permafrost, and repetition of the cycle.

In addition to the thermokarst ponds, tens of thousands of lakes were created primarily by glacial erosion and deposition. Most of the small lakes occur either where groundwater tables intersect hummocky dead-ice terrain or where glacial scour has locally deepened impermeable bedrock. These lakes are now slowly filling with aquatic organic material and detrital sediments. The largest lakes, such as Lake Iliamna, are usually dammed by moraines in glacial valleys just beyond mountain fronts. Most have one or more shorelines resulting from higher past lake stages (Detterman and others, 1965).

Dozens of mud volcanoes, the largest of which is about 100 m high and more than 1 km in diameter, penetrate the permanently frozen glaciolacustrine sediments of the Copper River basin, and issue warm saline water (Nichols and Yehle, 1961). All are late Pleistocene or younger, and apparently resulted from the effusive flow of mud, connate water, and possibly meteoric water from great depths.

A SIMPLIFIED MODEL

This section attempts to synthesize the geomorphic processes described separately above, and to present a simplified general

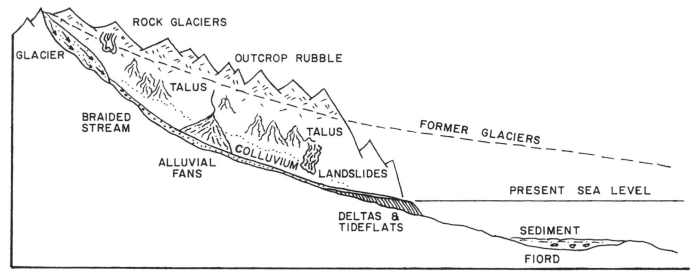

Figure 4. Schematic longitudinal profile of a representative valley in the Alaskan Pacific Coast and Mountain System showing common processes operating during interglacial conditions.

model for the geomorphology of the Alaska PCMS (Figs. 3 and 4). The two cornerstones for the model are southern Alaska's rapid rate of uplift along a convergent plate margin and its subarctic maritime to transitional climate. The tectonism is responsible for maintaining the tremendous relief in areas underlain to a large degree by relatively weak rocks, and to a lesser degree for accelerating slope processes through frequent ground shaking. The climate is responsible for creating and maintaining the largest system of glaciers in North America, and for producing an almost limitless quantity of mechanically fractured detritus through periglacial action. When combined, these salient aspects of southern Alaska geomorphology ensure the maintenance of extremely rugged topography, rapid denudation throughout the region, and rapid export of detrital material to the north Pacific.

In its simplest sense the landscape of the Alaska PCMS is almost entirely glacial in origin because it was nearly completely innundated and modified by the Cordilleran Ice Sheet as recently as about 15,000 years ago (strong structural control of glacial modification is clearly evident). Modern glaciers, however, are very restricted relative to Pleistocene times, occupying less than 5 percent of the total area today. The stratigrahic record indicates that alternating periods of widespread glaciation and ice withdrawal have occurred throughout Neogene time. Thus, a model based on alternating glacial and interglacial modes has logical appeal.

If Holocene time is representative of earlier periods of restricted glaciation, interglacial climates in southern Alaska were strongly periglacial in character over much of the area, with large glacier systems descending from the higher ranges. In general, these intervals of restricted glaciation are characterized by sediment accumulation and storage. Deglaciation exposes extensive upland areas of oversteepened rock slopes that were largely stripped of older sediment and regolith by the preceding glaciation. (If this were not the case, pre-Wisconsin sediments would be widespread.) The freshly exposed bedrock slopes are fragmented by exfoliation and frost riving, and large quantities of talus and colluvium are shed into glacial valleys occupied by underfit streams. The abundant clastic debris from reworked talus, alluvial fans, and recently deposited glacial drift, when combined with the steep slopes and flashy seasonal hydrology, maintains the braided character of most streams and their high sediment loads. Where these rivers meet the heads of fiords or other ice-scoured marine embayments, their loads are dropped, and they may either form prograding deltas or be reworked by shoreline and tidal processes. Fine-grained detrital sediments are either trapped in protected embayments as extensive salt marshes and tide flats, or accumulate in submarine ice-scoured basins. Much of the sediment accumulation may occur early in the postglacial period.

Upon the return to glacial intervals, upland sediment accumulations in glacial valleys are entrained and transported toward and beyond the coast. Coastal lowlands and fiords are excavated by tongues of ice that extend well beyond the present shoreline, and redeposit much of the entrained sediment as glaciomarine drift beyond the edge of the continental shelf. Another cycle is initiated when oversteepened uplands and reexcavated valleys and fiords emerge during rapid deglaciation.

Continued emergence of the land through tectonic processes allows the PCMS in Alaska to maintain its rugged alpine character. The buildup of Pleistocene glaciers on the continental shelf and the low glaciation thresholds for much of southern Alaska indicate that the Cordilleran Ice Sheet could have been extensive even if the bordering mountain ranges were much lower. Without strong uplift, the topography of the Alaskan PCMS might resemble that of the ice-sheet–covered northern Appalachians.

FUTURE RESEARCH PROBLEMS

(1) The classic geomorphologic problem of the age and evolution of the master drainage basins in southern Alaska is still unknown. Interpreting this history would greatly improve our understanding of the relative roles of fluvial incision, tectonic uplift, and asynchronous glacier advances in high-latitude drainage basin development.

(2) The presence of tremendous solid loads of large drainages in southern Alaska suggests that upland denudation rates are extremely high. Comparing reconstructed denudation rates with known rates of tectonic uplift would allow researchers to determine whether the existing relief is increasing or decreasing with time, or whether it is nearly in isostatic equilibrium.

(3) Finally, the causes, extent, and dynamics of ice sheets on the continental shelf bordering southern Alaska and the ice stream in the Cook Inlet-Susitna lowland are largely unknown. Interpreting these phenomena correctly would greatly enhance our understanding of the Cordilleran Ice Sheet, which is largely responsible for much of the geomorphology of the Alaskan PCMS.

GEOMORPHIC PROCESSES IN THE PACIFIC COAST AND MOUNTAIN SYSTEM OF BRITISH COLUMBIA

John J. Clague and W. H. Mathews

INTRODUCTION AND SETTING

Western British Columbia is a region of diverse topography, geology, climate, and vegetation. Rugged glacier-clad mountains reaching to 4,000 m elevation are within sight of fiords, island-studded inner coastal waterways, and the open Pacific Ocean (Figs. 5–7). The region extends north from the International Boundary along the forty-ninth parallel and Juan de Fuca Strait to southeastern Alaska, and east from the continental margin to the eastern slopes of the Coast Mountains. Its major physiographic subdivisions are: (1) the Insular Mountains comprising the Vancouver Island and Queen Charlotte ranges, with peaks up to 2,200 m and 1,130 m, respectively; (2) the Coastal Depressions that include the continental shelf and adjacent coastal lowlands; and (3) the Coast Mountains, a grand mountain battlement bordered on the west by the Pacific Ocean and on the east by rolling interior plateaus. The Insular Mountains are formed mainly of Paleozoic and Mesozoic volcanic and sedimentary rocks. In contrast, large parts of the Coastal Depressions are underlain by Cretaceous and Tertiary sedimentary strata, and the Coast Mountains by Jurassic to Tertiary granitic rocks and subordinate, variably metamorphosed, Paleozoic to early Tertiary, volcanic and sedimentary rocks (Tipper and others, 1981).

Western British Columbia is a collage of far-travelled crustal fragments ("terranes") that have become accreted to the craton in response to oblique convergence of North America and the Pa-

cific Ocean during the Mesozoic and early Cenozoic (Coney and others, 1980; Jones and others, 1982a; Monger and others, 1982; Chamberlain and Lambert, 1985). This process was accompanied by compression and by northwestward displacement of accreted rocks along strike-slip faults. As a result, the entire Canadian Cordillera, including western British Columbia, has a strong northwest-southeast structural and topographic grain.

The contemporary tectonic regime is controlled by the motions of the Pacific, America, Juan de Fuca, and Explorer plates, which are in contact in the northeast Pacific Ocean west of the Canadian continental margin (Fig. 5; Atwater, 1970; Riddihough, 1977). Northwest of Vancouver Island, the Pacific and America plates are bounded by the Queen Charlotte–Fairweather transform fault that extends to the Aleutian Trench (Chase and Tiffin, 1972; Keen and Hyndman, 1979). A system of spreading ridges and short transform faults separates the Pacific plate from the Juan de Fuca and Explorer plates (Barr and Chase, 1974; Riddihough and others, 1983). The boundary between the America plate and the Juan de Fuca and Explorer plates west of Vancouver Island is thought to be a zone of convergence and subduction (Riddihough and Hyndman, 1976).

The high mountains of western British Columbia today support cirque and valley glaciers and small ice caps. During Pleistocene cold periods, glaciers advanced from these mountains to bury lowland areas and parts of the continental shelf. In the process, the drainage network was repeatedly disrupted and rearranged, mountains were sculpted by glaciers, and prodigious quantities of sediment were deposited in valleys and on coastal lowlands and the sea floor. During the Holocene, the landscape was modified by fluvial, mass wasting, and marine processes, but these changes were minor in comparison to those produced by diastrophism, glaciation, and subaerial denudation during the Pleistocene.

GEOMORPHIC PROCESSES AND LANDSCAPE DEVELOPMENT

A variety of geomorphic processes have shaped the landscape of western British Columbia. The most significant of these are summarized below, with emphasis on the character of each process and its effects on the landscape.

Diastrophism

Diastrophism resulting from plate interactions in the northeast Pacific Ocean has created the framework in which other geomorphic processes, including glaciation, fluvial erosion, and mass wasting, have operated to produce the present landscape.

The present high relief of the Coast Mountains is a product of rapid late Cenozoic uplift. A mature erosion surface of moderate relief existed over most of southern British Columbia, including the southern Coast Mountains, in early to middle Miocene time (Mathews, 1968; Parrish, 1983). The Miocene landscape was characterized by broad low valleys separated by gentle slopes

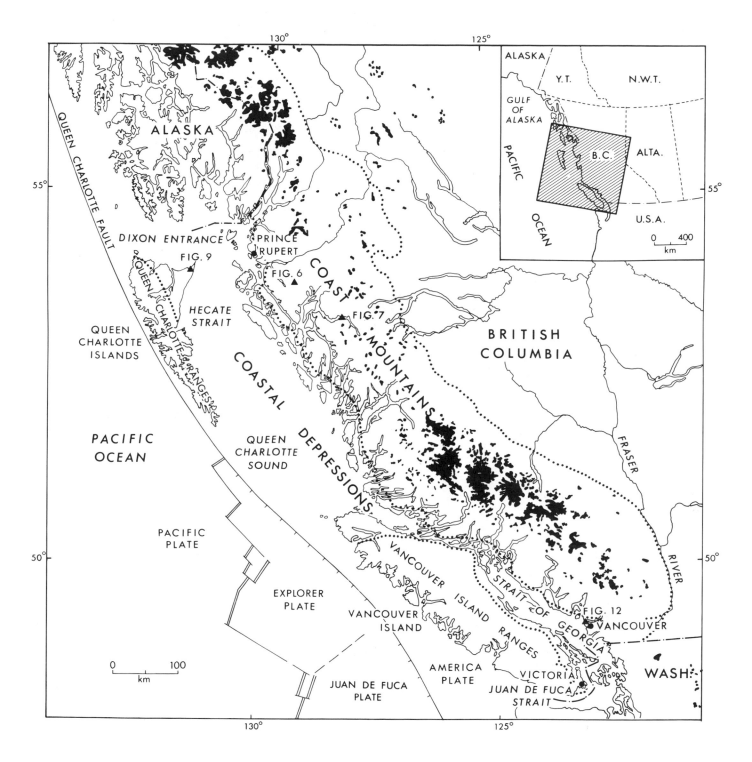

Figure 5. Index map of western British Columbia. Approximate boundaries of major physiographic regions, including the Insular Mountains (i.e., Queen Charlotte and Vancouver Island ranges), Coastal Depressions, and Coast Mountains, are indicated by dotted lines. Black areas are glaciers.

Figure 6. The Coast Mountains near Prince Rupert, British Columbia (location shown in Fig. 5). Note typical alpine glacial landforms including horns, arêtes, cirques, and U-shaped valleys. Province of British Columbia photo BC528-21.

Figure 7. Gardner Canal, a typical British Columbia fiord (location shown in Fig. 5). Province of British Columbia photo BC528-66.

rising to narrow divides and summits 300 to 500 m above the low ground (Mathews, 1987). Remnants of lavas erupted onto this surface when it was near sea level are found today on a few summits up to 2.5 km elevation (Mathews and Rouse, 1963; Tipper, 1969, 1978). Thus it appears that much of the height of the southern Coast Mountains was developed after the Miocene. In contrast, north of 52° N latitude, Miocene lavas or the erosion surface on which these lavas were deposited extend into valleys 1 to 1.5 km below mountain summits, suggesting that some of the present height of the central and northern Coast Mountains is pre-Miocene in age. Fission-track dates on apatite and zircon (Parrish, 1983) confirm these geologic observations and indicate that the Coast Mountains were raised 2 to 3 km in the last 10 m.y. (Fig. 8). Additional support for rapid recent uplift comes from an analysis of the present physiography of the Coast Moun-

tains. Over a large area in the south, summits form a broad plateaulike envelope at 2 to 3 km elevation that may represent the late Miocene erosion surface (Parrish, 1983). In contrast, north of 52°N, the summit surface is considerably lower and more irregular, suggesting greater antiquity of relief.

Uplift of the Coast and Insular mountains during late Cenozoic time was accompanied by fluvial and glacial erosion and mass wasting. Initially, erosion was localized along faults, fractures, and Tertiary drainage lines. However, the Tertiary setting exerted less and less control on denudation as the old Miocene land surface was destroyed and converted to mountainous topography. Most of the detritus eroded from the Coast and Insular mountains during this period was carried to the Pacific Ocean. Some of this material may form the upper part of the sedimentary prism underlying the continental shelf (Shouldice, 1971, 1973;

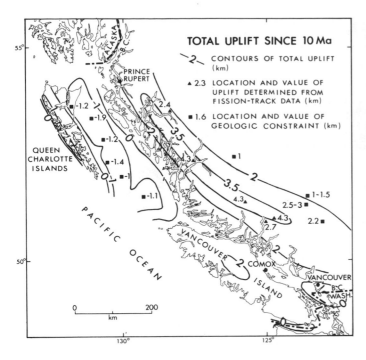

Figure 8. Total uplift of western British Columbia since 10 m.y. Adapted from Parrish (1983, Fig. 11).

Tiffin and others, 1972). However, the volume of Plio-Pleistocene sediments on the shelf is much smaller than the volume of rock eroded from the Coast Mountains since the Miocene. It thus seems likely that most of the material eroded from the mountains of western British Columbia during late Cenozoic time ultimately reached abyssal basins and plains beyond the continental margin.

Late Cenozoic uplift of the Coast and Insular mountains presumably is a product of plate interactions in the northeast Pacific Ocean. It is thought that the present plate tectonic regime is dominated by convergence and subduction of oceanic crust beneath western British Columbia south of 51° N latitude and right-lateral transcurrent faulting along the continental margin farther north (Riddihough, 1977). This pattern has not changed significantly since at least 10 Ma, and it is possible that the same basic plate configuration has persisted since the middle Cenozoic (Atwater, 1970; Coney, 1978). Yet, within this relatively stable plate tectonic setting, rapid uplift of the southern Coast Mountains began only at the end of the Tertiary. Also problematic is the fact that the central and northern Coast Mountains are located along a transform boundary; normally, sustained uplift would not be expected in such a tectonic setting.

Parrish (1983) suggested that rapid uplift in the south occurred in response to westward migration of a belt of arc-related volcanism and plutonism under the present Coast Mountains, accompanied by thermal expansion and thinning of formerly cool lithosphere. This shift may have occurred in response to steepen-

ing of the subducted edges of the Juan de Fuca and Explorer plates.

Less rapid uplift of the northern Coast Mountains may be a residual response to gross orogenic thickening of the crust along a destructive plate margin during the early Tertiary (Hollister, 1979). Early Tertiary orogenesis in this area terminated when the present strike-slip tectonic regime became established, but the mountains continued to rise through middle and late Cenozoic time to achieve isostatic equilibrium (Parrish, 1983). In the process, the large crustal root that supported the northern Coast Mountains during the early Tertiary was eliminated; today, crustal thicknesses in this area are similar to those in stable continental areas (Berry and Forsyth, 1975).

The landscape of western British Columbia continues to be shaped by diastrophism. The region has a high level of seismicity due to its location at the edge of the America plate (Milne and others, 1978). Several large (>6 Richter magnitude) earthquakes have occurred historically, the best known of which (M7.2 Vancouver Island earthquake, June 23, 1946) caused surface rupturing, vertical ground movements (Rogers and Hasegawa, 1978; Slawson and Savage, 1979), and widespread landsliding (Mathews, 1979). In addition, tidal records and levelling data indicate a complex pattern of seismic and aseismic deformation, with historical uplift of parts of the outer coast (i.e., Vancouver Island, Queen Charlotte Islands), and subsidence locally along the inner coast (Clague and others, 1982b; Riddihough, 1982). Raised beaches and elevated littoral sediments on western Vancouver Island and the Queen Charlotte Islands (Fig. 9), which have been dated by the radiocarbon method, provide further evidence for recent uplift of the outer coast (Clague and others, 1982b).

Given the tectonic setting and seismicity of western British Columbia, it is surprising that few active faults have been identified in this region. However, some large mapped faults probably have been active during the Holocene, and fresh linear scarps recently discovered in the Coast Mountains (Eisbacher, 1983) and in the Strait of Georgia (Hamilton and Luternauer, 1983) may have formed during earthquakes. There almost certainly are many small active faults in coastal British Columbia that have not yet been recognized by geologists.

Another indicator of recent and ongoing diastrophism is volcanism. There have been many volcanic eruptions in the Coast Mountains during the Holocene, the most recent less than 150 years ago (Souther, 1977). Most eruptive centers have been the locus of a single pulse of activity during which one or more small pyroclastic cones were built and a small volume of lava erupted to form thin blocky flows. In contrast, several centers are large composite volcanoes that have erupted repeatedly, often explosively, during Quaternary time.

Glaciation and Sea-Level Change

During the Pleistocene, coastal British Columbia was repeatedly enveloped by glaciers that flowed from the Coast and Insular mountains (Clague, 1981). At the climax of each major

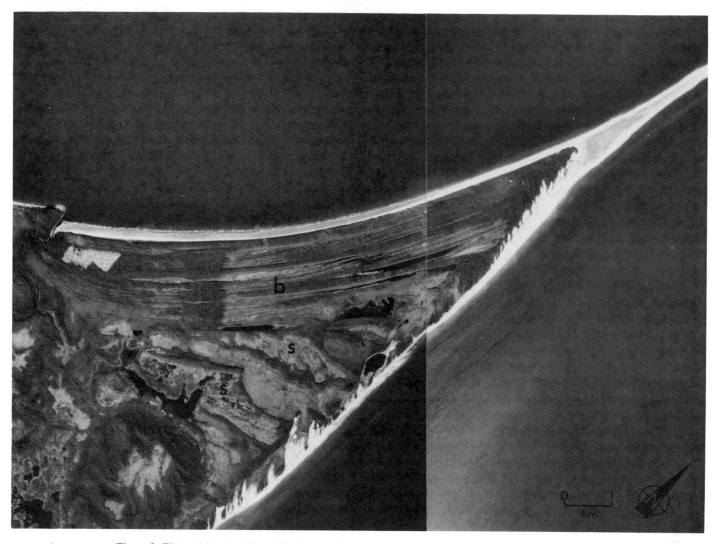

Figure 9. Elevated beach ridge plain (b) and spits (s), northeastern Queen Charlotte Islands (location shown in Fig. 5). These features formed and were uplifted above sea level during middle and late Holocene time. Province of British Columbia photos BC5630-197 and -199.

glaciation, valley glaciers coalesced over parts of the continental shelf and adjacent coastal lowlands to form a continuous sheet of ice, locally more than 2 km thick, representing the western fringe of the Cordilleran ice sheet. At times, the ice sheet was grounded to the outer edge of the continental shelf where it calved into deep water (Luternauer and Murray, 1983). However, it is unlikely that conditions were ever cold enough to support a floating ice shelf. Glaciers on the Queen Charlotte Islands and perhaps on parts of western Vancouver Island were independent of the Cordilleran ice sheet during most or all glaciations (Clague and others, 1982a).

During minor glaciations, the Cordilleran Ice Sheet did not form over coastal British Columbia. Instead, valley and piedmont glaciers occupied mountain valleys and fringing lowlands, leaving most of the continental shelf and coast ice free. A similar pattern prevailed in western British Columbia during the advance phase

of each major glaciation prior to the formation of the Cordilleran Ice Sheet.

Each glacial cycle terminated with climatic amelioration and rapid glacier decay. Deglaciation occurred by downwasting accompanied by complex frontal retreat. Eustatically rising seas destabilized the western margin of the Cordilleran Ice Sheet, causing rapid retreat back to the mouths of fiords and mountain valleys in most regions (Clague, 1981).

Pleistocene glaciers extensively modified the late Tertiary landscape of British Columbia. The Coast and Insular mountains are dominated by erosional glacial landforms, whereas coastal lowlands record both the erosional and depositional effects of glaciers. In high mountains, where summits and ridges stood above névé surfaces, classic alpine landforms were created, including cirques and overdeepened valley heads, horns, and comb ridges (Fig. 6). Most mountain valleys are typical glacial troughs with morphologic details reflecting intense glacial scour as well as local structural and lithologic control (Ryder, 1981). Many valleys are extensions of fiords which reach as far as 150 km inland and attain water depths of up to 755 m (Fig. 7; Peacock, 1935; Pickard, 1961). The major fiords are more or less perpendicular to the coast, but many are linked by structurally controlled, oblique and cross channels. Some fiords probably were eroded along late Tertiary river courses; others, however, appear to have been entirely excavated by glaciers.

The continental shelf between Vancouver Island and the Queen Charlotte Islands is crossed by three major, streamlined U-shaped troughs 20-30 km wide and up to 200 m deep (see Fig. 3 in Luternauer and Murray, 1983). The similarity in pattern and scale of these troughs to the fast-moving ice streams within the glacier apron draining westerly to the Ross Ice Shelf, Antarctica (Rose, 1979), supports the idea that they owe their origin to glacial scour. Juan de Fuca Strait, the Strait of Georgia, and Dixon Entrance also were formed or extensively modified by streams of ice at the western margin of the Cordilleran Ice Sheet. The Strait of Georgia, for example, was excavated from a Cretaceous-Tertiary sedimentary fill by piedmont glaciers flowing southeastward into Washington.

Much of the material eroded by Pleistocene glaciers was transported south into Washington and west to abyssal plains and basins of the northwest Pacific Ocean. Some, however, was deposited on the British Columbia continental shelf and in local mountain valleys and coastal lowlands. Thick Pleistocene sediments underlie much of the area around the Strait of Georgia and part of the lowland bordering Hecate Strait. These sediments were deposited in proglacial and ice-contact fluvial and marine environments, mainly during the advance and recessional phases of the last (late Wisconsin) glaciation (Clague, 1986). Deposits of older glaciations are less common, probably because they have been extensively eroded by late Wisconsin glaciers. However, even in areas where old deposits are present, they are poorly exposed because they are covered by late Wisconsin drift.

Growth and decay of glaciers in British Columbia triggered complex crustal movements that were dominantly isostatic in nature (Mathews and others, 1970; Clague, 1983). The crust was displaced downward during periods of glacier growth. Initially, this depression was localized beneath the major mountain ranges that served as loci of glacier growth. However, as glaciers advanced out of the mountains and into coastal lowlands, the area of crustal subsidence grew larger, and at the climax of each major glaciation, most of the province was displaced downward. The magnitude of isostatic depression at glacial maxima depended primarily on the thickness and extent of the Cordilleran Ice Sheet, the length of time during which it formed, and the structure and composition of the lithosphere. Presumably, isostatic depression was greatest beneath the center of the ice sheet and decreased west of the Coast Mountains and Strait of Georgia toward the continental margin.

Elevated glaciomarine sediments and shoreline features in coastal British Columbia provide evidence of isostatic depression

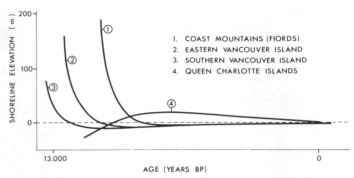

Figure 10. Generalized patterns of late Quaternary sea-level change on the British Columbia coast. Deglaciation and isostatic rebound occurred later in the Coast Mountains than on Vancouver Island. The Queen Charlotte Islands were weakly glaciated during late Wisconsin time and display a pattern of sea-level change different from that of most other areas. Curves are based on data presented in Mathews and others (1970) and Clague and others (1982b).

at the end of the last glaciation. The elevation of the late Wisconsin marine limit varies in relation to distance from former centers of ice accumulation and time of deglaciation. In general, the marine limit is highest (ca. 200 m) on the mainland coast and declines towards the west and southwest (Mathews and others, 1970; Clague, 1975, 1981). However, many mainland fiords were deglaciated after much of the local isostatic rebound had taken place; consequently, the marine limit in these areas is relatively low. Shorelines on the Queen Charlotte Islands at the close of the last glaciation were lower than they are today, indicating that glacio-isostatic depression there was less than the coeval eustatic lowering of sea level (Clague and others, 1982b).

Rapid deglaciation at the end of each glacial cycle triggered isostatic adjustments opposite in direction to those that occurred during periods of glacier growth (Clague, 1983). Material moved laterally in the asthenosphere from extraglacial regions toward the center of the decaying ice sheet. Areas at the periphery of the ice sheet, which were deglaciated earliest, rebounded first (Fig. 10). However, the total amount of uplift in these areas was less than at the center of the sheet where ice thicknesses generally were greater. As deglaciation progressed, the zone of rapid isostatic uplift migrated in step with receding glacier margins. The rate of uplift in each region decreased exponentially with time, and rebound was largely complete within about 5,000 years of deglaciation (Mathews and others, 1970).

Fluvial and Mass-Wasting Processes

Subaerial erosion accompanying and following uplift of the Coast and Insular mountains did much to develop the major valleys of western British Columbia. Fluvial incision during late Tertiary time probably formed V-shaped valleys, which subsequently were widened and deepened by glaciers. Although streams were instrumental in locating the major valleys, they did

SEDIMENT TRANSFER
MECHANISMS

LANDFORMS

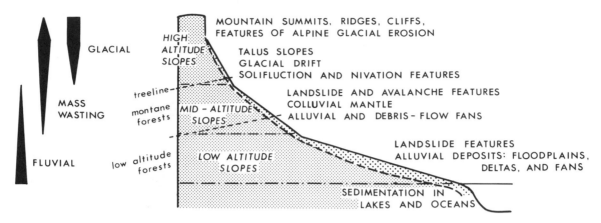

Figure 11. Zonation of major Holocene landforms and geomorphic processes in the Coast and Insular mountains. Modified from Ryder (1981, Fig. 9).

not necessarily flow along them in the same channels or even in the same directions as their modern counterparts. The mechanism of stream diversion can rarely be demonstrated, but rerouting due to blockage by glacier ice likely was very common in British Columbia during the Pleistocene. Damming and diversion of waters by glacial deposits, lava flows, and landslides also may have played a major role in modifying the drainage. On the other hand, piracy probably was very rare because persistently low hydraulic conductivity in most bedrock units would limit the supply of spring water to the head of a potential capturing stream (Mathews, 1987).

Mass-movement processes contributed to the widening of valleys and the denudation of mountains (Fig. 11). Debris flows, rockslides, and rockfalls are important today in moving material from oversteepened slopes to valley floors, and undoubtedly were important throughout the Quaternary on ice-free surfaces. Weathering and soil formation, accompanied by slopewash and solution transport of the alteration products, also contributed to denudation of valley walls and mountainsides (Slaymaker and McPherson, 1977).

Periglacial activity, limited today in British Columbia to high altitudes (Fig. 11), was more widespread during cooler parts of the Quaternary. It may have been instrumental in the formation of the British Columbia strandflat, a low coastal platform abutting inland against the rising western slopes of the Vancouver Island Ranges and the central Coast Mountains (Holland, 1964). Characterized by a small-scale local bedrock roughness, the strandflat extends up to 10 km inland and to about 150 m elevation. Locally, it comprises swarms of small islets and rocky reefs more or less modified by wave action. The origin of this feature is still debated, but accelerated Pleistocene frost weathering in the presence of abundant moisture and poor drainage likely contributed to its development (Holtedahl, 1960); marine planation and glaciation seem to have played a lesser role.

Fluvial and mass-wasting processes are agents of deposition as well as denudation. Some of the most prominent elements of the British Columbia landscape, including cones, fans, deltas, and floodplains, are products of these processes. However, these landforms are susceptible to removal during glaciation or as a result of long-sustained, regional base-level lowering.

Most existing alluvial and colluvial sediments in western British Columbia were deposited over the last 13,000 years during and following late Wisconsin deglaciation. During the early part of this period, thick alluvial and colluvial fills accumulated in many valleys as glacial sediments on slopes were mobilized by running water and gravity, and transported to valley floors. Such "paraglacial sedimentation" (Church and Ryder, 1972) commenced with deglaciation and was largely complete 2,000 to 3,000 years later. In most valleys, a period of fluvial incision followed as streams adjusted to reduced sediment supply and falling base levels. During the late Holocene, many British Columbia streams were relatively stable, although those with significant glacial drainage aggraded their floodplains and continue to carry substantial sediment loads to lakes and fiords (Gilbert, 1975).

Studies of alluvial, deltaic, and colluvial deposits in western British Columbia have provided insights into factors controlling Holocene sedimentation in the region (e.g., Ryder, 1971a, 1971b; Gilbert, 1975; Ashley, 1979; Church, 1983; Ryder and Church, 1986). They also have contributed to recent advances in the fields of process geomorphology and sedimentology. These studies have shown that the patterns and rates of sedimentation in this region are strongly influenced by climate, local geology and topography, and by antecedent events such as Quaternary glaciation.

Marine and Coastal Processes

Plate interactions in the northeast Pacific Ocean, coupled with Pleistocene glaciation and wave and current action, have

Figure 12. Quadra Sand: exposure (top) and schematic depositional model (bottom); location shown in Figure 5. This important stratigraphic unit was deposited at the margins of glaciers as they advanced down the Strait of Georgia and Puget Sound during the early part of the last glaciation.

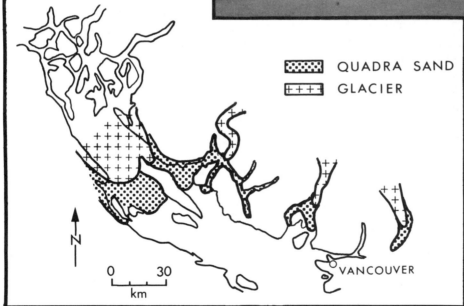

formed the British Columbia coast and continental shelf. Much of the continental shelf is underlain by thick sediments that were deposited in structural basins during late Cretaceous and Tertiary time. Up to 4 km of Upper Cretaceous to Miocene nonmarine and marine sediments underlie the Strait of Georgia and bordering lowlands (Mathews, 1972), and late Tertiary deposits of comparable thickness are present beneath Hecate Strait and Queen Charlotte Sound (Yorath and Chase, 1981). These sediments were derived from local sources to the east and west.

By the end of the Tertiary, the continental shelf had more or less its present areal extent. However, the morphology of the sea floor probably was very different from that of the present, and the coastline was less indented, looking perhaps somewhat like shores bordering the northern Gulf of Alaska today.

During the Pleistocene, the continental shelf was subject at times to erosion by grounded glaciers and at times to widespread

sedimentation. Sedimentation occurred mainly in ice-marginal subaqueous environments, which probably were similar to those fronting present-day tidewater glaciers in Alaska (Powell, 1983). Sediment was released from icebergs and from snouts of grounded glaciers; it also was carried to the sea by supraglacial, subglacial, and proglacial meltwater streams. Most coarse detritus accumulated near the fronts of glaciers; silt and clay, however, were carried in suspension far from glacier termini and rained out on the sea floor to form extensive deposits of mud. In areas of high sediment supply, focused meltwater inputs, and/or relatively stable glacier margins, large sandurs, deltas, and subaqueous fans were constructed. For example, during the early part of the last glaciation, large aprons of outwash sand (Quadra Sand) accumulated at the margins of glaciers in the Strait of Georgia and Puget Sound (Fig. 12; Clague, 1976). As these glaciers advanced southward, the locus of sedimentation gradually shifted; earlier deposited sand aprons were overridden and eroded to contribute to younger aprons farther south. Deposition probably occurred in braided sandur channels and on delta slopes where meltwater streams entered the sea. Sandurs and glaciomarine deltas were also constructed during periods of deglaciation. Deltas dating to the close of the last glaciation have been elevated above sea level and are conspicuous landforms in many coastal areas (Fyles, 1963; Armstrong, 1981; Clague, 1985). The deposits of these deltas commonly are associated with other elevated glaciomarine sediments, including silt and clay deposited from rain-out and turbidity flows, coarse outwash laid down as fans at the mouths of subglacial meltwater streams, and littoral sand and gravel. Presumably, similar sediments also are present below sea level on parts of the continental shelf.

Coastal and offshore erosion and sedimentation were more restricted during interglaciations than during glaciations. Significant interglacial erosion of the coastline has been limited to areas bordered by Pleistocene sediments and nonresistant bedrock. During the Holocene, for example, some shores on southern Vancouver Island and the eastern Queen Charlotte Islands experienced significant retreat due to wave and current attack (Clague and Bornhold, 1980). Most of the British Columbia coast, however, is situated on resistant rocks, and erosion in these areas has been negligible during the Holocene.

At present, energy levels over most of the continental shelf are too high to permit accumulation of thick fine sediments. In such areas, the sea floor is underlain by relict Pleistocene deposits, commonly covered by thin winnowed lags of sandy gravel or boulders (Carter, 1973; Luternauer and Murray, 1983). Thick Holocene sediments are restricted to semiprotected embayments and to deep-water troughs and basins adjacent to terrestrial sediment sources, for example, fiords and parts of the Strait of Georgia. Near the mouths of streams, marine sediments pass into deltaic deposits that may extend considerable distances inland beneath alluvial caps (Fig. 13).

SPATIAL AND TEMPORAL VARIATION OF PROCESSES

Although the geomorphic processes described above have all contributed to the formation of the British Columbia landscape, their efficacy has varied both in space and time. For example, while the Coast and Insular mountains were raised by diastrophism during the late Cenozoic, the adjacent continental shelf subsided. As the mountains rose, they became centers of subaerial erosion and mass wasting; the denudation products were deposited, in part, on the continental shelf.

Glaciation also has had a variable effect on the landscape. Late Tertiary glaciers, if present, were small and confined to mountain ranges; their geomorphic effects thus were limited. Likewise, during Quaternary interglaciations, glaciers in western British Columbia were restricted to the higher parts of the Coast and Insular mountains and did not directly affect lowland areas. On the other hand, during glaciations, ice extended into lowland

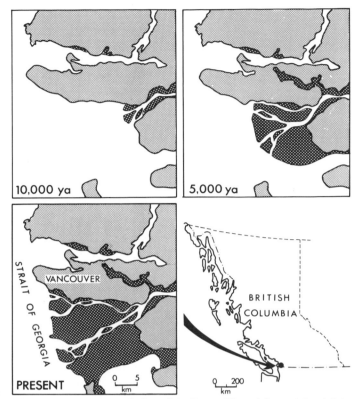

Figure 13. Holocene evolution of the Fraser River delta and floodplain. Dark stipple, Holocene floodplains, fans, and peat bogs; light stipple, preexisting land areas.

valleys and onto the continental shelf, and effected major changes in the landscape.

Erosion of mountainsides and valleys increased at the beginning of each glaciation. The products of glacial erosion and of related periglacial activity were carried downvalley by meltwater streams and deposited in outwash trains, fans, and deltas. Loci of outwash deposition gradually shifted away from the mountains and into coastal lowlands as glaciers advanced. Eventually, these lowlands and much of the continental shelf were enveloped in ice, whereupon they too were subject to glacial erosion. Subsequent deglaciation was accompanied by renewed deposition of stratified drift in glaciofluvial, glaciomarine, and glaciolacustrine settings.

The foregoing indicates that Quaternary sedimentation in western British Columbia was episodic (Clague, 1986). Most preserved Quaternary sediments in this region were deposited during the growth and decay phases of the last two glaciations. Sedimentation was more restricted and occurred at lower rates during interglaciations, and the stratigraphic record of these periods thus is meager. In most places, interglacial stratigraphic units are thin and discontinuous or are absent altogether. Commonly, an interglaciation is recorded only by an unconformity produced when streams incised valley and lowland fills shortly after the end of the preceding glaciation. This happened, for example, at the close of the last glaciation when isostatic rebound outstripped the eustatic sea-level rise on most parts of the coast. In conjunction with a reduction in the supply of sediment to streams, this fall in base level led to incision of deglacial and older sediment fills. Streams now flow, in part, in valleys cut into these fills, and large areas of coastal lowland are relict surfaces, little changed in the last 10,000 to 13,000 years.

LINKAGE OF PROCESSES

Geomorphic processes that have shaped the British Columbia landscape are linked to one another in a complex manner. The main independent element in this geomorphic system is diastrophism produced by North Pacific plate interactions. Diastrophism has created the relief in which the various agents of denudation have operated during the Quaternary. Major mountain ranges created by diastrophic processes were the source area of the Cordilleran Ice Sheet and its satellites, and thus indirectly controlled patterns of ice flow during Pleistocene glaciations.

Glaciation was also affected by other geomorphic processes. The first glaciers in the Coast Mountains occupied valleys eroded entirely by streams. These valleys exerted a strong control on ice-flow patterns during subsequent Pleistocene glaciations, although they became increasingly modified by the glaciers themselves. Sea-level changes, intricately linked with glaciation, also affected the distribution of glacier ice in western British Columbia. Eustatic lowerings of sea level facilitated the advance of glaciers across the continental shelf. However, this effect was offset by isostatic depression of the crust due to local ice loading. Where the isostatic factor was dominant, glaciers were unable to advance far into the relatively warm waters of the northeast Pacific Ocean. Catastrophic retreat of the western margin of the Cordilleran Ice Sheet at the close of the last glaciation, and also probably at the end of earlier glaciations, was likely triggered by a eustatic rise in sea level accompanying decay of the Laurentide and Fennoscandian ice sheets.

One of the most important geomorphic links in western British Columbia is that between glaciation and fluvial processes and mass wasting. As mentioned earlier, sediment production was markedly higher during glaciations than during interglaciations. Most of the increase in sediment production during glacial periods was a direct consequence of increased erosion by ice; some, however, may have resulted from expanded periglacial activity. At any rate, sediment supply to streams increased significantly during the early part of each glaciation, leading to local aggradation of valleys and coastal lowlands. Aggradation on a similar scale also took place during deglacial periods as meltwater from decaying glaciers deposited large amounts of sediment on floodplains and fans, and in lakes and the sea.

During interglaciations, most streams flowed more or less at grade in valleys that were partly incised into older sediment fills. These streams eroded laterally into the fills, and the sediment thus entrained was redeposited downstream on floodplains and in lakes and the sea. Other sources of sediment during interglacia-

tions were alpine glaciers and slope failures. Streams in heavily glacierized alpine basins aggraded their floodplains and carried substantial loads of sediment to lakes and fiords during interglacial periods.

Marine and coastal processes are also intimately linked with glaciation. Thick, heterogeneous glaciomarine sediments were deposited over large parts of the continental shelf during Pleistocene glaciations, whereas relatively fine deltaic and prodeltaic sediments accumulated more locally in fiords and inshore basins adjacent to the mouths of streams during interglaciations (Fig. 13). Eustatically and isostatically induced changes in water depth over the shelf during glaciations altered energy levels, water circulation, and thus patterns of sedimentation, and were accompanied by transgressions and regressions in coastal areas. Transgressions resulted from: (1) a eustatic rise in sea level without compensatory isostatic uplift; (2) isostatic loading by glaciers and sediments; and (3) tectonic subsidence. In contrast, regressions occurred in response to isostatic and tectonic uplift and eustatic lowering of sea level.

SUMMARY

The landscape of British Columbia is a product of several interdependent geologic processes, the most important of which are diastrophism, glaciation, fluvial erosion, and mass wasting. Rates of diastrophism in western British Columbia during the Cenozoic have been high, reflecting the proximity of this region to active lithospheric plate boundaries. Consequently, the landscape is very young—almost none of its elements predate the late Miocene. However, this landscape evolved within a geologic framework that became established in late Mesozoic and early Cenozoic time and that is responsible for both the northwesterly topographic grain of the region and for the distribution of the major mountain ranges and low-lying areas.

Elevation of the Coast and Insular mountains during the late Cenozoic was accompanied by subaerial denudation and episodic glaciation. These processes sculpted the mountains and created the valleys and fiords that are such conspicuous elements of the present landscape. Denudation products were in part deposited on the continental shelf and in part carried to abyssal plains and basins beyond the continental margin.

Erosion and sedimentation in western British Columbia during the Quaternary have been nonuniform both in time and space. Erosion rates were greatest during glaciations, when ice cover was extensive and periglacial activity widespread. Mountain areas were most severely eroded, although at glacial maxima, coastal lowlands and parts of the continental shelf were also scoured by glaciers. During glaciations, large amounts of sediment were deposited in valleys, fiords, coastal lowlands, and on the continental shelf. In contrast, lesser amounts of new sediment were generated during interglaciations; at these times, preexisting terrestrial glacial deposits were eroded by streams and redeposited on floodplains, deltas, fans, and in prodeltaic basins.

Quaternary geomorphic processes operating in western British Columbia are linked. The pattern of Pleistocene glaciation was controlled to a considerable degree by late Tertiary topography, which, in turn, was predetermined by the geologic framework of the region and by diastrophism and denudation. Among other things, episodic glaciation perturbed fluvial systems, induced isostatic deformation of the crust, and triggered sea-level changes. Changes in sea level, in turn, altered patterns of sedimentation and erosion on the continental shelf and along the coast.

GEOMORPHIC PROCESSES IN THE PACIFIC COAST AND MOUNTAIN SYSTEM OF OREGON AND WASHINGTON

Patricia F. McDowell

INTRODUCTION AND SETTING

The Pacific Coast and Mountain System in Oregon and Washington consists of three major regions with different structure and lithology: the Olympic Peninsula, Coast Ranges, and Klamath Mountains (Fig. 14; Baldwin, 1981; McKee, 1972; Tabor, 1975). The Olympic Peninsula has a central fenster exposing metamorphosed, complexly folded and disrupted basalts, and sedimentary rocks that were thrust beneath less disturbed Tertiary sedimentary and volcanic rocks. The underthrust rocks form the Olympic Mountains, with peaks of 2,000 m and above, and are surrounded on the north, east, and south by less disturbed peripheral rocks. Uplift occurred in the late Cenozoic, and a radial drainage pattern, deepened during Pleistocene glaciation, resulted. The southern and western flank of the Olympic Peninsula is a gently sloping, glacio-fluvial piedmont of Pleistocene age (Rau, 1973).

The Klamath Mountains of southwestern Oregon consist of four overlapping thrust sheets of Paleozoic and Mesozoic volcanic and sedimentary rocks, partly metamorphosed, that have been intruded by granitic and ultramafic rocks. The generally west-flowing rivers incised valleys and some deep canyons into the Klamath Mountains during late Cenozoic doming (Mortimer and Coleman, 1985), and present relief is 2,000 m for the region, and 600 to 900 m locally in the deeper canyons.

The Oregon and Washington Coast Ranges are lithologically and structurally simpler than the Olympic Peninsula and Klamath Mountains. Eocene and younger sedimentary and volcanic rocks have been broadly upwarped beginning in the Miocene; upwarping is continuing today. The rugged topography with moderate relief (1,200 m maximum) is largely the product of fluvial incision. Eocene and Oligocene intrusive rocks form the higher peaks. Pleistocene glaciation has occurred extensively in the Olympic Mountains (Tabor and Cady, 1978) and on a few high peaks in the Klamath Mountains (Moring, 1983), but was absent in the Coast Ranges.

INTERIOR ZONE

The interior landscape of the Pacific Coast and Mountain System of Oregon and Washington is a highly dissected one, characterized by deeply incised rivers, steep slopes, narrow sinuous ridgetops, and where not disturbed, heavy forest cover. Slopes of 30 percent or greater are common. In this humid and tectonically active region, lithology, intense weathering, and steep slopes combine to produce a landscape where denudation is dominated by dissolved load removal and mass-movement processes that deliver sediment to stream channels. The denudation rate (estimated to be 35 mm/1,000 years, for Rock Creek watershed in the Oregon Coast Range), is high, and 60 percent of the material leaves as dissolved load (Dietrich and Dunne, 1978). Recent research in the interior zone has focused on (1) the geologic setting and rates of mass-movement processes, (2) transfer of sediment through watersheds, and (3) impacts of timber harvesting on mass movement and sediment yield.

Mass Movement

At least five types of mass-movement processes are widely recognized in mountainous areas of the Pacific Northwest: shallow soil creep, deep creep, slump-earthflow, debris flow (debris avalanche), and debris torrent (Swanston and Swanson, 1976; Dietrich and Dunne, 1978). These types of movements are differentiated on the basis of velocity (Table 1), morphology, and landscape position. Debris flows (rapid, dry to wet movements on hillslopes) are associated mainly with resistant rocks, such as the basalts and well-consolidated sandstones of the Coast Ranges. On weaker rocks, especially sheared rocks of the Klamath Mountains and other rock types that weather to clay, creep and slump-earthflow dominate. Moring (1983), working in the Klamath Mountains, recognized two additional types of mass-movement processes. Large blockglides, with blocks up to 2 km^2 in size, occur in a few locations in ultramafic rocks. A catastrophic debris flow, of Pleistocene or Holocene age, carried sandstone blocks 5-10 m in size a distance of 15 km to form Blossom Bar, a famous rapid on the Rogue River. In the Olympic Mountains (Tabor and Cady, 1978), major landslide deposits occur mainly in the highly sheared sedimentary rocks of the eastern core (particularly along the sides of glacial troughs) and, less commonly, in basalts of the eastern peripheral rocks.

The occurrences and rates of both slow and rapid mass-movement processes are closely related to the soil moisture content (Swanston and Swanson, 1976). Slow mass-movement processes probably are related to longer-term moisture variations. Slump-earthflow movements are thought to have a long (thousands of years) history of somewhat episodic movement. Measurements of deep creep in the Pacific Northwest indicate that, while some sites show a distinct seasonal variation from rapid creep during the winter and spring to very slow creep during the summer, other sites show an even rate of movement year-round. Rapid mass-movement events often are triggered during in-

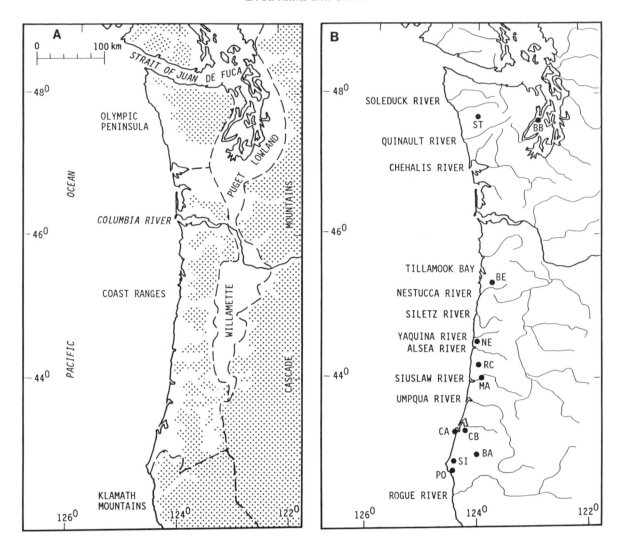

Figure 14. A: Geomorphic regions of western Oregon and Washington. Region boundaries are shown by dashed line. Shading shows areas above 300 m elevation. B: Drainage pattern of western Oregon and Washington, showing locations mentioned in text. BA, Baker Creek; BB, Big Beef Creek; BE, Bear Creek; CA, Cape Arago; CB, Coos Bay; MA, Mapleton; NE, Newport; PO, Port Orford; RC, Rock Creek; SI, Sixes River; ST, Stequaleho Creek.

tense precipitation events on saturated soil during winter. Debris flows are frequent after intense storms. Debris flows often originate from small, colluvium-filled hollows in bedrock on hillslopes (Dietrich and Dunne, 1978). The hollows, 3 m or more deep, periodically empty of soil material through debris flow, then refill over a period of 1,000 to 10,000 years. As filling progresses, the probability of slope failure increases, and failure may be initiated by a major winter rainstorm. Debris torrents (rapid movements of water, debris, and organic material down stream channels) are also associated with major storms. They are caused by debris flows that reach the channel, or by the mobilization of existing debris accumulations in the channel. Their recurrence interval is estimated to be 100 to 200 years in the western Cascades, and rates are probably similar in the Coast Ranges. Debris flows

involve several 1,000 m³ of material in each event, and debris torrents probably reach 10,000 m³ in size (Swanston and Swanson, 1976). Both colluvium-filled hollows and earthflows have been the subjects of much recent research in northern California (Kelsey, this chapter).

Sediment Transfer, Storage, and Discharge Processes

Sediment is transferred from hillslopes to valley floors and stream channels by a variety of processes, some operating more or less continuously, and others (e.g., debris flows) that operate episodically under the influence of climatic events. Over much of the Oregon and Washington Coast Ranges, the dominant sediment-transfer processes on hillslopes are probably creep into bedrock hollows and then creep or debris flow from the hollows. Slump-earthflows may be a major source of channel sediment in areas with susceptible rock types (Fredriksen and Harr, 1979), such as the Klamath Mountains and parts of the Olympic Mountains. Soil erosion by surface runoff and raindrop splash, on the other hand, is low because of protection by the vegetation and litter layer, and the contribution of organic matter and clays to soil aggregation (Fredricksen and Harr, 1979; Dietrich and others, 1982).

In forested areas, large organic debris (LOD) fragments (tree tops, limbs, root wads >10 cm diameter) are introduced into stream channels by natural processes, including mass movement, wind throw, and stream bank undercutting. LOD accumulations, also known as debris dams or log steps, often form in stream channels, and an individual LOD accumulation may remain relatively stable in the channel for 100 years or longer (Swanson and others, 1976; Swanson and Lienkaemper, 1978; Keller and Swanson, 1979). In the Oregon Coast Range, they are best developed in small streams of second to fifth order (Marston, 1982). LOD accumulations form a stepped stream profile, with sediment accumulation on the upstream side and scouring of a pool on the downstream side. These accumulations have a significant effect on sediment storage and movement through the stream system by acting as buffers on sediment yield. Under natural conditions, a series of LOD accumulations within a stream system will even out the release of sediment that may be introduced into the channel in pulses by mass movement (Swanson and Lienkaemper, 1978). Despite their morphological and sediment-routing impacts, Marston (1982) found that LOD accumulations do not have a statistically significant effect on stream equilibrium conditions.

The environmental effects of LOD accumulations are both positive (increased diversity of channel morphology and aquatic habitat) and negative (blockage of anadramous fish passage; Swanson and Lienkaemper, 1978; Beschta, 1979). Although LOD accumulations may be stable for long periods of time, they can be flushed downstream in violent debris torrents, which scour and destabilize the channel, resulting in subsequent higher sediment yields (Swanson and others, 1976; Swanson and Lienkaemper, 1978). Natural debris torrents usually are triggered by debris avalanches on hillslopes above the stream channel, but their erosive effectiveness is increased when they enter the channel and entrain more debris from LOD accumulations.

Sediment transport (bedload and suspended load) within stream channels also is highly variable, both temporally and spatially, and these processes have been a major focus of research in Oregon and Washington. Most sediment discharge occurs during flow peaks (of several days duration) during the winter rainy season. Short-term (within channel bed armor) and long-term (behind LOD accumulations) sediment storage areas within the channel exert a significant influence on short-term sediment discharge. Human impacts on stream systems cannot be assessed without considering the natural variability of sediment transport processes.

TABLE 1. RATES OF MASS-MOVEMENT PROCESSES IN OREGON AND WASHINGTON

Location	Rate of movement or discharge
Shallow Soil Creep	
Rock Creek, central Oregon Coast Range[1]	2.5 mm/yr (est.)
Deep Creep	
Baker Creek, southern Oregon Coast Range	10.4 mm/yr surface / 10.7 mm/yr subsurface maximum
Bear Creek, southern Oregon Coast Range	14.9 mm/yr surface / 11.7 mm/yr subsurface maximum
Debris Flow	
Stequaleho Creek, Olympic Peninsula	71.8 m³/km²/yr
Mapleton District, central Oregon Coast Range[2]	28 m³/km²/yr
Other Pacific Northwest sites	10-50 m³/km²/yr

[1]From Dietrich and Dunne (1978).

[2]From Swanson and others (1981); all other values from Swanston and Swanson (1976).

Bedload transport does not have a simple relationship with stream discharge or velocity, and Milhous argued that armored gravel beds influence both bedload and suspended transport (Milhous and Klingeman, 1973; Milhous, 1982). Early in a high-flow event, flow is incompetent to entrain the coarse gravel of the armor layer, and the armor limits entrainment of finer sediment within the stream bed. As streamflow increases to the hydrograph peak, the armor layer is mobilized, releasing large amounts of coarse and fine bed material stored beneath it. During the falling limb, first the armor particles come to rest, while transport of sand and fine gravel continues. For equivalent levels of discharge, therefore, bedload transport is lower during the rising limb, before the armor layer has been mobilized, than during the falling limb.

A related two-phase model for bedload transport proposed by Jackson and Beschta (1982) focuses on spatial variation of channel morphology, bed sediment, and transport processes in pool-and-riffle streams. During their phase II bedload transport, riffle armor is mobilized, bedload transport rates are high, and riffle scour and fill occurs. Bedload transport at a riffle is unsteady, and bedload transport among riffles is nonuniform, for several possible reasons. First, irregular channel geometry in the downstream direction causes succeeding riffles to respond differently to changes in discharge during a storm event. An increase in discharge may significantly increase velocity and therefore bedload transport out of one riffle, while having little effect on bedload transport out of the next riffle downstream. The result would be aggradation on the downstream riffle. With unsteady flow during a storm event, it is unlikely that equilibrium adjustment could be established among a series of riffles. Second, sediment inputs into the channel during a storm vary spatially and temporally, resulting in local perturbations in the sediment transport balance.

The armor layer also affects suspended sediment transport (Milhous and Klingeman, 1973; Milhous, 1982). A significant proportion of the annual suspended sediment load, as much as 30 percent, may be derived from fines stored in the channel bed (Paustian and Beschta, 1979). Voids between armor gravels serve as a reservoir for suspended sediment. Suspended sediment loads will vary for the same discharge level, depending on whether this reservoir is full and actively supplying suspended sediment, or empty. If a series of moderate to high flows occurs in succession, suspended sediment discharge will be lower later in the series, because the bed reservoir is depleted of fine sediment. The first high flow to follow the summer low period will have unusually high suspended sediment load because the reservoir is full (Milhous and Klingeman, 1973; Beschta, 1978; Paustian and Beschta, 1979). This suspended sediment reservoir acts as a buffer on episodic inputs of sediment by mass movement. Modelling studies indicate that supply-based models of suspended sediment concentration perform better than simple rating curve models (VanSickle and Beschta, 1983), adding further support to the conceptual models of Milhous, and Paustian and Beschta.

Over a longer interval, Brown and Krygier (1971) noted that in a small Oregon Coast Range watershed, the sediment concentration–water discharge relationship changed after the large flood of December 1964 to January 1965. During the flood, sediment from the hillslopes was deposited in the channel and served as a sediment source for several following years. In the Klamath Mountains, some stream channels aggraded and widened after the 1964 flood, and at some sites aggradation was followed by degradation (Lisle, 1981). Channel adjustment was asynchronous between upstream and downstream reaches. Adjustment to this large flood was still occurring 5-15 years after the flood.

Dietrich and Dunne (1978) constructed a sediment budget model (Fig. 15) and estimated some of the components for Rock Creek watershed (16 km²) in the Oregon Coast Range. The equilibrium through-flow of solid sediment is 31.2 tons/km²/year. They estimated that this material is introduced into the channel through shallow soil creep (45 percent) and creep from colluvium-filled hollows (55 percent), but the estimate of creep discharge from hollows is probably too large (Dietrich and others, 1982). They also identified several longer-term sediment storage areas and estimated mean sediment residence times (Table 2).

Innovative research in western Oregon and Washington has shown that the watershed sediment transport system consists of many different compartments which take in, hold, and release sediment at different rates and under different conditions. Many of the processes are threshold-controlled. The result is complex temporal and spatial patterns of sediment movement. Geomorphic processes in this heavily forested region are strongly influenced by biological processes; geomorphic processes also influence forest and aquatic ecosystem dynamics (Swanson and Lienkaemper, 1982).

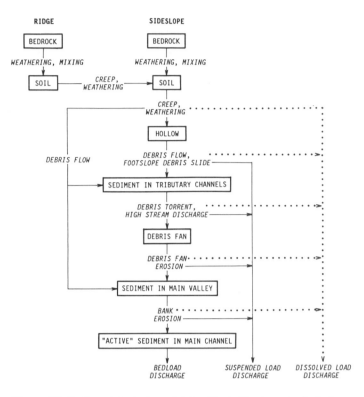

Figure 15. Sediment budget model for Rock Creek watershed, from Dietrich and Dunne (1978). Solid lines represent flow of sediments, and dotted lines represent flow of dissolved material. Rectangles show storage reservoirs. Processes are shown in italics.

Human Impacts on Sediment Yield

Road building associated with timber harvesting is the greatest source of increased mass movement, especially debris avalanches, and hence increased sediment yield relative to natural rates. Other aspects of timber harvesting also tend to increase sediment yield (Swanston and Swanson, 1976; Fredriksen and Harr, 1979). Several years after logging, mass movements (including creep and debris avalanches) increase on clear-cut areas, due to decay of roots and loss of soil shear strength (Swanston and Swanson, 1976; Beschta, 1978). For example, after clear-cutting, debris avalanche rates in the Pacific Northwest can be 2 to 12 times greater than natural rates, and after road building, they can be 25 to more than 300 times greater than natural rates. Slump-earthflow features may be reactivated by road construction and stream channel erosion resulting from logging (Swanston and Swanson, 1976). Slash burning removes organic cover from soil, increasing direct soil erosion, particularly if the burn is very hot. Heavily used forest roads can also produce a significant

TABLE 2. MEAN SEDIMENT RESIDENCE TIMES IN WATERSHED
STORAGE RESERVOIRS*

Storage Reservoir	Residence Time (years)
Soil on Hillslopes	20,000
Tributaries	20 to >200
Debris fans	30 to 90
Main channel valley floor	5000
Main channel active sediment	600
Main channel gravel bars	30

*From Dietrich and Dunne (1978).

amount of suspended sediment (Reid and Dunne, 1984). Increased sediment yields due to logging and road building activities are initially high, and decline to levels near normal after about 5 to 6 years (Brown and Krygier, 1971; Beschta, 1978), but delayed effects of mass movement may prolong the recovery. In the Big Beef watershed in the Puget Lowland, forest clearing increased suspended sediment input to the channel by a factor of ten and increased bedload input by a factor of six. Much of this input is temporarily stored in the channel, and it will take 20 to 40 years to remove it (Madej, 1982).

Timber harvesting also affects soil erosion and mass movement through its effects on the water budget of a watershed. After harvesting, transpiration is reduced and soil moisture remains high for a longer portion of the year than under forest cover. A greater annual rate of deep creep and slump-earthflow is thought to result (Swanston and Swanson, 1976). In addition, harvesting may increase peak flows by causing compaction and bare soils.

The effects of logging and road building on large organic debris in streams is complex. These activities may increase LOD loading in channels by (1) increasing mass-movement processes, and (2) directly introducing organic debris into streams (Swanson and others, 1976; Swanson and Lienkaemper, 1978). Perhaps the most damaging effect of logging and road building is to increase the frequency of debris torrents, which may destroy local aquatic habitats. Stream clearing of large organic debris accumulations was undertaken in the 1970s to increase fish passage. This practice has negative environmental effects in some cases. Beschta (1979) found significant channel scouring and increase in suspended sediment load after LOD removal in a third-order stream in the central Oregon Coast Range, but these effects were relatively short lived. Current practice in logging is to maintain, but not add to, existing stable LOD accumulations in the channel (Swanson and others, 1976).

Long-Term Landscape Development

The Quaternary stratigraphy of the interior area and the nature of long-term geomorphic changes there are very poorly known. Limited Quaternary mapping has been done by the Oregon Department of Geology and Mineral Industries (e.g.,

Baldwin and others, 1973; Schlicker and Deacon, 1974; Ramp, 1975) and the U.S. Geological Survey (Tabor and Cady, 1978; Moring, 1983). Diller (1914) recognized scattered bodies of weathered fluvial gravels on ridgetops in the Klamath Mountains, and he believed that they represented remnants of a Tertiary peneplain. The idea of the Klamath Mountains peneplain has largely been abandoned (Warhaftig and Birman, 1965), but some of these gravel deposits may represent Tertiary stream deposits, unrelated to the present drainage patterns, in a tectonically active landscape like the present one (Moring, 1983). Numerous examples of stream capture in the Coast Range (Baldwin, 1981) and the Klamath Mountains (Moring, 1983) attest to late Tertiary and Quaternary tectonic activity in this region, particularly eastward tilting (Reilinger and Adams, 1982).

Moring (1983) mapped four levels of terrace gravels, ranging up to 100 m above the present stream channel, along the Rogue, Applegate, and Illinois rivers and tributaries in the Klamath Mountains. The two upper terraces are mid-Pleistocene or older, and relict patterned ground is present on some of the surfaces (Parsons and Herriman, 1976). They have been tilted down to the east, and they merge upstream with the present stream profile. The lowest terrace is probably of late Wisconsin age. High fluvial terraces have also been identified within the valleys of some Coast Range rivers, but the number of terraces and their ages are not known (Schlicker and others, 1972; Schlicker and Deacon, 1974; Beaulieu and Hughes, 1975). These terraces are probably related to tectonic uplift and eustatic sea level changes during the Pleistocene.

COASTAL ZONE

The Pacific coastline of Washington and Oregon generally consists of straight or gently curved segments with sandy beaches, interrupted by rocky headlands. Along much of the coastline, straight segments dominate, punctuated by small headlands; however, in the Klamath Mountains and along the northwestern part of the Olympic Peninsula, the coastline is rocky and irregular, dominated by headlands and small coves formed in weaker rocks. The headlands are formed by harder Tertiary volcanic and sedimentary rocks, or metamorphic rocks of the Klamath Mountains. The straight segments are backed by Tertiary sedimentary rocks, Quaternary marine terrace deposits, or Holocene and historical sand dunes. A number of embayments, fronted by recent sand spits, occur within the straight segments; the embayments are occupied by estuaries and tidal flats today and are formed in deep Pleistocene river valleys.

The coastline of the Strait of Juan de Fuca is similar to the Pacific coastline, but the Puget Sound coastline offers a striking contrast. On the northeast and east sides of the Olympic Peninsula, Pleistocene glacial deposits form most of the coastline, the coastal outline has been partly shaped by glacial erosion, and wave erosion is less effective than on the more exposed coasts. Deltas, for example, are well developed in this area, although they are absent on the Pacific coastline except within estuaries.

The geomorphology of the coastal zone is shaped by tectonic uplift and warping, the effects of eustatic sea level changes, wave and wind action, and fluvial and tidal processes in estuaries. The nature of sediment movement into, within, and out of the coastal zone is less well known than in the upland zone, but available evidence indicates that there are some parallels with the upland zone. Coastal sediment is stored in a wide variety of landforms, including beaches, offshore bars, estuaries, marine terraces, and eolian dunes, in which sediment has different residence times.

Controls of Coastal Processes

During the Holocene, relative sea level along the Pacific coast of Washington and Oregon and along Puget Sound has been influenced by eustatic sea-level rise, spatially variable rates of tectonic uplift, and, in the Puget Sound area, postglacial isostatic rebound. Eustatic sea level at the Late Wisconsin glacial maximum was at 120 ± 60 m below its present level (Bloom, 1983). Regional sea level in the Pacific Northwest, independent of local tectonic influences, probably rose rapidly to about -10 m or less at 4,000 yr B.P. (7 to 10 m/ka), and rose slowly after 4,000 yr B.P. (<2.5 m/ka; Clark and Lingle, 1979). Superimposed on this change is a tectonic uplift rate of 2 m/ka in northern Oregon (Ando and Balasz, 1979) and 0.7 to 0.4 m/ka at Cape Arago, Oregon (Adams, 1984). These data indicate that, in the Pacific Northwest, the rate of eustatic sea-level rise outpaced tectonic uplift during the late Wisconsin and early Holocene, and that during the later Holocene, sea-level rise and tectonic uplift have been more nearly equal. These rates are supported by evidence at the Alsea River estuary on the central Oregon coast for rapid, deep-water sedimentation during the early Holocene, and slower, shallow-water sedimentation since 5,000 yr B.P. (see Table 3; Peterson and others, 1984a).

The most important source of sediment supply to the present beach zone is probably erosion of sea cliffs, especially cliffs cut into easily erodible Pleistocene marine terrace deposits and Holocene sand dunes (Komar, 1980). Most Oregon beaches are pocket beaches that are closed sediment cells, approximately in equilibrium and not receiving any significant amount of fluvial input. Several researchers have suggested or assumed that the heavy winter sediment load of rivers is the major beach source, as is the case in California (see Griggs and Hein, 1980), but recent studies have discouraged this viewpoint (Komar, 1983). Rivers empty into the littoral zone of Washington and Oregon through estuaries, which generally are effective sediment traps. The Columbia River, an exception, is a significant source of beach sand to the broad prograding beaches immediately north and south of it.

For most coastal rivers, estuary morphology determines sediment trapping efficiency and fluvial sediment supply to the beach. Peterson and others (1984b) studied six estuaries in Oregon and Washington and developed an estuary hydrographic ratio, an indicator of the amount of tidal inflow relative to fluvial

TABLE 3. SEDIMENTATION HISTORY OF ALSEA ESTUARY, OREGON*

Time (yr. B.P.)	Character of Sedimentation	Sedimentation rate
>10,000	High-energy fluvial	?
10,000-7,500	Low-energy, deep fresh water	0.4-0.7 cm/yr
7,500-5,000	Deep estuary	1.1 cm/yr
5,000-present	Shallow estuary	0.21 cm/yr

*After Peterson and others (1984a).

discharge into the estuary. Estuaries with a large hydrographic ratio (large estuary area with relatively small river inflow) have about 50 percent beach sand on their beds and are effective traps of both river and beach sand. Estuaries with a low hydrographic ratio, such as the Salmon, Siuslaw, Siletz, and Alsea river estuaries in Oregon, have about 30 percent or less beach sand, and they transport river silt and clay (and possibly some sand) to the littoral zone.

The Pacific coast of Oregon and Washington is a high-energy coast which has a strongly seasonal pattern of wave and wind orientation and strength. Summer winds are northerly to northwesterly. Winter winds are stronger and are southerly to southwesterly. Wave data from Newport, Oregon (Komar and others, 1976), indicate that the most destructive waves occur when, during the winter, a north Pacific storm passes over the coast, creating strong winds and locally generated wind waves that combine with swell waves. Longshore drift is generally to the north during the winter and to the south during summer. The Pacific coast of Oregon and Washington experiences a mixed tidal regime, with a moderate tidal range of about 4 m at spring tides and about 2 m at neap tides (Fox and Davis, 1976).

Coastal Landforms and Processes

More is known about the morphology, sedimentology, and dynamics of beaches in Oregon than in Washington, but generalizations based on Oregon beaches probably apply to beaches on the Pacific coast of Washington. Based on a study of 30 Oregon beaches, Dicken (1961) found that beach gradients vary from about 0.01 to 0.10, and fine-sand beaches have gentler slopes than coarser beaches. Berms are present on many beaches, although not universal, and slope on the berm face is as high as 0.33. On some beaches, the beach face consists of cobbles or a bedrock platform during the winter, but a layer of sand is deposited over the cobbles or bedrock during the summer.

The seasonal differences in wind and wave regimes are reflected in the morphology, slope, and width of beaches (Clifton and others, 1971; Komar and others, 1976; Fox and Davis, 1978;

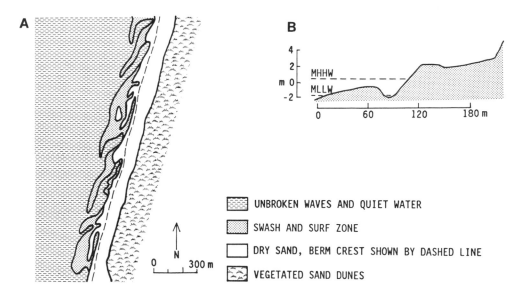

Figure 16. A: Map of oblique bar-trough systems, southern Oregon coast, summer condition. B: Profile of beach and oblique bar-trough system. MHHW, water level at mean high high water (highest of two daily high tides); MLLW, water level at mean low low water (lowest of two daily low tides). Both figures are based on Hunter and others (1979).

Hunter and others, 1979). During November through January, sand from the beach is moved offshore, resulting in the narrow storm beach typical of winter. Sand is returned to the beach face, beginning in the spring, and during the summer, beaches are characterized by a wide berm with a steep beach face.

The nearshore topography of beaches in this region is variable. Many Oregon beaches, particularly those on long, straight, sandy segments of the shoreline, have one or more submarine bars and troughs in the nearshore zone. On other beaches, nearshore bars are either entirely absent, or present only in winter, (Clifton and others, 1971). These bars can be considered temporary, renewable sediment-storage areas. Many Oregon beach faces are cusped, the result of erosion by rip currents (e.g., Komar and Rea, 1976; Aguilar-Tunon and Komar, 1978; Komar, 1983).

Several studies have focused on the morphology and dynamics of nearshore bars on Oregon beaches (Fox and Davis, 1976, 1978; Aguilar-Tunon and Komar, 1978; Hunter and others, 1979). Hunter and others (1979) described nearshore features, which they termed oblique bar-rip channel systems (Fig. 16). Largely formed by waves at higher tide levels, the bars and the separating troughs are oriented oblique to the beach face. The bar-trough unit migrates downdrift at about 0.5 to 1.0 m per day.

Although there is an overall seasonal pattern of onshore movement of sand onto the beach in the summer and offshore movement of sand into nearshore bars during the winter, at shorter time scales sand transport is more variable. Fox and Davis (1978) found periods of beach erosion in summer and periods of

net deposition on the beach face in winter. They considered the formation of bars and localized deposition during nonstorm periods in the winter to be evidence of partial beach recovery between winter storms. Seasonal variation in storm frequency, rather than wind and wave regimes, may thus be the ultimate control of beach profile morphology. As in the seasonal depletion of fine sediment from stream channel beds, high magnitude events early in the winter season are responsible for more of the seasonal geomorphic adjustment than equivalent events later in the winter season.

Rocky headlands, reefs, and sea stacks are effective barriers to longshore drift, and they divide the coastline into a series of littoral cells of variable length, each dependent on local sea cliffs (and in a few cases rivers) for sediment supply. Longshore drift presently is to the north in winter and to the south in summer. Most researchers have concluded that the net drift is approximately zero. Both northern and southern orientations are seen in sand spits located at the mouths of rivers and bays. River-mouth jetties effectively divide a littoral cell into two subcells, each with its own seasonally reversing drift. Lizarraga-Arciniega and Komar (1975) found that, at nine jetty systems on the Oregon coast, sand accretion and shore progradation occurs on both the north and south sides. Heavy-mineral composition of sands on the Oregon continental shelf indicates that over the last 18,000 years, however, net littoral drift has been to the north (Scheidegger and others, 1971).

Because they are composed of loose beach and dune sand, spits are highly susceptible to erosion, and they provide some dramatic examples of rapid erosion and sudden morphologic changes. Spits on the Oregon and Washington coasts typically are located at bay mouths and consist of a beach and a prominent foredune, with a deflation plain or smaller eolian dunes on the bay side. Erosion is caused by (1) winter narrowing of the beach profile, which puts the foredune within reach of wave attack, (2) the occurrence of storms producing high waves after the beach has narrowed, and (3) erosion of embayments on the beach by stationary rip currents, which allow waves to reach the foredune with less dissipation of energy (Terich and Komar, 1974).

Bayocean Spit, Tillamook County, Oregon, is a famous case of spit erosion (Dicken, 1961; Terich and Komar, 1974). The seasonally reversing littoral drift in the area of the spit was interrupted by jetty construction. The relatively short segment of the littoral cell south of the jetty was starved for sand during the summer, its normal time of replenishment. Severe erosion by winter storms occurred in 1932 to 1952, and the spit was breached in 1952.

In addition to the classic example of Bayocean Spit, there has been severe erosion of Siletz and Nestucca spits, Oregon, which are not influenced by jetties (Komar and Rea, 1976; Komar and McKinney, 1977; Komar, 1978, 1983). Foredune retreat of up to 30 m has resulted from a single storm (Komar and Rea, 1976). At Siletz and Nestucca spits, erosion exposed sawed drift logs buried several meters below the former surface of the spit, demonstrating the cyclic nature of severe spit erosion and breaching. The spits apparently undergo cycles of erosion and deposition, each 10 to 15 years long (Komar and Rea, 1976). A typical cycle involves (1) erosion of a vertical scarp on the seaward side of the foredune by wave action during one or more severe winter storms, (2) deposition of drift logs at the base of the scarp by poststorm high tides, (3) accumulation of sand around the logs by wind and wave action, (4) stabilization of new sand by beach grass, and (5) upward growth of the foredune over the logs. This process is strongly influenced by two factors associated with European-American settlement of Oregon since 1850: the introduction of European beach grass which holds sand on the foredune, and drift logs largely resulting from logging operations. Morphodynamics of coastal spits must have been significantly different during the prehistoric period.

Estuaries along the Pacific coast of Washington and Oregon are found in the presently drowned valleys of rivers that were deeply incised during glacial sea level minima, and in embayments presently closed off by sand spits. Narrow estuaries, such as that of the Rogue River in the Klamath Mountains, occur where rivers are incised in resistant metamorphic and igneous rocks and narrow, deep river valleys formed at the sea level minimum. Where rivers incised in less resistant sedimentary rocks, such as the Umpqua, Alsea, and Yaquina rivers, somewhat broader estuaries are found today, filling meandering river valleys. The head of tide is 40 or more river kilometers inland on major Coast Range rivers in Oregon, but only a few kilometers inland on the Rogue River and other rivers in the Klamath Mountains.

During the Holocene, estuaries along the Pacific coast have experienced varying water depths, sedimentation rates, and levels of salinity, due to the combined influences of eustatic sea level rise, tectonic uplift, and infilling of the incised river valleys. During the latest Wisconsin and early Holocene, estuaries were rapidly drowned as sea level rose and advanced eastward toward the present beach. During the late Holocene, estuaries have become shallower as sediment accumulation progressed and sea-level rise slowed or stabilized. Sediment-trapping efficiency of the estuaries was probably greatest during the early period. Peterson and others (1984a) identified four stages of sedimentation of the Holocene fill of Alsea estuary (Table 3). The spit that encloses the mouth of Alsea Bay was not established in its present position until 2,000 yr B.P. Tidal marshes within the bay have also become established only very recently, as evidenced by the absence of these deposits in cores from the estuary. In Tillamook Bay, Oregon, Glenn (1978) found a similar pattern of sedimentation.

Broad estuaries of the Pacific coast are efficient traps of both fluvial and beach sediment, although they discharge some suspended sediment to the littoral zone. At the Alsea River, Oregon, beach sand presently is carried 3 to 4 km into the estuary from the mouth, and river sand is transported down the estuary to within 1 km of the mouth during the winter high-flow season (Peterson and others, 1982). Similar patterns have been found for five other estuaries in Oregon and Washington (Peterson and others, 1984b). The narrower and steeper estuary of the Sixes River acts as a sediment sink for coastal and fluvial sediment during the summer, but is a sediment source for the coastal zone during high river discharge periods (Boggs and Jones, 1976).

Compared to the smaller estuaries on the Oregon and Washington coastline, the Columbia River estuary is much larger and more complex in its circulation and sediment transport (Fox and others, 1984). Near Portland, the Columbia River discharges an average of 13,000,000 tons of sediment per year, and probably most of this reaches the estuary. Sediment also enters the estuary from the ocean. The estuary is a sink for fluvial bedload sediment, but fluvial suspended sediment is discharged to the coastal zone, forming prograding beaches to the north and south. The floor of the estuary has well-developed, medium and large (up to 3 m high and up to 100 m in wavelength) subaqueous sand dune bedforms. These are generally oriented down-river in the upstream (eastern) part of the estuary, and upriver in deep channels in the central part of the estuary. The upstream (eastern) end of the central section is an area of net sediment deposition. In the downstream (western) section of the estuary, they reverse direction with the tidal cycle. Landforms of the Columbia River estuary include (1) a pattern of braided channels, commonly 7 to 12 m deep, (2) submerged sandbars, (3) tidal sand flats, both vegetated and unvegetated, and (4) vegetated islands.

Coastline Retreat

Coastline retrogradation or progradation during the Holocene has been controlled by the relative rates of (1) eustatic sea level rise, (2) tectonic uplift, and (3) sediment supply to the littoral zone. Because tectonic uplift is variable along the coast,

TABLE 4. GEOLOGIC MATERIALS, COASTLINE EROSION RATES AND PROCESSES, OREGON AND WASHINGTON

Coastline Lithology	Average Erosion Rate[1]	Processes
Pre-Tertiary bedrock	<5 cm/yr	Blockfalls; some slumps and earthflows; initiated by wave erosion.
Sheared pre-Tertiary bedrock[2]	5-30 cm/yr; highly episodic	Deep-seated earthflows reactivated by wave erosion at toe.
Tertiary sandstone	2-60 cm/yr[3]	Rockfalls, debris slides, erosion along bedding planes and joints, sea cave formation; if poorly consolidated, slumps, flows, debris shifts.
Tertiary siltstone	12 cm/yr[4]	Slumps, flows, shifts.
Tertiary basalt	very slow	Rock falls, debris slides.
Tertiary tectonic melange[5]	1 m/yr	Flows and slides.
Quaternary marine terrace deposits	15->220 cm/y; episodic	Direct wave erosion, slumps, shifts, slides on basal contact.
Holocene dunes	0->200 cm/yr; episodic or cyclic[6]	Direct wave erosion, wave overwash, slumping.

[1]Estimated average rates for the last 50-100 years, based mainly on historical maps and air photos. Short-term rates may be higher. Data from Beaulieu and Hughes (1975, 1976), North and Byrne (1965), Schlicker and others (1973), Komar and Rea (1976), and Byrne (1964).

[2]Occurs in Klamath Mountains, Curry County.

[3]Rau (1980) estimated average erosion rate for Tertiary marine rocks on the Olympic Peninsula at 91 cm/yr.

[4]Based on a single estimate (Beaulieu and Hughes, 1975). Range of rates is probably slightly higher than the range for sandstones.

[5]Hoh assemblage, southern Olympic Peninsula (Rau, 1973).

[6]Based on maximum short-term rates (at Siletz Spit) prorated over the length of a typical erosional cycle of 15 years (Komar and Rea, 1976).

retreat and stability are both found today. Where sediment supply from the uplands is high enough, progradation may be occurring. Cooper (1958) argued that stable, vegetated sea cliffs cut into late Holocene dunes indicated a slight regression of sea level and end of coastal retreat. Dicken (1961) in his inventory of erosion along the Oregon coast found sea cliffs stable at many sites. Rapid beach erosion rates occur where the back cliff experiences mass movement in Quaternary marine terrace deposits or weak Tertiary sedimentary rocks, and there is no major supply of sediment to the beach. Most of the coastline is stable, or possibly undergoing slow retrogradation. One exception to the general pattern of stability or slow retrogradation is a 70-km stretch of aggrading coastline centered on the mouth of the Columbia River (Cooper, 1958; Terich and Schwartz, 1981). A number of workers have investigated the rates of coastal erosion, using historic maps and air photos, and their results are summarized in Table 4.

Pleistocene Marine Terraces

Multiple elevated marine terraces, up to 500 m above present sea level, occur at many places along the Oregon and south-

ern Washington coast. Correlations of individual terraces along the coast are not well established, because terraces are discontinuous and differentially warped (Baldwin, 1981). Their elevations are largely controlled by the local rate and duration of tectonic uplift, and few radiometric dates are available. Adams (1984) summarized evidence for the ages of the terraces and demonstrated that they have been tilted down to the east as a result of subduction of the Juan de Fuca plate off the Oregon coast. The terraces are best developed and most numerous between Coos Bay and Port Orford, Oregon, where they have developed on sedimentary rocks of the southern Coast Range and Klamath Mountains. They occur as small discontinuous surfaces, however, from the southern Oregon boundary to the northwestern side of the Olympic Peninsula.

The terraces consist of a wave-cut platform topped by 3-15 m (locally up to 60 m) of sediment. The sediment includes beach sands and gravels, finer nearshore and offshore sediment, estuary deposits, old eolian dunes and colluvial deposits. Relict sea stacks are present on some of the terraces (Beaulieu and Hughes, 1976). Griggs (1945) discussed older sea cliffs buried by sediments of a younger marine terrace, presumably due to a eustatic sea level rise. Thus, the terrace deposits are stratigraphically complex, and some of the age assignments made by Adams (1984) may not be correct.

The marine terraces are relict landforms from interglacial high sea-level stands. Kennedy and others (1982) showed that the lowest terraces on the southern and central Oregon coast usually date from the last interglacial complex, based on aminostratigraphic methods. Thus, as is the case with California (see Muhs, this chapter), terraces form during high sea stands on a tectonically rising land mass. During sea-level rise, waves encroach upon older deposits, and sediment input by waves to the beach and littoral zone presumably is high. At the same time, sediment input by fluvial erosion of the interior zone is low, because estuaries are deep. Sediment stored in these landforms later is removed from the active coastal zone by tectonic uplift, and is unavailable to the modern geomorphic system, except by wave erosion into terrace deposits and by fluvial erosion. Equivalent relict landform assemblages from past low sea-level stands are not known, but presumably, during periods of falling sea level, sediment input from wave action is lower and sediment input from fluvial action is higher.

Coastal Sand Dunes

The Oregon and Washington coasts are marked by several dozen coastal sand dune bodies, each one a mosaic of deflation areas, presently active dunes and older dunes stabilized by forest cover. Dune sand is deflated from the beach, so active dunes are sinks for beach sediment. Dune formation is favored in sites where there is an abundant supply of sand, such as beaches near the mouths of the largest Coast Range rivers. Dunes act as temporary or long-term storage areas for sediment, with sediment recycled to the beach by wave or fluvial erosion. In the best-developed dune bodies, six major types of eolian landforms are present along a transect inland from the beach, shown in Figure 17.

Eolian processes are controlled by the distinctive seasonal wind regime described earlier. Dune building in summer, when the sand is dry and deflation is relatively effective, is controlled by north to northwesterly winds, while winter dune building, somewhat limited by cohesion due to moisture in the sand, is controlled largely by south to southwesterly winds. The dominant season and direction of coastal eolian activity in Oregon and Washington has been a subject of contention among researchers (Cooper, 1958; Lund, 1973). Hunter and others (1983) found that the net sand transport direction is N45°E, although wind data suggest that it should be N26°E. They concluded that the difference is due to moisture decreasing the erosional effectiveness of winter winds.

Cooper (1958) discussed the origin and seasonal regime of each of the dune types (Fig. 17). The foredunes occur only because of the ability of European beach grass (*Ammophila arenar-*

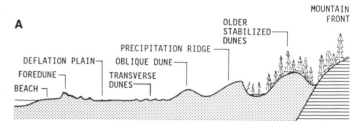

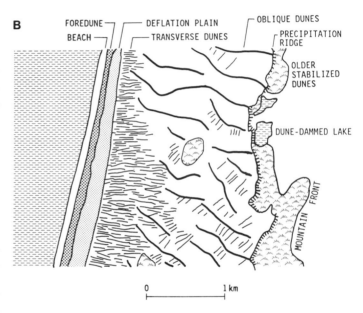

Figure 17. A: Generalized west-east profile of sand dunes, Oregon coast. Not to scale. B: Map of active dunes, south of Umpqua River, Oregon. Based on Cooper (1958).

ia) to colonize and stabilize sand immediately behind the beach, and foredunes apparently did not occur along the Oregon and Washington coasts before the introduction of this species during the twentieth century. Formation of transverse dunes, with crests oriented nearly east-west, occurs under the summer wind regime; they are nearly flattened during the winter (Hunter and others, 1983). The precipitation ridge is formed as the leading edge of an advancing sand sheet enters a forested area. Parabolic dunes form from blowouts on vegetated dune surfaces, where the site is sheltered from one of the seasonal wind directions, so that sand transport from the blowout is essentially unidirectional.

The most unusual and controversial dune type along the Oregon and Washington coasts is the oblique-ridge dune, described and named by Cooper (1958). These are large, east-southeasterly-trending ridges, with north-facing slipfaces during the winter, and smaller south-facing slipfaces during the summer. Cooper suggested that the oblique dunes form as tails trailing from an advancing precipitation ridge, but Hunter and others (1983) found several oblique dunes unrelated to a precipitation ridge. They concluded that the oblique dunes are the result of bimodal winds, with dune orientation controlled by winter winds, and net sand transport, controlled by both summer and winter winds, being oblique to the resultant sand-transport direction. Carson and MacLean (1985) argued that a significant amount of along-dune sand transport probably occurs on the Oregon oblique dunes, and that this dune form should be considered a hybrid form, which would likely develop into an east-southeasterly longitudinal dune if wind activity was not constrained by forest on the eastern edge of the dune field.

Most of the coastal dunes of Oregon and Washington are of Holocene age, although small bodies of dunes occur on, and presumably were contemporaneous with, the lower of the Pleistocene marine terraces. Cooper (1958) recognized three overlapping sheets of dune sand in many areas along the Oregon coast, representing three distinct periods of dune activity: episode I, occurring during the period of rapid postglacial sea level rise; episode II, during the middle or late Holocene; and episode III, beginning very recently and still in progress.

Cooper argued that dune formation along most of the Oregon coast occurred during periods of transgression, as rising sea level produced wave-eroded sediment for the beach zone, and that during regressive periods dune activity was absent or slight because the maritime climate allowed for rapid colonization of exposed land by vegetation. He envisaged a series of self-terminating episodes of dune initiation, inland advance, and stabilization. An episode of dune activity is initiated by a rise in relative sea level, or a climatic change, resulting in an increase in sand supply to the beach zone through wave attack on stabilized dunes or marine terrace deposits. Dune advance is maintained as long as the dunes remain in contact with a beach that is receiving sand, and the advancing dunes do not encounter any topographic barriers. A brief pause or reversal of sea-level rise, which decreases the supply of sand to the beach zone, may cause dune stabilization. Cooper's theory of dune initiation and stabilization has not been tested because data on the age of dunes are inadequate to establish correlations along the coastline and to compare the history of dune activity to possible triggering conditions.

Overview of the Coastal Zone

Active sediment in the coastal zone is transported among several compartments, including the beach, dunes, estuaries, and offshore zone. The beach is an area of sediment throughflow, although most beaches in the region appear to be relatively stable at present. Estuaries, dunes, and the offshore zone serve mainly as sediment sinks at present, although they may supply some sediment to the beach. Sediment transfer from the interior zone to the coastal zone apparently is relatively minor. Present conditions are typical of stable high-sea-level stands, and the coastal zone may be quite different under low sea level or changing sea level.

Landforms in the coastal zone are very dynamic, shifting position and morphology, and changing size. Sediment movement is variable at temporal scales ranging from the individual storm, to the annual cycle, to decadal cycles, to glacial-interglacial cycles. The main human impacts in the coastal zone have been the introduction of stabilizing vegetation, and interruption of littoral drift by jetty building.

SUMMARY

The Pacific Coast and Mountain System of Oregon and Washington is geomorphically active, due to its tectonic setting and climate. Mild, wet climatic conditions result in intense chemical weathering and dense forest cover that influence geomorphic processes. The strongly seasonal climate is felt both in the interior zone (through mass movement and fluvial processes) and in the coastal zone (through beach and dune processes). Active tectonic uplift has increased the rates of most interior-zone processes, particularly mass movement. In the coastal zone, uplift is less significant, but it removes sediment from the active beach zone.

Eustatic sea-level change has significantly influenced geomorphic development of the coastal zone. Present geomorphic conditions in the coastal zone are not representative of conditions during the past, and the coastal zone cannot be considered to be in long-term equilibrium. A change in eustatic sea level would greatly change sediment input to, and output from, the coastal zone, changing some landforms to relics (removing their sediment component from the active part of the system) or eroding some landforms (removing them as sediment storage areas and greatly increasing sediment input to the beach). In the interior zone, falls in eustatic sea level reactivate fluvial systems, but rises in eustatic sea level probably have little effect.

At present, linkages between the interior zone and the coastal zone do not appear to be strong, although the magnitude of sediment input to the beach from the interior is not well known. Quantitative rates of sediment input, transfer, and output are not well known for the coastal zone. Major contributions to geomorphic theory from this region have been made on the following topics: (1) conceptual models of mass-movement and stream-channel processes in mountainous, humid watersheds and quantification of their rates of sediment transfer, (2) episodic variation in sediment-transfer processes, (3) the interaction of biological and geomorphic processes, and (4) the dynamics of beach and nearshore landforms. The most important current research topics in this region are better quantification of fluvial and mass-movement processes in the interior, development of sediment budgets for the coastal zone, and chronologies of dune activity, marine terraces, and sea-level fluctuations. In addition, the structural geomorphology of the region, and its relation to tectonic processes, are not well known.

GEOMORPHIC PROCESSES IN THE RECENTLY UPLIFTED COAST RANGES OF NORTHERN CALIFORNIA

Harvey M. Kelsey

INTRODUCTION AND SETTING

This review paper focuses on geomorphic processes in a part of the Coast Ranges of northern California (Fig. 18), with an emphasis on channel- and slope-forming processes, and on the influence of tectonics. Most of the studies discussed below concern research in watersheds that drain the Franciscan Complex of the Coast Range Province (Blake and Jones, 1981) or the Klamath Mountain Province (Irwin, 1960, 1981) of northern California (Fig. 18). Rather than presenting a comprehensive review of all geomorphic process research in the area, the review concentrates on new research as well as those topics that have received an emphasis in this region.

The slopes of the northern Coast Ranges consist mainly of two physiographic types. The first consists of grasslands and grass-oak woodlands that range in morphology from smooth and undulating to hummocky, boulder strewn and in part poorly drained; these areas are generally underlain by incompetent rock types such as mélange or pervasively sheared fine-grained sediments. The other type consists primarily of straight, forested slopes with relatively sharp ridge crests and V-shaped canyons, underlain by competent sandstones or volcanic rocks. These two landforms can be referred to as "soft" and "hard" topography respectively, and they are sculpted by fundamentally different processes. In both topographic types, however, some sort of mass movement is the dominant transport process, and only in instances of extensive human disturbance has erosion by overland flow been of major importance.

The Coast Ranges have a Mediterranean climate, with high annual rainfalls occurring mostly from October through April. Rainfall amounts are in large measure orographically controlled and range from 125 to 300 cm/yr. Permanent winter snow cover occurs only at the highest elevations, and none of the major rivers have a spring snowmelt runoff peak. A major storm and flood in 1964, which caused considerable geomorphic change in Coast Range watersheds, subsequently became the focus of much process-oriented geomorphic research in the area. Through the use of alluvial stratigraphy, the dates and magnitudes of prehistoric floods (Helley and LeMarche, 1973; Kelsey, 1980; Zinke, 1981) can be compared with the deposits of the 1964 flood. Such a comparison suggests the 1964 flood has a recurrence interval of 60-80 years.

The California Coast Ranges are the product of the translational tectonics associated with the collision of the Pacific-Farallon Ridge with North America around 28 Ma (Atwater and Molnar, 1973), and the subsequent lateral plate motion along the San Andreas fault as the Mendocino triple junction migrated northward along the continental margin. An underlying context

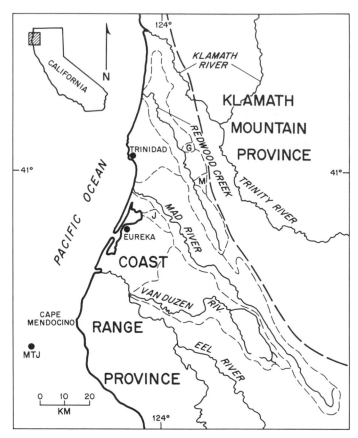

Figure 18. Location map of that part of the Coast Ranges in northern California that is discussed in this paper. MTJ, Mendocino triple junction; J, Jacoby Creek drainage basin; G, Garrett Creek drainage basin; M, Minor Creek drainage basin (site of Minor Creek earthflow). Heavy dashed line separates the Klamath Mountain Province from the Coast Range Province.

of all geomorphic processes in this region is Neogene uplift and deformation of the Coast Ranges associated with the San Andreas fault system, the migrating Mendocino triple junction, and the convergent margin north of Cape Mendocino. Uplift has varied in both time and space in northern California during the Neogene, and the present landscape records this variability. The triple junction was opposite San Francisco at approximately 7 Ma and has migrated to its present position opposite Cape Mendocino at a rate of approximately 3.2 cm/yr (Fox and others, 1985). The physiography of the northern California Coast Ranges reflects this migration history. The average elevation in the Coast Ranges increases progressively from San Francisco northward to the latitude of Cape Mendocino (Jachens and Griscom, 1983), as uplift appears to have migrated northward with the triple junction in a manner suggested by Glazner and Schubert (1985).

Calculated uplift rates for elevated marine terraces along the northern California coast vary from 1.0 to 4.0 m/ka (Lajoie and others, 1982; McLaughlin and others, 1983). Uplift-rate estimates farther inland are sparse (Mortimer and Coleman, 1985). Based on existing data, maximum uplift is 4 m/ka, areas of relative subsidence are common but are a small percent of land area (approximately 5 percent), and a regional averaged uplift rate may be on the order of 0.6 to 1.5 m/ka.

MASS-MOVEMENT PROCESSES

Due to the rugged relief of the Coast Ranges, both rapid and relatively slow mass movements are major hillslope-sculpting processes. Rapid mass movements in northern California entail speeds on the order of meters/second (Varnes, 1978) and include debris slides, debris avalanches, debris flows, and debris torrents (after the definitions of Swanston, 1971). Moderate (1.5 m/d) to very slow (.06 m/a) mass movements include earthflows and complex landslides, the latter being a combination of earthflow, translational and rotational sliding, and soil creep.

Earthflows and Complex Landslides

Earthflow landslides are common in incompetent rock units in northern California, occurring primarily in highly sheared Franciscan mélange. Earthflows move seasonally in the winter and early spring in response to prolonged rainfall of the wet season (Ziemer, 1984). The most visually obvious and fastest earthflows are unforested. They head in bowl-shaped depressions and most commonly their toes discharge sediment directly into a stream channel where it is carried downstream by high winter flows. Movement rates and sediment discharge of unforested earthflows have been reported by Kelsey (1978), Harden and others (1978), Iverson (1984), and Nolan and Janda (1987). In the Van Duzen River basin, Kelsey measured the downslope movement rate of 19 earthflows, both by seasonal monitoring of selected flows for three years and by aerial photographic measurements over a period of 35 years. Annual movement ranged from 0.6 to 26.0 m/yr, and average movement was 3.0 m/yr. Earthflows are highly gullied and typically have an axial gully that collects almost all runoff from the landslide surface area. Suspended sediment measurements in the axial gully show that sediment yield by fluvial transport off the highly disrupted earthflow surface is approximately equal to sediment yield through movement of the earthflow toe into the river channel (Kelsey, 1978).

Unforested earthflows in the Redwood Creek drainage show movement rates that are highly variable over time and in cross section across the flow due to sediment-flux imbalances within the landslides (Nolan and Janda, 1987; Iverson, 1986a, 1986b, 1986c). Portions of these earthflows underwent transient acceleration, accounting for annual movement rates ranging from 0.3 to 15.3 m/yr. In contrast to the Van Duzen earthflows, fluvial yield from gully erosion was far overshadowed by mass movement, being only about 10 percent of total sediment delivery to the Redwood Creek channel.

Borehole inclinometer surveys of the Minor Creek earthflow in Redwood Creek basin (Iverson, 1985) show that the zone of slide displacement in the main body of the earthflow averages 1 m thick and is 5 m below the surface. There is no single, discrete plane of shear. Drilling records and toe thicknesses for earthflows farther south in the Eel River drainage show the landslides to be between 30 and 35 m deep (California Department of Water Resources, 1970; Kelsey, 1978). Unforested earthflows are actually complex landslides with the major component of motion being earthflow, but also consisting of translational and rotational sliding, creep, and localized debris sliding (Iverson, 1985).

Forested earthflows and earth blockslides are a significant landscape-sculpting process as well. They generally have slower movement rates, which accounts for the ability of the slope to maintain a forest. Such mass movements are exemplified by numerous forested earth block glides (using the mass-movement classification of Varnes, 1978) in Redwood National Park (Sonnevil and others, 1985). These landslides are not easily distinguished topographically from surrounding areas. Forested conditions obscure slide boundaries, and the block glides appear to be only a faster-moving element of a generally mobile slope where the entire hillslope is subject to slow downslope translation. Evidence for this comes from borehole inclinometer surveys on schist hillslopes that do not exhibit block-glide morphology (Swanston and others, 1983). These surveys showed average movement rates of 16.4 mm/yr (for a 6-yr period), with shear planes of individual sites at 5.5, 6.4, and 12.6 m depth. In contrast, maximum movement observed on the active block slides (Sonnevil and others, 1985) is 7 m/yr, but more typical movement is 0.5-1.0 m/yr. Measured depths to the base of the failure planes for three active block glides was 6-7 m, with the thickness of the zone of failure being 1-1.5 m. Location of the failure surface was controlled at least in part by lithologic differences in the schist.

Iverson (1985) has developed a constitutive equation for complex mass movement such as unforested earthflows, using the Minor Creek earthflow as a field test model. Iverson's equation treats the deforming soil as "a nonlinear viscoplastic material with yield strength dependent on effective confining pressure and viscosity dependent on deformation rates" (Iverson, 1985, p. 146). Figure 19 shows rainfall data, changes in piezometric level, and landslide displacement across an extensometer on the Minor Creek landslide (Iverson, 1984) during two rainy seasons. The data show steady landslide movement following brief periods of acceleration. The acceleration is caused by gradual changes in pore pressure and effective stress that occur with the onset of the rainy season. The steady motion occurs in response to a steady stress state over time, which reflects a balance between driving and resisting forces. Iverson feels this dynamic force balance implies a viscous component to the deformation resistance. Another aspect of slow landslide movement is transient acceleration (Iverson, 1986a, 1986b, 1986c). A perturbation analysis, based on the constitutive model (Iverson, 1985), reveals that such localized perturbations in sediment flux in slow-moving landslides propagate slowly downslope as kinematic waves and spread rapidly outward by diffusion. All the above observations of earthflow behavior that are typical of the Minor Creek site are also documented elsewhere by Kelsey (1978) and Nolan and Janda (1987).

Rapid Mass Movements and Colluvium-Filled Bedrock Hollows

While various types of debris slides, avalanches, flows, and torrents are well documented in northern California (Furbish and Rice, 1983; LaHusen, 1984; Kelsey and others, 1987), the role of colluvium-filled bedrock hollows has only recently received due attention (Dietrich and Dunne, 1978; Lehre, 1982; Dietrich and Dorn, 1984; Marron, 1982, 1985; Wilson and Dietrich, 1985; Reneau and Dietrich, 1985; Reneau and others, 1984, 1986). These bedrock hollows typically occur in zero-order basins in upper and midslope positions where surface drainage is not developed. The following summary of research on these hollows is based on the references given above. Colluvial deposits may occupy V-shaped or U-shaped bedrock hollows that average 10–15 m wide and 3–4 m deep. Downslope extent of hollows is highly variable, but 60–100 m is a representative range of slope length. Both the geometry and the distribution of hollows indicate that they are formed by rapid evacuation of material by debris avalanches. Failure is most likely due to elevated piezometric levels that produce excessive pore pressures. Numerical experiments by Humphrey (1982) show that for conditions of failure during rainstorms, the hollow geometry dictates a greater saturated thickness in the downslope end of the fill. Slope-stability calculations and laboratory analyses suggest that failure can occur in saturated hollows only when hydrostatic pore pressures are exceeded by a significant amount, and this condition can only occur in the lower portion of the hollow if the flow depth is greatly expanded into weathered bedrock below the colluvium/ weathered bedrock contact (Wilson and others, 1984).

Landslides may evacuate hollows either completely to bedrock, possibly scouring out bedrock as well, or hollow evacuation may be only partial, leaving older basal colluvium remaining. Evacuation is followed by hollow infilling by creep, root throw, and fluvial transport. Frequently, a basal stone layer marks the surface of hollow evacuation where stones were transported into the bowl-shaped hollow by ravelling and concentrated flow of water prior to burial through the slower processes of creep, rainsplash, and periodic sloughing (Fig. 20).

Hollows are characteristic of geologically competent slopes made of sandstones, metasandstones, quartz-rich granites, and volcanic flow rocks. They are not characteristic of mélanges, highly foliated schists, and other rocks that tend to fail by earthflow. The recent interest in hollows stems from the evidence they provide for a denudational process operative on competent slopes throughout the late Neogene. Hollow evacuation was overlooked as a major process in earlier geomorphic studies in northern California because, historically, hollows had not failed in great

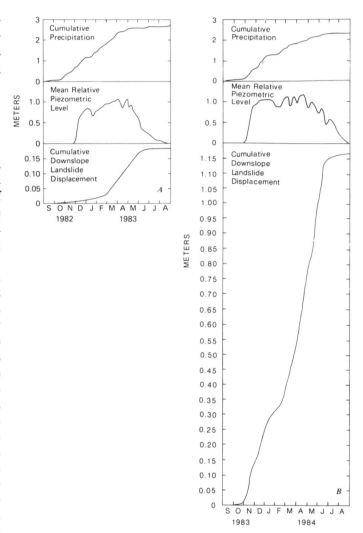

Figure 19. Temporal patterns of precipitation, piezometeric response, and surface displacement across a lateral shear zone at an extensometer site on the Minor Creek landslide (Redwood Creek basin) during two movement seasons. Most of the recorded shear displacement was distributed across a 1 to 2 m wide zone. Piezometeric data were collected weekly from two observation wells on the landslide complex about 100 m from the displacement gage. Piezometric levels are referenced to a datum 2.1 m below the ground surface. Increases in water content and pore water pressure and the resulting reduction of the ratio of static resisting forces to driving forces provide a simple explanation for seasonal landslide acceleration. Subsequent steady flow is suggestive of a dynamic balance between gravitational driving forces and viscous resisting forces in the deforming landslide (figure and caption from Iverson, 1984).

number during the major storms of the 1960s through 1970s. Recent storm-trigered shallow landsliding in the San Francisco Bay area (Reneau and others, 1984), helped bring attention to this process. Hollow evacuation is now recognized as perhaps being the most significant denudational process on competent lithologies in slope regions removed from the effects of bank undercutting on major channels. Support for this contention

A

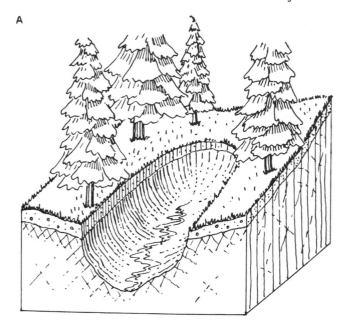

B

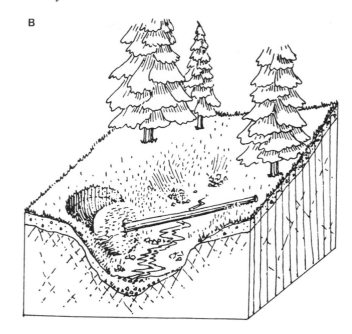

C

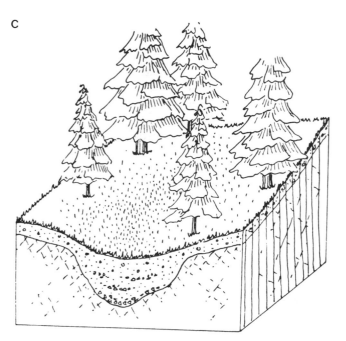

Figure 20. Evolution of a colluvial-filled bedrock hollow after a debris-avalanche failure on a hillslope. A: After the landslide, the exposed bedrock surface forms an impermeable horizon shedding rainwater and subsurface discharge into the depression as overland flow. B: Sediment eroded from the oversteepened soil perimeter into the depression is washed of its fine component, leaving a gravel-lag deposit covering the surface of failure. C: Continued soil creep and sloughing, as well as deposition by overland flow, leads to less frequent saturation overland flow and less surface transport. Eventually, the lack of surface wash causes the soil near the surface of the colluvial fill to become similar in texture to the surrounding material from which it was derived (after Dietrich and others, 1982).

comes from the fact that bedrock hollows containing colluvium appear to cover about 20–40 percent of zero-order basins (Dietrich and Dorn, 1984).

Pollen dates and ^{14}C dates on charcoal found in colluvial hollows on slopes near the San Francisco Bay area (Dietrich and Dorn, 1984; Reneau and others, 1986) and in Redwood National Park (Marron, 1982, 1985) have a range between 9,000 and 21,000 years B.P., with a clustering of dates of hollow evacuation between 11,000 and 14,500 B.P. These dates suggest a climatically induced, accelerated rate of landslide denudation on hillslopes in the late Pleistocene and early Holocene, though the limited number of dates so far leaves this interpretation open to question and verification.

SEDIMENT TRANSPORT AND STORAGE IN CHANNELS

Channel-Forming Processes in Nonalluvial Channels

Because of high uplift rates in the northern California Coast Ranges, most drainages flow on or near bedrock for most of their stream length. Nonalluvial channels (channels with bedrock boundaries) are therefore much more common than alluvial channels that are restricted to the lower valleys of the major rivers.

Large organic debris (logs, root wads, standing trees) is a significant determinant of channel morphology in small (fourth-order and less) Coast Range streams (Swanson and Lienkaemper, 1978), and there is an inverse relationship between the amount of organic debris loading and stream size. In steep mountain streams, large organic debris can form a stepped-bed profile. Keller and Tally (1979) found that for a second-order reach of a coastal stream in old-growth redwood forest, 60 percent of the total drop in stream elevation is caused by large organic debris. In such reaches, sediment stored behind this debris covers up to 40 percent of the active channel area and the residence time of the debris is frequently between 100 to 200 years, and can be greater. In lower-gradient streams bordered by large redwoods, both standing and downed trees influence the spacing and shape of more than half the pools in a reach (Keller and Tally, 1979).

Recent research has focused on the controls on riffle-pool spacing and bar formation in nonalluvial channels (Florsheim, 1985; Lisle, 1986). Lisle observed changes in Jacoby Creek (drainage area=35.5 km^2; Fig. 18) over a four year period, and noted the degree of thalweg migration in this fourth-order channel, which is incised into indurated Pleistocene siltstones and mudstones. Large streamside obstructions in the form of bedrock outcrops, large woody debris, and rooted bank projections appear to stabilize both the location and the form of gravel bars and adjacent pools. Obstructions are a stabilizing feature only if they are wider than approximately one-third the active channel width. Most of the pools (85 percent) were next to large obstructions or in bends, and 92 percent of large obstructions or bends had pools. Large obstructions form semi-stable flow vortices in scour holes, thereby fixing the location of the pool relative to the obstruction (Fig. 21) and also terminating upstream bars at fixed locations. To document the tendency for gravel channel stabilization by bedrock bends and streamside projections, projection width and bar volume were measured in channel-width long segments. These two measures are significantly cross-correlated at lags of –1, 3, and 4, relative to bank projections, indicating the tendency for large obstructions to form bars 3 to 4 bed-widths downstream and 1 bed-width upstream, thereby stabilizing the gravel channel (Lisle, 1986).

In Jacoby Creek, bedrock bends form the largest and most resistant bends and obstructions. Holocene terraces along the Jacoby Creek valley record lateral planation and migration of the creek and show that meanders migrate between major bedrock outcrops, but the channel bed has remained stationary where the channel is cut into bedrock. Large bedrock channel bends, therefore, tend to dictate long-term valley morphology in the lower reaches of this drainage.

Florsheim (1985) surveyed 15 gravel bed stream reaches in northern California to evaluate a slope threshold for bar formation. Study reaches excluded those affected by obstructions such as bedrock boulders, bedrock bends, or large organic debris. Surveys revealed that no riffle or storage bars form when the slope is greater than 0.02 except for bars downstream of flow obstructions, emphasizing the importance of obstructions for bar formation in steeper nonalluvial channels. Given slopes less than 0.02, bars form when channel width-to-depth ratio is greater than

eight. In the absence of obstructions, where slopes are greater than about 0.05, step-pool structure is common. In general, if reach slope is less than 0.02, bankfull flow does not build bars to heights equal to bankfull depth, as suggested in other places by previous workers (Simons and Richardson, 1966); storage bars reach a height only three-quarters of bankfull, and riffles attain a height that is a quarter of bankfull.

Sediment Storage in Channels

Another important physical component of nonalluvial channels of the northern California Coast Ranges is sediment storage. Stored sediment is a crucial link between sediment transported off the hillslopes and sediment transported out of a reach or past a point of measurement. Research on the movement of stored sediment in northern California (Madej, 1987; Pitlick, 1987; Kelsey and others, 1986, 1987) has focused on the tributaries and main stem of Redwood Creek. In order to quantify the movement of stored sediment, Madej (1987) classified alluvial material in the main channel of Redwood Creek as active, semi-active, inactive, or stable (Fig. 22), depending on its distance from and height above the thalweg, and the age and type of vegetation on the deposit. Important sediment storage compartments along the main channel include debris dams, lobate cobble bars, gravel flood berms, and alluvial fill terraces. Debris jams were only significant in the narrower, upper reach, and fill terraces were more prevalent in downstream reaches (Madej, 1987).

The volume of alluvium stored in each of the four relative mobility classes in the main stem of Redwood Creek was quantified as of 1980, based on aerial photographs and field measurements. Using these volume data in conjunction with bedload-transport rates from USGS gaging measurements, residence times for sediment in the four different sediment storage reservoirs were computed by the technique of Dietrich and Dunne (1978). Residence time ranged from 9 to 26 years for active sediment to 700 to 7,200 years for stable sediment (Madej, 1987).

Major erosion events on hillslopes produce significant temporal changes in channel-stored sediment. For instance, a significant amount of aggradation occurred in Redwood Creek during the 1964 flood, increasing the total volume of sediment stored on the valley floor by almost 1.5 times to 16×10^6 m^3. Streamside debris slides accounted for almost all the increase in channel-stored sediment, demonstrating the major role of streamside debris slides in determining temporal changes in storage volumes. Moderate to high flows (2–20 year recurrence intervals) in the 20 years after the 1964 event eroded stored sediment in the upper basin and redeposited it in lower reaches. Figure 23 shows that the distribution of stored sediment in Redwood Creek as of 1980 is best correlated with reaches having wide valley floors, which are also the reaches of relatively lower gradient. Stored sediment does not show a close correlation to sites of streamside debris sliding (Fig. 23) because of the relatively rapid movement of sediment from these sites to the wider, lower-gradient sites of optimum storage capacity. This migration of a slug of sediment

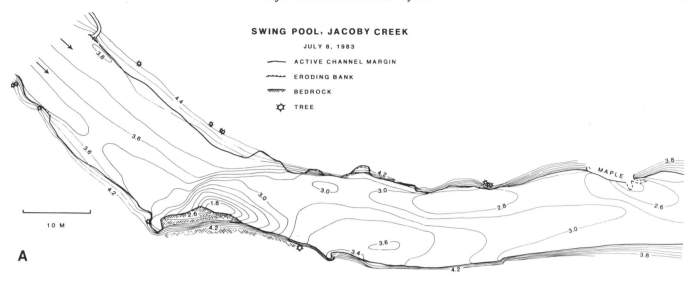

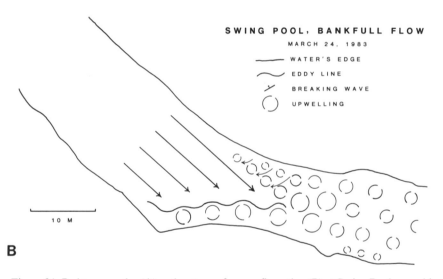

Figure 21. Bed topography (A) and water surface configuration (B) at Swing Pool, a pool in the Jacoby Creek channel adjacent to a bedrock bend. Turbulence and scour associated with this pool at the bedrock obstruction causes the truncation of the upstream channel bar. The bedrock bend therefore locally fixes the position of the pool as well as the associated upstream riffle and bar (after Lisle, 1986).

Figure 22. Schematic cross section of four sediment storage reservoirs in Redwood Creek: active, semi-active, inactive, and stable (after Madej, 1987).

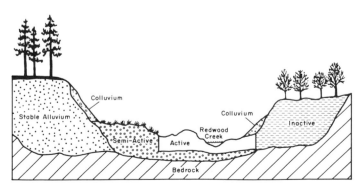

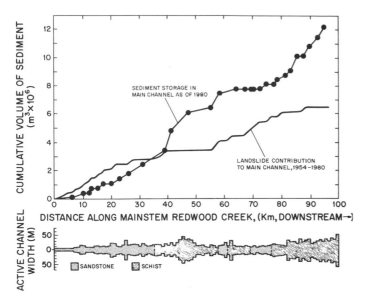

Figure 23. Streamside landsliding and sediment storage data for Redwood Creek: comparison of volume of sediment contributed by streamside landslides and volume of sediment stored in main channel, both cumulative through distance downstream along the main channel. Also shown on lower graph is the variation in valley floor width going down the main channel. Data in the figure show that sediment storage is well correlated with reaches of greatest valley width but is poorly correlated with sites of debris slide input, even though debris slides are the primary source of increases in channel sediment storage (after Madej, 1987; and Kelsey and others, 1987).

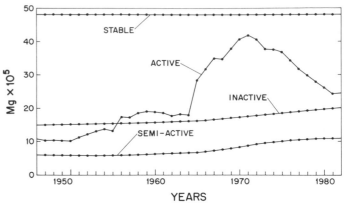

Figure 24. Mass of sediment (megagrams) in four different sediment storage reservoirs for the middle reach (33.3 km) of Redwood Creek for 1947 to 1982. For the active reservoir, the data show the movement of a slug of sediment through this reach as a consequence of major upstream sediment input from landsliding during the 1964 storm and flood. The data were generated from a first-order Markov chain model that simulated documented sediment movement in the active reservoir. Movement in the other three reservoirs was more variable (based on field measurements) than that shown in the model (after Kelsey and others, 1986).

down Redwood Creek was easy to follow on successive resurveys of channel cross sections along the main channel (Nolan and Janda, 1979).

Based on the above sediment storage data in Redwood Creek, Kelsey and others (1986) developed a statistical model using a first-order Markov chain that replicated the downstream transport of the slug of stored sediment in Redwood Creek. The model is based on the probability of transition of particles among the four sediment storage reservoirs, the probabilities being derived from the computed sediment residence times. Using initial volumes of channel-stored sediment in 1947 (derived from channel surveys and aerial photographic measurements) and incremental bedload transport volumes for each year through 1982, the model adequately characterized sediment storage changes in the middle reach of Redwood Creek for this 26-year period (Fig. 24), especially for the active reservoir. Two insights from the model are that total sediment flushing times appear to be highly dependent on the degree of interaction with the stable reservoir, and the median transit time for a particle traveling down a reach of Redwood Creek is an order of magnitude greater than the mean transit time. The first of these conclusions implies that the very infrequent, high-intensity storms (storms greater than that of 1964 and storms that mobilize the stable reservoir) are those events responsible for the long-term shifts in sediment storage.

In contrast to the main channel of Redwood Creek, first- and second-order tributaries are capable of transporting a high percentage of the sediment delivered to them, and total tributary storage in the basin amounts to only 6 percent of main channel storage volume (Pitlick, 1987). Main tributary storage sites are behind large accumulations of organic debris and in storage bars in the downstream-most, low-gradient reaches. Pitlick suggests that the tributary sediment transport regime is supply-limited, and therefore, residence times of stored sediment in tributaries are necessarily short. The general sediment routing scenario for Redwood Creek tributaries (similar to the situation described by Lehre, 1982), is that streamside debris slides and minor erosion by overland flow deliver sediment to the tributary channels where it is stored for a short time until more moderate flows (approximately 10 to 15 yr recurrence-interval events) redeposit the sediment in the wider, lower gradient reaches of the main channel. Once sediment reaches the main channel, similar size flow events move slugs of sediment down the channel in a fashion similar to that portrayed in Figure 24.

SEDIMENT BUDGETS

A sediment budget for a drainage basin is a quantitative statement of the rates of production, mobilization, and discharge of sediment in a watershed. To construct a sediment budget, both transport processes and sites of temporary sediment storage must be recognized and quantified as to rate and volume, respectively. Furthermore, the processes that link together different storage elements in the basin must be identified (Dietrich and

TABLE 5. SEDIMENT MOBILIZATION ON SLOPES AND PRODUCTION TO CHANNELS FROM INDIVIDUAL
PROCESSES, LONE TREE CREEK, CALIFORNIA*

Source	1971–1972		1972–1973	
	Mobilization (t/km^2)	Production (t/km^2)	Mobilization (t/km^2)	Production (t/km^2)
Landslides	0	0	1049	559
Slide scarp erosion	34	21	59	37
Slide scar sheet erosion	20	15	76	59
Headcut erosion	22	20	25	24
Soil creep	6**	included in gully	6**	included in gully
Hillslope sheet erosion	4**	scarp erosion	4**	scarp erosion
Gully scarp		82		194
Bank erosion		8		20
Bad erosion		2		92
Total	86	148	1219	985

Source	1973–1974		1971–1974 average	
	Mobilization (t/km^2)	Production (t/km^2)	Mobilization (t/km^2)	Production $(t/km^2/yr)$
Landslides	1795	1223	948	594
Slide scarp erosion	41	26	45	28
Slide scar sheet erosion	107	84	68	53
Headcut erosion	7	6	18	17
Soil creep	6**	included in gully	6**	included in gully
Hillslope sheet erosion	4**	scarp erosion	4**	scarp erosion
Gully scarp		201		159
Bank erosion		20		16
Bad erosion		15		37
Total	1960	1575	1089	904

*After Lehre (1982).

**Indicates effective mobilization. Effective mobilization, computed as **total length
of channel banks in catchment** multiplied by **mobilization rate per unit width of slope**,
gives rate at which sediment can be supplied to channel banks by processes acting
continuously over catchment hillslopes. In contrast, total mobilization (sediment moved
any distance) by such processes is given by **thickness of moving layer** multiplied by
total area affected by process, and is independent of downslope transport velocity.

others, 1982). In northern California, sediment budgets have been constructed for basins ranging in size from 1.74 km² to 575 km² (Kelsey, 1980; Lehre, 1982; Best and others, 1987).

For a 1.74 km² drainage basin 14 km northwest of San Francisco, Table 5 shows a summary sediment budget for 1971–1974 (Lehre, 1982). Debris slides and flows from colluvial-filled swales on hillslopes were the most important erosional processes in this basin, and accounted for most of the 3-yr sediment yield. After failure, these slide scars slowly fill through soil creep and scarp erosion, as discussed in more detail above. During the three years, only 53 percent of sediment mobilized in the basin was discharged past a gaging station at the basin mouth, the remainder being stored on slide scars, footslopes, and within the channel bed and banks. Based on a 15–20 year flood event in the basin during the study period (in January, 1973), sediment is removed from storage sites by storms with recurrence intervals greater than 10–15 years. In the intervening years of low to moderate storm activity, most of the mobilized sediment immediately returns to storage, mainly on the lower parts of slopes and in

the banks and the beds of gullies and channels. During major storm years, therefore, debris slides and flows are the major transporting mechanisms, whereas in drier years, sediment is mobilized by rainsplash and local sliding and spalling, and is only moved short distances.

A sediment budget for Garrett Creek, a 10.8-km² watershed tributary to Redwood Creek (Fig. 18), demonstrates the influence of intensive land use on sediment mobilization (Best and others, 1987). Unlike the largely grassland and brush-covered watershed of Lehre's (1982) study, roughly half of Garrett Creek has commercial timber, which was entirely logged and partially relogged again between 1947 and 1982. Based on a good aerial photographic record as well as field measurements, a sediment budget was constructed for 1956 to 1980 (Table 6), which includes the interval of widespread timber harvest and three major storm events in 1964, 1972, and 1975. Fluvial erosion contributed 62 percent of sediment to the main channel, and streamside landsliding contributed the remainder (Table 6). Almost all significant sources of erosion were due to road construction and logging. The

TABLE 6. SEDIMENT BUDGET FOR GARRET CREEK BASIN
(10.8 km² ; Tributary To Redwood Creek)
1956-1980*

Budget Component	Mass of Sediment (metric tons)	Percent of Total Input
Input		
Road-related gully erosion	51,000	16
Road-related debris torrents	43,000	14
Skidtrail-related fluvial erosion	29,000	9
Streamside landslides	121,000	38
Fluvial erosion from prairie and hardwood forest areas	74,000	23
Storage		
Additions to alluvial storage	19,000	
Output		
Inferred sediment yield 25-year budget period	299,000	

*After Best and others (1987).

two factors that accounted for 80 percent of road-related erosion were stream diversions at sites of plugged culverts and failure of road fills at the sites of stream crossings. In both cases, slumping or debris torrents initially deposited sediment in channels at a midslope position, and subsequently fluvial remobilization transported 20 to 80 percent of the sediment to the main channel. Of the total sediment delivered to the channel, only 6 percent remained in storage in lowermost Garrett Creek during the 25-year study period. Most sediment ended up in channel storage farther downstream in the wide, lower-gradient reaches of Redwood Creek. The dominant erosion processes in Garrett Creek are short-lived, event-related processes that occur in a matter of hours to a few days each year. Data from Garrett Creek, as well as other logged areas in California (Rice and Datzman, 1981; Weaver and others, 1987), strongly suggest that land management must be considered in the long term as another independent variable along with climate, tectonic processes, and geology in determining erosion rates.

In comparison to the more detailed views of small basins, Kelsey (1980) presents a sediment budget (Table 7) for a 575-km² basin drained by the upper half of the Van Duzen River. Although the budget spans 35 years (1941 to 1975), major changes in slope stability, channel morphology, and sediment transport are largely due to the December 1964 storm and flood. The vast majority of landslides are debris slides or debris avalanches that occur either on slopes adjacent to the main channels or on headwater slopes where debris moves directly into headwater channels. Virtually no instances of midslope, debris-avalanche failures of colluvial hollows occurred (with the exception of the large headwater avalanches that encompassed entire first-order channels), even though colluvium-filled hollows are abundant on competent, forested slopes. Fluvial hillslope erosion due to gullying is a highly significant process on mélange grassland and grass–oak woodland slopes. A major portion of the

sediment discharge (45 percent) presently comes from a small fraction of the basin area (5.5 percent) that consists of the most densely gullied grassland slopes, earthflow landslides in the mélange, and debris slides and avalanches in the competent sandstone units. Forested slopes not directly adjacent to major streams contribute a minimal amount of sediment, even though they constitute 57 percent of basin area. The 1964 flood, which lasted about three days, accounted for 7 percent of the sediment discharge during the 35-year study period, and mobilized, for the short flood period, almost as much sediment as moves out of the basin in a century. Channel-stored sediment increased dramatically due to storm-induced streamside landsliding, and almost one-fifth of the total sediment supplied to the channel during the study period remained as aggraded sediment in the main channels. Land use in the form of both logging and grazing increased sediment yields to channels in the Van Duzen, but the close spatial and temporal relations of storms, logging, and grazing precludes a definitive statement about land use, such as can be made for Garrett Creek.

RELATIVE EFFECTIVENESS OF PROCESSES OF DIFFERENT MAGNITUDES AND FREQUENCIES

The Coast Ranges of northern California are an appropriate site to discuss the relative effectiveness of processes of different magnitudes and frequencies, because infrequent, high-magnitude processes appear to perform the majority of geomorphic work in this region; this is contrary to the case for other less erosive temperate regions that do not have a pronounced Mediterranean climate. Based on suspended sediment data from the East Coast and the arid Southwest, Wolman and Miller (1960) concluded in their classic paper that the majority of the suspended load of rivers in these areas was carried by small to moderate flood flows and not by catastrophic floods. Suspended sediment discharge data for the Coast Ranges of northern California does not confirm this conclusion. Using five northern California gaging stations, Hawkins (1982) analyzed the effectiveness of suspended sediment transport for different recurrence streamflow events, using the identical procedure as Wolman and Miller. The highest 1 percent of the mean daily flows transported an average of 61 percent of the total suspended sediment, and the highest 10 percent of such flows transported an average of 92 percent. In contrast, in less erosive temperate regions, the highest 1 percent of the mean daily flows transported well less than 50 percent of the total suspended sediment (Wolman and Miller, 1960). The effectiveness of infrequent flows in north coastal California is due to the seasonal rainfall and high annual precipitation, which results in a kurtotic, highly positive skew to the distribution of daily streamflows. The consequence of this flow distribution, when combined with the highly erosive Coast Range terrain, is that infrequent flows, occurring on the average only one to three times a year, transport the majority of the suspended sediment. In less erosive temperate regions, flows occurring six or more times a year are required for equivalent transport (Wolman and Miller, 1960).

TABLE 7. SEDIMENT BUDGET FOR THE UPPER VAN DUZEN BASIN
(525 km^2 for 1941-1975)*

Budget Component	Mass of Sediment (metric tons)	Percent of Total Input
Input		
Fluvial sediment yield from hillslopes	45,509,000	73
Landsliding into main channel:		
Debris slides, debris avalanches	10,630,000	17
Earthflows	2,931,000	5
Streambank erosion, main channel:		
Melange bank erosion	426,000	1
Flood plain and fill terrace erosion	2,619,000	4
Storage		
Aggradation in main channel	10,601,000	
Output		
Total sediment discharge out of basin	51,036,000	

*After Kelsey (1980).

Brown and Ritter (1971) noted that for the Eel River basin near its mouth at Scotia, California (drainage area=8,063 km^2), sediment transport during the flood period from December 23, 1964, to January 23, 1965, accounted for 51 percent of the entire sediment transport for the 10-year period, 1958 to 1967.

The more infrequent processes associated with the unusual climatic events also appear to be those that deliver sediment from the hillslopes to perennial channels in competent geologic units. Debris avalanches are the major denudational processes that periodically evacuate colluvial fill and weathered bedrock from bedrock hollows. Recurrence at a site is on the order of thousands of years (Dietrich and Dorn, 1984; Reneau and others, 1986). Based on ^{14}C dates and alluvial stratigraphy, recurrence intervals for bedrock-sculpting avalanches in headwater basins in the Van Duzen basin range from 300 to 500 years (Kelsey, 1982). These slope failures are all triggered by infrequent storm events.

The most effective event for determining channel morphology is difficult to assess in northern California, because some channels have undergone prolonged, significant changes throughout the brief period of measurement. Wolman and Gerson (1978) suggest that the length of time for channel recovery compared to the recurrence interval of the event that created channel change may be the best time scale to measure effectiveness. The higher the ratio value in this measure, the more effective the infrequent event. They note that in temperate regions, channels widened by 50- to 200-year floods may recover original width in a matter of months or years. The December 1964 flood in northern California, a 50- to 100-year event, caused widespread aggradation (up to 4 m) and channel widening (up to 100 percent) on third-order and greater streams over a 3,000 km^2 area. Channel recovery from this flood, by degradation to the preflood level or to a stable but higher level, was as short as 5 years in some instances, but in other instances, recovery is still in progress at the time of this writing (Hickey, 1969; Nolan and Janda, 1979; Kelsey, 1980; Lisle, 1982; Kelsey and Savina, 1985). Recovery takes place

through sediment transport by moderate flow events. Therefore, using Wolman and Gerson's (1978) scale of effectiveness, the ratio of recovery time to recurrence interval of flood event is 0.05-0.6 for northern California, while in other temperate regions (Wolman and Gerson, 1978), it is 0.0015-0.06. These ratios suggest that infrequent flood events are more effective in determining channel morphology in the northern California Coast Ranges than in other temperate regions. Part of the reason for the effectiveness of extreme runoff events is the relative abundance of streamside debris slides in the narrow bedrock-confined canyons of northern California river channels (Nolan and Marron, 1985). These conditions can result in unusually large bed elevation changes (up to 4 m) during storms.

TECTONIC GEOMORPHOLOGY

The interplay of tectonic processes and hillslope mass movements have influenced landscape form in northern California. Two examples are the formation of inner gorges in downcutting stream valleys through the activity of streamside debris slides, and the effect of interbasin tectonic transport of sediment along thrust faults on hillslope processes and the sediment budget.

An inner gorge, after the morphologic definition of Dutton (1882), occurs along valley sideslopes adjacent to stream channels where slopes directly adjacent to the channel are steeper than those farther up the hillslope. A clearly defined break in slope separates the steeper inner gorge slopes from upper slopes. Though inner gorges are ubiquitous in this region of recent uplift, they are generally better developed in hard sandstone topography than in soft mélange topography. Farrington and Savina (1977) attributed inner gorge development in northern California to debris sliding during rapid downcutting. In a slope stability survey of part of the northern Coast Ranges, Furbish and Rice (1983) noted that 95 percent of debris slides occur on inner gorge hillslopes that are 30° or steeper. The inner gorge therefore appears to be formed by a coalescing of debris slide scars along inner gorge walls over time. Evidence for this process is relict, vegetated slide headscarps near the top of stable inner gorge slopes. Considering continuous uplift on the time span of 10^3 to 10^5 years, I estimate that inner gorge morphology would persist during slow retrogressive retreat of the slope break. An inner gorge with a 30° slope could retain that slope through a balance of debris-slide erosion and channel downcutting with uplift. If this balance is to be achieved during a constant uplift rate of 1.0 m/ka, debris slides must remove approximately 1 m of soil every 1,000 years. This streamside slope erosion rate compares favorably with short-term studies of streamside debris sliding (Kelsey, 1982; Kelsey and others, 1987), even if accelerated historic erosion due to land use is considered.

Lehre and Carver (1985) used simple sediment mass balance calculations to investigate and model the relations between tectonism and sediment yield in the Jacoby Creek basin (drainage area=35.5 km^2; Fig. 18), which is undergoing active thrust faulting (Carver and others, 1983) that is transporting sediment over the northern divide into the basin. Their model suggests that tectonism and erosion are not in equilibrium in Jacoby Creek, and that the northern flank of the basin is continuing to grow in relief and bulk. Earthflows that originate in the Franciscan mélange of the upper thrust plate redistribute the thrust material downslope. Their model of a growing divide being erosionally modified by earthflow can account for the observed sediment yield of the basin. This study shows the advantage of a sediment budget approach to the understanding of the interplay of tectonic activity and geomorphic processes.

SUMMARY

The focus of this section of the chapter is recent research on geomorphic processes particularly relevant to the landscape of the northern California Coast Ranges. Geomorphic processes operate in this region within the context of (1) active Holocene tectonism characterized by regional uplift rates on the order of 0.6 to 1.5 m/ka, (2) a diverse geologic substrate that nonetheless can be clearly divided into a "soft" topography typical of mélange and sheared fine-grained sediments and a "hard" topography typical of competent sandstones and volcanic rocks, and (3) a Mediterranean climate of dry summers, a pronounced wet season (October through April), and minimal winter snow accumulation that precludes spring snowmelt runoff events.

The above setting has led to research in the last few decades on a number of important topics. These fields of process-oriented geomorphic research, discussed in the above section, most notably include seven areas. These are (1) the behavior and dynamics of earthflows, block glides, and complex, slow-moving landslides; (2) the relative importance, history, and physical causes of rapid mass movements, particularly those that create colluvium-filled bedrock hollows and large debris-avalanche–sculpted bedrock bowls; (3) sediment transport, channel form, and bar formation in bedrock-controlled, nonalluvial channels; (4) the controls on and

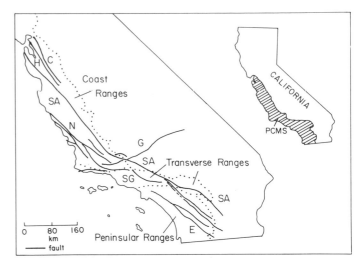

Figure 25. Location of the Pacific Coast and Mountain System (PCMS) in central and southern California and major faults. Abbreviations: SA, San Andreas Fault; H, Hayward fault; C, Calvareras fault; N, Nacimento fault; G, Garlock fault; SG, San Gabriel fault; E, Elsinore fault. Modified from Norris and Webb (1976).

the mobility of sediment stored in channels; (5) the development of and the insights gained from sediment budgets for basins ranging in size from 1.5 to 600 km^2; (6) the relative effectiveness of processes of different magnitudes and frequencies in northern California compared to other regions in the United States; and finally (7) the interplay of tectonics and geomorphic processes in creating distinctive landforms in the northern California Coast Ranges.

GEOMORPHIC PROCESSES IN THE PACIFIC COAST AND MOUNTAIN SYSTEM OF CENTRAL AND SOUTHERN CALIFORNIA

Daniel R. Muhs

INTRODUCTION AND SETTING

Central and southern coastal California in the Pacific Coast and Mountain System (PCMS) is an area of topographic, lithologic, and tectonic diversity. The area covered in this review extends from the United States–Mexico boundary to San Francisco Bay (Fig. 25) and encompasses the Peninsular Ranges, Transverse Ranges, and Coast Ranges as well as the coastal plain fronting these provinces.

The topography of the region is varied. Lowlands such as the Los Angeles Basin are rimmed by steeply-sloping mountains such as the San Gabriel and San Bernardino ranges, where the highest peaks are 3,000 to 3,500 m high and many parts of the ranges are over 2,000 m elevation. Much of the regional topography is controlled by the complex tectonic pattern. The most important single structure is the northwestward-trending San Andreas fault, a transform fault separating the North American and Pacific plates. This fault and many of the faults associated with it (Fig. 25) are highly active and are playing important roles in the geomorphic development of the region. The degree of tectonic activity and the tectonic style vary considerably from one place to another. In central California and extreme southern California, the San Andreas fault has a strike-slip character, and the Coast and Peninsular ranges tend to parallel the northwest strike of the fault. In contrast, northwest of Los Angeles the east-west–trending Transverse Ranges parallel the "big bend" in the San Andreas fault where intense compression is occurring, and very high north-south–shortening rates are estimated for this region (Yeats, 1983). Estimates of the relative motion along the Pacific and North American plates suggest an average of about 56 m/ka for the last 3 m.y. (Minster and Jordan, 1978). Sieh and Jahns (1984) estimated slip rates of about 34 m/ka along the San Andreas fault for the past 13,000 years, based on offsets of fluvial landforms of Wallace Creek in central California. Slip rates of approximately 24.5 m/ka for the past 14,000 years have been determined from offsets of terraces and other fluvial landforms for the San Andreas fault south of its junction with the San Jacinto fault (Weldon and Sieh, 1985). Along the San Gregorio–Hosgri fault system in coastal central California, slip rates of 6 to 13 m/ka have been reported (Weber and Lajoie, 1977). Coastal uplift rates vary with structural style (strike-slip versus compressive) and are as low as 0.2 m/ka (Muhs and Szabo, 1982) and as high as 10 m/ka (Lajoie and others, 1979). These estimates of horizontal slip rates and uplift rates serve to illustrate not only that the San Andreas and other faults are highly active, but also that they vary spatially in the degree of activity. Weldon and Sieh (1985) concluded that movement along the San Andreas fault system is currently accommodating about two-thirds of the plate motion in central and southern California; compression in the Transverse Ranges may account for some of the remainder.

The lithology of the region is varied and strongly influences geomorphic processes. The Peninsular Ranges and the eastern Transverse Ranges are composed of Paleozoic and Mesozoic granitic and metamorphic rocks. The western Transverse Ranges and most parts of the Coast Ranges, on the other hand, are composed of Tertiary sedimentary rocks, whereas the Los Angeles Basin is filled with Quaternary sediments. This lithologic diversity has important effects on geomorphology because different rock types have differing resistance to weathering and susceptibility to slope failure, and contribute differing particle sizes to fluvial systems. Lithologic variability also explains, in part, the irregular morphology of the coastline.

The climate of the region is Mediterranean, characterized by warm, dry summers and cool, wet winters. Within this region, however, temperature and precipitation can vary greatly. Mean annual precipitation values are as low as 100–200 mm/yr on the southern Channel Islands, whereas values are as high as 700-900 mm/yr at high altitudes in some of the mountain ranges. Generally, precipitation increases and temperature decreases with in-creasing elevation at a given latitude and northward at a given altitude. Vegetation tends to follow the moisture and temperature gradients.

An additional important factor for geomorphic processes, especially in southern California, is human activity. California's moderate climate and attractive coastline have resulted in large populations such that human activity must be regarded as a significant geomorphic element. The reverse is true as well: geomorphic events due to natural processes have had significant impacts on human populations.

MASS-MOVEMENT PROCESSES

Mass-movement processes in central and southern California coastal mountains deliver large quantities of sediment to valley bottoms. In fact, mass movements may be of greater importance to stream systems in this region than most others because of the effects of climate and lithology. Soil slips, debris flows, mudflows, and earthflows are the important mass-movement processes that have been studied in detail. Mass movements associated with sea cliffs will be discussed in the section on coastal processes.

Soil slips are common in California hills and mountains and have been studied in detail in both the San Gabriel Mountains (Rice and others, 1969; Rice and Foggin, 1971) and the Santa Monica Mountains (Campbell, 1974, 1975). These mass movements are shallow failures of soil, colluvium or ravine fill that occur only during heavy rainfall and only on steep slopes between 27° and 56° (Rice and others, 1969; Campbell, 1975). Steeper slopes do not have a mantle of soil or colluvium that can be mobilized, and motion is usually not initiated on gentler slopes; however, some movement has been reported on slopes as low as 15° (Campbell, 1975). Debris flows (discussed in more detail below) originate from soil slips when initial movement of slabs of soil and wedges of ravine fill cause remolding of the saturated moving mass into viscous debris-filled mud (Campbell, 1974, 1975). The result of this transformation is that the debris-flow mass is less resistant to shear and can accelerate down slopes. The debris flows often do not deposit material until they reach low gradients far from their sources and then they form fans at mouths of tributaries and debris trains in and along trunk streams (Campbell, 1974). Damage from soil slips is mainly due to inundation by, or high velocity impact of, the debris flows generated by the slips themselves.

Mudflows have been studied in detail in both southern and central California. Morton and Campbell (1974), Morton and Kennedy (1979), and Morton and others (1979) examined mudflows in the San Gabriel Mountains near Wrightwood and reported the hydrologic conditions under which mudflows occur. These workers hypothesized a spring mudflow cycle that seems to typify the area. Mudflows near Wrightwood have built up on coalescing fans of three creeks; the fans themselves are composed largely of earlier mudflow deposits. Mudflows here are associated with (1) short-period, high-intensity rainfalls from thunderstorms in summer (which are not common); (2) periods of heavy rainfall from winter rain, at which times they can flow for periods of a few hours to a few days; or (3) thaw of winter snowpacks, at which times flow can take place for as long as several weeks. The spring mudflow cycle that these workers identified consists of three stages: (1) a waxing stage, characterized by sporadic, short-duration mudflows, during which time the lower canyon floor is aggraded; (2) a climactic stage, which has longer and more frequent flows and during which a U-shaped channel is incised into the alluviated canyon floor; and (3) a waning stage, in which the meltwater supply decreases, and short-duration flows backfill the U-shaped channel.

In central California, Keefer and Johnson (1983) studied earthflows and debris flows in the Coast Ranges from San Francisco to Monterey Bay. They found that earthflows are tongue-shaped, composed predominantly of silt and clay, are bounded by slickensided shear surfaces, and can be up to several square kilometers in size. They are mobilized by increases in pore water pressure caused by water infiltration during and after rainstorms. Unlike debris flows, they are formed by multiple episodes of deposition, and at least 34 individual deposits were recognized at one locality.

Debris flows, in contrast to earthflows, have coarse-textured materials, move rapidly, are characterized by internal shear planes, and are not remobilized (Keefer and Johnson, 1983). They are found in California in many places; Scott (1971) has documented how the intense winter storms of 1969 generated one of these features near Glendora at the base of the San Gabriel Mountains. Three conditions for optimum debris flow initiation were found that winter: (1) the previous dry season had experienced a burn, (2) there was below-normal antecedent precipitation so that regrowth of vegetation was prevented after the burn, and (3) the below-normal antecedent precipitation period was followed by a major runoff-producing storm. Debris flows in the winter of 1969 mobilized channel bed material, scoured channels to bedrock, and then deposited sediment as fans at the mouths of canyons (Scott, 1971). Sediment yields of 59,000 m^3/km^2 were measured. The importance of fire in producing or enhancing water-repellent layers in soils and causing debris flow formation has also been discussed by Wells (1981, 1984). Scott (1971) concluded that debris flows are probably the most important mechanism for transporting coarse sediment in small basins of the San Gabriel Mountains.

Because central and southern California are seismically active regions, there is abundant evidence of initiation of mass movement due to earthquakes. After the 1971 San Fernando earthquake, Morton (1971) mapped more than 1,000 mass movements in a 250 km^2 area of the hills and mountains north of the San Fernando Valley. Because earthquake activity is variable both in space and time, it is difficult to make a quantitative estimate of the importance of earthquake-induced mass movements to the long-term sediment budget in California, but there is little question that they are significant.

Radiocarbon dating of large mass movements in southern California suggests that more humid paleoclimates enhanced landslide activity in the past (Stout, 1977). Most dated landslides cluster between 16 ka and 20 ka. Available paleoclimatic data suggest that this was a time of greater humidity than the present (Johnson, 1977); such conditions could have resulted in greater hillslope saturation, which in turn could have enhanced mass-movement processes.

FLUVIAL PROCESSES AND LANDFORMS

Erosion Rates and Sediment Transport

Erosion rates and sediment yields in drainage basins in central and southern California are functions of a large number of variables. Scott and Williams (1978) investigated a number of these variables for the eastern Transverse Ranges where sediment yields for the 1969 storms (representative of a 50-year recurrence interval storm) were known. They found sediment yield was largely a function of drainage area, basin elongation ratio, area (km^2) of slope failures, fire, mean annual precipitation, and total time of concentration (i.e., time of transit of water through the basin). Scott and Williams used a regression equation with these variables to estimate sediment yields from watersheds in the western Transverse Ranges in Ventura County, where measured sediment yield data were available. In the two studied basins, the predicted yields agreed closely with observed values.

Sediment yield data combined with drainage basin areas have been used to compute denudation rates for central and southern California (Scott and Williams, 1978; Griggs and Hein, 1980; Taylor, 1983). Although methods vary somewhat among workers, some regional trends emerge from these calculations. Taylor (1983) differentiated between mountainous, hilly, and plains areas in southern California from Point Conception to San Diego and calculated rates of 0.41–4.09 m/ka, 0.19–1.11 m/ka, and 0.06–0.07 m/ka, respectively. Denudation rate also tends to be a function of basin area (Fig. 26). Langbein and Schumm (1958) reported similar results for areas outside California and thought that dependence of denudation rate on basin size was due to reduced mean gradients in larger basins and reduced probability that a given short-term, high-intensity storm will affect the entire area of the larger catchment basins. Griggs and Hein (1980) reported denudation rates for central California drainage basins from Point Conception to San Francisco. These rates averaged 0.07 m/ka and, when plotted with Taylor's data, show that erosion rates for basins of a given size are generally lower in central California (Fig. 26). One factor that may explain this difference is tectonic uplift rate. Central California north of Point Conception is generally less active tectonically, and this could help account for the relatively low rates of erosion. Another factor may be a generally wetter climate and heavier vegetation cover in central California compared to southern California. Within the western Transverse Ranges of Ventura County, Scott and Williams (1978) did find that denudation rates are a function of the degree of tectonic activity. In the northwest part of their

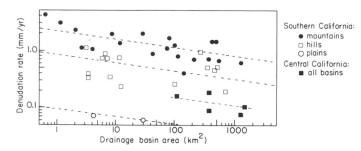

Figure 26. Denudation rate as a function of drainage basin area in central and southern California. Central California data from Griggs and Hein (1980); southern California data from Taylor (1983).

study area, where tectonic activity is greatest, average denudation rates were greater by a factor of 4 to 42, compared to the southeast part of their study area where tectonic activity is lowest. They calculated long-term denudation rates of approximately 2.3 m/ka for small (<.5 km^2) basins in tectonically active areas and noted that this was a small fraction of the uplift rates reported for this area. Marine terrace elevations suggest uplift rates of up to 10 m/ka near Ventura (Lajoie and others, 1979).

Taylor (1983) found that lithology and weathering processes strongly control particle size distributions of stream deposits in southern California drainage basins. The western Transverse Ranges have stream sediments that are about 80 percent silt and clay and only about 20 percent sand, reflecting the fine-grained sedimentary rocks predominant in that area. Streams in the chemically-weathered granitic basins of the Peninsular Ranges, on the other hand, yield an average of about 70 percent sand and about 30 percent silt and clay. In contrast, the tectonically-fractured granitic and metamorphic rocks of the San Gabriel Mountains yield stream sediments that are 10 to 25 percent silt and clay, about 50 percent sand, and 25 to 50 percent clasts coarser than sand. In the larger Santa Ana River system near Los Angeles, Haner (1984) was also able to identify sediment sources on the basis of particle-size distributions, but found better sorting and winnowing of coarse-grained sediments at greater distances from source areas. Sediments were also removed from floodplains during dry seasons by eolian transport.

The abundance of material coarser than sand size in small basins of the San Gabriel Mountains raises the question of the mechanism by which such material is transported. Observations made by Scott (1971) and Scott and Williams (1978) after the 1969 storms allowed generation of a model of coarse sediment transport via debris flows in small (<13 km^2) basins. In the dry season between major storms, there is a lateral supply of sediment to stream channels via dry sliding. In contrast, both mass movement and overland flow occur during the wet season so that channels experience more or less continuous fill. After major storms (such as those in 1969), channel bed material is mobilized and moved by debris flows with net scouring of the channel and

undercutting of the valley side slopes by bank erosion. This "cleaning out" process thus prepares the system for a new cycle of channel infilling.

Human activities have locally altered the natural sequences of erosion, sediment transport, and deposition in central and southern California. Scott (1973) studied the effects of the 1969 floods on scour and fill processes in Tujunga Wash in the San Fernando Valley. A large gravel pit upstream of a flood control basin caused a local reduction in base level; scour during the flow occurred upstream for more than 900 m and contributed to the destruction of three highway bridges, seven homes, and an entire residential street. Griggs and Walsh (1981) documented the effects of off-road vehicles (ORVs) on gullying, sediment discharge, and alluvial fan formation in Hungry Valley in the Transverse Ranges. Loss of vegetation contributed to severe soil erosion, gullying, and formation of alluvial fans at the bases of slopes. Griggs and Walsh estimated that gully development in heavily impacted areas contributed 800–12,122 t/km^2 during one winter season. Thus, in some ORV-affected areas, erosion rates are as much as three times greater than natural rates calculated for the Transverse Ranges by Scott and Williams (1978). Many other examples could be cited, but these two studies serve to demonstrate how human activities of very different types influence natural erosion and sedimentation processes, often with disastrous results.

Fluvial Landforms

Three of the most important fluvial landforms in coastal California are terraces, fans, and arroyos. Geomorphologists have traditionally considered stream terraces to form as a result of climatic change, changes in base level, tectonics, or combinations of the three. Bull (1964a, 1964b) studied stream terraces in valleys of the Diablo Range in western Fresno County as part of a larger study of alluvial fans. He found both low terraces and high terraces; the low terraces were associated with entrenchment that occurred from climatic changes of the nineteenth century. Periods of entrenchment coincided with times of above-normal daily and annual precipitation and preceded the introduction of livestock to the area. The best examples of high terraces were found along Little Panoche Creek; stream cutting and terrace formation there was due to uplift of the Diablo Range. At some distance from the mountain front, new deposition took place on preexisting fans. Longitudinal profiles showed that terrace divergence occurred upstream from the zone of maximum differential uplift; in contrast, terrace convergence was evident downstream. Thus, Bull found evidence of terraces resulting from both climatic change and tectonics. Lettis (1985), on the other hand, felt that regional climatic change was the dominant factor in high terrace formation along the Diablo Range. His evidence for this is similar ages of terraces (based on degree of soil development) in areas that have experienced uplift and areas that have experienced downwarping.

In areas that are highly active tectonically, terrace sequences can be complex, and it is not always possible to isolate climatic and tectonic effects. Weldon and Sieh (1985) studied fluvial landforms of Cajon Creek in order to determine slip rates of the San Andreas fault south of its junction with the San Jacinto fault. They found seven distinct terraces, which are all less than 13,000 years old. Both strath and fill terraces are present, and the sequence suggests a complex interaction of tectonic processes and climatic change during the late Pleistocene–Holocene.

In southern California, studies by Birkeland (1972), Keller and others (1982), and Rockwell and others (1984) have documented the combined effects of base-level change and tectonics leading to terrace formation. Birkeland worked in the Malibu area and mapped five Pleistocene stream terraces in valleys with mouths on the coast. The oldest four of these have tops that are graded to the shoreline angles of marine terraces, and they were interpreted to be fill terraces of eustatic origin. Filling took place during times of rising sea level and cutting took place during low stands of sea. The youngest Pleistocene stream terrace truncates the youngest marine terrace; Birkeland interpreted this stream terrace to be related to a mid-Wisconsin sea level high stand that was lower than present sea level. Alexander (1953) found similar relationships in central California, south of Santa Cruz. In both the Malibu and Santa Cruz areas, Birkeland (1972) and Alexander (1953) recognized that terraces had been uplifted tectonically. Keller and others (1982) and Rockwell and others (1984) studied stream terraces along the Ventura River from the coast to as much as 15 km inland. The two youngest terraces are of historic age. Age estimates of the older terraces are 0.5–5 ka, 8–12 ka, 15–20 ka, 30 ka, 38 ka, 44–64 ka, and 79–105 ka. The oldest four terraces were interpreted to be eustatic in origin, as global marine terrace studies indicate high sea-level stands at these times (Bloom and others, 1974). They also found that individual terraces were broken into segments by faults and warped over the axis of the Ventura Avenue anticline and were able to correlate these warped and faulted segments using degree of soil development. The importance of these studies to terrace theory is that these workers documented that there were far fewer terraces than originally mapped by Putnam (1942), who did not recognize the degree of terrace faulting that these later investigators found.

Alluvial fans are common landforms in both the coastal mountains and on the marine terraces that border the ocean. Fans in the mountainous regions have received far more attention than those overlying marine terraces on the coast. Fans derived from eastward-flowing streams heading in the Diablo Range were studied by Bull (1961, 1962, 1964a) and Lettis (1985). Bull identified some of the most important basic processes of fan formation from his detailed studies. He found that fan area is a function of both drainage area and lithology; fan area increases with drainage area as a power function (Fig. 27). Furthermore, fans derived from mudstone or shale are about twice as large as fans derived from sandstone basins of comparable size. The difference lies in the fact that the fine-grained rocks are more easily erodible. Finally, fan slope is inversely related to drainage basin

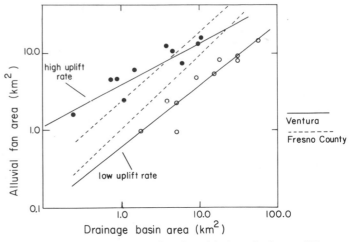

Figure 27. Alluvial fan area as a function of drainage basin area. Western Fresno County data are from Bull (1962). Ventura area data are from Rockwell and Keller (1985); individual points shown are for Ventura only.

area as a power function; the slope of the line defining this relationship for shale and mudstone is gentler than that for sandstone. Bull (1961, 1964a) also found that radial profiles of alluvial fans had tectonic significance. Fan radial profiles are not smooth lines, but have segments of differing slope, and each segment defines a different episode of fan deposition. In a sequence where the uppermost fan segment is the oldest and the lowermost segment is the youngest, incision rates greater than uplift rates are implied; streams incise the older fans and deposit material as new fans farther away from the mountain front. In contrast, when the youngest fans are found upslope, uplift rates greater than incision rates are implied.

Bull's observations on fan morphometry and relationships to tectonics have been closely confirmed by studies in the Ventura area by Rockwell and Keller (1985). They concluded that basinward tilting beyond a fan slope threshold of 6° to 8° caused fan entrenchment and shifted the locus of deposition downfan. In contrast, high rates of uplift in fan source areas caused fanhead deposition. Fans with coarser materials were found to have steeper slopes. Finally, they also found that fan area near Ventura is a power function of drainage basin area. However, in two of their study areas, uplift rates are so high (4–8 m/ka) that even small drainage basins produce large fans, so one of their regression lines plots, in part, above Bull's lines (Fig. 27).

Some alluvial fans in southern California may have regional paleoclimatic significance. Ponti (1985) worked in the Antelope Valley area and mapped fans derived from the Transverse Ranges and the Tehachapi Mountains. He used degree of soil development to date and correlate deposits. Similarity of ages of fans in most drainages led Ponti (1985) to conclude that there were several regionally synchronous episodes of alluviation that were probably climatically induced.

The problem of historic arroyo formation in the southwestern U.S. has fascinated geomorphologists for decades. The first serious study of these features in California was by Bull (1964b), who documented that arroyo formation in western Fresno County preceded livestock introduction and was related to above-normal periods of rainfall. A similar conclusion was reached by Scott and Williams (1978) for channel entrenchment in the Transverse Ranges. In contrast, both below-normal precipitation and overgrazing by sheep apparently caused arroyo formation on San Miguel Island (Johnson, 1980). The only other major study was the work of Cooke and Reeves (1976) who studied arroyo formation in Arizona and coastal California. They found little evidence for climatic control on nineteenth century arroyo formation on the mainland California coast, but, in contrast to Bull, concluded that grazing and cultivation were probably the major causes of entrenchment. These workers presented a model showing how grazing and cultivation brought about removal of vegetation on valley floors and alteration of valley bottom soils. Changes such as these in turn brought about increased erosiveness (velocity) of flows through valley bottoms, and increased erodibility of valley floor materials, all of which led to arroyo formation. Their study is important in that it suggests a very different cause for nineteenth century arroyo formation in California than that found in Arizona, where climatic factors were thought to be far more important; thus, similar geomorphic features can form by very different processes.

COASTAL PROCESSES AND LANDFORMS

The central and southern California coastline can be best described as a cliffed coast, the result of a generally high wave energy environment on a steep terrain. Headlands form where particularly resistant rocks are found, and because of differing lithologies or active tectonic processes, the coastline can be highly irregular in places. Barriers may constitute as much as 20 percent of the entire California coastline (Converse, 1982), but most areas are characterized by steep cliffs that are undergoing active erosion by waves. Delta and fan deposits derived from mountain streams form coastal plains in a few areas, such as the Los Angeles Basin and the Oxnard Plain. Some bays, such as Mission Bay in the San Diego area, are drowned river valleys related to the Holocene marine transgression. Most Holocene estuaries, however, are now filled with alluvium due to rapid deposition of sediment derived from streams heading in the coastal mountains. The generally southward longshore drift of sand causes barrier spits to develop across such features. Coastal dunes derived from either existing beaches or shelf sediments exposed during marine regressions are found at a number of localities.

Shore Platforms and Sea Cliffs

Shore platforms are common along much of the California coast, especially along rocky headlands. Bradley and Griggs (1976) studied modern platforms and ancient, uplifted shore platforms in the Santa Cruz area and observed that such features are composed of inshore segments with gradients of 0.02–0.04 and gentler offshore segments with gradients of 0.007–0.017. The modern inshore segments are 300–600 m wide and extend to depths of 9–13 m; major storm waves break in water 7–12 m deep. They concluded that inshore segments were related to storm-wave surf zones and offshore segments were associated with the zone of deep-water wave transformation. In their model, Bradley and Griggs suggested that slopes less than the minimums observed would so dissipate wave energy in offshore areas that waves in the surf zone would not be able to provide longshore transport of sediment; beach progradation would result. In the Santa Cruz area, the best platforms are developed where there is soft sandstone and where uplift rates are lowest. Bradley and Griggs suggested that where uplift rates are low and the rate of sea-level rise during the transition to an interglacial period matches the uplift rate, platforms should be widest.

Emery and Kuhn (1980) hypothesized that the two most important processes of platform lowering are solution and abrasion by grains in suspension. Basins develop in the intertidal segments of shore platforms due to solution of calcium carbonate cement; this occurs because nighttime respiration of organisms causes a decrease in pH, aiding dissolution. Emery and Kuhn (1980) photographed such features over a period of 35 years and found that rates of lowering from dissolution ranged from 0.003 to 0.06 cm/yr. They also estimated that rates of lowering by abrasion by sand in suspension were much higher, probably on the order of about 1 cm/yr.

Shore platforms widen as sea cliffs retreat. There is an enormous body of literature on sea cliff retreat processes and rates in California. The reason lies in the fact that the coastline is a preferred location for housing development, recreational areas, and critical structures such as nuclear power plants and liquified natural gas facilities. Numerous studies of sea cliffs on the California coast have identified the main processes of cliff retreat, and these have been summarized by Norris (1968), Kennedy (1973), Griggs and Johnson (1979), Emery and Kuhn (1980, 1982), and Kuhn and Shepard (1980, 1983). Marine processes include abrasion by sand and gravel in suspension, quarrying of blocks (especially where there are well-developed joints or faults), biological activities, and solution by ocean water. Terrestrial processes that contribute to sea cliff retreat include gullying, slopewash, wedging by tree growth, and mass movements. Mass movements have become increasingly important in recent years due to human activities such as heavy vehicle traffic, building construction, and saturation of cliffs by storm drain discharge or watering of lawns.

A number of variables modify the importance of various cliff retreat processes. In Santa Cruz County, Griggs and Johnson (1979) found that the most important environmental factors are available wave energy, degree of exposure to waves, presence or absence of a protective beach, cliff lithology, cliff structure (joints, faults, and folds), and height of the sea cliff. For example, they found that mudstones were more resistant to retreat than sandstones and siltstones, and these latter were in turn more resistant than unconsolidated marine terrace or alluvial deposits. Height of the sea cliff affects retreat rates because higher cliffs can potentially supply more material for erosion; this material may accumulate as talus at the base of the cliff and temporarily armor it against further erosion. This latter process may have limited effectiveness, however. Emery and Kuhn (1980) showed with photographic evidence that such talus accumulations are rapidly removed by boulder impacts and abrasion by sand in suspension.

Much sea-cliff retreat may take place by rather rapid mass movements after a threshold is crossed, followed by long periods of relative stability. The regional significance of rapid mass movement as a sea-cliff retreat process is suggested by the maps of landslides of the southern California coastal area by Emery (1967). He suggested that many coastal landslides were probably initiated in the last 5,000 yr, when sea level has been close to its present position and cliff undercutting would be more effective. A good example is the Portuguese Bend landslide in the Palos Verdes Hills, which experienced movement about 5,000 yr ago (Emery, 1967) as well as at present. This large landslide (1.4 km^2) has been studied by several workers (e.g., Merriam, 1960; Kerr and Drew, 1969) because modern movement began in 1956. Merriam (1960) showed that the "landslide" is actually a rotational slump in the upper regions, a series of independent large blocks in the middle part, and a complex earthflow in the lower part. Seaward-dipping Monterey Formation shales that compose much of the landslide have layers of tuff that have altered to clay. Kerr and Drew (1969) showed that this clay is predominantly smectite and is highly thixotropic. Movement on the slide takes place as shale beds slide over the smectite-lubricated layers. The modern movement was apparently initiated by a combination of above-average rainfalls, additions of water from cesspools, and possibly the peptizing effect on clays of organic matter additions from cesspools (Merriam, 1960; Kerr and Drew, 1969). The studies of the mechanisms of the Portuguese Bend landslide are important, because the Monterey Formation with its altered, interbedded tuffs crops out along much of the central and southern California coast.

A large body of data now exists on rates of sea-cliff retreat in California, and many of these rates have been summarized by Sunamura (1983). Rates vary from as low as 0.0–0.01 m/yr for some areas near Santa Cruz and San Diego (Griggs and Johnson, 1979; Kennedy, 1973; Emery and Kuhn, 1980), to as high as 1.38 m/yr for the period 1960–1980 near Half Moon Bay, after construction of a breakwater that interrupted littoral drift (Mathieson and Lajoie, 1982). Most rates are between 0.01 and 0.5 m/yr. It is difficult to make many generalizations about rates of retreat because of different methods of measurement, differing time scales, and differing sample sizes. In addition, long-term "average" rates of retreat, such as some of those given above, are misleading, because many cliffs experience little or no retreat for years and then may retreat several meters during one mass movement event.

A number of smaller landforms sometimes develop as a result of sea cliff retreat in California. The most spectacular of these are caves, arches, and stacks. Kennedy (1973) studied sea cliff erosion in the Sunset Cliffs area near San Diego and developed a cyclic model of cave development. He found that joints in the cliff bedrock first develop into surge channels; these then enlarge into sea caves. Partial collapse of the cave roof results in a blowhole and a natural bridge. Further cave roof collapse ultimately results in an embayment with a new cliff face. Shepard and Kuhn (1983) studied sea arches and stacks in detail near La Jolla, California and on a reconnaissance basis up the coast to as far as Point Arena. They examined old photographs of these features and compared them with the present-day morphology. One of the most important conclusions of their studies is that the time-honored hypothesis of Johnson (1938) of arch formation by cutting on two sides of a projecting point explained the origin of only one arch near La Jolla and one arch on Anacapa Island. Instead, they found that arches form by a variety of processes and are partly due to exposure of an irregular coastline (due to recent fault-block uplift) to Holocene sea-level rise. They also found that solution of calcium carbonate cement in sandstone plays a role, as do surge channels and embayment intersections of cave heads.

It is possible to study the subaerial processes acting on "inactive" sea cliffs as described by Emery and Kuhn (1982) by examining these features on uplifted marine terraces. Two such studies have been recently completed in California on flights of marine terraces. Hanks and others (1984) constructed a shore-normal profile of marine terraces and sea cliffs near Santa Cruz. Their profile shows that abandoned sea cliffs assume the general form predicted by Emery and Kuhn (1982) for homogeneous materials where subaerial processes are dominant. Furthermore, their profile shows that maximum slope angles on older cliffs are lower than on younger cliffs. A similar observation was made by Crittenden and Muhs (1986) on San Clemente Island. These latter workers found a log-linear relationship between cliff height and maximum slope angle, similar to, but with a weaker correlation than what Bucknam and Anderson (1979) found for fault scarps in unconsolidated materials in Utah.

Beaches and Coastal Dunes

Although most of the California coast is cliffed, some continuous stretches of beach sand do exist, and pocket beaches are common components of the cliffed segments of the coast. Fewer studies of beach processes have been made on the California coast than in other areas of the U.S., but good summaries of beach dynamics have been presented by Emery (1960), Norris (1964), Chamberlain (1964), and Inman and Brush (1973). The basic shore-parallel form of the beaches is a function of the "littoral cell" (Emery, 1960). The net littoral drift along much of the coast is southward. Narrow, pocket beaches are dominant in embayments of cliffed coasts, but at mouths of major streams heading in the coastal mountains, the southward littoral drift carries sediment supplied by a stream to form wide beaches to the south of the stream mouth. The southward extent of the wide beach is determined mainly by the location of submarine canyons, which lead to offshore basins and serve as important sediment sinks.

Thus, river mouths and submarine canyons define the limits of littoral cells. Heavy mineral analyses of beach and river sands show that leakage between littoral cells tends to occur only after major floods (Rice and others, 1976). Littoral drift rates are variable along the California coast. Rates of 3.6×10^5 m^3/yr have been reported for the coast around Santa Barbara (Norris, 1964). Higher rates of around 7.5–12.5×10^5 m^3/yr are found for the coast around the Oxnard Plain (Norris, 1964; Orme and Brown, 1983), and lower rates of 1.5–2.7×10^5 m^3/yr have been reported for parts of San Diego County (Chamberlain, 1964). The spatial variation in natural littoral drift rates is probably priimarily a function of the mass of sediment supplied to the coast by streams heading in the coastal mountains, since the amount of sediment produced by sea cliff erosion in California, unlike Oregon (see McDowell, this chapter), is one to two orders of magnitude less than that supplied by streams (Griggs and Hein, 1980). The higher rates around Oxnard are probably a function of the higher denudation rates found in the Ventura area, where tectonic activity is high.

Beach form and grain size change with the seasons. In summer, when wave energy is low, wide, sandy beaches are often built. In winter, when strong storm waves are generated, sand is removed to some distance offshore and often the beach narrows so that only lag gravels remain. The sand is replaced during the following summer when wave energy diminishes again. Locally, smaller landforms such as cusps are superimposed on beaches (Seymour and Aubrey, 1985). Although sediment is periodically supplied by rivers reaching the coasts, beaches are not prograding. This implies that beach sediment is lost somewhere along the coast (Norris, 1964). The major areas where beach sediments can be lost from the system are coastal dunes (Cooper, 1967), losses to deeper water during storms (Norris, 1964), and losses through submarine canyons to offshore basins as mentioned earlier (Emery, 1960; Chamberlain, 1964). Chamberlain (1964) estimated that about 2.0×10^5 m^3/yr of sediment were lost to the Scripps submarine canyon near San Diego. Because there are a number of submarine canyons and basins off central and southern California, these features constitute an important sediment sink. Chamberlain showed that during intense local storms, sediments are transported to the heads of submarine canyons by seaward-flowing rip currents. Loss of sediment from the canyon heads to deeper water is due to slope failure (either from earthquakes or thresholds of slope angle stability being exceeded) and is unrelated to waves, currents, or storms.

The simple picture sketched above has to be modified by recent studies of the effects of human activities on beach formation and erosion processes. Norris (1964) showed that by the mid-1960s a significant portion of the drainages of rivers that reach the ocean in southern California had been blocked by dams. He pointed out that dams affect sediment supply in three ways: (1) they reduce peak discharges, and thus reduce the ability of the stream to carry sediment, (2) they trap sediment in their reservoirs, and (3) water diversion via dams lowers surface flow. Kuhn and Shepard (1980, 1983) attribute the change from wide, sandy beaches found south of Oceanside in the nineteenth century to contemporary narrow, cobbly beaches in part to the construction of dams on the principal rivers in San Diego County. Converse (1982) showed that damming of the Tijuana River in Mexico (completed around 1940) had decreased the amount of northward-moving littoral drift that supplied sediment to Imperial Beach near San Diego; as a result, erosion is occurring on that beach and is expected to affect beaches farther north.

Beach formation processes are also modified by construction of jetties, groins, and breakwaters. One of the best examples of such influences is in northern Monterey Bay, where the effect of jetty construction on littoral drift and cliff erosion rates was studied by Griggs and Johnson (1976). They found that after jetty construction a wide, protective beach formed upcoast against a formerly retreating cliff. Cliff retreat rates there decreased from 12–42 cm/yr to 3–6 cm/yr. Downcoast beaches, however, became depleted in sand, and cliff retreat rates there increased by a factor of 2 to 3 because of the loss of protective beaches. Similar conclusions were reached by Mathieson and Lajoie (1982) who studied changes in cliff retreat rates due to breakwater construction at Half Moon Bay in central California.

Loss to coastal dunes constitutes one of the important beach sediment sinks, but dunes are also important landforms in their own right. The major study of these features in central and southern California is the comprehensive work of Cooper (1967). He found that both transverse ridge (barchanoid) and parabolic forms exist along much of the California coast. Dunes are developed on north-south–trending segments of the coast just south of locations where the coastline trends east-west. Dune ages increase as one moves inland. The youngest dunes are thought to be related to the Holocene transgression, and some Holocene dunes are still tied to their beach sources. Older dunes near Monterey Bay, Morro Bay, Santa Maria, and El Segundo underlie the Holocene dunes and are found up to several kilometers inland. They are characterized by subdued topography and well-developed soils. Cooper (1967) did not attempt to differentiate the older dunes, but noted that the presence of paleosols indicated eolian sands of at least three ages in some places.

Dunes on the California Channel Islands are very different from the mainland dunes. In contrast to the quartzose mainland sands, the insular dunes are highly calcareous due to large amounts of comminuted shell fragments (Vedder and Norris, 1963; Johnson, 1967, 1977; Muhs, 1983a). Meteoric waters have dissolved some of the primary carbonate, and it has been reprecipitated as secondary carbonate cement, weakly consolidating the dunes into eolianite. Importantly, the dunes overlie buried soils developed on marine terrace deposits of interglacial ages and are often found where no sand supply exists at present. This latter observation led Vedder and Norris (1963) and Johnson (1967) to speculate that the source for sand was the insular shelves that would have been exposed during glacio-eustatic sea-level regressions. This hypothesis was confirmed by Johnson (1977) and Muhs (1983a) who reported radiocarbon dates of around 20 ka for dunes on San Miguel and San Clemente Islands. Thus, while

Figure 28. The west side of San Clemente Island, looking south, showing a flight of marine terraces formed by high stands of sea superimposed on a tectonically rising land mass. The terraces here have been mapped and dated by Muhs and Szabo (1982) and Muhs (1983a).

dunes on the mainland have their sources in contemporary beaches during the present interglacial, insular dunes were largely derived from now-submerged shelves during glacial periods.

Marine Terraces

Early investigators (Lawson, 1893; Smith, 1898) studied the often spectacular flights of marine terraces on the California coast and the offshore islands (Fig. 28). These workers and, later, others (Davis, 1933; Woodring and others, 1946) recognized that marine terraces consist of shore platforms overlain by often fossiliferous marine sands and gravels, which in turn are overlain by terrestrial deposits of varying thickness. The terrestrial deposits can be alluvium, colluvium, or eolian sand and are often thick enough (on the order of 1–30 m) to obscure the terrace morphology and make accurate mapping difficult. Frequently, one "marine terrace" may consist of two or more shore platforms at different elevations, all overlain by a smooth cover of alluvium or colluvium (Bradley and Griggs, 1976; Kern, 1977).

In the early literature (e.g., Lawson, 1893), marine terraces were regarded as being mainly of tectonic origin, because terrace elevations were different from one terrace flight to the next. Later, based on mapping of both marine and stream terraces, workers such as Davis (1933) and Alexander (1953) recognized the importance of sea-level fluctuations (due to the growth and decay of continental glaciers), as well as tectonics. Birkeland (1972) mapped terraces in the Malibu area that had previously been

studied by Davis (1933). He recognized the similarities in age estimates between California terraces and those of coral reef terraces elsewhere in the world. Birkeland thought, however, that for an unambiguous correlation of a marine terrace with a eustatic high stand of sea, ideally there should be geomorphic evidence, such as a similarity of elevation of the shoreline angle of a marine terrace with the top of a valley-fill stream terrace. Not all California marine terraces exhibit such a relationship. By the early 1970s the mechanisms and timing of marine terrace formation were still not clearly understood. A bit later, however, Ku and Kern (1974) dated three corals from the Nestor terrace (at about 23 m above sea level) near San Diego and reported an average age of about 120 ka. This date was about the same as the dates reported by Ku and others (1974) for the Waimanalo shoreline (at +7.6 m elevation) on the tectonically stable island of Oahu. Ku and Kern (1974) thus concluded that the Nestor terrace was correlative with other terraces in California of about the same age (based on the few U-series dates on coral) and the Waimanalo shoreline on Oahu. They further concluded that all these terraces formed during a eustatic high stand of sea and that the Nestor terrace, because it was 13 to 17 m above the elevation of the tectonically stable Waimanalo shoreline, must have experienced tectonic uplift. Thus, for the first time, a reliably dated California marine terrace was recognized to have formed from a eustatic high stand of sea and then to have undergone subsequent uplift.

By the late 1970s, development of amino-acid geochronology allowed correlation of terraces between widely separated areas. The technique was first applied on a large scale in California by J. F. Wehmiller, K. R. Lajoie, and co-workers (Wehmiller and others, 1977). Since that time, there has been an intensification of effort; marine terraces for other parts of California and Baja California have been mapped, and new uranium-series and amino-acid age estimates have been generated (Kern, 1977; Wehmiller and Belknap, 1978; Wehmiller and others, 1978; Woods, 1980; Lajoie and others, 1979, 1982; Omura and others, 1979; Karrow and Bada, 1980; Szabo, 1980; Orme, 1980; Emerson and others, 1981; Kennedy and others, 1982; Muhs and Szabo, 1982; Muhs, 1983a, 1985; Muhs and Rosholt, 1984). In the tectonically active Santa Barbara–Ventura area, terraces dating at about 1.8–5.5 ka, 30–50 ka, and 80 ka have been mapped (Wehmiller and others, 1977; Lajoie and others, 1979, 1982). On other parts of the coast, terraces dating at 80 ka, 107 ka, 120 ka, and 200–250 ka have been mapped; still higher terraces are beyond the range of geochronologic methods currently in use. The dates for California terraces are similar to those of coral reef terraces that have been mapped in New Guinea, Barbados, Haiti, and Japan (Bloom and others, 1974; Mesolella and others, 1969; Dodge and others, 1983; Konishi and others, 1970), and thus reflect global high stands of sea superimposed on rising land masses (Fig. 29). Terraces of a given age may have different elevations at different locations that are reflective of their respective local uplift rates. Uplift rates are low (0.2–0.5 m/ka) on most parts of the coast as a result of predominantly strike-slip tectonics (Lajoie and others, 1979). In the Ventura and Santa Barbara areas, however, uplift rates are on the order of 1.5 to 10 m/ka due to a compressive tectonic style (Lajoie and others, 1979). Thus, these latter areas are the only places on the southern California coast where terraces as young or younger than about 45 ka are found. In lower uplift rate areas, one or both of the 80 ka and 107 ka terraces are usually not found (Kern, 1977; Emerson and others, 1981; Kennedy and others, 1982; Muhs, 1983a, 1985; Muhs and Rosholt, 1984). These "missing" terraces are the result of sea-cliff retreat keeping ahead of terrace emergence.

In addition to studies of global sea level, marine terraces are useful landforms for tectonic studies. Terrace elevations and ages combined with sea-level curves can yield uplift rates, as discussed above. In addition, because terraces begin as essentially horizontal surfaces, shore-parallel profiles can identify areas of warping or differential uplift and faulting (Alexander, 1953; Birkeland, 1972; Bradley and Griggs, 1976; Kern, 1977; Lajoie and others, 1979; Woods, 1980; Lajoie, 1986).

SOIL GEOMORPHOLOGY

One of the most recent developments in research on the southern and central California coastal region has been the application of geomorphology to studies of soil genesis. The numerous flights of stream and marine terraces described above form ideal frameworks for soil chronosequences, or arrays of landforms of different ages where the time of stabilization of the landform

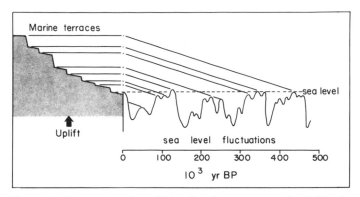

Figure 29. Diagram showing a flight of marine terraces on the California coast and their relationship to tectonic uplift and sea-level changes. Slope of diagonal lines indicates rate of uplift. Taken from Lajoie and others (1979).

marks time zero of pedogenesis (Jenny, 1941, 1980). Important in such studies is the assumption that the other soil-forming factors of climate, organisms, parent material, and relief have been relatively constant over time among members of the chronosequence.

Although at this stage of development, California soil genesis studies are dependent on geomorphic frameworks and their absolute-age control, the ultimate goal of many chronosequence studies is in fact the reverse, that is, to use soils as an aid in geomorphic studies. If it can be established that soil development shows time-dependent changes for certain properties in well-dated chronosequences, soils can be applied to assignment of relative ages to landforms in other areas where age control is lacking, following the approach outlined by Birkeland (1984). Such applications have already been alluded to in the studies of Keller and others (1982), Rockwell and others (1984), Lettis (1985), and Ponti (1985).

Detailed soil chronosequence studies have been carried out in four areas in southern California, two on marine terrace sequences at coastal locations, and two on river terrace sequences at inland locations. Muhs (1982, 1983b) studied soils on alluvial fans and marine terraces on San Clemente Island, which ranged in age from <3 ka to >1100 ka. Harden and Taylor (1983) and Harden and others (1986) examined soils on marine and stream terraces near Ventura that ranged in age from <1 ka to 60 to 105 ka. Inland of Harden's study area, soils on river terraces along the Ventura River were examined by Rockwell and others (1984, 1985). Their terraces ranged in age from 10–20 yr to about 79–105 ka. McFadden and Hendricks (1985) studied soils developed on alluvial fans and river terraces in the San Gabriel and San Bernardino mountains which ranged in age from 0–1 ka to greater than 700 ka.

These studies show that at least some properties change with time in all four areas, but not all properties behave similarly from area to area. In general, however, solum thickness and redness, structural development, clay content, clay mineralogy, soluble salt content, and Fe content appear to be useful potential relative age indicators for correlation of geomorphic surfaces in California. Rates of changes in soil properties differ among study areas, however. For example, when clay content (in g/cm^2-column) is plotted as a function of time for different areas, it is clear that clay accumulation rates vary significantly (Fig. 30). Ventura and San Clemente Island soils are accumulating clay much faster than soils in the Transverse Ranges, despite the fact that precipitation is lower in the former two areas than in most parts of the latter area. This can be explained by the fact that soils in both of the coastal study areas are in the paths of Santa Ana winds, which deliver eolian silt and clay to the coast from localities in the Mojave Desert several times each year (Muhs, 1983b). In contrast, clay in the Transverse Ranges soils builds up mainly by primary mineral weathering (McFadden and Hendricks, 1985). Despite the variability in importance of pedogenic processes, it is clear from the studies cited here that soils have considerable potential as relative-age indicators in geomorphological studies in California's mountains and coastal areas.

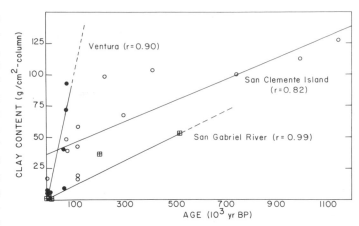

Figure 30. Clay buildup as a function of time in California coastal and Transverse Ranges soils. San Gabriel River data from McFadden and Hendricks (1985), Ventura data from Harden and others (1986), and San Clemente Island data from Muhs (1982). Midpoints of age estimates given in these references were used to plot points and compute best-fit regression lines.

CONCLUSIONS

Geomorphic studies in the last 30 years in central and southern California have added considerably to our understanding of mass movement, fluvial and coastal processes, and the relationships between soils and landforms. Perhaps one of the most important contributions of this region to geomorphology is a better understanding of the importance of tectonics to landform genesis. The California examples have wide application to geomorphic studies elsewhere where tectonic processes are active. The reverse is true as well: well-dated stream and marine terraces and alluvial fans that are offset by faults can yield slip rates and other tectonic information. The emerging discipline of tectonic geomorphology has its roots in many of the studies that have been conducted in California.

In addition, southern California, with its large population, provides for studies of the effects of human activities on geomorphic processes. The Los Angeles area in particular has presented an unusually good opportunity to study the modern impacts of humans on landforms and, in turn, the effects geomorphic processes have on urban dwellers (Cooke, 1984). As long as California continues to be an attractive place to live, human-landscape interaction will continue.

SUMMARY OF THE GEOMORPHOLOGY OF THE PACIFIC COAST AND MOUNTAIN SYSTEM

Daniel R. Muhs

This chapter has presented summaries of recent studies of geomorphic processes in the Pacific Coast and Mountain System (PCMS). It is clear that each subregion has made specific contributions to geomorphology. It is also apparent that the region as a whole has unique qualities that have made important impacts on the field of geomorphology.

In Alaska, R. M. Thorson points out the importance of the combination of climate, volcanism, tectonics, and glaciation in shaping the landscape of the PCMS. He presents a useful model that shows how geomorphic processes are linked and are conditioned by Alaska's tectonic regime and climate. Future geomorphic studies in Alaska should focus on the evolution of major drainage basins, rates of denudation versus uplift, and the dynamics of ice sheets on the continental shelf.

Linkage of geomorphic processes is also important in explaining the origin of landforms in the PCMS of British Columbia. J. J. Clague and W. H. Mathews show how tectonics, glaciation, fluvial processes, and mass movement are the most important landform-producing processes that have operated during the Cenozoic. Tectonic activity in the Tertiary set the topographic stage for the pattern of Quaternary glaciation. Glaciation, in turn, influenced fluvial processes and brought about isostatic adjustments of the crust and sea level changes which subsequently brought about changes in erosion and sedimentation processes on the coast and continental shelf.

The PCMS of Washington and Oregon has been the focus of detailed studies of coastal processes and erosion and sedimentation in coastal mountains, as summarized by P. F. McDowell. Here again, tectonic activity and climate have played important roles in the geomorphology of this subregion, resulting in high rates of denudation, solution, and mass movement. These processes have been strongly influenced by human activities as well, mainly in the form of timber harvesting. It is hoped that future research in this area will focus on the Quaternary fluvial history of the coastal mountains, sediment budgets both on the coast and in the mountains, and the ages of prominent marine terraces found on the coast.

In northern California, H. M. Kelsey shows how detailed

studies of mass movement and fluvial processes have yielded a wealth of information on sediment budgets in drainage basins. Sediment budget studies have probably been conducted in more detail here than in most parts of the PCMS, because of the impact of forestry practices. Results have shown the importance of tectonics and extreme climatic events in conditioning rates of erosion and sedimentation. Thus, the time-honored concept of most geomorphic work being done by high-frequency, low-magnitude processes may not apply in the PCMS of northern California.

Central and southern California, like the rest of the PCMS of North America, is tectonically active, and D. R. Muhs shows that tectonic activity has an important influence on mass movement and fluvial processes and in the genesis of fluvial and coastal landforms. Central and southern California have been the focus of several studies of the effects of eustatic sea-level changes on coastal and fluvial landforms as well. Again, linkage of processes is evident: marine terraces form by marine erosion processes and mass wasting during eustatic high stands of sea; these processes are superimposed on a tectonically rising land mass. Southern California is also a landscape that has been strongly influenced by human activity and will undoubtedly continue to be influenced by humans in the future.

Thus, each subregion of the PCMS has made its own special contributions to geomorphology. Not all geomorphic processes are important in all subregions, but tectonics and eustatic sea-level changes have influenced geomorphic processes over the entire PCMS region (Fig. 31). The proximity of the region to major plate boundaries results in high rates of uplift that produces relief, faulting and folding of landforms, and earthquakes, which induce mass movements. Superimposed on these effects are eustatic changes of sea level, which change the coastal geography, alter

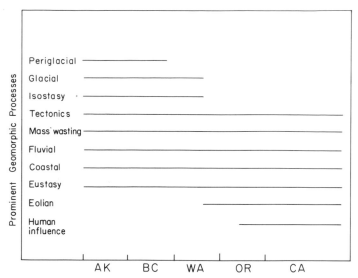

Figure 31. Importance of various geomorphic processes as a function of subregion within the Pacific Coast and Mountain System of North America.

coastal processes, influence stream systems by their effects on base level changes, and bring about alterations in sedimentation patterns. The effects of tectonics and eustatic sea-level changes can be seen in the landforms of the PCMS from California to Alaska. Thus, perhaps the most important contributions of studies in the PCMS are those that increase our understanding of the linkages between tectonics and eustatic sea-level changes, and geomorphic processes.

REFERENCES CITED

Ackerman, R. E., Hamilton, T. D., and Stuckenrath, R., 1979, Early culture complexes on the northern Northwest coast: Canadian Journal of Archaeology, no. 3, p. 195–209.

Adams, J., 1984, Active tectonic deformation of the Pacific Northwest continental margin: Tectonics, v. 3, p. 449–472.

Aguilar-Tunon, N. A., and Komar, P. D., 1978, The annual cycle of profile changes of two Oregon beaches: The Ore Bin, v. 40, p. 25–39.

Alexander, C. S., 1953, The marine and stream terraces of the Capitola-Watsonville area: University of California Publications in Geography, v. 10, p. 1–44.

Ando, M., and Balasz, E. I., 1979, Geodetic evidence for aseismic subduction of the Juan de Fuca plate: Journal of Geophysical Research, v. 84, p. 3023–3028.

Armentrout, J. J., 1983, Glacial lithofacies of the Neogene Yakataga Formation, Coast Range, Alaska, *in* Molnia, B. F., ed., Glacial-marine sedimentation: New York, Plenum Press, p. 629–665.

Armstrong, J. E., 1981, Post-Vashon Wisconsin glaciation, Fraser Lowland, British Columbia, Canada: Geological Survey of Canada Bulletin 322, 34 p.

Ashley, G. M., 1979, Sedimentology of a tidal lake, Pitt Lake, British Columbia, Canada, *in* Schlüchter, Ch., ed., Moraines and varves, origin/genesis/classification: Rotterdam, A. A. Balkema, p. 327–345.

Atwater, T., 1970, Implications of plate tectonics for the Cenozoic tectonic evolu-

tion of western North America: Geological Society of America Bulletin, v. 81, p. 3513–3536.

Atwater, T., and Molnar, P., 1973, Relative motion of the Pacific and North American plates deduced from sea-floor spreading in the Atlantic, Indian, and South Pacific oceans, *in* Kovach, R. L., and Nur, A., eds., Proceedings, Conference on tectonic problems of the San Andreas fault system: Stanford University Publication in Geological Science, v. 13, p. 136–148.

Baldwin, E. W., 1981, Geology of Oregon (third edition): Dubuque, Iowa, Kendall-Hunt Publishing Company, 170 p.

Baldwin, E. W., Beaulieu, J. D., Ramp, L., Gray, J., Newton, V. C., Jr., and Mason, R. S., 1973, Geology and mineral resources of Coos County, Oregon: Oregon Department of Geology and Mineral Industries Bulletin 80, 82 p.

Barr, S. M., and Chase, R. L., 1974, Geology of the northern end of Juan de Fuca Ridge and sea-floor spreading: Canadian Journal of Earth Sciences, v. 11, p. 1384–1406.

Bartsch-Winkler, S. R., and Garow, H. C., 1982, Depositional system reaching maturity at Portage Flats: U.S. Geological Survey Circular 844, p. 115–117.

Beaulieu, J. D., and Hughes, P. W., 1975, Environmental geology of western Coos and Douglas counties, Oregon: Oregon Department of Geology and Mineral Industries, Bulletin 87, 148 p.

—— , 1976, Land-use geology of western Curry County, Oregon: Oregon Department of Geology and Mineral Industries, Bulletin 90, 148 p.

Beikman, H. M., 1980, Geologic map of Alaska: U.S. Geological Survey, scale 1:2,500,000, 2 sheets.

Benson, C. S., and Motyka, R. J., 1979, Glacier-volcano interactions on Mt. Wrangell, Alaska: University of Alaska Geophysical Institute Contribution Series A-541, 25 p.

Berry, M. J., and Forsyth, D. A., 1975, Structure of the Canadian Cordillera from seismic refraction and other data: Canadian Journal of Earth Sciences, v. 12, p. 182–208.

Beschta, R. L., 1978, Long-term patterns of sediment production following road construction and logging in the Oregon Coast Range: Water Resources Research, v. 14, p. 1011–1016.

—— , 1979, Debris removal and its effects on sedimentation in an Oregon Coast Range stream: Northwest Science, v. 53, p. 71–77.

Best, D. W., Kelsey, H. M., Hagans, D. K., and Alpert, M., 1987, Role of fluvial hillslope erosion and road construction in the sediment budget of Garrett Creek, Humboldt County, California, in Nolan, K. M., Kelsey, H. M., and Marron, D. C., eds., Geomorphic processes and aquatic habitat in the Redwood Creek basin, northwestern California: U.S. Geological Survey Professional Paper 1454 (in press).

Birkeland, P. W., 1972, Late Quaternary eustatic sea-level changes along the Malibu coast, Los Angeles County, California: Journal of Geology, v. 80, p. 432–448.

—— , 1984, Soils and geomorphology: New York, Oxford University Press, 372 p.

Black, R. F., 1951, Eolian deposits of Alaska: Arctic, v. 4, p. 89–111.

—— , 1980, Isostatic, tectonic, and eustatic movements of sea level in the Aleutian Islans, Alaska, in Morner, N., ed., Earth rheology, isostasy, and eustasy: New York, John Wiley and Sons, p. 231–248.

—— , 1983, Glacial chronology of the Aleutian Islands, in Thorson, R. M., and Hamilton, T. D., eds., Glaciation in Alaska; Extended abstracts from a workshop: Fairbanks, Alaska, University of Alaska Museum Occasional Paper No. 2, p. 5–10.

Blake, M. C., and Jones, D. L., 1981, The Franciscan assemblage and related rocks in northern California; A reinterpretation, in Ernst, W. G., ed., The geotectonic development of California: Prentice-Hall, Incorporated, p. 306–328.

Bloom, A. L., 1983, Sea level and coastal morphology of the United States through the late Wisconsin glacial maximum, in Porter, S. C., ed., Late Quaternary environments of the United States, Volume 1, The Late Pleistocene: Minneapolis, University of Minnesota Press, p. 215–229.

Bloom, A. L., Broecker, W. S., Chappell, J.M.A., Matthews, R. K., and Mesolella, K. J., 1974, Quaternary sea level fluctuations on a tectonic coast; New ^{230}Th/^{234}U dates from the Huon Peninsula, New Guinea: Quaternary Research, v. 4, p. 185–205.

Boggs, S., Jr., and Jones, C. A., 1976, Seasonal reversal of flood-tide dominant sediment transport in a small Oregon estuary: Geological Society of America Bulletin, v. 87, p. 419–426.

Boothroyd, J. C., and Ashley, G. M., 1975, Processes, bar morphology, and sedimentary structures on braided outwash fans, northeastern Gulf of Alaska, in Jopling, A. V., and McDonald, B. C., eds., Glaciofluvial and glaciolacustrine sedimentation: Society of Economic Paleontologists and Mineralogists Special Publication 23, p. 193–222.

Bradley, W. C., and Griggs, G. B., 1976, Form, genesis, and deformation of central California wave-cut platforms: Geological Society of America Bulletin, v. 87, p. 433–449.

Bradley, W. C., Fahnestock, R. K., and Rowekamp, E. T., 1972, Coarse sediment transport by flood flows on Knik River, Alaska: Geological Society of America Bulletin, v. 83, p. 1261–1284.

Brown, C. S., Meier, M. F., and Post, A., 1982, Calving speed of Alaska tidewater glaciers, with application to Columbia Glacier: U.S. Geological Survey Professional Paper, 1258-C, 13 p.

Brown, G. W., and Krygier, J. T., 1971, Clear-cut logging and sediment production in the Oregon Coast Range: Water Resources Research, v. 7, p. 1189–1198.

Brown, W. M., and Ritter, J. R., 1971, Sediment transport and turbidity in the Eel River basin, California: U.S. Geological Survey Water Supply Paper 1986, 70 p.

Bruhn, R. L., 1978, Holocene displacements measured by trenching the Castle Mountain fault near Houston, Alaska: State of Alaska Division of Geological and Geophysical Surveys Report 61, p. 1–4.

Bruns, T. R., and Plafker, G., 1976, Structural style of part of the outer continental shelf in the Gulf of Alaska Tertiary Province: U.S. Geological Survey Circular 733, p. 13–14.

Bucknam, R. C., and Anderson, R. E., 1979, Estimation of fault-scarp ages from a scarp-height-slope-angle relationship: Geology, v. 7, p. 11–14.

Bull, W. B., 1961, Tectonic significance of radial profiles of alluvial fans in western Fresno County, California: U.S. Geological Survey Professional Paper 424-B, p. B182–B184.

—— , 1962, Relations of alluvial-fan size and slope to drainage-basin size and lithology in western Fresno County, California: U.S. Geological Survey Professional paper 450-B, p. B51–B53.

—— , 1964a, Geomorphology of segmented alluvial fans in western Fresno County, California: U.S. Geological Survey Professional Paper 352-E, p. 89–129.

—— , 1964b, History and causes of channel trenching in western Fresno County, California: American Journal of Science, v. 262, p. 259–258.

Byrne, J. V., 1964, An erosional classification for the northern Oregon coast: Annals of the Association of American Geographers, v. 54, p. 329–341.

California Department of Water Resources, 1970, Middle Fork Eel River landslide investigation: California Department of Water Resources Memorandum Report, 120 p.

Campbell, R. H., 1974, Debris flows originating from soil slips during rainstorms in southern California: Quarterly Journal of Engineering Geology, v. 7, p. 339–349.

—— , 1975, Soil slips, debris flows, and rainstorms in the Santa Monica Mountains and vicinity, southern California: U.S. Geological Survey Professional Paper 851, 51 p.

Carlson, P. R., Wheeler, M. C., Molnia, B. F., and Atwood, T. J., 1979, Neoglacial sedimentation in Glacier Bay, Alaska: U.S. Geological Survey Circular 804-B, p. B114–B116.

Carson, M. A., and MacLean, P. A., 1985, Storm-controlled oblique dunes of the Oregon coast, discussion: Geological Society of America Bulletin, v. 96, p. 409–410.

Carter, L., 1973, Surficial sediments of Barkley Sound and the adjacent continental shelf, west coast Vancouver Island: Canadian Journal of Earth Sciences, v. 10, p. 441–459.

Carver, G. A., Stephens, T. A., and Young, J. C., 1983, Quaternary thrust and reverse faulting on the Mad River fault zone, coastal northern California: Geological Society of America Abstracts with Programs, v. 15, p. 316.

Chamberlain, T. K., 1964, Mass transport of sediment in the heads of Scripps Submarine Canyon, California, in Miller, R. L., ed., Papers in marine geology, Shepard commemorative volume: New York, The MacMilliam Company, p. 42–64.

Chamberlain, V. E., and Lambert, R. St. J., 1985, Cordilleria, a newly defined Canadian microcontinent: Nature, v. 314, p. 707–713.

Chase, R. L., and Tiffin, D. L., 1972, Queen Charlotte fault-zone, British Columbia: International Geological Congress, 24th, Montreal, Proceedings, sec. 8, p. 17–27.

Church, M., 1983, Pattern of instability in a wandering gravel bed channel, in Collinson, J. D., and Lewin, J., eds., Modern and ancient fluvial systems: International Association of Sedimentologists Special Publication 6, p. 169–180.

Church, M., and Ryder, J. M., 1972, Paraglacial sedimentation; A consideration of fluvial processes conditioned by glaciation: Geological Society of America Bulletin, v. 83, p. 3059–3072.

Clague, J. J., 1975, Late Quaternary sea-level fluctuations, Pacific coast of Canada and adjacent areas: Geological Survey of Canada Paper 75-1C,

p. 17–21.

——, 1976, Quadra Sand and its relation to the late Wisconsin glaciation of southwest British Columbia: Canadian Journal of Earth Sciences, v. 13, p. 803–815.

——, 1981, Late Quaternary geology and geochronology of British Columbia, Part 2; Summary and discussion of radiocarbon-dated Quaternary history: Geological Survey of Canada Paper 80-35, 41 p.

——, 1983, Glacio-isostatic effects of the Cordilleran Ice Sheet, British Columbia, Canada, *in* Smith, D. E., and Dawson, A. G., eds., Shorelines and isostasy: London, Academic Press, p. 321–343.

——, 1985, Deglaciation of the Prince Rupert–Kitimat area, British Columbia: Canadian Journal of Earth Sciences, v. 22, p. 256–265.

——, 1986, The Quaternary stratigraphic record of British Columbia; Evidence for episodic sedimentation and erosion controlled by glaciation: Canadian Journal of Earth Sciences, v. 23, p. 885–894.

Clague, J. J., and Bornhold, B. D., 1980, Morphology and littoral processes of the Pacific coast of Canada, *in* McCann, S. B., ed., The coastline of Canada: Geological Survey of Canada Paper 80-10, p. 339–380.

Clague, J. J., Mathewes, R. W., and Warner, B. G., 1982a, Late Quaternary geology of eastern Graham Island, Queen Charlotte Islands, British Columbia: Canadian Journal of Earth Sciences, v. 19, p. 1786–1795.

Clague, J. J., Harper, J. R., Hebda, R. J., and Howes, D. E., 1982b, Late Quaternary sea levels and crustal movements, coastal British Columbia: Canadian Journal of Earth Sciences, v. 19, p. 597–618.

Clark, J. A., 1978a, An inverse problem in glacial geology; The reconstruction of glacier thinning in Glacier Bay, Alaska, between A.D. 1910 and 1960 from relative sea level data: Journal of Glaciology, v. 18, p. 481–503.

——, 1978b, Global changes in postglacial sea level; A numerical calculation: Quaternary Research 9, p. 265–287.

Clark, J. A., and Lingle, C. S., 1979, Predicted relative sea-level changes (18,000 years B.P., to present) caused by late glacial retreat of the Antarctic ice sheet: Quaternary Research, v. 11, p. 279–298.

Clifton, H. E., Hunter, R. E., and Phillips, R. L., 1971, Depositional structures and processes in the non-barred high-energy nearshore: Journal of Sedimentary Petrology, v. 41, p. 651–670.

Coats, R. R., 1950, Volcanic activity in the Aleutian Arc: U.S. Geological Survey Bulletin 974-B, p. 35–47.

Coney, P. J., 1978, Mesozoic-Cenozoic Cordilleran plate tectonics, *in* Smith, R. B., and Eaton, G. P., eds., Cenozoic tectonics and regional geophysics of the western Cordillera: Geological Society of America Memoir 152, p. 33–50.

Coney, P. J., Jones, D. L., and Monger, J.W.H., 1980, Cordilleran suspect terranes: Nature, v. 288, p. 329–333.

Converse, H., 1982, Barrier beach features of California, *in* Edge, B. L., ed., Proceedings, Eighteenth Coastal Engineering Conference, Volume 2: New York, American Society of Civil Engineers, p. 1008–1027.

Cooke, R. U., 1984, Geomorphological hazards in Los Angeles: London, George Allen and Unwin, 206 p.

Cooke, R. U., and Reeves, R. W., 1976, Arroyos and environmental change in the American South-west: Oxford, Clarendon Press, 213 p.

Coonrad, W. L., and Hoare, J. M., 1976, The Togiak Tuya: U.S. Geological Survey Circular 733, p. 44–45.

Cooper, W. S., 1958, Coastal sand dunes of Oregon and Washington: Geological Society of America Memoir 72, 169 p.

——, 1967, Coastal dunes of California: Geological Society of America Memoir 104, 131 p.

Coulter, H. W., Hopkins, D. M., Karlstrom, T.N.V., Péwé, T. L., Wahrhaftig, C., and Williams, J. R., 1965, Map showing extent of glaciations in Alaska: U.S. Geological Survey Miscellaneous Geological Investigations Map I-415, scale 1:2,500,000.

Cowan, D. S., and Boss, R. F., 1978, Tectonic framework of the southwestn Kenai Peninsula, Alaska: Geological Society of America Bulletin, v. 89, p. 155–158.

Crittenden, R., and Muhs, D. R., 1986, Cliff-height and slope-angle relationships in a chronosequence of Quaternary marine terraces, San Clemente Island, California: Zeitschrift für Geomorphologie, v. 30, p. 291–301.

Davis, W. M., 1933, Glacial epochs of the Santa Monica Mountains, California: Geological Society of America Bulletin, v. 44, p. 1041–1133.

Denton, G. H., and Armstrong, R. L., 1969, Miocene-Pliocene glaciations in southern Alaska: American Journal of Science, v. 267, p. 1121–1142.

Derksen, S. J., 1975, Raised marine terranes southeast of Lituya Bay, Alaska: Geological Society of America Abstracts with Programs, v. 7, p. 310.

Detterman, R. L., Reed, B. L., and Rubin, M., 1965, Radiocarbon dates from Iliamna Lake, Alaska: U.S. Geological Survey Professional Paper 525-D, p. D34–D36.

Detterman, R. L., Plafker, G., Hudson, T., Tysdal, R. G., and Pavoni, N., 1974, Surface geology and Holocene breaks along the Susitna segment of the Castle Mountain fault, Alaska: U.S. Geological Survey Miscellaneous Field Studies Map MF-738, scale 1:63,360.

Dicken, S. N., 1961, Some recent physical changes of the Oregon coast: U.S. Department of Navy, Office of Naval Research, Project Report NR 388-062, 151 p.

Dietrich, W. E., and Dorn, R., 1984, Significance of thick deposits of colluvium on hillslopes; A case study involving the use of pollen analysis in the coastal mountains of northern California: Journal of Geology, v. 92, p. 142–158.

Dietrich, W. E., and Dunne, T., 1978, Sediment budget for a small catchment in mountainous terrain: Zeitschrift für Geomorphologie, Suppementband 29, p. 191–206.

Dietrich, W. E., Dunne, T., Humphrey, N. F., and Reid, L., 1982, Construction of sediment budgets for drainage basins, *in* Swanson, F. J., Janda, R. J., Dunne, T., and Swanston, D. N., eds., Sediment budgets and routing in forested drainage basins: U.S. Department of Agriculture, Forest Service, Pacific Northwest Forest and Range Experiment Station, General Technical Report PNW-141, p. 5–23.

Diller, J. S., 1914, Mineral resources of southwestern Oregon: U.S. Geological Survey Bulletin 546, 147 p.

Dixon, J. C., Thorn, C. E., and Darmody, R. G., 1984, Chemical weathering processes on the Vantage Peak nunatak, Juneau Icefield, southern Alaska: Physical Geography, v. 5, p. 111–131.

Dodge, R. E., Fairbanks, R. G., Benninger, L. K., and Maurrasse, F., 1983, Pleistocene sea levels from raised coral reefs of Haiti: Science, v. 219, p. 1423–1425.

Driscoll, F. G., Jr., 1980a, Formation of the neoglacial surge moraines of the Klutlan Glacier, Yukon Territory, Canada: Quaternary Research, v. 14, p. 19–30.

——, 1980b, Wastage of the Klutlan ice-cored moraines, Yukon Territory, Canada: Quaternary Research, v. 14, p. 31–49.

Dutton, C. E., 1882, Tertiary history of the Grand Canyon district: U.S. Geological Survey Monograph, v. 2, 264 p.

Eckel, E. B., 1970, The Alaska earthquake March 27, 1964; Lessons and conclusions: U.S. Geological Survey Professional Paper 546, 57 p.

Eisbacher, G. H., 1983, Slope stability and mountain torrents, Fraser Lowlands and southern Coast Mountains, British Columbia: Geological Association of Canada, Mineralogical Association of Canada, Canadian Geophysical Union, Joint Annual Meeting, Victoria, Guidebook, Field Trip 15, 46 p.

Emerson, W. K., Kennedy, G. L., Wehmiller, J. F., and Keenan, E., 1981, Age relations and zoogeographic implications of late Pleistocene marine invertebrate faunas from Turtle Bay, Baja California Sur, Mexico: Nautilus, v. 95, p. 105–116.

Emery, K. O., 1960, The sea off southern California: New York, John Wiley, 366 p.

——, 1967, The activity of coastal landslides related to sea level: Revue de Géographie Physique et de Géologie Dynamique, v. 9, p. 177–180.

Emery, K. O., and Kuhn, G. G., 1980, Erosion of rock shores at La Jolla, California: Marine Geology, v. 37, p. 197–208.

——, 1982, Sea cliffs; Their processes, profiles, and classification: Geological Society of America Bulletin, v. 93, p. 644–654.

Evenson, E. B., and Clinch, J. M., 1986, The importance of fluvial systems in debris transport at alpine glacier margins: American Quaternary Association Program and Abstracts of the Ninth Biennial Meeting, Urbana, p. 58–60.

Farrington, R. L., and Savina, M. E., 1977, Off site effects of roads and clearcut units on slope stability and stream channels, Fox Planning Unit: U.S. Forest Service unpublished report, Six Rivers National Forest, Eureka, California, 76 p.

Ferrians, O. J., Jr., Nichols, D. R., and Williams, J. R., 1983, Copper River basin, *in* Péwé, T. L., and Reger, R. D., eds., Richardson and Glenn Highways, Alaska, Guidebook 1, IV International Conference on Permafrost: Fairbanks, Alaska Division of Geological and Geophysical Surveys, p. 137–175.

Field, W. O., ed., 1975, Mountain glaciers of the Northern Hemisphere, Volume 2: Hanover, New Hampshire, U.S. Army Cold Regions Research and Engineering Laboratory, v. 2, p. 3–650.

Florsheim, J. L., 1985, Fluvial requirements for gravel bar formation in northwestern California [M.S. thesis]: Arcata, California, Humboldt State Unviersity, 101 p.

Fox, D. S., Bell, S., Nehlsen, W., and Damron, J., 1984, The Columbia River Estuary atlas of physical and biological characteristics: Astoria, Oregon, Columbia River Estuary Data Development Program, 87 p.

Fox, K. F., Fleck, R. J., Curtis, G. H., and Meyer, C. E., 1985, Implications of the northwestwardly younger age of the volcanic rocks of west-central California: Geological Society of America Bulletin, v. 96, p. 647–654.

Fox, W. T., and Davis, R. A., Jr., 1976, Weather patterns and coastal erosion, *in* Davis, R. A., Jr., and Etherington, R. L., eds., Beach and nearshore sedimentation: Society of Economic Paleontologists and Mineralogists Special Publication no. 24, p. 1–23.

——— , 1978, Seasonal variation in beach erosion and sedimentation on the Oregon coast: Geological Society of America Bulletin, v. 89, p. 1541–1549.

Fredriksen, R. L., and Harr, R. D., 1979, Soil, vegetation, and watershed management of the douglas-fir region, *in* Heilman, P. E., Anderson, H. W., and Baumgartner, D. M., eds., Forest soils of the douglas-fir region: Pullman, Washington State University Cooperative Extension Service, p. 231–260.

Furbish, D. J., and Rice, R. M., 1983, Predicting landslides related to clearcut logging, northwestern California, U.S.A.: Mountain Research and Development, v. 3, p. 253–259.

Fyles, J. G., 1963, Surficial geology of Horne Lake and Parksville map-areas, Vancouver Island, British Columbia: Geological Survey of Canada Memoir 318, 142 p.

Galloway, W. E., 1975, Process framework for describing the morphological and stratigraphic evolution of deltaic depositional systems, *in* Broussard, M. L., ed., Deltas (second edition): Houston, Houston Geological Society, p. 87–98.

Gard, L. M., Jr., 1980, The Pleistocene geology of Amchitka Island, Aleutian Islands, Alaska: U.S. Geological Survey Bulletin 1478, 38 p.

Gilbert, R., 1975, Sedimentation in Lillooet Lake, British Columbia: Canadian Journal of Earth Sciences, v. 12, p. 1697–1711.

Glazner, A. F., and Schubert, G., 1985, Flexure of the North American lithosphere above the subducted Mendocino Fracture Zone and the formation of east-west faults in the Transverse Ranges: Journal of Geophysical Research, v. 90, p. 5405–5409.

Glenn, J. L., 1978, Sediment sources and Holocene sedimentation history in Tillamook Bay, Oregon: U.S. Geological Survey Open File Report 78-680, 64 p.

Goodwin, R. G., 1984, Neoglacial lacustrine sedimentation and ice advance, Glacier Bay, Alaska: Ohio State University, Institute of Polar Studies Report No. 79, 183 p.

Griggs, A. B., 1945, Chromite-bearing sands of the southern part of the coast of Oregon: U.S. Geological Survey Bulletin 945-E, p. 113–150.

Griggs, G. B., and Hein, J. R., 1980, Sources, dispersal, and clay mineral composition off central and northern California: Journal of Geology, v. 88, p. 541–566.

Griggs, G. B., and Johnson, R. E., 1976, Effects of the Santa Cruz Harbor on coastal processes of northern Monterey Bay, California: Environmental Geology, v. 1, p. 299–312.

——— , 1979, Coastal erosion Santa Cruz County: California Geology, v. 32, p. 67–76.

Griggs, G. B., and Walsh, B. L., 1981, The impact, control, and mitigation of off-road vehicle activity in Hungry Valley, California: Environmental Geology, v. 3, p. 229–243.

Gustavson, T. C., and Boothroyd, J. C., 1982, Subglacial fluvial erosion; A major source of stratified drift, Malaspina Glacier, Alaska, *in* Davidson-Arnott, R., Nickling, W., and Fahey, B. D., eds., Research in glacial, glaciofluvial, and glaciolacustrine systems: Proceedings of the 6th Guelph Symposium on Geomorphology, 1980, Norwich, Geo Books, p. 93–116.

Hackett, S. W., and Santeford, H. S., 1980, Avalanche zoning in Alaska, U.S.A.: Journal of Glaciology, v. 26, p. 377–391.

Hamilton, T. D., and Thorson, R. M., 1983, The Cordilleran Ice Sheet in Alaska, *in* Porter, S. C., ed., Late Quaternary environments of the United States: Minneapolis, University of Minnesota Press, p. 38–52.

Hamilton, T. D., Reed, K. M., and Thorson, R. M., eds., 1986, Glaciation in Alaska; The geologic record: Anchorage, Alaska Geological Society, 265 p.

Hamilton, T. S., and Luternauer, J. L., 1983, Evidence of seafloor instability in the south-central Strait of Georgia, British Columbia; A preliminary compilation: Geological Survey of Canada Paper 83-1A, p. 417–421.

Haner, B. E., 1984, Santa Ana River; An example of a sandy braided floodplain system showing sediment source area imprintation and selective sediment modification: Sedimentary Geology, v. 38, p. 247–261.

Hanks, T. C., Bucknam, R. C., Lajoie, K. R., and Wallace, R. E., 1984, Modification of wave-cut and faulting-controlled landforms: Journal of Geophysical Research, v. 89, p. 5771–5790.

Harden, D. R., Janda, R. J., and Nolan, K. M., 1978, Mass movement and storms in the drainage basin of Redwood Creek, Humboldt County, California; A progress report: U.S. Geological Survey Open-File Report 78-486, 161 p.

Harden, J. W., and Taylor, E. T., 1983, A quantitative comparison of soil development in four climatic regimes: Quaternary Research, v. 20, p. 342–359.

Harden, J. W., Sarna-Wojcicki, A. M., and Dembroff, G. R., 1986, Soils developed on coastal and fluvial terraces near Ventura, California: U.S. Geological Survey Bulletin 1590-B (in press).

Hawkins, R. H., 1982, Magnitude and frequency of sediment transport in three northern California coastal streams [M.S. thesis]: Arcata, California, Humboldt State University, 68 p.

Hayes, M. O., 1977, A modern depositional system; Southeast Alaskan coast, *in* Hayes, M. O., and Kana, T. W., eds., Terrigenous clastic depositional environments; Some modern examples: Columbia, South Carolina, Coastal Research Division, Department of Geology, University of South Carolina Technical Report no. 11-CRD, p. I-112–I-120.

Helley, E. J., and LeMarche, V. C., 1973, Historic flood information for northern California streams from geological and botanical evidence: U.S. Geological Survey Professional Paper 485-E, p. E1–E16.

Hickey, J. J., 1969, Variations in low-water streambed elevations at selected stream-gaging stations in northwestern California: U.S. Geological Survey Water Supply Paper 1879-E, 33 p.

Hicks, S. D., and Shofnos, W., 1965, The determination of land emergence from sea level observations in southeast Alaska: Journal of Geophysical Research, v. 70, p. 3315–3320.

Hogan, L. G., Scheidegger, K. F., Kulm, L. D., Dymond, J., and Mikkelsen, N., 1978, Biostratigraphic and tectonic implications of $^{40}Ar/^{39}Ar$ dates of ash layers from the northeast Gulf of Alaska: Geological Society of America Bulletin, v. 89, p. 1259–1264.

Holland, S. S., 1964, Landforms of British Columbia, a physiographic outline: British Columbia Department of Mines and Petroleum Resources Bulletin 48, 138 p.

Hollister, L. S., 1979, Metamorphism and crustal displacements; New insights: Episodes, v. 1979, no. 3, p. 3–8.

Holtedahl, H., 1960, Mountain, fiord, strandflat; Geomorphology and general geology of parts of western Norway: International Geological Congress, 21st, Norden, Guide to Excursions A6 and C3, 29 p.

Hudson, T., Plafker, G., and Rubin, M., 1976, Uplift rates of marine terrace sequences in the Gulf of Alaska: U.S. Geological Survey Circular 733, p. 11–12.

Hudson, T., Dixon, K., and Plafker, G., 1982, Regional uplift in southeastern Alaska: U.S. Geological Survey Circular 844, p. 132–135.

Humphrey, N. F., 1982, Pore pressures in debris failure initiation [M.S. thesis]: Seattle, University of Washington, 169 p.

Hunter, R. E., Clifton, H. E., and Phillips, R. L., 1979, Depositional processes, sedimentary structures, and predicted vertical sequences in barred nearshore systems, southern Oregon coast: Journal of Sedimentary Petrology, v. 49, p. 711–726.

Hunter, R. E., Richmond, B. M., and Alpha, T. R., 1983, Storm-controlled oblique dunes of the Oregon coast: Geological Society of America Bulletin, v. 94, p. 1450–1465.

Inman, D. L., and Brush, B. M., 1973, The coastal challenge: Science, v. 181, p. 20–32.

Irwin, W. P., 1960, Geologic reconnaissance of the northern Coast Ranges and Klamath Mountains, California: California Division of Mines and Geology Bulletin 179, 80 p.

—— , 1981, Tectonic accretion of the Klamath Mountains, in Ernst, W. G., ed., The geotectonic development of California: Prentice-Hall, Incorporated, p. 29–49.

Iverson, R. M., 1984, Unsteady, nonuniform landslide motion; Theory and measurement [Ph.D. thesis]: Stanford University, 303 p.

—— , 1985, A constitutive equation for mass movement behavior: Journal of Geology, v. 93, p. 143–160.

—— , 1986a, Dynamics of slow landslides; A theory for time-dependent behavior, in Abrahams, A. D., ed., Hillslope processes: Boston, Allen and Unwin, p. 297–318.

—— 1986b, Unsteady, nonuniform landslidemotion; 1. Theoretical dynamics and the steady datum state: Journal of Geology, v. 94, p. 1–15.

—— , 1986c, Unsteady, nonuniform landslide motion; 2. Linearized theory and the kinematics of transient response: Journal of Geology, v. 94, p. 349–364.

Jachens, R. C., and Griscom, A., 1983, Three-dimensional geometry of the Gorda Plate beneath northern California: Journal of Geophysical Research, v. 88, p. 9375–9392.

Jackson, W. L., and Beschta, R. L., 1982, A model of two phase bedload transport in an Oregon Coast Range stream: Earth Surface Processes and Landforms, v. 7, p. 517–527.

Jenny, H., 1941, Factors of soil formation: New York, McGraw-Hill, 281 p.

—— , 1980, The soil resource—origin and behavior: New York, Springer Verlag, 377 p.

Johnson, D. L., 1967, Caliche on the Channel Islands: (California Division of Mines and Geology) Mineral Information Service, v. 20, p. 151–158.

—— , 1977, The late Quaternary climate of coastal California; Evidence for an Ice Age refugium: Quaternary Research, v. 8, p. 154–179.

—— , 1980, Episodic vegetation stripping, soil erosion, and landscape modification in prehistoric and recent historic time, San Miguel Island, California, in Power, D. M., ed., The California islands; Proceedings of a multidisciplinary symposium: Santa Barbara, Santa Barbara Museum of Natural History, p. 103–121.

Johnson, D. W., 1938, Shore processes and shoreline development: New York, John Wiley, 584 p.

Johnson, P. R., and Hartman, C. W., 1969, Environmental atlas of Alaska: Fairbanks, University of Alaska Institute of Arctic Environmental Engineering, 111 p.

Jones, D. L., Howell, D. G., Coney, P. J., and Monger, J.W.H., 1982a, Recognition, character, and analysis of tectonostratigraphic terranes in western North America, in Hashimoto, M., and Uyeda, S., eds., Accretion tectonics in the circum-Pacific regions: Tokyo, Terra Scientific Publishing Company, p. 21–35.

Jones, D. L., Silberling, N. J., Berg, H. C., and Plafker, G., 1982b, Tectonostratigraphic terrane map of Alaska: U.S. Geological Survey Circular 844, p. 1–5.

Kamb, B., Raymond, C. F., Harrison, W. D., Engelhardt, H., Echelmeyer, K. A., Humphrey, N., Brugman, M. M., and Pfeffer, T., 1985, Glacier surge mech-

anism; 1982–1983 surge of Variegated Glacier, Alaska: Science, v. 227, p. 469–479.

Karlstrom, T.N.V., 1964, Quaternary geology of the Kenai Lowland and glacial history of the Cook Inlet region: U.S. Geological Survey Professional Paper 443, 69 p.

Karlstrom, T.M.V., and others, 1964, Surficial geology of Alaska: U.S. Geological Survey Miscellaneous Geological Investigations Map I-357, scale 1:1,584,000.

Karrow, P. F., and Bada, J. L., 1980, Amino acid dating of Quaternary raised marine terraces in San Diego County, California: Geology, v. 8, p. 200–204.

Keefer, D. K., and Johnson, A. M., 1983, Earthflows; Morphology, mobilization, and movement: U.S. Geological Survey Professional Paper 1264, 56 p.

Keen, C. E., and Hyndman, R. D., 1979, Geophysical review of the continental margins of eastern and western Canada: Canadian Journal of Earth Sciences, v. 16, p. 712–747.

Keller, E. A., and Swanson, F. J., 1979, Effects of large organic material on channel form and fluvial processes: Earth Surface Processes, v. 4, p. 361–386.

Keller, E. A., and Tally, T., 1979, Effects of large organic debris on channel form and fluvial process in the coastal redwood environment, in Rhodes, D. D., and Williams, G. P., eds., Adjustments of the fluvial system; Tenth Annual Geomorphology Symposia Series: Binghamton, New York, Kendall/Hunt Publishing Company, p. 169–198.

Keller, E. A., Rockwell, T. K., Clark, M. N., Dembroff, G. R., and Johnson, D. L., 1982, Tectonic geomorphology of the Ventura, Ojai, and Santa Paula areas, western Transverse Ranges, California, in Cooper, J. D., ed., Neotectonics in southern California: Geological Society of America Cordilleran Section Field Guidebook, p. 25–42.

Kelsey, H. M., 1978, Earthflows in Franciscan mélange, Van Duzen River basin: Geology, v. 6, p. 361–364.

—— , 1980, A sediment budget and an analysis of geomorphic process in the Van Duzen River basin, north coastal California, 1941–1975: Geological Society of America Bulletin, v. 91, p. 1119–1216.

—— , 1982, Hillslope evolution and sediment movement in a forested headwater basin, Van Duzen River, north coastal California; Workshop on sediment budgets and routing in forested drainage basins (proceedings): U.S. Forest Service General Technical Report PNW 141, p. 86–96.

Kelsey, H. M., and Savina, M. E., 1985, Van Duzen River basin, in Savina, M. E., ed., Redwood county: Guidebook for the American Geomorphological Field Group, p. 3–37.

Kelsey, H. M., Lamberson, R., and Madej, M. A., 1986, Modeling the transport of stored sediment in a gravel bed river, northwestern California, in Drainage basin sediment delivery: International Association of Hydrological Sciences Publication no. 159, p. 367–392.

Kelsey, H. M., Coghlan, M., Pitlick, J., and Best, D. W., 1987, Geomorphic analysis of streamside landsliding in the Redwood Creek basin, in Nolan, K. M., Kelsey, H. M., and Marron, D. C., eds., Geomorphic processes and aquatic habitat in the Redwood Creek basin, northwestern California: U.S. Geological Survey Professional Paper 1454 (in press).

Kennedy, G. L., Lajoie, K. R., and Wehmiller, J. F., 1982, Aminostratigraphy and faunal correlations of late Quaternary marine terraces Pacific coast, USA: Nature, v. 299, p. 545–547.

Kennedy, M. P., 1973, Sea-cliff erosion at Sunset Cliffs, San Diego: California Geology, v. 26, p. 27–31.

Kennett, J. P., and Thunell, R. C., 1975, Global increase in Quaternary explosive volcanism: Science, v. 187, p. 497–503.

Kern, J. P., 1977, Origin and history of upper Pleistocene marine terraces, San Diego, California: Geological Society of America Bulletin, v. 88, p. 1553–1566.

Kerr, P. F., and Drew, I. M., 1969, Clay mobility, Portuguese Bend, California: California Division of Mines and Geology Special Report 100, p. 3–16.

Komar, P. D., 1978, Wave conditions on the Oregon coast during the winter of 1977–78 and the resulting erosion of Nestucca Spit: Shore and Beach, v. 46, no. 4, p. 3–8.

—— , 1980, Beach processes and erosion problems on the Oregon coast, *in* Oles, K. F., Johnson, J. G., Niem, A. R., and Niem, W. A., eds., Geologic field trips in western Oregon and southwestern Washington: Oregon Department of Geology and Mineral Industries Bulletin 101, p. 169–176.

—— , 1983, The erosion of Siletz Spit, Oregon, *in* Komar, P. D., ed., Chemical Rubber Company handbook of coastal processes and erosion: Boca Raton, Florida, Chemical Rubber Company Press, p. 65–76.

Komar, P. D., and McKinney, B. A., 1977, The spring 1976 erosion of Siletz Spit, Oregon: Oregon State University Sea Grant College Program Publication ORESU-T-77-004, 23 p.

Komar, P. D., and Rea, C. C., 1976, Beach erosion on Siletz Spit, Oregon: The Ore Bin, v. 38, p. 119–134.

Komar, P. D., Quinn, W., Creech, C., Rea, C. C., and Lizarraga-Arciniega, J. R., 1976, Wave conditions and beach erosion on the Oregon coast: The Ore Bin, v. 38, p. 103–112.

Konishi, K., Schlanger, S. O., and Omura, A., 1970, Neotectonic rates in the central Ryukyu Islands derived from ^{230}Th coral ages: Marine Geology, v. 9, p. 225–240.

Ku, T.-L., and Kern, J. P., 1974, Uranium-series age of the upper Pleistocene Nestor terrace, San Diego, California: Geological Society of America Bulletin, v. 85, p. 1713–1716.

Ku, T.-L., Kimmel, M. A., Easton, W. H., and O'Neil, T. J., 1974, Eustatic sea level 120,000 years ago on Oahu, Hawaii: Science, v. 183, p. 959–962.

Kuhn, G. G., and Shepard, F. P., 1980, Coastal erosion in San Diego County, California, *in* Edge, B. L., ed., Proceedings, Coastal zone, 80, Volume III, Second Symposium on Coastal and Ocean Management: New York, American Society of Civil Engineers, p. 1899–1918.

—— , 1983, Beach processes and sea cliff erosion in San Diego County, California, *in* Komar, P. D., ed., Chemical Rubber Company handbook of coastal processes and erosion: Boca Raton, Florida, Chemical Rubber Company Press, p. 267–284.

LaHusen, R. G., 1984, Characteristics of management-related debris flows, northwestern California *in* O'Loughlin, C. L., and Pearc, A. J., eds., Proceedings, Symposium on effects of forest land use on erosion and slope stability, Honolulu, Hawaii, May 7–11, 1984: International Union of Forest Research Organizations, p. 139–147.

Lajoie, K. R., 1986, Coastal tectonics, *in* Active tectonics: Washington, D.C., National Research Council, National Academy Press, p. 95–125.

Lajoie, K. R., Kern, J. P., Wehmiller, J. F., Kennedy, G. L., Mathieson, S. A., Sarna-Wojcicki, A. M., Yerkes, R. F., and McCrory, P. F., 1979, Quaternary marine shorelines and crustal deformation, San Diego to Santa Barbara, California, *in* Abbott, P. L., ed., Geological excursions in the southern California area: San Diego, California, San Diego State University Department of Geological Sciences, p. 3–15.

Lajoie, K. R., Sarna-Wojcicki, A. M., and Ota, Y., 1982, Emergent Holocene marine terrces of Ventura and Cape Mendocino, California; Indicators of high tectonic uplift rates: Geological Society of America Abstracts with Programs, v. 14, p. 178.

Langbein, W. B., and Schumm, S. A., 1958, Yield of sediment in relation to mean annual precipitation: Transactions of the American Geophysical Union, v. 39, p. 1076–1084.

Lawson, A. C., 1893, The post-Pliocene diastrophism of the coast of southern California: University of California Department of Geological Sciences Bulletin, v. 1, p. 115–160.

Lawson, D. E., 1979, Sedimentological analysis of the western terminus region of the Matanuska Glacier, Alaska: U.S. Army Cold Regions Research and Engineering Laboratory Report 790-9, 122 p.

Lea, P. D., 1985, Eolian cover sands; A sedimentologic model and paleoenvironmental implications: Geological Society of America Abstracts with Programs, v. 17, p. 641.

Lehre, A. K., 1982, Sediment budget of a small Coast Range drainage basin in north central California, *in* Swanson, F. J., Janda, R. J., Dunne, T., and Swanston, D. N., eds., Sediment budgets and routing in forested drainage basins; Portland, Oregon, U.S.D.A. Forest Service General Technical Report

PNW-141, Pacific Northwest Forest and Range Experiment Station, p. 67–77.

Lehre, A. K., and Carver, G. A., 1985, Thrust faulting and earthflows; Speculations on the sediment budget of a tectonically active drainage basin, *in* Savina, M. E., ed., Redwood County: Guidebook for the American Geomorphological Field Group, p. 169–183.

Lettis, W. R., 1985, Late Cenozoic stratigraphy and structure of the west margin of the central San Joaquin Valley, California: Geological Society of America Special Paper 203, p. 97–114.

Lisle, T. E., 1981, The recovery of aggraded stream channels at gauging stations in northern California and southern Oregon, *in* Proceedings, Christchurch Symposium, Erosion and Sediment: International Association of Hydrological Sciences, IAHS-AISH Publication no. 132, p. 189–211.

—— , 1982, Effects of aggradation and degradation on riffle-pool morphology in natural gravel channels, northwestern California: Water Resources Research, v. 18, p. 1643–1651.

—— , 1986, Stabilization of a gravel channel by large streamside obstructions and bedrock bends, Jacoby Creek, northwestern California: Geological Society of America Bulletin, v. 97, p. 999–1011.

Lizarraga-Arciniega, J. R., and Komar, P. D., 1975, Shoreline changes due to jetty construction on the Oregon coast: Oregon State University Sea Grant College Program, Publication No. ORESU-T-75-004, 85 p.

Lund, E. H., 1973, Oregon coastal sand dunes between Coos Bay and Sea Lion Point: The Ore Bin, v. 35, p. 73–92.

Luternauer, J. L., and Murray, J. W., 1983, Late Quaternary morphologic development and sedimentation, central British Columbia continental shelf: Geological Survey of Canada Paper 83-21, 38 p.

Madej, M. A., 1982, Sediment transport and channel changes in an aggrading stream in the Puget Lowland, Washington, *in* Swanson, F. J., Janda, R. J., Dunne, T., and Swanston, D. N., eds., Sediment budgets and routing in forested drainage basins: Pacific Northwest Forest and Range Experiment Station, U.S.D.A. Forest Service General Technical Report PNW-141, p. 97–108.

—— , 1987, Recent changes in channel-stored sediment in Redwood Creek, California, *in* Nolan, H. M., and Marron, D. C., eds., Geomorphic processes and aquatic habitat in the Redwood Creek basin, northwestern California: U.S. Geological Survey Professional Paper 1454 (in press).

Mann, D. H., 1983, Glacial history near Lituya Bay, Alaska, *in* Thorson, R. M., and Hamilton, T. D., eds., Glaciation in Alaska, Extended abstracts from a workshop: University of Alaska Museum Occasional Paper no. 2, p. 62–66.

—— , 1986, Reliability of a fjord glacier's fluctuations for paleoclimatic reconstructions: Quaternary Research, v. 25, p. 10–24.

Marron, D. C., 1982, Hillslope evolution and the genesis of colluvium in Redwood National Park, northwestern California; The use of soil development in their analysis [Ph.D. thesis]: Berkeley, University of California, 187 p.

—— , 1985, Colluvium in bedrock hollows on steep slopes, Redwood Creek drainage basin, northwestern California, *in* Jungerius, P. D., ed., Catena Supplement 6 p. 59–68.

Marston, R. H., 1982, The geomorphic significance of log steps in forest streams: Annals of the Association of American Geographers, v. 72, no. 1, p. 99–108.

Mathews, W. H., 1968, Geomorphology, southwestern British Columbia, *in* Mathews, W. H., ed., Guidebook for geological field trips in southwestern British Columbia: University of British Columbia, Department of Geology Report 6, p. 18–24.

—— , 1972, Geology of Vancouver area of British Columbia: International Geological Congress, 24th, Montreal, Guidebook, Field Excursion A05–C05, 47 p.

—— , 1979, Landslides of central Vancouver Island and the 1946 earthquake: Seismological Societ yof America Bulletin, v. 69, p. 445–450.

—— , 1987, Development of Cordilleran landscapes during the Quaternary, *in* Fulton, R. J., Heginbottom, J. A., and Funder, S., eds., Quaternary geology of Canada and Greenland: Geological Survey of Canada (in press).

Mathews, W. H., and Rouse, G. E., 1963, Late Tertiary volcanic rocks and plant-bearing deposits in British Columbia: Geological Society of America

Bulletin, v. 74, p. 55–60.

Mathews, W. H., Fyles, J. G., and Nasmith, H. W., 1970, Postglacial crustal movements in southwestern British Columbia and adjacent Washington State: Canadian Journal of Earth Sciences, v. 7, p. 690–702.

Mathieson, S. A., and Lajoie, K. R., 1982, A drastic increase in sea-cliff-retreat rates due to insufficient geologic input into dynamic engineering practice; Half Moon Bay, San Mateo County, California: Geological Society of America Abstracts with Programs, v. 14, p. 558.

McFadden, L. D., and Hendricks, D. M., 1985, Changes in the content and composition of pedogenic iron oxyhydroxides in a chronosequence of soils in southern California: Quaternary Research, v. 23, p. 189–204.

McKee, B., 1972, Cascadia, The geological evolution of the Pacific Northwest: New York, McGraw-Hill, 394 p.

McLaughlin, R. J., Lajoie, K. R., Sorg, D. H., Morrison, S. D., and Wolfe, J. A., 1983, Tectonic uplift of a middle Wisconsin marine platform near the Mendocino triple junction, California: Geology, v. 11, p. 35–39.

Meier, M. F., Rasmussen, L. A., Post, A., Brown, C. S., Sikonia, W. G., Bindschadler, R. A., Mayo, L. R., and Trabant, D. C., 1980, Predicted timing of the disintegration of the lower reach of Columbia Glacier, Alaska: U.S. Geological Survey Open-File Report 80-582, 47 p.

Merriman, R., 1960, Portuguese Bend landslide, Palos Verdes Hills, California: Journal of Geology, v. 68, p. 140–153.

Mesolella, K. J., Matthews, R. K., Broecker, W. S., and Thurber, D. L., 1969, The astronomical theory of climatic change; Barbados data: Journal of Geology, v. 77, p. 250–274.

Mickelson, D. M., 1971, Glacial geology of the Burroughs Glacier area, southeastern Alaska: Ohio State University, Institute of Polar Studies, Report 40, 149 p.

Milhous, R. H., 1982, Effect of sediment transport and flow regulation on the ecology of gravel-bed rivers, *in* Hey, R. D., Bathurst, J. C., and Thorn, C. R., eds., Gravel-bed rivers, fluvial processes, engineering, and management: London, John Wiley and Sons, p. 819–842.

Milhous, R. T., and Klingeman, P. C., 1973, Sediment transport system in a gravel-bottomed stream, *in* Proceedings, Hydraulic Engineering and the Environment, 21st Annual Hydraulics Specialty Conference, Bozeman, Montana: AMerican Society of Civil Engineers, p. 293–303.

Miller, D. J., 1953, Late Cenozoic marine glacial sediments and marine terraces of Middleton Island, Alaska: Journal of Geology, v. 61, p. 17–40.

—— , 1958, Gulf of Alaska area, *in* Williams, H., ed., Landscapes of Alaska: Berkeley, Unviersity of California, p. 19–29.

—— 1960, Giant waves in Lituya Bay, Alaska: U.S. Geological Survey Professional Paper 354-C, 86 p.

Miller, R. D., 1973, Gastineau Channel Formation, a composite glaciomarine deposit near Juneau, Alaska: U.S. Geological Survey Bulletin 1394-C, p. C1–C20.

Miller, T. P., and Smith, R. L., 1976, "New" volcanoes in the Aleutian volcanic arc: U.S. Geological Survey Circular 733, p. 11.

—— , 1977, Spectacular mobility of ash flows around Aniakchak and Fisher calderas, Alaska: Geology, v. 5, p. 173–176.

Milne, W. G., Rogers, G. C., Riddihough, R. P., McMechan, G. A., and Hyndman, R. D., 1978, Seismicity of western Canada: Canadian Journal of Earth Sciences, v. 15, p. 1170–1193.

Minster, J. B., and Jordan, T. H., 1978, Present-day plate motions: Journal of Geophysical Research, v. 83, p. 5331–5354.

Molnia, B. F., 1980, Storm surge run-up in the northeastern Gulf of Alaska; Environmental implications: Geological Society of America Abstracts with Programs, v. 12, p. 142.

—— , 1983, Late Wisconsinian and Holocene glaciation of the Alaskan continental margin, *in* Thorson, R. M., and Hamilton, T. D., eds., Glaciation in Alaska, Extended abstracts from a workshop: University of Alaska Museum Occasional Paper no. 2, p. 67–70.

—— , 1986, Glacial history of the northeastern Gulf of Alaska; A synthesis, *in* Hamilton, T. D., Reed, K. M., and Thorson, R. M., eds., Glaciation in Alaska; The geologic record: Anchorage, Alaska Geological Society,

p. 219–236.

Monger, J.W.H., Price, R. A., and Tempelman-Kluit, D. J., 1982, Tectonic accretion and the origin of the two major metamorphic and plutonic welts in the Canadian Cordillera: Geology, v. 10, p. 70–75.

Moring, B., 1983, Reconnaissance surficial geologic map of the Medford 1° × 2° Quadrangle, Oregon–California: U.S. Geological Survey Miscellaneous Field Studies Map MF-1528.

Morris, R. H., and Bucknam, R. C., 1972, Gemorphic evidence of late Holocene vertical stability in the Aleutian Islands: Geological Society of America Abstracts with Programs, v. 4, p. 203–204.

Mortimer, N., and Coleman, R. G., 1985, A Neogene structural dome in the Klamath Mountains, California and Oregon: Geology, v. 13, p. 253–256.

Morton, D. M., 1971, Seismically triggered landslides on the area above the San Fernando Valley: U.S. Geological Survey Professional Paper 733, p. 99–104.

Morton, D. M., and Campbell, R. H., 1974, Spring mudflows at Wrightwood, southern California: Quarterly Journal of Engineering Geology, v. 7, p. 377–384.

Morton, D. M., and Kennedy, M. P., 1979, Landsliding and mudflows at Wrightwood, San Bernardino County, California; Part I, Wright Mountain landslide; Renewed movement in 1967: California Division of Mines and Geology Special Report 136, p. 1–5.

Morton, D. M., Campbell, R. H., Barrows, A. G., Jr., Kahle, J. E., and Yerkes, R. F., 1979, Landsliding and mudflows at Wrightwood, San Bernardino County, California Part II, Wright Mountain mudflows; Spring, 1969; California Division of Mines and Geology Special Report 136, p. 7–21.

Motyka, R. J., 1977, Katmai caldera; Glacier growth, lake rise, and geothermal activity: Alaska Division of Geological and Geophysical Surveys Report 55, p. 17–21.

Muhs, D. R., 1982, A soil chronosequence on Quaternary marine terraces, San Clemente Island, California: Geoderma, v. 28, p. 257–283.

—— , 1983a, Quaternary sea-level events on northern San Clemente Island, California: Quaternary Research, v. 20, p. 322–341.

—— , 1983b, Airborne dust fall on the California Channel Islands, U.S.A.: Journal of Arid Environments, v. 6, p. 223–238.

—— , 1985, Amino acid age estimates of marine terraces and sea levels on San Nicholas Island, California: Geology, v. 13, p. 58–61.

Muhs, D. R., and Rosholt, J. N., 1984, Ages of marine terraces on the Palos Verdes Hills, California, by amino acid and uranium-trend dating: Geological Society of America Abstracts with Programs, v. 16, p. 603.

Muhs, D. R., and Szabo, B. J., 1982, Uranium-series age of the Eel Point terrace, San Clemente Island, California: Geology, v. 10, p. 23–26.

Nelson, S. W., and Reed, B. L., 1978, Surficial deposits map of the Talkeetna Quadrangle, Alaska: U.S. Geological Survey Map MF 870-J, scale 1:250,000.

Nichols, D. R., and Yehle, L. A., 1961, Mud volcanoes in the Cooper River basin, Alaska, *in* Raasch, G. D., ed., Geology of the Arctic: Toronto, University of Toronto Press, v. 2, p. 1063–1087.

Nolan, K. M., and Janda, R. J., 1979, Recent history of the main channel of Redwood Creek, California, *in* A field trip to observe natural and management related erosion in Franciscan terrane of northern California: Geological Society of America, Cordilleran Section, p. X, 1–16.

—— , 1987, Sediment discharge and mass movement of two earthflows in Franciscan terrain, northwestern California, *in* Nolan, K. M., Delsey, H. M., and Marron, D. C., eds., Geomorphic processes and aquatic habitat in the Redwood Creek basin, northwestern California: U.S. Geological Survey Professional Paper 1454 (in press).

Nolan, K. M., and Marron, D. C., 1985, Contrast in stream-channel response to major storms in two moutainous areas of California: Geology, v. 13, p. 135–138.

Norris, R. M., 1964, Dams and beach-sand supply in southern California, *in* Miller, R. L., ed., Papers in marine geology; Shepard commemorative volume: New York, The MacMillian Company, p. 154–171.

—— , 1968, Sea cliff retreat near Santa Barbara, California: California Division of Mines and Geology, Mineral Information Service, v. 21, p. 87–91.

Norris, R. M., and Webb, R. W., 1976, Geology of California: New York, John Wiley and Sons, 365 p.

North, W. B., and Byrne, J. V., 1965, Coastal landslides of northern Oregon: The Ore Bin, v. 27, p. 217–241.

Oberlander, T. M., 1965, The Zagros streams: Syracuse Geographical Series, no. 1, 168 p.

Omura, A., Emerson, W. K., and Ku, T.-L., 1979, Uranium-series ages of echinoids and corals from the upper Pleistocene Magdalena terrace, Baja California Sur, Mexico: Nautilus, v. 94, p. 184–189.

Orme, A. R., 1980, Marine terraces and Quaternary tectonism, northwest Baja California, Mexico: Physical Geography, v. 1, p. 138–161.

Orme, A. R., and Brown, A. J., 1983, Variable sediment flux and beach management, Ventura county, California, *in* Magoon, O. T., ed., Proceedings, Coastal Zone '83, Volume III, Third Symposium on Coastal and Ocean Management: New York, American Society of Civil Engineers, p. 2328–2342.

Parrish, R. R., 1983, Cenozoic thermal evolution and tectonics of the Coast Mountains of British Columbia; 1. Fission-track dating, apparent uplift rates, and patterns of uplift: Tectonics, v. 2, p. 601–631.

Parsons, R. B., and Herriman, R. C., 1976, Geomorphic surfaces and soil development in the upper Rogue River valley, Oregon: Soil Science Society of America Journal, v. 40, p. 933–938.

Paustian, S. J., and Beschta, R. L., 1979, The suspended sediment regime of an Oregon coastal stream: Water Resources Bulletin, v. 15, p. 144–154.

Peacock, M. A., 1935, Fiord-land of British Columbia: Geological Society of America Bulletin, v. 46, p. 633–695.

Peterson, C. D., Scheidegger, K., and Komar, P., 1982, Sand dispersal patterns in an active margin estuary of the northwestern United States as indicated by sand composition, texture and bedforms: Marine Geology, v. 50, p. 77–96.

Peterson, C. D., Scheidegger, K. F., and Schrader, H. J., 1984a, Holocene depositional evolution of a small active-margin estuary of the northwestern United States: Marine Geology, v. 59, p. 51–83.

Peterson, C. D., Scheidegger, K., Komar, P., and Niem, W., 1984b, Sediment composition and hydrography in six high-gradient estuaries of the northwestern United States: Journal of Sedimentary Petrology, v. 54, p. 86–97.

Péwé, T. L., 1975, Quaternary geology of Alaska: U.S. Geological Survey Professional Paper 835, 145 p.

Pickard, G. L., 1961, Oceanographic features of inlets in the British Columbia mainland coast: Fisheries Research Board of Canada Journal, v. 18, p. 907–999.

Pitlick, J., 1987, Sediment routing in tributaries of the Redwood Creek basin, northern California, *in* Nolan, K. M., Kelsey, H. M., and Marron, D. C., eds., Geomorphic processes and aquatic habitat in the Redwood Creek basin, northwestern California: U.S. Geological Survey Professional Paper 1454 (in press).

Plafker, G., Ambos, E. L., Fuis, G. S., Mooney, W. D., Nokleberg, W. G., and Campbell, D. L., 1985, Accretion, subduction and, underplating along the southern Alaska continental margin: Geological Society of America Abstracts with Programs, v. 17, p. 690.

Ponti, D. J., 1985, The Quaternary alluvial sequence of the Antelope Valley, California: Geological Society of America Special Paper 203, p. 79–96.

Post, A., 1969, Distribution of surging glaciers in western North America: Journal of Glaciology, v. 8, p. 229–240.

Post, A., and Mayo, L. R., 1971, Glacier dammed lakes and outburst floods in Alaska: U.S. Geological Survey Hydrological Investigations Atlas HA-455.

Powell, R. D., 1983, Glacial-marine sedimentation processes and lithofacies of temperate tidewater glaciers, Glacier Bay, Alaska, *in* Molnia, B. F., ed., Glacial-marine sedimentation: New York, Plenum Publishing Company, p. 185–232.

Putnam, W. C., 1942, Geomorphology of the Ventura region, California: Geological Society of America Bulletin, v. 53, p. 691–754.

Ramp, L., 1975, Geology and mineral resources of the Upper Chetco drainage area, Oregon: Oregon Department of Geology and Mineral Industries Bul-

letin 88, 47 p.

Rau, W. W., 1973, Geology of the Washington coast between Point Grenville and the Hoh River: Washington Department of Natural Resources, Geology and Earth Resources Division Bulletin 66, 58 p.

——, 1980, Washington coastal geology between the Hoh and Quillayute rivers: Washington Department of Natural Resources, Geology and Earth Resources Division Bulletin 72, 57 p.

Reger, R. D., 1978, Bluff Point landslide, a massive ancient rock failure near Homer, Alaska: Alaska Division of Geological and Geophysical Surveys Report 61, p. 5–8.

Reger, R. D., and Updike, R. G., 1983, Upper Cook Inlet region and the Matanuska Valley, *in* Péwé, T. L., and Reger, R. D., eds., Guidebook 1, Richardson and Glen highways, Alaska; Fourth International Conference on Permafrost: State of Alaska Division of Geological and Geophysical Surveys, p. 185–263.

Reid, L. M., and Dunne, T., 1984, Sediment production from forest road surfaces: Water Resources Research, v. 20, no. 11, p. 1753–1761.

Reiger, S., Schoephorster, D. B., and Furbush, C. E., 1979, Exploratory soil survey of Alaska: Washington, D.C., U.S. Government Printing Office, Soil Conservation Service, 213 p.

Reilinger, R., and Adams, J., 1982, Geodetic evidence for active landward tilting of the Oregon and Washington Coastal Ranges: Geophysical Research Letters, v. 9, no. 4, p. 401–403.

Reneau, S. L., and Dietrich, W. E., 1985, Landslide recurrence intervals in colluvium-mantled hollows, Marin County, California: EOS Transactions of the American Geophysical Union, v. 66, p. 900.

Reneau, S. L., Dietrich, W. E., Wilson, C. J., and Rogers, J. D., 1984, Colluvial deposits and associated landslides in the northern San Francisco Bay area, California, U.S.A.: Proceedings of the IVth International Symposium on Landslides, International Society for Soil Mechanics and Foundation Engineering, Toronto, Ontario, Canada, p. 425–430.

Reneau, S. L., Dietrich, W. E., Dorn, R. I., Berger, C. R., and Rubin, M., 1986, Geomorphic and paleoclimatic implications of late Pleistocene radiocarbon dates from colluvium-mantled hollows, California: Geology, v. 14, p. 655–658.

Rice, R. M., and Datzman, P. A., 1981, Erosion associated with cable and tractor logging in northwestern California, *in* Davies, T., and Pearce, A. J., eds., Proceedings, Christchurch Symposium, Erosion and Sediment Transport in Pacific Rim Steeplands: International Association of Hydrological Sciences Publication no. 132, p. 362–374.

Rice, R. M., and Foggin, G. T., III, 1971, Effect of high intensity storms on soil slippage on mountainous watersheds in southern California: Water Resources Research, v. 7, p. 1485–1496.

Rice, R. M., Corbett, E. S., and Bailey, R. G., 1969, Soil slips related to vegetation, topography, and soil in southern California: Water Resources Research, v. 5, p. 647–659.

Rice, R. M., Gorsline, D. S., and Osborne, R. H., 1976, Relationship between sand input from rivers and the composition of sands from the beaches of southern California: Sedimentology, v. 23, p. 689–703.

Richter, D. H., 1976, Geologic map of the Nabesna Quadrangle, Alaska: U.S. Geological Survey Map I-932, scale 1:250,000.

Riddihough, R. P., 1977, A model for recent plate interactions off Canada's west coast: Canadian Journal of Earth Sciences, v. 14, p. 384–396.

——, 1982, Contemporary movements and tectonics on Canada's west coast; A discussion: Tectonophysics, v. 86, p. 319–341.

Riddihough, R. P., and Hyndman, R. D., 1976, Canada's active western margin; The case for subduction: Geoscience Canada, v. 3, p. 269–278.

Riddihough, R. P., Beck, M. E., Chase, R. L., Davis, E. E., Hyndman, R. D., Johnson, S. H., and Rogers, G. C., 1983, Geodynamics of the Juan de Fuca plate, *in* Cabre, R., ed., Geodynamics of the eastern Pacific region, Caribbean and Scotia arcs: American Geophysical Union, Geodynamics Series, v. 9, p. 5–21.

Riehle, J. R., Kienle, J., and Emmel, K. S., 1981, Lahars in Crescent River valley, lower Cook Inlet, Alaska: Alaska Division of Geological and Geophysical

Surveys, Geological Report 53, 10 p.

Ritter, D. F., 1982, Complex river terrace development in the Nenana Valley near Healy, Alaska: Geological Society of America Bulletin, v. 93, p. 346–356.

Rockwell, T. K., and Keller, E. A., 1985, Tectonic geomorphology of alluvial fans and mountain fronts near Ventura, California, *in* Morisawa, M. and Hack, J., eds., Tectonic geomorphology: Boston, George Allen and Unwin, p. 183–207.

Rockwell, T. K., Keller, E. A., Clark, M. N., and Johnson, D. L., 1984, Chronology and rates of faulting of Ventura River terraces, California: Geological Society of America Bulletin, v. 95, p. 1466–1474.

Rockwell, T. K., Johnson, D. L., Keller, E. A., and Dembroff, G. R., 1985, A late Pleistocene-Holocene soil chronosequence in the central Ventura Basin, southern California, *in* Richards, K. S., Arnett, R. R., and Ellis, S., eds., Geomorphology and soils: London, George Allen and Unwin, p. 209–327.

Rogers, G. C., and Hasegawa, H. S., 1978, A second look at the British Columbia earthquake of June 23, 1946: Seismological Society of America Bulletin, v. 68, p. 653–675.

Rose, K. E., 1979, Characteristics of ice flow in Marie Byrd Land, Antarctica [includes discussion]: Journal of Glaciology, v. 24, p. 63–75.

Ryder, J. M., 1971a, The stratigraphy and morphology of para-glacial alluvial fans in south-central British Columbia: Canadian Journal of Earth Sciences, v. 8, p. 279–298.

—— , 1971b, Some aspects of the morphometry of paraglacial alluvial fans in south-central British Columbia: Canadian Journal of Earth Sciences, v. 8, p. 1252–1264.

—— , 1981, Geomorphology of the southern part of the Coast Mountains of British Columbia: Zeitschrift für Geomorphologie, v. 37, p. 120–147.

Ryder, J. M., and Church, M., 1986, The Lillooet terraces of Fraser River; A palaeoenvironmental enquiry: Canadian Journal of Earth Sciences, v. 23, p. 869–884.

Scheidegger, K. F., Kulm, L. D., and Runge, E. J., 1971, Sediment sources and dispersal patterns of Oregon continental shelf sands: Journal of Sedimentary Petrology, v. 41, p. 1121–1125.

Schlicker, H. G., and Deacon, R. J., 1974, Environmental geology of coastal Lane County, Oregon: Oregon Department of Geology and Mineral Industries Bulletin 85, 116 p.

Schlicker, H. G., Deacon, R. J., Beaulieu, J. D., and Olcott, G. W., 1972, Environmental geology of the coastal region of Tillamook and Clatsop counties, Oregon: Oregon Department of Geology and Mineral Industries Bulletin 74, 164 p.

Schlicker, H. G., Deacon, R. J., Olcott, G. W., 1973, Environmental geology of Lincoln County, Oregon: Oregon Department of Geology and Mineral Industries Bulletin 81, 171 p.

Schmoll, H. R., and Yehle, L. A., 1986, Pleistocene glaciation of the upper Cook Inlet basin, *in* Hamilton, T. D., Reed, K. M., and Thorson, R. M., eds., Glaciation in Alaska; The geologic record: Anchorage, Alaska Geological Society, p. 193–218.

Schmoll, H. R., Szabo, B. J., Rubin, M., and Dobrovolny, E., 1972, Radiometric dating of marine shells from the Bootlegger Cove clay, Anchorage, Alaska: Geological Society of America Bulletin, v. 83, p. 1107–1114.

Scott, K. M., 1971, Origin and sedimentology of 1969 debris flows near Glendora, California: U.S. Geological Survey Professional Paper 750-C, p. 242–247.

—— , 1973, Scour and fill in Tujunga Wash; A fanhead valley in urban southern California, 1969: U.S. Geological Survey Professional Paper 732-B, 35 p.

Scott, K. M., and Williams, R. P., 1978, Erosion and sediment yields in the Transverse Ranges, southern California: U.S. Geological Survey Professional Paper 1030, 38 p.

Seymour, R. J., and Aubrey, D. G., 1985, Rhythmic beach cusp formation; A conceptual synthesis: Marine Geology, v. 65, p. 289–304.

Shepard, F. P.,and Kuhn, G. G., 1983, History of sea arches and remnant stacks of La Jolla, California, and their bearing on similar features elsewhere: Marine Geology, v. 51, p. 139–161.

Shepard, F. P., and Wanless, H. R., 1971, Our changing coastlines: New York, McGraw-Hill, 579 p.

Shouldice, D. H., 1971, Geology of the western Canadian continental shelf: Canadian Petroleum Geology Bulletin, v. 19, p. 405–436.

—— , 1973, Western Canadian continental shelf, *in* McCrossan, R. G., ed., Future petroleum provinces of Canada; their geology and potential: Canadian Society of Petroleum Geologists Memoir 1, p. 7–35.

Shreve, R. L., 1966, Sherman landslide, Alaska: Science, v. 154, p. 1639–1643.

Sieh, K. E., and Jahns, R., 1984, Holocene activity of the San Andreas fault at Wallace Creek, California: Geological Society of America Bulletin, v. 95, p. 883–896.

Simkin, E., Sieber, T. L., McClelland, L., Bridge, D., Newhall, C., and Latter, J., 1981, Volcanoes of the world, The Smithsonian Institution: Stroudburg, Pennsylvania, Hutchinson and Ross, 232 p.

Simons, D., and Richardson, E., 1966, Resistance to flow in alluvial channels: U.S. Geological Survey Professional Paper 422-J, 61 p.

Sirkin, L. A., Tuthill, S. J., Clayton, L. S., 1971, Late Pleistocene history of the lower Copper River valley, Alaska: Geological Society of America Abstracts with Programs, v. 3, p. 708.

Slawson, W. F., and Savage, J. C., 1979, Geodetic deformation associated with the 1946 Vancouver Island, Canada, earthquake: Seismological Society of America Bulletin, v. 69, p. 1487–1496.

Slaymaker, O., and McPherson, H. J., 1977, An overview of geomorphic processes in the Canadian Cordillera: Zietschrift für Geomorphologie, v. 21, p. 169–186.

Smith, W.S.T., 1898, A geological sketch of San Clemente Island: U.S. Geological Survey 18th Annual Report, Part II, p. 459–496.

Sonnevil, R., Klein, R., LaHusen, R., Short, D., and Weaver, W., 1985, Blocksliding on schist in lower Redwood Creek drainage, *in* Savina, M. E., ed., Redwood county: Guidebook for the 1985 American Geomorphological Field Group meetin, p. 119–130.

Souther, J. G., 1977, Volcanism and tectonic environments in the Canadian Cordillera; A second look, *in* Barager, W.R.A., and others, eds., Volcanic regimes in Canada: Geological Association of Canada Special Paper 16, p. 3–24.

Stephens, C. D., Fogleman, K. A., Lahr, J. C., and Page, R. A., 1984, Wrangell Benioff zone, southern Alaska: Geology, v. 12, p. 373–376.

Stephens, C. D., Fogleman, K. A., Page, R. A., and Lahr, J. C., 1985, Seismicity in southern Alaska, October 1982–September 1983: U.S. Geological Survey Circular 945, p. 83–86.

Stout, M. L., 1977, Radiocarbon dating of landslides in southern California: California Geology, v. 30, p. 99–105.

Sugden, D. E., and John, B. S., 1976, Glaciers and landscape: London, Edward Arnold Limited, 376 p.

Sunamura, T., 1983, Processes of sea cliff and platform erosion, *in* Komar, P. D., ed., Chemical Rubber Company Handbook of coastal processes and erosion: Boca Raton, Florida, Chemical Rubber Company Press, p. 233–265.

Swanson, F. J., and Lienkaemper, G. W., 1978, Physical consequences of large organic debris in Pacific Northwest streams: U.S. Forest Service General Technical Report PNW-69, 12 p.

—— , 1982, Interactions among fluvial processes, forest vegetation, and aquatic ecosystems, South Fork Hoh River, Olympic National Park, *in* Proceedings, Ecological research in national parks of the Pacific Northwest, Conference on Scientific Research in the National Parks, 2nd, 1979: Corvallis, Oregon State University Forest Research Laboratory, p. 30–34.

Swanson, F. J., Lienkaemper, G. W., and Sedell, J. R., 1976, History, physical effects, and management implications of large organic debris in western Oregon streams: Pacific Northwest Forest and Range Experiment Station, U.S.D.A. Forest Service General Technical Report PNW-56, 15 p.

Swanson, F. J., Swanson, M. M., and Woods, C., 1981, Analysis of debris-avalanche erosion in steep forest lands; An example from Mapleton, Oregon, USA, *in* Proceedings, Christchurch Symposium, Erosion and Sediment Transport in Pacific Rim Steeplands: International Association of Hydrological Sciences Publication no. 132, p. 67–75.

Swanston, D. N., 1971, Principle soil movement processes influenced by road-building, logging, and fire, *in* Proceedings, Symposium on forest land uses and stream environment Oct. 19–21, 1970: Corvallis, Oregon State University, Forestry Extension, p. 29–40.

Swanston, D. N., and Swanson, F. J., 1976, Timber harvesting, mass erosion, and steepland forest geomorphology in the Pacific Northwest, *in* Coastes, D. R., ed., Geomorphology and engineering: Pennsylvania, Dowden, Hutchinson and Ross, p. 199–221.

Swanston, D. N., Ziemer, R. R., and Janda, R. J., 1983, Influence of climate on progressive hillslope failure in Redwood Creek Valley, northwest California: U.S. Geological Survey Open-File Report 83-259, 43 p.

Szabo, B. J., 1980, ^{230}Th and ^{231}Pa dating of unrecrystallized fossil mollusks from marine terrace deposits in west-central California: Isochron/West, v. 27, p. 3–4.

Szabo, B. J., and Gard, M. L., 1975, Age of the South Bight II marine transgression at Amchitka Island, Aleutians: Geology, v. 3, p. 457–459.

Tabor, R. W., 1975, Guide to the geology of Olympic National Park: Seattle, University of Washington Press, 144 p.

Tabor, R. W., and Cady, W. M., 1978, Geologic map of the Olympic Peninsula, Washington: U.S. Geological Survey Map I-994, 2 sheets, scale 1:125,000.

Taylor, B. D., 1983, Sediment yields in coastal southern California: Journal of Hydraulic Engineering, v. 109, p. 71–85.

Terich, T. A., and Komar, P. D., 1974, Bayocean Spit, Oregon; History of development and erosional destruction: Shore and Beach, v. 42, p. 3–10.

Terich, T. A., and Schwartz, M. L., 1981, A geomorphic classification of Washington State's Pacific coast: Shore and Beach, v. 79, p. 21–27.

Thorson, R. M., 1978, Recurrent late Quaternary faulting near Healy, Alaska; Short notes on Alaskan geology: Alaska Division of Geological and Geophysical Surveys, Geology Report 61, p. 10–14.

Thorson, R. M., and Bender, G., 1985, Eolian deformation by ancient katabatic winds; A late Quaternary example from the north Alaska Range: Geological Society of America Bulletin, v. 96, p. 702–709.

Thorson, R. M., and Romick, J. D., 1984, The Alaska Tephrochronology Center; A status report, June, 1984: Unpublished report available at the University of Alaska, 130 p.

Thorson, R. M., Plaskett, D. C., and Dixon, E. J., 1980, A reported early man site adjacent to southern Alaska's continental shelf; A geologic solution to an archeologic enigma: Quaternary Research, v. 13, p. 259–273.

Tiffin, D. L., Cameron, B.E.B., and Murray, J. W., 1972, Tectonics and depositional history of the continental margin off Vancouver Island, British Columbia: Canadian Journal of Earth Sciences, v. 9, p. 280–296.

Tipper, H. W., 1969, Geology, Anahim Lake, British Columbia: Geological Survey of Canada Map 1202A, scale 1:253,440.

—— , 1978, Taseko Lakes (920) map-area: Geological Survey of Canada Open-File Report 534 (map), scale 1:125,000.

Tipper, H. W., Woodsworth, G. J., and Gabrielse, H., co-ordinators, 1981, Tectonic assemblage map of the Canadian Cordillera and adjacent parts of the United States of America: Geological Survey of Canada Map 1505A, scale 1:2,000,000, 1 sheet and legend.

Updike, R. G., 1982, Engineering-geologic facies of the Bootlegger Cove Formation, Anchorage, Alaska: Geological Society of America Abstracts with Programs, v. 14, p. 636.

VanSickle, J., and Beschta, R. L., 1983, Supply-based models of suspended sediment transport in streams: Water Resources Research, v. 19, p. 768–778.

Varnes, D. J., 1978, Slope movements and types and processes, *in* Landslides; Analysis and control: Transportation Research Board, National Academy of Sciences Special Report 176, p. 11–33.

Vedder, J. G., and Norris, R. M., 1963, Geology of San Nicolas Island, California: U.S. Geological Survey Professional Paper 369, 65 p.

von Huene, R., Crouch, J., and Larson, E., 1976, Glacial advance in the Gulf of Alaska implied by ice-rafted material, *in* Cline, R. M., and Hays, J. D., eds., Investigations of late Quaternary paleo-oceanography and paleoclimatology: Geological Society of America Memoir 145, p. 411–422.

Wahrhaftig, C., 1949, The frost-moved rubble of Jumbo Dome and their significance in the Pleistocene chronology of Alaska: Journal of Geology, v. 57, p. 216–231.

—— , 1965, Physiographic divisions of Alaska: U.S. Geological Survey Professional Paper 482, 52 p.

Wahrhaftig, C., and Birman, J. H., 1965, The Quaternary of the Pacific Mountain System in California, *in* Wright, H. E., Jr., and Frey, D. G., eds., The Quaternary of the United States: Princeton, Princeton University Press, p. 299–339.

Wahrhaftig, C., and Cox, A., 1959, Rock glaciers in the Alaska Range: Geological Society of America Bulletin, v. 70, p. 383–486.

Watson, R. A., 1980, Landform development on moraines of the Klutlan Glacier, Yukon Territory, Canada: Quaternary Research, v. 14, p. 50–59.

Weaver, W. E., Hagans, D. K., and Popenoe, J. H., 1987, Magnitude and causes of gully erosion in the lower Redwood Creek drainage basin, *in* Nolan, K. M., Kelsey, H. M., and Marron, D. C. eds., Geomorphic processes and aquatic habitat in the Redwood Creek basin, northwestern California: U.S. Geological Survey Professional Paper 1454 (in press).

Weber, G. E., and Lajoie, K. R., 1977, Late Pleistocene and Holocene tectonics of the San Gregorio fault zone between Moss Beach and Point Ano Nuevo, San Mateo County, California: Geological Society of America Abstracts with Programs, v. 9, p. 524.

Wehmiller, J. F., and Belknap, D. F., 1978, Alternative kinetic models for the interpretation of amino acid enantiomeric ratios in fossil mollusks; Examples from California, Washington, and Florida: Quaternary Research, v. 9, p. 330–348.

Wehmiller, J. F., Lajoie, K. R., Kvenvolden, K. A., Peterson, E., Belknap, D. F., Kennedy, G. L., Addicott, W. O., Vedder, J. G., and Wright, R. W., 1977, Correlation and chronology of Pacific coast marine terrace deposits of continental United States by fossil amino acid stereo chemistry; Technique evaluation, relative ages, kinetic model ages, and geologic implications: U.S. Geological Survey Open-File Report 77-680, 106 p.

Wehmiller, J. F., Lajoie, K. R., Sarna-Wojcicki, A. M., Yerkes, R. F., Kennedy, G. L., Stephens, T. A., and Kohn, R. F., 1978, Amino-acid racemization dating of Quaternary mollusks, Pacific Coast United States: U.S. Geological Survey Open-File Report 78-701, p. 445–448.

Weldon, R. J., II, and Sieh, K. E., 1985, Holocene rate of slip and tentative recurrence interval for large earthquakes on the San Andreas fault, Cajon Pass, southern California: Geological Society of America Bulletin, v. 96, p. 793–812.

Wells, W. G., 1981, Some effects of brushfires on erosion processes in coastal southern California, *in* Davies, T.R.H., and Pearce, A. J., eds., Erosion and sediment transport in Pacific Rim Steeplands: International Association of Hydrological Sciences, Publication no. 132, p. 305–342.

—— , 1984, Effects of brushfires on the generation of debris flows in southern California: Geological Society of America Abstracts with Programs, v. 16, p. 690.

Westgate, J. A., Hamilton, T. D., and Gorton, M. P., 1983, Old Crow Tephra, a new late Pleistocene stratigraphic marker across north-central Alaska and western Yukon Territory: Quaternary Research, v. 19, p. 38–54.

Williams, H., 1958, ed., Landscapes of Alaska, their geologic evolution: Berkeley, University of California Press, 147 p.

Williams, J. R., and Johnson, K., 1980, Map and description of late Tertiary and Quaternary deposits, Valdez Quadrangle, Alaska: U.S. Geological Survey Open-File Report 80-892-C, scale 1:250,000.

Wilson, C. J., and Dietrich, W. E., 1985, Lag in the saturate zone and pore pressure development after peak runoff in hollows: EOS Transactions of the American Geophysical Union, v. 66, p. 898.

Wilson, C. J., Reneau, S. L., Dietrich, W. E., and Narasimhan, T. N., 1984, Modelling the generation of excessive pore pressures in debris flow susceptible deposits: EOS Transactions of the American Geophysical Union, v. 65, p. 889.

Wolman, M. G., and Gerson, R., 1978, Relative scales of time and effectiveness of climate in watershed geomorphology: Earth Surface Processes, v. 3, p. 189–208.

Wolman, M. G., and Miller, J. P., 1960, Magnitude and frequency of forces in geomorphic processes: Journal of Geology, v. 68, p. 54–74.

Woodring, W. P., Bramlette, M. N., and Kew, W. W., 1946, Geology and paleontology of Palos Verdes Hills, California: U.S. Geological Survey Professional Paper 207, 145 p.

Woods, A. J., 1980, Geomorphology, deformation, and chronology of marine terraces along the Pacific coast of central Baja California, Mexico: Quaternary Research, v. 13, p. 346–364.

Yeats, R. S., 1983, Large scale Quaternary detachments in Ventura Basin, southern California: Journal of Geophysical Research, v. 88, p. 569–583.

Yehle, L. A., 1980, Preliminary surficial geologic map of the Valdez C-1 Quadrangle, Alaska: U.S. Geological Survey Map MF-1132, scale 1:63,360.

Yorath, C. J., and Chase, R. L., 1981, Tectonic history of the Queen Charlotte Islands and adjacent areas; A model: Canadian Journal of Earth Sciences, v. 18, p. 1717–1739.

Yount, M. E., Wilson, F. H., and Miller, J. W., 1985, Newly discovered Holocene volcanic vents, Port Moller and Stepovak Bay quadrangles: U.S. Geological Survey Circular 945, p. 60–62.

Ziemer, R. R., 1984, Response of progressive hillslope deformation to precipitation, *in* O'Loughlin, C. L., and Pearce, A. J., eds., Proceedings, Symposium on effects of forest land use on erosion and slope stability, Honolulu, Hawaii, May 7–11, 1984: International Union of Forest Research Organizations, p. 91–98.

Zinke, P. J., 1981, Floods, sedimentation, and alluvial soil formation as dynamic processes maintaining superlative Redwood groves, *in* Coats, R. N., ed., Watershed rehabilitation in Redwood National Park and other Pacific coastal areas: Center for Natural Resource Studies, John Muir Institute Incorporated, p. 26–49.

MANUSCRIPT ACCEPTED BY THE SOCIETY OCTOBER 10, 1986

ACKNOWLEDGMENTS

We wish to thank the following individuals who reviewed various parts of this paper and made valuable suggestions for its improvement: R. L. Beschta, S. Boggs, H. E. Clifton, and S. N. Dicken (Washington and Oregon); D. Iverson, T. Lisle, D. Marron, and M. Savina (northern California); and P. W. Birkeland and D. J. Ponti (central and southern California). In addition, two anonymous reviewers read the entire manuscript and we thank them for their comments. Finally, we express our sincere appreciation to Marjorie A. Henneck, U.S. Geological Survey, for typing the entire manuscript at its various stages of evolution.

Geological Society of America
Centennial Special Volume 2
1987

Chapter 14

Arctic Lowlands

L. David Carter
Branch of Alaskan Geology, U.S. Geological Survey, 4200 University Drive, Anchorage, Alaska 99508-4667
J. Alan Heginbottom
Geological Survey of Canada, 601 Booth Street, Ottawa, Ontario K1A 0E8, Canada
Ming-ko Woo
Department of Geography, McMaster University, Hamilton, Ontario L8S 4K1, Canada

INTRODUCTION

L. D. Carter

The Arctic Lowlands as defined for this chapter includes the Arctic Coastal Plains of Canada and Alaska and the western Arctic Islands of Canada (Fig. 1). In this chapter we discuss landforms and geomorphic processes that are characteristic of this part of North America, processes whose attributes are significantly different in the Arctic Lowlands than elsewhere, and landforms that occur elsewhere but have been studied most extensively in the Arctic Lowlands. Landforms and processes included are associated with the formation and degradation of ground ice, fluvial processes in the arctic environment, arctic coastal processes, and the formation of oriented lakes. We also discuss the sensitivity of the landscape of the Arctic Lowlands to human activity and to climatic change. Finally, we present recommendations for geomorphological research that will help us to better understand the evolution of the landscape of this region and to predict the effects on this landscape of natural or human-induced environmental perturbations. Brief descriptions of the physiography, climate, permafrost, and vegetation are presented as background for these discussions.

PHYSIOGRAPHY

As the name suggests, the Arctic Lowlands is an area of generally low relief, comprising plains, plateaus, and hills. The Arctic Lowlands as defined for this chapter includes five subdivisions (Fig. 1): (1) the Arctic Coastal Plain, (2) Sverdrup Lowland, (3) Parry Plateau, (4) Victoria Lowland, and (5) Shaler Mountains. This area differs from Bostock's (1970a) definition because the Sverdrup Lowland, the Parry Plateau, and the coastal plains of Canada and Alaska are included, and the Lancaster Plateau, Boothia Plain, and Foxe Plain of the eastern Arctic Islands are excluded. The Alaskan Arctic Coastal Plain was de-

fined and described by Wahrhaftig (1965), and the Canadian part of the Arctic Lowlands as defined herein was described by Bostock (1970b). Much of the following summary is derived from their reports.

The Arctic Coastal Plain is a coastal strip 15 to 150 km wide along the shores of the Arctic Ocean from Meighen Island to Alaska. In Alaska and the Yukon the coastal plain slopes gently from the foothills of the Brooks Range and Richardson Mountains to the Chukchi and Beaufort seas. Altitudes along the southern margin generally are less than 300 m. Coastal bluffs commonly are less than 10 m high except along the Chukchi Sea coast and the eastern part of the Yukon coast where bluffs 20 to 60 m high occur. A scarp locally as high as 125 m separates the coastal plain from the Mackenzie Delta. Local relief is principally due to thermokarst depressions, which are up to several meters deep; pingos, which are commonly 10 to 15 m high but may be as much as 60 m high; stream incisions, which reach a maximum of about 125 m near the Mackenzie Delta; and low hills, which rise as much as 150 m above the coastal plain in Alaska east of the Colville River. The Mackenzie and Colville rivers, the two largest rivers crossing the coastal plain, have built extensive deltas characterized by numerous lakes and distributary channels.

West of the Colville River, the Arctic Coastal Plain is a marine-abraded surface mantled by marine, fluvial, eolian, and lacustrine deposits (Dinter and others, 1987). Oriented lakes, which are described later in this chapter, are among the most striking aspects of the landscape in this region. Also present are stabilized Pleistocene linear dunes up to 30 m high and 20 km long (Carter, 1981) and smaller stabilized Holocene parabolic and longitudinal dunes (Black, 1951). East of the Colville River to near the Mackenzie Delta, the coastal plain is characterized by coalescing alluvial plains and fans, with pediments increasingly

Carter, L. D., Heginbottom, J. A., and Woo, M., 1987, Arctic Lowlands, *in* Graf, W. L., ed., Geomorphic systems of North America: Boulder, Colorado, Geological Society of America, Centennial Special Volume 2.

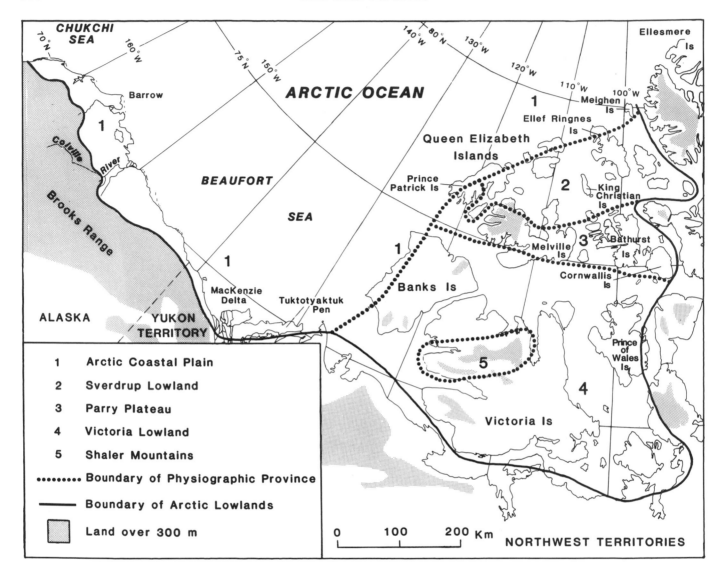

Figure 1. The Arctic Lowlands.

common eastward along the mountain front. Within a part of the coastal plain in Alaska east of the Colville River, however, streams have incised valleys across rising anticlines produced by active tectonism. Lobate moraines occur near the southern margin of the coastal plain along several rivers. East of 140°W longitude, most of the coastal plain is mantled by glacial drift and exhibits hummocky morainal morphology, except for the zone south of the glacial limits, which is characterized by alluvial fans and preglacial pediments (Rampton, 1982). Eolian silt and sand discontinuously mantles much of the coastal plain but is generally absent in areas of Wisconsin and Holocene glacial and fluvial deposits.

On the mainland east of the Mackenzie River, the coastal plain comprises the Tuktoyaktuk Coastlands (Rampton, in preparation) and Cape Bathurst. Most of this area is below an altitude

of 65 m with about 50 percent below 35 m. Pingos, thermokarst depressions, and coastal bluffs form the principal relief features. This area has been interpreted as consisting primarily of a Pleistocene fluvial and deltaic plain with morainal deposits in the southern Tuktoyaktuk Peninsula and marine deposits at Cape Bathurst (Mackay, 1963a; Rampton, 1974).

On Banks Island, the Arctic Coastal Plain is characterized by low rolling hills and well-organized drainage developed on till of the early Pleistocene or older Banks glaciation, and fluvial deposits of the upper Tertiary Beaufort Formation (Vincent, 1983, 1984). Coastal bluffs are as much as 70 m high. The Arctic Coastal Plain merges to the east with the Victoria Lowland at an altitude of about 100 m.

The Arctic Coastal Plain from Prince Patrick Island to Ellef Ringnes Island is low, flat, and uniform with a network of westward draining consequent streams. The inland boundary is at an elevation of about 30 to 40 m. On Meighen Island, the Arctic Coastal Plain is hilly terrain that supports an ice cap whose summit is at an altitude of more than 265 m. On all of these islands, the coastal plain is underlain by deposits of the Beaufort Formation.

The Sverdrup Lowland occupies the northern part of the Arctic Lowlands east of the Arctic Coastal Plain. This rolling, scarped lowland, which is mostly less than 165 m above sea level, developed on soft, poorly consolidated Mesozoic rocks that are moderately folded in the north and progressively more so to the south. Characteristic landforms include dissected domes and ring structures developed on evaporitic diapirs and igneous intrusions.

South of the Sverdrup Lowland is the Parry Plateau, which has an average elevation that is less than 250 m in the east and increases westward to more than 330 m. The Mesozoic sedimentary rocks that underlie this area are strongly deformed along the southern rim (Thorsteinsson and Tozer, 1970). Terrain in the eastern part of the Parry Plateau is characterized by wide, flat-topped ridges that are separated by broad, flat-floored valleys transected by incised cross valleys. To the west the plateau surface dominates the landscape and is surmounted by low hills, which support four small ice caps. The plateau surface is indented by fjordlike bays at the coast.

The Victoria Lowland lies between the Parry Plateau and the Canadian mainland. It consists of a smooth, undulating erosion surface that is formed on flat-lying Paleozoic sedimentary rocks and largely covered by glacial deposits, which over extensive areas are characterized by drumlinoid ridges. The Shaler Mountains are on Victoria Island and are bordered by the Victoria Lowland. The mountains are composed of Late Proterozoic stratified rocks intruded by cuesta-forming gabbro sills that are capped by flat-lying volcanic rocks.

CLIMATE

The effects of the cold climate of this region on geomorphic processes is the unifying theme of this chapter. This climate is characterized by low air temperatures, short cool summers, long cold winters, and low precipitation. Detailed descriptions of the climate of the Canadian Arctic Archipelago can be found in Brown (1972), and Maxwell (1980, 1981). Burns (1973, 1974) has published a climatological study that includes the Mackenzie Delta area, and details of the climate for the Alaskan Arctic Coastal Plain can be found in Dingman and others (1980), Selkregg (1975), and Searby and Hunter (1971).

Mean annual temperatures range from about –10 °C on the inner part of the Alaskan Arctic Coastal Plain to less than –17.5 °C in the northwesternmost part of the Arctic Lowlands. During July, which is the warmest month, mean temperatures range from 4° to 7 °C along the coast and throughout the western Arctic Islands and reach as high as 11° to 12 °C in the southernmost parts of the Arctic Lowlands in Alaska and the Yukon. Freezing temperatures may occur in any month of the summer. Mean daily temperatures drop below freezing in late August in the northwestern part of the Arctic Lowlands and in September in the southern part. Mean temperatures remain below 0 °C through May. January mean temperatures range from about –28° to –30 °C in Alaska and the Yukon to –37 °C north of 75°N latitude.

Precipitation is low throughout the Arctic Lowlands; as in other arid to semiarid areas, there are large variations in precipitation amounts from year to year. The annual total precipitation, as measured by standard precipitation gauges, ranges from 10 to 18 cm, of which about 35 to 50 percent falls as rain. Precipitation is greatest on the Alaska and Yukon coastal plains and southern Victoria Island, and the fraction that falls as rain decreases northward in the Arctic Archipelago. However, precipitation amounts, especially for autumn and winter, may have been underestimated by from 100 to 400 percent (Black, 1954; Benson, 1982; Woo and others, 1983) because of stormy conditions that are common during snowfall events. Rainfall is generally confined to the months June through September, and amounts are generally highest in late summer. Most rain is in the form of a persistent drizzle, but an occasional heavy rain during a single storm can account for more than a third of the summer's total. Snowfall is usually greatest in October, and more than 50 percent of mean annual snowfall has generally occurred by the end of December.

Snow covers essentially all of the landscape of the Alaskan and Yukon parts of the Arctic Lowlands from middle or late September through May. The maximum depth of snowcover is generally about 50 cm but varies because of drifting. In the northern part of the Arctic Lowlands, the mean depth of maximum snowcover is about 20 cm to 30 cm, and patches of ground blown free of snow become progressively more common northward.

PERMAFROST

Permafrost underlies the entire Arctic Lowlands. It is defined as rock or soil material, with or without included moisture, or organic matter that has remained colder than 0 °C continuously for two or more years (Muller, 1943). Permafrost is one of the manifestations of the cold climate that affects geomorphic processes, and it is associated with a characteristic suite of landforms related to the formation and destruction of ground ice. Ice in permafrost can cement materials that are otherwise unstable and subject to mass wasting and slopewash. When ice-rich permafrost is thawed, however, much sediment can be released from hillslopes and river banks. The presence of permafrost at shallow depths below the land surface restricts water storage in drainage basins, thus accelerating the runoff response to snowmelt and rainfall. In the Arctic Lowlands, permafrost has been constantly present for at least many tens of thousands of years, and may have been nearly constantly present for the past 2.4 m.y. (Carter and others, 1986a). All of the Arctic Lowlands is within the zone of continuous permfrost (Ferrians, 1965; Brown and Péwé, 1973; Brown, 1978; Péwé, 1983), which is underlain by permafrost everywhere beneath the land surface. Unfrozen zones (taliks) occur beneath water bodies that are sufficiently deep to prevent freezing to the bottom.

The surficial zone that thaws in summer and freezes in winter (the active layer) is generally less than 1 m thick and usually extends to the top of permafrost (the permafrost table). Thickness of the active layer is an important parameter for many geomorphic processes and is a principal control of surface drainage in areas of low relief. Active layer thickness depends upon climatic, topographic, and geologic conditions (Jahn and Walker, 1983), including slope aspect and inclination, thermal insulation of snow and vegetation, and material type and moisture content (Washburn, 1980). In general, the active layer is thicker on south-facing slopes than north-facing slopes; thicker where vegetation is absent than where there is a vegetation mat; thicker in dry, coarse-grained materials than in wet fine-grained materials; and thicker in bedrock than in soil. A rough correlation exists between thickness of the active layer and latitude, with the active layer becoming thinner northward (Brown, 1972). However, the variations caused by local conditions are of sufficient magnitude that a dry, unvegetated gravel site in the northern part of the Arctic Lowlands may have a thicker active layer than a tundra surface in the southern part of the Arctic Lowlands.

Permafrost thickness is imperfectly documented. This is because permafrost is defined solely on the basis of temperature, and therefore its base must be determined by observing ground temperatures in instrumented boreholes. Estimates of permafrost thickness, however, can be obtained from well logs and by geophysical methods. The available data indicate that permafrost is 300 to 600 m or more thick in most parts of the Arctic Lowlands, but may be considerably thinner near coastlines and near some areas that were glaciated during late Quaternary time (Judge and Taylor, 1985; Osterkamp and Payne, 1981).

VEGETATION

Vegetation is an important factor in geomorphic processes in the Arctic Lowlands because the type and density of plant cover influences the thickness of the active layer and the stability of ground ice. The flora of the Arctic Lowlands is now reasonably well known (Porsild, 1951, 1955, 1964; Anderson, 1959; Polunin, 1959; Wiggins and Thomas, 1962; Hultén, 1968), and the distribution of vegetation types has been extensively studied in many areas (Spetzman, 1959; Britton, 1967; Packer, 1969; Welsh and Rigby, 1971; Walker and others, 1980; Weber and others, 1980; Acevedo and others, 1982; Bliss and Richards, 1982; Edlund, 1982). Details of the relation of vegetation to permafrost were discussed by Bliss (1979).

The Arctic Lowlands is barren of trees except for the Mackenzie Delta, which supports stands of poplar and spruce (Mackay, 1963a). Elsewhere the vegetation is tundra, which can be described in terms of three vegetation belts distinguished on the basis of plant cover and vegetation type (Polunin, 1955). The southernmost belt includes the Arctic Coastal Plain of the main-

land and southern Banks and Victoria islands. In this region the vegetation is essentially continuous and comprises a variety of associations broadly grouped as: (1) wet tundra, which occurs on flat land and is composed predominantly of non-tussock-forming sedges, grasses, and forbs, with a minor component of shrubs such as dwarf willow, and (2) moist tundra, which occurs on better-drained land and consists primarily of tussocky sedges, grasses, forbs, and shrubs such as willow, dwarf birch, and Labrador tea. In the central zone, which occupies most of Banks Island and the central part of Victoria Island, vegetation is still conspicuous, but open plantless areas are increasingly common. Vegetation types are similar to those of the southern belt, but shrubs are less common, and cushion plant-lichen communities occupy upland sites. Barren ground predominates in the northern belt, with sedge and grass meadows and dwarf shrub tundra occurring only in the more favored localities.

OTHER GEOMORPHOLOGICAL RESEARCH

Considerable geomorphological research has been carried out in the Arctic Lowlands on topics other than those discussed here. Extensive work has been done on soil development and on the relation between soils, vegetation, and landforms (e.g., Tedrow, 1973, 1977; Tarnocai, 1976; Everett, 1979; Walker and others, 1980; Everett and Brown, 1982; Ugolini and others, 1982; Rieger, 1983). Washburn's (1980) thorough treatment of periglacial processes includes research carried out in the Arctic Lowlands. Papers dealing with periglacial sand dunes and eolian sand sheets were summarized by Niessen and others (1984) in a useful annotated bibliography. Research on glacial history, sea-level variations, climatic change, and landscape evolution is discussed in other Decade of North American Geology volumes (Dinter and others, 1987; Vincent, in preparation; Hodgson and others, in preparation), which also include references to research on periglacial processes carried out since Washburn's (1980) treatise.

ARCTIC FLUVIAL PROCESSES

M-k. Woo

In the Arctic environment, periglacial processes have received their well-deserved attention. Yet, as suggested by McCann and Cogley (1973, p. 118), "Viewed at a more general level, . . . it is the well developed drainage network . . . which holds the attention." There has been an increasing number of studies on fluvial processes in the Arctic Lowlands during the last two decades. Scott (1979) provided an annotated bibliography on the work done on arctic stream processes up to 1978. From the available evidence, there is little doubt that fluvial activities play an important role in shaping the arctic landscape. Fluvial process are not well understood, however, partly because most investigations involved only one field season, and partly because of the spatial variability of these processes, rendering it difficult to generalize results from small basin or channel segment studies to

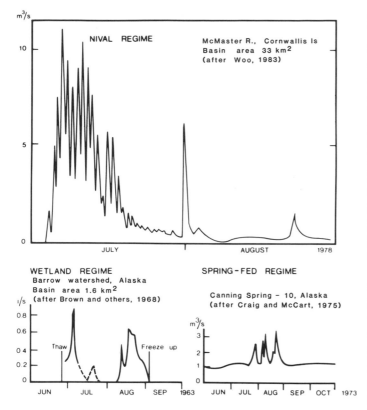

Figure 2. Three typical runoff regimes of the Arctic Lowlands.

(2) the wetland regime, and (3) the spring-fed regime (Fig. 2). The nival regime is dominated by snowmelt runoff. In spring, the frozen slopes and channels are largely impervious to meltwater; once the snow is saturated, runoff begins. Diurnal runoff rhythms are pronounced, corresponding to the daily pattern of snowmelt. Annual peak flows are often generated during this period. When the melt season is past, evaporation and continuous outflow deplete the basin water storage, and streamflow declines. Occasional summer rainstorms can quickly saturate the shallow active layer to produce flashy high flows. In rare instances, summer peaks nourished by exceptionally high rainfall can lead to a brief spell of intense fluvial activities (Cogley and McCann, 1976). As groundwater storage is limited, however, these peaks recede rapidly and streamflow resumes its summer low. Flow ceases completely after summer and does not recover until the following spring.

The wetland regime, also called the muskeg regime by Church (1974), is distinguishable from the nival regime mainly in the summer months. Like other areas in the Arctic Lowlands, wetlands are frozen completely in winter, and the frozen soil is quite impervious to meltwater infiltration. Spring often registers peak snowmelt runoff, but as thaw progresses, the surface organic mat or the peat layer allows more water to be stored. Vegetation growth also offers greater resistance to surface flow. Consequently, runoff response to rainfall is retarded, the peaks are lower than in nonwetland areas, and the recession flow is prolonged as water is gradually released to the stream. Brown and others (1968) showed that for a small basin (area 1.6 km²) near Barrow, Alaska, the recession flow after rainfall extends for about 50 hours, but can be as much as 160 hours. The peaks are delayed by only 3 to 10 hours, and they still exhibit a flashy appearance.

The spring-fed regime is often associated with carbonate terrains in Alaska and the Yukon, but thermal springs are also present. Streamflow is relatively stable because groundwater is the primary source of water supply. Hence, the discharge season is extended into early winter. The occurrence of flow during the freezeup period encourages ice formation in the channel, often as an icing, or aufeis, that builds up as groundwater or stream water floods the underlying ice layer in the channel.

Other than the spring-fed streams, rivers in the Arctic Lowlands have exceptionally high peak flow relative to mean annual runoff. Over half of the annual flow is concentrated within the snowmelt season, and the capacity for performing geomorphic work is potentially very high. In many instances, there is snow and ice in the channel and the ground is mostly frozen. These conditions provide partial protection for the beds and the banks against erosion and sediment entrainment.

the vast arctic domain. This section synthesizes the research findings obtained from the Arctic Lowlands and discusses the influence of permafrost and snow and ice on the fluvial processes in this cold region of northern North America.

Several large rivers, such as the Mackenzie, the Colville, and the Babbage, traverse the lowland region but have hydrologic characteristics acquired outside of the region and thus are not discussed here. It is to be noted, however, that studies have been carried out at the deltas of these large rivers, including their geomorphology, sediments, hydrologic, and hydraulic characteristics. Investigations on the Mackenzie River delta include those by Mackay (1963a) and the Mackenzie River Basin Committee (1981). The geomorphology of other deltas on the Yukon coast has been examined (e.g., Macdonald and Lewis, 1973; McCloy, 1970), and the Colville delta has also been studied (Arnborg and others, 1966, 1967; Walker, 1983, and references therein).

RUNOFF REGIMES

The runoff regime, or the seasonal pattern of streamflow, is controlled by the regional climate and the local storage and release of ground and surface water. Based on the classifications of Church (1974) and Craig and McCart (1975), three regimes can be distinguished for the Arctic Lowlands: (1) the nival regime,

CHANNEL PROCESSES

Many rivers in the Arctic Lowlands become dry before freezeup, and the channels are often bare of ice. Snow accumulates directly on the valley floors; drifting redistributes the snow

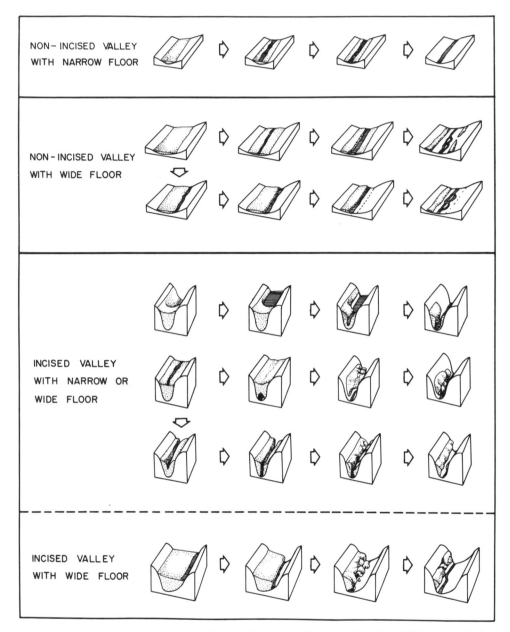

Figure 3. Break-up sequences in snow-filled valleys (after Woo and Sauriol 1980). In non-incised valleys rivers may downcut or downcut and shift laterally in the valley snow cover. In incised valleys, snow dams or snow tunnels may be formed and the rivers then downcut the snow cover, sometimes accompanied by lateral shifting.

into ridges and hollows along the valleys. The breakup sequence in spring has been described by Pissart (1967b) and by Woo and Sauriol (1980), a summary of which is given in Figure 3. The initial phase is a saturation of the valley snowpack by runoff from adjacent slopes. As the meltwater percolates through the valley snow cover to reach the streambed, intense coldness of the ground refreezes the water into basal ice (Woo and others, 1982). When stream flow commences, channels and occasionally tunnels are carved in the snow. These channels may initiate at the edge of the snow cover, and their ephemeral position can be high

above the streambed and on the hillslopes. Thus, fluvial activities are extended over a broad zone, reaching beyond the confines of the summer channels. The channels, partly or wholly cut in snow, are unstable, and they shift laterally and vertically in the snow and basal ice. Sometimes, when an ephemeral channel is abandoned, a cast remains of the sediments deposited by the stream onto its former bed in snow and ice (McCann and Cogley, 1973; Woo and Sauriol, 1981). Along segments of some valleys, large snow drifts accumulate across the valley to impound much of the meltwater until the snow dams are breached (Heginbottom,

1984). When this occurs, considerable discharge is released downstream. The position of the channels will not stabilize until the natural bed is reached; even so, lateral shifting can continue throughout the breakup season.

Some large rivers, such as those in Banks and Victoria islands and most streams on the Alaska coastal plain, may have sizeable pools left in their channels before freezing begins. Water in these pools will freeze up to 2.5 m deep, and the ice can be bottom fast if its thickness exceeds the water depth. Bottom-fast ice develops in rivers of the Alaska and Yukon North Slope because of negligible winter flow. This ice layer provides some protection against bed scour when initial breakup occurs, but the peak spring flood occurs after the disappearance of bottom-fast ice (Forbes, 1979). In spring, meltwater runoff floods the ice until the latter rots and rises from the bottom to float downstream as pans and chunks (Anderson and Durrant, 1976). Sometimes ice jams are created at narrower sections. The ice floes may be grounded or pushed to scour and groove the beds and the banks. Day and Anderson (1976) found that grooving is common where streamflow diverges, or where a variable river stage alternately lifts and grounds the ice. Ice-push on banks is found where flow converges, but both plucking and pushing are localized, though the activities are repeated at individual sites from year to year.

Another ice-forming process is icing caused by freezing of water that seeps from the banks or the bed to the channel. Icing can choke up segments of some rivers of the Alaska and Yukon coastal plains (Sloan and others, 1976). Breakup of this ice involves melting and mechanical erosion by running water. Sometimes icing can block the channel to such an extent that spring runoff is directed to the banks, thus causing serious bank erosion.

Where the stream flows over snow and ice-lined channels velocity increases, because the roughness diminishes. For a small stream on Cornwallis Island, Woo and Sauriol (1981) found that the Manning's roughness coefficient averaged 0.04 during the breakup period compared with 0.08 to 0.10 when the channel became ice free. The shear stress estimated for the creek also increased substantially in spring.

At the time of breakup, the power of the stream to erode and to transport varies rhythmically every day, according to the diurnal runoff cycles caused by daily melt. Although the potential for bed scour is high during the daily peaks, Pissart (1967b) argued that the presence of a snow and ice cover on the beds can protect the channels against erosion. Experiments using painted pebbles and painted lines on streambeds carried out in the Arctic Islands (Woo and Suariol, 1981) and on the Yukon Coastal Plain (MacDonald and Lewis, 1973) demonstrated that only a small fraction of the gravel was removed during breakup because it remained beneath the bottom ice. These experiments lend support to Pissart's argument. While protecting the beds, snow and ice may force incipient streams to impinge against the valley sides, extending the scouring action beyond the normal limits of natural channels.

After breakup, the floods recede. Thawing of the beds and the banks accelerates once the snow and ice have been removed

by the stream or by melting. Scott (1978) observed that before the peak of breakup flooding, erosion proceeds more slowly than the rate of thaw; but afterward, scour does not increase, even when the thalweg has thawed to greater depths. MacDonald and Lewis (1973) noted that bed scour is also complicated by the extent to which the bed materials are imbricated.

Besides bed scour, lateral migration of channels is another notable process in the Arctic Lowlands. According to Lawson (1983b), the factors affecting the erodibility of perennially frozen banks include exposure to currents and wind waves, texture and stratigraphy, ice content, slope aspect, and Coriolis effect acting on the banks of large rivers. Ground ice in streambanks has a significant effect on lateral erosion, and several types of bank failure associated with the presence of ice in bank materials have been described by Church and Miles (1982). Miles (1976) and Church and Miles (1982) observed that the presence of snow and ice in the channels tends to protect the banks against spring runoff so that bank erosion is relatively unimportant during the breakup period.

Lateral migration can be due to avulsion or preferential erosion along one bank (Lewis and MacDonald, 1973); the latter process being more rapid when thermoerosional niching is active (see discussion of thermoerosion in the Coastal Processes section of this chapter). Thermoerosion is especially effective on exposed ice wedges and on ice-rich bank material. Banks undercut by this process will subsequently fall, though Scott (1978) indicated that the slumping and niching are not necessarily synchronous, with the bank collapse delayed until the low flow period. Once collapsed, the slump blocks will protect the banks against further erosion until the blocks are removed by the river. Although ice-rich permafrost facilitates thermoerosion, the presence of permafrost increases the resistance to hydraulic action of streambanks composed of materials that are cohesionless in a thawed state (Cooper and Hollingshead, 1973).

Most studies on channel processes, carried out in a single field season, suffer from inadequate data to establish the representative nature of flood events. It is thus not possible to generalize upon the frequency of various magnitudes of events, nor is it possible to determine annual changes in channel forms. A comparison of aerial photographs taken at different times shows that many channels alter their courses, though the frequency of such occurrences is unknown. In one study carried out for over five years at the same reach of a river near Resolute, Cornwallis Island, Woo and Sauriol (1981) found that at-a-station hydraulic geometry manifested different sets of relationships amongst discharge, width, depth, and velocity each year (Fig. 4). Within the postbreakup period of each year, there was little scatter in the data points, suggesting that annual changes in the hydraulic geometry took place during breakup. This implies that channel processes are most effective in altering channel forms in the melt season.

SEDIMENT TRANSPORT PROCESSES

Arctic rivers entrain sediments derived from several sources.

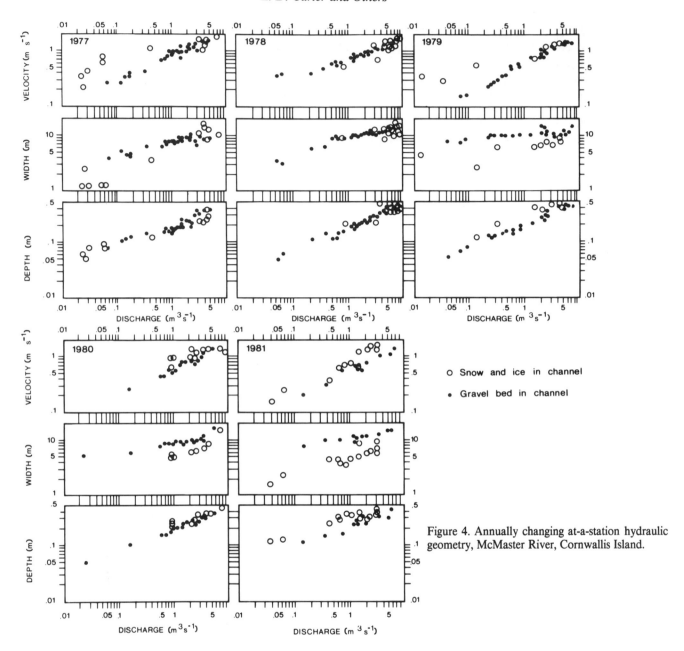

Figure 4. Annually changing at-a-station hydraulic geometry, McMaster River, Cornwallis Island.

One source is material left in the channel the previous fall, and another is aeolian and slope deposits that accumulated on the snow during winter. These unconsolidated sediments (Church, 1978) are easily entrained if they are not buried by snow and ice in spring. Another sediment source is mass wasting during and after snowmelt, but apart from the calving of streambanks, it is probably not as important as slope wash. Slope wash commences when meltwater removes materials from snow-free, thawing segments of slopes (Wilkinson and Bunting, 1975). Sometimes these sediments only reach the snowbanks adjacent to the channels and may not be entrained until the snowbanks have melted.

Arctic channel processes suggest that sediment transport is a function of the available power and the material supply. The latter may be restricted by a partial protective mantle of snow and ice, and limited by the degree to which the active layer has thawed. Despite these complications, the bulk of the sediment load is discharged during breakup, when high flows are maintained for a period of days. Summer rainstorms are short-lived and often of low intensity. Sediments are derived from the banks and beds. Hillslope processes are less effective in conveying materials to the streams unless intense rainfall reactivates slope wash.

Sediments are transported as dissolved load, suspended load, and bed load. The dissolved load of arctic rivers is highly dependent on the lithology of the bedrock and the surficial deposits in the drainage basin. Contrary to the low solute concentration found in the shield bedrock of the eastern Arctic (Church, 1972),

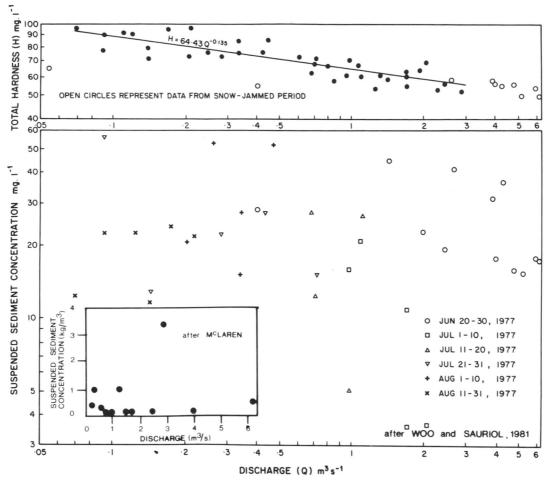

Figure 5. Relationship between discharge, total hardness, and concentration of suspended sediments, McMaster River, Cornwallis Island. Inset reproduces McLaren's data from eastern Bathurst Island (McLaren, 1981).

many rivers in carbonate terrains have high solute concentration (Cogley, 1972). Commonly, a strong relationship exists between solute concentration and discharge, with lower concentration at high flows. This is due to a dilution effect caused by a lesser portion of the water in contact with the bed and the banks at high water stage. The relative importance of dissolved load in terms of total stream load cannot be established for the Arctic Lowlands as a whole; however, for basins in carbonate areas, McCann and Cogley (1973) estimated that it exceeds suspended load by 50 to 100 percent.

Suspended sediment concentration in nonarctic rivers generally exhibits a statistical relationship with discharge, with the concentration increasing during the high stage. Depending on the availability of sediments during breakup, such relationships may not hold in arctic rivers. Woo and Sauriol (1981) reported that in spring, sporadic yet frequent waves of turbid water passed along the channel. One such wave was also described by McCann and Cogley (1973). These observations indicate that sediment is suddenly released, possibly due to the bursting of

snow dams or ice jams upstream (McCann and others, 1972). The supply of suspended sediments therefore is quite erratic and the sediments are poorly mixed in the water. Arnborg and others (1967) showed that the suspended load varies with discharge in a complex fashion in the Colville River, Alaska. McLaren (1981) suggested that two power functions be used for statistically fitting the sediment discharge rating curves: one for the breakup and the other for a postbreakup period. It is doubtful if such an approach is reliable because of the considerable scatter in the data (Fig. 5). In the absence of a simple, empirical relationship with discharge, most studies measured suspended sediments directly. It was found that the bulk of the suspended load is transported within the short breakup period.

Bed-load formulas cannot be used in the arctic because of the partial snow and ice cover in the channels and the frozen state of the materials. Evidence of significant bed-load transport can be found in the casts of coarse deposits on abandoned channels in the valley snow and ice cover (Fig. 6) and the considerable amount of material deposited on sea ice and on snow at the

Figure 6. Fluvial sediments stranded on snow, indicating that a broader stream was replaced by a narrow channel incised in snow. Photo by R. Heron on 12 July, 1978 when discharge was 4.9 m³/s.

mouths of streams (McCann and Cogley, 1973; Woo and Sauriol, 1981). Mobile beds have also been reported during stream gaging in the breakup period. After breakup, bedload transport is less marked.

The relative importance of bed load cannot be easily evaluated. Over the entire field season at Snowbird Creek, Bathurst Island, the 61 km² basin discharged 150 metric tons as bedload, 1,560 metric tons as suspended load, and 270 metric tons as dissolved load (Wedel and others, 1977). The magnitude of bed load was considered to have been underestimated.

PROCESSES AND FORMS

Cook (1967) suggested that fluvial action is ancilliary to periglacial processes, but McCann and Cogley (1973) contended that running water plays a major role in shaping the arctic landscape. Although subsequent research has confirmed the importance of fluvial processes, much of the evolution of the drainage network of the Arctic Lowlands probably occurred long before the present-day processes were activated. This regional drainage pattern is controlled by initial topography and geologic structure, Pleistocene history, and postglacial geomorphic processes that have varied significantly in intensity. A preliminary study of the river network, aided by information from various geological reports, permits a classification of the Arctic Lowlands drainage pattern into several groups outlined in Table 1.

Three broad regions are distinguished: (1) the continental margin, which includes the coastal plains of Alaska, Yukon, and the northern continental margin of the Northwest Territories; (2) the insular margin, which includes a 25 km strip of coastal zone for the Arctic Islands; and (3) the insular interior, which occupies the inland locations of the Arctic Islands. The choice of 25 km is entirely arbitrary, though it is felt that beyond this distance inland, the effect of coastal processes, both present and past, will be minimal.

The continental margin has mostly subparallel, low-order streams that are consequent to the regional slope (Fig. 7a). Two subgroups are identified; those controlled by topography and those controlled by coastal processes. Topographically controlled networks developed in a wetland environment, such as northern Alaska, are composed of streams that flow among lakes and are separated by poorly drained ground. In a nonwetland setting, such as the eastern Yukon Coastal Plain, the streams are often entrenched. In drainage patterns controlled by coastal processes, rivers run parallel rather than perpendicular to the coastline for a considerable distance. When the coast is breached, as at Horton River (70°N, 127°W), the stream finds a new exit to the sea.

The insular margin often has low-order streams, most of which are relatively short. Occasional large rivers from the interior area divulge to the sea, forming large deltas. For the smaller islands, a radial drainage is prominent if the islands are roughly circular, and where the islands are elongated, subparallel rivers drain from both sides of the central island ridge. Superimposed on the above general patterns are networks controlled by several factors. Structurally controlled networks are obvious at western Bathurst Island (Fig. 7f). There, a trellis pattern developed as

TABLE 1. MAJOR GROUPS OF DRAINAGE PATTERNS OF THE ARCTIC LOWLANDS

| Regions | Drainage Pattern | |
	Group	Subgroup
Continental margin	Topographically controlled	wetland environment nonwetland environment
	Coastal process controlled	
Insular margin	Structurally controlled Topographically controlled	low gradients moderate gradients high gradients
	Glaciation controlled Coastal process controlled	
Insular interior	Structurally controlled Glaciation controlled	main glaciated zone glacial margin
	Fluvial process controlled	

rivers exploited the limbs of folded sedimentary beds (Blake, 1964). Topography can also exert a marked control. Where the gradient is low, such as on northern Ellef Ringnes Island, distributaries are formed (Fig. 7b). In areas with moderate gradient, well-integrated dendritic patterns result (Fig. 7d). At coastal uplands, such as northeastern Banks Island (Fig. 7e) or northwestern Victoria Island (Thorsteinsson and Tozer, 1962), rivers carve canyons with over 30-m-high valley walls. Where glacial deposition dominates the landscape, drainage alignment is strongly influenced by linear features of the glacial deposits, as is seen at southwestern Stefansson Island (Fig. 7h). Finally, a stream may run parallel to the coast until an exit to the sea is found, as does Sachs River, Banks Island (Fig. 7c).

Rivers of the insular interior can reach high stream order, and the main streams are longer than the coastal rivers. Some drainage patterns are structurally controlled, such as the strike-oriented valleys of the Shaler Mountains, Victoria Island, where Precambrian sedimentary and volcanic rocks outcrop. Another group of drainage networks is glacially controlled. Where the land was within the main body of the glaciated zone, the present drainage network follows the linear features of the glacial deposits, as exemplified on the southern half of Victoria Island. Where the location was at the ice margin, the drainage pattern is deranged (Fig. 7g), an example of which is the Wollaston Peninsula of western Victoria Island (Sharpe, 1984). The third group of drainage networks has developed through most of the Pleistocene, producing well-integrated, often dendritic patterns. The best examples are found on Banks Island, where the late Wisconsin glaciers halted at various times at the eastern fringe of the island, sending meltwater that drained parallel to and away from the ice front. Many former valleys were also covered by glacial lakes or submerged by a sea (Vincent, 1982). At present, rivers of central Banks Island drain a broad plateau, whereas rivers in the northeast flow through a dissected upland.

Drainage patterns that are mainly inherited from the past are being actively modified by present-day fluvial processes. McCann and Cogley (1973) noted that valleys in much of the Arctic Lowlands have undergone considerable entrenchment since deglaciation, in part as a consequence of a falling base level due to isostatic recovery. In analyzing the longitudinal profile of a large river on Banks Island, Day and Lewis (1977) also suggested that the present river is attempting to degrade into a sediment wedge of Wisconsin materials whose extent appears to coincide with a relict outwash surface. Fluvial activities were probably more intense during deglaciation when the melting glaciers yielded high runoff to the sandur plains. This can be seen from a present-day analog in the glacierized area of the eastern Arctic (Church, 1972). There, rivers with a proglacial runoff regime provide a highly vigorous environment for erosion and sedimentation. It is clear that when the proglacial environment gave way to a periglacial setting, the large discharge was not maintained. Many abandoned and misfit channels are found in these formerly proglacial sites. The presence of such features in central Victoria Island attests to this (Fyles, 1963).

Valleys developed in the Arctic Lowlands are presently occupied by a single channel or by multiple channels. Single channel valleys are usually associated with narrow floors confined by bedrock or relatively cohesive bank materials. Knickpoints in bedrock are frequently encountered, and structure exerts a strong control over channel form. Examples include the canyons of Prince Albert Peninsula, northern Victoria Island, where rivers flow through the resistant Blue Fiord limestone, and the many bedrock knickpoints of the Allen River, Cornwallis Island. Another group of single channel valleys is associated with rivers of the coastal wetlands. These wetlands are covered with peat and tundra vegetation, and the streams tend to sweep across the low gradient area in meandering forms. Lateral erosion is conspicuous when there is active thermal niching and subsequent slumping of the banks.

Multiple channel valleys are more commonly found in the

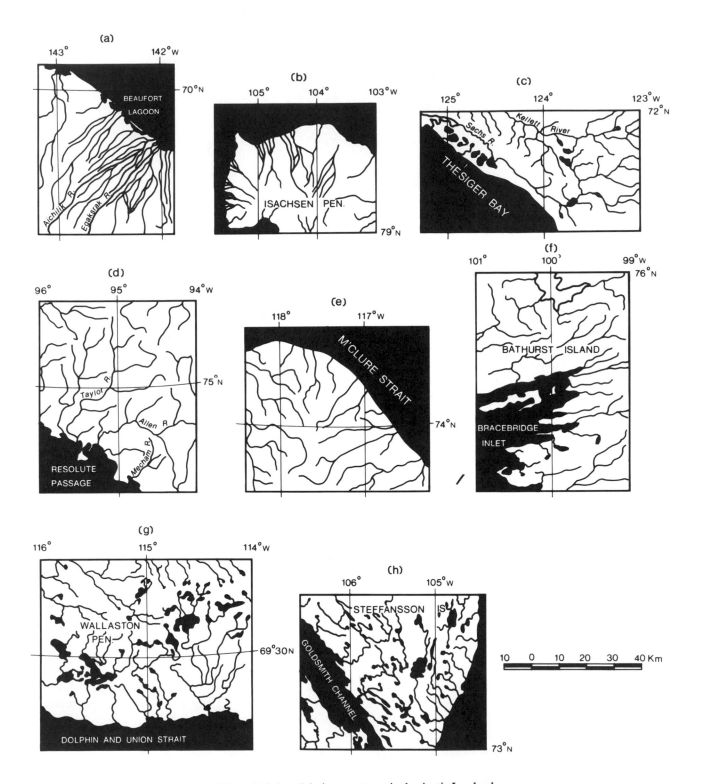

Figure 7. Selected drainage patterns in the Arctic Lowlands.

Figure 8. Multiple channel valley typical of a periglacial sandur, eastern Bathurst Island. Evidence of other processes active in the Arctic Lowlands (slope and periglacial processes) can be found as rills, solifluction lobes, and minor patterned grounds.

Arctic Islands (Fig. 8). Sometimes, a large part of the valley floor is covered by water at the high water stage to produce a single channel, and multiple channels become apparent when the stage subsides. French (1976) described the alluvial surfaces with multiple braided channels as periglacial sandurs, an analog to the proglacial outwash plains commonly found in the glacierized parts of the Arctic. Conditions favoring the formation of periglacial sandurs are rapidly varying discharge and high sediment load.

Sediments are easily eroded from the banks of braided rivers. These materials are noncohesive and seldom held together by vegetation. Lateral shifting of the braided channels leads to a continual rearrangement of the braid patterns, and is evidenced by the annual changes in the hydraulic geometry shown in Figure 4.

DISCUSSION

The landscape of the Arctic Lowlands is largely inherited from the past, developed under the control of structure and topography by processes that may no longer be active. There is sufficient evidence, however, to show that, together with periglacial and hillslope processes, fluvial processes are actively reshaping the landforms of the Arctic Lowlands (Fig. 8). In this regard, this section echos the viewpoint of McCann and Cogley (1973).

Two sets of conditions distinguish the arctic fluvial environment from other regions. One condition is the presence of permafrost at shallow depths, and the other is the preeminence of

snow and river ice. Scott (1978) provided an excellent discussion on the effects of permafrost on the behavior of stream channels in arctic Alaska. He concluded that the direct effect of permafrost is to retard channel erosion during breakup, and an indirect effect is to facilitate the erosion of cohesive banks after breakup by maintaining a high moisture content in the materials. While the presence of permafrost in naturally noncohesive materials can increase resistance to erosion, the abundance of ice can enhance thermal niching and thus facilitate bank erosion.

The severe arctic winters prevent any melting, and the snowfall of 8 to 10 months is stored and released rapidly during spring. This flashy supply of water produces an annually recurring peak flow period when the power to erode and transport is suddenly increased. On the other hand, the presence of snow and bottom ice can buffer the beds and the banks against fluvial activities during breakup. This apparent dichotomy between erosional and protectional tendencies leads to highly variable rates of fluvial action along most river segments, depending on a local rise in discharge (such as the bursting of a snow dam or ice jam) or a local increase in the channel snow and ice cover. The difference between the potential and the actual ability of the streams to work on their channels prevents the usage of conventional bedload formulas or empirical equations to estimate suspended load.

The above discussion does not suggest that permafrost or snow and ice dominate over other elements that affect the arctic fluvial landscape. The effects of summer rainstorms, the control of bedrock and Pleistocene geology, the presence of isostatic and

TABLE 2. SUMMARY OF GROUND ICE DESCRIPTIVE SYSTEM*

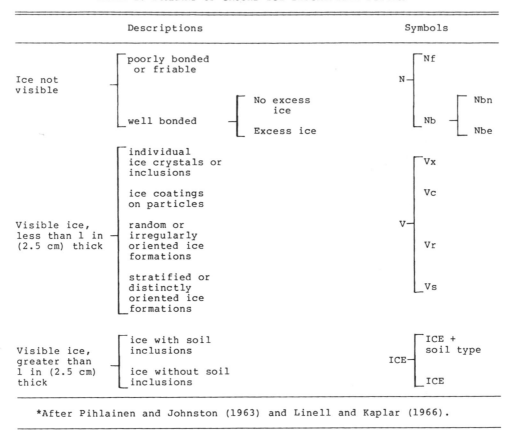

*After Pihlainen and Johnston (1963) and Linell and Kaplar (1966).

coastal processes, the variation in the regional and local topography and other factors such as vegetation cover, all play a geomorphic role in the arctic region. Similar to Scott's (1978) conclusion regarding the permafrost influences, it is concluded that both the permafrost and snow and ice factors should be considered as additional variables in the arctic fluvial system.

GROUND ICE

J. A. Heginbottom

INTRODUCTION

Because the Arctic Lowlands are entirely within the zone of continuous permafrost, most of the water in the ground occurs as ice; either on a seasonal basis within the active layer or year-round below the permafrost table. This discussion is concerned with perennial ground ice occurring within permafrost, which may include some glacier ice and ice that formed at the ground surface and subsequently was buried, or which formed within the active layer and later was incorporated into the permafrost.

Perennial ground ice occurs in three main forms: (1) crystals within the pores of sedimentary rocks and unconsolidated deposits (pore ice); (2) thin, lamelar lenses of ice (segregation ice); and (3) larger bodies of more-or-less pure ice, which can range from thin lenses and veins to large ice wedges and extensive sheets of massive ice, and form in several different ways.

Several schemes for the classification of ground ice have been proposed. In North America, the most widely used are the descriptive, engineering classification of Pihlainen and Johnston (1963; Table 2; see also Linell and Kaplar, 1966) and the genetic classification of Mackay (1972). Johnston (1981, p. 51) published a modification of Mackay's classification which related genetic ground ice forms or types to the ice descriptions of Pihlainen and Johnston (1963; Table 3). More recently, Heginbottom (1983) produced a morphological classification for the purpose of mapping the distribution of ground ice (Table 4).

Estimates of the quantity of ice in the ground have been made for two areas within the region: the Alaskan North Slope (Brown, 1967) and Richards Island, in the Pleistocene Mackenzie Delta (Pollard and French, 1980). Brown estimated that 175 km^3 of ice occurs in the upper 7.5 m of permafrost in a 50,000 km^2 sample area. When extrapolated to the whole Alaskan North Slope, the total ground ice volume is 1,500 km^3. On Richards Island, Pollard and French estimated that 10.27 km^3 of ground ice is present in the upper 10 m, over an area of 2,335 km^2 (of which 32 percent is lake covered). These estimates are both equivalent to about 45 percent of the total volume of the upper few meters of permafrost.

TABLE 3. GENETIC CLASSIFICATION OF GROUND ICE*

Water Source	Water Transfer	Water Freezing	Ice Types	Typical Ice Descriptions**
Atmospheric Water	Sublimation	Crack infilling — Vein ice	Ice wedges	Vertically foliated ice
Surface water	Gravity transfer	Dilation crack ice		
	Soil water expulsion	Water expulsion — Pore ice	Pore ice	Nf, Nbn, Vx, Vc
	Soil water gradient (pressure, gravity, thermal, matric, osmotic)	Water addition — Segregated ice	Ice lenses (downward) freezing	Nbe, Vx, Vc, Vr, Vs, ICE
Groundwater			Aggradational ice (upward) freezing	Nbe, Vx, Vc, Vr, Vs, ICE
			Reticulated vein ice	ICE + Nbn (for soil)
	Soil water pressure	Water intrustion — Intrusive ice (water)	Sill ice	ICE, pure or horizontally layered
			Pingo ice	ICE, pure or layered

*After Mackay (1972), Johnston (1981).

**See Table 2 for key to ice descriptions.

The volumetric estimates also show that pore ice is the greatest single component of ground ice, and that pore and segregation ice together constitute over 80 percent of the total ice volume. Wedge ice, while comprising 12 to 16 percent of the total ice volume in the upper 5 m of the ground, comprises 30 percent of all excess ice (ice in excess of the volume of natural voids), and, at some sites, may exceed 50 percent of the earth materials in the upper 1 to 2 m of the ground (Pollard and French, 1980).

Considerable quantities of ground ice are also known to occur at greater depths, but volumetric estimates have not been made. Rampton and Mackay (1971) and Mackay (1972, 1973) examined some 5,000 drill logs for the Canadian sector of the Arctic Coastal Plain, and found that several hundred penetrated deep, massive ground ice. The top of this ice was encountered at depths as great as 45 m; its thickness was typically less than 20 m, but greater thicknesses are fairly common. In some instances, "holes were drilled 125 feet [38 m] into solid ice, without bottoming through it" (Mackay, 1972, p. 15; Mackay, 1973, Fig. 4).

Ground ice is thus a significant component of the near-surface deposits of the Arctic Lowlands, although it is highly variable in its spatial distribution. In many respects, it can be considered similar to other earth materials, except that ice in the ground is near its melting point.

Most ground ice develops in place, within the ground, which generally requires the displacement of an equivalent volume of soil or rock material, either vertically or laterally. If this ice melts and the water drains away, the soil or rock material may be unsupported and so subside or collapse into the void. Buried surface ice, after melting, has similar effects. Thus the freezing or melting of ice in the ground changes the structure of the other earth materials enclosing the ice. These changes commonly result in alterations of the ground surface or landscape.

GROUND ICE AND THE LANDSCAPE

Over wide areas of the Arctic Lowlands, the growth and decay of ground ice have been major controls on the development of the present landscape. Ground ice is responsible for relief elements as small as 1 m or less, such as a soil hummock or the ridges either side of an individual ice wedge. At the other end of the scale, ground ice can produce individual hills several kilometers long and over 50 m in height. Relief elements within this scale range may occur in recurrent patterns, such as ice wedges in a polygon field, which influence large areas of the terrain. The thawing of ground ice, resulting in slumps, depressions and thaw lakes, similarly affects large areas of the landscape at scales rang-

TABLE 4. MORPHOLOGICAL CLASSIFICATION OF GROUND ICE*

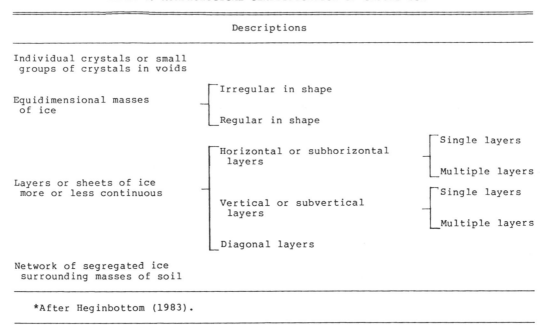

*After Heginbottom (1983).

ing from a few meters to several kilometers, and also in recurrent patterns.

Growth mechanisms and constructional landforms

Unfrozen soil is a complex mixture of mineral and organic particles, commonly exhibiting a range of sizes and mineralogies; liquid water, with dissolved salts; and air, including water vapor and other gases. During freezing, the properties of and interactions between these components change in a complex, temperature and pressure dependent manner. The growth of ground ice, particularly segregation ice, is one manifestation of the processes involved in the freezing of soil. Soil freezing and thawing, known collectively as frost action, are major geomorphological processes in the Arctic Lowlands (French, 1976, p. 12; Washburn, 1980, p. 62).

Initial freezing of soil normally progresses downward from the surface. As pore water freezes, it undergoes an initial volumetric expansion of 9 percent. This, however, is significant only in initially saturated soils. In unsaturated soil, the expansion is largely accommodated within the soil pores during the process of pore ice formation.

Ice segregation and the development of ice lenses are geomorphologically far more significant. Ice lenses form due to the movement of water from unfrozen soil toward the freezing plane, in response to suction created during freezing of soil pore water. Growth of an ice lens, however thin, results in a volume increase in the ground and an upward displacement of the ground, collectively known as frost heave. Heaving forces developed by the growth of ice lenses can be very large—in excess of 130 kN in a clay soil (Penner, 1970).

In fine-grained soils, partly because of the thermodynamic effects of fine capillaries on pore water, not all soil moisture freezes at 0°C. Rather, it freezes at some temperature less than 0°C, related to the specific diameter of the pores. Thus frozen, fine-grained soils can contain significant amounts of unfrozen, liquid water down to temperatures of several degrees below 0°C. This unfrozen moisture may continue to migrate deeper into the frozen ground, resulting in continued, secondary, frost heave, which has been demonstrated in the field at a site near Inuvik (Mackay and others, 1979). The thermodynamic nature of frost action and the mechanisms involved in soil freezing are reviewed in recent papers by Gold (1985) and Smith (1985).

Ice-cored terrain. Field observations throughout the Arctic Lowlands attest to the widespread occurrence of massive ground ice, ice-cored terrain, and high ice content soils. In Pleistocene sediments, many topographic high features owe a significant proportion of their elevation to ground ice. For example, at East Oumalik, 160 km south of Barrow, a sill of ice 11 to 13 m thick makes up about 50 percent of the vertical section above bedrock (Lawson, 1983a). Another example is "Involuted Hill," 16 km east of Tuktoyaktuk, where approximately 70 percent of the relief of a 30 m high hill is attributed to excess ice, including both segregated ice and massive ice (Rampton, 1973, p. 48; Rampton and Walcott, 1974, p. 118). Pleistocene sediments at other sites in the Tuktoyaktuk Coastlands, on Banks Island, and on southern Victoria Island contain similar proportions of ground ice (Rampton and Walcott, 1974; P. A. Egginton, personal communication, 1985).

Massive ground ice is also reported from other sites in the Arctic Archipelago, including Melville Island, King Christian Island, and Ellesmere Island (L. D. Dyke, J. Ballantyne, D. A.

Hodgson, personal communication, 1985). At these sites, the ground ice is occurring in bedrock or regolith rather than glaciogenic sediments. Because the sites have not been studied in detail, the full extent of the ground ice and its contribution to the form of the present landscape is not known.

For the Pleistocene sediments of the Arctic Coastal Plain, Mackay (1979, p. 52) suggested that ice-cored landforms form a genetic continuum ranging from tabular sheets of injection ice to areas of flat terrain with high ice-content soil, and including massive ice bodies and pingos. On this basis, pingos must be regarded as a special case of ice-cored terrain, rather than a peculiar feature in their own right. They are important, however, because much of our understanding of the process of ground ice development has resulted from detailed field observations of pingos, mainly by J. R. Mackay working in the Mackenzie Delta–Tuktoyaktuk Peninsula area (Mackay, 1979, 1983, 1985; Washburn, 1985).

Pingos. A pingo is a perennial frost mound, usually conical to rounded in shape, with a core of more-or-less massive ice that is covered with soil and vegetation. Pingos have commonly been divided into two groups, open-system pingos and closed-system pingos, depending on the source of the water that forms the ice core (Brown and Kupsch, 1974). Mackay (1979, p. 8–9) preferred to describe these groups as hydraulic pingos and hydrostatic pingos, respectively. He argued that in an area such as the Mackenzie Delta–Tuktoyaktuk Coastlands, "open-" and "closed-system" pingos were impossible to distinguish, and therefore the distinction should be based on the source of the force generating the water pressure that contributes to pingo growth. Based on current knowledge, all pingos within the Arctic Lowlands (Fig. 9) are "closed-system" or hydrostatic pingos. The greatest concentration of pingos in North America is found in the Tuktoyaktuk Coastlands (Mackay, 1962, Fig. 1), where over 1,450 have been counted. Pingos are also very numerous in the central portion of the Alaskan sector of the Arctic Coastal Plain (Carter and Galloway, 1979; Walker and others, 1985).

In the Arctic Lowlands, most pingos have developed in residual ponds remaining after rapid drainage of lakes formerly underlain by taliks (Fig. 10). Permafrost aggradation in the saturated sediments of the lake bottom creates the high pore-water pressures necessary for pingo growth. Mackay (1979) reports that these pressures can be great enough to lift the whole pingo and intrude a lens of water beneath it; this will gradually freeze unless the pingo ruptures, allowing partial or complete drainage of the water lens.

Most pingos are oval to circular in plan, reflecting the shape and size of the initial residual pond. A few pingos grow in abandoned river channels, taking the form of elongate ridges following the line of the old channel (Pissart and French, 1976).

The ice core of a pingo is not a single, massive, lens-shaped body of intrusive ice. Examination of the interior of a small pingo at Tuktoyaktuk (Rampton and Mackay, 1971) showed the core to contain segregated ice in silts and sands in numerous lenses, up to 2.5 cm thick in the silts and as much as 12.5 cm thick in sands,

separated by layers of frozen soil material. Ice wedges were also noted. Recently, Mackay (1985) compiled the results of many years of observations of pingo ice, concluding that pingos contain pore ice plus varying proportions of intrusive ice, segregated ice, dilation crack ice, and wedge ice. Ten years of growth measurements from one pingo suggest that pore ice and segregated ice contribute considerably more to pingo growth than does intrusive ice (Mackay, 1985, p. 1462). Dilation crack ice is a significant component in the upper part of the pingo where surface extension is greatest.

Ice wedges and ice-wedge polygons. The contraction crack or frost crack origin of ice wedge polygons, first proposed by Leffingwell (1915, 1919) following field observations in northern Alaska, is now generally accepted. The mechanics of the process were analyzed by Lachenbruch (1962, 1966), while Mackay (1974, 1975a, 1984) studied the frequency and distribution of cracking, and the direction and speed of crack propagation. The development of ice wedges in bedrock was considered by Dyke (1984).

Individual ice wedges in the Arctic Lowlands are seldom more than 5 to 8 m deep or greater than 2 to 3 m in maximum width (Harry, 1987). The majority appear to be epigenetic in origin, although large, syngenetic ice wedges occur in eolian silt at the inner edge of the Alaskan Arctic Coastal Plain (Carter and others, 1984), and several small, syngenetic ice wedges have been reported from Banks Island (French and others, 1982). The typical surface expression of an active ice wedge is a shallow trough—the width of which is indicative of the width of the ice wedge—bounded by a pair of low ridges. In cross section, the ice wedge itself is roughly wedge-shaped, and the enclosing sediments are deformed on either side of the wedge (Fig. 11). This deformation of the sediments extends well below the permafrost table; field experiments on Garry Island (Mackay, 1980) have confirmed Jahn's (1972) hypothesis that the deformation is a response to compressional stresses as the upper layers of the permafrost expand during the summer warm period. The stresses are relieved by upward bulging of the ground on either side of the ice wedge, resulting in the marginal ridges, which can form immediately after initiation of an ice wedge. Mackay (1980, Fig. 42.7) showed raised ridges that developed within six months of the formation of a frost crack.

Ice wedges and associated troughs and ridges seldom occur in isolation; rather, intersecting networks known as ice-wedge or frost polygons are formed. The two terms are important because not all large-scale polygonal forms of patterned ground are caused by ice wedges. In particularly arid environments, an open frost crack may fill with windblown sand, forming a sand vein. Repeated episodes of cracking and infill produce a sand wedge (Péwé, 1959). Sand wedges and wedges formed of both ice and sand are reported from Prince Patrick Island (Pissart, 1968). An extensive field of large sand wedges in a polygonal system developed over a broad area of the Alaskan Arctic Coastal Plain under desert conditions existing during Wisconsin time (Carter, 1983).

Ice-wedge polygon systems are among the most widespread

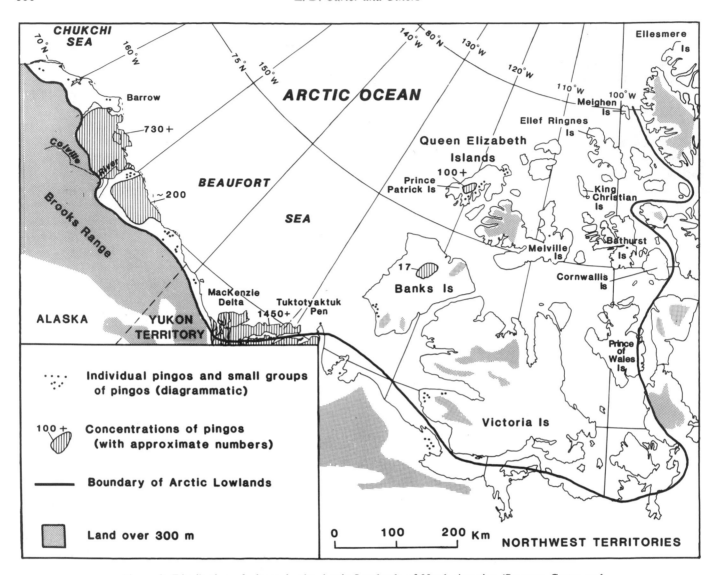

Figure 9. Distribution of pingos in the Arctic Lowlands of North America (Sources: Carter and Galloway, 1979, 1985; Galloway and Carter, 1978; French, 1975; Hodgson, 1982; Mackay, 1958, 1963a, 1966; Péwé, 1975; Pissart, 1967a; Pissart and French, 1976; Vincent, 1980; Williams, 1983).

and visible features of the landscape of the Arctic Lowlands, being found throughout the area, including Alaska (Leffingwell, 1919, Péwé, 1966a), northern Yukon (Rampton, 1982; Harry and others, 1985), the Mackenzie Delta and Tuktoyaktuk Coastlands area (Mackay, 1963a; Rampton, in preparation), Banks Island (French and others, 1982; P. A. Egginton, personal communication, 1985), Victoria Island (D. A. Sharpe, personal communications, 1985), and many sites in the Queen Elizabeth Islands (D. A. Hodgson, personal communication, 1985).

Two primary forms of ice-wedge polygons are distinguished: low-center and high-center polygons. Low-center polygons, up to 45 m in diameter, are characterized by distinct marginal ridges up to 60 cm in height and 2 to 3 m broad (Mackay, 1963a, Fig. 25). The low centers may retain standing water through much of the

summer. High-center polygons lack marginal ridges; they may develop by stabilization or degradation of low-center polygons or may develop directly in situations where the material being pushed up into ridges cannot maintain itself above ground level (Brown, 1973). The patterns exhibited within a polygon network were examined theoretically by Lachenbruch (1962, 1966), who concluded that ice wedges normally intersect at right angles. He recognized two main patterns, termed random orthogonal and oriented orthogonal. The former pattern is more widespread and develops on sites with no particular directional constraints. The oriented orthogonal pattern commonly develops adjacent to water bodies, with one set of frost cracks parallel to and one set perpendicular to the shoreline.

Although both field observations and theoretical considerations suggest that ice wedges should normally intersect at right angles, there are areas where polygons exhibit a generally hexagonal appearance. Washburn (1980, p. 166) was of the opinion that such sets initiated by permafrost cracking exist but are less common than the polygons described above.

Apart from the effects of water bodies on the pattern of polygon development, the geomorphic site conditions that favor ice-wedge and polygon development are not well known. In the Mackenzie Delta area, Mackay (1963a, p. 65–67) suggested that polygons are numerous in areas of sand and gravel terraces, Pleistocene sediments, and poorly drained areas, particularly former lake bottoms and sedgy depressions. Polygons are known to occur on hill tops and slopes, however, and on older sediments in the Arctic Islands. Rampton (1973, p. 45) proposed, "Ice wedges are probably present in all [unconsolidated] materials that are permanently frozen, even though their presence is not necessarily reflected at the ground surface by a polygonal pattern." The distribution of wedges with no surface expression is not known.

Relict ice wedges are reported from several localities in the Arctic Lowlands, including Alaska (Black, 1983; Carter and others, 1984), the northern Yukon (Harry and others, 1985), the outer Mackenzie Delta (Mackay, 1975b; Mackay and Matthews, 1983), and southern Banks Island (French and others, 1982). These wedges are commonly found at greater depths than currently active wedges, and in most cases their tops are truncated at a thaw unconformity. The Canadian wedges evidently formed during earlier cold phases, and the truncation occurred during a subsequent period of warmer climate. This warm phase has been interpreted variously as the early Holocene Hypsithermal event (French and others, 1982) or possibly an earlier warm period (Mackay, 1975b). The relict Alaskan wedges are associated with ancient tundra surfaces that have been buried by fluvial, eolian, lacustrine, marine, and colluvial deposits.

Inactive ice wedges are found in areas to the south of the zone of active ice wedges, where the present climate is not cold enough to cause cracking (Péwé, 1966b). Within the Arctic Lowlands, the only extensive area where most or all ice wedges are thought to be inactive is the forest-tundra area around the southern part of the Mackenzie Delta. Inactive ice wedges are also found throughout the zone of active ice wedges, primarily at sites where snow is trapped by vegetation or accumulates in drifts (Mackay, 1978, 1984). Inactive ice wedges may become reactivated following cooling of the ground (Mackay, 1976). This may result from a general climatic cooling, a decrease in snow accumulation, or erosion of the ground surface layers bringing relict wedges to within the depth range of active wedges. Several alternating episodes of rejuvenation and quiescence can lead to the formation of multiple wedge structures (Mackay, 1974).

Ground ice decay and destructional landforms

Melting of ground ice and the degradation of permafrost will occur with an increase of the mean annual ground temperature. This temperature change may result from either a general warm-

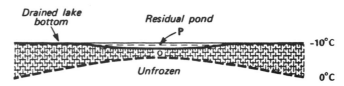

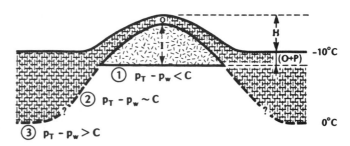

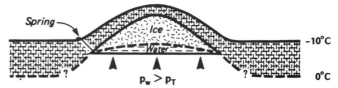

Figure 10. Diagram to illustrate the growth of a pingo in the residual pond of a drained lake (from Mackay, 1979, p. 36). (a) The residual pond has a depth of P in the center, the overburden thickness of permafrost is 0, and the permafrost is thinnest beneath the center of the residual pond. (b) Shows the growth of segregated ice. p_T is the total resistance to heaving and includes the lithostatic pressure and resistance to bending of the overburden. p_w is the pore water pressure, and C the soil constant. Ice lensing is favored at site 1; ice lenses and pore ice at site 2; and pore water expulsion at site 3. The values of p_T and p_w change from site 1 to site 3. (c) As the pore water pressure exceeds p_T, the total resistance to uplift, a subpingo water lens accumulates, and intrusive ice forms by freezing of bulk water. Peripheral failure may result in spring flow.

ing of the climate or from more localized disturbance of the ground surface such that a higher proportion of the available solar energy enters the ground. Common causes of such disturbance are human activity and forest or tundra fires (Heginbottom, 1973). Other causes include vegetation removal or burial, erosion or excavation of surface materials, or thicker snow accumulation. The response of the ground to thaw depends on the amount of ice in the ground and on the texture of the sediments or soil involved. A sediment containing no excess ice experiences little change in stability on thawing, except that it may be less able to resist other forces due to loss of bonding. Ice-rich permafrost, however, is subject to thaw consolidation upon melting. Where frozen soil containing considerable excess ice melts, the soil consolidates

Figure 11. Cross section view of an ice wedge, Maitland Point (70° 08′ N, 128° 18′ W) showing deformed sediments either side of the wedge (Jahn, 1972). The ice wedge, which is 6 to 7 m in height and over 2 m wide at the top, is developed in a sequence of alternating pond sediments and eolian deposits in a thermokarst basin (Photo by J-S. Vincent, Geological Survey of Canada no. 204375-I, 26 July 1985).

under its own weight. If thawing occurs more rapidly than melt-water can be expelled from the thaw zone by loading by the overlying soil, a positive pressure will develop within the intersti-tial water. This, in turn, can create a "quick" condition, with the water carrying part of the load. The result is a complete loss of strength within the soil mass, which then behaves as a viscous liquid and, even on a gentle slope, will flow rather readily (Taber, 1943; McRoberts and Morgenstern, 1974a; Carson and Bovis, in preparation).

Thermokarst terrain. Thermokarst terrain forms wherever ice-rich permafrost thaws and the ground surface subsides into the resulting voids. Thermokarst processes of thaw, ponding, drainage, surface subsidence, and related erosion, whether natural or man induced, are geomorphologically important, dynamic processes, capable of rapid and extensive modification of the landscape. Preventing or controlling man-induced thermokarst is a major challenge for northern development.

The extent of thermokarst resulting from local effects is

commonly confined to the site of occurrence and its immediate surroundings, whereas thermokarst resulting from climatic change is regional in its effects. Details of thermokarst landscapes are dependent primarily on the amount of ground ice initially present in the permafrost and the form in which it occurred. Thus, thaw of segregated ice, pingo ice, wedge ice, and massive ground ice may each result in different landforms.

In areas of ice-cored terrain, geophysical work and boreholes have shown that the sediments beneath most lake basins are generally free of excess ice, while the surrounding hills are ice cored (Rampton and Walcott, 1974). Widespread thaw modification of the landscape is reported from southern Victoria Island (Sharpe, 1985), Banks Island (French and Egginton, 1973), the Tuktoyaktuk Coastlands (Rampton, 1973, in preparation), Mackenzie Delta (Mackay and Matthews, 1983; Mackay, 1986a), and the northern Yukon (D. G. Harry, personal communication, 1985).

It was noted earlier that over much of the Arctic Lowlands, topographic high features owe much of their elevation to ground ice. On the other hand, much of the local relief of these areas is the result of thermokarst. In Siberia, such regional thermokarst of thick, very ice-rich permafrost produces a distinct landform, the "alas," which results from the coalescence of small thermokarst depressions. Alasses typically are large thermokarst depressions 5 to 20 m deep and of many square kilometers in extent (Czudek and Demek, 1970). Landscapes of this nature occur in loess in the southern part of the Alaskan Arctic Lowlands (Williams and Yeend, 1979), and Harry (1987) suggested that extensive lacustrine plains that appear to have evolved in response to the "thaw lake cycle" (Sellmann and others, 1975) are perhaps equivalent to the alasses of Siberia.

On level terrain, natural thermokarst typically begins as one or more small depressions or ponds, often occurring at the intersections of ice wedge polygons. In a polygon field, melting of ice wedges leads to subsidence of the trenches between the polygons that may be accelerated by thermal erosion of the wedge ice as meltwater drains through a polygon field. Low-center polygons may develop into high-center polygons, as the marginal ridges subside into the voids created where the ice wedges melt.

Thermokarst lakes. Small thermokarst ponds enlarge initially by thermal erosion (Mackay, 1970) of the surrounding banks; this process is particularly rapid where massive ground ice or wedge ice is exposed to the air or pond water. Larger thermokarst ponds continue to enlarge by thermal erosion, but other lacustrine processes may also become involved. Adjacent thermokarst ponds may enlarge until coalescence occurs, forming thermokarst lakes or thaw lakes (Hopkins, 1949). Over large areas of the Arctic Coastal Plain, thermokarst lakes are elongated with their long axes approximately parallel. The processes of formation of these "oriented lakes" are discussed in a later section.

Thermokarst lakes are prone to sudden drainage, which may occur when continued lake expansion intersects a river bank, coastal bluff, or area of lower terrain. Active retreat of a coast or erosion of a river bank will accelerate drainage. Mackay (1981) estimated that approximately one lake in the Tuktoyaktuk Peninsula and Richards Island region drains each year due to natural causes. Lake drainage is commonly initiated by water leaking along the troughs above ice wedges. Thermal erosion deepens the troughs by melting the wedge ice and drainage proceeds rapidly. Mackay (1981) described the induced drainage of "Illisarvik," a lake on northern Richards Island, that was achieved by digging a shallow ditch along an ice wedge and pumping lake water through it. When wedge ice was encountered, natural flow developed and the lake drained rapidly. Some 340,000 m^3 drained away in about seven hours. Following drainage of a lake, permafrost becomes reestablished in the sediments of the sublake talik.

Following regeneration of permafrost and the reactivation of old ice wedges or the development of new ones, polygons become reestablished. At "Illisarvik," frost cracking began the winter following lake drainage (Mackay, 1980), and a system of ice-wedge cracks is now developing (Mackay, 1984, Fig. 9). In northern Alaska, a "thaw-lake cycle," consisting of repeated development of thermokarst lakes, following lake drainage and regeneration of permafrost, has been identified (Britton, 1967; Billings and Peterson, 1980; Tedrow, 1969; Rawlinson, 1983). Repeated cycles of lake development, drainage, and redevelopment have not been reported from northern Canada. Rather, Harry and French (1983) working on southern Banks Island, suggest that the traditional thaw lake cycle has not developed there. Farther north on Banks Island and on Victoria Island, however, small thaw ponds are developing today within drained lake basins (P. A. Egginton, personal communication, 1985).

Thawing slopes. Tundra and forested slopes underlain by permafrost are generally stable. However, a rise in the ground surface temperature, leading to deeper thaw and an increase in the thickness of the active layer, can lead to slope instability and failure. Most slope failures in the Arctic Lowlands involve only thawed surficial materials. Two main classes of failures of thawing slopes are recognized: (1) detachment failures, and (2) thaw slumps. The amount and distribution of ground ice in the slope is a major factor determining which form of failure will occur and the progress of any failure event.

Detachment failures, also referred to as skin flows, occur when a thawed active layer and vegetation mat detach from the underlying permafrost table. They are common on colluvial slopes in areas of fine-grained, ice-rich soils, and occur more frequently in warm summers, particularly if associated with persistent summer precipitation, or following disturbance of the vegetation or the ground surface (Heginbottom, 1973). Typically, detachment failures in the low arctic exhibit an upper zone of detachment, the hopper; a middle zone of transportation, the chute; and a lower zone of deposition (Heginbottom, 1971). The failure plane is commonly in the ice-rich material immediately below the normal depth of the active layer. In the forested regions of the Arctic Lowlands, detachment failures occur most commonly on south to southwest-facing slopes. The initial failure is on steeper areas, where the slope angles are 12° to 14° (Heginbot-

Figure 12. Glacially deformed ground-ice bodies in a wall of the community ice cellar at Tuktoyaktuk. Layers of pale buff sand, 2 to 20 cm thick, are separated by dark layers of more or less pure ice, up to 25 cm thick. The icy sands have been deformed into horizontally overturned folds, visible in the upper part of the wall, as a result of being overridden by glacial ice; the ground ice therefore predates the last glaciation in this area (>40 ka). The floor of the ice cellar is seen at the lower left; hoar frost has built up on upper parts of the wall and ceiling (upper left). (Photo by J. A. Heginbottom, Geological Survey of Canada, no. 204375-G, 28 July 1983.)

tom, 1971; McRoberts and Morgenstern, 1973). North-facing slopes with steeper angles are more stable. In the Arctic Islands, particularly on ice-rich, poorly lithified Mesozoic shales, detachment failures can be initiated on slope angles as low as 3° and can propagate on even flatter slopes—as low as 0.5°.

Thaw slumps occur where massive ice or very ice-rich soil is exposed in areas of fine-grained soils; they result from a detachment failure, erosion of a river bank or lake shore, or coastal retreat. Thaw slumps, also known as bimodal flows, retrogressive thaw flow slides, ground-ice slumps, and thermocirques, consist

of a steep headwall, which retreats due to melting of the ice, and a debris flow formed from the mixing of thawed sediment and water, which slides down the face of the headwall and flows away. Headwall retreat is most rapid where debris is removed by, for example, a river. In coastal bluffs, the activity of thaw slumps tends to be cyclic, with periods of active retreat following storm erosion of the debris. Thaw slumps are common in icy glaciolacustrine sediments and fine-grained diamictons. These slumps have been described from all parts of the Arctic Lowlands. Other processes of slope failure involving ground ice, particularly thermoerosional falls, are peculiar to retreating coasts and river banks (Walker and Arnborg, 1966; Harry and others, 1983) and are described in a later section.

Deep-seated landslides involving frozen ground have been reported from the Mackenzie Valley (McRoberts and Morgenstern, 1974b). Thawing of the soil or ground ice is not thought to play a role in either the initiation or progression of such slope failures, even though the failure plane passes through frozen ground.

THE AGE AND ORIGINS OF GROUND ICE

In the Mackenzie-Beaufort region, three main periods of ground ice formation are recognized: (1) pre-Wisconsin, (2) early Wisconsin, and (3) Holocene, with the Holocene interval continuing today (Rampton, 1973; Mackay, 1986a).

Pre-Wisconsin ground ice may be seen in numerous exposures and some artificial excavations (Fig. 12) along some 400 km of the arctic coast of Canada, between Nicholson Peninsula (128°W) and Herschel Island (138°W). This ice and its enclosing sediments have been tilted and deformed by being overridden by glacier ice (Mackay, 1956a, 1959; Rampton and Mackay, 1971) of presumed early Wisconsin age (Mackay and others, 1972; Rampton, in preparation). This deformed, pre-Wisconsin ice is only known to exist within the area that was covered by ice of the Buckland glaciation but was not covered by later glacial advances (Rampton, 1986, in preparation). Rampton (1973) concluded that permafrost could only survive at or beyond the margins of the Laurentide ice sheet. Beneath the main body of the ice sheet, all permafrost disappeared due to geothermal heating, as the surface of the ground was insulated from the extreme glacial climate by the ice sheet itself (Fig. 13).

The major phase of ground ice formation in this region apparently occurred during the waning phases of the late Wisconsin ice sheet. Permafrost aggradation and ground ice formation proceeded rapidly during deglaciation (Rampton, 1973, Fig. 5d). The extensive bodies of massive ice reported from the area (Mackay, 1971) apparently formed in response to cold, late glacial climatic conditions coupled with the copious quantities of groundwater available from the melting ice sheet. Two mechanisms have been proposed for the origin of the massive ice bodies encountered in this area. Mackay (1971) theorized that the ice was of segregation origin, with water expelled during freezing of saturated sands as the source of excess water. This water was

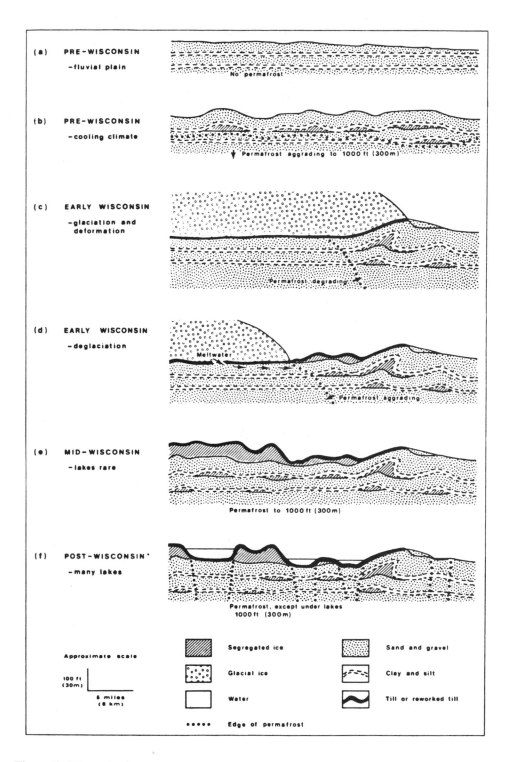

Figure 13. Schematic diagram of the development of ice-cored topography in the Mackenzie-Beaufort region (from Rampton, 1973, p. 55).

trapped beneath an aggrading, impermeable permafrost cover, creating high pore-water pressures, which are favorable to ice segregation. Rampton (1973; see also Fig. 13) proposed that the source of the excess water was meltwater from the waning ice sheet that was expelled from beneath the ice through permeable outwash sand and gravel by the weight of the ice sheet itself. Massive ice resulting from one or both of these processes occurs commonly near the base of tills of early Wisconsin age. Washburn (1985) favored Mackay's proposed model of the growth of segregation ice aided by pore water expulsion (Mackay, 1971). However, the two mechanisms (Mackay, 1971; Rampton, 1973) are not necessarily mutually exclusive.

Throughout postglacial time, degradation of permafrost and melting of ground ice have occurred locally, with a more regional impact during a latest Wisconsin–early Holocene warm interval. As the climate approached the conditions similar to those of today (ca. 11.5 ka) and became markedly warmer by 9 ka (Ritchie and Hare, 1971; Ritchie and others, 1983), the active layer thickened and thermokarst dramatically altered the landscape. Climatic deterioration about 5 ka (Ritchie and Hare, 1971) resulted in a sharp decrease in thermokarst activity. Thermokarst is still active, but the current rate of extension of thermokarst basins is not well known. Permafrost is aggrading in selected areas such as newly exposed estuarine and deltaic deposits and in the beds of drained lakes. Uplift of newly exposed, fine-grained sediments at Churchill, Manitoba, has been enhanced by the aggradation of permafrost and ice segregation (Hansell and others, 1983). The same mechanism is presumed to be operating today in aggrading coastal deposits of the Arctic Lowlands.

The age and origin of ground ice bodies in the Arctic Archipelago is poorly known. In the Alaskan part of the Arctic Lowlands, permafrost probably has been nearly continuously present for the last 2.4 m.y. (Carter and others, 1986a). Massive ice, such as is found in the Beaufort-Mackenzie region, is not widespread. Leffingwell (1915) concluded that in northern Alaska, ice wedges are the dominant form of distinct bodies of massive ice. In Alaska, epigenetic ice wedges that are early Wisconsin in age or older have been identified in coastal bluffs (Black, 1983), but late Wisconsin ground ice has been identified only as syngenetic ice wedges in loess in the eastern and southern Arctic Coastal Plain (Carter and others, 1984). Elsewhere, conditions were too dry for the formation of ice wedges, so sand wedges developed instead (Carter, 1983). General warming appears to have begun at about the same time as in Canada. Sellmann and Brown (1973, Table V) reported that in the Barrow area, warming began as early as 14 ka. Thermokarst probably was most active during the early and middle Holocene, just as in Canada. A general decrease in the size of thaw lakes near Barrow since the middle Holocene (Carson, 1968) may indicate that thermokarst activity has declined since that time.

Unlike bodies of massive ice and icy sediments, which often are amenable to stratigraphic evaluation, estimating the age of ice wedges is more difficult. Ice wedges cannot be older than their enclosing sediments, but may be much younger, with periods of active growth separated by periods of quiescence. Even in an area of active ice-wedge growth, many wedges fail to open in any given year (Mackay, 1974).

The extent to which burial of ice formed on the ground surface contributes to ground ice found in permafrost is a controversial topic. Buried surface ice may originate as stagnant glacier ice, icings, compacted snow, or river, lake, or sea ice. Burial may result from glacial overriding or colluvial or alluvial deposition of earth material over the ice. In the USSR, the origin of much ground ice, and particularly massive bodies of ice, has been ascribed to burial of surface ice, including Pleistocene glacial ice. In North America, however, this view has not been favoured and, until very recently, unequivocal agreement has not existed on the formation of significant bodies of ground ice in this way. Some massive ice on Victoria Island, however, may be buried glacier ice of early and/or late Wisconsin age (Lorrain and Demeur, 1985). This premise may also be applicable on Banks Island.

At present, attempts are underway to develop criteria that will determine the origins of bodies of massive ground ice. The promising lines of inquiry include examination of the isotope chemistry (^{18}O, ^2H, ^3H) of the ice, its petrography and crystallography, the nature and orientation of gas bubbles in the ice, and the mineralogy and chemistry of inclusions in the ice and the enclosing sediments.

COASTAL PROCESSES

L. D. Carter

The coastline of the Arctic Lowlands is diverse, ranging from low bluffs cut in unconsolidated deposits and fronted by barrier beaches (Hartwell, 1973; Walker, 1984) to cliffs 300 to 500 m high cut in rock (Bird, 1984). During the past three decades a great deal of research on coastal processes has been conducted along the Chukchi Sea and mainland Beaufort Sea coast, but comparatively little research has been accomplished along the coastline of the Canadian Arctic Islands. Nevertheless, it is clear that the intensity of coastal processes throughout the Arctic Lowlands is strongly influenced by the presence of ice. Ice covers the sea and prevents or inhibits wave action for most of the year, forms a protective coating on beaches, bonds nearshore subsea sediments, is responsible for nearshore erosive processes unique to polar regions, and locally composes as much as 75 percent of the materials that form mainland coastal bluffs of the Beaufort Sea. Processes influenced by ice can be conveniently discussed in terms of sea ice, ice-bonded beach and nearshore permafrost, ice in and on beaches, and ice-rich coastal bluffs.

SEA ICE

One of the most important effects of sea ice on coastal processes is that it prevents wave generation during most of the year, and limits it during the remainder (McCann, 1972, 1973). Sea-ice conditions are most severe in the western Sverdrup Is-

Figure 14. Sea ice along the Beaufort Sea coast on 1 August, 1976. The large river is the Colville River.

lands (Markham, 1981), where open water may occur for only a few days each year or not at all. In the eastern and southern Arctic Lowlands, the shore generally is free of fast ice by mid-July or early August, but freezeup and the formation of new sea ice commonly occurs in September or October (LaBelle and others, 1983; Markham, 1981). Wave generation is progressively inhibited as the ice canopy develops, and ceases when the ice cover is complete in October or November. Even during the brief open water season, wave action may be damped by ice floes, and fetch is limited by the presence of pack ice offshore (Fig. 14). Furthermore, pack ice may be blown against the shore, eliminating open water and wave generation. In spite of this, wave reworking and current transport dominate the sediment-transport regime in water depths shallower than 10 m along the mainland Beaufort Sea coast (Barnes and Reimnitz, 1985), and wave action and longshore drift play a major role in beach development and modification in this area (Rex, 1964; Short, 1979). Wave action is also important in the eastern Canadian Arctic Islands, but is less intense because ice conditions are more severe (Owens and McCann, 1970). In the western Sverdrup Islands, where ice conditions are most severe and wave activity may be limited to a few days, many beaches may be formed and altered primarily by ice push (Stefansson, 1913).

Sea ice also is an agent of sediment transport, and may contain sediment of two distinct types. One type consists of patchy concentrations of sediment, ranging in grain size from silt and clay to boulders (Fig. 15). These deposits are derived either directly from the seabed or the coast through mechanisms such as adfreezing, ice gouging, anchor ice flotation, seabed freezing in shallow water with subsequent flotation, deposition on the ice by rivers, deposition by wind, and mass-movement processes adjacent to steep coastal bluffs (Kindle, 1924; Reimnitz and Barnes, 1974). The second type of sediment consists of silt and clay-size particles in a zone of turbid ice that commonly occurs in the upper 1 m of fast ice and below a 10-cm-thick zone of relatively clear ice (Fig. 16; Barnes and others, 1982). This sediment may be entrained and concentrated by the interaction of sea ice with nearshore bottom sediment that was resuspended by storms during the early stages of growth of the ice canopy (Naidu, 1980; Barnes and others, 1982; Osterkamp and Gosink, 1984).

Sea ice can move sediment offshore or alongshore by rafting, and onshore by rafting or ice shove. Ice rafting is not thought to be a major mechanism of sediment transport west of the Mackenzie River for the first type of sediment described above (Reimnitz and Barnes, 1974). Ice rafting is important in transporting the second type of sediment, however, which sometimes oc-

Figure 15. Sediment-laden sea ice that has ridden up onto a barrier island on the Alaskan Beaufort Sea coast.

curs in sea ice in amounts that represent a significant proportion of the seasonal terrigenous influx of fine-grained material (Barnes and others, 1982). Coarse sediment can be carried or pushed up onto the beach during sea ice pile-up or ride-up (Fig. 15; Hume and Schalk, 1964; Taylor, 1978; Barnes, 1982; Kovacs, 1983, 1984). The movement of sea ice onshore along the Beaufort Sea coast can occur at any time of the year, but is most common in spring and fall (Kovacs, 1984). In the Canadian Arctic Islands, onshore movement has been observed primarily in the summer immediately following breakup (Taylor, 1978). Features formed by this process are described below in a discussion of beaches. Hume and Schalk (1964) estimated that coarse sediment supplied by sea ice typically made up from 1 to 2 percent of the beach deposits near Barrow and locally may have supplied 10 percent of the sediment. Ice shove during the winter of 1982/83 supplied sand and gravel to the beaches of some Beaufort Sea barrier islands in volumes estimated at >1 m^3/m of shoreline (Reimnitz and others, 1985). However, Reimnitz and others (1985) proposed that, over long periods, the contribution to beaches by ice shove is relatively small. Taylor and McCann (1976) pointed out that sea ice can also remove potential beach material from the littoral zone beyond reach of normal beach processes.

Interaction of sea ice with the seabed also results in ice gouges and "ice wallow" relief. An ice gouge is a furrow and associated ridges in the seabed produced when an ice keel scrapes along the bottom. Ice-gouge characteristics have been studied extensively on the Beaufort Sea shelf (e.g., Carsola, 1954; Rex, 1955; Kovacs and Mellor, 1974; Reimnitz and Barnes, 1974;

Figure 16. Fine-grained sediment entrained in sea ice. (Photo by P. W. Barnes, U.S. Geological Survey.)

Figure 17. Ice block about 3 m high stranded in a current-scour depression on a beach face. (Photo by E. Reimnitz, U.S. Geological Survey.)

Barnes and others, 1984). Barnes and others (1984) found that the furrows average nearly 8 m wide with a mean relief of slightly more than 1 m. They occur in all shelf water depths but are most numerous at depths of 20 to 40 m. In coastal waters, gouge intensity is less because ice motion is less and ice masses are small. Furthermore, nearshore gouges may be filled rapidly with sediment reworked by waves and currents (Reimnitz and Barnes, 1974). Reimnitz and Kempema (1982) applied the term "ice wallow" relief to closed depressions and mounds about 50 to 100 m in diameter and 2 to 3 m in relief that characterize most of the nearshore zone of the open coast of the Beaufort Sea. The depressions and mounds are thought to form as a result of erosion and deposition by currents that are accelerated as they pass around grounded ice (Fig. 17) and by pulsating currents produced by rocking motions or vertical oscillations of the grounded floes (Reimnitz and Kempema, 1982). Reimnitz and others (1985) pointed out that this irregular relief is attacked by normal waves, which results in increased bottom instability and sediment resuspension and transport. This in turn can cause steepening of the foreshore and accelerated coastal retreat.

Another effect of sea ice on coastal processes is its influence on patterns of meltwater erosion and deposition during breakup. River breakup and meltwater discharge commonly occur prior to the breakup of fast ice, causing flooding of the fast ice in the vicinity of river mouths (Kindle, 1924; Reimnitz and Bruder, 1972). This phenomenon results in minor sediment deposition on the fast ice (Walker, 1973) and erosion of the seabed where floodwater drains through cracks and seals' breathing holes in the

ice and flows seaward beneath it (Reimnitz and others, 1974). The drains, called strudel by Reimnitz and Bruder (1972), initially occur seaward of the 2 m isobath, which marks the general boundary between bottom fast ice and floating fast ice (Reimnitz and others, 1978). Later, as ice lifts off the bottom, strudel develop landward of the 2 m depth contour. "Strudel scours" developed below the drains correspond in shape to the strudel in the ice canopy and may be as much as 20 m across and 4 m deep. Reimnitz and Kempema (1983) found that strudel scours develop at a rate of about $2.5/km^2/yr$ and are filled in two to three years by bed load in the littoral sediment-transport system. They used this rate of filling to calculate an average bed load transport rate of 9 $m^3/yr/m$, which closely approximates the rate determined for nearby barrier island beaches. The rate of scouring and filling leads to the conclusion that arctic delta-front deposits should consist entirely of strudel-scour fill, which are characterized by dipping interbedded sand and lenses of organic material draped over steep erosional contacts, and an absence of horizontal continuity of strata. Reimnitz and Kempema (1983) proposed that these characteristics provide criteria that should uniquely identify high-latitude deltaic deposits.

BEACH AND NEARSHORE ICE-BONDED PERMAFROST

Ice-bonded permafrost generally is present beneath a thin active layer in Arctic Lowland beaches and barrier islands (Rex, 1964; McCann and Hannell, 1971; Taylor and McCann, 1974;

Carson and others, 1975; Owens and Harper, 1977; Morack and Rogers, 1981), and extends offshore at various depths below the seabed (Judge and Taylor, 1985; Neave and Sellmann, 1984). The influence of these ice-bonded materials on beach and near-shore morphology and erosive processes is controversial.

The depth to which beach sediments seasonally thaw and the depth below the seabed to ice-bonded sediments have been cited as providing a lower limit for sediment redistribution in the littoral zone (Owens and McCann, 1970; Harper and others, 1978). Because thaw depths on the upper beach face and berm are generally less than 1 m (Taylor and McCann, 1974; Owens and Harper, 1977), Harper and others (1978) proposed that sed-iment redistribution by wave action, storm surge washover, and ice push would be limited to this depth, except during periods of deeper thaw brought on by storm-induced high water levels. They argued that erosion or relocation of spits and barriers, breaching of barriers to form new inlets, and inlet migration commonly occur if the subsurface sediments are ice-free or if there is sufficient thaw during a storm or storms to melt the ice-bonded sediments. Maximum theoretical thaw rates for storm conditions are 50 to 70 cm/day (Harper and others, 1978; Tay-lor, 1980), however, which is insufficient to account for some of the large changes in beach and barrier configuration known to have occurred during storms. Harper and others (1978) assumed that areas undergoing changes more rapid than this rate of thaw-ing would allow must be underlain by sediments that are not ice-bonded. Morack and Rogers (1981) demonstrated that the cores of some barrier islands indeed are only sporadically ice-bonded. Reimnitz and others (1985) pointed out, though, that these islands do not exhibit the coastal irregularities that would be expected after storms if erosion were controlled by the distribu-tion of ice-bonded sediments. They concluded that ice-bonding of beaches and nearshore sediments does not appreciably affect rates of erosion and sediment transport of beach and nearshore deposits.

ICE IN AND ON BEACHES

Ice on the beach protects it from wave attack. Ice on and within beach deposits, including ice pile-ups and ride-ups, are responsible for morphologic features and sedimentary structures that are characteristic of arctic beaches.

Development of an ice cover on arctic beaches occurs dur-ing freezeup; it has been described by Short and Wiseman (1974) for the microtidal Beaufort and Chukchi sea environments and by McCann and Taylor (1975) for tidal conditions more typical of the island shores of the eastern Arctic Lowlands. Short and Wiseman (1974) note that freezeup processes and ice formation vary in intensity and chronology from place to place and year to year but include the following: (1) Freezing of the exposed beach surface between the prevailing swash zone and the vegetated backshore. (2) Formation of a snow cover, which insulates the frozen beach crust. During this period, swash accompanying storm surges may deposit thin layers of sediment that are incorpo-rated in the snow cover, while swash percolating through the snow freezes. (3) Deposition of cakes and boulders of sea ice that are generally less than 10 m in diameter. These may be deposited at any time during the open water season, and may be buried or nearly buried by subsequent sediment deposition (Short, 1976). (4) Deposition of ice slush berms derived from ice slush formed on nearshore water surfaces. Sediment is commonly scattered throughout the berm, and subsequent sediment deposition may bury the berm completely. (5) Freezing of swash, spray, and foam to the beach when air and water temperatures fall below 0°C, and freezing of the beach face within the swash zone. The frozen swash forms a continuous crust of frazil on the beach that is generally destroyed by the following swash, but may be preserved with falling sea level or decreasing wave energy.

Deposition of ice on the beach face in the form of ice boulders, ice slush, snow, frozen swash, spray, and foam can occur simultaneously with the deposition of sediment and freez-ing of the beach face. A flat-topped ice and sediment rampart up to 1.5 m thick (Fig. 18), called a storm-ice foot (Rex, 1964) or kaimoo (Moore, 1966), can develop through successive episodes of sediment deposition, ice slush deposition, and swash adfreezing.

Freezeup processes observed by McCann and Taylor (1975) in the eastern Arctic Islands are similar to those described above, but they pointed out that the kind of ice that forms in the beach zone is influenced by temperature, wind and wave conditions, and sea-ice conditions. The rate of freezeup is controlled primar-ily by the rate of temperature decline, but increased wave activity can delay the process of ice formation by disturbing sediments, eroding accumulated slush ice, and melting the ice glaze in the intertidal zone. Grounded ice floes large enough to prevent wave activity can inhibit the formation of an ice foot, which is of a different kind than kaimoo of the microtidal zone. In the eastern Arctic Lowlands, ice foot develops in the upper part of the inter-tidal zone of tidal beaches as layers of ice form there during successive low tides (McCann and Carlisle, 1972). Storm waves during the period of growth can result in the formation of ice above normal high-water line by freezing of swash and spray. The fact that a lack of wave activity can inhibit development of the ice foot suggests that waves, as well as tides, are important in its formation.

After freezeup, beach processes are inactive until breakup except for ice pile-up and ride-up along the coast of the Beaufort and Chukchi seas, which can occur during any season. Ice pile-up and ride-up produces ice-shove features when sediment is pushed or carried onto the beach by the ice (Hume and Schalk, 1964, 1976). When the ice melts, the bulldozed debris forms irregular ridges and mounds (Fig. 19). Boulder ridges as high as 3.5 m have been produced in this fashion (Barnes, 1982), and beaches can be significantly modified by scouring during these events to produce striations and boulder pavements (Kovacs, 1984). If snow and ice form a ramp at the base of a coastal bluff, sea ice can ride up it and deposit debris on top of bluffs as high as 6 m (Brower, 1942; Fig. 18). Sea ice that is thickened as it piles up on the beach may

Figure 18. Storm-ice foot (foreground) protecting a 4-m-high coastal bluff from wave attack on the Alaskan Beaufort Sea coast. Coarse debris on top of the bluff was deposited by ice shove prior to development of the ice foot. Coarse debris on the bluff face and ice foot has collapsed from the top of the bluff.

Figure 19. Ice-shove ridge (foreground) and mounds produced by sea ice riding up onto the beach. Helicopter in background for scale.

Figure 20. An ice pile-up along the Beaufort Sea coast insulating the beach from wave attack. Man in right middle-ground is about 1.7 m tall. Ice-shove deposits from a previous ice pile-up are in the foreground.

remain throughout the open water season and protect the beach from wave erosion (Fig. 20).

In the spring, waves cannot attack the beach until it has been cleared of fast ice. The period of time necessary to accomplish this varies from place to place depending on the thickness of the ice foot and exposure of the beach. It may require only a few days following the breakup of sea ice, but in some years, at favorable localities, an ice foot can remain on the beach through most or all of an open water season and thereby prevent wave erosion. Taylor and McCann (1976) considered tidal ice foot to be generally more effective at delaying wave action than kaimoo, and Reimnitz and others (1985) felt that, in general, kaimoo is relatively unimportant along the Alaskan Beaufort Sea coast.

Another type of fast ice that sometimes persists throughout the summer in areas of moderate tides occurs as a ridge at and below low tide level (Taylor and McCann, 1976). The ridge is the remnant of an ice foot that once covered the entire intertidal beach zone, and it forms if sufficient sediment is deposited in the first shore crack or lead to protect the underlying ice. When the remainder of the ice foot melts, a gravel-covered ice ridge remains to mark the position of the former shore crack. The ridge can protect the beach from grounding sea ice and inhibit waves from reworking sediment on the lower beach face.

Apart from the effects of ice shove, the distinctive geomorphic and stratigraphic features of arctic beaches form as ice in and on the beach melts during spring and summer (Short, 1976). The

amount of ice available to melt depends upon the amount formed during freezeup, and thus upon wave, sea ice, snow cover, and temperature conditions the previous fall. Melting of the ice results in subsidence or topographic inversion and produces features such as pits, sand and gravel piles and ridges, and distorted bedding. Buried ice masses may melt to form pits, and if the ice contained or was covered by a significant amount of debris, a conical pile may occur in the pit (Greene, 1970). When ice or snow interbedded with sediment melts, distorted bedding is produced. Also, the beach face is eroded by meltwater, and small alluvial fans are produced near the water line. These features are commonly destroyed by waves during fall storms, and so their preservation in the geologic record is expected to be rare. Reinson and Rosen (1982) reported the preservation of ice-formed features in a progradational beach sequence in Labrador, however, which they interpreted to represent a yearly cycle of beach accretion. Thus it may be possible to utilize the characteristics of modern arctic beaches to interpret the seasonal environments of some ancient beach deposits.

ICE-RICH COASTAL BLUFFS

Much of the mainland coast of the Beaufort Sea and southwest Banks Island is backed by low bluffs composed of ice-rich, unconsolidated deposits. Despite the limitations on wave action discussed above, these bluffs are undergoing rapid retreat

TABLE 5. COASTAL RETREAT RATES FOR THE ARCTIC LOWLANDS*

Location	Maxima (m/yr)	Mean (m/yr)	Source
Chukchi Sea coast			
Near Barrow	3.9	1.9	MacCarthy, 1953
Near Barrow	3.5	2.2	Hume and others, 1972
Barrow to Peard Bay	1.5	0.3	Harper, 1978a
Alaskan Beaufort Sea coast			
Near Barrow	4.5	2.1	MacCarthy, 1953
Oliktok Point	---	1.4	Dygas and Burrell, 1976
Near Prudhoe Bay	3	1-2	Barnes and others, 1977
Barrow to Harrison Bay	13.4	6.3	Hartz, 1978
Smith Bay to Prudhoe Bay	18	2.5	Reimnitz and others, 1985
Barrow to Demarcation Point	20.5	3.0	Lewellen, 1977
Canadian Beaufort Sea coast			
Kay Point	---	1.3	Forbes and Frobel, 1985
Yukon Territory	---	1	Mackay, 1963b
Yukon Territory	2	1	MacDonald and Lewis, 1973
Yukon Territory	5	---	Lewis and Forbes, 1974
Tent Island, Yukon Territory	27	15	Gill, 1972
Garry Island, Northwest Territory	7.3	2.3	Kerfoot and Mackay, 1972
	1.8	1.2	Forbes and Frobel, 1985
Pelly Island, Northwest Territory	13.2	6.3	Forbes and Frobel, 1985
Hooper Island, Northwest Territory	2.7	1.5	Forbes and Frobel, 1985
Pullen Island, Northwest Territory	---	9.2	Forbes and Frobel, 1985
Tuktoyaktuk Peninsula	5-8	---	Mackay, 1971
Tuktoyaktuk Peninsula	6-9	---	Rampton and Mackay, 1971
Tuktoyaktuk Peninsula	---	4	Rampton and Bouchard, 1975
Tuktoyaktuk Peninsula	8	---	Mackay, 1986b
Banks Island (Sachs Harbour), Northwest Territory	>4	---	Harry and others, 1983

*Modified from Harper (1978a), Hopkins and Hartz (1978), and Harper and others (1985b).

that locally is greater than 18 m per year (Reimnitz and others, 1985), and significant portions of the coast are retreating at rates of more than 2 m per year (Table 5). This rate exceeds that for the Gulf of Mexico, which has the highest rate of coastal erosion in the conterminous United States (May and others, 1983). Erosion along the Gulf coast, however, is occurring year-round, whereas erosion along Beaufort Sea shores is accomplished entirely within the three-month open-water period. When considered within this time frame, Beaufort Sea coastal erosion locally is several times more rapid than erosion along the Gulf coast. Such a rapid rate of coastal retreat is generally attributed primarily to the presence of ground ice in the coastal materials (MacCarthy, 1953; Mackay, 1963b, 1986b; Harper, 1978a), but rising sea level also has been suggested as a contributing factor (Harper and others, 1985a). Reimnitz and others (1985) suggested that nearshore processes of erosion are also important because a nearshore equilibrium profile appears to be maintained during coastal retreat.

Harper (1978a) described the degradational processes that affect ice-rich coastal bluffs. These processes include surface wash, debris slides, ground-ice slumps, and thermoerosional falls (Fig. 21). Surface wash is downslope sediment transport caused by snowmelt and summer rain runoff. Debris sliding occurs when the surficial thawed layer fails as a result of oversteepening at the cliff base. Ground-ice slumps were described earlier in the

Ground Ice section of this chapter. This process is especially important along rapidly receding parts of the Yukon coast and on the Mackenzie Delta and Tuktoyaktuk Peninsula where tabular sheets of massive ground ice occur (Mackay, 1963b, 1986b; Kerfoot and Mackay, 1972). Thermoerosional falls are also associated with some of the more rapid rates of coastal retreat (Lewellen, 1970, 1977) and are important in the erosion of ice-rich stream banks. The concept of thermoerosion was introduced in the Soviet literature (Gusev, 1959); it combines the melting of ground ice and the mechanical removal of thawed sediment by waves or turbulent currents. Waves or currents attacking a frozen bluff cut a thermoerosional niche into it (Fig. 22) that commonly is from 3 to 8 m deep and 1 m high (Walker and Arnborg, 1966), and may extend laterally for hundreds of meters (Leffingwell, 1919). The depth that the niche may attain prior to slope failure is determined by bluff height and frozen-ground material properties, including natural lines of weakness such as the margins of ice wedges. For banks of uniform composition, niching is deepest in fine-grained silt and clay (5 to 10 m) and less in sand and coarse-grained sediment (1 to 3 m; Miles, 1977). A common form of failure occurs when the niche intersects an ice wedge paralleling the shore (Fig. 22). The overhanging block detaches along the margin of the ice wedge, and the block falls (Walker and Arnborg, 1966).

The rate of bluff retreat varies spatially and temporally.

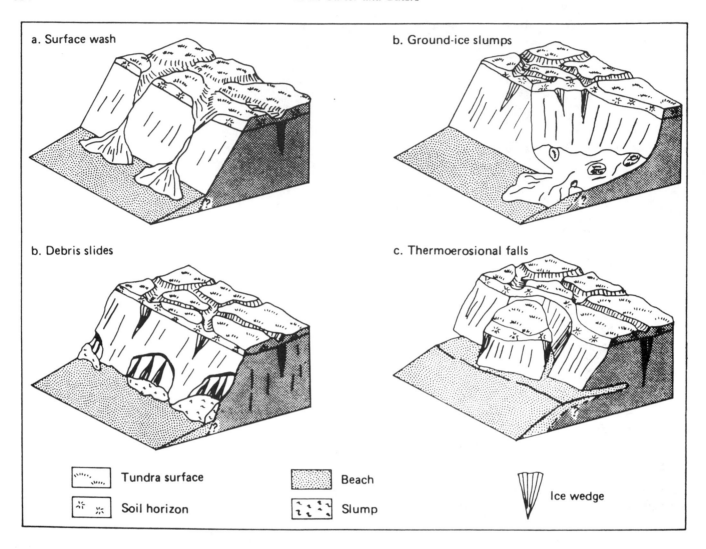

Figure 21. Degradational processes that affect ice-rich coastal bluffs (from Harper, 1978a).

Factors controlling spatial variations include bluff height and composition, which determine ice content. Bluff height is an important factor along all coasts, but is especially important along the Beaufort and Chukchi shores of Alaska and the western Yukon (Harper, 1978a) because the amount of excess ice in the ground decreases with depth. Measurements of the ice content of coastal materials near Barrow show that ground ice excluding ice wedges may compose 75 percent of the volume of the frozen sediments at a depth of 1 m, to zero at a depth of 8.5 m (Sellmann and others, 1975). In this area, ice wedges, which generally penetrate no more than 3 or 4 m below the surface, make up about 10 to 20 percent of the total amount of ground ice. As bluff height increases, the amount of sediment per unit volume of bluff materials rapidly increases, and larger volumes of sediment must be eroded for each meter of bluff retreat. Additionally, Harper (1978a) showed that thermoerosional falls are not likely to occur on bluffs higher than 6 m, because ice wedges rarely penetrate to

this depth, and thermoerosional niches deep enough to cause failure without ice wedges are not common. On parts of the Canadian Beaufort coast where tabular bodies of massive ice may occur at depth, however, the factor of bluff height is not as critical. Bluff composition is important because fine-grained sediments have more excess ice than coarse-grained sediments. Furthermore, debris from coarse-grained sediments, which is not as readily entrained by waves or currents as debris from fine-grained sediments, tends to accumulate and armor the base and face of the bluff.

Temporal variations in the rate of bluff retreat are largely controlled by climate, which determines the length of time each year that the bluffs are affected by the degradational processes described above, the effectiveness of marine processes in removing debris supplied by the bluffs to the beach, and the ability of the sea to produce thermoerosional niches. Waves and currents are most effective in transporting sediment and forming thermo-

Figure 22. Thermoerosional falls along the Beaufort sea coast localized along an ice wedge that parallels the coast. A thermoerosional niche undercuts the bluff beyond the fallen blocks. Bluff height is about 7 m.

erosional niches during late cyclonic summer storms that occur under conditions of maximum open water (Reimnitz and Maurer, 1979). The westerly winds that accompany these storms raise sea level as much as 3.5 m, allowing the sea to overtop beaches and directly attack coastal bluffs (Hume and Schalk, 1967). During such conditions, as much as 30 m of coastal retreat has occurred within a few days (Short and others, 1974). Thus, rates of coastal retreat vary dramatically from year to year, depending on the time of breakup of sea ice, variations in the size and positions of open water areas, and the timing and intensity of late summer and autumn storms (Hopkins and Hartz, 1978).

ORIENTED LAKES

L. D. Carter

Oriented lakes are groups of elongate lakes and lake basins that have a common long-axis orientation (Kaczorowski, 1977). In the Arctic Lowlands, oriented lakes (Fig. 23) are some of the most conspicuous features of the landscape over a large area of the Arctic Coastal Plain in Alaska (Sellmann and others, 1975), east of the Mackenzie Delta (Mackay, 1956b, 1963a), and on Banks Island (Harry and French, 1983; Fig. 24). Locally, from 20 to more than 50 percent of the surface is occupied by oriented lakes.

ORIGIN AND DEVELOPMENT

Oriented lakes of the Arctic Lowlands are initiated as ther-

mokarst lakes in unconsolidated, perennially frozen sand, sandy silt, and clayey silt. They range from a few tens of meters to a few kilometers in long dimension and are generally oval, triangular, or D-shaped. Sublittoral shelves form the lake bed adjacent to the long sides of the lakes and are separated by a central deep that commonly extends to both ends of the lake. Current bed forms on the shelves are clearly visible from the air (Fig. 23).

Cabot (1947) was the first to describe oriented lakes in the Arctic Lowlands. He discussed the lakes near Barrow, and attributed lake orientation to wave erosion generated by Pleistocene summer winds blowing from the Cordilleran ice cap and parallel to the long axes of the lakes. Black and Barksdale (1949) carried out detailed field investigations and also concluded that lake orientation was a relict Pleistocene feature produced by winds perpendicular to the modern prevailing winds, which blow ENE and WSW across the long axes of the lakes. Livingstone (1954) pointed out that the observed rapid rate of modification of the lake shores required that a change in wind direction such as that postulated by Cabot (1947) and Black and Barksdale (1949) must have occurred during the last few centuries; otherwise, evidence of this previous wind regime would have been obliterated by erosion. He suggested instead that lake elongation resulted from modern processes, and proposed that winds blowing across the lakes moved water to downwind shores, setting up a return flow around the lake ends where water would reach maximum velocity and thus maximum erosive power. Mackay (1956a) agreed with Livingstone's interpretation of the Alaskan lakes and proposed the same explanation for the oriented lakes east of the

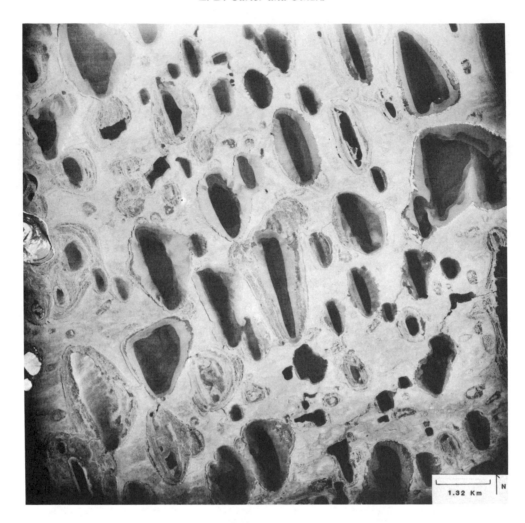

Figure 23. Vertical aerial photograph of oriented lakes on the Alaskan Arctic Coastal Plain. Location is approximately 90 km southeast of Barrow. Bedforms are visible on some of the sublittoral shelves that border the long axes of the lakes.

Mackenzie Delta, which are also oriented with their long axes perpendicular to the modern prevailing winds.

Rosenfeld and Hussey (1958), however, proposed that structural control is responsible for the orientation of the Alaskan lakes, based on the tendency of many small streams to consist of a series of reaches oriented at right angles to each other with one set parallel to the direction of lake elongation. Reasoning that permafrost would react competently to stresses, they thought it likely that a master fracture pattern had developed in the surficial deposits and controlled the maximum development of ice wedges. Thawing of these fracture-controlled ice-rich zones would give rise to oriented thaw depressions. This idea was incorporated by Carson and Hussey (1959) in a composite hypothesis for the origin of oriented lakes that included wave action and the effects of greater thaw in a north-south direction as a result of greater insolation at noon during the Arctic summer. Carson and Hussey (1962) later showed by potentiometric measurements that directional variations in insolation are of insufficient magnitude to be a factor in lake orientation. Structural control is now also thought to be unlikely because, in newly formed ponds, elongation initially proceeds parallel to the prevailing winds, and only as the lake grows does elongation perpendicular to the prevailing winds develop (Carson, 1968). Furthermore, zones of maximum ice-wedge development parallel to lake elongation have not been observed in the northern Alaskan lake district, and no fracture systems are known to occur in the unconsolidated deposits. In spite of these facts, and the research described below, some scientists have continued to propose that structural control is a factor in lake elongation (Cannon, 1976; Maurin and Lathram, 1977).

Rex (1961) applied hydrodynamic theory and the principles of shoreline erosion and sediment transport to the problem of elongation of the Alaskan lakes, drawing from a large body of observational and experimental work. In particular, he utilized the conclusions of Bruun (1953) that maximum littoral drift on a

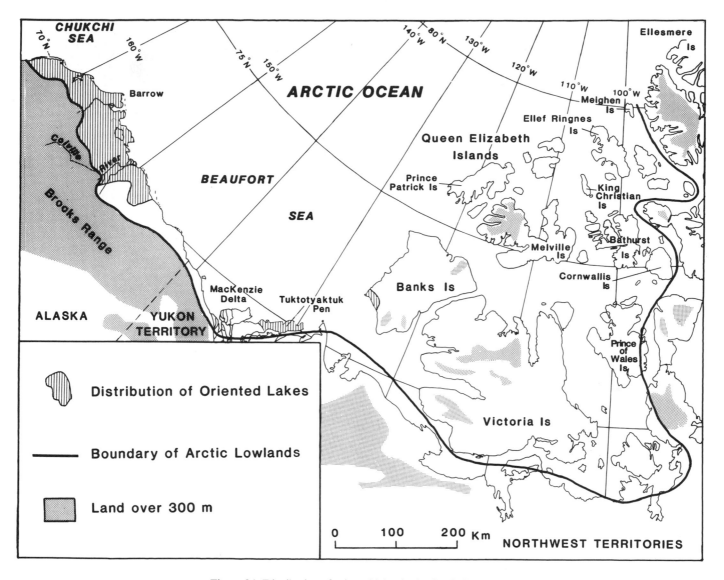

Figure 24. Distribution of oriented lakes in the Arctic Lowlands.

curved beach is at an angle of 50° between the deep water angle of the waves and a line normal to the shoreline, with minimum littoral drift occurring where the angle is 0° (the nodal point), or directly on the downwind side (Fig. 25). He concluded that elongation occurs as a result of the relative differences in rates of littoral drift, and thus rates of erosion, around the lake margin. The highest rates of erosion occur near the lake ends, where maximum rates of littoral drift create a deficit in sediment supply. For shorelines of finite length developed in readily moved unconsolidated deposits, the resulting equilibrium configuration is a cycloid. Lakes formed under the influence of mutually opposing winds of equal strength should have opposite shores of cycloid form, and thus be roughly elliptical. Rex (1961) explained the more rectangular configuration generally observed in the Alaskan lakes by considering the wave orthogonals not as unidirectional but rather as an array from one sector. In this way, for large shifts

in wind direction there are corresponding shifts in the locations of the nodal points, but little or no shift in the positions of the zones of maximum littoral drift.

Using these concepts, Mackay (1963a) developed a mathematical model to relate lake shape to resultant wind vectors for each of the 16 compass directions (Fig. 26). This model considered the oriented lake as a summation or integration form of 16 cycloids. Mackay (1963a) found that the parameters for lake shape computed from wind data for Nicholson Peninsula (Fig. 26B, Table 6) closely approximated those for mean lake shape determined from empirical equations developed from measurements of the lakes in the area near Nicholson Peninsula. He applied the model to Ikroavik Lake near Barrow, using data for shape and winds from Carson and Hussey (1960), and found good agreement between computed lake shape and actual lake shape (Fig. 26C). Harry and French (1983) showed that by using

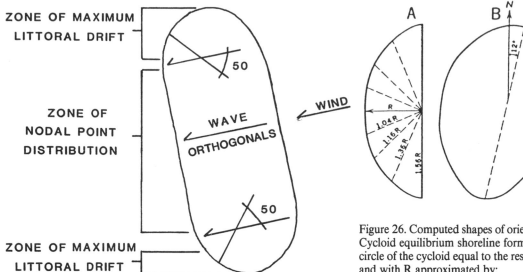

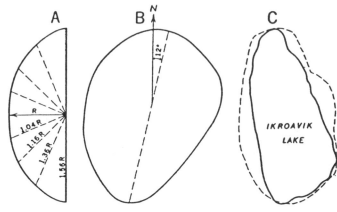

Figure 25. Schematic diagram of the relationship between zones of maximum littoral drift and wave orthogonals predicted by Rex (1961; after Carson and Hussey, 1962).

Figure 26. Computed shapes of oriented lakes (from Mackay, 1963a). A. Cycloid equilibrium shoreline form with the diameter of the generating circle of the cycloid equal to the resultant (R) for a given wind direction, and with R approximated by:

$$R = \sum_i f_i (V_i - v)^n$$

where i is the period of record, here taken on a monthly basis, f_i is the frequency for month 1, n is a positive exponent (see Landsberg, 1956), V is wind velocity, and v is the threshold wind velocity. Both n and v must be estimated. B. Computed lake shape based on n = 2 and v = 0. See Table 6 for shape parameters. C. Computed shape compared to real shape for Ikroavik Lake, Alaska.

data for storm winds, lake shapes calculated with this model closely simulate the D-shapes characteristic of oriented lakes on Banks Island.

Carson and Hussey (1960, 1962) carried out a detailed field study of lakes in the Barrow area, focusing on wind-controlled circulation patterns, wave action, morphology, sediment distribution, and thermal regime. They found that circulation patterns proposed by Livingstone (1954) occurred prominently only in small ponds and in the absence of large waves, and thus were not an effective orienting agent. They determined that the predictions of Rex (1961) for the zones of maximum current velocities and littoral drift are valid for lakes with a fetch of greater than 550 m, and noted that variations in lake form relative to calculated wind vectors agree with his interpretation of the lake shores as equilibrium forms adjusted to variations in the rate of littoral drift along their lengths. The mechanisms described by Rex (1961) are thus clearly important for lake elongation in large basins. They pointed out, however, that currents at the ends of lakes do not seem to have an erosive effect in small basins, and thus could not account for elongation during the early stages of lake growth. Moreover, they found that small lakes appeared to grow at the ends primarily by thaw, as indicated by an extensive cover of sunken tundra mat on banks and sublittoral surfaces at the ends of basins. They proposed that the most important factor in lake elongation during nearly all stages of lake growth is the development of sublittoral shelves, and showed that small, growing basins quickly establish wave-generated equilibrium bottom-profiles off downwind shores that rapidly adjust to changes in lake depth. In this way, sublittoral shelves are formed that insulate permafrost

and damp incoming storm waves. This serves to limit the rate of expansion parallel to prevailing winds, without restricting expansion perpendicular to prevailing winds. In the brief period prior to the development of sublittoral shelves, lake expansion occurs parallel to the prevailing winds (Carson, 1968).

More recent studies of current patterns carried out for the Alaskan lakes (Kaczorowski, 1977) support the results of Carson and Hussey (1962). Furthermore, Kaczorowski (1977) produced elongation of an initially circular model pond excavated in fine sand by generating waves with fans. Elongation occurred perpendicular to the direction of wind approach.

Hydrodynamic theory, field observations of lacustrine processes and deposits, mathematical models, experimental models, and records of modern winds thus support the mechanisms described by Rex (1961) and Carson and Hussey (1962) for lake elongation in the Arctic Lowlands. Mackay (1963a) has pointed out, however, that numerous details of lake development await explanation. These include the effect of lake ice on lake orientation, the preference for vegetation growth under favored microclimatic and topographic conditions, thermal effects, the aspect of two-cell circulation, and the amount of littoral transport.

SIGNIFICANCE FOR OTHER AREAS

Outside of the Arctic Lowlands, oriented lakes occur on Baffin Island (Bird, 1967), in the Old Crow Plain in northern

TABLE 6. LAKE SHAPE PARAMETERS CALCULATED FROM
EMPIRICAL EQUATIONS AND COMPUTED USING THE CYCLOID
EQUILIBRIUM SHORELINE MODEL*

Parameter	Mean Lake Shape (from equations)	Computed Lake Shape (from 16 cycloids)
Length (given)	1.00 mi	1.00 mi
Width d**	0.70 mi	0.73 mi
Width e	0.61 mi	0.62 mi
Axis	N11°E	N12°E

*From Mackay (1963a).

**Widths measured one-third the distance from either end of the lake.

Yukon Territory (Bostock, 1948), in Bolivia (Plafker, 1964), Chile (Kaczorowski, 1977), Texas (Price, 1968a), the northeastern USSR (Obruchev, 1938), and the Atlantic Coastal Plain of the United States, where they are referred to as Carolina Bays (Price, 1968b; Thom, 1970). The Carolina Bays were the first group of oriented lakes to be studied extensively, and numerous theories for their origin have been proposed, including meteorite showers, solution, upwelling of artesian springs, the action of spawning fish, and Pleistocene winds blowing parallel to the long axes (Price, 1968b).

The research described above on lake elongation in the Arctic Lowlands has shown that permafrost is not an essential criterion for the development of oriented lakes and demonstrates that the general mechanism of elongation—waves and currents generated by winds blowing at right angles to the direction of elongation—is azonal. Field studies of lake currents carried out by Kaczorowski (1977) indicated similar current patterns for the Alaskan oriented lakes and the oriented lakes of the Carolina Bays, Texas, and Chile. This work lends support to earlier suggestions that the Carolina Bays and the Alaskan oriented lakes developed through similar mechanisms (Thom, 1970), and Kaczorowski (1977) felt that the results of research on oriented lakes in the Arctic Lowlands are generally applicable elsewhere. Exceptions, however, are lakes in Bolivia and the Old Crow Plain, where orientations are thought to be structurally controlled (Plafker, 1964; J. A. Westgate, oral communication, 1986).

GEOMORPHOLOGY AND ENVIRONMENTAL CONCERNS

J. A. Heginbottom and L. D. Carter

Two types of environmental concerns, one an ongoing problem and the other an imminent problem, emphasize the importance of acquiring a better understanding of geomorphic processes and the origin and distribution of landforms and surfi-

cial deposits in the Arctic Lowlands. These concerns stem from ongoing engineering activity, and the prospect of climatic change due to a buildup of atmospheric greenhouse gases. Both concerns are relevant to other parts of North America, but involve distinctive problems in the Arctic Lowlands because of the presence of permafrost, and because global climatic changes are amplified in the Arctic by albedo feedback effects due to changes in snow-cover and in the extent of sea ice (Kellogg, 1975).

ENGINEERING ACTIVITY

Human activity in the Arctic Lowlands has intensified dramatically in the past three decades, principally as a result of the exploration for oil and natural gas. Development of these resources requires construction of producing and processing facilities and the means to move the resultant products to markets outside the Arctic Lowlands. Along with this goes the construction of an infrastructure of roads, airfields, port facilities, artificial islands, power and water supplies, settlements, and all the trappings of twentieth century life. This involves travel on the land, construction in the nearshore zone, excavation for construction resources, and drilling of deep and shallow wells and boreholes. These all present opportunities for disruption of normal fluvial and coastal processes, and of the thermal regime of permafrost, which can lead to pronounced geomorphic changes, especially those produced by frost heaving and by the degradation of ground ice (Ferrians and others, 1969; Johnston, 1981).

Recognition of these problems has resulted in several major programs in Canada and Alaska to study the potential impact of regional projects on permafrost and the land surface (Brown and Grave, 1979). These include the preparation of terrain and engineering-geologic maps at reconnaissance scales based on landform analysis (e.g., Rampton, 1974; Carter and others, 1986b). Numerous studies within these programs have generally shown that the character of the geomorphic changes that result from engineering activity depends upon the material properties of the terrain (including the amount and type of buried ground ice and its proximity to the surface), the thermal and hydrological regimes of the site, and the specific nature of the development being undertaken. For example, removal of surface vegetation during road construction results in a bare soil or ground surface, which leads to an increase in the mean annual ground surface temperature and the melting of shallow ground ice. In a fine-grained, ice-rich soil, this can result in a complete loss of soil strength, slope failure, ground surface settlement, and, if the meltwater cannot drain away, the development of a thermokarst pond. Conversely, in a coarse-grained, ice-poor soil, such as gravel, there may be no effective changes. Different problems arise with developments that lead to a cooling of the ground. In a fine-grained soil this can lead to renewed growth of segregated ice and heaving of the ground surface.

Many of the problems inherent in engineering developments in the Arctic Lowlands can be avoided or minimized by careful site selection and by appropriate designs for the geological environment. Permafrost and ground ice conditions beneath a building can be preserved unchanged, for example, providing the foundation is designed to insulate the ground from the heat of the building. Alternatively, the structure can be built in such a way as to accommodate changes in the ground resulting from the melting of some ground ice—a gravel road, for example. These approaches are referred to as the "Passive Method" and the "Active Method," respectively (Johnston, 1981, p. 249–258). For all except the smallest structures, however, even a passively designed foundation on an ice-rich soil may be subjected to stresses arising from deformation of the soil. A useful summary of the problems resulting from the particular rheological characteristics of frozen soil is given in Johnston (1981). Both the passive and active methods require a detailed knowledge of site conditions, including the geomorphic processes operating there, thermal regime, and material properties.

A good example of the use of geomorphology in a large engineering project that crossed the Arctic Lowlands, and of the level of detailed information necessary to ensure that such activities are carried out safely and economically, is afforded by the geotechnical investigations undertaken prior to construction of the Trans Alaska Pipeline System, which extends, 1,300 km from Prudhoe Bay to Valdez. After initial selection of a route corridor, early alignment changes and plans for detailed route investigations were based in part on reconnaissance maps of surficial deposits at a scale of 1:125,000 produced by the U.S. Geological Survey (e.g., Ferrians, 1971). These maps were prepared using landforms as a basis for extrapolating field observations of material type. Detailed investigations consisted of an integrated program of landform mapping (scale 1:24,000) and soil testing, with a computer-assisted data-index and storage system (Kreig and Reger, 1982). Soil properties such as texture, moisture, density, specific gravity, and thaw settlement were determined for the mapped landforms by testing of undisturbed soil samples obtained from borings. The results of the investigation were used for locating construction resources and evaluating slope stability, and to determine the construction mode—elevated or buried—of the pipeline, by considering the soil conditions and the anticipated effects of the hot oil line on the permafrost. The pipeline is buried where it crosses landforms composed of bedrock or thaw-stable soils such as gravel, and is elevated across landforms composed of thaw-unstable soils such as ice-rich silt (Liguori and others, 1979).

CLIMATIC CHANGE

The probable nature and magnitude of climatic change in arctic regions and the influence of increasing concentrations of carbon dioxide and other greenhouse gases in the atmosphere are controversial subjects. Recently, a consensus has appeared to the effect that over the next 50 to 100 years the global mean annual temperature will increase, and that this increase will be greatest toward the poles (Harvey, 1982; MacCracken and Luther, 1985). This warmer climate may result in warmer summers, or warmer winters, or both; it may also result in changes in precipitation, such as dryer summers, or changes in snow depths. The magnitude of these effects is still uncertain.

What are the implications of this? In qualitative terms, the effects can be predicted well enough (Mackay, 1975c; Smith and Riseborough, 1983; Judge and Pilon, 1984). Ground temperatures will rise, in proportion to the increase in the mean annual air temperature. The active layer will thicken, extensive thermokarst features may form from the melting of massive ice, and slopes will become less stable. In turn, these effects will result in widespread settlement of the ground surface and the formation of new thaw basins, many including lakes. Slope failures will lead to erosion; new drainage patterns will develop, and rivers will carry increased sediment loads. Rivers may change course more readily with their banks no longer frozen, and coastal erosion may increase. Subsidence of the ground surface in coastal and riparian areas is likely to result in extensive flooding.

A general degradation of permafrost, with melting of ground ice, may be accompanied by local instances of permafrost aggradation and ground ice growth. This is particularly likely at sites where existing lakes with subjacent taliks drain, and yet the climate is cold enough to maintain a permafrost environment. Permafrost will form beneath the newly exposed lake bed and, in favorable circumstances, pingo growth may be initiated.

The effects of differential settlement on human works—roads, airfields, buildings, pipelines—will also be considerable. The immediate results of climatic warming will occur in the upper layers of permafrost, where the effect on human activities will be greatest.

Even general quantitative predictions of the effects of climate warming on ground ice and permafrost conditions are much more difficult to make. This is because of (1) the very complex relationships between climatic elements (temperature, precipitation, snow cover), geology (rock properties), geomorphology (site conditions and their variability), hydrology, and permafrost conditions; and (2) a lack of widespread, long-term, base-line data (Smith, 1984). The complexity and extreme variability of local site conditions and thus of the response of permafrost to any climatic change is the most significant problem. An attempt to address this difficulty was made by Goodwin and others (1984), using a numerical energy balance modelling approach, with particular reference to sites at Fairbanks and Barrow. The results suggest that at Barrow, the depth of summer thaw is quite sensitive to an increase in summer air temperature and soil drying, less sensitive to increased winter temperatures, and rather insensitive to variations in snow depth.

Locally, the possibility of a cooler climate may exist. This will result in a thinner active layer, the aggradation of permafrost and, in favorable sites, the growth of ice lenses. Ice wedges may be reactivated, coastal advance may occur, and pingo growth may be accelerated.

Given the potential economic costs of either ignoring the probable effects of changes in permafrost conditions on structures and human activities, or overdesigning to resist such changes, research in this area of the effects of climatic change should receive considerable attention in the future.

CONCLUSION AND RECOMMENDATIONS

L. D. Carter and J. A. Heginbottom

Geomorphic processes in the Arctic Lowlands reflect the control of a cold climate and the permafrost and vegetation that are in equilibrium with it. Some processes, such as frost action, the processes responsible for the formation and destruction of ground ice, and certain coastal processes (for example, strudel scour, and ice-shove on beaches), are characteristic of this part of North America. Other processes, such as fluvial activity and wave action, occur around the globe; however, in the Arctic Lowlands they are modified by ice in and on the ground, snowcover, permafrost, and the presence or absence of tundra vegetation.

Much progress recently has been made in understanding the mechanisms of ground-ice formation, and in learning how fluvial and coastal processes in the Arctic Lowlands differ from these processes in other areas. However, quantitative relations between climatic parameters, vegetal cover, permafrost, and geomorphic processes in the Arctic Lowlands are still poorly known. These relations must be determined in order to accurately predict the consequences of any natural physical or biological change in the environment. Additionally, we need to know how geomorphic processes have responded to changes in the past. This knowledge will also help ensure that engineering, mining, and other human modifications of the land surface are carried out safely and economically. To this end, we should broaden and intensify research on modern geomorphic processes and on the origin of relict landforms in the Arctic Lowlands. Examples of the kinds of research necessary are as follows.

● Detailed mapping of landforms and surficial deposits should be carried out to provide base-line data for process-oriented studies.

● The quantity and type of sediment carried to the sea by representative rivers in the Arctic Lowlands should be monitored in order to relate rates of erosion and the character of weathering products to the modern climate and the physical characteristics of the drainage basins. These data will allow comparisons to be made between modern erosion rates in the Arctic Lowlands and those determined for other climatic regimes, which are necessary to predict how rates of soil loss in the Arctic will be affected by climatic change.

● The relations between modern climatic parameters and the rates of climate-controlled geologic processes such as frost action, ground-ice formation, ice-wedge growth, solifluction, and mass wasting should be determined through field studies to provide a basis for prediction of how these processes might be affected by climatic change or anthropogenic perturbations of the environment.

● Calculations based on the earth's orbital parameters show that at about 10 ka in high northern latitudes, insolation values were about 10 percent greater than today (Berger, 1978), and palyno-

logical data from the Yukon indicate that a thermal maximum did indeed occur at that time (Ritchie and others, 1983). Landscape-climate relations for that time period should be established through stratigraphic, geomorphic, and paleontologic studies as a possible analog for the predicted carbon dioxide-induced climatic warming.

● Elevated interglacial shorelines and their associated beach and nearshore marine deposits contain a record of arctic coastal processes that were active when the climate was warmer than now. In particular, deposits of the last interglacial interval (125 ka) are well preserved along much of the Arctic coast in Alaska (Hopkins, 1982), and a study of the deposits of this ancient coastal system would aid in understanding the changes in coastal processes to be expected during the predicted carbon dioxide–induced climatic warming.

● Measurement of oxygen and hydrogen isotopes and carbon dioxide should be made across fossil and active ice wedges. These measurements may provide a record of temperature and atmospheric carbon dioxide during the period of growth of the ice wedges, and so may possibly provide a basis for correlating regional changes in geomorphic processes with changes in climate and carbon dioxide.

● Sedimentary structures and other physical characteristics of sedimentary deposits in the Arctic Lowlands are important indicators of geomorphic processes. For example, debris flow deposits can be distinguished from fluvial deposits on the basis of sedimentary structures. Stratigraphic studies emphasizing these aspects of upper Cenozoic sedimentary deposits should be carried out in order to determine changes in the relative importance of the various climate-controlled geomorphic processes through time.

● Fossil dune fields, eolian sand sheets, and loess cover large areas of the Arctic Lowlands. Preliminary studies on the Arctic Coastal Plain indicate that some of these deposits formed in response to major changes in temperature, precipitation, and sediment supply over the past 40,000 years (Carter and others, 1984). Eolian deposits and their intercalated paleosols provide a sensitive record of climatic change and landscape evolution that can be dated by radiocarbon and thermoluminescence; they should be studied in detail.

● Glaciomarine deposits of the Arctic Coastal Plain formed during periods of high sea level that occurred while ice sheets in Canada were contributing icebergs to the Arctic Ocean (Hopkins, 1982). According to our present understanding of climate, glaciation, and sea level relations, however, when an ice sheet is present in Canada, eustatic sea level should be lower than it is today. Proposals to explain these enigmatic marine transgressions include rapid sea-level rises due to Antarctic ice surges, large volumes of floating ice in polar regions, and isostatic depression of the crust by Arctic Ocean ice sheets. These glaciomarine deposits should be studied in detail because they contain important information regarding regional and global paleoclimate, sea level change, and the stability of arctic coastlines.

● Glacial landforms and deposits provide direct evidence of climatic change. An effort should be made to relate the sequence of glacial advances to marine shorelines and deposits, nonglacial landforms and deposits, and the fossil record to allow the development of an integrated picture of climatic change and landscape evolution.

REFERENCES

Acevedo, W., Walker, D., Gaydos, L., and Wray, J., 1982, Vegetation and land cover Arctic National Wildlife Refuge coastal plain, Alaska: U.S. Geological Survey Miscellaneous Investigations Series Map I-1443, scale 1:250,000.

Anderson, J. C., and Durrant, R. C., 1976, Hydrologic reconnaissance, Thomsen River basin, Banks Island, District of Franklin: Geological Survey of Canada Paper 76-1A, p. 221–227.

Anderson, J. P., 1959, Flora of Alaska and adjacent parts of Canada: Iowa State University Press, 543 p.

Arnborg, L., Walker, H. J., and Peippo, J., 1966, Water discharge in the Colville River, 1962: Geografiska Annaler, v. 48A, p. 195–210.

—— , 1967, Suspended load in the Colville River, Alaska, 1962: Geografiska Annaler, v. 49A, p. 131–144.

Barnes, P. W., 1982, Marine ice-pushed boulder ridge, Beaufort Sea, Alaska: Arctic, v. 35, p. 312–316.

Barnes, P. W., and Reimnitz, E., 1985, Sediment reworking, transport, and deposition on the Alaskan Beaufort shelf; The role of ice, in relation to waves, currents, and infauna, in Arctic Land-Sea Interaction, 14th Arctic Workshop, November 6–8, 1985, Abstracts: Dartmouth, Nova Scotia, Bedford Institute of Oceanography, p. 37–40.

Barnes, P. W., Reimnitz, E., Smith, G., and Melchior, J., 1977, Bathymetric and shoreline changes, northwestern Prudhoe Bay, Alaska: U.S. Geological Survey Open-File Report 77-161, 10 p.

Barnes, P. W., Reimnitz, E., and Fox, D., 1982, Ice rafting of fine-grained sediment, a sorting and transport mechanism, Beaufort Sea, Alaska: Journal of Sedimentary Petrology, v. 52, p. 493–502.

Barnes, P. W., Rearic, D. M., and Reimnitz, E., 1984, Ice gouging characteristics and processes, in Barnes, P. W., Schell, D. M., and Reimnitz, E., eds., The Alaskan Beaufort Sea, Ecosystems and environments: Orlando, Academic Press, p. 185–212.

Benson, C. S., 1982, Reassessment of winter precipitation on Alaska's arctic slope and measurements on the flux of windblown snow: Geophysical Institute, University of Alaska Report UAG R-288, 26 p.

Berger, A. L., 1978, Long-term variations of caloric insolation resulting from the Earth's orbital elements: Quaternary Research, v. 9, p. 139–167.

Billings, W. D., and Peterson, K. M., 1980, Vegetational change and ice-wedge polygons through the thaw lake cycle in arctic Alaska: Arctic and Alpine Research, v. 12, p. 413–432.

Bird, J. B., 1967, The physiography of arctic Canada: Baltimore, The Johns Hopkins Press, 336 p.

—— , 1984, Arctic Canada, in Bird, E.C.F., and Schwartz, M. L., eds., The world's coastline: New York, Van Nostrand Reinhold Company, p. 241–251.

Black, R. F., 1951, Eolian deposits of Alaska: Arctic, v. 4, p. 89–111.

—— , 1954, Precipitation at Barrow, Alaska, greater than recorded: American Geophysical Union Transactions, v. 35, p. 203–206.

—— , 1983, Three superposed systems of ice wedges at McLeod Point, northern Alaska, may span most of the Wisconsin stage and Holocene, in Proceedings, Permafrost; Fourth International Conference: Washington, D.C., National Academy Press, p. 68–73.

Black, R. F., and Barksdale, W. L., 1949, Oriented lakes of northern Alaska: Journal of Geology, v. 57, p. 105–118.

Blake, W. J., 1964, Preliminary account of the glacial history of Bathurst Island, Arctic Archipelago: Geological Survey of Canada Paper 64-30, 7 p.

Bliss, L. C., 1979, Vegetation and revegetation within permafrost terrain, in Proceedings of the Third International Conference on Permafrost, v. 2: Ottawa, National Research Council of Canada, p. 31–50.

Bliss, L. C., and Richards, J. H., 1982, Present-day arctic vegetation and ecosystems as a predictive tool for the arctic-steppe mammoth biome, in Hopkins, D. M., Matthews, J. V., Jr., Schweger, C. E., and Young, S. B., eds., Paleoecology of Beringia: New York, Academic Press, p. 241–257.

Bostock, H. S., 1948, Physiography of the Canadian Cordillera, with special reference to the area north of the Fifty-fifth Parallel: Geological Survey of Canada Memoir 247, 106 p.

—— , 1970a, Physiographic regions of Canada: Geological Survey of Canada, Map 1254A, scale 1:5,000,000.

—— , 1970b, Physiographic subdivisions of Canada, in Douglas, R.J.W., ed., Geology and economic minerals of Canada (fifth edition): Geological Survey of Canada, Economic Geology Report no. 1, p. 9–30.

Britton, M. E., 1967, Vegetation of the arctic tundra, in Hansen, H. P., ed., Arctic biology (second edition): Oregon State University Press, p. 67–113.

Brower, C. D., 1942, Fifty years below zero: New York, Dodd, Mead and Company, 310 p.

Brown, J., 1967, An estimation of volume of ground ice, Coastal Plain, northern Alaska: U.S. Army Corps of Engineers Cold Regions Research and Engineering Laboratory Memorandum, 22 p.

Brown, J., and Grave, N. A., 1979, Physical and thermal disturbance and protection of permafrost, in Proceedings of the Third International Conference on Permafrost, Volume 2: Ottawa, National Research Council of Canada, p. 51–91.

Brown, J., Dingman, S. L., and Lewellen, R. I., 1968, Hydrology of a drainage basin on the Alaska coastal plain: U.S. Army Corps of Engineers Cold Regions Research and Engineering Laboratory, Research Report 240, 18 p.

Brown, R.J.E., 1972, Permafrost in the Canadian Arctic Archipelago: Zeitschrift für Geomorphologie Neue Folge, Supplementband 13, p. 102–130.

—— , 1973, Ground ice as an initiator of landforms in permafrost regions, in Fahey, B. D., and Thompson, R. D., eds., Proceedings, Research in Polar and Alpine Geomorphology, 3rd Guelph Symposium on Geomorphology, 1973: Norwich, United Kingdom, Geo Abstracts Limited, p. 43–49.

—— , 1978, Permafrost, in Hydrological atlas of Canada: Ottawa, Department of Fisheries and the Environment, Plate 32, scale 1:10,000,000.

Brown, R.J.E., and Kupsch, W. O., 1974, Permafrost terminology: National Research Council of Canada, Associate Committee on Geotechnical Research, Technical Memorandum no. 111, 62 p.

Brown, R.J.E., and Péwé, T. L., 1973, Distribution of permafrost in North America and its relation to the environment; A review, 1963–1973, in Permafrost; The North American Contribution to the Second International Conference on Permafrost, Yakutsk: Washington, D.C., National Academy of Sciences, p. 71–100.

Bruun, P., 1953, Forms of equilibrium coasts with a littoral drift: University of California Institute of Engineering Research, v. 347, p. 1–7.

Burns, B. M., 1973, The climate of the Mackenzie Valley–Beaufort Sea: Climatological Studies 24, Atmospheric Environment Service, Toronto, v. 1, 227 p.

—— , 1974, The climate of the Mackenzie Valley–Beaufort Sea: Climatological Studies 24, Atmospheric Environment Service, Toronto, v. 2, 239 p.

Cabot, E. C., 1947, The northern Alaskan coastal plain interpreted from aerial photographs: Geographical Review, v. 37, p. 639–648.

Cannon, P. J., 1976, Critical landform mapping of Alaska using radar imagery, in Proceedings, Mapping with remote sensing data, 2nd Annual Pecora Symposium, Sioux Falls, South Dakota, Oct. 25–29, 1976: American Society of Photogrammetry, p. 144–160.

Carsola, A. J., 1954, Microrelief on arctic sea floor: American Association of Petroleum Geologists Bulletin, v. 38, p. 1587–1601.

Carson, C. E., 1968, Radiocarbon dating of lacustrine strands in arctic Alaska: Arctic, v. 21, p. 12–26.

Carson, C. E., and Hussey, K. M., 1959, The multiple working hypothesis as applied to Alaska's oriented lakes: Proceedings Iowa Academy of Sciences, v. 66, p. 334–349.

—— , 1960, Hydrodynamics in three arctic lakes: Journal of Geology, v. 68,

p. 585–600.
——, 1962, The oriented lakes of arctic Alaska: Journal of Geology, v. 70, p. 417–439.

Carson, J. M., Hunter, J. A., and Lewis, C. P., 1975, Marine seismic refraction profiling, Kay Point, Yukon Territory: Geological Survey of Canada, Paper 75-1B, p. 9–12.

Carter, L. D., 1981, A Pleistocene sand sea on the Alaskan Arctic Coastal Plain: Science, v. 211, p. 381–383.

——, 1983, Fossil sand wedges on the Alaskan Arctic Coastal Plain and their paleoenvironmental significance, *in* Proceedings, Permafrost; Fourth International Conference, Washington, D.C., National Academy Press, p. 109–114.

Carter, L. D., and Galloway, J. P., 1979, Arctic Coastal Plain pingos in National Petroleum Reserve in Alaska, *in* Johnson, K. M., and Williams, J. R., eds., The U.S. Geological Survey in Alaska; Accomplishments during 1978: U.S. Geological Survey Circular 804-B, p. B33–B35.

——, 1985, Engineering-geologic maps of northern Alaska, Harrison Bay Quadrangle: U.S. Geological Survey Open-File Report 85-256, scale 1:250,000, 2 sheets, 47 p.

Carter, L. D., Forester, R. M., and Nelson, R. E., 1984, Mid-Wisconsin through early Holocene changes in seasonal climate in northern Alaska: American Quaternary Association Eighth Biennial Meeting, 13–15 August, 1984, Boulder, Colorado, Program and Abstracts, p. 20–22.

Carter, L. D., Brigham-Grette, J., Marincovich, L., Jr., Pease, V. L., and Hillhouse, J. W., 1986a, Arctic Ocean sea ice and terrestrial paleoclimate: Geology, v. 14, p. 675–678.

Carter, L. D., Ferrians, O. J., Jr., and Galloway, J. P., 1986b, Engineering-geologic maps of northern Alaska, coastal plain and foothills of the Arctic National Wildlife Refuge: U.S. Geological Survey Open-File Report 86-334, scale 1:250,000, 2 sheets, 9 p.

Church, M., 1972, Baffin Island sandurs; A study of arctic fluvial processes: Geological Survey of Canada Bulletin 261, 208 p.

——, 1974, Hydrology and permafrost with reference to northern North America, *in* Demers, J., ed., Proceedings, Workshop Seminar on Permafrost Hydrology: Canadian National Committee, International Hydrological Decade, Environment Canada, Ottawa, p. 7–20.

——, 1978, Palaeohydrological reconstruction from a Holocene valley fill, *in* Miall, A. D., ed., Fluvial sedimentology: Canadian Society of Petroleum Geologists Memoir 5, p. 743–772.

Church, M., and Miles, M., 1982, Discussion: Processes and mechanisms of river bank erosion, *in* Hey, R. D., Bathurst, J. C., and Thorne, C. R., eds., Gravel-bed rivers: New York, Wiley, p. 259–271.

Cogley, J. G., 1972, Process of solution in an Arctic limestone terrain: Institute of British Geographers Special Publication no. 4, p. 201–211.

Cogley, J. G., and McCann, S. B., 1976, An exceptional storm and its effects in the Canadian High Arctic: Arctic and Alpine Research, v. 8, p. 105–115.

Cook, F. A., 1967, Fluvial processes in the High Arctic: Geographical Bulletin, v. 9, p. 262–268.

Cooper, R. H., and Hollingshead, A. B., 1973, River bank erosion in regions of permafrost, *in* Fluvial processes and sedimentation: Proceedings of Hydrology Symposium Number 9, May 8 and 9, University of Alberta, Edmonton: Ottawa, Inland Waters Directorate, p. 272–283.

Craig, P. C., and McCart, P. J., 1975, Classification of stream types in Beaufort Sea drainages between Prudhoe Bay, Alaska, and the Mackenzie Delta, Northwest Territories, Canada: Arctic and Alpine Research, v. 7, p. 183–198.

Czudek, T., and Demek, J., 1970, Thermokarst in Siberia and its influence on the development of lowland relief: Quaternary Research, v. 1, p. 103–120.

Day, T. J., and Anderson, J. C., 1976, Observations on river ice, Thomsen River, Banks Island, District of Franklin: Geological Survey of Canada Paper 76-1B, p. 187–196.

Day, T. J., and Lewis, C. P., 1977, Reconnaissance studies of Big River, Banks Island, District of Franklin: Geological Survey of Canada Paper 77-1A, p. 75–86.

Dingman, S. L., Barry, R. G., Weller, G., Benson, C., LeDrew, E. F., and Goodwin, C. W., 1980, Climate, snow cover, microclimate, and hydrology, *in* Brown, J., Miller, P. C., Tieszen, L. L., and Bunnell, F. L., eds., An arctic ecosystem, the coastal tundra at Barrow, Alaska: Stroudsburg, Pennsylvania, Dowden, Hutchinson and Ross, p. 30–65.

Dinter, D. A., Carter, L. D., and Brigham-Grette, J., 1987, Late Cenozoic geologic evolution of the Alaskan North Slope and adjacent continental shelves, *in* Grantz, A., Johnson, L., and Sweeney, J., eds., The geology of the Arctic Ocean region: Boulder, Colorado, Geological Society of America, Geology of North America, American Geology, v. L (in press).

Dygas, J. A., and Burrell, D. C., 1976, Dynamic sedimentological processes along the Beaufort Sea coast of Alaska, *in* Assessment of the Arctic marine environment; Selected topics: Fairbanks, University of Alaska, Institute of Marine Science, p. 189–203.

Dyke, L. D., 1984, Frost heaving of bedrock in permafrost regions: Bulletin of the Association of Engineering Geologists, v. 21, p. 389–405.

Edlund, S. A., 1982, Vegetation of Melville Island, District of Franklin; eastern Melville Island and Dundas Peninsula: Geological Survey of Canada, Open File 852, 2 maps and explanation, scale 1:250,000.

Everett, K. R., 1979, Evolution of the soil landscape in the sand region of the Arctic Coastal Plain as exemplified at Atkasook, Alaska: Arctic, v. 32, p. 207–223.

Everett, K. R., and Brown, J., 1982, Some recent trends in the physical and chemical characterization and mapping of tundra soils, Arctic Slope of Alaska: Soil Science, v. 133, p. 264–280.

Ferrians, O. J., Jr., 1965, Permafrost map of Alaska: U.S. Geological Survey Miscellaneous Geological Investigations Map I-445, scale 1:2,500,000.

——, 1971, Preliminary engineering-geologic maps of the proposed trans-Alaska Pipeline route, Beechey Point and Sagavanirktok quadrangles: U.S. Geological Survey Open-File Report 71-101 (491), scale 1:125,000, 2 sheets.

Ferrians, O. J., Jr., Kachadoorian, R., and Greene, G. W., 1969, Permafrost and related engineering problems in Alaska: U.S. Geological Survey Professional Paper 678, 37 p.

Forbes, D. L., 1979, Bottom fast ice in northern rivers; Hydraulic effects and hydrometric implications: Proceedings, Canadian Hydrology Symposium: '79, Cold Climate Hydrology, Vancouver, B.C., p. 175–184.

Forbes, D. L., and Frobel, D., 1985, Coastal erosion and sedimentation in the Canadian Beaufort Sea, *in* Current Research, Part B: Geological Survey of Canada, Paper 85-1B, p. 69–80.

French, H. M., 1975, Pingo investigations and terrain disturbance studies, Banks Island, District of Franklin: Geological Survey of Canada, Paper 75-1A, p. 459–464.

——, 1976, The periglacial environment: London, Longman Group Limited, 309 p.

French, H. M., and Egginton, P. A., 1973, Thermokarst development, Banks Island, western Canadian Arctic, *in* The North American Contribution to the Second Internation Conference on Permafrost, Yakutsk: Washington, D.C., National Academy of Sciences, p. 203–212.

French, H. M., Harry, D. G., and Clark, M. J., 1982, Ground ice stratigraphy and late-Quaternary events, southwest Banks Island, Canadian Arctic, *in* Proceedings, Fourth Canadian Permafrost Conference: Ottawa, National Research Council, p. 81–90.

Fyles, J. G., 1963, Surficial geology of Victoria and Stefansson islands, District of Franklin: Geological Survey of Canada Bulletin 101, 38 p.

Galloway, J. P., and Carter, L. D., 1978, Preliminary map of pingos in National Petroleum Reserve in Alaska: U.S. Geological Survey Open-File Report 78-795, scale 1:500,000.

Gill, D., 1972, Environmental and ecological reconnaissance of the Pan Canadian Petroleum lease area, northeast coastal Yukon Territory: Edmonton, University of Alberta (Boreal Institute), unpublished manuscript.

Gold, L. W., 1985, The ice factor in frozen ground, *in* Church, M., and Slaymaker, O., eds., Field and theory; Lectures in geocryology: Vancouver, University of British Columbia Press, p. 74–95.

Goodwin, C. W., Brown, J., and Outcalt, S. I., 1984, Potential responses of permafrost, *in* McBeath, J. H., ed., Proceedings of a Conference on the

Potential Effects of Carbon-Dioxide Induced Climatic Changes in Alaska: University of Alaska, Miscellaneous Publication 83-1, p. 92–105.

Greene, H. G., 1970, Microrelief of an arctic beach: Journal of Sedimentary Petrology, v. 40, p. 419–427.

Gusev, A. I., 1959, K metokike kartirovanii Beregov Vdeltakh rek poliarnogo basseina: Leningrad Nauchno-issledovatelskii Institut Geodogii Arktiki, Trudy, Tome 107, p. 127–132.

Hansell, R.I.C., Scott, P. A., Staniforth, R., and Svoboda, J., 1983, Permafrost development in the intertidal zone at Churchill, Manitoba; A possible mechanism for accelerated beach uplift: Arctic, v. 36, p. 198–203.

Harper, J. R., 1978a, The physical processes affecting the stability of tundra cliff coasts [Ph.D. thesis]: Baton Rouge, Louisiana State University, 212 p.

—— , 1978b, Coastal erosion rates along the Chukchi Sea coast near Barrow, Alaska: Arctic, v. 31, p. 428–433.

Harper, J. R., Owens, E. H., and Wiseman, W. J., Jr., 1978, Arctic beach processes and the thaw of ice-bonded sediments in the littoral zone, in Proceedings of the Third International Conference on Permafrost, Volume 1: Ottawa, National Research Council of Canada, p. 195–199.

Harper, J. R., Collins, A., and Reimer, P. D., 1985a, Morphology and processes of the Canadian Beaufort Sea coast, in Arctic Land-Sea Interactions, 14th Arctic Workshop, November 6–8, 1985, Abstracts: Dartmouth, Nova Scotia, Bedford Institute of Oceanography, p. 110–111.

Harper, J. R., Reimer, P. D., and Collins, A. D., 1985b, Canadian Beaufort Sea physical shore-zone analysis, Final Report for Northern Oil and Gas Action Plan, Indian and Northern Affairs Canada: Sidney, British Columbia, Dobrocky Seatech, 105 p. and 2 appendices.

Harry, D. G. 1987, Ground ice and permafrost, in Clark, M. H., ed., International Perspectives in Periglacial Research: London, Wiley (in press).

Harry, D. G., and French, H. M., 1983, The orientation and evolution of thaw lakes, southwest Banks Island, Canadian Arctic, in Proceedings: Permafrost; Fourth International Conference: Washington, D.C., National Academy Press, p. 456–461.

Harry, D. G., French, H. M., and Clark, M. J., 1983, Coastal conditions and processes, Sachs Harbour, Banks Island, western Canadian Arctic: Zeitschrift für Geomorphologie Neue Folge, Supplementband 47, p. 1–26.

Harry, D. G., French, J. M., and Pollard, W. H., 1985, Ice wedges and permafrost conditions near King Point, Beaufort Sea coast, Yukon Territory: Geological Survey of Canada, Paper 85-1A, p. 111–116.

Hartwell, A. D., 1973, Classification and relief characteristics of northern Alaska's coastal zone: Arctic, v. 26, p. 244–252.

Hartz, R. W., 1978, Erosional hazards map of the arctic coast of the National Petroleum Reserve, Alaska: U.S. Geological Survey Open-File Report 78-406, 7 p. plus map, scale 1:1,000,000.

Harvey, R. C., 1982, The climate of arctic Canada in a 2 × CO$_2$ world: Canadian Climate Centre Report no. 82-5, 21 p.

Heginbottom, J. A., 1971, Some effects of a forest fire on the permafrost active layer at Inuvik, Northwest Territories, in Brown, R.J.E., ed., Proceedings of a seminar on the permafrost active layer 4 and 5 May, 1971: National Reseach Council of Canada, Associate Committee on Geotechnical Research, Technical Memorandum no. 103, p. 31–36.

—— , 1973, Some effects of surface disturbance on the permafrost active layer at Inuvik, Northwest Territories, Canada, in The North American Contribution to the Second International Conference on Permafrost, Yakutsk: Washington, D.C., National Academy of Sciences, p. 649–657.

—— , 1983, Problems in the cartography of ground ice; A pilot project for northwestern Canada, in Proceedings, Permafrost, Fourth International Conference: Washington, D.C., National Academy Press, p. 480–485.

—— , 1984, The bursting of a snow dam, Tingmisut Lake, Melville Island, Northwest Territories, in Current research, Part B: Geological Survey of Canada, Paper 84-1B, p. 187–192.

Hodgson, D. A., 1982, Surficial materials and geomorphological processes, western Sverdrup and adjacent islands, District of Franklin: Geological Survey of Canada, Paper 81-9, 44 p. and maps 1-1981 and 2-1981, scale 1:250,000.

Hopkins, D. M., 1949, Thaw lakes and thaw sinks in the Inuvik Lake area,

Seward Peninsula, Alaska: Journal of Geology, v. 57, p. 119–131.

—— , 1982, Aspects of the paleogeography of Beringia during the late Pleistocene, in Hopkins, D. M., Matthews, J. V., Jr., Schweger, C. E., and Yount, S. B., eds., Paleoecology of Beringia: New York, Academic Press, p. 3–28.

Hopkins, D. M., and Hartz, R. W., 1978, Coastal morphology, coastal erosion, and barrier islands of the Beaufort Sea, Alaska: U.S. Geological Survey Open-File Report 78-1063, 54 p.

Hultén, E., 1968, Flora of Alaska and neighboring territories: Palo Alto, Stanford University Press, 1008 p.

Hume, J. D., and Schalk, M., 1964, The effects of ice-push on arctic beaches: American Journal of Science, v. 262, p. 267–273.

—— , 1967, Shoreline processes near Barrow, Alaska; A comparison of the normal and the catastrophic: Arctic, v. 20, p. 86–103.

—— , 1976, The effects of ice on the beach and nearshore, Point Barrow, arctic Alaska: Revue Geographique Montreal, v. XXX, p. 105–114.

Hume, J. D., Schalk, M., and Hume, P. W., 1972, Short-term climate changes and coastal erosion, Barrow, Alaska: Arctic, v. 25, p. 272–278.

Jahn, A., 1972, Tundra polygons in the Mackenzie Delta area: Gottinger Geographische Abhandlungen 60, p. 285–292.

Jahn, A., and Walker, H. J., 1983, The active layer and climate: Zeitschrift für Geomorphologie Neue Folge, Supplementband 47, p. 97–108.

Johnston, G. H., ed., 1981, Permafrost engineering design and construction: Toronto, John Wiley and Sons, 540 p.

Judge, A., and Pilon, J., 1984, Climate change and geothermal regime; Introduction, in Final Proceedings, Permafrost, Fourth International Conference on Permafrost: Washington, D.C., National Academy Press, p.137–138.

Judge, A., and Taylor, A., 1985, Permafrost distribution in northern Canada: Interpretation of well logs, in Brown, J., Metz, M. C., and Hoekstra, P., eds., Workshop on permafrost geophysics, Golden, Colorado, 23–24 October 1984: U.S. Army Corps of Engineers Cold Regions Research and Engineering Laboratory Special Report 85-5, p. 19–25.

Kaczorowski, R. T., 1977, The Carolina Bays; A comparison with modern oriented lakes: Coastal Research Division, Department of Geology, University of South Carolina, Technical Report 13-CRD, 124 p.

Kellogg, W. W., 1975, Climatic feedback mechanisms involving the polar regions, in Weller, G., and Bowling, S. A., eds., Climate of the Arctic: Fairbanks, University of Alaska, Geophysical Institute, p. 111–116.

Kerfoot, D. E., and Mackay, J. R., 1972, Geomorphological process studies, Gary Island, Northwest Territories, in Kerfoot, D. E., ed., Mackenzie Delta area monograph: St. Catherines, Ontario, Brock University Press, p. 115–130.

Kindle, E. M., 1924, Observations on ice-borne sediments by the Canadian and other Arctic expeditions: American Journal of Science, v. 7, p. 249–286.

Kovacs, A., 1983, Shore ice ride-up and pile-up features, Part I; Alaska's Beaufort Sea coast: U.S. Army Engineer Cold Regions Research and Engineering Laboratory Report 83-9, 51 p.

—— , 1984, Shore ice ride-up and pile-up features, Part II; Alaska's Beaufort Sea Coast—1983 and 1984: U.S. Army Engineer Cold Regions Research and Engineering Laboratory Report 84-26, 29 p.

Kovacs, A., and Mellor, M., 1974, Sea ice morphology and ice as a geologic agent in the southern Beaufort Sea, in Reed, J. C., and Sater, J. E., eds., The coast and shelf of the Beaufort Sea: Arlington, Virginia, Arctic Institute of North America, p. 113–161.

Kreig, R. A., and Reger, R. D., 1982, Air-photo analysis and summary of landform soil properties along the route of the Trans-Alaska pipeline system: Alaska Division of Geological and Geophysical Surveys Geologic Report 66, 149 p.

LaBelle, J. C., Wise, J. L., Voelker, R. P., Schulze, R. H., and Wohl, G. M., 1983, Alaska marine ice atlas: Anchorage, Alaska, Arctic Environmental Information and Data Center, 302 p.

Lachenbruch, A. H., 1962, Mechanics of thermal contraction cracks and ice wedge polygons in permafrost: Geological Society of America Special Paper 70, 69 p.

—— , 1966, Contraction theory of ice wedge polygons; A qualitative discussion, in Proceedings, Permafrost; International Conference: Washington, D.C.,

National Academy of Sciences, National Research Council Publication 1287, p. 63–71.

Landsberg, S. Y., 1956, The orientation of dunes in Britain and Denmark in relation to wind: Geographical Journal, v. 122, p. 176–189.

Lawson, D. E., 1983a, Ground ice in perennially frozen sediments, northern Alaska, *in* Proceedings, Permafrost: Fourth International Conference: Washington, D.C., National Academy Press, p. 695–700.

—— , 1983b, Erosion of perennially frozen streambanks: U.S. Army Corps of Engineers Cold Regions Research and Engineering Laboratory Report 83-29, 22 p.

Leffingwell, E. de K., 1915, Ground ice-wedges, the dominant forms of ground ice on the north coast of Alaska: Journal of Geology, v. 23, p. 635–654.

—— , 1919, The Canning River region, northern Alaska: United States Geological Survey Professional Paper 109, 251 p.

Lewellen, R. I., 1970, Permafrost erosion along the Beaufort Sea coast: Published by the author (P. O. Box 2435, Littleton, CO 80161), 25 p.

—— , 1977, A study of Beaufort Sea coastal erosion, northern Alaska: U.S. National Oceanic and Atmospheric Administration, Environmental Assessment of the Alaskan Continental Shelf, Annual Reports of Principal Investigators for the year ending March, 1977, v. 15, p. 491–527.

Lewis, C. P., and Forbes, D. L., 1974, Sediments and sedimentary processes, Yukon Beaufort Sea coast: Environmental-Social Committee, Northern Pipelines Task Force on Northern Oil Development Report 74-29, 40 p.

Lewis, C. P., and MacDonald, B. D., 1973, Rivers of the Yukon North Slope: Proceedings, Hydrology Symposium Number 9, Edmonton, Alberta, p. 251–271.

Liguori, A., Maple, J. A., and Heuer, C. E., 1979, The design and construction of the Alyeska Pipeline, *in* Proceedings of the Third International Conference on Permafrost, Volume 2: Ottawa, National Research Council of Canada, p. 151–157.

Linell, K. A., and Kaplar, C. W., 1966, Description and classification of frozen soils, *in* Proceedings, Permafrost, International Conference: Washington, D.C., National Academy of Sciences, National Research Council Publication 1287, p. 481–487.

Livingstone, D. A., 1954, On the orientation of lake basins: American Journal of Science, v. 252, p. 547–554.

Lorrain, R. D., and Demeur, P., 1985, Isotopic evidence for relic Pleistocene glacier ice on Victoria Island, Canadian Arctic Archipelago: Arctic and Alpine Research, v. 17, p. 89–98.

MacCarthy, G. R., 1953, Recent changes in the shoreline near Point Barrow, Alaska: Arctic, v. 6, p. 44–51.

MacCracken, M. C., and Luther, F. M., 1985, Projecting the climatic effects of increasing carbon dioxide: U.S. Department of Energy, DOE/ER-0237, 381 p.

MacDonald, B. C., and Lewis, C. P., 1973, Geomorphic and sedimentologic processes of rivers and coast, Yukon Coastal Plain: Environmental-social committee, Northern pipelines, Task Force on Northern Oil Development Report no. 73-39, Information Canada, Ottawa, 245 p.

Mackay, J. R., 1956a, Deformation by glacier-ice at Nicholson Peninsula, Northwest Territories, Canada: Arctic, v. 9, p. 219–228.

—— , 1956b, Notes on oriented lakes of the Liverpool Bay area, Northwest Territories: Revue canadienne de geographie, v. 10, p. 169–173.

—— , 1958, The Anderson River map-area, Northwest Territories: Canada Department of Mines and Technical Surveys, Geographical Branch, Memoir 5, 137 p.

—— , 1959, Glacier ice-thrust features of the Yukon coast: Canada Department of Mines and Technical Surveys, Geographical Bulletin no. 13, p. 5–21.

—— , 1962, Pingos of the Pleistocene Mackenzie Delta area: Canada Department of Mines and Technical Surveys, Geographical Bulletin no. 15, p. 21–63.

—— , 1963a, The Mackenzie delta area, Northwest Territories: Canada Department of Mines and Technical Surveys Geographical Branch Memoir, 8, 202 p. (reprinted in 1974 as Geological Survey of Canada Miscellaneous Report 23).

—— , 1963b, Notes on the shoreline recession along the coast of the Yukon Territory: Arctic, v. 16, p. 195–197.

—— , 1966, Pingos in Canada, *in* Proceedings, Permafrost, International Conference: Washington, D.C., National Academy of Sciences, National Research Council Publication 1287, p. 71–76.

—— , 1970, Disturbances to the tundra and forest tundra environment of the western Arctic: Canadian Journal of Earth Sciences, v. 7, p. 420–432.

—— , 1971, The origin of massive icy beds in permafrost, western Arctic coast, Canada: Canadian Journal of Earth Sciences, v. 8, p. 397–422.

—— , 1972, The world of underground ice: Annals of the Association of American Geographers, v. 62, p. 1–22.

—— , 1973, Problems in the origin of massive icy beds, western Arctic, Canada, *in* The North American Contribution to the Second International Conference on Permafrost, Yakutsk: Washington, D.C., National Academy of Sciences, p. 223–228.

—— , 1974, Ice wedge cracks, Garry Island, Northwest Territories: Canadian Journal of Earth Sciences, v. 11, p. 1366–1383.

—— , 1975a, The closing of ice wedge cracks in permafrost, Garry Island, Northwest Territories: Canadian Journal of Earth Sciences, v. 12, p. 1668–1674.

—— , 1975b, Relict ice wedges, Pelly Island, Northwest Territories: Geological Survey of Canada, Paper 75-1A, p. 469–470.

—— , 1975c, The stability of permafrost and recent climatic change in the Mackenzie Valley, Northwest Territories: Geological Survey of Canada, Paper 75-1B, p. 173–176.

—— , 1976, Ice wedges as indicators of recent climatic change, western Arctic Coast: Geological Survey of Canada, Paper 76-1A, p. 233–234.

—— , 1978, The use of snow fences to reduce ice-wedge cracking, Garry Island, Northwest Territories: Geological Survey of Canada, Paper 78-1A, p. 523–524.

—— , 1979, Pingos of the Tuktoyaktuk Peninsula area, Northwest Territories: Geographie physique et Quaternaire, v. 33, p. 3–61.

—— , 1980, Deformation of ice-wedge polygons, Garry Island, Northwest Territories: Geological Survey of Canada, Paper 80-1A, p. 63–68.

—— , 1981, An experiment in lake drainage, Richards Island, Northwest Territories; a progress report: Geological Survey of Canada, Paper 81-1A, p. 63–68.

—— , 1983, Pingo growth and subpingo water lenses, *in* Proceedings: Permafrost, Fourth International Conference: Washington, D.C., National Academy Press, p. 762–765.

—— , 1984, The direction of ice wedge cracking in permafrost; Downward or upward?: Canadian Journal of Earth Sciences, v. 21, p. 516–524.

—— , 1985, Pingo ice of the western Arctic coast, Canada: Canadian Journal of Earth Sciences, v. 22, p. 1452–1464.

—— , 1986a, The permafrost record and Quaternary history of northwestern Canada, *in* Heginbottom, J. A., and Vincent, J-S., eds., Correlation of Quaternary deposits and events around the margin of the Beaufort Sea: Geological Survey of Canada Open File Report 1237, p. 38–40.

—— , 1986b, Fifty years (1935 to 1985) of coastal retreat west of Tuktoyaktuk, District of Mackenzie, *in* Current Research, Part A: Geological Survey of Canada, Paper 86-1A, p. 727–735.

Mackay, J. R., and Matthews, J. V., Jr., 1983, Pleistocene ice and sand wedges, Hooper Island, Northwest Territories: Canadian Journal of Earth Sciences, v. 20, p. 1087–1097.

Mackay, J. R., Rampton, V. N., and Fyles, J. G., 1972, Relic Pleistocene permafrost, western Arctic Canada: Science, v. 176, p. 1321–1323.

Mackay, J. R., Ostrick, J., Lewis, C. P., and Mackay, D. K., 1979, Frost heave at ground temperature below 0°C, Inuvik, Northwest Territories: Geological Survey of Canada, Paper 79-1A, p. 403–405.

Mackenzie River Basin Committee, 1981, Alluvial ecosystems: MacKenzie River Basin Study Report Supplement 2, 129 p.

Markham, W. E., 1981, Ice atlas, Canadian Arctic waterways: Ottawa, Atmospheric Environment Service, Environment Canada, 198 p.

Maurin, A. F., and Lathram, E. H., 1977, A deeper look at Landsat-1 images of Alaska, *in* Woll, P. W., and Fischer, W. A., eds., Proceedings of the First

Annual William T. Pecora Memorial Symposium, October 1975, Sioux Falls, South Dakota: U.S. Geological Survey Professional Paper 1015, p. 213–233.

Maxwell, J. B., 1980, The climate of the Canadian Arctic Islands and adjacent waters: Climatological Studies 30, Atmospheric Environment Service, Toronto, v. 1, 531 p.

—— , 1981, Climatic regions of the Canadian Arctic Islands: Arctic, v. 34, p. 225–240.

May, S. K., Dolan, R., and Hayden, B. P., 1983, Erosion of U.S. shorelines: EOS Transactions of the American Geophysical Union, p. 521–522.

McCann, S. B., 1972, Magnitude and frequency of processes operating on arctic beaches, Queen Elizabeth Islands, Northwest Territories, Canada, *in* Adams, W. P., and Helleiner, F. M., eds., International geography 1972, Papers submitted to the 22nd International Geographical Congress, Canada: Ottawa, University of Toronto Press, p. 41–43.

—— , 1973, Beach processes in an arctic environment, *in* Coates, D. R., Coastal Geomorphology: Binghamton, State University of New York, p. 141–155.

McCann, S. B., and Carlisle, R. J., 1972, The nature of the ice-foot on the beaches of Radstock Bay, south-west Devon Island, Northwest Territories, Canada in the spring and summer of 1970, *in* Polar geomorphology: London, Institute of British Geographers Special Publication no. 4, p. 175–186.

McCann, S. B., and Cogley, J. G., 1973, The geomorphic significance of fluvial activity at high latitudes, *in* Fahey, B. D., and Thompson, R. D., eds., Research in Polar and Alpine Geomorphology, Proceedings 3rd Guelph Symposium on Geomorphology: Norwich, England, GeoAbstracts, Limited, p. 118–135.

McCann, S. B., and Hannell, F. G., 1971, Depth of the "frost table" on arctic beaches, Cornwallis and Devon islands, Northwest Territories, Canada: Journal of Glaciology, v. 10, p. 155–157.

McCann, S. B., and Taylor, R. B., 1975, Beach freezeup sequence at Radstock Bay, Devon Island, Arctic Canada: Arctic and Alpine Research, v. 7, p. 379–386.

McCann, S. B., Howarth, P. J., and Cogley, J. G., 1972, Fluvial processes in a periglacial environment: Transactions Institute of British Geographers, v. 55, p. 69–82.

McCloy, J. M., 1970, Hydrometeorological relationships and their effects on the levees of a small arctic delta: Geografiska Annaler, v. 52A, p. 223–241.

McLaren, P., 1981, River and suspended discharge into Byam Channel, Queen Elizabeth Islands, Northwest Territories, Canada: Arctic, v. 34, p. 141–146.

McRoberts, E. C., and Morgenstern, N. R., 1973, A study of landslides in the vicinity of the Mackenzie River, Mile 205 to 660: Canada, Task Force on Northern Oil Development, Environmental-Social Committee, Report 73-35, 96 p.

—— , 1974a, The stability of thawing slopes: Canadian Geotechnical Journal, v. 11, p. 447–469.

—— , 1974b, Stability of slopes in frozen soil, Mackenzie Valley, Northwest Territories: Canadian Geotechnical Journal, v. 11, p. 554–573.

Miles, M., 1976, An investigation of river bank and coastal erosion, Banks Island, District of Franklin: Geological Survey of Canada Paper 76-1A, p. 195–200.

—— , 1977, Coastal and riverbank stability on Banks Island, Northwest Territories, Canada, *in* Proceedings, Third National Hydrotechnical Conference, May, Quebec: Ottawa, Canadian Society for Civil Engineering, p. 972–991.

Moore, G. W., 1966, Arctic beach sedimentation, *in* Wilimovsky, N. J., and Wolfe, J. N., eds., Environment of the Cape Thompson Region, Alaska: U.S. Atomic Energy Commission Report PNE-481, p. 587–608.

Morack, J. L., and Rogers, J. C., 1981, Seismic evidence of shallow permafrost beneath islands in the Beaufort Sea, Alaska: Arctic, v. 34, p. 169–74.

Muller, S. W., 1943, Permafrost or permanently frozen ground and related engineering problems: U.S. Army, Office Chief of Engineers, Military Intelligence Division Strategic Engineering Study 62, 231 p. (Also published 1947, Ann Arbor, Michigan, Edwards Brothers).

Naidu, A. S., 1980, An alternative conceptual model for sediment concentration in frazil sea ice of north arctic Alaska, *in* Schell, D. M., ed., Beaufort Sea Winter Watch: Fairbanks, University of Alaska, Outer Continental Shelf Environmental Assessment Program, Arctic Project Office, Special Bulletin no. 29, p. 28–30.

Neave, K. G., and Sellmann, P. V., 1984, Determining distribution patterns of ice-bonded permafrost in the U.S. Beaufort Sea from seismic data, *in* Barnes, P. W., Schell, D. M., and Reimnitz, E., eds., The Alaskan Beaufort Sea, Ecosystems and environments: Orlando, Florida, Academic Press, p. 237–258.

Niessen, A.C.H.M., Koster, E. A., and Galloway, J. P., 1984, Periglacial sand dunes and eolian sand sheets, an annotated bibliography: U.S. Geological Survey Open-File Report 84-167, 61 p.

Obruchev, S. V., 1938, Shakhmatnye (ortogonal'nye) formy v oblastiakh vechnoi merzloty (The checkerboard [orthogonal] forms in permafrost regions): Vsesoiuznoe geograficheskoe obshchestvo Izvestiya, v. 70, no. 6, p. 737–746.

Osterkamp, T. E., and Gosink, J. P., 1984, Observations and analyses of sediment-laden sea ice, *in* Barnes, P. W., Schell, D. M., and Reimnitz, E., eds., The Alaskan Beaufort Sea, Ecosystems and environments: Orlando, Florida, Academic Press, p. 73–93.

Osterkamp, T. E., and Payne, M. W., 1981, Estimates of permafrost thickness from well logs in northern Alaska: Cold Regions Science and Technology, v. 5, p. 13–27.

Owens, E. H., and Harper, J. R., 1977, Frost-table and thaw depths in the littoral zone near Peard Bay, Alaska: Arctic, v. 30, p. 155–168.

Owens, E. H., and McCann, S. B., 1970, The role of ice in the arctic beach environment with special reference to Cape Ricketts, southwest Devon Island, Northwest Territories, Canada: American Journal of Science, v. 268, p. 397–414.

Packer, J. G., 1969, Polyploidy in the Canadian Arctic Archipelago: Arctic and Alpine Research, v. 1, p. 15–27.

Penner, E., 1970, Frost heaving faces in Leda Clay: Canadian Geotechnical Journal, v. 7, p. 8–16.

Péwé, T. L., 1959, Sand-wedge polygons (tesselations) in the McMurdo Sound region, Antarctica; A progress report: American Journal of Science, v. 257, p. 545–552.

—— , 1966a, Paleoclimatic significance of fossil ice wedges: Biuletyn Peryglacjalny, v. 15, p. 65–73.

—— , 1966b, Ice wedges in Alaska; Classification, distribution, and climatic significance, *in* Proceedings, Permafrost, International Conference: Washington, D.C., National Academy of Sciences, National Research Council Publication 1287, p. 76–81.

—— , 1975, Quaternary geology of Alaska: U.S. Geological Survey Professional Paper 835, 145 p.

—— , 1983, Alpine permafrost in the contiguous United States; A review: Arctic and Alpine Research, v. 15, p. 145–146.

Pihlainen, J. A., and Johnston, G. H., 1963, Guide to a field description of permafrost: Canada, National Research Council, Associate Committee on Soil and Snow Mechanics, Technical Memorandum 79, 23 p.

Pissart, A., 1967a, Les pingos de l'Ile Prince-Patrick (76°N-120°W): Geographical Bulletin of Canada, v. 9, p. 189–217.

—— , 1967b, Les modalities de l'ecoulement de l'eau sur l'Ile Prince Patrick (76°Lat N., 120°Long O. Arctique Canadien): Biuletyn Peryglacialny, v. 16, p. 217–224.

—— , 1968, Les polygons de fente de gel d l'Ile Prince Patrick (Arctique Canadien, 76°Lat. N): Biuletyn Peryglacjalny, v. 17, p. 171–180.

Pissart, A., and French, H. M., 1976, Pingo investigations, north-central Banks Island, Canadian Arctic: Canadian Journal of Earth Sciences, v. 13, p. 937–946.

Plafker, G., 1964, Oriented lakes and lineaments of northeastern Bolivia: Geological Society of America Bulletin, v. 75, p. 503–522.

Pollard, W. H., and French, H. M., 1980, A first approximation of the volume of ground ice, Richards Island, Pleistocene Mackenzie Delta, Northwest Territories, Canada: Canadian Journal of Earth Sciences, v. 17, p. 509–516.

Polunin, N., 1955, Aspects of arctic botany: American Scientist, v. 43, p. 307–322.

—— , 1959, Circumpolar arctic flora: Oxford University Press, 514 p.

Porsild, A. E., 1951, Plant life in the Arctic: Canadian Geographical Journal, v. 42, p. 120–145.

——, 1955, The vascular plants of the western Canadian Arctic Archipelago: National Museum of Canada Bulletin no. 135, 226 p.

——, 1964, Illustrated flora of the Canadian Arctic Archipelago (second edition): National Museum of Canada Bulletin no. 146, 218 p.

Price, W. A., 1968a, Oriented lakes, *in* Fairbridge, R. W., ed., Encyclopedia of geomorphology: New York, Reinhold Book Corporation, p. 784–796.

——, 1968b, Carolina Bays, *in* Fairbridge, R. W., ed., Encyclopedia of geomorphology: New York, Reinhold Book Corporation, p. 102–108.

Rampton, V. N., 1973, The influences of ground ice and thermokarst upon the geomorphology of the Mackenzie-Beaufort region, *in* Fahey, B. D., and Thompson, R. D., eds., Proceedings, Research in Polar and Alpine Geomorphology, 3rd Guelph Symposium on Geomorphology, 1973: Norwich, United Kingdom, Geo Abstracts Limited, p. 43–59.

——, 1974, Terrain evaluation with respect to pipeline construction, Mackenzie transportation corridor, northern part, Lat. 68°N to coast: Environmental Committee Northern Pipelines, Task Force on Northern Oil Development Report no. 73-47, Information Canada Catalog no. R-72-9273, 44 p.

——, 1982, Quaternary geology of the Yukon Coastal Plain: Geological Survey of Canada, Bulletin 317, 49 p., and Map 1503A.

——, 1986, Quaternary history of the Arctic Coastal Plain in Canada, *in* Heginbottom, J. A., and Vincent, J-S., eds., Correlation of Quaternary deposits and events around the margin of the Beaufort Sea: Geological Survey of Canada, Open File Report 1237, p. 11–12.

Rampton, V. N., and Bouchard, M., 1975, Surficial geology of Tuktoyaktuk, District of Mackenzie: Geological Survey of Canada, Paper 74-53, 17 p.

Rampton, V. N., and Mackay, J. R., 1971, Massive ice and icy sediments throughout the Tuktoyaktuk Peninsula, Richards Island and nearby areas, District of Mackenzie: Geological Survey of Canada, Paper 71-21, 16 p.

Rampton, V. N., and Walcott, R. I., 1974, Gravity profiles across ice-cored topography: Canadian Journal of Earth Sciences, v. 11, p. 110–122.

Rawlinson, S. E., ed., 1983, Guidebook to permafrost and related features, Prudhoe Bay, Alaska; Guidebook 5, Fourth International Conference on Permafrost: Fairbanks, Alaska Division of Geological and Geophysical Surveys, 177 p.

Reimnitz, E., and Barnes, P. W., 1974, Sea ice as a geologic agent on the Beaufort Sea shelf of Alaska, *in* Reed, J. C., and Sater, J. E., The coast and shelf of the Beaufort Sea: Arlington, Virginia, The Arctic Institute of North America, p. 301–353.

Reimnitz, E., and Bruder, K. F., 1972, River discharge into an ice-covered ocean and related sediment dispersal, Beaufort Sea coast of Alaska: Geological Society of America Bulletin, v. 83, p. 861–866.

Reimnitz, E., and Kempema, E., 1982, Dynamic ice-wallow relief of northern Alaska's nearshore: Journal of Sedimentary Petrology, v. 52, p. 451–461.

——, 1983, High rates of bedload transport measured from infilling rate of large strudel-scour craters in the Beaufort Sea, Alaska: Continental Shelf Research, v. 1, no. 3, p. 237–251, plus erratum in next issue for missing page.

Reimnitz, E., and Maurer, D. K., 1979, Effects of storm surges on the Beaufort Sea coast: Arctic, v. 32, p. 329–344.

Reimnitz, E., Rodeick, C. A., and Wolf, S. C., 1974, Strudel scour; A unique arctic marine geologic phenomenon: Journal of Sedimentary Petrology, v. 44, p. 409–420.

Reimnitz, E., Toimil, L., and Barnes, P., 1978, Arctic continental shelf morphology related to sea ice zonation, Beaufort Sea, Alaska: Marine Geology, v. 28, p. 179–210.

Reimnitz, E., Graves, S. M., and Barnes, P. W., 1985, Beaufort Sea coastal erosion, shoreline evolution, and sediment flux: U.S. Geological Survey Open-File Report 85-380, 66 p.

Reinson, G. E., and Rosen, P. S., 1982, Preservation of ice-formed features in a subarctic sandy beach sequence; Geologic implications: Journal of Sedimentary Petrology, v. 52, p. 463–472.

Rex, R. W., 1955, Microrelief produced by sea ice grounding in the Chukchi Sea near Barrow, Alaska: Arctic, v. 8, p. 177–186.

——, 1961, Hydrodynamic analysis of circulation and orientation of lakes in northern Alaska, *in* Raasch, G. O., ed., Geology of the Arctic, Volume 2: Toronto, University of Toronto Press, p. 1021–1043.

——, 1964, Arctic beaches, Barrow, Alaska, *in* Miller, R. L., ed., Papers in marine geology, Shepard commemorative volume: New York, Macmillan Company, p. 384–400.

Rieger, S., 1983, The genesis and classification of cold soils: New York, Academic Press, 230 p.

Ritchie, J. C., and Hare, F. K., 1971, Late Quaternary vegetation and climate near the Arctic tree line of northwestern North America: Quaternary Research, v. 1, p. 331–342.

Ritchie, J. C., Cwynar, L. C., and Spear, R. W., 1983, Evidence from northwest Canada for an early Holocene Milankovitch thermal maximum: Nature, v. 305, p. 126–128.

Rosenfeld, G. A., and Hussey, K. M., 1958, A consideration of the problem of oriented lakes: Proceedings Iowa Academy of Sciences, v. 65, p. 279–287.

Scott, K. M., 1978, Effects of permafrost on stream channel behavior in arctic Alaska: U.S. Geological Survey Professional Paper 1068, 19 p.

——, 1979, Arctic stream processes; An annotated bibliography: U.S. Geological Survey Water-Supply Paper 2065, 77 p.

Searby, H. W., and Hunter, M., 1971, Climate of the North Slope, Alaska: National Oceanic and Atmospheric Administration Technical Memorandum AR-4, 54 p.

Selkregg, L., ed., 1975, Alaska regional profiles, Arctic region: University of Alaska, Arctic Environmental Information and Data Center, 218 p.

Sellmann, P. V., and Brown, J., 1973, Stratigraphy and diagenesis of perennially frozen sediments in the Barrow, Alaska region, *in* The North American Contribution to the Second International Conference on Permafrost, Yakutsk: Washington, D.C., National Academy of Sciences, p. 171–181.

Sellmann, P. V., Brown, J., Lewellen, R. I., McKim, H., and Merry, C., 1975, The classification and geomorphic implications of thaw lakes on the Arctic Coastal Plain, Alaska: U.S. Army Corps of Engineers Cold Regions Research and Engineering Laboratory Research Report 344, 21 p.

Sharpe, D. R., 1984, Late Wisconsinan glaciation and deglaciation of Wollaston Peninsula, Victoria Island, Northwest Territories: Geological Survey of Canada Paper 854-1A, p. 259–269.

——, 1985, The stratified nature of deposits in streamlined glacial landforms on southern Victoria Island, District of Franklin: Geological Survey of Canada, Paper 85-1A, p. 365–371.

Short, A. D., 1976, Observations on ice deposited by waves on Alaskan arctic beaches: Revue Geographique Montreal, v. XXX, p. 115–122.

——, 1979, Barrier island development along the Alaskan-Yukon coastal plains; Summary: Geological Society of America Bulletin, v. 90, p. 3–5.

Short, A. D., and Wiseman, W. J., Jr., 1974, Freezeup processes on arctic beaches: Arctic, v. 27, p. 215–224.

Short, A. D., Coleman, J. M., and Wright, L. D., 1974, Beach dynamics and nearshore morphology of the Beaufort Sea coast, Alaska, *in* Reed, J. C., and Sater, J. E., eds., The coast and shelf of the Beaufort Sea: Arlington, Virginia, The Arctic Institute of North America, p. 477–488.

Sloan, C. E., Zenone, C., and Mayo, L. R., 1976, Icings along the Trans-Alaska Pipeline route: U.S. Geological Survey Professional Paper 979, 31 p.

Smith, M. W., 1984, Climate change and other effects on permafrost, *in* Final Proceedings, Permafrost, Fourth International Conference: Washington, D.C., National Academy Press, p. 153–155.

——, 1985, Models of soil freezing, *in* Church, M., and Slaymaker, O., eds., Field and theory; Lectures in geocryology: Vancouver, University of British Columbia Press, p. 96–120.

Smith, M. W., and Riseborough, D. W., 1983, Permafrost sensitivity to climatic change, *in* Proceedings, Permafrost; Fourth International Conference: Washington, D.C., National Academy Press, p. 1178–1183.

Spetzman, L. A., 1959, Vegetation of the arctic slope of Alaska: U.S. Geological Survey Professional Paper 302-B, 58 p.

Stefansson, V., 1913, My life with the Eskimo: New York, The Macmillan Company, 384 p.

Taber, S., 1943, Perennially frozen ground in Alaska; Its origin and history: Geological Society of America Bulletin, v. 54, p. 1433–1548.

Tarnocai, C., 1976, Soils of Bathurst, Cornwallis, and adjacent islands, District of Franklin: Geological Survey of Canada Paper 76-1B, p. 137–141.

Taylor, R. B., 1978, The occurrence of grounded ice ridges and shore ice piling along the northern coast of Somerset Island, Northwest Territories: Arctic, v. 31, p. 133–149.

—— , 1980, Beach thaw depth and the effect of ice-bonded sediment on beach stability: Proceedings of the Canadian Coastal Conference, Burlington, Ontario, April 22–24, 1980, p. 103–121.

Taylor, R. B., and McCann, S. B., 1974, Depth of the "frost table" in the Canadian Arctic Archipelago: Journal of Glaciology, v. 13, p. 321–322.

—— , 1976, The effect of sea and nearshore ice on coastal processes *in* Canadian Arctic Archipelago: Revue Geographique Montreal, v. XXX, p. 123–132.

Tedrow, J.C.F., 1969, Thaw lakes, thaw sinks, and soils in northern Alaska: Biuletyn Peryglacjalny, v. 20, p. 337–345.

—— , 1973, Soils of the polar region of North America: Biuletyn Peryglacjalny, v. 23, p. 157–165.

—— , 1977, Soils of the polar landscapes: New Brunswick, New Jersey, Rutgers University Press, 638 p.

Thom, B. G., 1970, Carolina Bays in Horry and Marion counties, South Carolina: Geological Society of America Bulletin, v. 81, p. 783–814.

Thorsteinsson, R., and Tozer, E. T., 1962, Banks, Victoria, and Stefansson islands, Arctic Archipelago: Geological Survey of Canada Memoir 330, 83 p.

—— , 1970, Geology of the Arctic Archipelago, *in* Douglas, R.J.W., ed., Geology and economic minerals of Canada: Geological Survey of Canada, Economic Geology Report 1, p. 548–590.

Ugolini, F. C., Zachara, J. M., and Reanier, R. E., 1982, Dynamics of soil-forming processes in the Arctic, *in* French, H. M., ed., Proceedings of the Fourth Canadian Permafrost Conference: Ottawa, National Research Council of Canada, p. 103–115.

Vincent, J-S., 1980, Surficial geology, Banks Island, Northwest Territories: Geological Survey of Canada Map 16-1979 (north half) and Map 17-1979 (south half), scale 1:250,000.

—— , 1982, The Quaternary history of Banks Island, Northwest Territories, Canada: Geographie Physique et Quaternaire, v. 36, p. 209–232.

—— , 1983, La geologie du Quaternaire et la geomorphologie de l'Ile Banks, Arctique Canadien: Geological Survey of Canada, Memoir 405, 118 p.

—— , 1984, Quaternary stratigraphy of the western Canadian Arctic Archipelago, *in* Fulton, R. J., ed., Quaternary stratigraphy of Canada; A Canadian contribution to IGCP Project 24: Canadian Geological Survey, Paper 84-10, p. 87–100.

Wahrhaftig, C., 1965, Physiographic divisions of Alaska: U.S. Geological Survey Professional Paper 482, 52 p., 6 pl.

Walker, D. A., Everett, K. R., Webber, P. J., and Brown, J., 1980, Geobotanical atlas of the Prudhoe Bay region, Alaska: U.S. Army Corps of Engineers Cold Regions Research and Engineering Laboratory Report 80-14, 69 p.

Walker, D. A., Walker, M. D., Everett, K. R., and Webber, P. J., 1985, Pingos of the Prudhoe Bay region, Alaska: Arctic and Alpine Research, v. 17, p. 321–336.

Walker, H. J., 1973, Morphology of the North Slope, *in* Britton, M. E., ed., Alaskan arctic tundra: Washington, D.C., The Arctic Institute of North America, p. 49–92.

—— , 1983, Colville River Delta, Alaska, Guidebook to Permafrost and Related Features, Guidebook 2, Fourth International Conference on Permafrost: Fairbanks, Alaska Division of Geological and Geophysical Surveys, 34 p.

—— , 1984, Alaska, *in* Bird, E.C.F., and Schwartz, M. L., eds., The world's coastline: New York, Van Nostrand Reinhold Company, p. 1–10.

Walker, H. J., and Arnborg, L., 1966, Permafrost and ice-wedge effects on riverbank erosion, *in* Proceedings, Permafrost, International Conference: National Academy of Sciences, National Research Council Publication 1287, p. 164–171.

Washburn, A. L., 1980, Geocryology, a survey of periglacial processes and environments: New York, John Wiley and Sons, 406 p.

—— , 1985, Periglacial problems, *in* Church, M., and Slaymaker, O., eds., Field and theory; Lectures in Geocryology: Vancouver, University of British Columbia Press, p. 166–202.

Weber, P. J., Miller, P. C., Chapin, F. S., III, and McCown, B. H., 1980, The vegetation; Pattern and succession, *in* Brown, J., Miller, P. C., Tieszen, L. L., and Bunnell, F. L., eds., An arctic ecosystem, the coastal tundra at Barrow, Alaska: Stroudsburg, Pennsylvania, Dowden, Hutchinson, and Ross, p. 186–218.

Wedel, J. H., Thorne, G. A., and Baracos, P. C., 1977, Site-intensive hydrologic study of a small catchment of Bathurst Island, 1976: Hydrologic Regimes, Freshwater Project no. 1, Western and Northern Region, Environment Canada, Winnipeg, 123 p.

Welsh, S. L., and Rigby, J. K., 1971, Botanical and physiographic reconnaissance of northern Yukon: Brigham Young University, Science Bulletin, Biological Series, v. 14.

Wiggins, I. L., and Thomas, J. H., 1962, A flora of the Alaskan Arctic Slope: Arctic Institute of North America, Special Publication no. 4, 425 p.

Wilkinson, T. J., and Bunting, B. T., 1975, Overland transport of sediment by rill water in a periglacial environment in the Canadian High Arctic: Geografiska Annaler, v. 57A, p. 105–116.

Williams, J. R., 1983, Engineering-geologic maps of northern Alaska, Wainwright Quadrangle: U.S. Geological Survey Open-File Report 83-457, scale 1:250,000, 28 p.

Williams, J. R., and Yeend, W. E., 1979, Deep thaw lake basins of the inner Arctic Coastal Plain, Alaska, *in* Johnson, K. M., and Williams, J. R., eds., The U.S. Geological Survey in Alaska; Accomplishments during 1978: U.S. Geological Survey Circular 804-B, p. B35–B37.

Woo, M-k., and Sauriol, J., 1980, Channel development in snow-filled valleys, Resolute, Northwest Territories, Canada: Geografiska Annaler, v. 62A, p. 37–56.

—— , 1981, Effects of snow jams on fluvial activities in the High Arctic: Physical Geography, v. 2, p. 83–98.

Woo, M-k., Heron, R., and Marsh, P., 1982, Basal ice in high arctic snowpacks: Arctic and Alpine Research, v. 14, p. 251–260.

Woo, M-k., Heron, R., Marsh, P., and Steer, P., 1983, Comparison of weather station snowfall with winter snow accumulation in High Arctic basins: Atmosphere-Ocean, v. 21, p. 312–325.

MANUSCRIPT ACCEPTED BY THE SOCIETY DECEMBER 1, 1986

Index

[Italic page numbers indicate major references]

Typeset by WESType Publishing Services, Inc., Boulder, Colorado
Printed in U.S.A. by Malloy Lithographing, Inc., Ann Arbor, Michigan